The CABI Encyclopedia of Forest Trees

The CABI Encyclopedia of Forest Trees

Compilers:

**Andrew Praciak, Nick Pasiecznik, Douglas Sheil,
Miriam van Heist, Marieke Sassen, Cristina Sousa Correia,
Christopher Dixon, George E. Fyson, Keith Rushforth
and Claire Teeling**

www.cabi.org

CABI is a trading name of CAB International

CABI Head Office
Nosworthy Way
Wallingford
Oxfordshire OX10 8DE
UK

Tel: +44 (0)1491 832111
Fax: +44 (0)1491 833508
E-mail: info@cabi.org
Website: www.cabi.org

CABI
38 Chauncey Street
Suite 1002
Boston, MA 02111
USA

T: +1 800 552 3083 (toll free)
T: +1 (0)617 395 4051
E-mail: cabi-nao@cabi.org

A catalogue record for this book is available from the British Library, London, UK.

Library of Congress Cataloging-in-Publication Data

The CABI encyclopedia of forest trees.
p. cm.
Includes bibliographical references and index.
ISBN 978-1-78064-236-9 (alk. paper)
1. Trees--Encyclopedias. 2. Forest plants--Encyclopedias. I. C.A.B. International.

QK474.87C33 2013
582.1603--dc23

2013010068

ISBN-13: 978 1 78064 236 9

Commissioning editor: Victoria Bonham
Editorial assistant: Alexandra Lainsbury
Production editor: Shankari Wilford

Typeset by Columns Design XML Limited, Reading
Printed and bound in the UK by CPi Group (UK) Ltd, Croydon, CR0 4YY

Compilers

Christopher Dixon, University of Oxford, UK
George E. Fyson, Consultant, Auckland, New Zealand
Nick Pasiecznik, Agroforestry Enterprises, Cussy en Morvan, France
Andrew Praciak, Consultant, Carne, Co. Wexford, Republic of Ireland
Keith Rushforth, Chartered Arboriculturalist and Arboricultural Consultant, Cullompton, Devon, UK
Marieke Sassen, Plant Production Systems Group, Wageningen University, The Netherlands
Douglas Sheil, Institute of Tropical Forest Conservation, Kabale, Uganda; School of Environment, Science and Engineering, Southern Cross University, Lismore, New South Wales, Australia; and Center for International Forestry Research, Bogor, Indonesia.
Cristina Sousa Correia, University of Nottingham, UK
Claire Teeling, University of Birmingham, UK
Miriam van Heist, Institute of Tropical Forest Conservation, Kabale, Uganda; and School of Environment, Science and Engineering, Southern Cross University, Lismore, New South Wales, Australia

Picture credits

Special thanks go to the providers of the following images:

Abies amabilis, *Abies lasiocarpa*, *Acer macrophyllum*, *Alnus rubra*, *Chamaecyparis nootkatensis*, *Larix occidentalis*, *Pinus ponderosa*, *Populus tremuloides*, *Thuja plicata*, *Tsuga heterophylla*, *Tsuga mertensiana* © K. Klinka

Acacia melanoxylon, *Alstonia scholaris*, *Eucalyptus camaldulensis*, *Eucalyptus camaldulensis* var. *obtusa*, *Melaleuca leucadendra* © Australian Tree Seed Centre

Acacia nilotica, *Leucaena leucocephala*, *Prosopis juliflora*, *Schinus molle* © N. Pasiecznik

Adansonia digitata, *Baikiaea plurijuga*, *Pterocarpus angolensis* © D. Gwaze

Agathis australis, *Cupressus macrocarpa* © G. Fyson

Albizia lebbeck, *Eucalyptus robusta*, *Melaleuca cajaputi*, *Melaleuca quinquinervia* © D. Lea/CSIRO Forestry and Forest Products

Betula pendula © J. Barnett

Cedrus atlantica var. *glauca*, *Cedrus libani*, *Pinus radiata*, *Sequoiadendron giganteum* © V. Ordidge

Dalbergia nigra, *Peltophorum dubium* © S. Botelho

Enterolobium cyclocarpum, *Gliricidia sepium*, *Guazuma ulmifolia*, *Leucaena leucocephala* subsp. *glaberata*, *Peltophorum dubium*, *Prosopis juliflora*, *Robinia pseudoacacia*, *Samanea saman*, *Schizolobium parahyba* © C. Hughes

Fagus crenata © T. Kamitani

Juniperus procera © C. Couralet

Khaya senegalensis, *Parkia biglobosa*, *Terminalia superba* © D. Thies

Quercus ilex subsp. *rotundifolia*, *Quercus suber* © C. Sousa Correia

Paulownia tomentosa, *Picea abies*, *Pinus strobus*, *Terminalia catappa* © A. Pittaway

Prunus avium © C. Teeling

Quercus robur © P. Robakowski

Terminalia arjuna © D. K. N. G. Pushpakumara

Zelkova serrata © K. Yoda

A

Abies alba Mill.

European silver fir

Importance

Abies alba is important in Europe for its high-quality softwood timber and its ecological and historical value. It occurs naturally in mixed stands in montane forests, where it is found mixed with beech (*Fagus sylvatica*) or other broadleaves, and in subalpine forests, mixed with conifers such as Norway spruce (*Picea abies*), Scots pine (*Pinus sylvestris*) and larch (*Larix decidua*).

Botanical features

A. alba forms a monopodal tree, initially conical with spaced whorled branches but becoming columnar with age with a dense crown. Branches are generally short and horizontal. Forking is rare, although occasionally trees with a dominant side branch which becomes a competing stem may be seen. The lower branches will be retained where there is sufficient natural light but are lost in forests or plantations. In older trees the apical top is lost, forming a flat upper crown (known as 'stork's nest'). The crown colour is dark green with the silver underside of the foliage showing.

The average height at maturity is 25–35 m with a diameter at breast height (dbh) of 30–45 cm, although it has made 60 m in height, and old trees to 2 m diameter are recorded and can live for two or three centuries. The stem is straight and cylindrical, self-pruning in plantations with only a significant taper in isolated trees. Apart from in seedling trees, the tree has a lateral root system and only in shallow soils will the root system include surface roots. The bark in young trees is smooth and silver-grey, while in older ones the colour becomes darker and the bark surface is cracked into scaly plates with pinkish fissures. The young shoots are slender, fawn-brown, with dark brown hairs in shallow grooves. The buds are ovoid with a blunt point, not or scarcely resinous.

The upper surface of the needles is dark shiny green without any stomata or a few restricted to the groove near the tip, which is usually rounded or notched but sometimes acute, especially on coning shoots. The lower surface has two whitish-green or silvery-green waxy bands of 6–8 rows of stomata. Needles on branches in light shade are 20–30 mm long × 1.5–2.5 mm wide and set on shoots in a comb-like form ('pectinate') either side of the shoot; leaves in full sun are thicker and shorter and arranged in a brush-like form.

Male cones are greenish-yellow with purplish-red microsporophylls. The female cones (megasporophylls) are yellowish-green, sometimes with a purple tinge when they appear in late spring (May). The cones ripen through greenish-brown to brown or reddish-brown and at maturity are 10–18 cm long and 3–5 cm wide. Age of cone production is influenced by light, in isolated trees from about age 30 but in dense stands not until the trees are aged 60 years.

Seed ripening occurs in the autumn of the first year with the cones starting to fall apart sometime in the autumn. The seeds are 7–13 mm long, triangular, yellowish-brown, glossy, with resin blisters with a reddish-brown wing twice the length of seeds. Seed production is abundant every 3–4 years with only a few cones, or on a few trees in intervening years. Needle fall is continuous but with a maximum from August to November.

Distribution

A. alba has a natural distribution from the Pyrenees across Central Europe to the Carpathian and Transylvanian mountains. The distribution is discontinuous down the Apennines in Italy as far south as Calabria. There is a more continuous extension from the Dinaric Alps towards the Pindus and Rodopi/Rhodope mountains in Greece and Bulgaria, but this includes introgression with *Abies cephalonica* and is treated as a separate species, *Abies borisii-regis*. Isolated pockets of natural forest are found in Poland, central France (Massif Central) and Corsica. It is unclear whether the population in Normandy, France, is natural or the result of historic planting.

A. alba has scarcely been planted outside the natural area of distribution. However, plantations to replace degraded forests are often recorded in countries and regions within the natural range. Underplanting to replace low-quality coppice or to transform pure *P. abies* stands into mixed forests, is a practice often recorded in central Europe and – outside the natural area of distribution – in Ukraine.

Silviculture characteristics

A. alba is found over a wide range of altitudes, from sea level in Poland, to 1200–1700 (maximum 1900) m at continental sites in the Alps. The best growth occurs where mean annual rainfall exceeds 1500 mm and mean annual temperature is about 9°C. It will tolerate mean annual rainfall as low as 700 mm, but under these conditions growth is slow. Both regular uniform and seasonal (summer/winter/bimodal equinoctial) rainfall regimes are tolerated. Climatological studies suggest that *A. alba* is unsuitable for areas where the absolute minimum temperature falls below ca −25°C; the main limiting factors are a lack of summer heat and adequate moisture during the growing season, and frost damage, particularly to new growth.

A range of diverse soil types and substrata is tolerated, including on limestone and silicate substrates. Relatively fertile sites are preferred, the species demanding mainly deep and moist soils. Stagnant water, clay soils and a soil pH of less than 4 are not tolerated. Silver fir is not suited to growth on ridge tops because of wind damage to the foliage.

Silviculture practice

A. alba seedlings are not tolerant of spring forest and for natural regeneration a shelter wood system is necessary. In the nursery the seeds germinated better with a pretreatment of 30 days. Growth in the first year is slow, and planting stock is usually 2 + 2 transplants.

Current management of pure stands in parts of Europe is now often based on a close-to-nature type of silviculture. This involves creating a forest succession through the introduction of broadleaved species and is encouraged by thinning to develop a more natural forest.

Where top-quality timber is the primary objective, management is more intense, aiming for trees with a dbh of 45–60 cm on a rotation of up to 90 years. This will include five to seven thinnings with final stand density between 160 and 250 stems/ha. Mean annual increment (MAI) in natural forests is in the range of 10–12(–15) m^3/ha on good sites, but lower on less fertile or drought-prone sites. The management of stands affected by disease should be founded both on low plantation density and on early thinning. Studies for fibreboard found biomass production can reach 8 t/ha dry matter in stands to age 50 years, gradually reducing to about 6 tonnes between ages 50 and 100 years.

A. alba has not been widely planted in forestry outside the natural area of distribution but has become established in western Britain from historic plantings.

Diseases and pests

Butt rot and windthrow (uprooting) may occur in *A. alba* affected by the root fungus *Heterobasidion annosum*. Honey fungus *Armillaria mellea agg.* is an important source of infection. Breakage by wind and snow is more frequent in stands that are too dense. Dry years can result in heavy growth reductions, which may lead to particular pathological stresses.

Forest decline ('Tannensterben' in German) in *A. alba* has been recorded at various sites across Europe, with symptoms and damage attributed to an interaction of factors including air pollution, loss of genetic variability since the last glaciation and artificial silvicultural practices.

Needles and/or bark of both seedlings and adult trees are infected by phytophagous insects (*Mindarus abietinus* and *Dreyfusia nordmannianae*) which are sometimes responsible for serious damage, and have restricted past use in the UK and Ireland. *Epinotia nigricana* and *Cinara pectinatae* are insects which infect buds (serious damage only in young seedlings) and bark respectively. Of the xylophagous insects, *Cryphalus piceae* is considered to be the most detrimental to silver fir.

Natural regeneration (seedlings, advance growth) is often damaged or killed by ungulate browsing, and also by territory marking by deer.

Uses

Stands of *A. alba* are often important in terms of landscape or historical value, and their conservation as a part of a cultural landscape is sometimes emphasized. Conservation is a priority for mixed species stands too, and these stands are often rich in floristic diversity.

Natural regeneration in pure planted silver fir can be difficult. The introduction of mesophyll broadleaves (*Acer pseudoplatanus*, *Acer platanoides*, *Ulmus glabra*, *Fraxinus excelsior*, *Tilia platyphyllos*, *Castanea sativa*) in very old stands can be the basis for long-term natural regeneration of *A. alba*. Management of ungulate populations, especially deer, is necessary.

A. alba wood is utilized for similar products as Norway spruce (*P. abies*), including furniture, building timbers, framework, pulp, particle-, and blockboard. In the past, *A. alba* was in great demand for ship masts. The wood is less valuable than that of Norway spruce, because it is more difficult to work, having hard knots. Recorded wood densities range from 240 to 719 kg/m^3 at 12–15% moisture content.

Honey from *A. alba* has high commercial value. The tree was the source of Strasbourg turpentine used in the manufacture of paints and varnishes. Oil of turpentine, distilled from wood and needles, was used medicinally to treat orthopaedic problems, and aromatic spirits are still produced.

Further reading

Debreczy, Z. and Rácz, I. (2012) *Conifers Around the World*. Volume 1, DendroPress Ltd, Budapest.

Farjon, A. (2010) *A Handbook of the World's Conifers*. Volume 1, Brill, Leiden/Boston.

Abies amabilis Douglas ex J. Forbes

Pacific silver fir

Importance

Abies amabilis (Plate 1) is one of the major tree species of the Pacific North-west of North America, particular on high-elevation sites. Its large leaf biomass, shade tolerance, low crown spatial requirements, low taper and thin bark make it a productive species on a wide range of sites, except warm and water-deficient ones. It is a desirable component in mixed-species stands, mainly with *Tsuga heterophylla*, *Tsuga mertensiana*, *Thuja plicata* or *Cupressus* (*Callitropsis*/*Chamaecyparis*) *nootkatensis*. Natural regeneration, especially on montane and subalpine sites, is preferable to planting. It has been trialled on a minor scale in Europe.

Botanical features

A. amabilis ranges from a small tree in the subalpine parkland to 60–80 m at sea level with a diameter of 1 m; on high elevation sites it may live for more than 500 years. The trunk is straight, cylindrical and typically clear of branches for half its height. The crown is slender and compact with short initially horizontal branches. The branches are moderately persistent after they die. The bark is smooth, light grey, blotched with white patches, becoming scaly and grooved and reddish-brown at the base of mature trees. The twigs are yellowish-brown to greyish-brown, and initially densely covered with short brown hairs.

A. amabilis develops a good root system in well-drained soils but is flat and plate-like in poorly drained soils. Fine roots in the organic layers form a mycorrhizal association with *Cenococcum graniforme*.

The needles are 2–4 cm in length with a notched tip and retained for up to 8 years. The upper surface is shiny dark green and grooved with the underside bearing two silvery white bands of stomata. The needles spread beneath the shoot, curving upward at the sides and pointing forward above.

Pollen cones (micro-sporangiate strobili) are 1–1.5 cm and red, with the seed cones (mega-sporangiate strobili) purple. The cones ripen through purple to brown and are ovoid–cylindrical, 9–16 cm long, with scales as wide as long and the bract scales included. Seed dispersal is by wind between mid-September and October, with most seeds falling within 100 m.

Distribution

A. amabilis is a Pacific North-Western American species. Its native range extends from south-eastern Alaska to coastal British Columbia and Vancouver Island, along the western and upper eastern slopes of the Cascade Range in Washington and Oregon, throughout the Olympic Mountains and sporadically in the Coast Ranges of Washington and northern Oregon. Near Crater Lake, Oregon, it disappears from the Cascade Range but reappears at a few locations in the Klamath Mountains of north-western California. The major portion of its range lies between latitudes 43° and 56°N.

A. amabilis was included in a network of experimental plantations by the German Forest Research Institute in 1891; however, it has not become of real importance as a timber crop. In the British Isles, growth is best in cool, high-rainfall areas, notably western Scotland. It is grown in arboreta and parks in Western Europe.

Silviculture characteristics

Seed production begins at an age of 20–30 years with a low to moderate percentage of sound seed (7–51%) and a good crop generally every 3 years. In low light settings the potential for natural regeneration is high and it regenerates well on organic substrates. However, most seeds fall within 100 m, restricting the potential for natural regeneration in the open. The growth rate in the first 6 or more years is low to very low in heavy shade, and usually less than 10 cm/yr when planted. Advance regeneration responds well to release, with a 1–2 year delay in radial growth and 3–5 years in height growth.

It self-prunes in dense stands, with closely spaced, short, narrow crowns; it casts a dense shade, restricting understorey vegetation. Potential productivity is high but decreases with elevation. It is long-lived, up to 800 years on high elevation (subalpine) sites. It has a high tolerance to shade and snowpack, but does not tolerate frozen soils.

A. amabilis grows on a wide range of predominantly acidic soils in the Pacific North-west but growth is reduced on poorly drained or shallow, water-deficient soils, and on high elevation sites. It tolerates surface accumulations of organic matter, and moderate to extremely low base saturation, and often low available nitrogen.

A. amabilis grows at sea level along the coast from Alaska to the Olympic Peninsula but farther inland it is absent at lower elevations. In Alaska it occurs up to 300 m, in British Columbia to 1525 m but in the western Cascade Range of Washington it may be found from 240–1830 m. In the Olympic Mountains, it is the predominant montane species up to 1400 m, with lower limits at sea level on the west side and at 360 m in the central mountains. It is found between 610–1830 m in the Cascade Range in Oregon as far south as the divide between the Rogue and Umpqua Rivers. On the east side of the Cascade Range, it is confined to high elevations, down to 1000 m in Washington and 1160 m in Oregon.

A. amabilis does not tolerate a soil moisture water deficit, and frost resistance of roots is low, but it tolerates water surplus and nutrient- (nitrogen) deficient soil.

Silviculture practice

Limited clear-cutting, patch-cutting, strip shelterwood and group-selection systems are viable for the establishment and growth of *A. amabilis*.

Propagation by seed is the only contemporary method for regenerating *A. amabilis*. Natural regeneration occurs on a variety of seedbeds, and is preferred to planting when considering the slow initial growth on high elevation sites, and growth of released advance regeneration is more rapid than early growth from seeds or planting. Germination capacity averages 20–30% with 17–45,000 cleaned seeds/kg. The seeds are heavy and when dispersed by the wind, they carry <100 m from the parent tree.

A. amabilis germinates in spring after overwintering under snow. Seedlings in low light continue to grow very slowly, taking a century to make 1–2 m. Such advance regeneration has flat-topped crowns but resists being flattened by litter and heavy, wet snow.

In even-aged, pure or mixed stands the stocking can be more than 2470 stems/ha with understorey vegetation nearly eliminated by shade. Lower limbs become shaded and die, creating branch-free boles, but older stands will let more light into the understorey, allowing development of shrubs and advance regeneration.

Early re-spacing to 495–740 stems/ha is practised to increase radial growth, and such trees also respond to fertilization.

Site preparation is not required if advance regeneration is present; slash removal may be necessary in some situations to facilitate planting on high brush-hazard sites.

Growth rates vary with climates, elevation and latitude, with early height growth from seeds generally very slow.

Diseases and pests

A. amabilis is easily killed by fire because of its thin bark. It has a lower resistance to windthrow than *Pseudotsuga menziesii* and *T. plicata*, but is resistant to snow breakage with moderate resistance to frost damage.

It is susceptible to insect damage. Infestation of cones by seed chalcids (*Megastigmus pinus* and *Megastigmus lasiocarpae*) and cone maggots (*Earomyia abietum*) can be high during good seed years. Western hemlock looper (*Lambdina fiscellaria lugubrosa*) and western blackheaded budworm (*Acleris gloverana*) are serious defoliators in British Columbia, Canada. The silver fir beetle (*Pseudohylesinus serieceus*) and fir root bark beetle (*Pseudohylesinus granulatus*) can be very destructive together and in combination with the root-rotting fungi *Armillaria mellea agg.*, *Heterobasidion annosum*, *Inonotus weirii* and *Poria subacida*. Overmature trees are prone to heart rot, primarily by the Indian paint fungus (*Echinodontium tinctorium*) and the bleeding conk fungus (*Haematostereum sanguinolentum*).

The balsam woolly adelgid (*Dreyfusia piceae*) is the most devastating killer of *A. amabilis*. Attacks on the crown by this insect result in swelling or 'gouting' of branch nodes, loss of needles and reduced growth for many years, with attacks on the stem usually killing within 3 years. There is no effective direct control method for forest stands.

Uses

As a common tree on montane and subalpine sites, *A. amabilis* is a major component of watersheds, wilderness and recreational areas. Its beauty and ability to withstand or respond to human impact make it a suitable species for multiple-use management.

The wood is light, soft, weak, creamy-white to yellowish-brown, with no distinct heartwood. It has a density of 433 kg/m^3 at 12% moisture content. *A. amabilis* is marketed with *T. heterophylla* (often under the name hem-fir) and is typically used for construction framing, sub-flooring and sheathing. It is used for construction plywood although it is not as strong as *P. menziesii*. Because of its light colour and lack of odour, gum and resin, *A. amabilis* is well suited for container veneer and plywood, and can be used for interior finish or for poles. It gives a good yield of strong pulp from both mechanical and chemical processes.

A. amabilis is a minor Christmas tree species, and its boughs are occasionally used for decorative greenery.

Further reading

Crawford, P.D. and Oliver, C.D. (1990) *Abies amabilis*. In: Burns, R.M. and Honkala, B.H. (technical coordinators) *Silvics of North America, Volume 1*. USDA Agriculture Handbook 654. USDA, Washington, DC, pp. 17–25.

Farjon, A. (2010) *A Handbook of the World's Conifers, Volume 1*. Brill, Leiden/Boston.

Farjon, A. and Rushforth, K.D. (1989) A classification of *Abies* Miller (Pinaceae). *Notes from the Royal Botanic Garden, Edinburgh* 46, 59–79.

Klinka, K., Worrall, J., Skoda, L. and Varga, P. (1999) *The distribution and synopsis of ecological and silvical characteristics of tree species of British Columbia's forests*. Canadian Cartographics Ltd, Coquitlam, Canada.

Abies concolor **(Gordon & Glend.) Lindl. ex F.H. Hild.**

Rocky Mountain white fir

Importance

Once considered relatively less desirable for timber than other species, *Abies concolor* is now recognized as a productive and valuable species occurring as a stand dominant over large areas of montane forest habitat.

Botanical features

A. concolor is a large forest tree to 40–55 m or more with a dbh of 100 to >160 cm. In young trees the habit is conical with numerous horizontal tiers of branches. Mature trees form a more cylindrical crown with a rounded top.

The bark of young trees is smooth and grey, becoming divided into whitish ridges separated by deep furrows in mature trees, giving the tree its common name.

The needles are laxly set on the shoots and are curved or bowed. They are flattened in section, 3–9 cm in length, with a blunt to acute tip. They are grey-green to bluish-grey-green and similar on both sides (hence 'concolor'). Needles are usually retained for 5–8 years. In *A. concolor* var. *lowiana*, the leaves are spreading or widely parted on the shoot, and with fewer stomata on the upper surface, with two white bands of stomata separated by a green band on the lower surface.

Male cones shed pollen generally in May–June. The fruiting cones are cylindrical and 6–13 cm in length, and 3–5 cm wide. They are greenish-violet-purple when growing, maturing to brown. The seeds are about 15–20 mm in length, with a single broad wing.

Adequate to good cone crops occur every 2–5 years. Trees first produce cones at about age 40 years.

Distribution

The natural range of *A. concolor* var. *concolor* extends from the Sierra Nevada to the Sierra San Pedro Martir of Baja California, with scattered disjunct populations in the mountain ranges of the eastern Mojave Desert and Great Basin. It is common in the central and southern Rocky Mountains from extreme south-eastern Idaho, Utah, southern Colorado, Arizona, New Mexico and into northern Mexico. In the Cascade Range and coast ranges of southern Oregon south into California *A. concolor* var. *lowiana* is found.

A. concolor has been widely planted outside its natural range, particularly in Europe. In England, its use is largely decorative, but it has been tried in plantation forestry in other areas.

Silvicultural characteristics

A. concolor is relatively shade- and drought-tolerant. It thrives in the absence of ground fires, against which its thin bark provides no protection.

It inhabits a natural range characterized by long, cold winters with moderate to heavy snowpacks and short growing seasons. Summer temperatures are moderate, reaching highs of 30–35°C, tolerating winter low temperatures down to −30°C. In California and Oregon, where mean annual rainfall is from about 900 up to 1900 mm (mostly falling as snow) the majority of rainfall occurs from October to March, with 3–5 months of summer drought but in the Rocky Mountains populations of *A. concolor* receive only 500–900 mm of annual rainfall but with much of this falling in the summer. Years with low rainfall are often correlated with high growth rates, apparently because of early snow melt and thus a longer growing season.

A. concolor grows in montane habitats on a variety of soil types and the ecological range is more dependent on soil moisture conditions and temperature regimes than on soil pH or nutrient availability. Growth and development of *A. concolor* is best on moderately deep and well-drained sandy-loam to clay-loam soils, with higher altitude stands responding strongly to nitrogen fertilization. It can also be found growing successfully on steep and unstable slopes. Altitudinal ranges increase from north to south in response to temperature conditions, from about 600 m in central Oregon to 1200–2200 m in the Sierra Nevada and to 3000 m in the San Bernardino Mountains of southern California. In the Rocky Mountains, *A. concolor* is most common at 2100–2700 m, but reaches as high as 3400 m in central Colorado.

A. concolor is commonly a co-dominant climax forest species of mixed conifer forests, giving way on wet sites in the upper montane zone to *Pinus contorta*. In the Sierra Nevada of California it is often co-dominant with *Pinus ponderosa*, *Pinus jeffreyi*, *Pinus lambertiana* and *Calocedrus decurrens*, and it mixes with *Abies magnifica* at higher altitudes. Over its range in Arizona, New Mexico, and the central Rocky Mountains, *A. concolor* may occur as a climax component in a variety of major forest habitats, with associates including *Picea engelmannii*, *Picea pungens*, *P. ponderosa*, *Populus tremuloides*, and *Pseudotsuga menziesii*.

Silvicultural practice

Germination rates of *A. concolor* under standardized conditions are relatively low at 15–35%. Germination is generally increased by stratification for 30–90 days at about 5°C. Heavy seed crops are produced every 3–9 years where in an exceptional year seed production can exceed 1.5 million seeds/ha. *A. concolor* can be rooted from cuttings taken from juvenile trees.

Natural seed fall is commonly sufficient for regeneration after logging if site preparation is appropriate. In a nursery, seeds should be sown in the autumn in well-drained sandy loam or loam soil, mulching over for winter in cold climates or after stratification. Young seedlings are sensitive to strong light and do best with partial shading for their first six months of growth. Cutworms (Noctuidae) may be a major problem in nursery seedlings.

Natural regeneration following clear-cutting is usually successful so long as the maximum downwind width of openings does not exceed 1.5–2 times the height of trees left as a seed source. Although saplings are highly shade-tolerant, the species responds to release from suppression by rapid growth, with best growth in full sun. The shade tolerance makes uneven stand management possible but less productive.

Until the early 1960s *A. concolor* was considered relatively undesirable in managed forest stands, but its commercial value has since increased greatly. Stands in California, Oregon and Washington on good sites reach 27 m in height at age 100 years and produce 1066–1372 m^3/ha, with trees on average sites reaching 18 m and volumes of 633–805 m^3/ha.

Genetic studies show *A. concolor* is a highly variable species, with both inter- and intra-populational variation, with a correlation of characters with the latitude and altitude of the seed provenance.

Four major morphological/geographical races of *A. concolor* are found: (1) a central Oregon, north-western California race - indicating an origin from hybridization and introgression of race 2 morphologies with *Abies grandis*; (2) a Cascade Range race in south-central Oregon and north-eastern California; (3) a southern California, Baja California, Arizona race; and (4) an eastern Nevada, western Utah race. Geographic provenances have also been shown to vary in susceptibility to pathogen attack.

Diseases and pests

Root rot from *Heterobasidion annosum* is present in all stands and spreads from tree to tree by root contact; it only rarely kills *A. concolor* directly but damage increases water stress and reduces tree vigour and thus susceptibility to bark beetle infection and susceptibility to windthrow and stem breakage. Several other heart rots can cause significant losses of older growth trees.

Mistletoe infections are common throughout the range. Heavily infested trees have reduced growth rates and fungi may enter through open mistletoe stem cankers. Best silvicultural practice is to remove infected parts of trees and control stocking densities to allow the young trees to outgrow the spread of the mistletoe.

The most damaging insect pest on *A. concolor* is the fir engraver (*Scolytus ventralis*), which causes severe damage throughout the range of the species. Susceptibility to attack by fir engraver is increased by conditions which reduce tree vigour, including root disease fungi, mistletoes, drought stress, fire damage and overstocking. Other species of bark beetle (*Scolytus*), *Pseudohylesinus*, round-headed fir borer (*Tetropium abietis*), and the flat-headed fir borer (*Melanophila drummondi*) may attack and the only effective control is to maintain stand vigour.

The Douglas-fir tussock moth (*Orgyia pseudotsugata*), is a destructive defoliator. Western spruce budworm (*Choristoneura occidentalis*) is another defoliator but most outbreaks are short-lived, and the white fir needleminer (*Epinotia meritana*) may also be a significant foliage feeder.

Seed miners are the major cause of damage to seeds, periodically causing major damage.

A. concolor is less sensitive to drought than *A. magnifica* and *A. grandis*, but sustained drought conditions can make this species susceptible to bark beetle attack. Saplings and poles of *A. concolor* are killed by ground fires, but older trees become resistant with age. Large basal fire scars in old growth stands or logging scars can expose trees to a variety of pathogens. Young *A. concolor* may be sensitive to spring and autumn frosts but cold damage to mature trees is less common, but may cause frost cracks and ring shake.

Uses

A. concolor provides important forest cover for montane habitats throughout its range, where it commonly occurs in mixed conifer stands.

A. concolor produces a general purpose wood with widespread use for construction grade framing materials and plywood. Wood density is 370–390 kg/m^3.

Young trees are widely harvested from both plantations and natural regeneration for sale as Christmas trees.

Further reading

Blada, I. and Panea, T. (2012) Improvement of grafting procedures for the ornamental species: II. *Abies concolor* [(Gord. & Glend.) Lindl]. *Annals of Forest Research* 55, 25–31.

Furniss, R.L. and Carolin, V.M. (1977) *Western Forest Insects*. USDA Miscellaneous Publication No. 1339. US Department of Agriculture, Washington, DC.

Mason, R.R. and Wickman, B.E. (1991) Integrated pest management of the Douglas-fir tussock moth. *Forest Ecology and Management* 39, 119–130.

Wickman, B.E. (1988) *Tree Growth in Thinned and Unthinned White Fir Stands 20 Years after a Douglas-Fir Tussock Moth Outbreak*. Research Note No. PNW-RN-477. Pacific Northwest Research Station, USDA Forest Service, Portland, Oregon, USA, pp. 11.

Abies grandis (Douglas ex D. Don) Lindl.

Grand fir

Importance

Abies grandis is a fast-growing and very large fir of secondary and old-growth forests of western North America and is very productive on a wide range of sites. It has been introduced to Europe where it is grown in arboreta and on a limited scale in forestry.

Botanical features

A. grandis grows to 30–50(–81) m tall with dbh of 0.7 –>1.3 m, living for up to 300 years. The crown is cylindrical or oval, becoming irregular with a rounded top in old trees. Branches are horizontal with the lower ones drooping. The smooth, greyish-brown bark with resin blisters of young trees matures to thick, scaly and ridged. The twigs are slender, light brown to dark orange-brown and finely haired.

The flexible needles are 2–6 cm long by 1.5–2.5 mm wide, and set spreading either side of the shoot. Those in full sun are not arranged with a central parting. The cross section is flat with a groove but no stomata above with two whitish rows of stomata beneath; the apex is notched, sometimes with a few stomata. The resin canals are small and marginal.

The pendent pollen cones are ovoid or cylindrical and may be bluish-red, purple, orange, yellow or green.

The seed cones are erect on the uppermost branches. They are dark blue or purple before ripening brown and are cylindrical to narrowly ovoid, blunt and mostly 6–7 × 3–3.5 cm. The seed scales are 2.5–3 cm wide, with the bract scales concealed. Seeds are 6–8 × 3–4 mm with a wing about 10 mm long.

Flowering is in spring with the cones ripening by autumn when they break up to release the seeds.

Distribution

A. grandis is a western North American species occurring in both the Pacific (coastal or maritime) and Cordilleran (interior or continental) regions from latitude 39–51°N and at a longitude of 125–114°W. The coastal range is from southern British Columbia on the lee side of Vancouver Island and the adjacent mainland, the interior valleys of western Washington and Oregon, and in north-western California as far south as Sonoma County. Inland, its range extends from the Okanagan and Kootenay Lakes in British Columbia south through eastern Washington, northern Idaho, western Montana and north-eastern Oregon.

A. grandis is either an early or late successional species on different sites within its range. On dry sites and in dry climates, it is shade-tolerant and eventually dominant. In the wet climates of western British Columbia, Washington and Oregon, it grows with fast-growing, shade-tolerant and long-lived species such as *Abies amabilis*, *Tsuga heterophylla*, and *Thuja plicata*, initially overtopping them but dying out after 200–300 years.

Occasionally forming pure stands, it is much more common in mixed coniferous and broadleaf forests. West of the Coastal Mountains and Cascade crest in Washington and Oregon the main associates are *A. amabilis*, *Chamaecyparis lawsoniana*, *Picea sitchensis*, *Pseudotsuga menziesii*, *T. plicata* and *T. heterophylla*. In the interior forest to the east the principal associates are *Abies concolor*, *Abies magnifica*, *Calocedrus decurrens*, *Larix occidentalis*, *Pinus lambertiana*, *Pinus contorta*, *Pinus monticola*, *Pinus ponderosa*, *P. menziesii*, *Quercus garryana*, *T. plicata* and *T. heterophylla*.

A. grandis was introduced to Europe by David Douglas shortly after 1825. It was planted in arboreta, parks and on a minor scale in forests. In 1891, it was included into a network of experimental plantations by the German Forest Research Institute. Although not as important as *Picea sitchensis*, *P. contorta* or *P. menziesii*, it is considered a potentially productive species.

Silvicultural characteristics

A. grandis occurs in cool to temperate climates (maritime to drier, cool, warm mesothermal, continental) with a mean annual precipitation from 510 to 2540 mm (but down to 360 mm in the Blue Mountains of eastern Oregon). Most of the precipitation in continental areas of its range occurs during winter, generally with only 15–25% occurring during the growing season. The mean annual temperature range is 6–10°C and the average growing season temperature is 14–19°C. The frost-free season varies, from about 60 to >250 days, and frosts may occur in any month at inland locations within its range.

A. grandis grows predominantly on podzols derived from a variety of parent materials. It can tolerate a minor water deficit as well as water surplus and a fluctuating water table.

On Vancouver Island and the adjacent mainland of British Columbia it grows from sea level, but at around 460 m it is replaced by *A. amabilis*. In western Oregon, the upper elevation limit increases to 900 m, and in California to 1525 m. In the eastern Cascades, the range is 900–1525 m, from 460 m to 1830 m in the Blue Mountains of Oregon, and in central Idaho from 1220 to 1675 m.

Silvicultural practice

There are no recognized varieties of *A. grandis*, although five fairly distinct climatic forms have been identified. Seed sources west of the Cascade crest (Washington, Oregon) are preferred for planting in the UK and Europe. *A. grandis* hybridizes with *A. concolor* with a broad zone of introgression from the Klamath Mountains of northern California through south-western Oregon, and through the Oregon Cascade Range into north-eastern Oregon and west-central Idaho.

A. grandis is moderately windfirm, making it suitable for clear-cutting, patch-cutting, shelterwood or selection systems; its shade-tolerance makes it a desirable component in mixed-species stands including with shade-intolerant trees such as *P. ponderosa* and *L. occidentalis*, or shade-tolerant trees, such as *T. plicata* and *T. heterophylla*.

Seed production begins at about 20 years and increases with age. The yield of cleaned seed ranges from 26,200 to 63,100/kg. The seeds are large-winged and dispersed by wind and rodents, mainly in early autumn, remaining viable for only one winter period.

Seedlings are relatively resistant to drought and initial survival and growth is favoured by moderate shade. By age 20 the best growth is in the open. Understorey-grown saplings may respond positively to release where they have been managed to develop depth to the crown.

Early height growth from seeds is generally good and *A. grandis* ranks among the most productive species in the associations in which it grows, due to its large leaf biomass, shade-tolerance, moderate crown spatial requirements, low taper and thin bark. Yields at age 100 years range from 476 to 1330 m^3/ha. Using height as the site index, after 50 years the site index is about 40 m on productive sites.

Diseases and pests

A. grandis is rated medium in fire resistance, although this is influenced by habitat. Frost cracks, drought cracks and lightning scars appear more frequently than in associated vegetation and can contribute to the spread of fungal infections.

Indian paint fungus (*Echinodontium tinctorium*) is the most destructive fungus affecting this species in Washington and Oregon, but is rare where rapid growth rates close branch stubs quickly. *E. tinctorium* causes a brown stringy rot, but in early stages decay appears as a light brown or water-soaked stain in the heartwood. Decay is closely related to logging scars, frost cracks, broken tops and other mechanical injuries. *Armillaria* spp. and *Inonotus weirii* are important root-rot fungi.

Western spruce budworm (*Choristoneura occidentalis*) and Douglas fir tussock moth (*Orgyia pseudotsugata*) have caused widespread defoliation, top kill and mortality to grand fir. Early instar larvae of *C. occidentalis* mine and kill the buds, whilst late instar larvae are voracious and wasteful feeders, often consuming only parts of needles, chewing them off at their bases. Western balsam bark beetle (*Dryocoetes confusus*) and fir engraver (*Scolytus ventralis*) are the principal bark beetles. Fir cone moth (*Barbara* spp.), fir cone maggots (*Earomyia* spp.), and several seed chalcids destroy cones and seeds. The balsam woolly adelgid (*Adelges piceae*) is a serious pest of *A. grandis* in western Oregon, Washington and south-western British Columbia where feeding by this aphid causes twigs to swell or 'gout' at the nodes and the cambium produces wide, irregular annual growth rings consisting of reddish, highly lignified, brittle wood.

Uses

A. grandis is a large component of many municipal watersheds, wilderness areas and recreational areas where its beauty and tolerance of human impact make it a suitable species for multiple-use management.

A. grandis wood is light, soft, relatively weak, odourless, with a straight and even grain, and medium to somewhat coarse texture; density is 350–370 kg/m^3. The heartwood is indistinct from the sapwood and is usually whitish or light brown to yellowish-brown. The heartwood is reported to be rather difficult to treat with preservatives. It is weaker and more prone to decay than other Pacific conifers but it has good boring, gluing, moulding, nailing, screwing and planing properties, and is a valued pulpwood and timber.

The luxuriant foliage, crown symmetry and shiny deep green colour of the foliage make it one of the preferred species for Christmas trees grown in the Pacific North-west.

Further reading

Foiles, M.W. (1965) Grand fir, *Abies grandis* (Dougl.) Lindl. In: Fowells, H.A. (compiler) *Silvics of Forest Trees of the United States*. USDA Agriculture Handbook 271, Washington, DC, USA, pp. 19–24.

Foiles, M.W., Graham, R.T. and Olson Jr, D.F. (1990) *Abies grandis*. In: Burns, R.M. and Honkala, B.H. (technical coordinators) *Silvics of North America, Vol. 1*. USDA Agriculture Handbook 654, Washington, DC, USA, pp. 52–59.

Klinka, K. (2007) *Abies grandis* in Canada and in the United States of America. *Forst und Holz* 62, 10–13.

Klinka, K., Worrall, J., Skoda, L. and Varga, P. (1999) *The distribution and synopsis of ecological and silvical characteristics of tree species of British Columbia's forests*. Canadian Cartographics Ltd, Coquitlam, British Columbia, Canada.

Steinhoff, R.J. (1978) Distribution, ecology, silvicultural characteristics, and genetics of the *Abies grandis-Abies concolor* complex. In: Proceedings of the IUFRO Joint Meeting of Working Parties, Vol. 2: Lodgepole Pine, Sitka Spruce, and Abies Provenances. British Columbia Ministry of Forests, Vancouver, British Columbia, Canada, pp. 123–132.

Abies lasiocarpa (Hook.) Nutt.

Rocky Mountain fir

Importance

Abies lasiocarpa (Plate 2), the Rocky Mountain or subalpine fir, is the smallest of eight species of true fir indigenous to western North America. It is a common species in old growth, secondary growth, high elevation or continental forest. It is productive on a wide range of sites. Its light, soft, odourless wood is used mainly for lumber, and its seed is a favourite food of squirrels.

Botanical features

A. lasiocarpa ranges in size from prostrate shrub to large tree, occasionally over 30 m tall with a dbh of 60 cm, and living for about 250 years. The trunk is cylindrical, and the crown long, narrow, dense, conical and terminating in a conspicuous spike-like point. The branches are short and drooping. The bark is smooth, ash-grey, blotched with raised resin blisters when young, breaking into irregular greyish-brown scales when older (Plate 64). The twigs are stout, brownish, becoming ash-grey, remaining pubescent for several years.

In moderately deep, freely drained soils, *A. lasiocarpa* develops a relatively deep root system; a flatter, plate-like root system is formed in poorly drained soils. Fine roots occupy predominantly surface organic layers and are associated with mycorrhizae.

The needles are evergreen, 20–40 mm long, with a rounded or notched tip, greyish-green to light bluish-green, crowded, curving upward from the sides of the twig and standing almost erect. The resin ducts are large and remote from the surface.

A. lasiocarpa is monoecious, with male flowers borne in pendulous clusters from the axils of the needles on lower branchlets and female flowers borne erect and singly on the uppermost branchlets of the crown. Male flowers ripen, with pollen being wind disseminated, during late spring and early summer. The cones, which open during mid-August to mid-October, are erect, cylindrical, 5–13 cm long, purple to greyish-brown and resinous. They disintegrate when ripe, the scales falling away with the large, winged seeds, leaving only a central, spike-like axis. Nearly all seeds are wind dispersed.

Distribution

A. lasiocarpa is a major component of the high-elevation, continental forest in western North America. Its range extends from 32°N in Arizona and New Mexico, USA, to 64°N in Yukon Territory, Canada.

In the Pacific region, the range extends from south-eastern Alaska (the north-western limit); east to central Yukon Territory (the northern limit); south through British Columbia along the east slopes of the Coast Range to the Olympic Mountains of Washington, and along both slopes of the Cascades to southern Oregon. It also occurs on Vancouver Island and is found locally in north-eastern Nevada and north-western California.

In the Cordilleran region, *A. lasiocarpa* extends from the interior valleys of British Columbia west of the Continental Divide and south of the Peace, south along the high elevations of the Rocky Mountain system to southern New Mexico and Arizona. *A. lasiocarpa* var. *arizonica* is found mixed with var. *lasiocarpa* in south-western Colorado, northern, western and south-western New Mexico, and in the high mountains of Arizona.

A. lasiocarpa grows in pure stands or in various mixed-species stands. It is occasionally a pioneer species and present in early, intermediate and late stages of secondary succession; it is a minor component of old-growth in montane forests and a major component in subalpine forests.

In western USA, *A. lasiocarpa* is recognized as forming its own zone as well as forming a minor component in many other vegetation zones.

In the Pacific region, *Tsuga mertensiana* is its most common associate, while in the Cordilleran region, where *A. lasiocarpa* frequently extends to the timberline, its most constant associate is *Picea engelmannii*.

In contrast to other *Abies* species, *A. lasiocarpa* is grown mainly outside of its native range only in arboreta and parks.

Silviculture characteristics

A. lasiocarpa has an intermediate capacity for reproduction; seed production begins after 20 years. Vegetative reproduction by layering occurs on severe sites such as lava flows and talus slopes or near the timberline. The seed dissemination capacity is low, with a maximum dispersion of <100 m from the parent tree.

The potential for natural regeneration under low illumination is high, but much lower in the open, due to the short dispersal distance. The initial growth rate (up to 5 years) is low, particularly in low light conditions, but when planted in the open the potential growth rate is >10 cm/yr. The response of advance regeneration to release is high: 1–2 years delay in radial growth response, and 3–5 years delay in height growth response.

A. lasiocarpa has a high self-pruning capacity in dense stands and a narrow crown, producing dense secondary growth stands.

The potential productivity for this species is high; the site index after 50 years on intermediate quality sites decreases from about 30 m on montane sites to <10 m on upper subalpine sites. Depending on site and management objectives, clear-cutting, patch-cutting, strip shelterwood and group-selection systems are viable for establishment and growth. *A. lasiocarpa* can be grown in single- or mixed-species stands, preferably with less shade-tolerant species, such as *P. engelmannii*.

Silviculture practice

Propagation by seed is currently used to regenerate *A. lasiocarpa*. Seed viability is poor, with transient vitality and a mean germination capacity of 34%. Natural regeneration and/or planting containerized stock are also used to establish *A. lasiocarpa*. Natural regeneration, which occurs on various seedbeds, is preferred to planting, especially considering the slow initial growth on high elevation sites.

A. lasiocarpa begins producing cones when trees are over 1 m tall and 20 years old, but under closed-forest canopies, seed production is not significant until trees are older and taller. *A. lasiocarpa* var. *arizonica* may not begin to bear cones until about 50 years old. Maximum seed production occurs in dominant trees (150–200 years old), with good to heavy crops borne every 3 years. The seeds are fairly large, averaging 76,700/kg. Prevailing winds influence the dispersal pattern, with about half the seeds falling into openings within 30 m of the windward timber edge. Seedfall continues to diminish until about two-thirds the way across the opening, and then levels off before slightly increasing about 15 m from the leeward timber edge. Under natural conditions, seeds lie dormant under the snow and germinate the following spring. Although *A. lasiocarpa* grows under a wide range of light intensities, establishment and early survival are usually favoured by shade.

The growth of *A. lasiocarpa* is not rapid. Early shoot and root growth is very slow, with many first-year seedlings being less than 2.5 cm tall. Under favourable conditions, *A. lasiocarpa* can reach a height of 1.2–1.5 m in 20 to 40 years. Trees 25–51 cm in diameter are often 150–200 years old under closed-forest conditions. Trees older than 250 years are not uncommon, but, because the species suffers severely from heart rot, many trees either die or are complete culled at an early age. Under managed conditions, however, *A. lasiocarpa* can be as productive as its coastal subalpine counterpart, *Abies amabilis*. This productivity results from large leaf biomass, shade tolerance, very narrow crown spatial requirements (high basal area/ha), low taper and thin bark. The shade tolerance of *A. lasiocarpa* makes the species a desirable component in mixed-species stands including less shade-tolerant trees, such as *P. engelmannii* and *Picea glauca*.

Simulations for mixed *P. engelmannii–A. lasiocarpa* stands in the Rocky Mountains showed that periodic thinning to control stand density and maintain growth rates increases the yield and size of individual *A. lasiocarpa* trees. Furthermore, the growth rates for *A. lasiocarpa* are similar to those for *P. engelmannii* early in the life of the stand.

Diseases and pests

Fir broom rust (*Melampsorella caryophyllacearum*) and wood-rotting fungi are responsible for most disease losses. Important root and butt rots are *Gloeocystidiellum citrinum*, *Coniophora puteana*, *Armillaria mellea*, *Coniophorella olivaea*, *Polyporus tomentosus* var. *circinatus* and *Pholiota squarrosa*. Important trunk rots are *Stereum sanguinolentum*, *Phellinus pini* and *Amylostereum chailletii*.

The most important insect pests are the western balsam bark beetle (*Dryocoetes confusus*) and western spruce budworm (*Choristoneura occidentalis*). *A. lasiocarpa* is one of the most common hosts of *C. occidentalis*, the pest generally attacking low and middle elevation fir forests. *Adelges piceae*, introduced from Europe, has virtually eliminated *A. lasiocarpa* from some stands in the Cascades.

Uses

By providing cover, *A. lasiocarpa* assists in protecting watersheds and rehabilitating landscapes. The luxuriant foliage, long lived-crown, symmetrical form and deep bluish-green colour make the tree aesthetically valuable in recreation areas, urban plantings and landscapes.

The wood is light, soft, relatively weak and pale, with a density of 310–320 kg/m^3. It has no distinct heartwood and is low in bending and compressive strength although it is easy to work, glues well and holds nails or screws fairly well. The wood is primarily used for lumber for home construction and for prefabricated wood products, such as boxes, crates, doors and frames. The tree has excellent pulping properties, but has not been widely used for pulpwood because of its inaccessibility.

Resin produced by the tree is utilized as a cement for lenses and microscope slides in laboratories and in the optical industry. Various parts of the tree are used by Native Americans for medicinal and other purposes.

Further reading

Alexander, R.R. (1986) Silvicultural systems and cutting methods for old-growth spruce-fir forests in the central and southern Rocky Mountains. General Technical Report No. RM-126. Rocky Mountain Forest and Range Experiment Station, USDA Forest Service, Fort Collins, Colorado, USA.

Alexander, R.R., Shearer, R.C. and Shepperd, W.D. (1984) Silvical characteristics of subalpine fir. General Technical Report No. RM-115. Rocky Mountain Forest and Range Experiment Station, USDA Forest Service, Fort Collins, Colorado, USA.

Knudson, R.M., Chow, S., Lau, K.K., Lougheed, D.E. and Zaturecky, I. (2008) Extracting greater value from subalpine fir (*Abies lasiocarpa* (Hook.) Nutt.): some basic properties. *Forest Products Journal* 58, 31–39.

Abies magnifica A. Murray bis

Red fir

Importance

Abies magnifica is an important forest tree that forms stand dominants over large areas of upper montane forest habitats in the Sierra Nevada, Cascade Range, and Siskiyou Mountains of California and southern Oregon.

Botanical features

A. magnifica is a large forest tree, attaining up to 60 m with bole diameters to 2.5 m. The crown of young trees is conical, with numerous horizontal tiers of branches. Mature trees form a more cylindrical crown with a rounded top.

The bark of young trees is smooth and grey, becoming divided into reddish ridges separated by deep furrows in mature trees. It is this reddish bark in older trees that gives rise to the common name, red fir. Root systems of *A. magnifica* are highly variable, expressing an opportunistic approach to exploring available soil volume.

The foliage of *A. magnifica* consists of needles 20–35 mm in length, with an obscurely trapezoidal cross section of four sides. The needle apex is rounded to blunt or even lightly notched. Needles are glaucous-grey (almost bluish) in colour with several rows of stomata on the upper surface and two white bands of linearly arrayed stomata on the lower surface. These needles are arranged horizontally on the lower branch surface, whereas those on the upper surface distinctively curve upward, giving a branch the look of a large snowflake when viewed from below. Needle retention is commonly 6–10 years.

The large cylindrical female cones of *A. magnifica* are 15–25 cm in length and 7–13 cm wide. They are violet-purple when growing and brown when mature. As with other firs, the cone scales dehisce while still attached to the branches, shedding the wind-dispersed seeds. The seeds are about 12 mm in length, with a single broad wing.

Vegetative growth commonly begins in late May or early June in the Sierra Nevada, depending on altitude. Growth rates of trees near its lower altitudinal limits are restricted in late summer by drought. At upper altitudinal limits, however, drought may be less important than cool autumn temperatures in ending vegetative growth.

Distribution

The natural range extends from the Cascade Range of southern Oregon south into California through the Siskiyou Mountains to the southern Sierra Nevada. The variety *shastensis* occurs in southern Oregon and northern California where *A. magnifica* meets *Abies procera*, and also in the southern Sierra Nevada. The altitudinal ranges in the Cascade and Siskiyou Mountains are commonly 1620–2190 m, and 2130–2740 m in the southern Sierra Nevada.

At its northern limit in southern Oregon, *A. magnifica* hybridizes and merges with *A. procera* at 40°–44°N where there is a broad transition zone between these species. *A. magnifica* has not been widely planted outside of its natural range.

Silvicultural characteristics

A. magnifica is a climax forest species, forming relatively dense pure stands in upper montane habitats. It is co-dominant with *Abies concolor* at lower altitudes over much of its range. Near its upper limit in the Sierra Nevada it is often mixed with *Pinus jeffreyi*, *Pinus monticola*, and *Tsuga mertensiana*. In northern California, it occurs as a co-dominant with *A. procera* and *Picea breweriana* above 1850 m. Dense stands of *A. magnifica* on good sites produce such dense shade that no understorey vegetation is present.

The natural range is characterized by cool to cold and moist growing conditions with summer temperatures rarely reaching 30°C, and winter temperatures reported to −29°C. Rainfall is mainly in the winter from October to March, often as snow, with a mean annual range of 750–1500 mm, with 4–5 months of summer drought. Years with low rainfall are often correlated with high growth rates, apparently because of early snow melt and thus a longer growing season.

A. magnifica is intolerant of sites with seasonal saturation or poor drainage. Growth is best on deeper soils, which are often associated with glacial deposits or Pleistocene lake beds.

Where *A. magnifica* and *A. procera* meet they readily cross-pollinate with successful seedset as long as *A. magnifica* is the female parent, but viability is significantly reduced when *A. magnifica* is the male parent in such crosses.

Germination rates of *A. magnifica* are relatively low at about 40%, but heavy seed crops are produced every 1–4 years, and in an exceptional year seed production can exceed 1.4 million seeds/ha. The seeds germinate immediately after snow melt and are not tolerant of prolonged shading.

Silvicultural practice

It has not been used in silviculture outside its natural range. It grows well in Britain, quickly forming a narrow crowned tree with a stout bole.

The rates of growth and potential yield of *A. magnifica* are high, especially as the species will grow in relatively dense stands, and mean annual growth increment continues to increase for at least 140 years. On a good site, trees will reach 18 m in height at 50 years and 12 m in height on an average site. Yield tables for unmanaged stands indicate that a 160-year-old stand on a good site can carry 2320 m^3/ha of wood. An average site can carry 1470 m^3/ha at this age. Stand thinning can be extremely effective in increasing growth rates. Mixed stands with *A. concolor* 100 years old can be thinned to 50% of their basal area without significantly reducing overall stand growth increment. Site index equations for height and diameter increment in *A. magnifica* are available.

The ability of *A. magnifica* to support high growth rates in dense, even-aged stands makes clear-cutting the most common forestry practice. Where there are environmental and/or aesthetic limitations on large cuts, patch cuttings of 0.2–2.2 ha have been used with good success. Natural regeneration can be achieved using shelterwood and seed trees so long as the openings perpendicular to the wind direction do not exceed seed dispersal distances. Site preparation is important. Larger clear-cuts may require plantings of nursery-grown seedlings.

Diseases and pests

Root rot from *Heterobasidion annosum* is present in all stands and spreads from tree to tree by root contact; it only rarely kills directly but damage increases water stress and reduces tree vigour and thus susceptibility to bark beetle infection and susceptibility to windthrow and stem breakage. Several other heart rots can cause significant losses of older growth trees.

Mistletoe infections are common throughout the range. Heavily infested trees have reduced growth rates and fungi may enter through open mistletoe stem cankers. Best silvicultural practice is to remove infected overstorey parts and control stocking densities to allow the young trees to outgrow the spread of the mistletoe.

Fir engraver (*Scolytus ventralis*) is recorded on this species. Susceptibility to attack by fir engraver is increased by conditions which reduce tree vigour, including root disease fungi, mistletoes, drought stress, fire damage and overstocking. Other boring insects are also recorded. The Douglas-fir tussock moth (*Orgyia pseudotsugata*) is also recorded.

Seed miners are the major cause of damage to seeds, periodically causing major damage.

Uses

A. magnifica provides important forest cover for upper montane habitats throughout its range, where it commonly occurs in dense, pure stands.

The wood of *A. magnifica* ranges from nearly white to reddish-brown. The sapwood is indistinguishable from the heartwood. It has a medium to coarse texture and is generally straight grained. It is easy to work and is dimensionally stable when dried. The air dry density (12% moisture content) is 433 kg/m^3. It is moderate to moderately low in strength, stiffness, shock resistance and in nail withdrawal resistance. It dries easily, but may have problems with wet wood, a bacterial infection. The heartwood is not durable and is considered difficult to penetrate with preservatives. It is a general-purpose wood with widespread use for construction-grade framing material and plywood.

Young *A. magnifica* are widely harvested from both plantations and natural regeneration for sale as Christmas trees.

Further reading

Alden, H.A. (1997) Softwoods of North America. General Technical Report No. FPL-GTR-102. Forest Products Laboratory, USDA Forest Service, Madison, Wisconsin, USA, 151 pp.

Critchfield, W.B. (1988) Hybridization of the California firs. *Forest Science* 34, 139–151.

Dolph, K.L. (1992) A diameter increment model for red fir [*Abies magnifica*] in California and southern Oregon. Research Paper No. PSW-RP-210. Pacific Southwest Research Station, USDA Forest Service, Albany, California, USA, ii & 6 pp.

Zavarin, E., Critchfield, W.B. and Snajberk, K. (1978) Geographical differentiation of monoterpenes from *Abies procera* and *Abies magnifica*. *Biochemical Systematics and Ecology* 6, 267–278.

Zhang, J.W. and Oliver, W.W. (2006) Stand structure and growth of *Abies magnifica* responded to five thinning levels in northeastern California, USA. *Forest Ecology and Management* 223, 275–283.

Abies procera Rehder

Noble fir

Importance

Abies procera is the world's largest species of *Abies*, both in terms of height and wood volume, with desirable timber qualities. It grows in the subalpine forests of the Cascade Mountains where it often forms pure forests of large trees. It is not shade-tolerant and is self-pruning from an early age. *A. procera* is preferred to *Pseudotsuga menziesii* in areas which receive large amounts of snowfall; in productive sites within the subalpine forests it consistently outperforms the latter both in height growth and wood production.

Botanical features

Young trees are spire-shaped with tiered branching, but after crown closure side branches are lost, resulting in mature trees having a long clean bole with low taper and a short, rounded crown. Mature trees reach 45–60 m but 90 m is recorded, with diameters of 120–>150 cm. The bark is smooth and light grey with resin blisters; vertical fissures develop, eventually leading to a tight, checked or flaky bark, of a reddish-grey to purplish colour.

The twigs are finely pubescent and reddish-brown with small, non-resinous, ovoid buds. Leaves 10–30(–35) × 1.5–2 mm set with the proximal portion appressed to the twig for 2–3 mm before diverging. In cross section shade leaves are flat, with a prominent raised midrib on the lower surface, with or without a groove on the upper surface. Sun leaves have a four-sided cross section. The lower surface of leaves has 2–4 glaucous bands of stomata with 0–2 glaucous bands on the upper surface. Leaves are bluish-green, giving the tree a distinctive bluish blush. The leaf apex is rounded to notched and resin canals small and marginal. Leaves have a faint odour of turpentine.

Male and female cones are produced in the upper third of the tree in late spring and ripen in the autumn. Pollen cones can be purple, red, or reddish-brown. Seed cones are oblong–cylindric, 10–15 × 5–6.5 cm with the apex rounded. The cone scales are 2.5 × 3 cm, pubescent with exerted reflexed bracts, ripening from green, red or purple, overlaid with green bracts to brown. Seeds are 12 × 6 mm, reddish-brown with a paler longer wing.

Distribution

A. procera is restricted to a relatively small area of north-western Oregon and south-western Washington, USA, where it can be very dominant, especially in the Cascades north of the McKenzie River to Stevens Pass. There are discontinuous populations in the Coast Range of extreme south-west Washington and north-west Oregon.

The range becomes strongly discontinuous south of the McKenzie River south to Crater Lake, where trees have physical and ecological characteristics more typical of the *Abies magnifica* var. *shastensis* populations to the south.

A. procera is largely restricted to the montane (rainy – Coast Range) or subalpine (snowy – Cascades) forests influenced by a cool or cold maritime climate. Its most common associate is the very shade-tolerant *Abies amabilis*, which is nearly always present in the understorey of mature forests, as may be *Pseudotsuga menziesii*, *Tsuga heterophylla*, and occasionally *Thuja plicata*. In its more alpine exposures, *Abies lasiocarpa*, *Cupressus nootkatensis*, *Pinus contorta* and *Tsuga mertensiana* are common.

A. procera has not been widely planted outside of its natural range for forestry. It grows well in cool, moist valleys of Ireland and the UK, notably in Wales and Scotland, and has been planted elsewhere in Northern Europe. Its use as a Christmas tree and for greenery is widespread.

Silvicultural characteristics

It is shade-intolerant. As a pioneer species, especially in snowy situations, it often forms pure stands which grow into large to very large trees, which with its rapid growth rate and longevity allows it to dominate the canopy for centuries, especially at higher altitudes. Nevertheless *A. procera* often rapidly gives way to *A. amabilis* after 200–300 years of forest succession. Over time the stands are invaded by more shade-tolerant species. Its large seed allows the seedling to be stout, stiff and capable of bearing small debris loads often found following snow melt. *A. procera* is easily killed by fire because of its thin bark and tendency to grow on slopes. It has a high resistance to windthrow due to deep rooting; resistance to snow breakage is high, and to frost damage moderate.

Cones are generally produced after 20 years but commercial seed production is generally considered to begin at 50 years. A single tree has been recorded as bearing 3000 cones, suggesting 1.5 million seeds! Medium to good cone crops are on 3–6 year cycles. Seed viability is typically low, with results from one long-term study indicating a mean viability of only 10%, although higher in years with high cone production. Possible explanations include poor pollen dispersal, poor synchronicity between male and female cones, selfing, or insect damage in non-mast years. The seeds are heavier than those of other Pacific North-west conifers, often falling in pairs on the ovuliferous scales, with average seed weight of cleaned seeds being 29,750 seeds/kg. Dispersal is by wind, with most falling within two tree heights of the seed tree, but 600 m has been recorded, and up to 2 km for seeds blown around on the snow surface during winter.

Early seedling growth rate is slow to moderate, commonly taking more than a decade to reach breast height, but thereafter rapid with, on good sites, annual increments of 0.9–1.2 m, which can be maintained for decades.

Climate throughout the range of *A. procera* is distinctly moist and maritime, with cool summers having mean daily temperatures in July of 13.3–16.1°C. Winter temperatures are also relatively mild, with mean January temperatures ranging from −4.4 to −1.1°C. The mean number of frost-free days ranges from 55 near the tree line to more than 240 within the Oregon Coast Range. Mean annual precipitation varies greatly, ranging from 4740 mm in the Oregon Coast Range to only 1400 mm on ridges east of the Cascade Crest. Mean annual snowfall is in the range 1500–12,500 mm, with persistent snowpacks commonly 1–3 m deep. A summer dry season with ca 5% of the annual precipitation falling in July to September is usual.

A. procera grows on a wide variety of soils developed from various parent materials including successive deposits of volcanic tephra, and from ridge top to valley bottom, although slope forests are commonest. Best growth rates occur at low altitudes on lower slopes with fine-textured volcanic soils, reducing on poorly drained or shallow, water-deficient soils, and at higher altitude.

Silvicultural practice

A. procera is generally shade-intolerant, but seedlings have been seen to establish under partial shade in the Oregon Coast Range. Clear-cutting, patch-cutting, strip shelterwood, and group selection systems are acceptable management systems. Best results are achieved in pure stands, as these produce tall trees with high, clean trunks. Seeds germinate in the spring immediately after snow melt on any moist humic or mineral soil. On many upper-slope sites, stocking of *A. procera* is superior to that of *P. menziesii*, with 300–1800 seedlings/ha.

Sustained height growth, the tendency to grow in dense stands, high form factor and thin bark all contribute to the development of large stand volumes. On good sites it may make 28 m after 50 years with 20 m on an average site. One 110 year stand had 2100 m^3/ha (approximately 19 m^3/ha/yr) but 1200–1500 m^3/ha is more likely on an average site.

Stand thinning is extremely effective in increasing growth rates. The ability of *A. procera* to grow rapidly in dense, even-aged stands makes clear-cutting the most common forestry practice. It is the ideal tree for upper slope and subalpine reforestation efforts, both within its native range and in adjacent areas of Oregon and Washington, USA, and British Columbia, Canada, but should not be planted in areas with frost pockets or poor drainage, and is best on slopes greater than 20%.

A. procera × *A. magnifica* hybrids are present in wild populations between Crater Lake National Park and the McKenzie River in the Oregon Cascades (43°N–44°10'N), USA.

Diseases and pests

Root rot from *Heterobasidion annosum* is present in all stands and spreads from tree to tree by root contact; it only rarely kills directly but damage increases water stress and reduces tree vigour and thus susceptibility to bark beetle infection and susceptibility to windthrow and stem breakage. Several other heart rots can cause significant losses of older growth trees.

Mistletoe infections are common throughout the range. Heavily infested trees have reduced growth rates and fungi may enter through open mistletoe stem cankers. Best silvicultural practice is to remove infected overstorey parts, and control stocking densities, to allow the young trees to outgrow the spread of the mistletoe.

Seed miners are the major cause of damage to seeds, periodically causing major damage.

Uses

A. procera provides important forest cover for upper montane habitats throughout its range.

The wood is light and soft, creamy-white to yellowish-brown, with little or no distinction between sapwood and heartwood, and a density of 400–460 kg/m^3. It has a high strength to weight ratio and is suitable for light construction and pulping. High-quality *A. procera* wood is used for mouldings, window and door frames, and veneer. Lower-quality wood is often grouped with *T. heterophylla* and other *Abies* spp., often as 'hem-fir', and is typically used for construction wood. It can be used for construction plywood and produces good yields of strong pulp from both mechanical and chemical processes.

A. procera is a major Christmas tree species, with branches used for decorative greenery and plantations are sometimes grown for this purpose. The species is widely used for ornamental planting.

Further reading

Alden, H.A. (1997) Softwoods of North America. General Technical Report No. FPL-GTR-102, Forest Products Laboratory, USDA Forest Service, Madison, Wisconsin, USA, 151 pp.

Bega, R.V. (1978) Diseases of Pacific Coast conifers. Agriculture Handbook No. 521, US Department of Agriculture, Pacific Southwest Forest and Range Experiment Station, Berkeley, California, USA, iii & 206 pp.

Franklin, J.F. (1983) Ecology of noble fir. In: Oliver, C.D. and Kenady, R.M. (eds) Proceedings of the Biology and Management of True Fir in the Pacific Northwest Symposium, 24–26 February 1981. Contribution No. 45. College of Forest Resources, University of Washington, USA, pp. 59–69.

Pelz, S., Mehlin, I., Becker, G. and Bücking, M. (2003) Properties and utilization of *Abies grandis* and *Abies procera*. *Forst und Holz* 58, 290–296.

Zavarin, E., Critchfield, W.B. and Snajberk, K. (1978) Geographical differentiation of monoterpenes from *Abies procera* and *Abies magnifica*. *Biochemical Systematics and Ecology* 6, 267–278.

Abies sachalinensis (F. Schmidt) Mast.

Sakhalin fir

Importance

Abies sachalinensis forms pure stands within its natural distribution, and is a major timber tree of particular value for pulp production and hydrolysis (for production of sugar, ethanol, etc.). *A. sachalinensis* forest is also important for water-protection, water-regulation, and soil conservation functions. It has potential as an ornamental and amenity tree on suitable sites.

Botanical features

Some authorities take a broad concept of *A. sachalinensis* but others recognize up to four varieties, including the enigmatic *Abies gracilis* from a small stand on the Kamchatka peninsula.

Conical when young it develops a broad dense ovoid conic crown. Mature height is normally 25–30 m but occasionally to 40 m, with a bole diameter of 0.5–0.7 m. Young trees have a dark greyish bark tinged with green and numerous resinous blisters which in older trees is ash-grey or dark grey bark with vertical fissures. The shoots are buff or grey-brown with blackish hairs, darkening with age. The needles are 1.8–4 cm by 1.5–2 mm, sub-shiny mid green above with two white stomatal bands beneath, and arranged forwards above the shoot. The male cones are yellow and open in late May or June. The female cones are cylindrical or ellipsoidal, with shortly projecting reflexed bract scales. The cones are green or violet-blue, ripening to brown, and are 5–8 × 2–2.5 cm. The seeds are yellowish-brown, inversely wedge-shaped, 4.5–6 mm × 2.5–3 mm, with a 6–7 mm wing. The cones break apart in late September to October. Abundant seed crops occur every 3–4 years.

A. sachalinensis has a fairly typical Far Eastern phenological rhythm characterized by the growth buds flushing in late May or June with winter bud formation by July.

Distribution

The natural distribution of *A. sachalinensis* includes Sakhalin Island and the Kurile Islands in maritime (Far Eastern) Russia and Hokkaido Island, Japan. The northern limit of the distribution passes through the middle of Shmidt Peninsula on Sakhalin Island, where occasional young *A. sachalinensis* trees occur in mixed stands with *Picea jezoensis* in the River Pil'vo catchment. This evidence seems to indicate that the species is extending its range.

In the northern part of its range, *A. sachalinensis* ranges from an associate species to a dominant one, but forms pure stands south of 48°N. The montane *A. sachalinensis* biogeoclimatic zone in Hokkaido occupies mid altitudes and is characterized by *A. sachalinensis*, *P. jezoensis*, and *Picea glehnii*. In the northern part of its range, the *A. sachalinensis* understorey is characterized by green mosses but further south the understorey is characterized by *Sasa kurilensis*.

Experience with *A. sachalinensis* beyond the natural area is mainly limited to trials, and introductions in arboreta and botanical gardens.

Silvicultural characteristics

A. sachalinensis occurs in the temperate zone with a cool humid climate characterized by mean annual temperatures from −1 to 6°C, mean annual precipitation ranges from 570 to 1400 mm and annual potential evapotranspiration ranges from 350 to 550 mm. Snowfall is high, with maximum snow depths of 100–150 cm.

In nature, *A. sachalinensis* grows on a broad spectrum of habitats and on soils which include podzols, volcanic ash, and brown forest soils. In certain parts of its range *Sasa* (a genus of rampantly spreading dwarf bamboos) seems to play an important role in determining the soil type because it prevents podzolization by acting as a neutralizing agent to the abundant acidic conifer litter. Soil pH under *A. sachalinensis* ranges from 3.8 to 6.5.

On Hokkaido, *A. sachalinensis* occurs at mid to high altitudes, ranging approximately from 700 to 1500 m in the central region; from 400 to 1000 m in the northern region; from 100 to 700 m in the eastern region; from 500 to 1100 m in the southern region; and from 800 to 1300 in the south-western region, with an upper altitudinal limit on Hokkaido of ca 1650 m. The altitudinal limits are defined altitude and climatic factors including the influence of sea winds. On Sakhalin, it occurs up to 750–800 m.

The length of snow cover is the most important climatic factor affecting provenance characteristics. Provenances from along the Japan Sea coast and the northern part of the Okhotsk Sea coast have the largest seeds; they are affected less seriously by snow damage and more by winter desiccation, and show strong resistance to *Rhacodium* snow blight. The altitudinal gradient has a very important role in determining the suitability of seed sources for planting sites.

A. sachalinensis is a shade-tolerant species that regenerates well under closed canopies but needs high air humidity for good growth.

Silvicultural practice

A. sachalinensis is mainly propagated by seeds, which mature by September–October. The seeds require a period of low temperature stratification for optimal germination.

In Japan, there is long experience of managing plantations of *A. sachalinensis* with different thinning and fertilizer regimes used depending on the site quality. Young trees need shelter to protect them from desiccation, spring frost and from strong sunlight heating the forest soil surface to 40°C.

Non-indigenous silviculture is restricted to trial plots and the planting of trees in arboreta and botanical gardens. It was introduced into the British Isles by Maries in 1879, but it has suffered from spring frost damage although the few large trees have grown quite rapidly.

Planted stands of good site quality in Hokkaido attain sexual maturity at 35–40 years old and mean height of 20 m and mean dbh of 30 cm at age 53.

On Sakhalin and Kurile Islands, average volume of stands where *A. sachalinensis* is predominant is about 250 m^3/ha; however, in the better stands the volume may be as much as 450–800 m^3/ha. These stands are highly marketable but if the rotation age is longer the trees are often degraded by fungal butt rots. Naturally *A. sachalinensis* is usually found in mixed-age stands where there is no clear correlation between diameter and height; in one study 8-cm diameter trees were from 5.2 to 13.7 m in height, those with a dbh of 20 cm varied from 11 to 19 m, trees 32 cm in diameter varied from 11.5 to 24 m, and trees of 48 cm in diameter varied from 18.4 to 28.3 m. Bark volume ranges from 11 to 14% of wood volume, but decreases as bole diameter increases. Seed yield in Sakhalin averages 50 kg/ha but may be up to 80 kg/ha in the most fertile years.

Diseases and pests

The most harmful wood-attacking fungi are *Phellinus hartigii* (which is responsible for heart rot) and *Heterobasidion annosum*, affecting root systems and causing stump rot. *Armillaria* species often damage the root systems of weakened trees and those on dry sites. Many trees have some irregular wet wood in the heartwood, which may cause frost cracks. *A. sachalinensis* is affected by the canker *Gremmeniella abietina* in Hokkaido.

Repeated outbreaks of the Siberian moth (*Dendrolimus sibiricus*) have been recorded from Sakhalin and Kurile Islands and pose the most serious threat to *A. sachalinensis* in managed stands. *Calliteara abietis* and *Arbognophus amoenaria* usually infect trees of all ages and cause growth loss and mortality during severe outbreaks. The Sakhalin fir aphid (*Cinara todocola*) is a serious pest in young plantations in Hokkaido. Seeds and cones are injured by *Eupithecia abietaria*. Trunk-infesting insects include *Ips typographus*, *Polygraphus proximus* and *Monochamus urussovii*. Silvicultural control measures are the most efficient means of preventing outbreaks.

Uses

A. sachalinensis forests play an important ecological role in protection of water resources, regulating runoff and in soil conservation.

A. sachalinensis wood is used for sawn building timbers, containers, specially made chips, packing wood wool and other materials. It is suitable for making musical instruments. The wood is of particular value for pulp and paper production and hydrolysis (for production of sugars, ethanol, etc.). A basic wood density of 342 kg/m^3 has been reported.

A. sachalinensis needles and small twigs are distilled for their essential oils; 5–6 tonnes of needles and twigs can be collected from 1 ha of fir forest to produce 60–130 kg *A. sachalinensis* oil. The needles can be used to make carotene–chlorophyllic paste and coniferous-vitamin meal. The resin blisters in the bark can be used to produce turpentine, rosin and glue used in paper and textile production and in the optical, electrical and aerospace industries. The resin (fir balsam) is used to treat cuts and small wounds, burns, gastric ulcers and other conditions in native medicine; Far Eastern fir oil is used in medicine, perfumes and cosmetics.

Further reading

Abe, N. (1989) Studies on the system for thinning of *Abies sachalinensis* Mast. planted forest. *Bulletin of the Hokkaido Forest Experiment Station* 26, 1–95.

Ageenko, A. (1986) *Normative materials for inventory of forests of Sakhalin and Kamchatka*. Sakhalin Forest Experimental Station, Yuzhno-Sakhalinsk, USSR.

Goto, S., Iijima, H., Ogawa, H. and Ohya, K. (2011) Outbreeding depression caused by intraspecific hybridization between local and nonlocal genotypes in *Abies sachalinensis*. *Restoration Ecology* 19, 243–250.

Man'ko Yu I. (1976) The northern boundary of the distribution of *Abies sachalinensis*. *Botanicheskii Zhurnal* 61, 393–395.

Nikiforov, G. and Kalinin, A. (1977) *Production of fir oil*. Lesnaya promyshlennost, Moscow, USSR.

Acacia auriculiformis A. Cunn. ex Benth.

Northern black wattle

Importance

Few other species match the growth of *Acacia auriculiformis* on poor sites in the humid tropics. Rapid early growth, nitrogen fixation, tolerance of infertile, acid, alkaline, saline or seasonally waterlogged soils and short dry seasons make it useful for rehabilitating degraded land. It has been widely planted in South and South-east Asia, but also in Africa and South America for fuelwood, erosion control, and as an ornamental or shade tree. Stem form can be improved significantly by selection and breeding, further improving the prospects for plantations producing paper pulp and other wood products.

Botanical features

On favourable sites, *A. auriculiformis* grows 25–35 m tall with a long clear straight bole, more commonly 8–20 m, and occasionally a much-branched shrub to 3–5 m. Bark is grey or brown, sometimes blackened at the base, smooth in young trees, becoming rough and fissured with age. Like others in the subfamily Racosperma, bipinnate seedling leaves are soon replaced by entire phyllodes. In *A. auriculiformis*, phyllodes are falcate, 8–20 cm long and 1.0–4.5 cm wide, glabrous and greyish-green, with three prominent longitudinal veins and a distinct gland at the base. The tiny light-golden flowers are grouped in paired inflorescences up to 8.5 cm long. Mature pods are twisted and irregularly coiled, flat, flexible but hard and about 6.5 cm long by 1.5 cm wide, containing shiny black seeds, broadly ovate, 0.5 cm long by 0.35 cm wide, each encircled by a long red, yellow or orange funicle.

Distribution

The natural range of *A. auriculiformis* includes Australia, Papua New Guinea and Indonesia. In Australia, it occurs on Cape York Peninsula in Queensland, and the north of Northern Territory including several off-shore islands. In Papua New Guinea, it occurs in the Central and Western Provinces, extending into Irian Jaya [West Papua] and the Kai Islands of Indonesia.

It is a lowland tropical species often growing along river banks where it may become dominant, in depressions, or in open-forest dominated by various eucalypts and acacias. It may be common on floodplains and levees, or as scattered trees in riparian habitats, tall savannah woodland and in open monsoon forest. It is also found in swamp forests and is common in littoral forest behind either mangroves or coastal dunes, where it may be found with *Acacia mangium*, and other species of *Acacia*, *Eucalyptus* or *Melaleuca*.

It is been introduced to many humid tropical areas worldwide, and is especially common throughout Indonesia, South, East and South-east Asia, the Pacific and Indian Ocean islands, and parts of Africa and South America.

Silviculture characteristics

A. auriculiformis is a fast-growing species adapted to poor sites in the humid tropics. It fixes atmospheric nitrogen and is tolerant of infertile, shallow, acid, alkaline, saline or seasonally waterlogged soils and a moderate dry season. However, it does not tolerate shade, and branches break easily in strong winds. Poor stem form has been a major drawback limiting wider utilization, though some provenances appear to have straighter boles. Several countries have started genetic improvement programmes to enhance industrial uses for sawn timber and pulp production. Other limitations include the relative sensitivity of young trees under 2 years old to fire, weak coppicing ability and susceptibility of stressed trees to insect attack. It is has also proved invasive in some countries.

Seedlings grow quickly, reaching 30 cm in 3 months, 6 m in 2 years, and up to 12 m in 3 years under favourable conditions. Flowering usually begins 2 years after germination, and generally occurs between February and July in its native range. Flowers are insect-pollinated and large quantities of mature seed are available some 5 months later. Seeds retain viability well in storage, and immersion in boiling water for 1 min improves germination.

A. auriculiformis shows excellent growth under plantation conditions, with annual height increments of 2–4 m in the first few years being common, even on soils of low fertility. Under optimum conditions trees can reach 15–18 m tall and 15–20 cm diameter after 10–12 years, and 15–20 m^3/ha/yr is possible, reduced by half on less fertile or highly eroded sites, and much less where low rainfall or a prolonged dry season are limiting factors. On *Imperata* grasslands and very infertile soils *A. auriculiformis* will still usually grow faster than species of *Albizia*, *Eucalyptus*, *Leucaena* and *Pinus*, but provenance choice is also very important.

Diseases and pests

A number of diseases have been identified as potential threats to the future productivity of industrial plantations based on *A. auriculiformis*. These included stem cankers caused by a range of pathogens (*Botryodiplodia theobromae*, *Botryosphaeria* spp. and *Hendersonula* sp.) and are most often associated with stem borer damage, pink disease (*Corticium salmonicolor* also known as *Phanerochaete salmonicolor*) which is most prevalent in high rainfall areas, and phyllode rust (*Atelocauda digitata*) which has impaired the growth of *A. auriculiformis* in Australia and Indonesia. *Ganoderma* was observed to cause crown die-back and defoliation in plantations, and seedlings in the nursery can be infected by powdery mildew (*Oidium*), especially where there is heavy shading.

A. auriculiformis only rarely suffers significant damage from pests or diseases, though trees stressed by biophysical factors are more susceptible. Stressed trees are particularly susceptible to attacks by insect pests.

The beetle, *Sinoxylon* sp., can girdle small stems or branches. This beetle is of concern because damage causes the tree to develop multiple leaders and reduces the length of clear bole. The wood may also be attacked by borers and termites, and scale insects are prevalent on young trees.

Silviculture practice

Nursery seedlings reach planting size (25 cm) in 3–4 months. Inoculation with *Rhizobium* and mycorrhiza strains is rarely necessary but may be beneficial, especially when seedlings are raised in sterilized media or are for planting on highly degraded soils or mine spoil. Propagation of juvenile cuttings from pollarded trees is also a routine and simple operation. Typically, containerized planting material is used, giving higher survival, especially with heavy weed competition. Direct sowing is also possible but has achieved only limited success. Aerial seeding has sometimes been successful but appropriate site preparation prior to sowing is required. Profuse natural regeneration may appear after fire or on disturbed sites in the absence of severe weed competition.

The optimum spacing for *A. auriculiformis* depends on the end product and management considerations, from 1 × 1 m to 4 × 4 m, with 1 × 2 m and 1.5 × 1.5 m used for fuelwood and poles. Normal principles of site preparation apply, but with special attention to weed control in the first 2 years, as seedlings are shade-intolerant and are easily smothered.

A number of silvicultural systems are in use. One common practice is to clear fell at 10–15 years, followed by natural or artificial regeneration. It does not coppice well and careful management is required, with better results obtained when stumps are cut 60–100 cm from the ground. Tree age, stump diameter, and season of cutting also affect coppice regrowth. *A. auriculiformis* does, however, respond well to pollarding. It is sometimes included in mixed species systems in combination with native rainforest species or eucalypts that use its nitrogen-fixing ability for soil improvement, and its fast growth as a cover crop. It is also an income source to enhance the economic viability of planting the slower-growing species. Plantations in *Imperata* grasslands have survived fires, but are generally too severely damaged to be suitable for *Imperata* control.

Uses

A. auriculiformis is not widely used in agroforestry systems because of its spreading and competitive surface rooting habit, but intercropping of pruned trees has proved successful, and it has been used satisfactorily as a nurse tree in tea and sandalwood (*Santalum album*) plantations. The dense root system stabilizes eroding land and it is used widely in degraded lands, coastal areas, mine spoils, and for grassland rehabilitation. *A. auriculiformis* is also planted as an ornamental species due to its hardiness, dense foliage, and bright yellow flowers.

A. auriculiformis wood has a calorific value of around 4800 kcal/kg and does not smoke or spark, making it ideal for firewood and charcoal. The annual fall of leaves, twigs, and branches can amount to 4–6 t/ha, and is useful for household fuel even before harvesting.

The sapwood is yellow and the heartwood light brown to dark red. The timber is fine-grained, often attractively figured and finishes well. The basic density is 500–650 kg/m^3, and fibres are relatively short, about 0.85 mm in length. The heartwood is typically hard and durable, but the sapwood is highly susceptible to termite and borer attack. It makes attractive furniture, and is also suitable for construction work (e.g. framing, flooring), tools, wood turning and carving. However, the crooked stems which are a common feature of the species largely restrict its use as poles or sawn timber that requires a reasonable length.

The wood contains 59% cellulose, 24% lignin, 19% pentosan and 0.4% ash, with flavonoid substances also present. Plantation-grown trees appear very promising for the production of unbleached kraft pulp (for bags, wrapping paper, linerboard), and high-quality neutral sulphite pulp (for corrugating, medium, and higher-grade packaging-type products), but not for high-yield mechanical-type pulps, although there is significant variation in pulp-making properties between provenances.

Although *A. auriculiformis* is not recognized as a fodder species, trees are occasionally browsed by cattle, and foliage can meet minimum requirements for certain nutrients. The bark contains enough tannin (about 13%) for possible commercial exploitation, and a natural dye is extracted and used in the batik textile industry in Indonesia. Trees produce exudate gums that are commercially exploited, and flowers are a source of bee forage.

Further reading

Doran, C.J. and Turnbull, J.W. (eds) (1997) *Australian Trees and Shrubs: Species for Land Rehabilitation and Farm Planting in the Tropics*. Monograph No. 24. ACIAR, Canberra.

Kumar, J. and Yadav, R.S. (2009) Ecology of *Acacia auriculiformis*. *Plant Archives* 9, 487–494.

Lemmens, R.H.M.J., Soerianegara, I. and Wong, W.C. (eds) (1995) *Plant Resources of South-East Asia. No. 5(2). Timber trees: minor commercial timbers*. Backhuys Publishers, Leiden, The Netherlands.

NFTA (1996) *Acacia auriculiformis* - a multipurpose tropical wattle. NFTA 96–05. Nitrogen-Fixing Tree Association, Waimanalo, Hawaii, USA.

Acacia dealbata Link

Silver wattle, mimosa

Importance

Acacia dealbata is a fast-growing shrub or tree suited to cool to warm subhumid climatic zones. It is an excellent source of pulp for paper production, though is rarely raised in plantations for this purpose. It is planted on farms and hillsides as a windbreak or for erosion control and produces satisfactory firewood. The species is also often planted as an ornamental tree, being the common yellow-flowered 'mimosa' seen in many Mediterranean regions. It is a pioneer species which readily forms root suckers and produces prolific quantities of seed, making it a weed in some countries where introduced, such as in parts of South Africa and India.

Botanical features

A. dealbata has a variable habit, from a large spreading shrub to an erect tall tree with a rounded crown where space allows, usually 2–15 m tall but reaching 30 m in southern parts of its native range. The bark of mature trees is dark grey to brown-grey, often mottled white by lichen, though younger stems and newer growth are smoother and lighter in colour.

Leaves are bipinnate, 'feathery' and not large 'phyllodes' common in many other Australian *Acacia* species. The silvery-green to dark green leaves are densely hairy and 6–11 cm long. Each has 7–26 pairs of pinnae, 15–30 mm long, each with 20–50 pairs of very small leaflets 2–5 mm long. Glands are prominent, with one on the petiole and a raised jugary gland at the junction between each pair of pinnae. Like some other legumes, *A. dealbata* exhibits a diurnal rhythm of pinnule movement in which the leaves open by day and close by night.

Flowers are golden-yellow, lemon or pale yellow in colour, with 25–35 flowers in globular heads on peduncles some 6 mm long. The oblong pods are bluish in colour, commonly 6–8 cm long and 1 cm wide with undulated raised margins constricted between seeds. The seeds are black and oblong and some 5 mm long, with cap-like arils.

Distribution

A. dealbata is native to south-eastern Australia, on the tablelands and in the foothills of the Australian Alps from northern New South Wales to mid-western Victoria, at altitudes of 350–1000 m, and in Tasmania where it grows at elevations of 50–600 m. In its natural distribution it occurs mainly within forests and woodlands on a variety of soil types in cool to warm subhumid climatic zones. It is often a dominant shrub in eucalypt forests, becoming a small tree in clearings or on disturbed sites where it regenerates vigorously. *A. dealbata* was also introduced to South Australia, southern Queensland and New Zealand, where it has become naturalized.

It has also long been introduced to and continues to be planted in many other countries worldwide, and is still referred to in some of these by its old synonym, *Acacia mollissima*. It is especially common in parts of Chile and

Argentina; France and other southern European countries; North Africa; South Africa; East Africa; and in Asia, China, Japan, India and Sri Lanka. *A. dealbata* had already been introduced into Sri Lanka in 1870 to supply fuelwood for the tea estates and railways in the highlands, and later for the reforestation of marginal land. By 1900, it had also been introduced to many African countries, and was known to have become naturalized in parts of India before the 1920s.

Silviculture characteristics

In its native range, flowering occurs from July to November, and seed pods mature 5–6 months later. Seed are shed and remain viable in the ground for many years. *A. dealbata* may grow rapidly up to 16 m tall in 8 years, with a moderate lifespan of 20–40 years. It suckers extensively from roots and coppices readily from wounded stumps. It is a pioneer species, especially on cleared land, and prolific seed production and rapid growth also mean that it has become an invasive species in some countries.

Like other *Acacia* species, *A. dealbata* fixes atmospheric nitrogen through symbiotic bacteria living in nodules on its roots, and the top 40 cm of soil of an *A. dealbata* stand may contain 50% more nitrogen than under pine or eucalyptus plantations, and can significantly increase nitrogen cycling through litterfall. Foliage is less palatable than in many other *Acacia* species, and so the species suffers less browsing damage.

Although *A. dealbata* is known to grow rapidly in its native range, wood yields are highly variable. The highest growth rate in its native range has been recorded as 23 m^3/ha/yr after 10 years, but on other favourable sites yields were less than 1 m^3/ha/yr. Reports are equally variable where introduced, up to 46 m^3/ha/yr after 8 years, or 15–30 t/ha dry matter after 4–5 years.

Silviculture practice

Propagation of *A. dealbata* is by seed, with the common pretreatment immersion in boiling water for 1 min, but scarification is also an option. Germination rate averages 74% and there are approximately 50,000 viable seeds per kg. There are no standard nursery practices for *A. dealbata*, with seeds either sown directly into containers or into beds to produce bare-rooted seedlings. The ideal height for seedling establishment is around 30 cm with a collar diameter of 7 mm, usually achieved 4–5 months after sowing. Smaller seedlings tend to lose more foliage and have poorer survival and growth, and often develop multiple stems. Rapid root development may also be a problem as seedlings are sensitive to root wrenching, and several lighter root prunings are preferred rather than a single wrenching. Initial stocking rates should be increased if survival problems are anticipated. When introducing *A. dealbata* to new areas, seedlings are also likely to benefit from inoculation with *Rhizobium*.

Although *A. dealbata* can fix atmospheric nitrogen, fertilizer applications in the nursery and on planting will increase growth rates, especially on marginal sites. Form pruning is often recommended 1–2 years after planting when trees are 2–3 m tall. This is particularly important if foliage loss occurs after planting, as this can cause loss of apical dominance and the consequent production of numerous shoots from the base of the seedling. These shoots often persist and, without form pruning, are likely to produce trees with multiple leaders, reducing the quality and value of the timber. An important aspect of silviculture for *A. dealbata* is weed control, and survival and growth rates are improved when weed competition is eliminated. For pulpwood, rotations of 15–20 years are common.

Diseases and pests

Diseases such as wire-stem rot and fungal rot have been known to cause problems in trials, and various fungi have also been reported as causing serious losses to *A. dealbata* stock in a number of nurseries.

Little information is available on pests and diseases of *A. dealbata* in plantations, although within Australia, *A. dealbata* is susceptible to the fireblight beetle (*Pyrgoides orphana*) which defoliates the species and has a great impact on growth in natural stands in Tasmania; this is seen as a limiting factor in establishing *A. dealbata* as a plantation species. The beetle larvae can cause complete defoliation in winter and spring, and significantly lower growth rates. This allows secondary attack by other insects, such as wood borers, which cause major stem damage and sometimes death.

There are also some records relating to attacks on *A. dealbata* as an ornamental, such as cottony cushion scale (*Icerya purchasi*). However, much research has been conducted on organisms that could act as species-specific biological control agents.

Uses

A. dealbata used to be planted as a windbreak, and to control soil erosion in gullies and for hillside stabilization, especially in marginal and degraded upland areas. However, recognition of its invasive potential now limits such uses.

A. dealbata wood is used as a furniture timber and for poles, and occasionally for producing wood wool. It has good gluing properties, and makes a satisfactory fuelwood. Green wood has a density of about 800 kg/m^3, with an air-dry density of 540–720 kg/m^3.

A. dealbata produces a superior quality pulpwood and has a higher basic density than other fast-growing hardwoods. Its kraft pulping and papermaking properties make it suitable for a range of paper and paperboard products such as linerboards, bag and wrapping papers, white boards, writing, and printing paper. It has levels of brightness required for some high-grade paper products, and has lower alkali requirements than most eucalypts.

Exudate gum can be used as a substitute for gum arabic, and its bark is occasionally used for tanning but is lower yielding and of poorer quality when compared with *A. mearnsii*. Foliage can also be used as a yellow-fawn or green dye for wool, depending on the mordants used. The foliage is not widely known as a source of animal fodder but is used as such in some countries.

A. dealbata is valued as an ornamental tree in southern Europe where it is known as 'mimosa', and where it is also used in the cut-flower trade. The flowers are used in the production of more than 80 different perfumes, and manufacturers (especially in France) recognize the extract for its quality as a blender and smoothing agent for synthetics, and as a fixative in high-grade perfume. Flowers are also a valuable source of bee forage. It is also planted as a street tree in many other countries.

Further reading

Doran, C.J. and Turnbull, J.W. (eds) (1997) Australian trees and shrubs: species for land rehabilitation and farm planting in the tropics. Monograph No. 24. ACIAR, Canberra.

Kube, P.D. and Brooks, S.J. (eds) (1996) Evaluation of *Acacia dealbata* as a Plantation Species in Tasmania. Forestry Tasmania and Forests and Forest Industry Council, Hobart, Australia.

May, B.M. and Attiwill, P.M. (2003) Nitrogen-fixation by *Acacia dealbata* and changes in soil properties 5 years after mechanical disturbance or slash-burning following timber harvest. *Forest Ecology and Management* 181, 339–355.

Acacia mangium Willd.

Mangium, brown salwood

Importance

Acacia mangium is the most important plantation species in the humid tropical lowlands of Asia and one of the most important in the tropics, globally. Its success is due to its ease of establishment and extremely vigorous growth, even on very acid, low-nutrient soils. Importantly, it also outcompetes severe weed competition, and can be planted on the large areas of land that have been invaded by *Imperata* grass. Plantations in Indonesia and Malaysia are the main resource base for a large and growing pulp and paper industry, and are also used for fuelwood, building, and furniture timber and for the manufacture of particleboards.

Botanical features

A. mangium is a large tree to 30 m tall, with a long, straight, clear bole up to 60 cm in diameter, occasionally fluted at the base. On poorer sites, however, it may become only a small tree or large shrub of 7–10 m. The bark is pale grey-brown to brown, rough and longitudinally furrowed. Mature dark green phyllodes (leaves) are large, 11–27 cm long and 3–10 cm broad, characterized by four main longitudinal nerves (rarely three or five).Whitish (or cream) flowers are in rather loose spikes, 5–12 cm long, producing tightly coiled seed pods 7–8 cm long and 0.3–0.5 cm wide. Seeds are black and shiny, with a yellow or bright orange (rarely red) funicle, folded to form an oily, fleshy aril beneath.

Distribution

A. mangium is native to small areas of Australia, Papua New Guinea and Indonesia. The native range includes two restricted regions in northern Queensland, Australia; parts of Western Province in Papua New Guinea; and in Indonesia, in small areas of the provinces of Irian Jaya [West Papua] (Merauke District and Vogelkop Peninsula), and Maluku (Sula, Ceram, and the Aru Islands).

A. mangium grows on the margins of closed forest (rainforest), in tall and open forest and woodland alongside *Eucalyptus* and other *Acacia* species, especially where fires are common, or on better-drained sites within swampy coastal plains where *Melaleuca* species may be dominant.

It has been widely introduced to other parts of South, East and South-east Asia, with an increasing number of large-scale plantations especially in Indonesia, Malaysia, China, the Philippines, Thailand and Vietnam. It has also been planted in other humid lowland regions throughout the tropics, especially in Africa and South and Central America, and shows considerable potential, though it has proved to be invasive in some countries.

Silviculture characteristics

This moist tropical lowland species is fast-growing but relatively short-lived (30–50 years), is adapted to a wide range of acidic (pH 4.5–6.5) and infertile soils, but is shade-intolerant. Growth slows significantly when mean monthly temperatures fall below 17°C as in parts of China, and the species is sensitive to frost.

Smaller trees (stem diameter less than about 10 cm) are also sensitive to fire damage. Branches are persistent as the species does not naturally self-prune, and stem heart rot may develop from dead branch stubs. Stem straightness varies with site, being poorer on more fertile sites where growth is rapid, and it also has a tendency to form multiple stems, which appears to be related to soil fertility, competition, and planting of spindly, oversized seedlings. It is capable of outcompeting even the most persistent of weeds such as *Imperata* grass.

A. mangium can grow very rapidly, and can reach 10–15 m tall in 2–3 years, and 25 m after 5–15 years, when growth rates decline. However, on optimal sites, trees may grow at more than 5 m/yr. Annual production from Indonesia, cited in several reports, ranges from 20 to 40 m^3/ha/yr, though the best-growing provenances on better sites could produce more than three times these figures. Stumps coppice profusely, but the shoots lack vigour and are unsuitable for a second rotation. It has also become invasive in some countries.

Diseases and pests

In general, *A. mangium* is relatively free from serious diseases and pests, although heart rot resulting from branch stumps and other pruning wounds has proved to be a problem in timber plantations. Ganoderma root rot is also a serious pathogen in some areas.

However, researchers consider that where introduced, exotic *Acacia* species are likely to face an ever-increasing threat from new pests and diseases in the future. One example is a new defoliator (*Pelidnota filippiniae*) reported from Brazil in 2011.

Silviculture practice

Seed orchards have been established in Australia, Indonesia, Malaysia and Vietnam, but most seed for planting still comes from superior provenances in northern Queensland and Papua

New Guinea. Like many *Acacia* species and other legume trees, the seed has a hard coat and retains viability well in storage, and requires a pre-germination treatment such as immersion in just-boiled water for 1 min followed by soaking for 24 h. Newly emerged seedlings should receive 50% shade, with full sunlight once established, and reach planting size (from 25 cm high) in 3–4 months.

There is abundant natural regeneration after logging debris is burned, but generally logged areas are replanted with containerized seedlings. Direct seeding has been tried but results are variable. Initial spacing is governed by the need for rapid canopy closure and the end products required. For high-quality logs, sufficient numbers need to be planted to enable the selection of crop trees with excellent form, from 2 × 2 m spacing to 4 × 4 m, although 3 × 3 m spacing is common. Site preparation involves burning or spreading logging debris, or slashing or ploughing of *Imperata* grass with weed control applied during the first year, and occasionally in the second year.

A. mangium can fix nitrogen with a range of *Rhizobium* and *Bradyrhizobium* strains in many tropical soils, but is much more specific in its rhizobial affinities than *Acacia auriculiformis*. Plantations of *A. mangium* show large responses to fertilizer applications, especially phosphorus, and such additions may prove key to successful establishment on infertile sites.

On sites where *A. mangium* develops multiple stems from its base the trees need to be singled before they are 1.5 m tall. Later singling is more time-consuming and runs the risk of encouraging butt rot. Also, the more that the growth rate is increased by silvicultural treatments such as weed control, cultivation and fertilizer, the more important it is to prune to single stems very soon after planting.

Plantations for pulpwood, firewood or wood chips harvested at 6–7 years (when annual increments are at a maximum) are not usually thinned or pruned. When growing *A. mangium* on longer rotations for sawn timber, pruning is required to overcome the degradation caused by branches and associated heart rot, and regular heavy thinning is needed to maintain large deep crowns and fast diameter growth. One proposed management regime for *A. mangium* in Peninsular Malaysia involves singling at an average height of 1.5 m, two pruning operations after 6 and 9 months, and four thinnings at 9, 13, 22, and 27 months, with clear felling of the final 100 stems/ha at 15 years when average top height is approximately 28 m.

Uses

With its dense evergreen foliage, *A. mangium* makes a useful shade, screening, and soil-cover crop, and being a nitrogen-fixing species, it is also used in alley cropping systems, in mixed species plantations, and to ameliorate degraded soils.

Mature wood is dense and makes a good fuel, but young trees (2.5 years) make poor firewood. *A. mangium* produces charcoal of reasonable quality, and is suitable for the manufacture of wood pellets and activated carbon.

A. mangium has hard, pale yellow-brown heartwood with a narrow band of light coloured sapwood and close grain. It has a density of 450–690 kg/m^3 (at 15% moisture content), and it is classed as a light hardwood with low to moderate strength properties. Wood is easily sawn, polished, drilled and turned, producing timber that is fairly stable, with green to air-dry shrinkage of 6.4% tangentially and 2.7% radially; it also kiln-dries well without serious defects. It is very durable except when in contact with the ground. It is used for furniture and cabinet-making, light structural work, agricultural tools, boxes and crates. It has also been sliced for veneer and makes excellent particleboard.

Plantation-grown trees are used for the production of printing and writing paper from unbleached kraft pulp, and high-quality neutral sulphite semi-chemical pulp. The sulphate process with 12–15% alkali yields 47–57% screened pulp.

The use of *A. mangium* for non-wood products is relatively minor and incidental to its use for wood production. The leaves serve as occasional forage for livestock and as a soil mulch. Flowers are also a source of bee forage.

Further reading

Awang, K. and Taylor, D. (eds) (1993) *Acacia mangium* growing and utilization. MPTS Monograph series No 3. Jointly published with the FAO Forestry Research Support Programme for Asia and the Pacific (FORSPA) and Forest Tree Improvement Project (FORTIP), Winrock International, Arkansas, USA and FAO, Bangkok.

Buford Briscoe, C. (1995) Four *Acacia* species for acid soils. In: Evans, D.O. and Szott, L.T. (eds) *Nitrogen-Fixing Trees for Acid Soils*. Proceedings of a Workshop, 3–8 July 1994. *Nitrogen Fixing Tree Research Reports* Special Issue 96-05, 265–271.

Krisnawati, H., Kallio, M.H. and Kanninen, M. (2011) Acacia mangium *Willd.: Ecology, Silviculture and Productivity*. CIFOR, Bogor, Indonesia.

Acacia mearnsii De Wild.

Black wattle

Importance

Acacia mearnsii is a fast-growing, nitrogen-fixing tree, adapted to a wide range of sites from temperate and subtropical lowlands to tropical highlands. Its bark is the most widely used source of tannin in the world, and wood is used for paper pulp, cellulose for rayon, charcoal, and fuelwood. In addition, the tannin is being increasingly used for the production of waterproof wood adhesives for the reconstituted-wood industries. It is also planted for erosion control, soil improvement, shade, shelter, and as an ornamental, but it has proved very invasive in some regions, and more use could be made of this expanding resource.

Botanical features

A. mearnsii is a large shrub or small tree, typically 5–15 m tall with a dbh of 10–35 cm; however, it may grow to 20 m tall with dbh of 45 cm. Open-grown trees branch readily from near ground level with crooked main stems, but in denser stands the stem is straighter, without branches for at least half of the tree height. Bark on mature trees is brownish-black, hard and fissured, and is grey-brown and smooth on younger stems.

Foliage comprises olive to dark green, 'fern-like' bipinnate compound leaves, typically 8–12 cm long with many glands on the rachis and petiole. Each leaf has 8–21 pairs of pinnae, having 15–70 pairs of small crowded leaflets 1.5–4 mm long. Inflorescences consist of globular heads with 20–30 small, pale creamy-yellow flowers. Pods are more or less straight, 5–15 cm long and dark brown to blackish when ripe, often constricted between the seeds and finely hairy. They commonly contain 5–10 smooth black, ovoid seeds 3–5 mm long, each with a short, creamy coloured, terminal aril.

Distribution

The native range of *A. mearnsii* is south-eastern Australia; southern New South Wales, southern Victoria, south-eastern South Australia and Tasmania, from sea level to 1000 m. It occurs in the understorey of tall open-forest often dominated by eucalypts, on the margins of closed-forest, or rarely in woodland and coastal scrub. It can form dense, pure, even-aged thickets, especially where it has recolonized cleared land.

A. mearnsii is cultivated in many tropical, subtropical and even warm temperate regions throughout the world. Countries that first established extensive plantations include Brazil, China, India, Indonesia and in southern and eastern Africa, with many extensive, largely unrecorded plantings in agroforestry systems in many other countries.

Silviculture characteristics

A. mearnsii is a light-demanding but relatively short-lived pioneer species with a life-span of 10–20 years. It grows rapidly, up to 3 m/yr at 3–5 years old, when it reaches its maximum growth rate. Trees can reach 14–18 m in 10 years, with trunk diameters around 15 cm.

Trees usually start to flower when 2 years old, with prolific quantities of seed after the fifth year when yields may exceed 0.7 kg/tree. Flowering occurs in October–December in Australia, September–October in Brazil and August–October in South Africa. Seeds mature 12–14 months later, though are shed in 2–3 weeks, thus the timing of seed collection is critical.

Trees have shallow root systems and are susceptible to windthrow, and sensitive to fire when young. They are tolerant to some frosts and have survived temperatures to −12°C. *A. mearnsii* will tolerate relatively infertile sites but requires a good supply of phosphorus for rapid growth. Its coppicing ability is generally weak, although small stumps may coppice readily.

Wood volume yields reported from several different countries are in the range of 10–25 m^3/ha, assumed to be due to different site conditions. However, typical bark yields varied widely, from 15–20 tonnes of dry bark/ha in well-managed 10-year plantations in South Africa, to only 1–2 t/ha in Indonesia after 8–12 years.

Silviculture practice

Seed retains viability well in storage, and pre-germination treatments involving immersion in hot water for 1 min are common to ensure rapid germination. Vegetative propagation has had limited success. Seeds are mainly sown in containers, or in beds for bare-rooted seedlings, and reach plantable size (20 cm tall) in 4 months. *Rhizobium* inoculation may be beneficial, especially when seedlings are raised in sterilized media or are to be planted on highly degraded soils.

Establishment is usually in monoculture when managed for tannin production on a rotation length of 8–12 years, when sites are clear felled. Regeneration of cut-over areas is by replanting, spot or line sowing, or natural regeneration. The latter is improved by burning debris, and ploughing or ripping and weed control has been shown to improve establishment and growth. If replanting, reported initial spacings include 3 × 1.66 m (2100 trees/ha) and 3 × 1.33 m (2500 trees/ha) in Brazil, 2.7 × 1.5 m (2469 trees/ha) in Zimbabwe, and 2.75 × 1.2 m (3000 trees/ha) in South Africa. When sown directly in the field, the seed is sown in rows 1.8–2.7 m apart in well-cultivated weed-free ground at rates of 1.2–2.4 kg/ha, or five seeds per spot at various spacings, most commonly 3 × 1 m, thinning as required.

Intercropping is sometimes practised, with crops grown between the rows of trees in the first year, and *A. mearnsii* has also been used as a fallow crop in agricultural rotations. Significant increases in wood volume and tannin yield are observed with applications of phosphorus or combined NPK fertilizers at planting. Boron is also a major nutrient deficiency in several countries including China, Kenya and Tanzania. *A. mearnsii* is intolerant of weed competition when young, and plantations must be kept clean until the canopy closes.

One or two thinnings are common, to maintain rapid growth, increase bark thickness (directly correlated with stem diameter) and to obtain intermediate revenue streams. In Southern Africa, initial stocking of 2400 trees/ha may be reduced to 2000 trees/ha when tree height reaches 4 m, and to 1600 trees/ha when trees reach 7 m tall. Pruning is also sometimes carried out to reduce the incidence of forked trees and associated wind damage.

Bark stripping is best done when the sap is flowing during the warm summer months. If the bark must be stored, the strips should be carefully dried in the shade in a well-aerated area to minimize mould formation.

Diseases and pests

A. mearnsii is susceptible to various pathogens, some of which are responsible for economic losses in plantations. In several countries, significant problems and mortality are caused by gummosis, due to infection by various species of *Phytophthora*. Much research has been undertaken in the search for specific biocontrol agents, especially in South Africa.

The leaf-eating fireblight beetle (*Pyrgoides orphana*), is a serious pest in Australia. In young plantations in Zimbabwe, white grub (*Lepidiota mashona*), grasshoppers and cutworms are considered as problems in the nursery, and with the sap-sucking, the wattle mirid (*Lygidolon laevigatum*), are among the most serious pests. In southern Brazil, twig-girdling beetles and various stem borers also can cause severe damage.

Uses

A. mearnsii is planted in agroforestry systems and shelter belts, to improve soil fertility, control soil erosion on steep slopes, for soil and water conservation, and as a shade ornamental tree.

The wood is used for cabinet work, joinery, flooring, construction timber, poles, tool handles, matchwood and hardboard. The sapwood is very pale brown; the heartwood is light brown with reddish markings, and is fine textured with distinct growth rings. Wood density is 550–850 kg/m^3 (at 15% moisture content). The wood is hard and needs predrilling before nailing, but is moderately easy to work and takes a good polish. However, it must be seasoned slowly to avoid checking, and it is susceptible to attack by termites and *Lyctus* species.

Their mechanical, pulping and bleaching properties make *A. mearnsii* woodchips suitable for a wide range of paper and paperboard products, and 1 m^3 of wood produces about 320 kg of pulp. Plantation-grown *A. mearnsii* is used commercially and even exported to make kraft and soda-AQ (anthraquinone) pulp. *A. mearnsii* is also used to produce rayon.

The wood splits easily and makes fuelwood and charcoal of excellent quality. It is used as such extensively wherever it grows, especially in industrial kilns and for curing, and it is also used to produce activated carbon.

A. mearnsii produces the world's most important vegetable tannin, especially suited to the manufacture of heavy leather goods. The bark is very rich in condensed tannin (35–51% by dry weight), with yields influenced by several factors including genetic variability, age and environment. Tannin industries based on *A. mearnsii* have been developed in Brazil, China, Kenya, India, South Africa, Tanzania and Zimbabwe. The main exporting countries are South Africa, Kenya and Tanzania, and the main importers are the UK, Australia and the USA.

Powdered bark extract is also used to prepare tannin formaldehyde adhesives for exterior grade plywood, particle-board and laminated timber. It is used as an anti-corrosion agent of mild steel and cast iron, in insulation foams, mud fluidizing agents for drilling, calcite depressants in ore flotation, flocculants in water treatment, ion-exchange resins and polyurethane-type wood coatings. The sawdust is also an excellent medium for growing edible mushrooms. The leaves are rich in protein but high levels of tannin make them unpalatable, though they are used as a green manure, and plantations can produce more than 20 t/ha of fresh foliage annually.

Further reading

Brown, A.G. and Ho, C.K. (1997) *Black Wattle and its Utilisation.* Rural Industries Research and Development Corporation, Canberra.

Doran, C.J. and Turnbull, J.W. (eds) (1997) Australian trees and shrubs: species for land rehabilitation and farm planting in the tropics. Monograph No. 24. ACIAR, Canberra.

Jøker, D. (2000) *Acacia mearnsii* De Wild. Seed leaflet No. 4. Danida Forest Seed Centre, Humlebaek, Denmark.

Acacia melanoxylon R. Br.

Blackwood, Australian blackwood

Importance

Acacia melanoxylon wood is prized for use in cabinetmaking and as a source of veneer. It is also valued for shade, shelter and as an ornamental tree. Supplies from native Australian forests have dwindled to the point where all of the annual production of about 10,000 m^3 is now used within the country. However, it has been widely introduced as a plantation species in cool and warm humid climatic zones and warm subhumid climates throughout the world. To produce good timber, choice of provenance, site selection and careful management including protection from browsing are important. However, it has also become invasive in some countries.

Botanical features

A. melanoxylon commonly grows to 10–20 m tall and 0.5 m in diameter, but can also attain a height of 40 m and 1.5 m in diameter. In open situations, smaller and medium-sized trees are freely branched from close to ground level; larger trees have well-developed cylindrical trunks, although these may be shortly buttressed or flanged at the base. The hard rough bark is brownish-grey to very dark grey, furrowed and scaly, shedding in narrow vertical strips.

The dull green leaves (phyllodes) are falcate or almost straight, 8–13 cm long and 0.7–2 cm wide, commonly rounded at the tips, with 3–5 prominent longitudinal veins. Flowers are whitish to very pale yellow, gathered in groups of 30–50 in round (globular) heads (Plate 46). Inflorescences consist of 3–5 heads on short axillary racemes 2.5–5 cm long. Pods are flat, 6–10 cm long and 0.5 cm wide, irregularly twisted or openly coiled. Black, oval seeds have pink or red funicles half-encircling the seed in a double fold on each side, with a small aril.

Distribution

A. melanoxylon is native to south-eastern Australia, from south-east Queensland to southern Tasmania, including New South Wales and Victoria, extending westwards into South Australia. In its native range it is found in cool temperate rainforests (nanophyll moss forests) and as smaller trees in tall, open forest. It is predominantly an understorey species, a major early secondary tree species in disturbed areas, and a small shrub in mountain heath. It has a higher tolerance of saturated soil conditions than most eucalypt or rainforest overstorey species, and tends to become a more significant component of stands in swampy or seasonally flooded sites.

As an exotic, *A. melanoxylon* has been extensively grown in cool to mild climates with adequate rainfall, first in India, Sri Lanka, South Africa, Kenya and Tanzania, especially at higher altitudes. It is has since become a widespread plantation tree in many other parts of the world, in Europe (mainly Spain and Portugal), South America (mainly Argentina and Chile), China and New Zealand.

Silviculture characteristics

A. melanoxylon is a long-lived, hardy, nitrogen-fixing tree. It is tolerant of slightly saline soils and periodic waterlogging, even for several months continuously as long as the water is moving, however slowly. It is moderately resistant to drought and frost. It is known to survive temperatures as low as −7°C, and the roots may persist even during more severe frosts, although above-ground parts will die back. However, the best growth occurs on well-drained soils. It is a shallow-rooted species and is susceptible to windthrow, particularly in wet soils. It may grow well on some exposed sites, although trees will rarely attain sawlog sizes in such conditions.

The coppicing ability of *A. melanoxylon* is regarded as fair to good, best when cut at around 1 m above ground level. Trees also root sucker profusely. Trees may flower from as early as 2 years of age, but flowering times in the native range varies is variable. Seeds maintain viability well in storage, and can remain viable in the soil for more than 50 years. The tree also has characteristics that have seen it become an invasive species in several countries where introduced.

A. melanoxylon can be expected to produce trees 25 m tall on a 40-year rotation on favourable sites, and logs of 50 cm diameter with an acceptable amount of heartwood. Depending on the site, mean annual volume increments of 16–25 m^3/ha/yr after 20 years have been reported in plantations, and 15 m^3/ha/yr in natural stands. However, at these lower growth rates, it is considered uneconomical to grow *A. melanoxylon* unless combined with a commercial nurse crop such as pines or eucalypts.

Silviculture practice

Soaking in hot to boiling water is the most common seed pretreatment. After sowing into bags, seedlings can reach plantable size (20 cm in height) in less than 6 months, and inoculation with *Rhizobium* may be beneficial. Planting stock can also be raised and established successfully using open-rooted stock, stump plants and root cuttings and propagation by tissue culture has also been reported.

Although tolerant of a wide range of environmental conditions including low fertility soils, good site selection and silviculture are required to produce high-quality timber. Economic and silvicultural studies confirm that plantation management should be based on maximizing the production of clear heartwood for the high-quality furniture industry that is willing to pay a premium for defect-free *A. melanoxylon* timber.

Well-drained, sandy to silty loam soils with high humus content are preferred, and trees also require shelter from wind. Trees branch heavily at wide spacing, so close initial spacing of 2.5 × 2.5 m (1600 trees/ha) is often recommended in pure stands, or mixed with nurse trees such as *Eucalyptus* or *Pinus* species.

Intensive pruning and thinning are necessary to produce good timber trees, with for example three or four thinnings to a final density of 100–400 trees/ha after 10–15 years. Removal of branches exceeding 30 mm in diameter when trees are 3–7 years old is also recommended, and if in mixed stands, with regular and careful thinning of a nurse crop. Trees can also be pruned by about 50 cm in height each year or after each thinning, leading to final crop trees having a clear bole of 7–10 m. The current recommended regime in New Zealand is to have a stand with 200 trees/ha, pruned to 6 m with form pruning, clearwood pruning, and thinning completed after 10 years, with a rotation length of 35 years expected from fertile sites.

In natural stands in Tasmania, Australia, *A. melanoxylon* is managed on a 70-year rotation at a stocking of 200 trees/ha, producing logs above 50 cm diameter, with enrichment planting with nursery stock only when natural regeneration is inadequate. Natural regeneration can also be abundant in plantations, especially when logging residues are burnt.

Site preparation is reported to involve burning off all debris from the site, and cultivation. Eradication of all game from the site, erection of a game-proof fence, and effective game control and regular fence inspections should be undertaken for at least 3 years. Fertilizer application at planting or soon after is also considered important for fast early growth. Good weed control before planting and for the first 2 years after planting also leads to better survival and faster growth.

Diseases and pests

Many exotic species of *Acacia* including *A. melanoxylon* are attacked by an increasing number of pests and diseases. Susceptibility to fungal diseases such as *Armillaria* and *Phytophthora* appears to be minimal, although various fungal diseases have been reported on young plants of *A. melanoxylon*, including *Fusarium semitectum* which causes shoot die-back.

Significant pests in New Zealand are the ghost moth (*Aenetus virescens*) and pinhole borers (*Platypus* spp.). The sapwood is also susceptible to attack by *Lyctus* borer.

Seedlings may be defoliated by moths or grasshoppers, and larger trees attacked by wood borers, leaf eaters, psyllids and scale insects.

A. melanoxylon is subject to a substantial number of vertebrate pests in Australia, such as wallabies and rabbits.

Uses

Open-grown trees retain their lower branches for many years and form excellent single-row shelter belts. *A. melanoxylon* is also planted as a shade and ornamental species.

The wood is prized for cabinet work, panelling, inlays, bent work and staves. However, the availability of large, high-quality logs is increasingly limited, and today these tend to be used for sliced veneer, especially on particleboard for cabinet work and furniture. The wood has good acoustic qualities and is also used in instrument making.

A. melanoxylon sapwood is white and up to 10 cm wide. The heartwood is golden-brown to darker brown, sometimes with reddish tints and streaks. The grain is usually straight, but may sometimes be attractively figured with stripes or mottled, raindrop, bird's-eye and fiddleback patterns. The wood has a beautiful surface lustre, and takes a good polish. It is only moderately hard, however, with a basic density that is also moderately low (390–670 kg/m^3).

Small-diameter, fast-grown logs do not develop the growth stresses of some eucalyptus species, and good rates of recovery can be expected during sawmilling, from trees grown on

40–50-year rotations. *A. melanoxylon* has good pulpwood potential, giving acceptable pulp yields and paper properties favourable for fine papers.

The low-density wood makes a poor fuel, though it is known to be used as such in South Asia. Leaves are highly palatable for sheep and are occasionally harvested for cattle, though digestibility is low.

Further reading

Doran, C.J. and Turnbull, J.W. (eds) (1997) Australian trees and shrubs: species for land rehabilitation and farm planting in the tropics. Monograph No. 24. ACIAR, Canberra.

Hosagoudar, V.B., Archana, G.R. and Manojkumar, A. (2007) Diseases on wattles (*Acacia* spp.) - a review. *Journal of Mycopathological Research* 45, 219–223.

Nicholas, I. (2007) Understanding blackwood (*Acacia melanoxylon*) markets; an opportunity for improving blackwood plantation returns. *New Zealand Journal of Forestry* 52, 17–20.

Acacia nilotica (L.) Willd. ex Delile

Egyptian thorn, prickly acacia

Importance

Acacia nilotica (Plate 3) is an important timber tree in dryland Africa and Asia, and is also valued for fuelwood, fodder, gum, tannin and other products. It is a common tree of agroforestry systems, especially in northern India, and in natural stands, especially in riverine areas in the Sahel and East Africa. It is used extensively for rehabilitation of degraded saline and alkaline soils, in windbreaks, for shade, and as a street tree, and has a strategic value providing forage resources in silvopastoral systems in Africa. However, it has become invasive in many countries where introduced, particularly in more humid sites or where it is not heavily utilized.

Botanical features

The following is a general description, though *A. nilotica* is an extremely variable species divided into nine subspecies, six in Africa and three in South Asia. They are most easily distinguished by pod shape, size and degree of pubescence, but their habit and crown shape, habitat and geographic location may also be indicative.

The trees are often single-stemmed with 4–6 m of clear bole, generally large, to 15 m high but in riverine areas can reach 25 m or more with clean stems up to 8 m and with diameters of 50–80 cm. The species may also less commonly occur as small trees some 4 m tall with a short trunk. The crown can be upright and flattened, or rounded and spreading, and may vary from hemispherical to narrow and erect. Bark is grey or dark grey to brownish-black or black, rough and fissured, never powdery or peeling. Thorns are up to 10 cm, paired, straight, thin and light grey.

The alternate bipinnate leaves are green to blue-green, with rachis 1.2–8 cm long having a gland at the junction of each pinna pair or top few pairs only. Leaves have 2–14 pairs of pinnae, each with 7–36 pairs of leaflets 1.5–7 cm long and 0.5–1.5 mm wide, glabrous to pubescent. Flowers are bright or golden-yellow, sweet scented, in globose inflorescences on the current season's growth.

Pods are very variable depending on the subspecies, but all are indehiscent, grey to dark brown, straight or curved, glabrous or velvety, 4–22 cm long to 0.9–2.2 cm broad, the position of each seed clearly marked by a distinct raised bump in the pod valves, though margins may be entire or deeply constricted between each seed, such as the 'string of pearls' pods of subsp. *tomentosa*. Seeds are dark brown to brownish black, elliptic to subcircular, 7–9 mm in diameter, 6–17 per pod.

Distribution

A. nilotica is naturally widespread in the drier areas of Africa, from Senegal to Egypt and southwards to South Africa, and eastwards through Asia to India and Bangladesh. *A. nilotica* in Africa exhibits two very distinct ecological preferences: subsp. *subalata*, *leiocarpa* and *adstringens* occur in wooded grassland, savannah, and dry scrub forests on deep sandy loamy soils, and also on lateritic and calcareous sites; subsp. *kraussiana* also prefers dry grasslands and savannahs, especially on compacted sandy loam, shallow granite or clay soils along drainage lines and rivers, but away from flooding. On the other hand, subsp. *nilotica* and *tomentosa* are restricted to riverine habitats and seasonally flooded areas on clay alluvial soils. In the Indian subcontinent, subsp. *indica* forms low-altitude dry forests, usually on alluvium soils subject to flooding or black cotton soils, and has now been widely planted on farms throughout the plains. It will grow on saline, alkaline soils, even with calcareous pans.

A. nilotica has been widely introduced and cultivated in dry tropical and subtropical regions around the world, including areas in South Asia where it is not native, as well as being planted within its native range. However, its pioneer characteristics, especially on overgrazed lands, may lead to the formation of thorny thickets in its native range, though *A. nilotica* has also become a major invasive weed in some countries where introduced, notably Australia and Indonesia.

Silviculture characteristics

A. nilotica is a pioneering, light-demanding, nitrogen-fixing tree, regenerating well from seed, though seedlings are relatively shade-intolerant. Depending on the subspecies, it prefers a wide range of soil types, included degraded saline and alkaline soils up to pH 9, and may also tolerate both drought and flooded conditions for several months. *A. nilotica* forms a deep and extensive root system on dry sites, but on flooded sites the root system is largely lateral.

A. nilotica starts to flower after 3–4 years in ideal conditions and produce pods in abundance from 5–7 years. Flowering is prolific on the current season's growth during the rainy season, and can occur several times or almost continuously with available soil moisture, and mature trees can produce 2000–3000 pods or more in a year.

Growth is relatively fast, with reports of wood yields in the range of 3–9 m^3/ha/yr over a 15–20 year rotation, and 10–13 m^3/ha/yr in natural riverine forests over 20–30 years.

In addition, 12 t/ha of bark was harvested for tannin from plantations at 4 × 4 m spacing after 15 years, equivalent to 20 kg/tree. Pod production of 8–10 t/ha/yr is reported for use as fodder.

Silviculture practice

Seeds are extracted from the pods by pounding and winnowing, or collecting from animal pens after the pods have been eaten. There are 5000–16,000 seeds/kg, which store well, but require pretreatment, the easiest method being soaking in boiling water and allowing it to cool overnight. There are a number of reports of vegetative propagation, but regeneration by seed is preferred.

Plants can be sown directly, or into standard nursery bags with seedlings ready for outplanting after 4 months when 25–30 cm tall. Establishment varies considerably depending on site, though inoculating seed with improved rhizobia strains can increase survival and early growth, and salt-tolerant *Rhizobium* isolates have been identified. Seedlings require systematic weeding for the first 2–3 years for best results.

Spacing varies depending on the intended products. Dense initial stands are used for fuelwood plantations of 2 × 3 m or 3 × 3 m that can be grown on short 5–15 year rotations, and which are sometimes irrigated. These can also be thinned every 5 years and grown on longer 20–30 year rotations to yield sawlogs. Where tannin and gum are desired, 4 × 4 m spacing is usual with intercropping practised in the early years. In agroforestry systems in northern India with retained natural regeneration and some planting, intercropped trees may be 8–10 m apart, and are often pollarded for fodder.

Large timber trees have been exploited from the riverine forests along the Nile since the times of the ancient Egyptians. Remaining areas are currently managed on 20–30-year rotations. *A. nilotica* is also popular as a shelter belt, with special interest in subsp. *cupressiformis* because of its narrow crown. It is also a popular shade, ornamental and avenue tree throughout its native range.

Diseases and pests

A large number of pests and diseases are reported to attack *A. nilotica*, and some cause significant damage, including the scale insect *Anomalococcus indicus* that can kill trees in India.

Uses

The wood is nearly twice as hard as teak, strong, durable, termite resistant and very shock resistant, with an air-dry density of 650–830 kg/m^3. The heartwood is dark brown, and the sawn timber is of high quality and has a very wide range of uses, from cabinetry and construction, to rustic furniture and agricultural implements. It is also used for construction poles and fence posts, it good for carving and turnery, and its use for pulping has been investigated. The wood has a high calorific value of up to 5000 kcal/kg, making excellent fuelwood and high quality charcoal.

Trees are a valuable browse, with leaves and pods containing 12% crude protein and 8% digestible protein, and are rich in energy and minerals. Dried pods are relished by grazing livestock as they ripen towards the end of the dry season, and pods and occasional pod meal are used in mixed feed rations.

When harvesting for fuelwood or timber, the bark (containing 12–20% tannin) can be used or sold separately, and tannin is also extracted from the pods. The gum has some properties similar to those of true gum arabic (from *Acacia senegal*) and is used in calico printing, dyeing, sizing silk and cotton material, and in the manufacture of paints, paper and some medicines, marketed as 'Amravati gum' in India. The tannin contributes to many medicinal uses of *A. nilotica*, acting as a powerful astringent. Pods are a powerful molluscicide, and have killed both snail species which carry schistosomiasis when added to ponds in Sudan, without affecting the fish. There are many other reported uses, included sprouted seeds eaten as a vegetable, and well-roasted seeds mixed with coffee.

Further reading

Basavaraja, P.K., Sharma, S.D., Dhananjaya, B.N. and Badrinath, M.S. (2010) *Acacia nilotica*: a tree species for amelioration of sodic soils in Central dry zone of Karnataka, India. In: Gilkes, R.J. and Prakongkep, N. (eds) *Proceedings of the 19th World Congress of Soil Science: Soil solutions for a changing world.* IUSS, Brisbane, Australia, 1–6 August 2010, 73–76.

Brunck F. (1994) Pests and diseases of *Acacia nilotica*. *Bois et Forets des Tropiques* 239, 59–65.

Fagg, C.W. and Greaves, A. (1990) *Acacia nilotica 1869–1988. Annotated Bibliography*. CABI, Wallingford, UK and the Oxford Forestry Institute, Oxford, UK.

von Maydell, H.J. (1986) *Trees and Shrubs of the Sahel: Characteristics and Uses*. Schriftenreihe der GTZ, No. 147. GTZ, Eschborn, Germany.

Acacia saligna (Labill.) H.L. Wendl.

Blue-leaved wattle, Port Jackson wattle

Importance

Acacia saligna is the most widely planted non-timber *Acacia* species, with more than 600,000 ha established worldwide. It is a fast-growing, nitrogen-fixing large shrub or small tree, tolerant of calcareous, alkaline and saline soils in warm temperate climates. It has been planted extensively in North Africa, the Middle East, Southern Europe, South America, and elsewhere, principally for fodder production and soil conservation. It is, however, considered an invasive species in an increasing number of countries.

Botanical features

A. saligna is a dense and multi-stemmed, thornless, spreading shrub or single-stemmed small tree 2–6 m tall, but up to 9 m in height where introduced. Where it has become naturalized, however, it is commonly a dense bush which may be wider than the plant is high. The bark is smooth and grey to red-brown on branchlets, becoming dark grey and fissured with age.

Leaves are phyllodes, dark green to blue-green with conspicuous midribs, long and narrow to lanceolate and 8–25 cm long, straight or sickle-shaped and sometimes pendulous. Phyllodes may also be prominently broader at the base. There is a solitary gland on the upper margin of the phyllode, circular or oblong and 1–2 mm in diameter. Two distinct genetic types have been identified, one with an intricate bushy habit, long narrow phyllodes and a high leaf:stem ratio more suitable for forage production, and a second type with a more erect habit, broad phyllodes and low leaf:stem ratio more suitable for firewood production.

The inflorescence comprises globular flower heads, 5–10 mm in diameter, containing 25–55 (up to 78) bright yellow, five-parted flowers. The pods are narrow, 4–6 mm wide and usually 8–12 cm long, usually contracted between the seeds and with an undulate surface. The shiny seeds are 5–6 mm long × 3–3.5 mm wide, dark brown to black, with 14,000–25,000 seeds/kg.

Distribution

A. saligna is native to a relatively small part of south-western Western Australia, where it is very common on the poor sandy soils of the Swan Coastal Plain and on heavier clay soils further north. Where native, it grows best on deep sands and loams associated with watercourses, and in coastal dunes, more commonly in the hollows between dunes. It is also found on disturbed roadsides, and further inland in the wheat belt, with populations at the base of many of the large, granitic rock outcrops.

It has been very widely introduced throughout the world during the past 150 years, and is now sometimes a very common tree in arid and semi-arid sites with tropical, subtropical but especially Mediterranean climates, where it has often naturalized. It is particularly frequent in most countries of the Mediterranean basin (Southern Europe, North Africa, and the Middle East), South Africa, parts of South America, south-eastern Australia, and California, USA. It has escaped from cultivation in many countries, where it is observed to spread in a variety of habitats, but especially along watercourses. It is important to note that an old synonym, *Acacia cyanophylla*, is still widely but incorrectly used in parts of the introduced range to this day.

Silviculture characteristics

A. saligna reproduces almost entirely by seed and is an outcrossing species; however, it is also known to produce root suckers, and a single tree may form a clump or small colony. It produces seed in large numbers and builds up a significant seed bank, which can germinate rapidly after fire or sufficient rain. Like most legume seed, it can remain dormant in the soil for considerable periods of time. The species is relatively precocious, often producing seeds in the first 1–2 years after germination, depending on site.

A. saligna is relatively short-lived in cultivation (10–20 years), but tolerates alkaline and saline soils, waterlogging and salt winds. It is well-suited to heavy saline and waterlogged clay soils where it gives fast growth, excellent survival and large crown growth. It coppices well, especially after fires. In common with many other acacias, its roots are associated with nodules of certain strains of *Rhizobium* and fixes atmospheric nitrogen, but this ability is greatly reduced by drought, waterlogging, shading or defoliation. *A. saligna* exhibits frost sensitivity, and damage is likely to be severe if the temperature falls below −4°C. It generally grows poorly in tropical areas as compared to other *Acacia* species, except at high altitudes.

A. saligna grows quickly in favourable conditions, often reaching 8 m tall with a spread as great as its height in 4–5 years. Yields of *A. saligna* wood vary from 1.5–10 m^3/ha depending on site conditions. In arid parts of North Africa, plantations are reported to produce 3.5 tonnes of air dry wood/ha/yr on deep sandy loams. Foliage production for fodder can attain 12–13 t/ha in the first year and increase with age under irrigated conditions. The wide variability of production between trees suggested that useful gains are possible through selection and plant breeding. In subtropical areas, *A. saligna* trees can exceed 6 m in height after 3.5 years but growth was seen to slow after that age, with slower growth and increased mortality reported from tropical regions.

Silviculture practice

Seed stores well and dormancy is usually broken by soaking in hot or boiling water for 1 min. Planting of seedlings raised in nursery containers is the most common form of establishment, and vegetative propagation has shown only limited success. Direct sowing into well-cultivated rows is also used in some regions, mostly in North Africa and the Middle East, and natural regeneration may be possible for restocking clear felled stands.

It may be planted in dense monoculture plantations, especially on more favourable sites, but is also commonly planted at wider spacings in silvopastoral agroforestry systems in drier areas, allowing for herbaceous forage production between trees during the rainy season. The species is also commonly grown in rows, either planted or directly sown, where the ground has been ripped along contours to capture rain run-off, with rows often 8–20 m apart. On better sites, crops such as barley are also occasionally grown in between the tree rows. *A. saligna* is a successful farm tree in some areas, in agroforestry systems, improving soil fertility and reducing erosion, lowering water tables and mitigating soil salinity where this is a problem.

The clear felling of plantations for fuelwood production is usually undertaken at 10–15 years of age. For fodder production, regular harvesting of foliage every year is recommended, and regrowth can be rapid, or stands can be browsed directly. Trees can be stripped bare, particularly at the end of the dry season, and rotational grazing is suggested.

Diseases and pests

Older plants are susceptible to the gall rust *Uromycladium tepperianum* and various gall-exploiting insects. In parts of its native range, more than 90% of *A. saligna* trees bear conspicuous woody galls, and this gall rust has been used successfully as a biological control agent where the tree has proved invasive.

A number of other pests and diseases have been reported from other countries where it has been introduced. Termites can be a problem in tropical regions, and rodents may damage the bark at the base of the stems.

Uses

The species has been used for sand stabilization, and as a windbreak and ornamental, and is being increasingly planted in rangeland agroforestry systems especially in regions with dry Mediterranean climates.

A. saligna wood has an air-dry density of about 600 kg/m^3, but is not used as a sawn timber though it has been successfully tested for making particleboard. It is used for poles such as vine stakes, and for small agricultural implements. The wood is widely used as a fuel and for charcoal, though is of lower quality than many other *Acacia* species. It gives a charcoal yield of 24% with medium specific gravity and low tar yield.

The foliage, young shoots, pods, and seeds (fresh or dry) have high crude protein contents and are palatable to livestock, especially sheep and goats. They are especially valuable as a fodder resource during the dry season when other forage is scarce, and foliage on established trees can be completely grazed off in autumn without harming the plants. The phyllodes contain 50–55% dry matter, 12–16% crude protein, 20–24% crude fibre, 6–9% crude fat, and 10–12% ash. However, digestibility is reported to be moderately low, and leaves alone should not be fed to small ruminants due to low intake and negative nitrogen balance caused mainly by the high tannin content.

Trees were planted in the past for tannin production, with the bark containing 30% tannin. Leaves can also be used to dye wool a lemon-yellow colour using an alum mordant.

Further reading

El-Lakany, M.H. (1987) Use of Australian acacias in North Africa. In: Turnbull, J.W. (ed.) *Australian Acacias in Developing Countries*. ACIAR Proceedings No. 16. Australian Centre for International Agricultural Research, Canberra.

Fox, J.E.D. (1995) A review of the ecological characteristics of *Acacia saligna* (Labill) H. Wendl. *Mulga Research Centre Journal* 12, 39–56.

George, N., Byrne, M. and Yan, G. (2008) Mixed mating with preferential outcrossing in *Acacia saligna* (Labill.) H. Wendl. (Leguminosae: Mimosoideae). *Silvae Genetica* 57, 139–145.

Michaelides, E.D. (1979) Mini-monograph on *Acacia cyanophylla*. Technical Consultation on Fast-Growing Plantation Broadleaved Trees for Mediterranean and Temperate Zones. Lisbon, Portugal, 16–20 October 1979. FAO, Rome.

Acacia senegal (L.) Willd.

Gum arabic, gum acacia

Importance

Acacia senegal is a nitrogen-fixing tree of the dry tropics and subtropics of Africa, and the source of most of the internationally valued and traded 'gum arabic' used in the food, pharmaceutical, cosmetic and other industries. After a decline in production in recent decades due to desertification and over-exploitation for fuelwood, there is now a renewed interest and expansion in productivity and export from the centres of production in Sudan and other African countries. Secondary roles include the provision of fodder and fuelwood, restoring soil fertility in agricultural systems, as windbreaks, sand dune stabilization and desertification control. However, being a thorny, prolific seeder, it can become invasive and form impenetrable thickets.

Botanical features

A. senegal develops different growth forms most commonly 2–6 m tall; several distinct botanical varieties have been described to reflect these such as low shrubs with flat crowns, low bushes with tall whippy stems, small trees or larger trees to 12 m high. Bark on the main trunk is yellowish to grey or greyish-brown, rough and fissured, very scaly on older trees, or smooth and papery, peeling off in strips. Trees have distinctive thorns in groups of 3, the central thorn curved and 2–8 mm long, with two straighter thorns on either side.

The small leaves are bipinnate and grey-green. There are 2–12 pairs of pinnae, each with 5–25 pairs of leaflets 1–9 mm long and 0.5–3 mm wide. However, the common gum-producing variety in the Sahel region has 3–6 pairs of pinnae, each with 10–20 pairs of leaflets 3–6 mm long and 1–2 mm wide.

The very fragrant flowers are whitish or yellowish-white, on elongated spikes 3–8 cm long, sometimes up to 12 cm. The papery thin pods are yellowish or greyish-brown, 2–19 cm long (commonly 7–10 cm) and some 2 cm wide, and are dehiscent, splitting open when mature. Pods usually contain 3–6 light brown, flat round seeds, 2.5–6 cm in diameter.

Distribution

Its natural distribution includes the breadth of the African continent at its widest between the Sahara Desert and the more humid forest zones, continuing across the Arabian Peninsula to Pakistan and India, and extending south through East Africa with disjunct populations in Southern Africa as far south as Transvaal, South Africa. *A. senegal* var. *senegal*, the most widespread variety from which most of the internationally traded gum arabic is collected, dominates in the Sudano–Sahelian belt from Senegal to Somalia.

A. senegal is planted in and around its native range but has not been widely introduced elsewhere, though some ornamental plantings can be found, for example in Australia, Chile and some Caribbean islands.

Silviculture characteristics

A. senegal is a light-demanding, nitrogen-fixing, pioneer species that colonizes a wide range of dry sites, preferring well-drained sandy soils. It will also grow on many soil types including loamy sands and clays, and will tolerate poor stony or gravelly soils. It regenerates well from seed, providing the grazing pressure is low. Relatively slow-growing, it resprouts when coppiced and can be grown at close spacing for fuelwood production or at wider spacings in agroforestry systems.

A. senegal flowers at a relatively young age, at about 3 years in ideal conditions. Flowering varies between varieties and regions, occurring either before or soon after the rains, with pods maturing 3–4 months later. It has a relatively shallow root system, but can develop deep taproots depending on the site, with extended lateral roots under sandy conditions.

Growth estimates of *A. senegal* plantations and natural stands vary considerably, but are relatively slow, rarely exceeding 50 cm in height/yr. Wood production in gum-producing plantations is commonly less than 1 m^3/ha/yr, though higher-density fuelwood stands have been reported to produce up to 7 m^3/ha/yr.

Trees start to produce gum after 5 years, reaching a maximum when 10–15 years old, and are generally cut down after 15–25 years. Yields vary considerably between trees, sites, tapping dates and techniques. Average yields of gum per tree of 0.20–0.25 kg/yr were reported from Sudan, but only 0.1 kg/yr from Mauritania. However, in the main producing region of South Kordofan State in Sudan, another study found gum yields to be 0.3–1.0 kg/tree, with optimum tapping dates from mid-October to mid-November. Maximum yields in Kenya were found to be from June to September, indicating regional differences.

Silviculture practice

Fruits mature in the dry season, and pods are best collected (preferably before they have opened) by shaking the trees. Seed production is usually annual but is very variable between years and trees, with 8000–18,000 seeds/kg depending on the provenance. Seeds can remain viable for several years. When collected fresh they can be sown directly, but the seed hardens within a season, after which pretreatment by soaking in hot or cold water is usually recommended.

Seed are usually sown 1 cm deep in containers 30 cm long, with frequent root pruning recommended, before being ready for planting out after 4–6 months. In sandy soils, the minimum spacing is 4 × 4 m in plantations, or 10 × 10 m if agricultural crops such as millet, beans or groundnuts are to be intercropped. On clay soils, mechanical preparation by subsoiling is preferable, but this is costly, and planting in hand-dug pits is usual.

In Sudan, *A. senegal* is grown in traditional bush-fallow systems on approximately 20-year rotations. The land is cropped with agricultural crops for 5 years; this is followed by 5 years when the trees are established, and controlled grazing is gradually introduced; and then by 10 years of full gum production and grazing. Trees can then be coppiced, or cut and replanted. In the first 2 years, weed control and protection from fire and livestock are important, and manure or fertilizer application can improve establishment. In Asia, a coppice-with-standards system is often practised, sometimes on longer (30–40 year) rotations, and inoculation with mycorrhizae has been shown to improve growth on dry sites. Cultivation of *A. senegal* in traditional agroforestry and silvopastoral systems has increased in recent decades because of the need to improve soil fertility, with gum arabic proving to be a valuable extra bonus for subsistence farmers and pastoralists.

Tapping for the exudate gum occurs in the dry season after the leaves have fallen. Bark is peeled back for about 75 cm on younger stems using an axe or other tapping tools, and gum nodules form and harden in 3–8 weeks. The international market demands clean unbroken nodules, and these fetch the best price. Sudan dominates the world supply of high-quality gum arabic, with Chad and Nigeria rapidly expanding the production of lower-quality grades. One person can tap up to 100 trees/day and cover up to 30 ha per season, and may harvest 200–300 kg of gum/yr.

Diseases and pests

In Sudan, the tree locust (*Anacridium melanorhodon melanorhodon*) (Orthoptera, Acrididae) is the single most serious pest of *A. senegal*, and many species of longhorned beetle species are also known to cause considerable damage.

Uses

In Africa, wood is a secondary product from *A. senegal*, and is usually only harvested at the end of the gum production cycle. It has a basic density of 728 kg/m^3 (at 12% moisture content) and is used principally for fuelwood and charcoal, but also sometimes for construction poles and tool handles. In Asia, however, it is primarily grown for fuelwood and charcoal.

The most important product is gum arabic, a complex polysaccharide with a high solubility in water and a low viscosity, giving it emulsifying, stabilizing, thickening, and suspending properties. As defined by the Codex Committee for Food Additives, gum arabic also includes that from *A. seyal*, otherwise known as 'gum talha', though of less importance in terms of production and quality, and fetching only half the price of gum from *A. senegal*. In the pharmaceutical industry it is used for the preparation of tablets and syrups, and in the cosmetics industry as an adhesive and in lotions. It is also used in industrial production of adhesives, paints and coating papers. Other commercial gums are also produced by species of *Boswellia* and *Commiphora*, with lower-quality gums from other *Acacia* species, mostly *Acacia seyal*. There is scope for significant increases in gum production from the selection and breeding of superior *A. senegal* trees, and a number of programmes are under way in some countries.

In addition, *A. senegal* is an important source of fodder where it is abundant. The foliage and pods are palatable to livestock, both containing 10–15% digestible protein and 0.12–0.15% phosphorus. Dried *A. senegal* seeds are also used as a vegetable in parts of India, flowers produce nectar for bees and rope fibres are made from the roots.

Further reading

Fagg, C.W. and Allison, G.E. (2004) Acacia senegal *and the Gum Arabic Trade: Monograph and Annotated Bibliography*. Tropical Forestry Papers, No. 42. Oxford Forestry Institute, Oxford, UK.

Muller, D. and Okoro, C. (2004) *Production and Marketing of Gum Arabic*. NGARA Publication Series 2. Network for Natural Gums and Resins in Africa, KEFRI, Nairobi.

von Maydell, H.J. (1986) *Trees and Shrubs of the Sahel: Characteristics and Uses*. Schriftenreihe der GTZ, No. 147. GTZ, Eschborn, Germany.

Acacia seyal Delile

Whistling thorn

Importance

Acacia seyal is a nitrogen-fixing, drought-tolerant, and fire-resistant tree widespread in the dry lowland tropics and subtropics of Africa. It is principally used as a fuelwood and fodder tree, but is being increasingly exploited for the production of the valuable exudate gum sold in international markets. Natural stands are managed for these uses, but many plantations have also been established in its native range where it prefers heavy clay soils prone to waterlogging; it has rarely been introduced elsewhere.

Botanical features

A. seyal is a medium-sized tree around 10 m high with a flat-topped and spreading crown, occasionally up to 17 m. Mature trees have a stem diameter usually 20–30 cm but up to 60 cm under favourable conditions, with a very distinctive smooth and powdery bark that is white to greenish-yellow or orange-red, with a green layer beneath. Young branches are green, often bearing numerous reddish glands. The sharply pointed and paired thorns can reach 8 cm in length but are usually shorter, and are the main distinguishing feature between the two varieties. The thorns are separate in the more common type (var. *seyal*), but are fused at the base, forming inflated ‘pseudo-galls’, in var. *fistula*.

Alternate bipinnate leaves commonly have 4–7 pairs of pinnae, rarely 1–3, or 8–12, which are 1.5–4 cm long with glands at the junction of the top 1–2 pinnae pairs. Each pinna has 8–22 pairs of small close-set leaflets, 3–10 mm long and 0.75–1.5 mm wide.

Bright golden-yellow, scented flowers are in globose inflorescences 10–20 mm in diameter on stout peduncles 3–3.5 cm long, with 2–3 heads together at the leaf axils of the current year’s growth. Pods are flat, slightly curved and dehiscent, with the two thin papery walls splitting open when mature on the plant. They are reddish-brown, finely longitudinally veined, 7–22 cm long and 0.5–0.9 cm wide, margins discernibly constricted between the seeds. The olive to olive-brown seeds are elliptic and compressed, 7–9 mm long, 4–5 mm wide and approximately 2 mm thick.

Distribution

The native range of *A. seyal* stretches throughout the Sudano–Sahelian belt of Africa from Senegal to Somalia, south through East Africa to Malawi, Mozambique and Zambia, along the Red Sea coast of the Arabian Peninsula and up the Nile Valley into Egypt (the only part of the range outside the tropics). The range of *A. seyal* var. *fistula* overlaps almost entirely in the eastern part of this range, but is the only variety found south of Tanzania. It is commonly found along drainage lines or in basins and depressions on typically level or gently sloping sites. *A. seyal* has only rarely been introduced elsewhere, with reports from India, Iraq, Florida (USA), southern France and Portugal, mostly as an ornamental.

Silviculture characteristics

A. seyal is a drought-resistant nitrogen-fixing tree and will grow in areas with an annual rainfall as low as 250 mm, but is commonly found in areas receiving 500–1200 mm with a dry season of 6–8 months. It can withstand mean daily maximum temperatures up to 55°C, and is sensitive to frost. It is a light-demanding pioneer species that regenerates well from seed and colonizes a wide range of dry sites, flourishing on relatively fertile heavy clay soils prone to waterlogging. It has a relatively shallow root system with lateral roots mainly in the upper 50 cm of the soil profile, but can develop taproots to 6–7 m deep. Trees shed their leaves in the middle of the dry season, and remain leafless for 1–4 months, sometimes longer. Trees start to flower at 2–3 years under ideal conditions. Flowering times vary through the native range but generally occur in the early to mid-dry season, with fruits maturing 3–4 months later.

Reports of height growth in natural stands are 1–3 m after 4 years, and 4–5 m in irrigated plantations. Annual diameter increments as high as 1–1.5 cm have been recorded, although most estimates are much lower. Reports of wood production from natural woodlands are in the range of 0.3–3.2 m^3/ha/yr, more commonly 1–2 m^3/ha/yr. Gum production begins after 5 years, reaching a peak around 15 years, as with *A. senegal*.

Silviculture practice

Seeds are most easily collected by shaking trees before the pods have opened, though production is very variable between years and trees. Fruits mature in the dry season, and there are 13,000–22,000 seeds/kg. Seeds store well, and can be sown directly when fresh, but require pretreatment if not, by soaking in cold water or boiling water and leaving to cool for 24 h. Rooted cuttings can be used, but regeneration by seed poses no problems. Planting stock is usually raised in standard nursery containers, though regular root pruning is advised, and seedlings are ready for planting out after 4–6 months.

Direct seeding is also commonly used, either by broadcast or mechanized sowing. Usual spacings for outplanting or direct sowing are 2 × 2 m or 3 × 3 m, and thinning is often practised. One system uses an initial density of 1000 trees/ha, thinned to 675 stems/ha at 11 years and 450 stems/ha at 15 years before clear felling at 20 years. In plantations for fuelwood and charcoal, 15–20 year rotations are common, with older individuals retained as seed sources if restocking by natural regeneration.

Trees are also intercropped in early years with crops such as sorghum, millet or sesame, which reduce weed competition as well as providing valuable early returns. In the Sahel, gum as a secondary product is collected in the dry season after the leaves have fallen, with gum nodules hardening after 3–8 weeks where the bark has been damaged.

For fodder production, the beating of branches with sticks to bring down leaves and pods is well established, and lopping small branches is common, although coppicing and pollarding trials have not proved successful. It has shown value as a live hedge, where it may also serve as a compact fodder bank.

Diseases and pests

No pests or diseases are reported to significantly affect the growth of *A. seyal*, but a number of galls affecting the foliage and fruit have been identified.

Uses

The wood of *A. seyal* is heavy and shock resistant, pale yellow to medium-brown, some localized pinkish-brown patches, and sometime dark mahogany-red in sawn timber from older trees. Wood density is 730–800 kg/m^3 at 12% moisture content. The sapwood is perishable and susceptible to fungal attack, and the heartwood does not season well and is vulnerable to insect attack from wood-boring insects, which may account for its neglect as a source of timber. It is also difficult to work, rapidly blunts cutting tools, tends to split, and requires pre-boring before nailing.

It is used for construction poles, with durability greatly enhanced by soaking in water for 2 weeks and air drying, and for making stools and other small articles, but the wood is principally used for the high-quality fuelwood and charcoal it produces.

The exudate gum, a complex polysaccharide with high solubility in water, is an important secondary product. However, there are contrasting reports as to the quality and uses of 'gum talha' from *A. seyal* in comparison to 'gum arabic' from *A. senegal*. Several detailed studies state its suitability only for sizing textiles, as a binder in mould materials in foundry and ceramic industries and in the production of inks and lithographic work, and commenting that 'gum talha is not recognized as an acceptable food additive by any of the international regulatory authorities'. However, other researchers quote the 31st Codex Committee for Food Additives in 1999, that 'gum arabic is the dried exudate from the trunks and branches of *Acacia senegal* and *Acacia seyal* in the family Leguminosae', implying that both can be mixed and used for the same purposes.

A. seyal is also an excellent source of fodder where it is abundant. Leaves and pods are palatable to livestock (sheep, goats and camels), with leaves containing 8–13% digestible protein and 0.15–0.28% phosphorus, with slightly higher levels of both in the pods. The bark contains 20% tannin and is used as an astringent in traditional medicine, wood smoke is used to fumigate against insects in the home, pod extracts have been shown to have molluscicidal properties and flowers provide good bee forage.

Further reading

Hall, J.B. and McAllan, A. (1993) Acacia seyal: *a Monograph*. School of Agricultural and Forest Sciences Publication, No. 4. University of Wales, Bangor, UK.

Muller, D. and Okoro, C. (2004) *Production and Marketing of Gum Arabic*. NGARA Publication Series 2. Network for Natural Gums and Resins in Africa, KEFRI, Nairobi.

von Maydell, H.J. (1986) *Trees and Shrubs of the Sahel: Characteristics and Uses*. Schriftenreihe der GTZ, No. 147. GTZ, Eschborn, Germany.

Acacia tortilis (Forssk.) Hayne

Umbrella thorn

Importance

Acacia tortilis is one of the most drought- and heat-tolerant of tree species, found in some of the harshest desert environments where few other trees survive. It is dominant in many dry lowland, dry tropical, and subtropical zones of Africa and the Middle East, and is often dominant in savannahs. It is an important source of fuelwood and fodder and provides many other products. It was introduced to South Asia where it has proved very successful in fuelwood plantations in very arid areas. It is also planted for sand dune stabilization and in silvopastoral and other agroforestry systems.

Botanical features

A. tortilis is a variable species with several recognized subspecies. It can occur as a multi-stemmed shrub (subsp. *tortilis*) and as a tall, distinctive tree reaching heights of over 20 m with rounded (subsp. *raddiana*) or flat-topped (subsp. *heteracantha* and subsp. *spirocarpa*) crowns, but in very arid sites it may only develop into a shrub. Initially it forms a small, spreading shrub with many spines, although these become less frequent on larger trees. Over much of East and Southern Africa it is recognized by its characteristic spreading crown, giving rise to its popular name of umbrella thorn.

The bark is grey to grey-brown or black, rough and fissured. Younger stems are smoother, red-brown, with light grey lenticels, and can be glabrous to densely pubescent. It is distinguished from most African *Acacia* species in possessing both long straight (up to 2 cm long) and shorter paired hooked thorns (to 0.5 cm long) at each leaf axil.

The leaves are alternate, bipinnate and smaller than those of most other African acacias. Rachis are 0.2–4.6 cm long, usually with glands between the top 1–3 and lowest 1–2 pinnae pairs, with 2–10 pairs of pinnae, each with 6–22 pairs of leaflets 0.5–6 mm long and 0.5–1.5 mm wide.

Scented white-yellow-cream-coloured flowers are gathered in globose heads 0.5–1.1 cm in diameter. The pods typically spiralled or contorted, occasionally straight, 10–15 cm long and 0.5–1.3 cm wide, light brown when mature, indehiscent, containing 8–16 small oval light brown to red-brown seeds.

Distribution

A. tortilis occurs throughout the savannah and dry zones of Africa, from Senegal to Somalia, through East Africa to South Africa, in southern Arabia, and extending through the Middle East to as far as Iran. It is found in the countries that fringe the Sahara on both its latitudinal edges, and is often the species that extends furthest into the desert. *A. tortilis* occurs in a wide range of habitats in its native range, from sand dunes to rocky scarps and alluvial valley bottom soils with high water table levels, although it usually avoids seasonally waterlogged sites. The geographic ranges of the subspecies are generally distinctive, with subsp. *heteracantha* and subsp. *spirocarpa* found in Southern Africa, and subsp. *tortilis* and subsp. *raddiana* which overlap in East and West Africa.

It has not been widely introduced, with the exception of South Asia and parts of the Arabian Peninsula, where it has proved to be one of the most successful species for reforestation of the harshest sites, particularly in north-western India and parts of Pakistan. Where introduced, it favours alkaline soils, but will colonize a wide range of sites including saline, gypsic and sodic soils, and granitic and gravelly soils, but develops best on loamy valley soils, although large populations are also found on steep rocky slopes.

Silviculture characteristics

A. tortilis is a very drought-tolerant nitrogen-fixing tree that can be fast growing, regenerating by seed but also resprouting vigorously when coppiced or browsed. Older trees above 3 m tall can also withstand frosts and light grass fires. *A. tortilis* forms a deep taproot in sandy soils, and this has been known to extend to depths of 35 m in the southern Sahara. On shallower soils and in arid sites it also develops subsurface roots extending to over twice the width of the crown.

On favourable sites, *A. tortilis* can start flowering on the previous year's growth at 3–4 years old, just before and during the rainy season. Flowering is prolific, and usually broadly synchronous in the population, although occasional trees can flower out of season. Flowering to fruit maturity can take 3–4 months or more, being timed towards the end of the dry season. Mature trees can produce more than 6000 pods in a good year, and fruit drop can be spread over a number of months, with some pods retained for longer periods.

A. tortilis improves and conserves soils and stabilizes sand dunes on arid sites either as plantations, shelter belts, or in a range of agroforestry systems, where the tree can also provide valuable shade. In its native range, some of the most palatable grass species grow beneath *A. tortilis*, and the productivity and diversity of the herbaceous layer and its soil nutrients were found to be significantly greater under the canopy than away from it.

Silviculture practice

A. tortilis seeds are best collected by shaking pods down from the canopy, as naturally fallen indehiscent pods are soon consumed by animals. There are 10,000–50,000 seeds/kg depending on the subspecies. Seed is extracted by pounding and winnowing, or by collected from the dung of livestock fed on the pods.

The hard-coated seeds remain viable for several years, but require pretreatment to achieve high and uniform germination. Immersing in boiling water and allowing the water to cool is the simplest technique. Seedlings are most commonly raised in standard nursery bags, but they exhibit rapid taproot growth and frequent root pruning may be required, especially as they may take longer than other species before being ready for planting out, up to 8 months being reported.

Initial spacing depends on the end use and water availability, with 3 × 3 m to 5 × 5 m common on 10-year rotations for fuelwood production. For intercropping in agroforestry systems, 6 × 4 m spacing has been used, wider still in silvopastoral systems in degraded grasslands. Seedling growth can be slow at first on marginal sites, but later development is more rapid if roots can reach a water source. For optimal development, weeding and protection from browsing animals is required in the first 3 years.

Growth can be rapid, with trees in India reported to reach almost 4 m in height after 2.5 years without irrigation and only 320 mm mean annual rainfall, and 6–7 m after 10 years. Fuelwood production varies from 3.2 t/ha at 5 years to more than 50 t/ha after 12 years, although most estimates are in the range of 9–22 t/ha on a common 15-year rotation. The production of leaves for browse or fodder is considered to be around 1 t/ha/yr.

Diseases and pests

There are a number of reports of pests and diseases attacking *A. tortilis* in its native range, and particularly where introduced to India, though few appear to cause significant damage and only the bruchid and buprestid beetles are of economic importance.

Uses

The hardwood is red and dense (reported density 640 kg/m^3), and makes very good fuelwood and charcoal, burning slowly and producing practically no smoke, with a high calorific value of 4400 kcal/kg. Poles are commonly used in the construction of huts, fence posts, and for tools, though soaking in water increases durability. Although trees often have a short bole before forking, there are many reports of the use of sawn timber for making furniture and agricultural implements.

A. tortilis is an important fodder tree for arid areas, where it constitutes a staple browse. Leaves contain 12% digestible protein and are rich in energy and minerals. Seeds have high levels of phosphorus (usually low in grasslands) and are particularly high in crude protein (38%). The pods need to be milled prior to feeding to livestock, however, otherwise most seeds will not be digested. They can provide a low-cost concentrate supplement, and are found to increase milk yields in lactating animals.

They can provide a low-cost concentrate supplement, and given the dominance of the species, the tree is used by many pastoral societies. Flowers from *A. tortilis* provide a major source of good-quality honey in some regions. In parts of East Africa, porridge is made by mixing the pulp of the pods with milk, and the immature seeds are also eaten. The bark yields tannin, and the inner bark provides cordage. Thorny branches are used for enclosures and livestock pens. *A. tortilis* exudes a red gum which is used in many local medicines, as are the leaves, bark and seeds; the bark and roots can be used as a dyestuff. Ash of the dead wood is mixed with ground 'snuff' tobacco in West Asia.

Further reading

Fagg, C.W. and Greaves, A. (1990) Acacia tortilis *1925–1988*. Annotated Bibliography. CABI, Wallingford, UK and the Oxford Forestry Institute, Oxford, UK.

Luna, R.K. (1996) *Plantation Trees.* International Book Distributors, Dehra Dun, India.

Shabankareh, K.N.T. (2003) Investigation on some scientific characteristics and utilization of *Acacia tortilis* (Forssk.) Hayne. *Natural Sciences* 60, 69–79.

Acer campestre L.

Field maple

Importance

Acer campestre is distributed throughout Europe, North Africa, and West Asia, and has naturalized elsewhere. It grows to about 15 m height, and is mature at age 50 years. It is a later seral species, often associated with ash (*Fraxinus excelsior*), and is rarely found invading grasslands or the early stages of plantations. It is frost-resistant, and tolerates heavy soils, particularly calcareous ones.

A. campestre often constitutes quickset hedge or forest-border vegetation. It is rare in montane areas; for example, in France best growth is on sites below 600 m elevation. It is a common understorey component of mixed deciduous woodlands in lowland plains, lower hills, scrub areas, and along riverbanks. It is often thought of as a calcicole, but it grows equally well in acid soils. It rarely grows large enough to be a timber tree, but it coppices well and tolerates clipping, so is suitable for use in hedges and low screens. One of its principal uses is as an ornamental tree for landscaping in private and public gardens and parks, and for roadside, street and amenity planting. *A. campestre* also provides good honeybee forage.

Botanical features

A. campestre is a shrub or small tree which sometimes grows to 15 (20) m in height, but usually less. The stem is short, often sinuous and sometimes burred. The crown is rounded, with short side-branches; the branch ends droop then turn upward. There is a central primary root and numerous lateral secondary roots.

A. campestre is not long-lived, reaching maturity at 50 years, and rarely reaching 120 years old. It can be long-lived if coppiced or layered in hedges, but stem damage reduces its durability, allowing fungi to penetrate. It grows moderately fast to a height of about 9 m, but slower thereafter.

Young shoots are deep red-brown, usually with a lighter side, finely pubescent at first. The young bark is soft and corky, pale brown and uneven, with wide orange fissures or cracked into squares. Older trees have pale grey bark with fine cracks and pale ridges.

Buds are red-brown with grey, pubescent tips, 3 mm long. New foliage is pink-red, soon bright green, and then dark green. Leaves are deciduous, simple and palmate (3- to 5-lobed); the lobes have obtuse angles, and the middle lobe is often further lobed. The upper leaf surface is dull green, whereas the underside is paler and pubescent. The leaf petiole produces a milky sap if snapped. In autumn, the leaves are bright gold-yellow for a long period, some red or purple.

The flowers are greenish-yellow, appearing with the leaves in late April–May. They are small, with about 10 widely spaced in each head. The flowers have narrow petals and sepals dominated by 8 yellow stamens. Fruits develop from about 3–4 flowers in each inflorescence, and are 2.5–3 cm long, with horizontally spread wings which aid in wind dispersal. The fruits ripen in autumn.

Distribution

A. campestre has a large natural distribution area in central and Northern Europe, from northern England and southern Sweden eastwards to Russia and Western Asia, and is also found in North Africa. It is very common in mesophile stands, especially deciduous oak forests from sea level to the altitude at which beech (*Fagus sylvatica*) occurs. Distinguishing natural from naturalized distribution is difficult.

A. campestre has been widely planted within its natural distribution range yet not widely planted outside this range, except as an ornamental tree.

Silviculture characteristics

In natural forests, *A. campestre* is a later seral species; it is shade-tolerant, and will not invade grasslands or young forest plantations. It coppices well, and tolerates clipping in hedges. As it is not very tall, it is rarely blown down. It is frost-resistant, and in sub-Mediterranean climates, it can tolerate summer drought.

A. campestre is recommended for use and conservation in mixed deciduous woodlands, as it is a good host for a wide range of epiphytic lichens and bryophytes, and its presence in the understorey will encourage a wide range of insects, birds and mammal fauna.

Silviculture practice

A. campestre first produces seed after 10 years, and almost yearly thereafter. Seed should be harvested early – when samara wings are yellowish-brown – from standing trees, or collected by shaking or whipping the trees and collecting the fallen samaras on sheets spread on the ground. Samaras may also be collected from recently felled trees. Removal of empty samaras improves seed handling, storage, sowing and control of seedbed density. Freshly collected samaras have a moisture content of 30–160% of the dry weight. Seed should be air-dried before storage, to a recommended moisture content of about 10–15%. Seeds can be stored for 1 or 2 years in sealed containers at 3–5°C without appreciable loss of viability.

Seeds should be sown shortly after collection in the autumn, unless stratified; dried seeds will not germinate until the second year. Commercial dried seeds are usually stratified for a up to 1 year in sharp sand (at 4°C) before sowing in spring. To break dormancy, stratification at 1 month warm (20–30°C) followed by 3–6 months cold (3–6°C) is necessary.

Seed lots may vary considerably in the time required for after-ripening, so they should be checked regularly during the cold stratification period to determine whether germination has started.

Seeds are best sown in autumn in mulched seedbeds. Bare-rooted planting stock is usually 1+0 or 1+1. Spring-sown stratified seeds give poorer results than autumn-sown plants, and the seeds are usually sown 0.5–2.5 cm deep, broadcast or in drills. Seedbed densities of 15–30 per 30 × 30 cm are best for the production of vigorous seedlings. Shade is recommended during seed establishment.

The simplest and cheapest means of vegetative propagation is by layering. This is time-consuming and no significant quantity of new plants is produced unless many mother plants are used. Ornamental varieties may be bud-grafted in summer on to rootstocks of *A. campestre* or species of the same group.

Alternatively, softwood cuttings can be prepared by leaving two to three pairs of leaves and one pair of buds at the base of each cutting. They must be superficially wounded by removing a thin slice of bark at the base and dipped in rooting hormone before insertion into trays containing 2:1 peat and sharp sand. Once the cuttings have rooted, the trays can be moved for hardening off.

In an even-aged high forest, *A. campestre* is usually found as an understorey species. It may be included as a standard tree in coppice with standards, or it may be coppiced.

A. campestre may be managed as a coppice species, or as a component of mixed deciduous forests, where it usually forms small scattered groups. In natural forests it may regenerate vegetatively rather than by seedlings, although natural regeneration is usually abundant.

Diseases and pests

Generally, *A. campestre* has few serious pests or pathogens, either in the nursery or in the field. *Ceratocystis virescens* spreads extensively in sapwood and usually kills a tree within 4 years. Less serious is browning of leaf veins by *Apiognomonia* (*Gnomonia*) *veneta* or plane scorch – the commonest conidial stage being known as *Gloeosporium nervisequum*. The most conspicuous fungal pathogen is probably maple tar spot, caused by *Rhytisma acerinum*; this is widespread on sycamore maple (*Acer pseudoplatanus*) in Northern Europe, but is less common on *A. campestre*. Powdery mildew (caused by *Uncinula bicornis*) may be a serious pathogen in nurseries.

Xylophagous insects and defoliators generally cause more damage than fungi. Foliage may be damaged in spring and summer by leaf miners, defoliators, and weevils. A serious outbreak of *Erannis defoliaria* occurred in Toledo province (Spain) in 1981, which caused defoliation of oak and field maple.

Uses

A. campestre is used for landscaping plantings, in gardens and parks and for street and roadside plantings. It produces small trees, which blossom before leaves appear, handsome stem forms and beautiful colours, particularly in autumn. It is particularly recommended for cool, moist sites. In northern temperate Europe, it is an attractive component of mixed broadleaved woodlands, including protection forests on watershed slopes, as long as they are not too exposed. It is particularly valued for the shelter and food it affords to a wide variety of forest fauna.

Acer campestre wood is cream-coloured, fine-textured, close-grained, and slightly denser than *A. pseudoplatanus*, at about 690 kg/m^3. It is not decay-resistant in contact with the soil or exposed to weather. It is used to make tool handles, turnery, small items of furniture, cabinet work and for fuelwood. Wood cut from a burr or from behind a cluster of epicormic shoots produces the highly prized bird's-eye maple figure. Wavy grain also occurs in field maple.

A. campestre produces abundant honeybee forage (pollen and nectar); the typical honey has a pale amber colour with no strong flavour or aroma.

Further reading

Ducci, F., Proietti, R., Carone, G. and Apuzzo, S. (2009/2010) First surveys on genetic variability and structure of field maple (*Acer campestre* L.) in natural and managed populations in the landscape of central and southern Italy. *Annali dell'Istituto Sperimentale per la Selvicoltura* 36, 125–138.

Evans, J. (1984) Silviculture of broadleaved woodland. Bulletin, Forestry Commission, UK, No. 62, 232 pp.

White, J. (1995) *Forest and Woodland Trees in Britain*. Oxford University Press, Oxford, UK.

Acer macrophyllum Pursh

Broadleaf maple

Importance

Acer macrophyllum (Plate 4) is one of the few commercial deciduous hardwoods in the Pacific region of western North America. It is an excellent shade tree, its wood is used for furniture, especially piano frames, and the sap can be made into a syrup.

Botanical features

A. macrophyllum is a medium-sized tree occasionally over 30 m tall, with a dbh of up to 1 m. In the forest, the trunk may be unbranched for more than 50% of its length and has a narrow crown. In the open, the trunk soon divides into a few large spreading and ascending branches, and the crown is broad and rounded. The twigs are stout and reddish-brown, and the bark is greyish-brown and shallowly furrowed into narrow scaly ridges. *A. macrophyllum* has a shallow but wide-spreading root system, and the roots are associated with vesicular-arbuscular mycorrhizae. The leaves are deciduous, very large, 15–30 cm wide and almost as long, deeply notched into 5 lobes, with the upper surfaces dark green, but paler and hairless beneath.

A. macrophyllum begins to produce seed at about 10 years old and continues every year thereafter. Both staminate and perfect flowers are mixed in the same dense, cylindrical racemes. Flowers are greenish-yellow, scented, and appear before the leaves (from March at low elevations and in the south, to June at high elevations and in the north). Pollination by insects usually occurs within 2–4 weeks after the buds burst. Pubescent double samaras result, with wings 3.5–5.0 cm long. Seeds ripen in September and October. The initiation of phenological events varies greatly with latitude, altitude, aspect and weather.

Distribution

A. macrophyllum is a western North American species distributed almost exclusively in the Pacific region. The native range extends from 33° N to 51° N, always within 300 km of the Pacific Ocean – from near Sullivan Bay, British Columbia, to within 16 km of San Francisco Bay, California. Isolated groves are scattered along the southern Californian coast as far south as San Diego.

A. macrophyllum grows mainly in mixtures with mesothermal conifers or other broadleaved trees. It is most commonly present in the early and intermediate stages of secondary succession. It is a pioneer species (primary succession) on talus and floodplains, and is present in second-growth stands on upland sites. It may invade gaps created by windthrow of climax tree species such as *Thuja plicata* on floodplains and colluvial sites, and can persist in old-growth stands. It frequently grows in clumps of 3–5 trees, all originating from a single stump.

Common associates are: *Abies concolor*, *Abies grandis*, *Alnus rubra*, *Arbutus menziesii*, *Castanopsis chrysophylla*, *Chamaecyparis lawsoniana*, *Cornus nuttallii*, *Libocedrus decurrens*, *Lithocarpus densiflorus*, *Picea sitchensis*, *Picea lambertiana*, *Picea ponderosa*, *Populus trichocarpa*, *Pseudotsuga menziesii*, *Quercus chrysolepis*, *Quercus garryana*, *Quercus kelloggii*, *Sequoia sempervirens*, *Taxus brevifolia*, *Thuja plicata*, and *Umbellularia californica*.

In British Columbia, *A. macrophyllum* is a minor component in the Coastal Douglas-fir and Coastal Western Hemlock zones. In the western United States, it is a component of the Coastal Western Hemlock, Sitka spruce, Pacific Madrone–Douglas-fir, Coastal Ponderosa Pine, and Redwood zones. Introduction is mainly limited to arboreta and parks.

Silviculture characteristics

The reproduction capacity of *A. macrophyllum* is high, with a minimum age for seed crops of about 10 years. Its seed dissemination capacity is, however, low; seed dispersal by wind extends <100 m. There is little potential for natural regeneration in low light, but it can be higher where there is exposed mineral soil and if vegetative reproduction is also considered. The growth rate in the first 5 years can be high, with up to 5 m gained in one growing season from stump sprouts. Dense stands show an intermediate capacity for self-pruning. The mature crowns require a great deal of space, but few trees live more than 200 years. The productivity of *A. macrophylla* is probably similar to that of *A. rubra*.

A. macrophyllum does not aggressively invade clearcuts, but vigorous stump sprouting is a problem when it occurs in the harvested stand.

Silviculture practice

Seeds are abundant almost every year, but production by individual trees and stands can vary from year to year. Most of the seeds are dispersed by wind between October and January. The seeds are large, at 4–12 mm long and 4–9 mm thick. At field moisture content, filled seed weights are in the range 5200–7900 seeds/kg, with seed coat comprising 60%–70% of the seed weight. Seed collection and storage are best conducted when minimum moisture content is reached, before the start of the autumn rains.

Germination is epigeal and begins in late January or early February under field conditions; it is usually completed by April or May. Seed germination is excellent on mineral soil and organic substrates, and seedling establishment is best when those substrates do not dry excessively during the growing season. On average, 30–40% of viable seeds germinate if they are protected from predators, and seed lots occasionally attain 80% germination. All seeds germinate during the late winter and spring after seed dispersal.

A. macrophyllum seedlings have a high juvenile growth potential, exceeding that of *P. menziesii* and other conifers. When open-grown under conditions of adequate moisture and nutrients, seedlings reach heights of 1–2 m in one growing season. The morphology of young seedlings is strongly influenced by plant density. Temporary flooding is common on riparian sites, and the seedlings are able to survive short periods of inundation.

A. macrophyllum sprouts profusely after being cut. The large stumps produce more and taller sprouts, but all sizes regenerate vigorously. Sprout clumps have achieved heights of 5 m and crown diameters of 6.5 m in 3 years, with as many as 67 sprouts around a single stump. This sprouting vigour could be used in reproducing pure stands by coppicing.

Currently, silviculture of *A. macrophyllum* usually involves control rather than culture. Possible commercial thinning regimes, rotation ages and regeneration plans for managed stands are being planned.

Rapid height growth of *A. macrophyllum* continues through the sapling stage, but it slows as the trees grow from pole to saw timber size. Diameter growth is proportional to leaf area, and trees with large crowns develop more sapwood than trees with small crowns. Pure, 70-year-old stands have yielded about 315 m^3/ha. Under intensive management, rotations of 50 years or less could probably be used.

Diseases and pests

A. macrophyllum is attacked by a number of fungal and insect pests. Wood-rotting fungi including *Heterobasidion annosum*, *Fomitopsis pinicola*, *Polyporus berkeleyi* and *Inonotus dryadeus* can invade after injuries to the stem or branches. The tree may then 'weep' for up to 15 years before collapsing, usually in high winds.

Among the many insect pests, the most serious may be the larvae of the cossid moth *Prionoxystus robiniae*, known as the 'carpenter worm'.

Uses

A. macrophyllum is an excellent shade tree and is also planted as an ornamental.

Its wood is moderately hard, weak and light brown with a density of 440–480 kg/m^3. It is used in the furniture industry, but it is neither as hard nor as strong as the wood of *Acer saccharum*. It is a preferred wood for piano frames; it is also excellent for decorative face veneer and makes good container material, but is not suitable for flooring. However, the trees that originated from sprouts are of poor quality. Higher-quality trees may be produced by managing for stands that originate from seed or planted seedlings. The amount of *A. macrophyllum* being marketed for fuelwood is increasing as the use of wood stoves increases.

Like *A. saccharum*, *A. macrophyllum* has a sweet sap that can be made into syrup. The flow of sap is adequate for syrup production in January and February, but the syrup is of lower quality than that made from *A. saccharum*.

Further reading

Iddrisu, M.N. and Ritland, K. (2004) Genetic variation, population structure, and mating system in bigleaf maple (*Acer macrophyllum* Pursh). *Canadian Journal of Botany* 82, 1817–1825.

Kerbes, E.L. (1968) *Broadleaf maple in British Columbia*. Forestry Branch, Vancouver, Canada.

Minore, D. and Zasada, J.C. (1990) *Acer macrophyllum*. In: Burns, R.M. and Honkala, B.H. *Silvics of North America, Volume 2*. USDA Forestry Service, Washington DC, pp. 33–40.

Acer negundo L.

Box elder

Importance

Acer negundo is a small and variable tree of little economic value, and is native to much of North America. It has been widely planted mainly for amenity purposes in non-native parts of North America, and in Europe and Asia, where it is now regenerating freely and spreading, invading riparian zones and urban areas. In the temperate parts of the southern hemisphere, the spread of *A. negundo* is more limited. Although this pioneer species is invasive throughout much of its introduced range, actual impacts are not great, and this species has no important negative impacts on human activities. In natural habitats, it becomes an important component of the vegetation in riparian systems and increases siltation.

Botanical features

A. negundo is a tree reaching a height of no more than 20 m and a stem diameter of up to 1 m, often with multiple stems. In more open vegetation, the canopy usually exhibits a broad and open crown and may even become shrubby, whereas in the face of competition in a forest stand, the trunk tends to be single-stemmed and straighter. Shoots are initially green, turning violet in the second year. The bark is grey-brown, becoming dark grey, and is shallowly cracked. Buds are oppositely arranged, small, silky, white, 2–5 mm long and with two scales. The leaves are 15–35 cm long and pinnate, with 3–5(–7) leaflets. They are light green but paler below, lobed and serrated. Leaflet shape is variable; in *A. negundo* var. *interior*, for instance, the first pair of leaflets is 3-lobed. Male flowers are borne in corymbs with pendent stamens, whilst female flowers form small pendent racemes. Both types of flowers are small and pale yellowish-green in colour. There is much variation in the morphology of pistillate flowers, with the presence of stamens on a proportion of the flowers. The fruit consists of two fused winged samaras, up to 4 cm long and diverging at an angle of less than 60°. The samaras separate when shed and each contains a single wrinkled seed.

Distribution

The natural distribution of *A. negundo* ranges from southern Canada to the mountains of Mexico and as far south as Guatemala. It is absent from western North America but does occur along the Atlantic coast. The range of the species has greatly expanded in North America through planting and subsequent natural regeneration, and it is spreading in the western USA. In Canada, although only native to the southern tip of Ontario, *A. negundo* is now naturalized in eastern provinces from Ontario to Nova Scotia. The various recognized varieties appear to represent fairly distinct geographic races, but these often intergrade.

A. negundo is widespread in riparian and palustrine plant communities and has been reported from a variety of vegetation types in the USA. It is most commonly associated with deciduous forest communities but also occurs in scrub- and grass-dominated vegetation types; in Toronto, Canada, it is reported as invading all habitat types.

A. negundo is widespread in urban areas throughout much of Europe, following centuries of horticultural and landscape planting. It is probably more widespread in other temperate regions of the world but is absent from the tropics.

Silviculture characteristics

A. negundo is a fast-growing and relatively short-lived species with a maximum lifespan of 100 years, but more typically of about 75 years. Branches break easily during stormy weather and trees are readily injured by fire due to their thin bark.

Unlike other species of maple, which are all hermaphroditic with a wide array of sex expression, *A. negundo* is dioecious. The sex ratio is variable and often skewed towards males, especially on drier ground away from stream sides. The sexual dimorphism in *A. negundo* is well studied: male trees reach greater heights and stem diameters than female individuals, and males produce more leaf biomass and females more radial growth. The two sexes respond differently to water availability, leading to a segregation of sexes. In wet years, female individuals exhibit a higher growth rate and are more profligate water users, whereas both sexes show similar growth and physiological responses in dry years. Increasing precipitation would thus favour female trees and also alter the water flux from riparian ecosystems. Flowers of both male and female trees usually appear in spring, well before the leaves, and the fruits often persist on the tree well into the winter. In North America, large seed crops are produced annually.

A. negundo may start bearing fruit (around 30,000/kg) when it is just 5 years old. The tree is both wind-pollinated and wind-dispersed, although dispersal by birds and squirrels also occurs. Dispersal in maples is usually localized, no more than 100 m from the seed source, but may occasionally be spread over several kilometres if seeds are released when hard snow lies on the ground and very windy weather prevails. In riparian systems, samaras are probably also dispersed by water but the importance of this dispersal mechanism is unclear.

Vegetative reproduction is common on damaged individuals. Exposed or damaged roots will produce new shoots.

Shelter belt trees whose boles had died as a result of drought have been known to recover by producing root sprouts. Germination under test conditions varies widely from 0% to 96%. In the USA, northern populations require shorter stratification periods, germinate at colder temperatures and have a higher germination rate than southern populations. Under natural conditions, *A. negundo* regenerates readily on disturbed and moist soils and tolerates medium shade, but poor establishment or high seedling mortality occurs in dense ground cover or under closed canopies.

Silviculture practice

To date, little effort has been undertaken to control *A. negundo*, although its control has been advocated, for instance in Austria. No information is available for cultural or mechanical control, but cutting and stump treatments used on other tree species may prove effective. *A. negundo* has not been considered for biological control, although numerous insect pests are recorded on the species. Using only male trees in amenity planting would eliminate the risk of spread in the introduced range.

Diseases and pests

Fungi are much more damaging than insects to *A. negundo*, with species of *Fomitopsis*, *Perenniporia*, *Fomes*, *Inonotus* and *Ustulina* causing significant rotting, and *Eutypella parasitica* causing a stem canker. The only pathogen which kills significant numbers of *A. negundo* trees, however, is *Verticillium albo-atrum*, which causes verticillium wilt.

A. negundo is attacked by a wide variety of insects, including the boxelder bug (*Leptocoris trivittatus*) which is found across most of the host's range. Others include the boxelder aphid (*Periphyllus negundinis*), and the boxelder gall midge (*Contarinia negundifolia*). A number of scale insects also attack *A. negundo*.

Uses

A. negundo is planted as an ornamental, amenity, shade and shelter tree. It is a browse species of secondary importance to deer and the seeds are an important food source for birds and squirrels. In North America, the foliage of *A. negundo* has poor nutritional value for livestock. It has a good value as cover for both wildlife and livestock and has been used to revegetate riparian areas.

Little economic use is made of the species, as its timber has undesirable characteristics, being light, soft, close-grained and of low strength. Reported wood densities range from 416 to 540 kg/m^3. Locally, the wood may be used to make boxes and in rough construction, but it is only occasionally used to produce cheap furniture and other wooden products. Formerly, the wood was used for posts, fencing, and fuel, but it is a poor fuel as the wood is soft and spongy.

Further reading

Porté, A.J., Lamarque, L.J., Lortie, C.J., Michalet, R. and Delzon, S. (2011) Invasive *Acer negundo* outperforms native species in non-limiting resource environments due to its higher phenotypic plasticity. *BMC Ecology* 11, 28.

Rosario, L.C. (1988) *Acer negundo*. Fire effects information system. USDA Forest Service, Rocky Mountain Research Station, Fire Sciences Laboratory. Available at: http://www.fs.fed.us/database/feis/plants/tree/aceneg/introductory.html (accessed 5 March 2013).

Sachse, U. (1992) Invasion patterns of boxelder on sites with different levels of disturbance. *Verhandlungen der Gesellschaft für Ökologie* 21, 103–111.

Acer platanoides L.

Norway maple

Importance

Acer platanoides is a large, fast-growing tree, native to central and Northern Europe, Asia Minor, and the Caucasus as far east as northern Iran, and has been introduced into many countries, including Great Britain (in 1683). It is a popular ornamental tree and has been planted extensively in cities throughout Europe and North America as it is able to tolerate pollution. It thrives on both acid and calcareous soils, and grows very fast when young, usually faster than *Acer pseudoplatanus* (sycamore) for the first 30 years or so, but does not reach the size of the latter and does not live as long. It has attractive flowers and good autumn colour – mostly bright gold to orange, or in some cases bright red.

A. platanoides yields a good-quality hard wood, sometimes with attractive wavy grain or figure, which is often used for furniture and cabinet work, flooring, and tool handles. In the past, sugar was obtained from the sap in Norway and Sweden.

Botanical features

A. platanoides is a large tree, up to 25–30 m, with dbh at maturity of 60–80 cm, although exceptionally to 150 cm. The stem is straight erect, glabrous, with green to brown branches. The crown is tall-domed, and sometimes very broad, on a short stem. It is open in winter with rather short perpendicular shoots and often thin bunches of persistent fruits, but in summer has very dense foliage. The growth of *A. platanoides* is rapid during the first 30 years or so, by which time it can reach 18–22 m in height. Girth growth remains vigorous – up to 3 cm/yr – for 100 years. The stem is straight and cylindrical, and the bark is pale grey, smooth, and finely folded or shallowly ridged in a network.

The shoots are pinkish-brown or olive-brown. The buds are tall ovoid, with dark red or red-brown terminals and green appressed laterals. The leaves are simple, opposite, deciduous, generally five-lobed, 12 × 15 cm, with shallow, open indentations between short acuminate lobes, green on both sides and borne on a reddish stalk that emit a milky liquid when broken. The leaves vary greatly in size with the age of the tree and the vigour of the shoot. The petioles are 15 cm long (8–10 cm on old trees, up to 20 cm on young trees). The leaves' autumn colour is generally a good butter-yellow towards the end of October, but occasional trees may be scarlet in early October, and in some years most turn from yellow to orange.

Inflorescences are borne in erect pubescent panicles of about 30–40 flowers. Each flower is monoecious, 6–8 mm, bright acid-yellow-green with five oval petals and a large, green disc. The flowers open before the leaves, in late March, and last well into April, when the leaves are newly expanded. The fruits are double samaras, 6–10 cm across the pair, greenish-yellow when ripe; they have flattened carpels and broad wings, also flattened, set at a wide angle and not narrowing towards the base.

Distribution

The natural distribution of *A. platanoides* ranges from southern Sweden and Norway (as far north as Tromsø, where the tree is shrubby), Finland (to latitude 62° N), Belarus, Russia, the Baltic States as far north as Lake Lagoda and as far south as Ukraine (Crimea), to the Caucasus, but not east of the Urals. It also occurs in central Europe, but is not found in the UK, western France, Spain, the Netherlands or Denmark, although it has been introduced into these countries. *A. platanoides* subsp. *turkestanicum* is found in north-east Afghanistan and Turkestan in mountain forests, especially on the plateaux of the Tian Shan and Pamir–Alay.

It is commonly found throughout mainland Europe and North America in towns and villages, as an ornamental or shade tree.

Silviculture characteristics

A. platanoides is a light-demanding species and grows best on fertile sites. It has rapid early growth, with a height increment of about 1 m/yr at age 10 years. Thus, if planted in mixed stands, it is best planted in groups rather than intimate mixtures, because it tends to suppress other more slow-growing species. Growth is affected by competition from grasses when young, and weeding is important. It will not grow well on exposed sites, although it can tolerate smoke pollution and salt spray.

Silviculture practice

A. platanoides is very easily grown from seed. It will produce fertile seed from about age 25–30 years, with best seed crops at 1–3 year intervals between age 40 and 60 years, although some seeds are produced every year. Seeds are usually collected between August and October from standing trees. The mean weight of 1000 samaras ranges from 100 g to 140 g; there are about 7500 seeds/kg. Mature seeds are classified as orthodox. Seed can be stored at a moisture content of 8–11% and low temperature (−3 to −10°C) for 1–3 years.

Seed germination is 30–80%. If seeds are sown in autumn, they must be pre-chilled at 3°C for about 3 months. The recommended procedure in France is for seed to be sown in the spring (in nursery beds, or in fertilized peat) after stratification for 6–8 months at 3°C; bare-rooted planting stock (1+0 or 1+1) is utilized.

Propagation by cuttings from young material is easy. All cultivars can be bud-grafted on to the species, especially when whips are required. Side-grafting in the greenhouse is also easy. Grafting is better carried out in September than in February. *In vitro* regeneration of *A. platanoides* has been carried out.

There is interest in the use of *A. platanoides* as a forest plantation species, although it has not been extensively planted as a forest tree.

Although direct sowing is generally less reliable than use of planting stock, it is worth considering as a means of establishing Norway maple on lowland ex-agricultural sites, provided that appropriate protection and weed control is practised. Seed of *A. platanoides* has been directly sown in the UK.

Annual shoot increment is significantly improved by shoot pruning in the second and third years after transplanting, and shoot-pruned trees recovered better from the effects of transplanting.

Diseases and pests

A. platanoides can host unusually large numbers of aphids, and also suffers from verticillium wilt, powdery mildew and a bacterial infection known as wet wood. The introduction of the wood-boring beetle *Anoplophora glabripennis* to North America has forced the removal of many *A. platanoides* trees from urban locations.

Uses

The ability of *A. platanoides* to resprout vigorously after felling makes it suitable for use in live fences. It is extensively planted as a shade and street tree as it has attractive, colourful foliage, and a large, spreading crown.

A. platanoides wood, like that of other maple species, is fine-textured, white and has an average density at 15% moisture content slightly greater than that of *A. pseudoplatanus*, at about 660 kg/m^3. Wavy-grained maple is in great demand; this and other attractive wood figure effects, as well as good acoustic properties, have led to its use for musical instruments. It is also used for furniture, marquetry and tool handles.

Sugar has been obtained from the sap in Norway and Sweden. It is a prolific pollen- and nectar-producing tree, supplying valuable honeybee forage.

Further reading

Savill, P.S. (1991) *The Silviculture of Trees used in British Forestry*. CAB International, Wallingford, UK.

Webster, C.R. and Wangen, S.R. (2009) Spatial and temporal dynamics of exotic tree invasions: lessons from a shade-tolerant invader, *Acer platanoides*. In: Kohli, R.K. (ed.) *Invasive Plants and Forest Ecosystems*. CRC Press, Boca Raton, Florida, USA, pp. 71–85.

Acer pseudoplatanus L.

Sycamore

Importance

Acer pseudoplatanus is a large, handsome, fast-growing tree, attaining a height of 30–35 m in 50 years. It is adaptable to many site conditions, including those exposed to high winds or sea breezes; it also grows well in many sites suffering from air pollution or heavy metal contamination. Its dense foliage leads to it being widely used as an ornamental tree in European and North American cities. Many cultivars are appreciated for the shape and colour of their foliage. The timber, being soft, light and tough, is valued by saddlemakers and millwrights and is also used for musical instruments.

Botanical features

A. pseudoplatanus is a large tree with a straight trunk and erect branches forming a spreading, often irregular crown. The crown is often broader than tall, with massive low branches, although on suitable sites the crown may be very tall and less broad.

The average height of *A. pseudoplatanus* when mature is 30–35 m, although it may exceed 40 m; average dbh at maturity is 60–80 cm, although it may exceed 150 cm. The stem is straight and cylindrical. The bark is grey and smooth in young trees, but mature boles have pale orange-brown or pinkish-orange bark, broken into long scales which come away. The root system is much branched, providing stability for the tree in high winds. Deep roots penetrate to approximately 1.4 m depth, while the radial extent exceeds 9 m.

The shoots of *A. pseudoplatanus* are greenish-grey-brown, striated lengthwise with pale lenticels. The buds are 8–10 mm long, open, ovoid, with a few green scales and reddish margins; basal scales are red and decurved. The leaves are simple, opposite, deciduous, with five unequally toothed, pointed lobes at acute angles; their nerves are often pubescent, their bases cordate, and their margins denticulate. They are dark green above and glaucescent on the underside in young trees, but yellowish on pink petioles in old trees. There is rarely any worthwhile colour in the autumn, although the foliage may occasionally turn bright yellow. The leaves vary greatly with tree age and shoot vigour in the size and depth of the lobes. They are always borne on a long petiole, which does not emit milky sap when broken. The leaves open early, especially in shade, and lammas growth in July may emerge bright pink.

Flowers are in terminal, hanging racemes; they reach 6–12 cm long, and open in mid-April, attracting hordes of bumblebees. Each flower is monoecious, 4–6 mm across, pale green or yellowish, with narrow and very short petals; the stamens are whitish, and the anthers are bright yellow. Flowering occurs when the leaves are newly expanded. The fruits are double samaras, 5 cm long, with wings set in a 'V' at about 90°; they grow in clusters on short green stalks and appear abundantly when the tree is at least 20 years old.

Distribution

The native range of *A. pseudoplatanus* is broad, encompassing most of central and southern Europe: Germany (Harz Mountains); Russia (Caucasus); Ukraine (Crimea); Spain (Pyrenees); southern and central France; the Swiss Alps; Italy; Bosnia; northern Greece; and the Carpathian mountains in Poland, Slovakia and Romania. It is not native to the British Isles, Belgium, the Netherlands, north-west France, northern Germany or Scandinavia. All specimens of *A. pseudoplatanus* recorded from eastern Anatolia (Turkey) have proved to be *A. heldreichii* subsp. *trautvetteri*.

Silviculture characteristics

A. pseudoplatanus is a shade-tolerant species. In relation to light conditions under a closed canopy, it will perform better than *Fagus sylvatica* and *Quercus petraea*, and it is suitable for advance planting below the canopy of existing stands. *A. pseudoplatanus* produces suckers around the coppice base and may become relatively abundant in coppiced stands, while it remains isolated in mixed stands.

Silviculture practice

A. pseudoplatanus seeds are collected between August and October from standing trees. Good seed crops occur every 1–2 years. Mature seeds are classified as orthodox. The mean weight of 1000 samaras ranges from 60 g to 180 g. Seed can be stored for 1–2 years at moisture content 20–25% at low temperatures (−4 to −6°C). Stored seeds of *A. pseudoplatanus* will lose their viability if the moisture content is less than 15%. Germinative capacity is 50–70%. The sowing season is autumn, and a prior pre-chilling at 3–5°C for almost 2 months is needed, although mechanical rupture of the pericarp is not necessary. Samaras of *A. pseudoplatanus* containing dormant seeds can be treated before sowing by stratification (with no medium) with repeated soaking in cold water for 1 h every 2 weeks.

Shade is recommended during seedling establishment. Vegetative propagation by cuttings from young material is successful. All cultivars can be budded on to the species. Side-grafting in the greenhouse is also easy. Hardwood grafting is used for the widely grown cultivars 'Brillantissimum' and 'Prinz Handjery', both on 120–180 cm rootstock. *In vitro* regeneration of sycamore has been carried out from mature and juvenile trees.

In the wild, *A. pseudoplatanus* does not form forests, but occurs as fairly isolated groups of trees; only one uniform forest is known, in Thüringen, Germany. The tree usually forms a component of birch (*Betula*) and fir (*Abies*) forests. Some studies have shown that natural regeneration in Norway spruce (*Picea abies*) stands can be successful with a dominant proportion of *A. pseudoplatanus*, as long as gaps are sufficiently large to provide adequate light conditions. Stock planted in early spring (February–March) has the best chance of survival and above-ground development.

A. pseudoplatanus may reach a dbh of 55 cm at age 85 years (dominant height 38 m) on cool, mesotrophic sites. Wood yields of 350 m^3/ha have been recorded on good-quality sites (rotations of 80–100 years). In the UK, it is managed on rotations of 65–70 years, and can be planted in mixtures with European larch (*Larix decidua*), as the two species have similar growth rates.

Diseases and pests

A. pseudoplatanus frequently hosts a harmless leaf infection known as 'tar spot', caused by *Rhytisma* fungi. A number of other fungi can also attack *A. pseudoplatanus*, but are rarely serious.

Uses

A. pseudoplatanus is used in coastal areas and offshore islands of the western and northern UK, as windbreaks against strong Atlantic gales. Ramified adventitious roots, which often penetrate deeply, give these trees a solid hold, and the canopy significantly reduces wind speeds. Many cultivars are widely used in landscape planting because of their handsome foliage and crown shape.

The timber of *A. pseudoplatanus* is white, close-grained and moderately hard. It has an average density at 15% moisture content of about 630 kg/m^3. It was formerly much used in turnery – for cups, bowls, and pattern blocks – and it is still a highly regarded wood for saddlemakers and millwrights. It remains widely used for furniture manufacture, flooring, and joinery. Wavy-grained sycamore, which occasionally occurs, can fetch very high prices, and is in demand for musical instruments, especially violins. Veneer from *A. pseudoplatanus* is widely used in the furniture industry.

Further reading

Hein, S., Collet, C., Ammer, C., Le Goff, N., Skovsgaard, J.P. and Savill, P. (2009) A review of growth and stand dynamics of *Acer pseudoplatanus* L. in Europe: implications for silviculture. *Forestry* 82, 361–385.

Van Gelderen, D.M., de Jong, P.C. and Oterdoom, H.J. (1994) *Maples of the World*. Timber Press, Portland, Oregon, USA.

Acer saccharum Marshall

Sugar maple

Importance

Acer saccharum is one of the largest and longest-lived hardwood trees in eastern North America. It represents a large reserve of hardwood sawtimber and pulpwood, accounting for approximately 7% of all hardwood growing stock in the eastern USA. Following a century of agricultural abandonment, the hardwood forests of north-eastern North America are regenerating and maturing. In much of the northern portion of its range, current growth exceeds harvest because many stands are still considered to be young secondary forests. The wood of *A. saccharum* and *Acer nigrum* is known as 'hard maple'; it is a dense, fine-grained wood that machines, stains and finishes well and is widely used for furniture, cabinets, tool handles, toys and flooring. It continues to be an important short-fibre hardwood pulp, enjoys local use as charcoal for filtering whisky and contributes to local economies through 'fall foliage tours' to view the brilliant autumn colours. *A. saccharum* is also prized for its sap, the source of commercial maple syrup.

Botanical features

A. saccharum is a slow-growing but long-lived tree (up to 400 years), reaching heights of 27–37 m and diameters in excess of 90 cm. Forest-grown trees are usually well pruned and produce a straight bole, while open-grown trees often branch within 3 m of the ground and produce dense, ovoid or elliptical crowns.

A. saccharum has simple, opposite, medium-sized (7–12 cm), deciduous, palmately 5-lobed leaves with an entire margin. The leaf surfaces are glabrous, bright green above and pale green beneath. Mature shoots are slender, glabrous and brown-grey with paler lenticels. Terminal shoot buds are sharply pointed, 4–8 mm in length, with imbricate dark brown scales.

A. saccharum flowers are borne in corymbs on long pedicels, appearing with the leaves in March–June, depending on climate. The flowers appear perfect, but typically only one sex is functional within each flower. *A. saccharum* is primarily wind-pollinated, although small bees and flies also provide some pollination. The fruit, a double samara, typically has only one filled embryo. The period between good seed crops is 1–4 years in north-central USA, 2–5 years in other parts of the USA and 3–7 years in eastern Canada. Seed matures in August–September, with dispersal by wind over distances up to a few hundred metres. Excellent seed years can produce more than 11 million seeds/ha, often giving rise to populations in excess of 200,000 seedlings/ha.

The root system of *A. saccharum* is extensive, with strong lateral branching. Secondary roots on the upper side of the laterals grow into the humus layers, while those on the lower side grow downward. *A. saccharum* roots are extremely sensitive to flooding during the growing season. Seedlings develop root systems in leaf litter and humus, and require partial shade until the roots have extended into mineral soil. The roots of *A. saccharum* release allelopathic compounds that inhibit the growth of other northern hardwood species.

Distribution

A. saccharum is distributed throughout north-eastern North America, from Prince Edward Island south to central New Jersey, south-west across the Appalachians to Tennessee, then west to easternmost Kansas, and north to south-western Ontario and Manitoba. It is a major component of a number of northern forest types and occurs in mixed stands with species including yellow birch (*Betula alleghaniensis*), basswood (*Tilia americana*), beech (*Fagus grandifolia*) and hemlock (*Tsuga canadensis*).

A. saccharum has seldom been planted outside its natural range, probably because of its relatively slow growth, shade-tolerant habit and preference for cool, fertile sites. It is occasionally encountered as an ornamental tree in urban areas and parks outside its natural range, but not in plantations.

Silviculture characteristics

A. saccharum is considered a shade-tolerant species, capable of prolific reproduction in the shade of most other associates. Seedlings can form dense carpets even under heavy shade and can tolerate long periods of growth suppression. Seeds require a cold, wet stratification period and overwinter in leaf litter. The optimum temperature for germination is 1°C, with little or no germination above 10°C.

Its frequent and abundant seed crops, the ability of its seedlings to survive in shade and systems of partial cutting or small disturbances such as tree-fall gaps, appear to provide excellent conditions for perpetuating *A. saccharum* in northern hardwood forests. Seedlings are occasionally so abundant that regeneration of other preferred species (e.g. *Quercus rubra*) is precluded. In the absence of periodic catastrophic disturbance, *A. saccharum* can replace less shade-tolerant hardwoods such as yellow birch (*B. alleghaniensis*), but it is replaced in turn by beech (*F. grandifolia*).

A. saccharum is characterized by relatively slow growth as seedlings, saplings and pole-sized trees. Other, faster-growing species, such as white ash (*Fraxinus americana*), eastern white pine (*Pinus strobus*), red oak (*Q. rubra*), and yellow birch (*B. alleghaniensis*) occasionally overtop *A. saccharum* following partial cutting in second-growth stands. However, the development of dense sapling stands of *A. saccharum* under full or partial canopies provides a ready source of advance regeneration that responds well to release, even after long periods of suppression. Following release, even older (>75 years) *A. saccharum* trees can double their rate of diameter growth.

Silviculture practice

For *A. saccharum*, systems employing partial cutting are preferable to clear-cutting. Many even-aged northern hardwood stands originated following periods of destructive logging, but regeneration was probably from advance reproduction rather than seed. In general, greater canopy retention generally favours regeneration of shade-tolerant species like *A. saccharum* mixed with beech (*F. grandifolia*), basswood (*T. americana*) and yellow birch (*B. alleghaniensis*). Individual tree selection also promotes better natural pruning and reduced levels of epicormic sprouting, both of which are important for the development of high-quality stems.

Even-aged management of northern hardwoods has been advocated to retain a mixture of shade-tolerant and intolerant species desired by many resource managers. Clear-cutting and other forms of intensive management can retain trees characteristic of early and late successional series, but may result in poor regeneration and inferior stem quality. Early selection of crop trees for release, better control of logging damage and other silvicultural interventions can improve stem quality in even-aged stands, but may be costly and impractical.

Stocking levels of 16–21 m^2/ha are considered optimal for uneven-aged *A. saccharum* stands on average sites, with maximum annual sawtimber increments of 2 m^3/ha achieved at stocking levels of 20 m^2/ha. At these levels, stands can develop a balanced structure when harvested on a 10–15 year cycle. Mortality increases at higher stocking levels and sawtimber growth rates decline, although stem form improves. Maximum stand growth is achieved at slightly lower (12 m^2/ha) stocking levels, with total annual volume increments of 4–6 m^3/ha.

More emphasis has recently been placed on the creation or retention of old-growth characteristics in northern hardwood forests. Uneven-aged management using some form of selection has become popular, especially on public land. Converting even-aged stands to an uneven-aged condition is desirable, but can seldom be achieved in one or two cutting cycles. In many even-aged *A. saccharum* stands, thinning is required to develop adequate numbers of larger and smaller trees. Extremely dense pole-sized stands typically lack adequate regeneration or the size class diversity needed to create a balanced stocking condition, but management of stands for uneven-aged conditions is feasible, and some private landowners have enjoyed commercial success while maintaining some features of old-growth forests.

Diseases and pests

A. saccharum diseases tend to degrade or discolour stems, resulting in the loss of sawtimber value or volume, but seldom kill individual trees. Anthracnose damage during cool, wet springs can cause leaf scorch or defoliation. Stem cankers caused by *Eutypella parasitica* and *Nectria galligena* result in bole defects and occasional stem breakage, especially in the north.

A. saccharum suffers from a combination of stress-related pests that lead to a condition called maple decline. Typical symptoms of maple decline include reduced growth and a gradual deterioration of the crown (die-back) over several years. Leaf size generally declines and shows premature coloration. Attack by *Armillaria* root rot is often associated with these declines. Maintaining a healthy stand of trees by minimizing damage to roots from harvesting equipment or other disturbance is the primary management recommendation.

The most common insect pests of *A. saccharum* are defoliators, especially the gypsy moth (*Lymantria dispar*), other Lepidoptera including the forest tent caterpillar (*Malacosoma disstria*), several cankerworms and spanworms, and the pear thrip (*Taeniothrips inconsequens*). Two species of bud miners (*Proteoteras moffatiana* and *Obrussa ochrefasciella*) overwinter in the terminal bud, killing it and causing repeated forking and poor stem form, especially in even-aged second-growth stands.

Uses

A. saccharum is not widely planted even within its native range, primarily because natural regeneration is easy to obtain following minor disturbances or partial cutting. It is demanding in terms of site requirements and is rarely considered for reclamation or soil-improvement plantings.

The wood of *A. saccharum* and *A. nigrum* is known as hard maple – a dense, fine-grained wood that machines, stains and finishes well and is widely used for furniture, cabinets, tool handles, toys and flooring. The wood has a specific gravity of 0.56 (green) or 0.63 (dry). It is an important short-fibre hardwood pulp, and it is locally used as charcoal for filtering whisky. The wood is generally straight-grained but some *A. saccharum* trees produce wood with an unusual grain pattern called bird's-eye maple. Veneer logs from such trees command very high prices, and some log buyers have become adept at identifying such trees based on subtle bark characteristics.

The collection of *A. saccharum* sap for maple syrup is a cottage industry in many parts of the north-eastern USA and Canada. 'Sugaring off' is a late-winter activity for farmers, with best yields obtained during periods of sunny, warm days and cool nights. The commercial sap collection season extends from late February to mid-April, depending on climate, but early or dry springs typically reduce sap yields. About 30 l of sap is required to produce 1 l of syrup in large evaporation pans over wood or gas fires. Minimum diameters for 'tapping' trees are about 30–35 cm, with sap production roughly proportional to tree diameter and crown size.

Further reading

Godman, R.M., Yawney, H.W. and Tubbs, C.H. (1990) *Acer saccharum* Marsh. Sugar maple. In: Burns, R.M. and Honkala, B.H. *Silvics of North America. Vol. 2: Hardwoods.* Agriculture Handbook No. 654, USDA-Forest Service, Washington, DC, pp. 78–91.

Meiners, S.J. (2005) Seed and seedling ecology of *Acer saccharum* and *Acer platanoides*: a contrast between native and exotic congeners. *Northeastern Naturalist* 12, 23–32.

Tubbs, C.H. (1977) *Manager's handbook for Northern Hardwoods in the North Central States.* USDA Forest Service General Technical Report No. NC-39. North Central Forest Experiment Station, St Paul, Minnesota, USA, pp. 29.

Adansonia digitata L.

Baobab

Importance

In dryland Africa, *Adansonia digitata* (Plate 5) plays an important role in human nutrition, local culture and economies. The fruit, seeds and leaves provide protein and are rich in a variety of essential vitamins, minerals and amino acids. They are either used for fresh consumption, stored for use in the dry season, or put for sale. The tree is considered a 'wonder of nature' with its enormous, swollen and often oddly distorted trunk, and is linked to a wealth of ethnobotanical knowledge. It is characteristic of thorn woodlands in savannah ecosystems, though natural regeneration is poor and further planting is being promoted.

Botanical features

A. digitata is characterized by its massive size, reaching heights of 18–25 m, with a swollen trunk up to 10 m in basal diameter, and a rounded crown with irregularly distributed branches. Bark is smooth, soft and fibrous, grey to reddish-brown.

Leaves are alternate at the end of branches or on short spurs on the trunk, 2–3-foliate when young, 5–7(–9)-foliate when mature, and may reach a diameter of 20 cm. Elliptic to obovate–elliptic leaflets vary greatly in size, with medial leaflets 5–15 × 2–7 cm.

Large and showy flowers are pendulous, solitary or paired in leaf axils, on pedicels 15–90 cm long. The fruit is large, 8–54 × 8–20 cm, usually globose to ovoid but sometimes oblong–cylindrical, covered with velvety yellowish, sometimes greenish, hairs. The woody pericarp is 8–10 mm thick, green/yellow when immature becoming brown when ripe, containing a white, dry powdery pulp with dark brown to reddish-black seeds within, 10–13 × 8–10 × 4–5 mm.

Distribution

A. digitata is native to most of sub-Saharan Africa having a notable dry season, including the surrounding islands, and in south Yemen. It is completely absent only in Liberia, Uganda, Djibouti and Burundi, with distribution limited in South Africa to the Transvaal and to the eastern part of Chad. It is essentially associated with savannahs, but also in forest and coastal areas possibly associated with human settlements. Natural distribution appears to be in part determined by a minimum annual rainfall requirement of 300 mm.

Silviculture characteristics

Seedlings produce a prominent taproot, but mature trees have extensive systems of relatively shallow lateral roots ending in tubers. *A. digitata* has an outstanding ability to survive drought. The bark is remarkably fire-resistant, and trees can survive even if the core is completely burnt out. However, *A. digitata* appears not to regenerate well in natural conditions, with germination rates usually less than 20%. Trees are deciduous, being leafless for 2–5 months, depending on water availability.

Grafted trees can start bearing fruits after 3–4 years, and well-tended seedlings in approximately 9 years, but first flowering starts much later in most natural stands, after 16–23 years. Flowering occurs at different times within its range, lasting for up to 6 weeks, but can be sporadic or almost continuous. A single tree may have 1 or 2 to 10–50 sour-smelling flowers that open late in the afternoon and remain open throughout the night; they are pollinated mostly by fruit bats. Fruits develop 5–6 months later and tend to fall from the late rainy season onwards. There are 2000–3000 seeds/kg.

Seedlings can reach heights of 2 m in the first 1–2 years and 12 m after 15 years, young trees growing up to 3 cm/yr in diameter. The average fruit yield of mature trees is 200 kg per season, but varies greatly depending on site conditions, genotype and the amount of leaves harvested, though trees may not fruit at all for several years.

Silviculture practice

Healthy fruit is best picked from the tree and not the ground. *A. digitata* seeds can remain viable for several years if kept in a cool dry place. Pretreatment is needed to break dormancy. Cutting the seed coat and soaking in cold water for 3 days improves germination to 85–90%, which usually starts 4–6 days after sowing, with most seeds germinating within 18 days. Seedlings should be watered twice a day during germination and for the first 3 weeks, but not excessively as there is a danger of stem rot. They are ready for outplanting after 3–4 months, or when they reach 40–50 cm tall or have 2–3 pairs of leaves. Direct sowing is also possible with adequate tending and protection.

Vegetative propagation is also practised in the wet season, enabling desired phenotypes to be established. These include shorter stature, which aids management and harvesting and reducing time to first fruiting. The most commonly used methods are stem cuttings or veneer grafting, and also bud grafting and air-layering. Propagation by stem cuttings is the easiest, but the success rate is much lower (2–20%) than with the other methods (46–92%).

Trees are often deliberately planted near settlements, and naturally regenerated seedlings respond well to transplanting, though such seedlings are rare due to intensive browsing by livestock and wild animals. Tree spacing should be 0.5 × 0.2 m for intensive leaf production, 4 × 4 m or 5 × 5 m in home gardens, 13 × 13 m in small orchards for fruit production, or 3–4 m apart in single tree rows. Production is improved by applying 20 kg of farmyard manure or organic matter at planting, or mineral fertilizers can be used. Weeding is recommended, and plants must be protected from rodents and grazing animals, especially during the dry season.

It is better to allow the tree to establish before harvesting leaves, but if site conditions are favourable young leaves can be harvested within the first year. Harvesting is carried out constantly during the vegetative period and often stored for use in the dry season. Trees with good-tasting leaves are cut more regularly to improve quality and prevent the development of branches and fruits. Bark can be harvested at any time, but is often collected at the end of the rainy season.

Baobab is used in agroforestry systems, often intercropped with cereal and vegetable crops, and pollarding reduces competition but also facilitates resprouting, which increases leaf yields.

Diseases and Pests

It is often reported that there are no serious pests or diseases, but some fungal and viral diseases have been recorded and several insects attack the wood, fruit and young seeds (see Further reading).

Uses

The wood of *A. digitata* is light and spongy with a density of 320 kg/m^3 at 15% moisture content. It has no timber value and is unsuitable for fuel; however, it can be used to make canoes and fishing floats, and wood pulp is suitable for processing into writing paper.

In dryland Africa, the important role of *A. digitata* is in human nutrition. Fruit pulp, seeds and leaves are used in the preparation of traditional dishes with greater value during periods of food scarcity and famine. Young leaves are cut into pieces and cooked in sauces, and frequently dried, often powdered and used in sauces, porridges, thick gruels of grains and boiled rice. Dry leaves contain 13–15% protein, 60–70% carbohydrate, 4–10% fat, 11% fibre, and 16% ash; they score well for five of the eight essential amino acids; they are important sources of minerals, especially magnesium, manganese and iron; and they contain vitamin A, levels of which are much higher when not dried in full sunlight. The sale of fresh and especially dried leaves also provides substantial revenue to rural women in the dry season, when crop production is limited.

Fruit pulp is eaten fresh, added to gruels or ground and prepared into refreshing drinks. It can be stored for long periods in airtight containers, sun-dried or frozen. Dried pulp is also industrially processed and marketed as a health food. Ripe fruit contain 8.7% moisture, 2.7% protein, 0.2% fat, 73.7% carbohydrate, 8.9% fibre and 5.8% ash, a very high vitamin C content of 2.2 g/kg, and minerals such as calcium, phosphorus, iron, magnesium, zinc and manganese.

Seeds are used as a thickening agent in soups, fermented and used as a flavouring agent, roasted and eaten as snacks and a cooking oil can be extracted after pounding. Kernels contain 8.1% moisture, 33.7% protein, 30.6% fat, 4.8% carbohydrates, 16.9% fibre and 5.9% ash, are rich in lysine, vitamin B1 (thiamine), calcium and iron, and have an energy value of 1800 kJ/100 g (approximately 50% higher than leaves or fruit pulp).

Trees are regularly stripped for the inner bark, which produces strong fibres widely used for making ropes, fishing lines and nets, and for weaving into such items as baskets and mats. Other uses include the leaves as browse, but intake should be limited due to the high tannin content. The hard fruit shells are used for food and drink containers, or decorated and sold as art objects. All parts of the tree are used in traditional medicine. *A. digitata* plays a very important cultural role in many African societies, with many virtues attributed to it, and its presence is believed to be a good omen. Trees also reduce soil erosion, provide shade and shelter and are planted as ornamentals where introduced.

Further reading

Sidibé, M. and Hughes, A. (2004) *Baobab.* Adansonia digitata. *A Handbook for Extension Workers.* International Centre for Underutilised Crops, Southampton, UK.

Sidibé, M. and Williams, J.T. (2002) *Baobab.* Adansonia digitata. Fruits for the Future No. 4. International Centre for Underutilised Crops, Southampton, UK.

Wickens, G.E. (1982) The baobab – Africa's upside-down tree. *Kew Bulletin* 47, 173–209.

Adenanthera pavonina L.

Red-bead tree

Importance

Adenanthera pavonina is a multi-purpose species. It is a source of food, fodder, green manure, fuel, timber, dye, decoration, medicine and other products. In the South Pacific Islands, it is intercropped with spices, coffee and coconuts, or planted along borders as a windbreak in plantations or on agroforestry farms. Throughout the tropics and subtropics it is extensively cultivated as an ornamental tree along roadsides and in communal areas.

Botanical features

A. pavonina is a medium- to large-sized deciduous tree, 6–24 m high with a diameter of 45–240 cm depending on location. The bark is dark brown to greyish, the inner bark soft and pale brown, and the slash soft, white and fibrous. The crown is spreading and multiple stems are common. Trunks can be slightly buttressed in older trees and the upper bole is sometimes spirally fluted. Leaves are very large and bipinnate with a large swollen pulvinus. There are 2–6 opposite pairs of pinnae (5–6 on saplings), each with 8–21 leaflets on short stalks. The alternating leaflets are 2–2.5 × 3 cm, oval–oblong, with an asymmetric base and blunt apex, dull green on topside and blue-green underside. Leaves turn yellow with age and leave prominent scars on stems when fallen. Flowers are borne in narrow spike-like racemes, 12–15 cm long, at branch ends. The flowers are small, creamy-yellow and fragrant; each flower star-shaped with 5 petals, connate at the base, with 10 prominent stamen-bearing anthers tipped with minute glands. Pods are long and narrow, 15–22 × 2 cm with slight constrictions between seeds, dark brown, turning black upon ripening, leathery. The pods curve and twist upon dehiscence to reveal 8–12 hard-coated, lens-shaped, vivid scarlet seeds 7.5–9 mm in diameter, that adhere to the pod.

Distribution

A. pavonina is found between 20ºN and 15ºN and 300–1200 m altitude. It is endemic to south-east China and India but has been introduced throughout the humid tropics and become naturalized in Malaysia, Western, and Eastern Africa and most island nations of both the Pacific and the Caribbean. It is widely cultivated in tropical Asia.

It is found on coastland, in disturbed areas, and natural forest. *A. pavonina* invades intact as well as disturbed forests and can quickly form large colonies. The tree prefers neutral to slightly acidic soils but will grow on a variety of soils in moist and seasonally moist tropical climates. Optimal plant growth occurs in areas with precipitation ranging between 3000–5000 mm and with mean annual temperatures ranging from 28 to 32°C. It will tolerate a mean minimum of 14–22°C, but is not frost-tolerant.

Silviculture characteristics

After establishment, growth is initially slow but increases rapidly after the first year, during which average annual growth rates of 2.3–2.6 cm in diameter and 2–2.3 m in height can be attained. Trees planted 1 × 2 m apart for windbreaks and at 2 × 2 m in plantations can be thinned in 3–5 years to provide fuelwood and construction materials. For shade trees, spacing varies from 5 to 10 m, depending on the companion crop and site.

A. pavonina tolerates salt, wind and shade. It has the ability to self-prune and trees coppice easily and with good survival. Trees, and particularly the crowns, are susceptible to breakage in high winds.

Silviculture practice

A. pavonina can be propagated from seed (seed storage behaviour is orthodox) and the seedlings have epigeal germination. The seed coat is extremely hard and requires scarification. Manual scarification, immersing seeds in boiling water for 1 min, or treatment with sulphuric acid, significantly increases germination rates. Following treatment, seed can be directly sown in the field or in the nursery. Germination occurs within 10 days, with young seedlings reaching a height of 8–15 cm in approximately 3 months. Nursery stock outplants well. Seedlings are hardy and need minimal maintenance but are not able to suppress weeds. Propagation can also be carried out from large cuttings grafting and tissue culture.

Diseases and pests

A. pavonina is susceptible to the pest *Xylosandrus ater* and to sapstain. The sapwood is susceptible to drywood termites but the heartwood is resistant.

Uses

A. pavonina is valued as fuelwood and also yields very good charcoal. The timber is popular, being hard, durable and strong. It is easy to plane and takes a high finish. The wood is used for construction, flooring, paving blocks, and vehicle bodies, but also for furniture, cabinet work and turnery. The hardwood is medium to heavy with a density of 595–1100 kg/ m^3 at 15% moisture content. The heartwood is bright yellow when fresh, turning red, and is sharply demarcated from the light grey sapwood. The heartwood is closely and even grained, with a moderately fine to slightly coarse and even texture. Shrinkage during seasoning is variable, with only slight warping.

The seeds can be roasted and eaten. One-quarter of the seed weight is oil, with a high percentage of protein and fatty acids. Seeds require heating as they are poisonous when raw. Young leaves are eaten as a vegetable. The leaves are fairly high in digestible crude protein (17–22%) but low in mineral content, and can be used as a supplemental source of fodder. Nitrogen-fixing qualities (through association with *Rhizobium)* and fast-decaying leaves make the species a good green manure.

Medicinal uses include treating rheumatisms, migraines, dysentery and tonsillitis. *A. pavonina* is also used as an ornamental and for shade and shelter. It can be intercropped with most tropical field and tree crops, making it suitable to use in integrated production systems.

Further reading

Adkins, R. (1996) Adenanthera pavonina: *an underutilized tree of the subhumid tropics*. Forest, Farm, and Community Tree Network (FACT Net). Winrock International, Arkansas, USA.

Burkill, I.H. (1966) *A Dictionary of the Economic Products of the Malay Peninsula*. 2 ed., Volume 1, A–H. Government of Malaysia and Singapore, Kuala Lumpur, Malaysia, 1240 pp.

Clark, W.C. and Thama, R.R. (eds) (1993) *Agroforestry in the Pacific Islands: Systems for Sustainability*. United Nations University Press, Tokyo, Japan, 279 pp.

Slik, J.W.F. (2009 onwards) Plants of Southeast Asia. Available at: http://www.asianplant.net/Fabaceae/Adenanthera_pavonina.htm (accessed 5 March 2013).

Afzelia africana Sm. ex Pers.

Afzelia

Importance

Afzelia africana has a wide range of uses and is one of the most heavily used timber species in its natural region of distribution. It is nitrogen-fixing and can be used for soil conservation, and the leaves can be used as a mulch. The dense wood keeps its shape better than most other timbers during seasoning and does not warp or shrink.

Botanical features

A. africana is a large deciduous tree, growing up to 30 m in forests and 10–18 m in savannahs. The crown is dense in forests but more spreading in savannahs. The stem has unequal buttresses that are relatively thick and have a light concave profile; in general they are 1–1.5 m high and 1–2 m wide. The stem is generally straight, 15–20 m long with an average diameter of 1 m at 1.5–1.8 m above the buttresses.

The bark is reddish-grey, scaly, and about 2 cm thick. It exfoliates in rounded patches, protecting the tree against the frequent bush fires of the dry season. The slash is a rosy colour and exudes a dark yellow, highly aromatic liquid resin. *A. africana* has a principal taproot and superficial lateral roots. The leaves are bright green, paripinnate and up to 30 cm long, with 3–5 pairs of elliptic leaflets, which have a blunt apex and a rounded or wedge-shaped base. There are 7–9 principal pairs of fine secondary nerves.

The whitish flowers are 1.5 cm long, with a single, red-striped petal; they grow in terminal panicles up to 20 cm long. The fruits are flat pods, slightly rounded and dark brown to black. They are hard, hairless, 12–17 cm long, 5–8 cm wide and 3.5 cm thick and have a distinct beak at one end. Each pod contains several black seeds, each 2–3 cm long. The basal third of each seed is contained in a bright orange cup.

Distribution

The natural distribution of *A. africana* corresponds to the 'Sudanian Regional Centre of Endemism', the 'Guineo–Congolia/Sudania Regional Transition Zone' and the drier peripheral semi-evergreen Guineo–Congolian lowland rainforests in West Africa (Guineo–Congolian Regional Centre of Endemism). In the latter it grows with *Khaya grandifoliola*, *Triplochiton scleroxylon* and *Terminalia superba*. In the regional transition zone, the species is often found individually dispersed in natural dry forest and secondary forest, in association with *Daniellia oliveri*, *Khaya senegalensis* and *Erythrophleum guineense*. In the Sudan zone, it is found in gallery forests and savannah woodlands; however, regeneration is difficult due to the regular bush fires. Even if *A. africana* can tolerate early fires, in the long run it suffers and stands are decreased.

Silviculture characteristics

A. africana is a light-demanding species that tolerates light shade when young.

Silviculture practice

Natural regeneration is usually limited due to seed predation by wildlife and livestock, requiring seed collection to start soon after maturation. It may be necessary to fence off the chosen tree or to climb it to collect the seed. There are 350–450 seeds/kg. The germination capacity of the seeds can be conserved well at ambient temperature for at least 3–4 months. The orange cup has to be completely removed before sowing; this is facilitated by soaking seeds in water for 24–48 h at 20–30°C; scarification is also possible. Sowing has to be done with the placental end underneath. Germination will take place after 2 weeks. Treatment with concentrated sulphuric acid for 30 min followed by soaking in cold water for 3 h can increase germination to 100%. Under good conditions, the stem may reach a height of 30 cm after 2 months.

The minimum time recommended for bare-rooted seedlings is 6 months in the nursery, when plants will have a height of 30–40 cm. To obtain taller plants, plants have to stay in the nursery for 1 year, because from a height of 40 cm, an increase in growth of the root will be more important than that of the stem. Seedlings should stay in containers no longer than 4–6 months, otherwise the taproot will twist. The best size for planting is 30–40 cm. Direct sowing is possible but has the disadvantage that very young plants are highly appreciated by animals.

No special soil preparation is necessary. Planting holes of 40 × 40 × 40 cm should be dug early enough, particularly in savannah areas, to be able to plant if precipitation is favourable. Fertilizers can be used to promote growth.

Enrichment planting seems to be the most appropriate plantation technique in transition and deciduous forest. Spacing between lines that are 4 m wide should be 14–16 m, and spacing between the plants should be 2–3 m. This small spacing will force the plants to grow in height and will allow selection on thinning. In savannah, it is recommended to create a favourable microclimate before plantation of *A. africana*. This will also help to protect the plants against fire. A first plantation can be installed with *Gmelina arborea*, *Harungana madagascariensis*, and *Acacia mangium* at a spacing of 3 × 3 m, pure or mixed. After 2 or 3 years, when these trees reach a height of 2–3 m, some of them are harvested and *A. africana* can be installed in the free lines.

Diseases and pests

Seedlings of *A. africana* are attacked by fungi, grasshoppers, and small mammals. Its wood is largely resistant to attack from termites or fungi, but can be damaged by marine borers.

Uses

As a leguminous tree, *A. africana* can be used for soil conservation and improvement. The leaves can be used for mulching. In Guinea, it is increasingly planted by villagers for land demarcation and as a timber tree for future generations. In many villages in the Malinké region, *A. africana* is a sacred tree and is often situated near the middle of the sacred village forest.

A. africana has a yellow-white sapwood (3–8 cm thick) and a heartwood which is yellow-brown just after the tree has been felled, and reddish-brown once seasoned. The cross section clearly shows the vessels (pores) standing out, surrounded by 'eyes' formed by storage cells (parenchyma). The latter contain crystals (deposits of calcium oxalate) which are visible under the microscope and depend on the soil characteristics. A special feature of the wood is the white and yellow substances in the vessels and in other cells that can form light or hard layers.

The wood is strong, hard and dense (790 kg/m^3 at 12% humidity, 820 kg/m^3 when air-seasoned) and keeps its shape better than most other timbers during seasoning, not warping or shrinking to any degree. The heartwood is more resistant than other tropical timbers to some chemicals, particularly acids, and is not good for pulping or papermaking. The white or yellow substance in the vessels, afzelin, can cause the wood to stain textiles and other materials that come into contact with it when damp.

A. africana wood can be worked in many ways, but care must be taken when sawing to cut slowly and with specially hardened blades in order to get good timber and not to damage the saw. Because of its density, the wood takes a long time to season. Thicker planks should always be air-seasoned, in which case they must be carefully stacked and their ends protected to avoid warping and endshaking. The wood holds nails and screws well, although one must pre-bore to avoid the wood splitting. A grain filter should be used to achieve polished surfaces. When the wood is likely to come into direct contact with the ground or with water, an additional protective coating should be applied.

Its wide range of uses make it one of the most used timber species in its natural range. Because it is solid, durable and keeps its shape, it is suitable for exterior and interior construction purposes, joinery and for flooring and hard-wearing furniture in schools and public buildings. It can be used in shipbuiding and vehicle body construction, for various containers, wood carvings, mortars, poles, pilings and other traditional uses.

An infusion of the bark is used against paralysis, and a decoction against constipation, backache, stomach ache, gonorrhoea and, mixed with *Tamarindus* and *Ficus capensis*, sleeping-sickness. In Ghana, the pulp is used with *Pericopsis* and *Tamarindus* as a diuretic and febrifuge. The maceration is a remedy for leprosy. The crushed bark, mixed with honey, is used in veterinary medicine. The ash of the bark, prepared with Shea butter as soap, is used against lumbago. A leaf decoction mixed with *Syzygium guineensis* leaves and *Xylopia* fruit forms a drink to treat oedema. For veterinary purposes, the potassium-rich ash of the fruit is mixed with millet, and bark ash is mixed with honey.

Afzelia seed has a fat content of 31% and may be a source of seed oil for domestic and industrial use. Supplementing wheatflour-based foods by up to 30% with *A. africana* seed flour could improve the protein quality, reduce the cost of wheat imports and diversify the use of *A. africana* seed.

Further reading

Arkwright, P. (1962) Know your timber: No. 97. Afzelia [*Afzelia africana*]. *Mekhanicheskaya obrabotka drevesiny* 19, 415 pp.

Bolanle, A.O. (2010) Nutritional composition of *Afzelia africana*. *Biosciences, Biotechnology Research Asia* 7, 101–106.

Götz, E. (1983) *Timber trees of The Gambia*. Bundesforschungsanstalt fur Forst- und Holzwirtschaft/Stiftung Walderhaltung in Afrika, Hamburg, Germany, iii & 105 pp.

Agathis australis (D. Don) Steud.

Kauri

Importance

Agathis australis (kauri) is a resinous conifer renowned for the massive dimensions of the largest trees, excellence of stem form and heartwood quality; it was the main multi-purpose species widely used in New Zealand and exported during the first 100 years of European settlement.

For Maori, the indigenous people of New Zealand, *A. australis* was integrated into mythology, rituals, war, art and daily life.

Forests with *A. australis* in pure stands, as groves or scattered trees, once covered an area of about 1,000,000 ha or more. Destruction by fire, uncontrolled logging, and land clearance for farming has reduced the 'kauri estate' to some 7000 ha of old growth (virgin) forest, now mainly in reserves, and 60,000 ha of scrub and secondary forest with regeneration of *A. australis*.

Conservation of existing forests with *A. australis* is a main objective in management, but further aims are to sustainably manage second-growth forests outside public ownership and to establish intensively managed plantations on the most favourable sites available. Public and landowner interests in the heritage and cultural aspects of *A. australis* provide incentives for planting that go beyond purely economic constraints.

Botanical features

Mature trees have straight, untapered, cylindrical stems with flaking bark, free from branches or epiphytes for 12–25 m. Crowns are massive with upward-arching branches producing flat or slightly rounded tops. Total heights reach 30–50 m. Stem diameters of trees aged from 400–800 years are commonly 1–2 m but living giants are 3–5 m in diameter and often hollow, preventing accurate determination of age (Plate 65).

Trees with juvenile form have narrow, tall, tapering crowns with a strong monopodial habit which may persist until stems reach a diameter of about 50 cm at age 100–150 years in natural forests, and crowns start to spread. Branches of trees with the juvenile form abscise cleanly in natural forest but are more persistent in some plantations.

The root system of mature trees is extensive, with lateral roots often reaching beyond the crown spread. Deeply penetrating 'peg' roots descend from the laterals, giving firm anchorage so that trees are generally windfirm until senescence. Taproots disappear in juvenile trees. Fine feeding roots are superficial in layers of raw humus and litter, and bear nodules containing endophytic mycorrhizae.

Leaves are alternate to subopposite, sessile, thick, coriaceous, usually dull olive-green but occasionally with a bluish, glaucous appearance, 2–3.5 cm long, ovate–oblong for adults, 5–10 cm × 5–12 mm, lanceolate for juveniles.

Trees are monoecious with male cones borne in leaf axils above female cones, sometimes on the same branch. Male cones 2.5 cm × 8–12 mm. Female cones subglobose 5–7.5 cm diameter when ripe in late summer, producing single-winged seed.

Viable seeds are produced from cones on open-grown trees from an age of about 30 years. Seed crops are annual but variable in quantity and viability of seed. Seed-bearing cones break up on the tree when ripe in early autumn, some 2 years after cone initiation. The seeds are relatively heavy for distant wind dispersal, usually falling 20–100 m from tree crowns, or occasionally much further in freak winds.

Distribution

The natural distribution of *A. australis* is confined to the warm, humid, region of the northern part of the North Island of New Zealand, where frosts are infrequent and not severe.

A. australis has also been grown in number of warm temperate and tropical countries including Fiji, Malawi, South Africa, Zimbabwe, Kenya, Puerto Rico and Spain.

Silviculture characteristics

A. australis is shade-tolerant only in the small seedling stage, when shelter is required and seedlings are prone to desiccation. Full overhead light is required from the sapling stage for best growth.

Important characteristics making the species suitable for management are regular good seed crops, relatively fast growth on favourable sites after the seedling stage, tolerance of a wide range of soils and strong apical growth of trees with the juvenile form shedding lower branches to give clean, straight stems.

Silviculture practice

Seed should be collected by picking ripening cones directly from the crowns of selected trees before the cones disintegrate in late summer. Seed falls from cones that are air dried and is soaked before autumn sowing in containers kept under shade cloth for 1 year before hardening off to full light. Seedlings are planted out from 3 years of age when 30–50 cm tall.

Although freshly collected seeds are usually sown, seed has been successfully stored in airtight containers for 6 years at 6% moisture content and a temperature of 5°C.

Seedlings survive best after planting out in autumn and best growth has been recorded on soils of light texture on warm, moist fertile sites with control of grasses and release from overtopping vegetation. Interplanting in secondary forest on land that once supported mature *A. australis* forest has often resulted in slow early growth on infertile sites.

Thinning and fertilizer application in young stands of *A. australis* can produce a moderate growth response. Fertilizer application at planting time in tall scrub can also be beneficial.

Selective or partial logging of mature natural stands in the past, using tractors for log extraction, produced poor results in terms of damage to soil and residual trees. Most of these old-growth forests are now in reserves where logging is proscribed.

Annual height growth of unsuppressed *A. australis* seedlings under natural conditions is in the range of 10–25 cm, but with unfavourable lighting seedlings can stagnate or grow very slowly for 50 years or more. Seedlings planted on favourable sites can have annual height growth of 50 cm and in some plantations average annual diameter growth has exceeded 10 mm for periods of up to 40 years.

Plantations on favourable sites can have an annual productivity of up to 18 m^3/ha compared with 5–9 m^3 on untended second-crop stands.

The early growth rates of *A. australis* planted on favourable sites does not necessarily imply that the growth rates of more mature planted trees will differ from those in natural stands.

Under intensive management there are reasonable prospects of producing harvestable trees ca 60 cm diameter in plantations of *A. australis* in a period of 80–100 years, although a considerably longer period may be required to produce a significant proportion of heartwood.

Diseases and pests

The most significant pathogen and threat to *A. australis* is *Phytophthora* taxon *Agathis* (PTA) which causes kauri dieback, also known as kauri collar rot. First recorded in 1972, it was identified as a new species in 2008 and is assumed to be non-native to New Zealand. Symptoms include yellowing leaves, dead branches, lesions that bleed resin and tree death. The disease can kill trees of all ages. PTA spreads by soil and soil water movement, plant-to-plant transmission through underground root-to-root contact and human and animal vectors.

Another pathogen of concern is *Phytophthora cinnamomi* which causes a slow decline in canopy health and occasionally the death of stressed trees.

Mature trees are often affected by heart rot, leaving them hollow and causing a slow decline in health. A number of species of wood-rotting fungi may be responsible.

There are no significant insect pests of *A. australis*, but those causing local damage on living trees are the leaf miner (*Acrocercops leucocyma*), the leaf roller (*Planototrix excessana*), the New Zealand pin hole borer (*Platypus apicalis*), and the short-nosed kauri weevil (*Xenocnema spinipes*).

The roots of *A. australis* trees are easily damaged by feral goats and pigs, leading directly to a loss of health and potential transmission of disease.

Uses

In New Zealand, *A. australis* is a species regarded with national pride and interest, not only because of the remaining giant trees and mature forests which attract domestic and international tourists, but also because of the colourful history of logging and the timber trade, starting some 200 years ago and providing material for many books and a number of museums. Reasons for planting in regular plantations or groves, or for enrichment planting in scrub or secondary forest are not, therefore, confined to prospects for an economic return from harvesting, but also include aspects of culture and heritage. Such issues encourage use of some fertile farmland sites for planting to enlarge the greatly diminished kauri 'forest estate'. Public interest has also led to the reservation of the remaining state-owned forests; these contain old-growth trees or secondary forest with regenerating *A. australis*.

The heartwood from mature trees of *A. australis* had a deserved reputation as one of the finest softwoods of the world. Little heartwood is now available from harvesting of old trees, but wood of good quality is salvaged from old stumps, fallen trees and recycled timber. Wood recovered from ancient logs preserved in swamps is highly valued for craft work owing to its colours and lustre, but its mechanical properties have declined. The wood of second-growth *A. australis* has a high proportion of sapwood and its properties are not the same as those from old-growth trees. Reported wood densities range from 316 to 719 kg/m^3 at 12–15% moisture content.

The resin produced by old trees of *A. australis* once provided the basis for a thriving industry through the collection of the resin ('kauri gum') used for manufacture of varnishes and linoleum. The resin was collected from surface and subfossil deposits and from living trees. Damage caused by climbers and notching of stems has been associated with development of rots and decline of some trees.

Maori, the indigenous Polynesian people of New Zealand, used kauri timber for boatbuilding, carving and building houses. The resin was used as a fire starter, for chewing and was also burnt and mixed with fat to create the ink for moko (facial tattoos).

A. australis has potential for use in carbon forestry due to its high annual rate of carbon sequestration. The total carbon density of kauri forests can be nearly 1000 tonnes/ha.

Further reading

Bergin, D.O. and Steward, G. (2004) *Kauri: Ecology, Establishment, Growth and Management.* Forest Research, Rotorua, New Zealand.

Halkett, J. and Sale, E.V. (1986) *The World of the Kauri.* Reed Methuen, Auckland.

Orwin, J. (2004) *Kauri: Witness to a Nation's History.* New Holland Publishers, Auckland.

Stewart, K. (2008) *Kauri.* Penguin Books Ltd, London.

Agathis borneensis Warb.

Kauri

Importance

Agathis borneensis is an important source of kauri timber. As such it is highly valued and has been a subject of plantation forestry and natural enrichment plantings. The species also shows considerable ecological amplitude, making it extremely suitable for plantings from the lowlands to the mountains in a range of soil conditions. A disadvantage of the species is the difficulty in obtaining sufficient seed material for large-scale plantings.

More research is needed on the genetic variation of the species and to develop sustainable planting schemes using vegetative propagation.

Botanical features

This is a large tree reaching a height of 55 m, with a dbh of 70 cm and a deep conical crown. The bole is typically cylindrical without buttress and it is self-pruning. The bark is very variable, grey, blackish or fawn-brown, deeply fissured to thinly scaled.

The leaves are simple, opposite; usually more or less elliptic, bluntly pointed and very variable. Even on a single tree leaves can be between about 25–70 mm long, with many parallel veins, a short petiole and glabrous.

It is a monoecious tree with mature pollen cones, oblong, 4–7 cm × 20–25 mm, rounded at the apex, subtended by a 2–10 mm peduncle. Mature seed cones are oval, 6–8 × 5.5–6.5 cm. Seed 12 × 9 mm, blunt at one upper corner and a broadly rounded wing 20 × 16 mm at the other corner.

Seed cones are produced before pollen cones appear. Seed cone production appears to be continuous and cones take about 2 years to mature. In seedlings the leader shoot grows in flushes. In between growth, 4–6 scale leaves protect the resting bud. It is assumed that adult trees behave in a similar manner.

Distribution

A. borneensis is naturally distributed mainly in the submontane forests of Peninsular Malaysia, Sumatra and Kalimantan, and has also been planted in these areas. It is also found in kerangas forest (heath forest) at lower elevations.

Silviculture characteristics

Commercial plantations should be on gentle slopes, on well-drained soils with good aeration. Seedlings grow best under some shade, although heavy shade stunts growth. Growth and survival improves considerably when overhead light is given after the seedlings have been established. The species is best managed under selective felling systems, but with some manipulation of the canopy cover to promote initial growth of the seedlings.

Silviculture practice

Mature cones are difficult to collect, as they are produced in the upper parts of the branches and disintegrate at maturity. The seeds are winged and fallen seeds are not recommended for seedling production. Large quantities are easiest to obtain at the time of felling. Seeds must be sown as soon as possible, because viability drops rapidly, from 90–100% initially, to 0% after some weeks. However, quick drying using low temperatures can extend storage time when stored at temperatures below 0°C. After soaking for 24 h, seeds are directly sown on seedbeds and covered thinly with soil. Seedlings prefer shade and will only survive open planting if the roots are minimally damaged and soil is moist. Plants are more resistant to exposure when they are more than 1 year old.

Plantation trials on *A. borneensis* in Peninsular Malaysia, using spacing trials of 13.3 × 2.6 m, have shown good growth. Trees of *Agathis dammara* for tapping of kauri copal are planted at a wide spacing of about 10 × 5 m. On the other hand, close planting of *A. dammara* in Java, at 3 × 3 m, encouraged the formation of branchless boles.

Vegetative propagation has proved successful, either by root suckers from nursery seedlings, or by stem and leaf cuttings with auxin applications. Stem cuttings should preferably be taken from young plants or the lower branches of young trees.

Almost all of the commercial production of *A. borneensis* comes from natural stands in Peninsular Malaysia, Sumatra and Kalimantan. Standing volumes of 100–400 m/ha^3 of

A. borneensis have been exploited from large stands in heath forests in southern Kalimantan. Natural regeneration appears to be the only successful form of regeneration for the species at present. There is no published information on plantation yields of *A. borneensis*, but pulpwood plantations of *A. dammara* in Java on a rotation of 30 years have an annual wood production of 23–32 m^3/ha, and veneer production on a 50-year rotation of 22–28 m^3/ha. The total yield of pulpwood after 20 years is approximately 300 m^3/ha and total yield of timber after 40 years is 570 m^3/ha.

Diseases and pests

Seedling and young trees can host *Aecidium* rust, although it is not considered a serious problem.

A. borneensis has been recorded as a host of the rubber termite *Coptotermes curvignathus*.

Uses

The species has been used occasionally as an ornamental tree in urban planting.

A. borneensis wood is a general-purpose softwood and has many uses. The density of the wood is 380–550 kg/m^3 at 15% moisture content. It is excellent for joinery, boatbuilding (especially masts), construction, household utensils, panelling, turnery, mouldings, packaging and foundry pattern making. It is also used in drawing boards and rulers, matches, matchboxes, pencils, furniture, battery separators, piano parts and artificial limbs. The timber is odourless; it was formerly used for food containers such as tea chests and butter boxes. It makes a good peeled veneer, with an attractive colour and figure for decorative plywood panelling. It is very suitable for the manufacture of wrapping, writing and printing paper and rayon-grade pulp. The wood is also suitable for the manufacturing of charcoal and activated carbon.

The inner bark of *A. borneensis* exudes a translucent or clear resin known as copal or manila copal. This used to be an important component of varnish and was used in the manufacture of linoleum. There is still a steady export demand for varnishes for labels of food tins and photographic colour prints and reflector paint. There is potential for the species to be tapped for copal on a commercial scale.

There is some preliminary scientific evidence that extracts of *A. borneensis* have anti-protozoal properties.

Further reading

Becker, P. (2000) Measurement of sap flow in *Agathis borneensis* (Araucariaceae) with deep sapwood. *Journal of Tropical Forest Science* 12, 771–777.

Meijer-Drees, E. (1940) The genus *Agathis* in Malaysia. *Bulletin du Jardin Botanical Buitenzorg Series 3* 16, 455–474.

Whitmore, T.C. (1977) A first look at *Agathis*. Tropical Forestry Papers No. 11, University of Oxford Commonwealth Forestry Institute, Oxford, UK, 54 pp.

Whitmore, T.C. (1980) A monograph of *Agathis*. *Plant Systematics and Evolution* 135, 41–69.

Wong, T.M. (1981) Malaysian timbers – dammar minyak. Malaysian Forest Service Trade Leaflet No. 43. The Malaysian Timber Industry Board and Forest Research Institute Malaysia, Kuala Lumpur, Malaysia.

Agathis dammara (Lamb.) Rich.

Kauri

Importance

Agathis dammara is an important source of timber (kauri) and resin (manila copal). The timber is highly valued as a lightweight softwood; and the copal used to be an important component of varnish, and was used in the manufacture of linoleum.

Natural populations of *A. dammara* have been depleted in the Moluccas. The species is very suitable for large-scale plantations, particularly if vegetative propagation techniques can be developed and applied.

'Kauri' is a Maori word, applied by that people to *Agathis australis*, but generalized in modern usage to all species of *Agathis*. The nomenclature of the genus has been confused and confusing from its earliest descriptions and *A. dammara* in particular has many synonyms.

Botanical features

A. dammara is a medium to large evergreen tropical rainforest tree (30–65 m tall, with a diameter of up to 200 cm). Its bole is usually straight and cylindrical, often swollen at the base or with large, raised superficial roots. Its bark is light grey to reddish-brown, at first quite smooth with conspicuous lenticels, later peeling with large thin irregular flakes that gradually become thicker. These flakes leave irregular small depressions that create a rough, reddish-brown surface on larger trees. Stems sometimes bear accumulations of resin arising from stem wounds.

Leaves are tough, dark and leathery. Their size and shape vary, but most are elliptical, 6–8 × 2–3 cm, tapering towards the rounded apex. Adult leaves have solitary resin canals.

Female and male cones are on the same tree, but on different branches. Female cones are ellipsoid to globose, 6–8.5 × 5.5–6.5 (or even 13) cm; seed scales are 30–40 mm long, roughly triangular in shape with a well-rounded apical margin. Male cones are oblong, straight and continue to grow to 3–4 cm long, 10 mm across, sessile or with up to 4 mm-long peduncles. Male cones change from light green to brown when they are mature and pollen is released. The species is wind pollinated.

Mature seed cones are oval, 9–10.5 × 7.5–9.5 cm, and seed bracts are roughly obtriangular. Seeds are flattened and ovoid, 10–11 × 8 mm and with one lateral wing. There are 88–90 filled seeds per cone in a good season.

Distribution

A. dammara is native in the Philippines, on the Indonesian islands of Sulawesi, the Moluccas, and West Papua, as well as in Papua New Guinea. It grows at altitudes from 150 to 2200 m on a variety of soil types including shallow, infertile ones. It appears to favour well-drained sites.

Agathis is predominately a genus of tropical rainforests; most species thrive in frost-free sites that receive between 5000 and 10,000 mm of rain/yr. On these sites the trees may be solitary, but commonly occur in groves mixed with other rainforest trees.

Silviculture characteristics

A. dammara is cultivated as a plantation tree. In Java (Indonesia), it has been planted widely. It is also used as an enrichment crop in inadequately stocked areas in logged-over forest and for reforestation within its natural range, especially in Papua/Irian Jaya.

Stand establishment is normally through natural regeneration, planting stock and wildings. It needs partial shade in the early stages of its development. Growth of the seedlings is slow at first but later accelerates. The species tolerates wind, but is sensitive to waterlogging.

In plantations in Java, cone production begins when the trees are about 15 years old, but viable seeds are seldom produced before the age of 25. Trees produce cones throughout the year. The best 'fruiting season' occurs from August to October when the highest percentage of filled seeds is found in the cone. The cone breaks up upon maturity and the seeds are recalcitrant: these features represent possible disadvantages of the species, as it is difficult to obtain sufficient seed material for large-scale plantings.

Silviculture practice

To acquire viable seeds, cones are collected by climbing the tree when cones have become brownish-green. It is not recommended to collect seed from the ground. Mature cones are then placed in bags for 1–2 days, to release the seeds; there are 4000–5000 seeds/kg.

The seed cannot be stored for much more than 2 months in an air-conditioned room (18–20°C) at 30% moisture content. The seeds are sensitive to desiccation, which causes loss of viability.

Germination of *A. dammara* is relatively easy: no pre-treatment is needed before sowing. Seeds are buried superficially, with the wing pointing up. Germination may start only 6 days after sowing, and usually 90–100% of germination takes place within 10 days.

Vegetative propagation can be achieved by taking stem cuttings from seedlings aged 7 months (30 cm tall); these are dipped in indole-3-butyric acid (IBA, 10 ppm), and planted in a rooting bed of river sand. Roots will emerge 6–8 weeks after the stem cuttings have been placed in the rooting bed, and kept at a temperature of 28–32°C and relative humidity of 90–100%.

Plantations have generally been monocultures. The earliest involved the taungya system in Java, where the seedlings were grown among food crops for 1–2 years, and *Leucaena leucocephala* was planted for shade, fertilization, weed control and soil protection. Thus, at 30 years of age, the rotation length recommended for pulpwood, total volume, including thinnings, is estimated at 702–961 m^3/ha, or 23–32 m^3/ha/yr. A 50-year rotation is recommended for veneer, and total yield is predicted at 1099–1411 m^3/ha or 22–28 m^3/ha/yr.

In some regions resin tappers are said to be over-exploiting the species: the repeated deep wounding of the trees leads to infection and tree death. It is believed that strict controls, or a tenure-centred management system that encourages longer-term caring for the trees, is required.

Diseases and pests

Diseases encountered in large-scale plantations of *A. dammara* include rust caused by the fungus *Aecidium fragiforme* and other fungal infections such as *Fomitopsis pinicola* and *Phytophthora cinnamomi* (causing root rot).

It is attacked by insects such as the weevil *Aesiotes notabilis*. As noted above, heavy resin tapping can lead to stem rot and tree death. *Agathis* seeds are vulnerable to *Agathiphaga* moth caterpillars, which can severely impact seed production.

Uses

A. dammara produces a pale, high-quality timber, which is sought after for its attractiveness, straight grain and easy workability. The timber is used for joinery, boatbuilding (especially masts), utensils, mouldings, packaging, piano parts, veneers, plywood, matches and for pulpwood. It also makes a good charcoal, including activated charcoal. The wood is, however, not resistant to rot. Trees are often cut specifically to order, as extended storage readily leads to deterioration and reduced value. It has a density of 380–660 kg/m^3 at 15% moisture content.

Manila copal is the translucent or clear white resin exuded from the inner bark of *A. dammara*. It is soluble in alcohol and was formerly required for the production of many varnishes and of linoleum. Copal is still used in specialized fine-arts products and in traditional varnishes, lacquers, as a substitute for shellac, for paper sizing and for floor covering. Copal was used historically in sealing wax; as a liniment it is used to bind incense, as a fire starter and for (fire) torches. From 1920 to 1940, as much as 15–20,000 tonnes copal/yr was harvested from *A. dammara*. At present there is still a quite high local demand for copal, but less so for export, as it has been replaced by synthetic substitutes.

Further reading

Adams, J. and Erne, G. (1986) *Kauri: a King among Kings*. 2nd rev. ed. Wilson & Horton Ltd, Auckland.

Bowen, M.R. and Whitmore, T.C. (1980) A second look at *Agathis*. Occasional paper No. 13, Commonwealth Forestry Institute, Oxford, UK, 19 pp.

Eckenwalder, J.E. (2009) *Conifers of the World: The Complete Reference*. Timber Press, Portland, Oregon, USA, 720 pp.

Whitmore, T.C. (1980) A monograph of *Agathis*. *Plant Systematics and Evolution* 135, 41–69.

Agathis macrophylla (Lindl.) Mast.

Pacific kauri

Importance

Agathis macrophylla is a forest tree species of major ecological and economic importance in the south-west Pacific. It is a canopy-emergent species that constitutes a unique structural element in the forests in which it occurs. Its timber is of major commercial importance; a finely grained, pale, easily worked and uniform timber with various high-value end uses. From a silvicultural viewpoint, the most promising *Agathis* appears to be the southern Vanuatu provenances of *A. macrophylla* (formerly *Agathis obtusa*), which combine moderately fast growth, especially in more open-grown situations, with good form and self-pruning.

Botanical features

In the western parts of its range, *A. macrophylla* is a tall tree 30–40 m in height (rarely attaining heights of 45–55 m), while in Fiji the tree is more commonly 20–33 m tall. Bole diameters of older specimens are around 1.2–1.6 m in the Solomon Islands, but can be up to 3 m in Vanuatu and Fiji. Tree form may vary considerably depending on age, habitat and provenance.

The bole is slightly to strongly tapering, without buttresses, and clear for a variable length, approximately 9–20 m. The crown is at first monopodial and ultimately immense, then conical and massive, to 36 m diameter × 24 m deep, often asymmetrical and ragged in outline. Branches may be erect to horizontal and massive. Bark shows regional and local variation, typically grey to fawn and strongly scaly. Trees typically have a short but substantial taproot, and sometimes massive, far-reaching surface roots spreading up to 10 m.

Leaves are opposite decussate, but held in one plane except on vertical growing tips; blade simple, entire, elliptic to lanceolate, tapering to a more or less pointed tip, rounded at the base, the margins curved down at the edge; with many close inconspicuous parallel veins; size variable approximately 7–15 × 2–3.5 cm; dark green and shiny above and often glaucous below, leathery. Petioles short, almost sessile where the blade tapers down to the stalk base, to 5 mm long. The leaves of Solomon Islands plants are typically larger and broader than those from drier and cooler parts of its range in Fiji.

Male cones cylindrical, 8–12 mm diameter × 20–25 mm long; at anthesis have a cupule wider than cone base; green, drying to rufous-fawn; rather inconspicuous on short side branches. Microsporophylls, as seen in intact cone at anthesis, strongly imbricate, 1.5–2 mm across, margin thin, entire or erose; head in adaxial view 2(–2.5) mm across × 2 mm radially, thick centre thinning gradually to a narrow margin; stalk joining near head near abaxial edge; anthers 6–14.

Female cones are larger, globular, approximately 8–13 cm across, on thick woody stalks; green, slightly glaucous and becoming brownish when ripening. Each brown, winged seed is attached to a triangular cone scale about 2.5 cm across. About half the seeds are viable, these being located in the central part of the cone.

A. macrophylla is monoecious, having separate male and female flowers on the same tree. The first female cones are produced when trees are about 10 years old. These take 2 years to mature: at the end of the first year they are egg-shaped, about 5 cm long and 3 cm in diameter, and at the end of the second year they are more or less round and 8–10 cm in diameter.

Mature trees seed most years and some seeds are produced throughout the year. Seeds are wind dispersed, up to 10 km from the mother tree, following complete disintegration of the cone on the tree.

Distribution

A. macrophylla naturally occurs in the Solomon Islands, Vanuatu and Fiji. In the Solomon Islands, it is found on four islands in the Santa Cruz group: Ndende, Vanikolo, Tevai and Utupua.

It grows in lowland subtropical/tropical forests, as scattered trees and groves, being slightly more common on ridge crests. In Vanuatu, *A. macrophylla* is found on two southern islands: Erromango (central-eastern parts) and Aneityum. In Fiji, the species occurs on the islands of Viti Levu, Vanua Levu, Ovalau and Kadavu.

The species is recorded as being cultivated in Hawaii and has been included in *Agathis* species/provenance trials in many countries throughout the tropics.

Silviculture characteristics

As a gap opportunist, *A. macrophylla* has the ability to grow fairly rapidly in full sunlight. Seeds tend to germinate on the undisturbed forest floor but may also colonize open ground in abundance. Seedlings grow faster in brightly lit conditions but are able to endure shade for a long time. In partly shaded situations, plants grow slowly until they are exposed to fuller sunlight. In semi-shaded sites, the tree has good form and a self-pruning ability. In disturbed areas, *A. macrophylla* may either establish synchronously with pioneer tree species or invade such stands at a later date.

The species is capable of maintaining a viable population in undisturbed areas. It is also able to invade secondary forest and to establish itself as scattered trees and groves. Seeds are wind-dispersed up to a distance of 10 km which makes the species capable of colonizing remote localities. *A. macrophylla* is relatively resistant to strong winds and cyclones. Naturally regenerating *A. macrophylla* has the ability to grow fast in full light, especially if root/weed competition is reduced in early years.

Silviculture practice

A. macrophylla seed behaves in an orthodox manner at 9% and 13% moisture content, but is very sensitive to a range of conditions including chilling damage, carbon dioxide/oxygen balance, and the complex relationships between moisture content, temperature, and drying rate.

Precautions have to be taken in seed collection, extraction, and handling to avoid damage from lethal combinations of temperature or moisture content. Sowing of dewinged seed straight after collection avoids some of the risks. No pretreatment of seeds is required: fresh, undamaged seeds start to germinate at 26°C within a week and germination continues for 10–14 days.

For long-term storage, seed should be dried to around 9–13% moisture content and kept as cold as possible, preferably at −13°C.

Vegetative propagation using both main stem and lateral cuttings from 12-month-old nursery-raised seedlings has been successful; however, the latter tend to retain a plagiotropic growth habit, growing at an oblique or horizontal angle. Trials on decapitated seedlings showed that bigger shoots are produced by cutting at a top node rather than at a lower one, and roots are formed in 2 months.

In *A. macrophylla*, vescicular-arbuscular mycorrhizae are formed by infection of short and long roots by the phycomycete *Endogone*. Deliberate mycorrhizal inoculation of *Agathis* seedlings is unnecessary.

A. macrophylla trees generally display a good self-pruning habit; however, in open plantings the species forms persistent, rather heavy limbs which show no sign of self-pruning during the first 10 years; young trees have small branches which can be easily snapped by hand, mature plantations self-prune well and pruning operations are not practised. The Vanuatu provenances exhibit better self-pruning and branching characteristics than those in the Solomon Islands. Crown diameter studies indicate that for trees of 75 cm dbh, a spacing of 10 m between the planting lines is optimum for crown closure; in some areas of the Solomon Islands, a light thinning at age of 10–15 years may be necessary to achieve the desired stocking rate.

A. macrophylla is the fastest-growing *Agathis* species, with annual increments of up to 1.5 m in height and 2 cm in stem diameter over the first 10 years or so. It has been suggested that if selected planting stock is used, a 30-year rotation for enrichment plantings, and a spacing of 2–3 m in lines cut 9–10 m apart is suitable.

Diseases and pests

The species is seriously damaged by butt-rot fungi. On fine-textured soils with high moisture content, the species becomes susceptible to root fungi such as *Phytophthora*. The species has shown poor survival and suffered considerable die-back in the western Solomon Islands as a result of attack by *Amblypelta cocophaga* (a coreid bug). *Corticium salmonicolor* (pink disease) (*Phanerochaete salmonicolor*) has affected the species in the Solomon Islands and at sites in Fiji with waterlogged soils. Other diseases such as necrotic bark, canker (*Nectria cinnabarina*), hollow butt, *Phellinus noxius* and *Fomes annosus* have been reported.

The most serious insect pests are larvae of the primitive moth, *Agathiphaga vitiensis*, which can destroy up to 95% of seed in a cone. Wood may be damaged by pinhole borers, termites, and *Anobium* borers. *A. macrophylla* has relatively low susceptibility to termite attack.

Uses

Although *A. macrophylla* has not been widely cultivated outside its natural habitat (apart from its inclusion in trials), it exhibits some desirable silvicultural characteristics, including a fast growth rate, that could make it a valuable species for plantations or for enrichment planting.

A. macrophylla is a highly valued commercial timber species. It has finely textured, straight grained, pale, and uniform timber of an excellent quality. It is easily worked by hand and machine tools and has good gluing properties. It has good peeling and veneer properties and is highly sought after for veneer. The pale cream to gold-brown, lustrous heartwood and straw-yellow to pale brown sapwood is well known and highly appreciated. The wood is readily kiln dried with a medium shrinkage value. Air-dry density is 540 kg/m^3. In service the timber is very stable. The timber is suitable for a wide range of end uses including laboratory bench tops, vats, sauna baths, battery separators, weatherboards, bowls, novelties, handles, furniture, veneer and boatbuilding. Traditional wood uses in Aneityum, Vanuatu, include canoe-making and carvings. The species has also been recommended for production of high-quality veneer timber on a short rotation.

Manila copal produced from the living inner bark of *Agathis* was an important component of many varnishes and is still used, mixed with synthetics. *Agathis* resin is also traditionally used in Vanuatu for lighting and as canoe caulk, and soot from the resin is used for tattoos. In Fiji, the resin was formerly used for glazing pots and also for torches.

Further reading

Chaplin, G.E. (1993) *Silvicultural Manual for the Solomon Islands*. ODA Forestry Series, No. 1. Natural Resources Institute, Kent, UK, xv & 305 pp.

Thompson, L.J. (2006) *Agathis macrophylla* (Pacific kauri), ver. 1.2. *Species Profiles for Pacific Island Agroforestry*. In: Elevitch, C.R. (ed.) Permanent Agricultural Resources (PAR) Holualoa, Hawaii. Available at: http://www.traditionaltree.org (accessed 27 March 2013).

Whitmore, T.C. (1977) A first look at *Agathis*. Tropical Forestry Papers No. 11, Department of Forestry, Oxford University, Oxford, UK, 54 pp.

Agathis robusta (C. Moore ex F. Muell.) F.M. Bailey

Kauri pine

Importance

Agathis robusta has excellent form, moderate growth rate and produces a valuable knot-free timber on well-watered and drained, fertile sites in tropical and subtropical climates. The timber is suitable for a range of high-value uses such as for plywood and high-quality furniture and joinery. It is already widely cultivated as an ornamental species in the tropics and subtropics and has potential as a plantation species.

Botanical features

A. robusta is a tall forest tree to 50 m in height and 2.4 m in stem diameter, with a long, straight, clean bole without buttresses and with little taper. It has thick, smooth bark with scattered flakes, orange-brown, brown or grey in colour. The bark exudate is clear or somewhat milky and with a faint odour, slow to appear and generally rather meagre. Branchlets are smooth and glabrous. The root system is small and diffuse with fine roots and is endomycorrhizal.

Leaves are simple, alternate, opposite or whorled, almost linear to elliptic, flat, smooth and glabrous, 5–13 × 1–4 cm (laminae 2–7 times as long as broad). Petiole short, about 3–10 mm, apex acute or obtuse; base attenuate; margin entire.

A. robusta is monoecious. Male strobili are shortly pedunculate or almost sessile, usually axillary on slender leafy twigs. Cones 4–10 cm long × 0.7–0.9 cm broad with more than 500 scales per cone each with 2–8 pollen sacs on the underside. Male cones dry to a fawn colour at maturity.

Female cone globose to cylindrical, 9–15 cm long × 8–10.5 cm broad with 340 to 440 scales per cone, anthers 2–8. They are green, dry or woody and dehiscent at maturity. Seeds, excluding the wings, are narrowly cordate.

Cones develop and mature between 9 and 18 months depending on location.

Distribution

In Australia, *A. robusta* occurs in two disjunct regions of Queensland that are separated by ca 1000 km. One region takes in the Cook and North Kennedy Districts of far northern Queensland, while the other is in the south-east of the state in the Wide Bay District, including Fraser Island. *A. robusta* subsp. *nesophila* occurs in eastern New Guinea and New Britain.

In Queensland (Australia), *A. robusta* is an emergent species of a few rainforest types, but forms a dominant part of the stand in dry marginal rainforest types on the edge of the rainforest zone.

In Papua New Guinea, *A. robusta* subsp. *nesophila* occurs in the lowland tropical rainforests, which are exceedingly rich in angiosperm tree species. However, as in Queensland, it is only on limiting sites and never in mesic climax forest that *Agathis* becomes an abundant canopy-top tree and a major part of the forest biomass.

Silviculture characteristics

A. robusta is moderately fast growing when planted on well-watered, frost-free sites with well-drained, fertile soils. It is shade-tolerant and suitable for enrichment planting in gaps in the forest canopy. It has exceptional self-pruning characteristics, leading to high-quality, knot-free timber. Susceptibility to insect defoliation has led to a cessation of planting in Australia, but elsewhere it is relatively insect- and disease-free.

Silviculture practice

The cones of *A. robusta* should be collected when unopened and before they turn brown inside, since mature cones dehisce on the tree and the winged seeds are lost. The collected cones are dried on hessian trays under open-walled sheds and, after 1 or 2 days, open to release the seeds. Seed quality is best when the weather is dry during the drying period. The empty seeds are separated by winnowing and the unripe seeds are sorted by hand. If handled and processed correctly, seedlots give about a 90% germination rate if sown within 1 week after collection. *A. robusta* seed has a lower longevity than that of *Agathis australis* but is higher than that of *Agathis macrophylla*.

For longer-term storage of *A. robusta* seed, placement of dry seed (e.g. 5% moisture content) in airtight containers at sub-zero temperatures (e.g. −15°C) is recommended. Under sub-zero temperatures, seed retained its viability for at least 4 years compared to the refrigerator (3–5°C) where one-third of viability was lost in 12 months, and room temperature storage where viability was lost within a few months.

A. robusta can be propagated vegetatively by stem and root cuttings, root suckers and grafting. The species has plagiotropic growth habits and produces secondary leading shoots after loss or removal of the original leader. Stem cuttings, therefore, should always be taken from orthotropic shoots. *A. robusta* has been successfully propagated *in vitro* via axillary meristems from the leaf axils of cultured main stem segments taken from 2-year-old seedlings.

In the absence of the correct storage environment, it is important that seeds are sown as soon as possible after collection. A germination mixture of sieved nursery soil, sand and farmyard manure in the ratio of 2:1:1, respectively, with seeds spaced at 5 × 10 cm in the germination bed, is recommended. After sowing, the seeds are covered with a very thin layer of fine sand. Daily watering of sown seeds is necessary to facilitate rapid germination. Seed starts germinating within 1 week of sowing and germination is complete in about 3 weeks.

A. robusta was planted extensively in south-eastern Queensland until 1959, when the species was abandoned for plantation use following massive defoliation by the kauri coccid *Conifericoccus agathidis*. When planted on well-drained, fertile soils above the frost line, rotation lengths of 50–60 years were employed. Plantations were thinned to routine prescription and by 22 years of age had a stocking of 62 stems/ha. *A. robusta* can regenerate through coppicing. When planting *A. robusta* for ornamental and avenue planting, spacings of 12–18 m are recommended.

Manual pruning is unnecessary, as *A. robusta* naturally sheds its lower branches while still at fairly small girth, thus producing knot-free timber.

A. robusta is reasonably fast growing in correctly sited and managed plantations. At age 22, a plantation in southern Queensland thinned to routine prescription (919 stems/ha) had a mean annual increment of 13.8 m^3/ha with a predominant height (50 tallest/ha) of 18.9 m.

A mean annual increment of 8 m^3/ha has been estimated for recoverable sawlogs on a 30-year rotation.

Diseases and pests

An outbreak in 1959 of the defoliating coccid scale insect *Conifericoccus agathidis* led to the abandonment of *A. robusta* as a plantation species. After the age of 4 years, trees are prone to attack, but new leaves arising from buds of older trees, or scion wood from older trees, appear to be more susceptible than those of younger trees. Trees weakened by coccid attack are then susceptible to wood borers such as *Euthyrhinus meditabundus* and longicorn beetles.

A. robusta is also attacked by kauri thrip (*Oxythrips agathidis*) and by two leaf-attacking fungal pathogens (*Hendersonula agathi* and *Aecidium fragiforme*). Larvae of moths including *Agathiphaga queenslandensis* and an unidentified species of weevil have been observed in seeds of *A. robusta* in Australia. The larva of a moth tentatively identified as *Tarphyscelis* sp. has been observed in seed in Papua New Guinea.

Uses

A. robusta is a very fine tree for ornamental and avenue planting and is suitable for agroforestry and boundary planting.

The heartwood is pale cream to pale brown, the sapwood is indistinct, and growth rings are almost undetectable. The wood has no odour or taste. The timber of *A. robusta* is straight grained, finely and uniformly textured, soft, easy to work, machines well and readily takes stain and polish. Density (green density ca 720 kg/m^3 and air-dry density ca 480 kg/m^3) and durability are low. The timber is highly valued and uses include plywood, furniture, joinery and pattern-making.

A. robusta wood is suitable for wrapping, writing and printing papers and rayon-grade pulps. Wide, thin-walled tracheids are produced throughout the growth period with only a slight reduction in size and increase in wall thickness, followed by a band of narrow, thicker-walled tracheids that terminate the ring.

Trees are tapped for resin, which is also dug up at the base of the tree. However, the resin of *A. robusta* is reported to be slow to appear and generally rather meagre in quantity.

Further reading

Huth, J. and Holzworth, P. (2005) Araucariaceae in Queensland. In: Dargavel, J. (ed.) *Araucarian Forests.* Issue 2 of Australia and New Zealand forest histories. Australian Forest History Society Inc., Kingston, Canberra.

Whitmore, T.C. (1977) A first look at *Agathis.* Tropical Forestry Papers No. 11, Department of Forestry, Oxford University, Oxford, UK, 54 pp.

Whitmore, T.C. (1980) A monograph of *Agathis. Plant Systematics and Evolution* 135, 41–69.

Ailanthus altissima (Mill.) Swingle

Tree of heaven

Importance

In many warm temperate to subtropical climates worldwide, *Ailanthus altissima* is valued for its timber, its role as an amenity and ornamental tree and for shade and shelter. It is a pioneer species, characterized by rapid juvenile growth, prolific seed production, production of root suckers, and its ability to coppice readily if cut. It requires moist and well-aerated soils to produce good-quality, single-stemmed trees, but trees are also tolerant of drought, poor soils and pollution, and so can be grown in difficult urban locations. These characteristics also mean that it can become a weed, and it has spread widely in some places and is declared an invasive species in many countries where it was introduced.

Botanical features

A. altissima grows up to 30 m high or more, but is usually only 6–10 m tall, with gracefully curving branches. It often has a single clear bole for at least half of the tree height. The bark is smooth with pale stripes. Leaves are green, up to 90 cm long, pinnately compound, pubescent or nearly glabrous. Each has some 25 (10–40) pairs of large leaflets, 5–18 cm long and 2–5 cm wide, dark green above with whiter-green undersides, a tapering apex and 1–3 pairs of glandular teeth near the base; these readily distinguish it from some ornamental *Rhus* species with similar appearance.

Flowers are unisexual, on different trees, small and yellow, in large panicles, with male flowers having an unpleasant odour. Light reddish-brown seeds are contained within one-celled, one-seeded, oblong, thin, spirally twisted samaras, 3.0 × 0.8 cm in size, with no endosperm.

Distribution

A. altissima is native to northern and central China, from Liaoning and Hebei provinces in the north of its range to Guangxi and Fujian in the south, and from Zhejiang and Shandong in the east to Gansu in the west. From a relatively small natural distribution, it has been widely introduced and has become naturalized in temperate regions around the world, becoming invasive in parts.

It is present in many European countries, having been introduced in the early 1700s. It is now especially common and spreading in the Mediterranean region and south-eastern Europe. In central Europe, climate is the major factor affecting its distribution, whereas in Mediterranean regions its distribution is affected more by site fertility than by climatic factors.

A. altissima was introduced into the USA in the late 1700s and has become extensively naturalized across North America, from Canada to Mexico, and from coast to coast. It was reported to be already widespread and naturalized in some states in the late 1800s, and in some parts of the USA it is so well established that it appears to be a part of the native flora.

It is also a common tree in parts of Japan and the Korean Peninsula, northern India and Pakistan (where it is widely cultivated), Argentina, Chile, Australia, New Zealand and South Africa.

Silviculture characteristics

A. altissima is found in warm temperate to subtropical climates. It has become naturalized, however, in a broader range of climates, including cool temperature and tropical climates, but does not thrive in heavy monsoon rainfall areas. It is drought-resistant, and tolerates a 4–8 month dry season and annual rainfall as low as 400 mm. Preferred mean annual temperature is 7–18°C, but it can also tolerate heavy frosts, and survive absolute minimum temperatures as low as −35°C, but it is liable to snow damage during winter.

It grows best on loose and porous soils, but is found on a variety of soils from sandy or clayey loams to calcareous dry and shallow soils, and can even tolerate barren rocky hills with annual rainfall above 750 mm. *A. altissima* is light-demanding and grows rapidly in the open, and can withstand sooty and polluted environments in industrial or urban areas without needing tending. As it is competitive, it can be found on disturbed ground or wasteland. Its fire tolerance is low.

The tree produces a well-developed root system with a tapering root which penetrates deep into the soil and also produces abundant lateral roots. The roots and leaves also contain allelopathic and herbicidal compounds. It produces root suckers in abundance, coppices well and also regenerates prolifically by seed, with about 7000 seeds/kg. High growth rates of *A. altissima* have been recorded of more than 12 m high and with a diameter approaching 1 m in 20 years on favourable sites.

Silviculture practice

A. altissima produces prolific seed from an early age and abundant seed crops are produced every year. Fruits are picked from standing trees by hand or flailed or stripped on to canvas, then spread out to dry and then either macerated and fanned to remove impurities, or flailed or trampled in a burlap bag and run through a fanning mill. Seed can maintain viability for up to 2 years if stored in sealed containers at low moisture contents and temperatures of 1–4°C. Both autumn and spring sowings give good germination results, if seeds are soaked in cold water for 36 h or stratified.

Germination takes place after 12–15 days, and 1 kg of seed can yield about 1400 usable plants. One month-old seedlings are pricked out to transplant containers and can be planted out when they are 40 cm high. Excessive watering should be avoided. For bare-root planting, seeds can be sown into transplant beds at a spacing of 10 × 5 cm.

The tree propagates vegetatively with relative ease, through branch cuttings, root sections and air-layering. Branch cuttings 1 m long buried 0.6 m deep have given better results than shorter cuttings. Rooting can occur in almost half of all air-layers, but only 10–20% may withstand the shock of severing from the parent branch and planting out. Horizontally planted root sections give better results than vertically planted ones. *A. altissima* can also be propagated by tissue culture.

A. altissima establishes well following a variety of site preparation and establishment practices. Cultivation, fertilizer application, and adequate weeding greatly improve early growth. It has been planted successfully on dry, barren, and eroded hill slopes with shallow rocky soil, and on embankments and landslips, sometimes in mixtures with *Robinia pseudoacacia*, as the roots of both species have good soil-binding properties.

Diseases and pests

Seedlings mainly suffer from root rots (*Rhizoctonia* and *Fusarium* spp.), and are also susceptible to verticillium wilt, but plants are resistant to root-knot nematodes (*Meloidogyne* spp.) and *Anoplophora glabripennis*.

Two serious insect pests attacking *A. altissima* are *Eligma narcissus*, a defoliator of saplings, and *Lycorma delicatula*, a borer which attacks stems and branches and makes the bark darker in colour and susceptible to dry rot. *Eucryptorrhynchus chinensis* can also cause problems.

Uses

A. altissima is often valued more for its use as an amenity and ornamental tree, and for shade, in shelter belts and for erosion control, than for the direct tree products. It is particularly planted in cities where soils are poor and the atmosphere is smoky, as it tolerates pollution well. The ability to rapidly absorb and retain pollutants such as heavy metals, along with its rapid growth, means that the tree has also become a favoured species for planting in reclamation programmes on landfill sites and mine spoils. It also has important cultural values in its native range.

The wood is light and soft, with yellowish-white or cream-coloured sapwood and pale green to yellow or yellowish-brown heartwood with dark streaks, superficially resembling that of ash (*Fraxinus excelsior*). Specific gravity is 0.50–0.55 kg/m^3. It is flexible, easy to dry and process, is somewhat decay-resistant, though considered to be perishable and subject to staining in some regions. It produces a timber suitable for light construction and cabinetry, but is more commonly used in lower-grade furniture and packaging, and is often considered an inferior species in terms of wood quality. In parts of its native range, the wood is still used to manufacture traditional household items. It is used for processing into paper pulp, in the fibre industry, and as matchwood; it has also been identified as a potential fast-growing plantation species specifically for pulpwood production.

It is also used for fuelwood and charcoal in several countries. Other uses include feeding the leaves to silkworms, producing a lower quality silk than that obtained from mulberry (*Morus* spp.) trees. The roots can be used to treat epilepsy and asthma and have many other medicinal applications, especially in its native range in China. Seeds are a source of protein, and a bitter fatty oil that is used after refining.

Further reading

Feret, P.P. (1985) *Ailanthus*: variation, cultivation, and frustration. *Journal of Arboriculture* 11, 361–368.

Fryer, J.L. (2010) *Ailanthus altissima*. In: *Fire Effects Information System*. Rocky Mountain Research Station, USDA Forest Service, Fort Collins, Colorado.

van Valkenburg, J.L.C.H. (2001) *Ailanthus altissima* (Miller) Swingle. In: van Valkenburg, J.L.C.H. and Bunyapraphatsara, N. (eds) *Plant Resources of South-East Asia No. 12(2): Medicinal and poisonous plants*. Backhuys Publisher, Leiden, The Netherlands.

Ailanthus triphysa (Dennst.) Alston

White siris

Importance

Ailanthus triphysa is a large, fast-growing tree with a native range from South and South-east Asia to Australia. It produces a light-coloured, lightweight, comparatively soft wood, preferred as a matchwood, but also used for furniture parts, laminated wood assemblies and pulpwood; it also produces an incense resin popular in India. It is grown in monoculture plantations, but its compact crown and straight, self-pruning cylindrical bole also makes it suitable for mixed-species plantations and for agroforestry systems where a ready market and substantial revenues make it especially attractive to farmers. It is a common tree in the home gardens of southern India, and grows well in lateritic and degraded forests, and can tolerate water stress.

Botanical features

Trees to 30 m high but sometimes to 50 m and with diameters of 90 cm. The crown is ovate or oblong, compact, and the bole is clear for some two-thirds of tree height. The bark is pale brown, smooth, sometime mottled with grey. Branchlets are pubescent, thick, 1.5–2 cm across with prominent petiolar scars. Simple pinnate leaves are 20–70 cm long, each with 8–12 leaflet pairs. Leaflets are opposite or subopposite, oblong, lanceolate, acuminate with an oblique base, entire margins, glabrous above and glaucous below, with 8–11 pairs of lateral veins forking near the margins and with a sessile gland at the forks.

Sapwood and heartwood are not distinct, and growth rings are indistinct. The wood is diffuse porous, and resin ducts are occasionally present.

A. triphysa is unisexual, with many small flowers gathered into dense panicles 20–50 cm long. Fruits comprise 1–3 samaras (usually 2), yellowish to reddish-brown and prominently veined, 5–11 cm long and 2 cm wide, light in weight, winged and wind dispersed. Seeds are orbicular, compressed, in the middle of the samara, on fruiting pedicels 1–2.5 cm long.

Distribution

A. triphysa is native from South Asia, in Peninsular India and Sri Lanka, throughout South-east Asia in southern China, Myanmar, Thailand, Cambodia, Laos, Vietnam, Malaysia, the Philippines, and parts of Indonesia; and in western Australia as far as Queensland and northern New South Wales. It is generally found sporadically in tropical evergreen forests, monsoon forests, and seasonally dry forests, often at altitudes up to about 600 m, but to 900 m in Australia and to more than 1000 m in the Western Ghats of India. There are reports of *A. triphysa* being planted in Senegal and Yemen, and it is possible that this species has been more widely introduced.

Silviculture characteristics

A. triphysa is found mostly in the humid tropics, receiving an annual rainfall from 2200 mm to 4500 mm, though it will also tolerate a dry season of up to 5 months. It grows best on well-drained sandy loam soils with medium organic carbon contents, and in open grasslands the growth is poor. This species is also suited for planting on degraded and slightly acidic laterite soils. Trees are deciduous, though leaf shedding is rarely complete. In India, leaf shedding occurs in December–January with flowering in January–March and fruit set in February–April. Fruiting occurs every year, though a good seed year occurs only once every 2 or 3 years. It is a light-demanding species, with profuse natural regeneration only occurring in open conditions and rarely beneath the canopy of evergreen trees. However, nursery seedlings develop better when under partial shade. The taproot is prominent, woody and strong. *A. triphysa* is intolerant to waterlogging and weeds, however, and the presence of either will lead to very poor natural regeneration.

Silviculture practice

Fruits are usually collected by lopping branches and allowing them to dry under shade, and 1 kg of fruits will contain 8000 to 10,000 seeds. These remain viable for 3 months and are susceptible to insect damage during storage. Nursery-raised seedlings are commonly used, though seedlings are susceptible to root system damage during transplanting.

A common system involves the broadcasting of 0.5 kg of seeds over a 10 × 1 m shaded raised bed. Germination starts in 7–10 days and is complete by 20 days. Fresh seeds will often give 100% germination though 50% is more common, and only half of all germinated seeds may develop into seedlings. They are transplanted when 7–10 cm in height into polythene bags or transplanting beds at 8 cm spacing. Shade is removed when the seedlings have established, as those grown under shade become lean, tall and greenish in colour whereas those grown in full light are thicker, shorter and reddish in colour and grow well in the field. Micropropagation of *A. triphysa* has also been carried out successfully.

A. triphysa grows well in open, fertile, well-drained sites free from weeds. In monoculture plantations, a common spacing is 2 × 2 m, but it may be planted as densely as 1 × 1 m, and sometimes at wider spacings up to 4 × 4 m. Weeding is very important, and should be undertaken several times during the first 3 months after planting and again in the second and third year, and mulching is also sometimes used.

Early growth rates may reach 1 m in height/yr, reaching 10 m tall and 20 cm in diameter after 10 years with a volume of usable wood up to 0.1 m^3 per tree. After 20 years, tree heights can reach 15 m with diameters of 30 cm and tree volumes in excess of 0.3 m^3 per tree. In India, *A. triphysa* plantations are worked in 30-year rotations.

Diseases and pests

Reported plantation diseases in India include sooty mould caused by *Meliola ailanthii*, shot-hole leaf disease caused by the *Colletotrichum* state of *Glomerella cingulata*, pink disease caused by *Corticium salmonicolor (Phanerochaete salmonicolor)*, and stem canker and die-back caused by *Botryodiplodia theobromae*.

Many pests are recorded from southern India, of which the most severe are the Ailanthus defoliator or shoot webber *Atteva fabriciella* and the yellow hairy caterpillar *Eligma narcissus*. A high susceptibility to insect pests makes it unsuitable for monoculture plantations in some countries, leading to it being grown more in agroforestry systems or mixed plantations.

Uses

A. triphysa is suitable for use in agroforestry systems in the tropics, and is a common tree in the home gardens of southern India. It has been planted to enrich degraded patches of tropical rainforests and appears suitable for reforestation and improving wastelands and degraded soils. It is also used as a support tree for pepper, and has been tested as a host tree for sandalwood, *Santalum album*.

The wood is light, 350–465 kg/m^3, averaging 400 kg/m^3 in air-dry conditions, lustrous, yellowish-white in colour, straight grained with a coarse texture. It decays easily and is susceptible to sapstain infection and fungal attack, so green conversion and rapid seasoning are recommended, though the timber is easy to treat and easy to season. It is also easy to work, both by machine and by hand, with good nail-holding capacity. The wood is particularly suitable for use in match manufacture, being good for match splints but not very suitable for matchboxes; however, is usually preferred for this purpose because it gives a higher yield than other woods due to its clear cylindrical stem. It is suitable for light packing cases, boxes, fishing boats, spear sheaths, sword handles and toys. In addition, it is used for class III veneers, core cross bands, face veneers for blockboards, laminated wood assemblies, plywood cores and as pulpwood. Although not very suitable, it is also used as a fuelwood and for making charcoal.

The tree yields an aromatic exudate resin called 'mattipal' in India, which is used as incense and in making incense sticks. The bark, resin and leaves are used in indigenous medicine. The resin is used for treating dysentery and stomach disorders, and bark decoctions are used also for dysentery, and in the treatment of asthma, bronchitis, typhoid and constipation. Leaves and bark also yield a black dye used for dyeing silk and satin.

Further reading

Chandrashekara, U.M. (2007) Effects of pruning on radial growth and biomass increment of trees growing in homegardens of Kerala, India. *Agroforestry Systems* 69, 231–237.

Isaac, S.R. and Nair, M.A. (2004) Litter production and nutrients accretions in soil through *Ailanthus* leaf litter. *Indian Journal of Agroforestry* 6, 81–84.

Lemmens, R.H.M.J., Soerianegara, I. and Wong, W.C. (eds) (1995) *Plant resources of South-East Asia No. 5 (2). Timber trees: minor commercial timbers.* Prosea Foundation, Bogor, Indonesia. Backhuys Publishers, Leiden, The Netherlands, 655 pp.

Albizia lebbeck (L.) Benth.

Indian siris

Importance

Albizia lebbeck (Plate 6) is a medium- to fast-growing, drought-tolerant, nitrogen-fixing tree, widely introduced and now found throughout the tropics. It is adapted to a variety of climates from semi-arid to humid regions, and can tolerate saline, alkaline, and marginal soils. It is a valuable and versatile tree producing small timber, fuel and fodder, and is commonly used as an ornamental and shelter tree. Although commonly grown in plantations, it is also promoted as an agroforestry species, especially in dryland silvopastoral systems.

Botanical features

Trees can attain a height of 30 m and a diameter of up to 1 m, but more commonly 15–20 m tall and 50 cm in diameter, with grey fissured corky bark, somewhat flaky, and reddish inner bark. The distinct heartwood is a dark walnut colour streaked with darker bands, whereas the sapwood is pale yellow. It is multi-stemmed with only a short bole when grown in the open but can produce a single straight stem when grown in plantations. Bipinnate leaves have 2–4 pairs of pinnae, each with 2–11 pairs of oblique leaflets 15–65 cm long and 5–35 mm wide, initially bright green, maturing to a duller glaucous green and folding at night. It is fully deciduous but for only a brief period (4–6 weeks) in the dry season. The inflorescence consists of large clusters of fragrant, globular, yellow-green flower heads 6 cm in diameter. The indehiscent papery pods are pale straw-coloured to light brown, 12–35 × 3–6 cm wide. They are borne in large numbers, make a rattling noise in the wind, and distinctively, remain on the tree in the dry season when it loses its foliage. The seeds are brown, flat, elliptic, with 3–12 in each pod.

Distribution

A. lebbeck is indigenous to South Asia; parts of South-east Asia with a marked dry season; in parts of north-eastern Thailand, Malaysia, and some eastern islands of Indonesia; and in northern Australia, in the Kimberley region of Western Australia, the extreme north of the Northern Territory and offshore islands, and on Cape York Peninsula in Queensland.

It has been very widely introduced and today has a pantropical distribution. It is cultivated in many parts of the seasonally dry tropics, especially in Africa, Asia and South America as a shade tree, and has occasionally naturalized. The boundaries between indigenous and introduced populations have also become blurred in the native range, as in northern Australia for example where seed imported from Asia has been planted.

Silviculture characteristics

A. lebbeck is a nitrogen-fixing tree which can tolerate droughts, a wide range of soil types including those that are saline and alkaline, and even some light frost after the first year. It root suckers, coppices well, and responds to pruning and lopping. It has an extensive but fairly shallow root system, making it ideal for planting on degraded lands for erosion control, but is less successfully planted in combination with crops unless the spread of roots is controlled. Young plants can survive defoliation by fire and grazing. Mature trees may be vulnerable to strong winds.

Trees lose their leaves in the middle of the dry season, remain leafless for 1–2 months, with new growth appearing towards the end of the dry season and heavy flowering following shortly afterwards. Good seedset occurs into the wet season and pods remain on the tree for several months. In India, it is found in tropical evergreen, semi-evergreen, and deciduous forests with mean annual rainfall of 600–2500 mm. In Australia, it is dominant in semi-evergreen monsoon forest with a mean annual rainfall of 1300–1500 mm and a dry winter, and semi-deciduous closed canopy microphyll forests. It has also been recorded as an invasive species in some countries where introduced.

Silviculture practice

Containerized planting stock is preferred in dry regions, but bare-rooted seedlings, stumps and direct seeding are also occasionally practised. Seedlings are ready for planting after

4–5 months in the nursery. When direct seeding in the field, seeds are sown in pits or rip lines at the required spacing at the start of the wet season. The species can also be propagated vegetatively, by stump plants, stem cuttings, air layering, root explants, and by tissue culture.

Initial spacing may be as close as 1.5 × 1.5 m in fuelwood plantations, thinned to 3 × 3 m after 4–5 years and clear felled at 10 years. For timber production on 25–30-year rotations, plantations are thinned to 5 × 5 m, with very wide spacings of up to 15 × 15 m used in silvopastoral systems. However, *A. lebbeck* is also grown in hedgerows for fodder, when planting densities will be much higher, and can even approach 40,000 plants/ha. It generally grows well without artificial fertilizers, though supplementary watering may be required during dry periods in the first year, and protection from grass or weed competition will enhance establishment.

Thinning is a regular part of the silviculture, though coppicing is also practised in fuelwood plantations, with biomass production and green fodder increasing with coppicing height up to 1 m. Alternatively, trees can be left for 10 years and pollarded for fodder and fuel. Larger trees can be lopped annually and the entire green crown removed without loss of vigour; however, more than two cuts a year may adversely affect productivity.

Reports of timber, fuel, and fodder yields vary widely depending on the climate and site, which also reflects the wide variety of conditions in which *A. lebbeck* can grow, and depending on the management aims. In semi-arid areas, fuelwood plantations at 3 × 3 m spacing can yield about 5 m^3 of dry wood with an additional 100 kg of dry fodder/ha/yr, or 50 m^3/ha of fuelwood on a 10-year rotation. Under similar conditions, hedgerow stands grown only as a fodder bank yield 2.5 t/ha/yr when harvested twice a year. Fodder production from pollarding timber stands has been reported to yield 1700 kg/ha/yr.

In timber stands grown on 30-year rotations with adequate rainfall, trees can be expected to reach 10 m in height and 50 cm dbh and yield 300 m^3/ha, at an average of 10 m^3/ha/yr. Under semi-arid conditions and on shallow soils, however, annual increments can be as low as 2 m^3/ha/yr.

For incorporating *A. lebbeck* in existing pastoral systems, the following three management options were proposed in Australia:

- Grow trees intensively in rows and woodlots, as a protein supplement, for feeding weaners or as a drought reserve, lopping annually or as necessary.
- Establish trees at low density in open woodland with no management once established.
- Develop agroforestry regimes in which animal production benefits are combined with wood production.

Diseases and pests

A large number of pests have been recorded as attacking *A. lebbeck*, and experts consider that its value is critically dependent upon resistance to insect attack, which should be the first objective of any developmental work on the species.

Uses

A. lebbeck is a widespread ornamental tree common in many parts of the tropics worldwide. It is also used in alley farming systems in Africa and Asia, and in silvopastoral systems in northern Australia with benefits from forage production, soil improvement and shade. It is grown as a shade tree in coffee and tea gardens and its nitrogen-rich leaves are valuable as mulch and green manure, though the shallow, lateral roots may compete with crops and forage species. It is a good soil binder recommended for soil conservation, erosion control, land reclamation and revegetation of mine spoils.

A. lebbeck seasons and kiln dries well, and works and polishes easily. It is hard, strong and fairly durable, and moderately heavy with a specific gravity of 0.5–0.6 kg/m^3. The timber is used for furniture, panelling, flooring, turnery, and light construction, as pulpwood for writing and printing paper, and is suitable for decorative sliced veneer but not for plywood. India has a well-developed trade in the sawn timber where it is also used for tea boxes, sugar-cane crushers, rice pounders, wheels, agricultural implements and house posts. It is an excellent fuelwood and charcoal species, with air-dry heartwood having a calorific value of 5200 kcal/kg.

Green leaves, fallen flowers and dry leaves are palatable and valued as forage, with the foliage having *in vitro* digestibility of 45–70%. Leaves do not contain phenolics but pods do contain saponin. All fractions fed as supplements increased the digestibility of dry matter intake of a low-quality basal diet, and trees are often grown as livestock fodder. Flowers are a rich resource in the production of light-coloured honey. The bark contains 7–11% tannin as well as saponin and is used for tanning and for soap when dried and pounded. The reddish gum is a substitute for gum arabic. Extracts from the seed, flowers, fruit, bark and leaves have been used medicinally.

Further reading

Doran, C.J. and Turnbull, J.W. (eds) (1997) *Australian Trees and Shrubs: Species for Land Rehabilitation and Farm Planting in the Tropics.* Monograph No. 24. ACIAR, Canberra.

Jøker, D. (2000) *Albizia lebbeck* (L.) Benth. Seed Leaflet 7. Danida Forest Seed Centre, Humlebaek, Denmark, 2 pp.

Puri, D.B. and Gargya, G.R. (1995) Afforestation techniques for *Albizia lebbeck* Linn. (Benth.) on degraded lands. *Indian Forester* 121, 1102–1107.

von Maydell, H.J. (1986) *Trees and Shrubs of the Sahel: Characteristics and Uses.* Schriftenreihe der GTZ, No. 147, GTZ, Eschborn, Germany.

Alnus acuminata Kunth in Humboldt, Bonpland & Kunth

Andean alder

Importance

Alnus acuminata is a valuable component in forestry, agroforestry, and shelter belt plantings due to its rapid growth, soil improvement features, and its ability to act as a feedbank for stock. It is good as general-purpose timber, firewood, for soil protection and improvement, for increased livestock and crop yields under shelter, and is well adapted to cool–temperate to warm–temperate mountainous terrain.

Botanical features

A. acuminata is a deciduous or semi-evergreen tree up to 25 m tall and 40 cm dbh, although dimensions vary considerably with environment and genotype. The trunk is straight; under dense canopy, it is self-pruning and will lack branches for one-half to two-thirds of its height. The bole is cylindrical for most of the unbranched length. The bark is initially smooth, but becomes longitudinally furrowed with age. The canopy forms a rounded dome with robust lateral branches. The root system is deep and well developed, and superficial roots develop symbioses with actinomycetes such as *Frankia* or *Actinomycetes alni*, providing the tree with nitrogen-fixing and phosphate-releasing capabilities.

Leaves are coarse, oblong–elliptic, typically 10–15(–20) cm, with finely serrate to almost entire margins. They have 7–11 pairs of lateral veins and a 1–2 cm petiole. They are shortly acuminate at the tip and broad at the base, with a dark green, glabrous upper surface, whitish-silvery below with veins covered in yellowish pubescence.

A. acuminata may begin flowering at 4–5 years from seed (3–4 years from field planting). Male and female flowers are in separate catkins on the same tree. Clusters of male catkins are 10–12 cm long and pendent, opening in winter to release wind-borne pollen. Female catkins are borne in erect clusters, condensed into a 1.2–2 cm lignified cone. Approximately 100 winged seeds emerge from each cone when it dries in late summer. Near the equator, flowering and fruiting times depend more on the rainfall regime and altitude. In the south of its distribution, *A. acuminata* sheds its leaves towards the end of March–April and new leaves are sprouted in September–October, but those in the tropics shed their leaves progressively, and may never be bare.

Distribution

A. acuminata is native from Mexico to northern Argentina, following the mountains, and is still expanding in northern Argentina. In the tropics, it is found at 2000–3800 m altitude, usually associated with *Juglans*, *Cedrela*, *Tabebuia*, *Phoebe*, *Nectandra*, *Ocotea* and various Myrtaceae. Particularly further south, it forms monospecific stands at the treeline (2500–2800 m). *A. acuminata* has been introduced in many countries.

A. acuminata grows mainly in montane cloud forests, where it is often ruderal, colonizing gaps, landslides, river edges, forest margins, and previously disturbed areas. In the south, it can penetrate into grasslands at the forest–grassland ecotone.

Silviculture characteristics

When grown close together, *A. acuminata* forms tall, straight trees with narrow canopies, and is self-pruning. In more open situations, it forms stout trunks with large lateral branches. In unfavourable situations, trees may be dwarfed or twisted. Trees are tolerant of moderate to strong winds in shelter belts and plantations, and larger trees can tolerate fires in the undergrowth.

The important nitrogen-fixing symbiosis with actinomycetes requires soil aeration. The symbionts also transform unavailable phosphate into forms available to vascular plants. This deep-rooted species also 'mines' nutrients from deeper soil layers and returns them to the surface in the form of leaf and branch litter, which may explain how it can grow so well on poor soils and the reported yield increases of stock grazing under it and crops grown near *A. acuminata* hedges.

Silviculture practice

A. acuminata seeds are collected in late autumn, or whenever the fruit are ripe. The whole cones are collected and as they dry the seeds can be shaken out. The small seeds (less than 2 mm, 0.8–2.5 million seeds/kg) can be sown immediately or stored for several months in cool dry conditions, but lose viability relatively rapidly. Inoculation with the symbionts ensures vigorous growth (although this is often achieved naturally): soil or nodules from inoculated alders may be ground, mixed with water, and poured on to the seedlings. Inoculated seedlings grow rapidly and should be ready to plant out after 6–10 months.

Planting is effective during the wetter parts of the year. Where the tree is deciduous, winter planting allows establishment without leaf damage. Deep irrigation may be necessary in drier sites until roots can reach deeper levels. There is usually no need for fertilizers, but plants require protection from herbivorous mammals. Weeds must be kept under control in the first few years to achieve good growth rates. Preparations may require the removal of tall vegetation and either burning or slashing to reduce grass and weed competition before planting is intended.

In cooler and drier parts, growth rates of 0.5–1.0 m/yr are normal, while in more humid and temperate locations, over 2 m/yr is not uncommon. Although pines may grow faster than *A. acuminata*, they do not shed their leaves in winter or fix nitrogen, and are unpalatable to livestock. *A. acuminata* is larger, and has faster growth and reduced water dependence than most other *Alnus* species. In the first decade, its growth rates may be as high as managed, improved *Pinus radiata* plantations. Further growth rates in unmanaged, unimproved stands quickly fall back, but timber volumes and growth rates in managed plantations can continue for the full rotation cycle.

Timber quality is related to trunk diameter, with larger trees producing better wood, making thinning important. Management for agroforestry has demonstrated improved growth and protein content of pasture grasses and a 33% increase in weight gain of cattle grazing under *A. acuminata* compared to open pastures. At the end of its cycle, *A. acuminata* produces a better timber than that of *P. radiata*.

Diseases and pests

There are few serious diseases or pests of *A. acuminata*. Outside its native range, leaf spot caused by *Septoria* and stem canker are reported to cause significant damage.

The most serious are leaf-eating insects, such as the beetle *Nodonota* and soil fungi such as *Rosellinia*.

Uses

From an environmental and social point of view, *A. acuminata* is a most useful species for revegetation. It shows rapid growth, soil-improvement capability, tolerance of a wide range of soils and conditions, and the production of a variety of economically valuable products.

The deep-rooting system and fertility enhancement favours whatever crops it is desirable to shelter. The final height required is important, as the larger and faster-growing types may be inconvenient. Deciduousness is an advantage in cold climates where sun can penetrate the shelter during the winter. Shelter belt plantings are made at 1–1.5 m spacing and branches can be pruned back to avoid excessive width and shading.

Possibly the tree's most important 'product' is environmental friendliness or ecosystem services (i.e. soil improvement, erosion control, and shelter). In addition to soil and companion plant effects, its deciduous and palatable foliage attracts birds and other fauna. Fast growth of large volumes of biomass make *A. acuminata* a good option as a carbon sink.

A. acuminata wood is used locally, and its wood characteristics suggest it could have a market worldwide. The wood is light, at 230–470 kg/m^3; the strongest (densest) woods are regularly used for construction in Argentina. It is considered durable outdoors under water but not highly durable above ground. The wood of *A. acuminata* has good dimensional stability and can be dried rapidly without cracking or twisting. It is easy to work and has numerous applications in construction, carpentry, furniture, carving, turnery and packaging. It was considered to have excellent performance in the fabrication of pencils, matches and matchboxes. *A. acuminata* is also appropriate for shingles and lightweight particleboards, and can be laminated easily. Its fast growth and good burning properties make *A. acuminata* a good source of firewood which is widely used in the Andes.

Pulp can be produced under appropriate techniques and a good-quality paper may be obtained by mixing with *Podocarpus parlatorei* fibres.

The abundant wind-borne pollen is a valuable bee food supplement, coming in winter when there are few other sources available. The palatable, nitrogen-rich leaves make a useful source of emergency fodder. The bark is rich in tannin which can be extracted to tan leather. Macerated leaves are used for joint and muscular pains, rheumatism and skin infections, and an infusion is used to treat prostatitis.

Further reading

National Research Council (US) Advisory Committee on Technology Innovation (1980) *Firewood crops: shrub and tree species for energy production*. Report of an ad hoc panel of the Advisory Committee on Technology Innovation, Board on Science and Technology for International Development, Commission on International Relations. National Academy of Sciences, Washington, DC.

Rojas, F., Torres, G., Arnáez, E. and Moreira, I. (1991) *Alnus acuminata* ssp. *arguta*. *Especies Forestales Tropicales* 1, 9 pp.

Sachsse, H. and Schulte, A. (1991) Wood properties of the Andes alder (*Alnus acuminata*). *Forstarchiv* 62, 196–199.

Salazar, R. and Jøker, D. (2000) *Alnus acuminata* spp. *argutta* (Schlecht.) Farlow. Seed Leaflet 1. Danida Forest Seed Centre, Humlebaek, Denmark, 2 pp.

Alnus cordata (Loisel.) Duby

Italian alder

Importance

Alnus cordata is a species of *Alnus* that tolerates limited soil humidity, and so is more widely used in plantations for high-quality wood than other species in the same genus. Its wood is suitable for many different uses, including carpentry, turning, carving, rotary cutting and pulp. Its good characteristics are mostly recognized and appreciated in Italy, within its natural range. Besides its importance in timber production, *A. cordata* is also used in landslide reconsolidation. It is also valued for its leaf litter, which is very rich in nitrogen (low C:N ratio) and so improves soils.

Botanical features

A. cordata is a tree with a straight stem and a pyramidal, compact, dark green crown. The bark is smooth, darkish-red when young, later becoming thicker and deeply fissured. The root system is superficial. On the secondary roots, nodules up to several centimetres in diameter can frequently be found where the nitrogen-fixing bacterium *Frankia alni* is present. In its natural area of distribution (in particular in southern Italy), *A. cordata* achieves a height of 20(–26) m and dbh of 50 cm.

The leaves of *A. cordata* are simple, alternate, ovate and (2.5–)5–8(–12) cm long; their edges are crenate–dentate, and the tips are slightly pointed, and the base is typically cordate. The leaves are dark, bright green and smooth on the upper surface, lighter beneath, with 6–8 clearly visible secondary ribs, and reddish axillary hairs. Trees from Corsica typically have leaves with more pointed tips.

Both male and female inflorescences are usually present on the same shoot. Male flowers are borne in pendulous terminal catkins, each 7–11 cm long. These are bright yellow at anthesis and contain (1–)4–6 flowers. Female inflorescences are subterminal and consist of 1–4 catkins. They are much shorter than the male catkins, at 0.9–1.5 cm, with the red stigmas being clearly visible at anthesis.

The fruits of *A. cordata* are pseudostrobili, reaching 25–30 × 15–20 mm. When they ripen, each scale becomes greatly lignified and holds two small achenes with short marginal wings.

A. cordata retains its leaves for a very long period: leaves appear in March–April, and fall in November–December. At the beginning of summer, inflorescences become visible: first female, then male. At leaf fall, catkins are one-third of their final size. The catkins are dormant until the end of February. Anthesis occurs before new leaves sprout at the beginning of March. Seeds are usually collected in September–October, but the collecting period can be longer.

Distribution

A. cordata is endemic to southern Italy and Corsica, and is considered to be a Tertiary relict because of its resemblance to eastern species, in particular *Alnus subcordata* from the Caucasus. In southern Italy, the natural area of distribution of *A. cordata* runs along the mountain chains on the western side of the peninsula, from Avellino to Cosenza, and it is also reported on the island of Ischia. In Corsica, *A. cordata* is mainly present on riparian sites and is concentrated in the south-central area.

In Italy, *A. cordata* has been introduced to the northern Apennines to consolidate landslides. It has also been used in mixed-tree plantations for high-quality wood in different parts of the peninsula. Plantations have also been established in southern France, Spain and Portugal. In England, *A. cordata* is recommended as a fast-growing species that can be coppiced profitably to produce biomass. The species has also been introduced into New Zealand.

Silviculture characteristics

A. cordata is a very light-demanding species, although it can sometimes grow under a rather sparse tree canopy, and becomes shade-tolerant under favourable rainfall regimes. It can be managed as either a high forest species or a coppice species, as it suckers easily. It naturally establishes in both pure and mixed stands, but pure *A. cordata* woods are seldom larger than 50 ha. It is a fast-growing species: in mixed plantations with *Pseudotsuga menziesii* in the central Apennines, it has grown faster than the Douglas fir. Similarly, in coppiced stands where it mixes with chestnut in the Sorrento Peninsula and in Calabria, *A. cordata* has the greater increment.

Silviculture practice

A. cordata seeds can be collected from the end of summer until autumn. If collected too early, seeds are more difficult to store, while a late harvest can result in poor seed quantity because of natural dispersal. The seed can be considered recalcitrant. Pseudostrobili in the upper part of the crown produce more and larger seeds than those in the lower parts. Thinning interventions in tree seed orchards can influence seed production, resulting in a greater number of seeds in each pseudostrobilus and a greater seed weight, without altering germinative capacity and empty seed percentage. A 4-week pre-chilling treatment at 3–5°C on wet filter paper doubles the germination percentage to 98% and improves the speed of germination from 16 to 9 days.

Cuttings can be used; the best results are obtained using 1-year-old hardwood cuttings harvested in March and grown in the greenhouse under mist conditions. Grafting is easy, using both cleft grafting and spliced-graft techniques.

A. cordata can be grown both in open-ground nursery beds and in containers. Enough material suitable for planting can usually be obtained after 5 or 6 months of nursery growth. When environmental conditions are unfavourable, nursery growth may be extended for a further year.

A. cordata stands can be managed either as high forest or as coppice. The small area clear-cutting system is the most widely used management system for high-forest stands. When mixed stands are present, or the stand is intended to be converted towards a mixed stand, the strip-and-group shelterwood cutting system is suggested, starting from existing regeneration groups. The suggested rotation for coppice stands is 25–30 years. Longer rotations can result in a great reduction in the ability to produce new coppice shoots. Coppicing of shoots with a dbh of 20 cm is preferred. After thinning, *A. cordata* reacts quickly, with increased growth in both its crown and stem diameter. Research carried out in a 15-year-old even-aged stand indicated that after a selective thinning, diameter growth showed a swift increase, but after 4 years this effect was reduced, returning to the level before thinning. Thus, the species needs fairly strong and frequent interventions because of its fast growth. New stands are established using 1-year-old (or more rarely 2-year-old) planting stock, bare-rooted or grown in containers. Stocking density can vary according to different purposes: when planting for erosion control and soil conservation, high densities are used (1100–1600 plants/ha), usually in square planting designs. *A. cordata* has more recently been proved to be very useful in stimulating the growth of associated species. *A. cordata* is therefore used in tree plantations for high-quality wood, together with valuable broadleaved species such as *Juglans regia*, *Prunus avium* and *Quercus robur*. In this case, planting density is usually lower, and the geometric arrangement can be much more complex (quincunx planting, etc.).

The greatest yield occurs in Alta Irpinia, in southern Italy. There, mean annual increment in coppice stands with a rotation age of 15–20 years can reach 16 m^3/ha, with an average of 12–13 m^3/ha. The largest specimens can also be found in Alta Irpinia, where some trees reach up to 26 m in height, 65 cm dbh and 4 m^3 volume.

Diseases and pests

Damage is more frequently observed in plantations established outside of the natural area of distribution, where imperfect environmental adaptation has to be taken into consideration among the predisposing causes. In general, it can be said that pathogens affecting *A. cordata* are only of limited importance. Damage to the root system, stem and leaves are reported, due to fungi of the genera *Armillaria*, *Phytophthora*, *Cryphonectria*, *Fomes* and *Microsphaera*. *A. cordata* is susceptible to infection from an unidentified *Phytophthora* species, causing chlorosis of the foliage and bark necrosis. This particular pathogen limits production across Northern Europe, and is spreading into the Mediterranean area.

Of insect pests, *Xyleborus*, *Cossus cossus*, *Zeuzera pyrina* and *Saperda scalaris* can attack the cortical and subcortical zones in plants living under precarious health conditions. A disease ascribed to a bacterium belonging to the genus *Erwinia* has been identified, causing wide cankers frequently spreading out to the heartwood.

Uses

A. cordata is planted to protect against landslides and to consolidate road slopes. It is suitable in these situations because it has low requirements for soil fertility, can improve soil quality, its wood is relatively light, and it keeps its leaves for a longer period than species such as *Robinia pseudoacacia*.

Because of its crown shape and easily degradable, nitrogen-rich litter, *A. cordata* can be profitably associated with other forest tree species. It has been introduced around the world in windbreak plantations, producing better results than poplars or willows.

Italian alder wood is suitable for many uses, such as carpentry, turning and carving. *A. cordata* can also be used to manufacture plywood, particleboard and fibreboard, as well as in rotary cutting for veneers. Although its wood has good combustibility, live *A. cordata* is fairly resistant to fire.

Further reading

Barghchi, M. (1988) Micropropagation of *Alnus cordata* (Loisel.) Loisel. *Plant Cell, Tissue and Organ Culture* 15, 233–244.

Ducci, F. and Tani, A. (2009). *EUFORGEN Technical Guidelines for Genetic Conservation and Use of Italian Alder* (Alnus cordata). Bioversity International, Rome, 6 pp.

Lumini, E., Bosco, M., Puppi, G., Isopi, R., Frattegiani, M., Buresti E. and Favilli, F. (1994) Field performance of *Alnus cordata* Loisel. (Italian alder) inoculated with *Frankia* and VA-mycorrhizal strains in mine-spoil afforestation plots. *Soil Biology and Biochemistry* 26, 659–661.

Santini, A., Barzanti, G.P. and Capretti, P. (2010) A new *Phytophthora* root disease of alder in Italy. *Disease Notes* 85, 560. doi: 10.1094/PDIS.2001.85.5.560A (accessed 6 March 2013).

Alnus glutinosa (L.) Gaertn.

European alder

Importance

Alnus glutinosa is a dominant structural tree in European lowland alder forests in moist and wet habitats, which are important habitats for many rare plants and animals. The nodulated root system of *A. glutinosa* contains symbiotic bacteria that assimilate nitrogen from the atmosphere, which explains the tree's wide tolerance of infertile soils. This pioneer species is therefore very useful in the recultivation of waste ground. It is also an excellent agroforestry tree, especially when planted along ditches. The wood is durable when stored under water, and is thus important for constructions sited in water.

Botanical features

A. glutinosa is usually a medium-sized tree with a pyramidal or narrowly oval, sparse crown, with a low base when the tree grows in isolation; in forests, most individuals have straight stems with an umbrella-shaped crown. *A. glutinosa* rarely grows as a shrub unless in extreme habitats or if the original stem has been cut at the base. The average height at maturity is 30–35 m (occasionally 40 m). The stem rarely achieves a dbh >50 cm, but may exceptionally reach 80–100 cm.

The root system is rarely deep and can be quite shallow when groundwater levels are high. It possesses root nodules containing the bacterium *Schinzia mollis*, which assimilate atmospheric nitrogen. In boggy conditions, horizontal roots bear the nodules while the stronger near-vertical roots contain aerenchymatic tissue. These vertical roots grow into the soil and lift the tree up, forming characteristic mounds.

The bark of young trees is dark brown and rather smooth; later it becomes almost black with linear crevices. The leaves are alternate, simple, oval or transversely ovate, usually with a retuse apex. They are dark green above and lighter green below, 4–11(–14) cm long and 3.5–11 cm broad, with doubly serrate margins. Young leaves are sticky, as reflected in the Latin epithet 'glutinosa'.

A. glutinosa is monoecious; male catkins are 8–12(–16) cm long, reddish-brown with red anthers, loose, and protrude from a 7 mm-long peduncle. The female catkins are about 3.5 mm long, reddish or brownish, and clustered in groups of 3–5, each on a separate peduncle. The small, winged seeds are grouped in cone-like green strobili, and turn dark brown or black in late autumn.

In the core of its range, *A. glutinosa* flowers at the beginning of spring (i.e. in March and April), fruits ripen in early autumn (September), and leaves appear at the end of April (or beginning of May) and drop in October and November.

Distribution

The natural distribution of *A. glutinosa* ranges from ca 63°N in Scandinavia (with isolated populations to 66°N), to ca 36° N in Algeria. It is present in almost the whole of Europe, and also grows spontaneously in the Atlas Mountains and in some places in northern Turkey, Georgia, and on the southern shore of the Caspian Sea. The eastern limit of the species' continuous range is near the Russia–Kazakhstan border, where some isolated localities have been found near 76°E.

The altitudinal distribution ranges from below sea level to high-altitude sites, especially in the mountains in the south of its range: *A. glutinosa* subsp. *barbata* has been observed growing at sites at up to 2000 m altitude. In central Europe, *A. glutinosa* occurs only sporadically at higher-altitude sites, being replaced by *Alnus incana*, and can thus be regarded as a lowland species.

A. glutinosa subsp. *glutinosa* is the widespread subspecies occurring across almost the entire range. In north-eastern Turkey, the Caucasus, and northern Iran, it is replaced by *A. glutinosa* subsp. *barbata*.

It has been introduced to North America, India, South Korea, New Zealand and possibly elsewhere.

Silviculture characteristics

A. glutinosa is a pioneer, light-demanding, short-lived species. It is a shade-intolerant species and seedlings do not usually survive in the shade of mature trees. It may live up to ca 100 years, but may only live for 20–25 years in poorer sites. *A. glutinosa* coppices well when young. It grows well on various wet and moist soils in temperate climates. Due to its ability to fix nitrogen, it can be used for soil improvement and land reclamation. It is resistant to frost and waterlogging. The species is important for silviculture because it grows well on wet sites which are unsuitable for other species.

A. glutinosa is relatively tolerant to frost, wind and pathogenic fungi and insects. However, it is sensitive to changes in water balance. A decrease of groundwater level (beyond natural seasonal amplitude) reduces its vigour significantly.

Silviculture practice

A. glutinosa can be naturally regenerated, but it is difficult to obtain enough seedlings because the seeds often fall on water surfaces or into a dense herbaceous layer. The 'cones' of *A. glutinosa* are collected in November and December. The seeds are orthodox and can be stored in dry, cold conditions. They can also be collected from the water surface, but must then be sown immediately.

A. glutinosa is a light-demanding species, and a clear-cutting system is often suggested on sites where it dominates. *A. glutinosa* should be allowed to regenerate naturally, since a complete artificial regeneration is often extremely difficult on wet sites after the clear-cutting.

Naturally regenerated seedlings are suitable for transplanting at 1–2 years old. Nursery stock (2+0) is most often used for planting. The trees can be planted in holes, or on wet sites on mounds, at spacings of 1.2–1.5 × 1.2–1.5 m. Young plantations of *A. glutinosa* need no particular nursing or control, but coppice shoots must be cut systematically.

In coppice systems, seedlings should be planted to replace old stumps with weak or deformed sprouts. In the thicket phase, the same cleanings are made as in naturally regenerated stands and only the 2–4 best coppice shoots per stump are left to grow on. The thinnings in coppice stands should start at 10–15 years; in stands established from seedlings, thinnings should be carried out at about 15–20 years. In the first case, the number of stems has to be reduced as fast as possible to obtain one stem per stump; in the second case about 300–400 trees/ha should be selected. The early thinnings finish at an age of 30–35 years.

The rapid early growth of *A. glutinosa* makes its use as a nurse species difficult because the trees being nursed can easily be suppressed and killed.

The most important problem in alder swamp forests is regulating the water balance. The stand should be accessible for all forestry measures and yet the groundwater level should be high all year. In particular, soil drying should be avoided to protect the local flora.

The most intensive growth of *A. glutinosa* is in the first 5 years of cultivation; this rate decreases to 75% within 20 years to reach 33% at 30 years and only 17% of the maximum growth rate at an age of 40+ years.

There is some interest in growing *A. glutinosa* on short rotations for biomass, as it can produce more volume per tree than species such as *Platanus occidentalis*.

Diseases and pests

The most serious disease affecting *A. glutinosa* is the oomycete *Phytophthora alni*, discovered in 1993. It causes necrosis of the stem bark, often manifesting as a rusty or tarry exudate from the lesions, and by the unusually small, sparse and yellowed leaves. The disease is now found in many countries, and is causing severe mortality in some areas. The alder bracket *Inonotus radiatus* may also be regarded as an important pathogen causing die-back.

Examples of insects feeding on *A. glutinosa* foliage are the alder leaf beetle (*Agelastica alni*) and the blue leaf beetle (*Melasoma aenea*), which are both closely connected with alder and sometimes may considerably decrease the growth of young trees or even cause their death. A number of leaf-mining caterpillars also feed on the foliage, and the mite *Eriophyes laevis* is a common cause of galls on leaves. There are also numerous bark and wood borers that attack *A. glutinosa*; the most important of these is the poplar-and-willow borer (*Cryptorhynchus lapathi*) Minor pests of *A. glutinosa* include sawflies and tent caterpillars.

Uses

A. glutinosa stands are important habitats for many rare plants and animals: one reason is that its cones open gradually and release seed throughout the winter, and are therefore a reliable source of food. Its rapid growth, tolerance of acid soils, and nitrogen-fixing ability mean that *A. glutinosa* is a desirable species for shelter belts, mine-spoil rehabilitation, biomass production and landscapes.

The wood has a uniform structure with a density of 400–439 kg/m^3 and is used for turnery and the production of plywood. Generally, its wood is considered to be of low quality, although wood of trees grown from seedlings is better than that grown from coppice shoots. It is often used by farmers for fuel. The wood is easily worked and so is used for furniture, banisters, piles, toys, coat racks, boxes and shoes. Due to its elasticity, softness and relative lightness, it is suitable for cigar boxes, pumps, carvings, slippers and moulds for glass manufacture. *A. glutinosa* wood is as durable in water as oak, and has therefore been much used for water-based constructions such as bridges. It is often used for wood pulp and semi-chemical pulp, and can also be impregnated or modified with polystyrene for various uses.

Further reading

Boratynski, A. (1980) Systematics and geographic distribution of alders. In: *Nasze drzewa lesne. Monografie popularnonauk tom 8 (Olsze – Alnus Mill.)*. Institute of Dendrology, Polish Academy of Sciences, Warszawa-Poznan, Poland (in Polish with English summary), pp. 35–71.

Fennessy, J. (2004) Common alder (*Alnus glutinosa*) as a forest tree in Ireland. Reproductive Material no. 8. Coford Connect, Dublin, 4 pp.

McVean, D.V. (1953) *Alnus glutinosa* (L.) Gaertn. *Journal of Ecology* 41, 447–466.

Savill, P.S. (1991) *The Silviculture of Trees used in British Forestry*. CAB International, Wallingford, UK.

Alnus hirsuta (Spach) Turcz. ex Rupr.

Siberian alder

Importance

Alnus hirsuta is an important broadleaved tree species in eastern Russia, north-east China, Japan, and the Korean Peninsula. It produces high rates of lumber due to its straight trunk which has a small taper. It is an ideal tree to plant with conifers, as it is nitrogen-fixing and produces large amounts of leaf litter, both of which tend to improve soil fertility. It has long-lasting foliage that flushes early in the year, and is therefore planted as an amenity tree in urban areas. It also plays an important role in soil and water conservation as it has a well-developed root system, and is used for fuelwood because it is productive and easy to coppice.

Botanical features

A. hirsuta is a deciduous tree, living for 80–90 years. It grows to 20 m high with a mostly straight trunk up to 80 cm in diameter. The bark is purplish-brown and smooth with prominent transverse greyish lenticels. The shoots are dull grey, mostly villous at first, later glabrous. Winter buds are stipitate and are covered with two purplish-brown pubescent scales.

The leaves are alternately arranged, almost round or ovate–round. They are 4–9 cm long, 2.5–9.0 cm across, with a round or cuneate base and pointed at the tip. There are 5–8 pairs of lateral veins, and the leaves are coarsely doubly serrate and shallowly lobed at the margin. They are glaucous and rufous-pubescent or subglabrous below, except for hirsute veins and axillary tufts of hairs, but dark green and nearly glabrous above. The petiole is 1.0–2.5 cm long and subglabrous or nearly glabrous.

A. hirsuta is monoecious. Staminate (male) catkins hang in groups of 3–4 from the ends of the shoots. Pistillate (female) catkins occur in groups of 3–5 in axillary racemes underneath the staminate catkins. The strobili are ellipsoid to oblong–ovoid; their fruiting scales are persistent, woody and five-lobed, and they consist of one outer bract and two inner pairs of compressed, narrowly winged bracteoles, all completely connate.

Flowering occurs in early April before the leaves appear. The strobili are produced prolifically and evenly across years; they ripen and turn brown in October and November.

Distribution

A. hirsuta is widely distributed in north-eastern China, Japan, Korea and Siberia. In China, it occurs naturally in Helongjiang, Jilin and Liaoning provinces, and in the Lao mountain area of Shandong Province (700–1500 m elevation). In Liaoning Province, it occurs mainly in the southern peninsula or in lowlands or along valleys with a high water table or which are seasonally waterlogged.

In Russia, it occurs in the central part of eastern Siberia and the Far East (Primorsky Krai, Amur Oblast, Sakhalin Island, Kuril Islands, Kamchatka, Magadan basin). In Japan, it is distributed in central and northern areas (Honshu and Hokkaido).

Silviculture characteristics

A. hirsuta is a light-demanding species, but requires some shade at the juvenile stage. It is fast-growing, beginning to fruit after 10 years and reaching maximum fruit production at 20 years old. It is frost-tolerant, withstanding minimum temperatures as low as −40°C in the Changbai Mountains area. It is also flood-tolerant, the root system being able to grow in flowing water, although stagnant or polluted water can eventually cause the roots to rot.

As an early successional species, *A. hirsuta* can improve wet or waterlogged land, being replaced by *Picea jezoensis*, *Abies nephrolepis*, *Fraxinus mandshurica* and *Ulmus japonica*. It does not grow well in dry soils. It develops a short root system which penetrates 30–40 cm into the soil, with 3–4 larger lateral roots.

A. hirsuta is a nitrogen-fixing tree which can gradually improve soil fertility. The leaves contain higher levels of nitrogen, potassium, phosphorus and calcium than *Betula platyphylla*, *Pinus densiflora*, *Larix gmelinii*, *Quercus acutissima* or *Zelkova schneideriana*.

An endophytic actinomycete, *Frankia* strain AsI1, has been isolated from nodules of *A. hirsuta*. This has similar hyphae and sporangia to other *Frankia* strains isolated from *Alnus* species. The optimum carbon source is propionic acid and the optimum nitrogen source is casamino acid. The optimum conditions for the growth of AsI1 are a pH of 6.8 and a temperature of 30°C.

Silviculture practice

A. hirsuta seeds are small (0.6–0.8 g/1000 seeds), usually ripen in autumn, and persist throughout the winter. Germination rates can be low, especially if stored in dry conditions for a prolonged period, although pretreatment prior to sowing will give better germination. Treating seed with high temperatures (90°C) for a short time prior to sowing also enhances germination rates. Sowing is usually performed in late April to early May and seeds are covered with fine soil 0.2–0.3 cm deep. Sowing quantities of about 2.9 g/m^2 results in final seedling stocks of 47–53/m^2.

Site preparation along streams, rivers and hill bottoms is performed by digging planting holes in autumn. Seedling stocks are planted out at 2 years old (50 cm high) at a spacing of 1.5 × 1.5 m or 1.75 × 1.75 m. Juvenile stands must be tended for the first 3–4 years.

A. hirsuta is fast-growing: stands at 17 years old reach about 15 m with a dbh of 18.7 cm, representing an individual stand volume of 0.117 m^3. According to data from sampled trees, the peak growth in height occurs when 4–10 years old, over which time an annual increase of 0.8–1.5 m can be achieved. The standing volume reaches a maximum after 6–20 years, which is similar to *Robinia pseudoacacia*. It is better to thin the trees initially between 8 and 9 years old, and *A. hirsuta* is mainly managed with a rotation of 25–30 years. Coppiced stands usually grow faster than seedling stands.

Diseases and pests

A. hirsuta is favoured for its immunity to pests and diseases, although it is affected by *Phyllactinia corylea*, a powdery mildew disease.

Taphrina epiphylla (witches' broom) has been reported as affecting the species in Japan, and in parts of East Asia it is one of a number of species attacked by the polyphagous scale insect *Lepidosaphes ussuriensis*.

A. hirsuta also suffers from damage by pests such as the chrysomelid *Linaeidea aenea*, the sawfly *Pamphilius histrio* and the linden spider mite *Eotetranychus tiliarius*.

Uses

A. hirsuta can improve soil fertility and is flood-tolerant, so has great potential for soil conservation in wet areas.

The wood is soft and light (specific gravity 0.45 kg/m^3), but is tolerant of wet conditions and resistant to decay. It is easy to dry and polish, with good properties for painting and gluing, but only moderate properties for nailing. It is used for furniture, light construction, tools and packaging, plywood, paper pulp, mine poles, railway sleepers, wood carvings, matchwood, small items and fuelwood.

The bark and cone-like fruit can be used for dyeing purposes. The bark contains 10% tannin and can be used for extracting tanstuffs. The flowers are regarded as good bee forage.

Further reading

Eom, H.J., Chang, K.S., Kim, H. and Chang, C.-S. (2011) Notes on a new overlooked taxon of *Alnus* (Betulaceae) in Korea. *Forest Science and Technology* 7, 42–46.

Xing, S.-P. (1988) *Forestry of Jilin*. Jilin Science and Technology, Jilin, China.

Alnus nepalensis D. Don

Indian alder

Importance

Alnus nepalensis is a fast-growing nitrogen-fixing tree which can grow at altitudes up to 3000 m in both temperate and tropical zones, and can also tolerate poorly drained soils. It is a typical multi-purpose tree, grown for timber, fuelwood, pulp, tannin, fodder, green manure, soil improvement and erosion control. As a vigorous pioneer species, it is particularly useful for rapid colonization of landslides, road cuttings and other areas of bare mineral soil in full light.

Botanical features

A. nepalensis typically achieves heights of 28–35 m with a dbh of up to 2 m on good sites. It is self-pruning, resulting in a good clean bole, which is also typically tall and straight.

Its leaves are simple, alternate and 8–20 cm long; they are ovate–elliptical, with prominent parallel veins, usually with slightly toothed edges. Their petioles are 2–3 cm long. A white powder usually covers the undersides of the leaves, particularly along the veins.

Male catkins grow in drooping bunches at the ends of branchlets and are 10–25 cm long; female catkins are 1–2 cm long and arise from the leaf axils. The flowers are yellow. Fruits are black and cone-like, about 1.5–2 cm long and 0.7 cm wide, ovoid or cylindrical, short-peduncled in lateral panicles. The nut has a narrow membranous wing.

The catkins are produced in September and October in India and China, and from September to November in Nepal. The new cones appear between December and March (Nepal) and the seeds ripen in December in Yunnan (China).

Distribution

A. nepalensis is native to the middle hills of the Himalayas, occurring at altitudes of 500–3000 m in Pakistan, eastern Nepal, Bhutan, northern India (Darjeeling hills, Mishmi hills, Sikkim), Xizhang (Tibet) in China, upper Myanmar and parts of Indochina.

A. nepalensis is used in afforestation programmes in upper Myanmar and has been planted in Hawaii and Costa Rica. It has also been experimentally included in species trials in Burundi, Uganda and the highlands of Java, and in agroforestry trials in Bolivia.

Silviculture characteristics

A. nepalensis is a pioneer species, which grows well in full light and is thus moderately light-demanding. It only tolerates shaded conditions in its early growth stages. It does not require high soil fertility, but prefers permeable soils. Natural regeneration is good on open wasteland sites with sparse vegetation cover. *A. nepalensis* successfully competes with weeds, and can even stay ahead of sugar-cane ratoon crops. It is slightly susceptible to wind damage.

A. nepalensis is sometimes reported to coppice well, although some authorities state that it will not coppice – these differences may be due to site conditions, the age and size of the trees, or the time of year when coppiced. Successful regrowth seems to depend on the season of felling.

A. nepalensis has a symbiotic relationship with nitrogen-fixing actinomycetes of the genus *Frankia*, located in root nodules.

Silviculture practice

A. nepalensis seeds are collected between November and March, depending on locality, when the cone-like fruits turn yellowish-brown and begin to open, but before the seeds have been dispersed by the wind. Some authorities state that the seeds have a limited viability of only 3 months, although others claim seeds are orthodox and retain viability for at least 1 year. The germination rate is only 22% after 1 year's storage at 0°C. If the seeds are dried thoroughly and stored in sealed containers in a refrigerator at 4–5°C, they will retain viability for 1 year or more. The species is readily propagated from fresh seed with no pretreatment. Tissue culture of *A. nepalensis* using the shoot tips of aseptically germinated seedlings has been shown to be successful.

The nursery bed needs to be carefully prepared, incorporating soil collected from mature *A. nepalensis* forests (to ensure suitable symbionts are present), and made level. Seeds are broadcast at 15 kg/ha, or sown in lines in March (or in the autumn in areas not prone to frosts), and then covered with a thin layer of fine soil with a further covering of rice straw, pine needles or a plastic film. Seedlings are thinned out during the rainy season when they have produced six leaves and attained a height of 2–3 cm, at which stage they need to be watered and tended regularly. When seedlings are 12–18 months old, they are transplanted, usually during the spring or summer (south-east China), or during the autumn/winter in damp and misty locations in India.

Below altitudes of 1200 m, seedlings should reach plantable size (25–35 cm high) in 4–5 months, but above this altitude they may take as long as 11 months.

Trials were undertaken in Nepal into the possibility of using bare-rooted plants, mainly in order to avoid difficulties in transporting large numbers of potted plants using human labour; generally, trial results have been variable, depending on site and climatic conditions. Intensive site preparation is a common practice in China, where the soil is cultivated to a depth of 20–30 cm and a planting hole (40 × 40 × 30 cm) prepared.

Planting is usually carried out at a spacing of 1.5 × 2 m or 2 × 2 m in China. Loosening of the soil and tending is undertaken once or twice a year during the early growth stages. Superphosphate (25 g/plant) and micronutrients (Mn, B, Zn, and Cu at 1 g/plant) enhance nodulation and tree growth.

A. nepalensis grows very rapidly under favourable conditions. In the Darjeeling region of northern India, *A. nepalensis* is reported to have attained a height of 15.2 m and a dbh of 28.6 cm in 3 years, giving a mean annual increment (MAI) of 9 cm. In Kashmir (Himalayas), a tree with a height of 33 m and a dbh of 80 cm was recorded, but trees with a dbh of 30–50 cm are more common.

In Darjeeling, estimates were made of the biomass distribution and accumulation in *A. nepalensis* plantations 7–56 years old; total biomass increased from 106 t/ha to 606 t/ha.

In Nepal, a 6-year-old plantation on a good site produced a mean annual yield of wood of 4.3 t/ha, and it is estimated that a fully stocked pure *A. nepalensis* plantation could yield 6 t/ha.

In Hawaii (USA), an MAI of 10.9 m^3/ha at the age of 29 years was recorded, while in India, MAI of 16.27 m^3/ha by the age of 22 years was reported. Coppice growth is fast during the early stages, and *A. nepalensis* coppice can attain heights of 14–17 m in 7 years under favourable conditions.

In south-east China, plantations can reach heights of 20 m with an average dbh of 23 cm in 20 years. In natural forests on mountain red soils in the same area, *A. nepalensis* can attain average heights of 13.8 m with an average dbh of 13.6 cm in 10 years.

Diseases and pests

Two species of fungi, *Melampsoridium hiratsukanum* and *Passalora alni*, have been recorded on *A. nepalensis* leaves in Nepal.

A. nepalensis leaves may frequently be eaten by insects and so present a ragged appearance. Pests found on *A. nepalensis* in China are mainly omnivorous insects belonging to the Scarabaeidae, *Stylotermes* or Cerambycidae, as well as *Chrysomela adamsi*.

Uses

As a pioneer, nitrogen-fixing species, *A. nepalensis* can grow on sites with low soil fertility, and so is very useful for soil improvement on eroded soils, or the rehabilitation of degraded land, or for recolonizing new landslides or road cuttings. *A. nepalensis* is used for the revegetation of 'jhumias' in north-eastern India, which are areas affected by shifting cultivation. Other sites which benefit from *A. nepalensis* planting are landslides and steep valley sides. It gives partial shade, thus allowing luxuriant undergrowth to develop, further reducing rates of soil erosion.

A. nepalensis is a very popular tree among farmers in eastern Nepal, as it supplies fuelwood, small timber, shade for certain crops, fodder and can be grown in ravines or areas unsuitable for crops. The cultivation of *Cinchona* spp. (grown for quinine production from the bark) or large cardamon (*Amomum subulatum*) in agroforestry systems with *A. nepalensis* is a common practice in the central Himalayas.

The wood is moderately soft but tough, with a specific gravity of 0.32–0.37. The density of the wood varies from 320–370 kg/m^3 to 480–590 kg/m^3. Wood calorific value is estimated to be between 18,230 kJ/kg and 20,480 kJ/kg. It is an important source of fuelwood, and the wood dries and burns easily.

The wood has an even grain, seasons fairly well, and is easy to saw and finish by hand or machine. The timber is used for the manufacture of tea chests, matches, splints, frames, inner partitions, furniture, farm implements and posts. It is useful for all interior or unexposed conditions. *A. nepalensis* pulpwood can be mixed with that of conifers to produce good-quality newsprint or paper pulp.

The bark has occasionally been used for the production of dyes and tannins in India. The bark contains approximately 21% tannin. Although the leaves contain only 3–4% nitrogen, 0.4% phosphorus, and 1.1% potassium, making it a low- to moderate-value fodder, it is an important fodder source in the Himalayan area. Mature leaves are eaten by sheep and goats, but not by cattle, although the foliage is palatable to cattle when mixed with other feeds. Fallen leaves are also used as a green manure and as a biofertilizer.

Further reading

Gupta, R.K. (1993) *Multipurpose Trees for Agroforestry and Wasteland Utilisation*. Oxford and IBH Publishing Co Pvt Ltd, New Delhi, 562 pp.

Jackson, J.K. (1994) *Manual of Afforestation in Nepal*. 2nd edition. Forest Research and Survey Centre, Kathmandu, Nepal.

Jøker, D. (2000) *Alnus nepalensis* D. Don. Seed Leaflet 8. Danida Forest Seed Centre, Humlebaek, Denmark, 2 pp.

Neil, P.E. (1990) *Alnus nepalensis*: a multipurpose tree for the tropical highlands. *NFT Highlights* 90–06, 2 pp.

Alnus rubra Bong.

Red alder

Importance

Alnus rubra (Plate 7) is a common deciduous broadleaved species of commercial importance in the Pacific region of western North America. It is the largest alder species and the only one regularly exploited for timber and pulpwood. It is a relatively short-lived pioneer species that exhibits rapid juvenile growth and favours disturbed areas. The wood is easily worked and used in woodcarving, in making furniture and for firewood. Because it fixes nitrogen, *A. rubra* is under consideration for forest management systems.

Botanical features

A. rubra is a medium-sized tree, large specimens reaching heights of 30 m. The trunk is straight and slightly tapered in forests, with a narrow, rounded crown. In open situations, the crown is broadly conical and may extend to the ground with spreading branches. The bark is smooth, light grey and sometimes shallowly furrowed.

A. rubra has extensive, fibrous root systems, which are commonly ectomycorrhizal, bearing nodules that fix atmospheric nitrogen through symbiosis with the actinomycete *Frankia*. Nodulation occurs soon after germination; root systems of seedlings a few months old have dozens of visible nodules up to 25 mm in diameter.

The leaves are deciduous, oval to rhombic in shape, 7–13 cm long, thick, double-toothed and with the margins rolled under. There is considerable leaf-fall during the summer.

A. rubra reaches sexual maturity after 3–4 years for solitary trees and 6–8 years for most dominant trees in a stand. It is generally monoecious, with separate male and female catkins developing on the previous year's young branches. Pistillate (female) catkins also occur in clumps, but are borne upright. The seeds are small winged nuts (9 × 18 mm) borne in pairs on the bracts of woody, cone-like strobili. *A. rubra* is a prolific and consistent producer of seed. Moderate seed crops are produced almost annually and large crops occur every 3–5 years. The seeds are very light (800–3000/g) and wind dissemination is effective.

Seed dispersal begins in late September in the middle of the species' range, somewhat earlier in Alaska, and several weeks later in California. Most seeds are shed during the late autumn and winter. Flowering occurs in late winter or early spring, and the peak shedding of pollen generally precedes peak receptivity by only a few days.

Distribution

A. rubra is a submontane or maritime, mesothermal species. Its range extends from southern California (34°N) to southeastern Alaska (60°N). It is generally found at altitudes below 750 m, and within 200 km of the ocean, although several isolated populations exist in northern Idaho.

A. rubra grows in even-aged, pure stands or with shade-tolerant conifers in later successional stages. It is a pioneer species on floodplains, and is present in early and intermediate stages of secondary succession on cleared, cut-over and burnt coastal and upland sites.

A. rubra is commonly associated with *Pseudotsuga menziesii*, *Tsuga heterophylla*, *Thuja plicata*, *Abies grandis*, *Picea sitchensis*, *Populus trichocarpa*, *Acer macrophyllum* and *Salix* spp. In British Columbia, *A. rubra* is a minor component in the coastal Douglas fir and coastal western hemlock zones. In the western United States, it is a component of the coastal western hemlock, Sitka spruce, redwood, and Pacific madrone–Douglas fir zones.

Introduction of *A. rubra* outside its natural range is largely limited to arboreta and parks. Extensive trials as a timber species in England and Scotland have been disappointing due to unsustained rapid early growth, decline and die-back.

Silviculture characteristics

A. rubra reproduces rapidly, with seeds produced as early as 3 years old. The seeds are readily disseminated by wind and water. *A. rubra* also reproduces vegetatively from stump sprouts.

A. rubra is very intolerant of shade and hardly regenerates in low light. In the open, natural regeneration occurs favourably on mineral soil, and direct seeding yields good results.

Initial growth rates are high, and exceptionally rapid (>1 m) in the first growing season. The potential productivity is therefore high, representing a site index (50 years old at breast height) of about 40 m on productive sites, albeit limited by the species' short lifespan.

A. rubra is self-pruning in dense stands, and it has moderate spatial requirements for the crown. Moderate or high light levels are found beneath closed-canopy, mature stands, and they are associated with well-developed understorey vegetation.

Silviculture practice

Clear-cutting and patch-cutting are both viable silvicultural systems for growing *A. rubra*, but due to its rapid growth, it is best grown in single-species stands.

A. rubra germinates and grows well on moist mineral soils in full light. Seedlings tolerate partial shade for several years, but full sun is required afterwards for normal development. *A. rubra* is an aggressive pioneer of areas where mineral soil has been freshly exposed to seed fall. Artificial regeneration can be accomplished with either bare-root or containerized seedlings. If the soil is sterilized, it may be necessary to re-inoculate it to encourage the formation of root nodules. Planted seedlings usually survive and grow well.

A. rubra grows in height exceptionally fast: on favourable sites, seedlings can grow 1 m or more in the first year, and on all but the poorest sites, seedlings surpass breast height in the second year. The highest annual height growths (>3 m a year) can be achieved by 2–5-year-old seedlings. Height growth begins slightly later in the season than radial growth, and continues throughout the growing season until soil moisture, temperature or light conditions become unfavourable.

A. rubra sprouts vigorously from stumps when young and can be repeatedly coppiced on short cycles, although rootstock mortality increases with each harvest. Stumps sprout best when trees are cut in the winter and when the stump is >10 cm high. Older trees rarely sprout and coppice regeneration cannot be expected after pole-sized or sawlog-sized material is harvested.

A. rubra is a relatively short-lived species, maturing after 40–70 years and with a maximum age of about 100 years. On favourable sites, trees can be 30–40 m tall with a dbh of 55–75 cm in diameter. The maximum volume (709 m^3/ha) is attained after 50 years. In pure stands on good sites, it has been estimated that *A. rubra* can achieve annual volume increment rates of 21 m^3/ha in pulpwood rotations of 10–12 years, and 14 m^3/ha in sawlog rotations of 30–32 years.

Early control of spacing is necessary to keep live crown ratios high enough to maintain good growth beyond the juvenile phase. Sawlog yields can be maximized on short rotations by combining early spacing control with pulpwood thinnings. In previously unthinned stands, thinnings are most effective in stimulating growth of residual trees if done before height growth slows after about age 15–20 years. Thinning in older stands can reduce mortality and help maintain the vigour of residual trees, but usually without accelerating diameter growth.

Diseases and pests

A. rubra is fairly free from insect and disease problems, especially when young and uninjured. *Phellinus igniarius*, a white heart rot, is probably the major cause of death in older trees. Ice storms and unseasonable frosts can, however, cause damage and mortality.

Uses

The ability of *A. rubra* to fix atmospheric nitrogen can increase both its nitrogen content and the nitrogen availability in the soil. Fixation rates vary with time, site and stand age, but annual fixation rates up to 320 kg/ha have been reported in pure stands and 130 kg/ha in mixed stands. *A. rubra* also increases the organic matter content in the soil and concomitantly decreases the soil bulk density and pH.

A. rubra has been proposed for use in mixtures with other species because of its soil-improving characteristics. Its rapid juvenile growth means it has been experimentally planted in various situations: as a nitrogen source for other species; on coal mine spoils, landslides, and other eroded or low fertility areas; for stream bank or roadside protection; in areas of poor drainage; as a firebreak or windbreak; and for wildlife areas.

A. rubra has also been tested as a replacement for conifers susceptible to root pathogens such as *Phellinus weirii*, which can survive for many years in soil organic matter.

A. rubra wood is diffuse-porous, moderately dense (370–410 kg/m^3) and uniformly textured, typically with a straight grain. Cut heartwood is whitish, turning light reddish-brown, but is not distinct from the sapwood. The timber is susceptible to decay and to attack by wood borers and the beetle *Anobium punctatum*. Rapid conversion into lumber after felling is recommended. The wood has excellent gluing characteristics, fairly good mortising characteristics, responds well to carving and machining (planing) but does not hold nails or screws well. The wood dries easily and seasons fairly rapidly with little degradation and is used in the production of solid wood products (such as furniture, case goods, pallets and novelties), composite products (plywood and flakeboard), fibre-based products and as a veneer.

A. rubra is a common fuelwood and is burned domestically and in mills that use residues to produce heat for drying and other processes. Because of its rapid juvenile growth and ability to coppice, *A. rubra* has also been evaluated for biomass production.

Further reading

Harrington, C.A. (1990) *Alnus rubra* Bong. Red Alder. In: Burns, R.M. and Honkala, B.H. (eds) *Silvics of North America. Volume 2: Hardwoods*. United States Department of Agriculture Handbook 654, Washington, DC, pp. 116–123.

Peterson, E.B., Ahren, G.R. and Peterson, N.M. (1996) *Red Alder Manager's Handbook for British Columbia*. FRDA Report 240, Ministry of Forests, Victoria, British Columbia.

Vanden Heuvel, B.D. (2011) *Alnus*. In: Kole, C. (ed.) *Wild Crop Relatives: Genomic and Breeding Resources: Forest Trees*. Springer, Heidelberg, Germany, pp. 1–14.

Alstonia scholaris (L.) R. Br.

White cheesewood

Importance

Alstonia scholaris (Plate 8) is a common, medium to large, tropical hardwood tree with an extensive occurrence in the Asia–Pacific region. It has a moderate to fast growth rate. Its timber is a non-durable, lightweight hardwood which can be used for a range of indoor purposes and for pulp and paper production. The bark contains useful medicinal alkaloids.

Botanical features

A. scholaris is a medium to large tree, reaching about 40 m high and 125 cm dbh, with somewhat tessellated corky grey-white bark. The boles of larger trees are strongly fluted to 10 m, tapering and somewhat twisted. The outer blaze is cream to yellowish in colour with an abundant, milky latex that flows rapidly when cut. The crown is narrow-domed, deep and dense, or eventually diffuse with large spreading limbs, small in relation to size of the bole.

The leaves are in whorls of 4–8. The leaf stalks are 1.0–1.5 cm long, the lamina obovate to elliptical or elliptical–lanceolate, tapering towards the base, 11.5–23 cm long by 4–7.5 cm wide. The leaves of *A. scholaris*. var. *scholaris* are glabrous, whereas those of *A. scholaris*. var. *velutina* have a strigillose–pilose undersurface. The upper surface is dark green, the lower green-white with 25–40 pairs of lateral veins on each side of the midrib and 2–6 mm apart. The tip of the leaf is rounded or shortly pointed, tapering towards the base.

The inflorescence is a much-branched terminal panicle, up to 120 cm long. The flowers are 7–10 mm long, white, cream, or green, the tube hairy, lobes sparse or densely pubescent and 1.5–4.0 mm long, the left margins overlapping, strongly perfumed. The fruit is a pendulous, two-lobed dry-dehiscent follicle, brown or green, dry or woody, spindle-shaped, 15–32 cm long, 4–6 mm diameter. *A. scholaris* is wind-dispersed: each fruit contains numerous flat, oblong brown seeds, 4–5 mm by 0.9–1.2 mm, with a tuft of hairs 7–13 mm long at each end. In contrast to several of its close relatives, the seed of *A. scholaris* does not taper to a point at either end.

A. scholaris is an evergreen species. In Australia, the flowering period is from October to December. In Sri Lanka, there are two flowering events per year – April–June and October–November – and mature follicles are collected in February for drying. Seeds are collected after release. Delay means loss of all seeds by dehiscence of the follicles on the tree.

Distribution

A. scholaris occurs from India and Sri Lanka through mainland South-east Asia and southern China (Yunnan Province), throughout Malesia to northern Australia, the Bismarck Archipelago and the Solomon Islands. In Australia, the species is widespread in coastal areas of northern Queensland including the Cape York Peninsula, and south to central Queensland. Disjunct populations occur away from the coast in the Northern Territory. *A. scholaris* var. *velutina* is found in Peninsular Malaysia, Borneo and New Guinea.

A. scholaris is most abundant in lowland rainforest, but it tolerates a variety of habitats including drier monsoon forests. In the Solomon Islands, *A. scholaris* is mainly a lowland, coastal species. It is common as a small tree on coral but can also be found inland in primary and secondary forest, attaining heights of 46 m. In Vietnam, it occurs on forest margins or along streams and rivers mixed with species such as *Bischofia javanica*, *Litsea cubeba* and *Randia oxydonta*.

In Australia, *A. scholaris* is a dominant canopy species found in mesophyll vine forest with a canopy height of 35–45 m. On soils with impeded drainage derived from granite or basalt, *A. scholaris* occurs in mesophyll vine forest dominated by palms with a dense closed canopy to about 20 m. It is also a dominant species in notophyll vine forests, with a canopy of 25–45 m, associated with *Argyrodendron peralatum*, *Castanospermum australe*, *Ceratopetalum succirubrum*, *Darlingia ferruginea* and various *Ficus* species. *A. scholaris* is found with other typical rainforest species in forests dominated by eucalypts including *Eucalyptus tereticornis*, *Eucalyptus tessellaris* and *Eucalyptus intermedia* and in a mosaic of vegetation types on swampy coastal plains. These are characterized by palm species such as *Archonthophoenix alexandrae* and *Licuala ramsayi*.

Outside its native area, *A. scholaris* is planted for its ornamental value.

Silviculture characteristics

There is little published information on the silviculture of *A. scholaris*.

Silviculture practice

Propagation of *A. scholaris* is mainly by seed which dehisces naturally from the mature follicles on drying. There are 37–87 seeds/g and an average germination rate of 50%. The storage life of the seeds is unknown but may be as brief as 2 months. About 90% of germinants will provide healthy, vigorous seedlings suitable for planting out the field. *A. scholaris* has been grafted successfully: cleft grafting and inverted-T grafting were found to be the most appropriate. There are no special requirements for the sowing of seed, but it does best in full sunlight. With a shallow covering of sowing mix and light and regular watering, germination starts within about 12 days and continues for 3 months. Plantable seedlings of about 30 cm tall are produced in 9–12 months. Stumps of 6 mm diameter at the collar may also be produced.

In most countries of South-east Asia, *A. scholaris* is harvested selectively from natural forest with a diameter limit of 40, 50 or 60 cm. Natural regeneration of *A. scholaris* is often scarce, so enrichment planting may be employed, using the strip system. However, sufficient opening of the canopy is essential for optimum growth of seedlings. Commercial plantations are best set at a square spacing of 1.8 m at the start of the wet season and may require regular thinning. Young trees coppice well. Plantation areas should be cleared and burnt. Clean weeding is essential after the first month and twice more during the first year. After 1 year of growth, when the trees may have reached 1.5 m tall, climbers should be removed by cutting.

In Darwin, Australia, urban plantings of *A. scholaris* receiving regular dry season watering and deep mulching at establishment have shown fast early growth and exhibited good form.

The species reached an average height of 3.6 m and diameter of 10 cm at 3.5 years in mixed-species social forestry plantings in West Bengal, India. This performance was considered poor among the various species tested.

In plantations in Taiwan, *A. scholaris* reached an average of 23.5 m in height and 51 cm dbh in 18 years. A maximum of 35 m in height and 109 cm dbh was attained at 41 years of age.

Diseases and pests

A fungal disease, caused by *Colletotrichum gloeosporioides*, and manifesting as leaf spotting, has been described from India.

As the culture of *A. scholaris* has only a short history, it is unclear if it will suffer from insect pests. In parts of its range, its leaves are consumed by several caterpillars, including *Hyblaea puera*, *Eutectona machaeralis* and *Parotis marginata*. The timber is readily attacked by boring insects.

Uses

A. scholaris was included in a species trial in the Philippines which was intercropped with corn for three croppings. While trees of *A. scholaris* were healthy, their growth performance failed to impress as they reached only 1.5 m dbh and 1.7 m tall in 2 years. The species is recommended for roadside planting for shade and decorative purposes in Vietnam and also in southern Florida and California, USA, and in northern Australia.

The heartwood of *A. scholaris* is cream to pale yellow in colour, often with a strong odour and a bitter taste. The sapwood is visually indistinct from the heartwood. The air-dry density of the wood is approximately 400 kg/m^3. The heartwood is not sufficiently durable for external use but can be readily impregnated with preservatives. It is used for pattern-making, corestock, plywood, carving and mouldings. It is not a structural timber. In Sri Lanka, the wood is used for making coffins, because of its lightness. It is regarded as suitable for pulp and paper production.

A. scholaris was recommended as a fuelwood species for the Patana lands of Sri Lanka, managed under a short coppice rotation (6–8 years). The wood has a relatively high calorific value of 11.38 kcal/g. Due to its low density, however, the species makes a poor-quality charcoal.

In a study of the ethnobotany of the Nagas of Nagaland in north-east India, *A. scholaris* was among the native plants used in magico-religious beliefs. Australian Aboriginal people used the bark for treatment of abdominal pains, dysentery, and fevers and the latex for neuralgia and toothache. In India, the bark is used to treat bowel complaints and has proved a valuable remedy for chronic diarrhoea and the advanced stages of dysentery. The active constituents are alkaloids found in the bark and latex. The hill tribes of northern Thailand boil the bark in water, which is then drunk to combat the effects of malaria; the bark and leaves are used to treat a variety of ailments.

Further reading

Baliga, M.S. (2010) *Alstonia scholaris* Linn. R.Br. in the treatment and prevention of cancer: past, present, and future. *Integrative Cancer Therapies* 9, 261–269.

Dey, A. (2011) *Alstonia scholaris* R.Br. (Apocynaceae): phytochemistry and pharmacology: a concise review. *Journal of Applied Pharmaceutical Science* 1, 51–57.

Rudjiman, Gintings, N., Martawijaya, A. and Ilic, J. (1993) *Alstonia* R.Br. In: Soerianegara, I. and Lemmens, R.H.M.J. (eds) *Timber trees: major commercial timbers. Plant Resources of South-East Asia* 5, 82–90. Pudoc Scientific Publishers, Wageningen, The Netherlands.

Aniba rosaeodora Ducke

Rosewood

Importance

Rosewood essential oil was introduced to the world market in 1926, but its industrialization began in France in 1975 using wood from French Guiana, following an initiative by M. Samain. The largest use of rosewood essence is now in the perfume industry; rosewood essence contains approximately 91% linalol (the main alcohol representative of this essence). It is also used in the drinks industry. Due to the medicinal action of the oil, people in the Amazon region frequently use it against rheumatism and arthritis by rubbing it into the affected parts.

For many years, rosewood extraction stood out as one of the most prosperous industries in the Amazonian region, and was the third most important export from the Amazon. By the end of the 1950s, the production and exportation of essential rosewood oil reached approximately 450 t annually, from 53,000 t of wood processed. Due to the high price the product reached on the international market, abundant raw material near the distilleries and easy wood extraction, there was a great demand for the species. Approximately 53 distillation plants were known in the 1960s: three in the Brazilian state of Pará and 50 in Amazonas. However, as a result of the intensive exploitation of natural populations and competition from synthetic products, only five factories remain active in Amazonas, working only a few months a year due to a lack of raw material. Today, the species is on the Red List of Threatened Species of the International Union for Conservation of Nature (IUCN).

The threat of species extinction led to regulated cultivation and control by governmental organizations in Brazil. Some of the means used to make the production of oil sustainable included raising awareness, providing technical assistance in oil extraction, using branches and leaves instead of felling the tree, studies on other plants containing linalol and reforesting for cultivation.

Botanical features

Aniba rosaeodora trees are 30 m tall and 30–70 cm in diameter, though they can reach 2 m in diameter. All parts of the tree are aromatic. The stem is cylindrical and rectilinear, with thin bark that desquamates in large scales. The leaves are alternate and regularly distributed along branches. The petioles are 1.5–2 cm long, canaliculate, glabrous, while the leaves are 14–17 × 4–6 cm, papyraceous and sub-coriaceous, glabrous, with an acute base, acuminate apex and revolute margin. They have 7–9 equal and prominent secondary ribs, each forming an angle of 55–65° with the principal rib.

The inflorescences are sub-terminal panicles in the junctions of falling bracts and persistent leaves, 4–12 cm long, comprising many flowers. The flowers are pubescent, 2 mm long and 2 mm in diameter, on a pubescent pedicel, angular and 0.5–1.5 mm long. The stamens are external (series I and II), 0.5–1.5 mm long, with hair filaments wider than the anthers, glabrous and with ventral locules; the interior stamens (series III) with biglandulous filaments, narrower than the anther, and glabrous; stamens (series IV) absent, 2 mm long and tomentose except for the glabrous base. Fruits are ellipsoidal, 3.5 cm long and 3.8 cm in diameter. Cupule is 2 cm tall and 3.8 cm in diameter, sub-hemispheric, with some ferruginous stains in the double margin.

The species has perennial leaves. Bearing in mind its characteristics of slow development, high-density wood, shade tolerance in the juvenile phase, medium-sized seeds and abundant regeneration with high mortality, *A. rosaeodora* could be considered a secondary species. However, the species is heliophilous and regeneration occurs only in clearings in the forest.

Distribution

A. rosaeodora is found in northern South America: in the basins of the Tapanahoni and Gonini rivers in Suriname, in the Oiapoque river basin in French Guiana, near Iquitos along the Napo and Pucalpa rivers in Peru, in the Negro river in Venezuela, in the Brazilian Amazon and in Colombia and Ecuador.

In Brazil, it occurs on the border of Amapá with French Guiana. It has its largest habitat in the upper and middle reaches of the Amazon, especially in areas formed in the Upper Carboniferous. The areas of larger concentration of the species occur in the southern part, extending from the river Curuá-Una to Peru, and in the north from the Trombetas river to Colombia. The species is usually found in areas of well-drained higher lands, mainly in pluvial non-flooded forest along the river banks.

Silviculture characteristics

The species reproduces annually, albeit rather irregularly. Flowering occurs between October and February, and fruiting between November and March.

Silviculture practice

One of the alternatives to assure a constant supply of raw material for distillation work and to reduce or avoid the random exploitation of the species is the establishment of plantations or other reforestation systems. However, there is a lack of information about the silviculture and ecology of the species and also about how improvements can be made to the distillation system, and the use of foliar biomass for the production of essential oil.

In the case of wide-scale plantings for the production of essential rosewood oil, propagation through seeds has only been moderately successful so far, due to the difficulty in finding a significant number of adult trees in the seed production phase. This is due to intensive exploitation in the past, and a rapid loss of viability. However, some work has been carried out on techniques for raising seedlings from seed.

The most recommended method for mass production of planting stock is through the vegetative propagation of cuttings. Cuttings of juvenile material removed from resprouts of trees gave a high rate of survival and superior rooting, whereas cuttings from adult material had lower rooting capacity.

The survival of stumps 6 years after logging was 100% in all tested treatments, showing it to be an excellent coppicing species. The number of surviving resprouts on stumps varied, however, and the diameter of the stump affected the diameter growth of the resprouts. It is preferable to use stumps with diameters of 8–10 cm due to good resprouting capacity. *A. rosaeodora* can also be grafted.

The establishment of plantations has not been greatly successful and continues at a small scale; as the species is on the Red List of species threatened with extinction, determining sustainable management methods is urgent.

The problems to be overcome before domestication can be considered a realistic economic option should not be underestimated. Unlike some other Amazonian species that have attracted attention because of their much larger potential market (as sources of fruits or oil seeds, for example), comparatively little research has been carried out on *A. rosaeodora*.

Diseases and pests

The main insect pest of *A. rosaeodora* appears to be the beetle *Heilipus odoratus*. This weevil has a spermophagous life cycle; eggs are laid in the seeds and the hatchling completes its entire development within the seeds, feeding on the endosperm while sheltered from potential predators inside the thick shell of the seed.

It is likely that attack by ants could be a problem in pure plantations.

Uses

A. rosaeodora is suggested for agroforestry systems where the climate is favourable to the species. In the Amazon region of Brazil, owners of small properties are starting to combine the cultivation of rosewood with annual crops. There is a project in Manaus, Amazonas, providing support to traditional rosewood oil producers for the experimental cultivation of rosewood and extraction of the essential oil from the leaves. It is recommended that each producer should plant an area of 30 ha with 10,000 rosewood seedlings, interspersed with cassava and 2500 seedlings of other aromatic species.

The wood is moderately heavy (with a density between 600 and 700 kg/m^3), and hard. However, it is easy to work, with a medium texture, fine grain and good natural durability. The wood is brown-yellowish, and light or dark. It contains 9–15 vessels/mm^2, approximately 60% of which are solitary; the mean length of vascular elements is 0.38 mm; rays are up to 0.75 mm in height, with a maximum width of 0.06 mm, and fibres up to 0.15 mm long.

The principal product is the high-value essential oil. To produce rosewood oil, trees are felled, cut into 1 m sections, and chipped and distilled on arrival at the distillery (usually in 1000-kg chip capacity equipment). Yield is typically in the region of 1%. Overexploitation has meant that rosewood stands are now becoming more remote from established distilleries. Annual production was estimated at 300–400 t in the 1960s, twice the then official figure, but data from the entire range of Peru, Colombia, Ecuador, Suriname and French Guiana, as well as from parts in Brazil, leads to current estimations of annual production of less than one-tenth of this figure, 20–30 t.

The enantiomeric composition of oil from different trees varies according to inherited genetic factors. Some analysts take the eremophilene content as an index of quality of the wood oil, but because rosewood leaf oil contains relatively higher amounts of this compound, this may lead to the possibility of the leaf oil being mixed with synthetic linalol and sold as rosewood oil. Sales of adulterated oil have occurred; however, adulterated oils are being termed 'US quality' in trade, as the only unadulterated oil is shipped directly from Manaus, Brazil.

Further reading

Gonçalves, J.F. de C., de S. Barretol, D.C., dos S. Junior, U.M., Fernandes, A.V., de T.B. Sampaio, P. and Buckeridge, M.S. (2005) Growth, photosynthesis and stress indicators in young rosewood plants (*Aniba rosaeodora* Ducke) under different light intensities. *Brazilian Journal of Plant Physiology* 17, 325–334.

Pereira Santos, R., da Silva Ângelo, P.C., de Tarso Barbosa Sampaio, P., Quisen, R.C., Leite, Â.M.C. and de Oliveira, C.L. (2008) Geographic pattern of genetic diversity in natural populations of Rosewood (*Aniba rosaeodora*), in the Central Amazonia. *Acta Amazonica* 38, 459–466.

Souza, R.C.Z., Eiras, M.M., Cabral, E.C., Barata, L.E.S., Eberlin, M.N. and Catharino, R.R. (2011) The famous Amazonian rosewood essential oil: characterization and adulteration monitoring by electrospray ionization mass spectrometry fingerprinting. *Analytical Letters* 44, 2417–2422.

Anisoptera thurifera (Blanco) Blume

Palosapis

Importance

Anisoptera thurifera wood, known for its beauty and durability, is suited for use in general construction work, for plywood, furniture and in the manufacture of pulp and paper. *A. thurifera* is planted in the Philippines as an enrichment crop or under other species in tree plantations as it needs partial shade in its early stages of development. The seeds lose viability rapidly, and the wood is hard to saw and needs careful seasoning.

Botanical features

A. thurifera reaches 40–45 m tall with a dbh of 80–120 cm (a tree 200 cm in diameter has been recorded). It has a straight unbuttressed regular bole that is 60–67% of the tree's height. The crown is dense, especially during the rainy season, but is leafless for a short period during the dry season. The bark is thick, smooth and yellowish-grey when young; mature tree bark is fissured and flaking in elongated plates at the tree base. Beneath the surface, the bark is reddish-brown and leathery.

The leaves are elliptical–lanceolate, shortly pointed at the apex and rounded at the base, 6–15 cm long and 2.5–6.5 cm wide, on a 3–4 cm petiole. Their texture varies from thinly papery to coriaceous (leathery).

The flowers are borne in racemes. The flower buds are acuminate and about 7–8 mm long; their pedicels are about 2 mm long, tomentose, with tufts of stellate hairs. The flower is yellow, fragrant and about 1.5 cm across when fully open. The sepals are lanceolate, acute to obtuse, the two longer sepals about 2 mm long, acute at apex, with stellate hairs outside and pubescent inside, 0.6 mm wide at the base. The petals are lanceolate, 8–9 mm wide with about 15 principal longitudinal nerves, recurved at the apex, and twisted in bud. The part exposed in bud is pubescent, whereas the rest is glabrous. There are 45–47 subequal stamens, with filaments scarcely widened at base, 0.1–0.3 mm long; the anthers are ellipsoid, one pair twice as long as the other, 0.2–0.4 mm long, 0.2 mm wide. The ovary is partly sunken in the receptacle, pubescent or tomentose, hemispherical, about 1 × 0.6–0.7 mm. The stigmas are minute and usually closely appressed.

The fruit is a globose samara, slightly constricted at the top, 4–17 mm in diameter. The two long fruiting calyx wings are 5–15 cm long, 1–5 cm wide, spathulate, narrowly obtuse, about 4 mm wide at the base with three prominent longitudinal nerves and numerous roughly oblique transverse nerves. The three short wings are linear to linear–oblong, 2–3 cm long and 2–3 mm wide. The presence of wings helps disseminate the seed and reduces the impact of falling fruit, bringing the fruit gently to the ground and preventing injury to the embryo.

Flowers of *A. thurifera* form in the latter part of the dry season. Fruit may be produced annually, with fruit maturation taking about 3–4 months from the first appearance of the flowers. The racemose character of the inflorescence suggests that fruit development and maturity are equal.

Distribution

A. thurifera is the commonest *Anisoptera* species in the Philippines, and is found throughout the country, except at high altitudes. It is also found in Papua New Guinea, Myanmar, Indonesia, Malaysia, Brunei, Thailand and Cambodia. It occurs frequently, but scattered, in lowland forest.

Silviculture characteristics

A. thurifera tolerates shade and wind. It can regenerate rapidly and is self-pruning.

Silviculture practice

The change in colour of the fruit from greenish-yellow to brownish is not an indication of maturity but rather an indication that the fruit has attained full size. Seeds should only be collected from healthy mother trees (i.e. with straight bole, balanced crown and free from defects). *A. thurifera* seeds can be collected by climbing the tree or by using a bamboo pole with scythe to cut the branchlets. For seed collection on the ground, clear the basal portion of the tree using the crown circumference as boundary guide; seeds that fall within this area must be randomly checked for viability prior to mass collection. The collected seeds should be placed in a well-ventilated container during transport. Seeds are recalcitrant, with about 830–900/kg. A simple viability test is done by opening the seed to determine if it still has a green embryo. Moisture content after harvest is 32–51%.

It is highly recommended that immediately after collection, viable seeds are sown in germination beds. However, seeds that are not sown must be stored in perforated bags/sacks to maintain viability for up to 2 weeks. Seeds can also be stored in finely ground charcoal at 25–30°C and these may remain viable for 5 weeks.

Remove wings before sowing in seedbeds with forest topsoil mixed with garden soil at a ratio of 1:1. Sow small end first or where the wings were attached, so that the radicle emerges and the roots grow directly downward and deep into the soil. This prevents the development of crooked and bent roots. The depth of sowing should be about half the diameter of the seeds. Cover lightly with mulch to prevent moisture loss, and water once or twice a day during sunny days. Germination begins 5 days after sowing. Only germinating seeds should be potted, preferably in 16 × 20 cm plastic bags filled with forest soil. Potted seedlings should be placed under 75% shade and arranged in rows for easy inventory. The seedlings should be left to harden for at least 2 months prior to planting out; do this when the seedlings reach 30 cm tall.

The plantation site should have enough trees or brush (about 75% shade) to protect seedlings from extreme temperatures and soil moisture loss. Open areas such as grasslands are not suitable. Recommended sites are inadequately stocked forests or brushlands.

For site preparation in inadequately stocked forest, circular brushing of 1 m diameter is recommended. When brushing, unwanted species should be cleared, leaving only premium species to serve as nurse trees for the seedlings. Likewise, for brushlands, strip brushing of 1 m is preferred.

Planting holes should be staked at a spacing of 3 × 3 m. Holes 30 cm deep and 20 cm wide are large enough to accommodate the seedlings.

Seedlings should be planted out at the start of the rainy season when soil has been thoroughly drenched by rain. When planting, remove the plastic bag carefully and refill the bottom of the planting hole with topsoil. Then place the potted seedling in the centre of the hole and push the subsoil to the sides and on top. Tamp the side-refilled soil lightly around the stem to keep the seedling erect.

Weed around the plant three times during the first year and twice during the second and third years. Weed at the peak of the rainy season when abundant growth of weeds occurs. Pile the cut grasses at the base of the plant to serve as mulch to conserve moisture and keep the soil surface cool.

Apply 15 g of complete fertilizer (14:14:14) around the seedlings only if no endomycorrhizal inoculation has been done. Replace all dead seedlings 1–2 months after planting, if there is still enough rainfall. Visit the plantation twice a month to detect insect infestation, animal intrusion and fire occurrence, and to monitor growth. For seedlings to reach sapling and pole stages, they should be released from competing vegetation to enhance growth and development.

In released *A. thurifera* stands, studies have shown that growth in diameter and height is more than twice that of untreated stands. In the case of natural stands, the best cutting cycle varies from 20 to 30 years, depending on site quality.

Diseases and pests

A. thurifera is affected by a canker caused by the fungus *Nectria haematococca*. Although this is not known to cause significant mortality, it may have an effect on yields. The occurrence of the seed-borne fungus *Botryodiplodia theobromae* has been reported on seeds of *A. thurifera*.

The species is vulnerable to powder-post beetles, especially in cases where the trees are weak.

Uses

A. thurifera is used for soil improvement, soil conservation, and erosion control.

A. thurifera wood is good for interior finish, ship planking, vehicle works and automobile bodies. It is also used in furniture making, veneer and plywood, general construction, wooden tanks and containers. The timber can be worked to a good finish but there is considerable dulling of cutters due to silica content; carbide-tipped tools are recommended. The wood needs careful seasoning as it is inclined to shrink excessively and remain so once dried; it is also hard to saw compared to other woods of equivalent density (615–820 kg/m^3 at 15% moisture content).

A. thurifera bark yields an abundant oily resin when freshly cut, which dries white, and often covers stumps. This resin is very similar to 'balau', which is extracted from *Dipterocarpus grandiflorus* in the Philippines.

After cooking, the nuts can be eaten and have a high edible oil content.

Further reading

de Guzman, E., Umali, R.M. and Sotalbo, E.D. (1986) *Guide to Philippine Flora and Fauna*. Natural Resources Management Center, Ministry of Natural Resources and University of the Philippines. JMC Press Inc., Manila, Philippines.

PCARRD (1985) *Philippines recommends for dipterocarp production*. Plantation Development and Management Technical Bulletin Series No. 58, PCARRD, Los Baños, Laguna, Philippines.

Schulte, A. (1996) *Dipterocarp Forest Ecosystems: Towards Sustainable Management*. WSPC, Singapore, 680 pp.

Araucaria angustifolia (Bertol.) Kuntze

Paraná pine

Importance

Araucaria angustifolia is a tree with a straight, nearly cylindrical stem that reaches up to 35 m in height with a dbh of 1 m. With a soft wood, this species is ideal for milling and use in carpentry, furniture and buildings. During the first half of the 20th century, exploitation of *A. angustifolia* in natural forests led to a significant economy in southern Brazil, and many towns were born around sawmills in the forest. Forests where *A. angustifolia* is a dominant species originally covered 183,900 km^2 in southern Brazil and north-eastern Argentina. Intensive harvesting and conversion to agriculture nearly eliminated the tree from the landscape. As the seeds are an important food supply for animals in the forest, natural regeneration cannot compete with deforestation. Not all details of regeneration ecology and forest dynamics are known yet. Although establishment of plantations is not difficult, poor site conditions can severely affect production, even leading to failure of the plantation.

Botanical features

At maturity, the tree reaches 30–35 m tall, with a dbh of 100–130 cm. The stem is straight, with first-order branches arranged in pseudo-whorls, horizontally spreading, and up to 10 m long. Second-order branches are short-lived (4–5 years), falling to the ground once dead. There is a deep taproot with a shallow secondary root system. The thick bark (5–10 cm) is fire-resistant when mature. The leaves are triangular and end in a thorn.

A. angustifolia is dioecious. Male strobili are 10–15 cm long and 2–3 cm in diameter; each female strobilus (known locally as a *pinha*) is spherical when mature, and 20–30 cm in diameter, weighing 2–4 kg. The seeds are conical, 5–7 cm long and 2–2.5 cm in diameter, weighing 5–10 g each, and with a high starch content. Pollination occurs in September–October (spring); seeds mature 18 months later in April–June (autumn). Depending on the pollination conditions, each *pinha* may produce up to 150 seeds and 700 sterile scales. The strobili open when mature, letting the seeds fall to the ground.

Distribution

A. angustifolia is distributed in southern and south-eastern Brazil (Minas Gerais, Rio de Janeiro, São Paulo, Paraná, Santa Catarina, and Rio Grande do Sul provinces), north-eastern Argentina (Misiones) and eastern Paraguay (Alto Paraná). *A. angustifolia* has been planted outside its natural range but usually only as isolated trees in gardens or parks.

Silviculture characteristics

A. angustifolia tolerates shade and frost, but grows optimally in full sunlight. Poor weed management may damage the apical meristem through early frost or sunburn. The recovery after damage is rapid, but can result in the main stem forking. The apical meristem normally grows orthotropically, but branched apical meristems grow plagiotropically. *A. angustifolia* is self-pruning when grown in the forest, but self-pruning is poor in open situations.

Silviculture practice

Seeds are dispersed by birds (especially the jays *Cyanocorax caeruleus* and *Cyanocorax chrysops*) and rodents (such as squirrels and mice). Natural regeneration is frequent in secondary forests or openings in the forest, as long as seeds are available and growing conditions are favourable. Seeds may be stored in plastic bags at low temperatures (5°C), and should not be dried or stored at room temperature. The cold treatment also kills the larvae of the moth *Cydia araucariae*, which may already be present in some seeds. Seeds can be stored for nearly 1 year with some loss of viability.

Seeds germinate without any treatment; the best results are obtained by pre-germinating the seeds in the plastic bags at room temperature and then positioning the seeds horizontally in the soil after the radicle emerges and reaches approximately 10 mm long. The seed must always be positioned horizontally in the soil, covered with 10 mm of soil and preferably a 10 mm thick mulch of *A. angustifolia* needles. Germination is hypogeous, and the two cotyledons are haustorial, transferring nutrients from the endosperm to the seedling over 3–4 months.

Good seedlings may be raised in 100–200 ml containers, avoiding excessive moisture or containers with poor drainage. The seedlings are ready for planting out 3–4 months after germination, when they are 20–30 cm high. Containers may be filled with any growing medium; good results are obtained with a 1:1 mixture of soil and composted *A. angustifolia* bark. Fertilizer should not be added, as nutrition for the seedling is supplied by the endosperm.

Vegetative propagation trials show that due to the strong differentiation of orthotropic and plagiotropic growth of stem and branch meristems, only stem meristems should be used. Grafting using apical meristems is also possible.

If the area has been mechanically prepared, plantation may be established by direct planting using 3–4 seeds at each seed-spot, (2000–3000 seed-spots/ha) or in lines, with one seed every 50 cm, and lines 2.5–3.0 m apart. After 2 years, the best plant at each spot or per 2 m of line is selected, to achieve a stand of 2000–2500 plants/ha. Direct sowing on unprepared sites or near forests carries some risks, such as seed consumption by rodents, and should be avoided.

Planting is best when associated with a crop, such as corn or beans: the seedlings profit from the light shade and *A. angustifolia* provides fertilizer for the crops. If the correct spacing is used, the crops may be grown for at least 4 years, with mechanical soil preparation between the *Araucaria* rows but manual harvesting of the crop.

Because of the high variability of the trees, dense stands should be established; spacings of 1.6 × 2.5 m, 2.0 × 2.5 m, 3 × 3 m and 2 × 4 m are common for young stands. Although the species is shade-tolerant, the best growth is achieved in full sunlight, as long as the soil is sufficiently humid. Light shade normally reduces evapotranspiration and improves survival rates. If necessary, supplementing nitrogen (e.g. 25 g urea per plant in spring during the first 2 years), or the use of mulch, can be a good practice.

Enrichment planting of *A. angustifolia* is feasible. Nurse trees should be less than 10 m high and removed after a few years. Seedlings are planted in lines, 1 m apart, in open brush vegetation. Every 6 months, competing vegetation is cleared away, as well as some of the shading trees, and after 5–10 years the overstorey should be removed. Avoid enrichment planting when long-lived tree species constitute the canopy.

Growth of *A. angustifolia* in the first few years is slow, but after 10 years, increments of up to 1.5 m in height may be achieved. Because of the typical crown architecture, this species needs more space than *Pinus* or *Eucalyptus* spp. for the same increment in diameter. The trees should be pruned when the stem diameter reaches 10 cm.

Diseases and pests

Fungi reported from *A. angustifolia* include *Armillaria mellea*, *Cylindrocladium* sp., *Sphaeropsis sapinea*, *Rosellinia bunodes*, and an unidentified wood decay fungus of the family Mucronoporaceae (Basidiomycetes). *Uleiella paradoxa* parasitizes and modifies the male flowers. *Phomopsis araucariae* has been isolated from *A. angustifolia* trees that appeared sickly.

Natural regeneration or plantations of *A. angustifolia* can be severely damaged by the ants *Atta* sp. and *Acromyrmex niger*. *Cydia araucariae* (Lepidoptera: Olethreutidae) is considered the most important insect pest of shoots, seeds and cones of *A. angustifolia*. *Phrasterothrips conducans* (Thysanoptera: Phlaeothripidae) attacks the base of the foliage in the terminal shoots of the whorls. Other defoliators include *Dirphia araucariae* (Lepidoptera: Attacidae) (which can cause complete defoliation and damage wood yields), *Fulgurodes sartinaria* (Lepidoptera: Geometridae), *Elasmopalpus lignosellus* (Lepidoptera: Phycitidae) and *Eupithecia* spp. (Lepidoptera: Geometridae).

Woodborers of the family Cerambycidae have been reported from *A. angustifolia* and Ambrosia beetles (Coleoptera: Scolytidae) have been associated with the tree. Several weevils of the genus *Araucarius* (Coleoptera: Curculionidae) breed in the stems of *A. angustifolia*.

Uses

A. angustifolia is commonly used in pastures and around houses in southern Brazil. Sometimes these trees remain from forest clearance; in other cases, the species is planted as a shade or ornamental tree. *A. angustifolia* grows well on grassland and pasture where fire is not used for removing brush, and is suitable for silvopastoral systems, and plantations where an agroforestry system is used in the establishment phase. The ideal stocking levels for trees on pastures are not yet established.

The wood of *A. angustifolia* may be classified as soft, with a density of 500–610 kg/m^3. It is white-yellow, with a reddish heartwood and a straight grain. The timber is used for structural purposes, panelling, furniture, plywood and charcoal. It is also excellent for fibre production, but it has low increments and production levels in comparison with *Pinus taeda*.

There is both a local and an international market for *Araucaria* wood. Regional regulations restrict the trade, although the sale of logs from plantations is permitted. Unfortunately, neither the dimensions nor the quality of plantation timber are comparable with wood from natural forest. *A. angustifolia* is listed as 'Critically Endangered' by the IUCN because of the rapid losses it has suffered through logging.

Resin was once collected from branch knots of *A. angustifolia*, but this is no longer viable following the introduction of *Pinus elliottii*. For farmers, seed production can be an important source of fodder for cattle, sheep or pigs. Maturation of the seeds coincides with a shortage in grasses in the winter, so this is an important feed. Seeds are also collected for human consumption.

Further reading

Farjon, A. (2006) *Araucaria angustifolia*. IUCN Red List of Threatened Species, version 2011.2. Available at: http://www.iucnredlist.org (accessed 4 May 2012).

Mecke, R. (2002) *Insetos do Pinheiro Brasiliero – Insekten der Brasilianischen Araukarie – Insects of the Brazilian Pine*. Attempto Service GmbH, Tübingen, Germany.

Silva, L.C.R., Anand, M., Oliveira, J.M. and Pillar, V.D. (2009) Past century changes in *Araucaria angustifolia* (Bertol.) Kuntze water use efficiency and growth in forest and grassland ecosystems of southern Brazil: implications for forest expansion. *Global Change Biology* 15, 2387–2396.

Wrege, M.S., Higa, R.C.V., Miranda Britez, R., Cordeiro Garrastazu, M., de Sousa, V.A., Caramori, P.H., Radin, B. and Braga, H.J. (2009) Climate change and conservation of *Araucaria angustifolia* in Brazil. *Unasylva* 60 (231–232), 30–33.

Astronium urundeuva (Allemão) Engl.

Urunday

Importance

Astronium urundeuva is a medium to large tree, 6–33 m tall, with a dbh of 40–60 cm. It is a potentially important economic species and abundant in dry forests of central South America, especially in caatinga, chaco, and cerrado, where it can be the dominant tree species, sometimes growing in almost pure stands. The wood is extremely dense and resistant to rot because of its high tannin content and is often used for posts and poles. However, the high density of the timber often imposes high transport costs.

It is regarded by the IUCN as being of 'Endangered' status in Argentina, Bolivia and São Paulo state, Brazil, and is classed as vulnerable in Argentina (Jujuy and Salta provinces) and other parts of Brazil (Bahia, Goias, Mato Grosso and Piaui).

Botanical features

A. urundeuva (also called *Myracrodruon urundeuva*) is a medium to large tree with an irregular, narrow and open crown. Its branches are thick and tortuous. The mean height is 6–11 m in caatinga forests, 20–25 m in cerrado forests, and up to 33 m in closed semi-deciduous forest. The average dbh is 40–60 cm, but it can reach a maximum of 120 cm. *A. urundeuva* has a straight, cylindrical trunk, which is somewhat fluted at the base, with clear bole lengths of up to 12 m. The grey external bark is scaly with irregular fissures, often with the appearance of small blocks on older trees, and can reach 5.5 cm thick. The internal bark is sticky, white and purple, which changes to a chestnut colour after oxidation. Small quantities of white latex are exuded from bark cuts.

Leaves are pinnately compound (5–11 leaflets) and arranged in spiral phyllotaxy, ranging in size from 10 × 7 cm to 25 × 15 cm (length × width). The leaflets are each 4–5 cm long, pubescent to glabrous on both sides, and deep green in colour.

Inflorescences are displayed in clusters (5–20 cm) that grow out of the leaf axils. *A. urundeuva* is a dioecious species, and its small, fragrant flowers are bright white and have five petals. The small, compound fruits are ovoid to oblong drupes with a smooth and resinous exocarp that shrivels when dry. The fruits are borne in hanging racemes and are effectively winged, due to the persistent attachment of the five sepals which are bright green and up to 6 mm in length. There are around 47,500 seeds/kg.

A. urundeuva is deciduous in the dry season between June and October, with leaves being renewed at the beginning of the rainy season. Leaf fall varies with the intensity of the dry season and flowering coincides with this leafless period, which is normally between June and August. Male trees usually flower 2–3 weeks before female trees. Fruiting usually occurs between August and October, but may be prolonged to as late as December.

Distribution

A. urundeuva is restricted to northern and central South America. Its distribution extends from the sub-Andean belt of northern Argentina and Bolivia to northern Paraguay, and the north-eastern part of Brazil. It is part of the caatinga, cerrado, chaco, and deciduous closed forest vegetation types within these countries, as well as the more humid Atlantic forests of Brazil.

Silviculture characteristics

A. urundeuva is a shade-intolerant tree species that germinates and grows best in high light environments, particularly where there has been some light soil disturbance. Adult growth rates are often slow. Because of its intolerance to competition, it is often found in highest abundance on poor or shallow soils.

Silviculture practice

Cryopreservation has been used for seed storage: seeds are dried for 120 h and immersed in liquid nitrogen for 15 days. Seeds are then ready to germinate without treatment. Sand, enriched with organic matter, gives a good seed bed for germination. The highest germination rates were obtained with seed that had been stored on damp filter paper for 6 days at 4–5°C.

Germination begins in 3–18 days after planting with a germination capacity that can exceed 80%.

The growth of *A. urundeuva* in mixtures with *Pinus caribaea* has been successful as well as with other tree species. Other studies found no significant differences in growth for species in pure stands or mixtures with other hardwoods. In young pure stands, a spacing of 1 × 1.33 m provided the best self-pruning and stem form.

Natural forest management using *A. urundeuva* has been employed. The species is regenerated mostly by seed. Abundant seed crops are produced annually. The species is strictly shade-intolerant and its lightweight wind-dispersed seeds germinate best on disturbed soils. Whereas high seedling densities are often found on old logging roads and log landings, very little regeneration is found in logging gaps or undisturbed forest. Abundant regeneration is also found in areas burned by wild fires and in abandoned slashed and burned clearings.

Establishment of the species is likely to be best accomplished through intensive logging, perhaps with some follow-up competition control. Leaving seed trees is an important requirement, especially along logging roads and other areas of disturbed soil. Since the species is dioecious, it is important to leave female trees in these areas or enough trees so that the retention of female trees is highly probable. Some male trees also need to be retained to ensure successful pollination, which is mostly performed by small bees. Since the species yields excellent honey, similar silvicultural strategies are recommended to produce high local tree densities for honey production. Males can therefore be culled from stands in favour of females, except for a small percentage of males to ensure pollination.

Although locally valuable for some uses, *A. urundeuva* is best managed in natural, multi-species stands as a secondary species. The species is rarely more valuable than most other tropical timber species with which it commonly occurs, and is therefore not a primary focus for timber management.

Diseases and pests

The timber of *A. urundeuva* is naturally resistant to fungal and insect attack, and few pests or diseases have been reported. A cerambycid beetle, *Brasilianus* sp., has been observed boring into the trunk of the species and may have minor effects on wood quality.

Uses

Since *A. urundeuva* establishes exceedingly well on disturbed soils, it has high potential for use in erosion control. Although growth rates of adult trees are relatively slow, the species is used in low densities in some agroforestry systems with perennial crops or as shade for cattle. It is also used in shelter belt systems. Because of its crown form and flowering, it is planted as an ornamental in gardens and street plantings.

A. urundeuva sapwood is yellow-pink and extends for 30 mm. The heartwood is brown-pink to violet when fresh, turning reddish-brown after exposure to air. It has a fine, uneven texture, a wavy, sometimes interlocked grain, and indistinct growth rings.

The wood of *A. urundeuva* is very hard, with a density of 750–950 kg/m^3. It is highly resistant to rot and insect attack, mostly due to its high tannin content (17%). The timber is therefore excellent for heavy exterior construction or for use as railway sleepers. It is also commonly used for posts and poles, although care must be taken to avoid excessive warping during seasoning. The wood is difficult to split and to work with in general, but finishes well. It is used more than any species for the elaborately sculptured posts in the Jesuit Missions in eastern Bolivia. It also makes very good firewood and charcoal.

The abundant and fragrant flowers of the species are attractive to bees, particularly *Apis mellifera*, and it is therefore valued for honey production.

Indigenous inhabitants of eastern Bolivia use the species for a number of medicinal purposes. A jelly is made from the bark to mould cloth in the form of a cast for bone fractures. A tonic made from the leaves is used to control dysentery or as an antidote for snake bites. An alcoholic extract of the bark has anti-ulcerogenic properties.

Further reading

Caetano, S., Prado, D., Pennington, R.T., Becj, S., Oliveira-Filho, A., Spichinger, R. and Naciri, Y. (2008) The history of seasonally dry tropical forests in eastern South America: inferences from the genetic structure of the tree *Astronium urundeuva* (Anacardiaceae). *Molecular Ecology* 17, 3147–3159.

Sá, R.A., Gomes, F.S., Napoleão, T.H., Santos, N.D.L., Melo, C.M.L., Gusmão, N.B., Coelho, L.C.B.B., Paiva, P.M.G. and Bieber, L.W. (2009) Antibacterial and antifungal activities of *Myracrodruon urundeuva* heartwood. *Wood Science and Technology* 43, 85–95.

Silva, L.M. de M., Rodrigues, T. de J.D. and de Aguiar, I.B. (2002) Efeito da luz da temperatura na germinação de sementes de aroeira (*Myracrodruon urundeuva* Allemão). *Revista Árvore* 26, 691–697.

Aucoumea klaineana **Pierre**

Okoume

Importance

Aucoumea klaineana is a long-lived, medium–large pioneer tree, which can form pure natural stands in natural or artificial forest clearings. These pure stands, covering large areas in Gabon and Equatorial Guinea, offer the potential for intensive natural forest management. It is one of the best West African timber species for producing plywood. It is therefore a major commercial timber species in Gabon and Equatorial Guinea, where it represents more than 70% of the timber production, and to a lesser extent in Congo (Brazzaville).

Botanical features

A. klaineana reaches 50 m in height and 1.2–1.5 m dbh, and has light foliage and a globular crown. When mature, the base of the trunk has large buttresses up to 2–3 m; the trunk is rarely perfectly straight, and is often slightly bent, grooved, or somewhat swollen immediately above the buttresses. The bark is dark orange-grey, and detaches in roughly rectangular scales, often spotted by white, yellow or orange bands of encrusted lichen. The root system is shallow with virtually no taproot, even on deep soils, and with extensive lateral roots.

The roots of neighbouring individuals are often joined below the ground by root grafts.

The leaves are alternate, pinnate with 3–6 pairs of opposite leaflets plus one terminal leaflet (10–30 × 4–7 cm), with long, thin petioles. New leaves are bright red for 1 week and appear in September–December.

A. klaineana is dioecious; its small (5–7 mm), whitish flowers form densely ramified racemes. Flowering starts in August and lasts for approximately 1 month, with any single flower in full bloom for only a few days. The fruit is a dry capsule that opens into five sections, with a single winged seed in each section. Fruiting starts in September, fruits reach their final size in 40 days and are mature in about 80 days.

Distribution

The natural distribution of *A. klaineana* covers about two-thirds of Gabon and four-fifths of Río Muni (the continental part of Equatorial Guinea). In Congo (Brazzaville), the species is restricted to the Massif du Chaillu and to the Mayombe. A few small natural stands also exist in Cameroon, near the border with Equatorial Guinea. *A. klaineana* timber is produced principally in Gabon, Equatorial Guinea and Congo (Brazzaville).

A. klaineana is a component of evergreen rainforests. As a long-lived pioneer species, it is found in most successional stages: colonizing, secondary (where it often forms pure stands) and mature forests.

It has been planted for timber outside its natural range in Cameroon and, to a lesser extent, in Côte d'Ivoire. It also has been tested on a small scale in several other countries.

Silviculture characteristics

A. klaineana requires full sunlight to grow satisfactorily. It is easily propagated by seed, and seedlings and saplings can survive under shade, although they will not grow properly. There is no natural regeneration in tree-fall gaps, either natural ones or those created by logging operations.

Unlike most rainforest species, *A. klaineana* can form important monospecific or semi-specific stands. Pure stands derived from the colonization of open areas are part of the succession to mature forests and are of great economic importance.

At 12–15 years old, both in natural stands and plantations, populations segregate into dominant and suppressed individuals, the latter having their growth seriously hindered. At this stage, the population becomes organized into 'biological cells' via the development of root grafts (anastomoses) between dominant and suppressed trees.

Silviculture practice

A healthy dominant mature tree can produce up to 15,000–20,000 seeds; a suppressed individual will only produce 1000–2000 seeds. Seeds are wind dispersed up to 50–80 m from the mother tree, and can be collected on the ground within 24 h. They lose their germinating ability within 1 month in the field, but can be stored at 4°C in airtight containers for up to 3 years, after being dried to a moisture content of 8%. Although regeneration by seeds is usually unproblematic, grafting and propagation has been achieved with cuttings.

Seed can be directly drilled in polythene bags (20–30 cm depth, 10–15 cm diameter) filled with a mixture of sand and clay (two seeds per bag, thinned to one after 3–4 weeks). After 2.5 months, the 20–25 cm seedling has 5–7 simple leaves and is ready for planting.

Natural regeneration requires a sufficient number of seed trees, fruiting every 2 or 3 years, a large canopy opening or clearing (>2500 m^2) and clean soil during the fruiting season. There is little or no regeneration in canopy gaps resulting from natural tree-fall or selective logging, but in appropriate conditions, *A. klaineana* regenerates in almost pure stands.

Two main methods are used in order to produce land with enough light for *A. klaineana* cultivation:

- Mechanized clear-felling: the existing woody vegetation is bulldozed, and the remnants are put into windrows and burned; seedlings are then planted in lines.
- Regrowth method: existing woody vegetation is cut to 50 cm above ground level, in order to allow resprouting and regrowth. Larger individuals are devitalized and there is no windrowing or burning of the remnants. Planting lines are then manually opened in the regrowth.

Protection is needed for up to 5 years against herbaceous climbers, especially *Mikania scandens* and *Musanga cecropioides*, which are the main species competing with *A. klaineana* in open areas.

Thinning operations are generally needed in both natural and artificial stands, although sudden intense lateral illumination of the stem may increase the sensitivity to black canker attacks. In almost pure natural stands, thinning aids lateral growth, and suppressed trees are much more reactive than dominant individuals (for which the thinning has been done). In these stands, thinning should be restricted to young stands (<15 years old), as the dominant individuals are already in good growing conditions in older stands and the thinning reduces the volume. Thinning should be carried out by girdling *A. klaineana* trees or poisoning other species, taking care to avoid intense bole illumination of lower-strata individuals.

Mean individual height growth increases from 1.5 m after 1 year, to 15–50 m after 40–50 years. From 10 years old onwards, the appearance of dominant and suppressed sub-populations explains the large increase in the range of mean height, whereas height growth is similar in plantations and pure natural stands.

Diameter growth is variable and depends on various parameters such as age, stand density, site fertility, hierarchical position of the tree and thinning rates. Studies in the coastal area of Gabon showed values of 30–80 cm dbh after 50–60 years for dominant trees and 10–40 cm dbh for suppressed individuals. The mean annual increment in diameter ranged from 1.4 cm (after 7–12 years) to 0.7 cm (after 50–60 years) for dominant trees and from 0.6 cm (after 5–20 years) to 0.1 cm (10–40 years) for suppressed trees. Other studies in different sites of inland Gabon indicate mean annual diameter increases of 0.5–0.7 cm, without separating dominant and suppressed individuals.

In properly established plantations, the estimated potential yield varies from 7 m^3/ha/yr at 43 years (moderately fertile site, no thinning) to 11 m^3/ha/yr at 32 years (fertile site, thinned). However, production will probably be lower since 20–30% of the trees above the minimum felling diameter of 60 cm are unusable for timber due to their poor shape.

Diseases and pests

The most serious disease of *A. klaineana* is black canker. This begins as an infestation of scale insects such as *Asterolecanium pustulans*, spread by ants, with the fungus *Botryodiplodia theobromae* then entering through injuries in the bark. The bark becomes blackened by smut, and the resin is affected.

Less significant pests include psyllids and centipedes, which attack young plants, and damage to larger trees by elephants. Caterpillars of the moth *Pleuroptya balteata* can occasionally defoliate entire stands, but this has little lasting effect on growth.

The heartwood is resistant to *Lyctus* beetles, but fairly susceptible to termite, marine borer and fungal attacks. Preservation treatment is therefore necessary in external or humid conditions.

Uses

A. klaineana is planted and managed solely for timber. It has a soft, light wood, with a density of 400–500 kg/m^3 at 12% moisture content, which dries easily with low to medium shrinkage. The heartwood is salmon-pink with a shiny or pearly aspect. The sapwood is well differentiated, white or greyish, and annual growth rings are distinct.

From 1898 to 1955, approximately 14 million m^3 were exported from French Equatorial Africa (Gabon and Congo). Recent production data (1991–1997) are available for Gabon (10 million m^3) and Equatorial Guinea (1.5 million m^3).

Most *A. klaineana* timber is stripped for plywood production, for which it is considered one of the best woods. Controlled steaming or seasoning appears to increase the overall quality. Sawn wood is becoming an important timber use; good-quality logs are exported or processed locally for plywood, and lower grades are now used for sawn timber. The silica-rich (3%) wood blunts saw teeth rather quickly, and so sawing requires appropriately tipped cutters.

A. klaineana produces a resin widely used to make torches and oil lamps in Gabon and Equatorial Guinea. Other recorded uses include firewood, pulp and the construction of traditional canoes.

Further reading

Born, C., Kjellberg, F., Chevallier, M.-H., Vignes, H., Dikangadissi, J.-T., Sanquié, J., Wickings, E.J. and Hossaert-McKey, M. (2008) Colonization process and the maintenance of genetic diversity: insights form a pioneer rainforest tree, *Aucoumea klaineana*. *Proceedings of the Royal Society B: Biological Sciences* 22, 2171–2179.

Brunck, F., Grison, F. and Maître, F.H. (1990) *L'Okoumé Aucoumea klaineana Pierre: Monographie*. Centre Technique Forestier Tropical, CIRAD, Nogent-sur-Marne, France.

Fuhr, M., Nasi, R. and Delegue, M.-A. (2001) Vegetation structure, floristic composition and growth characteristics of *Aucoumea klaineana* Pierre stands as influenced by stand age and thinning. *Forest Ecology and Management* 140, 117–132.

Azadirachta indica A. Juss.

Neem

Importance

Azadirachta indica can be grown under a range of tropical and subtropical conditions but tends to prefer semi-arid climates. It produces valuable timber, fuelwood and fodder, though most of the interest in this tree surrounds its use for medicines, pesticides, and oil produced from its seeds, fruit and leaves. It also a common street tree with shade, ornamental and even religious value in addition to its economic and medicinal benefits. Nearly every part of the tree can be used for some purpose.

Botanical features

A. indica is a medium to large, deep-rooted tree. It is evergreen and has wide spreading branches forming a dense rounded or oval crown. It grows to 12–15 m, rarely to 30 m, with a straight bole of 3–8 m and a diameter of 0.4–0.8 m. Bark is dark grey and reddish inside. Leaves are compound, imparipinnate, 20–38 cm long, each with 4–7 pairs of lanceolate coarsely serrate leaflets, 2.5–7.5 cm long and 1.2–4.0 cm wide. The inflorescence consists of many-flowered panicles with white flowers. The flowers are small, bisexual and male on the same tree, with a honey fragrance, in dense bunches. The oval fruit is yellowish-green to dark yellow, and 12–18 mm long.

Distribution

A. indica is now found in the tropics and subtropics worldwide. Due its value it has been introduced since prehistory and its exact origins are obscure. However, it is considered to be originally native to dry areas of South Asia (India, Pakistan, Bangladesh and Sri Lanka) and possibly neighbouring countries (Afghanistan, Myanmar and China).

Silviculture characteristics

A. indica is light-demanding but it also tolerates fairly heavy shade during early development. It can grow with other crops, but cannot compete with weeds and grasses. It is drought-hardy but cannot withstand waterlogging and is highly susceptible to frost damage when young. It regenerates naturally and is propagated mainly through seed, but regenerates by coppicing and pollarding and can produce root suckers, especially in dry localities.

It improves soil fertility and water-holding capacity, and can neutralize acidic soils, and is therefore useful for wasteland reclamation. It can be grown on saline and alkaline soils with pH up to 9.8 and with soluble salt content up to 0.45% in the subsoil. It is very suitable for growing along roadsides and canal banks. With a deeply taprooted habit in suitable soils, it competes little with agricultural crops for soil moisture, and crops such as sorghum grow well close to the trunk.

Silviculture practice

A. indica is propagated primarily by seed, though germination is about 50% and seeds only have a short viability of 3–4 weeks. Viability is enhanced by fruits being depulped, though stones give better germination rates than hulled seeds. The most popular and economic methods of propagation are by direct sowing, transplanting seedlings in plastic containers and by stumps, though it can also regenerate from root suckers, stem cuttings, and air-layering. Seedlings in polythene bags with a height of 20–25 cm are planted out at the beginning of the rainy season in pits at a spacing of 3 × 3 m, although other spacings are used depending on the management aims. Trimming of leaves and roots also improves survival.

When direct seeding, two or three weedings are usually required in the first year, with the first soon after germination is complete and a second 1 month later. First thinning is done at 3–4 years when direct sowing, and after 5 years for transplanted seedlings.

Silvicultural systems for wood products are usually coppice or clear felling, and replanting or use of natural regeneration. Under dry conditions without irrigation, *A. indica* may still attain a height of 4 m in 5 years, and up to 10 m after 25 years, though the height is often 6–7 m. Mean annual increments are commonly reported in the range of 5–18 m^3/ha, but may sometimes be as low as 2–3 m^3/ha; MAIs of more than 20 m^3/ha have been recorded. The best economic rotation for wood production is considered to be 20–25 years.

Trees start fruiting at about 5 years but economic yield of fruit is obtained after 10 years. A medium-sized tree produces 37–55 kg of fruit/yr, with 3300–4500 seeds/kg. The financial benefits of using *A. indica* products for controlling pests on agricultural crops are reported to be high.

Diseases and pests

Due to the presence of chemicals in plant parts with pesticidal qualities and its uses as a natural pesticide, *A. indica* is not attacked significantly by any pests.

Uses

The hard to very hard heartwood of *A. indica* is reddish to reddish-brown, somewhat lustrous, darkening on exposure to air, whereas sapwood is yellow to yellowish-grey, turning pale yellowish-brown on exposure. The wood is usually heavy with an specific gravity of 0.72–0.83 when air-dried. It has interlocked grain, sometimes exhibiting a ribbon-grain effect on longitudinal surfaces, usually medium to somewhat coarse-textured, and occasionally a little finer in texture when fast grown.

As the wood is hard and heavy, sawing is difficult, but it is generally reported to work well with hard machine tools including lathes, showing excellent machining properties except for planing. A good pattern is noticeable in plain sawn boards due to parenchyma bands as well as small knots. The timber is durable, is resistant to termites, borers and fungi, and seasons well with little deformation during air drying. The wood peels well and is considered suitable for general-purpose plywood, fire-retardant plywood and plywood for blockboard on the basis of tests for glue adhesion, tensile, bending and compressive strength, and amenability to preservative treatment and fire proofing.

A. indica timber has satisfactory strength and is used for furniture, doors, windows, agricultural implements, carts, fence posts and in ship- and boatbuilding. It was not a preferred timber for furniture manufacture due to its rough and interlocked grain which does not allow a good finish, but more recently, painted furniture made from *A. indica* wood has become fashionable. It makes good fuelwood with a calorific value of 7000 kcal/kg, and good quality charcoal, and is the main source of fuel in parts of West Africa.

Leaves are used as fodder, mulch and green manure, and are sometimes mixed with other vegetables in the preparation of soups and curries. The fodder contains 12–18% crude protein, 11–23% crude fibre, 8–19% total ash, 1–2% calcium, 0.2–0.4% phosphorus and 50–55% total digestible nutrients.

Leaves also contain active medicinal and pesticidal chemicals, though these are more concentrated in the fruits and especially the seeds. The most active anti-feedant in the seeds is azadirachtin, at concentrations of 0.2–3.5 g/kg. Stem and root bark have astringent, tonic, anti-periodic and other medicinal properties. Stem bark contains 12–16% tannin and 8–11% non-tannin polyphenolics.

A. indica produces natural pesticides, and extracts prepared from *A. indica* seed are reported to affect more than 200 insect species as well as some mites, nematodes, fungi, bacteria and viruses. They have a controlling effect on more than 80% of the major insect pests in stored grain, reducing grain damage for up to 6 months and remaining effective for up to 1 year. Leaves can also be simply be mixed with grain, though will have a more limited effect than seed extracts.

The trees have been well known for their medicinal properties since antiquity, and *A. indica* has a long history of use dating back to the Vedic period of India. In Ayurvedic medicine, the bark and leaves are used for skin diseases, flowers as a tonic and stomachic, and fruits as a purgative and emollient. Fruits are used as a purgative and an emollient, and are useful in the control of intestinal worms, urinary tract diseases and piles, but it is the seed oil that is the most important medicinal product. The oil saponifies readily and is used in the manufacture of a medicinal soap because of its antiseptic properties, which is very effective for washing sores and for general uses similar to those of carbolic soap. Twigs are also used extensively as tooth brushes. The bark exudes a clean, bright amber-coloured gum which is used as a stimulant and as an adhesive in silk dyeing.

Further reading

National Research Council (2002) *Neem: A Tree for Solving Global Problems*. BOSTID, Washington DC, 144 pp.

Singh, K.K., Phogat, S., Tomar, A. and Dhillon, R.S. (2008) *Neem, a Treatise*. IK International, New Delhi, 568 pp.

Suri, R.K. (2001) *Neem, the Virtuous Tree*. Society for Forest and Environmental Management, New Delhi, 133 pp.

B

Bagassa guianensis Aubl.

Tatajuba

Importance

Bagassa guianensis occurs in tropical forest in north-eastern parts of South America (northern Brazil and the Guiana Shield). It produces a widely used high-density wood which has a natural resistance against fungi and termites. This species demonstrates high mean annual increments (MAI) in open plantations, and also shows potential for use in agroforestry. Research on genetic variation within the species is needed to identify the best genotypes.

Botanical features

B. guianensis is a tropical tree which reaches a height of 27–30 m with a dbh of 60–120 cm, although taller and larger specimens can also be found. The bole is cylindrical and straight, and is 18–21 m high, with an open crown. The bark is thick and fibrous, and produces a latex which coagulates rapidly.

The leaves are simple and borne oppositely. They may be entire and ovate, or divided into three lobes with sharp tips. They are 15–20 cm long, and are borne on petioles 6 cm long. They are cordate or emarginate at the base, and pointed at the tip. The upper leaf surface is glabrous. The stipules are 10–15 mm long and are deciduous.

B. guianensis is dioecious, with cymose inflorescences. The male flowers are small and sessile, with a 4-segmented perianth. The female flowers are smaller, compacted into spikes about 5 cm long.

The fruit is an globose, orange achene, 2.5–3.5 cm in diameter on a peduncle 1.3–1.5 cm long. It is edible and slightly astringent. The seeds are small and numerous, with up to 400,000 seeds/kg; they are disseminated by local fauna.

B. guianensis is leafless in July and August, produces flowers between April and September, and fruits develop from September to January, being more intense in October and November. This phenology does vary, however, according to the prevailing climate variations.

Distribution

The natural distribution of *B. guianensis* ranges from several areas of northern Brazil (states of Acre, Amapá, Amazonas, Maranhão, Pará, Rondônia) to French Guiana, Guyana and Suriname. It grows in low upland forests in Amazonia as scattered trees. In Suriname, *B. guianensis* is a component of specific vegetation types – namely, mesophytic forests – and has a lower frequency in dry forests. *B. guianensis* is found at 0–160 m altitude in Suriname, and up to 650 m in Brazil.

Silviculture characteristics

B. guianensis is a light-demanding species, and light is a limiting factor for growth in Amazonian forests. The species is characteristic and exclusive to Amazonian alluvial forests, occurring at moderate or irregular frequency. In Suriname, *B. guianensis* grows at densities such that there are four trees with a dbh of 10–15 cm/ha, and two trees with dbh >40 cm/ha. This species requires a large quantity of light for seed germination, and seedlings are rare around the adult trees, except after any disturbance. Natural regeneration of *B. guianensis* is significantly affected by the competition of secondary vegetation, which is more aggressive. Birds have a preference for the fruit and seed of this species, and they enhance its dispersal. *B. guianensis* produces a large quantity of seeds annually, and they remain viable for a long period, with germination only occurring in sufficient levels of light.

Silviculture practice

B. guianensis fruits are collected from the ground from early September until February or March. The peak period of fruit production is October and November. In the nursery, seed germination occurs in a few weeks. After 45–60 days, seedlings 5–7 cm tall are transplanted to plastic containers, and may reach a plantable size in 4–5 months. Bare-root seedlings 20–25 cm tall gave good results in experimental plantations; satisfactory results have also been obtained using stumps.

B. guianensis has potential for reforestation and agroforestry programmes, due to its high survival rates, MAI in open plantations, and wood quality and utilization potential.

B. guianensis grown in Brazil in open-managed plantations showed average heights of 16 m and dbh of 20 cm after 14 years. Natural regeneration of this species in deforested areas reached a growth rate average of 3 m in height/yr.

Diseases and pests

B. guianensis may be attacked by shoot borers in open and full light plantations. The wood is naturally resistant to termites and fungi.

Uses

B. guianensis is suitable for planting in degraded areas and for the enrichment of disturbed forests. It is also recommended for use in agroforestry systems with crops such as beans.

The timber is of medium density (800–900 kg/m^3 at 12% moisture content), and is hard and heavy. The heartwood is yellow when freshly cut, sometimes with dark streaks; after drying, it becomes olive-yellow and golden-brown on exposure to light. The sapwood is thin, yellow to yellowish-white and is well differentiated from the heartwood. Despite having very hard timber, *B. guianensis* is easily worked with manual and mechanical tools. It produces an excellent finish, but is not easy to nail or screw; it receives good finishing when painted, varnished or just polished. The timber is very durable and resistant to white and brown fungi and termites, and moderately resistant to marine borers.

B. guianensis is used in general construction, bridges, railway sleepers, furniture, picture frames, panels, door and window frames, floors, parquetry, carpentry, cabinet work, tool handles, flooring, sliced veneers, boats and ships and sports equipment.

B. guianensis produces edible fruits. The latex is used by indigenous populations.

Further reading

Kazuya, M., Takumi, T. and Tsuyoshi, Y. (1999) Contribution of extractives to fungal and termite resistance of Brazilian wood *Bagassa guianensis* Aubl. *Wood Preservation* 25, 164–171.

Silva, M.B., Kanashiro, M., Ciampi, A.Y., Thompson, I. and Sebbenn, A.M. (2008) Genetic effects of selective logging and pollen gene flow in a low-density population of the dioecious tropical tree *Bagassa guianensis* in the Brazilian Amazon. *Forest Ecology and Management* 255, 1548–1558.

Simmenn, B. (1992) Competitive utilization of *Bagassa* fruits by cympatric howler and spider monkeys. *Folia Primatologica* 58, 155–160.

Baikiaea plurijuga Harms

Rhodesian teak

Importance

Baikiaea plurijuga (Plate 9) is mainly distributed in Southern Africa and has been tested in plantation trials and systematically cut on a sustained-yield basis in Botswana, the Caprivi region of Namibia, Zimbabwe and Zambia. The wood is hard, strong and durable, and has a beautiful red-brown colour; it is rated as one of the world's finest commercial timbers.

B. plurijuga has been overexploited, and there is a need to encourage regrowth, as well as to encourage the use of other species and to determine sustainable harvest levels.

Botanical features

B. plurijuga is a medium to large deciduous tree, 8–16 m tall and up to 1 m in diameter, with a large, dense, spreading crown. The tree is often multi-stemmed, with a clear bole varying from 3 to 4 m in height. The bark is smooth and pale on young trees, whereas in adult trees it is vertically fissured, cracked and brown to grey in colour.

The leaves are alternate, compound – with 4–5 pairs of opposite leaflets – and up to 12 cm long. The leaflets are oblong to elliptic, 3.5–7 × 2–2.5 cm; the apex is broadly tapering to rounded, often notched, and bristle-tipped. The margins are entire, and are inclined to be wavy and finely rolled under. The swelling on the margin near the base of each leaflet is a distinguishing feature.

The flowers are large, and are borne in axillary racemes up to 30 cm long. The buds are dark brown or golden-brown, with a dense, velvety pubescence. The flowers have four sepals, adorned with dark brown velvety hairs, five pinkish-mauve petals, each 2–3.0 × 1–1.5 cm, and ten stamens (Plate 47).

The fruits are flattened woody pods, up to 13 × 5 cm, broadest near the apex and tapering to the base, with a dark brown velvety pubescence, which is frequently rubbed off to reveal golden-yellow patches. The fruit splits explosively, dispersing seeds a considerable distance. The seeds are hard, smooth and shiny brown.

Flowering occurs between December and March, and fruits mature between June and September.

Distribution

B. plurijuga is confined to deep Kalahari sandy soils, occurring at latitudes of 13–20°S, usually in open deciduous woodland. It occurs in pure stands or in association with *Pterocarpus angolensis*, *Schinziophyton rautanenii* and *Guibourtia coleosperma*. It is mainly found in Angola, Botswana, Namibia, Zambia and Zimbabwe, and is restricted to the Zambezi floral region. It is occasionally found on sites with seasonally impeded drainage, for instance near swamp margins, where it grows as a dwarf form.

Silviculture characteristics

Propagation of *B. plurijuga* from seed is difficult in areas other than the deep Kalahari sands where it occurs naturally. In the natural environment, seedlings require shady conditions for optimum growth and survival. Seedlings exhibit remarkable development of the taproot, growing very slowly for the first 3 years, during which time the taproot rapidly penetrates the soil. The shoot does not respond noticeably to watering under nursery conditions, but the root does, reaching a depth of 6–9 m at maturity.

Seedlings of *B. plurijuga* are susceptible to fire, drought, animal damage and frost, and mature trees remain susceptible to fire. Young trees can be coppiced after harvest or stem damage.

Silviculture practice

B. plurijuga seed is orthodox in its storage characteristics and can be collected between June and September. The seeds retain their viability for least 1 year in open storage at room temperature, and can remain viable in cold storage for more than 3 years. Suitable pre-treatment of seed requires soaking in cold water for 24 h or soaking in hot water for 2 min. Seeds germinate easily and grow best in sandy soils under both natural and nursery conditions.

Attempts at artificial regeneration, by enrichment planting and establishment of trial plantations, have so far met with little success. Considerable losses are caused by the removal and destruction of directly sown seeds and newly planted seedlings by animals, such as rodents or duikers (*Sylvicapra grimmia*) in Zambia. The seedlings and saplings regularly suffer severe die-back caused by frost damage.

The management of *Baikiaea* forests is based solely on natural regeneration. Logging is selective and on a sustained-yield basis, although the basis at arriving at the sustainable cut may be questionable in many cases. In Zimbabwe, for example, trees above 35 cm can be harvested and the cutting cycle is 40 years. Rotation length is estimated to be between 80–100 years if grown from seed. In its natural environment, *B. plurijuga* is very slow growing, representing an increase of approximately 2.5 mm in diameter. However, its growth rate was found to be much lower (1.0–1.5 mm in diameter/yr) in Zimbabwe. The species has a lengthy establishment phase during which the taproot develops very extensively, reaching 2 m in length within 6 months, while the stem attains no more than 15 cm in height for several years.

Multiple land use is commonly practised, because the forests where *B. plurijuga* occurs are important for wildlife, soil and water protection. Due to the fire sensitivity of the species, some of the forests are protected from fire, and grazing of domestic animals is encouraged to reduce fuel loads. In some areas, prescribed light fires are used to reduce fuel load and to enable the viewing of wild game.

Diseases and pests

B. plurijuga is virtually free from specific diseases, and does not suffer from any of the leaf-spot infections that are common in the region. Some fungal attack can affect germinating seeds, in the form of damping off in the nursery. This can be avoided by following standard nursery management practices. The only minor fungal pathogen encountered so far in young plantations is a sooty mould on twigs.

Frost damage occurs in young seedlings, causing severe die-back, retarding growth and affecting stem form. Young seedlings are also affected by fire, drought and animal damage.

Sawn wood can be infested by a number of wood borers during seasoning. The most harmful class of timber insects are the powder-post beetles (Lyctidae and Bostrichidae), especially *Heterobostrychus brunneus*.

Uses

B. plurijuga is managed not only for timber production, but also for wildlife, soil and water protection. *Baikiaea* forests are important for providing employment, particularly for rural people. They cover some of the most important national parks in the region, and are therefore important for biodiversity conservation. The forests are of international and local importance with respect to tourism and scientific studies.

B. plurijuga produces one of the world's finest commercial timbers. The wood is hard, resistant to abrasion, strong and durable, and has a beautiful dark red to brown colour. Its density at 12% moisture content is 815–955 kg/m^3. It is suitable for painting, varnishing and gluing purposes. The wood is resistant to termites, but may be susceptible to attack by longhorn beetles. It is hard-wearing and ideal for all outdoor work. The grain is straight to slightly interlocked and the texture is fine and even, giving a smooth, hard surface. The wood is difficult to saw, blunts tools severely and pre-boring is necessary before nailing, to prevent splitting.

B. plurijuga has been used for furniture, high-quality flooring, railway sleepers, mine props, carving, structural work, firewood and the building of canoes along the Zambezi River in Caprivi and Zambia. The timber is sold regionally and internationally, especially in Europe, as round logs and for finished products, such as floor tiles. A high level of wastage of the wood is reported both in the field and at sawmills, and more research needs to be done to improve recovery and minimize the wastage. Forests dominated by *B. plurijuga* provide extensive grazing for cattle and opportunities for wildlife management. A decoction of the bark is drunk to give strength and to cure eye diseases. Decoctions of the bark are also used in the treatment of syphilis.

Further reading

Gambiza, J., Campbell, B.M., Moe, S.R. and Frost, P.G.H. (2005) Fire behaviour in a semi-arid *Baikiaea plurijuga* savanna woodland on Kalahari sands in western Zimbabwe. *South African Journal of Science* 101, 239–244.

Martin, J.D. (1940) The *Baikiaea* forests of Northern Rhodesia. *Empire Forestry Journal* 19, 8–18.

Piearce, G.D. (ed.) (1986) *The Zambezi Teak Forests: Proceedings of the First International Conference on the Teak Forests of Southern Africa, March 1984, Livingstone, Zambia*. Zambia Forest Department, Lusaka.

***Baillonella toxisperma* Pierre**

Moabi

Importance

Baillonella toxisperma is the most important tree species of Central African moist tropical forests. It dominates the canopy and remains a popular species among loggers for its wood quality, and among local populations, mainly for its fruit. The fleshy fruits are eaten fresh and oil is extracted from the seeds.

Botanical features

B. toxisperma is 45–70 m high when mature, with a dbh of 70–280 cm. Trunks of mature trees are remarkably straight and cylindrical, with up to 30 m of clear bole before the first branches appear, the trunk base being buttress-free but having flares. The bark is deeply furrowed, and copious white gum exudes from any wounds.

The spreading crown is formed from large, characteristically tortuous branches, bearing heavy and dark deciduous foliage. The leaves are simple, entire, alternate, stipulate and verticillate at the end of the branches. Leaves are obovate and very elongate, 20–30 cm long and 6–10 cm wide, abruptly acuminate at the apex and cuneate at the base; the petiole is 3–4 cm long.

The hermaphroditic flowers are grouped together in heavy fascicles with pedicels 2–3 cm long. The flowers have eight 1-cm-long sepals, an eight-lobed corolla, eight stamens and eight staminodes. The fruit is a spherical berry, about 9 cm in diameter, weighing around 250 g. It contains up to three oleaginous seeds which are among the largest found in Central African moist tropical forests.

The native range of *B. toxisperma* crosses the equator, and its phenology varies accordingly; fruiting occurs in June–August in Cameroon, but in December–February in Gabon. Fruiting accompanies the total loss of leaves, although sterile trees retain their foliage. In Cameroon, fruiting starts at the end of the long dry season and lasts for about 3 months, each tree fruiting for about 4 weeks. For dominant trees (dbh ≥70 cm) fruiting occurs annually, but massive fruiting only takes place every 3 years, corresponding to a production of 6000 seeds per tree over a 3-year period.

Distribution

The native range of *B. toxisperma* covers southern Nigeria, western and southern Cameroon, Equatorial Guinea, Gabon, the Mayombe region of the Democratic Republic of the Congo and Congo (Brazzaville). *B. toxisperma* also occurs disjunctly in the Cabinda region of Angola, but not in the central Congo Basin.

Silviculture characteristics

The seeds of *B. toxisperma* are widely dispersed by animals but are not densely distributed; they show no dormancy and do not contribute to the seed bank. Fresh seeds initially germinate well (over 85%) but lose viability with the rapid degradation of fatty acids. Seeds need not be digested by animals before germinating, which can occur in complete darkness. The seed has a fine, lignified tegument, marked by a large ventral scar. The endosperm is absent and the large cotyledons contain fat reserves. Seedlings germinate epigeally, with exposed fleshy cotyledons and a short (1 cm) hypocotyl.

Removing the terminal bud is generally fatal to seedlings, although an adventitious bud, axillary to the cotyledons, occasionally develops into a robust stem. The seedlings become autotrophic 6 weeks after germination. Regeneration under the crown of the producing tree is almost non-existent despite a huge number of seeds, but a few seedlings remain alive at least 50 m from the producing trees, despite survival rates of one in a million.

B. toxisperma occurs within a wide range of light exposure conditions. Seedlings tolerate half-light, but do not develop their maximum growth potential there. In well-lit sites, *B. toxisperma* seedlings have to deal with more competitive light-demanding species and ideal growth conditions are met around small gaps, especially where *B. toxisperma* already grows. It seldom finds its optimal light exposure conditions until it reaches the dominant stage at a dbh of 40–70 cm.

B. toxisperma has an annual phenological rhythm, but massive and general fructification takes place every 3 years. Annual mortality rates of 6% are recorded for seedlings with a height of less than 100 cm, falling to less than 1% mortality for trees with a dbh of 2 cm. These seedlings grow 6.9 cm each year, corresponding to a relative growth rate of 4.9%. Populations of *B. toxisperma* seedlings show a relatively slow but regular growth pattern in the undergrowth, although individuals seem to undergo successive periods of fast and slow growth related to the light conditions.

Initial annual diameter increments are very low. Trees with a dbh of 40–70 cm have easy access to light and their growth rates increase initially. Mortality rates decrease progressively with age, from 0.3% at a dbh of 10–40 cm, becoming insignificant thereafter until senescence (dbh >190 cm), when mortality rates are 1–3%. The highest observed dbh was 280 cm. The majority of felled trees (dbh ≥100 cm) are 200–500 years old.

Silviculture practice

In natural conditions, seeds are disseminated by elephants, rodents and humans. *B. toxisperma* produces massive amounts of fruit during a short period – especially when most tree species are also in fruit (before the long rainy season) – which saturates the capacity of destruction by predators and which takes advantage of the ideal ecological conditions for the early development of seedlings. Germination rates are high when seeds are planted immediately after collection. Planting seeds in logging gaps also proved to be quite successful, at 2 m spacing after clearing; 8 months later, the average height of the seedlings was 50 cm. Propagation via seeds can be successful, but there is no information on vegetative propagation.

Only a few experimental plantations of *B. toxisperma* have been established, and no commercial plantations. Instead, trees are still exploited in natural forest areas. Logging companies must abide by minimum exploitation diameters of 70–100 cm. In natural conditions, there is approximately one mature, exploitable tree every 20 ha. In Cameroon – the main exporting country for the species – *B. toxisperma* is particularly important in the Dja Reserve, with a density of 0.6 trees/ha with a dbh of >70 cm. Its population structure is sigmoidal. In Cameroon, cutting cycle periods cover 30 years, at least in cases of forests with a management plan. After 30 years of exploitation, it takes 300 years to restore the initial population structure. In order to increase regeneration rates, selective thinning in favour of young stems with a dbh of 10–70 cm has been proposed, but not tested.

The IUCN considers *B. toxisperma* to be a 'Vulnerable' species because of over-exploitation, habitat fragmentation and increasing disturbance through human activities.

Diseases and pests

In moist rain forests, different frugivorous animals act as predators as well as disseminators of seeds. The most recurrent seed parasites are the larvae of *Carpophilus* beetles and *Mussidia* moths, which attack the fruit before it falls from the tree. Affected seeds do not germinate when the embryo is damaged, and seedling growth is reduced in other cases. In some cases, up to 53% of the seeds can be infected.

Uses

The wood is heavy, with a density of 860 kg/m^3 at 12% moisture content, and is hard or moderately hard. The sapwood is dark grey or pinkish-grey, and rather well demarcated from the heartwood, which is pinkish-brown or red-brown. Due to the silica content, tools are rapidly dulled during sawing. Growth rings are often visible, giving the wood a finely veined aspect. The wood has outstanding mechanical properties and is resistant to fungi, beetles and termites. The timber dries slowly and does not normally warp, but other defects may occur, such as checks. During milling, dust may affect the workers' mucous membranes. The wood tolerates various weather conditions, making it suitable for exterior carpentry and boat construction, although its capacity for impregnation by preservatives is very poor. The fruits of *B. toxisperma* are eaten by many animals. During periods of abundant fruiting of a large tree, animals are attracted by the smell or the sound of falling fruit. Total fruit production has been estimated to be on the order of 2.5 tonnes per tree, representing an important energetic effort as well as an important but seasonal nutritional contribution for many animals. *B. toxisperma* is an important source of non-timber forest products. Local people collect the fruit for home consumption or to extract an oil from the seeds that is rich in palmatic, stearic and oleic acids. For this, the fruits are usually left to rot first, aiding the extraction of seeds, which are then dried over a fire for about 1 month. They are then broken, their kernels crushed in a mortar and the resulting paste is kneaded and pressed. The oil is boiled before purification; it has therapeutic qualities and is used to treat rheumatic pain. The press-cake by-product from oil processing is toxic and can be used as poison for fishing. The plant's latex is used to heal wounds and a bark decoction is used to treat kidney pain.

The Baka tribe believes the bark, which they use while chasing elephants, makes them invisible. Judicial courts take place near *B. toxisperma* trees, and the tree is mentioned in numerous folk stories and Bantu songs.

Further reading

Aubréville, A. (1961) Sapotacées in: *Flore du Gabon. Volume 1.* Muséum National d'Histoire Naturelle, Paris, 162 pp.

Debroux, L. (1998) L'aménagement des forêts tropicales fondé sur la gestion des populations d'arbres: l'exemple du moabi (*Baillonella toxisperma* Pierre) dans la forêt du Dja, Cameroun. PhD thesis, Faculté des sciences agronomiques de Gembloux, Belgium.

Debroux, L., Delvingi, W., Mbolo M. and Amougou, A. (1998) Régénération du moabi et du mukulungu au Cameroun, perspectives pour l'aménagement. *Bois et Forêts des Tropiques* 225, 1–17.

Mapaga, D., Ingueza, D. and Louppe, D. (2002) Moabi. CIRAD, Département Forestier, Montpellier, France, 4 pp.

Betula papyrifera **Marshall**

Paper birch

Importance

Betula papyrifera is one of the few North American transcontinental species which is distributed mostly in Canada. It is a fast-growing but short-lived tree that may form nearly pure stands on disturbed sites. Its wood is used for veneer, pulpwood and many speciality items. The foliage and showy white bark make the trees attractive for landscaping. The foliage provides an important browse for animals. The seeds, buds and bark are also eaten by wildlife.

Botanical features

B. papyrifera is a medium-sized, deciduous, broadleaved tree, occasionally reaching over 30 m tall, 40 cm dbh and 200 years old. The trunk is slender and often curved. The crown is narrowly oval and open, with ascending branches. Twigs are slender, dark reddish-brown, with sparse resin glands. The bark is thin and smooth, dark red on young stems and bright creamy-white on mature stems. Roots are shallow and associated with ecto- and endomycorrhizae.

Leaves are deciduous, ovate to triangular, 5–10 cm long, tip pointed, base broadly wedge-shaped, margin double-toothed, upper surface dull green, lighter beneath.

B. papyrifera is monoecious and flowers from mid-April through early June depending on location. In the late summer, staminate flowers appear in aments (catkins) 2–2.5 cm long at the ends of twigs and lateral shoots. These mature and grow in length to 4–10 cm by the following spring. Pistillate flowers are borne in cylindrical aments 2.5–5 cm long and 0.8 cm in diameter. Fruits are winged nutlets 1.5 mm long by 0.8 mm wide with styles 0.8 mm long. Bract shape in the pistillate catkins is characteristic for the species, and variations are useful in distinguishing varieties. The seeds ripen from early August until mid-September.

Initiation of phenological events varies greatly with latitude, longitude, altitude, aspect and weather.

B. papyrifera is shade-intolerant to moderately shade-tolerant, with a rapid initial growth rate and a wide climatic and edaphic amplitude.

Distribution

As a transcontinental species, the range of *B. papyrifera* closely follows the northern limit of tree growth from Newfoundland and Labrador west across the continent into north-west Alaska; south-east from Kodiak Island in Alaska to British Columbia and Washington; east in the mountains of north-east Oregon, northern Idaho and western Montana, with scattered outliers in the northern Great Plains of Canada, Montana, North Dakota, the Black Hills of South Dakota, Wyoming, Nebraska, and the Front Range of Colorado; east to Minnesota and Iowa, through the Great Lakes region into New England. *B. papyrifera* also extends down the Appalachian Mountains from central New York to western North Carolina. Introductions are mainly limited to arboreta and parks.

B. papyrifera grows in even-aged, pure stands or mixed with shade-tolerant conifers mainly in second-growth stands. It may be a pioneer on fragmental sites (talus), is present in early and intermediate (infrequently in late seral) stages of secondary succession on floodplains and upland sites, and is a minor component in old-growth stands on some montane boreal sites. *B. papyrifera* invades gaps on floodplains created by windthrow of climax species such as *Picea glauca*, and retains a long-term presence.

B. papyrifera associates with both hardwoods and softwoods throughout boreal and cool temperate climates; the common associates are *Abies balsamea*, *Abies lasiocarpa*, *Picea glauca*, *Picea mariana*, *Picea rubens*, *Pinus banksiana*, *Pinus contorta*, *Populus balsamifera*, *Populus tremuloides*, *Pseudotsuga menziesii*, *Thuja plicata* and *Tsuga heterophylla*.

Silviculture characteristics

Reproduction capacity is high. *B. papyrifera* reproduces vegetatively from stump sprouts. Minimum age for a seed crop is about 15 years, the optimum seed-bearing age being 40–70 years. The light, winged seeds (3 million/kg) are dispersed readily by the wind, and some seeds travel great distances, particularly when blown across the surface of snow. Potential for natural regeneration in low light is low, practically nil in the absence of adequate seedbeds. Potential for natural regeneration in the open is high in the presence of exposed mineral soil or burnt forest floor.

Potential initial growth rate (up to 5 years) is intermediate; the best growth (about 40 cm/yr) is in partial shade and on forest floor seedbeds. Self-pruning capacity in dense stands is intermediate. Crown spatial requirements are intermediate.

Light conditions beneath closed-canopy, mature stands are high and associated with well-developed understorey vegetation. Potential productivity is intermediate. Longevity is low; maturing at about 70 years, few trees live longer than 200 years.

Silviculture practice

Depending on site and management objectives, clear-cutting, patch-cutting, strip shelterwood and seed tree are viable silvicultural systems. Considering its intermediate growth rate, *B. papyrifera* may also be grown in mixed species stands with either shade-intolerant or shade-tolerant softwoods.

Propagation by seed is the only method practised. Seed years vary with locality. Nearly 2.5 million seeds/ha are produced in average seed years, rising to 86 million in bumper years. Discoloured or empty seeds make up 14–47% of the crop.

Most seeds are dispersed during September–November. Time of dispersal does not depend on seed crop size, but varies among stands and from year to year depending on weather conditions. Seeds that fall in late autumn and winter have a higher germination capacity than those that fall in early spring.

B. papyrifera seed may be stored for at least 2 years at room temperature if the moisture content is maintained at less than 5%.

Germination is epigeal. Because of the small seed size, newly germinated seedlings are very fragile and sensitive to moisture, temperature, light and seedbed conditions. Best germination occurs on mineral soil, with germination on humus being lower by 50% and on undisturbed litter by 90%. Shaded sites produce about twice as many germinants as full-sun sites. Early survival of seedlings follows similar patterns, but initial height growth is better on humus than on undisturbed sites.

Following disturbance, the bulk of *B. papyrifera* regeneration becomes established during the first growing season from seed that fell the previous autumn and winter. Some seeds may lie dormant in the forest floor for 1 year or more.

Even though natural regeneration of *B. papyrifera* is obtained readily, planting of bare-root or container-grown stock may be needed in old fields. Site preparation to remove sod is required for satisfactory survival and growth. Protection from girdling by rodents and browsing by deer may be required in some locations.

B. papyrifera can regenerate from sprouts following cutting or fire. Prolific sprouting usually occurs when young, vigorous trees have been cut in the spring to stump heights of 15–30 cm. Whereas sprouts are seldom abundant enough to reproduce mature stands, they can be valuable supplements to seedlings, particularly on difficult sites. *B. papyrifera* can be propagated by grafting, air-layering, cuttings or tissue culture. Cuttings from seedlings root sooner and at higher frequency than cuttings from mature trees.

About 370 trees/ha should be selected for crop trees. Individual trees often have a diameter of 20 cm after 30 years. With age, the growth rate declines, and in old age becomes almost negligible. Trees in mature stands average 25–30 cm in diameter and 21 m in height. On the best sites, an occasional tree in an old stand may exceed 76 cm dbh and 30 m in height. Yields at maturity range from 230 to 270m^3/ha on good sites, and from about 100 to 185m^3/ha on poor sites. Site index ranges from 11 to 24 m at base age 50 years, depending on location.

Diseases and pests

Principal decay-causing fungi include *Inonotus obliquus*, *Phellinus igniarius* and *Pholiota* spp. The root-rotting fungus *Armillaria mellea* causes cracks at the base of the stem; attacks can result in uprooting by the wind.

White-tailed deer, porcupines, moose and hares cause browsing damage. The bronze birch borer (*Agrilus anxius*) is the most serious insect pest of *B. papyrifera*. It usually attacks over-mature or weakened trees. The most serious defoliators are *Malacosoma disstria*, *Bucculatrix canadensisella*, *Fenusa pusilla*, *Heterarthrus nemoratus*, *Profenusa thomsoni* and *Coleophora serratella*, as well as the general forest defoliators *Heterocampa guttivitta*, *Lymantria dispar* and *Rheumaptera hastata*. Defoliation alone seldom causes mortality of otherwise healthy trees. Fire is one of the most serious threats to established stands; as the bark is thin and highly flammable, even large trees may be killed by moderate fires. *B. papyrifera* is very susceptible to mechanical logging damage during partial harvesting.

Uses

Young stands of *B. papyrifera* provide prime browse and cover for deer and moose, and are also an important source of food for birds. The graceful form and attractive white bark make *B. papyrifera* a prized species for ornamental planting and landscaping around homes and public buildings. The species is a prime hardwood for revegetating acid coal mine spoils and other drastically disturbed sites.

The wood is moderately dense (480–620 kg/m^3) and is used for veneers, woodware, musical instruments, carvings and other speciality items. Full tree chips can be used in the manufacture of pulp, paper and other reconstituted products, as well as for fuel. The bark's fuel value is high: 5740 cal/g or 3209 cal/cm^3.

B. papyrifera can be tapped in the spring to obtain sap from which syrup, wine, beer and medicinal tonics can be made. The carbohydrate content of about 0.9% consists of glucose, fructose and sucrose.

Further reading

Ashburner, K. and McAllister, H. (2012) *The genus* Betula. Royal Botanic Gardens, Kew, UK.

Peterson, E.B., Peterson, N.M., Simard, S.W. and Wang, J.R. (1997) Paper birch managers' handbook for British Columbia. Pacific Forestry Centre, Victoria, British Columbia, Canada.

Safford, L.O. (1983) Silvicultural guide for paper birch in the Northeast. Research Paper No. NE-535. Northeastern Forest Experiment Station, USDA Forest Service, Broomall, Pennsylvania, USA.

Betula pendula Roth

Silver birch

Importance

Betula pendula is a fast-growing deciduous tree, yet it rarely dominates forest stands. In Northern Europe, birch is the most commercially important broadleaved species, especially in Scandinavia and Russia, and the wood, bark and sap have many uses. *B. pendula* is an important pioneer species, and is tolerant of low temperatures, water deficits and infertile soils. *B. pendula* often grows together with *B. pubescens* (downy birch), although the latter requires higher humidity and cannot grow on dry soils which are suitable for silver birch. It can be an important component in coniferous plantations, particularly Scots pine (*Pinus sylvestris*) or Norway spruce (*Picea abies*).

Botanical features

B. pendula is recognized by its pyramidal, graceful crown and characteristic white stem. *B. pendula* can be distinguished from *Betula pubescens* as in the latter the branches are upward pointing or spread horizontally, but never pendulous. The bark of young trees is brown, becoming white with age due to an accumulation of a chemical compound which refracts light (Plate 66). The bark of mature tree buttresses is darker, thicker and more deeply fissured. The main trunk is upright; branches are long and pendulous, hairless and covered with resinous warts. Buds are hairless, sticky, brown-black, larger at the base and pointed at the apex. Leaves are alternate, simple, 3–7 cm long, broadly ovate, triangular or rhomboidal, doubly and unequally serrate. The upper surface is dark green, the underside is brighter green. Petioles are up to 3 cm long, stipules up to 8 mm long.

B. pendula can reach an age of 90–100 years, exceptionally 120–150 years. A typical maximum height is 23 m, with a dbh of 18–25 cm.

B. pendula is monoecious and begins to flower at 10 years of age if open-grown, and 20–25 years of age in a stand. It fruits every year; however, full seed years only occur every 2–3 years. Male catkins appear on the tips of elongated shoots; each has sticky scales and numerous flowers. Flowers consist of 3 perianth segments and many stamens; anthers are divided at the base. Cylindrical female catkins also consist of many flowers clustered in threes within scales. The pistil has two stigmas and two ovaries; only one of the two ovules develops. Many small winged fruits (achenes) are produced, containing at least one seed, without an endosperm.

Buds open from early March, and leaves and flowers usually develop in April. Male catkins appear 1 year before flowering. Female catkins appear and develop in spring, together with leaves. Seed crops depend on tree age and site conditions. Fruits ripen at the end of summer, most often in August.

Distribution

B. pendula is native to Europe, and to Central and South Asia. It has been introduced to other temperate regions especially in North America. It is mainly a lowland species, but it occurs also in the mountains, especially in Asia. The natural distribution of *B. pendula* is very similar to that of the closely related *B. pubescens*. However, the range of *B. pubescens* extends further north to Arctic Europe and Iceland, whereas *B. pendula* stretches further south.

Silviculture characteristics

B. pendula is a light-demanding and fast-growing pioneer species. It is not particularly resistant to wind. *B. pendula* prefers moderately moist sites but it can also grow on dry sands. It is tolerant of low and high temperatures, and is frost-resistant. It does not tolerate high groundwater levels. *B. pubescens* is generally more tolerant to waterlogging and cold climates than *B. pendula*.

Silviculture practice

Silver birch reproduces by natural seeding. Seeds germinate better after pre-treatment: 30–60 days at temperatures slightly above 0°C. Afterwards, the seeds should be at a temperature above 20°C and experience long daylight conditions. Seeds can be sown in summer immediately after harvest and extraction, or after storing. It is possible to sow in autumn or winter on snow-covered surfaces, as well as in spring, to get the optimal light, temperature and humidity conditions for breaking dormancy.

Two-year-old seedlings are used for regeneration and reforestation by planting. The height of seedlings should be 25–50 cm. Suggested spacing is 1.3 × 1.3 m to 1.6 × 1.6 m, and seedlings should be planted in autumn or in spring. It may be necessary to use plastic tree guards to protect the seedlings from small herbivores such as voles.

B. pendula can also be planted for use as a protective stand for plantations of *Fagus sylvatica*, *Abies alba*, *P. abies* and other forest tree species.

On optimal sites, *B. pendula* will begin to achieve higher increment rates earlier (at an age of 10–15 years). On other sites, higher increment rates will not occur until the trees are older (at 20–25 years). After 60 years, trees will only increase in diameter, and not in height.

Diseases and pests

B. pendula is host to an exceptionally high number of pathogens, in particular fungi, which attack birch at all stages of development – from the seeds, which are damaged by

Ciboria betulae, to a host of other fungi affecting different parts of the tree.

Seedling damping off caused by many fungi in nurseries is dangerous for germinating seeds and young seedlings. For *B. pendula*, the most serious are: *Fusarium* sp., *Rhizoctonia solani*, *Botrytis cinerea* and *Pythium debaryanum*.

Birch is also affected by the so-called 'witches' brooms' and foliar deformations caused by *Taphrina betulina* and *Taphrina nana*. Birch rust (*Melampsoridium betulinum*) is a very common disease of the leaves and is associated with retarded height and increased mortality. *Godronia multispora*, a damaging fungus isolated from the bark of birch root suckers, causes necrosis and bark injuries.

The honey fungus *Armillaria mellea* is often found on birches. Birch is also affected by wood rots which may be caused by *Piptoporus betulinus*, *Fomes fomentarius* or *Phellinus igniarius* f. *betulae*. These fungi reduce significantly the wood quality and eventually kill the trees. Birches are also attacked by the peculiar pathogen *Inonotus obliquus* which forms enormous sporophores.

Birch also suffers from die-back which affects crown health; fungi associated with such die-back are *Anisogramma virgultorum*, *Discula betulina* and *Marssonina betulae*. Stem lesions and top dying caused by *Phytophthora cactorum*, and leaf lesions caused by a number of different fungi, have also been reported.

Numerous forest insects as well as other species typically occurring in agricultural land and orchards often attack silver birch. Larvae of *Semudobia betulae* (Diptera) damage seeds of birch, and those of *Argyresthia goedartella* (Lepidoptera) attack the male flowers.

Many pests attack the leaves and shoots. These include the leaf miners *Lyonetia clerkella*, *Phylloporia bistrigella*, *Agromyza alnibetulae* and *Scolineura betulae*. Birches infested by leaf miners are weakened and become highly susceptible to secondary invaders. *Cimbex femorata* (Hymenoptera), *Byctiscus betulae* (Coleoptera), *Deporaus betulae* (Coleoptera), *Ocneria dispar*, *Amphidasis betularia*, and others seriously damage the leaves. *Melolontha melolontha* and *Melontha hippocastani* feed on leaves of birch species, and their white grubs cause damage to the roots of seedlings. *Polyphylla fullo*, *Amphimallon solstitialis*, *Euxoa segetum* and *Gryllotalpa gryllotalpa* also attack the roots.

Infestation by the larvae of the birch cambium borer, *Phytobia betulae*, which tunnel through the wood, can considerably reduce the value for veneer due to discolouration which may run for many metres through the wood.

Viruses, bacteria, and Acarina also cause some of the less important diseases of *B. pendula*.

Uses

B. pendula is tolerant of a wide range of environmental conditions and is therefore used for revegetation, land reclamation and erosion control. The species can grow on mine spoil or bare rock piles. It can be used in afforestation of former mining areas and farmland to prepare the soil for other broadleaved species and conifers.

B. pendula is planted not only for its many wood uses but also for its ornamental value. Its straight silvery-white trunk and pendulous branches make it one of the most decorative trees in Europe. There are a number of attractive birch varieties in cultivation, and these are often used as street and park trees and also as specimen trees in gardens.

The diffuse porous wood of *B. pendula* is light in colour, smooth and fine grained with a density of 670 kg/m^3 (at 15% moisture). It has a number of uses including pulp (for paper), plywood, veneer, timber, particleboard, fibreboard, furniture and firewood. The wood of *B. pubescens* is used in industry in a similar way to that of *B. pendula*. However, its chemical and mechanical properties are somewhat different and it is less suitable for the production of plywood.

Birch is also used, to a lesser extent, in the production of tool handles, barrels, toys and musical instruments. The highly figured wood of Karelian or curly birch (*B. pendula* var. *carelica*) and of birch burls is used for turnery and carvings, for souvenirs, caskets, sculptures and plates, and also for wall panelling.

Birch also makes a very good fuelwood due to the high levels of fatty acids both in the bark and wood.

Non-wood products of birch include a kind of tar from the bark which has antiseptic properties; birch sap that is used to produce beverages; and tannins, which are extracted to tan hides.

The medicinal properties of birch have been noted since at least the medieval period in Europe, and it has been well known as a diuretic, anti-rheumatic, anti-fever purgative and wound-healing agent. A tea is also made from the leaves or a decoction of the buds. *B. pendula* used to be considered a holy tree, revered by Celtic and Germanic tribes, and was considered to have sacred powers of renewal and purification.

Further reading

Cameron, A.D. (1996) Managing birch woodlands for the production of quality timber. *Forestry (Oxford)* 69, 357–371.

Hynynen, J., Niemistö, P., Viherä-Aarnio, A., Brunner, A., Hein, S. and Velling, P. (2010) Silviculture of birch (*Betula pendula* Roth and *Betula pubescens* Ehrh.) in northern Europe. *Forestry (Oxford)* 83, 103–119.

Lilja, A., Poteri, M., Petäistö, R.-L., Rikala, R., Kurkela, T. and Kasanen, R. (2010) Fungal diseases in forest nurseries in Finland. *Silva Fennica* 44, 525–545.

Bischofia javanica Blume

Bishop wood

Importance

Bischofia javanica is a fast-growing, fairly large, deciduous or evergreen tree, found scattered in primary or old secondary dry and deciduous forest or monsoon forest, and sometimes in open, savannah woodland tracts. The wood is not always suitable for sawing, but can be used for the manufacture of plywood and composite board, and is a source of long-fibre pulp. It is a potential tree-plantation species, and an excellent shade tree in coffee and cardamom plantations and in other agroforestry situations.

Various parts of the tree are widely used in the Pacific region for traditional herbal medicine, and a red-brown dye prepared from the bark is used for dyeing tapa cloth in Polynesia.

A major disadvantage of the species is that it is susceptible to top and twig borers on less favourable sites, which can cause plantations to fail. Research is needed to find a provenance resistant to insect attack that can also produce a high yield. There is also concern that the species has the potential to invade native plant communities when planted outside its natural range.

Botanical features

B. javanica is a medium-sized to large tree, typically 20–35 m tall (maximally 50 m), with a straight or poorly shaped bole up to 80 cm diameter (exceptionally up to 170 cm). It sometimes has buttresses up to 3 m high. The outer bark is dark reddish-brown, fibrous and flaky; the inner bark is reddish-brown and exudes a red sap. The crown is dense, rounded and far-spreading. *B. javanica* is usually deciduous in habit. Its root growth is rapid; the taproot is tuberous for a short length and tapers quickly. Thin but long secondary roots are present below the collar.

The leaves are trifoliolate, alternately arranged, with a rachis 9–10 cm long. The leaflets are ovate, 9–13 cm long, 4.5–6.5 cm wide, with an acuminate apex, rounded base and crenate margins; they are leathery in texture with glabrous upper and lower faces. They have 6–8 lateral nerves, occasional intermediate veins, and tiny domatia in the axils of lateral veins and midrib; tertiary veins are very faint. The petiole of the central leaflet is 30–35 mm, while those of the side leaflets are only 6–12 mm.

The inflorescence is paniculate, arising from the upper leaf axils, equal to or shorter than the foliage, and greenish. The numerous flowers are small, dioecious and apetalous. The male flowers are scattered or clustered; their five concave, imbricate sepals conceal the five anthers. There is no disc, and the rudimentary ovary is short and broad. Female flowers have caducous sepals, five small or absent staminodes, a 3- or 4-celled ovary with two ovules in each cell. The fruits are globose, fleshy, brown or reddish, 10–15 mm in diameter, with 3–4 cells; the 3–4 seeds (oblong, 5 mm long) are embedded in a colourless, edible pulp.

Leaves are shed during the summer months, and flowering takes place after a new set of leaves is produced. *B. javanica* flowers annually from the age of 8 years onwards. In west Java, flowering usually takes place from August to November (exceptionally also December) and fruits mature from February to June (exceptionally from January), with a peak in March. In central Java, fruiting is from May to November, and in east Java it lasts from November to December.

In India, the tree will remain leafless for a short period over winter, particularly in areas with low rainfall. In the Ganges basin, bright green leaves appear in February–March, followed by minute flowers in pendulous clusters. The fruits are evident during colder weather, and ripen by the following February. The fruits ripen in Karnataka in September–October, in West Bengal during December–January, and in Assam in February–March.

Distribution

B. javanica has a vast natural range in the Asia–Pacific region, where it is found scattered in primary or mature secondary forests from India and the Himalayas in the west; China, Taiwan, southern Japan and Indo-China, Thailand and throughout the Malesian area (although rare in Peninsular Malaya and large parts of Borneo); to north-eastern Australia and the Pacific east to Samoa and Tonga.

It may have been introduced in ancient times to the Cook and Society Islands in the South Pacific. It is also found in dry, open parang formation areas in many parts of the Philippines.

B. javanica is planted locally as a timber-plantation species, and as an ornamental tree within the natural area of distribution. It has been introduced as a fast-growing ornamental or shade tree in East Africa and also in the USA.

Silviculture characteristics

B. javanica can tolerate moderate shade, and ordinary frosts, but not drought. It tends to become evergreen in moist localities and deciduous in dry ones.

Seedlings can grow through the weeds characteristic of moist locations. The species has the ability to self-prune once the canopy has closed (which, at a spacing of 2 × 3 m, will be 5 years after planting). Pruning wounds heal very well, and the tree may survive girdling, involving the removal of a strip of bark 30 cm wide.

The tree may develop forks as a result of attacks of top and twig-boring insects. Roots spread out superficially, and the tree also sprouts vigorously following coppicing.

Silviculture practice

B. javanica can be propagated by seed, wildings and stem cuttings. Seeds may be collected in large quantities but should not be collected from early or late crops. Seed rapidly loses viability after 3 months, although it can be stored for up to 6 months. One kilogram of fruit yields about 37 g of clean seed.

Seeds may be sown under shade or in full sunlight, provided adequate water is supplied. Germination starts 1–3 weeks after sowing, and after 5–6 weeks, about 70% of the seed will have germinated. Young plants need plenty of water, so direct sowing in the field is not appropriate.

Bare-rooted seedlings aged 1 or 2 years can be transplanted quite successfully. Stump plants should be robust, at least 2.5 cm in diameter, with a recommended shoot length of 20 cm and root length of 30 cm; these dimensions should ensure a survival rate close to 100%. Stumps can be stored for over 3 weeks if kept moist. If dry, they only remain viable for 3–4 days. Air layering, and root and shoot cuttings have all been successfully used to propagate the species in India.

In Java, *B. javanica* has been planted in pure stands at spacings of 2 × 3 m, and in mixed plantations in alternating rows with *Calophyllum inophyllum* and *Bombax ceiba* at 1 × 3 m spacing, and with *Acacia mearnsii* at 2.5 × 5 m spacing.

In India, *B. javanica* is often planted in mixtures as a subsidiary species, with light-canopied species such as *Trewia nudiflora*, in order to keep weeds and climbers down. It is removed when it begins to interfere with the principal species.

In Java, an 8-year-old pure plantation of *B. javanica* planted on a moderately fertile soil and with a spacing of 2 × 3 m yielded 12 m^3/ha of clear-bole wood. A mean annual increment of 13–22 m^3/ha at age 7 years has been reported.

Diseases and pests

In Indonesia, fungi reported to attack the trees are *Corticium salmonicolor* (*Phanerochaete salmonicolor*) and *Glomerella cingulata* (conidial state: *Colletotrichum gloeosporioides*).

In southern China, where *B. javanica* is used for afforestation and ornamental planting, the species has suffered severely from witches' broom, caused by a mycoplasma-like organism.

In Java, young trees of *B. javanica* are heavily attacked by top and twig borers, causing failure of plantations in less suitable locations. On favourable sites, trees can grow rapidly and survive such pest attacks. Caterpillars of *Metanastria hyrtaca* and *Selepa celtis* are found feeding on its foliage.

Uses

In India, *B. javanica* is considered to be an excellent shade tree in coffee, tea and cardamom plantations. It is also used as a shade tree in the Philippines. It is recommended for use in agroforestry and social forestry plantations.

B. javanica wood is of medium texture, with an interlocking grain and indistinct growth rings, and vessels scarcely visible to the naked eye. Density varies from 520 to 1010 kg/m^3 at 15% moisture content. The sapwood is light, cream or reddish-brown, whereas the heartwood is red to dark red-brown. The wood is moderately heavy, soft, and liable to warp and crack. It is durable, however, particularly under water.

It is used for many general construction purposes (beams, posts, bridges, decking, sleepers, mine timber) as well as for joinery, and also for agricultural implements, carving, pencils and billiard cue butts. It is also used for fuelwood (calorific value 5142 (heartwood) to 5316 kcal/kg), and to produce charcoal. It is a potential source of long fibres for pulp and paper production. It is also suitable for plywood and veneer production.

In Polynesia, a red dye is extracted from the bark, which also contains tannins, used for toughening nets and ropes. A leaf astringent, used for toothache and eye diseases, contains 18% crude protein. The seed yields a drying oil, useful for surface-coating products.

Further reading

Airy Shaw, H.K. (1967) Notes on the genus *Bischofia* Bl. (Bischofiaceae). *Kew Bulletin* 21, 327–329.

Lemmens, R.H.M.J., Soerianegara, I. and Wong, W.C. (eds) (1995) *Timber Trees: Minor Commercial Timbers (Plant Resources of South-east Asia 5)*. Backhuys, Leiden, The Netherlands.

Tanimoto, T. and Toyoda, T. (1996) Survivorship and growth of akagi (*Bischofia javanica* Bl.) seedlings under the forest canopy and different temperature conditions. *Bulletin of the Forestry and Forest Products Research Institute, Ibaraki* 370, 1–19.

Yamashita, N. (2008) Physiological acclimation and plasticity of an invasive tree species, *Bischofia javanica*, in the subtropical pacific Bonin Islands. *Research Bulletin of the Hokkaido University Forests* 65, 19–59.

Byrsonima crassifolia (L.) Kunth

Wild cherry

Importance

Byrsonima crassifolia is a fast-growing tree that thrives in bright sunlight and produces edible fruits, which are consumed fresh or may be processed. The wood is a good source of charcoal and is used for carpentry and construction. The tree is especially used for the recuperation of degraded areas and in agroforestry systems, because of its rapid growth and its fruits, which are attractive to wild animals.

Botanical features

B. crassifolia is typically a small tree around 2–5 m high, and also occurs less frequently as a shrub. It is commonly sclerophyllous, with a deep and intensive root system, an open crown and a tortuous trunk and branches.

The main trunk is cylindrical, usually 30–40 cm dbh, with grey, rugose bark. The leaves are simple, with a short (1 cm) petiole; they are borne oppositely, and are concentrated at the tips of the branches. Each leaf is 7–15 cm long, 3–7 cm wide, elliptical, ovate or obovate and leathery in texture. The upper leaf surface is smooth, but the lower surface is covered with greyish or reddish hairs.

The flowers are borne in racemes up to 12 cm long and 1.5–4 cm wide. They are initially yellow, becoming reddish as they mature. Each flower is about 1.5 cm in diameter, with two elongated glands in each sepal, and a corolla formed of 5 free petals. The flowers produce abundant lipids instead of nectar. The fruit is a globose, slightly acid, fleshy yellow drupe of 1.5–2 cm diameter, which usually contains a single viable seed.

B. crassifolia is evergreen and perennial, and produces flowers throughout the year, although there is a maximum flowering period from October to December. Fruiting occurs between October and March.

Distribution

B. crassifolia is a widespread species in the Neotropics, occurring naturally from Central America to Paraguay. It is found in almost all Amazonian regions and is one of the most common trees in Venezuelan savannahs, and is also frequent in coastal plant communities (dunes and coastal plain forests, or restingas) and caatingas in north-eastern Brazil. It is cultivated in southern California and along the western coast of Mexico.

Silviculture characteristics

B. crassifolia is a fire-resistant tree, occurring where burning is frequently undertaken towards the end of the dry season. The tree is considered to have an intermediate to high resistance to drought. Although it is tolerant of infertile soils, experiments have shown that *B. crassifolia* has rapid growth in plantations that are established on rich organic matter, on loose soils and in weed-free areas, which can result in fruiting within 1 year of planting.

Silviculture practice

Despite its great potential, *B. crassifolia* is of limited interest for use in plantations, and its edible fruit is the main reason for the importance of the species. *B. crassifolia* is propagated by seeds collected from mature fruits, and vegetative propagation has not been reported. The species can compete well against the natural vegetation where it occurs, but it has shown improved performance under cultivation techniques that use increased organic matter and weeding.

B. crassifolia is presently grown on a limited commercial scale for the production of fruit, and in some cases its importance is restricted to a subsistence level within its native range. However, the species is considered to have good potential for increased economic importance through cultivation, as observed in Veracruz, Mexico, where cultivated fields of *B. crassifolia* cover 236 ha.

Diseases and pests

A fungus of the genus *Meliola* caused a leaf spot known as 'fumagina' in Mexico and Brazil. Trees grown in glasshouses can also be subject to attack from scale insects and spider mites.

B. crassifolia is largely free from serious pests, although in Brazil various parts of the tree are attacked by a number of beetles. These include *Macraspis festiva*, which eats the fruits, and *Costalimaita ferruginea*, *Oncideres dejeani* and *Orthezia insignis*, which all make incisions into the branches.

Uses

Phytosociological studies have shown that *B. crassifolia* is one of the most common species in savannahs, and that it is important silviculturally and ecologically. As with other savannah species, *B. crassifolia* naturally builds the heterogeneous vegetational structure needed for sustained ecological development. The species can be used for live fences and has been used successfully to restore the agricultural productivity of old fields in tree-based agricultural systems. It can also be cultivated in association with other trees; for example, mixed plantations of *Spondias* sp. and *B. crassifolia* have been observed in Mexico. *B. crassifolia* restores the regenerative capacity of the forest, since the tree grows rapidly in full sun and produces fleshy fruits – which attract seed-carrying animals – soon after planting. Chemical studies conducted in Venezuelan savannahs that included *B. crassifolia* have shown it induces a significant accumulation of organic matter in the soil.

The wood of *B. crassifolia* is not used very intensively, being essentially limited to light construction and carpentry. It is more generally used as fuelwood, and also provides good-quality charcoal. Wood density is 520–630 kg/m^3.

B. crassifolia is a highly prized species due to its edible, slightly acid fruits, which are consumed fresh or processed into jam, alcoholic drinks, ice cream, juice and other products. It is an important human food resource in northern and north-eastern Brazil, being highly nutritious, and a good source of both vitamin C and fats. The fruits can be conserved for up to 15 days in water and up to 40 days with sugar. The bark contains 20% tannins and is used to tan hides. It is also considered to have medicinal properties, being widely used in folk medicine to treat coughs, fungal diseases of the skin, diarrhoea, fever, indigestion, tuberculosis and snakebites, and to promote menstruation. Extracts of the bark show notable activity against a range of pathogens, including fungi, several enterobacteria and the Chagas disease parasite *Trypanosoma cruzi*.

Further reading

Cavalcante, P.B. (1976) *Frutas comestíveis da Amazônia*. Museu Paraense Emilio Goeldi, Belém, Brazil.

Duarte, O. and Vernon, R. (2002) Floral and reproductive biology of nance [*Byrsonima crassifolia* (L.) H.B.K.]. *Proceedings of the Inter-American Society for Tropical Horticulture*, Tegucigalpa, Honduras, pp. 41–42.

Moreno, E.M., Torres, T.C., García, E.A., González, A.M.C., Terrazas, T., León, M.T.B.C., de La Cruz Lázaro, E. and Torres, R.M. (2010) Caracterización morfológica de hojas de nanche (*Byrsonima crassifolia* (L.) H. B. K.). *Revista Fitotecnia Mexicana* 33, 15–19.

C

Calocedrus decurrens (Torr.) Florin

Incense cedar

Importance

Calocedrus decurrens forms a distinctive component of mixed conifer forests of the Sierra Nevada and Cascade Ranges of the western USA. These large trees reach 50 m in height and 3 m or more in diameter. The durability and resistance to decay of lumber from *C. decurrens* makes it an ideal wood for exterior use. Most of the top-grade lumber is used for exterior siding and manufacture of pencils. Due to its value as a lumber tree, old-growth trees of *C. decurrens* have been extensively logged over much of its range outside of National Parks and other protected areas.

Botanical features

C. decurrens is a large tree with pyramidal canopy form, becoming more columnar with age, typically reaching 20–50 m in height (occasionally 70 m). Diameters of mature trees are commonly 1–2.5 m, occasionally 3.5 m. Growth form is regular, with the primary bole gradually tapering with height. Large trees show some degree of buttressing at the base.

The bark is brick-red in colour and fibrous, forming a thick protective layer around the trunk. Root systems, which are spreading and relatively extensive compared with many associated conifer species, begin growth slowly in the first year but extend rapidly in older seedlings.

The foliage of *C. decurrens* consists of elongate, flattened scale leaves, arranged in four rows surrounding and adherent to the branchlets. The foliage is aromatic when crushed.

Staminate cones are 5–7 mm in length and light yellow. Ovulate catkins are borne singly at the tip of branchlets; the female cones are red-brown and 18–25 mm in length, with two seed-bearing scales separated by a sterile scale. Seeds are 8–12 mm in length, with two unequal wings. Embryos have two cotyledons.

Radial growth commonly begins in April with stem elongation, followed around June by vegetative branch growth. Unlike typical temperate conifers, *C. decurrens* lacks a distinctive spring flush of foliage growth. Successive years of stem growth are not readily separable along the branchlets. Male cones shed pollen from late winter to early spring, and female cones mature in late summer.

Distribution

In North America, *C. decurrens* is found as a distinctive component of mixed conifer forests from the Cascade Range of Oregon through the Siskiyou north coast ranges, Sierra Nevada and Transverse Ranges of California to the Sierra San Pedro Martir in Baja California, Mexico. It barely crosses into Nevada in the Washoe Range east of the Sierra Nevada. Over this range it may be found from around 50 m to nearly 3000 m altitude. *C. decurrens* reaches its best development in the Sierra Nevada at 600–2100 m altitude in the lower and drier portions of the mixed conifer forest.

C. decurrens is a characteristic co-dominant to subdominant species in drier associations of mixed conifer forest over its range. Common associated species are *Abies concolor*, *Abies grandis*, *Pinus ponderosa*, *Pinus jeffreyi*, *Pinus lambertiana* and *Pseudotsuga menziesii*.

It has been widely planted as a specimen and ornamental tree in Europe, and has been found to be very tolerant of low winter temperatures in southern Russia and the Ukraine.

Silviculture characteristics

C. decurrens is a long-lived conifer, reaching more than 500 years in age. It is relatively tolerant of drought and infertile soils in comparison with most associated conifer species. Only rarely is *C. decurrens* a stand dominant, and generally it accounts for about 5–10% of stand volume. In stands of moderate tree density, *C. decurrens* typically grows more slowly than associated species, but on open sites with poor growing conditions it often exceeds most other conifers in rates of growth. The size of mature trees varies greatly over its range, depending on climatic and local growing conditions. In the Sierra Nevada, old-growth *C. decurrens* frequently reaches heights of 45 m and diameters of 2 m or more. In southern California and Baja California under greater seasonal drought, mature trees commonly reach only 18–24 m in height and 90–120 cm diameter. Individual trees at high altitudes or on dry, exposed sites are often small and shrubby in growth form. Seedlings are strongly shade-tolerant, but saplings require higher light exposure for growth. In this respect, *C. decurrens* is more shade-tolerant than associated *Pinus* species and *P. menziesii*, but less shade-tolerant than *A. concolor* and *A. grandis*. Although *C. decurrens* is a prolific seed producer, medium to heavy seed crops are borne at intervals of 3–6 years, with light or sometimes no seed crop in intervening years.

Silviculture practice

C. decurrens germinates with 20–40% success with no treatment, indicating that at least a proportion of seeds lack dormancy. Germination rates can be doubled by stratification at 35–40°C for 2–3 months. In good seed years, production of as many as 961,500 seeds/ha has been reported. Fresh seed loses its viability rapidly and needs to be collected annually. Under natural conditions, wind dispersal allows seed to be carried great distances. This high seed production, good dispersal and high germination make *C. decurrens* an effective colonizer of open forest sites after harvesting.

C. decurrens does not propagate vegetatively under natural conditions, but cuttings can be induced to root under greenhouse conditions. Tissue culture is also possible. For silvicultural applications, autumn sowing of seeds is recommended instead of vegetative propagation, with inoculation with mycorrhizae effectively increasing early seedling growth.

Well-drained soils in the open provide the best seedbeds. Seed drills have shown good success, but broadcast sowing may also be used. If seeds are sown in spring, light mulching is recommended to reduce drought stress, as seedlings are particularly sensitive in their first year of growth.

Diseases and pests

The most significant pathogen impacting *C. decurrens* is pocket dry rot (*Tyromyces amarus*), which may infect as many as 75–100% of mature trees in some parts of the Sierra Nevada. This dry rot enters trees through fire scars or other open wounds. Once it invades woody tissues, it appears to be remarkably resistant to the heartwood extractives that are toxic to most other heart rot fungi. Under good growing conditions, pocket dry rot only becomes a significant problem in trees older than about 220 years, or 165 years in suppressed trees.

Root disease is the greatest cause of death for *C. decurrens*, with the most destructive root rot species being *Heterobasidion annosum*, although *Armillaria* sp. and *Inonotus weirii* are also involved. Foliage diseases are generally of only minor significance. *Gymnosporangium libocedri* may infect trees of all ages and cause witches' brooms, but does not cause death. However, it may reduce rates of growth and cause infections in the main stem that may reduce lumber value. Mistletoe infects *C. decurrens* throughout its range and may suppress growth rates.

Few insects cause major problems for *C. decurrens*. Cone sawfly (*Augomonoctenus libocedrii*) sometimes infects cones and significantly reduces seed production. Incense cedar tip moth (*Argyresthia libocedrella*) feeds on cones through the summer and may virtually destroy an entire crop. Juniper scale (*Carulaspis juniperi*), an introduced European species, is distributed throughout the range of *C. decurrens* and attacks young growing tissues and turns foliage yellow. Only in rare cases does it cause tree mortality. Although several species of bark beetle may be found in association with *C. decurrens*, these are usually inconsequential as a threat, except to trees in very poor health.

Fire plays a significant role in the health and abundance of *C. decurrens*. Saplings are readily killed by fire, but the thick bark of mature trees gives them strong protection against the heat of ground fires. Intense fires that open up fire scars at the base of mature trees serve to promote infections of pocket dry rot. In comparison with associated pines, *C. decurrens* is relatively tolerant of oxidant air pollution that is a chronic stress in mixed conifer forests of the southern Sierra Nevada and Transverse Ranges of southern California.

Uses

C. decurrens is often favoured in reforestation of mixed conifer stands after harvesting because it is a prolific natural seeder producing seedlings and saplings that are sufficiently shade-tolerant to persist for long periods in the understorey. Once established, *C. decurrens* is highly drought-resistant, and its relative tolerance of air pollution increases its value in plantings in southern California. *C. decurrens* is widely planted as an ornamental in Europe and the eastern USA, as well as in its natural range.

Wood specific gravity is 0.35 (green wood) and 0.37 (at 12% moisture content). *C. decurrens* is valuable for exterior construction use because of its durability and resistance to decay when exposed to moisture. In addition, it is dimensionally stable and holds paint well, making it relatively low in maintenance. It is widely used for exterior panelling, window sashes, mud sills, sheathing under stucco or brick veneer, fencing, trellises and poles set into the ground. The wood's rich colour, pleasant fragrance and hard knots make it also useful for interior panelling. Low-quality pecky cedar (sawn timber from trees infected with pocket dry rot) is widely used for fencing and interior panelling. Because of its straight grain, workability and dimensional stability, *C. decurrens* is preferable to other species, including hardwoods, for the manufacture of pencils.

Further reading

Eckenwalder, J.E. (2009) *Conifers of the World: The Complete Reference*. Timber Press, Portland, Oregon, USA, 720 pp.
McDonald, P.M. (1973) Incense-cedar (an American wood). USDA - Forest Service No. FS-226, Washington, DC.
Schubert, G.H. (1957) Silvical characteristics of Incense-Cedar. Technical Papers. *California Forest and Range Experiment Station* 14, 18 pp.

Calophyllum brasiliense Cambess.

Santa Maria

Importance

Calophyllum brasiliense is a medium or large tree that occurs from Mexico to south-eastern Brazil, although its distribution is discontinuous. The wood is medium in weight, easy to work, and durable; it has been used for the construction of boats since the 17th century. The wood of *C. brasiliense* is one of the most used of all species of Neotropical timber. It is used for construction beams, interior mouldings, cabinetmaking, decorative features, parquet, musical instruments and general carpentry. It is amenable to plantation forestry, reforestation, ornamental plantings, agroforestry and silvopastoral systems. The species also produces an abundance of fruit, which makes it an important food source for wildlife.

Botanical features

C. brasiliense is a medium to large evergreen tree with a dense, rounded crown. The height of mature trees typically averages

20–30 m, but can reach up to 40 m. The dbh averages 40–60 cm, but diameters of up to 130 cm have been recorded. The trunk is straight and cylindrical to conical, without buttresses, and the trunk ranges from 8 to 12 m tall. The branches are thick and twisted.

The external bark is dark maroon to grey. The trunk is deeply fissured longitudinally in some individuals. The external bark is lacking in spines and has no characteristic odour. The thickness of the external bark ranges from 5 to 12 mm. The fibrous, internal bark is pink to white, bitter to the taste, and ranges from 10 to 20 mm thick. An orange-coloured substance exudes from any wounds to the trunk or branches.

The simple, opposite leaves are dark green, smooth and lustrous, and are elliptical or oblong to ovate in outline. They are typically around 10–15 cm long and 2.5–5 cm wide, and are borne on the end of short petioles, each around 6–19 mm long. The margins of the leaves are entire and the leaves' undersides are light green. The central vein is prominent, and is flanked by multiple parallel veins.

C. brasiliense is polygamodioecious – some individuals have a mixture of bisexual and female flowers, while others have a mixture of bisexual and male flowers. The small, cream-coloured flowers are displayed in axillary panicles, which are 4–8 cm long. On male flowers, there are two ovate sepals, three ovate petals with rounded apices and numerous stamens. Female flowers are similar to male flowers, and have one style and a superior ovary. Flowering usually occurs at the beginning of the rainy season.

Fruiting occurs from the middle to the end of the rainy season. The fruits are greenish-yellow drupes, 25–30 mm in diameter and ovoid to spherical in shape. The endocarp is tough and the pulp is resinous with a fragrant odour. Each fruit contains a globose seed, which is 10–15 mm in diameter. The fruits are dispersed either by water or by animals; bats are especially important seed dispersers.

Distribution

C. brasiliense is found naturally from central Mexico and the Caribbean region south to lowland parts of Bolivia and Brazil. It is present in the majority of humid forests throughout its range. *C. brasiliense* can also be found occasionally in successional forests.

It was introduced into the Hawaiian islands during the 1950s for reforestation, ornamental planting and timber production. It has also been introduced as an ornamental tree throughout the Caribbean, where it is used for shelter belts and coffee shade.

Silviculture characteristics

C. brasiliense is tolerant to shade, and a relative abundance of advance regeneration can be found in shaded forest understoreys. It is capable of rapid growth, especially when growing in full sun. Diameter and height growth were among the best of all species in natural stands in one Brazilian study. *C. brasiliense* is often considered an indicator of mature or late-successional forest. Plantations using direct seeding are often successful, and germination rates can approach 100%.

Silviculture practice

C. brasiliense seeds can germinate – and seedlings develop – in full shade, but rates of initial growth are best when grown in only partial shade. Shade should be retained over seedlings until they reach a height of 30–40 cm. For plantations, direct seeding without scarification can be effective. The germination capacity of seeds is almost always greater than 50% and is usually 100%; germination usually begins 40–50 days after planting. In order to maintain viability, seeds should not be stored for more than 90 days. Regeneration by resprouting is not an important establishment mechanism.

The species was considered one of the most promising of 12 species tested for plantation potential in Costa Rica, with 80% survival after 4 years and height growth of >6 m. In natural management systems, the species may adapt well to single-tree selection because of its shade tolerance. However, in one study in Bolivia, regeneration did not vary under differing harvest intensities, perhaps indicating adaptability to a broad range of silvicultural systems.

Financial returns for selective cutting in natural forest on a 50-year rotation of *C. brasiliense* and other very valuable timber species were considered highly profitable in a study in Bolivia. The annual biomass increment of a 41-year-old plantation in Puerto Rico was 4 t/ha.

Diseases and pests

There are few significant pests and diseases of *C. brasiliense*. Among those reported are a wilt caused by a species of *Cephalosporium* which has rapidly killed trees in Central America, and a number of other fungal pathogens.

The wood of *C. brasiliense* is fairly resistant to decay, but is not resistant to termites or marine borers.

Uses

C. brasiliense has a high potential for use in windbreaks and is also used in other agroforestry systems as shade for coffee, cacao, palm hearts and 'Rheedia' (*Garcinia* spp.). It has been used to stabilize soils and to relieve soil compaction in degraded pastures. Its evergreen habit makes it suitable for use in ornamental and shade tree plantings.

The sapwood of *C. brasiliense* is pinkish and paler than the heartwood; there is a transition to the pink, reddish or purple heartwood, with some sparse dark striations. The wood is moderately dense (in the range 550–720 kg/m^3) and is durable even under conditions of high humidity. The wood is easy to saw and work, and is one of the most used woods in the tropics of the Americas, with uses ranging from veneer and musical instruments to fuelwood.

The heartwood is highly resistant to fungi and moderately resistant to termites. A high content of xanthones provides resistance against wood-rotting fungi. The wood produces a resin that reduces inflammation and is also used as an astringent for skin conditions. Oil from its seeds is used as lamp oil, and the fruits are used to feed livestock.

Further reading

Devall, M.S. and O'Rourke, T.M. (1998) *Calophyllum brasiliense* Camb. In: Burns, R.M., Mosquera, M. and Whitmore, J.L. (eds) *Useful Trees of the Tropical Region of North America*. North American Forestry Commission, Washington, DC, pp. 53–62.

Lorenzi, H. (1992) *Árvores Brasileiras: Manual de Identificaçao e Cultivo de Plantas Arbóreas Nativas do Brasil*. Editora Plantarum, Nova Odessa, São Paulo, Brazil.

Oliveira, V. de and Joly, C.A. (2010) Flooding tolerance of *Calophyllum brasiliense* Camb. (Clusiaceae): morphological, physiological and growth responses. *Trees – Structure and Function* 24, 185–193.

Calophyllum inophyllum L.

Alexandrian laurel

Importance

Calophyllum inophyllum is a medium-sized tree which produces a decorative timber used for cabinet work, parquet flooring, construction, musical instruments, veneers, plywood, canoes and boats. The seeds are edible, and an oil from the seeds is used for illumination, or, when mixed with other plant oils, for medicinal use. The latex and bark also have medicinal properties. *C. inophyllum* is often planted as an ornamental, avenue or shade tree, or on sandy beaches for dune stabilization. It is also used in afforestation and reforestation projects. *C. inophyllum* is mainly found in coastal locations around the Indo-West Pacific, from East Africa, including Mauritius and Madagascar, through India, Sri Lanka and Malesia to northern Australia and Oceania. It has also been planted in tropical America.

One disadvantage of this species is that its wood is difficult to saw, and the sawn surfaces tend to be woolly. Research into the handling of the timber after harvesting, as well as into drying, wood preservative treatments and wood mechanical properties, is needed in order to enhance the usability of the timber.

Botanical features

C. inophyllum is a medium-sized tree (up to 25 m, or sometimes 35 m, tall) with a spreading crown of large branches, usually with a twisted or leaning bole up to 150 cm in diameter and lacking buttresses. The twigs are four-angled or rounded, with a plump terminal bud, which is 4–9 mm long.

The leaves are simple, opposite, without stipules, shiny and glabrous, with very numerous parallel nerves at right angles to the midrib. The leaves are oblong, elliptic or obovate–oblong in outline, obtuse or emarginate at the tip, and acute at the base, 8–20 cm long × 6–9 cm wide. The petiole is broad and around 2 cm long.

The inflorescences are usually unbranched but occasionally with 3-flowered branches, and are borne in the axils of leaves. The showy, fragrant flowers each have 8 (–13) tepals, and are white with yellow stamens. The fruit is spherical to obovoid, smooth, glabrous, 3–4 cm in diameter and 25–50 mm long; it has a fairly thin, compact outer layer, which is greyish-green and pulpy on the outside. The fruits may be carried by ocean currents, but are usually dispersed by bats.

C. inophyllum can flower throughout the year, although flowering generally occurs from March to August in the Philippines. For the flowers that open in May–June, fruits will mature in August. In the Indian states of Tamil Nadu and Mysore, flowers usually appear in the cold season and fruits ripen in March. In Kerala, flowers appear in March–April and fruits ripen in May–June, although both flowers and fruits can be found at other times of the year. In Orissa, there are two seasons, with flowering during May–June and October–November. In the Andaman Islands, the tree will flower profusely during the rainy season and, to a lesser extent, at other times of the year, with fruiting from June to August.

Distribution

C. inophyllum is found in East Africa, Madagascar, the islands of the Indian Ocean, coastal India, Sri Lanka, Myanmar, throughout South-east Asia, to Taiwan, the Ryukyu Islands (Japan), to northern Australia and islands in the Pacific. It is thought to be indigenous to India, Malaysia, Indonesia and the Philippines.

C. inophyllum is found in coastal forests and flourishes on sandy or xeric soils. It tolerates salt winds and spray, exposed positions, and sites where the soil water may be slightly brackish. It is also found along river margins further inland.

Silviculture characteristics

C. inophyllum is not a light-demanding species, and can survive some shade. It grows rapidly on beaches or on sandy soils, and can tolerate xeric conditions when the water table is high, but cannot withstand waterlogging, fire or frost; it is resistant to strong winds and salt spray. It regenerates rapidly, and natural regeneration is usually good; it also coppices moderately well.

Silviculture practice

A selective cutting system and removal of undesirable trees can enhance the natural regeneration of *C. inophyllum*.

C. inophyllum produces seeds throughout the year; in India, seeds are usually collected in March. Seeds should be collected only from mature fruits. Only one seed is produced in each fruit. Seeds remain viable for more than 1 year, but the thick seedcoat can delay germination. Complete removal of the seedcoat has been shown to be very effective in accelerating germination and increasing the germination percentage.

Seedlings do not require much care in nurseries beyond the provision of adequate shade and water. Nursery plants can be prepared either as bare-rooted stock or as container plants. Seedlings can be transplanted when they are about 50 cm tall. Direct sowing can also be carried out, provided the seed is fresh. Once planted out, the seedlings should be weeded at least twice a year.

If direct sowing is used to establish a plantation, the spacing should be 2 × 2 m (in Indonesia, trial plantations usually use a spacing of 2 × 3 m). If, however, planting stock is used, plants should be spaced in rows 20 cm apart. *C. inophyllum* can also be propagated from wildings, which may be abundant beneath the canopy of mother trees. The wildings, having hard, thick leaves, have a high survival rate.

The growth of *C. inophyllum* is generally fairly slow. In Orissa, India, a height of 1.2–1.8 m is reported for irrigated trees 6–7 years old. Data from a few sample plots in Peninsular Malaysia indicated that trees may take about 70 years to attain a diameter of 50 cm.

Diseases and pests

In the Seychelles, *C. inophyllum* has recently started to succumb to a vascular wilt disease associated with the fungus *Verticillium calophylli*. A leaf wilt disease caused by *Pestalotia* sp. was recorded damaging seedlings in forest nurseries in Bangladesh.

No pests cause serious damage to *C. inophyllum* in the Philippines. The psyllid *Leptynoptera sulfurea* has been recorded as a pest of *C. inophyllum* in Indonesia, and in the Mariana and Caroline Islands. In Orissa, an insect known locally as 'sambalua' has been reported to cause damage to coastal plantations. Laboratory tests of tree species grown in Hawaii have indicated that the timber may be resistant to termite (*Coptotermes formosanus*) attack.

Uses

C. inophyllum is planted as an ornamental, shade and avenue tree throughout its range, and for sand dune stabilization. It is also a good species for inclusion in windbreaks, as it has dense foliage.

The timber of *C. inophyllum* is generally slightly heavier, stronger and more durable than that of other *Calophyllum* species, and the wood is often fine textured and lustrous, with an interlocked grain and no annual growth rings. The heartwood is reddish-brown with darker streaks, and the sapwood is pale reddish-white. The wood has a density of 560–800 kg/m^3 at 15% moisture content; in Hawaii, it has been tested and found to be resistant to termite attack. It is said to be durable, particularly under water, and in Kerala (southern India), it is preferred over many other species for boat construction. The wood can be obtained in many parts of South-east Asia in fairly large quantities, and is used for many purposes, including construction, furniture and cabinet work, cartwheel hubs, vessels, musical instruments, canoes and boats.

The seeds yield a viscous, dark green oil, known in India under the names *domba*, *laurel nut*, *dillo*, *pinnay* or *poonseed oil*. It is used for lighting, soap-making and for medicinal purposes, and research is being carried out into its potential use in biodiesel. The oil contains about 10–30% resinous substances, which may make it useful as a varnish. Mixed with the resin from *Vatica indica*, the oil is used for caulking boats in India.

The latex and pounded bark are also used medicinally. A decoction of the bark and the latex of *C. inophyllum* is used internally after childbirth and against diarrhoea, and externally against skin and eye diseases and rheumatism. A bark decoction is sometimes used to toughen and dye fishing nets. The leaves contain saponins and other substances which are poisonous to fish. The seed oil has been tested as an insect pest antifeedant, for example against the brown planthopper (*Nilaparvata lugens*) and the green leafhopper (*Nephotettix virescens*), which are pests of rice in India. The fruit, although sour, is edible.

Further reading

Ong, H.C., Mahlia, T.M.I., Masjuki, H.H. and Norhasyima, R.S. (2011) Comparison of palm oil, *Jatropha curcas* and *Calophyllum inophyllum* for biodiesel: a review. *Renewable and Sustainable Energy Reviews* 15, 3501–3515.

Soerianegara, I. and Lemmens, R.H.M.J. (eds) (1993) *Timber Trees: Major Commercial Timbers*. Plant Resources of South-East Asia No. 5 (1). Pudoc Scientific Publishers, Wageningen, The Netherlands.

Troup, R.S. and Forest Research Institute and Colleges (India) (1975) *Troup's The Silviculture of Indian Trees, volume I*. Controller of Publications, Government of India, Delhi.

Campnosperma brevipetiolatum Volkens

Campnosperma

Importance

Campnosperma brevipetiolatum is a major commercial timber species in Papua New Guinea and the Solomon Islands, and forms almost pure stands in swamp forests both there and in lowland forests in Micronesia. The wood is used for structural-grade plywood, and light general-purpose applications. Approximately 9,300 ha of plantations have been established in the Solomon Islands, but planting of *C. brevipetiolatum* ceased in 1988 due to difficulties in producing quality nursery stock, which resulted in poor stem form and short bole lengths. Research is needed into silvicultural practices, thinning regimes, monitoring for pests and diseases and evaluation of plantation material, and further work is needed into growth and yield models.

Botanical features

C. brevipetiolatum develops into a large tree, typically 35–40 m in height and 100 cm in diameter above the buttresses, with the biggest trees attaining diameters of 160 cm. The clear bole length can reach 25 m, but is usually 13–20 m. The trunk commonly divides into several large, slightly tapering and sometimes leaning or bent, upwardly pointing candelabra-like limbs, each with a sub-crown. The crowns are large and more or less flat-topped. The buttresses are low and rounded, spreading along the ground some distance from the tree, and are creamy-yellow with large pale lenticels. The bark is cream, fawn or grey and usually smooth.

The foliage is diffuse, with leaves borne alternately, and arranged in tight spiral clusters near the ends of branches. Each blade is obovate–oblanceolate, leathery, glabrous, medium glossy green above and yellow-green below, up to 60 × 15 cm. They have a very short winged petiole with a distinct small lobe on each side of the base.

The species has male, female and hermaphroditic flowers on the same plant. The flowers are very small, greenish-yellow, tetramerous, and borne in large axillary panicles. They produce nectar, and the pollinators are thought to be small, generalist insect pollinators. The fruit are small, round, fleshy drupes, approximately 5 mm across, which begin green and turn dark reddish-purple when fully ripe, and contain a large stone. The seeds are dispersed by birds and possibly by bats.

C. brevipetiolatum can flower at any time; mature fruits have been collected between October and March.

Distribution

C. brevipetiolatum has a wide distribution in the Western Pacific. It occurs naturally in the Moluccas, Sulawesi, New Guinea, the Bismarck archipelago, the Solomon Islands and Micronesia. It is widely distributed in the Solomon Islands, with the Santa Cruz group forming the south-eastern limit of its range. *C. brevipetiolatum* is widespread in south-west Micronesia, including Palau and the islands of Yap, Pohnpei and Kosrae in the Federated States of Micronesia.

C. brevipetiolatum may form near-monotypic stands in swampy, lowland forests in Papua New Guinea and the Solomon Islands, and also in lowland forests in Micronesia. On alluvial sites and in swamps of the Solomon Islands, the species is often mixed with *Terminalia brassii*. In other areas, it is found in association with *Calophyllum vitiense*, *Calophyllum peekelii* and *Dillenia salomonensis*, or with *Agathis macrophylla* (Santa Cruz Islands). The species regenerates following disturbance in gaps in high forest – in the open, or in association with pioneer species such as *Macaranga* – often becoming the predominant canopy species in such secondary regrowth forests.

Silviculture characteristics

C. brevipetiolatum is strongly light-demanding, although it is considered more shade-tolerant than many other pioneer species. Plants may regenerate in small forest gaps such as those created by an occasional windfall. The species regenerates both within the forest matrix and also in highly disturbed areas such as roadsides, and at the edge of forest clearings, after logging or widespread cyclone damage.

C. brevipetiolatum seedlings occur as scattered individuals (ca 120/ha) or densely concentrated in small patches (equivalent to up to 200,000/ha) in logged forests, either where mature trees are present, or below obvious bird perches. The species cannot regenerate after coppicing.

Silviculture practice

Collecting seed from the progressively ripening crop has proven tedious. Only purple berries should be collected, and care should be taken to remove the pericarp around the seed without damaging its viability. *C. brevipetiolatum* seed appears to behave in an orthodox manner, with seeds being storable for long periods under cool storage (3°C) at 8% moisture content.

Fully developed seed has been successfully germinated when broadcast on a seed bed of coarsely sieved soil. Seed is sown on sieved topsoil and remains uncovered in a germination bed. Germination starts 5–15 days after sowing and may continue for 3 weeks. Seedlings are ready for pricking out when they have developed two or more true leaves. During the nursery phase, it is important to provide adequate shade and water. Seedlings are ready for planting out when they are 20–30 cm high or about 4 months old. They benefit from routine fertilizer applications at 1, 6 and 10 weeks, and roots should be regularly pruned once seedlings have reached 15 cm in height. Watering should be progressively reduced, and root pruning conducted in the 2 weeks' hardening period prior to planting. Small wildings can also be collected and potted, but they must be lifted from logged-over areas and planted out the same day.

C. brevipetiolatum is strongly light-demanding and benefits from full site clearance. In the Solomon Islands, the species was originally line-planted, usually at 10 × 3 m spacing, but trials showed that plants grew equally well when line-planted at close spacing, and later plantings were undertaken at 5 × 3 m spacing on cleared sites. Rigorous weeding is required, especially during the first 6 months after establishment and through to canopy closure.

Early growth is rapid in plantations (e.g. 3 m height and 4 cm dbh per year during the first 3–4 years), with plants gaining about 20 m height and 30 cm dbh per year at 11 years of age. Height growth slows considerably thereafter, and annual dbh increment reduces to about 1 cm between 11 and 23 years of age.

In the Solomon Islands, two thinnings of up to 50% (to give a final crop stocking rate of 150–250 stems/ha) appears a suitable regime for sawlog production. The first selective thinning is undertaken at canopy closure on stands with more than 500 stems/ha. A second thinning is carried out, again at canopy closure, to give the final density. The final crop yield at 30 years is estimated to be 200–300 m^3/ha.

Measurements of trees that have regenerated naturally after clearings showed that free-growing trees are capable of annual girth growth of 64–76 mm, reaching 1.8 m girth in 25 years or less; trees in high forest on Kolombangara (Solomon Islands) grew 51–64 mm over a range of girths up to 2.4 m.

Diseases and pests

The coreid bug *Amblypelta cocophaga* is a serious pest of young plants (6–18 months old), causing die-back of the leading shoot and young leaves. Establishment of the species under shade and opening up of lines when trees are 4–5 m tall would be beneficial, but difficult to manage silviculturally. Other insect pests on *Campnosperma* at Kolombangara (Solomon Islands) include *Eucerocoris* sp., *Dromaeolus* sp. and *Pterolophia* sp. Borer holes (unidentified species) have also been seen in the pith of young seedlings in the Solomon Islands.

Trees up to 10 m height are considered to be wind-firm, but *C. brevipetiolatum* is one of the most susceptible trees in mature, natural forest to windthrow.

Uses

The species has good potential for replanting and environmental rehabilitation of deforested swampy sites in the humid tropics. No important non-timber uses of *C. brevipetiolatum* have been reported in the Solomon Islands.

The wood is pale pink-brown, fine straight grained to slightly interlocked (frequently incorporating tension wood), rather light and soft, odourless and lacking annual rings. It is susceptible to blue stain and pinhole borer and is not durable. It seasons readily, though there is a tendency to end-split. The timber saws well but the resulting sawdust can cause skin irritation. The timber kiln-dries rapidly without distortion. Plantation-grown timber is rather lighter and softer than mature timber. A sample of plantation-grown timber had a basic density of 225–330 kg/m^3 with an average of 310 kg/m^3, whereas trees from native forest have a density of around 325–360 kg/m^3.

C. brevipetiolatum is an excellent source of veneer for a wide variety of uses. Its easy working properties and stability, together with fine bending and gluing properties, make it particularly suitable as a mouldable plywood. Because of the timber's soft nature, chiselling needs care, but the wood can be turned and bored. There is a tendency for checking to occur, and fibrous material may prove difficult in turning. The timber is also used for other light general purposes include siding, sheeting, furniture, boxes, turnery, instrument cases, splints, mouldings, packing cases and blackboards. The tree has been used traditionally for making canoes in Micronesia.

Pulping tests give a reasonable yield with good strength characteristics, but the wood requires moderately severe pulping conditions, and may only be acceptable in a mixture with other species.

Further reading

Chaplin, G.E. (1988) The collection, processing and storage of *Campnosperma brevipetiolata* seed. *Forest Research Note No. 54-22/88.* Forestry Division, Munda, Solomon Islands, 13 pp.

Chaplin, G.E. (1993) *Silvicultural Manual for the Solomon Islands.* ODA Forestry Services, Natural Resources Institute, University of Greenwich, UK, xv & 305 pp.

Hosokawa, T. (1954a) On the structure and composition of the *Camnosperma* forests in Palau, Micronesia. *Memoirs of the Faculty of Science, Kyushu University. Series E Biology* 1, 199–217.

Hosokawa, T. (1954b) On the *Campnosperma* forests of Yap, Ponape and Kusaie in Micronesia. *Memoirs of the Faculty of Science, Kyushu University. Series E Biology* 1, 219–243.

Whitmore, T.C. (1966) *Guide to the Forests of the British Solomon Islands.* British Solomon Islands Protectorate Forestry Department Forest Record 2, Oxford University Press, London.

Canarium schweinfurthii Engl.

African elemi, bush candle tree

Importance

Canarium schweinfurthii is a large tree of tropical Africa cultivated principally for its fruits which are marketed mainly in West Africa. It produces a low-durability timber with a good appearance, similar to mahogany.

It makes an attractive avenue tree and farmers will often leave the tree standing while clearing land, as it provides shade as well as a wind break.

C. schweinfurthii appears not to compete with crops and thus has great potential for agroforestry applications. It has been planted for reforestation in Uganda.

Botanical features

C. schweinfurthii is a large dioecious canopy tree, with a clean, straight and cylindrical bole often exceeding 50 m tall. The diameter above the heavy root swellings can be up to 4.5 m. It has a thick bark, smooth on young trees but increasingly scaly and fissured on older ones. The slash is reddish- or light brown and smells like turpentine. A sticky oleoresin from the slash turns a sulphurous-yellow and becomes solid. Leaves are pinnate, clustered at the end of the branches, and may be 15–65 cm long, with 8–12 pairs of leaflets, mostly opposite, oblong, cordate at base, 5–20 cm long and 3–6 cm broad, with prominent and pubescent lateral nerves beneath. The lower leaflets are bigger than the upper ones. The lower part of the petiole is winged on the upper side.

Long inflorescences (up to 28 cm) growing in the axils of the leaves consist of many creamy-white unisexual flowers about 1 cm long. The fruit is a small drupe, bluish-purple, glabrous, 3–4 cm long and 1–2 cm thick. The calyx is persistent and remains attached to the fruit. The fruit contains a hard, spindle-shaped, trigonous brown stone that eventually splits, releasing 3 seeds.

Distribution

C. schweinfurthii is distributed throughout tropical Africa in rainforest, gallery forest and transitional forest in Angola, Cameroon, Ethiopia, Ghana, Guinea-Bissau, Liberia, Mali, Nigeria, Senegal, Sierra Leone, Sudan, Tanzania, Togo, Uganda and Zambia. It grows at low altitudes up to 1000 m with mean annual rainfall of 900–1400 mm.

There are trial plantings of this tree in Java, Indonesia.

Silviculture characteristics

C. schweinfurthii is amenable to intercropping and its cultivation has therefore received attention from agroforestry research. In forest settings the seeds are mainly dispersed by hornbills and elephants. The tree is artificially propagated through seedlings, wildings and direct sowing on site.

Silviculture practice

The ripe fruits should be collected when they fall to the ground and allowed to decompose; the stones should then be separated from the outer fruit coats. Seed germination is considered poor, due to the hard endocarp. The seeds can be stored for a long time.

Good germination rates (90% and above) have been obtained by immersing seeds in hot water, allowing them to cool and then soaking for another 24 h in cold water before sowing them in nursery beds.

Others recommend cutting through the seedcoat with a sharp scalpel and piercing the strophiolar plug or filing with sandpaper, and state that heat treatments were ineffective. With such methods the best seed germination was 70%.

After 3 months the seedlings may be transplanted, and 12-year-old trees reached an average total height of 6.8 m, with trunks of about 8.6 cm mean diameter. Floral analyses of an experimental stand revealed a sex ratio of 38% of male individuals for 62% of female trees. Female trees started producing fruits from around 8 years old.

The use of vegetative multiplication techniques offers the potential to fruit producers to work mainly with female trees.

Diseases and pests

Little information is available on pests or diseases of *C. schweinfurthii*.

Uses

The timber is valuable, and has a recorded green density of 360 kg/m^3. The sapwood, often up to 15 cm thick, is white with pinkish reflections; the heartwood is pinkish when freshly cut, but darkens to a light brown mahogany colour. It is used as a substitute for true mahogany, and seasons slowly but fairly well, works easily, and stains and polishes well. However, end-splitting may occur during the drying process. The timber is not durable and considered sensitive to termites and fungi. The wood is used as core veneer, for decorative panelling, parquetry, furniture, flooring and for general utility purposes. Locally, the wood is used for mortars, planks and canoes. The tree makes a good fuelwood, igniting readily and burning with a lot of heat.

Fruits are sold seasonally in African markets. The fleshy greenish pulp of the fruit is oily. Fruits are rarely eaten raw and most consumers first soften them in hot water. According to research in Cameroon, fruits – depending on their origin – can be softened at 40–50°C for 35–55 min. Fruits start drying quickly after harvesting and seldom soften after 6 days. Oil extracted from the fruit pulp, varying from 36 to 58% of dry weight, is sometimes used for cooking, and the kernel is oily and edible. These are cooked, and in Nigeria they are sometimes prepared into a vegetable butter and eaten as a substitute for shea butter.

The pulp has antioxidant activity. However, during the usual soaking, this declined from 45.3 to 5.31 ± 3.36 mmol/100 g.

The resin burns readily and is used as a bush candle and as incense (it releases a lavender-like smell). The resin is used as a fumigant against mosquitoes. The flame is smoky, and in Liberia the soot is collected from the outside of pots held over the flames, for use in tattooing and to make ink.

The resin is used locally to repair broken pottery, for caulking boats and as a gum for fastening arrowheads to shafts. In the past, it was exported to Europe for pharmaceutical use, and in World War II it was used as a substitute for gum mastic in wound dressings.

Traditional medicinal applications include a bark decoction against dysentery, gonorrhoea, coughs, chest pains, pulmonary affections, stomach complaints and food poisoning. The resin is used against roundworm infections and other intestinal parasites. It is an emollient, stimulant, diuretic and has action on skin affections and eczema. The pounded bark is used against leprosy and ulcers. Root scrapings made into a poultice are used for ganglion inflammation.

The bark of the tree is reported to have been used for preparing varnishes and printing inks, and in Gabon, people make boxes of the bark.

The light brown seed stones polish well and are strung into necklaces or attached to traditional instruments.

Further reading

Abayeh, O.J., Abdulrazaq, A.K. and Olaogun, R. (1999) Quality characteristics of *Canarium schweinfurthii* Engl. *Oil Plant Foods for Human Nutrition* 54, 43–48.

Katende, A.B., Birnie, A. and Tengnäs, B. (1995) Useful trees and shrubs for Uganda. Identification, propagation and management for agricultural and pastoral communities. Technical handbook No. 10. Regional Soil Conservation Unit, Regional Soil Conservation Unit (RSCU), Swedish International Development Authority (SIDA), Nairobi.

Njoukam, R. and Peltier, R. (2002) *Canarium schweinfurthii*: a first attempt at its domestication in the West of Cameroon. *Fruits (Paris)* 57, 239–248.

Nwosu, M.O. (2000) Studies on the germination of seeds of *Canarium schweinfurthii* L. (Burseraceae). *Journal of Economic and Taxonomic Botany* 24, 651–654.

Carapa guianensis Aubl.

Andiroba

Importance

Carapa guianensis is a popular timber tree with fragrant flowers found in South American rainforests. The wood is hard and resistant and is appreciated for its rich texture and favourable grain. It can be used as a substitute for mahogany (*Swietenia*). The seed oil (andiroba), bark and leaves have medicinal uses. The oil is also used in the cosmetic industry.

Botanical features

C. guianensis is a deciduous or semi-evergreen, monoecious tree reaching a height of up to 35–55 m and a diameter of 100 cm (maximum 200 cm), with a straight cylindrical bole and short buttresses up to 2 m high. The bole is sometimes fluted and branchless up to 20–30 m. The bark has characteristic wide fissures with horizontal splits and is light grey to greyish-brown or dark brown, sometimes reddish. The inner bark is pink-red or pinkish-brown and fibrous. Young plants produce taproots but the trees tend to become surface rooted. The leaves are alternate, paripinnate with a dormant glandular leaflet at the apex, exstipulate. Leaflets are opposite and entire. The tree shows gigantic leaves in the monocaulous juvenile stage, decreasing in size when branching is initiated. Flowers are small, white and borne in a large, axillary or subterminal thyrsi. They are unisexual but with well-developed vestiges of the opposite sex and tetramerous to pentamerous (maximum sextamerous). The calyx is lobed almost to the base and the petals are slightly contorted. The fruit is dehiscent, 4-lobed, pendulous and subglobose. It forms a woody capsule (nut) containing 2–4 seeds in each lobe. Seeds are smooth, pale brown, angular, with woody sarcotesta. They have oil-rich kernels.

Distribution

C. guianensis is distributed from Honduras to Colombia, Venezuela, Peru, Brazil and the Guianas, including the West Indies, Cuba and Trinidad. It is an exotic in Indonesia, Malaysia and Singapore. The species can be common in South American evergreen to semi-evergreen rainforest, as part of the canopy or sub-canopy, but sometimes also as a dominant tree or in almost pure stands. *C. guianensis* is able to tolerate shade. It is found both on higher ground and along rivers and periodically flooded areas or in swamps. The tree grows in light-, medium- or heavy-textured soils. *C. guianensis* grows between 0–1200 m altitude with mean annual precipitation between 1500 and 3200 mm. It does not tolerate frost.

Silviculture characteristics

Early growth in the nursery is moderately fast, and seedlings may attain 0.5 m in the first year. In South America, *C. guianensis* showed a mean annual diameter increment of 1.6–2 m in 25-year-old plantations. A large tree may produce 750–4000 seeds per year, but almost none in unfavourable years. The flowering period is usually concentrated in one short period per year, but this depends on the climate. Pollination is probably by insects. Usually only 1 or 2 fruits in an inflorescence mature in 8–12 months. In swamp forest, trees reach felling size in 20–25 years, and at higher elevations probably in 40–60 years. *C. guianensis* can tolerate shade, but requires full overhead light for fast growth. It coppices freely and is resistant to fire.

Silviculture practice

C. guianensis can be propagated from seed. Seed storage behaviour is recalcitrant. Seeds are viable only for a short period and cannot be dried well. When sown on the surface or shallowly in moist soil, all fresh seeds germinate. Non-scarified seeds complete germination in 19–21 days, while scarified seeds take 6–7 days. The roots of 1-year-old seedlings are pruned at about 15 cm while the plants are still in the nursery bed; when new rootlets begin to develop, the plants are uprooted and planted out. Propagation can also be done from large cuttings planted in swampy soil.

Diseases and pests

The sapwood of *C. guianensis* is susceptible to *Lyctus* and pinhole borer attack. Logs are susceptible to severe ambrosia beetle attack. *C. guianensis* is attacked by shoot borer (*Hypsipyla grandella*) that causes malformation of the stem. Young trees are browsed by deer.

Uses

C. guianensis is planted as an ornamental in the Caribbean and can be used for enrichment planting.

The timber is reddish-brown, hard, of medium density (560–750 kg/m^3 at 12% moisture content), durable, strong and easy to work. The heartwood is moderately durable and resistant to termites. The wood is used in high-quality furniture, medium and light construction, veneers, flooring, shipbuilding, plywood and turnery. Its colour darkens with time and exposure to the sun. The grain sands easily and accepts finishes well. It is not recommended for exterior uses where it makes direct ground contact. *C. guianensis* is moderately easy to air or kiln dry, but it can develop warping and checking. The wood is also suitable for the production of pulp and paper.

Traditionally, *C. guianensis* is used by the Amerindians to treat skin problems (rashes, boils and ulcers) and as an insect repellent. In Suriname, the tree is used against fever, diarrhoea, rash and gout. The bark contains an alkaloid, carapina, which is a febrifuge. Boiled leaves are applied to itchy skin; fever and intestinal worms are treated with a fruit rind decoction; the seed oil, known as andiroba, is used in the treatment of hepatitis, tetanus and ringworm. The oil has anti-inflammatory properties and is rich in essential fatty acids. It is also used as lamp oil and for making soap and candles. It is a natural emollient, used in cosmetics. The nut yield of *C. guianensis* is about 300 lbs per tree/year.

Further reading

Embrapa (2005–2007) Andiroba - Características Gerais da Espécie *Carapa guianensis*. Empresa Brasileira de Pesquisa Agropecuária – Embrapa. Available at: http://dendro.cnptia.embrapa.br/Agencia1/AG01/arvore/AG01_32_309200411812.html (accessed 1 December 2012).

Forget, P.-M. and Kenfack, D. (2008) A website dedicated to trees in the genus *Carapa* (Meliaceae). Taxonomy, biology, ecology and uses. Muséum National d'Histoire Naturelle, Brunoy, France. Available at: http://www.carapa.org (accessed 13 March 2013).

Fournier, L.A. (unda) *Carapa guianensis* Aubl. In: Tropical Tree Seed Manual. Part II – Species Descriptions, 360–362.

RNGR, USDA Forest Services, USA. Available at: http://www.mgr.net/publications/ttsm.

Sosef, M.S.M., Hong, L.T. and Prawirohatmodjo, S. (eds) (1998) *PROSEA 5(3) Timber Trees: Lesser Known Species.* Backhuys Publishers, Leiden, The Netherlands.

Carapa procera DC

Crabwood, carapa

Importance

Carapa procera is an evergreen subcanopy tree species of tropical Africa and America, where it is locally abundant in some gallery, deciduous and dry forests.

Its reddish-brown timber resembles true mahogany (and is in the same family) and is valued for furniture and interior fittings. In Latin America, the tree is highly valued for the multi-purpose 'andiroba' oil extracted from the kernels.

C. procera can be planted for erosion control, hedges and windbreaks and is considered suitable for revegetation of seasonally waterlogged sites. Although little is known about how the species behaves in agroforestry systems, it is considered to have considerable potential as a multi-purpose tree.

Botanical features

C. procera is typically a medium-sized, and occasionally large, tree with wide spreading and arching branches, reaching 15 m high in swamp forest. It can grow into a tall tree up to 30 m in lowland rainforests in Africa and even reach 40 m in such forests in tropical America. The bark is reddish and smooth. When adult, the species has compound odd-pinnate leaves, as long as 1.5 m and often clustered at the end of branchlets. Each leaf consists of 6–18 pairs of (sub) opposite leaflets, bright glossy green, 40 × 16 cm and oblong-elliptic. However, the first leaves of seedlings are simple. The inflorescence is a pyramidal panicle of white and pink fragrant flowers, much branched and up to 70 cm long. The flowers have 5 (or 6) petals that are 5–6 mm in length. Pollination is probably by small insects. Flowers appear early in the rainy season and fruits in the mid- to late rainy season.

The fruits are large (12–15 cm in diameter), oblong-ellipsoid, long-beaked, or truncate and short-beaked, with ridges along each valve. They contain 1–3 capsules with 3–7 seeds each. The capsules open simultaneously at the apex and the base. The seeds (typically 3 cm long) are shiny, dark brown and are dispersed by rodents (in South America by the red acouchy (*Myoprocta acouchy*) and the red-rumped agouti (*Dasyprocta leporine*)).

Note: the species currently known as *C. procera* is variable – the type specimen is African and at least one recent treatment implies that the American species should be considered distinct.

Distribution

C. procera occurs in West and Central Africa, and South and Central America (but see Note above). Its area of distribution is characterized by annual rainfall of 800–3000 mm with a dry season lasting less than 7 months and mean annual temperatures in the range of 20–35°C. The species does not tolerate temperatures below 5°C.

It is a common, but localized species of lakeshores, riparian and mid-altitude (1100–1800 m) forest, especially where drainage is impeded. It has also frequently been recorded on sandy soils, generally at sea level.

Silviculture characteristics

C. procera is easy to germinate, and seedlings grow best on medium to heavy, acid to neutral soils, and tolerate seasonal waterlogging and shade. It regenerates rapidly and is suited for coppicing. Young trees are relatively fast growing.

Silviculture practice

Fruits should be picked from the trees when dehiscence of the capsules occurs, and seeds are extracted after the fruits have dried. Alternatively, the seeds are picked from the ground, soon after falling from the tree.

Seed weight ranges from 30 to 80 seeds/kg and the kernels contain ca 55% oil. Seeds quickly lose viability when dried and should be kept in moist sawdust/vermiculite to maintain moisture content, even for short-term storage.

Direct planting is recommended for locations near the seed production area. Though some authors recommend scarification, others state that no pretreatment is required. Germination success is usually high (85–95%) and starts within 6–10 days after being sown in sand covered by sawdust. Germination is complete within 2–3 months and the seedlings can be transplanted.

Seedlings tolerate some shade but perform better in light gaps than in the understorey. If grown for its timber *C. procera* could be harvested on a 18–23-year cycle, when grown on yellow latosols and planted at 3 × 4 m.

Regeneration of *C. procera* in natural forest is from seeds cached by scatter-hoarding rodents (which bury individual seeds as food reserves in shallow, scattered caches in the topsoil). Seed-predating insects and mammals destroy most non-hoarded seeds. Exceptions are seeds falling in treefall gaps: these high-light environments permit seedling establishment, even from infested seeds.

Diseases and pests

Seeds are predated by insects and mammals. As with most timber species in the mahogany family, seedlings are susceptible to shoot borer (*Hypsipyla grandella*) attack.

Uses

C. procera has been recommended in Brazil for enrichment planting and/or as an agroforestry species. The species has great potential to help recuperate the large degraded areas in the more humid parts of Amazonia.

The heartwood is reddish-brown, with a regular grain. The wood has a density of 560–670 kg/m^3 and works well; it is popular on local African markets for furniture and cabinetry and can also be used for light construction and veneers. The wood makes a charcoal with high calorific content. In Benin, carapa wood is used for the production of toothpicks; this led to overexploitation and restrictions on cutting the trees have been enforced.

C. procera is also important for its oil, particularly in tropical America. Andiroba oil is extremely bitter, whether pressed cold or hot. Small quantities of oil are used to sooth arthritis and muscular distentions, skin tumors, superficial skin ailments and herpes. The Amerindians use it as an insect repellent. When obtained by hot pressing, it becomes cloudy when left standing, condensing out a solid white fat composed principally of palmitin. This solid fraction is used for making sodium soaps in northern Brazil. The oil is also used in West Africa to prepare 'black soap'.

Traditional oil extraction in Brazil is a process requiring up to 2 months. The seeds are boiled to produce a seed cake which is stored for 20 days. The cake is then kneaded and placed on an inclined plate to collect the oil dripping from it. Extraction in the shade takes longer than in the sun, but the oil quality is considered better than the oil extracted under the sun. Presses may be used to express the oil more efficiently.

The leaves can be used for mulching. The bark, root, leaves and fruit are used medicinally, and the flowers attract honeybees.

Further reading

Ferraz, I.D.K. and Sampaio, R. de T.B. (1996) Simple storage methods for seeds of andiroba (*Carapa guianensis* and *Carapa procera*). *Acta Amazonica* 26, 137–144.

Ferraz, I.D.K., Camargo, J.L.C. and de Sampaio, P. de T.B. (2002) Roba-mahogany (*Carapa guianensis* Aubl. and *Carapa procera* D. C.): ecological, botanical and technological aspects of its seeds and seedlings. *Acta Amazonica* 32, 647–661.

Forget, P.M., Baraloto, C., Blanc, L., Petronelli, P. and Gourlet-Fleury, S. (2004) Sustainable harvesting of Crabwood (*Carapa procera*) seeds, dispersal, predation and regeneration. In: Report of the Technical Workshop on Crabwood, Iwokrama International Centre, Georgetown, Guyana.

Mendonça, A.P. and Ferraz, I.D.K. (2007) Crabwood oil: traditional extraction, use and social aspects in the state of Amazonas, Brasil. *Acta Amazonica* 37, 353–364.

Carpinus betulus L.

Hornbeam

Importance

Carpinus betulus is a common tree in many broadleaved forests of central Europe. It produces a very hard, dense, white timber sometimes referred to as ironwood, used for making small hard-wearing items and flooring. It is also a valuable fuelwood commonly managed by coppicing, and often in mixed woodlands as an understorey with other more valuable species such as oak, ash, beech or planted softwoods such as Scots pine (*Pinus sylvestris*) or Norway spruce (*Picea abies*). It is also planted as a hedgerow and can be pruned to different forms. The foliage can be used as a valuable fodder.

Botanical features

C. betulus grows to 20–25 m with dbh to 40–50 cm when open-grown, having a spreading crown when free from lateral competition. The stem is straight or slightly crooked, tapering, often forked, distinctly ribbed or fluted, and sometimes buttressed from the base up to half of stem height. Bark is smooth and light grey-green. There is a strong shallow lateral root system to a depth 50 cm. Leaves are glabrous, oval-oblong, rarely up to 12 cm long, rounded, slightly cordate at the base, finely dentate, with 12–14 pairs of veins with no tertiary veins. Flowers are unisexual, borne in pendulous catkins with both sexes occurring on the same tree. Fruits are small ribbed nutlets with a 3-lobed bract, in clusters. Tree form and foliage can appear similar to the common beech (*Fagus sylvatica*).

Distribution

C. betulus has a broad native range from the Atlantic Ocean to the Caspian Sea, and from the Baltic to the Mediterranean. The range includes an area from the Pyrenees to southern Sweden including southern England, the whole of central Europe, the Balkans and Eastern Europe. The northern extent includes some areas of Russia east of the Dnieper river, and southwards to southern Italy and the Peloponnese in Greece. Eastwards, *C. betulus* is found in northern Turkey, and the Caucasus as far as northern Iran.

In the post-glacial period, *C. betulus* moved relatively late from its refuges in south-eastern Europe and the present range was reached only in historical times. During the last 1500 years the frequency of the species increased steadily in hilly and lowland forests. The further spread in oak (*Quercus* spp.) and beech forests is probably due to forest clearance. That it is not present in the Iberian Peninsula or along the French Atlantic coasts is attributed to a still incomplete post-glacial spread, or unsuitability to too strongly oceanic climates. Oak–hornbeam forest is the classic forest type on fertile soils in temperate Europe, and it is also found in some beech forests. Pure hornbeam stands are rare and may be an accidental consequence of human activities. In Southern Europe, *C. betulus* is a minor species, often in relict stands located in mountain zones where annual rainfall is higher, or on lowland plains on moist alluvial soils.

C. betulus has been introduced to parts of North America and on a small scale in some other countries.

Silviculture characteristics

C. betulus is a typical temperate-climate species, requiring fairly abundant moisture. In sub-continental climates, high summer temperatures can be a limiting factor. It is known to be very hardy and is found even in frost hollows, though it can be more damaged by early autumn frosts than by late spring frosts.

Although found in many different types of soil, it grows best on bottomland sites with deep, moist and well-drained sandy and loamy soils. It is, however, also found on wet, heavy clay soils and on light, dry sandy soils. Nitrogen availability can enhance shade tolerance, so that the species becomes more competitive with light-demanding species. On clay soils it may replace beech. It can tolerate calcareous stony soils provided rainfall is abundant, but is not found on acid soils.

It is a moderately to strongly shade-tolerant but not long-lived species, often found in the final phases of secondary forest succession as an understorey tree. In upland sites it is sometimes found as a colonizer of bare soil, and on sites which are not too infertile, it provides a fertile humus.

C. betulus can play an ambiguous role in mixed oak or beech forests, however, as it can be seen as a dangerous invader because of its shade tolerance, or can be valued as an accessory understorey species in stands of more light-demanding and valuable timber species such as oak, ash (*Fraxinus excelsior*) or Scots pine. *C. betulus* coppices and pollards well. Natural seed regeneration is easy, and sometimes too easy. In young mixed regenerating stands, timely and repeated tending operations are required, or stands can otherwise evolve towards an undesired conversion towards pure *C. betulus*, and more valuable species could disappear.

Silviculture practice

It is often managed with the same silvicultural systems designed for other broadleaved species with which it mixes, as its understorey position helps to suppress epicormic branching. Several aspects must be considered, including succession in natural or semi-natural stands; natural regeneration; and the role of *C. betulus* in the evolution of coppice with standards.

When raising nursery stock, mature, brown fruits are collected in October–November. Fruit bracts are removed by hand or machine. The weight of 1000 seeds is 3–10 g, and percentage viability is 50–60%. Seeds can be stored up to 2 years, though seed dormancy is marked and seeds need to be stratified before sowing. One report states that the best results were obtained with 6 months of dry storage at −5°C followed by 1 month of warm stratification and 6 months of cold stratification. Another suggests warm stratification (8–21°C) for 8 weeks, followed by cold stratification (0.5–1°C) for 8–12 weeks, before sowing in spring at 150–250 fruits/m^2. Alternatively, to avoid seed dormancy fruits can be collected early in September–October when they are still green, and then immediately sown and kept moist throughout the whole winter.

The best planting stock is 2-year-old bare-rooted transplants, either transplanted after the first year (1+1) or undercut at the end of the first season. *C. betulus* is planted as an accessory understorey species in stands of light-demanding species such as oak or ash, or during the conversion of conifer plantations of species such as Scots pine or Norway spruce. It is also planted with oak in afforestation plantations. It is planted on bare sloping land for soil protection and maintained in a bushy form when needed. When used in hedgerows, it can be pruned to retain various shapes.

In mixed high forests, a *C. betulus* understorey can be left to grow until it becomes shaded and eventually dies as the main canopy species develops. Where dominant in mixed coppice stands, a 30–40 year rotation gives a mean annual volume increment of 4–6 m^3/ha.

Diseases and pests

As a species generally considered of minor importance, the diseases and pests of *C. betulus* have not been exhaustively studied. Cankers due to a pathogenic fungus of the genus *Nectria* are also sometimes reported. It is also known to be damaged by the same non-specialized fungi (e.g. *Armillaria mellea*) and insects (e.g. Lymantridae) which attack tree species with which it is associated.

Uses

The wood is dense, usually about 770 kg/m^3 at 15% moisture content. It is strong and very hard, one of the hardest of temperate timbers and sometimes given the name ironwood. The wood is white with an occasional yellow or greenish tinge, has a fine and uniform texture, and finishes very smoothly. It is most commonly used for small items where hardness is a key required quality such as mallet heads, tool handles, chopping blocks, piano key movements, drum sticks, axles and in block and tackles. It can be used as a satisfactory alternative to maple (*Acer* spp.) for flooring, either as boards or as end-grain blocks. In the past, the wood was stained black and used as an alternative to ebony (*Diospyros* spp.). It is occasionally manufactured into veneer. However, stems are often fluted and have poor form, and are considered difficult to work. The wood has a high calorific value and makes an excellent fuel and charcoal, and in Western Europe it is mainly used as such.

C. betulus is a good hedgerow tree, and can be pollarded for fodder. Recent studies have indicated that it was an important fodder species especially for use as 'tree hay', cut and stored over winter in Neolithic farming systems. It can be planted to help soil conservation and erosion control, particularly on slopes and to stabilize landslides. It can be used in improvement planting and land reclamation. As an ornamental species, it can be planted along garden paths to provide shade and amenity value.

Further reading

Kiaei, M. (2012) Effect of site and elevation on wood density and shrinkage and their relationships in *Carpinus betulus*. *Forestry Studies in China* 14, 229–234.

Szwagrzyk, J., Szewczyk, J. and Maciejewski, Z. (2012) Shade-tolerant tree species from temperate forests differ in their competitive abilities: a case study from Roztocze, south-eastern Poland. *Forest Ecology and Management* 282, 28–35.

Vacik, H., Rahman, M.M., Ruprecht, H. and Frank, G. (2009) Dynamics and structural changes of an oak dominated Natural Forest Reserve in Austria. *Botanica Helvetica* 119, 23–29.

Castanea sativa **Mill.**

Sweet chestnut

Importance

Castanea sativa is an important, vigorous, long-lived multipurpose tree and a valuable source of food, fuelwood and timber. Nuts are nutritionally rich in starch, oils and vitamins, and chestnut supplied a great variety of wood products to many rural people in upland Mediterranean Europe and Western Asia until the beginning of the 20th century. The wood is similar to oak (*Quercus*) in colour and texture, but lacks the broad rays of oak. Wood from old trees tends to contain shakes and have spiral grain, especially from trees grown on light soils liable to drying out. *C. sativa* was commonly managed in coppice woodlands, many being converted to uneven-aged high forest stands in the 20th century. It is also valued as an ornamental, street and parkland tree.

Botanical features

These large deciduous trees grow to 30–40 m with diameters up to 150 cm, although exceptionally old specimens may have diameters in excess of 3 m. Young trees have conic, open and whorled crowns, becoming dense, domed or widely spreading with age, and may eventually develop low crowns with huge low branches and twisting upper branches. Bark is smooth, at first purplish-grey, becoming silvery-grey with vertical or spiral fissures. Leaves are simple, 10–20 cm long (up to 32 cm long on coppice shoots), 5–10 cm wide, shiny upper surface, paler beneath, oblong-lanceolate, acute or acuminate, with distinctive serrated margins. Autumn leaf colour pale yellow in October turning rich brown in some years. Trees are monoecious, with long, slender, creamy-white male catkins 25–32 cm in length, having a pungent smell. The fruit is a red-brown edible nut 2–3.5 cm long, in bunches of 2−3, encased in a yellow-green husk covered in sharp spines.

Distribution

C. sativa is native to the Mediterranean basin, through the effects of the last glaciation and early movements by man since prehistory have obscured its true origins, and it is difficult to distinguish between the geographical range of natural versus naturalized populations. It is however, considered native to Italy and south-eastern Europe, north-eastern Turkey and the Caucasus, from where it was then spread throughout mainland Europe and North Africa. It is now found throughout Europe up to 57°N in Scotland (UK), Denmark and southern Sweden, and plantations for nut production have also been established in many other temperate regions of the world such as in Argentina and Chile, Canada and the USA, and Australia and New Zealand.

Silviculture characteristics

C. sativa is light-demanding and heat-tolerant, but frost tender. Although windfirm due to a deep taproot and extensive root system, it will not flourish on exposed sites. Naturally, it occurs only sporadically in mixed broadleaved forests and is not considered suitable in mixed broadleaved stands. *C. sativa* requires more acid soils than many temperate broadleaved species will tolerate. Once established, its heavy shade and tough leaf litter will inhibit the growth of most ground vegetation.

C. sativa is suitable for coppicing since the stump is long-lived and vigorous. The stem is straight and self-pruning only if grown in dense coppice or high forest. In orchards, only a small part of the bole is clear of branches.

It prefers a mean annual rainfall of over 600 mm with no dry season or only a very short one, and less than 700 mm can be tolerated only on deep, fertile soils or in moist atmospheric conditions. Good fruit production requires a mean monthly temperature above 10°C for at least 6 months of the year, and north of 49°N (Paris, France), cooler climates result in irregular fruiting. Trees begin fruiting at 30–40 years old, with peak production after 50 years.

C. sativa is a highly variable species and there are a large number of local varieties and provenances. Even after considerable efforts in *in situ* and *ex situ* conservation, much is thought to remain undocumented. Nonetheless, significant breeding programmes have been undertaken, particularly to overcome the problems posed by disease infection. These have involved intervarietal crosses and interspecific hybridization because of the compatibility among different *Castanea* species.

Silviculture practice

Seeds, collected when fruit has fallen, are considered recalcitrant and should be sown immediately in a moist soil–peat mixture, protected from rodents and other predators. Storage for several years is only possible in a cold chamber at 3°C. After 1–2 years in the nursery, seedlings are usually transplanted as bare-rooted plants (1+0 or 2+0) in January–March. Trees can also be propagated by cuttings and grafting. Alternatively, natural regeneration presents no problem if light is sufficient.

C. sativa is mainly grown in orchards for nut production with stand height varying from 5 to 18 m, or in coppices (simple coppice or coppice with standards) for production of fuelwood, poles and small timber. Spacing depends on management goals, for example 6 × 12 m for fruit production, 3 × 4 m for general-purpose timber, and 2 × 3 m in coppice stands.

Rotation length varies depending on final assortments required. Coppice rotations of 10–15 years are used for posts and fuelwood, and 35-year rotations on good sites produce logs suitable for carpentry and general construction (2.1 m long × 25 cm diameter), or smaller assortments suitable for weatherboarding or shingles (1.1 m long × 15 cm diameter).

Only on the very best sites, rotations of 45–50 years produce larger roundwood (30–40 cm diameter), with small logs used for poles and stakes. On a rotation of 50 years in a high-forest system, stands may be thinned at 15–20 years, although thinning is rarely carried out today. Guidelines for sustainable silvicultural systems for chestnut coppice with the ability to produce quality timber suggest a rotation period of up to 50 years with thinning every 7–10 years, starting from the 6th to 10th year. However, pure stands for high-quality timber are rare, though some coppice stands are undergoing silvicultural conversion to high forest. Mixed broadleaved high-forest stands are more common in the eastern part of its range.

Trees grow relatively quickly when young, though yield data is highly variable. Many reports of annual volume increment are in the range of 8–12 m^3/ha/yr, though yields of more than 20 m^3/ha/yr have been recorded on very good sites.

Diseases and pests

The most significant chestnut disease causing important economic losses is a root crown canker known as 'ink disease', caused by *Phytophthora cambivora* and *Phytophthora cinnamomi*. Wet, heavy soils favour disease, and infection causes root die-back, wilting and eventual death of the tree. The diagnostic black xylem is revealed by cutting into the trunk at ground level. One of the most destructive plant pathogens is chestnut blight fungus (*Cryphonectria parasitica*), a wound parasite attacking the phloem, forming a girdling canker on the main trunk, with trees typically dying within 3 years of infection. Other diseases include *Mycosphaerella maculiformis* rust that can cause complete loss of leaves on nursery plants, and *Sclerotinia pseudotuberosa* which causes black spots on harvested nuts. There are many viruses, the most important being Chestnut line pattern virus but none are economically important.

The most damaging insect pests that feed on the nuts are the weevil *Curculio elephas* and the shuckworms *Cydia splendana* and *Cydia fagiglandana*, which can cause severe commercial losses.

Uses

C. sativa is common along rural roadsides and in parks and gardens, being a very attractive landscape species.

Large trees are made into beams used in the construction of houses and barns, and sometimes sawn for planks. However, it is not often used for furniture or interior carpentry as ring shake enhanced by variable stem diameter increments is a frequent problem. It is commonly used for fence posts, stakes, vine supports, barrel staves, roofing shingles and similar products, as coppice poles can be split easily, and as the wood is naturally very durable due to its high tannin content. The wood is also valued as a fuel and for making charcoal. Reported wood densities range from 430 to 520 kg/m^3.

However, trees tend to be valued more for their fruits, which unlike most other nuts are high in carbohydrate (50%) and moisture (45%), but low in protein (4%) and fat (1%). This makes chestnuts more suitable for use as a staple item of diet, and they are eaten fresh after roasting or boiling, or are easily conserved. Records of their use and of selection of varieties with larger fruit date back to Greek literature, and during some periods in European history its cultivation was so dominant and essential for human survival that some authors called them 'chestnut civilizations'. Dried, they are ground into a sweet, gluten-free flour used for making various traditional dishes. They are also an important forage for livestock and wild animals.

Tannins extracted from chestnut wood are still used in the manufacture of leather and in wood-preservation products, with a content of 10–13% reached when trees are at least 30 years old. The bark is not used because it gives a dark colour and has a high sugar content. The copious flowers are also a valuable bee forage and 'chestnut honey' is a sought-after product with a distinctive flavour.

Further reading

Bourgeoi, C., Sevrin, E. and Lemaire, J. (2004) *Le Châtaignier, un Arbre, un Bois [The Chestnut, a Tree, a Wood]*. 2nd ed. IDF, Paris, France, 352 pp.

Breisch, H. (1995) *Châtaignes et Marrons [Sweet Chestnuts]*. CTIFL, Paris, France, 239 pp.

Conedera, M., Krebs, P., Tinner, W., Pradella, M. and Torriani, D. (2004) The cultivation of *Castanea sativa* (Mill.) in Europe, from its origin to its diffusion on a continental scale. *Vegetation History and Archaeobotany* 13, 161–179.

Castanopsis purpurella (Miq.) N.P. Balakr.

Chinese evergreen chinkapin

Importance

Castanopsis purpurella has a wide natural and planted distribution in China and elsewhere in Asia. Under suitable conditions, it is a fast-growing tree. It produces high-quality timber widely used for furniture, building construction, boat-building and agricultural implements. Its bark is used as a source of tannin. The fruit is edible and sold locally.

Botanical features

A large evergreen tree, it can grow up to 30 m tall and 1.5 m dbh. Branches are covered with soft hairs, and apertures in branches are not apparent. There are plank-like roots on the trunks of mature trees. The bark of mature trees has shallow fissures and falls in chips; the cortex is greyish-white, becoming deeply reddish-brown, 0.6–0.8 cm thick. *C. purpurella* has a deep root system with a well-developed primary root, short secondary roots and few fine roots.

Leaves are thin, leathery, ovate, oval or elliptical, 5–12 cm long, 1.5–2.6 cm wide, with a sharp apex and broad base; margin glabrous or calloso-serrate at the apex. The central vein is concave, with 10–14 pairs of side veins, and branch veins not apparent on the underside of leaves; the petiole is 0.5–1 cm long at 2 years old. The female inflorescence is 10–18 cm long, and the fruit is 15 cm long, with one nut in the fruit shell. Trees flower from April to June and fruits form from August to November of the following year, maturing in November and December.

Distribution

C. purpurella is distributed between latitudes 29° N and 18° N, in typically warm and sometimes hot climates. It grows in China in south Fujian and Hunan, Guangdong, Guangxi, south-western Guizhou, south Yunnan, south-eastern Xizhang, also Taiwan, Hong Kong, Hainan and Jiangxi. It also grows in Vietnam, Cambodia, Laos, Myanmar, north-eastern India, Bhutan and Nepal, and also Thailand. It is found in broadleaved evergreen forest; mixed forests; secondary forests; subtropical rainforests; coniferous forests; mountain forests; broadleaved forests and tropical rainforests in the climax communities.

Silviculture characteristics

C. purpurella grows well in both pure and mixed forests; however, it prefers semi-shade and grows better in shade at the juvenile stage. Saplings survive under low light intensity, growing well at the forest edge, due to effective utilization of scattered and/or transmitted light. Because of this, the choice of silvicultural model may affect its early growth, and *C. purpurella* is better planted in uneven-aged mixed forest. In mixed forest with *Pinus elliottii*, the pine offers shading for *C. purpurella* during the early stage after afforestation. The percentage of different species in a mixture also affects the growth of *C. purpurella*, and the major species used in China are *Castanopsis fargesii*, *Cunninghamia lanceolata*, *Elaeocarpus sylvestris*, *Phoebe zhennan*, *Pinus elliottii*, *Pinus massoniana*, *Sassafras tsumu* and *Schima superba*. Trees at 15 years old begin to set fruit, and produce abundantly during the period from 25 to 70 years old, although the quantity of fruits produced varies greatly from to year to year. Mature seeds are attractive to rats, birds and insects, and so protection is necessary. As an intermediate successional species, *C. purpurella* shows fewer negative effects of high nitrogen deposition, low irradiance and increased ambient temperature than late successional species. Exposing trees to ambient and increased levels of carbon dioxide and nitrogen deposition showed that *C. purpurella* is a good species for carbon fixation under conditions of global climatic change.

Silviculture practice

C. purpurella may be propagated by grafting and cuttings, but is mainly carried out by sowing seeds.

Grafting is usually carried out at the end of winter or in the early spring (mid- or late January to late March) using *C. purpurella* or other *Castanopsis* spp. as rootstocks; however, the crowns of grafted trees easily become asymmetrical.

C. purpurella can be established by direct seeding, artificial planting of containerized or bare-root stock, or natural regeneration. This species naturally regenerates well when the stand canopy density is maintained at about 0.4 and at 15–30 mother trees/ha retained in the felled stand. *C. purpurella* has good germination ability, and trees from naturally germinating seeds grow faster than trees from planted seedlings. The aim of natural regeneration is to have approximately one young tree up to 1 m tall/m^2 (9700 trees/ha).

C. purpurella is sensitive to growth conditions such as water, fertilizer and temperature. Initial planting density depends on site quality, with 2250–3000 seedlings/ha in medium-quality land, and 3750–4500 seedlings/ha on poor land.

Mycorrhizal fungi are associated with the roots of *C. purpurella* and form symbiotic relationships which can offer nutrition for the tree's growth and development. Mycorrhizal inoculation can promote seedling growth, benefit the development of roots and affect the physiological activities of seedlings. Some fungal strains which can accelerate tree growth have been selected.

As *C. purpurella* generally grows slowly in natural stands at the juvenile stage, it is recommended that applying fertilizers, loosening soil, controlling weeds and tending management should last 3–5 years. Good silvicultural practices could improve growth during early establishment, and trees might attain rapid growth rates after only 2 years. High productivity forests of *C. purpurella* will form when the stand canopy is almost closed at 5 years old, and associated trees and stressed trees should be gradually removed, leaving 900–1200 trees/ha.

Diseases and pests

The main seedling disease is root decay.

C. purpurella seeds are easily attacked by weevils such as *Curculio davidi*, but by soaking in water, seeds can be stored or sown after drying. Scarab beetles (e.g. *Adoretus corpulenta*) attack leaves and roots of seedlings in nurseries. Stick insects such as the common pest *Diapheromera femorata* may damage both the young trees and mature forests in March, with extreme damage in May to July when complete defoliation may occur. Mature trees and marginal trees are easily damaged by a number of pests including termites (*Reticulitermes* spp.).

Uses

C. purpurella has high ability to improve soil, and can prevent soil erosion. Its presence in mixed stands benefits the forest through its ability to resist attack by pests and diseases and its tolerance of fire. It can tolerate shade and has a high capacity for natural regeneration. *C. purpurella* is a useful ornamental species. In addition, there are many fungi associated with soil in *C. purpurella* forests, which not only improve growth, but which are also collected and used as delicious and highly sought-after edible mushrooms with a high market value.

Wood is light to moderately heavy, and moderately hard to moderately soft. The wood produced in Hainan, China is, however, generally heavy, hard and strong, and resistant to corrosion and termite attack. The heartwood is reddish-brown or red; sapwood is light reddish-brown and distinct, somewhat distinct or not distinct from the heartwood. Wood density is 584 kg/m^3 at 15% moisture content and air-dry density is 733 kg/m^3. The timber is used in making ship parts, furniture, stationery and craft works such as sculpture, apparatus boxes and dishes. It can be used in construction, for pillars, shutters, door-frames and larger construction such as bridges. In addition, it is used for such items as stakes, sleepers, poles, vehicle parts and sports apparatus. It can also be made into veneers. The branch wood, sapwood and shavings are good materials for making fibre products, pulp, composite boards and plywood.

The bark and fruit shells contain up to 15% tannin and are a good source of tannin for industrial use. The nut is rich in starch and is used as a minor food source.

Further reading

Chen, Y.-L., Gong, M.-Q., Chen, Y. and Wang, F.-Z. (2001) Effects of inoculation with 11 ectomycorrhizal fungal isolates on growth and photosynthesis of *Castanopsis hystrix* saplings. *Forest Research, Beijing* 14, 515–522.

Huag, P.Q. and Wang, L.X. (1983) *Silviculture of Major Plantation Trees in Guangdong Province, Guangdong, China*. Guangdong Forest Science and Technology Press, Guangdong, China, pp. 108–111.

Peng, Y.-H., Hao, H.-K., He, Q.-F., Cao, Y.-Y., Huang, Z.-L., Wei, J.-M. and Lu, G.-D. (2011) Nutrition effects on the growth of *Castanopsis hystrix* container bred seedlings. *Journal of Southwest Forestry University* 31, 27–30.

Qin, L., He, Y.-J., Li, Z.-Y., Shao, M.-X., Liang, X.-Y. and Tan, L. (2011) Allocation pattern of biomass and productivity for three plantations of *Castanopsis hystrix*, *Pinus massoniana* and their mixture in south subtropical area of Guangxi, China. *Scientia Silvae Sinica* 47, 17–21.

Casuarina equisetifolia L.

Casuarina

Importance

Casuarina equisetifolia is a nitrogen-fixing tree of considerable social, economic and environmental importance in many tropical areas of the world. It is widely planted for the reclamation of unstable coastal sites in the tropics and subtropics. It is salt-tolerant and can grow on pure sand, so is especially useful for erosion control and windbreaks along coastlines and estuaries. The wood has many uses but is renowned as a fuel. It is a truly multi-purpose species, providing a range of products and services for industrial and local end-users.

Botanical features

C. equisetifolia is commonly 8–16 m tall, but in some populations may attain heights of 35 m with diameters to 50 cm. Form in wild populations is very variable, from crooked, low-branching trees on exposed seashores, to straight-stemmed forest trees in more sheltered situations. Crowns are finely branched. Bark is light grey-brown, smooth on small trunks, with prominent encircling bands of lenticels, becoming rough and thickly furrowed on older trees. Inner bark is reddish and astringent. Twigs are drooping, needle-like, furrowed, 1–2.5 mm in diameter and 23–38 cm long, with minute teeth-like reduced leaves in whorls, 7–8 per node. Male flowers occur on simple terminal, elongated spikes 7–40 mm long, arranged in whorls, with female flowers borne on lateral woody branches. The 'cones' are globose or somewhat cylindrical, 10–35 mm long and 9–15 mm in diameter, with dull brown samaras 6–8 mm long.

Distribution

The more common *C. equisetifolia* subsp. *equisetifolia* occurs naturally between 20°N and 20°S, along the tropical coastlines from northern Queensland and Northern Territory in Australia, throughout the whole Malesian region to the Kra Isthmus of Thailand and adjoining coastal areas of the Andaman Sea in southern Myanmar, which is the western-most distribution range. To the east its natural occurrence extends throughout Melanesia and Polynesia. The subsp. *incana* has a much narrower distribution, extending from the coastlines of central New South Wales to north Queensland in Australia, and in Vanuatu and New Caledonia.

Silviculture characteristics

C. equisetifolia is relatively fast growing on poor soils and tolerates salt-laden winds. It can withstand gusty winds making it suitable as a windbreak, and young trees can even resist hurricane-force winds with little or no damage. Although a coastal species, it can be grown up to 1500 m altitude. It is not a legume, but fixes atmospheric nitrogen by nodulation with the actinorhizal symbiont *Frankia*. It is shade-intolerant and smaller seedlings are quickly suppressed by more vigorous seedlings. The ability to coppice is limited, usually restricted to young trees up to 4 years old, though trees respond reasonably well to hedging. Trees are sensitive to fire and to waterlogging.

C. equisetifolia has a lifespan of 40–50 years and shows rapid early growth up to 2–3 m in height per year, and good form in cultivation. Although it is evergreen, it sheds large amounts of twigs throughout the year. Shoot growth is generally active for much of the year except during cool and dry months. Where there is no distinct wet or dry season, flowering and fruiting tend to be irregular and may occur throughout the year, becoming more regular and occurring once or twice per year with more pronounced wet and dry seasons. Female cones mature about 18–20 weeks after anthesis, and will release small samaras (winged seeds) within 3 days when dried under full sun. There are about 250 cones/kg, yielding 20–60 g of seed of which less than half is viable. The fruits on a single tree do not all mature at the same time, often presenting a problem for seed collection.

C. equisetifolia may form pure stands on coastal dunes, growing over a ground cover of dune grasses and broadleaved herbs, or can be part of a richer association of trees and shrubs collectively termed the Indo-Pacific strand flora. It can also grow in narrow belts adjacent to mangrove forests or scattered in open woodlands dominated by *Eucalyptus*.

Silviculture practice

Plantations are established using containerized seedlings, rooted cuttings or bare-root seedlings. Mature seed is orthodox and maintains satisfactory viability for extended periods in air-tight storage in cool temperatures. Seed does not require pretreatment. Germination is epigeal and takes up to 2–3 weeks. It is also easily propagated by rooting stem cuttings, lateral or side shoots, terminal twigs, heel cuttings or basal sprouts, with suitable cutting material being 5–15 cm long × 1–2 mm in diameter. When propagating clonal material from old trees, air-layering on 1–2 cm diameter branches yields better results than cuttings.

Seed is sown in beds or trays containing sand or sand/peat mixtures, with seedlings transferred to containers when they are 3–5 cm tall. Seedlings reach plantable size (25–30 cm) in 3–4 months, using 50% shade, though excessive watering may cause damping off. Seedlings 5–10 cm tall in germination beds can also be transplanted to open beds at densities of 100–400 seedlings/m^2 for raising bare-rooted planting stock.

Inoculation of seedlings and cuttings with a pure culture of effective *Frankia* is recommended when *C. equisetifolia* is introduced to a new area, by applying a water suspension of the inoculant. Applying a solution of crushed nodules works less well. Where the species is already planted, adding soil collected from established stands to the potting medium will ensure the availability of mycorrhizal fungi.

When planting on moving sands, however, seedlings 1–1.2 m tall with strong root balls are preferred, planted 40 cm deep in order to withstand strong winds. A planting density of 2500 stems/ha is commonly used, or up to 10,000 stems/ha for production of fuelwood and small poles. Plantations are grown on 5–15-year rotations with several thinnings. Young trees compete poorly with weeds, so weeding is important during the first 2 years. With poor self-pruning capacity, pruning of lower branches needs to be carried out within 2 years to make plantations accessible for maintenance, with branches used for fuel.

On favourable sites trees can grow 4–5 m per year during the first 2 years, and annual increments of 15 m^3/ha over 10 years are possible. Other reported yields include mean annual increments of 8–12 m^3/ha reported from plantations at 1.5 × 1 m spacing on 4–7 year rotations, and 50–200 t/ha from plantations at 1 × 1 m or 2 × 2 m spacing on 6–15 year rotations, with dry weights per tree of 15–25 kg after 3 years, depending on site quality.

Diseases and pests

The tree is attacked by a number of pathogens, of which the most serious is blister bark disease associated with the fungus *Trichosporum vesiculosum*, now present in much of South and South-east Asia and on Indian Ocean islands. Infected trees die rapidly after exhibiting foliar wilt and cracking of the bark where blisters develop, enclosing a black powdery mass of spores. Bacterial wilt associated with *Pseudomonas solanacearum* causes foliage to yellow, and wilting and death has been reported in China and India. Other serious recorded diseases include stem cankers and die-back caused by *Phomopsis casuarinae*, *Botryosphaeria ribis* and pink disease (*Corticium salmonicolor* or *Phanerochaete salmonicolor*)).

More than 50 species of insects are known to feed on the species, but serious pest problems have not occurred.

Uses

C. equisetifolia is commonly planted as shelter belts, for sand dune stabilization, land reclamation, and erosion control. Many areas of occurrence are susceptible to tropical cyclones or typhoons and the general tolerance to strong winds has encouraged the use in protective plantings against coastal erosion. First-year plantations may also be underplanted with crops such as peanuts, sesame, pulses, cucumbers or melons. *C. equisetifolia* is also used for revegetating tailings following open-cast mining.

The wood is hard and heavy, with an air-dry density of 900–1000 kg/m^3. The heartwood is dull reddish-brown and not easily distinguished from the pinkish sapwood. Rays are prominent on the radial faces of sawn timber. It is difficult to use for fine carpentry. Seasoning is accompanied by heavy and relatively uneven shrinkage. The logs are very difficult to saw in small circular saw mills, and air-dried timber is difficult to machine because of its density and hardness. Poles are popular as masts for fishing boats, boat oars, piles, posts and tool handles. The wood is susceptible to attack by the dry-wood termite *Cryptotermes brevis* and has only limited durability. The wood is also a source of wood fibre for production of paper pulp and as a raw material for rayon fibres. The wood is highly regarded as a fuel and produces quality charcoal. The energy value of the wood is 24,000 kJ/kg and that of the charcoal exceeds 33,500 kJ/kg.

Bark contains 6–18% tannin and was used extensively in some countries for tanning purposes. It is also used for dyeing fishing nets. Leaf litter is also often removed from plantations and used as fuel. The species has medicinal uses, with root extracts reportedly used for the treatment of dysentery, diarrhoea and stomach-ache; bark and twigs provide a treatment for beriberi.

Further reading

Midgley, S.J., Turnbull, J.W. and Johnston, R.D. (eds) (1983) *Casuarina* ecology, management and utilization. Proceedings of the 1st International Casuarina Workshop, 17–21 August 1981, CSIRO, Melbourne, Australia.

Pinyopusarerk, K., Turnbull, J.W. and Midgley, S.J. (eds) (1996) Recent *Casuarina* research and development. Proceedings of the 3rd International Casuarina Workshop, 4–7 March 1996. Da Nang, Vietnam. CSIRO, Canberra.

Zhong, C., Pinyopusarerk, K., Kalinganire, A. and Franche, C. (eds) (2011) Improving smallholder livelihoods through improved *Casuarina* productivity. Proceedings of the 4th International Casuarina Workshop, 22–25 March 2010. Chinese Academy of Forestry, Haikou, China.

Cedrela angustifolia Moc. & Sessé ex DC

Cedar

Importance

The timber of *Cedrela angustifolia* is easy to shape, and has great durability, even when exposed to weathering. It is therefore greatly appreciated in the regions where it occurs naturally, and is widely used. A number of scientific papers have discussed the problems of *Cedrela* silviculture in the American tropics, and the technical details of its silvicultural and technological characteristics are now known. Cultivation is, however, complicated by the incorrect application of ecological principles, and by attack from the shoot borer *Hypsipyla grandella*. Despite these problems, cultivation is possible if the species is managed correctly and is not planted in pure stands.

Botanical features

C. angustifolia is a large tree, reaching 30–35 m tall when mature. Its stem is thick and cylindrical, with an angular base and forming large tabular roots (buttresses), which are always present in large individuals. The bark is wrinkled, longitudinally furrowed and dark grey in colour with lighter spots. The branches are thick, cylindrical or angular, with glabrous or pubescent branchlets with conspicuous lenticels. The leaves are paripinnate, alternate, polymorphic and typically 28–40 cm long. The 6–10 pairs of leaflets are subopposite; each leaflet is sub-sessile, 8–14 cm long, 2.5–4.5 cm broad, lanceolate or ovate–lanceolate, sub-coriaceous or membranaceous, with an asymmetrical base. The main vein of each leaflet is densely covered with silky hairs, and the secondary nerves have tufts of hair in their axils. The leaflets each have a short, cylindrical, pubescent petiolule.

The inflorescence is a large terminal panicle with few branches, reaching around 20 cm long in total. Its peduncle has a glabrous base, a pubescent tertiary axis, triangular-lanceolate bracts, and persistent tiny flowers, which are 5–10 mm long and borne on short pedicels. The calyx is shortly campanulate, 5-dentate, 1.5–2 mm long, and with a pubescent external face. The corolla comprises 5 elliptical petals, which are pubescent, 6–8 mm long and 1.5–2.5 mm wide. There are 5 stamens with cylindrical filaments and glabrous, elliptical anthers. The gynophore is glabrous and elongate, at around 3–5 mm long. The ovary is ovoid and glabrous, with a glabrous style which is 1–2 mm long and has a capitate head. The fruit capsule is oblong-ovoid, 3–6 cm long and thick, with the valves opening up to the apex. Its pericarp is woody, lenticellate, and dark brown when ripe, with grey lenticels. The seeds are winged, elliptical, 4 cm long and 1 cm broad. Flowers are produced from December to February.

Distribution

C. angustifolia occurs in rainforests from Central America to the middle of South America, including parts of countries from Mexico to Argentina. However, as the species is not well defined taxonomically, its natural distribution is also somewhat uncertain. *C. angustifolia* is cultivated in Mapane, Suriname, where it occurs naturally, and is also present in the northern states of Brazil including Pará, Rondônia, Acre and Amapá. *C. angustifolia* has been reported in Rio Grande do Sul, although this is doubtful, being south of the Tropic of Capricorn. It is also present in Paraguay and Argentina, but is limited to the northern parts of both countries.

Silviculture characteristics

In natural forests, *C. angustifolia* has a tendency to develop a straight trunk until it surpasses the canopy and reaches direct sunlight above the forest. However, when in open areas, the height at which it opens its crown is much lower. Thus, when planting this species, special care must be taken when pruning. Periodic inspections must also be carried out to look for signs of pest attacks.

Silviculture practice

Sowing is carried out in seed beds containing previously prepared soil (mixed with wood ash). Seedlings are transferred to individual nursery tubes 15 days after germination or when they are approximately 5 cm high. During the first few weeks, they must be protected by moderate shade cover. After that, the seedlings only need regular water and fertilization. The plants usually remain in the nursery for 6–8 months and are planted out when 20–30 cm high.

Planting can be carried out in secondary forests in the initial phase of development, where the tree species are not over 10 cm dbh, and not over 5 m in height. In these cases, lines are opened which are 1.5–2 m wide, and the seedlings are planted 2 m apart. The distance from line to line can vary from 5 m to 10 m, and planting must be carried out in the rainy season. Intensive management is necessary during the first 4 years, until the plants are in a condition to develop on their own. It is necessary to remove undesirable plants that are established due to the light conditions during the first year and to control any weeds.

This species has mostly been exploited in its natural forest habitat, not being produced in commercial plantations except in experimental areas. There is no specific information on management either in natural or planted stands, but monoculture planting is not generally advisable, mainly because of pest attacks. Thus, if planting in an open area, it is best to plant mixed crops rather than a single-species stand, preferably alternating species. These can be planted 1.5–2 m from each other. It is important to include fast-growing species in the mixture to provide some shade and stimulate the growth in height of *C. angustifolia*. With this technique, it is prudent to use other commercially important species, to optimize the resulting products and add commercial value.

Diseases and pests

All species of *Cedrela* are severely attacked by the larvae of shoot borer moths of the genus *Hypsipyla*, which has somewhat limited their usefulness as plantation crops within their natural area of distribution. Nevertheless, a few *Cedrela* species are being tested on a large scale throughout the tropics and are likely to succeed if they are planted on better sites with more fertile soils and good drainage.

The largest restriction to production from plantations of *C. angustifolia* is attack by *H. grandella*, and the best way to avoid attacks is to have healthy plants, which can be achieved through correct silvicultural practices and soils. Homogeneous plantations in open areas are not recommended.

Uses

C. angustifolia has several uses, including as an ornamental due to its dense crown, and in agroforestry projects, as long as there is careful management to control *H. grandella* attack. It can also be used for the restoration of disturbed areas.

The greatest importance of *C. angustifolia*, however, is for its wood. Its timber is in general use due to its various desirable properties, such as easy shaping and great durability even when exposed to weathering. Wood density is 360 kg/m^3. No record of non-wood products or other uses for this species was found.

Further reading

Smith, C. E. Jr (1960) A revision of *Cedrela* (Meliaceae). *Fieldiana: Botany* 29, 295–341.

Villalba, R., Boninsegna, J.A. and Holmes, R.L. (1985) *Cedrela angustifolia* and *Juglans australis*: two new tropical species useful for dendrochronology. *Tree-Ring Bulletin* 45, 25–35.

Zapata, M.A., del Castillo, E.M. and Pennington, T.D. (2004) El género *Cedrela* (Meliaceae) en la Argentina. *Darwiniana* 42, 347–356.

Cedrela fissilis Vell.

Spanish cedar

Importance

Cedrela fissilis is one the most important commercial timber species in South America and an important component of forest exports in many countries, particularly in Bolivia and some Brazilian states. It is noted for its lightness, resistance to insects and versatility. Its major silvicultural problems include over-harvesting to the point of commercial extinction in many regions. Plantation efforts have been hindered by the shoot borer *Hypsipyla grandella*.

Botanical features

C. fissilis is a large tree with an open, round, parasol-shaped or umbrella-shaped crown, and numerous basal branches. Trees average approximately 25 m high, but individuals can reach 40 m. Average dbh at maturity is 90 cm, but can reach 1.6 m. The stem is long, cylindrical and straight with no buttressing at the base. The dark brown outer bark is rough and deeply fissured; the soft, fibrous, reddish internal bark is 1.2–2 cm wide. The wood is aromatic.

The alternate, imparipinnate leaves are 25–60 cm long, on petioles 6–8 long. There are 6–20 pairs of leaflets, each 3–10 cm in length and 1.5–4 cm wide. The leaflets are pubescent, pinnately veined, aromatic and obovate with acuminate apices and asymmetrical or oblique bases.

C. fissilis is monoecious with unisexual flowers in terminal panicles 30–80 cm long and 20 cm wide. The flowers are pubescent, actinomorphic, yellow-white and approximately 1 cm long; they have five elliptical, fleshy, pubescent petals.

The five stamens are attached to the base of the gynoecium, and the ovary is superior. The brownish-black fruits are oblong, woody capsules up to 10 cm in length, with conspicuous scabrous lenticels. They dehisce in five parts, releasing numerous winged seeds.

C. fissilis is dry-season deciduous. In Bolivian dry forests, leaflessness can last 4–6 months with leaves flushing at the beginning of the rainy season, but may vary with the length of the dry season. Flowering usually occurs at the end of the dry season or beginning of the rainy season, in October–December in Bolivia, but extending to February elsewhere. Fruits mature 8–9 months after flowering.

Distribution

C. fissilis is strictly Neotropical, but has the widest distribution of any *Cedrela* species. Its range extends from northern Argentina to north-eastern Paraguay, west to the base of the Andes, and into Central America, where it occurs patchily in Panama and Costa Rica.

C. fissilis appears in various vegetation types, but is mostly associated with seasonally dry lowland forests with a subtropical or tropical climate. It has an irregular distribution in the Chaco region of Bolivia and Paraguay at the foothills of the Andes, in humid Amazonian forests, and the rainforests in the Brazilian Atlantic. *C. fissilis* is largely replaced in Amazonia by *Cedrela odorata*, and in mountainous forests by *Cedrela lilloi*.

In New South Wales (Australia), *C. fissilis* has been introduced for wood production in plantations and agroforestry systems. It has also been introduced in Mexico and Cuba for forestry plantations.

Silviculture characteristics

C. fissilis is a shade-intolerant species that requires relative freedom from competing vegetation to regenerate. Through its small, wind-dispersed seeds, it naturally colonizes disturbed sites, provided there are seed trees nearby. Trees are long-lived and diameter growth can be substantial even for older trees. The species can be grown in mixed plantation systems, but rarely in monocultures because of problems with shoot borer attack.

Silviculture practice

C. fissilis is noted for its abundant production of seeds, most of which remain viable for more than 4 months if stored cold, although it is recommended that seeds be used within 1 year of collection. Seeds can be sown in germination beds and then transplanted to nursery beds, or directly planted in bags or nursery beds. Germination occurs within 12–18 days. The seedlings grow rapidly, but height growth is dependent on light availability. Seedlings should be well watered at least twice daily. Mixing seedlings in beds among other species may help reduce damage from shoot borers. Sewage sludge has been used as a bedding material in nurseries. Seedlings may be planted out once they have developed woody shoots (2–3 months) and are 25–50 cm tall. The most efficient method of transplanting uses bare rootstock.

Plantation establishment should consider the seedlings' growth requirements and protection from *H. grandella*. Seedlings need high light availability overhead and do not tolerate competition from weeds, although clean weeding causes greater exposure of seedlings to shoot borers. Likewise, mixed planting of *C. fissilis* and other species at wider spacings helps reduce shoot borer infestations. Spacing should not be too wide (>2 m), because the trees seem to grow best with some lateral shade.

Grafts of *C. fissilis* have been made using rootstock of *Toona ciliata*, which may reduce susceptibility to damage by the shoot borer *Hypsipyla robusta*.

C. fissilis has been successfully planted under widely-spaced stands of *Pinus elliottii* and *Araucaria angustifolia*. Seedlings should be planted out at the beginning of the rainy season, as drought can cause mortality or reduced growth.

The species has good production of seeds and subsequent germination, but often encounters natural regeneration problems due to pests and competition. Best regeneration is observed in soils with sparse vegetation, such as on the border of rock outcrops or in areas cleared by wildfire. Silvicultural systems should be used that simulate these conditions, such as group selection or strip clear-cutting. Since *C. fissilis* establishes mostly from seed, it is important to ensure that seed trees are left near cleared areas. Some post-establishment release of seedlings may be necessary where weed infestations are severe.

Culmination of mean annual increment averaged 152 years in the Misiones province of Argentina (range 61–180 years). Actual cutting cycles in natural stands of this species are much shorter, typically 30–50 years. In Bolivia, trees produce 1.5–3.5 m^3 of wood per stem, with greatest wood values in size classes above 120 cm. Standing dead trees have been harvested for timber in salvage operations in Bolivia.

Because of high-grading and over-exploitation, *C. fissilis* is in danger of commercial extinction in natural stands over much of its range. In cases where insufficient or only poorly formed seed trees exist, enrichment planting may be needed to restore the species to its original densities in natural stands.

Diseases and pests

The major pest of *C. fissilis* is the shoot borer *H. grandella*, a pest that has caused major problems with other species of *Cedrela* and with *Swietenia macrophylla*. It causes deformations of the stem and growth reductions when larvae bore into terminal shoots, resulting in their death. Shoot-borer attack has prevented the planting of *Cedrela* species in pure stands in many parts of tropical America, although *C. fissilis* is much more resistant than *C. odorata*. Attack by shoot borers is greatest at the beginning of the rainy season when new growth of *Cedrela* commences. Control is best achieved through mixed planting, which reduces the chance that shoot borers will encounter their host and decreases the probability of outbreaks. Natural biological control agents have been introduced, including various Hymenoptera and nematode species.

The lepidopteran larva *Antaeotricha dissimilis* feeds on the leaves of *C. fissilis* and was considered a potentially dangerous pest in Brazil. In Bolivian dry forests, many seedlings and saplings of *C. fissilis* are infested with an aphid that may reduce tree growth.

Uses

C. fissilis is suggested for use in shelter belts for middle and upper crown positions. Because of its rapid growth and ability to tolerate poorer soils, it can also be used in soil stabilization plantings. When grown as an agroforestry species, its flower production and pollen quality make it valuable in honey production. Because it is important to plant *C. fissilis* in mixtures to avoid shoot borer attack, it is amenable to multi-cropping systems in agroforestry. It is also used as an ornamental tree for gardens and street plantings.

All *Cedrela* species have similar wood properties, and are marketed in Brazil under the general name 'cedar' or 'cedro', although some authors consider the wood quality of *C. fissilis* to be lower than that of *C. odorata*. The wood of *C. fissilis* is fine grained, lightweight (density 340–550 kg/m^3), easy to work and has good finishing properties. It is highly regarded for furniture, general carpentry, decorative mouldings, doors, windows, decorative woods, musical instruments, naval construction, cooperage, cigar boxes, wood turning and toys. Because of its light weight, it is not recommended for construction. It is moderately resistant to rot, except at constant high humidity or when in contact with soil. It has low permeability to wood preservatives. It is exported to Chile, USA, Europe, Mexico and Uruguay. The importance of *C. fissilis* as a timber tree has increased due to the commercial restrictions imposed on *Swietenia macrophylla*, which it is slowly replacing in commercial markets.

Indigenous peoples in South America have often used the wood of *C. fissilis* for making canoes. It has been used for sacred art in Brazil, as well as in the Jesuit Missions of eastern Bolivia. Its bark is astringent and is used in traditional medicine in South America as a febrifuge and abortifacient.

Further reading

Garcia, M.G., Silva, R.S., Carniello, M.A., Veldman, J.W., Rossi, A.A.B. and de Oliveira, L.O. (2011) Molecular evidence of cryptic speciation, historical range expansion, and recent intraspecific hybridization in the Neotropical seasonal forest tree *Cedrela fissilis* (Meliaceae). *Molecular Phylogenetics and Evolution* 61, 639–649.

Nunes, E. da C., de Castilho, C.V., Moreno, F.N. and Viana, A.M. (2002) *In vitro* culture of *Cedrela fissilis* Vellozo (Meliaceae). *Plant Cell, Tissue and Organ Culture* 70, 259–268.

Pennington, T.D., Styles, B.T. and Taylor, D.A.H. (1981) *Meliaceae.* Flora Neotropica Monograph 28. New York Botanical Garden, New York, New York, USA.

Cedrela odorata L.

Spanish cedar

Importance

Cedrela odorata is a large tree, up to 40 m tall and 2 m in diameter, which produces a lightweight timber. Its natural range is confined to the New World, extending from northern Mexico to Argentina, and including the islands of the Caribbean. It is widely planted throughout the tropics and its timber is well known for its use in cigar boxes and a broad range of other products, including musical instruments. It is also occasionally planted for shade and used as an ornamental tree on roadsides and in parks. *C. odorata* has great potential as a plantation species, due to its fast growth and timber-producing characteristics. It is also used as an agroforestry species in cocoa and coffee plantations. *C. odorata* is highly vulnerable to attack by shoot borers (*Hypsipyla* spp.), whose larvae damage seedlings and saplings. There is some evidence that infestation can be controlled by planting *C. odorata* with other species, such as *Leucaena leucocephala*, *Neolamarckia cadamba* or (under light shade) with *Eucalyptus deglupta*.

Botanical features

The trunk of *C. odorata* reaches 20 m tall with a diameter that can exceed 2 m. One characteristic of this tree is its aroma: the leaves smell strongly of garlic or onions, and the flowers have a strong malty smell. The leaves are very long, up to 80 cm, with 6–7 pairs of leaflets (exceptionally as few as 5 or as many as 14); the leaflets are ovate to lanceolate, acute to rounded at the base, acute, acuminate or obtuse at the tip, 8–20 cm long × 2.5–5.5 cm broad (occasionally up to 8 cm), and generally glabrous. The large and much-branched inflorescences bear numerous small, five-parted, symmetrical greenish-white flowers. The flowers are 6–9 mm long; their petals are greenish-cream in bud, becoming white after opening. The fruits are 2.5–4.5 cm long, winged and septicidally 5-valved. The seeds are flat, chestnut-brown, 20–25 mm long and 6 mm broad, including the wing.

Distribution

The natural distribution range of *C. odorata* is confined to the New World, extending from northern Mexico to Argentina, including the Caribbean, approximately between 26°N and 28°S. It is widespread but never very common throughout moist tropical American forests; its numbers are continuing to be reduced by exploitation without successful regeneration.

Silviculture characteristics

C. odorata tolerates drought, fire, frost, shade, termites and wind, and can be pollarded. It is a climatic generalist, found over a wide geographical range of warm latitudinal belts, from subtropical dry forests through subtropical moist forest to subtropical wet forest, and into tropical moist and wet zones and tropical premontane moist and wet zones in equatorial regions. It is most abundant in the lowlands and foothills, and is replaced by other species such as *Cedrela montana* and *Cedrela lilloi* at higher elevations. Its distribution is within the frost-free tropics for the most part, although it has been collected at latitudes 26°N and 28°S where occasional light frosts can be expected. Mean temperatures of 23–26°C are found in the Caribbean part of its range; in tropical South America, the mean temperature is slightly higher at 28°C, with a mean minimum of 23°C and a mean maximum of 32°C. At the southern limit of its range in Argentina, the mean temperature is 24°C, mean maximum temperature is 30°C and mean minimum is 18°C.

C. odorata prefers seasonally dry climates, as reflected in its deciduous habit and formation of (presumably annual) growth rings. It prefers an annual rainfall of 1200–2400 mm, with a dry season of 2–5 months; in areas with only 1000 mm of rainfall, its growth is often stunted and slow. It occurs only sporadically in areas with up to 3500 mm annual rainfall, and then only on very well-drained sites.

C. odorata is always found naturally on well-drained soils – often but not exclusively on limestone – or on sites with heavy or waterlogged soils. It may be exacting in its soil requirements, but these are still imperfectly understood. In the West Indies, it is most commonly found on limestone-derived clay soils but it also grows in well-drained sites on acid soils derived from volcanic rock (ultisols). The common factors appear to be the drainage and aeration of the soil, rather than its pH, although no definitive studies of nutrient requirements in the forest have been performed.

Silviculture practice

The seeds of *C. odorata* are orthodox, and the species can also be propagated vegetatively by air-layering, with cuttings or from stump plants. Stands can be established through direct sowing, natural regeneration, planting stock or wildings.

Diseases and pests

By far the most serious insect pest of *C. odorata* is the mahogany shoot borer (*Hypsipyla grandella*). The larvae of this moth eat the pith just behind growing shoot tips, causing death of the apical meristem, which slows the plant's growth and reduces the tree form by promoting multiple leaders or causing seedling mortality. The tree is also attacked by termites and a number of other generalist boring insects and fungal pathogens.

Uses

C. odorata is an important street tree both in its native and introduced ranges, and has also been used successfully as a shade tree in cacao and coffee plantations.

The timber is attractive and moderately light (with a specific gravity of around 0.4), and remains in demand wherever it is available in its native American range. Its main uses are in household articles used to store clothing, as the heartwood contains an aromatic and insect-repelling resin that is the source of its popular name 'Spanish cedar', as it resembles the small or true cedars (*Cedrus* spp.). The wood is easy to work and is used for furniture, windows and doors, and is reported to be occasionally used for a wide range of other products. Also, being resistant to fungal decay and termites means it is particularly suitable for exterior construction, and it makes excellent plywood and veneer. It would be more widely used if it could be successfully grown in plantations. *C. odorata* also has a number of uses in traditional medicine.

Further reading

Cavers, S., Navarro, C. and Lowe, A.J. (2003) A combination of molecular markers identifies evolutionarily significant units in *Cedrela odorata* L. (Meliaceae) in Costa Rica. *Conservation Genetics* 4, 571–580.

Cintron, B.B. (1990) *Cedrela odorata* L. Cedro Hembra, Spanish Cedar, Meliaceae. Mahogany family. In: *Silvics of North America. Hardwoods.* Agriculture Handbook 654. Vol. 2. US Department of Agriculture, Washington, DC, pp. 250–257.

Muellner, A.N., Pennington, T.D., Koecke, A.V and Renner, S.R. (2010) Biogeography of *Cedrela* (Meliaceae, Sapindales) in Central and South America. *American Journal of Botany* 97, 511–518.

Cedrelinga cateniformis (Ducke) Ducke

Tornillo

Importance

Cedrelinga cateniformis is an abundant leguminous tree species of the tropical humid forests of Amazonia. It has various common names throughout its range, including tornillo and cedrorana. Economically valuable for wood production and also culturally important to local communities, *C. cateniformis* is a canopy tree which in some areas has one of the highest natural volumes of any timber species, reaching more than 10 m^3/ha. Its reddish wood is durable, lightweight, easy to work, and is utilized in boatbuilding, construction and carpentry.

It has been planted commercially in Colombia, Ecuador and Peru. Introductions outside its native range are unknown.

Botanical features

C. cateniformis is a large canopy tree with a rounded, uniform and compact crown. It can reach 25–45 m in height with a dbh of 80–150 cm. The stem is cylindrical with prominent buttresses up to 1.5 m. The greyish-brown external bark is fissured, while the internal bark is dark red with brownish latex.

Leaves are bipinnate, opposite and 30–40 cm long, with 1–2 pairs of pinnae with 2–4 pairs of leaflets, 10–20 cm long and 6–10 cm wide. The petiole has small, interfoliar glands.

The compound, capitulate inflorescences form terminal or axillary panicles 15 cm long. The hermaphrodite flowers range from greenish-brown to creamy-white. The yellow or green fruits are twisted, indehiscent legumes, 40–70 cm long and 3–6 cm wide, with 1–8 seeds.

C. cateniformis is evergreen, but loses a large percentage of its leaves during the dry season. The timing and regularity of fruiting and flowering are very variable. Generally, flowering occurs during the peak of the rainy season (December–January), and fruits mature within 2 months.

Distribution

C. cateniformis is a strictly South American tropical forest species. Its range extends from northern Bolivia north to Venezuela, Colombia, Suriname and French Guiana, and from Peru to eastern Brazil in low- to mid-elevation forest.

Silviculture characteristics

C. cateniformis is moderately shade-tolerant and has the ability to fix nitrogen. It can outcompete other commercial tree species in height growth under shaded conditions. As an aggressive colonizer in logging gaps, it responds well to release treatments. However, it is gap-dependent, and seedling mortality is high in the absence of gap formation. *C. cateniformis* also regenerates well in disturbed areas. Although relatively drought tolerant, it prefers sites with a dry season of less than 4 months, and high annual precipitation.

C. cateniformis commonly grows on nutrient-poor soils with high acidity (such as oxisols, cambisols and ultisols) and can also tolerate a wide range of soil and physiographic conditions, from dry, rocky, upland soils to alluvial flats, and soil textures ranging from sandy–loam to clay–loam. It tolerates shallow soils, often occupying hillsides and slopes.

Silviculture practice

Natural forest management of *C. cateniformis* is relatively simple because of its abundant natural regeneration from seed. However, seeds require significant rainfall on dispersal in order to establish successfully. Seedlings often become established in abundance under parent trees.

Seedlings can be produced in the nursery or collected from natural or wilding stocks. Seed viability is greatly reduced after 50–60 days and can significantly decrease after only 1 week. Ideal seed storage conditions are 12% moisture content and 5°C, but seeds can rarely be stored for more than 100 days under any conditions. Seed germination capacity is generally high and seeds do not require fungicidal treatment. Successful regeneration is variable because of its dependence on adequate moisture for germination and retention of seed viability. Wilding reproduction is also abundant, particularly when understorey clearing is conducted. Cuttings, grafting and micropropagation are also possible.

Plantation establishment is best carried out using bare-root stock. Topsoil removal and soil compaction caused by forestry machinery can reduce the diameter growth of planted seedlings.

C. cateniformis performs well in both monocultures and mixed species plantations. Mean annual increments of 1.5 cm in diameter and 1.5 m in height have been recorded in a 10-year-old plantation, while in pure-block plantations the species attained a mean dbh of 35.8 cm, mean height of 26.4 m, basal area of 30.3 m^2/ha and a mean volume of 380.2 m^3/ha after 20 years. The species is also utilized in enrichment planting; seedlings planted in 5 × 200 m strips within a climax forest can reach 14–16 m in 6 years. In a strip clear-cutting system, tornillo regenerating in two strips (30 × 150 m) clear-cut in the Peruvian Amazon forest reached commercial size (≥30 cm dbh) 40 years after the initial cutting. When used in agroforestry systems, trees are planted at 10 × 10 m spacings.

Diseases and pests

Of concern is a canker disease caused by three species of *Fusarium* with similar levels of pathogenicity; this serious disease was observed for the first time in 2002 on the main stems and side branches of trees in plantations in Ecuador, and is now becoming more widespread. Symptoms include severe cracking of the bark, kino exudation and wood discoloration in the region of the bark cracks, as well as die-back and growth reduction.

C. cateniformis is relatively pest free, although parrots (Psittacidae) are important seed predators and bark beetles (Scolytidae) can attack green wood. The wood is susceptible to attack by fungi, termites and boring insects.

Uses

C. cateniformis is used in agroforestry systems in Peru with bananas, citrus, coffee, and cacao. As the species has the ability to fix nitrogen, it has the potential to be used in soil improvement and reforestation of degraded lands. *C. cateniformis* is also suitable for use as a shade tree in silvopastoral systems.

The heartwood of *C. cateniformis* is pale brown with a golden lustre, is prominently marked with dark red vessel lines and merges gradually into the lighter-coloured sapwood. Wood texture is coarse, and freshly cut timber is reported to emit an unpleasant odour when worked, although this is absent in dried wood. Basic specific gravity of the wood (oven dry weight/green volume) varies from 0.41 to 0.53 depending on provenance. The timber is resistant to compression and is excellent for construction and structural uses. However, compression failures on live trees are common sources of wood defects. Because of its light weight and workability, it is also used for boats, furniture, woodware and in turnery. *C. cateniformis* is also used locally as fuelwood.

Sawdust from *C. cateniformis* is one of the best media for shiitake mushroom (*Lentinula edodes*) production.

The bark is used as a purging admixture in *ayahuasca* psychoactive infusions, prepared by indigenous people in Amazonia from the vine *Banisteriopsis caapi*. It is also used medicinally for gastrointestinal ailments.

Further reading

Barefoot, A.C. and Traywick, J.D. (1971) Mechanical and related properties of tornillo (*Cedrelinga catenaeformis*). *Wood Science* 3, 245–253.

Farias, V.C.C., Varela, V.P., Costa, S. de S. and Batalha, L.F.P. (1997) Análise de crescimento de mudas de cedrorana (*Cedrelinga catenaeformis* (Ducke) Ducke) cultivadas em condições de viveiro [Growth analysis of cedrorana (*Cedrelinga catenaeformis* (Ducke) Ducke) seedlings cultivated in nursery conditions]. *Revista Brasileira de Sementes* 19, 193–200 [in Portuguese with English summary].

Salazar, R. and Soihet, C. (2001) *Cedrelinga catenaeformis* Ducke. In: Salazar, R. and Soihet, C. (eds) *Manejo de Semillas de 75 Especies Forestales de América Latina*. Volumen II, CATIE, Turrialba, Costa Rica [in Spanish], pp. 105–106.

Cedrus atlantica (Endl.) G. Manetti ex Carrière

Atlas cedar

Importance

Cedrus atlantica (Plate 10) is an important species of the Mediterranean forest zone both for economic as well as for environmental reasons. Resistant to drought, it is a valuable timber tree, growing quite quickly despite the unfavourable conditions. This species has been successfully introduced for afforestation since 1860 in coastal locations in northern Mediterranean countries, to restore degraded areas and for the production of high-quality timber. It regenerates easily and provides a valuable aesthetic element to many Mediterranean landscapes, and has also been introduced in parks and gardens for ornamental purposes. Due to the diversity in form and colour, several horticultural varieties have been selected.

Botanical features

C. atlantica can reach 50 m height and more than 2 m dbh. In its native range, trees can live to be several hundred years old. Young trees have a conical crown, whereas mature trees are characterized by a typical flat-topped crown. The main branches are located in pseudo-whorls at the top of each annual growth unit. Smaller branches are located all along the annual shoot. If *C. atlantica* grows alone or at low density, the main lateral branches develop and can become very thick, giving a wide range of forms. The axillary branches of the genus *Cedrus* comprise two types of axes: long pubescent shoots and short shoots. The short shoots form a cluster of needles each year at the base of the terminal bud. *C. atlantica* is an evergreen species and needle length, often less than 25 mm, depends on growing conditions. Needles are sharp-pointed, green in colour, bluish-green in some trees. Taproots form if soil depth allows, and in shallow soils a plate-like root system develops. A range of fungi has been identified to provide potential mycorrhizal symbiosis with *C. atlantica*, some of which are specific to the species.

C. atlantica is monoecious even if the trees are in general either male or female; the male and female flowers are never on the same branches. The phenology of reproduction lasts over 2 years. In the first year, male flowers appear in July and female cones appear in August, both at the top of short shoots. Pollination occurs in September but fertilization is not effective before the spring of the second year. The cones grow to 5–8 cm long until September of the same year before disarticulating while on the trees; most seeds fall in November of the third year, though some remaining seeds fall in the subsequent winter and spring. The period of bud break varies greatly depending on climatic conditions and seed provenance.

Distribution

As a native species, *C. atlantica* covers a total of about 150,000 ha in the mountainous regions of Morocco (120,000 ha) and Algeria (25,000 ha). Its natural area is discontinuous in the Rif and Atlas Mountains. Mainly in pure stands, it is often associated with *Quercus rotundifolia* and sometimes on acid soils with *Abies maroccana* in the Rif Mountains and with *Abies numidica* in Algeria, where rainfall is a more important factor.

Silviculture characteristics

The litter is not easily flammable and fire does not readily spread beneath the closed canopy, where the light level is insufficient for undergrowth and brush development. Except in very dense stands, the self-pruning capacity of *C. atlantica* is low because of its tendency to produce large branches at wide spacing. Successful natural regeneration depends on site quality, with reduced regeneration on better sites due to increased weed competition, and fire favouring regeneration by reducing this competition. *C. atlantica* begins to produce seed after 25 years but production is not sufficient for natural regeneration until trees are 40 years old.

Silviculture practice

Seed production is irregular, with one heavy and one medium crop every 7 years on average. Closed cones can be soaked in water for 1–2 days before disarticulation, which can also be made easier by freezing followed by defrosting. Seeds (10,000–15,000/kg) are stratified at 4°C for 5 weeks to break dormancy. Removing seed wings can have a negative influence on germination rate. After drying to a moisture content of 8–10%, seeds can be stored for a long time at temperatures between −15°C and −5°C.

Seeds are generally sown in containers with an optimal capacity of 400 cm^3 in a mixture of equal parts peat and bark. Seedlings are generally planted out after 1 year's growth in the nursery. Site preparation often involves subsoiling on calcareous rocky soils, with pit planting recommended on other sites. *C. atlantica* is very sensitive to most types of herbicide. Artificial pruning is highly recommended to avoid the growth of thick branches and subsequent large knots.

General silvicultural practice recommends a reduction of stand density to 1200 trees/ha when stand top height reaches 10 m. Subsequent thinning operations are then carried out to obtain a density of 100–250 trees/ha by the time of clear felling which, depending on the site index, may be at 25–35 m top height. The silvicultural objective when marking a stand for thinning is to concentrate the yield on crop trees that have been selected previously. To promote natural regeneration, the overstorey is progressively removed after a heavy seed production, depending on occurrence of seedlings. Cedar forests are often managed by even-aged methods even if the stands can be managed by uneven-aged methods. Generally, without thinning or because of lack of regeneration, cedar forests are unable to maintain the reverse-J-shaped diameter structure presumed necessary for sustaining an uneven-aged state.

On the best sites, early height growth (at 15–30 years old) can exceed 80 cm per year and annual mean volume increment at 100 years old can be as high as 15 m^3/ha.

Diseases and pests

In the natural range of *C. atlantica* in North Africa, significant die-back has been observed in stands closest to the desert in the Great Atlas (Morocco) and in Teniet el Had (Algeria), accompanied by an outbreak of Scolytidae insects causing high mortality, particularly in Morocco. In these same areas, cedar stands are periodically attacked by the pine processionary moth (*Thaumetopoea pityocampa*) and population levels are regularly surveyed and infested zones aerially treated with a preparation based on *Bacillus thuringiensis*. Between 1980 and 1990, much defoliation by the cedar processionary moth, (*Thaumetopoea bonjeani*) was observed in Algeria. In the

Moroccan Middle Atlas, an outbreak of the budworm *Acleris undulana* was observed from 1983 to 1990.

Stands in France, Italy and Spain were affected by die-back from 1967 following the introduction of the aphid *Cedrobium laportei*, but introduction of *Pauesia cedrobii*, its specific parasite, has reduced aphid population levels significantly since 1980. Every 5–6 years, the budworm *Epinotia cedricida* causes important defoliation and can greatly reduce natural regeneration. Since 1994, a seed insect, *Megastigmus schimitscheki*, has also been observed. It originates from Asia Minor where it is a major pest.

Uses

C. atlantica is often used in parks because of its desirable and aesthetically pleasant shape, and there are horticultural varieties with a wide range of foliage colour or growth form. In Europe, it is often used to enrich coppice stands and to improve the local forest productivity. In North Africa, cedar forests are grazed with livestock in silvopastoral systems.

The wood is well adapted to most uses. It is highly appreciated for its very high quality and natural durability, particularly its resistance to decay fungi. The heartwood is often used for exterior fittings such as fences, posts, poles, outside shutters and doors, rustic furniture and outdoor sports equipment. The wood has high stability and remarkably low shrinkage values. Although ring width is large, the density of wood remains high (540–620 kg/m^3). Despite its high density, the moduli of elasticity and rupture are low, making the wood a little brittle compared to other softwood species. The sawn or hewn timber must be oversized for building use. Due to its smell, the wood cannot be used for food containers, though it is well known for its mothproofing properties.

The wood is commonly used in Morocco and Algeria and is one of the most valuable timbers in the region. In the south of France, the market in *C. atlantica* is beginning to develop and the market price for standing trees is now one of the highest of any species in the country.

Essential oils from *C. atlantica* and other *Cedrus* spp. have a range of uses, and also exhibit potential molluscicidal activity. Epiphytic lichens are collected in Morocco to produce medicines.

Further reading

Pijut, P.M. (2000) *Cedrus* - the true cedars. *Journal of Arboriculture* 26, 218–224.

Toth, J. (2005) *Le cèdre de France: étude approfondie de l'espèce [The cedar of France: in-depth study of the species]*. Editions L'Harmattan, Paris.

Cedrus deodara **(Roxb. ex D. Don) G. Don**

Deodar cedar

Importance

Cedrus deodara is the most important timber tree of the western Himalayan region, where it occurs at 1100–3000 m altitude, usually on silicate mother rocks. Its wood is highly valued and extensively used for construction, railway sleepers and wherever durability is required, as well as for furniture, general carpentry and many other purposes. It is widely planted as an ornamental, and numerous cultivars exist differing in needle and shoot colour, and in habit.

Botanical features

A very large evergreen tree, up to 65 m in height and more than 4 m in stem diameter, with horizontal or slightly ascending or descending branches, which are not whorled but arise irregularly from the stem. Up to middle age, the trees have conical crowns with definite leading shoots, but in later life, their crowns become rounded or broad and flat with horizontal branches. The bark is dark greyish-brown, with vertical and diagonal cracks dividing it into irregular oblong scales. Shoots are dimorphic.

Foliage is silvery or silvery-blue. Needles are acuminate, solitary, stiff, sharp-pointed, 25–37 mm long, arranged spirally on normal long shoots and in pseudowhorls on short arrested shoots.

C. deodara is dioecious. Male flowers are pale green to yellowish-green with a purplish tinge, oblong–ovoid, and 2.5–4.6 cm long by 1–1.5 cm in diameter. On opening they elongate rapidly to 5–7.5 cm and become yellow with pollen. The female flowers are solitary and erect at the end of arrested branchlets, and at the time of pollination are oblong–ovoid, 1.2–2.0 cm long and 0.6 cm in diameter, and pale glaucous green. The scales occur in a spiral of 8 × 5; at the time of pollination they stand perpendicular to the axis, exposing ovules, and close after pollination.

The ripe cones (solitary or in pairs) are erect, brown, ovoid or ellipsoidal, 7.5–12 cm long and 5–8.7 cm in diameter with numerous fan-shaped scales arranged in spirals of 8 × 5 on a persistent woody central axis. On each scale rests a pair of winged seeds. The seed with wing is 2.5–3.7 cm long, triangular; wings with rounded corners, 2–2.5 cm wide; seed without wing 2–2.5 cm wide, irregularly triangular. There are about 8000 seeds/kg. Maximum cone weight (226.5 g), seed weight/cone (16.86 mg) and seed yield (7.44%) occur in the 41–50 cm tree diameter class. Seeds are oily and soon lose their viability, but good fresh seed is usually highly viable. The periodicity of good seed years is about one in three.

New shoots appear in March or early April. Old leaves are shed during the hot season in May; leaf persistence varies with age and growth vigour. Solitary and erect male flowers or catkins appear in June, ripen and shed pollen from the middle of September to the middle of October according to locality and altitude, at which time female flowers also ripen for pollination. Female flowers or cones appear in August, are inconspicuous and partly hidden by the rosette of leaves. There is no growth in young fertilized cones until the following spring, and by early May they are large enough to become visible. They become full-sized by the end of June or July and are pale bluish-green. Cones ripen from the end of September to November, but may not fall until the early part of December at higher altitudes. Thus the time occupied from the first appearance of the female flowers to the ripening of cones is 12.5–13.5 months.

Distribution

C. deodara forests are found naturally throughout the western Himalayan ranges in Afghanistan, Pakistan and India at an elevation of 1200–3300 m, but most commonly at 1800–2550 m. They generally occur as pure stands of considerable extent on moderate to precipitous slopes as well as on level ground, dominating cool northern slopes in particular.

Silviculture characteristics

C. deodara is typically gregarious and light-demanding. It tolerates drought when mature, but not at the seedling stage. Trees are resistant to atmospheric pollution. A mild climate and an altitude of around 2000 m is ideal for this species, as frost injury and inhibited growth occur above 3200 m and at temperatures below −25°C. It does not grow in heavy rainfall areas. Ectomycorrhizae are associated with roots.

Silviculture practice

Where conditions are favourable, the natural tendency of *C. deodara* is to regenerate in groups of more or less even age in the neighbourhood of seed bearers, either in gaps of sufficient size or in completely isolated situations. The techniques for artificial establishment of plantations are generally patch sowing, line sowing or planting of nursery stock (2¼ years old, pricked out at 3 months of age), this last being commonly recommended as the cheapest method of securing established plantations in the shortest time.

Seed is generally sown in the nursery in March. Seed germination rate is best in seeds soaked in water for 24 h. The type of growing medium also has a significant and varying impact on germination, length and weight of shoots and roots, collar diameter, shoot/root and sturdiness ratios, and quality index. Maximum germination is observed in soil from deodar forest floor. Treatment of seed with mycorrhizal fungi significantly increases plant biomass. Seeds stratified at 5°C in moist sand for 30 days also give a much higher percentage germination (44.7%) than those stratified for 15 days (16%) or not at all (11.3%).

C. deodara can be propagated vegetatively, with enhanced rooting and survival of plantlets obtained by treating greenwood cuttings with plant growth regulators.

In young crops, whether natural or artificial, tending operations consist of weeding for 3–4 years, freeing trees from suppression by blue pine (*Pinus wallichiana*) and thinning, which should begin early and be repeated at intervals of about 10 years for gradual opening of the crop, to make it resistant to snow damage and to remove suppressed and damaged trees.

C. deodara forests have for a long time been managed through a system under which selection fellings are carried out with a definite felling cycle, and with a minimum exploitable girth or diameter of selected trees. Improvement fellings are also made to remove unsound or badly shaped trees as well as interfering accessory species. Introduction of intensive forest management based on a combination of a shelterwood system with clear felling of trees down to an advanced growth of up to 30 cm diameter, with artificial regeneration mainly by planting, has proved successful. This system allows areas to be assigned for natural regeneration assisted by artificial regeneration, followed by felling. These blocks have a rotation period of 120 years, an exploitable diameter of 55 cm and a regeneration period of 30 years.

Forests respond well to regeneration fellings and conversion to a uniform silvicultural system, with such forests giving an estimated growing stock of 228.73 m^3/ha, 331.28 trees/ha and average annual increment of 4.33 m^3/ha.

Diseases and pests

The parasitic fungi *Fomes annosus* and *Peridermium cedri* are injurious, causing mortality or the formation of witches' brooms. *Pestalotiopsis cryptomeriae* causes leaf blight on young, 5–8-year-old trees. *Ploioderma cedri* has caused foliar infection and premature defoliation on 60-year-old plantations in Himachal Pradesh, India.

In China, 3-year-old *C. deodara* seedlings were observed to develop root rot, crown rot, chlorosis, and defoliation, often resulting in death from attack by *Phytophthora cinnamomi* and *Phytophthora drechsleri*. A study of 115 trees in Pakistan, including *C. deodara*, showed that 80% of the total volume of decay was by the fungus *Phellinus pini* but this was found only on old trees (>90 years).

Other pests reported on *C. deodara* include *Ectropis deodarae* causing complete defoliation, *Dioryctria abietella* damaging cones and seeds, and borers causing widespread mortality of trees of all ages, but particularly those with a diameter of 20–29 cm.

Trees are damaged by snow, fire, and browsing by goats. Among wild animals, bears, porcupines and monkeys are the most injurious. *Rosa moschata* scrambles into the crowns of young trees and can suppress them.

Uses

C. deodara shelterwoods increase dry land pasture productivity in New Zealand. The species is widely planted as an ornamental tree, and in the USA has shown promise as an alternative Christmas tree species.

The wood is white to light yellowish-brown, with a characteristic odour and oily feel. It is straight grained, medium fine and somewhat uneven textured. Its average density is 560 kg/m^3. The heartwood is very durable, but insects and fungi attack sapwood. The wood is easy to saw and work to a smooth finish. Due to the presence of volatile oleoresins in the wood, to which the heartwood owes its characteristic odour and durability, it is not suitable for high paint and polish work.

Deodar is an important structural timber, and is highly suitable for railway sleepers, carriage and railway wagon work, housebuilding, beams, floorboards, door and window frames, light furniture, ordnance boxes and pattern making. Plywood of satisfactory quality can be made using hot-setting urea formaldehyde (UF) glue or water-soluble phenol formaldehyde (PF) glue.

Oil and extracts are used as insecticides, molluscicides and in herbal medicines for the treatment of animal diseases in India.

Further reading

Maheshwari, P. and Biswas, C. (1970) *Cedrus*. CSIR Botanical Monograph No. 5. CSIR, New Delhi.

Shamet, G.S., Bhardwaj, D.R. and Parmar, Y.S. (2000) Vegetative propagation of Deodar (*Cedrus deodara*) through cuttings. *Advances in Forestry Research in India* 23, 118–129.

Tewari, D.N. (1994) A monograph on deodar (*Cedrus deodara* (Roxb.) G. Don). International Book Distributors, Dehra Dun, India.

Cedrus libani A. Rich.

Cedar of Lebanon

Importance

Cedar of Lebanon (*Cedrus libani*) (Plate 11), the national emblem of Lebanon and displayed on its flag, is mentioned in the oldest recorded legend, the Epic of Gilgamesh, as well as in several books of the Old Testament. It is a large, magnificent tree, with a beautiful shape, originating in Asia Minor and extensively planted elsewhere for ornamental purposes in parks and gardens. Its wood is used for carpentry and in the past was highly esteemed and widely used, so that today natural forests have almost completely disappeared in some parts of its natural area of distribution, such as Lebanon, where only a few monumental trees survive. *C. libani* is not much used in silviculture in Europe, and at present is only used for reforestation or afforestation in Turkey.

Botanical features

A tall tree to 40 m, living up to 700 years. Main bole often branched, dividing near ground level to give stout, near-horizontal branches, which may extend a long way, or may curve upwards to ascend vertically. Hence, stem circumference may attain as much as 15 m. Horizontal branches irregular, divided into numerous densely arranged branchlets (denser than those of *Cedrus atlantica*), short, glabrous or irregularly pubescent. Thick main bole with blackish-grey bark, split or divided into narrow plates. The broad crown is pyramidal when young, but may develop a flattened, umbrella shape on old trees. Growth is rapid until age 45–50 years, then the increase slows progressively until approximately 70 years, after which growth is extremely slow.

Needles and buds are solitary on young twigs, joined together in bundles of 30–40 needles on older twigs, 15–35 mm long, about 1 mm wide, stiff, 4-sided, broader in width than in height, sharpened apex, dark or bright green. Needles remain on trees for several years.

Male microsporophylls 3–5 cm long, green-grey at the early stage, becoming yellow when pollen is ripe. Pollen is dispersed by wind. Male flowers, after pollen has been discharged, become light or dark brown, and fall 1–2 months later. Female flowers, 1 cm long, are green at first, becoming purple when mature. When female flowers are fertile, they develop an ovoid form and a globose cone. Female cones are carried on a stalk and are solitary, upright, barrel-shaped, 8–10 cm long, 6–7 cm across, impressed at apex, sometimes resinous; scales are numerous, densely imbricated, narrowed to base, about 5 cm across, outside slightly decurved, entire, dorsal side short downy pubescent. When cones are 1 year old they are green, when they are 2 years old they become light or dark brown. Trees produce cones from about age 40 years. Cones open from top to bottom of the crown, releasing their seeds with the combined effects of sun and rain; only the rachids remain on the branch.

Seeds are oblong–ovoid, triangular, narrowed to base, compressed, 15–18 mm long; wing thinly membranous, half fan-shaped, straight truncate above, 25 mm long, brown-red. Seeds have a protective resinous blister that, if broken, makes germination impossible, because the embryo is very fragile. Seeds are relatively heavy and they tend to fall close to the mother tree, although they may be dispersed by wind, birds or rodents.

Cedrus is the only genus of the Pinaceae that flowers in autumn, with male flowers appearing in September and female ones at the end of September or the beginning of October; fertilization occurs in autumn. Seeds ripen from late summer to October, and are shed until the following spring. Sometimes cones remain on the tree for several years without opening.

Distribution

In Lebanon, the large cedar forests have practically disappeared due to over-exploitation and forest degradation, so that today *C. libani* is found in only 15 fragmented populations covering a total of about 2300 ha on the western slopes of the western mountain range, between altitudes 1050 m and 1925 m. These populations are either pure *C. libani* or mixed forest with pine, oak and juniper. There is one small isolated population in north-west Syria. The main populations of *C. libani* are found in eastern Turkey, in the Taurus mountain region, covering about 100,000 ha between altitudes 530 m and 2000 m. Many of these forests are pure *C. libani*.

Silviculture characteristics

C. libani can grow rapidly when young on deep, fertile and light-textured soils. It is a light-demanding, calciphile coniferous species that is most vigorous when planted in montane regions.

Silviculture practice

Silvicultural systems incorporate natural regeneration, although this may be enhanced by additional direct sowing.

Both shelterwood and strip clear-cut methods are in use in Turkey, in order to encourage natural regeneration. The shelterwood method may be more suitable for use on south-facing, shallow karstic soils where root competition from seed trees is not severe, as the shelterwood retained would protect regenerated seedlings from late frosts and from direct sunlight during the summer. In contrast, on north-facing, shady and/or more sheltered aspects, the strip clear-cut method could be applied.

For the shelterwood method, a seeding cutting is made in a medium or good seed year, to be completed by the end of September; about 20–25 seed trees/ha are sufficient. If prescribed fire is applied before seed dispersal, care should be taken to remove slash from residual seed trees, as *C. libani* bark is not especially fire-resistant. Direct sowing, at a suggested rate of 15–20 kg/ha of additional seed, is recommended for some sites in Turkey; seed should be sown in November–December, just before first snow falls, although seed can be scattered on snow.

Under certain ecological conditions, the strip clear-cut method is used to promote natural regeneration. For successful regeneration, felling is completed by September. The first strip is 35–40 m wide, with subsequent strips 20–25 m wide. Because medium or good seed years occur about every 3 years, subsequent strips can be opened every third year or so, after it has been ascertained that it is a good seed year.

On calcareous or limestone montane sites, soil tillage is not usually necessary, as *C. libani* seedling roots can penetrate cracks in the bedrock. Prescribed fire may be used to reduce grass competition.

During the seedling stage, weeding, stocking level and beating up (replanting) may be necessary. Weeding is generally not needed where prescribed fire has been used, but may be necessary for the first 3 years in order to eliminate or reduce weed competition. From plantation age 5–6 years, when height differentiation between seedlings becomes obvious, and if seedling densities are high (15,000–20,000 plants/ha), some stocking level control can be carried out.

At the sapling stage, 6–8 years after stand establishment, height growth of *C. libani* increases notably, and young plantations will exhibit a more pronounced layered structure as regards height, especially on karstic soils. Cleaning should begin after crown closure, by the removal of slow-growing saplings.

At the pole stage, low thinning should be carried out, to promote growth of good-quality trees. Studies have indicated that, in stands aged 40–50 years, diameter and basal area growth of released trees did not increase dramatically 4–5 years after seed trees were removed.

Afforestation trials on karst areas in Turkey since the 1980s have indicated that direct sowing is the most appropriate method for reforestation. Planting stock is, however, used in areas with medium or deep soils, on lower slopes and in basins. In such cases, 1+1, 1+2 or 2+0 containerized stock should be planted in either November or April.

C. libani stands subjected to selective and partial fellings have an increased stand volume increment rate up to stand age 30 years, but this rate decreases after age 40 years.

Diseases and pests

Some fungal pathogens attack seedlings, and *Botrytis cinerea* may cause needle discoloration and loss. *Armillaria mellea* may attack roots in wet soils.

In Lebanon, *C. libani* relict forest stands are attacked by the cedar shoot moth (*Parasyndemis cedricola*), and by *Cephalcia* sp., causing necrosis of shoots, particularly young ones. The damage caused is considered serious in view of the threatened status of these forests.

Uses

C. libani is planted in warm temperate Europe as an ornamental tree; it was the first cedar species to be so utilized. Mature specimens, with their massive trunks and symmetrical crowns of widely spreading branches, give great dignity and beauty to many parks and gardens.

The wood of *C. libani* grown in Asia Minor is held in high regard, is highly fragrant and has a dark yellow colour with red speckled marks. Wood density is 560 kg/m^3. It is extremely durable and resistant to insect attack. Wood of trees grown in Europe is of poor quality. In the Middle East, the wood is traditionally used in construction, as well as for furniture and wood carvings, although it lacks figure. It is also valued as firewood as it gives off a fragrant odour.

It is possible to extract a resin called *cedrie* and an important essential oil called *cedrum* from the wood and fruits.

Further reading

Boydak, M. (1996) *Ecology and Silviculture of Cedar of Lebanon (*Cedrus libani *A. Rich.) and Conservation of its Natural Forests.* Ministry of Forestry, Ankara, Turkey.

Boydak, M. and Çalıkoğlu, M. (2008) *Biology and Silviculture of Lebanon cedar (*Cedrus libani *A. Rich.).* Foundation for Supporting the Mission of Forestry Development and Combating Forest Fires, Ankara, Turkey.

Shackley, M. (2004) Managing the cedars of Lebanon: botanical gardens or living forests? In: Harrison, D. and Hitchcock, M. (eds) *The Politics of World Heritage: Negotiating Tourism and Conservation.* Channel View Publications, Clevedon, UK, pp. 137–145.

Ceiba pentandra (L.) Gaertn.

Kapok

Importance

Ceiba pentandra is an important forest tree in the tropical Americas, West Africa and South-east Asia. It is a moderate to very large pioneer species but is also established as a plantation tree or to provide shade for coffee, cocoa or other crops. The timber is lightweight but easy to work, and is widely used. However, it is most commonly raised for the production of kapok, a cotton-like material found in the seed pods. The seeds also produce an acceptable cooking oil, the leaves are eaten and the trees have very many other uses.

Botanical features

C. pentandra is a majestic, deciduous tree which can reach 70 m high and a diameter of 2.4 m, although it usually grows to 40–50 m and diameters of 130–180 cm. The bole is straight and can be branch-free for up to two-thirds of the total height. The lower trunk is distinctly bottle-shaped with well-developed plank buttresses up to 8 m above the ground. Thick branches grow out horizontally with a low degree of rebranching giving rise to the characteristic, widely spreading crown that may be more than 50–60 m in diameter. The bark is greenish-grey, covered with spines, thin and pointed in young trees and becoming conical when mature.

Leaves are alternate, bright green, palmately compound, each with 7–9 distinctly veined leaflets, 10–20 cm long and

2.5–4 cm wide. The five-petalled flowers are white, yellowish-white or rose-pink-coloured, appearing at the end of branchlets after the tree has shed its leaves. Hanging fruits are leathery, elliptic, greenish, 10–15 cm long and up to 6 cm thick, flattened at both ends, containing 120–175 small, black seeds, embedded in a whitish-grey cotton-like floss.

Distribution

The true native range of *C. pentandra* probably includes Central America, the Caribbean, northern South America and West Africa, but its exact history is difficult to ascertain. It has been widely introduced, and is now has a pantropical distribution, in various types of evergreen and deciduous moist forest, as well as in dry forest, gallery forests and secondary forests.

Silviculture characteristics

The main distribution corresponds to tropical climates with an annual precipitation of 1400–3500 mm and a dry season of no more than 6 months, and a mean annual temperature of about 23–27°C. It is a light-demanding pioneer species and an indicator of human disturbance. It is drought-tolerant and withstands a certain amount of frost, although fruit set fails if temperatures drop to below 15°C during and after flowering. It is very sensitive to fire. It grows in a variety of soils but prefers light and medium slightly acid to neutral soils on alluvial sites, and does not tolerate waterlogged sites.

Silviculture practice

Propagation is generally by seed, root or shoot cuttings, 2–3-year-old stakes, budding, air-layering, grafting and sets. Natural regeneration is abundant, with 90–95% germination of the lightweight seeds, though regeneration can be prevented by the inhibiting effect of ground flora. Seed viability diminishes rapidly and it has not yet been possible to preserve them for extended periods, even with specialized storage techniques. Reports of improved germination include soaking in cold to lukewarm water for 24 h, or soaking in boiling water for 5 min.

Seeds are sown at 1–2 cm depth, germination taking place after 8–21 days. Young plants are grown under lightly shaded conditions until they are 12–15 cm tall, then exposed to full sunlight until they are ready to be transplanted in the field when 45–60 cm tall. Bareroot planting stock can be used, as well as container plants. Seedlings that are left for 8–10 months in the nursery should be cut back to about 1.2 m. No special soil preparation is recommended, with plantation holes of 40 × 40 × 40 cm dug in sufficient time to take advantage of precipitation being preferred in savannah areas. Application of fertilizer will enhance growth, but is not essential.

Pure stands for the production of kapok can be planted at 3 × 3 m spacing and selectively thinned out as early as the third year resulting in a final density of about 8 × 8 m. Plantations may also be underplanted with crops in the early years. Enrichment planting is performed in lines or dense, spot plantings to create a forest microclimate for the plantation of coffee, cocoa or kola nut trees (*Cola nitida*, *Cola acuminata*). Many reports of growth increments during the first 10 years are more than 1 m in height and 3–4 cm in diameter per year. Generally, rotations of 30–45 years are used, although kapok production can continue for up to 60 years. Typical annual kapok production figures are 150–300 kg/ha after 3 years; 800–1600 kg/ha at 8 years; and 2000–4000 kg/ha for mature trees.

Diseases and pests

C. pentandra is generally free of diseases and pests, though the stem-boring beetle (*Plocaederus obesus*), the kapok pod borer (*Mudaria variabilis*) and parasites from the Loranthaceae can cause significant losses.

Uses

C. pentandra is a multi-purpose tree and is grown in agroforestry systems in conjunction with coffee, cocoa or as a support tree for pepper, and also as an ornamental, shelter or shade tree and for soil protection.

The timber is greyish-white to brown, but can range from pale red to yellow and occasionally has a beige vein running through it, and the sapwood and heartwood are almost identical in colour. The wood has an unpleasant odour when cut or newly sawn. The wood is soft, light and easy to work, with a density of 450–480 kg/m^3 when green or 300–400 kg/m^3 when air seasoned, which is quick and easy without distortion or warping. It withstands strain and stress relatively well. Insects and fungal attack can destroy the wood.

Due to its light weight and softness, the wood is easy to work and peels to give good veneers. However, tools should be kept as sharp as possible as sawing produces ‘woolly’ or fuzzy surfaces and any shavings produced tend to stick together quickly. It is easy to nail and screw but it is best to use glue to obtain sound, stable joints. The wood is easy to paint when first treated with a grain filler, but clear varnish should not be used as it cannot easily penetrate the rough surface. The wood is non-durable, but because of its stability it is suitable for making canoes, tables, doors, packing cases, boxes and wooden sandals; it is also used for plywood, particleboard, pulping purposes, plywood cross-bands and machine patterns. It is excellent for radio cabinets and for sound isolation purposes. Buttresses, which can be 15 cm thick in older trees, are used as house doors in West Africa.

However, the main commercial product is the silky cotton that surrounds the seeds, marketed under the names of ceiba, Java cotton and kapok, with around 200 fruits yielding 1 kg of floss. It is water-repellent and supports 30 times its own weight in water (compared to cork, which supports only five times its own weight), making it an excellent material for life jackets. It is also used in textile production and for making insulation materials (thermal, electrical and acoustical). It can also be dry-sterilized. Specialized uses include stuffing for hospital mattresses and padded surgical dressings. It is used locally for stuffing cushions, pillows, quilts, and saddles, although if the seeds are not extracted from the floss, rats will eat the stuffing. The floss is being steadily replaced by synthetic fibres but it is still collected for local use.

Seeds contain up to 25% oil, composed of largely unsaturated fatty acids (43% oleic acid, 31% linoleic acid) and 26% protein which is well balanced in amino acids but lacks tryptophan; and is a good source of calcium and phosphorus.

It is comparable to groundnut oil and has an agreeable flavour. However, it has to be purified and contains cyclo-propenoid acids which may be harmful if ingested in large quantities. Pounded and ground to a meal, seeds are used in soaps or eaten after having been roasted. Young fruit can be eaten raw or cooked and used to flavour spicy dishes. Young leaves are used as a substitute for baobab (*Adansonia digitata*) leaves, and cooked and used as a spice. Flowers are eaten raw or cooked and added to groundnut dishes. Seed cake contains about 26% protein and is used as livestock feed and as a manure. Seeds and leaves produce a suitable fodder, and leaves may also be used as manure or as a mulching material. Leaves, bark and roots have a wide range of medicinal uses. In Africa, it is often considered as sacred and preserved while clearing land for agriculture, whereas in Mexico it was the 'World tree', sacred to the Mayas.

Further reading

Neves, E.J.M., Martins, E.G. and dos Santos, Á.F. (2003) *Potencialidade de* Ceiba pentandra *(L.) Gaertn. para Plantios na Amazônia Brasileira [The Potential of* Ceiba pentandra *(L.) Gaertn. for Planting in Brazilian Amazonia].* Documento No. 89, Embrapa Florestas. EMBRAPA, Combo, Brazil, 26 pp.

Poorter, L., Bongers, F., Kouamé, F.Y.N. and Hawthorne, W.D. (2004) *Biodiversity of West African Forests: An Ecological Atlas of Woody Plant Species*. CABI, Wallingford, UK, 521 pp.

Sahid, M. and Zeven, A.C. (2003) *Ceiba pentandra* (L.) Gaertn. In: Brink, M. and Escobin, R.P. (eds) *Plant Resources of South-East Asia, No 17. Fibre Plants*. Backhuys Publishers, Leiden, The Netherlands, pp. 99–103.

Celtis mildbraedii Engl.

Red-fruited white-stinkwood

Importance

This common African species has a strong, light-coloured timber. Its main importance is in local markets. The species is traditionally used for poles in housebuilding and for pestles, tool handles and spoons, and it provides a high-quality, slow-burning fuelwood. It is also favoured for charcoal production.

Celtis mildbraedii is judged less valuable than many other hardwoods that grow in the same forests, reflecting poorer durability. The timber is vulnerable to rot and to insects; problems with shrinkage and splitting have also been noted.

The fruits are eaten by primates and birds and can thus, in some contexts, possess a significant conservation value.

Botanical features

C. mildbraedii is a large, evergreen or deciduous monoecious tree with a cylindrical bole, has buttresses and can reach 45 m tall. The crown is compact, with thin horizontal boughs. The bark is smooth and pale brown, scaling in small discs. The stem and branches often show conspicuous pale lenticels and young twigs are tawny pubescent. The freshly cut bark exudes a pungent aroma – the reason for its name, 'stinkwood'.

Leaves are oblong, 5–17 cm long and 4.5–6 cm wide, glossy green and obscurely 3-veined from the base which is slightly asymmetrical.

Flowers are borne in greenish clusters. The fleshy red fruits are ovoid–ellipsoid, 7–12 mm long and 5–7 mm across and conspicuously 4-ribbed when dried.

Distribution

C. mildbraedii is a widely distributed and locally common, shade-tolerating canopy-tree species found throughout the African moist tropics, including the Guineo–Congolian zone and the East African coastal forests, at altitudes ranging from 200 to 1600 m above sea level. It has been recorded from Angola, Benin, Cameroon, Central African Republic, Democratic Republic of the Congo, Gabon, Ghana, Guinea, Ivory Coast, Kenya, Nigeria, Mozambique, Madagascar, Liberia, Republic of the Congo, São Tomé and Príncipe, South Africa, Swaziland, Tanzania, Togo, Uganda, Zambia and Zimbabwe.

Much research has been carried out in Ghana, where the species is especially common and can locally make up 20–25% of all trees over 10 cm dbh. In Uganda, the species may be the main canopy former of the forest, forming 30% of its basal area. Its abundance declines as the landscape becomes wetter, both at regional, climatic and local scales, and it appears absent from Africa's wettest forests. In southern Africa and Madagascar, it is less common and occurs in somewhat drier conditions.

Silviculture characteristics

C. mildbraedii trees of 10 cm diameter may already produce fruits. These are eaten by birds, including hornbills, and primates, which disperse the seeds. Trees generally regenerate well under normal forest conditions, though germination is depressed in larger gaps. Seeds germinated well (75–83% success within 21–23 days) under an experimental range of 2–50% irradiance. Under full irradiance, the success rate drops to 44% on average after 34 days.

Seedlings may be abundant near mother trees. Established saplings grow well in forest gaps. Young trees may grow 1–2 m per year. In Ghana, seedlings reached 1–3 m tall when 4 years old.

C. mildbraedii has been widely harvested in natural high forests. It frequently sprouts from stumps.

Silviculture practice

So-called 'fruit stones' are usually collected from the ground; they are produced in large quantities in the fruiting season. Once cleaned there are about 20,000 seeds/kg; germination takes 2–4 weeks. Wildings are sometimes collected for planting.

Experience with plantations is limited, though trials have been established in Ghana. Results suggest that (heavy) thinning had beneficial effects on recruitment and increment (diameter, basal area).

Diseases and pests

General practice requires that logs are extracted from the forest soon after felling to avoid attacks by blue-stain fungi and pinhole borers. However, an experiment in exposing logs of three different species for 6 weeks in the dry season showed that *C. mildbraedii* had a comparatively low infestation of fungi colonies and wood borers.

Leaves often have dark brown galls 3–5 mm in diameter on both surfaces. In Zimbabwe, larvae of *Libythea labdaca laius* (African snout butterfly) eat the leaves.

Uses

In Côte d'Ivoire and Uganda, *C. mildbraedii* trees are left for shade in plantations of banana, cocoa, coffee and tea. The tree is planted as an ornamental in large gardens, parks and along streets, mainly in South Africa.

The wood of *C. mildbraedii* is valued as a building material, for poles and general carpentry and for carving household, domestic and personal items such as musical instruments, games and toys. However, the wood is not durable under damp or wet conditions as it is prone to rot. It is also an excellent firewood, burning slowly, and makes a good-quality charcoal. Studies show that the wood density is highly variable among individuals with published values for basic density from two comparable studies in Ghana ranging from 490 to 812 kg/m^3.

In an attempt to test lesser-used timber species for strength, Ugandan researchers found that the properties of *C. mildbraedii* were comparable with those of *Milicia excelsa* and *Khaya anthotheca*, both well-known and commercially utilized species. *C. mildbraedii* also had good values of work to shear strength, maximum load, and total work, implying that it could be used for structural purposes where such toughness is desired. They concluded that this species should be promoted as a substitute for other, more valued timbers.

A wide variety of medicinal uses are reported across Africa. For example, in Cameroon, bark decoctions are used as a wash to invigorate seriously weakened babies; they are also taken with *Solanum anguivi* fruits to treat venereal diseases; and root ash mixed with palm oil is applied to scarifications as a treatment for headache. In Congo, bark decoctions are used as enemas to treat menstrual problems. In Angola, a decoction of the root bark is drunk to treat malaria, and a tea of the roots is given to children against constipation, diarrhoea, cough, urinary complaints and heart problems; a maceration of leafy twigs is applied as a bath or lotion to treat headaches and as a vermifuge; hot-water extracts of the flowers are mixed with ash and rubbed on to the belly to promote childbirth; and flowers are also used in a rub to treat hernias. The pounded leaves are used as a fish poison in Angola.

Further reading

Burkill, H.M. (2000) *The Useful Plants of West Tropical Africa. 2nd edition. Volume 5, Families S–Z, Addenda*. Royal Botanic Gardens, Kew, Richmond, UK, 686 pp.

Coates Palgrave, K. (2002) *Trees of Southern Africa*. 3rd ed. Struik Publishers, Cape Town, South Africa, 1212 pp.

Durrieu de Madron, L., Favrichon, V., Dupuy, B., Bar Hen, A., Houde, L. and Maître, H.F. (2003) *Growth and Productivity In Dense Tropical Wet Forest: Results from the Experimental Site at Mopri, Côte d'Ivoire (1978–92)*. CIRAD, Montpellier, France, 73 pp.

Ofori, J. and Obese-Jecty, E.K.B. (2001) Some physical and drying characteristics of the wood of *Celtis mildbraedii* (Engl.). *Ghana Journal of Forestry* 10, 1–16.

Chamaecyparis lawsoniana (A. Murray bis) Parl.

Port Orford cedar

Importance

Chamaecyparis lawsoniana is a shade-tolerant, slow-growing forest tree which maintains a continuous diameter increment even at an old age. International trade and the spread of the introduced pathogen *Phytophthora lateralis* have put enormous pressure on remaining old-growth stands. Because of limited supply and its excellent wood properties, the timber is highly valued in the USA. Most of the timber harvested in native stands is exported to Japan where it is used as a substitute for *Chamaecyparis obtusa*. In Europe, North America and elsewhere the species is commercially important as an ornamental. Because *C. lawsoniana* grows relatively slowly and suffers from several diseases, its silvicultural and commercial importance is rather small.

Botanical features

C. lawsoniana grows to 30–50 m tall (occasionally up to 70 m) and 200 cm in diameter (occasionally up to 300 cm or more). In its native stands it lives up to about 600 years. Stem form varies greatly from very straight single stems to multi-stemmed trees. Young trees are narrowly conical. Older trees are frequently forked and have a columnar crown with a flat top. The main stem is usually straight and the diameter tapers very slowly from the bottom to the top. The bark is smooth dark grey-green when young, later turning dark purplish-brown or red-brown, becoming deeply fissured vertically and forming thick scaly or spongy plates. Self-pruning progresses slowly. *C. lawsoniana* forms a rather horizontal rooting system in the surface soil layers with no taproot, but it produces vertical sinkers. It forms endomycorrhizal associations with *Acaulospora*, *Gigaspora*, *Sclerocystis* and *Glomus* species.

Leaves are evergreen, small and scale-like, with four of them forming two unequal pairs. Each scale is about 1.5–2 mm long, flattened, closely appressed to the twig; each scale with a translucent central oval spot, commonly with hypodermal fibres on the adaxial leaf surface. Lower surfaces are dusted white at the base of each scale. The white edges of four scales form an X. Upper sides of the leaves are dark green to bluish-green.

Ovoid male flowers, 1–2 mm, are located at the top of terminal shoots, their black scales becoming red in early spring. Female flowers occur on terminal shoots near the axes of the sprays, blue to black, each scale white edged. Female and male flowers are found on the same branches, protruding beyond surrounding leaves. Cones are small, about 10 mm, have 7–10 scales with a short mid-thorn, and usually bear 2–4 seeds.

Flowering begins in late spring or summer; the development of reproductive organs continues during the summer. Pollen is formed in winter and is shed in March–April. Fertilization occurs 1–2 months later. Seed maturation continues in September or October of the same season. Lower scales of cones open partially in winter, with complete opening in March and April. Seedfall takes place in late autumn and spring. Large seed crops occur every 4 or 5 years.

Distribution

The natural distribution of *C. lawsoniana* is limited to a 350-km strip, mainly within 65 km of the Pacific Coast, between south-western Oregon and north-western California, a latitude range of only 3°. It is found from just above sea level to 1500 m. There are disjunct inland populations south-west of Mount Shasta, California, where it is found up to 1950 m altitude.

C. lawsoniana is found with an extremely wide range of associated plants and vegetation types. It occurs mainly within *Picea sitchensis*, *Tsuga heterophylla*, *Abies concolor*, *Abies magnifica*, *Sequoia sempervirens*, mixed-pine and mixed-evergreen forests.

Plantations of *C. lawsoniana* have been established in the USA, Germany, France, The Netherlands, Denmark, UK, Australia, South Africa, Kenya, New Zealand and Sri Lanka; however, it is of minor importance as a forest tree outside its natural range.

Silviculture characteristics

The slow to moderate growth rate of *C. lawsoniana* and losses caused by *P. lateralis* root rot have diminished the silvicultural and commercial interest in this species; however, the demand for its valuable timber is still high. The most valuable silvicultural property of this species is its shade-tolerance. Its moderate growth rates beyond the sapling stage result in its being overtopped; however, it tolerates competition, maintains growth rates in the understorey, and retains into old age the ability to respond to increases in light availability. Therefore, it is well suited to mixed stands with species such as *Pseudotsuga menziesii*, where the faster-growing species would be removed part-way through the rotation. The potential damage by root rot can be minimized by keeping the proportion of *C. lawsoniana* low in mixed forests.

Young trees are easily killed by fire, but mature trees are very resistant due to the thick bark. Seed size varies considerably, ranging from 130,000 to 1,300,000 seeds/kg.

Silviculture practice

Propagation from cuttings and seed is easy. Open-grown container seedlings produce cones 3–4 years after planting; seed production of *C. lawsoniana* in stands is usually expected by 20 years old. The seed is classified as intermediate and can be stored frozen at less than 10% moisture in a sealed container for more than 10 years. Shading of sowings in the nursery and the establishment of 1+1 seedlings in plantations are recommended. Seed is sown in spring, and seedling densities of 322–530/m^2 are ideal. The seed is covered with 0.3–0.6 cm soil. Bare-root stock is usually grown as 1+1 or 2+0 seedlings; container-grown 1-season-old stock can also be used successfully. For vegetative propagation, cuttings should be taken from tips of major branches located in the lower crown of young trees, preferably from December to February.

Plantations of *C. lawsoniana* can be planted with 2+0 stock at 2.4 m spacing in rows mixed with *P. menziesii* (alternate rows) after clear felling. Container stocks (20–30 cm) are planted at 6–8 m spacings. The narrow crowns of *C. lawsoniana* allow high stocking densities. Thinning increases the risk of windthrow as well as the risk of introducing root rot.

Mean annual increments of 13.7 and 16.9 m^3/ha have been recorded in Oregon at ages 61 and 57 years, respectively, while in Germany 12–14 m^3/ha were obtained on a medium to poor site. The growing stock in this stand had increased by 8.6 m^3/ha/yr between the ages of 75 and 89 years.

The diameter increment of *C. lawsoniana* is very consistent. The difference in its diameter compared with the faster-growing *P. menziesii* decreases with age. The height increment may be more rapid in younger stands; *C. lawsoniana* achieves only 86% of the height growth rate of *P. menziesii*. In mixed stands, *C. lawsoniana* is usually overtopped by 20–25 years.

Diseases and pests

Port Orford cedar root disease, caused by the oomycete *P. lateralis*, is a major cause of *C. lawsoniana* tree mortality. This introduced disease, of unknown origin, was first reported on ornamental trees near Seattle, Washington, USA, in 1923. It was not until 1942, however, that the causal agent was discovered (in Oregon) and named. Losses became so severe among ornamental *C. lawsoniana* trees in north-west Oregon and western Washington that, within a decade, production of valuable horticultural varieties was virtually abandoned. In 1952, the disease was detected in south-western Oregon, within the geographic range of *C. lawsoniana*. As spores are carried by water, one introduction can infect all trees in streamside zones below the point of introduction. Humans assist spread via vehicles, shoes, equipment and transport of contaminated soil. The disease causes rapid death of infected trees.

The oomycete *Phytophthora cinnamomi* causes major losses to *C. lawsoniana* in some nurseries, and also to cultivated trees.

Several insects have been reported from *C. lawsoniana* but cause relatively little damage. The midge *Janetiella siskiyou* damages the seeds, while the moths *Argyresthia cupressella* and *Argyresthia libocedrella* infest twigs and leaflets and sometimes cause damage to ornamental trees. *C. lawsoniana* is one of the hosts of redwood bark beetle (*Phloeosinus sequoiae*), which confines its attacks to weakened, felled or fire-scorched trees. Among borers, *Semanotus amethystinus* attacks the bole and large branches, while *Serropalpus substriatus* lives under the bark and in the wood of dead trees.

Uses

The main use of *C. lawsoniana* outside its natural range is as an ornamental tree, when it is known by the common name of Lawson cypress. There are over 200 horticultural varieties in cultivation. Although commonly planted as a hedge or shelter belt due to its dense, narrow crown, it is often planted as a solitary specimen tree.

C. lawsoniana wood is characterized by uniform texture and density, good machining properties, moderate shrinkage, lightness, excellent paint-holding properties and resistance to decay. It is easy to machine and is suitable for a variety of uses. Wood density is low to medium (465–495 kg/m^3 at 12% moisture content).

The available volume of *C. lawsoniana* from stands in its native range is almost completely exported to Japan, where it is used for woodware and toys, as well as in shrines and temples. Old-growth *C. lawsoniana* timber, which is clear, fine-grained and of large diameter, commands high prices. The wood is also used for arrow shafts.

Cut branches are used in floral displays and wreaths.

Further reading

Kitzmiller, J.H., Sniezko, R.A., Mitchell, A.K., Puttonen, P., Stoehr, M. and Hawkins, B.J. (2000) Range-wide genetic variation in Port Orford cedar (*Chamaecyparis lawsoniana* [A. Murr.] Parl.): I. early height growth at coastal and inland nurseries. *Journal of Sustainable Forestry* 10, 57–67.

Zobel, D.B. (1990) *Chamaecyparis lawsoniana* (A. Murr.) Parl. Port-Orford-Cedar. In: Burns, R.M. and Honkala, B.H. (eds) *Silvics of North America: 1. Conifers*. Agriculture Handbook 654, Vol. 1. US Department of Agriculture, Forest Service, Washington, DC.

Zobel, D.B. and Hawk, G.M. (1980) The environment of *Chamaecyparis lawsoniana*. *American Midland Naturalist* 103, 280–297.

Zobel, D.B., Roth, L.F. and Hawk, G.M. (1985) Ecology, pathology, and management of Port-Orford-Cedar (*Chamaecyparis lawsoniana*). General Technical Report PNW-184, USDA Forest Service, Portland, Oregon, 161 pp.

Chamaecyparis obtusa (Siebold & Zucc.) Endl.

Hinoki cypress

Importance

Chamaecyparis obtusa exists as var. *obtusa* (hinoki) in Japan and as var. *formosana* (Taiwan cypress) in Taiwan. Both are long-lived, evergreen conifers producing uniform, relatively strong, highly durable, lustrous, scented sawnwood of high quality on a range of sites. The smaller size reported for hinoki compared to Taiwan cypress is probably due to centuries of felling of the largest hinoki trees for construction purposes.

Popular as an ornamental, *C. obtusa* is often planted as a specimen tree.

There are few plantations of *C. obtusa* var. *obtusa* outside the 2.5 million ha in Japan and eastern Asia, where annual growth rates are moderate and not more than 6 m^3/ha at the maximum current annual increment. Plantations of var. *formosana* in Taiwan are restricted by competition from dwarf bamboo.

Botanical features

C. obtusa var. *obtusa* is a long-lived, straight-stemmed, monopodial tree, reaching over 40 m tall and up to 2 m diameter.

The crown is pyramidal in exposed trees, but spire-like or conical in mature forest trees. Branches persist for years but are eventually shed in mature trees, leaving clear boles up to 20 m. In 20–50-year-old trees, 70–80% of live branches are in the 3–4 m of the top, crown length is nearly constant regardless of stand density and upper diameters can be estimated from dbh.

Bark is fissured, reddish-brown and comes off in long strips. Roots are shallow and the root system is very fine, lacking heavy laterals. Leaves are closely appressed, of two sizes, the lateral pair much the larger, boat-shaped, 2 mm long, blunt at the apex or with a minute point; the smaller pairs about 1 mm long, triangular with a thickened apex, all prominently lined below with white X-shaped markings produced by a coating of wax. Leaves on the branchlets are oblong, unequal, the lateral 6 mm long, the facial 3 mm long with spreading, blunt points. The bright green foliage emits a resinous scent when crushed.

Male flowers are terminal, minute, on the smallest branchlets. Cones often densely borne, globular, 2 cm across, green but ripening to orange-brown, with few flat scales. Seeds, 2–5 on each scale, are 3 mm long and winged.

Trees are monoecious and bear cones from about age 10 years in most areas. Cone crops can be heavy. Seeding is annual, with a heavy seed crop occurring every second year. Shoot elongation occurs from April to October, with shoots in the upper canopy growing more than those in the middle and lower canopy.

C. obtusa var. *formosana* trees grow up to 40 m tall and up to 3 m in diameter, and differ from var. *obtusa* in producing smaller leaves, cones and seeds, and being longer lived.

Distribution

The natural distribution of hinoki in Japan is confused, both by the extent of new plantings and by the two millennia of exploitation. There are numerous scattered relict stands, from Sado Island in the Sea of Japan at 38°N to Yaku Island at nearly 30°N. The main natural stands are from Central Honshu south to Yamaguchi Prefecture in the extreme south of Honshu, where it becomes scattered. It grows throughout Shikoku. There are a few scattered natural occurrences in Kyushu. Its southern limit is in Yaku Island in a climate of extreme rainfall (>6000 mm annually). It extends to the colder east coast prefectures in central and south Honshu, but not further north on the Sea of Japan side.

The range of latitudes in Japan has been extended by plantations established up to 40° N, as it is often the preferred species for private growers. When *C. obtusa* var. *formosana* is taken into account, the southern distribution of the species is extended to the Tropic of Cancer. It occurs throughout most of the Taiwan mountains.

C. obtusa var. *obtusa* occurs in Japan in small pure stands, but more commonly in mixture with other conifers in cool temperate conifer or conifer–broadleaf forests, while var. *formosana* occurs in mixed softwood forest growing at 2000–3000 m altitude in Taiwan.

Silviculture characteristics

The tree tolerates moderate to heavy shade, eventually developing into a climax canopy species. This ability means it can regenerate successfully where the canopy is dense enough to suppress competing vegetation (such as dense stands of dwarf bamboo). Newly exposed mineral soil sites resulting from landslides are often colonized, as bamboo is usually absent.

Crown breakage due to ice-glaze is a serious problem, especially at 1100–1200 m altitude. Heavy snowfalls cause butt sweep and often major stem curvature. Consequently the trees have compression wood. This is one of the most serious defects of plantation wood in Japan.

Silviculture practice

C. obtusa regenerates well from seed. It can also be grown from cuttings and can be grafted successfully, with survival rates of up to 90%; timing and humidity control appear critical. Tissue culture using cotyledon and plumule explants is also possible.

Although some attempts have been made to regenerate *C. obtusa* naturally, the majority of productive stands are from direct planting, followed up by 10–15 years of tending.

In Japan and Taiwan, weeding of plantations is intensive, and is usually carried out several times during each of the first 3 years. In addition to bamboo, re-establishment of tree crops in Japan is hindered by vigorous growth of climbers, such as kudzu vine (*Pueraria thunbergiana*), and cutting these back is necessary for the first 15 years in many plantations. Competition from bamboo in high-altitude var. *formosana* forests in Taiwan is even more severe than in Japan.

Pre-existing seedlings are an important factor in regeneration, which in a mixed conifer stand is best on mineral soil beneath a canopy gap.

Height increment is slightly increased, and diameter increased more after release felling. Regeneration is best when soil, understorey and canopy are all disturbed.

Pruning is often carried out with the aim of producing knot-free faces on sawnwood. Branches persist in open-grown hinoki trees. The branches are at a right angle or moderately angled up from the stem. This branch angle results in good knot characteristics in the sawnwood. The optimum ages for pruning are 7, 9, 12 and 15 years (i.e. four lifts), with harvesting at age 42 years. The dbh and height growth of pruned stands are increased by fertilizers. Pruning reduces both dbh and height growth, while thinning does not affect height growth. Thinning or removal of the overstorey can damage the remaining crop.

The first hinoki plantations in Japan were established about 500 years ago, initially in the Yoshino area of central Honshu. However, it has often been suggested that, following the initial successful establishment of these early plantations, Japanese silviculture remains frozen in a past when labour was very cheap and plentiful and wood was worth more than it is now, the consequence being that current stands are very heavily overstocked. As thinning prescriptions have not been carried out, current plantation standing volume in Japan is about 387 million m^3, with the largest volumes per 5-year age class concentrated in the 30–40-year age classes.

There is a considerable movement to convert uniform stands to two- or three-level stands and a type of selection forest. With its rich tree flora, this should be relatively easy in Japan.

Diseases and pests

Rooshi pitch canker is frequent in areas where heavy snowfall causes damage, wounds subsequently becoming infected with fungi such as *Sarea resinae* and *Pezicula livida*. *Seiridium unicorne* causes lesions, leading to broken or dead stems and branches, though its frequency reduces as trees age. *Chlorosypha* needle blight occurs following salt wind injury.

The bark beetle *Semanotus japonicus* attacks *C. obtusa* and has become a major pest in recent years. Males and females live in both dead and live trees. *Xeris spectrum* and *Urocerus japonicus* are the main siricids on *C. obtusa* and are parasitized by *Rhyssa persuasoria* and *Megarhyssa praecellens*. *Pseudorhyssa sternata* is a cleptoparasite fairly common in logs of *C. obtusa*.

The main mammals causing localized damage by browsing and gnawing include serows (*Capricornis crispus*), Japanese deer (*Cervus nippon*), black bears (*Ursus thibetanus japonicus*), hares (*Lepus brachyurus*) and giant flying squirrels (*Petaurista leucogenys*). Voles (*Microtus montebelli*) are probably the most important mammal pests, as numbers can rise rapidly; outbreaks usually follow the flowering of bamboo (*Sasa* and *Pleioblastus* spp.) and cause serious damage.

Tokkuri disease is the abnormal growth of the stem base, which thickens and extends 0.8–1.2 m above the ground. Symptoms start at about age 10.

Uses

Forest grazing is practised but is not widespread in Japan, and has little effect on tree growth rates. *C. obtusa* makes thick shelter belts.

With attractive foliage and form, numerous, possibly hundreds, of horticultural varieties, including dwarf forms under 30 cm tall, are grown in Japan, Europe and North America. It is also a popular species for bonsai.

C. obtusa wood is usually rated as light and strong, and has the same basic strengths as other softwoods of its density range. Its strength in service is largely dependent on stand history and consequent branch size and condition. Superior wood qualities include surfacing quality, stability, durability, and texture, as well as scent and lustre. Basic specific gravity of the wood ranges from 0.35 to 0.488. The heartwood is difficult to penetrate with preservatives.

C. obtusa wood has been used in construction from at least as early as 300 BC in Japan, and was the favoured species for temple building. Pagodas were built on unique earthquake-resistant principles, based on durable *C. obtusa* sawnwood and use of a solid trunk of up to 65 m in length for the shinbashira or central column. In Taiwan, *C. obtusa* var. *formosana* was also traditionally used in construction, particularly for columns, horizontal timbers and rafters.

Other traditional uses for the wood in Japan include wooden bathtubs, coffins and as a basis for lacquer work.

To utilize the large volume of wood available from over-stocked plantations, zephyr wood has been developed. It is made by roller-crushing small logs and sawmill slabs. The material has a continuous grain running in one direction. Boards are then prepared by gluing a number of zephyr sheets together, like plywood, with the grain at right angles in adjoining sheets. Bending strengths are comparable to plywood, but shear perpendicular to the surface is poor.

Another product being developed is 9 mm 5-ply laminated veneer board for use as floor sheathing. Oriented strand board is made from very thin strands of *C. obtusa*, whereas small-diameter wood and branchwood is used for particleboard. *C. obtusa* is used for decorative sliced veneer, but volumes are very small. The high extractive content makes it an unattractive pulpwood.

C. obtusa bark is durable, lasting up to 60 years when used as roofing. It is also used as cladding.

The bark contains large amounts of growth-inhibiting substances, while the wood contains two termiticidal compounds.

Wood oils extracted from both *C. obtusa* varieties show good control of house mites (*Dermatophagoides pteronyssinus*) and have potential in perfumery.

Further reading

Igarashi, T. and Kiyono, Y. (2008) The potential of hinoki (*Chamaecyparis obtusa* [Sieb. et Zucc.] Endlicher) plantation forests for the restoration of the original plant community in Japan. *Forest Ecology and Management* 255, 183–192.

Sato, K. (1974) Chamaecyparis obtusa *in Japan*. Oregon State University, Corvallis, Oregon, USA.

Yamamoto, S.I. (1998) Regeneration ecology of *Chamaecyparis obtusa* and *Chamaecyparis pisifera* (Hinoki and Sawara Cypress), Japan. In: Laderman, A.D. (ed.) *Coastally Restricted Forests*, Oxford University Press, New York, New York, USA, pp. 101–110.

Chlorocardium rodiei* (M.R. Schomb.) Rohwer *et al.

Greenheart

Importance

Chlorocardium rodiei is a tall tree with a localized distribution centred on Guyana. It has long been known for its very heavy, hard timber ('greenheart'), suitable for use under exacting conditions, highly resistant to marine organisms and fire, and immune to decay and termites. Commercial harvesting, particularly from Guyana, dates back to the late 18th century, but has sped up since the introduction of chainsaws in 1967. At one time greenheart constituted 70% of the timber exports from Guyana.

The sustainable management of this species has been a subject of controversy. The species was added to the International Union for Conservation of Nature (IUCN) Red List of Threatened Species (1996) as 'Vulnerable' and its trade was restricted. In 2007, its status reverted to 'Data Deficient' due to controversy over the reliability of data and claims. The IUCN website notes that:

> This revised assessment is a ruling made by the Red List Standards and Petitions Working Group (SPWG) of the IUCN-SSC Biodiversity Assessments Subcommittee after considering material provided by the Guyana Forestry Commission who petitioned against the 1998 assessment of this species.

Elsewhere the IUCN reports that:

> Allegations that production of Greenheart timber has declined significantly over the past 50 years as reported in *The World List of Threatened Trees* appear to be unfounded. Hammond *in litt.* has provided evidence that Greenheart timber production in 1998 was close to 80,000 m^3, a similar level of output to that achieved in 1953. Hammond *in litt* agrees that there was a decline in production between 1963 and 1991, but this was not steady or sustained. It appears this was due to the loss of human capacity and capital resources from the country during that period.

Available at: http://www.iucnredlist.org/details/34688/0 (accessed 27 March 2013)

Regeneration in natural stands is very slow and poor germination success inhibits the establishment of plantations. The low growth rates would require much longer rotations than have typically been contemplated in practice.

The ships used by the famous Antarctic explorers Amundsen and Shackleton for their polar expeditions in the early 20th century (the *Fram* and the *Endurance*) were sheathed in wood of this species to prevent the ships from being crushed by ice. They are considered to have been the two strongest wooden ships ever constructed.

Botanical features

C. rodiei (known previously as *Nectandra rodiei* or *Ocotea rodiei*) is an evergreen tree with a small dense conical or oval crown growing to 40 and occasionally even 50 m tall with a trunk diameter of 50–60 or rarely 100 cm. Stem base swollen or with buttresses to 1 or 2 m. Bark is light or creamy-brown with large vertical irregular flakes. The stiff leathery leaves are opposite, simple, with an entire margin. The twigs are angular and when young are distinctly puberulous, and end in a typical 5–10 mm apical bud. The inflorescence is an axillary panicle to 5 cm long, the flowers are creamy-white, jasmine scented and 5–10 mm in diameter. The fruit is a large grey or brown single-seeded ellipsoid to ovoid–globose drupe (6–7 cm long and 5 cm diameter) with white speckles. It is held in a distinctive woody cup or cupule (2–2.5 cm diameter) attached to the twigs.

Distribution

C. rodiei has a very localized distribution within the Guiana Shield. It occurs primarily in Guyana, and a small area in neighbouring Suriname and Venezuela. The species dominates a restricted belt of lowland rainforest on brown sand and is also found occasionally in other forest types.

No *C. rodiei* plantations have been established, due to poor germination success.

Silviculture characteristics

C. rodiei is selectively logged from natural forests, but the resulting canopy opening is locally high due to the species' tendency to occur in clumps.

Fruiting is reported to be intermittent with a masting event every 12–15 years. There are reports of the seeds being dispersed by mammals, but most simply fall under gravity, and seed and seedling densities below the crown are significantly higher than beyond the crown area. It is unclear to what extent these patterns may reflect hunting.

Germination success of *C. rodiei* in large gaps is low. Once established, however, high light levels increase seedling survival.

The tree seems able to coppice effectively and sprouting of cut stumps may have the potential to contribute to the maintenance of *C. rodiei* in logged sites.

Silviculture practice

Conventional logging is considered unsustainable: logging intensities appear to be low (e.g. 5 m^3/ha) when reported as averages over extensive areas, but local damage can be extensive as harvestable stems often occur in clumps. Logging therefore creates large gaps, while skidder movements destroy seedlings and compact the soil, further affecting forest recovery. Reduced-impact logging systems have been devised to redress these environmental problems, reducing skidding damage by 65% and average size of felling gaps by 40%. Long cutting cycles would be needed to reflect the slow growth of this species.

Diseases and pests

Because there are no plantations of *C. rodiei*, little is known or reported on pests or diseases of the species. The seeds may be attacked by infesting scolytid beetles and may be predated by rodents.

One study looked into infestation by *Stenoma catenifer*, a Neotropical pest of avocado, which is a member of the same family of Lauraceae. However, infestation levels were always less than 10%, in comparison with up to 80% found for avocado.

Uses

The wood of *C. rodiei* tends to be a pale olive-green colour with darker streaks. It has an air-dry density of 1058 kg/m^3. Its yellowish-green sapwood is poorly distinguishable from heartwood. The wood is outstanding for its strength and durability and is especially prized for marine construction due to its resistance to attack by marine borers. As it is extremely hard and strong, it requires specialized tools to work it and there are records of respiratory problems associated with prolonged exposure to the sawdust.

The timber is available in very large sizes and long lengths, and is suitable for pilings, piers, lock gates, and dock and harbour works. It is useful for pier decking and hand rails, flooring and in the engineering industry as bearers for engines. It also gives good service in chemical plants for vats, filter press plates and frames. Furthermore, it is suitable for joinery in both exterior and interior situations, and it is also used for fishing rods.

In times of food shortages, the seeds of *C. rodiei* were used by local people in a way similar to cassava: they grated the seeds, and the starch was repeatedly washed to remove the bitter alkaloids. Bark alkaloids have been indicated to have anti-malarial properties.

Further reading

Gerard, J., Miller, R.B. and ter Welle, B.J.H. (1996) *Major Timber Trees of Guyana: Timber Characteristics and Utilisation.* Tropenbos Series III. The Tropenbos Foundation, Wageningen, The Netherlands.

Hout, P. van der (2000) Testing the applicability of reduced impact logging in greenheart forest in Guyana. *International Forestry Review* 2, 24–32.

Ter Steege, H., Bokdam, C., Boland, M., Dobbelsteen, J. and Verburg, I. (1994) The effects of man-made gaps on germination, early survival and morphology of *Chlorocardium rodiei* in Guyana. *Journal of Tropical Ecology* 10, 245–260.

Ter Steege, H., Welch, I. and Zagt, R. (2002) Long-term effect of timber harvesting in the Bartica Triangle, Central Guyana. *Forest Ecology and Management* 170, 127–144.

Chukrasia tabularis A. Juss.

Chittagong wood

Importance

Chukrasia tabularis is a handsome tree with a straight bole and high-quality wood. Its distribution has become localized due to over-exploitation in the past; however, its valuable wood has resulted in the creation of *C. tabularis* plantations in some areas of the tropics, despite their susceptibility to attack by *Hypsipyla robusta*.

Botanical features

C. tabularis is a very large tree with a tall, straight bole and spreading crown. On good sites it generally attains heights of up to 40 m, with a clear bole to 25 m and a dbh of 1.2 m. There are sometimes convex buttresses up to 1.5 m high. The bark is dark brown, fissured vertically, scaling or cracking in rectangular blocks; the inner bark is red-brown or pinkish.

Leaves are compound, paripinnate and 30–50 cm long when mature, with incised or lobed leaflets in juveniles (rarely retained at maturity), with 6–12 leaflets on each side; leaflets ovate to oblong, more or less asymmetric or even falcate; subapical leaflet the largest (10–11.5 × 3.5–6.5 cm), the most proximal one the smallest (4 × 2.2 cm); subglabrous to finely scattered pubescent on the lower surface, subglabrous on the upper surface except axils and veins; veins 9–11 on each side, bifurcating near the margin. Leaf margin smooth. Petioles 4–9 cm long and swollen at base.

Flowers are borne in axillary panicles. The panicled inflorescence develops from the axils of the youngest leaves, which are not always apparent, so that often the inflorescence can appear terminal. Panicles 10–30 cm long, primary branches to 16 cm, secondary to 4 cm, bearing fascicles of flowers, axes short-pubescent. Flowers creamy-green or yellowish, sweet-scented. Anthers are united into a glabrous staminal tube. Ovary is densely appressed-pubescent. Fruit a capsule, 2.8–5.0 cm long, 2.5–4.0 cm in diameter, dark brown, lenticellate. Seed about 1.2 cm long, brown, flat, with a brown membranous wing twice the length of the remaining portion of the seed (1.2–2.0 × 0.6–0.8 cm including wing).

Distribution

C. tabularis is widely distributed in South Asia, occurring in Bangladesh, India, Myanmar, Nepal, Sri Lanka, Thailand, Peninsular Malaysia (not in the southern regions), Sarawak, Kalimantan (limestone regions), Sumatra (northern regions but rare), Vietnam and southern China. Due to its high wood value, *C. tabularis* has been grown in plantations in Australia, the western regions of Nigeria and South Africa.

Silviculture characteristics

Although *C. tabularis* is a light-demanding species, it can tolerate shade in the early stages of its growth; however, open conditions are recommended for any plantings. It is relatively windfirm in tropical cyclone zones.

C. tabularis grows in a diverse range of soil types, including limestone soils, brown forest soils and alluvial soils. It prefers soils which are freely drained, have a pH range 4.0–7.5 and occur on level to steeply sloping sites, preferably with a south-facing aspect. It remains stunted on infertile limestone soils.

C. tabularis produces coppice shoots. It has a tendency to branch and fork, usually as a result of infestation by the shoot borer *H. robusta*.

Silviculture practice

Natural regeneration can be generally satisfactory, but many seedlings fail to become properly established. For gap regeneration in wet evergreen forests of Karnataka (India), *C. tabularis* performed well when planted in areas logged in the current year.

Ripe fruits, collected from trees during March and April, are spread out in the sun for 2–3 days to dehisce, after which the seeds are separated by gentle thrashing, care being taken to protect them from being blown away. Seeds can be stored in gunny bags for about 5 months, being recalcitrant to intermediate.

Fresh seeds should be sown in porous soil, shaded from direct sunlight. The soil must not be allowed to dry out since the seedlings are sensitive to drought, although seeds can germinate under certain levels of drought stress, depending on provenance. Germination generally takes from 10 to 40 days. Best results are achieved by sowing seeds in germination trays and pricking out into polythene bags. *C. tabularis* can also be propagated vegetatively from stem and root cuttings.

For plantations, sites are generally prepared by burning after clear felling; however, since seedlings are sensitive to drought, efforts should be made to reduce the danger of early mortality, such as providing partial shade with faster-growing species.

At least three weedings during the rainy season are necessary for the first 3 years. The seedlings are liable to attack by shoot borers, hence mixing with other, non-meliaceous species is desirable. As the species has a tendency to branch and fork, thinnings, especially the early ones, should be comparatively light. *C. tabularis* can be badly browsed and debarked by deer, and therefore fencing around plantations is necessary.

Performance of *C. tabularis* in provenance and plantation trials in the Northern Territory of Australia has been good, with significant differences observed between provenances for some growth traits, indicating that a domestication programme for the species would be feasible.

Diseases and pests

Phytophthora nicotianae, a new disease infecting *C. tabularis* with an incidence of 30–80%, has been observed in forest nurseries in India.

Throughout its natural range and in plantations, *C. tabularis* can become heavily infested with the shoot borer *H. robusta*. Light to moderate infestation by *Pagiophleus longiclavis* (a collar borer) has been recorded in Bangladesh.

Uses

The straight bole and self-pruning ability of *C. tabularis* make it a suitable tree for growing in combination with crops, such as banana, *Citrus* spp. and guava. Its coppicing and pollarding ability make it particularly suitable for home gardens. Its fire resistance and high dust-retention capacity make it suitable for urban forestry and landscaping.

C. tabularis has a moderately heavy and strong wood, often with interlocked grain. At 15% moisture content the wood density is 625–880 kg/m^3. It is easy to saw and work, can be turned and carved easily, and brought to an excellent lustrous finish. It can be air dried, although this takes a long time, and kiln drying with care produces good results. The heartwood is durable under cover; however, it is classified as a non-durable timber.

The wood is used for furniture, flooring and interior construction. Selected wood is used for ornamental figures and for decorative panelling, carvings and other ornamental articles. Some speciality articles, such as rifle butts and shuttles for jute and textile mills, are produced using this wood as an alternative to walnut (*Juglans regia*). The wood peels well on a rotary machine and face veneer can be used for ornamental plywood using a slicing veneer cutter.

C. tabularis bark is a powerful astringent, and a red/yellow dye can be produced from the flowers. It also produces a gum which is apparently not used.

Different parts (leaves, bark, fruits) have ethnobotanical, medicinal and biopesticidal uses due to the presence of an abundance of phenolic compounds, including terpenoids and limonoids. The young leaves and bark contain up to 22% tannins, and leaf extracts exhibit antimicrobial activities. Seeds have a high fatty oil content (31.2%), the oil being rich in linoleic and linolenic acids, with no oxygenated or epoxy fatty acid glycerides, indicating its potential as an edible oil. Leaf oils are rich in oxygenated monoterpenes and phenyl-propanoids.

Further reading

Gunn, B.V., Aken, K. and Pinyopusarerk, K. (2006) Provenance performance of *Chukrasia* in a five-year-old field trial in the Northern Territory, Australia. *Australian Forestry* 69, 122–127.

Jøker, D. (2000) *Chukrasia tabularis* A. Juss. Seed Leaflet 42. Danida Forest Seed Centre, Humlebaek, Denmark, 2 pp.

Kaur, R. and Arora, S. (2009) Chemical constituents and biological activities of *Chukrasia tabularis* A. Juss. - a review. *Journal of Medicinal Plants Research* 3, 196–216.

Troup, R.S. and Joshi, H.B. (1981) *Troup's The Silviculture of Indian Trees. Volume III*. Controller of Publications, Government of India, Delhi.

Colophospermum mopane (Kirk ex Benth.) J. Leonard

Mopane

Importance

Colophospermum mopane is one of the best-known and most valuable trees in Southern Africa, where it forms dominant deciduous woodlands over some of the driest areas. It is important for the production of fodder and a very dense timber, but also for the production of edible caterpillars which feed on the leaves, as well as being a key browse species for both domestic and wild herbivores. Trees are capable of growing on poor, compacted soils and have been promoted for agroforestry systems in some of the driest parts of South and South-east Asia.

Botanical features

C. mopane is a deciduous shrub or medium-sized to tall tree from 4 m to 18 m tall, commonly 10 m, with single or multiple stems from the base and sparse branches growing upwards. Trunks in large individuals have diameters up to 1 m, with bark dark grey to blackish, deeply and vertically fissured, flaking in narrow strips. Leaves are alternate, bifoliate with two stalkless leaflets resembling a pair of butterfly wings, and vestigial remains of a third terminal leaflet forming a very small appendage between the leaflets pair. Leaflets are glabrous, ovate, 4.5–10 cm long and 1.5–5 cm wide, margins entire, petiole 2–4 cm long. The base is asymmetric, slightly lobed on one side with several veins arising from it. Flowers are greenish, small and inconspicuous, borne in short axillary racemes or sprays. There are four greenish sepals; petals are absent. Flowers are hermaphrodite with 20–25 stamens and have enlarged stigmas. The fruit is an indehiscent, leathery and flat pod, kidney-shaped or almost semi-circular, up to 5 cm long and 2.5 cm wide, enclosing a single seed, which is flat, wrinkled, sticky and dotted with resin glands.

Distribution

The natural distribution of *C. mopane* is in Southern Africa between latitudes 10°S and 26°S, between Zambia, Malawi and Angola in the north, with its southernmost limit just south of the Olifants River in the Kruger National Park, South Africa. *C. mopane* is one of the dominant trees of the hot and low-lying areas of Southern tropical Africa and is the principal species in one of the most distinctive vegetation groups, usually forming pure stands known as mopane or *Colophospermum* woodland. Two distinct kinds of mopane stands are recognized: 'cathedral mopane', consisting of very tall trees; and 'scrub mopane', referring to vegetation where either soil conditions are unfavourable or the plants have been stunted by heavy elephant browsing. It has been planted in desert regions in Rajasthan, India, where it performs well in silvopastoral and silvoarable agroforestry systems, and has been promoted for South-east Asia.

Silviculture characteristics

C. mopane is deciduous and trees are leafless for up to 5 months during the dry season, often growing in dense natural stands. Leaf flush occurs in spring, independent of rainfall. Flowering begins at about 5 years of age. Seed production is highly variable between years. *C. mopane* is wind pollinated with seed dispersed by wind, water and animals. In India, abundant seed production is observed even during the driest years, with carpets of young seedlings covering the ground under stands of trees, indicating the potential for natural regeneration but also the potential for it to become an invasive weed. Trees produce suckers, coppice, and respond well to lopping, with rapid regrowth of new shoots.

Trees grow in areas with a mean annual temperature of between 20°C and 28°C, and are susceptible to frost. The native range is a summer rainfall area receiving between 100 mm and 800 mm per year and a dry season ranging from 5 to 10 months, though droughts have been observed to kill large trees. *C. mopane* has a wide altitudinal range (200–1200 m) and grows on a wide range of soils, but is often found on alluvial soils and on alkaline and poorly drained soils, which it tolerates better than many other species.

Flowering occurs between October and March, although this can be erratic, with trees in a given area producing no flowers for several years. Fruiting occurs between March and June. There is considerable variation in seed traits among provenances, differing significantly in seed length, width and weight, attributed to more favourable conditions such as higher rainfall in more northern latitudes.

Silviculture practice

C. mopane regenerates rapidly both vegetatively and by seed. Seed storage behaviour is orthodox but seeds are seldom viable for more than 1 year under natural conditions. They germinate readily and do not require pretreatment, but soaking in water for 24 h speeds germination. Seeds can be sown directly or into polythene bags, though seedlings grow slowly in open-air nurseries and may only reach 5 cm high after 40 days. Plant spacing partly depends on the silvicultural system used, with 5 × 5 m in agroforestry systems or 10 × 4 m for rehabilitating degraded rangeland. Young trees are susceptible to competition, and weeding is required for the first few years after planting.

Cutting height is a major determinant of the number of coppice shoots produced and regrowth rates. After 3 years, plants cut at 1 m above ground level produce about twice the number of coppice shoots compared to trees cut at 10 cm high, and had shoots more than ten times longer. However, established seedlings grow slowly, and may have height increments of around 10 cm per year.

In coppicing cycles of 10–20 years in mixed acacia and mopane woodlands, biomass production was 5–10 t/ha/yr, but more long-term research is needed on coppice management in mopane woodlands. Thinning of naturally occurring dense stands reduced inter-tree competition and results in significant increases in flowering and fruiting of the remaining trees. Fruits and seeds from trees in thinned stands were larger and heavier than those from trees in dense stands. However, there were no differences in the germination of seeds from thinned and dense stands.

Diseases and pests

The main insect that feeds on *C. mopane* is the leaf-eating caterpillar of the grey emperor moth (*Gonimbrasia belina*), but as this is a protein-rich food source for local people it is an important resource, and is not treated as a damaging pest. There are no reported diseases.

Uses

C. mopane is planted as a multi-purpose tree for soil improvement, fuelwood and fodder in agroforestry systems. Trees are also planted to provide shade and shelter, and it is also a useful species for land rehabilitation and has been used for stabilizing dry, alkaline soils and sand dunes.

It produces dark grey wood that is hard, heavy (density ranges from 960–1200 kg/m^3 at 12–15% moisture content), durable and resistant to termites. It is used for sawn or hewn building timbers, for heavy and light construction, flooring, exterior fittings, fences, railway sleepers and furniture, as well as fuelwood, roundwood, posts, pit props, building poles and tool handles.

However, trees are often better known as an indirect source of food for local people, as caterpillars of a brownish-coloured grey emperor moth (*G. belina*) (known as 'mopane worms') feed on the leaves and are an important part of the diet. The crude protein content of the larvae is estimated at 48%, and they are collected, eaten at home, sold in local markets, and are even processed and sold further afield, often conserved in tins or jars.

Burnt *C. mopane* wood produces an ash that is used to make whitewash, and being rich in calcium and phosphorus, the ash is also a useful fertilizer. Trees also produce gum or resin. An extract of the bark is used as a remedy for syphilis and as an application to inflamed eyes; a deep red decoction of the bark is drunk to cure stomach pains. It is a valuable browse species for both domestic and wild herbivores. Fresh leaves have a crude protein content of about 11% but are high in tannins and are therefore less palatable, whereas fallen leaves have a crude protein content of about 4% but are readily eaten during the dry season.

Further reading

Gondo, T., Frost, P., Kozanayi, W., Stack, J. and Mushongahande, M. (2010) Linking knowledge and practice: assessing options for sustainable use of mopane worms (*Imbrasia belina*) in Southern Zimbabwe. *Journal of Sustainable Development in Africa* 12, 127–145.

Musvoto, C., Mapaure, I., Gondo, T., Ndeinoma, A. and Mujawo, T. (2006) Reality and preferences in community mopane (*Colophospermum mopane*) woodland management in Zimbabwe and Namibia. *International Journal of Human and Social Sciences* 1, 173–176.

Timberlake, J.R. (1995) Colophospermum mopane, *Annotated Bibliography and Review*. The Zimbabwe Bulletin of Forestry Research 11, Forestry Commission, Harare, 49 pp.

Cordia alliodora (Ruiz & Pav.) Oken

Ecuador laurel

Importance

Cordia alliodora is an important tree, combining fast growth on fertile sites with the production of high-quality timber. The wood is valued throughout its natural range in the tropical Americas, both for local use by farmers and as a commercial timber in national markets. Farmers favour natural regeneration within agroforestry systems, which are appropriate for this species with its light crown and self-pruning habit. Given the pressures on the fertile land to which it is most suited, *C. alliodora* is likely to be increasingly planted in pure plantations.

Botanical features

C. alliodora is a tall, thin tree commonly growing to 15–20 m tall and 30–50 cm diameter, but may reach more than 40 m and diameters of 1 m, with a single stem, minimal forking and a narrow, open crown. Buttressing to 1–1.5 m is limited to larger trees on shallow soils. Bark is light grey to brown, and smooth, fissured in drier regions. Leaves are simple, alternate, petiolate, up to 18 cm long and 5 cm wide, pointed at the base. Large, conspicuous auxiliary or terminal inflorescences 10–30 cm across contain 50–3000 hermaphrodite, unspecialized flowers about 1 cm long. The mature fruit is shed approximately 10 weeks after pollination, complete with calyx and corolla, the latter acting as a parachute aiding dispersal by wind.

Distribution

C. alliodora has a wide natural distribution from 27°N to 25°S, from northern Mexico through Central and South America to Paraguay, southern Brazil and northern Argentina. It is also found on most Caribbean islands from Cuba to Trinidad, but is almost certainly not native to Jamaica. As a successful colonizer of disturbed sites after forest clearance, its numbers have undoubtedly increased under human influence. *C. alliodora* has been introduced to a large number of countries as part of species and provenance trials, but it has only been planted outside of its native range to any extent on Pacific islands.

Silviculture characteristics

C. alliodora seeds prolifically and regenerates easily, readily colonizing exposed fertile soil, often found as pure stands of varying densities following forest clearance. It will survive under light shade but full overhead light is necessary for full vigour, and on fertile sites the species is capable of rapid early growth. It is a long lived, shade-intolerant, and variously referred to as a late secondary species, small gap specialist, or big pioneer. It is not typical in mature forests, occurring very infrequently in undisturbed tropical lowland rainforest, but is common in undisturbed dry forest. It coppices readily, even from large mature trees. A strong taproot is complemented by a spreading root system, but trees only have average susceptibility to windthrow, and stem snap may occur in mature trees even in areas not prone to strong winds.

Mature trees are deciduous even in aseasonal climates, losing foliage for 1–2 months following seed production. *C. alliodora* starts flowering at 2 years old, but reaches sexual maturity at 4–5 years and will seed profusely when mature. The flowering season varies throughout the natural distribution, though length and abundance of flowering varies both between individuals of the same population in any given year, and between years for any given individual. In wet climates, a proportion of trees do not flower in any given year, while in seasonally dry climates all trees of a sexually-mature size flower every year.

Good growth is effectively limited to wide spacing on weed-free, fertile sites. The tree is self-pruned even in open conditions and often up to 50–60% of total height, although the degree of pruning is variable and natural regeneration shows great variation in form. Some trees have pronounced nodal swellings where branches have been shed, and incomplete healing of such areas may represent suitable sites for entry of pathogens.

Throughout its native range, *C. alliodora* occurs under a wide variety of ecological conditions from very wet, up to 6000 mm of rainfall per year, to seasonally dry with as low as 600 mm and a 7-month dry season; and from sea level to 2000 m. It grows larger in lowland humid tropical regions, whereas in seasonally dry deciduous and semi-deciduous forest it is smaller and more poorly formed, with the best growth on well-drained, medium-textured soils with more than 2000 mm annual rainfall and mean annual temperature of about 24°C. Trees will survive under a range of nutrient conditions and occur on degraded and abandoned areas once used for pasture or shifting cultivation, coastal lowlands with deep infertile sands and little organic matter and very dissected mountainous uplands. Trees do not tolerate poor drainage, soil compaction, waterlogging or very acidic soils.

Silviculture practice

Seed should be collected from a tree 2 weeks after the last flowers have opened, and is mature when the embryo is hard (like a grain of rice) when squeezed. Seed can be stored for several years at low temperature (<5°C), after drying in the shade. No pretreatment is required, and germination occurs 10–28 days after sowing. Transplanting to containers or beds is carried out after the first true leaves appear. Planting of *C. alliodora* generally uses nursery seedlings in containers after 3–4 months, or stump plants 6–8 months old. Farmers may also transplant wildings or seed directly. Vegetative propagation has only been used to establish seed orchards by grafting and cuttings.

In pure plantations, spacings of 3 × 3 m to 4 × 4 m are recommended, and timely thinning is required if the plantation is not to stagnate and fail. Effective control of rapid weed growth is essential, but the cost of this operation can be countered by intercropping during the early years of establishment. In agroforestry systems with perennial crops such as coffee or cacao, the shade of mature trees cannot be manipulated in the same way as the 'service' shade trees, since pollarding is not an option and even thinning is undesirable unless carried out at a very early stage because of damage to the underlying crops. In these circumstances, *C. alliodora* should be planted at wide spacing (e.g. 10 × 10 m) that will be maintained until the end of the rotation. In fence rows, trees are often planted at a spacing of 2.5 m, but must be thinned to every 5 m for final spacing.

On suitable sites with good management, annual growth of 2 m in height and 2 cm in diameter may be obtained during the first 10 years, with higher diameter increments at wider spacing. At 10 × 10 m on fertile sites, trees may reach 30 m tall and 50 cm diameter after 20 years, with a final stem volume of 2 m^3, and financial analysis showed better and more stable financial indicators at 100 trees/ha. Yields in plantations are very variable, reported to be 5–20 m^3/ha/yr on 20–30 year rotations, depending on site conditions.

Diseases and pests

Despite rapid growth, introduced plantations have suffered from diseases to the point that further planting has been discontinued in some countries. In the native range, a number of pests and diseases may cause infestations, including a canker disease caused by *Puccinia cordiae* and parasitism by several species of Loranthaceae, but none are considered very important.

Uses

Most coffee and cacao in Latin America is grown under shade trees which provide little direct economic benefit, with valuable timber trees maintained at low densities (e.g. 100 trees/ha) as a strata above the shade trees, and *C. alliodora* is particularly suited to inclusion in such systems.

Heartwood is light greenish-brown to olive-brown with dark streaks when freshly cut, turning a pale golden-brown to brown on seasoning, with the streaks becoming more conspicuous, and having a slightly mottled appearance when quarter-sawn. The boundary between the darker heartwood and the creamy-yellow sapwood is, however, not always sharply defined. There is seldom any odour or taste, but heartwood from very old trees may have a spicy aroma. Heartwood is described as outstandingly resistant to termite attack, very resistant to white- and brown-rot fungi but with only moderate to no resistance to marine borers. Reported wood densities range from 270 to 624 kg/m^3.

C. alliodora is renowned for the quality of its timber, having wood which is usually straight grained with a fine to medium texture and a medium to high lustre. It seasons well, is dimensionally stable when dry, has satisfactory strength and durability properties, works easily with little dulling of cutting edges giving a very smooth attractive finish, and polishes and stains well. The wood is used for furniture, turnery, panelling, veneer manufacture, mouldings, light internal and external construction, boat timbers and carving.

Flowers are known to bee keepers as a major source of nectar that give a viscous extra-white honey, and some medicinal properties are reported.

Further reading

Boshier, D.H. and Lamb, A.T. (1997) Cordia alliodora: *Genetics and Tree Improvement*. Tropical Forestry Paper No. 36. Oxford Forestry Institute, Oxford, UK, 96 pp.

Greaves, A. and McCarter, P.S. (eds) (1988) Cordia alliodora *1922–1987. Annotated Bibliography*. CAB International and Oxford Forestry Institute, Oxford, UK, 28 pp.

Somarriba, E., Valdivieso, R., Vásquez, W. and Galloway, G. (2001) Survival, growth, timber productivity and site index of *Cordia alliodora* in forestry and agroforestry systems. *Agroforestry Systems* 51, 111–118.

Cordia subcordata Lam.

Beach cordia

Importance

The fine-grained, dark-coloured, durable wood of *Cordia subcordata* is highly prized and has many local uses where it is native, along tropical coastlines in the Indian Ocean, South-east Asia and the Pacific Ocean. It is especially favoured for carvings and curios which provide a significant source of local cash income on Pacific islands. It provides coastal protection and is planted as a windbreak, for amenity and shade. It has not proved very amenable to plantation management to date, and continuous removal means that it is becoming rare in parts of its natural range.

Botanical features

C. subcordata is typically 5–10 m tall and 40 cm in diameter, but can grow to 15 m in height and 70 cm in diameter, and up to 20 m tall and 1 m diameter in favourable conditions. The bole is usually short but sometimes branchless for up to 8 m, crooked or leaning, without buttresses, with a dense, round to irregular crown. Bark is mid- to dark brown, usually weathered light grey, smooth when young, developing evenly spaced vertical fissures and lesser horizontal cracks forming a grid pattern, peeling in square to rectangular scales along these lines with age. Inner bark light brown in distinct layers, grading to very light brownish-white on the wood; with a small amount of clear sap, turning slightly darker greenish-brown after exposure to the air for some time.

Leaves are alternate, simple, broadly ovate to elliptic, 8–25 cm in length and 5–15 cm in width, on a long (2–8 cm), thin petiole. Leaves are discolourous, shiny dark green above with light yellow-green venation, and dull, light green with fine white hairs in the axils of the main veins below. The leaf base is cuneate or obtuse and the leaf margin is entire or crenulate. Very young leaves and shoots may be covered with a mixture of fine hairs and scattered coarse hairs, but most surfaces are glabrous.

Its showy, pale orange or bright orange-red, bisexual flowers are arranged in short terminal cymes, 6–20 flowered, 2–5 cm long (without flowers). The fruit is a hard brown drupe, subglobose to ellipsoid, 2–3 cm in diameter and 2–3.3 cm long with a pointed tip and slight longitudinal ridges, completely and tightly enveloped within the enlarged calyx tube. The mesocarp is spongy or corky when fresh and the endocarp is hard, angled, with four locules. Fruit hangs in clusters, green at first, turning yellow-green to dark brown or black when mature.

Distribution

C. subcordata is a common constituent along tropical coastlines in the Indian Ocean, South-east Asia and the Pacific Ocean. The natural distribution ranges from Tanzania and Kenya in East Africa, Madagascar and Indian Ocean islands, South-east Asia including Myanmar and Thailand to Hainan Island (China), Indonesia, northern Australia and New Guinea. It is widespread in most Pacific archipelagos in the South Pacific Ocean including Hawaii and as far east as the Society Islands; however, as it was an important cultural plant of the early Pacific islanders, it may have been introduced to the most eastern part of its range.

Throughout its natural range, *C. subcordata* is a fairly common component of strand vegetation, on sandy shores, lagoons and in coastal secondary forests and thickets, low deciduous forest, littoral forest and thickets, but is rarely found very far inland.

Silviculture characteristics

C. subcordata grows in maritime areas up to 20–30 m altitude in tropical humid and subhumid climates, sometimes to 150 m. The species is adapted to strong, salt-laden winds, including tropical cyclones over much of its range. It is a pioneer coastal species which grows in full sunlight or partial shade, but plantations in the open result in trees poorly suited for timber production, and it is recommended that trees be replanted into native forest with scattered trees or bush.

Trees mainly occur with 1000–4000 mm annual rainfall and a short dry season of 3–4 months, though can tolerate 500–700 mm with a pronounced dry season of 4–6 months with access to groundwater. Rainfall may be distributed more or less uniformly or irregularly throughout the year, bimodally, or with a peak over 4–6 months usually during the hottest period of the year. Temperatures are warm to hot throughout the year with little seasonal and diurnal variation. Mean annual temperatures are typically 24–28°C, with mean maximums for the hottest month of 28–36°C and mean minimums for the coolest month of 17–25°C. The entire range is frost-free with absolute minimal temperature recorded of 12–22°C. The corky fruits float, and may often be found in the shore drift, and seeds are readily dispersed along ocean currents in common with other widespread, strandline species.

Silviculture practice

No seed pretreatment is required. Fruits are soaked in a fungicidal solution and then sown in germination boxes filled with river/beach sand or in soil mixed with humus, germinating 4–10 weeks after sowing. Seedlings should be grown under well-ventilated and aerated conditions to minimize risk of fungal attack. They are ready for planting after 6–8 months when they reach about 40–50 cm high, and are often planted at a spacing of 1.5–2 m. Tending operations in native forest and dense bush include frequent weeding, thinning and pruning. On small islands it is perhaps best suited to replanting in small pre-existing gaps, where tending will be required less than when planted into dense bush. As an amenity or shade tree, the often crooked habit means that regular pruning is required to achieve a desirable form.

Height growth rates of 1 m/yr are common, and up to 1.5 m/yr on the most favourable sites in the first few years, with corresponding diameter increments of 1.5–3 cm/yr. However, on a very dry clay soil receiving only 500–700 mm annual rainfall, a height increment of only 36 cm was reported after 6 years. Trees take 60–70 years to reach large enough diameters, and whereas thinning may provide earlier cash returns, even these are unlikely to occur before 20–30 years.

Diseases and pests

Seedlings are susceptible to attack by several pathogenic fungi, including *Pythium*, *Phytophthora* and *Fusarium*. Susceptibility to other diseases is also reported if the species is planted in excessively wet sites.

Trees are very vulnerable to the kou leaf worm (*Ethmia nigroapicella*) which may cause significant defoliation and even kill trees, and this has led to its substitution with other trees on some Pacific islands.

Uses

C. subcordata is important for coastal protection and as a windbreak, being resistant to strong, salt-laden winds. It is also widely used in amenity purposes and as a shade tree, and is common in gardens, parks and as a street tree. It is sometimes planted around farms and fields in rows and as a live fencepost, and has been tried in alley farming systems.

The wood is hard, moderately heavy (470–650 kg/m^3), durable and easy to work and dress. The heartwood is light brown, prominently marked with dark brown or black streaks, and the sapwood is pale yellowish-brown. The fine-grained and attractively marked wood is among the most favoured carving woods in the South Pacific. It is also has many other uses, including for plank canoes, paddles, trays, beams, posts, wharves, cabinets, bowls, furniture, air-tight reef boxes for fishing equipment, slit gongs, drums and other musical instruments, scabbards, tools, veneer, fancy articles and other carved objects. It is favoured for kitchen utensils because it does not impart any flavour to foods. It is used in general and light construction where it is recorded as lasting more than 100 years.

The wood is fast burning, and is very easily ignited, even by vigorously rubbing two pieces of dry wood together, though it is sometimes considered too valuable to burn.

The showy flowers are included in floral garlands, the seeds are edible, and leaves are sometimes chewed with betel nut. Leaves may also be fed to livestock and various plant parts are used medicinally. In parts of Polynesia, the inner bark is soaked in seawater and made into dance skirts, hats, fans, baskets and garlands.

Further reading

Elevitch, C.R. and Wilkinson, K.M. (2003) Propagation protocol for production of container *Cordia subcordata* Lam. plants. Permanent Agriculture Resources, Hōlualoa, Hawaii.

Friday, J.B. and Okano, D. (2006) *Cordia subcordata* (kou), ver. 3.1. In: Elevitch, C.R. (ed.) *Species Profiles for Pacific Island Agroforestry*. Permanent Agriculture Resources (PAR), Hōlualoa, Hawaii, 15 pp.

Lemmens, R.H.M.J., Soerianegara, I. and Wong, W.C. (eds) (1995) *Timber Trees: Minor Commercial Timbers*. Plant Resources of South-East Asia 5. Prosea Foundation, Bogor, Indonesia. Backhuys Publishers, Leiden, The Netherlands, 655 pp.

Croton megalocarpus Hutch.

Croton

Importance

Croton megalocarpus is a hardy, fast-growing East African multipurpose tree planted for the production of timber and fuelwood, for shade, as an ornamental, for boundary marking and as a live fence. It is also important for charcoal production, as bee forage, for mulch and has potential as a bioenergy crop.

Botanical features

C. megalocarpus is a large monoecious or occasionally dioecious tree up to 35 m tall, though commonly 15–25 m, with a clear bole up to 20 m high, a diameter of 1 m and a flat, spreading crown with distinctive layering of branches. The bark is pale grey-brown to dark grey, and closely longitudinally fissured.

Young twigs are densely fulvous lepidote; older twigs are greyish-brown, tardily glabrescent. Petioles 2–8 or up to 11 cm long, densely fulvous lepidote to fulvous or ferruginous stellate–lepidote.

Leaf blades are ovate, elliptic–ovate, elliptic–lanceolate or oblong–lanceolate, generally 7–14 cm long and 3–7 cm wide, but with extremes of 4 and 19 cm in length and 1.5 and 12 cm in width; they are shortly and acutely or subacutely acuminate, cuneate–rounded, rounded or shallowly cordate, with 2–4 sessile or stalked basal glands visible or not from above, entire, firmly chartaceous to subcoriaceous, lateral nerves 15–25 pairs, not or slightly impressed above, slightly prominent beneath, midrib impressed above, prominent beneath, evenly minutely stellate–puberulous and medium to dark green above, densely silvery lepidote or rarely stellate–lepidote and brown flecked beneath, the scales fimbriate, stipules linear–filiform, sometimes somewhat laciniate towards the base, 5–10 mm long, sparingly scurfily stellate–lepidote and sub-persistent.

Terminal racemes are up to 25 cm long, all male or with a few female flowers at the base, axis angular, densely fulvous lepidote to fulvous or ferruginous stellate–lepidote. Bracts are linear lanceolate to elliptic–lanceolate, 2–4.5 mm long, densely stellate–lepidote, often brown flecked. The calyx is five-lobed, the lobes triangular-ovate, 4.5 × 3.5 mm, subacutely acuminate, densely silvery lepidote and brown flecked without, densely pubescent within, somewhat villous on the margin at the apex, and greenish-yellow.

Male flowers are short-lived, but conspicuous, with five petals, obovate, 5 × 3.5 mm, densely silvery lepidote and sparingly light brown flecked without, pubescent within, somewhat villous on the margin, pale straw-yellow; disc-glands free and truncate; stamens 25–30 (exceptionally 40), filaments 5–8 mm long, glabrous above, sparingly pilose below, anthers 1.5–2 mm long, receptacle pubescent. Female flowers have pedicels 4–12 mm long, up to 2 cm long in fruit, stouter than in the male; calyx-lobes 3.5 × 2 mm, otherwise as in the male; five petals, linear, 5 × 0.5 mm, lepidote without in the lower part, otherwise subglabrous, pubescent within and villous on the margin; disc shallowly five-lobed or more or less annular; ovary subglobose–subtrilobed, 4 mm in diameter, densely silvery lepidote and brown flecked; three styles, spreading, 1.5 mm long, subglabrous above, densely stellate–lepidote beneath, shortly bipartite, the segments two- to four-lobulate. Trees can begin flowering at 4 years old.

Fruits are numerous grey woody capsules, ellipsoid–ovoid to subglobose, not lobed, 3–4.5 cm long, 2.5–4 cm in diameter, loculicidal from the apex, lepidote; endocarp woody, 3–5.5 mm thick, with three double rows of pits outside corresponding to the septa. Each fruit contains 3 seeds, ellipsoid–ovoid or oblong–ellipsoid, 2.2–2.4 cm long, 1.2–1.4 cm wide, somewhat ridged and rugulose, white when fresh, grey-brown when dry, with a minute caruncle.

After pollination by insects, fruit development takes 5 months. In Kenya, seeds mature during October–November in central regions and from January to March in western regions.

Distribution

C. megalocarpus has a native distribution in Eastern Africa from Uganda, Kenya and Tanzania in the north, through Rwanda, Burundi and eastern parts of the Democratic Republic of Congo to Malawi, Mozambique and Zambia in the south.

It occurs in evergreen and semi-deciduous forests, where it can be a dominant overstorey tree, in a wide range of areas at altitudes of 900 to 2500 m, and is particularly widespread in montane semi-evergreen forests with a mean annual rainfall under 1020 mm.

Silviculture characteristics

C. megalocarpus is a pioneer species and grows in cleared parts of natural forests, at forest margins or as a canopy tree. It regenerates well through seedlings; under favourable climatic conditions it can sometimes become invasive. It is fast growing in areas with rainfall above 900 mm, but slower in drier areas. It tolerates a dry season duration of up to 4 months. Although preferring light, deep and well-drained soils, it can be planted on heavy and medium-textured soils.

Trees can be trimmed, pollarded, lopped or coppiced.

Silviculture practice

Direct sowing is a viable propagation method. Mature fruits are collected from the ground and seeds are extracted by cracking open the shell. On average there are 1700 seeds/kg. The seeds are dried in the sun to approximately 5–9% moisture content and can be stored up to 1 year at 3°C, retaining up to 80% germination. After sowing, the seeds germinate at a rate of 95% within 35–45 days without any pretreatment. Traditional farmer sowing practice involves dropping ripe fruits into 8-cm deep furrows at spacings of 15–20 cm. Seedlings raised in seed beds require light shade and watering twice a day at first. *C. megalocarpus* is sometimes propagated by wildings and stump planting. Vegetative propagation is possible using cuttings, tissue culture or grafting.

Depending on growing conditions, seedlings can reach 1.7 m in height in 1 year, 3 m in 2 years and 11.5 m in 5 years. In Burundis planted trees reached on average only 3.6 m tall 7 years after planting, and 15 m with a bole diameter of 24 cm after 32 years.

Young trees coppice well after pruning, but fruiting is unlikely with intensive pruning as practised in hedgerow management. In agroforestry systems, *C. megalocarpus* is sometimes managed as scattered trees in farmland because of its open canopy and usefulness for mulching.

Diseases and pests

In Kenya, trees of *C. megalocarpus* growing at altitudes between 1300 and 2100 m have been reported to be damaged by ambrosia beetles. Sporadic outbreaks of the defoliator *Amyna punctum* can cause up to 100% foliage loss in 1–2 weeks, more so in older than younger trees. Such defoliation can lead to tree mortality if other stress conditions, such as a prolonged dry spell, prevail at the same time.

Uses

C. megalocarpus trees are planted widely as shelter belts and windbreaks, and also as live hedges, although these usually do not become very dense, but are not browsed by livestock. Trees are widely retained as shade trees in arable and plantation crop lands and in pastures. Although not particularly recommended for intercropping due to competition and shade, as is the case with maize cultivation, an agroforestry trial showed croton trees had little effect on sugarcane yields.

C. megalocarpus wood is of medium weight, with a density of 700–750 kg/m^3 at 12% moisture content. It is liable to splitting and moderate distortion during air drying. Once dry, it is often not stable in service, especially when used in larger sizes. The wood is easy to saw and work with hand tools, but moderately difficult to machine. It usually planes to a smooth and lustrous surface. It is resistant to abrasion. Nailing, screwing, gluing, varnishing, painting and jointing properties are all satisfactory. The wood is moderately durable, being only slightly susceptible to termite and borer attacks, but is liable to attacks by blue-stain fungi. The heartwood is yellowish-white to brownish-grey, often with irregular dark brown streaks, and not distinctly demarcated from the 2.5–5 cm wide sapwood; it has a bad odour when freshly sawn.

The main uses of the timber are for construction poles, flooring, stools, mortars, beehives, veneer and plywood. It is suitable for joinery, interior trim, shipbuilding, vehicle bodies, furniture, cabinet work, railway sleepers and agricultural implements.

The wood is an excellent fuel and can be used directly or made into charcoal.

Decoctions of the bark are used in traditional medicine to treat stomach ache and whooping cough, remove intestinal worms and relieve pneumonia. Chiromodine, a clerodane diterpene from the bark, is reported to have anti-ulcer, anti-cancer, and co-carcinogenic properties. Aqueous extracts of the species have given positive antibiotic tests against *Staphylococcus aureus*, but negative tests with *Escherichia coli*. Seeds are traditionally fed to poultry in Kenya as a treatment for coryza sinusitis, and the seed oil can be used as a strong purgative.

Seeds contain 50% protein, and with an oil content of over 30%, interest has grown recently in *C. megalocarpus* as an alternative to *Jatropha curcas* for producing biodiesel from the seed oil, croton having several economic advantages over *Jatropha* including a much lower water requirement. Wild-growing croton trees can yield about 0.35 l oil/kg of nuts.

C. megalocarpus is a good forage plant for bees, the honey produced from it being dark amber in colour with a strong flavour. Its conspicuous flowers also make it popular for ornamental plantings.

Further reading

Aliyu, B., Agnew, B. and Douglas, S. (2010) *Croton megalocarpus* (musine) seeds as a potential source of bio-diesel. *Biomass and Bioenergy* 34, 1495–1499.

Dharani, N. (2002) *Field Guide to Common Trees & Shrubs of East Africa*. Struik Publishers, Cape Town, South Africa, p. 75.

Smith, A.R. (1987) 8. *C. megalocarpus* Hutch. In: Polhill, R.M. (ed.) *Flora of Tropical East Africa. Euphorbiaceae (Part 1)*. A.A. Balkema, Rotterdam, The Netherlands, pp. 144–145.

Cryptomeria japonica (L. f.) D. Don

Japanese cedar

Importance

Cryptomeria japonica, a native of China (var. *sinensis*) and Japan (var. *japonica*), is an extremely long-lived, evergreen conifer producing uniform, moderately strong, variably durable, scented sawnwood. It exists in a large number of forms and is tolerant of a wide range of sites and climates. It has been widely planted as a specimen or shelter tree in both its native range, and in warm and cool temperate climates around the world. In Japan, *C. japonica* is used for cladding temples, shrines and sacred places generally.

Botanical features

C. japonica var. *japonica* is a straight-stemmed, monopodial tree, up to 65 m tall and up to 2.5 m in diameter. It is very long-lived, with ancient specimens on Yakushima Island in Japan being well over 1000 years old. Crown pyramidal in exposed trees, but spire-like in mature forest trees. Branches persist for years but are eventually shed in mature trees; clear boles up to 20 m; in most strains, branches leave stem horizontally, then turn upwards.

Bark is fissured on old trees, light brown. Bark is thin, only 12–18 cm thick on 200 year-old trees.

Leaves spirally arranged in 5 rows, decurrent, linear–subulate, ascending and slightly incurved, slightly flattened laterally, anterior-posteriorly keeled, acuminate, stomatal bands on both sides. Leaves turn reddish-brown in winter, especially if exposed to wind.

Cones are globose, woody, the scales 20–30, 2- to 5-seeded. Seeds narrowly winged; irregularly triangular; 6 × 3 mm. Cones ripen (colour brown) in the first year and persist 1–2 years longer, with branchlet growth often temporarily continuing through cone. Seeding is heavy and starts at 5–10 years old.

Monoecious; cone crops can be heavy. Seed is usually available annually. Shoot elongation occurs from April to October in the northern hemisphere.

C. japonica var. *sinensis* is diffuse in habit with deflexed branches and long slender branchlets; it has more slender and drooping branches, longer and thinner needles, and cones with fewer scales (rarely more than 20) than var. *japonica*. There are usually 2 seeds to each scale. The cones are much less spiny than in var. *japonica*.

Distribution

The natural distribution of *C. japonica* var. *japonica* in Japan is confused both by the extent of new plantings and the millennia of use. The range of latitudes of natural stands is from 40°12’ to 30°15’ N in Japan; plantings in Japan now extend to nearly 43°N in Hokkaido to 27°N in Okinawa.

Var. *japonica* is native only to Japan, where it occurs naturally in pure and mixed stands from Kyushu to northern Honshu at elevations to 400 m. The variety has also been widely introduced for forestry in Taiwan and many provinces of mainland China. Var. *sinensis* is native to China, in Fujian (Nanping Shi), Jiangxi (Lu Shan), Sichuan, Yunnan and north-west Zhejiang (Tianmu Shan), and has also been widely introduced for forestry in other provinces of China. It grows in forests on deep, well-drained soils subject to warm, moist conditions at elevations from below 1100 m to 2500 m.

C. japonica var. *sinensis* is found in Fujian, Jiangxi, Zhejiang, Sichuan and Yunnan, China. This represents a distribution from about 25 to 29°N.

C. japonica has been introduced to Australia, Brazil, Canada, Europe, India, Indonesia, Iran, Malawi, Mauritius,

New Zealand, the Philippines, Russia, South Africa, Tanzania, the UK, the USA and Zimbabwe.

Silviculture characteristics

C. japonica tolerates moderate to heavy shade, eventually developing into a climax canopy species. This means it can regenerate successfully where the canopy is dense enough to suppress competing vegetation. It can be used in underplanting. Newly exposed mineral soil sites resulting from landslides are often colonized.

Heavy snowfalls can cause butt sweep, and often major stem curvature. Affected trees have compression wood which is among the worst defects of *C. japonica* var. *japonica* as a plantation species. A variety has been bred, *okinoyama-sugi*, which withstands the snow effects better; it develops less compression wood. Snow damage is lessened if trees grow above the susceptible height of 2 m as soon as possible. Frost cracks, usually found 2–3 m above the ground, can affect trees older than 40 years of age.

Silviculture practice

C. japonica regenerates well from seed. It can also be grown from cuttings and it layers readily. Unusually for a conifer, *C. japonica* will coppice from cut stumps.

Propagation is mainly by cuttings from mature trees and seed. Seeds are broadcast into shaded beds mixed with wood ash and fine earth, which require daily watering. Germination occurs after 14–28 days. Seedlings are ready for transplanting after 12 months. Seed are sown in early spring in a cold frame. Higher germination rates are achieved with a short cold stratification for 2–3 weeks at 4°C. Germination usually takes place within 3–9 weeks at 15°C. Seedlings should be pricked out at a spacing of 90–120 cm in early summer. Shade is removed after 3 months, and seedlings are transplanted to the field when 3 years old at a spacing of 5.4 m in early summer or mid-winter. *C. japonica* needs full sun, sheltered from harsh winds and where the air circulation is good, especially in summer, to prevent leaf blight. Plants may grow 60–120 cm/yr when young.

Pruning has been carried out in Japanese forests (with the aim of producing knot-free faces on timber for specialized products) for several centuries, but was not widespread until recently. There are usually two to four pruning steps. It is usual to apply fertilizer to the stands to compensate for reduced vigour.

In Japan, *C. japonica* has been underexploited in plantations, as the thinning prescriptions have not been carried out. Most forests in Japan belong to a large number of small-scale forest owners. Further, because the costs for growing forests are very high, forest owners tend to be reluctant to engage in forestry practices.

Pollinosis caused by *C. japonica* is acknowledged as a nationwide problem in Japan, where efforts are being made to develop and plant lower-pollen *C. japonica* varieties.

Mature trees can grow very slowly; ring widths of 1–2 mm at 1.3 m above ground were recorded in Niigata. Ring widths of younger trees were 3–4 mm, but diameter increments are often rapid. Japanese *C. japonica* trees used in forest products tests had annual ring widths of 3 mm.

Foliage mass increases exponentially until age 14, when this levels off.

The expected rotation length for maximum volume is 40–50 years for *C. japonica* plantations. Plantations have shown an above-ground biomass of 158 t/ha at age 40.

Diseases and pests

Cercospora sequoiae causes needle blight and stem canker in *C. japonica* and is an important disease in Japanese nurseries.

Damping-off caused by *Rhizoctonia* spp. and *Fusarium* spp. has been observed in nurseries. Three root rots, *Rosellinia necatrix*, *Helicobasidium mompa* and *Armillaria mellea* are considered the most important soil-borne diseases in natural forests and plantations.

The main insect pests of *C. japonica* include the bark beetle *Semanotus japonicas*, the wood wasps *Xeris spectrum* and *Urocerus japonicus*, the defoliating sugi tussock moth (*Dasychira abietis argentata*), sugi web worm (*Homona issikii*), sugi leaf beetle (*Basilepta pallidulum*) and chafer beetles (*Anomala* spp.).

Voles, field mice, rabbits, hares, wild boar, deer, black bears, and monkeys are recorded as causing damage to *C. japonica*.

Uses

C. japonica has been used extensively for shelter belts in South Korea, where it forms windbreaks in 65% of citrus orchards. In New Zealand, single-row belts of trees planted 1.5m apart and 10 m high, often of one clone from cuttings, are common around fruit crops and on farms, except for those in very dry regions. These are trimmed annually and, in effect, become tall, thin, hedges.

C. japonica wood is coarse grained, with thick layers of annual growth, dark-reddish heartwood and thick, pale sapwood; it is easily worked, durable and strong enough for most purposes. Wood density is 310–460 kg/m^3. The wood turns dark green when it is buried in the ground; the Japanese regard such wood as being very precious, and use it to decorate objects such as cabinets.

In Japan, *C. japonica* was the most widespread building wood in the period from ca 300 BC to AD 300, and was the major species used in the lowlands, which were being developed for rice cultivation. Some earlier records relate to its use around 10,000 BC. From AD 400 to 800 it became even more widely used. It was occasionally used for coffins in the period from AD 400 to 800.

It was also used for carved wooden ships in central Honshu in 1500 BC. By 500 BC it was used for tools and some buildings. This use increased considerably from AD 500 to AD 1400.

The wood is fissile, and is used for saké barrels. The wood contains the oil sugiol which gives the saké a peculiar, pleasant fragrance.

C. japonica bark is used for roofing. The bark is moderately durable, but not as good as that from other Japanese conifers. The species also yields a turpentine or semi-solid resin known as *sugi-no-jamai*, a viscous, greenish liquid which is used for incense.

Further reading

Kitahara, R. (2011) Usages of the wood in *Cryptomeria japonica*. *Bulletin of the Faculty of Agriculture, Miyazaki University* 57, 103–112.

Vidakovic, M. (1991) *Conifers: Morphology and Variation.* Translated from Croatian by Soljan, M. Graficki Zavod Hrvatske, Croatia/CAB International, Wallingford, UK.

Wilson, E.H. (1916) *The Conifers and Taxads of Japan.* Publications of the Arnold Arboretum No. 8. Cambridge, Massachusetts, USA.

Cunninghamia lanceolata (Lamb.) Hook.

Chinese fir

Importance

Cunninghamia lanceolata has been cultivated for more than 8000 years in China, with archaeological evidence existing for large *C. lanceolata* trees in Hunan some 2000 years ago. The species has good form and fast growth. It is one of the main plantation tree species in China, with estimates giving the total plantation area as 7,953,000 ha and standing volume as 27,070,000 m^3. *C. lanceolata* accounts for 20–30% of the total commercial timber production in China; recently, about 400,000 ha of *C. lanceolata* plantations have been planted annually, accounting for about 10% of the total area of new forest plantations.

The main use outside China is as a landscape and ornamental species.

The wood of *C. lanceolata* is durable and widely used for construction, boatbuilding, transmission poles, windows, doors, furniture and veneers. It also has good pulping properties.

Botanical features

An evergreen tree with good form, up to 30 m tall and 2.5–3.0 m in diameter.

Bark of old trees is grey-brown, shed in long irregular strips, showing the young reddish bark beneath. Branches of young trees more or less whorled in groups of 5 or 6; branches of old trees irregular, spreading, pendulous at the base. Lateral branchlets deciduous. Young shoots without pubescence, densely covered with leaves. Winter buds round, flattened and covered by leafy scales, the flower buds larger than the others.

Leaves are spirally arranged, those on the main axis standing out from all sides of the stem, those on the underside of the branches turning upwards by a basal twist so that all appear to spring from the sides and surface of the shoot; persisting alive for 5 or more years and remaining dry and dead on the branches for several years more; linear–lanceolate or lanceolate, curving backwards, 3–6 cm long, 3–5 mm wide at the base, green or glaucous-green; margins finely toothed, apex a long, slender point; stomata in a broad band on each side of the midrib on the underside.

Trees are hermaphrodite, unisexual. Male flowers small, subterminal in shoot catkins, about 12 of which form a dense cluster at the end of the shoot. They form in autumn and open in April to early May. Female flowers terminal, form in autumn near the ends of shoots, enclosed in leafy scales, open in April. Cones terminal, ovoid or rounded, 2.5–5 cm long and 3–4 cm wide, solitary or several together; scales brown, stiff, margins toothed, apex narrowed abruptly into a long spine. Seeds 3 to each scale, thin, about 7–8 mm long and 5 mm wide, surrounded by a thin membranous wing.

The terminal buds of *C. lanceolata* start swelling and needles start to flush from mid-March to early April in the central zone of its distribution, and about 1 month later at the southern margin of its distribution. Male flowers bloom for 5–10 days, in mid- to late March when mean monthly temperatures reach 10–13°C. Female flowers close scales tightly after successful pollination and form young cones 10 days later. Seeds ripen late October to November, with the optimum period for harvesting cones occurring 2 or 3 weeks later when cones are brown.

Distribution

C. lanceolata is widely distributed in the Chinese subtropics. The limits of its natural distribution are marked in the north by the southern part of the Qinling, Tongbai Shan, Dabie Shan and Ning-Zhen Ranges; in the east by Taiwan, Fujian and Zhejiang; in the south by Hainan Island, Guangdong and Guangxi; and in the west by south-west Yunnan, the Anning River, and the middle and lower reaches of the Dadu River. It occurs at altitudes from sea level up to 700 m in the eastern part of the distribution, up to 1800 m in the western part of the distribution, and up to 2600 m on Mount Cangshan in Dali, Yunnan.

The natural distribution of *C. lanceolata* can be divided into three zones (northern, central and southern) according to soil type and climatic characteristics. The distribution of *C. lanceolata* vegetation types is closely related to these distribution zones: evergreen and deciduous broadleaved mixed forests occur in the northern zone; evergreen broadleaved forest, and evergreen and deciduous broadleaved mixed forest occur in the central zone; and partially moist monsoon evergreen broadleaved forest occurs in the southern zone.

Most *C. lanceolata* occurs in plantations; however, natural secondary forests are occasionally found in the central zone.

Silviculture characteristics

C. lanceolata is slightly tolerant of shade when young, but grows best in full sunlight during later growth stages. It is intolerant of drought, waterlogging and strong winds. Provenances from the north are more frost-tolerant than provenances from the south. Propagation is easily achieved by coppicing (unusual for conifers) and from root cuttings.

Silviculture practice

Flowering and seed set first occur in 6–8-year-old trees. Mature trees flower from late February to late March, and produce mature seeds during late October and early November. Seeds account for about 3–5% of the dry cone weight. Seeds dried to 8–10% moisture content retain a high germination rate following storage in airtight plastic containers for 1 year at low temperatures or under normal temperature conditions.

C. lanceolata is relatively easy to propagate using vegetative methods. Before the 1950s, planting stock was produced by coppicing basal stumps after felling, and then collecting the coppice sprouts when they had grown to a height of 30–50 cm. The sprouts were then planted directly in the soil. This regeneration method has been replaced by planting of nursery-grown seedlings, which have faster early growth and better survival rates. Since the 1980s, techniques have been developed for producing planting stock from root cuttings. Cuttings are collected from the root collar when they reach a height of about 10 cm and are placed in rooting beds; these cuttings have a survival rate of 90–95%. The quality of rooting stocks is usually similar to that of seedling stocks. Cuttings from lateral shoots have lower survival rates and produce heavily tapered stems, and are therefore not used in practice.

C. lanceolata is easy to graft using the cleft grafting method, in which a seedling is used as a rootstock and the terminal part of a 1-year-old shoot is used as a scion; the survival rate of grafts produced by this method is typically >90%.

Small areas of mature *C. lanceolata* plantations are usually clear felled, and then regenerated using 1-year-old seedlings or cuttings. In some montane areas, coppice regeneration is frequently used. However, with felling stool age, the size of the timber decreases and taper becomes more pronounced; therefore coppice regeneration is usually not used for more than two rotations.

Planting densities in *C. lanceolata* plantations are usually quite high (2500–3600 stems/ha); on marginal, lower-quality sites, planting densities are usually at the higher end of this range.

Tending should be carefully carried out 2 or 3 times during the growing season in each of the first 3 years after planting. Pruning is not usually practised. In forests, the lower branches will eventually self-prune when tree canopies close. Thinning 2 or 3 times is recommended according to initial density and growth, with about 1500 trees/ha retained as the final crop.

C. lanceolata produces good height growth from 3 to 10 years old, with mean annual values of 0.5–1.0 m. At 30 to 35 years old, the mean annual height increment decreases to 0.2–0.3 m. Mean annual diameter growth from 3 to 10 years old can be up to 3 cm, but is usually less than 1 cm. Maximum volume growth occurs at 30 to 35 years old, with individual annual volume increments reaching up to 0.5–0.6 m^3.

Rotation lengths usually range from 20 to 30 years. In the central zone in China, mean height at 20 years old is estimated at 15–18 m, and annual volume increment is estimated at 20–30 m^3/ha. Total harvest volume is estimated at 500–800 m^3/ha.

Diseases and pests

The main diseases of *C. lanceolata* are caused by *Rhizoctonia solani*, *Glomerella cingulata*, *Pseudomonas cunninghamiae*, and *Lophodermium uncinatum*. They frequently occur on weak trees or on lower-quality plantation sites.

Three insects are serious pests of *C. lanceolata*. *Semanotus sinoauster*, a cerambycid borer, is controlled by artificial capture or by physical removal of the pests (by scraping the bark); *Lobesia cunninghamiacola*, which eats the terminal buds on new shoots, is controlled by the release of parasitoids such as *Trichogramma dendrolimi* during egg production; and the third pest is the borer *Callidium villosulum*.

The main pests affecting *C. lanceolata* in nurseries are *Macrotermes barneyi*, *Odontotermes formosanus*, *Gryllotalpa africana* and *Agrotis ipsilon*.

Uses

C. lanceolata is widely used in agroforestry systems in China. In some montane areas it is frequently interplanted with crops or mixed with other tree species. It is a popular species for landscaping in public gardens, parks and temples because of its fast growth, beautiful crown shape and resistance to pests and diseases.

The wood is characterized by straight grain, uniform texture, and only minor differences between early- and latewood. Wood density is 260–370 kg/m^3. The wood is resistant to warp, splitting, and decay, and its properties include weight by volume 0.39, latewood percentage 22.0%, coefficient of shrinkage (volume) 0.386, crushing strength along the grain 358 kg/cm^2, ultimate bending strength 661 kg/cm^2, and end section strength 285 kg/cm^2.

C. lanceolata wood is widely used for house construction, engineering structures, transmission poles, boat- and vehicle building, agricultural tools and furniture. The wood has a strong fragrance and is therefore not suitable for use in food containers.

The older and larger branches are used in turnery, as the branchwood is characterized by beautiful grain and closely spaced annual rings. Young branches can be made into water bucket handles.

The bark is a source of tannin. It is also used as a raw material for papermaking and traditional textiles, and as an insulating material. Chinese traditional medicine uses the white resin beneath the bark (which can also be processed into furfural), leaves, wood dust and burned bark for the treatment of various ailments. *C. lanceolata* stems are a good source of essential oils for use in the perfumery industry.

Further reading

Anon. (1980) *Ecological studies on the artificial* Cunninghamia lanceolata *forests*. In Chinese with English summaries. Institute of Forestry and Pedology, Academia Sinica, Beijing, 268 pp.

Jøker, D. (2000) *Cunninghamia lanceolata* (Lamb.) Hook. Seed Leaflet 43, Danida Forest Seed Centre, Humlebaek, Denmark, 2 pp.

Sheng, W.T. (1991) The management of *Cunninghamia lanceolata* plantations in China. In: Shi, K.S., Deng, J.F., Ding, Y.Y., Li, W.C. and Chen, Z.D. (eds) *Development of Forestry Science and Technology in China*, China Science and Technology Press, Beijing, pp. 108–114.

Cupressus arizonica **Greene**

Arizona cypress

Importance

Cupressus arizonica is widely cultivated as an ornamental tree, particularly in Europe where it is used for landscaping, erosion control, windbreaks and to a minor extent for lumber.

Botanical features

C. arizonica var. *arizonica* (Arizona cypress): a dense to somewhat openly branched tree, 20–25 m high, straight central leader with a trunk diameter less than 75 cm. On trees over

30 cm in diameter, the bark of the main trunk is grey, dark brown, split with longitudinal furrows, thick, fibrous and does not exfoliate.

C. arizonica var. *glabra* (smooth cypress): a dense somewhat open tree, 7–15 m high, straight central leader with a trunk diameter usually 40–60 cm. Bark of the main trunk is smooth, cherry-red, exfoliating annually in thin layers or plates, non-fibrous and afterwards grey.

C. arizonica var. *arizonica* leaves about 2 mm long, sharply acute, keeled or ridged on the back, glaucous in colour, dorsal glands are very variable. On most cypresses, the glands are mainly inconspicuous, closed and inactive, but some tree leaves have active open glands with resin flow. Crown colour is very variable: from pale grey to more or less bright and dark green, to reach a crown colour with bluish tints.

C. arizonica var. *glabra* leaves about 1.5–2 mm, sharply acute, keeled or ridged on the back, with resin pits. Dorsal glands are well developed, visible and active, and exude resin which turns dark or whitish as it dries on nearly all the leaves. Crown colour is generally grey-green or grey, due to the presence either of a glaucous flowering and/or of resin exuded by glands.

Both varieties are monoecious and have many aspects of flowering and fructification in common.

The microsporophylls (male or staminate inflorescences), borne in the lower part of the crown on short branchlets usually less than 8–10 mm long, are about 3–5 mm long and 2 mm in diameter, tapering to the tip, composed of 12–20 scales for var. *arizonica* and 10–18 for var. *glabra*. There are 4–6 pollen sacs per scale.

The macrosporophylls (female or ovulate cones) are produced at the end of short branchlets in the upper part of crown (prior to pollination: 3 mm long), with 6–8 scales.

In early summer, the fertilized cones (8–10 mm long) turn from green to yellow to brownish and glaucous-grey. When seeds are ripe (18 months after pollination), the mean size of cones is 20–25 mm long and 20–22 mm wide. Shortly after seed maturity, trees shed seeds, but it is common for the cones to remain unopened for several years in clusters (serotinous cones) and they frequently hang on for many years after shedding their seeds. Var. *arizonic* cones produce an average of 90–120 seeds (10–30% filled or sound seeds), and var. *glabra* cones of 90–100 seeds (1–50% filled or sound seeds). The weight of 1000 seeds is 8–9 g.

Distribution

C. arizonica is native from the western USA to north-western Mexico (dry mountains in south-east Arizona, New Mexico and Texas between 1000 and 1500 m as well as in the mountains of northern Mexico – Chihuahua and Coahuila regions – up to 2200 m). The Arizona cypress complex is confined to the upper Sonoran life zone and to the lower borders of the arid transition zone. At lower elevations, it generally grows in canyon bottoms and ravines, while at higher elevations it occupies ridges and slopes, particularly those with a north-east to north-west exposure.

Silviculture characteristics

C. arizonica vars. *arizonica* and *glabra* are quiescent taxa and thus are able to tolerate shallow, rocky, clayey soils which become very dry in the summer. Both are tolerant of winter and spring frosts. Their frost hardiness depends on the physiological state of the trees at the time of the frost event. Both varieties planted near industrial areas (brick kilns or glassworks) do not show any damage due to SO_2 and hydrogen fluoride (HF) pollutants. They are tolerant to the surfactants of sea aerosols. The varieties replace the common cypress (*Cupressus sempervirens*) in the cold areas of Italy, France and Greece, and where canker disease causes heavy damage.

Silviculture practice

C. arizonica seed is collected from cones in autumn. Pre-germination treatment of seed is not required. Usually seeds are placed in water for 48 h before sowing. *C. arizonica* seedlings have 3–5 cotyledons. Average annual height growth is 30–40 cm in 1-year-old seedlings and reaches 70–90 cm in 2 year old seedlings. Vegetative multiplication is used commercially only for some selected clones and for particular purposes (ornamental plantations and Christmas tree plantings). Both varieties have a considerable diffusion in Europe as ornamental trees, and recently also as forest trees, used especially to prevent soil erosion and in areas subject to winter and spring frosts; also in dry, barren soils.

Diseases and pests

Seiridium cardinale, cypress canker disease, is a pathogenic fungus on *C. arizonica* var. *arizonica*, but not so much on *C. arizonica* var. *glabra*, which has a high percentage of resistant individuals in its populations.

Kabatina thujae causes leaf tip death, especially in areas with high relative humidity and fairly low temperatures.

C. arizonica is very susceptible to the cypress aphid, *Cinara cupressi* sensu lato (Hemiptera: Aphididae). The eriophyid mite, *Trisetacus juniperinus* (Acari: Phytopidae), causes stunting of the plant and the development of numerous side-shoots, giving them a bushy appearance.

Uses

C. arizonica was and is planted in many European countries with winter and spring frosts, to reduce the frequency of fire and wind damage to crop production. Due to its dense foliage and acidic litter, var. *arizonica* and var. *glabra* cover the soil completely, reducing weeds and, indirectly, the risks of fire. The species is also cultivated as a Christmas tree.

C. arizonica wood has properties similar to that of *C. sempervirens*, but with some disadvantages. *C. arizonica* wood is less durable and has larger knots that are less adherent and often split during drying. Wood specific gravity is 0.44 at 12% moisture content.

Further reading

Farjon, A. (2005) *A Monograph of Cupressaceae and Sciadopitys*. Royal Botanic Gardens, Kew, London.

Farjon, A. (2010) *A Handbook of the World's Conifers*. Brill Academic Publishers, Leiden, The Netherlands.

Cupressus lusitanica Mill.

Mexican cypress

Importance

Cupressus lusitanica is used as an ornamental (in living hedges, because of its low sensitivity to pruning) as well as in forestry. It is a fast-growing species, especially when young, and has been introduced all over the world.

Botanical features

C. lusitanica is a medium-sized tree up to 25–30 m high, with a broadly pyramidal crown that, in older trees, is broadened with pendulous branches. Crowns are fairly open in young trees, being dark and dense in older ones. The diameter at maturity may reach 120 cm. This species has a straight stem, and a thick, reddish-brown bark with longitudinal fissures.

Young plants have fine dark green, grey-green or glaucous foliage on pink to dark red shoots. Older trees have dark grey-green foliage with the shoots turning orange-brown then dark purplish-brown. The leaves are scale-like, narrowly acute to a short spine, free for most of their length and pointing forwards, having almost no scent when crushed.

Male flowers are very small and ovoid on the tips of the smallest shoots. The cones are small, globose, 12 mm across, having a bluish-green colour in the juvenile stage, turning dark brown when they ripen. Each cone is composed of 6–8 scales with a central hooked spine, erect on the upper scales. Flowering takes place at the driest time of year.

Distribution

C. lusitanica is considered by some to be native between the parallels 15–28°N and 89–108°W, including only Mexico, Guatemala and Honduras. However, many also consider it native to the mountains of El Salvador, Nicaragua and Costa Rica.

C. lusitanica is an adaptable species with a wide geographical and altitudinal range. It is planted in Portugal, England, California, East Africa, New Zealand, Australia and Brazil, mainly for wood production, erosion control and as an ornamental. It has been introduced worldwide as an ornamental tree. The specific name is derived from Lusitania, Portugal, where the tree was introduced in the 17th century.

Silviculture characteristics

C. lusitanica is a fast-growing species of moderate longevity. Because it is a shade-intolerant species, it should be managed in even-aged stands. *C. lusitanica* may have abundant natural regeneration and it is normally used in a pure stand configuration, although it may also grow well in mixed stands.

Silviculture practice

C. lusitanica commences flowering at approximately 15 years old. Seed collection should only be conducted in mast years. The cones should be collected manually on standing trees, by the end of the summer when they turn brown. After being collected, the cones must be immediately packed in porous bags and stored in a ventilated place. Cones are dried in the sun or in ovens at a temperature of approximately 35°C until they open. Seeds may be stored up to 20 years when kept in containers at 2–4°C, and must be stratified for 21 days at a temperature of 2–3°C before sowing. Germination rates can be low due to a high percentage of empty seeds, as well as pests.

Grafting is the most successful method for vegetative propagation of *C. lusitanica*. Cuttings must be taken from the base of plants at the beginning of the growing period. These are then packed in plastic bags and stored at 2–3°C until required. After the removal of the lower leaves, the cuttings (between 10–15 cm long) are treated with rooting hormones and kept warm (22–24°C) on rooting benches for 4 months, until the roots have developed. The plants are gradually acclimatized for a few weeks in order to achieve sufficient height to withstand the transfer to an open-air nursery.

Seeds are sown in beds covered in a thin layer of sand and peat. The seedlings should be kept moist and under shade until the end of summer.

After one growing season in the nursery, the seedlings, measuring 20–30 cm tall, are usually adequate for afforestation purposes. For other uses, requiring taller seedlings, they must be taken away from the containers or seed beds and the taproot undercut before planting in pots with higher volume.

The silvicultural system used for *C. lusitanica* is clear-cutting, with its different alternatives: in alternate strips, in progressive strips and in patches.

The most frequent method for stand establishment is using nursery seedlings which are one growing season old. The recommended initial densities range between 1333 and 1667 trees/ha (at spacings of 2.5 × 2.5 m, 3 × 2.5 m, 3 × 2 m or 3.5 × 2 m).

In order to improve height growth, fertilization is advisable with nitrogen, phosphorus and potassium, 15:15:15 or 12:24:12, using 208 g/tree or 130 g/tree, respectively.

Weeding is necessary in the first few years after planting.

A low thinning system is usually applied for this species in Portugal, which is performed on three occasions during the rotation: between the 13th and 18th year, between the 20th and 25th year, and between the 30th and 35th year. The purpose of this thinning plan is to provide an intermediate income, as well as to select the best trees for timber production. For the final felling, it is recommended to have a density of 300–500 trees/ha. Current recommendations are that pruning should be done below a 12 cm stem diameter (or to retain at least 5 m green crown).

Pruning of trees with the best form to leave 5 m green crown and thinning to maintain diameter growth is recommended for timber production. For the final felling, it is recommended to have a density of 300–500 trees/ha.

Growth rates and rotation length vary depending on where in the world *C. lusitanica* is grown. Generally, under good conditions the mean annual increment for *C. lusitanica* should reach 20–30 m^3/ha/yr. Lower figures indicate unsuitable sites, poor genotypes or inadequate management.

Diseases and pests

Few records exist of diseases that affect *C. lusitanica*. While fungi of the genus *Seiridium* can cause severe branch killing of some species of *Cupressus* (e.g. *Cupressus macrocarpa* and *Cupressus sempervirens*), *C. lusitanica* appears to be relatively resistant to infection except on wet sites.

Two species of bark beetles, *Phloeosinus baumanni* and *Phloeosinus tacubayae* (Coleoptera: Scolytidae), attack this species in Mexico. They confine their attacks to weakened or stressed trees and are more of a pest in ornamental plantings than in natural forests. Several indigenous defoliators have found the foliage of *C. lusitanica* to be a favourable host in plantations established in Colombia, and have caused damage to this species. These include *Glena bisulca*, *Melanolophia commotaria* and *Oxydia trychiata* (Lepidoptera: Geometridae). Several species of ants, including *Acromyrmex* sp. and *Atta* sp. (Hymenoptera: Formicidae) are also known to defoliate this species in Colombia.

The cypress aphid, *Cinara cupressi* sensu lato (Hemiptera: Aphididae), has become a major pest of *C. lusitanica* in Eastern and Southern Africa, where this species is widely used for industrial forest plantations and as an ornamental hedge. This insect was first discovered in Malawi in 1986 and spread rapidly to neighbouring countries. This insect has also been introduced into Colombia. The aphids feed in colonies and suck plant juices from twigs and branches. Heavy feeding can kill trees.

Uses

C. lusitanica is very well suited for the establishment of hedges and windbreaks due to its low sensitivity to pruning. Its characteristics of fast growth, abundant natural regeneration, and tolerance of limestone soils make it suitable for revegetation of degraded soils, where the direct establishment of indigenous trees may be difficult. It is also used for land reclamation, and its hardiness and tolerance of salt wind and to limestone soils make it well suited for erosion control and coastal protection. *C. lusitanica* is a beautiful tree, and is widely used as an ornamental.

The wood is extremely resistant to degradation agents such as water and moist air. It also has general mechanical properties similar to tropical broadleaved trees of half-hard wood. On the other hand, it shows the following disadvantages: tendency to have many nodes, relatively low diameter and a sapwood that is non-durable, as well as relatively thick (5 cm average). Wood density ranges from 360 to 599 kg/m^3 at 12–15% moisture content.

It is utilized for poles and props and for exterior doors and mouldings. For this reason, it has been used for centuries for making cabinets and chests. It is one of the few temperate-zone woods resistant to the attack of marine mollusc borers, and is used in naval construction. It is also well suited for sculptures, as it is easily worked. Due to its good natural durability and aesthetic qualities, it can be used in adverse conditions, such as bathrooms, and in framework of beams, furniture and garden flooring.

Further reading

Hocking, D., Milne, P., Tantrum, D., Brann, G., Low, C., Aimers-Halliday, J., Palmer, H. *et al.* (2006) Special feature on cypress. *New Zealand Tree Grower* 27, 3–14.

Nicholas, I. (ed.) undated. Best Practice with Farm Forestry Timber Species, No. 1: CYPRESSES. NZFFA Electronic Handbook Series. New Zealand Farm Forestry Association. Available at: http://www.scionresearch.com/general/publications/technical-reports/forest-management/other-species2, (accessed 14 March 2013).

Teissier du Cros, E., Ducrey, M., Barthelemy, D., Pichot, C., Giannini, R., Raddi, P., Roques, A., Sales Luis, J. and Thibaut, B. (1999) *Cypress: A Practical Handbook*. Studio Leonardo, Florence, Italy, 139 pp.

Cupressus macrocarpa Hartw. ex Gordon

Monterey cypress

Importance

Cupressus macrocarpa, despite its highly restricted natural range in California, is often planted as an ornamental tree and windbreak along temperate coastal areas throughout the world. It is of conservation concern because of its rarity in its natural range.

Botanical features

C. macrocarpa is a moderate-sized tree, reaching heights up to 20–25 m. Diameters of mature trees may reach as much as 200 cm. The canopy growth form of *C. macrocarpa* is often irregular, due in part to strong coastal winds where it occurs. Under protected conditions, the branches are ascending and form a conical crown. Under windy conditions, the crown is often flat topped and wind trimmed in form.

The bark of *C. macrocarpa* is thick and reddish-brown on younger trees, but becomes nearly white on older, exposed trunks. Trunk cross-sectional shape is commonly highly irregular, with fluted buttresses common in exposed trees. The root systems are spreading and relatively extensive, but shallow.

The foliage of *C. macrocarpa* is bright green and consists of small, triangular scale leaves closely appressed to the branchlets.

Staminate cones are about 6 mm in length, light yellow. Ovulate cones are red-brown and spherical to elliptic, with a diameter of 17–25 mm (Plate 61). Female cones have woody shield-like scales that open rapidly after fires and shed seeds. Without fire, the cones eventually open, but more slowly, and shed seeds gradually. The dark brown seeds are 5–6 mm in length and 2–3 mm thick. Cones mature from February to March and female cones ripen in August.

Distribution

The natural range of *C. macrocarpa* is highly restricted to two small groves along the coast of central California near Monterey. These groves occur on the Monterey Peninsula and a few kilometres to the south at Point Lobos. The stands of *C. macrocarpa* are relatively pure along the immediate coast and grade gradually into forests of *Pinus radiata*.

C. macrocarpa has been much planted in mild coastal climates in both the northern and southern hemispheres. It is notably common along the coasts of England, countries around the Mediterranean basin, Chile, Australia and New Zealand. It is intolerant of low temperature extremes.

Silviculture characteristics

C. macrocarpa has a notable resistance to wind and salt wind.

Silviculture practice

There has been extensive silvicultural work carried out with *C. macrocarpa* in several areas of the world, most notably in New Zealand and the Mediterranean basin.

Green cones may be collected from *C. macrocarpa* several years after seed maturity, as the closed cones remain vascularized. Cones can then be opened by drying in the sun or by application of moderate temperature (up to about 40°C). Seeds appear to retain partial germination for 10 years or more with dry storage at room temperature. Although *C. macrocarpa* is commonly germinated without pretreatment, some dormancy appears to be present. Chilling for 30 days has been shown to significantly increase the rate and speed of germination of *C. macrocarpa* seed.

C. macrocarpa propagates well from cuttings. Cuttings from second-year stockplant growth rooted as easily as cuttings originating from first-year stockplant growth. Rooted cuttings are best planted out in mid-winter in Mediterranean climate regions, where frost is not a problem. Protection from rabbits may be important in the first few years of establishment. Propagation by grafting is possible but not widely practised.

Autumn sowing of seeds is recommended. Seeds may be broadcast sown or planted in drills, covering them with 6–12 mm of sand. Seedbeds should be kept in moistened but well-drained sandy soil for the 2–3 weeks before germination takes place.

Although natural seedlings are relatively rare in the species' habitat in California, seedlings may be common in plantations. Direct sowing of plantations has been used, but field planting of 2–3-year-old seedlings or rooted cuttings is more desirable. In areas with mild temperatures and winter rainfall, young plants are planted out in autumn to allow them to become well established before the following summer drought.

C. macrocarpa is shade-tolerant, and health and form tend to be better on sheltered sites. *C. macrocarpa* can stand severe pruning down to a trunk diameter of less than 5 cm without seriously affecting growth. Up to 15–20 years the best 400 trees/ha will grow at much the same rate regardless of stocking rate, so it is important not to thin too early.

Short rotations at high stocking rates (600 trees/ha plus), or production thinning, are realistic options. However, if the target is clear, knot-free heart wood, especially in larger dimensions, then lower final stocking rates, 200–300 trees/ha and longer rotations of 35–40 years are required.

Diseases and pests

The most significant pathogen affecting *C. macrocarpa* is the coryneum or cypress canker, caused by the fungus *Seiridium cardinale*. This fungus invades the bark through wounds and the bark is killed as the fungus spreads. Patches of dead foliage appear in the crown as the branches die. This fungus was first discovered on ornamental Monterrey cypress in California in 1928, where it was causing severe damage. It is now found throughout the world and affects several species of *Cupressus*, especially *Cupressus sempervirens* in the Mediterranean basin. Once infected by *S. cardinale*, other fungal pathogens are often able to invade infected trees. It is the main reason that *C. macrocarpa* is not planted more widely as a valuable plantation species. The disease is managed by sanitation and use of genetically resistant strains. Thus far, this pathogen has not caused significant damage to the relict native stands of *C. macrocarpa* along the California coast.

Another stem canker, caused by the fungus *Lepteutypa cupressi*, was responsible for major damage to plantations of *C. macrocarpa* in East Africa during the 1940s and resulted in the suspension of further planting of this species in Kenya in 1952. This fungus also has a worldwide distribution.

C. macrocarpa is attacked by the root pathogens *Phytophthora cinnamomi* and *Phytophthora cactorum* in New Zealand, where this species has been used in windbreak plantings.

A number of insects are known to attack *C. macrocarpa* but most are of minor importance. The larvae of the Cypress bark moth *Cydia cupressana* (Lepidoptera: Olethreutidae) bore into cones, branch nodes, and injured branches of *C. macrocarpa*. The moths are attracted to cankers caused by *S. cardinale* where they lay eggs. *C. cupressana* may have a role in the dissemination of this fungus.

The larval stage of *Henricus macrocarpara* (Lepidoptera: Cochylidae) feeds on the seeds in young cones of *C. macrocarpa*.

Epinotia hopkinsana cupressi (Lepidoptera: Olethreutidae) feeds on the foliage, cones and under the bark of limbs of *C. macrocarpa* in California. Two bark beetles of the genus *Phloeosinus*, *Phloeosinus cristatus* and *Phloeosinus cupressi* (Coleoptera: Scolytidae), attack the boles and branches of several species of *Cupressus* in western North America, including *C. macrocarpa*. They can hasten the death of trees infected by *S. cardinale* and may function as a vector of this fungus.

In Kenya, *C. macrocarpa* is rated as highly susceptible to damage by the cypress aphid, *Cinara cupressivora*. In New Zealand, *Ambeodontus tristis* (Coleoptera: Cerambycidae) attacks *C. macrocarpa* through dead branch stubs or wounded parts of living trees.

Uses

C. macrocarpa is a useful species for use in coastal windbreaks and shelter belts, as it can tolerate salt winds. It is planted as a windbreak around plantations of other tree species such as *Pinus radiata*, or along roads or dwellings in windy coastal habitats.

C. macrocarpa yields a medium- to lower-density (360–599 kg/m^3) softwood of moderate strength and stiffness. It has good working and finishing characteristics and quite uniform wood properties. It is very stable, and easily sawn with minimal reaction wood. The wood can experience some collapse when kiln dried. The sapwood is very light brown and normally occupies about five growth rings. The heartwood is darker yellow-brown in colour. Heartwood is not generally regarded as suitable for in-ground use, and nor is it durable in situations where it remains moist.

Uses of *C. macrocarpa* include for ceiling sarking, exposed beams, flooring, wall panelling, framing, furniture, solid wood bench tops, architraves and skirtings. It may also be used for a limited amount of short-fibre pulp: the latter use is limited to kraft paper production with the addition of fibre from other species.

Further reading

Hocking, D., Milne, P., Tantrum, D., Brann, G., Low, C., Aimers-Halliday, J., Palmer, H. *et al.* (2006) Special feature on cypress. *New Zealand Tree Grower* 27, 3–14.

Miller, J.T. and Knowles, F.B. (1990) *Introduced Forest Trees in New Zealand: Recognition, Role and Seed Source. 9. The Cypresses.* Cupressus *spp. and* Chamaecyparis *spp.* FRI Bulletin 124, Forest Research Institute (N.Z.), New Zealand Forest Service, Wellington, New Zealand, 33 pp.

Nicholas, I. (ed.) undated. Best Practice with Farm Forestry Timber Species, No. 1: CYPRESSES. NZFFA Electronic Handbook Series. New Zealand Farm Forestry Association. Available at: http://www.scionresearch.com/general/publications/technical-reports/forest-management/other-species2 (accessed 14 March 2013).

Somerville, A. (1993) Growth and utilisation of young *Cupressus macrocarpa*. *New Zealand Journal of Forestry Science* 23, 163–178.

Cupressus nootkatensis (D. Don) Spach

Alaska cedar

Importance

The taxonomy of *Cupressus nootkatensis* (Plate 12), commonly known as Alaska cedar (not a true cedar and also known as yellow or Alaska cypress), has undergone several revisions in recent years. At one time or another the species has been assigned to the genera *Cupressus*, *Chamaecyparis*, *Callitropsis*, *Xanthocyparis* and back to *Cupressus*. Taxonomy of *C. nootkatensis* aside, it is an important timber species of western North America. It is confined to the cool, humid climate of the Pacific coast in Alaska and British Columbia, and of the Cascade Range in Oregon and Washington. It is a slow-growing conifer, producing wood which is extremely durable and excellent for speciality uses. Little effort has been made to manage the species to assure a continuing supply.

Botanical features

Generally a medium-sized tree, occasionally over 35 m tall, 100 cm in dbh, and up to 1000 years old. The trunk is tapered, often with a broadly buttressed and fluted base. The crown is sharply conical. The branches are spreading, drooping and bear loosely hanging branchlets. The leading shoot is slender, flexible and curved. Old trees may have one or more dead spire-like tops. The bark is thin, greyish-brown and scaly when young, separating into narrow intersecting ridges with age. The wood is light, hard, strong, close grained, even textured and pale yellow; sapwood is not clearly distinct from heartwood.

In freely drained soils, roots are extensive. The taproots are poorly defined or non-existent, but fine roots develop a profuse, dense network. Where a thick forest floor layer is present, many roots occur in it rather than in the underlying soil. In poorly drained soils, roots tend to be shallow and develop in complex patterns associated with a long history of branch layering. Root systems of krummholz – apparently the result of root sprouting and layering – have been observed to extend a distance of 30 m. Roots have mycorrhizae of the vesicular-arbuscular type.

Leaves are evergreen, scale-like, small, up to 3 mm long and dull bluish-green.

C. nootkatensis is monoecious. Flowering occurs from April to June depending on latitude. The tiny inconspicuous yellow or reddish male pollen-bearing strobili and green female cones are borne on the tips of branchlets. Pollination occurs from mid-April to late May in cones that were initiated the previous summer. Cones generally mature in 2 years, but in the southern part of the range they may mature in 1 year. Both first- and second-year cones occur on the same branch. Mature and immature cones are nearly the same size, about 12 mm in diameter and globe-shaped, so care must be taken to collect only mature cones for seed. Immature cones are green and soft, often with purple markings, and are borne near the tips of branchlets. Mature cones are yellow-green and hard, often with brown markings, and are borne farther from the branch tips.

Initiation of phenological events varies greatly with latitude, altitude, aspect, weather and snowpack.

C. nootkatensis exhibits a high tolerance of shade and water surplus.

Distribution

C. nootkatensis grows predominantly in the Pacific region, its range extending from northern California to Prince William Sound in Alaska. Except for a few isolated stands, it is found within 160 km of the Pacific Coast.

C. nootkatensis grows in even- and uneven-aged, mixed-species stands, rarely in pure stands. It is present in early, mid- and late stages of secondary succession and forms a major component in old-growth stands. Major associates include *Abies amabilis*, *Abies magnifica*, *Abies procera*, *Picea sitchensis*, *Pinus contorta*, *Pseudotsuga menziesii*, *Taxus brevifolia*, *Thuja plicata*, *Tsuga heterophylla* and *Tsuga mertensiana*.

C. nootkatensis is a minor or major component of a number of forest communities.

Silviculture characteristics

Reproduction capacity is low; the proportion of filled seeds in mature cones is low. Seed dissemination capacity is medium, with seed dispersed <150 m from the parent tree. Reproduction by layering is frequent.

Potential for natural regeneration is medium in low light, less than that of *A. amabilis*, but high in the open. Most natural regeneration occurs as ingress in the open. The regeneration period on high elevation sites is long.

Potential initial growth rate (up to age 5 years) is low to medium: <10 cm per year, higher on low elevation sites. Response of advance regeneration to release is high; advance regeneration is rarely present in forest understories. Self-pruning capacity in dense stands is high, providing that initial stand density is high. Dense stands are very infrequent. Crown spatial requirements are low. Light conditions beneath closed-canopy, mature stands are medium; closed-canopy stands are very infrequent.

Potential productivity is medium, decreasing with increasing elevation, and is comparable to that of *T. mertensiana* on high elevation sites, and to *T. plicata* on low elevation sites.

Longevity is high, frequently >1000 years, possibly up to 2000 years.

C. nootkatensis survives heavy snow loads because of its narrow, flexible crown and drooping branches, and its flexibility allows it to survive on avalanche tracks.

Silviculture practice

Depending on site and management objectives, clear-cutting, patch-cutting, shelterwood and selection systems are all appropriate for the establishment and growth of *C. nootkatensis*. It can be grown in mixed species stands where uneven-aged management is practised. When sawtimber is to be produced, even-aged management regimes are more suitable. Disturbed mineral soil seedbeds seem to be a major requirement for regeneration from seed.

Propagation by seed is the only contemporary method for regeneration. Large crops of seed occur at intervals of 4 or more years. The proportion of filled seeds from mature cones is generally low and extremely variable. Cleaned seeds average 240,000/kg. Seed orchards offer the opportunity for treatment and thereby provide a practical means of increasing cone production.

Germination is epigeal, and germination rate tends to be low. Warm stratification followed by cold stratification greatly improves germination, but optimum stratification schedules have not been developed. In British Columbia and Alaska, seeds ripen from mid- to late September and are shed during dry periods in the autumn and early winter.

C. nootkatensis reproduces vegetatively under various natural conditions from low-elevation bogs to krummholz at tree line. In south-east Alaska, layering is common on low elevation bog sites and less common on better-drained sites. The species can also be reproduced from cuttings. Container stock suitable for planting can be produced in the greenhouse in 1 year.

In common with *T. plicata*, *C. nootkatensis* should be grown at high densities (over 2000 trees/ha). Maintaining a nearly closed canopy at all times will improve tree form, because open-grown trees tend to develop excessively large limbs and multiple tops.

In Washington, dominant trees on productive sites are typically 30–38 m tall; in British Columbia they are 90 cm in dbh and 23–30 m tall; and in Alaska they are often 60 cm in dbh and 24 m tall.

Interest in management of *C. nootkatensis* is relatively new, and information on growth and yield of young stands is not available. Most *C. nootkatensis* timber comes from logging mixed-species old-growth stands. Because of its slow rate of growth in comparison with other more commercially important species, there has been little interest in managing the species for timber on the more productive sites.

Diseases and pests

A total of 78 taxa of fungi has been reported throughout its range, including 50 in Alaska. The wood, however, is very durable and resistant to fungal attack, partly because the naturally occurring chemicals (nootkatin, chamic acid and chaminic acid) in the heartwood inhibit fungal growth. Certain black-stain fungi are capable of degrading nootkatin, thereby increasing the susceptibility of the heartwood to decay. Living trees often attain great age, and over time heart-rotting fungi cause considerable loss and defect in standing trees.

C. nootkatensis is relatively free from damage by insects, and no infestations of defoliating insects are known.

Since at least 1880, *C. nootkatensis* has suffered advancing decline and mortality on more than 100,000 ha of wetlands in south-east Alaska. Abiotic factors appear to be partly responsible, but the primary cause remains unknown.

Uses

Its drooping branchlets give *C. nootkatensis* a graceful weeping appearance, making it an attractive specimen tree for parks and open spaces. It is also used for hedging.

Special attributes of *C. nootkatensis* wood include durability, freedom from splitting and checking, resistance to acid, smooth-wearing qualities and excellent characteristics for milling. The wood has a density of about 500 kg/m^3 (at 12% moisture content). Heartwood qualities make it highly desirable as firewood.

C. nootkatensis wood is extensively used by Native Americans of the Pacific North-west coast. It is suitable for boatbuilding, utility poles, heavy flooring, framing, bridge and dock decking, marine pilings, window boxes, stadium seats, water and chemical tanks, cooling towers, bedding for heavy machinery, furniture, patterns, moulding, sash, doors, panelling, toys, musical instruments and carving. The wood is highly regarded in Japan, and most high-quality logs are exported there from North America.

Further reading

Harrington, C. (ed.) (2010) *A Tale of Two Cedars*. International symposium on western redcedar and yellow-cedar. Pacific Northwest Research Station General Technical Report No. 828. US Department of Agriculture Forest Service, Portland, Oregon.

Harris, A.S. (1990) *Chamaecyparis nootkatensis* (D. Don) Spach), Alaska-cedar. US Department of Agriculture Forest Service, Washington, DC.

Stewart, H. (1995) *Cedar: Tree of Life to the Northwest Coast Indians*. Douglas & McIntyre Ltd, Vancouver, British Columbia, Canada.

Cupressus sempervirens L.

Mediterranean cypress

Importance

Cupressus sempervirens is a multi-purpose tree species, which may reach up to 30 m in height with a dbh of 50–60 cm. This species produces high-quality and very durable wood at low yields and is utilized more frequently as windbreaks, soil conservation, protection forest and as an ornamental. It is suitable for plantation programmes in unfavourable conditions in many countries, including for afforestation on dry, shallow, clay soils. Since the beginning of the 1950s, *Seiridium cardinale* (cypress canker) has decimated *C. sempervirens* plantations in Italy, Greece and France.

Botanical features

C. sempervirens is a medium-sized tree which may reach up to 30 m in height, with a dbh of 50–60 cm. *C. sempervirens* var. *horizontalis* has spreading branches and a broad conical crown, the most common crown shape found in its natural range. The bole is straight with bark that is usually grey, thin and fibrous with fine longitudinal ridges. Trees with a columnar crown with erect branches are sometimes treated as *C. sempervirens* var. *pyramidalis* (syn. *stricta*, *fastigiata*).

The leaves are usually scale-like, decussate, imbricated, tightly pressed against the axis, partly grown together with the branchlets and exhibiting large, more or less visible resin ducts in the central-dorsal part.

C. sempervirens is a monoecious species with oblong microsporophylls ('male' inflorescence) (3–5 mm long, approximately 2 mm wide), which appear at the tip of the brachyblasts and are very numerous, already visible in winter, and generally distributed through the lower part of the crown. The macrosporophylls ('female' inflorescence) are subglobose (2.5 mm), composed of 6–12 scales, each bearing 6–20 ovules, generally located in the upper portion of the crown.

Pollen dispersal occurs from mid-winter to early spring, and is dependent on location. Pollination is anemophilous. In early summer the fertilized female cones are visible (approximately 1 cm) and green. In early winter they turn pale yellow, and the seed gradually reaches maturation in the second year after pollination, between summer and late autumn. The cones (average 2.5 cm long, 2 cm wide) turn silver-grey and their scales tend to open slightly. In the autumn, or more usually in winter, the scales open out further and the seeds are released. Each of the 9–12 scales contains about 8–15 seeds (5–6 mm long, 3–4 mm wide); 1000 seeds weigh approximately 6–8 g.

Distribution

The natural range of *C. sempervirens* is predominantly in the Mediterranean, including Greece, the Aegean islands, Crete, Cyprus, Syria, Israel, Cilicia (Turkey) and Lebanon. *C. sempervirens* var. *indica* reaches as far as south-western Iran.

C. sempervirens has become naturalized in many Mediterranean countries, especially Greece, Italy, France, Portugal and Spain, where extensive plantations have been established for afforestation, windbreaks and ornamental purposes. The heartland of cypress in Italy is Tuscany, where it is utilized extensively as a forest and ornamental tree; in France its heartland is the Val du Rhône, where it is commonly used as a windbreak.

C. sempervirens is planted as an ornamental in suitable locations around the world.

Silviculture characteristics

C. sempervirens is a quiescent species; it is able to tolerate shallow, rocky, or clayey soils, which become very dry in the summer. It regenerates quickly when young, showing rapid height growth and a slow increase in diameter. It is suitable for coppicing, is able to tolerate salt winds, has a well-developed root system and is also a very good pioneer species on abandoned soils. This species is generally frost-hardy.

Silviculture practice

The cones are harvested towards the end of summer and in the autumn. Any old cones should be avoided; these can remain on the tree for as long as 20 years. The cones open when dried in the sun or in kilns at an average temperature of 35–40°C. The seeds are moistened with water for 2–3 days and are kept at a constant temperature of 19°C; germination requires light and occurs in 20–30 days. Sowing takes place in spring. One-year-old seedlings are removed from the seedbed and are transplanted into pots which are filled with a mulch–sand–perlite substrate (3:1:1). Seedlings for planting out are generally grown in containers (4 seeds per container, but only one seedling is left per container). At the end of the growing season, the seedlings are 20 cm in height. For reforestation, 1+1 container seedlings are normally used. Generally, for rootstocks, seedlings less than 1 year old are used.

Vegetative propagation is used for multiplication of clones for windbreaks and ornamental trees. Grafting is the most successful method used in nurseries, and has a 90% success if correctly executed. Side-veneer grafting is the most successful grafting technique; cleft grafting can also be applied. Propagation of cuttings by rooting is more problematic than grafting, and more expensive. *C. sempervirens* roots form mycorrhizal associations.

C. sempervirens is planted at various densities depending on the use, for example as a shade tree, windbreak or in protection forest (where planting densities of 600 trees/ha are recommended). For timber production, a density of 1100 trees/ha is planted, and for firebreaks a dense stand of 2500 trees/ha is used. *C. sempervirens* is suitable for pruning and coppicing; however, all the wounds should be disinfected to prevent attack from *S. cardinale*.

Timber production of *C. sempervirens* is low, at around 1–7 m^3/ha/yr.

Diseases and pests

The most important disease of *C. sempervirens* is cypress canker, caused by the fungus *S. cardinale*. This fungus invades the bark through wounds and the bark is killed as the fungus spreads. Patches of dead foliage appear in the crown as the branches die. This disease is especially damaging to plantings of *C. sempervirens* in Italy, Greece and southern France, and is managed by sanitation and use of genetically resistant strains of *C. sempervirens*. *Diplodia pinea* f. sp. *cupressi* causes a disease common in Israel, but is also reported from Greece, Morocco and Tunisia. This fungus can be damaging wherever summer drought is a stress factor for *C. sempervirens*.

The fungus *Phomopsis occulta* causes a twig canker and die-back in the nursery, but can also infect young and mature trees.

In the Mediterranean basin, the bark beetles *Phloeosinus aubei*, *Phloeosinus thujae*, and *Phloeosinus armatus* (Coleoptera: Scolytidae) are vectors of the fungus *S. cardinale*, the cause of cypress canker. Removal of infested branches and/or infested trees is one of the pest management tactics used to reduce losses by this disease.

The cypress aphid, *Cinara cupressi* sensu lato (Hemiptera: Aphididae) is an important pest of several species of Cupressus, including *C. sempervirens*. This insect probably has its origins in the Near East and Western Asia and its most likely natural host is *C. sempervirens*, to which it occasionally causes damage. This insect has been introduced into portions of Europe where *C. sempervirens* is widely planted and into Eastern and Southern Africa and South America, where it is damaging to other species of *Cupressus*.

The eriophyid mite, *Trisetacus juniperinus* (Acari: Phytopidae), colonizes the foliage at the tips of twigs of several species of *Cupressus* and *Juniperus*, including *C. sempervirens*, and causes swelling of the leaf bases and destruction of the apical bud. This results in stunting of the plant and the development of numerous side-shoots, giving them a bushy appearance. Cones of *C. sempervirens* may be also be killed by this mite.

Uses

C. sempervirens is a multi-purpose species which is utilized in protection forest, soil conservation, to maintain the beauty of the landscape, as an ornamental, as a windbreak and to re-establish vegetation on abandoned land. *C. sempervirens* var. *fastigiata* is more suitable as an ornamental and as a windbreak, and *C. sempervirens* var. *horizontalis* is more suitable for timber production. The spindle-shaped *C. sempervirens* var. *pyramidalis* also has a social function in Mediterranean villages, churches and cemeteries because common cypress was once considered a protective and sacred tree. *C. sempervirens* has a low flammability and may be planted as a firebreak.

C. sempervirens has a yellow-pink timber with fine grain and irregular growth rings. Wood density is 380–490 kg/m^3. It is very durable, especially in water, does not require preservation and is not attacked by termites. The heartwood is successfully used for posts. Its high stability and mechanical properties make *C. sempervirens* very suitable for small carpentry pieces which are required to withstand stress, and exterior woodwork (e.g. front doors, windows, shutters and even garden furniture), even in humid surroundings. It is one of the few temperate timbers that is resistant to attack by marine molluscs and is widely used in ship building.

The wood's odour repels numerous harmful insects (particularly mites) and for this reason it has long been used for chests and wardrobes to store linen and food, as well as for coffins. It is easy to glue, producing glued and laminated frame elements, laminated wood panelling and glued window framing. *C. sempervirens* is an easy wood to work, except for problems with spiral grain knots. It is also very suitable for wooden sculptures. It is well adapted to machining and finishing.

The cones of *C. sempervirens* contain a tannin which has a constricting effect on the vascular system. The oil is used by the perfume industry.

Further reading

Farjon, A. (2005) *A Monograph of Cupressaceae and Sciadopitys*. Royal Botanic Gardens, Kew, London.

Farjon, A. (2010) *A Handbook of the World's Conifers*. Brill Academic Publishers, Leiden, The Netherlands.

D

Dalbergia latifolia Roxb.

East Indian rosewood, rosewood

Importance

Dalbergia latifolia is a multipurpose timber tree in South and South-east Asia, which produces one of the highest-value timbers in the region, used for furniture, cabinetmaking and decorative work. *D. latifolia* is managed within natural forests and is being increasingly established in plantations, although the slow growth rate and long rotations restrict its potential when compared with other tropical plantation species such as teak. It is being promoted for use as an agroforestry tree, but more research is required.

Botanical features

D. latifolia grows to 25–40 m high with a diameter of up to 180 cm, with a dense, rounded crown. The bole is more or less straight, cylindrical, and clear for about half of total tree height or up to 12 m. Leaves are compound, petiolate, alternate, 10–30 cm long, with leaflets entire, obtuse, rounded or cuneate at base and rounded, obtuse, emarginate or rarely subacute at apex. Lax, axillary, corymbose cymes or panicles, 5–20 cm long, with white or slightly yellowish flowers, having a papilonaceous corolla almost 7 mm long and 4 mm wide. Fruits are strap-shaped pods, 4–15 cm long and 1–2 cm broad, reticulate-veined throughout, with 1–4 reniform, compressed seeds.

Distribution

D. latifolia is indigenous to South and South-east Asia. In India, it is widely distributed from the sub-Himalayan tract to southern India at altitudes up 1350 m. It is also reported as native to Nepal and Malaysia, and may also be native in some neighbouring countries. It was introduced into Sri Lanka in the 1800s and plantations were established in Indonesia in the 1900s, especially in Java. It has also been introduced into African countries, including Nigeria, Kenya and Tanzania.

Silviculture characteristics

Natural reproduction by seed is generally satisfactory, and seeds germinate early in the rainy season with drought and weed competition being the main cause of suppression and mortality. *D. latifolia* also regenerates well by root suckers, improved by wounding and in situations exposed to light. *D. latifolia* is moderately light demanding, benefiting from overhead light but tolerant to some shade, particularly when young.

The mature tree is drought-resistant, but seedlings may be susceptible to drought in exposed situations. Trees can tolerate a dry season of about 6 months with a high saturation deficit. It is only moderately fire-resistant and is often killed outright by crown fires, but withstands annual ground fires in moist deciduous forests. It coppices and pollards well, and produces numerous root suckers, particularly when the roots are exposed or wounded. It is not a gregarious tree and occurs scattered in mixed deciduous forests. In moist conditions trees remain evergreen, but in drier habitats, trees may lose some or all of their leaves by the end of January, and fresh foliage appears in April–May. Flowering begins by December and normally continues to March, or rarely to October.

It grows naturally in tropical and subtropical climates, with mean annual rainfall varying from 750 to 5000 mm, available mostly from May to September. In its natural habitat in India, the absolute maximum shade temperature varies from 37 to 50°C and absolute minimum from 0 to 15°C. The mean daily maximum temperature in May, which is generally the hottest month, varies from 33 to 43°C, and the mean daily minimum temperature in January, the coldest month, fluctuates from 7 to 21°C. The mean relative humidity in July varies from 70 to 100% and in January from 40 to 80%. *D. latifolia* occurs on deep loam and clayey soils as well as on black cotton soil, but best development is reported where the soil is deep and moist; trees do not tolerate poor drainage.

Silviculture practice

Ripe, dark brown pods are collected from trees by lopping branches; they are dried in the sun and broken into fruit segments. Extraction of seeds is not necessary for sowing. Seeds remain viable for about 6 months and lose viability appreciably when kept for more than 1 year. There is considerable variation in seed weight (18,500–40,000 seeds/kg). No pre-sowing treatment is required, and germination takes 7–25 days with germination rates of 45–80%.

In the nursery, seeds are sown in well-raised seedbeds of porous sandy loam, and seedlings are ready for outplanting 3 months after sowing. Containerized seedlings raised in polythene bags are also successfully used for planting, though stump planting is considered the best method for raising *D. latifolia*. Irrespective of the age of seedlings from which stumps are produced, collar diameter for stump plants should be 5–15 mm with root and shoot lengths of approximately 3.5–4.5 cm. Usually 2–3-year-old seedlings produce stumps of this standard, and planting pre-sprouted stumps in polythene containers during pre-monsoon showers gives best results.

Usual initial spacing varies from 1.2 × 1.2 m to 1.8 × 1.8 m; to obtain straight stems with a clear bole, close spacing in either mixed or pure plantations is desirable. As side shade is beneficial, planting can be done under the broken shade of miscellaneous species left standing after felling. Regular weeding and soil working is required, as plants benefit considerably from soil aeration.

As *D. latifolia* accounts for only a small proportion of the exploitable timber in natural forests, it is usually worked according to the requirements of other important species present, usually teak (*Tectona grandis*). High forests are worked by using selection systems with a prescribed exploitable diameter, or improvement fellings. Pure plantations are also raised, however, by clear felling and artificial regeneration, although the slow growth rate and long rotation has limited the use of large-scale plantations. Mean annual diameter increments rarely exceed 1 cm even during early growth, and will decrease significantly as the tree ages.

Diseases and pests

Tar spot caused by *Phyllachora dalbergiae* and *Phyllachora spissa* is a major problem at the nursery stage. Other pests and diseases that are known to attack other *Dalbergia* spp. may also cause problems. *Phytophthora* sp. is not specific to *D. latifolia*, but can lead to tree death.

The termites *Macrotermes gilvus* and *Odontotermes grandiceps* are not specific, but can cause also tree mortality.

Uses

D. latifolia is primarily a furniture and cabinet wood, but due to its strength and durability it is used for many other purposes. It should not be confused, however, with the true Indian rosewood, *Dalbergia sissoo*. The wood is hard and heavy, with an air-dry density of about 880 kg/m^3, with a medium-coarse texture and straight to shallowly interlocked grain. The yellowish or pinkish-white sapwood is perishable but is readily treatable, whereas the heartwood, which varies in colour from light purplish-brown to dark purple with darker streaks, is very attractive, highly durable and refractory to treatment. The timber is fairly easy to season and offers no difficulty to kiln drying.

In India, the wood produced is one of the most valuable timbers with an average price higher than that of teak (*T. grandis*). It is used mainly for high-class furniture, cabinets, panelling, carving, turnery and other decorative applications. Wherever available, it is also used for building construction as posts, rafters, doors, windows, flooring and for cart and carriage building. The timber is also in great demand for veneers of decorative plywood and blockboard, and is used for marine- and aircraft-grade plywood. Although technically suitable for railway sleepers, mining and constructional timber, the timber is too valuable to be used for such purposes. It is one of the best timbers for railway wagon-building, vehicle bodies, boatbuilding, tool handles, agricultural implements and bentwood furniture, and is also commonly used such items as musical instruments, novelty items, woodware, toys, brush backs and sports goods. The timber has also been tested and found suitable for hardboard.

As *D. latifolia* is relatively slow growing, it is not widely used as a plantation species at present. However, its timber value and flexibility of ecological requirements makes it suitable for a range of agroforestry systems and afforestation of eroded soils. Tree extracts are used for tanning and medicinal products, and is locally lopped for fodder.

Further reading

Soerianegara, I. and Lemmens, R.H.M.J. (eds) (1993) *Plant Resources of South-East Asia No. 5(1). Timber trees: major commercial timbers.* PROSEA Foundation, Bogor, Indonesia and Pudoc Scientific Publishers, Wageningen, The Netherlands, 610 pp.

Tewari, D.N. (1995) *A Monograph on Rosewood (*Dalbergia latifolia *Roxb.).* International Book Distributors, Dehra Dun, India, 74 pp.

Westley, S.B. and Roshetko, J.M. (eds) (1994) *Dalbergia*: Proceedings of an International Workshop. Nitrogen Fixing Tree Research Reports Special Issue. NFTA, Waimanalo, Hawaii, USA, 207 pp.

Dalbergia melanoxylon Guill. & Perr.

African blackwood

Importance

Dalbergia melanoxylon produces one of the world's most valuable timbers, with few parallels for woodwind instruments and the finest turned and carved craft work. The native range extends throughout much of sub-Saharan Africa. Once abundant, commercially exploited wild stocks in Eastern Africa have suffered from bushfires and over-harvesting: in Kenya to the point of commercial extinction, yet commercial supply persists from Tanzania and Mozambique. The decline in the quality of traded timber and the increasing scarcity of trees of suitable dimensions have raised concerns from within native-range countries, as well as from importing governments and international conservation agencies, and the species is protected, being listed in Appendix II of the Convention on International Trade in Endangered Species (CITES).

Botanical features

D. melanoxylon is a deciduous shrub or tree, usually 4–10 m tall and 40 cm in diameter, attaining 15 m in height and 1 m in diameter on better sites with age. It has a narrow, irregular, open crown, increasingly rounded and heavier, characteristically multi-stemmed and heavily branched. The bole is highly fluted, clear for 1.2–1.8 m, occasionally to 3.6 m. Bark is pale grey to greyish-brown, papery, smooth and flaking in long, narrow strips. Twigs are whitish, with terminal thorns 0.5–5.0 cm long, though branches and boles of younger trees carry straight, conical spines at the nodes which may bear both leaves and flowers. Leaves alternate, imparipinnate, 5–10 cm long, with 9–13 opposite or sub-alternate leaflets, obovate to elliptic, 1–3.5 cm long and 1–2 cm wide, with a truncate or emarginate tip, and pinnate or more or less palmate nerves. White, scented flowers are 0.5 cm long, in dense clusters of axillary panicles up to 10 cm long. Pods are indehiscent, papery, pale brown to dark, oblong to elliptic–lanceolate, 3–6 cm long by 1–1.5 cm wide, with 1–4 small, reniform seeds.

Distribution

D. melanoxylon occurs naturally in much of dryland sub-Saharan Africa from Senegal to northern Ethiopia, and south to Angola, Mozambique and the Transvaal, South Africa, but absent in more humid regions such as coastal West Africa and the Congo Basin. Distribution in Tanzania is patchy, in densities of 9–90 trees/ha often growing in clusters, in most vegetation types except in montane forest and desert scrub, and frequent in mixed forest and coastal savannahs. In Uganda, it occurs in the dry regions in the north and east of the country. In Sudan, it is common on clay plains or tall grass savannah, often on rocky places or valleys with impeded drainage. In West Africa, *D. melanoxylon* is present in the eastern regions of Senegal, northern and eastern Guinea, and the northern regions of Côte d'Ivoire, Ghana, Togo, Benin, Nigeria and Cameroon. In Namibia, it is only reported on Impalila Island near the Botswana border.

D. melanoxylon has been introduced to southern India. It was imported to Australia as a promising timber species but soon became an aggressive weed and was quickly eradicated.

Silviculture characteristics

D. melanoxylon grows with annual rainfalls of 400–1200 mm, and is a common component in dry savannah woodland and an understorey species in the open miombo woodlands of Southern Africa. Trees are nitrogen-fixing, and develop an extensive root system which sustains them during the long dry season. Bark is thin and soft, making trees susceptible to damage by bruising and fire. Natural regeneration occurs through seeds, suckers and coppicing, although the ability to coppice seems to be lost as the tree matures. Trees are semi-deciduous, shedding leaves during the dry season, when many branches are covered with scented flowers. *D. melanoxylon* is water- and light-demanding, and juveniles are much more likely to be found in open areas than under tree canopies, supporting the hypothesis that it is a pioneer species. Trees do not compete with most agricultural crops and could be maintained on farms.

Silviculture practice

Seeds are orthodox and germinate readily without pre-treatment, but have a short viability period and should be planted within 6 months after collection. Alternatively, pods are broken into one-seeded pieces and soaked in water for 6 h prior to sowing, with a germination of 50–60% in 8–20 days. Seedlings are transplanted to tubes or pots after 4–5 weeks, and planted after 6–7 months, but 2-year-old stumps (12 cm root + 2 cm stem) are also recommended. It is recommended that *D. melanoxylon* is planted in pure stands, mixed with fast-growing trees which do not cast too much shade. Initial spacing of 2 × 2 m promotes good branching, and later thinning improves growth. Intensive weeding is crucial until root collar diameter has reached 5 cm, and pruning improves the bole length.

Plantation management of *D. melanoxylon* has not yet been shown to be economically viable, given the long estimated rotation length of 70–100 years for 40-cm diameter trees. However, well-tended planted or wild trees had annual increments of 0.3–1 m in height and 1–2 cm in diameter during the first 5 years. For the first 5–8 years in the wild, annual increment in height was 0.5–0.7 m/yr. Levels of commercial harvesting are not thought to be sustainable, and although there are formal controls that license the harvesting of *D. melanoxylon* in some countries, illegal felling continues.

Diseases and pests

There are no significant pests or diseases reported for *D. melanoxylon*. However, logs may exhibit heart rot caused by fungal infection after fire damage, and the species may be damaged by tunnel-boring larvae of cerambycid beetles, especially in the sapwood.

Uses

Farmers recognize the species' importance in agroforestry systems, with increased growth of associated crops. As solitary trees they are valued for shade and shelter, promoted for boundary planting and restoration of degraded lands.

The heartwood is close, even-grained and oily, and exceptionally hard, heavy and durable. Density is 1300–1400 kg/m^3 in the best-quality fresh-cut timber, and 900–1200 kg/m^3 when air dry. Sapwood is white to yellowish-white, 1–2 cm thick, with a density of 750 kg/m^3. The heartwood darkens on exposure to air from a purplish colour to a variety of shades from jet-black through purplish-brown to brown. The heartwood makes up 83% of total log volume, the remaining 17% is sapwood and bark. The wood dries slowly over 2–5 years, and needs careful handling to minimize degradation and checking, with common drying defects being heart shake and end splitting, but it becomes very stable after seasoning. Wood has a severe blunting effect on cutting tools, and metal working tools and tungsten carbide-tipped saws are often used. Pre-boring is needed for nailing and can be tapped for screw threads.

D. melanoxylon is the raw material for the finest objects of local art and tourist souvenirs, and was also the 'ebony' of ancient Egypt. The wood is valued throughout the world as a raw material for carving and turned objects, in particular woodwind musical instruments such as clarinets, oboes, Scottish bagpipes, Irish pipes, flutes and piccolos, and it is also used in bowed stringed instrument fittings for tuning pegs, tailpieces and chin rests. It is almost without parallel in its qualities for use in woodwind instruments, being unusually stable when subject to the stresses of being turned, hollow bored and drilled to take metal key-work, while retaining precise dimensions when subject to frequent changes in moisture and temperature as the instruments are played. Its density and stability provide excellent tonal qualities and it takes an outstanding finish, and high oil content prevents moisture uptake from breath that otherwise results in dimensional changes and inconsistent musical pitch. It is also exported for parquet flooring and blanks for turnery and other craft work.

Principal exporting countries are Mozambique, Tanzania, Kenya, Senegal, Malawi and South Africa, with imports going to Europe (70%), Asia (20%) and North America (10%). Although it has a wide distribution, it is only in East Africa that the tree reaches harvestable size in sufficient abundance for it to be a significant commercial proposition. In recent years, Kenya's stocks have been completely exhausted, such that carvers there must now obtain their wood from Tanzania, and most *D. melanoxylon* comes from Mozambique and southern Tanzania. The most standardized commodities are the semi-processed billets supplied to musical instrument manufacturers.

Pods and leaves are used as fodder, mulch and green manure. The heartwood yields a red dye and flowers provide a good nectar source for bees. All tree parts have medicinal uses.

Further reading

Ball, S.M.J. (2004) Stocks and exploitation of East African blackwood *Dalbergia melanoxylon*: a flagship species for Tanzania's miombo woodlands? *Oryx* 38, 266–272.

Jenkins, M., Oldfield, S. and Aylett, T. (2002) *International Trade in African Blackwood*. Fauna and Flora International, Cambridge, UK, 32 pp.

Malimbwi, R.E., Luoga, E.J., Hofstad, O., Mugasha, A.G. and Valen, J.S. (2000) Prevalence and standing volume of *Dalbergia melanoxylon* in coastal and inland sites of southern Tanzania. *Journal of Tropical Forest Science* 12, 336–347.

Dalbergia nigra (Vell.) Allemão ex Benth.

Brazilian rosewood

Importance

Dalbergia nigra (Plate 13) or Brazilian rosewood is one of the most prized timbers in Brazil. The wood has been used globally since the 1600s for luxury furniture and musical instruments, and over-exploitation of trees from Atlantic forests over the centuries led to fears for its extinction. It became so scarce that international trade was banned, and it was the first tree species to be listed in Appendix I of the Convention on International Trade in Endangered Species (CITES) in 1992. *D. nigra* is, however, being increasingly planted for reforestation or conservation of degraded regions and riparian forests, land reclamation, timber production, and as a shade and ornamental tree. It is recommended for use in mixed plantations in full sun in association with pioneer species, and in enrichment planting in secondary forests.

Botanical features

D. nigra is an evergreen to semi-deciduous tree that is typically 10–25 m in height with a diameter of 15–80 cm, and in exceptional cases reaching 50 m in height and more than 150 cm in diameter. The trunk is twisted and irregular, and the bole can reach 10 m. Branches are dark, having a distinct zig-zag appearance. Leaves are paripinnately compound, alternate, 5–13 cm long, with 10–20 pairs of alternate leaflets, 7–25 mm long, 4–10 mm wide, pilose when young and glabrous later. Fragrant white-yellowish flowers are 5–10 mm long, in axillary bunches up to 6 cm long, in panicles up to 20 cm long. The fruit is a samara, elliptical or oblong, flat, membranaceous, 3–8 cm long and 2 cm wide, with one seed (rarely two) centrally located.

Distribution

The natural distribution of *D. nigra* is restricted to the Atlantic forest zone of Brazil, from southern Bahia (13° S), through Espírito Santo, Minas Gerais and Rio de Janeiro, to northern São Paulo (23°S). *D. nigra* has been planted in some other states of Brazil including Amazonas, Paraná and Pernambuco, and has also been introduced to Colombia and Peru.

Silviculture characteristics

In Brazil, *D. nigra* grows in tropical lowland forest to sub-montane forest up to 1700 m altitude, with mean annual rainfall in the range of 1000–2100 mm. It is moderately resistant to drought, able to tolerate a dry season of up to 4 months. *D. nigra* grows naturally even in low-fertility soils with a pH greater than 5.2. It is not demanding of phosphorus, and prefers soils with low aluminium content. In southern Bahia, it grows mainly in mountainous areas, occupying peaks and slopes with clay and sandy clay–loam soils that are deep and well drained. *D. nigra* can grow in a moderately shaded environment. It is a nitrogen-fixing tree and forms endomycorrhizal associations. Flowering time ranges from September to May depending on location, with fruits ripening from May to January, though seed may only be produced every second year. Seeds are dispersed by wind, though the species can also regenerate from stumps and by producing sprouts.

Silviculture practice

Seeds of *D. nigra* are collected when the fruits are brown, and seeds should be harvested between 335 and 350 days after anthesis. Fruits are sun-dried to enhance opening, after which they are individually opened by hand, with 17,300 seeds/kg. Seeds with 12% moisture content in semi-permeable bags stored at 10°C with 65% relative humidity maintained their viability for 105 days. A germination rate of 65% was also achieved after 2 years when stored in semi-permeable containers at 3–5°C and 92% relative humidity. Seedling development in the nursery requires up to 50% full light, and seedlings reach a plantable size after 4–6 months. Seedlings can also be produced from cuttings.

D. nigra is grown in mixed plantation for land reclamation, restoration of riparian forests and small-scale wood production. Pure stands are not recommended due to problems with forking and susceptibility to the wood borer *Stenodontes spinibarbis*. *D. nigra* is therefore generally recommended for use in mixed plantations in full sun, and also in enrichment planting in secondary forests. It has also been planted with other plantation species and a cover crop of *Desmodium ovalifolium*.

Due to its tendency to form a very branched trunk, this species should be planted at close spacing such as 2 × 2 m and 3 × 2 m, to favour self pruning and reduce forking of the bole, assisted by artificial pruning where required. The economic rotation for wood production is estimated at 40 years, predicting a productivity of 100–150 m^3/ha.

In its native range in Minas Gerias, *D. nigra* planted in a mixed understorey of *Eucalyptus grandis* established to restore disturbed land showed the highest growth rate among the 15 native species planted, reaching 1.8 m high and a diameter of 2.6 cm after 3 years. *D. nigra* has also showed good growth and survival in plantations outside of its native range. Survival rates were 94% in a pure stand in Amazonas after 19 years, and 100% survival was recorded in Paraná after 14 years, even after being subjected to a light frost. Trees in a 5-year plantation in Amazonas had a mean height of 11 m and a diameter of 13 cm. In cooler temperatures in Paraná, trees reached more than 7 m high and 7 cm in diameter after 5 years, more than 9 m high and 12 cm in diameter after 10 years, and more than 13 m high and 15 cm in diameter after 14 years, at spacings of 2 × 2 m.

Diseases and pests

Several seedborne fungal pathogens reduce seedling emergence and the quality of seedlings, including *Cladosporium cladosporioides*, *Colletotrichum crassipes*, *Fusarium semitectum* (*Fusarium pallidoroseum*), *Phomopsis dalbergiae* and *Pestalotiopsis* sp. All except *C. cladosporioides* are highly pathogenic to seeds and seedlings, causing seed rot, root rot and reduced seedling height and survival. *Pestalotiopsis* sp. induces severe chlorosis and defoliation; the leaf spots of *P. dalbergiae* cause die-back, whereas those caused by *C. crassipes* remain restricted to the leaves.

D. nigra has particular susceptibility to the cerambycid wood borer *S. spinibarbis*, which bores into living trunks. The coleopteran *Scolytus submarginatus* has also been identified, and seeds within the fruits are damaged by the beetle *Troezon championi*.

Uses

D. nigra is sometimes planted in association with pastures in south-central Brazil and also with annual crops. In Colombia, this species has been planted experimentally with guaraná (*Paullinia cupana*). It can be used in windbreaks, and is grown in mixed plantations for environmental restoration and widely used as an ornamental.

The common name 'rosewood' relates to the fragrant oil in the heartwood, which has the scent of roses. The wood perfumes the air when cut, machined or handled. The heartwood is prized for its rich, deep red colour, but the colour ranges from deep purple-black through dark chocolate-red with ebony-coloured streaks. *D. nigra* wood density ranges from 750 to 1220 kg/m^3, most frequently 800–1000 kg/m^3. The sapwood is white to yellowish and is well distinguished. The timber has many fine characteristics, being hard, shock resistant, flexible, stable and strong. The wood is fairly easy to work, bend and glue, machines and veneers well, and polishes to a high sheen due to its oily nature. It stands up to years of constant handling, and is very durable and resistant to the attack of decomposition organisms and termites.

In the past, rosewood was used in fine panelling, cabinetry and furniture. It was used in marquetry and inlay and as detailed design elements in billiard cue butts, knife handles, brush backs and other fine tools and implements. It has often been used in carpentry tools such as spirit levels and plane handles. In furniture, rosewood was used in table tops and cutlery cabinets and boxes. One of its main traditional uses was also in the manufacture of some of the world's finest musical instruments, especially guitars and violins, and also piano cases and piano keys. The timber is still much sought after in local markets as building material for use in flooring, and for structural beams and wall panelling/lining. The oil produced from the timber is also used as a fragrance in cosmetics and for medicinal purposes.

Further reading

Fonseca, C.E.L. da, Bueno, D.M. and Sperândio, J.P. (1990) Performance of five-year-old Brazilian rosewood [*Dalbergia nigra*] under four different spacings in Manaus, Amazonas, Brazil. [Comportamento do Jacaranda-da-baia aos cinco anos de idade em quatro diferentes espacamentos em Manaus - AM.] *Revista Arvore* 14, 78–84.

Ribeiro, R.A., Lemos Filho, J.P., Ramos, A.C.S. and Lovato, M.B. (2011) Phylogeography of the endangered rosewood *Dalbergia nigra* (Fabaceae): insights into the evolutionary history and conservation of the Brazilian Atlantic Forest. *Heredity* 106, 46–57.

Westley, S.B. and Roshetko, J.M. (eds) (1994) *Dalbergia*: Proceedings of an International Workshop. Nitrogen Fixing Tree Research Reports Special Issue. NFTA, Waimanalo, Hawaii, USA, 207 pp.

Dalbergia sissoo Roxb. ex DC.

Indian rosewood

Importance

Dalbergia sissoo is a very useful multipurpose tree, extensively planted as single trees and in linear or block plantations, especially in South Asia. It is a fast-growing nitrogen-fixing tree, easily propagated from seed, seedlings, stumps and root or shoot cuttings. It can be grown on well-drained soils, even on pure sand. The heartwood is hard, heavy, strong and durable, and the species provides among the finest general-use timbers in South Asia, commonly used for high-class furniture, cabinetmaking and marine-grade plywood. It is also suitable for all kinds of construction, is an excellent fuel and leaves are useful as livestock fodder.

Botanical features

D. sissoo is a medium to large, deciduous, long-lived tree with a spreading crown and thick branches, up to 30 m and a diameter of 75 cm. The bole is often crooked and the bark is thick, rough and grey, with shallow, broad, longitudinal fissures exfoliating in irregular woody strips and scales. Leaves are compound, imparipinnate and alternate, with rachis 3.5–8 cm long, swollen at the base. There are alternate 3–5 leaflets, 3.5–9 cm long and 3–7 cm wide, broadly ovate and entire, with 8–12 pairs of veins and terminal leaflets larger than the others. The inflorescence is an axillary panicle 3.5–7.5 cm long, with small flowers 7–9 mm long, white to yellowish-white with a pervasive fragrance, papilionaceous and hermaphrodite. Pods are 4.5–10 cm long, linear–oblong, indehiscent, with 1–4 kidney-shaped seeds, 1 cm by 0.5 cm, pale brown to brownish-black, compressed with papery testa.

Distribution

D. sissoo is native to South Asia and neighbouring countries, from Iraq and Iran to Myanmar. It is particularly common in India and Pakistan in the sub-Himalayan tract and outer Himalayan valleys over a large area from the River Indus to Assam, except in the coldest, wettest and driest areas. It has been introduced with varying degrees of success into many other tropical and subtropical countries in Asia including China, Indonesia, Malaysia, Thailand, Sri Lanka and Australia, and is widespread in Africa. There are occasional reports from some Pacific, Caribbean and Mediterranean islands, the Middle East, and in North and South America, though it is likely to have been more widely introduced.

Silviculture characteristics

D. sissoo is a strong light-demander, and only vigorous trees attain large size in dense stands, which tend to suppress weaker trees. Seedlings are sensitive to drought, but mature trees may withstand dry seasons of up to 9 months. Annual rainfall in the native range is 750–4500 mm, but plantations are most common when the rainfall is 1000–17000 mm, though trees will survive with 400 mm. It is frost-hardy and grows up to 1000 m altitude, occasionally to 1500 m. It is a gregarious colonizer, growing in tropical dry mixed deciduous and dry deciduous and thorn scrub forests, prefers free-draining soils and does not tolerate heavy clays or saline soils.

Seedlings produce a long taproot at an early age, which becomes thick during the sapling stage. In trees, the taproot is sometimes thicker than the bole itself and lateral roots may also attain considerable size. Trees are generally windfirm but windthrow is not uncommon in irrigated plantations, when only superficial roots may develop. In hot, arid areas, trees can die in large numbers if irrigation ceases for any length of time. Trees coppice and sucker vigorously and profusely, withstanding frequent cutting and lopping. *D. sissoo* is not fire-resistant.

Silviculture practice

In South Asia, *D. sissoo* is generally grown in block plantations established under irrigation or in floodplains. Silvicultural systems used are clear felling with artificial regeneration generally using stump plants, and coppice with standards. Pods are collected in late January or February. Although very young trees bear seed, collections should only be made from mature trees. Fresh seed retains its viability for 1 year with 70–90% germination. No pre-sowing treatment is required; germination starts after 7–10 days and is completed in about 21 days.

The most common method of artificial regeneration uses stump plants consisting of 5.0–7.5 cm of shoot and some 25 cm of root from 1–2-year-old nursery seedlings, generally at a spacing of 2 × 3 m. The frequency of irrigation during the first year is high, and is reduced in subsequent years as plants establish themselves. Pruning of lower branches in the early years helps produce a clear bole. Thinning is commonly carried out at 5–6, 10–11 and 14–16 years. Weeding and cleaning are carried out when necessary. Rotations of 10–22 years are used for fuelwood and small timber in irrigated plantations, with rotations of 40–60 years for production of larger timber. Annual growth rates in plantations are 10–22 m^3/ha, and biomass yields of 8–18 t/ha per year have been reported.

Diseases and pests

By far the most severe disease is the wilt *Nectria haematococca*, favoured by heavy soils and high soil moisture. It attacks through dying or weakened roots, causing yellowing and death of leaves and even mass mortality in severe infestations. Other root rot diseases include *Ganoderma lucidum*, *Oxyporus latemarginatus* and *Phellinus gilvus*; a butt rot is also caused by *P. gilvus*. Many pathogens are reported attacking leaves, though are less significant, including *Cercospora sissoo*, *Colletogloeum sissoo*, *Maravalia achroa*, *Mycosphaerella dalbergiae*, *Phyllachora viventis*, *Phyllactinia dalbergiae* and *Rhizoctonia solani*.

D. sissoo is attacked by more than 125 insect pests, of which only about ten are considered important. *Plecoptera reflexa* causes defoliation resulting in reduced growth and mortality, caterpillars of *Leucoptera sphenograpta* mine into the leaves, *Cladobrostis melitricha* bores into the twigs causing heavy branch loss, and *Drosicha mangiferae* and *Drosicha stebbingi* are the most important sapsuckers. However, the most damaging are the heartwood borers *Aristobia horridula* and *Crossotarsus externedentatus*. Among the several reported parasites, the mistletoe *Loranthus longiflorus* can cause severe damage and tree mortality.

Uses

Tree planting on farmlands is economically feasible, and use of *D. sissoo* as an agroforestry component with crops is financially superior to agriculture alone. It is widely grown in field, along field boundaries, around fruit orchards, as windbreaks and shelter belts, and as scattered trees. Every year before the onset of or during winter, farmers carry out pollarding, lopping, individual branch pruning, trimming or cutting of trees to promote coppicing as well as for fuelwood and fodder. Trees can reduce crop production, so trenches may be dug to cut lateral tree roots.

Growing *D. sissoo* on farms yields numerous environmental and economic benefits including ameliorating the microclimate, controlling soil erosion and improving soil fertility, as well as producing timber, fuelwood, fodder and other minor products (e.g. honey) of high economic value. The strong development of root suckers and runners favours its use as a living barrier against soil movement, and for this reason it is planted across the bottom of eroding gullies. *D. sissoo* is also valued as a shade tree, street tree and as an ornamental.

The heartwood is golden-brown to dark brown with deep brown streaks which become dull soon after sawing, whereas sapwood is white to pale brownish-white. The depth of colour is highly variable, with lighter colours often coming from irrigated plantations on sandy soils, with dark forms with heavy figures from higher altitudes in Himalayan foothills. Wood density is 770–800 kg/m^3 at 12% moisture content. The wood is moderately coarse textured with fairly straight to somewhat interlocked grain and semi-ring-porous with inconspicuous growth rings. The wood is hard, strong and very elastic, dimensionally stable and not liable to cracking or splitting. The heartwood of *D. sissoo* is extremely durable. Strength properties vary, with wood from the Himalayan foothills being superior to that from the plains.

Timber can be air- or kiln-seasoned with little degrade. Air seasoning may take 2 years to attain a moisture content of 10% in 9 cm-thick planks. Kiln seasoning of 2.5-cm thick timber will take about 12–15 days to reduce moisture from initial green to 12%.

Timber is easy to work and saw, machines well and takes an excellent polish. Veneer is suitable for high-class work, commercial plywood and tea chests. Timber is preferred for door and window shutters and frames, cabinetmaking, flooring, panelling, construction, tool handles, vehicle bodies, farm implements, boatbuilding, bentwood articles, superior-quality felloes for wheels, turnery, laminated skis and a wide range of other uses such as fence posts and poles. The sapwood gives satisfactory bleached and unbleached pulp. It also makes an excellent fuelwood and charcoal, with a calorific value of around 5000 kcal/kg.

Leaves contain 12–24% crude protein on a dry-weight basis, and palatability and nutrient content of leaves is highest in April and May when the need for green fodder is greatest owing to scarcity of other sources. Green pods are also used as a fodder. Roots, bark and leaves are used for medicinal purposes. The heartwood also yields light brown, non-drying highly viscous fixed oil, which is suitable as a lubricant for heavy machinery.

Further reading

Pande, P.K. and Rawat, L. (2012) Intra, inter-tree and inter-location variations in wood anatomical traits and specific gravity in seed raised plantations of *Dalbergia sissoo* Roxb. *Indian Forester* 138, 555–561.

Roshetko, J.M. and Westley, S.B. (1994) Dalbergia sissoo *Production and Use: A Field Manual.* NFTA/Winrock, Morrilton, Arkansas, USA, 48 pp.

Tewari, D.R. (1994) Monograph on *Dalbergia sissoo* Roxb. International Book Publishers, Dehra Dun, India, 316 pp.

Diospyros digyna Jacq.

Black sapote

Importance

Diospyros digyna is a persimmon from Central America. It is a small to medium evergreen, broad-topped tree about 10–25 m in height. It is best known in Mexico and Guatemala where it is native. The tree has medicinal properties and the edible fruit is also known as 'chocolate pudding fruit'.

Botanical features

D. digyna is an evergreen tree of 10–25 m in height with a dominant trunk. It bears pseudowhorled tiers of slender spreading applanate branches. Leaves are alternate, elliptic–oblong to oblong–lanceolate and 10–30 cm long. They are dark green, glossy and leathery. Flowers are axillary, unisexual or hermaphrodite. Male flowers are usually in clusters of 3(–7) and female flowers usually solitary. Different types of flowers are mixed on individual trees. The flowers are white, tubular, with a persistent green calyx. The ovary is 8-, 10- or 12-celled with 1 ovule per cell. The fruit is a flattened globose to slightly lobed berry, 5–15 cm in diameter, dark olive-green, and seated on a persistent 4–6-lobed calyx which reflexes at maturity. The pulp is soft, white when unripe and brown to black when ripe. Seeds 0–12 are flat, smooth and brown, 2–2.5 cm long. The fruit from cultivated trees is often seedless.

Distribution

D. digyna is probably native to Mexico and Guatemala. It grows between 0 and 1800 m altitude, but normally below 600 m. *D. digyna* was brought to the Philippines by the Spanish in the late 1960s and then spread to the Moluccas and Sulawesi where it became naturalized, and to Mauritius, Hawaii, Brazil, Cuba, Puerto Rico and the Dominican Republic. Its most important area of cultivation is Mexico and Guatemala.

D. digyna adapts to different types of soil, and is found both in dry forests and on alluvial clay near streams. It tolerates flooding, but is sensitive to drought. It prefers moist sandy loam, on well-drained sand or even limestone with very little topsoil.

Silviculture characteristics

D. digyna is a perennial. After establishment, growth is initially slow, but the tree becomes a vigorous grower later on. The juvenile phase can be as short as 3–4 years.

Young plants of *D. digyna* do not tolerate frost, while older plants are more frost-hardy and may reportedly tolerate temperatures of −1 to −2°C. However most reports suggest it does not tolerate frost and is successful only in the tropics and subtropics. Seedlings normally begin to fruit within only 3–4 years. Flowers are borne on the new shoots, hence the crop cycle is linked to flushing. Flowering takes place either in the spring or the summer flush. In Mexican lowlands and in the Philippines, the tree flowers in March. Fruiting is reportedly abundant, but information on yields is unavailable.

Silviculture practice

D. digyna is usually propagated from seed or by grafting. The seeds remain viable for several months if stored dry. The seeds probably do not show recalcitrant seed storage behaviour and germinate in about 30 days. Seedless types can be propagated by budding or grafting on seedling rootstocks. As some trees bear flowers that are self-incompatible, it is not advisable to plant a solitary tree. Trees grow fast and should be spaced 10–12 m apart. There is little information on husbandry or crop protection.

Diseases and pests

The fruit flies *Anastrepha obliqua*, *Anastrepha serpentina* and *Anastrepha striata* are reported pests. Damage from a noctuid, *Hypocala andremona*, was reported in Florida, USA.

Uses

D. digyna is planted as an ornamental.

The wood is yellowish to deep yellow with black markings near the heart of old trunks; it is compact and suitable for cabinet work, but is little used. Wood density is 790 kg/m^3.

The main use of *D. digyna* is for its edible fruit, which is eaten when ripe and soft, it is also used in drinks, ice-cream, cakes and liqueurs.

Unripe fruits are inedible; they have a high tannin content and have been used to poison fish in Central America, the Philippines and the West Indies. The bark and leaves have been used medicinally against fever and skin disease. The fruits of *D. digyna* reportedly have high vitamin C content and provide a good source of calcium and phosphorus.

Further reading

Janick, J. and Paull, R.E. (eds) (2006) *Encyclopedia of Fruit and Nuts*. CABI, Wallingford, UK, 954 pp.

Ricker, M., Siebe, C., Shimada, K., Larson, B.C., Martínez-Ramos, M. and Montagnini, F. (2000) Optimising seedling management: *Pouteria sapota*, *Diospyros digyna* and *Cedrela odorata* in a Mexican rainforest. *Forest Ecology and Management* 139, 63–77.

Verheij, E.W.M. and Coronel, R.E. (eds) (1991) *Plant Resources of South-East Asia No 2. Edible Fruits and Nuts.* Pudoc Scientific Publishers, Wageningen, The Netherlands, 447 pp.

Diospyros melanoxylon Roxb.

Coromandel ebony, East Indian ebony

Importance

Diospyros melanoxylon is a tree native to India, Pakistan and Nepal. It is valued locally for its timber quality and for its leaves, which are used in making 'bidi' cigarettes, as well as for a range of other non-timber uses. It is a member of the ebony group, although its wood is not as black as 'true' ebony (*Diospyros reticulata* and *Diospyros ebenum*). The wood is dark, hard and heavy and is used for posts, rafters, shafts, textile-shuttles, toys and mine props.

The principal commercial value of this species is currently its leaves, known as 'tendu patta' or 'bidi/beedi/biri' leaves. They are unrivalled in terms of flavour, decay resistance and easy workability for wrapping the cigarettes/cigarillos locally termed 'bidis'. Bidi rolling alone can account for as much as 60% of a household's total cash income.

Due to the size and importance of the bidi trade and exploitation of local collectors – mostly indigenous people – Indian state and central government sought to establish ownership rights of the collectors to these non-timber products. This culminated in a constitutional amendment in 1996, giving such ownership to local groups or entities. There is an established network of traders composed of wholesalers and retailers, and some companies export the bidis to neighbouring countries.

Botanical features

D. melanoxylon is a medium-sized deciduous tree up to 25 m tall and 2 m in girth, though is often coppiced or pollarded to produce a shorter, heavily branched shrubby form. The bark is pinkish, with rectangular scales. Leaves can be opposite or alternate and coriaceous, up to 35 cm long, tomentose on both sides when young, becoming glabrous above when fully grown. Flowers are mauve, often in threes; the male 1–1.5 cm long (smaller than the female), 5 or 6-merous, (nearly) sessile in short peduncles; the female flowers are extra-axillary, generally in pairs but sometimes solitary. Fruits are olive-green berries, ovoid or globose, 3–4 cm across; they contain a yellow soft sweet pulp and 1–8 oblong shiny seeds.

Distribution

D. melanoxylon is native to India, Pakistan and Nepal and has also been planted widely in the region. It is drought and frost-hardy. *D. melanoxylon* can tolerate a wide range of soil conditions but is sensitive to waterlogging. It is normally found among other tree species in dry deciduous forest (teak, sal and mixed forests of *Acacia leucophloea*, *Boswellia serrata*, *Butea monosperma*, *Lannea coromandelica* and *Terminalia tomentosa*) at altitudes ranging from 0–900 m and annual rainfall from 500–1500 mm.

Silviculture characteristics

It is a light-demanding, drought- and frost-hardy species and it coppices and pollards well. A characteristic of the species is its prolific root suckering. It can also be propagated from seed and nursery-raised stock.

D. melanoxylon grows best on loose, porous soils in cool and moist sheltered valleys. However, it can tolerate poor denuded soils; hot and dry hill slopes; stony soils with quartzite, shale and sandstone; and even heavy clays.

The deep taproot is believed to aid nutrient cycling and to reduce competition with agricultural crops. Coupled with the deciduous habit, fodder value and tolerance to coppicing, it is considered a notable multi-value agroforestry species.

Silviculture practice

Cultivation of *D. melanoxylon* as boundary trees or scattered in fields of oilseed or cereals under an agroforestry system may be feasible and attractive. Competition with annual crops is limited, due to their deep taproots.

The species regenerates naturally through seedlings, from coppice and from root suckers. Despite its good timber the tree is usually reserved and is not felled except in clear felling coupes.

Plantation growing has been successful by direct seeding as well as by planting out nursery-raised seedlings or stumps. The yield from 5 kg of ripe fruit is 1 kg of seeds (1100–2000 seeds/kg). Seeds are sun-dried and remain viable at room temperature for up to 1 year. Germination success increases after soaking in cold water for 12 h. Propagation is also possible by rooting shoot sucker pieces.

Seedlings are planted at 2 × 2 m. If production of leaves is the main objective, heavy pruning is required to promote vegetative growth, with relatively large and thin leaves. To maximize the number of leaves, the tree should be regenerated regularly through root suckers from damaged root points, pollarding-cum-trenching, pruning or coppicing. Pruning too frequently may result in stunted growth.

Most bidi leaves come from the low root suckers; leaves of bigger trees are too stiff and brittle and therefore not capable of producing trade leaves for wrapping bidis.

Stems up to 15 cm diameter should be cut at ground level to encourage coppicing, and good-quality leaves are obtained after 40–50 days. To obtain the best-quality leaves for the bidi trade it is advised to coppice flush to the ground level twice, 2 weeks apart, rather than just once. The tree also pollards well, but shoot growth is slow. Pollarding yields a good flush of tender leaves. Larger plants that are over 15 cm dbh should be cut at 60–90 cm height.

For small timber and pole production from *D. melanoxylon* plantations, managers should consider a coppice rotation of 30 years. A much longer rotation would be needed to produce the more valuable ebony heartwood.

Diseases and pests

Severe outbreaks of blight disease caused by *Colletotrichum gloeosporioides* have been reported. The disease is characterised by dark brown necrotic lesions which increase rapidly during the rainy season, causing decay and death of entire growing tips and newly formed leaves. The disease accounted for considerable losses in the regeneration of the species in the early 1990s. The fungus *Stereum lobatum* causes white spongy rot in felled timber, while *Cercospora kaki* causes leaf spot.

The Tendu gall psyllid (*Trioza obsoleta*), is a sap feeder and forms reddish-yellow galls on the leaves. Caterpillars of *Miresa albipuncta*, *Euthalia laudabilis*, *Euthalia nais* and *Hypocala rostrata* are important defoliators. The larvae of the longicorn beetle *Plocaederus ferrugineus* bore between the bark and the sapwood.

Uses

D. melanoxylon is a fodder species and is reportedly a good fuelwood. It has a place in traditional agroforestry approaches as a boundary and shade tree, and is also a focus of current research.

The dark heartwood is very heavy and is valued for carving and other ornamental work. The sapwood is whitish-pink, tough, fairly durable and used for building, shoulder poles, mine props and carriage shafts. *D. melanoxylon* has a wood density of 680 kg/m^3.

This species is the primary source of 'tendu patta' or 'bidi/beedi/biri' leaves used for wrapping the local 'bidi' cigarettes. The fruits and powdered seeds are edible and sold in local markets.

Many medicinal applications are locally important; the seeds are prescribed in India as a cure for mental disorders, nervous breakdowns and palpitations of the heart. They can be intoxicating. The fruits have a cooling and astringent effect. Dried flowers are used to cure urinary, skin and blood diseases. The bark is astringent and a decoction of it is used to treat diarrhoea.

Further reading

Boaz, A. and Boaz, O. (2003) Community-based sustainable management of tendu leaves (*Diospyros melanoxylon* Roxb.): A case study of Harda district of Madhya Pradesh, India. USDA Forest Service General Technical Report Pacific Northwest. *Sustainable Production of Wood and Non-Wood Forest Products* 604, 51–61.

Hunter, J.R. (1981) Tendu (*Diospyros melanoxylon*) leaves, bidi cigarettes, and resource-management. *Economic Botany* 35, 450–459.

Ghosh, R.C., Mathur, N.K. and Singh, R.P. (1976) *Diospyros melanoxylon* - its problems of cultivation. *Indian Forester* 102, 326–336.

Gupta, T. and Guleria, A. (1980) *Some Economic and Management Aspects of a Non-Wood Forest Product in India: Tendu Leaves.* Centre for Management in Agriculture Monograph No. 88, 184 pp.

Dipterocarpus grandiflorus (Blanco) Blanco

Apitong

Importance

Dipterocarpus grandiflorus is a medium to large tree of seasonal primary semi-evergreen or evergreen forests of lowland South-east Asia up to 600 m altitude.

It has a high quality, strong, dense wood of many uses from construction and paper pulp to fine furniture. Tapped trees yield an oleoresin ('balau' or 'minyak keruing') with numerous industrial applications. Because of its quality wood and resin production, it has high potential for commercial plantation. Once established, it tends to aggressively colonize an area in almost monoculture stands which tend to be attacked by bark beetles and other pests. More research is needed on artificial propagation and on the protection aspects of infestations affecting natural stands.

Botanical features

A medium to very large tree attaining a diameter of 100–180 cm and a height of 25–45 m. It usually has a straight cylindrical bole, branchless up to 25–30 m. The crown is short, compact, roughly flat-conical or irregular and partly open. Buttresses low to moderate or sometimes none at all. Twigs to 12 mm, stout, becoming grey-brown; internodes short, leaves clustered around twig endings.

D. grandiflorus has a relatively deep taproot to reach the water table, as it usually thrives on the upper slopes of mountains.

Bark is light grey, smooth, shed in large rectangular outwardly curved plates, surface covered with numerous corky pustules; the flakes leave faint jigsaw-puzzle marks on the bark surface. The inner bark is reddish.

Leaves are alternate, ovate, 15–20 cm long, 9–12 cm wide, apex acuminate, base rounded or slightly cordate, margin more or less undulate, texture leathery, both faces glabrous, shiny above, dull below; lateral nerves 12–20 pairs, very prominent beneath, parallel veins curve towards the apex shortly before reaching the margin; petiole 5–7 cm long, slender, woody, rough, thickened at the base of the blade.

Inflorescence very slender, glabrous, up to 18 cm long, axillary, borne in dense groups behind the leaves, unbranched bearing up to 3 remote distichous flowers. Floral buds 35 × 13 mm, fusiform. Stamens 30, exceeding style at anthesis; filaments broad, compressed, tapering. Ovary tapering into the short stoutly columnar style; ovary and style puberulent in the basal half. Tube of the fruiting calyx 7 × 3.5 cm, ellipsoid, with 5 pruinose, to 1.5 cm wide, prominent coriaceous wings continuous from base to apex, 2 longer lobes 22 × 3 cm, oblanceolate, spatulate, obtuse, tapering abruptly to about 1.3 cm wide at base; 3 shorter lobes to 2 × 1.5 cm elliptic, subrevolute.

Distribution

D. grandiflorus is widely distributed in almost all parts of the Andaman Islands (India), Indonesia (Kalimantan and Sumatra), Myanmar, Peninsular Malaysia, the Philippines (except at higher elevations), Thailand and Vietnam.

Silviculture characteristics

Like most dipterocarps, *D. grandiflorus* needs partial shade in the early stages of development. However, when seedlings reach sapling and pole stages they should be released from competing vegetation to enhance their growth and development. Diameter and height growth under such management increases by twice or more compared with untreated stands.

Heavy precipitation and strong winds cause serious damage not only to mature trees but also to young seedlings. The trees may be uprooted, especially on top of narrow ridges, steep slopes, precipitous banks of ravines, gullies or canyons, or be carried away by landslides and other forms of soil denudation. Wide openings in the forests formed in this manner are taken over by grass and fast-growing weed species.

Openings made in the forest due to logging operations cause damage to remaining seedlings and immature trees. This may be due to several factors such as intense insolation, change of temperature and pressure, sun scorch, unfavourable soil conditions and drought.

Silviculture practice

Seeds of *D. grandiflorus* are generally collected from the ground after they fall, because of the difficulty of climbing the tall trees. The seeds have very short viability (a few days) which can be extended up to 8 weeks with cold storage in sealed containers under nitrogen.

Nursery-sown potted stocks or balled wildings may be used to establish plantations. The technique is to first establish a suitable stand of nurse trees such as *Leucaena leucocephala*, *Gmelina arborea*, *Paraserianthes falcataria*, *Acacia* spp. or any other adaptable pioneer/secondary species to provide shade to the seedlings/wildings.

As soon as the nurse trees are tall enough to provide the necessary partial shade, potted or balled stock is planted between nurse tree rows. However, in order to attain straight boles, a spacing of 3 × 3 m, 4 × 4 m or 2 × 4 m is required. In strip planting, spacing is 2–3 m in the strip and 6–10 m between strips. The most suitable size/age of seedlings for transplanting is about 40–60 cm tall or about 1 year old. Progressive thinning should be done at a later stage until the favoured spacing is reached. Gradual release from the shade trees should be done when signs of severe light competition are observed. This can be done by pruning branches and successive thinning of nurse trees until the shade is no longer necessary.

In plantations, weeding is necessary during the first 3 years. Thinning should be carried out after 5, 10, 15 and 25 years. In the first 2–3 years, shade trees may be used as a nurse crop.

Diseases and pests

The prevalent indicator of heart rot in *D. grandiflorus* is the presence of sporophores or fruiting bodies of *Ganoderma applanatum*. The sporophores are readily seen on scars on exposed woody tissue resulting from undetermined causes. Broken branches or some mechanical injuries may provide avenues for the entry of spores or other forms of inoculum. Affected parts are cut and removed, or, if not badly affected, trees are immediately utilized for other purposes such as pulp and paper before heart rot and decay take over the whole tree. Wilding blight caused by *Lasiodiplodia theobromae* and apitong wilt caused by *Polyporus* sp. are common problems for the natural reproduction of *D. grandiflorus* in the Philippines.

D. grandiflorus is one of the host trees of the bark engraver *Sphaerotrypes philippinensis* and the bark beetle *Dryocoetiops laevis*. The fruit is attacked by the seed weevil *Alcidodes crassus*.

Termites, in association with fungi, simultaneously attack and inhabit the central core of the butt, resulting in hollow core. Normally, termites infest only old, decaying, sickly, injured or partly burned trees.

Uses

D. grandiflorus is a component of dipterocarp forest which is generally uneven-aged, similar in structure and reproductive character and hence self-perpetuating, so continuously benefiting human and wildlife communities, unless adversely disturbed by human activities.

D. grandiflorus can make up the biggest forest cover component of watersheds. It can store much of the rainwater and regulates its flow on the slopes to streams, lakes and reservoirs for the irrigation of food crops and the generation of electricity to provide energy for homes and industry. It minimizes soil erosion on slopes and resultant sedimentation of streams, lakes and reservoirs. It also improves soil conditions through its fast rate of litter deposition and organic matter decomposition.

This tree has been utilized commercially because of its high wood quality and wood extracts that command good prices in local and international markets. Known as apitong in the Philippines and keruing belimbing in Malaysia, it is an important source of timber which is used mostly for building construction, especially for the posts and frames of houses and for planking in lighters and ships. It is used for flooring when well-seasoned, and for car body construction and for other purposes where moderately hard and comparatively heavy wood is required. Wood density ranges from 650 to 945 kg/m^3 at 15% moisture content. *D. grandiflorus* is a good substitute for most of the uses of Oregon pine (*Pseudotsuga menziesii*). It is a standard wood for any type of piling. It is also used in the manufacture of veneer and plywood, pulp and paper, hardboards, furniture, boats, crates, boxes and wooden shoes.

Large quantities of oleoresin (called balau or minyak keruing) can be obtained from the wood. The oleoresin is obtained by cutting a semicircular niche about 1 m from ground level. The niche is freshened frequently to ensure regular flow. The exudation has a thick honey-like consistency, a reddish-brown colour and a characteristic balsamic odour. It consists of about 85% volatile oil and a hard yellow lustrous resin. The resin, when processed or dissolved in linseed oil and turpentine, gives varnish, lacquer and thinner. A tannin formaldehyde adhesive can be produced from bark extracts. In the Philippines, substituting apitong oil for diesel fuel gave positive results but needs further research.

The bark is considered to be a tonic, and a hot decoction of it is used to treat rheumatism and liver complaints. The nuts may be boiled and eaten.

Further reading

Ashton, P.S. (1982) Dipterocarpaceae. In: van Steenis, C.G.G.J. (ed.) *Flora Malesiana, I, Spermatophyta.* Martinus Nijhoff/Dr W. Junk Publishers, The Hague, The Netherlands, 9, 237–552.

Cadiz, R.T. and Modino, R.C. (1991) Apitong, *Dipterocarpus grandiflorus* Blanco. ERDB, DENR, College, Laguna. *RISE – Research Information Series on Ecosystems* 3, vi & 16 pp.

Prasad, P.R.C. (2011) Ecological analysis of Dipterocarpaceae of North Andaman Forest, India. *Journal of Plant Development* 18, 135–149.

Dipterocarpus turbinatus C.F. Gaertn.

Gurjan

Importance

Dipterocarpus turbinatus is an important timber tree, and is used extensively in construction as rafters, scantlings and planks, and in boatbuilding. Due to high demand for its heavy timber, this tree is found in plantations in many countries, especially Bangladesh where it is the most important indigenous tree species grown in long-term plantations. It is also the principal source of an oleoresin known internationally as East Indian copaiba balsam.

Botanical features

D. turbinatus is a lofty evergreen tree reaching a height of 35–50 m, with a dbh of at least 1.7 m, and sometimes 4.6 m. It has an elevated crown. The bole is long, clean, cylindrical, self pruning and mostly clear of branches for more than 70% of the total height. The bark is light grey, hard, rough, fibrous, fissured longitudinally and exfoliates in stripes. Young branchlets bear stipule scars. *D. turbinatus* is characteristically known by its smooth ovate leaves, which dry to a dark coppery-brown.

Leaves are simple, stipulate, petiolate, 20–30 × 8–13 cm, ovate or ovate–lanceolate to elliptic; lateral nerves distinct, 14–24 pairs, glabrous on both surfaces, petiole 2.5–4.0 cm in length, stipule 1.5–9 cm in length (or even as long as 12 cm), densely greyish to buff.

Flowers are borne on a few flowered racemes, which develop from the axils of fallen leaves. Flowers are 3.0–3.5 cm in length, white or pinkish. Fruit is in the form of a nut, spindle shaped, 2–3 cm in diameter, smooth ovoid, supported by 2 thickened winged calyxes; wings 11–18 × 2.5–4.0 cm, strongly reticulate with 3 veins, with central vein prominent. Each fruit contains one seed, and the fruit coat strongly adheres to the seed coat.

Generally the older leaves are shed in March, and flowers appear from March to April. Fruits tend to ripen approximately from late April to May in Bangladesh, and in May–June in upper Myanmar. In May–June fruit fall occurs as the rains begin. Fruiting may be irregular in certain localities, and good seed years often occur in alternate years.

Distribution

D. turbinatus is distributed in evergreen and semi-evergreen forests in India, including Assam, Manipur, Nagaland, Tripura, Meghalaya, Mizoram and the Andaman Islands. It is also found in Sylhet, Chittagong and Chittagong Hill Tracts (Bangladesh), and in the greater part of Myanmar extending as far as Thailand, through Cambodia and Laos to Vietnam. It has also been reported from Cochinchina and Indonesia. *D. turbinatus* has been planted in plantations in many countries including those within its natural range in Bangladesh, Myanmar and Karnataka (India).

Silviculture characteristics

D. turbinatus is shade-tolerant during its juvenile stages, and becomes more light demanding as it matures and in the latter stages of development. The tree is able to self-prune, but will not coppice or produce root suckers. It is sensitive to fire, and any poles which have been severely damaged by fire may fail to recover. The species has proven relatively windfirm in tropical cyclone zones.

Silviculture practice

The seeds are recalcitrant with poor viability, having a germination rate of 82% and a plant survival rate of 50%. *D. turbinatus* stands may be established using direct sowing, nursery stock or natural regeneration. With natural regeneration, seeds germinate in moist soils or partially decomposed litter. Using a shelterwood system, some seed trees are left after felling, for seed dispersal. Clearing the soil surface may aid recruitment of young seedlings; the ground is prepared by brushwood cutting. In Bangladesh, the middle canopy is retained until the seedlings are established, and later opening of the canopy may encourage further development. In the forest, germination is initiated within 3–4 days after fruit fall, and is completed within 3 weeks.

Direct sowing in the field is more or less the uniform method used to artificially regenerate *D. turbinatus* over its entire range. Ripe fruits are collected from beneath mother trees, each fruit containing one seed, and the whole fruit is treated as a seed for germination. Seeds should be stored in the shade and used preferably within 1 week. The seeds are planted in holes 2.5 cm deep at 1 × 1 m spacing, covered with soil with wings protruding and the soil pressed lightly. During the rains intensive weeding and hand-pulling of creepers is required. Weeding, fire-protection, and cutting back climbers are repeated for approximately 3–4 years, until the trees are established and the shelterwood is removed.

Seedlings may also be raised relatively easily in the nursery, using shaded beds or plastic bags or tubes. The seeds are sown within 2–3 days after collection and covered with soil, leaving the wings protruding. Watering and weeding are necessary, and overhead shading induces vertical growth and reduces branching during the summer. *D. turbinatus* may also be propagated vegetatively using air-layering, which is carried out during the monsoon (June–July).

Generally, plantations are established after 4 years. At least three weedings are required during the rainy period for the first 3 years, which aids stand establishment. After 5 years the stand is thinned to 1.2 × 1.2 m spacing, and selection thinning is conducted after 10 years of age. A 10-year-old plantation produces an average crop height of 11.9 m and a dbh of 10.4 cm. At 45–60 years rotation the mean annual increment (MAI) recorded from Bangladesh is 10–12 m^3/ha.

Diseases and pests

No major pest or disease has been recorded for *D. turbinatus*. Some Lepidoptera larvae may bore into the seeds and fruits, and cyclones may sometimes result in the destruction of seed before it matures.

Uses

D. turbinatus forests are of ecological importance, supporting high species diversity. Within its natural range, *D. turbinatus* is now being planted in preference to exotic species, in order to enhance conservation. *D. turbinatus* is also being used in revegetation and has successfully been established in taungya agroforestry systems.

In the light of the current interest in carbon trading and sequestration, *D. turbinatus* was one of five species examined in Bangladesh, and showed a high ability to sequester organic carbon from the atmosphere.

Due to its wide distribution and high availability, *D. turbinatus* has become an important timber in the Indian subcontinent. It is widely used for various purposes, and several countries in Asia such as Myanmar export it in substantial amounts to countries such as Indonesia and Thailand. The wood is heavy, strong and hard, with a specific gravity of 0.655, and is often used for heavy construction in ship-, boat- and bridge-building. It has also been used for railway sleepers and transmission poles. *D. turbinatus* may also be considered as a substitute for European oak, and can be used in flooring. It is used to manufacture plywood, although it has a tendency to peel due to the resin content of the wood. It also makes a good fuelwood (sapwood 5300 kcal/kg, and heartwood 5050 kcal/kg). Sawdust of *D. turbinatus* may be mixed with rice husks to produce high-quality fuel briquettes.

Despite its durable properties, wood preservative treatments are essential for outdoor uses and for indoor applications if the products come into contact with the ground. It seasons quite slowly, and is moderately refractory, but may be successfully kiln dried. The heartwood is moderately durable against fungal attack, but is susceptible to attack by termites and borers, despite the fact that standing trees are reasonably tolerant.

D. turbinatus is the principal source of a commercial oleoresin oil, which is known locally as kangin oil in Myanmar, garjan oil in Bangladesh, gurjan oil in India and internationally as East Indian copaiba balsam. The essential oil consists of two sesquiterpenes, alpha and beta gurjunene. The resin remaining after distillation of the volatile oil contains gurjunic acid. The oleoresin is an important ingredient of lithographic ink and an anti-corrosive coating composition for iron. It also has various medicinal uses, such as the treatment of ulcers, ringworm and other cutaneous infections. It is a stimulant to mucous surfaces and a diuretic.

Further reading

Haque, S.M.S. and Osman, K.T. (1993) Performance of gurjan (*Dipterocarpus turbinatus*) and teak (*Tectona grandis*) in pure and mixed plantations at Kaptai, Bangladesh. *Indian Forester* 119, 738–743.

Mohammed Alamgir and Al-Amin, M. (2007) Organic carbon storage in trees within different geopositions of Chittagong (South) forest division. *Journal of Forestry Research* 18, 174–180.

Rai, S.N. (1983) Notes on nursery and regeneration techniques of some species occurring in southern tropical wet evergreen and semi-evergreen forests of Karnataka (India). *Indian Forester* 109, 127–136.

Troup, R.S. and Joshi, H.B. (1980) *Troup's The Silviculture of Indian Trees. Volume II. Dipterocarpaceae*. Controller of Publications, Delhi, 471 pp.

Dipteryx panamensis (Pittier) Record & Mell

Tonka bean tree

Importance

Dipteryx panamensis is one of the most impressive rainforest trees in Central America. It is an important leguminous timber species, producing hard, durable timber which is utilized in construction and as fence posts. *D. panamensis*, along with other species of the genus, is commonly known as the almendro or tonka bean tree, the latter name applying more appropriately to species such as *Dipteryx odorata* which are the main sources of the coumarin-containing tonka beans. The seeds of *D. panamensis* are roasted and consumed in Panama; germinated cotyledons are also eaten. The species is ecologically important, with many species using it for shelter, nesting and food, but has been overexploited; in Costa Rica in 2008 the government introduced restrictions on its felling and offered compensation for its conservation. It has great potential as a tropical hardwood by management in systems such as mixed plantation or enrichment planting.

Botanical features

D. panamensis is a large, emergent, slow-growing, deciduous tree attaining heights of 30–60 m and a dbh of 0.8–1.2 m, occasionally 2 m. The bole is straight and clear to about 20–25 m. The bark is smooth, whitish-brown in colour and has vertical lenticels. Mature trees are buttressed and have a semispherical crown and ascendant branches.

The leaves are alternate, pinnate, 30–50 cm long, with leaflets 8–20 jugate, oblong, obtuse or acute, almost glabrous. The rachis is noticeably winged, with an apical prolongation that makes the species easily identifiable. Leaves are shed after the dry season.

The inflorescence is a panicle, 30–50 cm long, racemose, terminal or axillary. Flowers are pink, bisexual, zygomorphic, with a short pubescent pedicel 0.5–0.8 cm long. The calyx is gamosepalous, tubular, pink, pubescent and glandular, pentamerous, with the upper two lobes coloured, enlarged and petaloid, as long as the petals. The corolla has 5 petals, with the two bottom petals joined to form the keel of the typical papilionate corolla. The adaxial member outside forms the vexillium or standard, pink to purple, reaching 1–1.4 cm. The wings are falcate, emarginate, dark violet and deciduous. The androecium has 10 monadelphous stamens. The anthers are ovate and versatile. The gynoecium is monomerous (one carpel), unilocular, stipitate, linear and glabrous. The style is recurved and narrow, and the stigma is truncated and glabrous. Placentation is submarginal and suspended. The ovule is amphitropous, bitegmic and crassinucellate.

The fruit is an ellipsoid, flattened pod, 6–8 cm long, 3–5 cm wide, 2–3 cm thick, single-seeded, 1-loculate, indehiscent, and with a membranaceous exocarp, a pulpy mesocarp and a woody endocarp. There are 29–80 (on average about 50) fruits/kg. The single seed is only 14% of the total weight of the fruit, and the average moisture content of fresh seeds is about 37%. In good years, trees can produce 20 or more fruits/m^2 of crown; individual trees have alternate good and poor fruiting years.

Trees begin to flower and set fruit at 11–12 years of age. The purple-pink flowers bloom at the end of branches in early summer, from May to July, after the onset of the rainy season and are cross-pollinated by bees. Flowers remain open only 1 day. Seed dispersal is aided by fruit bats, which after consuming the fruit pulp drop seeds far from the parent tree, and by agoutis, which subsequently collect the seeds, burying some of them in the ground.

Distribution

The native range of *D. panamensis* extends along the lowlands of the Atlantic coast of Central America, from southern Nicaragua to Panama and into Colombia. Commercial plantations and demonstration plots have been established on sites on the Atlantic lowlands of Costa Rica.

It grows in humid and very humid tropical forests, where annual temperatures vary between 24 and 30°C and annual rainfall is more than 3500 mm, and at elevations ranging from 20 to 1000 m above sea level.

Silviculture characteristics

Although *D. panamensis* is a non-nodulating leguminous tree, it can fix nitrogen, albeit at much lower rates than nodulating legumes. The seeds are recalcitrant. Germination is epigeal and the seedling is phanerocotylar. Its ecology and micro-habitat distribution indicate that *D. panamensis* possesses several characteristics typical of a shade-tolerant non-gap species, in contrast to previous descriptions of it as a shade-intolerant gap species. It grows best in clearings and well-illuminated forest edges, and also does well in grasslands and ranchlands.

Silviculture practice

Pods are usually collected from the ground, given that the tree's height discourages climbing for manual collection. The seeds may be soaked for 1 day to aid germination, and then planted either in boxes filled with washed sand or directly in nursery beds with shade and frequent watering. Germination is gradual and the radicle emerges after 15–17 days. Seedling survival rate is approximately 90% in the nursery. After 1 month, seedlings can reach 17.7 cm in height and 1.66 cm in diameter. After 6–8 weeks, seedlings are transplanted into plastic bag containers; initially seedlings require shade, which should be gradually reduced to full light conditions.

Nursery stock is transferred to the field 6–8 months after germination, using either potted seedlings or stump plants if the root collar thickness is over 1 cm. An initial spacing of 3 × 3 m (1111 trees/ha) has been utilized in Costa Rica.

A first thinning of 50% is carried out after 3–4 years when the crowns are touching, resulting in a density of 500–550 trees/ha. Other thinning rates may be suitable. In a rotation length of at least 30 years, the final recommended density for a dbh of 45–60 cm is approximately 150–200 trees/ha. Pruning is suggested on basal branches to obtain knot-free timber.

In Costa Rica, 7-year-old stands in the Atlantic lowlands reached 15.9 m in height and 16.7 cm in diameter at final densities of 538 trees/ha, while an 11-year-old commercial plantation initially planted at a 3 × 3 m spacing (740 trees/ha) reached 17 m in height and 15.3 cm in diameter. A mean annual increment of 10–13 m^3/ha after 7 years was achieved in experimental plots and some commercial plantations, giving an expected yield at 30 years of approximately 300 m^3/ha.

Diseases and pests

D. panamensis appears fairly resistant to pests, and no phytosanitary problems have been consistently reported. Occasionally *Colletotrichum* sp. (anthracnose), which produces tip die-back, has been reported. Deer eat the apical meristems and may cause bifurcation during resprouting in plantations.

Plantings should be checked regularly for pests and diseases, and for early detection of bifurcation or multiple sprouting.

Uses

D. panamensis is an ecologically very valuable species, offering habitats, shelter and food to almost 1000 different organisms, with more than 60 animal species alone depending on its fruits for sustenance. The endangered great green macaw (*Ara ambiguus*) is almost exclusively dependent on *D. panamensis* for nesting sites, food and drinking water, which collects in cavities in the tree. At least 13 species of bee visit the flowers. With its profuse and attractive flowers, *D. panamensis* is considered to have potential ornamental use.

The fresh sapwood is whitish, turning brownish-yellow when dry, while fresh heartwood is yellowish, turning brownish-red. The transition from heartwood to sapwood is very gradual and difficult to detect, making the wood very attractive. The wood is very hard and heavy, with a basic specific gravity (oven-dry weight/green volume) of 800–900 kg/m^3. It is considered one of the heaviest woods in the world and because it was so difficult to saw and work it was not utilized until the mid-1980s, when advances in sawing and woodworking technologies were introduced. The timber is very durable and is used in heavy construction and for railway sleepers, flooring, cogs and shafts, dock fenders, cattle yards, corrals, tool handles and turnery. The timber is resistant to attack by fungi and termites, difficult to work but fair to good for machining, including planing, turning, boring, moulding, mortising and routing. Sanding qualities are also good. Its high mechanical resistance makes it valuable for the manufacture of sporting goods.

Seeds can be roasted and consumed, and germinated cotyledons are also eaten. Raw seeds are used as candlenuts.

Further reading

Hanson, T.R., Brunsfeld, S.J., Finegan, B. and Waits, L.P. (2008) Pollen dispersal and genetic structure of the tropical tree *Dipteryx panamensis* in a fragmented Costa Rican landscape. *Molecular Ecology* 17, 2060–2073.

Leigh, E.G. Jr (1999) *Dipteryx panamensis* (Leguminosae, Papilionoideae). In: Leigh, E.G. Jr (ed.) *Tropical Forest Ecology: A View from Barro Colorado Island*. Oxford University Press, New York, New York, USA, pp. 15–16.

Zuchowski, W. (2007) *Dipteryx panamensis*. Tonka bean, almendro. In: Zuchowski, W. (ed.) *Tropical Plants of Costa Rica*. Cornell University Press, Ithaca, New York, USA, pp. 240–242.

Dryobalanops aromatica C.F. Gaertn.

Borneo camphorwood

Importance

Dryobalanops aromatica is a fairly fast-growing timber tree and among the dipterocarps it is perhaps the one with all the necessary features to make an excellent plantation species. These include excellent form; almost conical and compact crown, allowing many trees to be packed within an area; and the ability of its seedlings to survive in shade for many years, making it possible to raise future crops through natural regeneration. Also, unlike most dipterocarps, *D. aromatica* fruits annually and so planting material is easy to procure, making it possible to plan annual planting programmes. Despite the potential of the species, no major plantings have been undertaken on a commercial scale, although experimental and small-scale plantations have been established in Malaysia since the early 1900s. In the past, *D. aromatica* has been an important source of camphor, which was traded between South-east Asia and China as early as the 4th century.

Botanical features

D. aromatica is a very large tree, reaching heights of 60–67 m with girths of 2–3.4 m. The bole is straight, cylindrical and branchless for up to 40 m. The bark is smooth in young trees, but turns shaggy, with large, irregular, thin flakes in mature specimens. The slash smells of camphor, with the outer bark dark purple-brown, the inner bark red-brown and slightly resinous, and the sapwood dull yellow-brown. Buttresses are short and stout, sometimes up to 5 m high.

Branches very slender, glabrous when mature, marked with elevated decurrent leaf traces; stipules linear, about 1.2 cm long, falling so early that they are to be seen only on very young shoots. Leaves simple, petiolate, alternate, coriaceous, 4–6 × 2–4 cm, broadly ovate, base cuneate, acumen to 1.5 cm long, nervation indistinct, midrib depressed above, prominent beneath, petiole slender, 0.5 cm long.

Panicle to 7 cm long, terminal or axillary, singly or doubly branched. Flower bud 9 × 4 mm, fusiform. Calyx glabrous, lobes subequal, lanceolate, obtuse. Corolla white, petals glabrous, oblong, acute. Approximately 30 stamens, subequal, shorter than the style, filaments slender, tapering, united at base; anthers linear; appendage to connective short. Ovary small, ovoid, glabrous; style filiform, approximately twice the length of the ovary.

Fruit calyx base 6–8 mm deep, 8–15 mm diameter, cup shaped, constricted at the rim; lobes enlarged into 5 equal, broad, spathulate, about 9-nerved wings, each about 10 cm long. Nut 3 × 1.5 cm, ovoid, glabrous, constricted at the apex of the calyx cup, acute with short style remnant.

Distribution

D. aromatica is strictly a west Malesian species, found in Peninsular Malaysia, north-west Sumatra, East Sumatra, Musala, Lingga, Singkep and Kalimantan. Where it occurs, it is usually localized.

Silviculture characteristics

D. aromatica is a light-demanding species which can, however, tolerate some shade as a seedling. Seedlings and advanced saplings have the ability to coppice, a characteristic which is lost in older plants. Since *D. aromatica* fruits are available almost every year, planning annual planting programmes is relatively easy. The fruits mature after approximately 4 months, and each weighs about 10 g. If the fruits do not fall in a wet place, they dry up. The seeds are recalcitrant, with germination starting immediately upon fruit fall, although when the weather is wet the germination process can start while the fruit is still on the tree.

D. aromatica is found mainly on sandy soils where gap formation is more frequent and predictable than on clay soils (where *Dryobalanops lanceolata* predominates); here, its ability to allocate more above-ground biomass to stems gives it a competitive advantage in maximizing its height growth.

Silviculture practice

Natural stands of *D. aromatica* can easily be managed using the Malaysian Uniform System. Regeneration is usually very good, and if adult trees are harvested without damaging the regeneration, a dense, almost pure, second crop results. Stands are usually managed under rotations of 50 years. Selective felling is usually practised, since smaller residual trees are usually present in adequate numbers, and the second cut can be made after only 30 years. Selective felling also leads to the opening up of the canopy, favouring the dense young regeneration. Established seedlings can survive for many years in the shade, and gradually overtake the secondary growth. When light becomes available they respond with rapid height and diameter growth. Where there are gaps, enrichment planting may be practised with a high success rate because of the drought tolerance of the seedlings.

At a temperature of 16°C and moisture content of 30%, it is possible to store the fruits for a short period of about 2 weeks. Germination frequency is usually very high, with a success rate of over 80%, and the process is also rapid, taking only 2 weeks. Fruit collections should, therefore, be started as soon as the wings of the fruits start to turn brown. In the nursery, germination is usually high and simultaneous.

Wildings are also comparatively easy to obtain, another feature that makes *D. aromatica* attractive for plantations. *D. aromatica* seedlings, and even advanced saplings which can coppice, also make suitable material for planting up an area. However, a living branch on the stump is necessary for successful establishment. Single-node juvenile stem cuttings can be used for propagation, but with less than 50% rooting their use is currently limited. Tissue culture experiments on *D. aromatica* have proved successful and hold potential for the creation of uniform planting material.

In trials in degraded secondary forest in Sarawak, nutrient addition consistently increased seedling growth by up to 50%, while responses to mulching and planting hole size were variable and inconsistent. *D. aromatica*, like all dipterocarps, is associated with mycorrhizae and infection is necessary for optimum growth.

A seedling mortality rate of 20%/year has been recorded in the forest understorey, although this is lower in lighter parts of the forest. This early mortality rate may be high, but is greatly reduced once the seedlings reach the six-leaf stage, resulting in a rich seedling bank in natural stands and plantations.

For enrichment planting, seedlings from nurseries or wildings should be at least 30 cm tall. The recommended spacings for planting are 3 × 2 m or 4 × 3 m, and survival can be very good in plantation conditions. In plantations, height growth is rapid, and early on young plants form a long conical crown, with slender side branches. The conical form of the crown is kept until the tree reaches a height of 30 m. Branches are shed to give a smooth bole.

D. aromatica is one of the fastest-growing indigenous timber species in the regions in which it occurs. A unique feature of *D. aromatica* is that a comparatively high basal area can be retained without affecting girth increment, due mainly to the conical crown, resembling that of conifers. Such a form is not only suitable for plantations, but the early thinnings can also be utilized for roundwood poles.

D. aromatica has been used in single-species plantations for rehabilitation of degraded forest in Indonesia, and for the same purpose in checkerboard plantations in Sarawak.

Diseases and pests

A leaf disease caused by brown pinhead spot *Colletotrichum gloeosporioides* (*Glomerella cingulata*) has been reported from Malaysia.

Insects are the principal seed predators in the canopy, with vertebrates playing a minor role. Fruits can be heavily infested by dipterocarp weevils of the genera *Alcidodes* and *Nanophyes*. Other fruit pests include squirrels, mice, fruit bats and parakeets. The shoots of *D. aromatica* seedlings are frequently bored by the scolytid beetle *Hypothenemus cuneolus*. Malayan porcupines (*Hystrix brachyura*) have also been reported to damage seedlings. In nurseries, seedlings can be damaged by termites.

The felled wood is affected by an ambrosia beetle, and trees are infested by a minute lymexylonid beetle. Neither the beetle nor the borer survives seasoning of the wood. Quick extraction of logs from the felling areas may be important.

Uses

D. aromatica has potential as a species for land reclamation, revegetation, soil conservation and ornamental plantings, since it grows fast and planting material is readily available.

D. aromatica produces a moderately heavy hardwood timber (density 680 kg/m^3) with reddish-brown heartwood. The wood is suitable for use in heavy construction, such as posts, beams, joists and railway sleepers. It is peeled for making plywood, but the frequent presence of pinholes lowers its grade as a veneer. The timber seasons easily, but is somewhat prone to splitting, warping and surface checking during kiln drying. Green timber is easier to saw than dry timber. Planing, boring and turning are easy, giving a smooth surface, although nailing properties are poor.

D. aromatica is believed to be highly durable, but in the tropics it may be rapidly destroyed by termites when in contact with the ground. It is, however, resistant to fungi. The sapwood is liable to attack by powder-post beetles.

D. aromatica was a major source of Borneo camphor but because camphor is now synthesized cheaply from pinene, the importance of natural camphors has declined. *D. aromatica* camphor has very similar properties to true camphor (from *Cinnamomum camphora*) and is used in the same manner in medicines, perfumery and in the production of varnish, lacquer and plastics.

D. aromatica is also the source of an oleoresin called camphor oil or kapur oil. This oil is aromatic and volatile, and is used in medicine.

Further reading

Afzal Ata, M. and Zulkifli Tukiman, D. (1992) Production and utilization of sawlogs, roundpoles, fuelwood and posts from a 36-year-old kapur (*Dryobalanops aromatica* Gaerth F.) plantation stand: with particular reference to small-diameter logs. *Malaysian Forester* 55, 89–107.

Hamzah, M.Z., Arifin, A., Zaidey, A.K., Azirim, A.N., Zahari, I., Hazandy, A.H., Affendy, H., Wasli, M.E., Shamshuddin, J. and Nik Muhamad, M. (2009) Characterizing soil nutrient status and growth performance of planted dipterocap and non-dipterocarp species on degraded forest land in Peninsular Malaysia. *Journal of Applied Sciences* 9, 4215–4223.

Lim, L.S., Wickneswari, R., Lee, S.L. and Latiff, A. (2001) Genetic structure of natural populations of *Dryobalanops aromatica* Gaertn. f. (Dipterocarpaceae) in Peninsular Malaysia using microsatellite DNA markers. International Conference on In situ and Ex situ Conservation of Commercial Tropical Trees, Yogyakarta, Indonesia, 11–13 June 2001, pp. 1–12.

Dyera costulata (Miq.) Hook. f.

Jelutong

Importance

Dyera costulata is an emergent canopy tree in the lowland forests of South-east Asia.

The wood is a pale-coloured light hardwood timber of great workability but limited durability and strength. Jelutong latex (generally called simply 'Jelutong') was once valued as a component of chewing gum; this latex was harvested from *D. costulata* as well as from *Dyera lowii* ('Swamp jelutong').

In the first half of the 20th century there was considerable investment in establishing plantations for latex production in South-east Asia, but *Hevea brasiliensis* latex and synthetic gums later replaced most of this use.

It is still a protected species in parts of Malaysia and Thailand and is grown commercially for timber. It performs very well in cultivation.

Botanical features

D. costulata is a large deciduous tree up to 60 m tall (occasionally even 80 m, according to some accounts), with diameters up to 2–3 m. It has a straight and clear bole up to 15–30 m, without buttresses. Its bark is smooth, dark grey with small square scales; the inner bark is thick, with a milky latex. Leaves grow in whorls, are glabrous, 12–25 cm long and 6–11 cm wide. They are usually shed once a year. Flowers are small and white, glabrous, and occur in 5–18 cm long panicles.

The pod-like fruits grow at the end of branches, in pairs. Fruits are 30–40 cm long and 2–3 cm wide. They contain 20–40 seeds, which are released upon maturity and are dispersed by wind.

Distribution

D. costulata occurs naturally in Peninsular Thailand and Malaysia, Singapore, Sumatra and Borneo islands, where it is widespread but occurs at low density (1 tree/4.5 ha). Its main habitat is primary lowland rainforest at altitudes up to 300 m, occasionally up to 800 m. Typically it grows in heavy rainfall areas on soil with good drainage (e.g. on ridges or alluvial soil, and on lateritic clay soil) and not in swamp forest.

Silviculture characteristics

D. costulata is an excellent plantation tree. It is easy to establish as it grows well under high light conditions and requires little maintenance. It may be suitable for planting under large canopy gaps in the secondary forest. Once a young tree is well established in full light, it tends to spread its crown. Under heavy shade it can persist but growth becomes stunted.

Growth in the early stage is not very rapid, 6-month-old seedlings being only 7–10 cm high, and developing at an uneven rate under identical conditions. Heights of 4 m at 4 years have been recorded, but the average of a group of transplants set out was only slightly over 1 m. It seems that seedlings require a protracted period of 3 or 4 years to become established and do not make appreciable growth until the root system is well developed.

In the past, latex harvesting used to damage some trees severely and increased mortality. The tree can generally survive high levels of damage, and seedlings grow quickly.

Silviculture practice

Seed collection is challenging and requires timely collection of the fruit pods from the high crowns of mother trees. Pods must be mature but not yet split to release the wind-borne seeds. Fruits do not all mature at the same time but pods that are light brown, more flat than round, and with wrinkles, can be collected together with the fully mature fruits. Long-handled pruners are used, combined with climbing. Shaking does not work as the fruits will remain on the tree until after seed dispersal and shaking can cause the fruits to open.

The fruits are dried in the sun for about 1 week, sheltered or covered against wind blow, until they release the seeds. Contradictory experiences with seed storage have been reported: in Malaysia, seed viability remained good after 8 months' storage in a well-ventilated and cool room (20–24°C) with 60% humidity. Others, however, found it difficult to store the seed for more than 3 months. It seems that the seeds do not tolerate drying or low temperatures.

The seeds do not show dormancy, but soaking in water can enhance germination. Mature seeds normally germinate well (80–90%). The seeds are sown in a mixture of peat or soil and sand (1:1) in shady conditions. It is recommended to sow them in a seedbed with a light cover of sand. The seedlings can be transplanted after the first pair of leaves has emerged.

Propagation by cuttings is also possible. The cuttings should be 0.5–1.5 cm in diameter and 30–40 cm in length. The cutting bed should be covered with transparent material (e.g. plastic) under shady conditions. By this method, 30% of rooted cuttings are ready for planting after 3 months.

The latex used to be tapped at intervals of 2–3 days during 4 months or so each season. Coagulation was obtained either by allowing the latex to stand, or by evaporation, with or without the addition of a coagulant. The most effective method was to add diluted phosphoric acid to fresh latex, which was then heated for a few minutes.

Diseases and pests

D. costulata plantations support a characteristic fauna dominated by four Coleopteran species of *Acicnemia*, *Batocera*, *Platypus* and *Schizotrachelus*. The most prevalent of these insects is *Batocera rubus*, which forms galleries in the bark and is not necessarily confined to the tapping panel. *Platypus spretus* attacked trees through tapping wounds, especially in the days of intensive latex harvesting; this facilitated termite attack that ultimately damaged or even killed the tree. Such infestations were most prevalent in dense shade and were likely to be related to reduced vigour of the tree brought about by age, suppression, disease and injury. Sound trees of almost any age apparently resisted infestation well, but tapping, whether good or bad, often brought them within the range of susceptibility.

Various measures for the control of *Batocera* have been investigated. Exposure of trees to the sun appears effective, as is the manual collection of eggs. Consideration has also been given to the development of more resistant varieties of trees.

Uses

The wood of *D. costulata* is a yellow, lightweight hardwood with no distinction between heartwood and sapwood. The wood density has been reported as (220–)390–435(–560) kg/m^3 at 15% moisture content. The texture is fine and straight, and the surface is glossy. It is not durable or suitable for construction purposes, but due to its workability it is liked for pattern carving, picture frames, drawing boards, blackboards, toys, packing cases and wooden shoe heels. It is also a favourite wood for pencil manufacture.

Until recently the principal product of *Dyera* was its latex. Before *Hevea* plantations in South-east Asia came into commercial production and ousted all other sources of rubber, jelutong was exported for the manufacture of inferior rubber articles, in which elasticity was not a prime consideration. Factories were constructed to extract and purify the materials needed to make a higher quality rubber, but low yield (20% of dry material compared with 70% or more from *Hevea*) made this too costly, and interest in jelutong as a source of rubber soon waned. In the 1920s, jelutong regained importance as a basic ingredient in the manufacture of chewing gum, to supplement other tree-derived gums. Since then, this has been practically its exclusive use. Apart from that, jelutong was also used in paints and as a priming for concrete.

Reports from Sarawak in the early 20th century state that the roots were used locally as a substitute for cork and the follicles as torches, since they burned with a steady flame and kept mosquitoes at bay by the pungent smoke they emit.

One source mentions that the burned fruits yield a resin effective as an emollient to heal sores.

Further reading

Bauer, P.J. (2007) Plantation forestry at Bukit Cheeding Estate 1990–2005. *Planter* 83, 621–626.

Lemmens, R.H.M.J., Soerianegara, I. and Wong, W.C. (eds) (1995) *Plant Resources of South-East Asia No. 5(2): Timber Trees: Minor Commercial Timbers*. Blackhuys Printers, Leiden, The Netherlands, 655 pp.

Williams, L. (1963) Laticiferous plants of economic importance. IV. Jelutong (*Dyera* spp.). *Economic Botany* 17, 110–126.

E

Endospermum diadenum (Miq.) Airy Shaw

Sesendok

Importance

Endospermum diadenum is a fast-growing tree with a light-coloured timber, and has potential to be further exploited in plantations. In Malaysia, it is famous for its use in the match industry. This species has a good market in the East, especially in Japan and Taiwan. The wood is used as a general utility timber and for industrial purposes.

Botanical features

E. diadenum is a medium to large tree, reaching up to 40 m in height with a dbh of up to 150 cm. The bole is stout and columnar, with thick equal to steep fluted buttresses. The bark is grey-fawn, smooth, hooped and becoming regose to scaly in patches. Inner bark is thick and coarsely granular, cream with orange flecks. The wood is fawn coloured. The species has large, spreading limbs with a diffused dome-shaped crown.

The leaves appear clustered at the end of the twigs. Leaf blades are non-peltate (sometimes peltate in saplings) rounded to strongly heart-shaped at the base, cordate, 7–25 × 4–22 cm, palmately 3–9 veined and the mid-rib with 3–5 pairs of lateral veins. Leaf stalks are 13 cm long with two prominently raised glands at the junction of the stalk to the leaf-blade.

The flower spikes are erect, 15 cm in length, protruding above the leaves. The male flowers have short side axes and 9–11 stamens. Female flowers with a 2–3 celled ovary and a 1.5 mm-wide stigma. Fruits are rounded or occasionally bi-lobed with a width of 13 cm. They are finely velvety blue-green in colour with walls containing white-coloured sap.

Generally, flowering occurs twice a year between February and April and August–October. *E. diadenum* is pollinated by small insects such as bees and flies.

Distribution

E. diadenum is found in Brunei Darussalam, Indonesia (Kalimantan and Sumatra), Malaysia (including Peninsular Malaysia, Sabah and Sarawak), Singapore and Thailand. It occurs in lowland forest to lower montane forest up to 1000 m in latitude. This species is found in primary forest and especially in secondary forest. *E. diadenum* is an opportunist species and is often found in former shifting cultivation sites. It has been planted in Peninsular Malaysia and trialled as part of enrichment plantings in eastern Sumatra.

Silviculture characteristics

E. diadenum is a pioneer species with fast growth, and has the ability to compete with other species. It regenerates rapidly, is able to self-prune and is suitable for coppicing.

Silviculture practice

E. diadenum bears fruit at an early age, normally once 2–3 years old. The fruit is collected approximately 2–3 months after flowering, once it has turned slightly yellow. In a heavy fruiting season, a 13-year-old tree can produce 2–3 kg of fruits, or 9000–15,000 fresh seeds. Once clean and dry, the seeds may be cold stored for up to 6 months; 1 kg of dried seeds contains about 40,000 seeds.

E. diadenum has a low germination rate, and a high incidence of insect predation of the seeds has been reported. However, the seeds collected from Gunong Tampin Forest Reserve, Negeri Sembilan, had a germination success of up to 80%. *E. diadenum* seedlings are usually potted 3 months after sowing, and are transplanted in the field approximately 9–12 months after potting.

Wildings are a useful immediate source of planting material for this species. Approximately 2 months after the fruiting season, the ground beneath the mother tree is cleared to enable germination of fallen seeds. The wildings are collected 1–2 months later, and seedlings with 2–4 leaves are suitable for transplanting. During collection, the wildings must be wrapped with moist tissue paper to prevent them from wilting, and may be kept for a maximum of 3 days before replanting in the rooting bed. Wildings are left in the rooting bed for at least 2 weeks before potting; once potted they are kept in the nursery for another 6–9 months before planting out.

Larger wildings can be collected using the stumping method. The stem and root are cut at approximately 5–7 cm above and below the collar region, to produce stumps of 10–14 cm in length. All lateral roots are also severed. Stumps can be kept for 1–2 days under moist conditions before planting in polybags or directly in the ground.

Being a light-demanding species, *E. diadenum* can be open-planted. The method of establishing *E. diadenum* in plantation is similar to that of other plantation species. The area should be clear-felled and burnt, which is usually carried out during the dry period. Once the area is prepared, transplanting can be conducted during the rainy season at the end of the year.

In Peninsular Malaysia, it has been reported that 2-year-old seedlings planted in open sites at the Sungai Buloh Forest Reserve have a high survival rate (>86%) and good growth (mean height 5 m). Close planting at 3 × 3 m is recommended in order to ensure fast growth, reduction of weeds and enhancement of self-pruning. The first thinning of the crop is recommended after 5 years' growth, and a second thinning at 13–15 years.

E. diadenum is a fast-growing species, with an estimated rotation age of 18–25 years. Mean annual increments (MAI) have been reported as 2.5 cm in dbh in a 13-year-old tree, and 4.6 cm in 3-year-old trees. On a degraded site at the Forestry Research Institute of Malaysia (FRIM), the largest tree with a dbh of 38.4 cm exhibited an MAI of 1.92 cm after 20 years' growth. MAI declined to 1.18 cm once the tree reached 36 years of age.

Diseases and pests

Urapteroides astheniata larvae may attack *E. diadenum*. Generally the damage is localized and limited to mature leaves only, because this insect occurs only during the dry spell before the flushing of new leaves. Repeated attacks on young plantations can reduce the diameter increment of the trees by more than 50%. Fruits of this tree may be eaten by insects.

Uses

E. diadenum is used in reforestation and as a shade tree, or may be used in enrichment planting. The species is used in agroforestry, such as interplanting with rubber in an alley-cropping system. After 25 years, both the rubber and *E. diadenum* are harvested together. *E. diadenum* may also be interplanted with short-term crops, but only for the 3 years before the canopy closes above.

E. diadenum is classified as a light hardwood timber with an average density of 300–650 kg/m^3 at 12% moisture content. The wood is bright yellow, and it is difficult to differentiate between the sapwood and the heartwood. It is not very durable. The timber is suitable for a wide range of general utility products such matches, drawing boards, blackboards, toys, trays and other small articles. It is reputed to be one of the favourite timbers for making clogs, pattern making, furniture parts, plywood chests, low-grade coffins and disposable chopsticks. The thinning material from *E. diadenum* can be used as chip which is sold to local fibreboard industries.

E. diadenum has medicinal properties; the bark is used to treat dropsy and the roots help to heal injuries.

Further reading

Aminah, H. (2003) Vegetative propagation of *Endospermum malaccense* by leafy stem cuttings: effects of indole butyric acid (IBA) concentrations and propagation systems (mist and non-mist). *Journal of Tropical Forest Science* 15, 249–258.

Darus Ahmad, Hashim Md. Noor, Ab Rasip Ab Ghani, and Lok, E.H. (1991) *Khaya ivorensis* and *Endospermum malaccense* as potential species for future reforestation programmes. In: Appanah, S., Ng, F.S.P. and Ismail, R. (eds) Proceedings of a conference on Malaysian forestry and forest products research. FRIM, Kepong, Malaysia, 3–4 October 1990, pp. 60–65.

Muuss, U. (1996) Enrichment plantings in the tropical rainforest of Sumatra - a silvicultural challenge. [Anreicherungspflanzungen im tropischen Feuchtwald Sumatras - eine waldbauliche Herausforderung.] *Forstarchiv* 67, 65–70.

Zuhaidi, Y.A., Zakaria, I., Rosdi, K. and Krishnapillay, B. (2002) Species for timber plantations. *A manual for forest plantation establishment in Malaysia*. Forest Research Institute Malaysia (FRIM), Kuala Lumpur, Malaysia, pp. 13–24.

Endospermum peltatum Merr.

Sesendok

Importance

Endospermum peltatum is a valuable, fast-growing, tropical timber tree species with a native distribution in South-east Asia. It is primarily planted in pure or mixed plantations, and is also used as a nurse crop and for reforestation.

The species is an important source of sesendok timber, which is generally used for manufacturing small and speciality wooden products such matches and pencils, and also for veneer and pulp. *E. peltatum* also has some traditional medicinal uses.

Botanical features

E. peltatum is a moderately large tree up to 30–45 m in height, with a diameter of 80 cm. The bole is generally straight, branchless, cylindrical and 8–12 m in length. It has a flat open crown with light green foliage and very slight buttresses. The bark is golden-green, smooth and has a disagreeable odour. The inner bark is yellow-brown and hard.

The leaves are simple, grouped at the ends of twigs, 11–25 cm long and 10–18 cm wide, elliptical to ovate–oblong in shape, rounded to deeply cordate at the base, and palmately veined, with 7–9 veins, and with secondary nerves and cross bars. The petioles are 10–15 cm in length, glandular at the distal end, coriaceous, soft and pubescent, especially on the upper surface.

The inflorescence is axillary, nearly as long as the petiole, densely pubescent. The male flowers are whitish, densely clustered along the short paniculate branches, with calyx subtruncate; they have approximately 10 stamens. The female flowers are borne in spicate racemes; the ovary is 4-celled with connate styles. The fruit is ovoid, glabrous, fleshy, nearly 1 cm thick, stipulate, crowded by the stigma and subtended by the calyx, with 1-4 seeds.

The flowering season is in May and the fruits ripen in August, although this varies with climate.

Distribution

E. peltatum is a tropical species with a distribution ranging from the Andaman Islands, through peninsular Thailand to Borneo, northern Sulawesi and the Philippines. It is commonly found in logged-over areas and in secondary forests at low and medium altitudes (100–300 m, but up to 1000 m), and prefers regions with evenly distributed rainfall rather than distinct wet and dry seasons.

Plantations have been established throughout its range, especially in the Philippines (where it is known commonly as gubas), as well as Peninsular Malaysia.

Silviculture characteristics

E. peltatum is a prolific seeder, and natural dispersal is mediated by wind or birds. It regenerates well in open areas created after logging or forest fires, but tolerates shade, and will self-prune and coppice.

Natural regeneration of *E. peltatum* has been observed 2–3 months after logging, yet only 36% survival was noted after 6 months. This is indicative of the relative intolerance of *E. peltatum* in a residual dipterocarp forest.

Silviculture practice

Plantations can be established by direct seeding or by using nursery stock. Mature fruits are collected from healthy dominant mother trees, and are then macerated using hands or feet. The fruit is left for 1–2 days in order to enhance separation of the seed. The pulp is then removed mechanically or by washing in water. Soaking the seed overnight until the seedcoat becomes soft reduces germination time. The seeds, which are orthodox, are sown on a seed bed by broadcasting, covered with about 2 cm of soil, mulched and well watered but not waterlogged. Germination begins about 24 days after sowing. As soon as the cotyledons have fully emerged from the seeds, approximately 2–6 weeks after germination, the seedlings are potted on.

Once they are 0.5 m in height, the seedlings are ready for planting out. This is usually done at the beginning of the rainy season or when the soil is moist to a depth of 18 cm. Hardening or root pruning may be conducted prior to planting. Vegetation is removed from the area to be planted by strip or spot clearing, and seedlings are planted at distances of 2 × 2 m, 3 × 3 m or 4 × 4 m, with thinning being carried out later.

Fertilizer application may improve growth; for example, adding 0.3 g NPK fertilizer per pot has been shown to increase growth rates by 129% in comparison with controls. After 1 year's growth, 14:14:14 NPK fertilizer can be applied at a rate of 125 kg/ha in order to enhance tree survival and development in the plantation.

The most preferred silvicultural system is uniform clear-cutting with a series of compartments. Where logging is by diameter limit, this is usually about 30 cm for matchwood and pulpwood. The rotation for clear-cutting is generally 15–20 years.

The mean annual increment of *E. peltatum* in the Philippines, where it is widely planted, is reported as 2.1–2.2 m^3/ha, in terms of stocked volume. On average sites, the yields of plantations at 10, 15 and 20 years old were 67, 140 and 236 m^3/ha, respectively.

Diseases and pests

At the seedling stage *E. peltatum* is susceptible to certain fungal diseases, such as *Lasiodiplodia theobromae* and *Athelia rolfsii*. Control measures include sterilization of the growing medium using heat or chemicals, avoiding both overcrowding and too much watering, gradual exposure of seedlings to sunlight, and the removal and burning of any infected plants. No common diseases have been observed in older trees.

A blue stain of the wood is caused by attacks of *Diplodia* and *Ceratocystis* species.

Uses

E. peltatum is planted commercially for timber production and as a nurse tree for species which require shade and protection at the early stages of development; these include species of important timber trees such as *Agathis* and *Dipterocarpus*, as well as rattans such as *Calamus* spp. *E. peltatum* is suitable for use in mixtures with agricultural crops, and has also been used in reforestation programmes.

The timber has a density of 335–460 kg/m^3 at 15% moisture content. The wood is generally used in the manufacture of matches, matchboxes, lollipop sticks, toothpicks, pencils and clogs, as well as for veneer and pulp. Shavings are used in the manufacture of wood panels known as placarol in the Philippines; these are used mainly for making doors.

The seeds are edible. Leaves are used to make a poultice to relieve diarrhoea, while a decoction made from the bark is considered an effective depurative and alterative.

Further reading

Generalao, M.L. and Torrenueva, E.T. (1972) *Silvical Characteristics of Gubas (*Endospermum peltatum *Merr.)*. Department of Agriculture and Natural Resources, Manila, Philippines.

Ramos, V.J.A. (1977) Yield and growth production for gubas in natural stands. MS thesis, University of the Philippines, Diliman, Philippines.

Teow Chuan Ti and Tangau, W.M. (1991) *Endospermum peltatum*. In: Teow Chuan Ti and Tangau, W.M. (eds) *Cultivated and Potential Forest Plantation Tree Species, with Special Reference to Sabah*. Institute for Development Studies (Sabah), Kota Kinabalu, Malaysia, pp. 41–43.

Entandrophragma angolense (Welw.) C. DC.

Tiama mahogany

Importance

Entandrophragma angolense is probably the most important of the highly valued 'African mahoganies'. Its beautiful dark timber is much prized for many uses. Exploitation has resulted in reduced production throughout its range. Significant genetic erosion has been reported in some countries and its status on the International Union for Conservation of Nature (IUCN) Red list is 'Vulnerable'.

Botanical features

E. angolense is a large dioecious rainforest tree up to 50 or even 60 m tall bearing a dome-shaped deciduous crown. Its bole is usually straight and cylindrical up to 30(40) m, with diameter up to 200 cm and with buttresses up to 6 m high, often extending into surface roots. The bark is pale greyish to orange-brown, smooth but later scaly with irregular scales up to 20 cm diameter leaving concave scars of pinkish-red inner bark, with whitish streaks.

The leaves are large and compound with (8)14–20(22)

oblong-shaped leaflets of variable size (7–28 × 3–8.5 cm). Flowers are axillary or terminal panicles, up to 40 cm long, individual flowers 5-merous, greenish-white and 4–5 mm long. Fruits are pendulous, cylindrical capsules with many wind-dispersed seeds.

Distribution

E. angolense is widespread in tropical Africa and occurs from eastern Guinea to southern Sudan, Uganda and western Kenya, and south to the Democratic Republic of the Congo and Angola.

In West Africa, *E. angolense* is most common in moist semi-deciduous forest (annual rainfall range 1600–1800 mm). It may also grow in evergreen forest, but its abundance strongly declines in regions with an annual rainfall of more than 2300 mm. In East Africa, it occurs in lowland and mid-altitude rainforest, but sometimes also in gallery forest and thickets, up to 1800 m altitude. It requires well-drained soils with good water-holding capacity.

Silviculture characteristics

E. angolense is a slow-growing canopy emergent, though under optimal conditions seedlings may grow as fast as 1–2 m per year over the first 2 years. *Entandrophragma* is related to the American 'true' mahoganies (*Swietenia*). It is a non-pioneer light-demander favouring shade for initial establishment but needing improved light conditions in order to grow well. In natural forest, saplings are most common in gaps. It regenerates well after logging but not after burning. The seed does not appear to disperse over great distances and regeneration is rare away from mother trees. Natural regeneration is often scarce in natural forest, but logging operations that create gaps may promote regeneration. Tests with seedlings indicate that growth is good under conditions simulating the light conditions in small forest gaps, and that growth was still fair under the light conditions of medium-sized gaps. Seedlings need some shade for the first few years after germination and perform poorly under full-light conditions. These results suggest that the best way to favour recruitment of *Entandrophragma* seedlings in natural forest would be through a series of small- to intermediate-sized gaps.

Silviculture practice

Seeds can be stored for some time in sealed containers in a cool place. There are about 2000–3000 seeds/kg. These are very susceptible to insect damage, but adding ash can prevent this. Fresh seeds may have a high germination rate, more than 75%, but this decreases to about 25% after 3 weeks' storage in open air. Germination of fresh seeds starts 1 week after sowing, but germination of seeds stored for 10–15 days may take 30–45 days. Soaking of the seeds for one night is reported to accelerate germination.

Seeds should be planted in seed beds with the wing pointing upward and three-quarters of the seed (without wing) buried. Some overhead shade is required for young seedlings. It is recommended to plant out under a light cover of a young secondary forest or in enrichment lines in forest paths (after 3–4 years some clearing around the young trees is necessary). Only on exceptionally fertile soils or on dead termite nests may it be possible to plant in full sun, but mortality after some years may be as high as 50%. Because of their exceptional value, enrichment planting of *Entandrophragma* trees may be economically viable, both as part of forest management and agroforestry systems. Seedlings of *E. angolense* can be combined in mixed plantings in forest gaps, avoiding periods of drought.

In many regions *E. angolense* is overexploited. The rather low growth rates under natural conditions, the long time needed to reach maturity in terms of fruit production and poor dispersal of the seed can all lead to poor and inadequate regeneration. On the other hand, *E. angolense* is capable of fair regeneration in many logged-over forests if seed sources are present. More research is needed on appropriate management measures such as sowing of seeds in suitable localities in logged-over forest. Rotation cycles needed to ensure sustainable exploitation in natural forest are longer than current practices allow.

Shoot borers have discouraged the planting of all the African mahogany species as monoculture plantations.

Diseases and pests

Entandrophragma species are susceptible to the mahogany shoot borer *Hypsipyla robusta*, but have lower susceptibility than the related *Khaya* species.

Uses

The tree is planted as roadside tree, and occasionally as a shade tree in banana, coffee and tea plantations.

The wood, usually traded as 'gedu nohor' or 'tiama', is highly valued for exterior and interior joinery, furniture, cabinet work, veneer and plywood, and is used for flooring, interior trim, panelling, stairs, shipbuilding, vehicle bodies and coffins. It is also suitable for light construction, musical instruments, toys, novelties, boxes, crates, carvings and turnery. Wood that is not suitable as timber is used as firewood and for charcoal production. Wood density is 440–585 kg/m^3.

The bark is used in traditional medicine: a decoction is drunk to treat fever, including malaria, while an anodyne can be applied externally against stomach ache, peptic ulcers, earache and kidney, rheumatic or arthritic pains. Some pharmacological studies have confirmed moderate *in vitro* anti-malarial activities against *P. falciparum* parasites. Further studies are necessary to evaluate the pharmaceutical potentials.

Further reading

Hawthorne, W.D. (1995) Ecological profiles of Ghanaian forest trees. *Tropical Forestry Papers* 29, Oxford Forestry Institute, Oxford, UK.

Katende, A.B., Birnie, A. and Tengnäs, B. (1995) *Useful Trees and Shrubs for Uganda: Identification, Propagation and Management for Agricultural and Pastoral Communities*. Technical Handbook 10. Regional Soil Conservation Unit, Nairobi, 710 pp.

Styles, B.T. and White, F. (1991) Meliaceae. In: Polhill, R.M. (ed.) *Flora of Tropical East Africa*. A.A. Balkema, Rotterdam, The Netherlands, 68 pp.

Ward, S.E., Boshier, D. and Grogan, J. (eds) (2008) Special issue: Sustainable management of high-value timber species of the Meliaceae. *Forest Ecology and Management* 255(2), 265–365.

Entandrophragma candollei Harms

Kosipo

Importance

Entandrophragma candollei is a very large timber tree of the African semi-deciduous forest zone. It is widely exploited and classified as vulnerable on the IUCN Red List of Threatened Species. It was a valued timber species for export from Côte d'Ivoire in the 1970s and later from Gabon, Cameroon, the Central African Republic and Ghana.

Botanical features

E. candollei is a large deciduous, dioecious tree of up to 50–70 m high. The bole is usually straight, cylindrical and up to 200–400 cm in diameter. It is usually branchless up to 30 m height and often has blunt buttresses up to 4 m high that sometimes extend into large, rounded surface roots. The bark is pale greyish-brown to dark grey, irregularly scaly with round scales leaving shallow, pinkish pits. The slash is pinkish light brown, granular with orange inclusions. The crown is dome-shaped, widely spreading and dense. Young twigs are brownish with short hairs; older twigs show scars of fallen leaves. Leaves are alternate, clustered near the ends of twigs, paripinnately compound with 10–20 leaflets. Stipules are absent. The petiole is 7–20 cm long, flattened and often slightly winged at the base. Petiolules are 1–7 mm long. Leaflets are (nearly) opposite, oblong–elliptical to oblong–obovate and 5–18 × 2.5–6.5 cm in size. The leaflets are cuneate to rounded at the base, and obtuse to shortly acuminate at the apex. They are thick-papery to thinly leathery, almost glabrous, and pinnately veined with 15–25 pairs of puberulous lateral veins. Flowers are borne in a panicle up to 30 cm long. They are unisexual, regular, 5-merous; the pedicel is 1–3 mm long. The calyx is cup-shaped, lobed to about the middle and 2–2.5 mm long, with short hairs on the outside. The flower petals are free, greenish-white and hairy, oblong–oblanceolate, with stamens fused into a cup-shaped tube 5 mm long, and with 10 slips 3–4 mm long, each with an anther at the apex. The disk is cushion-shaped, the ovary superior, conical, 5-celled, with a thick style that is slightly longer than the stamens. The stigma is disk-shaped, with 5 lobes. The fruit is a pendulous, cylindrical capsule 17–23 × 3–5 cm in size, with a 2.5 cm long stipe, purplish- to dark brown and dehiscing from the top with 5 thin valves. The many yellowish-brown seeds are attached to the upper part of the central column, 5–12 cm long and winged (12 cm apical wing).

Distribution

E. candollei is widely distributed throughout the African semi-deciduous forest zone. It is most common in moist semi-deciduous forest, although also found in evergreen rainforest. The trees are never particularly abundant and they can be locally uncommon. They do relatively well on acidic soils compared to other *Entandrophragma* species from the same region. *E. candollei* occurs in damp or less damp sites, in areas with an annual rainfall of about 1800 mm.

Silviculture characteristics

An average annual diameter increment of 5.1 mm has been recorded under natural conditions. In plantations in Guinea, trees reached a mean height of 2.7 m after 6 years. Harvesting trees of less than 50 cm bole diameter seriously reduces fruit production. Fruits mature about 5 months after flowering. Seed production can be somewhat erratic. In experiments with seedlings *E. candollei* performed well on infertile soils. Under natural forest, saplings are most common in gaps, and following logging operations. Long-rotation cycles are probably needed to ensure sustainable exploitation in natural forest. In southern Cameroon, the average density is one tree of more than 60 cm bole diameter/15–30 ha, and the average wood volume is up to 0.7 m^3/ha. On average, a tree 1 m in diameter yields 10.3 m^3 of commercial timber and a tree 2 m in diameter yields 41.8 m^3.

Silviculture practice

E. candollei is propagated by seed, and the seedlings have epigeal germination. Fresh seeds germinate 1 week after sowing with a germination rate of 60–75%. The growth of seedlings is slow; they reach 13 cm in height after 5 months and 20–30 cm after 1 year. The seedlings are very sensitive to sun-scald and have to stay under permanent shade in the nursery; they will also need slight shade for at least 2–3 years after being planted out. Later, more light is needed for good growth.

Diseases and pests

E. candollei is sensitive to powder-post beetle, pinhole borer and marine borer attacks. It is moderately resistant to termites.

Uses

Wood from *E. candollei* is used for construction, joinery, boatbuilding, furniture, cabinet work, veneer and plywood. It is suitable for flooring, vehicle bodies, toys, boxes, crates and turnery. The heartwood is dull brown to dark purplish-brown, slightly darkening upon exposure. It is distinctly demarcated from the sapwood, which is up to 8 cm wide and greyish-white to pale brown,. The grain is straight to slightly interlocked, texture coarse and even. Wood from *E. candollei* saws fairly easily. It usually contains silica (up to 0.5%) and has moderate blunting effects on cutting edges. It can be finished to a good surface, but the use of a filler may be required. It is not liable to splitting on nailing and screwing, and has good holding properties. Steam-bending properties are poor. The wood is medium-weight (density of 570–810 kg/m^3 at 12% moisture content). Proper drying is needed to avoid warping.

Bark sap is applied externally as an analgesic in traditional medicine. The sap of the root bark is used to treat snake bites and venomous insect stings. It can also be used as an anti-feedant against stored-product insect pests. The root bark is also locally used in traditional religious practices.

Further reading

Bolza, E. and Keating, W.G. (1972) *African Timbers: The Properties, Uses and Characteristics of 700 Species*. Division of Building Research, CSIRO, Melbourne, Australia, 710 pp.

CTFT (Centre Technique Forestier Tropical) (1978) Kosipo. *Bois et Forêts des Tropiques* 181, 29–40.

Hall, J.S., Ashton, P.M.S. and Berlyn, G.P. (2003) Seedling performance of four sympatric *Entandrophragma* species (Meliaceae) under simulated fertility and moisture regimes of a Central African rain forest. *Journal of Tropical Ecology* 19, 55–66.

Hall, J.S., McKenna, J.J., Ashton, P.M.S. and Gregoire, T.G. (2004) Habitat characterizations underestimate the role of edaphic factors controlling the distribution of *Entandrophragma*. *Ecology* 85, 2171–2183.

Entandrophragma cylindricum (Sprague) Sprague

Sapele

Importance

Entandrophragma cylindricum, the West African cedar or sapele, a large deciduous tree producing timber with very scented slash, is a common timber tree which is heavily exploited in its range in tropical Africa. As a result, the IUCN Red List of Threatened Species classes it as vulnerable, and there are protected subpopulations and felling limits in various countries. Sapele is one of the most important export timbers of tropical Africa in terms of both quantities produced and wood quality.

Small-scale timber plantations of *E. cylindricum* have been established in Côte d'Ivoire and Cameroon, while line planting in the forest is practised in Cameroon.

Botanical features

E. cylindricum is a dominant tree in the high forest with a crown that is rounded and not spreading. Trees range in height from 46 to 94 m, the average at maturity being 60 m, with an average girth at breast height of 5.4 m (~1.7 m dbh), but reaching 6.4 m (~2.0 m dbh). Trees can be over 500 years old.

The bole is tall, cylindrical and straight, and branchless for up to 40 m. Buttresses are generally short, but can reach 3 m.

The bark is usually a mustard-grey or dull silvery colour, smooth and with a diamond pattern on young trees less than 2.1 m girth. Later it becomes scaly and shaggy brown. It has many scattered lenticels. The slash is pink, discolouring to saffron-brown, moderately thick and scented. There are ripple marks on the white sapwood and heartwood darkens from red to reddish-brown.

The leaves are about 25 cm long on mature trees, alternate, exstipulate and paripinnate, with about 6–7 pairs of sub-opposite, glabrous leaflets. The leaflet is oblong, about 10 cm long and 3.8 cm broad, entire, broadly apiculate, more or less rounded at the base but with equal sides. The midrib and nerves are raised below. There are about 8 pairs of fine and widely spaced nerves. The petiolule is very short. The rachis is pubescent towards the base.

Flowers are small, yellow-green and in lax panicles. The fruit is an elongated woody capsule, 7–10 cm long and pendulous. The valves are about 1.9 cm broad, open from the apex and bend backwards. The valve is smooth and dark on the outside but inside it is light brown and shiny, with darker flecks. The columella is sessile and broadest at the base. On its 5 faces are the seeds, up to 6 on each. The seeds are brown and 6.2–8.7 cm long, with a terminal wing on which are the impressions of the other seeds. These impressions are due to the seeds being very tightly packed in the small capsule.

Trees are deciduous in September and October, and crowns are complete with their new leaves by the end of November. Flowering, though irregular for individual trees, is principally from October to January. Pollen flow extends over large distances. Fruiting is from May to August. The capsules open on the trees and the seed is dispersed by wind but does not travel far. Seed production is erratic. Monkeys eat the mature fruits.

Distribution

The natural distribution of *E. cylindricum* ranges from Sierra Leone in the west, east to the Democratic Republic of the Congo and Uganda and as far south as Angola.

The species is most commonly found scattered in semi-deciduous forests.

Silviculture characteristics

E. cylindricum is a non-pioneer and a light-demander. It requires shade as a seedling for about 2 years, but strong overhead light for fast growth when older. Growth rates are amongst the slowest in the genus. Trees start flowering when 35–45 years old, and fruit production starts when trees have reached bole diameters over 50 cm.

E. cylindricum can grow in drier habitats than other species of *Entandrophragma*, including abandoned fields, but it does not respond well to burning. It prefers regions with an annual rainfall of about 1750 mm, a dry period of 2–4 months and a mean annual temperature of 24–26°C.

Silviculture practice

E. cylindricum is generally raised from mature, brown, hinged seeds. Mature seed is classified as orthodox, but has notoriously short viability. When sown soon after collection, a germination of about 90% can be obtained. Germination takes place within 14–26 days; the seedlings require shade.

Under natural conditions, seeds germinate abundantly, but mortality of seedlings is high, less than 1% reaching 10 cm stem diameter. Seedling growth is slow, 20–40 cm/yr, due to slow root development. Height growth increment is about 17.7–40.6 cm in the first year. Vegetative propagation can be carried out using cuttings about 90–110 cm long. Natural regeneration is often scarce in natural forest, but logging operations creating gaps may promote regeneration, larger gaps appearing more favourable.

Height growth in silviculturally treated forests may be

about 1 m after 4 years, and it may take 81 years to reach 2.4 m girth. Mean annual diameter increment for trees in the Central African Republic has been established at 3.9 mm, but the variability is large. For trees planted in lines in forest in Cameroon, the average annual height growth over 40 years was 30–50 cm and average annual diameter growth 4–8 mm. Trees planted in the open in Côte d'Ivoire reached an average height of 5.4 m and an average stem diameter of 10 cm after 7 years, with a survival rate of 74%.

In forests in the Central African Republic there are seven *E. cylindricum* stems >10 cm diameter/ha and a wood volume of 25 m^3/ha. In southern Cameroon, the average density is up to one tree of >60 cm bole diameter/ha, and the average wood volume is up to 11.5 m^3/ha. Depending on the country, minimum felling diameters range from 60 to 110 cm. Minimum felling diameters should be well above 50 cm to allow natural regeneration.

In many regions, *E. cylindricum* is still not exploited on a sustainable basis. It has been suggested that intensive silviculture, possibly involving the use of shifting cultivation in a taungya-like system, is needed to achieve sustainable management.

Diseases and pests

Hypsipyla robusta attacks *E. cylindricum*, feeding on the soft tissue inside the terminal stem. In heavy infestations as many as 20–24 wounds may occur on a stem, resulting in heavy exudation. Attack is usually most severe on trees growing in full sunlight. Larvae can bore through seeds, and mature larvae move from fruit to fruit to feed. Trees are attacked in their second or third year. Generally, shoot borer attack weakens the tree and predisposes it to other pests and pathogens. With repeated infestations, mortality can occur. Economic loss is usually due to reduced height growth and poorer tree form. Silvicultural control methods are used to increase shade and isolate infested trees.

Gyroptera robertsi attacks old trees. *Catopyla dysorphnaea* lays eggs on fruit; larvae at first feed within individual seeds and then move from fruit to fruit.

Mussidia nigrivenella breeds in green and dried fruits. Larvae feed on the seeds and fruit wall.

Uses

E. cylindricum is used in taungya and in enrichment planting of disturbed areas. It is also used as a roadside tree and as an ornamental shade tree. Caterpillars of the butterfly *Imbrasia oyemensis* are commonly found on the leaves; they are edible and in East Africa much sought after for human consumption.

E. cylindricum produces a heavy and scented timber with persistent fragrance. The wood is whitish or pale yellow, and the heartwood is pinkish at first, darkening to a typical mahogany colour (red-brown) with age. Logs are sometimes beautifully figured, mottled or with a regular and marked stripe or sheen, which shows well on quarter-cut timber. Sapele is known for a wide variety of figured grain patterns, such as ribbon, pomelle, quilted, mottled, wavy, beeswing and fiddleback.

The wood is tough and resembles Honduran mahogany (*Swietenia macrophylla*) but with less tendency to split if straight-sawn. It is, however, also liable to split and distort and is therefore best kiln-dried gently and should be quarter-sawn to prevent warping and splitting. It is of close and heavy texture, and fairly resistant to decay. It is sometimes damaged by pinhole borers and is not resistant to termite attack. It is medium-heavy, with a density of 650 kg/m^3. It is also of moderate durability and medium shrinkage. While being worked, sapele has a distinct, cedar-like scent. It nails and polishes well and, as it glues easily, is suitable for sliced and rotary veneers for use in cabinet work, furniture, piano cases, panelling and interior decoration. The wood itself is also suitable for good-quality furniture, joinery, cabinet work, doors, fitments, frames, flooring, boat construction, plywood, blockboard, carvings, turnery and musical instruments.

The bark is used in traditional medicine, with decoctions or macerations used to treat a range of ailments. Bark pulp is also applied externally to wounds. Bark extracts have insecticidal properties and have been used to protect stored maize grain against pests.

Further reading

Anon (1974) Sapelli. *Bois et Forêts des Tropiques* 154, 27–40 [in French].

Phongphaew, P. (2003) Sapelli/sapele (*Entandrophragma cylindricum*). In: Phongphaew, P. (ed.) *The Commercial Woods of Africa*. Linden Publishing Inc., Fresno, California, USA, pp. 170–171.

Siepel, A., Poorter, L. and Hawthorne, W.D. (2004) *Entandrophragma cylindricum* (Sprague) Sprague, Meliaceae. In: Poorter, L., Bongers, F., Kouamé, F.N. and Hawthorne, W.D. (eds) *Biodiversity of West African Forests: An Ecological Atlas of Woody Plant Species*. CABI Publishing, Wallingford, UK, 412 pp.

Styles, B.T. and White, F. (1991) 13. Entandrophragma. In: *Flora of Tropical East Africa. Meliaceae*. A.A. Balkema, Rotterdam, The Netherlands, pp. 49–55.

Entandrophragma utile (Dawe & Sprague) Sprague

Sipo

Importance

Entandrophragma utile is a large and impressive tree, the largest and most light-demanding species of its genus. It is an important timber species in tropical Africa, producing a mahogany-like timber known as sipo, utile or ogipogo mahogany. Commercial interest in this valuable timber has resulted in over-extraction of large individuals from the forest throughout its range, in some regions even leading to its near extinction. It is classed in the IUCN Red List of Threatened Species as vulnerable.

Low growth rates under natural conditions, the long time needed to reach maturity in terms of fruit production and poor dispersal ability of the seed seem to be serious drawbacks for the sustainable management of populations, making very long rotation cycles inevitable. Intensive silviculture, possibly involving the use of shifting cultivation in a taungya-like system, is needed to achieve sustainable management. Although the potential for plantation development is hampered by susceptibility to insect, rodent and antelope damage, timber plantations of *Entandrophragma cylindricum* have been established in Côte d'Ivoire.

Botanical features

E. utile is a large, deciduous, dioecious tree of the high forest, with an average height of 57 m at maturity (range 40–65 m) and an average girth at breast height at maturity of 6.5 m (~2.1 m dbh), the range being from 5.7 to 7.6 m (~1.8–2.4 m dbh). The crown, which consists of large spreading branches, is not dense as the leaves are grouped at the ends of the branches and leaflets are shed.

The bole is straight and cylindrical, bifurcating into two huge branches at the top, with characteristic light silvery-grey bark and long regular vertical scales 2.5 cm or more in width. Blunt buttresses reach up to 4.5 m or more but do not spread from the tree. The slash is brown in the outer layer and bright pink-red below. It is slightly scented. Vague ripple marks may sometimes be seen in the white sapwood. The heart wood is reddish-brown and slightly scented when fresh.

The leaves are alternate, paripinnate, existipulate and about 38 cm long (more on young plants), with about 8–12 pairs of subopposite leaflets. The leaflet is simple, about 8.8 cm long and 3.8 cm broad, oblong–ovate, gradually acuminate, more or less rounded at the base and with unequal sides. The petiole is very short. The midrib and nerves are raised below and pubescent. The lamina is dull green on the upper side. There are white villose hairs on the rachis and it is flattened on the upper side at the base. The terminal buds are covered with brown tomentose hairs.

Flowers are white, small and scented. The fruit is an elongated, woody, pendulous capsule about 20.3 cm long, the largest among *Entandrophragma* species. It is broadest at the apex, and on the outside is dark brown with small warts. The 5 thick valves open at the apex and remain attached to the columella at the base. On the inside, they are shiny brown. The columella is sessile and bears about 6 winged seeds on each of its 5 faces. The wing is about 12.7 cm long and about 2.5 cm broad, brown and papery. The seed is not rectangular and is borne towards the straighter side of the wing, near the oblique base. The 1000-seed weight is about 500 g.

E. utile trees are deciduous from November to February. Flowering takes place from about December–March. Young capsules may be seen in April, but ripen from December to March. Thus, flowering and fruiting take place during the same period but 1 year apart in age. The seeds are wind distributed but do not travel far from the mother trees.

Distribution

E. utile ranges from 10°S to 10°N in Africa, from Guinea and Sierra Leone in the west to Uganda in the east, and south to Angola. There is widespread occurrence in Liberia, Côte d'Ivoire and Ghana. In other regions, such as Nigeria, Cameroon and Central African Republic, it is either naturally rare or has become rare due to extensive over-exploitation.

It is predominantly a tropical lowland species found at altitudes of up to 500 m.

Silviculture characteristics

E. utile is a non-pioneer and a light-demander that will survive shade when a seedling, but demands a great deal of light by the time it reaches sapling stage. Seedlings are physiologically well suited to the deep shade of the forest floor, and the healthiest seedlings usually are found under at least some shade.

Tree growth is best where mean annual rainfall exceeds 1400 mm and mean annual temperature is about 24–26°C. *E. utile* can tolerate drought and dry periods of 2–4 months. It also tolerates a range of soil types, although it grows best in well-drained deep soils. It is generally undemanding with regard to soil quality, but does not tolerate waterlogged or swampy sites.

Silviculture practice

When seed is sown fresh, the germination period is about 13–19 days and germination rate is about 75%. However, the rate drops to about 60% when seed has been stored for about 3 months. Seedlings should be raised under screens in the nursery to prevent the mite and insect attacks which are common in full sunlight; irradiance with 25% full sunlight is optimum for seedling growth, full sunlight tending to lead to 100% mortality. Seedlings can attain a height of 38 cm in 6 months and about 73 cm in 1 year.

Vegetative propagation is by stumps or striplings, although stump planting is not practicable. Propagation by cuttings, grafting or tissue culture is also possible.

Stand establishment is by natural regeneration, direct sowing or planting stock. Damage by rodents and antelopes tends to obscure any influences of silvicultural management on regeneration. Mortality rates are fairly high for the first 100 days, with a slight decline thereafter.

E. utile grows relatively slowly. Diameter growth rates of 1.0–1.5 cm per annum can be expected. Trees can reach a height of 1.5 m at age 7–8 and 14.4 m at age 15 years.

Selective logging in a rotation of 30–40 years has been practised. Minimum felling diameters vary between countries, ranging from 60 cm to 110 cm. On average a tree 1 m in diameter yields 10–13 m^3 of commercial timber, and a tree 1.5 m in diameter 23–30 m^3. Freshly harvested logs float in water and can thus be transported by river.

Diseases and pests

Under conditions of strong sunlight, seedlings become infested with mites and insects. *Xylosandrus compactus* attacks young shoots and *Catopyla dysorphnaea* lays eggs on fruit, the larvae at first feeding within individual seeds, then moving from fruit to fruit. *Mussidia nigrivenella* attacks green and dried fruits; larvae feed on seeds and the fruit wall. Attacks of *Hypsipyla robusta* on seedlings are very frequent.

Browsing by antelopes and rodents frequently causes damage to young seedlings.

Uses

E. utile is considered ideal for enrichment planting.

It produces a dark timber which is unscented or has a slightly cedar-like scent, and has a moderately fine texture. The heartwood is rich reddish- to purplish-brown and distinctly demarcated from the pinkish-white to light brown sapwood. The sapwood is liable to insect attack, but the heartwood is moderately resistant to termites and other borers. The wood is moderately resistant to decay. It can be worked fairly easily with all tools, and finishes and polishes well, though requiring filler. It takes nails and stains easily. It is of medium density (675 kg/m^3 at 12% moisture, with a range of 400–740 kg/m^3) and moderate natural durability, with medium shrinkage. It has easy sawing and satisfactory fixing properties. It air dries rather slowly, and may be liable to splitting and distortion. However, there seems to be considerable variation in drying characteristics.

Sipo wood is used for high-quality exterior/interior joinery, doors, window frames, furniture, cabinet work, flooring, sliced veneer, rotary veneer, plywood and boatbuilding. It is also suitable for construction, vehicle bodies, boxes, crates, carvings and turnery. The bole is traditionally used for dug-out canoes. Wood that cannot be used as timber is used as firewood and for charcoal production.

The bark of *E. utile* is used in traditional medicine, including as an anti-malarial. Bark extracts have shown fungicidal activity against *Pyricularia oryzae*, a major fungal pathogen of rice.

Seeds contain limonoids and steroids, but no tannins or saponins. Their fatty acid composition is characterized by the presence of about 30% *cis*-vaccenic acid, a rare isomer of oleic acid that can be used in the industrial production of nylon-11. The main fatty acid is linolenic acid (46%).

Further reading

Mujuni, D.B. (2008) *Entandrophragma utile* (Dawe & Sprague) Sprague. In: Soerjanegara, I. and Lemmens, R.H.M.J. (eds) *Timbers 1 (Plant Resources of Tropical Africa, volume 7)*. PROTA Foundation, Wageningen, The Netherlands, pp. 245–249.

Phongphaew, P. (2003) Sipo/utile (*Entandrophragma utile*). In: *The Commercial Woods of Africa*. Linden Publishing Inc., Fresno, California, USA, pp. 172–173.

Siepel, A., Poorter, L. and Hawthorne, W.D. (2004) *Entandrophragma utile* (Dawe & Sprague) Sprague, Meliaceae. In: Poorter, L., Bongers, F., Kouamé, F.N. and Hawthorne, W.D. (eds) *Biodiversity of West African Forests: An Ecological Atlas of Woody Plant Species*. CABI Publishing, Wallingford, UK, 413 pp.

Enterolobium contortisiliquum **(Vell.) Morong**

Tamboril

Importance

Enterolobium contortisiliquum is a medium- to large-sized tree occurring in eastern-central South America. It displays rapid growth and is managed for timber both in plantations and natural forest; it is also utilized for soil improvement and as an ornamental. Its timber is light, easy to work and is resistant to rot even in conditions of high humidity.

Botanical features

E. contortisiliquum reaches 15–40 m in height with a dbh of 60–300 cm. It is shade-intolerant, and deciduous in the dry season. The crown is wide, flat and open. The bole is straight and cylindrical, approximately 8–15 m in height; buttressing is not prominent. The external bark is grey, rough to scaly, and has shallow longitudinal striations at the base of the trunk. On the upper stem and branches, the bark is smooth with abundant and conspicuous lenticels. Bark thickness ranges from 5 to 20 mm. The internal bark is fibrous, orange to brown, and is 15–20 mm thick.

Leaves are bipinnately compound, alternate to spiral in arrangement, and 8–14 cm long. The bipinnate leaves are made up of pinnae (5–7 pairs per leaf); each pinna has 12–20 pairs of leaflets, 1–2 cm in length. They are asymmetrical at the base and have a sharp-pointed apex.

Inflorescences are pedunculate heads (10–24 flowers) displayed in short axillary or apical racemes. The small hermaphroditic flowers have an adnate calyx. Flower petals are united. Up to 70 stamens, with small anthers. The fruit is a woody indehiscent legume, initially green and later turning brownish-black. It is wide and curved in the form of an ear or kidney. Fruit size ranges from 5–8 cm long by 5–7 cm wide. Each fruit contains 5–15 oval, brownish-tan seeds about the size of a kidney bean.

Distribution

E. contortisiliquum occurs infrequently in the humid forests of Venezuela, Amazonian Brazil and northern Bolivia, and is more common to the south in the dry forests of south-eastern Bolivia, south-eastern Brazil, Paraguay, northern Argentina and Uruguay. This species is found in both forest (especially 'matta atlantica') and 'cerrado', as well as shrubby 'restinga'. In Argentina it occurs in humid subtropical forest.

E. contortisiliquum has also been planted in Argentina, Bolivia and Brazil, and introduced in the Dadra and Uttar Pradesh regions of India for panelling and boat construction, as well as in Guandong province and the island of Hainan in China. It has also been introduced in Australia.

Silviculture characteristics

E. contortisiliquum is shade-intolerant; it regenerates and grows best in high-light environments. It is typically uncommon in natural tropical forests and is restricted to sites with relatively high soil moisture retention. Therefore, timber production for this species is generally low under natural forest management. Abundance is higher in secondary forests and the species responds favourably to disturbance. *E. contortisiliquum* can regenerate from seed, but also has the ability to resprout from roots and cut stumps. Due to a proclivity for lateral branching, close spacing (2 × 2 m) may be required for good stem form.

Silviculture practice

Mature seeds should be collected rapidly to avoid predation by bruchid beetles. The hard seeds have long dormancy periods and store easily without need of refrigeration. Some type of scarification is often desirable to accelerate germination. Soaking seeds in hot water yields rapid germination in 5–20 days, with germination of 80–90%. Placing seeds in damp sand for 4 days at 4–5°C has been found to break seed dormancy. Clipping the seedcoat to allow imbibition of water also accelerates germination. Numerous studies have shown the efficacy of sulphuric acid and other treatments in breaking the dormancy of fresh or stored seeds and improving germination frequency. However, seedcoat abrasion by sanding has proved a more practical method for use under forest nursery conditions. Seed germination rates increase with temperature, but germination capacity does not differ between 18.2 and 38.8°C. In addition to establishment by seed, the species can be propagated vegetatively by root or branch cuttings; these are effective for establishing plantations for fuelwood and posts.

Upon germination, seedlings grow rapidly from their large seed reserves. Usually 60 days is sufficient for nursery seedlings to reach 30 cm, when they are ready for planting out, although full light (or only up to 20% shade) must be provided for best seedling development and survival. Soil mixed with 80% composted urban waste has proved effective in enhancing seedling growth. *E. contortisiliquum* is prone to prolific branching and therefore benefits from close spacing and pruning. Weed control is important to obtain adequate growth in plantations. Although the species is a nitrogen fixer, seedlings have been found to respond positively to nitrogen fertilization up to rates of 100 kg/ha; lagoon sludge at various concentrations has been used successfully as a fertilizer for young plants.

The species is amenable to both natural forest management and plantation systems. In Argentina, *E. contortisiliquum* has been established in mixtures with other species at 3 × 3 m spacing, and has been found suitable for intercropping with *Eucalyptus grandis* in reforestation and agroforestry systems in Brazil. *E. contortisiliquum* has been recommended for enrichment planting in regenerating and over-exploited forests in Argentina and Brazil.

The species has rapid growth rates on favourable sites and can grow 5 cm in diameter and up to 2.5 m in height per year. Planted seedlings can reach a mean height of 6.5 m with a dbh of 11 cm in 7 years. However, in this particular case, variation in growth was high and survival was only 80%.

Diseases and pests

A range of seedborne and storage fungi has been identified infecting seeds. Seedling damping off due to infection by *Sclerotium rolfsii* is a problem, whereas the fungus *Colletotrichum dematium* f. *truncatum* causes damage to young seedlings in nurseries. Nematode problems have been reported in nurseries in China.

Seed predation by bruchids such as *Merobruchus bicoloripes* and *Caryedes* sp., and by larvae of the cerambycid beetle *Lophopoeum timbouvae*, can reduce seed viability by up to 75%. Another cerambycid, *Lochmaeocles sladeni*, does minor damage to branches by cutting sections for oviposition, whereas larvae of *Cryptophlebia carpophagoides* cause damage to fruits. *E. contortisiliquum* is a preferred species of the leaf-cutter ant *Atta sexdens rubropilosa*.

Uses

In Bolivia, *E. contortisiliquum* is used for shade for cacao and in mixed plantings on small farms, and in Argentina it is used in agroforestry systems with yerba maté (*Ilex paraguariensis*). It is also recommended for wood-pasture systems. The species has proved useful for reclamation and restoration planting. *E. contortisiliquum* is tolerant of copper-contaminated soil. Because of its rapid growth and large, spreading crown, the species is also an excellent shade tree for street plantings and gardens. Its rapid growth and nitrogen-fixing ability makes it suitable for revegetation, soil stabilization, land reclamation and soil improvement.

Uses of the wood are varied and include boxes, carpentry, furniture, boats, beehives and floors. Because it is easy to work, the wood is also used for turnery, utensils and toys. It is also suitable for the manufacture of particleboard. The yellowish-white sapwood and the maroon heartwood are resistant to rot. The wood is light, with a specific gravity of 300–400 kg/m^3 and is flexible and porous. It is considered to be very similar in its properties to teak (*Tectona grandis*).

Fruits of *E. contortisiliquum* are eaten by large mammals, and cattle will eat the seeds; however, the saponins they contain can cause abortion of calves, while ingestion of pods is also associated with hepatogenous photosensitization. Although the foliage is apparently somewhat toxic, cattle do not avoid it, and *E. contortisiliquum* is actually grown as a fodder legume in pastures due to its tolerance of browsing and good fodder quality and feed value.

The abundance of flowers in October and November provides a good pollen source for bees. Saponins in the seeds form a soap, and seed essential oils have exhibited anti-microbial activity.

Further reading

Araújo, A.P. de and Paiva Sobrinho, S. de (2011) Germination and production of seedlings of tamboril (*Enterolobium contortisiliquum* (Vell.) Morong) on different substrates. *Revista Árvore* 35, 581–588.

Carvalho, P.E.R. (1994) Brazilian forest tree species: silvicultural recommendations, potential and use of wood. Especies florestais brasileiras: recomendacaes silviculturais, potencialidades e uso da madeira, CNPFlorestas-EMBRAPA, Brasilia, Brazil, 639 pp.

Shahat, A.A., El-Barouty, G., Hassan, R.A., Hammouda, F.M., Abdel-Rahman, F.H. and Saleh, M.A. (2008) Chemical composition and antimicrobial activities of the essential oil from the seeds of *Enterolobium contortisiliquum* (Leguminosae). *Journal of Environmental Science and Health. Part B, Pesticides, Food Contaminants, and Agricultural Wastes* 43, 519–525.

Enterolobium cyclocarpum (Jacq.) Griseb.

Ear pod tree

Importance

Enterolobium cyclocarpum (Plate 14) is one of the largest and most conspicuous trees in the dry forests of Mexico and Central America, reaching up to 40 m in height and 2–3 m dbh, with a huge, spreading crown. It is well known for its distinctive, thickened, contorted indehiscent pods which resemble a human ear in form. Although introduced throughout the tropics as a shade and ornamental tree, it is not widely planted today. *E. cyclocarpum* can tolerate a wide range of soil types but is intolerant of frosts or waterlogging. Propagation is mainly by seed, containerized seedlings or stump plants.

It is as a pasture tree in silvopastoral systems that *E. cyclocarpum* is particularly important. Large-crowned trees scattered in extensive cattle-rearing pastures are a common and distinctive feature of the landscape throughout the seasonally dry Pacific zone of Central America and the Caribbean, providing shade, a high-quality dry season fodder from the pods, and a valuable timber from older trees. However, silvopastoral use has so far been important only within the native range, where adult trees occurred naturally and were retained when forest was cleared.

The wood is relatively light and of medium quality, used for fuelwood, posts and sawn or hewn timber for building construction. The dust from sawmilling *E. cyclocarpum* wood is a pungent irritant to mucous membranes and can produce an allergic reaction in workers. *E. cyclocarpum* is the national tree of Costa Rica.

Botanical features

A giant, thornless canopy tree to 30–40 m tall and 2–3 m stem diameter with a short, clear bole, often supported by small buttresses, heavy branching and a huge spreading rounded crown. The bark is smooth to granular, sometimes lightly fissured, pale grey-brown with many elongated lenticels arranged in longitudinal lines. The inner bark is light brown, bitter and astringent and exudes a brownish gum.

The leaves are 15–40 cm long, opposite, bipinnate and finely divided into 5–15 pairs of pinnae and 20–30 pairs of small leaflets per pinna, with a small sessile dome-shaped gland on the petiole. The leaflets are linear–oblong, mostly 8–15 mm long, acute, inequilateral, the mid-vein distinctly eccentric, glabrous or sparsely hairy below.

The flowers are arranged in densely packed heads 1–1.5 cm in diameter when the flowers are fully open, with a small densely hairy greenish calyx 2.5 mm long, a glabrous pale whitish-green corolla 5 mm long and creamy-white stamen filaments which are basally united into a staminal tube (Plate 48).

The ear-shaped pod is quite unlike that of any other leguminous plant and is perhaps the most immediately notable feature of *Enterolobium*, providing the basis for many of its widely used common names (Plate 56). The pods are 3–4 cm wide, curved to form an almost complete circle 8–10 cm in diameter, and dark, lustrous brown. The seeds are extremely hard, dark brown, compressed ovoid–ellipsoid, 13–20 × 11 × 8 mm, with a pale pleurogram and are arranged in two rows with 10–15 and occasionally up to 22 seeds per pod.

In its native range, flowering occurs in March–April at the time when the new leaves are appearing, 4–8 weeks before the beginning of the rainy season on branches that have been bare for 1–3 months of the mid-dry season. At about the same time, the full-sized fruits from the previous year's flowering begin to mature and turn brown. After flowering, the minute green pods remain less than 2 cm in length for 9 months, then expand rapidly to a green fruit 10–14 cm in diameter. The mature, dark brown indehiscent fruits fall in March–April and in pastures they are eaten avidly by cattle and horses.

Distribution

E. cyclocarpum occurs from latitude 23° N in central Mexico, south and east through the seasonally dry tropical zone of southern Mexico and Central America to about 7° N in northern South America, where it occurs as far south as the extreme north of Brazil.

Silviculture characteristics

It is a strongly light-demanding species and must receive adequate sunlight as a dominant or co-dominant from the sapling stage onwards. These traits make it particularly suitable for extensive cattle ranching pasture systems, and it is in seasonally dry silvopastoral systems with scattered widely spaced *Enterolobium* trees (often mixed with *Samanea saman*) that it is most widely used.

It is known to be nitrogen fixing and to resprout vigorously after cutting, coppicing or ring-barking.

Silviculture practice

Propagation is mainly by seed, containerized seedlings or stump plants. Seed production can start at 15–25 years of age but may be delayed in plantation conditions. Seed collection is straightforward by collecting the indehiscent pods from the ground during the late dry season before it rains, provided that livestock have been excluded from the area before pods fall. Trees in Central America produce seed crops in most years. Given that the pods are indehiscent, hard and not easily broken up, seed extraction is more problematic. The pods can be pounded until well broken then winnowed and sieved to separate the seed from pod fragments and residue.

The seeds of *E. cyclocarpum* are large, heavy and extremely hard with a thick seedcoat which comprises 63% of the dry weight of the seed. Hard, dormant seeds do not germinate unless the seedcoat is scarified by wear, passage through the gut of an animal or mechanical scarification. This means that seeds are orthodox, can be readily stored for long periods under conventional seed-storage conditions (less than 10% moisture content and 4°C in sealed containers) and require pretreatment. If the seedcoat is pierced, high germination rates are usually obtained. Mechanical, hot water and sulphuric acid pretreatment methods have all been used successfully. For small seedlots, manual nicking is optimal.

Seed should be sown at a depth of 1–2 cm with the micropyle pointing downwards, as the emerging root is not strongly geotropic and may come up out of the soil if the seed is planted upside down. Seed can be sown directly into a seedbed, although sowing into biodegradable or polyethylene containers gives very good results in terms of germination and seedling development. Seeds germinate rapidly within 3–4 days. The early growth of the seedling is very rapid and vigorous, probably fuelled by the very large seeds. This advantage over smaller-seeded species can continue for several months after outplanting, but thereafter its growth rate, though still vigorous, is no longer exceptional relative to other fast-growing species. Seedlings require little shade in the nursery and are ready for outplanting to the field within 6 months.

Although trees are known to resprout vigorously after cutting, pollarding or ring-barking, there are no reports of successful vegetative propagation by grafting, although stump planting is successful. A protocol for the micropropagation of *E. cyclocarpum* resulting in the production of rooted shoots has also been developed.

Enterolobium is naturally a spreading tree with heavy branching and a wide crown and is not readily adapted to plantation conditions. If used in plantations, wide spacings of 3 × 3 m or 4 × 4 m must be used. Bole length and form are improved in plantation conditions and can be further improved by pruning.

Seedlings are susceptible to fire, desiccation, grazing and competition from grasses, so careful protection and weeding during the early years are essential.

Diseases and pests

No serious or particularly damaging pathogens or pests of *E. cyclocarpum* have so far been identified. However, trees are damaged by a number of minor pathogens and pests in particular areas.

In Puerto Rico, a variety of root rots, butt rots and burls develop after mechanical damage. The fungus *Fusarium oxysporum* can cause copious yellow exudation from bark fissures on infected trunks or branches, later attracting various wood-boring insects which can subsequently lead to branch break or loss.

Entire leaf crops are commonly eaten off by the larvae of the moth *Coenipita bibitrix* in Guanacaste in Costa Rica. The mite *Tetranychus urticae* has been observed on *E. cyclocarpum* in Venezuela.

In forests in its native range, the gall-forming fly *Hemiasphondylia enterolobii* destroys flowers. The full-sized green pods are heavily preyed on by *Amazona* parrots which remove the green seed contents, and spiny pocket mice, *Liomys salvini*, may consume up to 93% of dispersed seeds.

Uses

The most important use of *E. cyclocarpum* is for cattle, horse or goat shade and fodder in pasture lands, where it makes an excellent shade tree because of its broad crown. *Enterolobium* is also cultivated as a shade tree over coffee. However, its large size and spreading crown make shade management over plantation crops difficult and other species are generally preferred and more widely used. It is also planted as an ornamental for its majestic appearance.

The wood is light, of medium density, with a specific gravity of 0.35–0.60 g/cm^3. The heartwood of *E. cyclocarpum* is deep walnut-brown, often with various shadings and sometimes a reddish tinge and a high lustre, and its appearance has been compared with walnut. The sapwood is dull white and merges into the darker heartwood. The heartwood is fairly durable, resistant to dry-wood termites, and may be used for boatbuilding because of its durability in water. In contrast, the sapwood is highly susceptible to insect attack.

Some *E. cyclocarpum* timber used to reach the USA; however, commercial use has subsequently declined. In Central America the wood is used for many construction purposes: furniture, boats (including dug-out canoes and modern boatbuilding), water troughs, general carpentry, panelling and for fuel. In Nicaragua, its timber was identified as having potential for local craft work, with the seeds being used to make necklaces, curtains and earrings. The dust from sawmilling *Enterolobium* wood is a pungent irritant to mucous membranes, can produce an allergic reaction in workers and can kill fish if dumped into rivers.

In silvopastoral systems, *E. cyclocarpum* trees produce three to ten times more fruit than other tree species. The fruit is of high quality and is used for cattle feeding in seasons with limited fodder availability. Although not as nutritious as the seeds, pods and whole fruits, the leaves can also be used as fodder.

The seeds have food uses after being toasted and ground, and immature pods and seeds are cooked as a vegetable. Heat treatment of seeds is important as it reduces the levels of cyanogenic glycosides. *E. cyclocarpum* is also a source of starches and gums for use in food manufacture.

Other minor uses include using the pulp from the pods as a soap substitute; using bark extracts as traditional medicines against colds and bronchitis; and obtaining tannin from the fruits and bark. The flowers are a rich source of nectar for honey production, and leaf extracts have proved to be highly effective attractants for control of the pest *Cosmopolites sordidus* in banana plantations.

Further reading

Hughes, C.E. and Stewart, J.L. (1990) *Enterolobium cyclocarpum*: the ear pod tree for pasture, fodder and wood. *NFT Highlights* 90–05, 2 pp.

Martinez Pacheco, M.M., Rio, R.E. del, Flores Garcia, A., Martinez Muñoz, R.E., Ron Echeverria, O.A. and Raya Gonzalez, D. (2012) *Enterolobium cyclocarpum* (Jacq.) Griseb.: the biotechnological profile of a tropical tree. *Boletín Latinoamericano y del Caribe de Plantas Medicinales y Aromáticas* 11, 385–399.

Thirunavoukkarasu, M., Siddartha Parida, Shiba Prasad Rath and Asmita Behera (2006) Micropropagation of *Enterolobium cyclocarpum* (Jacq.) Griseb. *Journal of Sustainable Forestry* 23, 1–12.

Eucalyptus camaldulensis Dehnh.

Red gum

Importance

Eucalyptus camaldulensis (Plate 15) is a medium to large tree. It has the widest geographical range of any eucalypt and is extensively planted outside Australia. Its tolerance of extreme drought and high temperature, combined with rapid growth when water is available, deep penetration of roots, tolerance of periodic waterlogging and soil salinity, some tolerance of frost, good coppicing ability and usefulness of the wood, are the keys to its success.

Botanical features

In Australia, *E. camaldulensis* commonly grows up to 20 m tall and rarely exceeds 50 m, while stem dbh can reach 1–2 m or more. In open woodlands it usually has a short, thick bole which supports a large, spreading crown. In plantations, it can have a clear bole of up to 20 m with an erect, lightly branched crown. The bark is smooth white, grey, yellow-green, grey-green, or pinkish-grey, shedding in strips or irregular flakes.

Juvenile leaves are petiolate, ovate to broadly lanceolate, up to 26 × 8 cm, green, grey-green, or blue-green, slightly discolorous. Adult leaves lanceolate to narrowly lanceolate, acuminate, lamina 8–30 cm long × 0.7–2 cm wide, green or grey-green, concolorous; petioles terete or channelled, 1.2–1.5 cm long.

Inflorescence axillary, 7–11(–13)-flowered; flowers white, peduncles slender, terete or quadrangular, 6–15 mm long; pedicels slender, 5–12 mm long. Buds pedicellate; hypanthium hemispherical, 2–3 mm long × 3–6 mm wide, operculum globular-rostrate (typical) (Plate 49). Fruits are hemispherical or ovoid, 5–8 mm long and wide; disc broad, ascending; 3-5 exserted valves.

Eucalypts do not develop resting buds and grow whenever conditions are favourable. Early growth may exceed 3 m per year for well-adapted provenances on favourable sites.

Distribution

E. camaldulensis is the most widely distributed of all eucalypts. It is native to Australia and has been introduced around the world.

Silviculture characteristics

The success of *E. camaldulensis* as an exotic is attributed to its superiority over other trees in the production of wood on infertile dry sites, its tolerance of extreme drought and high temperature combined with rapid growth when water is available, deep penetration of roots, tolerance of periodic waterlogging and soil salinity, some tolerance of frost, good coppicing ability and usefulness of the wood.

Silviculture practice

E. camaldulensis is usually propagated from seed. There are about 700,000 viable seed/kg of seed and chaff mixture. As a rule of thumb, 1 kg of *E. camaldulensis* seed is sufficient to provide plants for 100 ha of plantation at routine spacings (3 × 2 m) and typical seedling survival rates (25%). Viability of seed stored dry (5–8% moisture content) in refrigerated, airtight containers (3–5°C) will be maintained for several years. No pre-sowing treatment is required.

E. camaldulensis is one of several eucalypt species suited to mass vegetative propagation through stem cuttings. Protocols for successful tissue culturing have also been documented.

E. camaldulensis seed should be sown under shade (optimum temperature for germination is 32°C) on a free-draining and sterilized medium and covered very sparingly with inert material (e.g. sand). Germination should be complete after 7 days and then shade can be reduced. At the second leaf-pair stage germinants are transplanted to containers filled with sterilized potting mix. Shade cover is needed for the first week after transplanting, after which time plants should be exposed fully to direct sunlight. Growth is fast under tropical conditions and plants could reach plantable size (30 cm) in 6 weeks, although 12 weeks is more usual.

Common mistakes in propagation are over-watering (leading to disease problems), over-shading (allowing the germinants to become too large for easy transplanting, resulting in malformed taproots or root curling) and allowing roots from the containers to grow into the soil. Occurrence of disease problems in the nursery can be limited by strict attention to hygiene, reducing watering and shade and allowing good ventilation.

E. camaldulensis is usually grown on short rotations and clear felled at an age that maximizes production of logs of optimum size for a particular end use. This is usually small-diameter material suitable for pulpwood, mining timber, charcoal or fuelwood. The species coppices well for five or more rotations. The season of felling affects coppice regeneration. Felling during the dry season delays sprouting and increases the risk of the stump drying out. Felling by saw to give a cleanly-cut short stump with minimum bark damage is best. Reduction of the number of coppice shoots on a stool is a very important, if time-consuming, operation in coppice management.

While clear felling is usual, the use of coppice with standards to supply a range of wood products is also recommended. A thinning from below of 700 stems/ha at half rotation age (5 years) provides posts, poles, fuelwood and pulpwood, leaving the better trees to grow on to 10 years for sawn timber and other products.

Spacing and cropping systems are very variable, from community plantings around homes, villages and along roads to closely spaced commercial plantations, and depend on the end products required. When pulpwood is the principal objective, a spacing of 3 × 2 m (1667 stems/ha) is often used. Wider spacings of 4 × 2 m (1250 stems/ha) or 5 × 2 m (1000 stems/ha) are recommended when larger trees are required. Wider spacing between rows facilitates mechanical cultivation for weed control.

On old agricultural land, intensive site preparation by ploughing to a depth of 10 cm when the soil is moist, followed by disc harrowing 3 months later, is beneficial; mouldboard ploughing can give effective weed control by burying weed seed.

On compacted sites, deep ripping of the planting lines (1 m with a bulldozer) is recommended. Ripping in combination with mounding improves survival and growth on most sites, mounding being essential on wet or saline soils. On stony or very steep sites, complete cultivation may be impracticable. It is then necessary to prepare planting pits with a 1-m surround cleared of weeds.

E. camaldulensis roots form symbiotic mycorrhizal associations with various fungi. Many exotic plantations are devoid of such mycorrhizal associations since compatible fungi are not always present in the soil. While this has seldom affected the successful establishment of *E. camaldulensis* in exotic environments, there is a growing pool of evidence to suggest that a tree's growth rate and resistance to environmental stresses may not be optimal unless suitable mycorrhizal fungi are present.

Application of fertilizer to each tree at planting to assist establishment and growth is common. *E. camaldulensis* plantations are sometimes irrigated. The species' ability to compete with weeds is poor and the area within at least 1 m of the tree should be kept free of grasses.

Thinning may be carried out; *E. camaldulensis* naturally sheds its moribund branches, so artificial pruning is seldom necessary.

Diseases and pests

In Australia, natural stands and plantations of *E. camaldulensis* are affected by many insects and fungi.

In the nursery, it is susceptible to a diverse range of fungi causing damping off, collar rot and leaf diseases (*Pythium* spp., *Phytophthora* spp., *Rhizoctonia* spp. and *Cylindrocladium* spp.).

Disease is most common where the species is planted off-site, or where inappropriate provenances are used, resulting in stressed, unthrifty trees which are more susceptible to disease than healthy ones. Stem cankers and leaf diseases proliferate where rainfall and humidity are much higher than normally encountered in the natural habitat (e.g. in parts of India). In humid regions of Vietnam and Thailand many *E. camaldulensis* plants are defoliated by fungi, causing severe reductions in growth rate and deformed crowns and stems. Several pathogens are associated with these characteristics in these countries with *Cylindrocladium quinqueseptatum* an important causal agent. The most susceptible provenances suffer mortality and general decline, but some provenances (e.g. Katherine) are little affected, allowing selection and breeding for resistance to the disease.

In Indonesia, prominent leaf spot symptoms on *E. camaldulensis* have been attributed to an unidentified *Mycosphaerella* species. *Colletotrichum gloeosporioides* has been found responsible for leaf spot and twig blight diseases of young *E. camaldulensis* plantations in Bangladesh.

Leaves are often attacked by leaf-chewing insects, particularly of the Chrysomelidae and Curculionidae families (such as *Paropsis* spp., *Chrysophtharta* spp., *Gonipterus* spp. and *Oxyops* spp.).

Where the tree is well adapted outside Australia, it is relatively free from problems. In parts of Africa and Asia termites affect young trees. In Africa, the eucalyptus snout beetle (*Gonipterus scutellatus*), of Australian origin, feeds on young shoots; moribund or newly felled trees may become infested with the Australian stem borer or longicorn beetle (*Phoracantha semipunctata*). The longicorn beetle has caused major damage to plantations of *E. camaldulensis* in parts of North Africa and the Middle East. The sapwood is susceptible to attack by *Lyctus* borers.

The bronze bug *Thaumastocoris peregrinus*, originating in Australia, has been rapidly spreading in eucalyptus plantations in the southern hemisphere, and is found on *E. camaldulensis* in South Africa, Argentina and Brazil. Infested trees show leaf silvering, ranging from chlorosis to bronzing. Heavy infestations cause leaves to become red/brown and defoliation occurs.

The red gum lerp psyllid (*Glycaspis brimblecombei*) was recently introduced from Australia into North America, where it causes severe defoliation and some tree mortality. It is a pest on *Eucalyptus* species, and is mainly found on *E. camaludulensis*. The eucalyptus gall wasp (*Leptocybe invasa*), is also found on *E. camaludulensis*.

Uses

E. camaldulensis is planted extensively for shade, shelter belts, agroforestry and amenity purposes. *E. camaldulensis* is useful for mine site rehabilitation and the reclamation of degraded lands.

The wood burns well and makes a good fuel, and is used in several countries for the large-scale production of charcoal for iron and steel industries. Its dense wood and coppicing ability make it an excellent species for fuelwood production. The wood is used mainly for poles, posts and paper pulp. Logs may be sawn for construction timber, furniture and packing cases, although the quality is sometimes poor.

The timber has a handsome red colour, a fine texture, and interlocking wavy grain. It is hard, durable, resistant to termites and has many uses. Correctly handled, the wood is useful for speciality furniture, construction timber, pulpwood and roundwood. Post-harvest procedures can be employed to ameliorate unfavourable wood characteristics of small diameter plantation-grown *E. camaldulensis*, such as growth stresses, shrinkage on drying, collapse, spiral grain and starch in the sapwood.

Green densities of 500–700 kg/m^3 for wood from young trees and 1130 kg/m^3 for old trees are reported. Density is positively correlated with charcoal and pulp yield.

Some provenances of *E. camaldulensis* are potential sources of medicinal-grade *Eucalyptus* oils. Substantial variability within and between provenances for commercial oil traits and their high heritabilities indicates significant potential for improvement through selection and breeding.

E. camaldulensis is of major importance in Australia as a source of honey, producing heavy yields of nectar in good seasons and also provides the bees with an important source of good-quality pollen.

Further reading

Coppen, J.W.J. (2002) *Eucalyptus*: The Genus *Eucalyptus* (Medicinal & Aromatic Plants). CRC Press, Boca Raton, Florida, USA, 464 pp.

Ishiguri, F., Iizuka, K., Tanabe, J., Wedatama, S., Yokota, S., Yoshizawa, N. and Diloksumpun, S. (2012) Technical note: solid

wood properties of *Eucalyptus camaldulensis* planted for pulpwood production in Thailand. *Wood and Fiber Science* 44, 108–110.
Wei, R.P. and Xu, D. (eds) (2003) *Eucalyptus* Plantations: Research, Management and Development - Proceedings of the International Symposium. World Scientific Publishing Co. Pte Ltd, Singapore, 432 pp.

Eucalyptus citriodora Hook.

Lemon-scented gum

Importance

Eucalyptus citriodora is a handsome tree of excellent form, with a well-shaped but sparsely foliaged crown. It is widely adaptable, drought tolerant and capable of reasonable growth rates in both summer and winter rainfall areas. This species is one of the better eucalypts for cultivation at lower latitudes of the subtropics and tropics, but not in equatorial regions. It produces strong, hard and moderately durable wood suitable for fuelwood, charcoal, posts and poles, household products, tool handles, sawn timber and general construction materials. *E. citriodora* is also used as a source of non-wood forest products, most notably a lemon-scented essential oil.

Botanical features

A medium-sized to large, straight-stemmed tree, 25–40 m tall, of handsome appearance, with smooth, pale grey, cream or pink, powdery bark throughout, decorticating in flakes. Foliage is fine, somewhat sparse in the crown and emits a strong lemon scent after rain or when abraded. The lemon-scented leaves are very useful in identifying the species. There is only one other species in the genus with this characteristic, the ironbark (*Eucalyptus staigeriana*), which is easily separated from *E. citriodora* based on morphology, oil chemistry and partly on natural distribution.

All leaves are strongly lemon scented when crushed. Seedling leaves opposite for a few pairs then alternate, petiolate, peltate, ovate, 6.5–17 × 2.3–7.5 cm, pale green, slightly discolorous, setose. Juvenile leaves alternate, petiolate, ovate to broad-lanceolate, 14–21 × 4.5–8 cm, some remaining setose, pale green and peltate for many pairs, others becoming glabrous and glossy green soon after the seedling stage. Stems, petioles and leaves (particularly veins) are setose with bristle glands up to 0.5 cm long, in both the seedling and juvenile leaf stages. Intermediate leaves alternate, petiolate, broad-lanceolate to lanceolate, 13–30 × 2–5 cm, green, concolorous. Adult leaves on petioles 1.3–2.2 cm long, alternating, lanceolate to narrow-lanceolate, 7–22 × 0.5–2.2 cm, concolorous, glossy, green; reticulation very dense with numerous large island oil glands, lateral veins just visible, intramarginal vein distinct, up to 0.1 cm from margin.

Inflorescences clustered on short, leafless shoots in axils of leaves towards ends of branchlets, 3–flowered; peduncles 0.3–0.7 cm long. Buds pedicellate, clavate, to 1 × 0.6 cm, no scar; operculum conical to slightly beaked, 0.3–0.4 cm long, 0.4–0.5 cm wide; flowers creamy-white; fruit on pedicels 0.1–0.6 cm long, truncate–ovoid to urceolate, often warty, 0.8–1.5 cm long, 0.7–1.1 cm diameter; 3–locular; disc approximately 0.2 cm wide, seeds glossy red-black, dorsiventrally compressed, keeled on dorsal side, hilum ventral, minute cracks in the seedcoat, 0.2–0.3 mm long, 1.5–2.5 mm wide.

Distribution

E. citriodora is endemic to Queensland, Australia. It has been extensively planted as an ornamental tree in many regions of the world, and also for commercial purposes in many countries.

Silviculture characteristics

E. citriodora is a useful multipurpose tree for wood and oil that can be grown profitably by farmers with smallholdings in countries such as India. Although not a species of outstanding growth rate, it gives acceptable growth on a wide variety of sites. *E. citriodora* is a lignotuberous species which is resistant to damage by fire. It can survive light frosts and is moderately drought-tolerant; a minimum annual rainfall of 600 mm is generally required, and for rapid growth 900 mm is desirable. The climate range suggests that there may be provenance differences in the ability of this species to tolerate frost and drought. The species is susceptible to damage by typhoons. Variable stem form, from straight to sinuous and frequent forking, and sparse foliage are less satisfactory characteristics of this species.

Silviculture practice

E. citriodora is usually propagated from seed. It can also be regenerated vegetatively by micropropagation; however, propagation by cuttings is very difficult.

Containerized planting stock are preferred, as problems have been experienced with poor lateral root development, and losses due to transplanting shock using open-rooted stock. The seed are relatively large for a eucalypt and can be sown directly into containers without pretreatment, and covered with a light sprinkling of fine sand. Germination percentage should be checked first and sowing rate gauged to give at least one germinant per container. A successful potting medium is a freely draining mixture of loam and sand that has been sterilized before use. Lignotubers develop early in the life of seedlings. *E. citriodora*, in common with other species in the genus, is highly susceptible to damping off and other fungal pathogens during the nursery phase. Occurrence of disease problems in the nursery can be limited by strict attention to hygiene, reducing watering and shade and allowing good ventilation. Seedlings are planted out in the field when they reach a height of about 25 cm, 3 months after sowing. This should coincide with the onset of the wet season in tropical countries. Direct seeding of *E. citriodora* into carefully prepared ground has been successful in some areas, but it is unreliable, as favourable weather conditions and freedom from weeds are critical to success, and it is not generally recommended.

Optimum rotation length for plantations of this species will depend on site and the end products required.

Spacing varies with the purpose of the plantation. When grown solely for oil production with frequent coppicing, 3 × 1.5 m (2222 plants/ha) is appropriate. For fuelwood, poles or charcoal production with initial harvesting at 7 years, a spacing of 3 × 2 m (1667 plants/ha) is commonly used.

Intensive site preparation by ploughing is beneficial. On compacted sites deep ripping may also be used.

Eucalypt roots form symbiotic mycorrhizal associations with various fungi, such as ectomycorrhizal or vesicular-arbuscular mycorrhizal (VAM) associations can occur in the genus. *Pisolithus tinctorius*, an ectomycorrhizal fungus, has been isolated from *E. citriodora* in Brazil. The fungi are associated with the tree roots, and enable the tree to derive certain nutrients (especially phosphorus); in turn the fungi benefit from other nutrients made available to them by the tree. Many exotic plantations are devoid of mycorrhizal associations since compatible fungi are not always present in the soil. While this has seldom affected the successful establishment of eucalypts in exotic environments, there is a growing pool of evidence to suggest that a tree's growth rate and resistance to environmental stresses may not be optimal unless suitable mycorrhizal fungi are present. Nurseries growing eucalypts, especially where the soils are deficient in phosphorus, are recommended to introduce appropriate mycorrhizae to the nursery soil.

Although the species is often cultivated without addition of organic or chemical fertilizers, on sites of very low soil fertility, fertilizer application may be necessary to increase productivity. Plantations of this species are not normally irrigated.

The ability of the species to compete with weeds is poor, so inadequate weed control may lead to the complete failure of a planting. Mechanical or manual cultivation is the main means of controlling weeds.

E. citriodora is very light-demanding; frequent and regular thinning is a prerequisite for healthy, vigorous plantations. *E. citriodora* is strongly self-pruning and is only hand pruned when leaves are harvested for oil production.

Diseases and pests

E. citriodora has been damaged by a range of diseases including damping off and leaf spots caused by *Cylindrocladium* spp., *Puccinia psidii* (a rust) and *Cryphonectria cubensis* (a stem canker). Gummosis induced by *Cytospora* sp. and *Macrophoma* sp., and cankers from infection by *Endothia havanensis* have also been noted. The species is also susceptible to *Melampsora* sp. (a rust), *Corticium salmonicolor* (pink disease) (*Phanerochaete salmonicolor*), and by *Ganoderma* sp. and *Pseudophaeolus baudonii* (both root rots).

E. citriodora is very susceptible to termites. A range of defoliating insects and a stem borer *Apate indistincta* have been noted as causing occasional damage to plantations.

Uses

E. citriodora is widely used in park and avenue plantings as an attractive, large-sized ornamental tree noted for its bark colour, branching habit and glossy leaves. However, its crown is too sparse for shelter belts.

The wood of *E. citriodora* has a white or cream sapwood up to 60 mm wide and is very susceptible to *Lyctus* borer attack; the heartwood is pale grey-brown to dark brown. The sapwood is permeable, but heartwood is extremely resistant to preservative treatment. The grain is straight or interlocked and occasionally wavy, and the texture is open and coarse. The timber is hard, strong and tough; it saws easily and planes well, but is rather difficult to nail, and is prone to checking and collapse in drying. It is also liable to marine borer and termite attack. The air-dry density of wood from natural stands is about 1000 kg/m^3 with a basic density of 800 kg/m^3.

The timber of *E. citriodora* is used for general and heavy construction such as framing and bridge construction, flooring, cladding, tool handles and case manufacture. Reasonable chemical pulp qualities have been obtained from wood of young trees, but wood from old trees is generally unsuitable, due to high extractive content and high density. *E. citriodora* is also grown for charcoal production.

The pleasant, lemon-smelling foliar essential oil of *E. citriodora* is widely used in less expensive perfumes. Its citronellal-rich (65–91%) oil is a preferred natural source of hydroxycitronellal, and is obtained by fractionation of the crude oil. Hydroxycitronellal is one of the most widely used of all perfumery materials, and is also used for the manufacture of menthol. Other minor constituents recovered during fractionation of the oil, such as citronellol, are also used by the fragrance industry. The oil is also used in soaps and disinfectants; it has anti-bacterial, anti-fungal and insecticidal activity.

E. citriodora is an important source of nectar for bees and yields a light amber honey.

Further reading

Koriesh, E.M., El-Fattah, Y.M.A., El-Dayem, M.A., El-Etriby, M.A., Hammerschlag, F.A. and Saxena, P., (2003) Micropropagation of juvenile *Eucalyptus citriodora*. *Acta Horticulturae* 625, 283–288.

Shashikala, S. and Rao, R.V. (2009) Radial and axial variation in specific gravity and anatomical properties of plantation grown *Eucalyptus citriodora* Hook. *Journal of the Institute of Wood Science* 19, 84–90.

Singh, H.P., Kaur, S., Negi, K., Kumari, S., Saini, V., Batish, D.R. and Kohli, R.K. (2012) Assessment of *in vitro* antioxidant activity of essential oil of *Eucalyptus citriodora* (lemon-scented Eucalypt; Myrtaceae) and its major constituents. *LWT - Food Science and Technology* 48, 237–241.

Eucalyptus deglupta Blume

Kamarere

Importance

Eucalyptus deglupta has been planted throughout the humid tropics and is ranked among the world's fastest-growing trees. It is also capable of growing to enormous size. Mature individuals have been recorded up to 80 m tall, with a dbh of

3 m. The main use for plantation wood grown in rotations of 5–12 years is for production of strong sulphate pulp that can be readily bleached to a high brightness. However, *E. deglupta* planting has been constrained by the fact that it requires relatively fertile, alluvial sites for optimum growth, and these sites are being increasingly taken up for agricultural crops. Also, the wood from fast-grown, short-rotation plantations has a relatively low density. On poorer-quality, humid, lowland sites in the tropics, acacias such as *Acacia mangium* are now preferred over *E. deglupta* for pulp production. *E. deglupta* hybrid combinations have proved successful with species such as *Eucalyptus urophylla* and *Eucalyptus wetarensis* which tolerate a diverse range of sites, have a higher wood density, and some disease tolerance; and also *Eucalyptus pellita* in Papua New Guinea which has higher wood density, and good disease tolerance. *E. deglupta* × *E. pellita* has been very successful in Mindanao, Philippines. The best hybrids are being propagated clonally using cuttings.

Botanical features

E. deglupta is a large tree, commonly 35–60 m high, 0.5–2.0 m diameter, occasionally reaching 80 m height and 3 m diameter. The bole is typically straight, cylindrical and self-pruning, often clear of branches for more than 75% of the total height. Buttressing to 3–4 m high is common on individuals growing on unstable soils.

Bark smooth, 3–8 mm thick, exfoliating in strips of varying shape and size, leaving a smooth, white to pale green surface which ages through light green, green, grey, pink, red and orange to a deep purple, all colours mentioned being visible on different parts of the trunk simultaneously.

Seedling leaves opposite, thin, petiolate, ovate, ovate–lanceolate or oblong–lanceolate, first 1–3 pairs 2–4 × 0.5–1.5 cm then 6–10 × 2.5–4.0 cm; lateral veins visible, very few, initially 40–50° to the midrib then curving to form the remote intra-marginal vein; upper surface light green to green, lower pale green to purplish.

Juvenile leaves opposite, thin, shortly petiolate, broadly ovate, ovate–lanceolate, oblong–lanceolate or acuminate, 7–11 × 5–7 cm; lateral veins visible, few, curved to form an irregular, remote intra-marginal vein; upper surface green to dark green and slightly shiny; lower surface light green to greenish-purple and dull.

Adult leaves opposite to subopposite, rarely alternate, shortly petiolate, ovate to ovate–lanceolate or acuminate, thicker than juvenile leaves, 10–20 × 6–10 cm, occasionally larger; lateral veins visible, few, initially inclined about 60° to the mid-vein, then curving to form the intra-marginal vein, upper surface green to dark green and shiny, lower surface pale green and dull.

Leaves and terminal branches held horizontally, terminal branchlets and twigs mainly square or flattened in cross-section, often with four longitudinal keels.

Terminal or axillary panicles 5–20 × 5–18 cm, umbels 5–7 flowered; peduncles terete or slightly angular, 0.5–1 cm long; pedicels about 5 mm long, young buds small, green, with double opercula, developed buds pale green or cream, globular, apiculate, 0.2–0.4 × 0.2–0.5 mm, operculum hemispherical, apiculate and wider than long; flowers with many white to pale yellow or cream stamens 2–10 mm long, strongly reflexed in the unopened bud, anther dehiscing by separate slits.

Fruit pedicellate, hemispherical, valves 3–4, thin, deltoid, exserted to 2 mm making the capsule appear globular, 3–5 × 3–5 mm, disc very narrow. Mature fruits brown to dark brown, usually with 4 valves, each containing 3–12 well-formed seeds. Seeds minute, brown, flattened with a small terminal wing.

Distribution

E. deglupta occurs between tropical latitudes 9° N and 11° S in a markedly discontinuous distribution through Mindanao (Philippines), Sulawesi, Ceram and Irian Jaya (Indonesia) and Papua New Guinea including New Britain. It has been planted as an exotic throughout the lowland humid tropics.

Silviculture characteristics

Seed production is often profuse. Capsules may be picked when they are greenish-brown and will open to release the seed after 2–3 days of drying in the sun. After drying, seeds may be stored for at least 2 years in refrigerated airtight containers.

For natural regeneration, water is the most important vehicle for dispersal, continually leading to new riverine stands on fresh alluvial deposits. In addition, the seeds have a small wing which, combined with their small size and light weight, enables them to drift into nearby bare areas.

This eucalypt is shade-intolerant. Full overhead light is required for development past the seedling stage. The species will not reproduce under shade, only in natural or artificial openings. Where it finds suitable conditions it regenerates in huge numbers. This accounts for the species typically forming almost pure stands.

E. deglupta is a site-sensitive species; when planted on poor sites it may grow so slowly it will never produce useable timber.

This eucalypt does not coppice well enough for practical regeneration of plantations using coppice. Coppicing can be induced by partial girdling low down on the stem or by application of hormones to cut stumps, in order to produce physiological juvenile shoot materials for cuttings.

Silviculture practice

Sowing seeds in shaded and glass or plastic-covered germination trays is typically the first step in production of seedlings. Seedlings are transplanted to tubes or pots when they have two to three pairs of leaves above the cotyledons. Seedlings reach a plantable size (20–30 cm tall) 3–4 months after germination. Shade should gradually be reduced from full during germination, to one-quarter for the first month then to none (full sunlight) by the end of the nursery phase.

Successful site preparation procedures include removal of slash, control of weeds and soil cultivation. As seedlings of this eucalypt are prone to wilting in dry weather, planting has to be done during rainy periods, and seasonal conditions are important in timing planting. After 1 month in the field, seedlings will usually be able to withstand short periods of moisture stress. Fertilizer is often applied at the time of planting to boost early growth.

The greatest annual height increment usually takes place during the first year, and gradually decreases over the next 25–30 years. If warranted, thinning for stakes may begin at the end of the first year and continue through the life of the stand successively for posts, pulpwood, poles and sawlogs. The rotation age for pulpwood, the most important product from plantations of *E. deglupta*, is from 5 to 12 years.

Diseases and pests

Heart rot affects this eucalypt in many parts of the world, particularly where the plantations are established on poorly drained sites, although generally the amounts of decay present in each tree are insufficient to affect pulpwood quality. An unidentified canker disease is commonly associated with wounds caused by machete cuts to the stem made during vine removal. Both this and heartwood decay are minimized by avoiding damage to trees during silvicultural operations.

Wood borers have been reported on this species; these significantly reduce the value of trees as sawlogs. The larvae of a cossid moth *Zeuzera coffeae* bores into the cambium and sapwood of young trees, leaving them susceptible to breakage by wind. Losses of seedlings in unprotected plantations have been inflicted by the giant snail (*Achatina fulica*) that eats the bark; seedlings may snap off under the snail's weight. Termites are also severe pests of *E. deglupta*. They attack the trees from the seedling stage through maturity, occasionally consuming all the heartwood. Trees are also susceptible to the stem girdler *Agrilus opulentus* and the related varicose borer *Agrilus sexsignatus*. Problems with leaf-cutting ants (*Atta* sp.) have been reported. Tip necrosis of *E. deglupta* caused by the coried beetle (*Amblypelta cocophaga*) was reduced by clearing vegetation between the rows of trees, and by introducing the green tree ant (*Oecophylla smaragdina*).

Uses

This species provides a measure of stream-bank stabilization in areas prone to flooding. It does not fix nitrogen. Because of its attractive bark, the species is frequently planted as an ornamental and as a shade tree.

The heartwood is reddish-brown and coarse grained, more like a rainforest wood than a typical eucalypt. It has a typical ribbon grain and good lustre. The sapwood is white to pale pink. Of moderate strength, it saws and machines easily, except for some splitting on quarter-sawn faces, and polishes and glues well. Specific gravity is greater for old-growth natural forest trees (450–650 kg/m^3) than for young plantation wood (350–400 kg/m^3). The heartwood is resistant to *Lyctus* sp. (powder post beetles), but is not termite resistant. The wood is not durable in contact with the ground.

The wood is used for furniture (including veneer and plywood), joinery, mouldings, flooring, wood-based panels, heavy and light construction and boatbuilding. Most plantation-grown wood is destined for short-fibre pulp production. *E. deglupta* is also used for firewood and charcoal.

This eucalypt supports production of a good-quality honey. Leaf oils are present, but not in sufficient quantities to be of commercial interest.

Further reading

Harries, E.D., Phillips, F.H. and Logan, A.F. (1997) The pulping and papermaking potential of plantation-grown *Eucalyptus deglupta* from Papua New Guinea. Part 1. Pulpwood characteristics. *Journal of Tropical Forest Products* 2, 199–215.

Phillips, F.H., Logan, A.F. and Harries, E.D. (1997) The pulping and papermaking potential of plantation–grown *Eucalyptus deglupta* from Papua New Guinea. Part 2. Sulphate and NSSC pulps from wood of various ages. *Journal of Tropical Forest Products* 3, 20–42.

Schaller, M., Schroth, G., Beer, J. and Jiménez, F. (2003) Species and site characteristics that permit the association of fast-growing trees with crops: the case of *Eucalyptus deglupta* as coffee shade in Costa Rica. *Forest Ecology and Management* 175, 205–215.

Takizawa, T. (1999) Wood quality and working properties of tropical fast-growing trees (2) *Eucalyptus deglupta* and *E. robusta*. *Tropical Forestry* 44, 81–85.

Eucalyptus globulus **Labill.**

Tasmanian blue gum

Importance

Eucalyptus globulus varies from a multi-stemmed shrub on exposed sites to a medium-sized woodland tree of 20 m to a very tall forest tree of 70–80 m with a large open crown. The wood is strong and moderately durable, and is used for light and heavy construction; it is regarded as being excellent for cellulose and paper manufacture. It is slightly to moderately salt tolerant. Production from plantation-grown trees is mainly for pulpwood, but also for other timber products. It is an important plantation species in Australia and has experienced outstanding success as an exotic in many countries. The essential oil of *E. globulus* has dominated the market for cineole-rich oils since soon after the inception of the industry in 1852.

Botanical features

The forest-tree form can grow as high as 80 m with a tall, straight trunk; its branches retained for less than half the total tree height. The lower bark is rough, greyish or brownish, as a stocking at the base; the upper bark smooth pale often with a bluish or yellowish tinge; decorticating into long strips. The woodland form is a medium-sized tree, 15–20 m tall with a compact crown. The single trunk is much branched, the first branch usually fairly high up.

Seedling leaves are opposite, sessile, the base clasping the stem (amplexicaul), ovate, bluish-green, glaucous, strongly discolorous; juvenile leaves opposite, sessile, amplexicaul, elliptic-ovate, bluish-green, glaucous, discolorous. The stems of both seedling and juvenile leaves are square in section, flanged and glaucous. Adult leaves are petiolate, falcate, lanceolate or narrow–lanceolate, green, (a similar shade of green on upper and lower surfaces).

The inflorescence is simple, usually single flowered (occasionally 3), and axillary. The stamens are usually white or cream. The pedicel is sometimes absent or very short and stout, up to 4 mm long. Buds are top-shaped (turbinate) with 4 distinct ribs (sometimes more), extremely glaucous. The bud cap or operculum is flattened, very warty with a very distinct, central knob. The ribbed capsules are sessile, sub-globular to hemispherical with 4 distinct ribs (sometimes more), glaucous on the enlarged receptacle (hypanthium); disc broad with slight lobes; valves 4 or 5, more or less horizontal or slightly exserted. The capsules have a distinct concave, calycine ring. The flowers are predominantly insect pollinated although birds and small mammals may also assist in pollination.

Distribution

Although naturally found mainly in Tasmania, *E. globulus* also occurs in southern Victoria. It has been widely planted around the world in areas of mild climate free of severe frosts.

Silviculture characteristics

E. globulus is a fast-growing species which has been widely planted as a short-rotation pulp species. Plantations of *E. globulus* are generally more successful on sandy soils than on clay; on soils that retain moisture without becoming water-logged; on acid rather than alkaline; and on deep rather than shallow soils. Seedlings are not particularly tolerant of frosts, and temperatures of below about −5°C will usually kill them. Mature trees may survive lower temperatures.

E. globulus has a low to moderate salt tolerance; reduced growth is expected with increasing soil salinity.

Silviculture practice

Propagation is usually by seed, which germinates readily in warm, moist conditions with no pretreatment. Seedlings are usually transplanted from the trays in which seed has been sown to tubes or pots at the four-leaf stage. Seed can also be sown in specially prepared seedbeds, transplanted into nursery beds at the four-leaf stage and left there until required for planting out. Pure seed can be also be sown successfully by machine directly into tubes or pots for later field planting. Seedlings should be protected from severe frosts.

E. globulus can also be propagated vegetatively, although this has not proved to be particularly successful on a commercial scale. In fact, many commercially important selections have proved very difficult to propagate from cuttings because of poor rooting ability. Recent advances have increased the success of vegetative propagation, and since 1995 many companies have started planting extensive areas of *E. globulus* cuttings in Portugal.

Seedlings can be planted by hand or by machinery; the inter-row width should be wide enough to allow machinery to be used, with between-plant spacings of 3–5 m. Machine planters have been developed to plant two rows at a time, and incorporate electronic programming for spacing and fertilizer application. Planting rates of approximately 10 ha per machine day at stockings of 1100 stems/ha are possible. Commonly used initial spacings in plantation forestry are 3 × 2 m (1667 trees/ha), 3 × 3 m (1111 trees/ha) and 4 × 2.5 m (1000 trees/ha). Thinning is not usually undertaken when the species is grown for pulp; however, the final stocking may be reduced to as low as 100 stems/ha when grown for sawlogs.

The planting area is often deep ripped to a depth of up to 1.2 m in heavy soils where a hardpan may be present at depth. The area may also be ploughed. Slashing weed growth to ground level is usually sufficient once the seedlings are established. Seedlings may need protection from browsing by rabbits and other small animals and sometimes goats, but are generally unpalatable to other domestic stock and deer.

Survival and growth can be enhanced with the addition of fertilizer, particularly nitrogen and phosphorus, at establishment. Deficiencies in numerous other nutrients and trace elements may cause a variety of disorders in this species, and visual symptoms should be verified by chemical analysis of affected plant tissue. Inoculation with ectomycorrhizal fungi may have the effect of increasing phosphorus uptake and growth of young seedlings.

E. globulus is considered a fast-growing species, and height growth of 2–3 m per year is commonly achieved on favourable sites. It is grown primarily for pulpwood, firewood and mine timbers, and hence rotations are relatively short. Plantations are often cut on a coppice rotation of about 8–10 years, with 2–3 rotations before replanting. Some trees are allowed to grow to a large size and are used as sawn timber, but because of growth stresses the output of high-quality boards is low. On the most favourable sites the mean annual increments are more than 20 m^3/ha.

Diseases and pests

Penicillium sp. and *Fusarium* sp. are important seed pathogens; the latter is highly destructive to stored seed. Seedlings are highly susceptible to *Botrytis cinerea* (and other types of damping off diseases) in the nursery. *E. globulus* is susceptible to *Mycosphaerella* leaf disease. Other diseases, including *Diplodia* cankers and *Armillaria* root disease, have been reported in various countries, but are generally of lesser importance.

E. globulus foliage is susceptible to attack by autumn gum moth (*Mnesampela privata*), Christmas beetle (*Anoplognathus* spp.), leafblister sawfly (*Phylacteophaga froggatti*), leaf beetle (Chrysomelid beetles – various genera), various scale insects including *Eriococcus coriaceus* and the bluegum psyllid (*Ctenarytaina eucalypti*). It is also susceptible to autumn gum moth and leafblister sawfly.

Snout beetles of the genus *Gonipterus* in both larval and adult stages feed on leaves and have caused severe damage to plantations around the world. The scale insect *Eriococcus coriaceus* has greatly impacted *E. globulus*, and on drought-prone sites the species is liable to attack by the stem-boring insect *Phoracantha semipunctata*.

Uses

E. globulus has been planted for a variety of uses which include erosion control, amenity and roadside plantings, windbreaks and shelter belts for pasture and livestock and around orchards, and other watershed and environmental purposes. It has also been planted as part of salinity control measures; in the rehabilitation of mining sites; for drying out marshy areas; and in plantations to re-use sewerage and chemical effluent.

E. globulus has also shown a degree of resistance to high levels of ozone, which may be useful in regions with the potential for photochemical smog formation, such as some of the rapidly developing industrializing areas where tolerance to ozone may need to be considered in selecting plantation trees.

A number of horticultural varieties showing dwarf characteristics or fastigiate growth are utilized for ornamental plantings.

The heartwood of *E. globulus* is light yellowish-brown, open-textured, has distinct growth rings and is moderately durable. Mature trees from natural stands have an air-dry density of about 900 kg/m^3. It is used for light and heavy construction, poles, piles and railway sleepers. It is a good firewood and carbonizes easily for good charcoal production.

Sawn timber needs care when sawing and drying to minimize checking of the tangential surface. Quarter-sawing is desirable. The wood has a high incidence of spiral grain and can be difficult to nail. Other important wood products include fibreboard, particleboard, parquetry flooring, cooperage, low-grade veneer and furniture.

Mature timber produces poor pulp, but young plantation-grown wood is less dense (air-dry density of around 650 kg/m^3) and is suitable for paper and rayon pulps. It is a major component of large paper-making industries around the world.

Although *E. globulus* is primarily grown for wood products, it is also utilized for a variety of non-wood products, usually a by-product or additional use for trees grown for wood products. For example, the leaves from felled trees may be distilled for their oil, which is high in cineole.

Crude oil concentrations obtained from *E. globulus* are not high compared to eucalypt species grown specifically for oil; however, *E. globulus* produces relatively large amounts of leaf biomass which helps to increase oil production per unit land area and to compensate for modest concentrations. The oil is used for household utility products, perfumery, pharmaceuticals (pastilles, lozenges and syrups for coughs and colds) and general flavouring.

Compounds extracted from the leaves of *E. globulus* have shown anti-viral, anti-microbial, and anti-tumour activities. *E. globulus* flowers produce pollen and nectar, and natural populations and plantations are used for the production of honey.

Further reading

Hardel, D.K. and Sahoo, L. (2011) A review on phytochemical and pharmacological of *Eucalyptus globulus*: a multipurpose tree. *International Journal of Research in Ayurveda and Pharmacy (IJRAP)* 2, 1527–1530.

Ong, S.H. (1988) *Notes on Some Properties of Eucalyptus globulus*. FRC Publication 44. Forest Research Centre, Sandakan, Sabah, i & 7 pp.

Stackpole, D.J., Vaillancourt, R.E., Downes, G.M., Harwood, C.E. and Potts, B.M. (2010) Genetic control of kraft pulp yield in *Eucalyptus globulus*. *Canadian Journal of Forest Research* 40, 917–927.

***Eucalyptus gomphocephala* DC**

Tuart

Importance

Eucalyptus gomphocephala is a large hardwood with a limited natural distribution in coastal south-western Western Australia. It is a fast-growing exotic that has been planted in many countries with a Mediterranean climate on sandy soils with a high percentage of lime. The wood from exotic plantations is mainly used in the round for poles, stakes and fuelwood; it is also used for environmental purposes such as shade and shelter (windbreaks), and for ornamental purposes.

Botanical features

A medium-sized to tall tree that can grow to 25–40 m in height and 1–2 m dbh with a well-developed crown and fairly dense, spreading branches. Open-grown trees can produce individuals with large branches.

In natural stands the trunk is often short and one-third to under a half of the tree height. The species forms large branches and the crown can become umbrageous.

It has a deep taproot and in Western Australia unusual stalagmiform roots on *E. gomphocephala* have been recorded. These specialized roots consist of stalagmite-like aggregations of roots and rootlets which protrude upright from the floor of certain limestone caves. These roots have ectomycorrhizal associations and non-pathogenic endophytes are present.

This species has typical box-type bark, rough up to the small branches and dark grey in colour. The dead bark is retained, short-fibred, moderately thin, firm or hard with narrow longitudinal fissures. The outer branchlets are yellow and smooth.

Juvenile leaves are petiolate, alternate, ovate or cordate, 9–15 × 5–10 cm, discolorous, light green, thin. Adult foliage, petiolate, alternate, lanceolate or falcate, 9–16 × 1.6–2.5 cm, slightly discolorous or the same colour on both sides, glossy, light green.

Inflorescences axillary, unbranched, 7-flowered; the peduncles are broad and flattened, 13–27 mm long. The buds are sessile, or on short, stout stalks, mushroom-shaped, 14–23 × 8–15 mm, scar present; operculum hemispherical to slightly conical, wider than the hypanthium. The flowers are white. Fruit are sessile, bell-shaped to cupular, sometimes winged or ribbed, 13–22 × 13–17 mm, with a thick rim and disc level, valves 4 to rim level or slightly exserted. The seeds are black, flattish to saucer-shaped, often flanged with a distinct net-like pattern on the surface.

Distribution

E. gomphocephala is a species that is endemic to south-west Western Australia found on limestone, coastal dunes and on sub-coastal plains. It has been introduced into a number of countries.

Silviculture characteristics

E. gomphocephala is one of very few eucalypts capable of forming large trees on neutral to alkaline soils. It coppices strongly and plantations can be managed on this basis for several rotations.

Silviculture practice

Seed can easily be stored in sealed, airtight containers at ambient temperatures for long periods of time. No special seed pretreatment is required to get a good germination response. Seedlings can be raised without difficulty in the nursery, and seedlings suitable for planting in the field can be produced within 3 months of sowing.

Good silvicultural management of mature-aged stands involves encouraging soil-stored seed to germinate on the forest floor, and protection of subsequent seedling regeneration from fire, invasive weeds and cattle grazing.

Cattle were once used as a management tool to control grasses in some parts of the tuart forest, lowering the chances of bushfires. Unfortunately the cattle ate the tuart regeneration, leaving the unpalatable species *Agonis flexuosa*, large dense thickets of which have replaced once-open areas of natural grasses and shrubs.

As a result of weed invasion and grazing damage, cattle were removed from areas where conservation values for flora and fauna values were maximized in Ludlow forest (Western Australia), and are restricted to buffer areas. The declaration of the Tuart Forest National Park in 1987 stopped many of these forest practices: a new prescription uses naturally occurring small gaps in the forest as the focus of stand establishment. Into these gaps dead material is gathered, heaped and burnt to provide an ash bed into which either seed is sown or seedlings planted. Standing dead or decaying material is retained, and only the *A. flexuosa* understorey within the gap and its immediate surrounds is cleared. This favours the Western Australia ringtail possum *Pseudocheirus occidentalis*, a threatened species that makes extensive use of tree hollows for nest sites and has *A. flexuosa* foliage as the main constituent of its diet.

Exotic tuart plantations can coppice vigorously and are usually managed by this means after the first rotation. Pruning is not undertaken as tuart self-prunes.

Initial spacing is reported to vary greatly from 2 × 2 m to 4 × 5 m, no doubt influenced by rainfall and soil moisture-holding capacity where plantations are established. Where mechanical operations are intended a spacing of 2 × 6 m is sometimes used.

Seedlings should be raised in containers. Direct sowing of seed for farm plantings has been tried experimentally in Western Australia, with some success. This method is not recommended for intensively managed industrial plantations.

Control of weeds in the first few years after establishment is often critical to the survival and growth of eucalypts.

For sawlog production, rotations vary from 10 to 15 years; thinnings are carried out at 6, 9 and 11 years in Morocco. Rotations of 7–10 years are used in unthinned coppice crops for production of smallwood.

Diseases and pests

The butt-rot fungus *Armillaria luteobubalina* has been identified on *E. gomphocephala*. The eucalyptus cerambycid borer *Phoracantha semipunctata* is the major insect pest of tuart in exotic plantations. It causes severe damage, especially when the trees are under drought stress, and can kill them. *Phoracantha impavida* also attacks tuart, and can cause substantial crown decline and tree death.

A taxonomic revision of the Australian cerambycid genus *Atesta* resulted in the description of a new species, *Atesta patula* sp. nov., which attacks *E. gomphocephala*. *Atesta* was not split away from *Phoracantha* but some specimens now considered *Atesta* were originally misplaced in *Phoracantha*. It is unclear how seriously this new species damages tuart.

The tuart bud weevil (*Haplonyx tibialis*) is a major pest of eucalypt flowers but this has little effect on the overall productivity of the species. A vinegar fly *Drosophila flavohirta* lays its eggs in the flowers of several eucalypts, including *E. gomphocephala*. The larvae which develop feed on nectar and pollen, and the adults feed on pollen. Feeding by *D. flavohirta* reduces nectar availability for honeybees (*Apis mellifera*).

The Australian tortoise beetle (*Trachymela tincticollis*) is a chrysomelid defoliator that has been found on *E. gomphocephala*.

Uses

E. gomphocephala is used in natural silvopastoral systems where cattle graze natural stands of tuart, with their dense understorey of grasses. *E. gomphocephala* is also used as a boundary tree in farming systems.

E. gomphocephala has been used successfully to stabilize sand dunes in a number of countries, and has also been planted as a roadside ornamental and for shelter, windbreaks and land reclamation. The heartwood is pale yellow-brown, texture fine and even with a very interlocked grain. Air-dry density is about 1020–1030 kg/m^3. The sapwood is susceptible to lyctid borer attack. The timber is slow to dry and prone to checking. It is very hard, durable, somewhat termite-resistant and although rich in tannins (7%) is said to have little corrosive effect on metals.

In Western Australia, *E. gomphocephala* was once harvested commercially for sawn timber. The wood was used for purposes where high strength was required, such as wagon wheels (rims, spokes and hubs), railway carriage frames, decking for wagons, telegraph pegs (to support insulators) and tool handles. In Australia it is little used today, largely because of its relative rarity.

Wood from exotic plantations has been sawn for use as flooring blocks, and for boxwood and framing timber. Smaller material is used for particle- and fibreboard and for agricultural use as fence posts, poles and stakes. *E. gomphocephala* has been found to be moderately suitable for wood cement composites; it is also used for pulp.

E. gomphocephala flowers yield a profuse, good-quality nectar and the honey is light in colour, of a pleasing flavour and fine grained texture.

E. gomphocephala leaves yield very low quantities of essential oils which are high mainly in alpha-pinene; however, the species is not harvested for its essential oils as yields are too low and the chemical composition is not attractive for commercial exploitation.

Further reading

Archibald, R.D., Bradshaw, J., Bowen, B.J., Close, D.C., McCaw, L., Drake, P.L. and Hardy, G.E. St J. (2010) Understorey thinning and burning trials are needed in conservation reserves: the case of Tuart (*Eucalyptus gomphocephala* D.C.). *Ecological Management & Restoration* 11, 108–112.

Boland, D.J., Brooker, M.I.H., Chippendale, G.M., Hall, N., Hyland, B.P.M., Johnston, R.D., Kleinig, D.A., McDonald M.W. and Turner, J.D. (2006) *Forest Trees of Australia*. (5th ed.). CSIRO Publishing, Victoria, Australia, 736 pp.

Close, D.C., Davidson, N.J. and Swanborough, P.W. (2011) Fire history and understorey vegetation: water and nutrient relations of *Eucalyptus gomphocephala* and *E. delegatensis* overstorey trees. *Forest Ecology and Management* 262, 208–214.

Eucalyptus grandis W. Hill ex Maiden

Saligna gum

Importance

Eucalyptus grandis can grow rapidly under favourable conditions and can tolerate short-term flooding. It has been a favoured species for industrial plantation forestry in subtropical regions because of its ease of nursery care, rapid early growth, good form and versatile wood properties including excellent qualities for pulp and papermaking. It is reputedly one of the best native species available from Australia for establishment of fast-growing timber plantations over widely differing sites. Given a favourable environment and good silviculture, *E. grandis* plantations commonly yield timber volumes in excess of 25 m^3 ha/yr. However, if conditions are unfavourable and appropriate silvicultural inputs are not supplied, the species will grow slowly or even fail totally.

Botanical features

In its natural habitat, *E. grandis* is a medium-sized to very tall tree, 45–55(–75) m high with a short basal stocking of rough bark then smooth, thin, fibrous or flaking, grey to grey-brown bark above. Branches are mostly smooth white or grey-white. Newly exposed bark may be bright pink. Form is generally excellent with a clear, straight bole up to three-quarters of the total height.

Juvenile leaves are stalked, opposite for several pairs then alternating, ovate to 4–9.5 × 2–4 cm, discolorous, green to dark green. Adult leaves are alternate, lanceolate to broad–lanceolate to 10–16 × 2–3 cm, glossy dark green above, pale green below.

The inflorescence is axillary, 7–11-flowered with flattened stalks 0.8–1.8 cm long. The buds are sessile or shortly stalked, pear shaped with a conical or slightly beaked operculum, 8 × 5 mm. The flowers are white. Fruit are woody capsules, shortly stalked or sessile, conical to slightly pear shaped, often glaucous, to 8 × 7 mm, often contracted towards the top; valves 4 or 5, exserted, broad and incurved. The seed is brown.

Distribution

The southerly limit of natural stands of *E. grandis* is near Newcastle in New South Wales. Its distribution then stretches northwards almost continuously in the coastal regions and subcoastal ranges until just east of Gympie in south-eastern Queensland. Further northwards it occurs in disjunct populations.

The greatest area of plantations of *E. grandis* and its hybrids are those established in Brazil, and several other Central and South American countries contain major plantation areas. It has been planted extensively in India, South Africa, Zambia, Zimbabwe, Tanzania, Uganda and Sri Lanka. Smaller areas of plantations have also been grown in the USA (California, Florida and Hawaii).

Silviculture characteristics

E. grandis is a light-demanding, shade-intolerant species which responds well to heavy thinning. In the first 1–2 years after planting it can present a bushy appearance. After that, height growth is faster and it develops tall columnar stems with moderately dense crowns that can shade out competition from an early age.

Plantations of the species are usually grown on short rotations and clear felled at an age that maximizes production of logs for a particular end use. This is usually small-diameter material suitable for poles, mining timber or pulpwood. Stands to be managed for production of larger-size logs for solid timber applications usually require one or more thinnings from an early age.

Silviculture practice

E. grandis is most commonly propagated from seed and can also be reproduced vegetatively with relative ease. Successful techniques reported include grafting, rooted cuttings and a wide range of micropropagation techniques.

E. grandis is generally raised as potted stock either in single containers or trays containing a number of individual cells. Supporting cells/containers clear of the ground to allow air to circulate freely around their bases encourages both air pruning of vertical roots which emerge from containers and the development of superior root systems.

Associations between *E. grandis* roots and appropriate ectomycorrhizae may increase both tree growth and resistance to environmental stresses in some environments, especially on less fertile soils. In some situations ectomycorrhizae can also increase the efficiency of soil water and nutrient uptake by the trees.

Frost protection may be required in nurseries in cooler areas. *E. grandis* nursery plants can be frost-sensitive when young, but will withstand some frost if appropriately hardened.

Container-raised stock is generally preferred for *E. grandis* establishment. Direct sowing of seed into the plantation has generally provided variable and unreliable results. Establishment spacings for plantations have varied around the world from 1.8 × 1.8 m to as wide as 7.3 × 7.3 m.

E. grandis plantations have frequently been established on sites that were either naturally infertile or have become so through over-exploitation; consequently, many positive responses to fertilizer applications within the first 18 months after planting have been recorded both in Australia and elsewhere. The effects, however, do vary greatly according to tree age, application rates, climatic and soil factors, and nutrient deficiencies.

E. grandis stands managed on medium- to long rotations (i.e. >20 years) for production of solid timber are usually thinned from an early age (as early as 3 years or less), but those for short-rotation crops such as pulpwood or small posts are not normally thinned.

The larger proportion of plantations of *E. grandis* and its hybrids worldwide are grown on short rotations for either production of pulp or charcoal, particularly in Brazil where some of the largest areas have been established. Smaller areas worldwide are grown on longer rotations for solid wood production.

Management practices for *E. grandis* vary widely, being influenced by size and quality objectives, growing environment and genetic quality of the tree stock available.

Diseases and pests

In the nursery, *E. grandis* is susceptible to various fungi causing damping off and collar rot including *Phytophthora* and *Pythium* spp. In nurseries, the species may also be affected by seedling blight caused by *Cylindrocladium quinqueseptatum* and leaf spot disease caused by *Cylindrocladium ovatum*. The former is one of the most significant widespread pathogens in *E. grandis* plantations.

Other significant pathogens that have impacted *E. grandis* plantations to varying degrees include *Cylindrocladium ilicicola*, *Cryphonectria cubensis*, *Puccinia psidii*, 'Mal do Doce' (a disease of uncertain origin), *Corticium salmonicolor* (*Phanerochaete salmonicolor*), *Chrysoporthe cubensis* and *Ceratocystis* spp.

The leaf disease *Cylindrocladium pteridis* has also been reported to cause massive defoliation in some plantations. In recent times, serious new stem canker diseases affecting *E. grandis*, which manifest as longitudinal stem cracks from the butt and slightly swollen areas on the bark, have also emerged in South Africa. A well-known pathogen of woody plants in Southern Africa, *Endothia gyrosa*, is now thought to be one of the major causes of this disease. The pathogen *Botryosphaeria dothidea*, which infects leaves, is also associated with the die-back and canker diseases.

E. grandis is moderately to highly susceptible to insect attack. The more predominant insect pests include leaf beetles (*Paropsis atomaria* and *Paropsisterna cloelia*), Christmas beetles (*Anoplognathus* spp.), leaf blister sawfly (*Phylacteophaga froggatti*) and psyllids and lerps (Psyllidae family). Others which are less important but can cause significant problems in some circumstances include the gum tree scales (*Eriococcus* spp.), the wingless grasshopper (*Phaulacridium vittatum*) and stem-boring insects including the wood moth (*Endoxyla* spp.) and longhorn beetles (*Phoracantha* spp.). Autumn gum moths (*Mnesampela privata*) have on occasions caused massive defoliation in some plantations, even though *E. grandis* is only considered a secondary host for the species.

Infestation of stressed or felled *E. grandis* trees by the wood-boring longhorn beetle (*Phoracatha semipunctata*) has been reported. Occasional epidemics of the Eucalyptus snout beetle (*Gonipterus scutellatus*) have been recorded, and a range of Lepidoptera species (caterpillars) can be significant predators on *E. grandis* leaves.

Leaf-cutting ants (*Atta* spp. and *Acromyrmex* spp.) can cause massive defoliation and even death of *E. grandis*. In parts of Southern Africa and South America the species is very susceptible to attack by termites on some sites, and these can cause heavy early mortality, particularly when trees are young.

Uses

E. grandis has been used for shade, and although sometimes used in shelter belts, its tall growth habit and lighter branching make it an inferior shelter species relative to other eucalypts. Plantations of the species on some acidic soils were found to improve fertility in upper soil horizons and to decrease soil acidity. Pot trials have also shown it to be suitable for revegetation of mined areas. It has been used in irrigated plantations established for disposal of effluent waters. It is also suitable for use in some agroforestry systems.

Sapwood of *E. grandis* is a very pale pink, and is generally resistant to attack by *Lyctus* borers. The moderately strong heartwood varies from almost white to pink, or dark red-brown with a pink tinge. The timber has moderate strength and durability, a straight grain but can have a coarse texture. It has an air dry density of 600–750 kg/m^3.

Some of the earliest plantings of the species as an exotic were for production of pole and mining timbers in South Africa. In Brazil, large areas of the early plantations were established to produce wood for charcoal to use in their steel industry. It is now very widely grown for pulp as its fibres are considered to have excellent paper making properties.

Plantations of *E. grandis* are also grown in many countries for a wide variety of other applications including fuelwood, cellulose, sawn timber, posts, poles and veneers, particleboard production, rayon and it has also been used for production of railway sleepers. The wood is not naturally durable.

Applications for sawn timber of *E. grandis* range from packing cases and house construction to flooring and furniture.

E. grandis is an important nectar source for honeybees in many countries. The leaves of *E. grandis* also yield essential oils.

Further reading

Forrester, D.I. and Baker, T.G. (2012) Growth responses to thinning and pruning in *Eucalyptus globulus*, *Eucalyptus nitens*, and *Eucalyptus grandis* plantations in southeastern Australia. *Canadian Journal of Forest Research* 42, 75–87.

Ramos, L.M.A., Latorraca, J.V. de F., Pastro, M.S., Souza, M.T. de, Garcia, R.A. and Carvalho, A.M. de (2011) Radial variation of wood anatomical characters of *Eucalyptus grandis* W. Hill ex Maiden and age of transition between adult and juvenile wood. *Scientia Forestalis* 39, 411–418.

Vivian, M.A., Santini, E.J., Modes, K.S., Carvalho, D.E., Morais, W.W.C., Souza, J.T. de and Susin, F. (2012) Estimation of strength and stiffness of *Eucalyptus grandis* wood by ultrasound waves. *Pesquisa Florestal Brasileira* 32, 109–113.

Eucalyptus occidentalis Endl.

Swamp yate

Importance

Eucalyptus occidentalis has been extensively planted in many semi-arid and arid temperate regions around the world. Planting is mostly for firewood, shade, windbreaks and catchment afforestation rather than for industrial wood production, and is nowhere on a very large scale. The species displays excellent survival and moderate growth rates in arid, semi-arid and sub-humid winter rainfall climates, and can tolerate soil salinity and seasonal waterlogging. The wood can be used for firewood and to produce paper pulp of fair quality.

Botanical features

A small to medium-sized tree, usually with a fairly dense, flat-topped canopy. There is considerable provenance variation in habit and stem form within the species. The trunk is usually short and forks at or below half the total height, resulting in a tree with several stout primary branches and a flat-topped crown. On the most favourable sites in its natural range *E. occidentalis* attains heights of 25–30 m and a dbh up to 80 cm. On poor or fire-prone sites, it may grow as a mallee no more than 5–10 m in height with several stems that arise from a basal lignotuber. The bark is dark grey-brown to black, rough and hard, fibrous or flaky on the lower trunk, becoming loose and ribbon-like above. The bark of the upper trunk and branches is smooth white or pink.

Juvenile leaves are petiolate, alternate, ovate, 6–14 × 3–7 cm, discolorous, dull, blue-green. Adult leaves are petiolate, alternate, lanceolate, 6–16 × 1–3.3 cm, concolorous, glossy, green, with prominent side veins. The branchlets are red-brown.

The inflorescences are axillary, unbranched, and 7–9-flowered, borne on down-curved, flattened peduncles 1–4 cm long. The buds are 1.6–3.3 × 0.5–0.7 cm, pedicellate, pendulous and elongated with a broad horn-shaped operculum. After the operculum falls at anthesis, the exposed stamens are creamy white in colour. The fruits are bell-shaped, woody capsules, 0.8–1.5 × 0.6–1.2 cm, with a thick rim, level disc and four exserted valves, often outward-curved.

Distribution

The species is endemic to south-western Western Australia. It has been successfully introduced to many countries with sub-humid and semi-arid winter rainfall climates.

Silviculture characteristics

Because of its adaptability to dry temperate climates and somewhat saline soils, *E. occidentalis* has been extensively planted in such environments which are unsuitable for the growth of more productive temperate *Eucalyptus* species such as *Eucalyptus globulus* and *Eucalyptus grandis*. It is notable for its use in reafforestation of bare hillsides for erosion control, shade and shelter in countries such as Italy, and its drought resistance in Israel and North Africa.

Some of the earliest plantings of this species near saltlands were established by the Forests Department of Western Australia in 1950.

Silviculture practice

Seeds may be sown into sand in germination trays and seedlings pricked out into individual containers when they have developed two leaf pairs, or sown directly into containers (polybags or plastic root trainers) and thinned down to the best plant per container. A friable, well-drained potting mix is recommended, with complete fertilizer incorporated in the mix at a rate of 2 kg/m^3 or applied regularly in aqueous solution. The species is generally planted in the spring in temperate latitudes, and growth will be relatively slow in the nursery during the winter months, unless supplementary heating is used. In southern Australia, seedlings are sown in January–February and raised over an 8-month period, for spring planting in September (or July–August in warmer districts where frost is not a problem and soils dry out earlier).

Seedlings are planted out in the spring, when conditions are ideal with rising temperatures and reduced risk of frost, but while the soil is still wet from winter rains. Soil ripping to a depth of at least 70 cm and weed control will improve survival and early growth rates. Control of grasses and other weeds is critical to reduce competition for moisture during the dry summer months. On seasonally waterlogged and saline sites, mounding with a disc plough is recommended to provide a better-drained environment for early growth.

Spacing between trees varies from country to country, but is typically quite wide, to give each individual tree access to reasonable soil moisture resources in the dry climates where the species is typically planted. Spacings from 3 × 3 m on the most productive sites, and up to 4 × 5 m in drier conditions have been used.

Many plantings are managed on a clear felling rotation of about 12 years, for production of firewood or sometimes pulpwood. Plantings which are primarily for windbreaks or shelter belts could be managed on a longer rotation.

After harvesting, the plantation may be regenerated by coppicing. Regeneration by coppice is very good in many countries. Coppice regrowth should be cut back to the best one or two stems per tree, once it reaches a height of about 3 m. If planting stock is of good genetic quality, yielding predominantly single-stemmed, straight trees, it would be possible to thin the stand and manage it for pole, or possibly sawlog, production.

Diseases and pests

E. occidentalis is susceptible to Armillaria root disease. The species is reported to be susceptible to damage by lerps (*Cardiaspina* spp.), gumleaf skeletonizer (*Uraba lugens*), sawflies (Pergidae) and termites in its natural range in Western Australia. Minor damage from leaf-eating beetles has been reported and also further lerp attack by *Cardia jerramungae*. The eucalypt borer *Phoracantha semipunctata* has also been reported on the species, leading to tree death, especially when it is subjected to moisture stress.

In the natural range, hot fires occurring under dry conditions with heavy fuel loads will kill the stems of adult trees, although the trees will usually recover by coppice regrowth.

Uses

Plantations in Italy play an important role in soil conservation in hilly areas of Calabria and Sicily, whereas roadside and shelter belt plantings are valuable for amenity and protective purposes in many countries. Catchments in Calabria, southern Italy, which had been afforested with *E. occidentalis*, were found to have lower water outputs and erosion rates than those covered with natural herbaceous vegetation. Water output and erosion were greater in recently coppiced stands of *E. occidentalis* than in mature stands of the species.

E. occidentalis is now being planted in southern Australia in situations where rising semi-saline water tables are threatening agricultural land. Its ability to grow and transpire water at a reasonable rate under somewhat saline conditions makes it particularly valuable for this application. It is also one of the recommended species for planting near salt seeps in south-western Western Australia, to improve water usage and reduce water and salt accumulation in surface soils.

Groves of *E. occidentalis* irrigated by run-off water are planted in the arid region of Israel for amenity purposes. Low earth dams are used to trap run-off water to increase soil moisture reserves in the plantation areas. *E. occidentalis* has an attractive appearance and has been used for ornamental planting.

The heartwood is pale, hard, somewhat straight grained and durable in damp soils. It has been used for building poles, piling, posts, heavy construction, firewood and charcoal production. Air-dry density of wood from mature natural stands is 850–900 kg/m^3.

E. occidentalis was considered to show some promise for pulpwood production, its only serious fault being low paper-tearing resistance due to short fibre length.

The flowers have some value for honey production, and the bark is reported to have high levels of tannin.

Further reading

Elliott, C. and Byrne, M. (2003) Genetic diversity within and between natural populations of *Eucalyptus occidentalis* (Myrtaceae). *Silvae Genetica* 52, 169–173.

Jeddi, K., Cortina, J. and Chaieb, M. (2009) *Acacia salicina*, *Pinus halepensis* and *Eucalyptus occidentalis* improve soil surface conditions in arid southern Tunisia. *Journal of Arid Environments* 73, 1005–1013.

Shariat, A., Assareh, M.H. and Ghamari-Zare, A. (2010) Effects of cadmium on some physiological characteristics of *Eucalyptus occidentalis*. *Journal of Science and Technology of Agriculture and Natural Resources* 14, 145–154.

Eucalyptus paniculata Sm.

Grey ironbark

Importance

Eucalyptus paniculata is a species of subtropical, warm humid to sub-humid climates. It has proven itself to be a good to very good plantation species, combining quick growth and good coppicing ability with adaptability to a wide range of soil conditions, and some tolerance of drought and light frost. It is a good species for shade and shelter belts, and for production of poles, and an excellent fuelwood and charcoal timber; it is also useful in the production of good honey.

In Australian dry schlerophyll woodland, *Eucalyptus paniculata* may be associated with *Eucalyptus punctata*, *Eucalyptus tereticornis* and *Syncarpia glomulifera*, and it is one of the food plants of the koala (*Phascolarctos cinereus*).

In South Africa the species has been declared a category 2 invader under the Conservation of Agricultural Resources Act (1983).

Botanical features

E. paniculata is a medium-sized to tall tree, 15–30 m high and with a dbh of up to 1 m, but on favourable sites, heights of up to 50 m and diameters of 1.5 m have been recorded. Trunks are typically long and straight with a furrowed, hard or corky, light grey 'ironbark' persistent to the small branches. It has been described as a tall evergreen tree, with leaves dark green and discolorous at all stages of growth, 95–100 mm long when mature and pendulous; whereas young leaves are shorter and broader; cream flowers with long exserted stamens, buds with conical lids and up to 11 mm long. The fruits are hemispherical capsules, 6–10 mm long, tapering to a short stalk and with enclosed or slightly protruding valves (frequently five).

Distribution

E. paniculata originates in eastern Australia. It has now been planted in a number of countries in Europe, Africa, Asia and the Americas.

In its native Australia, *E. paniculata* occurs as a component of dry sclerophyllous woodland (e.g. in New South Wales). In one of its exotic locations, South Africa, it is an invader of watercourses.

Silviculture characteristics

E. paniculata grows rapidly and is able to tolerate a range of soil conditions, as well as drought and light frost.

Silviculture practice

In South Africa this species flowers from May to January, and reproduces from seed.

Diseases and pests

The pathogen *Armillaria mellea* (Armillaria root rot) and a cerambycid herbivore *Phoracantha semipunctata* (blue gum borer, or Eucalyptus longhorn borer) are known to attack *E. paniculata*.

Uses

E. paniculata is a good shade and shelter belt tree.

It is a good pole species and excellent for fuelwood and charcoal timber. The dark to red-brown wood is very hard, strong and durable, and is used principally for heavy-engineering construction. Due to its high density (over 1000 kg/m^3), the wood is difficult to work. Wood products include transmission poles, posts, piles, building poles, building timbers and for boatbuilding.

The species is also used in the production of good-quality honey.

Further reading

Florence, R.G. (1996) *Ecology and silviculture of eucalypt forests.* CSIRO Publishing, Melbourne.

Hamza, K.F.S. (1999) Basic density and some anatomical properties of *Eucalyptus camaldulensis* Dehnhr., *E. citriodora* Hook, and *E. paniculata* Sm. grown at Ruvu, Tanzania. *Annals of Forestry* 7, 221–226.

Mattei, V.L. and Longhi, S.J. (2001) Evaluation of *Eucalyptus paniculata* Smith natural regeneration. *Ciência Florestal* 11, 55–65.

***Eucalyptus regnans* F. Muell.**

Mountain ash

Importance

Eucalyptus regnans is capable of fast growth, giving trees of excellent form and producing valuable timber and very high-quality material for pulp and paper manufacture, once the demanding site requirements for good growth of this species are met. Successful plantation sites are restricted to areas with well-drained, fertile soils and with rainfall of more than 1000 mm evenly distributed throughout the year. The species is intolerant of drought and severe cold.

Botanical features

E. regnans is one of the tallest hardwood trees in the world occasionally reaching up to 100 m in height. Its open crown occupies the top one-third or less of tree height. Typically, forests of *E. regnans* are pure, even-aged stands with mean heights of 70–80 m and with exceptional trees attaining more than 95 m. The typical range in diameters of large, mature trees is 2–3 m. The stem form of this species is usually excellent with straight, clean boles reaching for two-thirds of the total height of the tree.

E. regnans lacks lignotubers. The bark is rough only at the base or to about 15 m up the trunk, thin, fibrous, brown. The upper trunk and branches are smooth, white, cream, grey or greenish, thin, shed in long ribbons, often conspicuously hanging from the branches.

The juvenile leaves are alternate, petiolate, ovate, oblique, tapering suddenly to a pointed tip, 11–17 × 4.5–8 cm, green, glossy, at first discolorous, soon becoming concolorous. The petioles, midribs and leaf edges of juvenile leaves may have numerous raised oil glands. Adult leaves are alternate, petiolate, broad–lanceolate to lanceolate, falcate, oblique at the base of the leaf, 9–14 × 1.5–2.7 cm, green, concolorous.

The inflorescence is simple, axillary usually paired, 9–15-flowered; peduncles angular. The buds are club-shaped (clavate), 5–7 × 3–4 mm. The opercula are hemispherical–apiculate or conical. The fruits are stalked, obconical or pear-shaped, 5–9 × 4–7 mm with a broad disc, valves 3 to rim level or slightly enclosed, or occasionally slightly exserted.

Distribution

The natural forests are in Victoria and Tasmania (Australia). *E. regnans* is grown in commercial plantations in New Zealand and Chile. It has been grown on a small scale in in South Africa and Zimbabwe, and has also been trialled in other countries including Sri Lanka, India and Kenya.

Silviculture characteristics

E. regnans is a fast-growing species reasonably demanding in its site requirements for good growth: it prefers well-drained, fertile soils and a rainfall greater than 1000 mm evenly distributed throughout the year. This eucalypt will not tolerate long periods of water stress and on shallow soils young trees may die of drought and excessive insolation. Although *E. regnans* is intolerant of severe cold (below −7°C), it is quite frost hardy, provided suitable provenances are selected for planting.

Silviculture practice

E. regnans is typically propagated by seed. The small, brown seed is orthodox and will remain viable in storage for several years if kept dry (5–8% moisture content) in refrigerated, airtight containers at 3–5°C. There are, on average, 181,000 viable seeds/kg of seed and chaff mix. Cold, moist stratification of seed at 3–5°C for 3 weeks is recommended to break dormancy and promote rapid and even germination on sowing. Germination takes place over 3 weeks under moist, mild to warm conditions (15°C or 20°C are optimal in the laboratory) in the presence of light. Germination is epigeal.

Stem cuttings taken from seedlings of *E. regnans* have been propagated successfully. However, the regular technique of vegetatively propagating a wide range of eucalypts by taking stem cuttings from basal epicormic shoots of selected trees is not viable with this species, as it does not coppice readily. Juvenile *E. regnans* has been successfully micropropagated, and the species lends itself to grafting.

E. regnans is generally raised in containers, and plantable size is reached at 8–9 months of age. It can be more troublesome to raise in the nursery than some eucalypts, and the species is particularly susceptible to damping off.

Natural forests of *E. regnans* are usually regenerated through a system of clear felling, burning and aerial sowing of pelleted seed, supplemented by later spot-sowing to fill the gaps. Regeneration of plantations of *E. regnans* also relies on clear felling followed by replanting, as this species does not coppice well enough to ensure a coppice rotation. Rotation lengths are commonly 15–25 years.

Spacings of 2.5–3.0 × 2.5–3.0 m (1100–1600 stems/ha) are recommended for eucalypt woodlots.

Control of weed competition in the first few years after establishment is often critical to the survival and growth of eucalypts. Their roots form symbiotic mycorrhizal associations with various fungi. Ectomycorrhizal or dual ectomycorrhizal and endomycorrhizal (vesicular-arbuscular or VA) associations can occur in the genus. Many exotic plantations lack mycorrhizal associations, since compatible fungi are not always present in the soil.

As a general practice, nitrogen, phosphorus, potassium and trace elements in normal- and slow-release forms are added to potting mixes in Australian eucalypt nurseries; plantation soils are also fertilized.

E. regnans is sensitive to stem damage because of its thin bark. Thinning causes injury to remaining stems and this allows entry of organisms that cause rapid spread of brown staining and a slower spread of decay. However, losses anticipated due to decay and other silvicultural problems which may result from thinning, such as increased branching, coppicing, epicormic shooting, wind throw or pest and disease attack, are likely to be offset by better diameter growth of residual trees.

Diseases and pests

Seedlings of *E. regnans* are highly susceptible to attack by *Phytophthora cinnamomi* and to leaf spot diseases; *Phytophthora nicotianae* has also been reported to attack nursery seedlings. *Mycosphaerella* spp. are associated with severe leaf-blight epidemics on mature *E. regnans* in Australia, and *Mycosphaerella cryptica* has been found to cause stunting and malformation of trees in New Zealand. *Armillaria luteobubalina* occurs widely in native eucalypt forests and was reported to cause the death of a small area of trees in a vigorous plantation of *E. regnans* in eastern Victoria. In Tasmania, a die-back known as 'regrowth die-back', of uncertain origin, caused economically significant damage to regrowth forests.

Plantations of *E. regnans* are very susceptible to attack by the eucalypt leaf or tortoise beetle (*Paropsis atomaria*). The species is also highly susceptible to attack by *Gonipterus* sp. beetles (eucalypt weevils). The longhorn beetle (*Phoracantha semipunctata*) bores into the trunks of *E. regnans* trees that are under stress, leading to red kino stains exuding on to the outer bark.

Large populations of the defoliating stick insect *Didymuria violescens* have caused severe damage to natural stands of *E. regnans*. Defoliation by the leaf beetle *Chrysophtharta bimaculata* causes reduction in height and basal area increment.

Conspicuous damage by a complex of wood-boring moths, including two species of *Xyleutes* and an unidentified xyloryctid species, has also been reported.

Heavy and continuous browsing by wildlife (wallabies, possums and rabbits) causes significant reductions in growth of newly established plantations of *E. regnans* and in extreme cases may cause them to fail completely.

Uses

E. regnans is recommended for growing in farm woodlots in New Zealand, for slope and gully stability planting and as a windbreak.

The heartwood is pale pink or pale straw in colour, with a straight grain and often prominent growth rings, whereas the sapwood is not always clearly differentiated. The sapwood is resistant to lyctid borer attack and the air-dry density is about 600–800 kg/m^3. Collapse occurs during drying, so reconditioning is standard practice for dressed products. The timber is relatively easy to work, good for steam bending, and glues satisfactorily. When fumed with ammonia the colour changes to pale walnut. The heartwood is not durable. The timber is used for engineering and building construction, marine craft, flooring, furniture, sliced veneer and plywood. The timber of *E. regnans* is similar in appearance to two closely related species, *Eucalyptus delegatensis* and *Eucalyptus obliqua*, to the extent that all three species are interchangeable for many uses and are often sold in mixed batches under the trade name of Tasmanian Oak, although there are important physical and anatomical differences in their timbers.

Multi-aged forests of *E. regnans*, with their substantial numbers of hollow-bearing trees, provide important habitats for a suite of arboreal marsupials including Leadbeater's possum (*Gymnobelideus leadbeateri*), and for birds such as the yellow-tailed black cockatoo (*Calyptorhynchus funereus*) and other hollow-dependent wildlife. There are concerns that current harvest methods in natural stands, based on clear felling, even-aged regrowth and relatively short rotations of 50–80 years, will substantially reduce the abundance of trees with cavities and put at risk hollow-dependent fauna; this has led to calls for modification of current harvesting strategies to ensure that wildlife habitat is maintained.

E. regnans has a reputation as one of the best of the eucalypts for pulp yield and quality for all types of mechanical and chemical pulp.

Bees thrive on the pollen of this species and produce a strongly flavoured honey. The foliar essential oils of *E. regnans* have been studied but are not of interest for commercial production.

Further reading

Haslett, A.N. (1988) Properties and utilisation of exotic speciality timbers grown in New Zealand. Part V: Ash eucalypts and *Eucalyptus nitens. Eucalyptus delegatensis* R. Baker; *Eucalyptus fastigata* Deane et Maiden; *Eucalyptus fraxinoides* Deane et Maiden; *Eucalyptus obliqua* L'Her, *Eucalyptus regnans* F. Muell, *Eucalyptus sieberi* L. Johnson; *Eucalyptus nitens* (Deane et Maiden) Maiden. *FRI Bulletin*, 119, 20.

Loyn, R.H., Kennedy, S.J., Beadle, C., Duff, G. and Richardson, A. (2009) Designing old forest for the future: old trees as habitat for birds in forests of Mountain Ash *Eucalyptus regnans*. *Forest Ecology and Management* 258, 504–515.

Waters, D.A., Burrows, G.E. and Harper, J.D.I. (2010) *Eucalyptus regnans* (Myrtaceae): a fire-sensitive eucalypt with a resprouter epicormic structure. *American Journal of Botany* 97, 545–556.

Eucalyptus robusta **Sm.**

Swamp mahogany

Importance

Eucalyptus robusta (Plate 16) is a medium to large tree capable of rapid early growth in suitable environments. This species has a broad environmental amplitude, and is planted widely outside Australia in equatorial to cool temperate regions. The species may tolerate adverse conditions and is especially useful on sites subject to prolonged flooding. It is often used as a shade tree, an ornamental, in shelter belts and in water catchment rehabilitation. The wood of *E. robusta* is a good fuel and is commonly used for charcoal production. It is durable and is used in the round for posts and poles. Sawn timber can be used for general construction, but requires kiln seasoning to avoid degradation during drying. It is possible to use *E. robusta* as a source of pulpwood for papermaking, but other eucalypts are usually preferred.

Botanical features

E. robusta has a dense crown and long, spreading branches when open grown. It regularly attains heights of 20–30 m, with a dbh of up to 1 m. The trunk is usually straight and extends to about one-half the height of the tree, or to two-thirds of the tree height in dense stands on favourable sites. The bark is rough and persistent to the small branches, thick, held in coarse, soft, spongy, elongated slabs with deep longitudinal furrows, grey or reddish-grey-brown. Trees growing on wet sites in open stands often form both small buttresses and aerial roots in major branch crotches. Trunks may become encased by aerial roots, some reaching 20 cm in diameter, under wet tropical conditions.

The juvenile leaves are petiolate, ovate, up to 19 × 8 cm, strongly discolorous, green, opposite for several pairs, then alternate. Adult leaves are petiolate, leathery, discolorous, glossy and dark green above, pale green below, broad–lanceolate, up to 17 × 4.5 cm.

The inflorescence is axillary, 9–15 flowered, the peduncles are strongly flattened, up to 3 cm in length; buds with prominent stalks (pedicels) to 0.9 cm in length, rarely sessile, to 2.4 × 0.8 cm, scar present, operculum long, beaked; flowers white. The fruit is a woody capsule, with prominent stalk, cylindrical, 1.8 × 1.1 cm; valves 3 or 4, usually joined across the orifice, to rim level or slightly exserted. The seed is brown.

Distribution

E. robusta occurs naturally in Australia, mainly within a narrow coastal strip from southern New South Wales to coastal south-eastern Queensland (and offshore islands).

E. robusta as one of the eucalypt species most commonly planted outside Australia. It is best known as a plantation species in the Madagascar highlands, but also in the USA (Hawaii, Florida, California) and Central America (Puerto Rico).

Silviculture characteristics

E. robusta is a moderately fast-growing tree with reasonable form and excellent coppicing ability. It grows well on a wide range of soil types in the presence of adequate moisture, including degraded and exposed lateritic soils in the tropics. It is sensitive to frosts when young. This species recovers well after fire, is highly tolerant to waterlogging, and will withstand severe coastal exposure. These characteristics facilitate its use in afforestation of coastal swamps and sand dunes. However, older trees may be severely damaged in high winds and may shed large limbs. *E. robusta* is slightly to moderately salt-tolerant.

Silviculture practice

Seed is the main means of propagating *E. robusta*. Seedlots may be cleaned to a high level of purity by sieving. Viability of seed stored dry (5–8% moisture content) in refrigerated, airtight containers at 3–5°C may be maintained for several years. No pre-sowing treatment is required. Rapid and complete germination is achieved under moist, warm conditions in the presence of light (25°C is optimal in the laboratory). Germination is epigeal.

E. robusta may also be regenerated vegetatively by stem cuttings and micropropagation.

Containerized seedlings are now preferred to bare-rooted planting stock because of enhanced survival and growth, although bare-rooted planting of *E. robusta* has been successfully applied in the past. Seed may be sown under shade either directly in nursery containers, or first sown in germination beds with seedlings and later transferred to containers once the seedlings are at the second leaf-pair stage (about 6 weeks from sowing). The seedlings can be grown in a variety of potting mixes including peat/coarse sand/sandy loam; peat/vermiculite; and peat/vermiculite/perlite. Slow-release fertilizers may be added to the potting mixture at the rate of 1.75–4.3 kg/m^3, or liquid fertilizer applied as the seedlings grow in the nursery. During the nursery phase, as with other species in the genus, *E. robusta* is highly susceptible to damping off and other fungal pathogens. Occurrence of disease problems in the nursery can be limited by strict attention to hygiene, reduced watering and shade, and good ventilation. After 2–6 months following sowing, or once the seedlings reach a height of about 25 cm, they are planted out in the field. This should coincide with the onset of the wet season in tropical countries.

E. robusta is grown in plantations with rotations of approximately 4 or 5 years for fuelwood, 8–10 years for pulpwood and 15–20 years for poles. Longer rotations of 30–60 years are used to produce sawlogs. Coppicing is the main means of regeneration after felling.

As with other species in the genus, *E. robusta* is intolerant of shade and does not compete well with grasses for water and nutrients. Intensive site preparation by ploughing is beneficial; on compacted sites deep ripping may also be used, and on wet sites mounding improves root aeration and provides well-drained conditions which facilitate planting.

Eucalypt roots form symbiotic mycorrhizal associations with various fungi. Both ectomycorrhizal and VAM associations may occur within this genus. In Brazil, the ectomycorrhizal fungi *Pisolithus tinctorius* has been isolated from *E. robusta*. The fungi are attached to the roots and benefit from other nutrients available from the tree, and in return the tree derives certain nutrients (especially phosphorus) from the fungi. Many exotic plantations are devoid of mycorrhizal associations since compatible fungi are not always present in the soil. While this seldom affects the successful establishment of eucalypts in exotic environments, evidence suggests that growth rates and resistance to environmental stress may not be optimal unless suitable mycorrhizal fungi are present. Nurseries growing eucalypts, especially where the soils are deficient in phosphorus, are recommended to introduce appropriate mycorrhizae to the nursery soil. Various inoculation systems include soil, spores, sporocarps and vegetative mycelium.

Although this species is often cultivated without organic or chemical fertilizers, on sites of very low soil fertility, fertilizer application may be necessary to increase productivity. *E. robusta* may be grown under irrigation and is highly tolerant to waterlogging. Some plantations are weeded, and light grazing by cattle may be used to implement this.

Diseases and pests

Leaf spots on *E. robusta* have been attributed to the fungal pathogens *Harknessia hawaiiensis*, *Harknessia insueta*, *Colletotrichum gloeosporioides*, *Cylindrocladium ovatum*, *Cercosporella theae* and *Readeriella mirabilis*. In the past, the fungus *Cylindrocladium scoparium* has caused serious seedling losses. *E. robusta* is susceptible to gummosis and trunk rots caused by *Polyporus schweinitzii* and *Fomes* sp., cankers caused by *Botryosphaeria ribis* and root rots. *E. robusta* has been attacked by the bacterium *Phytomonas tumifaciens*.

The juvenile foliage is attacked by leaf-blister sawfly (*Phylacteophaga froggatti*), autumn gum moth (*Mnesampela privata*) and the leaf-eating beetle (*Maecolaspis favosa*), whereas adult foliage is susceptible to lerp (*Cardiaspina* sp.), autumn gum moth, leaf beetle (*Paropsis* sp. and *Chrysophtharta* sp.), leafblister sawfly, gumtree hoppers (*Eurymela* sp. and *Eurymeloides* sp.), scale (*Eriococcus coriaceus* and *Eriococcus confusus*) and Christmas beetle (*Anoplognathus* sp.). Older trees under stress by drought are attacked by eucalypt borers (*Phoracantha* sp. and *Epithora dorsalis*), bullseye borer (*Phoracantha acanthocera*, synonym *Tryphocaria acanthocera*) and cossid borer (*Endoxyla* spp.). *E. robusta* is susceptible to eucalyptus snout beetle (*Gonipterus scutellatus*). Young trees are susceptible to termites.

Uses

E. robusta has a dense, deep crown of dark green leaves and seasonally abundant creamy-white flowers, and is suitable for shade, shelter and as an ornamental tree. The tendency of older trees to shed large limbs in strong winds makes it an undesirable selection for areas affected by typhoons. It has been used for watershed protection, mine site reclamation and dune stabilization.

E. robusta provides a moderately durable hardwood with an air-dry density of 805–900 kg/m^3 for wood from natural stands in Australia and 725–800 kg/m^3 for plantation timber in Hawaii. The heartwood is pale red when freshly cut, but turning orange-red or red-brown with age. The sapwood is pale brown, 40 mm wide and susceptible to attack by lyctid borers. The logs are prone to sap-stain and pinhole borer attack, yet are moderately resistant to termite attack. The heartwood is moderately durable, having an approximate life in the ground of 8–25 years. It may split and distort during seasoning and is not stable in fluctuating atmospheric conditions. This can be avoided if the timber is air-dried to less than 30% moisture content before kiln-seasoning. It is not favoured for general building purposes because of degradation during drying, or for steam-bending. The wood is resistant to marine borers.

The timber is used in general construction and for underground pilings, utility poles and fence posts, pallets and furniture. However, high humidity causes shrinkage and expansion which makes it unsuitable for furniture construction in many areas. The wood of *E. robusta* forms excellent stakes used in irrigation systems, and for conveyor-belt slats used in the sugar industry. Truck beds, weatherboards for houses, flooring, interior trim and panelling are made from *E. robusta*, and it may also be used to produce rotary-cut, construction-grade veneer and pulpwood. *E. robusta* is commonly used for fuelwood in many parts of the world, and also for charcoal. Studies of its use in plantations as a potential source of woody biomass to generate electricity have also been conducted.

Bees foraging on the nectar of *E. robusta* flowers produce a dark amber, highly aromatic honey of acceptable flavour.

Further reading

Eldridge, K., Davidson, J., Harwood, C. and van Wyk, G. (1994) *Eucalypt Domestication and Breeding*. Clarendon Press, Oxford, UK, 308 pp.

Hernández-Hernández, R.M., Ramírez, E., Castro, I. and Cano, S. (2008) Changes in quality indicators of hillside soils reforested with pines (*Pinus caribaea*) and eucalyptus (*Eucalyptus robusta*). *Agrociencia (Montecillo)* 42, 253–266.

Skolmen, R.G. (1963) Robusta Eucalyptus wood: Its properties and uses (U.S. Forest Service research paper PSW). Pacific Southwest Forest and Range Experiment Station, Forest Service, U.S. Dept. of Agriculture, Berkeley, California, 12 pp.

Eucalyptus saligna Sm.

Sydney blue gum

Importance

Eucalyptus saligna is a tree of inherently good form, capable of fast growth, amenable to management as a short rotation coppice crop and provides a useful, general-purpose hardwood. It has proven successful as a reforestation species in many countries.

Botanical features

A medium to tall tree, usually 30–55 m in height with a dbh up to 2 m. Exceptional trees may attain 65 m in height with a dbh of 2.5 m. The bark is rough at the base or higher, persistent, moderately thick, short-fibred, sometimes tessellated, brownish-grey, becoming thin and finally shed in long thin flakes or strips; rest of trunk and branches smooth grey-green or bluish-green. The trunk is generally of good form, straight and clear of branches for one-half to two-thirds of the total tree height. The seedlings have lignotubers.

The juvenile leaves are petiolate, opposite for several pairs then alternate, ovate to broad lanceolate 12 × 5 cm, discolorous, green above, pale green below. Adult leaves are alternate, stalked, lanceolate to broad lanceolate, to 16 × 3 cm, discolorous (green above, pale green below).

The inflorescences are axillary, unbranched, supported by flattened peduncles 1.8 cm long, with 7–11 flowers. Buds are sessile to shortly stalked, tapering into the base of each bud, pyriform to 9 × 4 mm; scar present from shed outer operculum, operculum conical or slightly beaked; flowers white. The fruit are woody capsules, shortly stalked or almost sessile, obconical to slightly pyriform, to 8 × 7 mm; valves 3 or 4, erect with thin pointed tips, either just protruding above the rim level or strongly exserted, usually conspicuously outcurved; seed brown.

Distribution

In Australia, *E. saligna* is a common species from the south coast of New South Wales to southern Queensland, mostly within 120 km of the coast (with some isolated populations further north). It has been widely planted outside Australia, particularly in countries in South America and southern Africa, and in India and Sri Lanka.

Silviculture characteristics

E. saligna is a lignotuberous species of good coppicing ability and will tolerate a wide range of soil types from clays to sands. Although best growth is achieved on sites with a high rainfall (more than 900 mm) well distributed throughout the year, it can survive moderate droughts and has been successfully established where there are 4 summer months with rainfall of less than 25 mm/month. Young stands are moderately sensitive to fire damage, but in older trees their basal stocking of fibrous bark gives some protection from the effects of fire. It is moderately frost-tolerant and will withstand winter temperatures down to about −8°C, when sufficiently hardened by night temperatures lower than 5°C over several weeks prior to the onset of frosts. It shows intolerance of winds, both strong and salt-laden, and possums.

Silviculture practice

E. saligna is easily raised from seed in the nursery. *E. saligna* is also suited to mass vegetative propagation through stem cuttings, and can be propagated by air-layering and grafting.

E. saligna planting stock is usually raised in containers in the nursery. Seed should be sown under shade (optimum temperature for germination is 25°C) on a free-draining and sterilized medium, and covered very sparingly with inert material (e.g. sand). Germination should be complete after 2 weeks and then shade can be reduced. At the second leaf-pair state germinants are transplanted to containers filled with sterilized potting mix. Shade cover is needed for the first week after transplanting, after which time plants should be fully exposed to direct sunlight. Seedlings should reach a size suitable for planting out in little as 12 weeks. Seedling survival may be reduced by damping off fungi such as *Pythium* sp. and *Rhizoctonia* sp. Common mistakes in propagation are over-watering (leading to disease problems), over-shading, allowing the germinants to become too large for easy transplanting (resulting in malformed taproots or root curling) and allowing roots from the containers to grow into the soil. Occurrence of disease problems in the nursery can be limited by strict attention to hygiene, reducing watering and shade and allowing good ventilation.

In the cool, moist conditions of New Zealand, *E. saligna* can be raised successfully in open nursery beds and established as open-rooted planting stock. Regular root pruning is a critical part of the process. Planting is best done under moist, cloudy, frost-free conditions.

E. saligna is capable of producing very high volumes on short rotation. Spacings to produce small-diameter material suitable for pulpwood, mining timber, charcoal or fuelwood are usually 2.1 × 2.1 m (2.4 × 2.4 m in drier areas) with rotations of 9 years; second and subsequent rotations are regenerated by coppice thinned to two or three stems per stump.

Successful growth of eucalypts requires good establishment practices. On old agricultural land, intensive site preparation by ploughing to a depth of 10 cm when the soil is moist, followed by disc harrowing 3 months later, is beneficial; mould-boarding can give effective weed control by burying weed seed. On compacted sites, deep ripping of the planting lines (to 1 m depth with a bulldozer) is recommended. Ripping in combination with mounding improves survival and growth on most sites, mounding being essential on wet or saline areas. On stony or very steep sites, complete cultivation may be impracticable. It is then necessary to prepare planting pits with a 1 m surround cleared of weeds.

Fertilizer application in the nursery and in the field is often necessary to ensure the good growth of eucalypts, and *E. saligna* is no exception. Weed control is especially important to the successful establishment of eucalypts both before and after planting.

Pruning of *E. saligna*, irrespective of whether it is established in pure plantations, mixed with *Pinus radiata* or interplanted into scrub, is standard practice in New Zealand to reduce knots which downgrade sawn timber. Branches should be removed before their diameters exceed 2.5 cm, as large stubs increase the risk of fungal infection and stem decay.

Diseases and pests

Die-back disease has killed *E. saligna* in patches throughout coastal New South Wales; the cause is unknown. *Armillaria luteobubalina* has been associated with deaths of young trees of *E. saligna* in rehabilitation plantings in forests of *Eucalyptus marginata* in south-western Australia. Pink disease, *Corticium salmonicolor* (*Phanerochaete salmonicolor*), causes stem and branch cankers and die-back of tops in *E. saligna* plantations in humid climates.

A stem-decay fungus *Chondrosterum purpureum* associated with decay around pruned branch stubs, has been noted in stands of *E. saligna* in New Zealand, and can cause the death of young trees. Other pathogens observed on *E. saligna* in New Zealand but recorded as causing only minor damage are *Kirramyces epicoccoides* (causing leaf lesions), *Botryosphaeria dothidea* (associated with branch cankers), *Phytophthora* sp. (causing defoliation of some *E. saligna* stands in the Bay of Plenty) and *Peniophora sacrata* (a root disease).

Diseases of *E. saligna* in South Africa include *Cryphonectria cubensis* (black stem canker), which is limited to the warmer (sub-tropical) areas; *Endothia gyrosa* (yellow stem canker), which is most prevalent in cooler areas and affects moisture-stressed, suppressed individuals; and leaf spots such as *Aulographina eucalypti* and *Phaeoseptoria eucalypti*. *E. gryosa* is a major cause of timber degradation in South Africa.

Cryphonectria cubensis has also caused significant damage to *E. saligna* plantations in Brazil, and *Endothia havanensis* has caused serious cankers to 3 year-old plantations of *E. saligna* in Suriname. *Mycosphaerella parkii* has been associated with leaf spots of *E. saligna* in Indonesia.

In a species/provenance trial of susceptibility to *Puccinia psidii* (guava rust) in Brazil, the susceptibility of *E. saligna* to the disease was found to vary among provenances.

Natural forests and plantations of *E. saligna* in New South Wales may be severely damaged by a wide range of insect pests including a large and diverse suite of phytophagous insects, among which are at least 16 species of psyllid. Debilitated trees are then predisposed to attack by the eucalypt keyhole borer *Xyleborus truncatus* and to an associated brown-staining fungus, a combination often leading to tree death. Principal insect defoliators of juvenile *E. saligna* include *Phaulacridium vittatum* (wingless grasshopper) and *Pylacteophaga froggatti* (leaf blister sawfly). A gall wasp thought to be closely related to *Ophelimus eucalypti* has been found on *E. saligna* in New Zealand. Defoliation can be so severe that young trees die, and significant growth loss occurs in older trees.

The wood borer *Phoracantha semipunctata* is a serious pest of both natural and planted stands of *E. saligna* both in Australia and in other countries such as South Africa and the USA (California). Stressed trees are most susceptible to attack. A range of wood-boring Lepidoptera may severely degrade logs of plantation-grown *E. saligna*.

Uses

E. saligna is used in shelter belts and makes a fine avenue tree. Coffee/tree agroforestry systems (including *E. saligna*) are being promoted in Costa Rica as a means of providing protective buffer zones to biosphere reserve core zones.

The sapwood is pale pink and susceptible to attack by *Lyctus* borers; heartwood pink to red, usually straight grained but occasionally interlocked, causing ribbon figure on the quarter-cut face, moderately coarse textured, relatively easy to work. Green density is about 1070 kg/m^3 whereas air-dry density is about 850 kg/m^3. However, the density of rapidly grown wood from plantations is usually significantly less. The heartwood is moderately durable. When grown slowly the heartwood may present problems for semi-chemical pulp, but not when the trees are young and fast grown.

The wood is suitable for use for general construction, joinery, decking, decorative flooring, panelling, cladding, boatbuilding, heavy furniture, medium-density fibreboard and plywood, reconstituted wood products and pulpwood. In addition it may be used for fuelwood, chipwood and boxes.

E. saligna produces a strongly flavoured, rather dark honey which frequently lacks density and does not store well.

Further reading

Haslett, A.N. (1990) Properties and utilisation of exotic speciality timbers grown in New Zealand. Part VI: Eastern blue gums and stringy barks. *Eucalyptus botryoides* Sm. *Eucalyptus saligna* Sm. *Eucalyptus globoidea* Blakey. *Eucalyptus muellerana* Howitt. *Eucalyptus pilularis* Sm. *FRI Bulletin* 119, 21 pp.

McKenzie, H. and Hay, E. (1996) Performance of *Eucalyptus saligna* in New Zealand. *FRI Bulletin* 195, 31 pp.

Onofre, F.F., Engel, V.L. and Cassola, H. (2010) Natural regeneration of Atlantic Forest species in the understory of *Eucalyptus saligna* Smith. in a former forest production unit at the parque das Neblinas, Bertioga, SP. *Scientia Forestalis* 38, 39–52.

Eucalyptus sideroxylon A. Cunn. ex Woolls

Red ironbark

Importance

Eucalyptus sideroxylon, commonly known as ironbark or red ironbark, is native to Australia.

E. sideroxylon is a relatively slow growing, coppicing species of indifferent form. Its tolerance of drought, some frost and poor shallow soils has seen it widely planted in Northern Africa. Its timber, although difficult to work, is very heavy, very hard and very durable and is used in general construction and particularly where durability is important. The tree has a number of other uses and is also planted as an ornamental.

In its alien range, *E. sideroxylon* has been reported to invade some semi-natural habitats, such as timber plantations in Namibia and Northern Cape Province, South Africa, and it has also been described as an invader of watercourses in South Africa.

E. sideroxylon subsp. *tricarpa* is now considered to be the separate species, *Eucalyptus tricarpa*.

Botanical features

E. sideroxylon is a small to medium-sized woodland tree, commonly 10–25 m tall with exceptional specimens reaching 35 m, and with stem dbh up to 1.0 m. The form of the trunk is often rather poor, and the length does not usually exceed one-half of the tree height. The 'ironbark' is persistent on larger branches, hard and deeply furrowed, dark brown to black, with upper limbs covered in a smooth, whitish bark. It has been described as an evergreen tree with the blackest bark of all eucalypts (although some authors state that the bark is dark red), leaves dark grey-green, 60–110 mm long, lance-shaped, pendulous, young leaves much more variable in shape. Flowers cream, pink or deep rose-red with exserted stamens. The buds are pendulous with conical or beaked lids, up to 12 mm long. The fruits are brown pendulous capsules, round to oval, 8–10 mm long with deeply enclosed valves, which may be covered by a sub-persistent staminal ring.

Distribution

E. sideroxylon is native to Australia, occurring on the western slopes and plains of New South Wales and also west of Sydney to the Blue Mountains. It is also widespread in south-eastern

Queensland west of the Great Dividing Range, with a small extension into northern Victoria.

The ability to withstand droughts, frosts and poor soils has led to the introduction of this species to regions outside its native Australia, including countries in Asia, Africa, the Americas and Europe.

E. sideroxylon is considered an invasive species in South Africa where it has been widely planted for various uses including timber, fuelwood, in shelter belts, for honey production and as an ornamental plant; however, it has been reported to invade riparian habitats.

Silviculture characteristics

E. sideroxylon is a woodland species, typically found on infertile, shallow soils, including sands, gravels, ironstones and clays.

Provenance variation can be anticipated as the species has a very wide ecological range. However, few comprehensive provenance trials appear to have been established as a basis for properly assessing the potential of the species. *E. sideroxylon* is a relatively slow growing, coppicing species. It reproduces from wind-dispersed seed.

The climate in the native range of *E. sideroxylon* is largely warm sub-humid, but the species extends to the warm humid and warm semi-arid zones, with annual rainfall of 420–1000 mm, with a winter or summer maximum and extended dry season depending on location. It can tolerate a range of soil types, including shallow, saline and infertile sites, though prefers free-draining soils, and can be grown at altitudes up to 2000 m.

Silviculture practice

E. sideroxylon may be grown from seed. It has the ability to produce vigorous shoots after coppicing.

Diseases and pests

Two herbivores, *Anoplognathus* sp. and *Gonipterus* sp., are noted as occasional pests; limited information is available. *E. sideroxylon* is also host to the red gum lerp psyllid (*Glycaspis brimblecombei*) and the leafblister sawfly (*Phylacteophaga froggatti*).

Uses

The tree has ornamental value, a pink-or red-flowering form (*E. sideroxylon* var. *rosea*) being used in horticulture. It also makes a good shade and shelter belt tree.

The timber of *E. sideroxylon*, although difficult to work, is very heavy, very hard and very durable. It has a density of 1040–1105 kg/m^3 at 12% moisture content. It is used in general construction and particularly where durability is important. Wood products made from this tree include posts, building timbers, beams, exterior fittings, bridges, railway sleepers, boats, pulp, fuelwood and charcoal.

Essential oils and plant extracts from *E. sideroxylon* are reported to have antimicrobial and anti-tumour properties; it is also an excellent species for use in honey production.

Further reading

Ashour, H.M. (2008) Antibacterial, antifungal, and anticancer activities of volatile oils and extracts from stems, leaves, and flowers of *Eucalyptus sideroxylon* and *Eucalyptus torquata*. *Cancer Biology and Therapy* 7, 399–403.

Boland, D.J., Brooker, M.I.H., Chippendale, G.M., Hall, N., Hyland, B.P.M., Johnston, R.D., Kleinig, D.A., McDonald, M.W. and Turner, J.D. (2006) *Forest Trees of Australia*. 5th ed. CSIRO Publishing, Collingwood, Victoria, Australia.

Craze, B. and Salmon, J. (2004) *A Review of Literature on Ironbark Ridges and Associated Lands*. Dept. of Infrastructure, Planning and Natural Resources, Temora, New South Wales, Australia.

Eucalyptus tereticornis Sm.

Forest red gum

Importance

Eucalyptus tereticornis is a versatile, fast-growing and widely planted tree species. It has been especially successful in areas with summer rainfall followed by a moderate to severe dry season, although fast-growing provenances adapted to a winter rainfall are also available from its extensive natural distribution.

The species is marginally less drought-tolerant than its close relative *Eucalyptus camaldulensis*, but appears to be less susceptible than the latter to leaf pathogens and stem cankers when planted in areas of seasonally high humidity. Provenances have been identified from the southern, temperate part of the species distribution that are moderately frost and salt-tolerant.

E. tereticornis is proving a valuable partner in various hybrid combinations with other fast-growing eucalypts, including *Eucalyptus grandis* and *Eucalyptus urophylla*. The wood is used for a wide range of purposes including fuel, charcoal, paper, poles, posts, mining timber, hardboard and particleboard, while the tree is used for shelter belts, shade and in apiculture.

Botanical features

A medium-sized to tall forest tree attaining 20–50 m in height and up to 2 m in diameter. The trunk is usually straight and clear for more than half of the total height, with the major branches steeply inclined. The bark is shed in irregular plates leaving a smooth or granular, mottled surface of white, grey, dark grey and bluish patches over the trunk. The trunk may carry a short stocking of rough, dark grey to black dead bark.

Juvenile leaves are on stalks, opposite for a few pairs then alternating, ovate, slightly discolorous, bluish-green, to 21 × 10 cm. The adult leaves are stalked, alternating, narrow–lanceolate to lanceolate, to 20 × 2.7 cm, concolorous, green.

The inflorescences are axillary and unbranched, 7–11 flowered, on stalks 2.5 cm long. The buds are to 2 × 0.5 cm, elongated, with a distinctive scar left after the first operculum has been shed during early bud development; the second operculum is distinctively horn-shaped, widening at the base, up to 7 times as long as the hypanthium. The flowers are white, rarely pink. Fruit are woody capsules, stalked, hemispherical or ovoid including the disc, to 0.6 × 0.8 cm; disc broad, often broader than the hypanthium, steeply ascending; valves 4(5), strongly exserted. The seed is usually black with pitted surface and toothed edges.

Distribution

E. tereticornis has the most extensive latitudinal range of any species in the genus. It occurs along the coast and on adjacent hills and plains of eastern Australia, from eastern Victoria in the south to about Cooktown in far northern Queensland. It is in widespread use in eucalypt afforestation projects in many parts of the world, but particularly in the seasonally dry tropics.

Silviculture characteristics

E. tereticornis is a fast-growing, versatile and strongly coppicing species. It is noted for its drought hardiness, although an annual rainfall of over 800 mm is preferred.

Regarded generally as fairly sensitive to frost, in Uruguay *E. tereticornis* been grown with reasonable success in areas where unseasonal frosts are likely to occur. This no doubt depends on the provenance used, as selection of the correct provenance for a particular planting site is paramount to planting success with this species. *E. tereticornis* is a moderately salt-tolerant species and it is also relatively fire-resistant.

Silviculture practice

E. tereticornis is usually propagated from seed, and it is relatively easy to propagate vegetatively. The species may also be grafted, and micropropagation of the tissues from bud, shoot or apex of stem has been carried out in China and India.

Seed may be sown under shade either directly to nursery containers or first in germination beds with seedlings transferred to containers at the second leaf-pair stage, or about 6 weeks from sowing. The seedlings can be grown in a variety of potting mixes including peat/coarse sand/sandy loam; peat/vermiculite; and peat/vermiculite/perlite.

Slow-release fertilizers may be added to the potting mixture, or liquid fertilizer applied as the seedlings grow in the nursery. In common with other species in the genus, *E. tereticornis* is highly susceptible to damping off and other fungal pathogens during the nursery phase. Occurrence of disease problems in the nursery can be limited by strict attention to hygiene, reducing watering and shade, and allowing good ventilation. Seedlings are planted out in the field when they reach a height of about 25 cm, 3–5 months after sowing. This should coincide with the onset of the wet season in tropical countries.

Rotation varies with site quality and management objectives. *E. tereticornis* is usually grown on a short rotation and clear felled at an age that maximizes production of logs of optimum size for a particular end use. This is usually small-diameter material suitable for pulpwood, mining timber, charcoal or fuelwood.

Plantations of *E. tereticornis* are expected to regenerate from coppice at least three times. The season of felling affects coppice regeneration. Felling during the dry season delays sprouting and increases the risk of the stump drying out. Felling by saw to give a cleanly cut short stump with minimum bark damage is best.

Spacing and cropping systems are very variable – from community plantings around homes, villages and roads to closely spaced commercial plantations – and depend on the end products required. When pulpwood and firewood are the principal objectives on a rotation of 5–6 years, a spacing of 1.5 × 1.5 m is considered desirable in India, where wider spacings of 4 × 2 m or 6 × 1.5 m are recommended when crops are to be grown between the tree rows during the first 3 years. Otherwise, plantings are done at closer spacings such as 2 × 2 m, 2.6 × 2.6 m and 3 × 1.8 m. Wider spacings between rows also facilitate mechanical cultivation for weed control.

Intensive site preparation by ploughing is beneficial; on compacted sites deep ripping may also be used and on wet sites mounding improves root aeration and provides well-drained conditions that facilitate planting. Application of fertilizer to each tree at planting to assist establishment and growth is common.

As with other species in the genus, *E. tereticornis* is intolerant of shade and does not compete well with grasses for water and nutrients. In India, the usual practice is to carry out three weedings in the first year, two in the second year and one in the third year during the rainy season. In addition to weed control, it is important to protect plantations against fire.

Diseases and pests

One of the most serious diseases of *E. tereticornis* is pink disease caused by the fungus *Corticium salmonicolor* (*Phanerochaete salmonicolor*). Other fungal pathogens recorded as causing damage to *E. tereticornis* include *Ganoderma lucidum*, *Endothia gyrosa* and *Cylindrocladium*/*Calonectria* spp., of which *Calonectria quinqueseptata* is dominant. *Cylindrocladium clavatum* was recorded as causing seed rot, seedling blight and seedling wilt of *E. tereticornis* in Punjab, India.

Other potentially serious diseases were web blight (*Rhizoctonia solani*) in the nursery, and stem cankers caused by *Cryphonectria cubensis* and *Cytospora eucalypticola*, which can result in heavy mortality. The following fungal diseases have been recorded in plantations of *E. tereticornis* and *Eucalyptus camaldulensis* in Andhra Pradesh, India: anthracnose caused by the colletotrichum state of *Glomerella cingulata*; red rust caused by *Cephaleuros virescens* and leaf spots caused by *Pestalotiopsis versicolor*, *Pestalotiopsis disseminata*, *Robillarda sessilis*, *Pseudocercospora eucalyptorum* and *Coniella castaenicola*.

In South America, the fungus *Mycosphaerella molleriana* has been reported as attacking *E. tereticornis*. Seedlings of *Eucalyptus deglupta*, *E. tereticornis* and *Eucalyptus urophylla* in the Solomon Islands suffered from shoot-tip necrosis (die-back), foliar wilt and stem canker shortly after establishment. The fungal diseases followed primary attack by the coreid bug *Amblypelta cocophaga*.

In Australia, *E. tereticornis* has a medium to high susceptibility to attack by a range of insect pests, such as Christmas beetles (*Anoplognathus* spp. and *Repsimus aeneus*), leaf beetles (*Paropsis* spp. and *Chrysophtharta* spp.), lerps (psyllids, *Cardiaspina* spp.) sawflies (*Perga*, *Pergagrapta* and

Pseudoperga spp.), gum tree scales (*Eriococcus coriaceus* and *Eriococcus confusus*), gumleaf skeletonizer (*Uraba lugens*), eucalypt weevils (*Gonipterus* spp. and *Oxyops* spp.), leaf-tying moth (*Agrotera amathealis*) and the defoliating, tessellated phasmatid (*Cternomorphodes tessulatus*).

In Brazil, the leaf-cutting ant (*Atta sexdens rubripilosa*) prefers the leaves of *E. tereticornis* to *Eucalyptus grandis*, and can cause severe defoliation of the tree. In India, a cerambycid beetle *Cerosterna scabrator* is reported as a pest of young *E. tereticornis* plantations in several states.

In Tamil Nadu, India, the scarabaeid beetle *Adoretus versutus* has also been found on a single clone of *E. tereticornis*. Subterranean termites are reported to damage seedlings and young plants of *E. tereticornis* in Asia.

Uses

E. tereticornis has been trialled in alley-cropping systems and used in wind breaks/shelter belts. This species has been used successfully to plant eroded gullies and to reclaim low-lying areas and sand dunes. It has been used in combination with other species to rehabilitate fly ash dumps in the area surrounding a hydroelectric power station in India. It is a large ornamental tree suitable for parks.

E. tereticornis is a major source of fuelwood, charcoal and timber for local use. It is also used for light and heavy construction, railway sleepers, bridges, wharves, piles, poles, mining timber, pulpwood, hardboard and particleboard. *E. tereticornis* has been used in parts of India as raw material for the production of rayon-grade pulp.

Sapwood is pale, yellowish, and sometimes susceptible to attack by borers (*Lyctus* sp.). The heartwood is red, with a moderately fine texture and interlocked grain, and is hard, strong and durable. Density varies with tree age and environment. In areas of natural occurrence, green density is about 1200 kg/m^3 and air-dry density about 1050 kg/m^3. The air-dry density of wood from fast-growing plantations is less and ranges from about 600 to 800 kg/m^3.

E. tereticornis leaf extracts have been tested as insecticides against the grain moth *Sitotroga cerealella*, and as a soil drench against the root knot nematode *Meloidogyne javanica*.

The leaves of *E. tereticornis* also provide a good source of essential oils; however, it appears that, despite the widespread utilization of the species for wood products in countries such as India and China, production of essential oil as a secondary product takes place only rarely.

E. tereticornis is a major source of pollen in apiculture and produces a medium amber honey of distinctive flavour. The wood and bark of the tree have a tannin content of 6–12% and 3–15%, respectively, although these are not used commercially.

The species is also a very important koala food tree. Open eucalypt woodland comprising *E. tereticornis* is a suitable habitat for the common striped possum (*Dactylopsila trivirgata*) in far northern Queensland.

Further reading

Dhanda, R.S., Davinder Singh and Gill, R.I.S. (2005) Timber volume and weight tables of *Eucalyptus tereticornis* Sm. in alluvial plain of India. *Indian Journal of Agroforestry* 7, 30–39.

Rakesh Kumar and Bangarwa, K.S. (2006) Clonal evaluation in *Eucalyptus tereticornis* Sm. *Environment and Ecology* 24S, 1188–1191.

Shalinder Kaur, Singh, H.P., Batish, D.R. and Kohli, R.K. (2011) Chemical characterization, antioxidant and antifungal activity of essential oil from *Eucalyptus tereticornis*. *Journal of Medicinal Plants Research* 5, 4788–4793.

Eucalyptus urophylla S.T. Blake

Timor mountain gum

Importance

Eucalyptus urophylla has an important role in afforestation in some countries, and has assumed importance in wood production in plantations in low-latitude, seasonally dry tropics. *E. urophylla* is grown as either a pure species or in hybrid combinations, and has the potential to be much more widely used in humid and sub-humid tropical regions. Its fast growth, coppicing ability, adaptability to a range of environments, relative resistance to fire, pests and diseases, and suitability for a variety of wood products, place it among the most useful tropical trees.

Botanical features

Over most of its natural range *E. urophylla* is a forest tree usually reaching 25–45 m in height with a dbh of 1 m; it has a straight bole for half to two-thirds of tree height. Exceptional specimens attain 55 m in height with a dbh of over 2 m.

There is extreme variation in bark characteristics, apparently associated with differences in moisture conditions and altitude. Below 1000 m on Alor and Flores (Indonesia) a large proportion of trees exhibit relatively smooth boles with a variable (0–0.5 tree height) stocking of rough bark at the base. Between 1000 m and 2000 m on Timor (Indonesia), where conditions are moister, the trees are usually covered with sub-fibrous, shallowly and closely fissured longitudinally, red-brown bark extending to the smaller branches. A wide range of bark types is also found within some stands on other islands. In some localities *Eucalyptus alba* occurs in association with *E. urophylla*, and the inter-specific hybrids which are produced occasionally may have a range of intermediate bark characters.

There is considerable variation in adult leaf characters such as length, width and expression of a 'drip-tip'. There are differences between the seedling, juvenile, intermediate and adult leaves. The adult leaves are dark green above and paler green below, subopposite to alternate on the branch, stalked, broad–lanceolate, 12–20 × 2–5 cm, narrowing abruptly to a short point, or lanceolate and tapering to a narrow point or 'drip-tip'. The leaves have visible lateral veins 45–50° to the midrib, and an intra-marginal vein.

There is also variation in fruit size, fruit shape and bud characteristics. The white-cream flowers are in clusters of 7–11, forming an umbel which has a flattened stalk. The buds are globular to ovoid and the fruits cup-shaped with a wide, flat or oblique disc and 3–5 valves more or less at rim level.

E. urophylla flowers in January–March and mature seed is collected August–September.

Distribution

E. urophylla is one of the two species of eucalypts that do not occur naturally in Australia, the other being *Eucalyptus deglupta*. There are extensive stands of *E. urophylla* on the large Indonesian island of Timor above 1000 m, and throughout Wetar. Scattered stands are found on the nearby islands of Adonara, Alor, Flores, Lomblen and Pantar in the south-eastern end of the Sunda Archipelago (Indonesia). Much of the range is poorly accessible and it has not been possible to confirm the presence or absence of natural occurrences on islands other than the seven listed above.

In addition to its natural range, *E. urophylla* has also been planted in countries across South America, Africa, Asia and Oceania.

Silviculture characteristics

E. urophylla is similar to other eucalypts in that it is a vigorous, light-demanding species, and growth is curtailed with increasing weed competition. Under favourable conditions the species is capable of annual growth rates in excess of 30 m^3/ha mean annual increment after provenance selection. The species has some level of drought and frost-tolerance but is generally grown where neither of these factors are limiting.

Silviculture practice

The species is usually propagated by seed and established using containerized stock. Flowering usually starts within 2 years and seeds are produced abundantly within 4 years' growth. Summer flowering in the natural habitat and insect pollination results in mature seed 6 months later. *E. urophylla* has an average of 450,000 viable seeds/kg. Seed is orthodox and can be stored for extended periods at low humidity and temperature. It requires no germination pretreatment.

Nursery establishment is generally by sowing untreated seeds in germination beds. The resulting seedlings are transplanted into containers when they have two pairs of leaves. The potting medium is usually a free-draining loam–sand mix. Seedlings usually reach planting-out size (25 cm) in 10–12 weeks.

The species coppices readily and is easily propagated by rooted cuttings or in tissue culture.

Intensive site preparation by ploughing is beneficial, and on compacted sites deep ripping may also be used. Fertilizer (NPK) is applied in each planting hole. Boron deficiency has been identified as one of the more important causes of poor stem form and malformed leaves in *E. urophylla* plantings. Spacing between planting spots varies, and is commonly 3 × 2 m in pulpwood and charcoal plantations, but may be closer in fuelwood or pole plantations. The tree is highly sensitive to competition in the early stages, and the plantations must be kept weed-free for 6–12 months after planting. After that time, the dense crown inhibits competing weeds. Mechanical cultivation is advisable to remove imperata grass or other dominant vegetation prior to planting.

Fertilizer is an essential requirement for fast growth rates. Although the species is relatively tolerant of fire, control measures are often undertaken to protect commercial plantations.

E. urophylla grows fast and under average conditions produces 20–30 m^3/ha/yr of wood with bark at 5–10 years of age. The better provenances can yield up to 50 m^3/ha/yr on favourable sites. High yields have been recorded in South America (e.g. Brazil) and Africa (e.g. Cameroon, Congo and Ivory Coast) with good silviculture (especially weed control and fertilizer) and genetic improvement. The species is particularly productive in hybrid combinations.

Diseases and pests

A canker disease caused by *Cryphonectria cubensis* is found on *E. urophylla* in West Africa and South America. Although it is much more resistant than *Eucalyptus grandis* or *Eucalyptus saligna*, some provenances are quite susceptible, especially in humid tropical lowland conditions without a marked dry season. In Indonesia, the death of seedlings about 2 months after planting has been attributed to attack by root fungi such as *Botryodiplodia* sp., *Fusarium* sp. and *Helminthosporium* sp. *E. urophylla* is moderately susceptible to guava rust (*Puccinia psidii*).

Seedlings and small trees of *E. urophylla* are susceptible to termite attack, and wood borers such as *Zeuzera coffeae* cause damage to older trees. In parts of South America the leaf-cutter ants *Atta* spp. and *Acromyrmex* spp. can destroy the trees, and control measures are essential where they are present.

E. urophylla is susceptible to attack by leaf-eating caterpillar in Brazil. Numerous pathogens reduce the productivity of trial plantings of eucalypts in the Sumatran lowlands, among these being leaf spot fungi (e.g. *Cylindrocladium* sp., *Pestalotia* sp.), root rot fungi (e.g. *Botryodiplodia* sp., *Fusarium* sp. and *Helminthosporium* sp.), leaf roller insects, ring borer insects and the stem borer (*Z. coffeae*).

Uses

E. urophylla has an important role in afforestation in only a few countries. Its wood properties are very suitable for a wide range of purposes, the young trees providing satisfactory firewood and charcoal.

The wood of *E. urophylla* is less dense than most eucalypts. The basic density is in the range 540–570 kg/m^3. The heartwood is pinkish-brown to red-brown and contains little gum. Fibres are relatively short, about 1.0 mm in length. The wood is very suitable for producing bleached chemical pulp and has good pulp yield (49.5%). The timber of older trees can be sawn and used for general construction purposes. In the round, the wood is suitable for building poles and fence posts. The average annual volume increment is 20–30 m^3/ha.

The leaves yield essential oils of commercial use. The bark has a 10% tannin content.

Further reading

Evans, J. and Turnbull, J.W. (2004) *Plantation Forestry in the Tropics: The Role, Silviculture and Use of Planted Forests for Industrial, Social, Environmental and Agroforestry Purposes.* 3rd ed. OUP, Oxford, UK, 482 pp.

Sein, C.C. and Mitlöhner, R. (2011) *Eucalyptus urophylla S.T. Blake: Ecology and silviculture in Vietnam.* Center for International Forestry Research (CIFOR), Jakarta, Indonesia, 16 pp.

Vercoe, T. and Clarke, B. (1994) Trial growth performance of *Eucalyptus urophylla* S.T. Blake. Report to FAO Forestry Division on the Growth of *Eucalyptus urophylla* in International Provenance Trials and Other Comparative Growth Trials. Australian Tree Seed Centre, Canberra, 56 pp.

Eusideroxylon zwageri Teijsm. & Binn.

Billian

Importance

Eusideroxylon zwageri is one of the most renowned timbers of Borneo and Indonesia, a species of great commercial importance producing an exceptionally durable timber, favoured both for local use and for export. The most valuable characteristic is its resistance to termites and other ubiquitous, tropical wood-eating insects and fungi, and the wood is therefore in great demand for construction throughout Indonesia. Over-exploitation, together with forest clearance, has, however, led to the decline of this slow-growing timber species, and being slow growing it has generally been considered unsuitable for large-scale plantations. Problems of uncontrollable exploitation caused by concessionaires (logging operators with valid contracts) opening forest roads in Kalimantan, shifting cultivation, the introduction of chainsaws and logging and land clearance for plantations and agriculture have increased, especially since the 1998–1999 economic crisis in the lowlands east of the Barisan Range. Extensive stands of *E. zwageri* have been almost entirely destroyed in southern Sumatra and it is now categorized as a threatened tree species. Although *E. zwageri* is afforded official protection through legislation in many areas, illegal logging is widespread.

Botanical features

E. zwageri is a tall tree, up to 40(–50) m, bole straight, branchless up to 20 m, but usually less, sometimes slightly fluted at the base, up to 1.5 m in diameter. The largest recorded tree had a girth of 5.69 m (approximately 1.8 m diameter). The tree has small, rounded buttresses, sometimes many, often few, giving the base an elephant-foot like appearance. The bark surface is red or grey-brown with thin cracks, debarking in small sub-quadrangular pieces, turned up at the lower side. The crown is dense and globular with strong horizontal branches. As formation of the crown takes place relatively early, it occupies most of the tree height. Branches are few in number but are thick and heavy, with smooth twigs, slightly angular and tomentose. *E. zwageri* has a strong and extensive root system, and forms numerous superficial roots similar to the 'hunger roots' found in the root collar of old trees in pine forests on poor soils in Europe. These hunger roots look like a thin carpet, strong with many branches, and are more commonly found in poor, sandy soil.

The leaves are arranged spirally and are evenly separated, elliptical, without stipules, apex obtuse to shortly acuminate and the base rounded-subcordate, (12–)14–18(–30) cm long and (5–)6–10(–11) cm wide. Young leaves are reddish-green when mature, upper surface glabrous, lower surface hairy on the larger veins; petiole 6–15 mm long.

The flowers are small (about 1 mm), in aggregated groups 5–10 cm long, and with a pleasant fragrance. Flowers are bisexual, actinomorphic, on a 3–11 mm long pedicel, perianth tube shallow, tepals 6, in 2 whorls, imbricate, caducous, 3–3.5 mm long, greenish, yellow or purplish, puberulous outside. Stamens without glands, in 4 whorls, in outer 2 whorls staminodial and petaloid, 1.5 mm long, yellowish with purple tip, ciliate, those of the third whorl fertile, thick, with minute red or white anthers with a pink hue. Anther cells 4, in 1 horizontal row, stamens of the inner whorl staminodial subulate, small. Ovary superior, sessile, unilocular, with a single ovule, tapering into the cubulate style; stigma small, discoid. Fruit completely enclosed in the accrescent flower tube, the latter assuming the function of the exocarp and the mesocarp; the exocarp of the fruit becomes woody. Fruits mature 3 months after flowering, a 1-seeded berry generally oval in shape, 8–18 cm long and approximately 7 cm (4–10 cm) in diameter, with a horny seedcoat, furrowed, brittle, pale with a very small embryo.

Trees begin to flower at 15–20 years. They bear fruits at irregular intervals, often every 2–3 years, sometimes annually. The mass flowering of *E. zwageri* is generally in the middle or end of the dry season, but it does not occur every year. In specific trees, flowering can occur for long periods but in irregular seasons. In Indonesia, the flowering season of *E. zwageri* in Palembang (south Sumatra) is August–November, while in Jambi, it is in July.

Distribution

E. zwageri is found in lowland forest in tropical humid Southeast Asia. In Indonesia, it is found in Kalimantan, south and south-east Sumatra, Lampung, the southern part of west Sumatra and the west coast residency. It is also found in the Philippines, Sabah, Sarawak and Brunei Darussalam.

Silviculture characteristics

E. zwageri is a slow-growing species which is shade-tolerant, requiring heavy shade during the seedling stage, although its light requirement gradually increases with age. It resprouts well after cutting, and is fire and drought-resistant. Trees may begin to flower at an age of 15–20 years. They bear fruits at irregular intervals, often every 2–3 years, sometimes annually. Seeds tend to fall to the ground and roll down slopes, and may also be spread by mammals, but large numbers of *E. zwageri* trees are often found along rivers at places where the fruits wash ashore in large quantities, suggesting that water is an important dispersal agent.

Silviculture practice

The problem of natural regeneration of *E. zwageri* is that seedling distribution is uneven due to reliance on gravity, slopes and growing in lowland areas and creeks. To ensure that seedling distribution is more evenly spread over a site, enrichment planting is required. Although *E. zwageri* does not bear fruit every year, in good years trees can produce hundreds of fruits; a single mother tree with a diameter of 20 cm was observed to produce 761 young fruits. *E. zwageri* seedlings can be obtained from wildings or from nursery-raised stock. Seeds require pretreatment, soaking in water for 30–60 minutes and then sun-drying, the seedcoat breaking some 2–4 h later. Seeds can be sown in seedbeds containing sand, or directly to standard containers with equal parts of sand, topsoil and organic matter, and will germinate after 1–3 months. Transplanting may be carried out when seedlings reach 40 cm high or 6 months old. *E. zwageri* can also be direct sown in the field; however, planting seedlings raised in containers gives the best results. *E. zwageri* may also be propagated vegetatively by stem cuttings.

As almost all felled trees coppice, the management of regrowth offers another silvicultural practice, as the increment growth of shoots is better than that of seedlings due to the root systems already developed. The number of coppice shoots per stump varies widely, from 2 to 64, and thus the most important treatment is thinning, which should be done at an early stage when the shoot diameter is 1–5 cm, with thinning to one shoot per stump, and selecting for vigour, good health and straightness. Shoots emerging close to the soil surface are preferable, as this improves resistance against wind or other mechanical disturbances and these shoots form new root systems easily. Thinning should be carried out regularly, every 6–12 months until the coppice shoots reach 5–8 years old.

E. zwageri is principally a shade-tolerant tree which grows well on damp soil; this must be taken into consideration when planning plantations. When planted in an open area with inadequate shade in Sabah, all *E. zwageri* plants had died within 10 years. Best results are obtained by planting out trees in shaded rows under old trees or high virgin forest in north–south rows 1.5–2 m wide, 5 m between rows and 2 m between seedlings within rows. The trees forming the shade should be removed gradually, and secondary forest regrowth is to be cleared (weeded) after 12 years.

Diseases and pests

Standing trees are colonized by termites and brown rot, and there are reports of the twig borer *Xyleborus morstaeti* (*Xylosandrus compactus*) causing die-back. However, few of these pests cause serious problems.

In Indonesia, plantations of *E. zwageri* in east Java suffered from more insect attacks than those in other parts of its range. Trees planted at 600 m altitude in the 'Zuidergebergte' near Malang showed 'feathered' leaves, where the leaf blade is divided into parallel strips of tissue in which the side nerves run longitudinally.

Uses

E. zwageri can be used as a component of community-based forestry in southern Sumatra which includes timber and fruit trees, perennials and food crops. The main trees include rubber (*Hevea brasiliensis*), damar (*Shorea javanica*) and kayu ulin (*E. zwageri*), while fruit trees include durian (*Durio zibethinus*), duku (*Lansium domesticum*) and petai (*Parkia speciosa*). The aims of this activity are to obtain additional income, fulfil the wood and fuel needs of local people and to provide environmental and ecological benefits. Some *E. zwageri* trees had been planted by local people in tembawang (referring to forests that include fruit trees and other valuable flora cultivated by the Dayak over generations) under *Shorea* spp. in Embaong, Sanjan and dan Gok Tanjung. The problem here is that the growth rate is very slow and the seed supply limited, due to a limited number of mother trees. If the seeds are provided, the people would want to plant more *E. zwageri* in their tembawang.

In Indonesia, *E. zwageri* and other local tree species have been planted for the purpose of mine site reclamation.

E. zwageri is one of the most important construction woods in Indonesia. Wood density is 830–1190 kg/m^3 at 15% moisture content. It is used for making furniture, window and door frames, doors, heavy construction, bridges, railway sleepers, marine piling, harbour structures, boat constructions, fenceposts, heavy-duty industrial flooring, shingles and vehicle bodywork. The most valuable characteristic of *E. zwageri* wood is that it is resistant to attack by termites and other wood-eating insects and fungi. For this reason, it is in great demand for construction throughout Indonesia.

Although the seeds are poisonous, the fruit and seeds have medicinal uses, and are used locally to treat swellings. Seed extract was shown to have anti-melanogenetic properties, indicating a potential for use in skin-whitening cosmetics.

Further reading

Irawan, B. (2002) Ironwood (*Eusideroxylon zwageri* T. et B.): Present conditions and future development in Jambi, Indonesia. In: Birner, R., Nurrochmat, D. and Rosyadi, S. (eds) *Sustainable Development: Socio-economic and Environmental Problems*. Cuvillier Verlag, Goettingen, Germany.

Irawan, B. (2005) *Ironwood (Eusideroxylon zwageri Teijsm. & Binn.) and Its Varieties in Jambi, Indonesia*. Cuvillier Verlag, Goettingen, Germany.

Haryati, J.R., Azizah, I.N. and Arisoesilaningsih, E. (2010) *Eusideroxylon zwageri* (Ulin) as key species in two zones of Sangkima rain forest, Kutai National Park, East Kalimantan. *Journal of Tropical Life Science* 1, 47–50.

F

Fagus crenata Blume

Japanese beech

Importance

Fagus crenata dominates mesic sites in cool temperate Japan. Due to extensive forest management, primary forests are restricted to only a few areas of the national forests, one of the largest consisting of 16,971 ha in the Shirakami Mountains in northern Honshu. This area was registered as a UNESCO World Heritage Site in 1993.

Botanical features

A large deciduous tree up to 25 m high, with a mostly straight trunk to 1.5 m in diameter; bark smooth, greyish, usually with various patterns due to epiphytic lichens (Plate 67); open crown shape; shallow root system.

Leaves alternate, distichous, with hirsute or glabrous petioles, wavy on the margin, covered with long silky hairs when young but glabrous when mature, except for hirsute nerves on both surfaces, with 7–11 pairs of lateral veins terminating at the sinus on the margin.

Flowers monoecious, opening April–May. Staminate flowers in globose heads on drooping slender peduncles axillary on the basal part of new shoots; pedicels long, hirsute; perianth campanulate, 5–8 lobed, hirsute; stamens 7–10, exserted with filiform filaments. Pistillate flowers in 2-flower clusters in the axils of leaves on the middle part of new shoots, surrounded by numerous bracts adnate to an involucre. Nuts ripen in October, with 3 winged angles; cupules woody, dehiscent into 4 valves revealing 2 nuts.

Distribution

Naturally distributed throughout Japan. On the Japan Sea side *F. crenata* extends from 34°N in south Honshu to 42°49'N at the Kuromatsunai depression in south-west Hokkaido. On the Pacific side, it extends from 31°29'N on Mt Takakuma in south Kyushu to 37°N in east Honshu. In south Kyushu, in its southernmost distribution area, *F. crenata* forest is found above 1000 m altitude. From about 39°N in north Honshu to the northernmost distribution limit in the Kuromatsunai depression, *F. crenata* forest can be found at sea level.

Silviculture characteristics

To promote regeneration, shelterwood logging has been a common practice in primary *F. crenata* forests. In this system, seed trees are left standing (30–70% by volume) until sufficient regeneration has taken place, after which they are logged.

In natural forests, the synchronous flowering and then withering of dwarf bamboo (*Sasa*) favour tree establishment, causing a wave of recruitment in the forest canopy. In order to regenerate successfully, both a tree seedling bank and a subsequent canopy gap are required. The interval between simultaneous *Sasa* death events and the duration of the *Sasa* recovery period are important factors controlling *F. crenata* forest dynamics.

Silviculture practice

A combination of leaving adequate seed trees and ground preparation are necessary for natural regeneration. As only 20% of fallen nuts germinate following a mast year, more than 50 seed trees/ha are needed for successful regeneration. The actual number of seed trees to be retained is determined on the basis of dispersal range in an average crop year, but as the interval between good crop years is long and irregular, a management plan is difficult to establish. The period for retaining seed trees is determined with a standard of over 10 years, during which several average crop years and at least one good crop year may occur. The basis of ground preparation is clearing, usually carried out three times or more; it should be carried out according to recovery of forest floor vegetation, but not regularly. In places where recovery is rapid, the clearing must be increased and in the places where it is delayed, it should be reduced. Complete clearing is desirable, particularly initially.

Nuts dried to moisture content as low as 8% maintain viability and germinate after 2 years of storage.

The estimated annual production in primary *F. crenata* forests is ca 40 t/ha at the lower altitudinal limit (600 m) and ca 20 t/ha at the upper limit (1600 m). Stem volume at 900 m altitude may be up to 550 m^3/ha. Annual biomass increment is about 2.3 t/ha at low altitude (550 and 900 m) and less than 1.3 t/ha above 1300 m.

Stem volume in planted stands about 50 years old at 500 m altitude in central Honshu island is about 400 m^3/ha, and the net production 17–19 t/ha/yr, which is higher than that of deciduous broadleaved forests in the temperate zone. The oldest planted stand (130 years old), named Gartoner's beech stand, is situated in southern Hokkaido, and has an average tree height and diameter of 21 m and 70 cm, respectively.

The Forestry Agency of Japan supplies timber from primary forests situated in national forests. The yield has been decreasing in recent years because of the shortage of growing stock and the need to conserve remaining primary forests.

Diseases and pests

Damping off caused by *Colletotrichum dematium* and *Cylindrocarpon* sp. is common in current-year seedlings, with mortality particularly high under the canopy; survival is higher in gaps and at forest edges due to light-controlled defence mechanisms.

Didymella fagi sp. nov. (anamorph *Ascochyta fagi*) was identified as causing yellow leaf spot disease of *F. crenata* in Japan in 1995. *Apiognomonia errabunda* (anthracnose), *Fomes fomentarius* (hoof fungus) and *Fomitopsis pinicola* (brown crumbly rot) also affect *F. crenata*.

Mikiola bicornis and *Mikiola glandaria* (Diptera: Cecidomyiidae) inducing acorn-shaped galls on *F. crenata* leaves were described as new species in 2008.

Populations of beech caterpillars, *Quadricalcarifera punctatella* (formerly *Syntypistis punctatella*), regularly increase explosively every 8–11 years and cause major damage to leaves. Outbreaks occur at the same time in different areas.

Trees planted in lowlands are susceptible to damage by the longicorn beetle *Apriona japonica*.

The problem with gnawing damage by rodents to seedlings is most likely to occur under *Sasa* cover. Deep snow cover in winter prevents excessive nut predation by wood mice. *Pseudopammene fagivora* is an important insect species related to seed survival of *F. crenata*. Other pests recorded include *Anoplophora chinensis* (black and white citrus longhorn) and *Xylosandrus mutilatus* (camphor shoot beetle).

Tropospheric ozone is considered to be the main air pollutant causing the decline of *F. crenata* forest in Japan.

Uses

Cattle and horses graze in *F. crenata* forests in northern Honshu. The trees are planted in natural parks and rural areas in Japan; the species is seen as a national symbol of environmental conservation. It is also a favourite subject for bonsai, with seedlings from around Mt Fuji being particularly prized.

F. crenata wood has a density of 551 kg/m^3 with a compact texture and is easily bent by heat treatment. It is used in the production of sawn or hewn building timbers, railway sleepers and industrial or domestic woodware, including flooring, toys and utensils; and furniture such as tables, chairs and bed frames. The wood warps easily and/or decays unless it is fully seasoned. Artificial seasoning is, therefore, required before processing.

F. crenata smallwood is utilized for wood pulp and as a medium for edible mushroom culture.

Steamed *F. crenata* wood has been used in experimental diets for ruminants, its dietary energy value being nearly equal to that of rice straw.

Further reading

Koyama, Y., Takahashi, M., Murauchi, Y., Fukatsu, E., Watanabe, A. and Tomaru, N. (2012) Japanese beech (*Fagus crenata*) plantations established from seedlings of non-native genetic lineages. *Journal of Forest Research* 17, 116–120.

Shimano, K. (2002) Regeneration dynamics, causal factors, and characteristics of Pacific Ocean-type beech (*Fagus crenata*) forests in Japan: a review. *Folia Geobotanica* 37, 275–296.

Tanaka, N. and Taoda, H. (1996) Expansion of elevational distribution of beech (*Fagus crenata* Blume) along the climatic gradient from the Pacific Ocean to the sea of Japan in Honshu, Japan. In: Omasa, K., Kai, K., Taoda, H., Uchijima, Z. and Yoshino, M. (eds) *Climate Change and Plants in East Asia*. Springer-Verlag Tokyo, Tokyo, pp. 175–184.

Wagner, S., Collet, C., Madsen, P., Nakashizuka, T., Nyland, R.D. and Sagheb-Talebi, K. (2010) Beech regeneration research: from ecological to silvicultural aspects. *Forest Ecology and Management* 259, 2172–2182.

***Fagus grandifolia* Ehrh.**

American beech

Importance

Fagus grandifolia is a primary constituent of the eastern hardwood forests of the USA and Canada. It is the only species of *Fagus* native to North America, and occurs in forest stands occupying approximately 6 million ha. American beech is a principal component of three cover types (sugar maple–beech–yellow birch; red spruce–sugar maple–beech; and beech–sugar maple), and occurs in an additional 17 types. As such, *F. grandifolia* has major ecological and economic significance throughout its wide range.

Botanical features

F. grandifolia may grow to a height of 18–22 m and have a dbh of 45–75 cm at maturity. The tree usually develops a broad, compact crown. Branch shedding is slower than for most other associated hardwood species, but when grown under forest conditions it can shed branches readily, achieving a straight smooth bole for one-half to two-thirds of its length. *F. grandifolia* has a shallow root system when compared with other deciduous species in the north temperate region. It can develop prolifically from root sprouts, especially after stand disturbances. These sprouts can be long-lived and develop into mature trees. Sprouts originating from cut stumps may also develop, but are usually short-lived.

The bark of *F. grandifolia* is light to blue-grey in colour, and very smooth on both young and old trees. This characteristic makes it distinctive among other species of the temperate forest. Variations in bark appearance occur only rarely. The oval leaves are 7.5–15 cm long and 2.5–6.5 cm wide, and coarsely toothed. Leaves are glabrous on both upper and lower surfaces. Winter buds are long, slender and sharply pointed or lanceolate.

Flowers are monoecious, and appear immediately after bud break early in the spring. Staminate flowers, which persist for only a few days, contain 8–16 stamens. Pistillate flowers contain 2–4 flowered spikes. The fruit consists of a spiny burr containing 2–3 triangular nuts. Nut production begins at about age 40. The nuts are edible and used by a wide range of wildlife, as well as by humans. Seeds are produced abundantly every 2–3 years.

Near the more northerly limits of its range, at about 45°N, bud break occurs during the first week in May. Stem elongation develops rapidly, and is generally complete by early July, when bud set occurs. Fruit ripening occurs by mid-October, and seeds are dispersed after frost. Autumn foliage coloration first develops as a pale or light yellow in mid- to late September, turning to a bronze-brown by mid-October.

Distribution

In the USA, *F. grandifolia* occurs from northern Michigan south to eastern Texas and eastward to the Atlantic Ocean, with the exception of the Florida panhandle. In Canada, it occurs from Nova Scotia and Prince Edward Island throughout New Brunswick, southern Quebec and south-eastern Ontario. A disjunct varietal population also occurs in the mountains of north-eastern Mexico. A great deal of phenotypic variation is shown by *F. grandifolia* across its broad range.

Silviculture characteristics

F. grandifolia is one of the most shade-tolerant forest tree species in eastern North America. Perhaps because of this trait, height growth tends to be comparatively slow, especially during the early phases of stand development. Additionally, it is recognized as growing more slowly in diameter than many associated tree species, although on good sites with little competition, growth rates can match those of other northern hardwoods. Establishment of *F. grandifolia* after stand disturbances, such as intensive harvesting, can be aggressive due to its ability to develop sprouts prolifically, and this can lead to a drastic alteration of stand species composition, with a large portion of affected stands occupied by dense beech thickets.

F. grandifolia develops as a principal component of the cover types mentioned above, or in hardwood and conifer mixtures. Pure or nearly pure natural stands are not common, except where successional trends have been interrupted by destructive harvesting, by the occurrence of beech bark disease, or by both.

F. grandifolia is very sensitive to flooding, but tolerates salt winds and a considerably wide range of temperatures. In the northern limits of its range, temperatures below –42°C occur, while in the southern limits, summer temperatures may routinely reach above 38°C for extended periods. It is not well-adapted to rapid changes in temperature, however, and is subject to sunscald and frost cracks, conditions thought to be exacerbated by its bark anatomy. *F. grandifolia* can grow on a wide range of soils, but reaches its best development on well-drained alluvial soils.

Silviculture practice

Hardwood stands in which *F. grandifolia* occurs may be managed using either even age or uneven age methods of silviculture. Silvicultural methods that result in higher shade levels, such as individual tree selection or dense shelterwoods, are especially conducive to regenerating *F. grandifolia*. Regeneration can be in the form of root sprouts or seedlings. Experience in the north-eastern USA suggests that nearly equal numbers of individuals arise from seeds and sprouts, although individual stand history and harvesting practices can strongly influence this ratio. Because *F. grandifolia* regeneration can develop under various conditions and by both sexual and asexual strategies, silvicultural methods which limit the proportion of beech in a stand are often the preferred goal. This is especially true in the north-eastern USA, where low-quality stems develop as a result of beech bark disease. However, because beech nuts provide a high-value food source for wildlife, managers will maintain at least some component of beech in stands if other sources of hard mast are unavailable.

Seeds can be sown in the autumn, or in the spring following cold stratification. Nursery recommendations follow those used for germination of *Fagus sylvatica*: chilling for 42 days at 2.5–5°C, followed by 28 days at 20°C. A sowing density of approximately 485 seeds/m^2 is recommended.

Diseases and pests

The most serious disease of *F. grandifolia* is beech bark disease. This disease results in a complex decline and is initiated by the feeding of a scale insect, *Cryptococcus fagisuga*, on phloem tissues. The feeding is then followed by bark infection by any of several fungi of the genus *Nectria*. When scale populations become high, the *Nectria* infections will girdle the stem, resulting in tree mortality. Root sprouts of killed or salvaged trees can develop into dense thickets, but remain susceptible to the disease complex, accumulate extensive cankering and grow very slowly. Introduced into North America from Europe approximately a century ago, the disease has slowly spread through eastern Canada and the north-eastern USA as far west as western Pennsylvania and as far south as Tennessee. Numerous fungi occur on the foliage of beech, but few cause any damage. *F. grandifolia* is subject to many internal decays and discolorations. There are over 80 trunk rots represented by species in the genera *Fomes*, *Hydnum*, *Polyporus*, *Stereum*, *Poria* and *Daedalea*. In addition, several canker-causing fungi are known, including species of *Strumella*, *Endothia* and *Nectria*.

Principal insect defoliators include the saddled prominent moth (*Heterocampa guttivitta*), spring and fall cankerworms (*Paleacrita vernata* and *Alsophila pometaria*, respectively) and the gypsy moth (*Lymantria dispar*). Scale insects, including *C. fagisuga* and the oystershell scale (*Lepidosaphes ulmi*), are important sucking pests. Beech nuts are especially susceptible to damage from insects, with losses being as high as 100%. Filbertworm (*Melissopus latiferreanus*) is an important and range-wide pest of beech nuts. *F. grandifolia* is very sensitive to fire damage, owing to its smooth bark with an extremely thin phellem. The flowers are very susceptible to late spring frosts.

Uses

F. grandifolia is rarely used as an ornamental around homes and parks, most individuals in these settings being the result of natural occurrence rather than horticultural plan. Difficulty in transplanting, susceptibility to stem decay fungi, and the ease with which bark is damaged and vandalized make the tree unattractive in this regard.

F. grandifolia has been widely used for a variety of wood products, but is best known for its use in small-piece products. Small tool handles, clothes pins, spools, small wooden toys, and other woodenware are traditional uses, because the wood is easily turned on the lathe, and the small piece size eliminates some drying-related defects. Beech was also used extensively as railroad ties in the past, but less so today. High-quality logs can be cut into veneer products or used in furniture or as flooring. Low-quality material is used as pulp for papermaking, or as pallets. The wood is dense (560–660 kg/m^3), and makes a highly valued fuelwood.

F. grandifolia beech nuts are valuable as a wildlife food for numerous species of birds and mammals, and have been used in the production of certain types of creosote.

Further reading

Busby, P.E., Motzkin, G. and Foster, D.R. (2008) Multiple and interacting disturbances lead to *Fagus grandifolia* dominance in coastal New England. *Journal of the Torrey Botanical Society* 135, 346–359.

Takahashi, K., Arii, K. and Lechowicz, M.J. (2010) Codominance of *Acer saccharum* and *Fagus grandifolia*: the role of *Fagus* root sprouts along a slope gradient in an old-growth forest. *Journal of Plant Research* 123, 665–674.

Williams, C.E. (2007) Why are there so few insect predators of nuts of American beech (*Fagus grandifolia*)? *Great Lakes Entomologist* 40, 140–153.

Fagus sylvatica L.

Common beech

Importance

Fagus sylvatica is a deciduous broadleaved tree found in temperate and warm temperate climates of western Eurasia, where it dominates the natural development of forest vegetation. Although individual trees may attain an age of 900 years, in natural forests a 300-year lifespan is more common. *F. sylvatica* is able to grow well in a wide range of site conditions provided it has a satisfactory supply of nutrients; it does not tolerate waterlogging. Although cold tolerant, it may be damaged by spring frosts. It is the most shade-tolerant broadleaved tree in its range.

Botanical features

Deciduous, monoecious, mono- and polycormic trees with a characteristic smooth, silver-grey bark. Isolated open-grown trees develop an extremely wide, spreading crown with branches springing from low down the stem. Trees growing in a forest stand can develop cylindrical clear boles up to 25 m. On good sites, mature beech will attain an average height of 35 m and mean dbh of 45 cm.

Twigs are green after flushing and become reddish-brown during lignification. Vegetative structures are monopodial, alternate and acrotonic. Terminal shoots develop from secondary growth of basically plagiotropic-oriented shoots, when light conditions are sufficient. Only the terminal shoot of seedlings and secondary shoots (prolepsis) are orthotropic. Two types of shoots may be distinguished: long shoots with alternate buds which branch out regularly, and short shoots which only develop a terminal bud, do not branch out and are frequently located at the twig base.

Buds are brown, fusiform and acute, >20 mm lanceolate, <10 mm spherical, and frequently covered by trichomes.

The bark of young trees is dull coloured, turning silvery-grey to grey-white with age. Because primary cork cambium activity is maintained, the bark remains smooth throughout the whole life of the tree and branch scars remain on the bark. Bud scales also leave annual scars.

Leaves are simple, petiolate, stipulate and alternate. Leaf shape is ovate, elliptic or obovate, symmetric or asymmetric. Leaf margin is more or less sinuate to slightly crenate. Leaf length ranges from 3 to 14 cm, and width from 2 to 8 cm. Petioles are up to 12 mm long. Tertiary veins are obliquely orientated. On the abaxial surface major veins are raised, interveinal areas flat, stomata 30 μm long and 22 μm wide. On the adaxial surface, major veins are level with the leaf surface, interveinal areas are flat. Primary leaves are similar to secondary leaves.

Male flowers are numerous, in pendant heads on peduncles up to 45 mm long. Calyx 4–7-lobed. Stamens 8–16. Pollen with spheroidal grains, tricolporate with pronounced pores. Female flowers usually 2, surrounded by a 4-partite, pedunculate, pubescent and scaly cupule becoming woody in fruit. Perianth 4–6-lobed. Ovary inferior, 3-celled, styles 3. Fruit a brown, triquetrous nut up to 20 mm long; 2 nuts are enclosed in the woody cupule.

Flowers bloom before and during leaf-bud break in April–May. Fruits fall in October, pass the winter in leaf litter and germinate, depending on temperature and moisture, between January and May.

The macroscopic differences between *F. sylvatica* subsp. *sylvatica* and subsp. *orientalis* are, respectively: cupule length 20–25 mm versus 15–20 mm; cupule peduncle 5–25 mm versus 20–45 mm; secondary veins 6–10 versus 6–11; and cupule with acute bristles versus cupule with foliar appendages.

Distribution

The southern limit of the natural distribution of *F. sylvatica* in Europe stretches from the Pindos Mountains in Greece to the Cantabrian Mountains in northern Spain. The western distribution limit is the coast of continental Europe, although beech is indigenous to southern England and southern Scandinavia (beech was successfully introduced as a forestry species into Ireland from the 17th century and to Scotland during the 19th century). The eastern limit is roughly a line from the Baltic to the Black Sea.

There is a transition zone in the eastern Balkans and the north-west Black Sea coast, from which *F. sylvatica* subsp. *orientalis* spreads via northern Asia Minor to the Crimea, the Caucasus Mountains, the Caspian Sea and the Elburz Mountains. In this transition zone, subsp. *sylvatica* colonizes mountainous areas, whereas subsp. *orientalis* is found at lower altitudes. The altitudinal limits of *F. sylvatica* vary from close to sea level in Scandinavia to 2700 m in the Elburz Mountains.

In the oceanic west, the southern Alps, Mediterranean montane zones and the northwest Caucasus, beech forests occur at the tree line.

F. sylvatica forms pure or mixed forests in its range. Within closed forest canopies, it can establish almost pure regeneration due to its high shade-tolerance; this regeneration will subsequently develop into beech-dominated stands. Such forests occupy a very large area under widely varying site conditions and ground vegetation associations. Most areas of beech forest today are assumed to be secondary forest.

Silviculture characteristics

Although *F. sylvatica* is very shade-tolerant, initially it has a low height growth rate. However, it can adapt to light conditions with much greater increment rates when older. It can create pure stands of high-quality trees. In mixed stands on good sites it will outgrow and dominate most species if silvicultural measures are not undertaken.

F. sylvatica grows well on a wide range of sites, with a preference for base-rich soils. Windthrow appears mainly on gley and pseudogley soils. As *F. sylvatica* is susceptible to damage by late spring frosts and to competition from ground vegetation, it is difficult to establish on open, exposed sites without overhead shelter. It has a moderate capacity for natural pruning, and is prone to stem forking or twisting and epicormic growth; to avoid these developments, beech should be established at a minimum density of around 5000–6000 plants/ha.

Silviculture practice

Natural regeneration of *F. sylvatica* is widely applied, based mainly on seedlings arising from fruits and seeds dispersed by mammals and birds.

Artificial propagation can be successful using shoot cuttings from seedlings; 2–3-year-old seedlings are used as stock plants, which are grafted with 2 year-old twigs, mainly by splice grafts.

Artificial regeneration is mainly carried out by sowing and planting. Seeds are collected after fruit-fall in November. Seed weight is 150–300 g/1000 nuts. After harvesting, nuts are spread out in 10 cm deep layers in a dry, well-aired location and turned every second day. During this process, seeds ripen fully and germination inhibition is reduced. The moisture content of the fruits should decrease to 30%. In January, nuts are spread out in enclosed chilled rooms to a depth of 30 cm and covered with jute tarpaulins or straw. Storage in sand (30 cm depth covered with 20 cm of leaf litter) is an alternate method of stratification. In approximately the second half of March, fruits are stratified in moist sand, and are sown as soon as the first germinating tops are visible in April and May.

Fruits can be stored for 6 months without special treatment, for 2 years at reduced moisture content (25–30%) and 4–5°C, and long term at 3–10% moisture content and temperatures between −3 and −10°C.

Sowing is in strips at least 30 cm wide in prepared seedbeds. Approximately 200–300 kg nuts/ha are required. Seedlings are covered to prevent late spring frost damage. In the autumn seedlings are transplanted to nursery beds at 10 × 25 cm spacing. Transplants are undercut in October of the first year (or in the following spring). Overall, 1000 seedlings and 900 transplanted plants will be obtained for each kg of fruits. Planting stock is mainly 2–3 years old (2+0; 1+1; 1+2), and 30–100 cm in height.

The silvicultural system most widely applied to beech is high forest. Coppices with standards (middle forest) are present over small areas, frequently as the result of traditional forest utilization. *F. sylvatica* was managed in the past as simple coppice forest (low forest), but this was never an important system, as beech does not produce sufficient numbers of stump shoots. Within the high forest system, even-aged *F. sylvatica* forest is the predominant type. Selection forest is mostly restricted to mixed forests. *F. sylvatica* high forest is regenerated by shelterwood and selective felling.

The success of *F. sylvatica* regeneration is highly dependent on microclimatic conditions. As with many climax species, *F. sylvatica* has relatively slow initial growth. Therefore, weeds need to be controlled in situations where radiation is high and *F. sylvatica* loses its advantage of high shade tolerance.

Even-aged *F. sylvatica* stands with an optimum density need careful tending to remove predominant, multi-branched individuals and to encourage the development of an understorey layer. When the self-pruning process has achieved a defined branchless stem length, pre-commercial selection thinning is applied to promote 150–200 evenly-spaced trees/ha, which are of high vigour and quality. As soon as these trees dominate the future stand development, the long-term growth potential of beech is utilized and stand increment is directed to crop trees by commercial increment thinnings.

On good sites, merchantable volume achieves a maximum annual increment of 10–14 m^3/ha between age 40 and 80 years, while mean annual increment gradually increases with age, to reach 8–9 m^3/ha at 120 years old. Corresponding merchantable gross volume is 900–1100 m^3/ha and standing volume 500–600 m^3/ha, depending on thinning.

The production goal of *F. sylvatica* forest management should be to produce high-quality wood of a size suitable for veneer and furniture manufacture. This involves the production of straight, cylindrical, branch-free boles, 8 m or more in length and 50–60 cm diameter at mid-point. Consequently, crop trees should have a dbh of >60 cm. On good sites this objective can be achieved over a rotation age of 80–120 years, depending on silvicultural treatment.

Diseases and pests

F. sylvatica is less affected by serious diseases than other tree species in western Eurasia. Germinative capacity can be reduced by infection of beech nuts with the pathogen *Thanatephorus cucumeris*. Beech bark disease, the first symptoms of which are brown-black, moist, hand-span sized stains on the bark, is the only serious disease of beech. It tends to occur on trees previously subjected to stress due to prolonged drought or other causes. *Cryptococcus fagisuga* attacks the tree, followed by *Nectria coccinea* infection. This combined attack of insect and fungus can result in tree death due to moisture stress. In some cases, the succession of secondary pathogens may include wood-boring insects (*Xyloterus domesticus* and *Hylecoetus dermestoides*). Eventually white rot (*Fomes fomentarius*, *Fomitopsis pinicola*, *Polyporus* spp.) destroys the wood.

Protection against biotic damage can be achieved by the immediate removal, through felling, of infected trees. Abiotic damage from solar radiation can be avoided through carefully planned forest edges. Late frost damage to young trees can be reduced by overhead shelter.

Rodents, especially mice, can damage and destroy beech nuts; 26 bird species and 17 mammals are known to consume high amounts of beech nuts in the winter. With regard to seeds and fruits, the insects *Laspeyresia fagiglandona* and *Rhynchaenus fagi* can kill 40% of beech mast.

Elateridae larvae, snails (*Agriolima agrestis*, *Arion* spp.), *Agrotis segetum* larvae and fungi (*Phytophthora cactorum*, *Phytophthora cinnamomi*, *Pythium debaryanum*, *Pythium ultimum*) cause high seedling mortality. Correct soil preparation and fertilization can minimize biotic damage.

Air pollution from ozone, fluorine and sulphur dioxide directly affects foliar vigour.

Uses

F. sylvatica forests perform many environmental functions, such as soil conservation through nutrient cycling and control of soil erosion by the extensive root system. Including *F. sylvatica* in coniferous forests can improve soils through the acceleration of leaf litter decay in humus layers, reducing interception and increasing ground water rates and water quality. About 1800 animal species are adapted to *F. sylvatica* forest ecosystems. Many ornamental varieties and forms of *F. sylvatica* are planted in parks and gardens.

Wood of *F. sylvatica* is heavy, hard, dense (585–650 kg/m^3) and not elastic. It is easy to saw and split, and, with the exception of red heartwood, takes wood preservatives and waterproofing well. It is in great demand for veneers. It can be turned and polished, and it absorbs stains, lacquers and paints well.

Industrial wood has a wide range of applications and is also used in the production of charcoal, wood gas and fuelwood. Hardwood pulp may be processed for paper or synthetic fibres. Chemicals such as calcium acetate, acetic acid, methanol, acetone and tar can be extracted from beech wood by distillation.

F. sylvatica roundwood is used for building timbers, in general carpentry and the manufacture of parquet flooring, floorboards, tools, toys, utensils and other products. Peeled veneers are used to produce plywood and blockboard, while sliced veneers are produced for the furniture industry.

As a by-product of *F. sylvatica* forests, *F. sylvatica* nuts can be pressed to produce a bright yellow, odourless, non-toxic oil that is durable under regular storage conditions, and has nutritional and pharmaceutical uses; 1 l of oil can be produced from 4 kg beech nuts. A flour can be produced from the residue of oil pressing for use in food production. Intoxication can occur if large amounts of unpeeled nuts are eaten, due to the presence of the saponin fagin. Tannins can be extracted from the bark.

Further reading

Brumme, R. and Khanna, P.K. (eds) (2009) *Functioning and Management of European Beech Ecosystems*. Springer-Verlag GmbH, Heidelberg, Germany.

Falusi, M. and Calamassi, R. (1996) Geographic variation and bud dormancy in beech seedlings (*Fagus sylvatica* L.). *Annales des Sciences Forestières* 53, 967–979.

Meilby, H. and Nord-Larsen, T. (2012) Spatially explicit determination of individual tree target diameters in beech. *Forest Ecology and Management* 270, 291–301.

Peters, R. (1997) *Beech Forests*. Springer-Verlag GmbH, Heidelberg, Germany.

Faidherbia albida (Delile) A. Chev.

Apple ring thorn

Importance

Faidherbia albida is an important nitrogen-fixing tree in African savannahs, parkland and some dry forests, and is particularly suited to agroforestry under arid conditions. It is well known for its unusual phenology where it sheds leaves with the onset of the rainy season and is green throughout the dry season. Due to its ability to improve soil, its presence reduces the need for fallow periods on poorer soils, and it has recently been promoted as a key component in sustainable and robust 'evergreen agriculture' systems.

Botanical features

F. albida is a large deciduous tree up to 15 m tall in Western Africa or reaching 25 m or more in Southern Africa and a diameter of 1 m. Mature trees have a large rounded crown, spreading branches, with a straight bole that may fork after 4–5 m. The bark is dark brown to grey or greenish-grey, rough with vertical fissures. Young stems are distinctly whitish, with paired thorns at each leaf axil. Leaves are bipinnate, blue to grey-green, 2–10 cm long, with 2–12 pairs of pinnae each having 4–23 leaflet pairs. Leaflets are small, 2.5–14 mm long and 0.7–5 mm wide. Stipules are spiny and up to 2 cm long. Soon after the leaves develop, flower buds appear on the current season's growth, with approximately 100 creamy-white flowers occurring on spicate inflorescences up to 16 cm long and 1.5 cm wide. The majority of flowers are aborted and usually only 5 or fewer are pollinated, developing into mature pods 3–4 months later. Pods are 11–30 cm long, 1.4–6.7 cm wide, orange to reddish-brown, often coiled, twisted or falcate, containing up to 30 seeds.

Distribution

F. albida has a broad native range, extending throughout dry tropical Africa from the Sahel in the north, from the Atlantic Ocean to the Red Sea, and south through Eastern Africa to Southern Africa, as well as in the Middle East from Arabia to Iran. This species is found at altitudes from 270 m below sea level in on the shores of the Dead Sea in Jordan to 2500 m on Jebel Marra in Sudan. It has been introduced to South Asia. In Western Africa, the largest populations are found in areas of high human population density, as a result of livestock dispersing seeds in their dung in regions of cultivation.

Silviculture characteristics

It is a light-demanding pioneer species which regularly colonizes disturbed fertile habitats such as riverine sites and floodplains, but suffers from competition and does not colonize undisturbed woodland. Once established at wide spacings, *F. albida* may survive extremes of drought and flooding. It may also sucker and produce coppice shoots. It fixes nitrogen, and in sandy Sahelian soils, the highest densities of *Rhizobium* are found at the water table some 30–35 m below the surface, whereas on moister sites abundant nodules are found near the surface. *F. albida* is well known for its unusual phenology, shedding leaves with the rains, contrary to most tropical deciduous trees, and remaining in leaf throughout the dry season, with ripe fruit falling late in the dry season. Fruit production is highly variable between trees and between years, with some trees producing more consistently than others.

Silviculture practice

Seedlings are prone to competition by grasses, and successful natural regeneration requires cultivated areas, abundant seed and the presence of animals in the dry season to control grass growth. Good extension work is necessary to increase the awareness of local farmers to the benefits of maintaining *F. albida*, and to identify, mark and select regenerating trees so that they are well-spaced and not weeded out. Marker stakes can be used to identify seedlings in fields at least in the first couple of years, as seedlings require protection from livestock in dry seasons until they have grown out of reach. Pruning will also promote vertical growth.

Individual mature trees may produce more than 1 kg orthodox seed each, which stores well. Pretreatment is required for rapid uniform germination, and covering with boiling water and leaving for 24 h is simple and effective. Seeds may be sown directly or planted in nurseries in polythene bags. They require frequent root pruning and may be transplanted 3–6 months later at 10 × 10 m spacing in cultivated or pasture land depending on availability of moisture and local farming traditions. Fertilizers are not required, although planting outside the natural range may require inoculation with appropriate bacteria. Establishment in farmers' fields provides desirable protection and weed control, and tractor ploughing between mature trees may promote coppicing from damaged roots. Clonal propagation from shoot and root cuttings and from callus is possible but is not widely practised, and elite stock for propagation requires identification.

Adaptation to arid sites, soil improvement and fodder have been the major factors that have prompted the interest in *F. albida*, and high wood yields have been less important. On average, growth rates of 1–1.5 m/yr have been recorded on favourable sites, though there are few systematic data on wood production. Economic analyses on managing natural stands give high-income gains as compared to treeless fields. Reports of production leaf fodder may be 50–100 kg/tree/ha at 10 trees/ha, with trees often lopped at the end of the dry season when little other forage is available, and pods also fall at the end of the dry season. Pod production from unlopped trees may yield an average of 50–150 kg pods/yr with a maximum of 340 kg recorded from a single tree, but most trees are lopped and yield only around 5–20 kg of pods/yr.

Diseases and pests

F. albida is more resistant that many *Acacia* species to gall nematodes (*Meloidogyne* sp.), though this may be due to phenolic chemicals in the roots. Pods are attacked by a number of seed-feeding insects, mostly Bruchidae, but there are few reports of any significant damage by any pests or diseases.

Uses

F. albida is principally utilized as a soil improver in subsistence farming systems with cereals such as millet, sorghum and maize, especially where fertilizers are not used, and as a fodder tree using the nutritious leaves and pods. The mulch created by falling leaf litter and canopy shade at time of planting can increase crop yields due to the improvements to the soil, as small leaflets rapidly decompose and increase the soil organic matter. Trees also affect the microclimate by increasing rainfall infiltration and reducing evapotranspiration and temperature extremes. However, improvement of some soils may be due in part to trees being planted on more fertile microsites, rather than creating fertility. On dry sites, the species is an important shade tree, and is valued in tribal customs and traditional law.

The heartwood is soft and easy to work, with a smooth finish, and relatively light with a specific gravity of 0.6–0.7 and an air-dry density of 560 kg/m^3. It seasons and planes well, but is prone to attack by borers and termites. Due to the relatively large size of the trees, the wood is used locally for dugout canoes, mortars, doors and some light carpentry. However, anatomical and chemical studies indicate that the wood is not suitable for pulping or particleboard manufacture. It is used for fuel but is less suitable than many other acacias. Thorny branches are cut for fencing in many regions.

Pods, leaves and seeds contain on average 15%, 20% and 25% total protein/kg of dry matter, respectively, and total protein digestibility may reach 73%. Tannins limit digestibility, but incorporating pods into low-quality fodder enhances ingestion without reducing digestibility. Pods may also be collected and fed to animals, or sold in markets or at roadsides. Milling the pods increases digestion of the seeds when fed to cattle, otherwise the seeds pass through the gut undigested.

Cooked seeds are eaten as a human famine food both in Western and Southern Africa. *F. albida* flowers later than most species and is a useful source of pollen or nectar for honey bees. Log beehives are made from its bark; the bark is removed from a large trunk and sealed in the form of a tube. *F. albida* is widely used for local medicines; tannins (29% in bark) have astringent properties and are used to treat toothache; and bark is also used as a toothbrush. Seed extracts have antibiotic properties.

Further reading

Barnes, R.D. and Fagg, C.W. (2003) Faidherbia albida *Monograph and Annotated Bibliography*. Tropical Forestry Papers No. 41. Oxford Forestry Institute (OFI), Oxford, UK, 267 pp.

CTFT (1989) Faidherbia albida, *A Monograph*. Translated into English by Wood, P.J. Centre Technique Forestier (Tropical and Technical Centre for Agricultural and Rural Cooperation), Paris.

Garrity, D.P., Akinnifesi, F.K., Ajayi, O.C., Weldesemayat, S.G., Mowo, J.G., Kalinganire, A., Larwanou, M. and Bayala, J. (2010) Evergreen agriculture: a robust approach to sustainable food security in Africa. *Food Security* 2, 197–214.

Falcataria moluccana (Miq.) Barneby & J.W. Grimes

Batai wood

Importance

Falcataria moluccana, previously known as *Paraserianthes falcataria*, is one of the world's fastest growing multipurpose tree species. This fact, coupled with other positive attributes, makes it a suitable species for plantation programmes and agroforestry applications in the humid tropics. It is commonly planted as an ornamental and shade tree. Other uses of the species being tested include alley farming, intercropping in forest plantations and reforestation. As a nitrogen-fixing tree it helps improve soil fertility.

Botanical features

F. moluccana is a medium- to fairly large-sized tree up to 40 m high with a small buttress. The bole is branchless up to 20 m and up to 100 cm or more in girth. In dense stands the bole is generally straight and cylindrical. When grown in the open, trees form a large umbrella-shaped canopy. In plantations of 1000–2000 trees/ha the crowns become narrow. The bark is white, grey or greenish, smooth or slightly warty.

Leaves alternate, bipinnately compound and 20–40 cm long with 4–15 pairs of pinnae, each of the pinnae containing 8–25 falcate leaflets. There is a large nectary below the lowermost pair of pinnae and smaller ones between or below most pairs of pinnae.

It has paniculate racemes, which are axillary. The flowers are bisexual, regular and 5-merous. The corolla is creamy-white and sericeous. The stamens are numerous. Pods chartaceous, narrow and flat, 10–13 cm long and 2 cm wide, winged along ventral suture with many seeds.

Two flowering periods per year have been observed. Mature fruits appear 2 months later. Seeds are wind dispersed.

Distribution

F. moluccana is native to the Moluccas, Papua New Guinea, the Bismarck Archipelago and the Solomon Islands from 10° S to 30°N. The species probably originated in the East Malesian region and has been widely planted in the tropics, becoming an invasive and problematic weed in some areas. It occurs more frequently in secondary forest than primary forest and on river flood terraces.

Silviculture characteristics

F. moluccana is a strongly light-demanding species which regenerates naturally when the soil is exposed to sunlight. The abundance of growth of wildings in the forest only occurs when the soil is cleared from the undergrowth and the canopy opened. *F. moluccana* thrives on comparatively poor soils without fertilizers as long as they are well drained, and growth of young trees is promoted by inoculation with mycorrhizal fungi. The presence of *F. moluccana* in mixtures with eucalypts gives a better growth increment in girth, height and biomass than pure eucalypt stands. Trees coppice fairly well.

Silviculture practice

Seeds are usually collected by cutting branches bearing matured brown pods. As germination of untreated seeds is irregular and may be delayed for up to 4 weeks after sowing, seeds can be treated by soaking in boiling water for 1–3 min or immersing in concentrated sulphuric acid for 10 min, with subsequent washing and soaking in water for 18 h. For storage, seeds are air dried for 24 h, packed in polyethylene bags, then stored at 4–8°C; germination rate is 70–90% after 18 months. Sowing is usually by broadcasting, with seeds then pressed gently into the soil and covered with about 1.5 cm of fine sand. The soil in seedbeds should be well drained and loosened, the surface mulched and excessive shading avoided.

Transplanting of the seedlings to the field can be carried out at 2–2.5 months when the seedlings are about 20–25 cm high and have developed good fibrous root systems and woody stems. Container-grown plants are transplanted in the field at 4–5 months old. Seedlings are planted at a spacing of 2 × 2 m to 4 × 4 m. Only one complete and three spot weedings in the field are required in the first year, as initial growth is very rapid. *In vitro* micropropagation has been successful using seed tissue explants.

F. moluccana can reach 7 m in height in 1 year, 15 m in 3 years and 30 m in 10 years. Volume wood yields are on average 39 m^3/ha/yr in a 10-year rotation, and may reach 50 m^3/ha/yr on better soils. Timber production stands can be thinned at age 4–5 years to a density of 250 stems/ha and after 10 years to 150 stems/ha.

Pruning is required, as there is a tendency for stems to fork. Trees grown for timber have a harvest cycle of 12–15 years, while for pulp production the cutting cycle is approximately 8 years. In agroforestry systems with annual crops in the first year and grazing animals in subsequent years, the cutting cycle for *F. moluccana* is 10–15 years. *F. moluccana* plantations must be kept weed-free in the first few years and fertilizers may be applied to improve the yield.

Diseases and pests

Seedlings in nurseries are prone to damping off caused by *Sclerotium*, *Rhizoctonia*, *Fusarium*, *Phytophthora* and *Pythium* fungi. Sterilization of the soil may control the disease. Gall rust caused by the fungus *Uromycladium tepperianum* has become a serious disease of *F. moluccana* plantations in some countries since the 1990s. Infection causes a decrease in relative growth rate at all stages of tree development and is promoted by high relative humidity (≥90%) and slow wind speed (≤80 km/h). Pruning, thinning and clear felling reduce disease incidence and severity. Other fungal diseases include pink canker caused by *Corticium salmonicolor* (*Phanerochaete salmonicolor*) and red root caused by *Ganoderma pseudoferreum*.

Leaf-eating caterpillars and aphids are also occasional problems of seedlings and trees in plantations. The stem borer *Xystrocera festiva* (longicorn beetle) and the red borer *Zeuzera coffeae* (cossid moth) are also found in plantations. *Coptotermes curvignathus* (rubber termite), *Euwallacea fornicatus* (tea shot-hole borer), *Helicotylenchus dihystera* (common spiral nematode), *Homona coffearia* (tea tortrix), *Hypomeces squamosus* (green weevil), *Indarbela quadrinotata* (bark-eating caterpillar), *Rastrococcus iceryoides* (mango mealy bug), *Xylosandrus morigerus* (brown twig beetle) and *Xyleborus similis* have also been recorded as pests of *F. moluccana*.

Rapid extraction, conversion and seasoning of harvested wood are crucial to prevent infection by fungi, insect attack and sap-stain attack.

Uses

The thin crown of *F. moluccana* provides partial shade to tea, cacao and coffee crops. However, the importance of this species as a shade tree is rather limited, as solitary trees are prone to wind damage. When planted in rows it is suitable as a windbreak for bananas. When managed as hedges, *F. moluccana* is able to produce 2–3 dry tonnes of green leaf manure/ha/yr. Because of its fast growth, *F. moluccana* is often used in reforestation and afforestation of denuded lands, for firewood and charcoal production. Pure stands give a good protective cover in preventing erosion on slopes. The ability to coppice fairly well makes it suitable for pulpwood production. It is also commonly used in agroforestry systems throughout its range. It is widely planted as an ornamental tree due to its attractive form, grey bark and feathery leaves and flowers.

The wood is soft with a density of 430 kg/m^3, but is not durable. Due to its excellent pulping characteristics, it is widely used to supply pulp for the manufacture of paper. Being pale in colour, it requires only limited bleaching to obtain high-quality white paper. The wood is easy to work and suitable for matchsticks and for general utility purposes such as light-weight packing materials and construction, boats, panelling, cabinets and furniture. As the wood is very easy to cut, it is also suitable for making toys, wooden shoes, musical instruments and general turnery. The wood is also an important source of light-weight veneer and plywood, and proves very suitable for the production of low- and medium-density particleboard, hardboard, wood wool board and blockboard. In some countries, *F. moluccana* is an important source of firewood.

The leaves of *F. moluccana* are used as a green manure, and as feed for chicken and goats. The bark yields the gum kino, which has tanning properties and is also sometimes used as a soap. The edible pods are used as substitutes for *Parkia speciosa* in cooking.

F. moluccana has shown potential as a biofuel crop.

Further reading

Krisnawati, H., Varis, E., Kallio, M.H. and Kanninen, M. (2011) Paraserianthes falcataria *(L.) Nielsen: Ecology, Silviculture and Productivity*. Center for International Forestry Research (CIFOR), Jakarta, Indonesia.

Roshetko, J.M. (1998) Albizia *and* Paraserianthes *Production and Use: A Field Manual*. Forest, Farm, and Community Tree Network (FACT Net), Morrilton, Arkansas, USA.

Yan, S., Hu, D.-H., Wei, R.-P., Wang, R.-H., Liu, J., He, H.-B. and Zeng, J.-X. (2011) Criteria for selecting superior trees of *Paraserianthes falcataria*. *Forest Research (Beijing)* 24, 272–276.

Flueggea flexuosa Muell. Arg.

Poumuli

Importance

Flueggea flexuosa is highly regarded for production of very durable timber, being especially favoured for posts and poles. Its moderately rapid growth, good bole form, wood and non-wood uses make it a promising tree for agroforestry plantings, especially on infertile soils. It grows rapidly, straight and without large branches and is widely exploited for local usage. *F. flexuosa* is planted in rural gardens and house-yards to provide poles and has potential for intercropping with high-value timber species. Further provenance selection and refinement of silvicultural practices are needed.

Botanical features

F. flexuosa is a deciduous, small to medium tree, 10–16 m tall with a diameter to 30 cm and a clear bole up to 6 m, rarely attaining 25–30 m with a diameter of 50 cm, sometimes with indistinct buttresses. The crown is narrowly columnar to conical with many small straight, radiating, horizontal branches. With age, branching develops a less regular appearance and the canopy may exhibit an oblong or more spreading form. Branchlets have a characteristic zig-zag appearance with slightly pendant tips. The bark is rough, light grey, light sandy to mid-brown, with close evenly spaced vertical fissures to 4 mm deep becoming scaly with age, peeling in small thin strips along these lines. Inner bark is light pinkish-brown grading to a very light pink.

Leaves are simple, alternate, oblong–elliptic, with a rounded or tapered base, and prominent pointed, often recurved tips, shiny dark green above and light green below, usually 8–14 cm long and 3–5 cm. There is a mid-vein with 5–7 pairs of side veins, curving forward steeply, almost meeting well inside the margin. Petioles are 5–10 mm long, and are either red or green, distinguishing two morphological types (along with different venation). The masses of small, light greenish-yellow flowers are arranged in short axillary clusters all along the twigs. Both male and female flowers, found on separate trees, have 5 sepals but no petals. Fruits are in clusters on slender pedicels, 5–8 mm long. Fruits are small, globose berries, 3–5 mm in diameter, which ripen from green to red or purple or black, each containing 4–6 small angular seeds 1 mm long.

Distribution

The natural range of *F. flexuosa* extends from the Philippines, eastern Indonesia to the Solomon Islands and Vanuatu. It is rare in Indonesia but widespread in the Solomon Islands, localized in Irian Jaya (West Papua) but with no records from

Papua New Guinea. It is considered to be either an ancient (usually Polynesian) or else quite modern introduction in many parts of the South Pacific, including Vanuatu, New Caledonia, Fiji, the Wallis and Futuna Islands, Tonga and Samoa.

F. flexuosa occurs in primary forest and dense shrub savannah at low altitudes, and is often a pioneer growing in river floodplains, fallow fields and abandoned coconut plantations. In the Solomon Islands, the species has a restricted habitat in lowland coastal forests on coralline soils and beside rivers on poorly drained sites. In South Pacific countries where it has been widely planted, such as Vanuatu and Samoa, *F. flexuosa* readily colonizes adjacent disturbed sites, but only infrequently escapes into secondary and primary forest habitats.

Silviculture characteristics

F. flexuosa is adapted to the lowland, humid tropics, and its native range is uniformly warm to hot throughout the year with mean annual temperatures of 22–28°C, with no frost. Rainfall is high, between 1500 mm and 5000 mm and uniformly distributed; however, the months during which higher rainfall is experienced can vary markedly between locations and at different altitudes, up to 350 m. It can tolerate dry seasons of up to 4 months.

It grows well on a wide range of soils including coralline soils, red clays, nutrient-poor soils and poorly drained sites, though prefers freely draining, light-textured soils. It may be dominant in the floodplains of river and abandoned coconut plantations, but will not tolerate waterlogging for more than a few weeks. It is a light-demanding species, although it can become established under a light canopy of other trees such as coconut. Regeneration is often plentiful within natural stands and plantations. It is tolerant of light salt spray but not to exposed coastal locations. It is not tolerant of fire.

In the Philippines, flowering has been observed in January and May–June and fruiting in January and May, though flowering and fruiting appear to occur all year round in Vanuatu and Samoa.

Silviculture practice

Planting of seedlings is the most common method of propagation. Ripe fruits should be soaked in water overnight, and then depulped by rubbing and washing over a fine mesh sieve. Germination occurs within 12–20 days after sowing and the nursery phase is about 4 months, although seed is sometimes broadcast over newly planted food crop gardens to supplement natural regeneration. No work has been carried out on vegetative propagation.

Trees are established on cleared sites, the most appropriate planting technique being close-line planting, because at close spacing, seedlings are able to rapidly dominate the site. This reduces weeding requirements and the need for thinning, and will provide roofing poles or rafters. Otherwise, regular weeding will be required. For plantations, whether pure or mixed, various spacings between 2 × 2 m and 4 × 4 m are reported in pure plantation or in mixtures.

Singling has not proved to be necessary, although pruning to 3 m height at an age of 3–4 years old is recommended for the production of posts. Appropriate thinning regimes need to be enforced to ensure the required size of the end product is achieved. For pole production, a final density of 1000 stems/ha at age 6 years is found to be appropriate, with trees of 15 m height after 7 years.

Annual timber increments are estimated to be around 4–6 m^3/ha/yr. Annual increments can be more than 1.5 m during the first 5–10 years, but may decline sharply as trees reach 20 m. On some sites, trees were reported to grow more than 3 m in height/yr and 3–4 cm in diameter in the first 3–4 years.

Diseases and pests

The species is not very susceptibility to pests and diseases, with the only reported disease being brown root rot (*Phellinus noxius*); newly planted seedlings may be damaged by termites.

Uses

F. flexuosa is well suited to planting on farms and in agroforestry systems, including boundary plantings and property demarcation; in planted fallows or in plantations mixed with other tree and food crops; as a shade crop for cacao or other crops; or as a nurse tree for other valuable but long-rotation timber species. As a locally common lowland pioneer species, it also plays an important environmental role in the revegetation of various categories of disturbed site.

The heartwood is heavy, with a density of 810–935 kg/m^3 at 15% moisture content. It has a pale yellowish-brown or reddish-brown colour, sometimes hardly distinguishable from the pale sapwood. The grain is straight and the texture moderately fine. The wood is very hard and strong, but timber is easily worked, resistant to dry-wood termites and fungi, and the heartwood is very well suited for uses that involve contact with the ground, whereas the sapwood is not durable in the ground, being susceptible to *Lyctus* attack.

Uses include house and fence posts, ground posts, rafters, piles and other smaller parts in house construction, being especially favoured for the poles in traditional 'fale' houses in Samoa. Poles can be split easily. Although usually only available in small dimensions, its timber is suitable for heavy construction, bridging, piling and for buildings that stand in the sea, because the wood is durable in salt water. The wood makes an excellent, hot-burning fuel. It may also be suitable for wooden buttons and other uses where a fine-grained strong timber is required.

Leaves of the red-veined form are used to stain *Pandanus* leaves a charcoal colour. A medicinal drink is made from the rasped bark and can be used to treat fever and the shredded root is used medicinally. The purple juice from the fruit is also used as a dye. The tree has a moderate importance for bee keepers for both pollen and nectar production.

Further reading

Chaplin, P.E. and Ngoro, M.L. (1988) *The Status of* Flueggea flexuosa *in Solomon Islands—An Appropriate Species for Small Scale Community Forestry*. Forestry Research Note 46–14–88. Forestry Division, Ministry of Natural Resources, Solomon Islands.

Sosef, M.S.M., Hong, L.T. and Prawirohatmodjo, S. (eds) (1998) *Plant Resources of South-East Asia. No 5(3). Timber Trees: Lesser-Known Timbers.* Backhuys Publishers, Leiden, The Netherlands.
Thomson, L.A.J. (2006) *Flueggea flexuosa* (poumuli), ver. 2.1. In: Elevitch, C.R. (ed.) *Species Profiles for Pacific Island Agroforestry.* Permanent Agriculture Resources (PAR), Hōlualoa, Hawaii, 13 pp.

Fraxinus excelsior L.

Ash

Importance

Fraxinus excelsior, common ash, is widely distributed in Europe and South-west Asia, and is an important commercial broadleaved tree that is able to grow rapidly and produce high-quality wood. Extensive stands of pure ash are rare. It is regarded as a useful component in mixed natural forests with other broadleaved species, but is not suitable for large-scale planting; it is only worth persisting with on ideal sites where it will grow to large sizes. Ash is one of the most important timbers because of its elasticity and toughness after seasoning.

Botanical features

Crown oval-shaped to spherical, with steeply angled branches, sparsely but regular branched on all sides. Monopodial growth. The tree can reach heights of more than 40 m although 30 m is the norm. Average dbh at maturity is 60–80 cm. The bole is straight, cylindrical, with a slight tendency to eccentricity and can be clear of branches for 40–50% of the total height. Forking of shoots is common.

The root system is characterized as a sinker root system consisting of a taproot and horizontally wide-spreading roots from which vertical or sinker roots develop. The bark is olive-green and smooth in young trees but becoming grey, rough and broad-furrowed with age.

F. excelsior has sooty-black winter buds set in opposite pairs, and smooth grey twigs. The leaves have 7–15 stalkless, broad or narrow, long-pointed leaflets which are 5–12 × 1.5–4.5 cm, oblong–ovate to oblong–lanceolate, long–acuminate, tapering to a round base, sessile, and crenate–serrulate.

Panicles of purplish flowers develop from side buds in April before the leaves flush. The flowers are wind pollinated, trioecious, with male, female and bisexual flowers in different combinations on different trees. Petals and sepals are absent. Male flowers have two anthers, and female flowers one ovary and two sterile anthers.

Fruits are samaras (dry, indehiscent, winged, fruity), 2.5–5 cm long, 4–6 mm wide, lanceolate or broader towards the tip. Fructification starts at around 30 years. Mast years occur every 2–3 years. Seeds are flat, lanceolate and acuminate, with endosperm and flat embryo. They ripen on the tree during July–October and are dispersed by the wind over the winter and early spring. They germinate in the following spring. Germination is epigeic.

Distribution

F. excelsior has a wide natural distribution throughout Europe, from the Mediterranean to central Sweden and from the Atlantic to central Russia. It extends eastwards to the Caspian Sea and is also found in North Africa. It is a component in different types of mixed forest. Plantations are found throughout Europe; countries of importance are Belgium, Denmark, France, Germany, Ireland and the UK.

Silviculture characteristics

In common with other pioneer species, ash is a strong light-demander in mature-growth phases and does not grow well with overhead cover or strong crown competition. Seedlings are shade-tolerant for approximately 7 years. Conditions for natural regeneration are best in woodlands with a dense canopy. Young trees show a very rapid growth in the early years. Full overhead light is necessary for full vigour and for high-quality timber production. Thinnings should be early, heavy and rather frequent to keep crowns entirely free. Tending and pruning improve timber quality. Thinning at 10 m top height encourages crown development and stimulates diameter growth; potential final crop trees are selected at 15 m top height.

Silviculture practice

Seeds are usually harvested when they turn brown in October/November, but can also be collected as soon as they fill and should be sown immediately. A high proportion will germinate in the following spring. Seeds are sown from mid-March onwards. No special site preparation is needed. Seeds can be stored for up to 10 years.

Afforestation uses 1+1 and 1+2 year old transplants with a minimum root collar diameter of 7 mm and a maximum height of 60 cm. A minimum root collar diameter of 5 mm and 20 cm height are recommended for bare-rooted plants, 6 mm and 60 cm height for container-grown stock. An area of 1 m^2 around the plant should be kept weed-free for at least 3 years to maximize water and nutrient availability.

For timber production, a maximum initial spacing of 2 × 2 m is recommended for planting bare land (2500 plants/ha) and 2.5 × 2.5 m for restocking (2000 plants/ha). Side shelter is necessary when pure ash is planted on exposed ground. Individual protection or fencing against damage by browsing is essential.

With natural regeneration, sufficient overhead light is necessary for rapid seedling development, so canopies have to be opened up and small gaps need to be enlarged to more than 400 m^2 in area. The advantages of natural regeneration are a high seedling density which favours the development of straight stems with small branches, low costs and the benefit of genetic diversity. However, several early cleanings are necessary to avoid the occurrence of canker.

In mixed stands, other broadleaves play a role in natural pruning of ash and in producing conditions suitable for natural regeneration. In naturally regenerated stands, the first tending operation should take place when the crop reaches a top height of 5 m, to eliminate coarse-branched, deformed and cankered trees (4000 stems/ha). A second intervention follows at a top height of 7–8 m (2000 stems/ha). At this stage naturally regenerated associated tree species are favoured to encourage species diversity. One tending operation is considered sufficient in plantations with a density of 3300 plants/ha.

Shaping and pruning operations are well established. Forking of the stem occurs frequently in young trees (approximately 60% of all trees in a plantation). These, and competitive branches, should be cut when the crop reaches a height of 6–7 m to favour the development of a straight bole. Repetitive shaping operations are recommended on a designated number of trees (250 stems/ha). In plantations of heterogeneous and low stocking densities (1000 stems/ha), artificial pruning is necessary to obtain a branch-free bole of a minimum length of 6 m.

F. excelsior is a strong light-demander and crowns should be kept free to grow quickly in order to stimulate diameter growth. Thinning begins at 10 m and is repeated for every 2 or 3 m increase in top height, according to site quality. A strong reduction in stem numbers at the first thinning facilitates crown development at this stage. Thinning intensity is 35% for very good sites and 30% for medium sites. Delaying thinning has to be avoided as it will retard crown and diameter development and produce only small stems at a high density, making the stand sensitive to windthrow following thinning, and encouraging the dissemination of canker. In such cases a very strong thinning has to be undertaken to favour the growth of selected dominant healthy trees. Furthermore, seed production of *F. excelsior* in close stands is poor, and ash will not regenerate naturally.

The objective of silviculture aimed at high-quality timber production is to produce a straight, defect- and branch-free bole at a minimum length of 6–7 m, with a dbh of 50–60 cm at age 50–65 years. This corresponds to 60–80 stems/ha. *F. excelsior* has a relatively short economic rotation of 60–80 years, after which the wood tends to become discoloured.

A dbh increment of 1 cm/year is adequate, and should not fall below 0.6 cm/year if quality timber is required. The common dimensions for sawing are logs with a minimum length of 2.5 m and minimum diameter of 40 cm.

Diseases and pests

The most frequent stem defect of *F. excelsior* is canker. Bacterial canker, caused by *Pseudomonas syringae* subsp. *savastanoi*, is characterized by a brown, corky swelling with wart-like excrescences on the stem. Fungal canker (*Nectria galligena*) is characterized by tissue death and some swelling at the edges of the canker. Cankered trees should be removed by thinning at the earliest opportunity.

Ash die-back, causing progressive death of branches and eventually death of the top of the crown, is among the most important new, invasive tree diseases in Europe. It is caused by the new fungal species *Hymenoscyphus pseudoalbidus* (anamorph *Chalara fraxinea*) and has affected large numbers of ash trees since the mid-1990s, particularly in eastern and northern Europe, although it has been observed across much of *F. excelsior*'s range.

Forks on the trees are produced by wind, late spring frosts, birds or the ash bud moth (*Prays fraxinella*) killing the buds. The ambrosia beetle *Leperisinus fraxini* tunnels in the wood at the base of small branches, producing galls and black swollen strips on the stem. This disease occurs when the tree is grown on inadequate sites or at too high a density. As a precaution, it is recommended that dead branches that still hang on the tree be eliminated. Trees and logs are liable to attack by a range of beetles, including members of the Cerambycidae and Buprestidae.

Rabbits can damage the bark, while hares snip off the stems. Deer browse on young plants and should be excluded by fencing.

Uses

F. excelsior produces a ring-porous wood with very large earlywood and strong, hard, dense latewood that gives it toughness after drying. Density at 12% moisture content ranges from about 510 to 830 kg/m^3. The timber is perishable and fairly easy to treat.

The timber is elastic and tough after seasoning. High-quality timber logs for the production of sawtimber and veneer for the furniture industry should be from trees of 50–60 cm dbh. The wood is used for furniture, agricultural implements and in boatbuilding. The bending ability and flexibility (coming from large ring width) make the wood very suitable for sports goods and handles. The name 'olive ash' refers to wood of a brownish colour, often streaked, and hence of high value for decorative veneers.

Leaves and bark have traditional medicinal attributes.

Further reading

Asche, N. (1995) Ash - ecological and silvicultural aspects. *Allgemeine Forst Zeitschrift* 50, 1087–1089.

Dobrowolska, D., Hein, S., Oosterbaan, A., Wagner, S., Clark, J. and Skovsgaard, J.P. (2011) A review of European ash (*Fraxinus excelsior* L.): implications for silviculture. *Forestry (Oxford)* 84, 133–148.

Forestry Commission (2003) *The Management of Semi-Natural Woodlands. 2: Lowland Beech-Ash Woods*. Forestry Commission Publications, Stockport, UK.

Opstal, N.A. van (ed.) (2011) EPPO Workshop on *Chalara fraxinea*, a major threat for ash trees in Europe, Oslo, Norway, 30 June–2 July 2010. *Bulletin OEPP/EPPO Bulletin* 41, 1–26.

G

Garuga floribunda Decne.

Garuga

Importance

Garuga floribunda is an important timber and multipurpose tree from South-east Asia and the South Pacific. The medium-heavy, reddish-brown heartwood has many uses including general construction and interior uses. The tree offers considerable potential for reforestation or enrichment planting in degraded, logged-over or cyclone-damaged forests. *G. floribunda* is commonly used as a live fence and for boundary plantings in the South Pacific.

Botanical features

G. floribunda is a medium to large tree to 15–35 m tall. In favourable situations the bole may attain 1.3–1.5(–2.2) m dbh above buttresses. The crown is spreading, with a feathery appearance, and may be irregularly shaped, domed or flat-topped.

In forest situations the bole is generally long, straight and cylindrical, but in open situations the trees may exhibit a short and crooked bole. When present, buttresses are variable, steep to even, straight to arched, thick to thin, and sometimes branching above the ground. The bark is light grey-brown to grey, smooth with wide, evenly spaced conspicuous lenticel protrusions on young trees. With age, the bark on the lower bole becomes rough and scaly. The inner bark is pink- to red-brown in distinct layers grading to white on the wood. Following damage to the bark a slight clear to milky, slightly resinous-scented exudate is produced, which becomes sticky on exposure to air. Branchlets minutely puberulous when young, becoming glabrous, and often rough due to many leaf-scars.

The leaves are large, pinnately compound, and crowded in dense spirals at the ends of the branches. Leaflets are arranged in 4–15 pairs along the rachis, typically ca 10–12 pairs, with or without a terminal leaflet. The leaf rachis is slightly swollen at the base and up to 35 cm long. Each leaflet is shortly petiolate, somewhat obliquely narrowly ovate to ovate, 38–100 × 11–31(−38) mm, lanceolate around the base, margins entire or crenate, gradually tapered to a long–acuminate apex, mid to dark green and shiny above; dull light green and velvety hairy below. Stipules are present at the base of each leaflet, up to 10 × 3 mm, pointing towards the base of the leaf, and usually falling as the leaf matures. Trees are deciduous, losing all or most of their leaves once a year: leaflets turn red and fall singly. The new red to light reddish-green leaf flush appears shortly after the onset of flowering.

The flowers are held in terminal branched erect panicles, up to 15 cm long and are conspicuous, being borne on otherwise bare branches. Each flower is bisexual with functional male and female parts. The flowers are 5-merous, ca 7 mm across, with light pink and yellow petals. The ovary glabrous to pilose, style equal to stamens, stigma 5-lobed. Large clusters of fruits are born in drooping panicles, held clear of the foliage. The fruits are somewhat globular to obovoid, 10–15 mm across, green, ripening to greenish-blue or purple-black.

Flowers appear before the new leaves. In general, flowering is at the end of the dry season or the beginning of the rainy season. Flowering has been recorded from June to November (Java, Indonesia), September to October (Sulawesi, Indonesia), March to October (Philippines), June to October (Vanuatu), and August to September (Samoa). Fruiting has been recorded from October to April (Sulawesi), March to October (Philippines), October to February (Vanuatu), and October to December (Samoa). The fruits are eaten and seed dispersed by birds, including Pacific pigeons and fruit doves, and by flying foxes.

Distribution

G. floribunda is found distributed from South Asia (India, including the Himalayas and Bangladesh); through East Asia (south-west China and Hainan Island) and South-east Asia, including the Philippines, Indonesia (Central and East Java, Lesser Sunda Islands, Celebes, Moluccas, Irian Jaya (West Papua), but absent from Sumatra); Malaysia (where it is very rare in Sabah and Peninsula Malaysia); Papua New Guinea; northern Australia; the Solomon Islands (where it is restricted to drier parts in the north of Guadalcanal); to South Pacific islands including Vanuatu, Samoa, Tonga but not Fiji.

In Vanuatu, the species is most common in the central islands in areas with a pronounced dry season and is recorded from Tanna, Efate, Epi, Paama, Malekula, Espiritu Santo, Pentecost and the Banks group. It is also reported from Futuna, Erromango, the Shepherds group and Ambae. In Australia the species occurs in vine thickets and rainforest in the Kimberley region (Western Australia), often in near-coastal areas, from port Warrender to Walcott Inlet, on islands of the Bonaparte Archipelago and in Queensland.

G. floribunda occurs in seasonal climates in primary and secondary forests, often periodically dry or very dry monsoon forests and vine thickets and in lower montane rainforest, up to 1200 m altitude. It is also found in coastal forests and teak-dominated associations.

In the South Pacific, the species occurs in primary and mature secondary forest in Samoa and Vanuatu, especially in associations with *Gyrocarpus americanus*. In Papua New Guinea, *G. floribunda* occurs in lowland and lower montane rainforests; in Samoa, the species is most common in the driest areas and in coastal and lowland forest.

Silviculture characteristics

The species is tolerant of semi-shaded conditions in seedling and juvenile stages, but mature trees are light-demanding.

Silviculture practice

There are about 1,300 fresh fruit/kg. It is preferable to collect only fully mature, blackish fruits, although slightly immature fruits may be sun ripened for a few days and soaked overnight in water for easy seed extraction. There are 15,500–23,000 seeds/kg. No seed pretreatment is required, and fresh seed is sown in sawdust or potting media in germination trays.

Seedlings exhibit epigeal germination and appear after about 6 weeks. Seedlings may be pricked out at the two-leaf stage, and the total duration of the nursery phase is about 4 months. The species grows best with uniform, light shade during the first 5 years and cover should then be progressively opened. In Vanuatu, it has also been observed that growth of young plants is much better in more sheltered, partly shaded sites than in the open.

Planting at close spacing is recommended in order to avoid production of large low branches.

Vegetative propagation using large woody branch pieces is commonly used in Tonga for the development of live fences.

There is very limited published information concerning management of this species in native forest or plantations. In Vanuatu (Espiritu Santo and Malekula), the recommended minimum diameter cutting limit is 60 cm dbh.

Diseases and pests

In Vanuatu, young plants may be severely defoliated by an unnamed white hairy caterpillar. In Samoa, the bole may be attacked by an unidentified species of small beetle. The species is reported to be susceptible to fire, but trees are moderately resistant to cyclones and drought.

Uses

The species is found in Samoa near garden areas, as live fences around plantations and as a protected species near settlements or as live fences and boundary markers in Tonga. It may also be planted as a shade tree. In Samoa, fruits attract pigeons and for that reason the tree is frequented by pigeon hunters; it is also a habitat for edible beetles. In Palawan, in the Philippines, stands of *G. floribunda* var. *floribunda* serve as nesting sites for the critically endangered Philippine cockatoo *Cacatua haematuropygia*.

The heartwood varies in colour from brownish- to reddish-brown, but with considerable variation, whereas the sapwood is greyish. The wood exhibits an interlocked to wavy grain and moderately coarse texture. It is moderately hard and moderately strong. Machining is easy; however, gluing can create problems. Wood density is quite variable, typically in the medium to high range (e.g. densities have been reported as 680–690 kg/m^3 at 12% moisture content or 500–900 kg/m^3 at 15% moisture content).

The wood is very slow to dry, and shrinkage of the wood upon seasoning is reported to vary from low to moderately high, although with little degradation and only slight end splitting. Careful stacking to avoid localized warping is required during the lengthy air-drying phase. Reports of timber durability vary from low to durable; it is suitable for exposed exterior uses and moderate service in ground contact. The sapwood is easy to saw and work with hand and machine tools.

It is susceptible to dry-wood termites and liable to stain. The wood is also susceptible to *Lyctus* (powder-post beetle) attack.

In the South Pacific, the timber is mainly used for high-quality furniture, construction, canoes and in the timber industry. It is also used for general construction, bridge-building, posts, light-duty flooring, cabinet work, interior trim, mouldings, shelving, skirting, sporting goods, agricultural implements, boxes and crates, carving, toys, novelties and turnery. It also provides veneer and plywood. It is appreciated as a hot-burning fuelwood in Vanuatu.

The species is widely used in traditional medicine in Vanuatu where the inner bark is used to treat stomach ache and infected cuts. The leaves are used for animal fodder, and the fruit is edible. A decoction of the bark has been used as a black dye to dye mats made from *Corypha* leaves.

Further reading

Jalonen, R., Choo, K.Y., Hong, L.T. and Sim, H.C. (eds) (2009) Forest genetic resources conservation and management: status in seven South and Southeast Asian countries. FRIM, Kuala Lumpur, Malaysia; Bioversity International, Selangor, Malaysia; and APAFRI, Kepong, Malaysia.

Sopsop, L.B. (2011) Floral survey in the coastal forest of rasa wildlife sanctuary, narra, Palawan, Philippines. *Journal of Environmental Science and Management* 14, 71–76.

Wardani, M. (1994) Description and key of commercial wildings in Kintap forest, South Kalimantan. *Buletin Penelitian Hutan* 560, 1–42.

Gleditsia triacanthos L.

Honey locust

Importance

Gleditsia triacanthos is a hardy, adaptable and useful tree, native to hardwood forests of eastern, central, and southern USA. It is tolerant of drought and frost and grows in all types of soil, thriving from cold-temperate to subtropical climates and even in tropical conditions where introduced. It is an ornamental, shelter belt, shade and erosion-control tree. The timber is hard, heavy, strong and durable, and is used for various purposes including structural timbers. *G. triacanthos* can easily be grown from seed, suckers or cuttings, but has a tendency to form dense thickets and become invasive.

Botanical features

G. triacanthos grows to 21–25 m with a diameter of 60–100 cm, up to 45 m tall and 180 cm diameter. It has attractively arranged branches ascending in the lower part of the stem and horizontal in the upper part, resulting in an umbrella-like crown form. The crown is variable in shape, from narrow and plume-like to moderately wide. The stem form is straight, with a round cross section, often short and divided near the ground. The bark of mature trunks is usually 0.6–3.5 cm thick with narrow ridges divided by fissures, peeling in strips. Branches are densely armed with reddish, simple or compound stout thorns, flattened at the base. Trees have strong taproots with deep, widely spreading and profusely branched root systems. Leaves are pinnate (20–30 leaflets) or bipinnate in the same tree. Leaflets are sessile acute, ovate, 2–3 cm long, finely crenate and bright green, golden-yellow before leaf fall. *G. tricanthos* is polygamo-dioecious. Flowers are small, greenish, actinomorphic, in axillary racemes (5–7 cm long) with 4–5 petals. Fruits are flat, sickle-shaped, and twisted pods, 30–40 cm long and glossy dark brown or purplish-brown with sweet, edible pulp. Seeds are oval, dark brown and 7–8 mm long.

Distribution

The natural range of *G. triacanthos* covers much of central USA, from northern Florida and eastern Texas, north to southern Ontario, Canada, and from Nebraska east to Pennsylvania, with scattered native populations in Mexico. From the 1600s, *G. triacanthos* was widely introduced elsewhere in North America, and later to Europe and temperate South America; it can now be found throughout Africa, South and South-east Asia, Australia and the Pacific, and Europe.

Silviculture characteristics

G. triacanthos is a light-demanding species, colonizing bare soils by seed propagation or suckers. Seed propagation is reliable, as good seed crops occur nearly every year, the fruits falling gradually throughout the whole winter. Trees bear seed from about 10 years until 100 years of age, with optimum production occurring between 25 and 75 years. In mixed natural stands it behaves as a dominant tree. When coppiced, *G. tricanthos* resprouts abundantly and forms dense thickets. Growth is rapid and maturity is reached at about 120 years. *G. triacanthos* is easily injured by fire due to its thin bark and appears to be excluded from prairies or rangelands by frequent fire. Damage to trees by poor air quality has also been recorded, but they are highly resistant to ice damage and average in resistance to flooding.

The generalized root system is adaptable to environmental conditions and enables the tree to grow on both upland and lowlands sites. It is commonly found on rich alluvial soils in mixed forests associated with others broadleaved trees, but is generally only a minor component of natural forest stands. *G. triacanthos* is deciduous, beginning to flower when its leaves are nearly full grown, from around early May in the southern parts of its range to late June in the northern parts. The pods ripen from September to October, usually falling after ripening but sometimes remaining on the tree until February.

G. triacanthos thrives in its natural range in climates ranging from cold-temperate to subtropical. In the western portion of its range, it grows in a subhumid climate, whereas in the middle and eastern portions the climate is humid. Typical annual precipitation is about 510 mm in South Dakota and Texas to >1520 mm in southern Louisiana, Mississippi and Alabama. The mean annual snowfall can be up to 102 cm. In the north and north-east of its range, the length of the growing season is about 150 days, increasing to >300 days in the south. Honey locust is tolerant of low temperatures and in the northern parts of its distribution it is hardy at −29° to −34° C, but outside of its native habitat, it has also been successfully planted in tropical climates. It is drought and frost tolerant. The natural range of *G. tricanthos* is generally below 760 m elevation, although the upper limit appears to be 2200 m.

Silviculture practice

Seed propagation is easy and employed when no particular cultivar is required. Good seed crops occur nearly every year. There are about 8000 seeds/kg and the viability percentage is high after pretreatment. Either soaking in hot water, mechanical scarification or concentrated sulphuric acid treatment is required, as the seed coat is hard.

Sowing in the nursery is advisable in late spring during warmer conditions; a suitable density is 150–200 seeds/m^2. Nursery-grown seedlings attain a suitable size for planting in the field (30 cm) after 1 year. Dormant nursery-grown seedlings can be stored at 0°C, bare-rooted, for several weeks before outplanting.

Vegetative propagation of *G. triacanthos* is necessary when selecting cultivars and it is generally performed by bud grafting on young green material that is still herbaceous. Cutting and *in vitro* propagation are possible. Root cuttings appear to be the best and most practical method of reproducing desirable lines or strains in large quantities. Hardwood cuttings for propagation taken from thornless areas of mature trees (upper and outer crown growth) generally produce scions that remain thornless.

In natural stands the silvicultural management of *G. triacanthos* usually follows the same system adopted for the more profitable tree species with which it is growing. Since the outbreak of Dutch elm disease, the artificial spread and planting of *G. triacanthos* has increased as a substitute for elm as roadside trees, in agroforestry and especially as a fodder tree. It can be pollarded profitably, and when wood products are also required, *G. triacanthos* can be managed as a short-rotation coppice species. The thornless variety *G. triacanthos* f. *inermis* is now generally preferred for plantation use. Many of the cultivars selected for roadside plantations are derived from this variety and are both thornless and fruitless, or they bear only a small amount of fruit. Average annual reported growth rates are approximately 0.5 m in height and 0.5–1.5 cm in diameter.

Diseases and pests

G. triacanthos does not suffer from any serious diseases. Stem cankers caused by *Nectria cinnabarina* and *Thyronectria austroamericana* may cause significant damage, with honey locust knot of unspecified aetiology being a new and potentially damaging disease.

In general, attacks by insects or pathogens are not fatal, but they do weaken the tree and slow growth. The mimosa webworm (*Homadaula anisocentra*), is a severe and widely distributed defoliator. Heavy infestations of the spider mite *Platytetranychus multidigituli* can occur in hot dry weather and severely defoliate trees. Other important defoliators include the white-marked tussock moth (*Orgyia leucostigma*), the honey-locust plant bug (*Diaphnocoris chlorionis*) and the very similar *Orthotylus robiniae*, the walkingstick *Diapheromera femorata* and the leaf hopper (*Empoasca pergandei*). The gall midge pest (*Dasineura gleditchiae*) appears to cause substantive damage in some regions. Another defoliator, the white flannel moth caterpillar (*Norape ovina*), is an additional problem on amenity trees as the caterpillars can also 'sting' humans. A number of seed-feeding insects are also known to attack *G. triacanthos*, notably members of the Bruchidae family, common pests of legume seeds.

Uses

G. triacanthos has excellent characteristics for use as a windbreak, and when cut, it resprouts strongly producing thick, thorny and impassable hedges. As a fast-growing species with an attractive crown, it is also suitable for ornamental planting.

The heartwood is reddish-brown with an attractive grain, in contrast to the wide, yellowish sapwood. The wood is very heavy, dense (700–800 kg/m^3), very hard, strong in bending, stiff, resistant to shock, and durable when in contact with the soil. Although it does not have high value, it is used locally for fuelwood, fence posts, and also as lumber for pallets, crating and general construction.

Pods have 30–40% total sugar content, and young leaves have a forage value similar to lucerne with 15–20% crude protein, 25–30% crude fibre and about 70% dry matter digestibility. Pods can be fermented into alcohol, and seeds have been roasted and used as a coffee substitute. Many plant extracts have been traditionally used for folk medicine.

The high levels of adaptation to extreme climatic conditions, rapid growth and early fruit production make it suitable for use in agrosilvopastoral systems as cattle forage, particularly in semiarid regions. It allows sustained and diverse forage production during periods of fodder scarcity, although thornless cultivars are recommended. When pods are still greenish and immature, they contain a sweetish substance which makes them appetizing to cattle, but after maturation they become too hard and bitter to be eaten.

Further reading

Benedetti, R.S., Cáceres, G.V., Delard, R.C. and González, O.M. (2000) *Monograph of the honeylocust:* Gleditsia triacanthos [*Monografía de* árbol *de las tres espinas* Gleditsia triacanthos]. Instituto Forestal (INFOR), Santiago, Chile, 31 pp.

Blair, R.M. (1990) *Gleditsia triancanthos* L. In: Burns, R.M. and Honkalla, B.H. (eds) *Silvics of North American Hardwoods*. Agricultural Handbook 654, Hardwoods, Vol. 2. USDA Forest Service, Washington DC, pp. 391–399.

Gold, M.A. and Hanover, J.W. (1993) Honeylocust (*Gleditsia triacanthos*), a multipurpose tree for the temperate zone. *International Tree Crops Journal* 7, 189–207.

Gliricidia sepium (Jacq.) Kunth ex Walp.

Mother of cocoa

Importance

Gliricidia sepium (Plate 17) is an extremely versatile, fast-growing, nitrogen-fixing, multipurpose agroforestry tree that is adaptable to a wide range of humid and sub-humid climates and soil conditions. It is extremely easy to propagate and manage in a wide diversity of smallholder farming systems, providing many high-quality products including fuelwood, construction poles, crop supports, green manure, animal fodder and bee forage. In addition, it is used in living fences, to shade plantation crops, to stabilize soils, as an ornamental, a rodent poison and in traditional medicine. It is one of the most common and well-known trees in Central America. It has been very widely introduced, and now has a pantropical distribution; it is probably the most widely cultivated agroforestry tree after *Leucaena leucocephala*.

Botanical features

G. sepium is a small- to medium-sized, thornless, single- or multiple-stemmed tree. It is 2–15 m tall and 5–30 cm in diameter, occasionally to 20 m tall and 1 m diameter, with an open rounded crown, often greatly modified by lopping. Young bark is grey-brown or pale whitish-grey, smooth with raised pale brown lenticels, becoming fissured with age. Leaves are alternate, sometimes subopposite, pinnate, 15–35 cm long with an odd terminal leaflet, with 6–24 opposite elliptic leaflets, 44–83 mm long, 17–48 mm wide, with characteristic dark purplish tannin patches, especially on lower surfaces. Flowers are borne on erect racemes, 2–15 cm long, with almost synchronous maturation of 30–100 flowers on a single inflorescence. The pink flowers are 2 cm long with a five-lobed bell-shaped calyx and a typical pea-shaped corolla with five strongly unequal petals (Plate 50). Yellow-brown pods are 10–17 cm long, 14–22 mm wide, strongly compressed, opening explosively when dry.

Distribution

The natural distribution of *G. sepium* is considered to be Mexico, Central, and South America, but has been greatly altered and extended by a long history of cultivation and introduction alongside massive habitat disturbance. Although it is difficult to discern the true extent of its native distribution, molecular studies suggest a centre of diversity in seasonally dry zones of southern Mexico and northern Central America. Many outlying populations exhibit unusually low levels of genetic diversity, suggestive of introduction and subsequent naturalization. It has been very widely introduced and is now present in tropical humid and sub-humid regions worldwide.

Silviculture characteristics

Trees are generally deciduous, losing some or all of their leaves during the dry season. The species flowers and fruits while leafless, and leaves flush as seed is shed, usually about

1 month prior to the first rains. In non-seasonal, humid areas, trees may be evergreen with continuous shoot growth, and in these areas flowering is often sporadic with little seed set. Ease of propagation and management, along with many positive associative traits including nitrogen fixation, a light crown, and limited root competition mean that *G. sepium* competes less aggressively with associated crops than with most other woody legumes. It is nitrogen-fixing, with nodulation occurring widely in its native range and where introduced. It withstands fire well by quickly resprouting after damage.

Silviculture practice

G. sepium can be propagated easily by seed or by cuttings. In seasonally dry areas it produces abundant crops of seed from as early as 1–3 years of age. Pods can be collected when they are on the point of opening. Pod opening is explosive and seeds can be catapulted up to 25 m, so pods should be covered with mesh or netting. Seeds are orthodox, can be stored without loss of viability for at least 5 years, and no pre-treatment is required.

Direct sowing is sometimes used to establish hedgerows but container stock or bare-rooted seedlings are more frequently used. Nursery stock can be outplanted within 10–12 weeks. Seedlings are susceptible to competition from grasses during the first year, but once established, *G. sepium* is an aggressive competitor and has been cultivated in West Africa and Indonesia to reclaim grasslands dominated by spear grass (*Imperata cylindrica*). Vegetative propagation by shoot cuttings is very easy, commonly with large branches at least 6 months old, 0.5–2 m long, 3–6 cm thick, and with the bark incised to promote rooting. Rooting occurs within 6 weeks, but trees established from cuttings have shallower roots and are less hardy than trees grown from seed.

Tree spacing depends on the particular agroforestry combination. For hedgerows, trees are generally spaced 5–100 cm apart in rows 4–10 m apart. In fodder plots, spacings of 0.25 × 1 m may be used and leaf yields apparently vary little even over a range of spacings from 5000 to 40,000 stems/ha. In woodlots, 2 × 2 m (2500 stems/ha) spacings are common and are managed on 2–4 year coppicing cycles with mean annual increments of 2–5 t/ha dry weight. When grown for leaf production, the first harvest can be as early as 6–8 months in wetter areas from plants grown from stakes and at 12–16 months from seedlings. Harvesting should be limited to once or twice per year during the first 2 years, and thereafter, every 3–4 months during the dry season and every 2–3 months during the rainy season. Regrowth should be 1–2 m in height before harvest, with cutting heights from 0.5 m in contour hedgerows and fodder banks to 2 m in living fences or shade trees. Under average conditions, yields of 3–4 kg dry matter/tree/harvest may be obtained, translating into annual yields of 9–16 t/ha leaf dry matter, or up to 43 t/ha fresh leaves. Sometimes trees are planted at wider spacings (e.g. 10 × 10 m) in pastures.

Diseases and pests

G. sepium has so far remained free of serious diseases throughout its extensive cultivated range. However, a new virus-like disease and a 'little leaf' disease thought to be caused by a phytoplasma adversely affect trees within the native range, with the potential to cause serious problems if introduced elsewhere. Other foliage diseases include *Cercosporidium gliricidiasis*, *Cladosporium* sp. and *Sirosporium gliricidiae* causing defoliation, leaf spot caused by *Colletotrichum gloeosporioides*, and scab caused by *Sphaceloma* sp.

There are a number of reports of insects attacking *G. sepium*, but no serious pests have been encountered to date. One of the reasons for the popularity of *G. sepium* is its resistance to the Leucaena psyllid defoliator *Heteropsylla cubana*.

Uses

G. sepium is used to restore soil fertility as an important and sometimes dominant element of managed secondary bush fallows.

The wood is light to dark olive-brown, very hard and heavy, strong, coarse textured with an irregular grain. It seasons well and although difficult to work, takes a good polish, is highly durable, resistant to termites and fungus and valued for house construction and fence posts. Specific gravity is moderate to high at 500–800 kg/m^3. Timber of large dimensions is rarely available, however, and wood is little used for sawn timber and does not enter commerce. Heartwood burns slowly, giving good embers and producing little smoke, and older wood makes a good fuelwood.

G. sepium is an important forage crop in cut-and-carry systems in many parts of the tropics. The nutritive value of leaves is high, with 18–30% crude protein, 13–30% crude fibre, 48–77% digestibility and relatively low tannin content. However, palatability may be extremely problematic in some areas, possibly due to anti-nutritional factors such as flavonols and phenols, or the odour of the leaves possibly due to presence of coumarin or other volatile substances. Some toxicity has also been documented, possibly caused by conversion of coumarin to dicoumarol, a haemorrhagic compound, during fermentation. Despite this, it remains an important high-quality ruminant fodder and there are many reports of improvements in animal production from feeding it as a supplement. Leaves are usually consumed as a high-quality supplement to low-quality grass, straw or other residues, though results with non-ruminants have generally been poor.

Leaves of *G. sepium* are widely used as a green manure where used as an understorey or shade tree, and more recently in formal intercropping systems such as alley farming. Prunings have a low C/N ratio, high nutrient levels, and decompose quickly providing a rapid influx of nutrients; however, they provide minimal mulching benefits in terms of weed control and soil moisture conservation. Very fast growth, although an advantage in producing plenty of green manure, can be a problem in alley farming systems by demanding high labour inputs to carry out the frequent pruning required to avoid shading. Flowers are cooked locally for human food and are a useful source of nectar for bees. Dried bark or leaves, ground and mixed with cooked maize, are used as a rat poison.

Further reading

Kpikpi, W.M. and Sackey, I. (2012) *Gliricidia sepium* (Jacq.) Walp.: hardwood with potential for pulp and paper-making. *Canadian Journal of Pure and Applied Sciences* 6, 1961–1966.

Nyoka, B.I., Simons, A.J., Akinnifesi, F.K., Baidu-Forson, J.J., Jones, M. and Hodgkin, T. (2012) Genotype-environment interaction in *Gliricidia sepium*: phenotypic stability of provenances for leaf biomass yield. *Agriculture, Ecosystems and Environment* 157, 87–93.

Stewart, J.L., Allison, G.E. and Simons, A.J. (eds) (1996) *Gliricidia sepium*. Genetic Resources for Farmers. Tropical Forestry Paper 33. Oxford Forestry Institute, Oxford, UK, 125 pp.

Glyptostrobus lineatus (Poir.) Druce

Canton water pine

Importance

Glyptostrobus lineatus, recognized by some as *Glyptostrobus pensilis*, is indigenous to China and Vietnam, and is the sole species in the genus. It is a fast-growing, multipurpose, coniferous tree producing good-quality timber. It is widely used in shelter belts and windbreaks, and as an ornamental in wet coastal plain areas.

Botanical features

G. lineatus is a deciduous or semi-evergreen tree with a conical crown, spreading lower branches and more erect upper ones, and two types of shoot. It ranges in size from 8 to 10 m tall in China, but in Vietnam it grows to 20–30 m tall and its dbh is commonly 60–70 cm and sometimes to 120 cm. Trees aged about 400 or 500 years old have been found in several small populations scattered throughout the southern provinces of China, with average heights from 35 to 41 m and diameters from 72 to 110 cm.

G. lineatus develops rhizophores from lateral roots, similar to those of *Taxodium* species, on wet sites, which can spread up to 6–7 m from the base of the tree. The bark is thick, greyish-brown and longitudinally fissured.

Two kinds of leaves are produced on the two different types of shoot. On short, terminal, barren branchlets the leaves are needle-like, sickle-shaped, enlarged at the base, quadrangular in cross section, 2- or 3-ranked, 1–3 cm long and 1.5–4 mm wide, light green when young, red-brown in autumn before falling, with stoma belts on both sides of the middle vein on the lower surface of the leaf. On the long, terminal shoots, which represent fruiting branchlets and mature persistent vegetative branchlets, leaves are overlapping and scale-like, thicker, 2 mm long, persistent for 2 or 3 years, delicate green when young, rich brown before falling in autumn, with white stomata spots on the leaf surface. As one type of leaf falls every autumn and the other type is persistent for 2 or 3 years and remains green in winter, *G. lineatus* has been referred to as a semi-evergreen tree.

Male flowers are terminal on short shoots, and female flowers terminal on lateral shoots. The male strobili resemble those of *Taxodium*, but the cones differ in being pear-shaped or ovate and borne on stalks 1.3–1.9 cm long. Cones are 1.2–1.8 cm long, disintegrating, with rather thin scales bearing wart-like, sub-apical teeth and a small, triangular mucro above their centre. Seeds are oval or oblong, 0.45–0.63 cm long, thin coated and terminated by a hatchet-shaped wing about 0.3 cm long.

G. lineatus usually flowers in February or March and develops mature cones by October or November. It generally begins flowering and bearing cones when 5 or 6 years old.

Distribution

In China, *G. lineatus* occurs mainly in the Pearl River Delta area in the south-eastern part of Guangdong, usually on low, damp sites; however, small populations have been found in the northern part of Fujian, south-eastern Guangxi, eastern Jiangxi, Yunnan, Sichuan and Guizhou.

G. lineatus was formerly common in many provinces in Vietnam as the dominant species in permanently inundated swamp forest. However, it is now confined to Dac Lac province where fewer than 100 individuals exist. It is one of the most threatened tree species in Vietnam and is classed by the International Union for the Conservation of Nature (IUCN) as 'Critically endangered' there.

Silviculture characteristics

G. lineatus is a strongly light-demanding species. It has proven to be strongly windfirm in tropical cyclone zones. It is also well suited to coppicing.

Silviculture practice

G. lineatus seed is collected from mature, yellow-brown cones with slightly open scales. Seedling size for planting is dependent on the site. Generally, vigorous 2-year-old stock is planted in deep waterlogged soils, and 1-year-old or 1.5-year-old seedlings on shallow wet sites. Containerized seedlings have a good survival rate.

G. lineatus is increasingly being used as a plantation species, and much experience in afforestation has been gained in southern China. Proper site choice is essential for good survival, particularly on coastal plains waterlogged throughout the year. Site preparation and weeding enhance survival and early growth. Planting containers 40 cm wide, 40 cm long and 30 cm deep is suggested. Generally a planting density of 2 × 2 m or 2 × 2.5 m is practised, with higher densities for windbreak forests. Small bamboo culms or poles are required for support soon after planting to ensure good survival rates; this is particularly important on wet sites susceptible to erosion from strong river waves or tidal flow.

G. lineatus stands are fast growing, with average heights of 12 m and dbh of 22 cm recorded at 18 years old on good sites. Biomass in a *G. lineatus* stand at 11 years old has been measured at up to 110.6 t/ha, which is even greater than for *Cunninghamia lanceolata* grown on valley soils with class I site quality (106.6 t/ha).

Diseases and pests

Although little research has been conducted on pests and diseases, *Pestalotia* and *Cercospora* fungi, and *Cryptothelea* insect pests have been found on *G. lineatus*, with only a moderate influence on tree growth.

Uses

G. lineatus is widely used in windbreak forests, for soil conservation on river banks, and as an ornamental tree in warmer coastal waterlogged areas. It is suited to these sites because of its good adaptability (particularly through its rhizophores in wet environments), fast growth, straight stem form, and semi-evergreen nature.

G. lineatus wood has straight grain, is soft, with low strength, and is susceptible to slight shrinkage. It has an air-dry specific density of 578 kg/m^3, coefficient of shrinkage (volume) of 0.445, bending strength of 700 kg/cm^2, sidewise compressive strength of 349 kg/cm^2 and an end section strength of 301 kg/cm^2. It is lustrous with a sweet odour, easy to cut and polish, and resistant to termites and other wood-boring insects.

G. lineatus wood is used for various purposes, including small fishing boats, the side-boards of trucks, agricultural tools, construction, packaging board and in the pulp industry.

Specific gravity in the roots of *G. lineatus* is so low (0.094 g/m^3) that this soft, spongy material is widely used in safety devices in water sports, for buoyancy in fishing nets, for making corks and as an insulating material.

The leaves and cones are used in Chinese traditional medicine, providing relief from back pain. Tannin materials are also extracted from the leaves, cones and bark.

Further reading

Butala, J.R. (1973) Investigations of *Glyptostrobus*, a Taxodiaceous conifer. PhD thesis, Washington State University, Pullman, Washington, DC.

Eckenwalder, J.E. (2009) *Glyptostrobus*. In: Eckenwalder, J.E. (ed.) *Conifers of the World*, Timber Press, Portland, Oregon, USA, pp. 267–269.

Gmelina arborea Roxb.

Candahar

Importance

Gmelina arborea produces an important and valuable creamy-white timber which is sometimes called white teak. It is a fast-growing species that demands good site conditions. Native to Asia, the species has great plantation potential and is widely planted, especially in Africa. Large-scale plantations of *G. arborea* are found in Nigeria, Sierra Leone and Malawi. Compared with other high-quality timber species, *G. arborea* can be managed for relatively short rotations of 15–20 years. With recent developments in wood-processing technology for utilizing smaller-diameter logs, *G. arborea* plantations are often managed on rotations of less than 15 years; this is especially the case when fuelwood is the main end product.

G. arborea is also valuable as a multipurpose tree and is economically important in various agroforestry systems. Different parts of the tree are used in traditional medicine.

Botanical features

G. arborea is a medium-sized tree which produces a clear, but frequently bent, bole. It reaches 30 m tall on favourable sites. However, it can become shrubby in drier environments. *G. arborea* has low, obscure buttresses and a yellowish-grey, smooth bole. The branches are numerous and spread to form a large shading crown. The corky bark is yellow inside and greyish-white outside when young, turning brown with age. The old bark then sheds, producing patches of paler bark on the bole.

Leaves are simple, normally cordate, broad–ovate acuminate, about 10–25 cm long and 5–18 cm broad, denticulate, with about 3–5 primary veins originating from the base and 4–6 pairs of secondary veins. The petiole is about 7.5–12.0 cm long. A pair of shining glands is located at the intersection of the blade and petiole, along the lower half of the midrib.

The flowers are numerous and arranged in a panicle about 30 cm long. Corolla brown-yellow, upper lip shortly bifid, larger than the lower. The fruit is a drupe with sweet flesh; fruit turning yellow while ripening, 2.0–2.5 cm long and with a bony endocarp (stone) in 1–4 cells, though it is rare for all 4 cells to be fertile and each fruit usually has only 2–3 seeds.

G. arborea sheds its leaves from the end of January to March. The flowers and new leaves are seen together in April–May after the leaf-shedding period. Flowering time varies according to location, individual tree, and even between inflorescences within trees. The fruits ripen from the end of April to June or July. Correlations of flowering time and fruit ripening with latitude or forest type have not been detected.

Distribution

G. arborea is indigenous to India, Pakistan, Bangladesh, Myanmar, Sri Lanka, Thailand, Laos, Cambodia, Vietnam and Yunnan and Guangxi provinces in China.

Silviculture characteristics

G. arborea is rarely gregarious, although it grows well in pure plantations. It is a pioneer species and prefers full sunlight, although it can withstand partial shade; it usually colonizes gap openings in disturbed forests. It is moderately drought- and frost-tolerant, but severe drought and frost may kill seedlings. The shedding of leaves after drought is also common in this species. After severe drought damage to the main stem, *G. arborea* can still produce shoots which will grow vigorously in subsequent favourable periods. The permanent wilting point has been suggested as a criterion for assigning clones to sites differing in water availability. The ability to coppice is an added advantage for *G. arborea* as a timber plantation species.

G. arborea can flourish on wasteland, roadsides and occasionally in patches of urban secondary forest.

Silviculture practice

G. arborea planting stock can consist of seedlings, stump plants cuttings or coppice shoots.

Germination of *G. arborea* seeds occurs 7–15 days after sowing; the sowing medium should be kept wet and well-aerated throughout the period. Seedlings should be transplanted to a good potting mixture (one part organic material, three parts sand and two parts topsoil), to provide good aeration and drainage. Potted seedlings are kept in the nursery until they reach 25–30 cm tall. Application of fertilizer at monthly intervals is necessary to ensure good growth of the potted seedlings.

Stump plants are a popular planting stock because of their relatively high survival rates in the field. The quality of stump plants is dependent on their seed quality; 1-year-old seedlings growing on germination beds are kept for producing stump plants.

Vegetative cuttings have also been successfully propagated. Stem cuttings are preferred over other vegetative parts because they produce better planting stock. Protocols for *in vitro* micropropagation from nodal tissue have also been documented. In addition, *G. arborea* coppice is a potential source of planting stock.

Planting using stump plants or seedlings requires good site preparation. Establishment of *G. arborea* necessitates removal of existing vegetation and a clean site void of *Imperata cylindrica*. Planting must be done in the early rainy season to ensure good survival and establishment of transplanted seedlings. Stump plants are popular for plantation establishment due to their drought tolerance, relatively low cost, and ease of planting. Stump plants require shoot pruning after sprouting to maintain a single leader.

Initial spacing of *G. arborea* stands varies according to management objective. For chip or pulpwood production, *G. arborea* can be planted at 1.2 × 1.2 m to 1.8 × 1.8 m. For sawlog production wider spacing is anticipated; an initial spacing of 3 × 3 m is recommended for sawlog production in Malaysia. Spacing of 2.5 × 2.5 m ensures early suppression of weed growth and 4.5 × 4.5 m decreases the form quality of stands.

Weeding is a very important silvicultural practice for enhancing the survival and growth of *G. arborea*. Intensive care and frequent fertilizer application will help the establishment of *G. arborea* seedlings. Thinning is usually undertaken only in plantations producing veneer and sawlog timber.

Diseases and pests

Severe pest and disease problems are frequent in plantations within the natural distribution of *G. arborea* and have even led to the failure of plantations. Fungal attacks are less prevalent, but can cause damage to *G. arborea* stands. Fungal species found to be causing disease include *Gnomonia* sp., various *Coniella* species, *Poria* species, a powdery mildew *Phyllactinia suffulta* var. *gmelinae* and *Fusarium* wilt.

Outside the natural distribution of *G. arborea*, fungal diseases are mainly root diseases found in Africa and Latin America. Root diseases of nursery seedlings are normally caused by *Gibberella fujikuroi* in Gambia and *Sclerotium rolfsii* (*Athelia rolfsii*) in Sierra Leone, Gambia and Nigeria. *Thanatephorus cucumeris*, *Chaetophoma* sp., *Polyporus* and *Armillaria mellea* are also causal agents of root diseases in Nigeria and Côte d'Ivoire. In Amazonia, *Ceratocystis fimbriata*, which is an important pathogen of rubber (*Hevea brasiliensis*), mango (*Mangifera indica*), coffee and cocoa, also infects *G. arborea*; the vectors of the fungus are insects of the genera *Scolytus* and *Platypus*. In Malaysia, diseases such as leaf spot, collar-rot and wilt have been identified, caused respectively by the fungi *Colletotrichum gloeosporioides*, *Rhizoctonia solani* and *Pythium* spp. Stem rot and anthracnose are common seedling diseases in Malaysia.

Insect pests that are common in *G. arborea* plantations within the natural distribution area are stem borers (including *Dihammus cervinus*), leaf defoliators (including the larvae of *Calopepla leayana* and nymphs and adults of *Tingis beesoni*), and leaf or shoot cutters (such as the larvae of *Alcidodes gmelinae* which cause die-back). *Alcidodes ludificator* is also a serious pest in nursery and young plantations.

Although *G. arborea* planted outside its natural range suffers similar pest problems, on the whole these are less serious. Defoliators are very common and include *Empacamenta calabrica*, *Apophylia nigricollis*, *Zonocerus variegatus*, *Achaea lienardi*, *Chrysodeixis chalcites*, *Acherontia lachesis*, *Ozola minor* and *Attacus* spp. In Latin America, leaf-cutting ants are a major problem for stem quality and growth of plantations. In Malaysia, the beehole borer *Duomitus ceramica* is a damaging insect pest and the termite *Coptotermes curvignathus* severely damages *G. arborea* plantations in Malaysia.

Uses

G. arborea is much used in agroforestry systems involving intercropping with arable crops, coffee or fruit trees, where it has been shown to give better growth and yield than in sole plantations. *G. arborea* has also been shown to have some potential for the phytoremediation of polluted soils.

The species produces a highly-prized cream-coloured timber known for its easy working and good finishing properties. Wood density is 400–560 kg/m^3. When the timber is adequately protected, it is excellent for house posts, boat decks, furniture, boxes, musical instruments, cars, railway carriages, matches, curing tobacco, building, panels and general carpentry work. It has also been used to manufacture plywood and wood wool cement board with high resistance to subterranean termite attack. *G. arborea* can easily be pulped, and produces paper for various uses, including wrapping, writing and printing. *G. arborea* wood can be pyrolysed to produce bio-oil and bio-char.

Plantation-grown *G. arborea* in Malaysia had superior strength properties when compared with *G. arborea* grown in Myanmar and India.

Various parts of the tree are used in Ayurvedic and traditional medicine, in both its native range and introduced range.

Bark extracts have also been used to control crop pests, while wood ash was equally as effective as commercial products in controlling nematodes (*Meloidogyne* spp.) in cowpea crops. Leaves have potential as feed for goats.

Further reading

Nayak, B.S., Jena, P.K., Dinda, S.C. and Ellaiah, P. (2011) An overview on silviculture and traditional therapeutic uses of *Gmelina arborea* Roxb. *Journal of Pharmacy Research* 4, 1329–1330.

Wee, A.K.S., Li, C.-H., Dvorak, W.S. and Hong, Y. (2012) Genetic diversity in natural populations of *Gmelina arborea*: implications for breeding and conservation. *New Forests*, 43, 411–428.

Wong, W.C. and Khoo, K.C. (1980) *Gmelina arborea* - a literature review. FRIM Reports No. 14. Forest Research Institute Malaysia (FRIM), Kuala Lumpur.

Gonystylus bancanus (Miq.) Kurz

Ramin

Importance

Ramin (*Gonystylus* spp.) is a tropical hardwood from the swamp forests of South-east Asia. The pale timber is highly prized as a decorative and workable timber. Of the 32 species in the genus, six (*Gonystylus affinis*, *Gonystylus bancanus*, *Gonystylus forbesii*, *Gonystylus macrophyllus*, *Gonystylus maingayi* and *Gonystylus velutinus*) are commercially valuable. *G. bancanus* is the most heavily traded species; it also appears the best known and is judged to be relatively typical of the genus.

Gonystylus has been widely depleted and annual production figures have plummeted since the 1990s (Indonesia reported a decline of more than 80% in 6 years). Over-exploitation has led to international concern and since 2005 all ramin is listed under the Convention on International Trade in Endangered Species (CITES). This regulates the international trade in species threatened through trade, though there are concerns about poor controls in exporting countries. Fifteen *Gonystylus* species, including *G. bancanus*, are classified as 'Vulnerable' in the IUCN's Red List of Threatened Species due to habitat loss and over-exploitation. Timber cutting – both selective- and clear-cutting – is recognized as the major threat to these species, though conversion of forest to other uses such as oil-palm also contributes a significant threat.

Within its natural range, regeneration and reforestation are generally rather poor. Only in Sarawak, Malaysian Borneo, has replanting or promotion of natural regeneration been attempted.

Botanical features

G. bancanus is a canopy tree with a straight and cylindrical bole up to 40–45 m tall and 120 cm in diameter. The stem is sometimes fluted at the base, is free of branches up to 20 m, and typically has knee-roots (pneumatophores). Its bark surface is greyish- to red-brown and is often cracked. The inner bark is yellow and fibrous, with pale cream or white sapwood. Its leaves are elliptical, oblong–oblanceolate or obovate, 4–14.5 × 2–7 cm, broadly cuneate to rounded at the base, suddenly narrowing into a point; petioles range from 8–18 mm. The inflorescence is up to 9 cm long with short hairs. Flowers have pedicels 8–14 mm long, and narrowly lanceolate, glabrous petals of 13–20 cm. The fruit is a capsule up to 4.5 cm long, with 3–4 valves, minutely rough but not rugose. There are 1–3 seeds per fruit, which are compressed ovoid, black, 28 × 22 × 6 mm.

Distribution

G. bancanus is native to the freshwater coastal peat swamp forests of Brunei Darussalam, Malaysia, Indonesia (western and central Kalimantan and south-eastern Sumatra and Bangka), Singapore, the Solomon Islands and the Philippines. It is a lowland species that is rarely found above 100 m altitude and occasionally forms pure stands.

Apart from targeted logging of *Gonystylus* trees, drainage of its habitat for new land uses such as rice cultivation, oil palm plantations or shrimp ponds has led to its threatened status.

Silviculture characteristics

G. bancanus has high potential as a plantation species since planting material can be easily obtained by either growing from seeds or through vegetative propagation. However, plantation establishment in peat swamp forest has been difficult and costly because of poor accessibility and low soil productivity. In Malaysia, a decade-long trial of growing ramin outside peat swamp, in well-drained mineral soils, resulted in 52% survival rates, with average increments of 70 cm height and 1 cm dbh/yr. However, based on good physical and mechanical properties, this small-sized timber may already be used for furniture manufacturing, internal joinery, holders and wooden toys.

G. bancanus has not been widely cultivated, and the age when it begins to flower and set seed is not well known (observations in Kalimantan and Sumatra showed that 8-year-old trees had not started flowering yet). Pollinators are not known. Under natural conditions, the time that the tree flowers differs from year to year and place to place. It may not flower every year.

Silviculture practice

Seed collection is generally from around tracks, as access in swamp forest is often difficult. The seeds should be collected when freshly fallen (40–45% moisture content) and should be protected from drying out during transportation by covering them with a moist gunny sack and placing them in the shade.

The seeds are washed and must be stored at high moisture content. They can be stored temporarily with moist sawdust in sealed polyethylene bags in an air-conditioned room at 18°C. With this method, viability can be kept above 80% for 2 weeks. If stored too long the seeds will begin to germinate.

Soaking the seeds in water for 24 h before sowing enhances germination. The seeds are sown in sand or a mixture of sawdust and soil (2:1). The seedbed should be covered with mulch to prevent the seeds drying out. After about 5 days the radicle should start emerging, and after about 30 days the seedlings may be transferred to containers with peat or peat with charcoal. Again, the seedling containers should be placed under shade. After 8–11 months, the seedlings are about 20 cm tall and ready for planting out.

Experiments have shown that best seedling growth is achieved in partial sunlight. Also, seedlings grew much better on flooded soil compared with those grown on drained soil. The results of this experiment suggested that drainage of peat swamp forests would hamper regeneration. It also suggests that selective logging operations which produce moderate canopy gaps should improve regeneration in otherwise undisturbed swamp forests.

Diseases and pests

The seeds are vulnerable to rats during germination in the nursery. Ambrosia beetle infestation of the logs of *G. bancanus* is reported (more so when debarked), and logs should not generally be stored for long in field conditions.

Uses

Ramin wood is prized for its suitability for many high-value products that depend on its appearance and its easy workability. Wood density varies from 460 to 840 kg/m^3 at 15% moisture content. The main traded ramin products are decorative cabinets and other furniture, picture frames, venetian blind slats, pool cues, tool handles, rulers, toys and drawing boards. It is also very suitable for veneer and plywood. Uses for general light construction include door and window frames, moulding, skirting, ceilings and partitions. The major importing countries are China, the USA and Italy.

Further reading

Kartiko, H.D.P. (1998) Simple storage technology for recalcitrant seeds of an endangered species: ramin (*Gonystylus bancanus*). *Seed Technology Bulletin* 5, 1–8 (in Bahasa Indonesia).

Kartiko, H.D.P. (2002) *Gonystylus bancanus* (Miq.) Kurz. Seed leaflet no 58. Danida Forest Seed Centre, Humlebaek, Denmark, 2 pp.

Shamsudin, I. (1996) *Gonystylus bancanus*: Some observations on its flowering, fruiting, seed predation and germination. *Journal of Tropical Forest Science* 8, 424–426.

Soerianegara, I. and Lemmens, R.H.M.J. (eds) (1993) *Plant Resources of South-East Asia No. 5(1). Timber Trees: Major Commercial Timbers.* Pudoc Scientific Publishers, Wageningen, The Netherlands.

Grevillea robusta A. Cunn. ex R. Br.

Silky oak

Importance

Grevillea robusta is a popular tree in warm temperate, subtropical and tropical highland regions. Originally a shade tree for tea and coffee and later an agroforestry tree for small farms, it provides economically valuable products including sawn timber, poles, firewood, fodder and mulch. It is easy to propagate and establishes and grows well even in low-fertility soils. Trees tolerate heavy branch and root pruning, do not compete strongly with adjacent crops, and are relatively free of pests and diseases. In addition, its fern-like pinnate leaves and attractive, orange flowers make it a popular ornamental tree.

Botanical features

G. robusta is an erect, single-stemmed tree typically reaching 20–30 m in height and 80 cm in diameter. The crown is conical and symmetrical with branches at intervals of about 1 m, projecting upwards at angles of 45°. Bark is dark grey and furrowed into a lace-like pattern. The fern-like foliage is very distinctive, with leaves 10–34 cm long, 9–15 cm wide, variably pinnate to bipinnate, green but silvery underneath, on petioles 1.5–6.5 cm long. The bright golden-yellow to orange flowers are about 2 cm long, and borne in numerous pairs along flower spikes, which are themselves in groups of one to six panicles, borne on older wood. Fruits are two-seeded follicles 2 cm long with a slender persistent style. The flat seeds are 1–2 cm long and 1 cm wide, with a papery wing around the brown, ovate central seed body.

Distribution

The natural range of *G. robusta* is eastern Australia, from southern Queensland to northern New South Wales, from approximately 30°S to 25°S, and from sea level to 1120 m in the Bunya Mountains at the western extreme. In this area, stands are found along rivers and streams sometimes associated with *Casuarina cunninghamiana*, and in vine forests and thickets dominated by *Araucaria cunninghamii*, but not in the fire-prone *Eucalyptus* forests and grasslands that occupy much of its natural range. It has been widely introduced to most warm temperate, subtropical and tropical highland regions from the 1800s and is commonly planted throughout parts of Africa, South Asia and Central and South America.

Silviculture characteristics

G. robusta is semi-deciduous in its natural range, being almost leafless shortly before flowering. Proteoid roots (sections of the secondary roots which develop as dense cylindrical clusters of rootlets) develop in conditions of low phosphorus availability, and are thought to increase the plant's ability to take up nutrients. The species is relatively deep rooted and regarded by farmers as being non-competitive with crops, compared to other tree species such as *Eucalyptus*. It tolerates repeated heavy pruning and pollarding, also enabling farmers to regulate the degree of competition with adjacent crops. It coppices well only when very young, up to 2 years old, and coppicing ability declines sharply thereafter, so management on a coppicing rotation is not feasible.

The distribution is largely in the warm humid to warm subhumid climatic zones. Climate varies widely within the natural range, because of the substantial altitudinal range and rainfall gradients created by prevailing weather systems interacting with rugged topography. It performs poorly in lowland tropical environments, and growth is also reduced at high altitudes, above 2300 m.

The species is more common on more fertile soils such as those derived from river alluvia or basalts but will grow on shallower, less fertile soils derived from sedimentary material. Best growth is obtained on neutral or slightly acid, sandy loam, loam and clay loam soils. It will not grow on poorly drained, swampy sites where waterlogging occurs for more than a few days at a time, and performs poorly on heavy clays. Trees are also vulnerable to fire.

Silviculture practice

Propagation is usually from seed, with about 40,000 viable seeds/kg. Seed will retain viability for at least 5 years if dried to below 8% moisture content and stored dry and cool (<20°C). No pretreatment is required, with seeds usually germinated on loamy soil under a shallow covering of sand. Seedlings are pricked out when their second leaf-pair starts to develop, into polythene bags filled with a fertile, loamy potting mix. Seedlings are grown on for around 4–6 months in the nursery until planting out at a height of 20–40 cm during the rainy season. In many countries, demand for seed currently exceeds supply, so farmers also obtain planting stock by digging up wildings. Cuttings can also be easily struck using shoots of seedlings or saplings, which can also be air-layered.

The species is usually planted in rows or small woodlots, or as scattered individual trees. Mechanical site preparation is seldom used, but some weeding is usual in the first few years. Trees are normally planted in rows with 2–2.5 m between trees, or at between 1.5 × 1.5 m and 2.5 × 2.5 m in plantations and woodlots. Thinning of inferior trees is often carried out after 4–5 years to yield poles and firewood for local use or sale. The branches can start to be cut for around the same time and every few years subsequently, to regulate shading and competition with crops, and are used for firewood, fodder and small poles, and the tree grows back well after heavy pollarding and pruning. Some farmers also cut all surface roots around the tree base to reduce competition with adjacent crops.

When grown at close spacing in plantations and woodlots, growth rate is relatively modest, and mean annual wood yield may be only 10–12 m^3/ha over 10–15 years at stocking rates of 800–1200 trees/ha. When planted singly or in lines, however, annual growth rates can be 2 m in height and 2 cm in diameter over the first 5 years, and more than 3 m height and 3 cm diameter are possible on the most favourable sites. Sawlog sizes are achieved after 15–25 years, or up to 40-year rotations on poorer sites.

Diseases and pests

Whereas a number of pests and diseases have been reported on *G. robusta*, such as a canker and die-back disease caused by an unidentified *Botryosphaeria* sp., none are considered significant over widespread areas. Trees may, however, act as hosts for a number of pests and diseases of associated crop plants, such as the nematode *Helicotylenchus agricola*.

Uses

G. robusta is one of the most important agroforestry trees in the tropical highlands of East and Central Africa. It is commonly planted as a boundary tree around the perimeter of fields and small farms, and as scattered individuals over crops such as coffee and maize. From the 1800s, trees have been planted extensively as high shade for tea and coffee plantations, often pollarded to produce a spreading crown, with a typical working life of 40–50 years before they become senescent and need replacing. Shade trees provide effective protection against frost to open-grown coffee plants at planting densities of 70–120 trees/ha without reducing crop yields due to competition.

It is used a nurse crop for commercial timber species. The dense, brilliant golden-yellow to orange flowers, silvery fern-like leaves and symmetrical crown encourage its use as an attractive ornamental. It is also planted as a component of contour rows to conserve soil on sloping lands, further aided by the use of leaves as a mulch.

The heartwood is pale pink on cutting, darkening to red-brown after drying, whereas the sapwood is cream-coloured. Growth rings are not visible, although broad rays give the wood a distinctive and attractive appearance on both the quarter-sawn and back-sawn faces. The air-dry density of heartwood is 550–650 kg/m^3. Sawn timber is of medium strength and is used for furniture, packing cases, flooring, panelling, plywood and the manufacture of small wooden items such as pencils. The wood produces short-fibre pulp of acceptable quality, but has not been used for pulp production on a commercial scale. Poles are used for house construction in rural areas, scaffolding, rafters and branches and twigs are used for firewood. The gum produced from stressed or wounded trunks may have commercial value.

Leaves are used by some farmers as a fodder supplement in the dry season when other fodder sources are scarce, and as bedding in livestock stalls. The leaves contain a number of useful chemical compounds, in particular rutin, which has pharmacological applications.

Further reading

Harwood, C.E. (1989) *Grevillea robusta*: An Annotated Bibliography. International Council for Research in Agroforestry (ICRAF), Nairobi, 123 pp.

Harwood, C.E. (1992) *Grevillea robusta* in agroforestry and forestry. Proceedings of an International Workshop. International Council for Research in Agroforestry (ICRAF), Nairobi, 190 pp.

Muchiri, M.N. (2004) *Grevillea robusta* in agroforestry systems in Kenya. *Journal of Tropical Forest Science* 16, 396–401.

Guaiacum officinale L.

Lignum vitae

Importance

Guaiacum officinale is a small lowland tree of dry Neotropical forests and thickets. It produces extremely hard and heavy wood commonly known as lignum vitae. This name, meaning 'wood of life', derives from the fact that during the 16th and 17th centuries the wood and resin were believed to have extraordinary healing powers over serious diseases and were introduced from Spain to medical practitioners in Europe. The species has a long history of use in traditional herbal medicine in the West Indies.

The wood is highly valued and has a wide range of specialized and technical uses. Most modern use, however, is for decorative applications.

The heartwood is greenish-brown and very hard due to its dense diagonally crossing layers of fibres. *G. officinale* is listed as 'Endangered' on the IUCN Red List of Threatened Species and was placed on Appendix II of CITES in 1992.

It is the national flower of Jamaica.

Botanical features

G. officinale is an evergreen, often multi-stemmed, tree bearing a dense rounded crown of dark green leaves. It reaches 10 m tall and 60 cm diameter. The stem is often crooked, with knotty branches. The bark is light brown, mottled and peeling. Each leaf is composed of 2 or 3 pairs of smooth, stalkless 6–13 cm leaflets arranged on a slender mid-rib. Masses of light to dark blue flowers grow at the end of its branches, and different parts of the same tree bloom at various times during the year. The fruits are small, flat, yellow heart-shaped capsules, containing up to 5 cells, each with one or two seeds covered in a bright red aril, which are released at maturity.

Distribution

G. officinale is native to dry neotropical lowland forests. It occurs in the West Indies, the Florida Keys of the USA, and Central America (Mexico to Colombia and Venezuela), where it grows in dry lowland forests and thickets. The species is especially associated with coastal plains with mild climates.

G. officinale is planted as an ornamental through much of tropical America and has also been introduced to India, Pakistan, Ghana and Thailand.

Silviculture characteristics

G. officinale readily self-propagates in the wild, grows well under shade, and is extremely resistant to drought. Seedlings can be raised from seeds but growth is slow.

Seeds (3200–3300 seeds/kg) are available throughout the year and are collected from the ground around the tree or taken from the tree once the seed is released from the fruit.

Silviculture practice

Seeds should be cleaned of their fleshy arils by briefly soaking and then rinsing under running water.

Once germinated, growth itself is slow and seedlings may need a long time in nurseries; to address this, research has been carried out into several treatments to stimulate growth. Trees can be propagated from cuttings and by air-layering.

Diseases and pests

There is little published information on pests or diseases of *G. officinale*.

Uses

G. officinale wood is dark, resinous and extremely hard and heavy (it sinks in water). Wood density is 960–1200 kg/m^3 at 12–15% moisture content. The resinous timber was once favoured for propeller-shaft bearings in ships and windmills, due to its extreme hardiness, durability and self-lubricating properties. Currently it is mostly used for decorative items such as wood carvings but also for cutting boards, mortars and pestles. It is an important fuelwood and charcoal feedstock in the Dominican Republic. It is also planted as an ornamental in the various parts of the drier tropics due to its attractive appearance especially when in flower. It is also suited for certain underwater use. Lignum vitae is used in increasing amounts in steel and tube mills to replace brass and babbit metal for bearings in roller mills and pumps, as the initial cost is less than metal, the life is several times longer, and lubrication is unnecessary.

Other uses for the timber are mallets, pulley sheaves, caster wheels, bowling balls, masthead trucks, stencil and chisel blocks, cable dressers and turned novelties; it is employed to a limited extent for brush backs. West Indian musicians use lignum vitae sticks, and when they are knocked together they sound like metal.

The resin known as 'guaiac' or 'guaiaci resin' is used by pharmaceutical companies. It sometimes flows spontaneously from the stem, but can also be tapped. There are a number of ways to do this: by cutting the stem and allowing the exuding juice to harden; by boring holes in logs of the wood and then placing them on a fire so that the resin is melted and runs through the hole; by boiling wood chips in salt and water, when the resin floats on the surface of the water; it is also extracted with solvents from sawdust and other wood waste.

Resin thus obtained is used in pharmacies to prepare powders, pills and tinctures. The resin is a stimulant and sudorific and appears effective in various treatments. It is considered a valuable remedy and applied externally against toothache and chronic rheumatism, and taken orally against gout, abnormal blood pressure and arteriosclerosis. In Jamaica, it is soaked in rum as a gargle for a sore throat.

The resin's colour changes from red-brown to a blue or blue-green with oxidations, a property used for differential staining. This is important in forensic medicine, where a solution of the resin in alcohol is used to detect bloodstains: it turns blue in contact with blood.

Further reading

Basilis, G. (1978) Lignum vitae [*Guaiacum officinale*]. *Forestal* 2, 21–23.

Berni, C.A., Bolza, E. and Christensen, F.J. (1979) *South American Timbers: The Characteristics, Properties and Uses of 190 Species.* Division of Building Research, Commonwealth Scientific and Industrial Research Organization (Australia), Melbourne, 229 pp.

Francis, J.K. (2002) *Guaiacum officinale* L. In: Vozzo, J.A. (ed.) *Tropical Tree Seed Manual.* Agriculture Handbook 712. Forest Service, Washington, DC, pp. 483–485.

Guazuma ulmifolia Lam.

Bastard cedar

Importance

Guazuma ulmifolia (Plate 18) is widely distributed on the American continent from Mexico and the Caribbean to Paraguay where it is used for firewood, as posts and for charcoal production. The species has great potential for different industrial uses such as interior carpentry, boxes, crates, shoe horns, tool handles and light construction, but it has been little used as such. It is planted for ornamental purposes, as human food, is an important source of livestock fodder, and is important in traditional medicine. It has been little studied, however, regarding its forest management in natural stands, or in forest plantations for commercial purposes with different rotation periods.

Botanical features

G. ulmifolia is a deciduous tree with an irregular, rounded crown that reaches 25–30 m in height with diameters of 40–70 cm. It has a straight trunk with grey or grey-brown bark (5–10 mm thick) which peels off in thin flakes to expose a yellowish layer underneath which changes to reddish-brown. It presents branches from its stem base. Leaves are simple and alternate in a flattened arrangement in 2 rows, oblong–lanceolate, ovate–lanceolate or lanceolate, 3–13 cm long, 2.5–6 cm wide, with a saw-toothed margin, an acuminate or pointed apex, and an asymmetric base. The leaves are thin, with three or sometimes five main veins from a rounded or notched base, nearly hairless, with a dark green upper surface with paler undersides. The leaf stalks are slender, 5–15 mm long, and covered with tiny stellate hairs.

The sweet-smelling flowers occur in panicles 2–5 cm long and covered with tiny stellate hairs. The actinomorphic flowers are 5 mm in diameter, on a pedicel 2–4 mm long. The calyx is green with 2–3 lobes, with five yellow petals 3–4 mm long. The fruits are round to elliptical black warty capsules retained for a long time on the tree, 3–4 cm long and five-celled, opening at the tip or irregularly by pores. Each contains 40–60 egg-shaped seeds 2–2.5 mm long, grey, with a sweet odour and flavour.

Distribution

G. ulmifolia is native to a wide area from north Mexico to Paraguay, and also most of the Caribbean islands in the Greater and Lesser Antilles. In Mexico, it is found from southern Tamaulipas to Quintana Roo and from southern Sonora to Chiapas. *G. ulmifolia* has been introduced to several countries in Africa and Asia. In India, it has been cultivated for more than 100 years and is, for example, a common street tree in Chennai and other cities in Tamil Nadu. It is also reported in Pakistan, Myanmar, Indonesia, the Seychelles, Mozambique, Kenya, Ghana and the Gambia.

Silviculture characteristics

G. ulmifolia has a rapid regeneration capability, good growth rate, tolerance to both wet and dry growing conditions, an ability to compete with weeds, and the capability to sprout and regenerate asexually. Trees also tend to lose their leaves during the dry season. It is a pioneer species that grows best in full sunlight, colonizing recently disturbed areas, and is also commonly found growing along stream banks and in pastures. In Mexico, it is a common species in secondary forest.

G. ulmifolia requires a subtropical or tropical climates with annual rainfall of 700–1500 mm but occasionally up to 2500 mm. It is common in warm lowlands below 500 m altitude with mean annual temperatures of more than 24°C. However, it can also be found at higher latitudes up to 1200 m altitude.

On the Atlantic coast of North and Central America, it grows in tropical subhumid climates with summer rainfall from May to October, although the entire area is affected by hurricanes which occur most frequently in September. During November and December, the area is affected by boreal air masses from north-eastern Canada and the USA which cause a violent drop in temperature and, frequently, persistent light rain and fog. *G. ulmifolia* grows in various soil types varying from fertile to barren limestone, although it grows best in rich lowland soils and in alluvial and clay soils.

It flowers all year long, especially from April to October. Seeds are mature almost all year long, especially from September to April. The number of seeds/kg ranges from 100,000 to 225,000 and averages 187,500.

Silviculture practice

Seeds collected from standing trees are sun dried, stored at ambient temperatures, and are viable for 5 months. Seed germination can be improved using scarification or by soaking in boiling water for 30s. With fresh seed, 60–80% germination occurs in 7–14 days, whereas with stored seed, germination may be as low as 17% at 48 days after sowing.

Seedlings are ready for outplanting after about 15 weeks, when they reach 30–40 cm. With root stumps, plants are left in the nursery for 5–8 months or until they reach a stem diameter of 1.5–2.5 cm. Forest plantations propagated from seed are not very common, due to its easy natural regeneration by seed under natural growing conditions in its natural range. When it has been planted, regular pruning increases the fodder yield, and trees can produce 10 kg of dry matter when pruned four times/yr.

There are few research studies available on forest management for this species. Research conducted in Morelos, Mexico at 800–1600 m altitude, with 800–900 mm annual rainfall and mean annual temperature of 23–24°C, showed that the tropical dry forest was highly disturbed due to land use changes from forest to agriculture, and that *G. ulmifolia* was one of the dominant forest species. In non-disturbed tropical dry forest in the same region, there were five *G. ulmifolia* trees/ha out of a total of 304 trees/ha and 31 species, which accounted for a total basal area of 10.8 m^2/ha. These individuals had diameters of 11–40 cm, heights of 5–7 m and the total volume was 1.35 m^3/ha, out of a total of 74.5 m^3/ha. In Mexico, there is no formal management of this species.

Diseases and pests

Few pests and diseases have been reported. A bacterial leaf spot is caused by *Xanthomonas axonopodis*.

In Costa Rica a weevil, *Phelypera distigma*, has been observed feeding exclusively on new foliage, and *Amblycerus cistelinus* may attack up to 75% of seed pods in a particular site, along with insects of the family Anobiidae that enter ripe fruits and feed on the seeds.

Uses

G. ulmifolia is considered a potentially valuable species for agroforestry systems because of its great capacity for regeneration after cutting, although it is not used extensively for this purpose. Forest management in natural stands includes its selective harvesting in Mexico with the aim of conserving the soil. It is also been tested as a companion tree for raising high-value timber trees in silvopastoral systems. It is planted as a landscape tree, as a shade tree on streets and as an ornamental in gardens and parks. Throughout its natural range of distribution *G. ulmifolia* is used for firewood and charcoal.

The sapwood of *G. ulmifolia* is light brown and the heartwood pinkish to brownish, with large vessels and conspicuous rays. The texture of the wood is medium and the grain is interlaced. The wood has no characteristic odour and has a density of 550–570 kg/m^3. The wood is easily worked; it saws, planes and polishes very well, but it is not used commercially.

Leaves and immature fruits are cut to provide fodder, especially for cattle, and fruits are fed to pigs, usually during the dry season. The crude protein content of young leaves is 16–23% and stems have 7–8%. *In vitro* dry matter digestibility for young leaves is 56–58% and of stems is 31–36%. The seeds are edible, either raw or cooked, and this species can be used for apiculture. Condensed tannins are present in the bark, which can be used for tanning. Formerly, a decoction of the bark was used to polish cocoa beans. The seeds have many medicinal uses.

Further reading

Carvalho, P.E.R. (2007) Mutamba Guazuma *ulmifolia*. Circular Técnica - Embrapa Florestas No. 141. Embrapa Florestas, Colombo, Brazil, 13 pp.

Giraldo, V.L.A., Sánchez, M.D. and Rosales Méndez, M. (1999) Potential of *Guazuma ulmifolia* in silvopastoral systems. FAO Animal Production and Health Paper No. 143. Rome, pp. 295–308.

Manríquez-Mendoza, L.Y., López-Ortíz, S., Pérez-Hernández, P., Ortega-Jiménez, E., López-Tecpoyotl, Z.G. and Villaruel-Fuentes, M. (2011) Agronomic and forage characteristics of *Guazuma ulmifolia*, Lam. *Tropical and Subtropical Agroecosystems* 14, 453–463.

Guibourtia coleosperma (Benth.) J. Leonard

Large false mopane

Importance

Guibourtia coleosperma is an economically important hardwood timber species in Southern Africa and is exploited commercially with *Baikiaea plurijuga*, *Pterocarpus angolensis* and *Afzelia quanzensis*. The red skin of *G. coleosperma* seeds is an important food for local people during droughts. *G. coleosperma* is also an important canopy species of *Baikiaea* woodlands that grow on infertile Kalahari sands. These woodlands are important for soil and water conservation because they reduce erosion and are also a key habitat for wildlife.

Botanical features

G. coleosperma is a medium-sized tree 6–20 m tall. The crown is large, rounded and drooping. The trunk is slightly buttressed; bark grey to red-brown or black, smooth when young, becoming rougher with age. Young branches are conspicuously red. Leaves are compound with two wing-like leaflets up to 10 cm long and 4 cm wide, sickle-shaped, hairless, dark glossy green above and paler below, covered with fine transparent dots and with one main vein. Leaflets are not hinged together and are carried on very short stalks, petiole 1.5–3.5 cm long. Flowers are small, creamy-white, star-shaped and about 10 mm in diameter produced in panicles up to 16 cm long. The fruit is a small woody pod that is almost circular, 2–3 cm long, brown to dark brown and dehiscent, containing one brown seed covered in a scarlet aril. When the fruit ripens, the pod opens and the bright seed protrudes and remains hanging below the pod. The seed is up to 10 mm in diameter and shiny reddish-brown. *G. coleosperma* is almost evergreen and seldom leafless, new leaves appearing in spring. Flowering occurs from December to March and fruits ripen from May to July, but the time can extend either way by a month or more.

G. coleosperma is very similar to *Colophospermum mopane*, but can be distinguished using several characters. *C. mopane* has a rough fissured bark whereas that of *G. coleosperma* is smoother. Leaflets in *C. mopane* have no stalks, are joined together and between them is a vestigial leaflet about 3 mm long; in contrast, *G. coleosperma* leaflets have short stalks, are not hinged and the vestigial leaflet is absent. *C. mopane* pods are indehiscent, whereas those of *G. coleosperma* are dehiscent; *C. mopane* seeds are covered with red resinous dots, whereas those of *G. coleosperma* are not; and *C. mopane* flowers have 20–25 stamens of equal length, whereas *G. coleosperma* has a maximum of 12 stamens of variable length.

Distribution

G. coleosperma is native to Angola, Botswana, Namibia, Zambia and Zimbabwe, in open woodlands or dry forests, and has also been described from Zambezian dry deciduous forests and scrub forests or broad-leaved woodlands. *G. coleosperma* usually grows in association with *B. plurijuga*, *P. angolensis*, *A. quanzensis*, *Combretum collinum* and *Schinziophyton rautanenii*.

Silviculture characteristics

G. coleosperma is difficult to establish and slow growing. It regenerates both vegetatively and by seed; however, attempts to establish the species in plantations have failed, and this has been attributed to poor germination and establishment of seedlings due to desiccation, frost and herbivory. For example, in a 16-year planting trial in Zimbabwe, annual height increments were 0.14–0.21 m but survival was very poor, only 4–5%. The annual diameter increment of mature trees is also low (1.02–2.37 mm). It has therefore been concluded that natural regeneration is the key to the conservation of this economically important timber species. Seed production varies from year to year and is largely influenced by the amount and distribution of annual rainfall; seed is dispersed by birds. However, *G. coleosperma* grows slowly leading to a long establishment phase.

Silviculture practice

Seed storage behaviour is orthodox. Seeds germinate readily although manual scarification improves germination. Survival of *G. coleosperma* seedlings is low and erratic, and seedling survival after the first growing season ranged between 13% and 52% in north-western Zimbabwe. Seeds can be sown direct or into polythene tubes, and germination usually occurs within 10 days. Root pruning is necessary when seedlings are kept in the nursery for too long (i.e. 8 months). *G. coleosperma* can produce root suckers and coppices when cut or when damaged by elephants or fire, and tree stumps with diameters of 30–45 cm produced more sprouts than stumps with diameters of 50–80 cm.

Tree spacing ranging from 2 × 2 m to 5 × 5 m has been used in trials to establish *G. coleosperma* under plantation conditions. Even so, all trials gave very low survival rates even when there was a good soil environment (moisture, nutrients, and mycorrhizae), and weeding is required during the first 3 years after planting to reduce competition from weeds. *G. coleosperma* seedlings and saplings tolerate shade.

G. coleosperma together with *B. plurijuga*, *P. angolensis* and *A. quanzensis* are managed for timber production from natural woodlands. Individuals with a minimum dbh of 35 cm are harvested in a 40-year cutting cycle. Logging results in increased grass growth leading to increased fire hazards during the late dry season, and protection of logged woodlands from fire is a major challenge.

Human-induced fires are a major problem in *B. plurijuga* woodlands that contain *G. coleosperma*. Early burning and grazing to reduce fuel loads are two management strategies that have been proposed to reduce the incidence of intense fires during the late dry season. However, fire usually does not kill *G. coleosperma* saplings, and trees and saplings generally coppice. There is limited information on the growth rates of coppice shoots.

Diseases and pests

G. coleosperma is susceptible to damage by drought, fire, frost and elephants. Other damage agents include small rodents such as Peter's gerbil (*Tatera leucogaster*) and springhare (*Pedetes capensis*). Large herbivores such as duiker, kudu and eland also damage trees.

Uses

G. coleosperma and other hardwood species in *B. plurijuga* woodlands are managed for timber production, but these woodlands are also important for conserving soil and moisture and are valuable habitats for wildlife.

G. coleosperma produces attractive wood (also known as Rhodesian mahogany) that is hard and heavy. The density of dry wood is 800–960 kg/m^3. The heartwood is reddish-pink, often with dark and light veins. It is used for railway sleepers, building, joinery, furniture, parquet flooring and for canoes.

G. coleosperma is an important source of food for local people during drought periods. The red arils enclosing the seed are edible, eaten raw, mixed in porridge or made into a drink. The seeds are roasted and ground to make a meal. A red dye is also made from the aril and is used to stain furniture. *G. coleosperma* leaves and roots are used medicinally, roots are applied to wounds and leaves are used for treating coughs.

Further reading

Burke, A. (2006) Savanna trees in Namibia - factors controlling their distribution at the arid end of the spectrum. *Flora (Jena)* 201, 189–201.

Coates-Palgrave, M. (2002) *Trees of Southern Africa*. Struik Publishers, Cape Town, South Africa.

Phongphaew, P. (2003) *The Commercial Woods of Africa: A Descriptive Full-Color Guide*. Linden Publishing Co. Inc., Fresno, California.

H

Heritiera simplicifolia (Mast.) Kosterm.

Mengkulang

Importance

Heritiera simplicifolia is the most important timber species of the genus *Heritiera* and is the major species producing mengkulang timber. It has local commercial importance, especially in Peninsular Malaysia.

As it produces a superior timber for general utility purposes and for veneer and plywood, *H. simplicifolia* is worth considering as a forest plantation species.

Botanical features

H. simplicifolia is a large, monoecious, emergent tree 50 m tall, reaching 64 m in some cases, and 135 cm in diameter. It has a straight, columnar trunk with tall buttresses. The crown is rather open, made up of a few large branches and spreading to 36 m. The bark is fibrous, greyish- to reddish-brown, and shallow fissured with small rectangular scales. The inner bark is pinkish and laminated. Twigs are slender, with prominently raised leaf scars, often with clustered or stellate hairs and scales.

The leaves are alternate, simple, densely scaly beneath, with entire margins. In young trees the leaves are palmately compound. The petiole is 2.5–6 cm long, with apex thickened and minutely scaly. The leaf blade is broadly elliptic to obovate–elliptic, 5.5–17 × 3–10 cm in size, with apex rounded or notched, the base rounded, thickly leathery, drying to a coppery colour, midrib raised above, secondary veins 9–16 pairs, raised below, and tertiary veins scalariform, visible below.

The inflorescence is in axillary panicles, densely reddish hairy. Flowers are unisexual, yellow-pink, with a 4–5, sometimes 6, lobed bell-shaped calyx; petals absent. Male flowers are more numerous than female flowers, having an androgynophore bearing 8–10 sessile anthers with or without minute sterile flowers. Female flowers are slightly larger than male flowers, with 4–5, sometimes 6, sessile ovaries with short styles alternating with small groups of sterile anthers. The fruit is a non-splitting, ellipsoid nut, with a woody inner wall and a well-developed longitudinal ridge enlarging into a large wing (i.e., a samara) for wind dispersal. It is glabrous, ovoid to 1 cm long, with the wing up to 10 × 3 cm.

H. simplicifolia flowers at the beginning of the rainy season but not every year.

Distribution

H. simplicifolia is found throughout Malaysia, as well as on the islands of Sumatra and Borneo, where it occurs in undisturbed mixed dipterocarp forests up to 600 m altitude. In secondary forest it occurs mainly as a pre-disturbance remnant. Trees are widely distributed but not generally abundant, except in a few localized areas.

Silviculture characteristics

H. simplicifolia is a shade-tolerant, self-pruning species. Although preferring flat or undulating land with well-drained clay soil, it can grow on hillsides and ridges with sandy soils.

Silviculture practice

H. simplicifolia can be propagated by sowing seeds in a nursery. Germination is hypogeal, with seeds germinating in anything from 5 to 31 days. Stem cuttings can also be used for propagation. Seedlings are transplanted into the field when 30–50 cm high, and planted at a spacing of 3 × 4 m. The survival rate is 60% from planted stem cuttings. Propagation is also possible using grafting and tissue culture.

Natural regeneration is generally poor, although natural regeneration techniques used with regular management of mixed dipterocarp forests can be successful. It is reported that generally there is only one tree of exploitable size/ha. However, in a 50 ha plot in lowland forest in Malaya there were 160 stems with a diameter of 10 cm or more, of which 24 stems had a diameter of 30 cm and above.

The maximum tree diameter recorded in a 40 year-old plantation was 30 cm.

Diseases and pests

Resistance to wood rotting fungi varies. Incidence of dry wood termite attack on the sapwood has been reported. *H. simplicifolia* wood is also prone to marine borer attack, but is not particularly susceptible to powder-post beetle attack.

Uses

The wood of *H. simplicifolia* is of medium weight, moderate hardness and coarse texture. The heartwood is reddish-brown to dark brown, whereas the sapwood is pale yellow-brown to reddish. The density is 520–890 (but sometimes up to 990) kg/m^3 at 15% moisture content. The rate of shrinkage is moderate to fairly high. The wood seasons rapidly but has a slight tendency to warp and to exhibit surface checking. It is difficult to work due to a high silica content. Nail-holding properties are fairly good.

Its good, uniform colour and freedom from defects makes it a superior timber for general utility purposes. It is suitable for plywood and veneer. The timber is used for interior construction, flooring, furniture, ship masts and other ship construction above the water line. It is also used for ceilings, joists, window frames, staircases and in particleboard manufacture. Treated wood can be used for foundation piling. The fibre is used in various fibreboards with good results.

Further reading

Kostermans, A.J.G.H. (1959) *A Monograph of the Genus* Heritiera *Dry: (Sterculiaceae) including* Argyrodendron *F.v.M. and* Tarrietia *Bl.* Madjelis Ilmu Pengetahuan Indonesia, Jakarta.

Soerianegara, I. and Lemmens, R.H.M.J. (1994) *Heritiera simplicifolia* (Masters) Kosterm. In: Lemmens, R.H.M.J. and Soerianegara, I. (eds) *Plant Resources of South-East Asia No. 5 (1) Timber Trees: Major Commercial Timbers.* Backhuys Publishers, Leiden, The Netherlands, pp. 237–238.

Hevea brasiliensis (Willd. ex A. Juss.) Muell. Arg.

Rubber

Importance

Hevea brasiliensis is a large, deciduous tree native to South America but grown in lowland, humid tropics worldwide, known for the latex collected by tapping the bark, which is made into natural rubber. Extensive plantations are grown primarily for tapping, but the wood is increasingly important and plantations are now being grown exclusively for timber production.

Botanical features

H. brasiliensis naturally grows up to 30–40 m tall, but only about 15 m when cultivated. The tree has a very large root system, with a taproot 1–2 m long and laterals spreading to about 10 m. The trunk is cylindrical, bark smooth to slightly corky, pale to dark brown, and branching varies considerably. Leaves are alternate or subopposite at apex of shoot, trifoliolate, long petioles with apical glands and deciduous stipules. Leaflets are elliptic or obovate, 4–50 cm long and 1.5–15 cm wide, entire and pinnately veined. Flowers in axillary panicles on basal part of new flush, unisexual, with bell-shaped, 5-lobed perianths, bright yellow on ripening. Fruit a 3-lobed capsule, 3–5 cm in diameter, light brown when mature; one seed per carpel, ovoid, about 1 × 2 cm.

Distribution

The centre of origin covers part of the Amazon Basin, the upper Orinoco Basin and the Guianas, and wild and semi-wild members of the genus *Hevea* are found throughout the northern part of South America from Brazil to Venezuela, and from Colombia to Peru and Bolivia. Natural rubber was first introduced into South-east Asia in 1876 and grew into a valuable plantation species with the expansion of the motor car industry. Today, it is a very important crop in Malaysia, Indonesia, Thailand, Sri Lanka, Vietnam and China in Asia, as well as Côte d'Ivoire, Nigeria, Cameroon, Liberia and Gabon in Africa.

Silviculture characteristics

H. brasiliensis is a tropical tree that grows best at temperatures of 20–28°C with an annual rainfall of 1500–3500 mm. It is light-demanding, requiring moist but well-drained soil, and severe damage is caused by waterlogging. It is relatively insensitive to soil type though is sensitive to salinity, but higher yields and disease resistance can be expected on highly fertile soils. Trees grow satisfactorily up to 600 m altitude, but tend to be damaged by high winds in exposed areas.

Silviculture practice

Seed from vigorous, high-yielding parents are used to produce rootstock. As seeds are viable for only a short time, planting must be done soon after harvesting, germinated on shaded beds, and transferred to nurseries. Improved planting material is often preferred, however, and this is easily obtained by bud-grafting rootstock. Budwood is grown in special nurseries in which trees budded with the desired clone are closely spaced. Brown-budding was the traditional method, later superseded by the green-budding technique. Crown budding is another method producing a three-component tree which, for example, combines a disease-resistant crown with a high-yielding production trunk budded on a seedling rootstock.

After budgrafting, planting material can be nursed as bare-root stumps (e.g. budded stumps, stumped buddings, or mini-stumps), or as polybag-raised buddings. Raising bare-root planting material requires a suitable and well-prepared soil, while for polybag plants only the potting medium and a good water supply matters. Lifting and root pruning bare-root plants are time consuming, but once done the material is easy to handle and transport. Polybag plants need constant attention. They are ready for transplanting immediately but need great care during transport to prevent root damage, whereas the large polybag plants are difficult to handle. Polybag plants, however, develop more quickly after planting. Soil-core buddings have much the same advantages and disadvantages as polybag plants, but as they are raised in the ground they are less susceptible to drought in the nursery.

For the production of seedling trees, 'clonal' seed is used, obtained from monoclonal or polyclonal plantings which are known to produce high-yielding families. These 'clonal' seedlings are cheaper to produce, may have greater wind resistance, may reach maturity earlier than brown-budded rubber, but are more variable and seedling plantings usually give lower yields. Germination and nursery procedures are essentially the same as for raising stock. All planting material, buddings and seedlings, are pruned to restrict development to one single stem free from any branches up to 3 m to ensure enough tappable bark for high panel tapping.

In smallholdings, temporary intercropping of young rubber with food crops is a common practice to provide cash income when the trees are still immature. On flat or undulating land, intercropping can be carried out during the first 1–3 years after planting without adversely affecting rubber plants.

Budded stumps with bare roots are planted in pre-dug holes 45 cm across, and rock phosphate is added at 100 g/hole during refilling. A similar procedure is used for planting advanced planting material (maxi-stumps) with bare roots, except that larger planting holes are used. Due to the plants' susceptibility to drought, cylinders of polythene sheeting (sarongs) are temporarily placed in the planting hole around the upper half of the taproot, filled with a mixture of good soil and rock phosphate, and covered with grass after watering. When the first leaves are properly hardened, the sarongs are removed. For polybag plants, planting holes are made at the time of planting.

The preference for planting patterns has varied over the years, between square planting at 5 m spacing, to avenue planting with 8–10 m between rows and 2–3 m in the row. The former has the advantage of optimal use of soil and space, early closure of the canopy, and less wind damage; the latter ensures cheaper maintenance, lower tapping costs, and more space for temporary intercropping. The current recommendation from the Rubber Research Institute of Malaysia for smallholders practising intercropping is to plant rubber in east–west rows at distances of 9 × 2.7–3 m. High planting densities give the highest yields/ha but trees take longer to reach a tappable size and give lower yields per tree and per tapper. This is why smallholders, who are usually interested in maximization of yield/ha, plant at higher densities (500–600 trees/ha) than are used in estates (400–450 trees/ha) which are interested in maximization of net financial returns.

Cover crop establishment before planting is a standard practice both in new planting and when replanting on estates. The most common leguminous species used are *Pueraria phaseoloides*, *Centrosema pubescens*, and *Calopogonium mucunoides*, and although the cover crops compete rather strongly with the rubber in the first year of establishment, their overall effect on the production is beneficial and may extend over a 20-year period. Drainage is required in waterlogged areas.

Diseases and pests

The most damaging leaf disease of *H. brasiliensis* is South American leaf blight caused by *Microcyclus ulei*. Infected trees lose their leaves after every new flush resulting in die-back and ultimately, tree mortality. Other fungal leaf diseases include *Glomerella cingulata*, *Corynespora*, and *Phytophthora*. Important root pathogens include *Rigidoporus microporus*, *Ganoderma philippii*, and *Phellinus noxius*, giving rise to white, red, and brown root disease, respectively, causing significant total tree losses in some new plantings.

Important fungal leaf diseases are *Colletotrichum* and *Oidium* spp., causing secondary leaf fall, leaf spot (*Corynespora cassiicola*), and phytophthora leaf fall (*Phytophthora palmivora* and *Phytophthora botryosa*). Bird's eye spot disease caused by *Bipolaris heveae* is also common but is confined to the nursery. Pink disease caused by *Corticium salmonicolor* (*Phanerochaete salmonicolor*) attacks trunks and branches causing branch snap, but is easily controlled.

Underground pests are important and require attention at all stages of tree growth. These include termites (*Coptotermes curvignathus*) and grubs of certain *Melolontha* beetles. Among the above-ground pests, yellow tea mite (*Polyphagotarsonemus latus*) and thrips (*Scirtothrips dorsalis*) are common in nurseries, where they cause defoliation of tender leaflets. Other common sap-sucking insects, such as scale insects and mealybugs may occasionally cause sufficient damage to warrant treatment.

Uses

The main product is rubber made from the milky latex, which is first processed into a concentrate, sheet rubber or block rubber, and marketed to manufacturers as natural raw rubber. The main users are tyre manufacturers, who consume 60–70% of the total world production. Other uses include rubber car components such as engine mountings, engineering components such as anti-vibration mounts and flooring, and many consumer products such as footwear, sports goods, toys, gloves, swimming caps and condoms.

When felled for replanting, rubber trees are also sawn to produce rubber wood, and with proper treatment, the timber is used for high-value-added furniture, particle board, parquet flooring and many other wood products. It is a light-coloured and medium-dense timber with an air-dry density of 560–650 kg/m^3, with a wide variety of applications. It can be easily steam-bent, and stained to resemble other timbers depending on consumer demand.

Plantations are estimated to cover almost 9 million ha, 80% in South-east Asia. Until recently, the wood was considered a by-product of rubber plantations and was used largely for industrial fuelwood and charcoal, but rubberwood is increasingly important and plantations are now being grown exclusively for timber production.

Further reading

Oktavia, F., Lasminingsih, M. and Kuswanhadi (2011) Genetic diversity of wild germplasm and cultivated clones of *Hevea brasiliensis* Muell. Arg. detected by RAPD analysis. *Journal of Rubber Research* 14, 241–251.

Priyadarshan, P.M. (2011) *Biology of Hevea Rubber*. CAB International, Wallingford, UK, 240 pp.

Yi Peng, T., Mat Don, M. and Ujang, S. (2011) Assessment of the properties, utilization, and preservation of rubberwood (*Hevea brasiliensis*): a case study in Malaysia. *Journal of Wood Science* 57, 255–266.

Hibiscus elatus Sw.

Blue mahoe

Importance

The Caribbean tree *Hibiscus elatus*, known as blue mahoe or majagua, is a highly valued, fast-growing timber species that has adapted to a wide range of sites and is often used in reforestation. It produces excellent timber for high-grade furniture, but can also be used as a windbreak. Ease of germination results in successful plantation establishment. It could be grown on a larger scale as a plantation crop but suffers from epicormic branching and sweep. *H. elatus* is also grown for ornamental and shade purposes and is the national tree of Jamaica.

Botanical features

H. elatus commonly grows to a height of 20–25 m and a diameter of 30–45 cm, although on fertile sites it may reach a diameter of 90 cm. The bole is straight and long. The bark is soft in texture and consists of many layers that can be separated after beating.

The leaves are broad, simple, entire or lightly toothed, and arranged spirally. The petiole is up to 14 cm long, slender, greenish and densely covered with white hairs. The blade is cordate, up to 20 × 20 cm in size, acuminate at the apex, and palmately 7–9 nerved at the base. The nerves are depressed above and raised below. The blade is glabrous above and densely whitish-haired below.

Typical hibiscus-like flowers are borne in terminal panicles or laterally near the end of twigs. They are trumpet shaped, 7.5–10 cm long and 8–13 cm broad, with yellow petals that change colour as they mature to orange, red, and finally crimson. The calyx is tubular with five narrow pointed lobes. Flowers last 1 day and flowering occurs irregularly throughout the year.

Fruits are grey-green, hairy capsules, 2.5–3 cm long, containing numerous blackish-brown seeds.

Distribution

H. elatus is native to Cuba and Jamaica. It has been widely planted on both islands and has naturalized in southern Florida (USA), Mexico, Peru, Brazil and throughout the West Indies.

Silviculture characteristics

H. elatus is a fast-growing timber species that can adapt to a wide variety of sites. Drought can be a limiting factor on well-drained sites where annual rainfall is less than 1500 mm; even so, it will survive on alluvial plains receiving an annual rainfall of 1000 mm or less. In wetter areas with annual rainfall ranging from 1500 to 3800 mm it will grow under a wide range of soil conditions and at altitudes of up to 1200 m. *H. elatus* grows best near mountain bases and on lower mountain slopes. It performs poorly on shallow soils over bedrock, on very poorly drained soils and on severely eroded or nutrient-depleted sites, but tolerates acid clay soils. It is intolerant of exposed conditions, such as on exposed ridges. It suffers from epicormic branching and sweep.

Silviculture practice

In the West Indies, seeds ripen in March and April when the capsules are picked and laid in the sun to dry. Seed weight averages 1.8–1.9 g/100 seeds in Jamaica but slightly less in Puerto Rico. Germination percentage is high in Jamaica (80%), but poor in Puerto Rico. *H. elatus* can also be propagated by cuttings, grafting and tissue culture.

Seedlings can be transplanted after 1 year when they range in height from 45 to 60 cm, and rapid growth commences about 6 months after outplanting.

In order to avoid the development of epicormic branches, planting densities of at least 2.5 × 2.5 m and thinning to 3.5 × 3.5 m in mature plantations are recommended.

The volume of stands aged from 16 to 27 years ranges from 97 to 979 m^3/ha, and mean annual increment ranges from 4.5 to 36 m^3/ha.

Diseases and pests

H. elatus is relatively free from diseases and pests, and there are only occasional reports of damage. In Jamaica, fungal leaf spots caused by species of *Septoria* and by *Pestalotia heterocornis* have been recorded, while in Cuba, leaf spots caused by bacteria have been observed.

Anomis illita is an important defoliator. *H. elatus* is also a host to cotton stainers (*Desdercus* spp.) and to the pink hibiscus mealybug (*Maconellicoccus hirsutus*).

Uses

H. elatus is often planted as an ornamental and shade tree, and can also be used as a windbreak, despite its tendency to lose leaves during dry spells and susceptibility to limb breakage. The species has shown promise when used in agroforestry trials in Cuba.

The fairly straight grain of *H. elatus* heartwood is richly variegated with shades of steely-blues, metal-greys, deep purples, olive-greens, and browns, with an elegant chatoyance in the wood. The narrow sapwood is pale white and subtly flecked, creating an attractive contrast with the heartwood. The wood is hard, elastic, very durable, highly resistant to attack by decay fungi, and resistant to subterranean termites. It has a basic specific gravity (oven dry weight/green volume) of 0.58 to 0.62, while the average density has been recorded as 500 kg/m^3. The timber works easily and is generally easy to saw, plane, rout, mould, mortise, carve, glue, nail, screw, sand and turn, with a natural gloss in the wood when finely finished. It responds very well to both hand and machine tools in all woodworking operations.

The wood is used for high-quality furniture, house interior trim and craft items, and various other purposes including light construction, railway sleepers, containers, agricultural implements, turnery, musical instruments, flooring, inlays, carvings and boatbuilding.

The inner bark of young trees is used for making rope and cords, which are reported to be very durable in salt and brackish water.

The mucilaginous young leaves and twigs are traditionally used to treat dysentery, and when boiled with *Cissus sicyoides* can be used to treat colds.

Further reading

Adams, C.D. (1971) *The Blue Mahoe & other Bush: An Introduction to Plant Life in Jamaica*. McGraw-Hill Far Eastern Press, Singapore.

Leiva Sánchez, A.T. (2007) Majagua. In: Leiva Sánchez, A.T. (ed.) *Trees of Cuba*. Macmillan Education, Oxford, UK, p. 64.

Hopea odorata Roxb.

Merawan

Importance

Hopea odorata is a medium to large evergreen tree with a straight, cylindrical bole up to 45 m tall and 120 cm dbh and with prominent buttresses. Indigenous to much of South-east Asia, it is essentially a riparian species, rarely occurring far from streams. It grows easily in a wide range of habitats and is easy to handle as a plantation species. One major disadvantage is the short viability of the seeds, making storage and transport of seed difficult. However, the species can be propagated artificially by cuttings.

The timber is a fine, light hardwood known in the trade as merawan, suitable for a wide range of uses from rollers in the textile industry to piles and bridge construction. The species is also widely planted as a shade and ornamental tree. Tannin is sometimes extracted from the bark to tan leather. The species also produces an inferior resin (dammar), which is used in Myanmar as a varnish over paint. It is also used to caulk boats.

Botanical features

A medium to large evergreen tree up to 45 m tall, bole straight, cylindrical, with a diameter of up to 120 cm and prominent buttresses; bark surface with irregular longitudinal shallow fissures in immature trees but becoming scaly with age. Crown dense, dark green, pyramidal or conical in young trees, spreading in mature trees. Twigs slender, glabrous except when young. Stipules minute and caducous.

Leaves simple, alternate; petiole rather slender, approximately 15 cm long, drying black; blade usually ovate–oblong, unequal-sided, 7–14 × 3–7 cm, venation scalariform, approximately 11-nerved with pore-like domatia in the axils of the lower pairs, drying a pale yellow-green colour.

Inflorescence paniculate, slender to 12 cm long, terminal or axillary, twice-branched, with many up to 205 cm-long branchlets bearing up to 9 secund flowers. Flowers with petals pale yellow, fimbriate at the apex; style as long as the ovary. Fruits two-winged, nut ovate–conical, approximately 1 cm long, free from the calyx lobes; wings approximately 6.4 cm long, green when immature. *H. odorata* flowers at more or less regular intervals, usually once every 2 years.

Distribution

H. odorata is distributed in the Andaman Islands (India), Burma, Thailand, and Indochina, with its southernmost limit in north Peninsular Malaysia (in North Perak and Terengganu). This is a common tree of the evergreen forests, although it is essentially a riparian species, rarely occurring far from the streams.

Silviculture characteristics

H. odorata is a light-demanding species but is not invasive. In Peninsular Malaysia, it regenerates freely along rivers. Its tolerance of disturbed habitats has made the species popular in urban plantings. This tolerance also makes the species suitable for plantation forestry.

H. odorata is propagated mostly by seeds. During fruiting years, fruits are produced in great quantities. However, there is high incidence of seed predation from weevils, particularly in seeds collected from the forest floor. For mass collection purposes, the fruits are often collected by shaking branches to cause the fruits to drop. A canvas is spread under the tree to facilitate collection.

Silviculture practice

Seeds of *H. odorata* are recalcitrant and do not store well for long periods. Germination can be obtained at 75% after 14 days storage at 14°C. With longer storage, viability is often lost and germination percentage decreases even after storage at low temperatures, particularly at temperatures below 10°C; storage at low temperature causes chilling damage. Storage in nitrogen gas did not slow the deterioration of *H. odorata* seeds.

H. odorata can be easily rooted from stem cuttings, and orthotropic shoots must be used in such vegetative propagation. Shoots from juveniles are often used, and perform best. A good rooting percentage is obtained with juvenile material from seedlings of 3–12 months old and coppice shoots from older seedlings. In general, juvenile *H. odorata* is easily rooted with the appropriate hormonal treatment. In trees, reservoir grafting of mature branches onto seedlings is another possible source for stem cuttings.

Fruits collected from the field should be sown as soon as possible. After germination, the seedlings from the sowing beds can be transplanted into polythene bags. Seedlings perform best under 50% shade for 6–9 months, by which time they reach a height of 40–60 cm and are suitable for planting in the field. Ectomycorrhizal infection appeared to improve growth rates of seedlings of *H. odorata*.

Growth data are lacking, but records of arboretum trees at the Forest Research Institute Malaysia (FRIM) showed growth rates of about 1.8 cm diameter per year for trees at 46 years old. These trees at FRIM were about 80 cm dbh at about 46 years and were large enough to be harvested. Rotation cycles have not been calculated or developed for the species, but based on growth rates of a limited number of trees, a cycle of 30–40 years is feasible for the species on suitable sites.

The species also performs favourably in planting trials in degraded land and in logged-over forests in log landings and skid trails. Little tending is required once the trees are established after about 3 years. Initial tending such as weed control and the removal of competition from surrounding trees is essential during the establishment phase.

Diseases and pests

In nurseries in Vietnam the psyllid *Trioza hopeae* causes serious damage to young plants of *H. odorata* by inducing leaf galls. In Bangladesh, the larvae of the tasser silkworm *Antheraea paphia* has been observed feeding on the foliage of *H. odorata*. Another lepidopteran larva *Calliteara cerigoides* is a serious defoliator of several species including *H. odorata*. Damage to seeds by the weevil *Nanophyes shoreae* has also been recorded in *H. odorata*.

Uses

Planting experiments have shown that *H. odorata* is suitable for planting in degraded sites such as log yards, skid trails from logging activities, and ex-mining land. The species is now widely planted as an ornamental tree in Malaysia. For this purpose sufficient space must be provided for the tree to grow, as mature trees even in the urban environment can reach large sizes. *H. odorata* is planted as a street tree in Ho Chi Minh City, Vietnam.

H. odorata, together with other species of light hardwood *Hopea*, are commercially grouped together as 'merawan' in Malaysia and elsewhere. Merawan has a density of 495–980 kg/m^3 when air dried. The timber is strong, and texture is moderately fine and even, with interlocked grains. The timber is moderately easy to re-saw and crosscut. Planing is fairly easy and the surface produced is smooth. Its nailing properties are rated as very poor.

The timber seasons slowly with slight cupping as the only defect; 15-mm boards take approximately 4 months to air dry, while 40-mm boards take 6 months. For kiln drying, Schedule H is recommended; 25-mm boards take approximately 5 days to kiln dry from 50 to 10% moisture content. The timber is moderately durable.

Tannin is sometimes extracted from the bark to tan leather. The species also produces an inferior resin known as dammar, which is used as a varnish over paint on pictures. It is also used to caulk boats.

Further reading

Chiu, S.B. and Siow, A. (2003) Planting *Hopea odorata* (merawan siput jantan), a fast growing timber species. *Planter* 79, 779–783.

Jøker, D. (2000) *Hopea odorata* Roxb. Seed Leaflet 49. Danida Forest Seed Centre, Humlebaek, Denmark, 2 pp.

Mohd Afendi, H. and Ang, L.H. (1994) *Nursery techniques for dipterocarps - Part I. Planting stock production of Shorea assamica (meranti pipit), Shorea parvifolia (meranti sarang punai), Shorea leprosula (meranti tembaga), Dryobalanops aromatica (kapur) and Hopea odorata (merawan siput jantan).* Forest Research Institute Malaysia (FRIM) (Institut Penyeli dikan Perhutanan) Technical Information 49, Kuala Lumpur, 4 pp.

Noor, M.M., Aminah, H., Zaki, A.M. and Krishnapillay, B. (2002) *Merawan siput jantan. A manual for forest plantation establishment in Malaysia.* Forest Research Institute Malaysia (FRIM), Kuala Lumpur, pp. 189–197.

Sakai, A., Visaratana, T., Vacharangkura, T., Thai-ngam, R. Tanaka, N., Ishizuka, M. and Nakamura, S. (2009) Effect of species and spacing of fast-growing nurse trees on growth of an indigenous tree, *Hopea odorata* Roxb., in northeast Thailand. *Forest Ecology and Management* 257, 644–652.

Hyeronima alchorneoides Allemao

Bully tree

Importance

As a timber tree, *Hyeronima alchorneoides* is highly prized for its dense, durable wood. The name pilón, as it is commonly known in parts of its range, refers to its putative use as a pestle for pounding maize in traditional Amerindian culture. The wood is used for a number of purposes that include heavy construction, railway ties, marine work, furniture, decorative veneers and joinery. *H. alchorneoides* is a food source for various forest birds and animals that consume its fruits.

Botanical features

H. alchorneoides is a massive emergent tree of lowland tropical forests. It exhibits sympodial growth, with multiple, orthotropic branches that form a dense crown. Mature forest trees can attain a height of up to 50 m (typically 30–45 m) and can grow to over 1 m in diameter. Boles tend to be straight and robust, with very little taper, and unbranched up to a height of 20 m or more. The tree forms rounded, sinuous buttresses that can be taller than 1 m, and extend 10 m or more from the base.

The bark is smooth, and is a distinctive deep reddish-brown colour on the outside, pinkish on the inside and bitter-tasting. The tree has a taproot system. A relatively high proportion of biomass is allocated to fine roots.

Leaves are simple, alternate, petiolate, and stipulate. Leaves have abundant peltate hairs on both surfaces. The petioles are long, about half the length of the leaf lamina, and are distinguishable by a pulvinus at either end. Stipules are conspicuous and leafy, but vary in size and shape among individuals and even on the same individual. In juveniles the leaves are large, but the tree produces progressively smaller leaves as it ages.

The tree is reproductively mature by the time it reaches a diameter of 30 cm, and is dioecious, with male and female flowers borne on separate individuals. Flowers are small and inconspicuous, pale yellow, zygomorphic, and arranged on axillary branched panicles 5 cm long. They are pollinated by a diverse suite of small insects. The single-seeded drupaceous fruit is small and berry-like. At maturity, fruits turn red or dark purple. They have a fleshy, sweet mesocarp, and are dispersed by monkeys and birds.

The tree is evergreen and leaves emerge and abscise continuously. *H. alchorneoides* is readily recognized in the forest due to the presence of red, pre-senescent leaves in the canopy all year round. There are some reports that the tree has two flowering periods during the year, with peak flowering from May to July, and a smaller episode in November–December. Others report only a single flowering event per year, in January–February or in July–October.

Distribution

H. alchorneoides has a wide distribution, occurring from Belize at the northern extent of its range, to the Brazilian Amazon region in the south, and extending to the islands of the Caribbean. It is also reported as occurring in southern Mexico.

Silviculture characteristics

H. alchorneoides is known to occur naturally mainly on old alluvial, clayey, or acid soils. In plantations it has also been successfully grown on older, more weathered ultisols. In areas prone to frosts *H. alchorneoides* shows very high mortality.

In studies of the demography of large canopy trees, an association of small sapling and pole stages of *H. alchorneoides* with large forest gaps has been demonstrated, suggesting a requirement of the species for high light at the seedling stage, and then again later, to reach the canopy. Nevertheless, the studies also demonstrate the ability of juveniles to persist for long periods of time under low light, with very low mortality. Thus, it may take an individual several hundred years to reach its full size in the forest. Plantation-grown individuals in high light show remarkably rapid growth (as much as 3 m/yr) on fertile alluvial soils.

The tree forms a dense canopy with a high leaf-area index. In experimental plantations, monocultures of the species achieved 90% interception of incoming solar radiation at 2 years old. This high light capture is associated with the ability of *H. alchorneoides* to suppress understorey regeneration. Furthermore, a high leaf-area index is hypothesized to indicate a high shade tolerance of lower leaves. This may make the species compatible for planting in mixed species plantations, where *H. alchorneoides* forms the lower stratum in a stratified canopy, in combination with species that have taller, more open crowns.

In addition to high light capture, *H. alchorneoides* also has high biomass allocation to fine roots, which may increase its effectiveness at soil nutrient capture. This is likely to confer an ability to tolerate relative nutrient deficiencies. Initial experiments with *H. alchorneoides* grown on soils of varying fertility further support this hypothesis, making it a potentially suitable candidate for revegetation of sites that have been degraded due to previous land-use practices.

Thinned stumps are known to resprout, but the resulting root systems tend to be weak. Trees show good growth and form, and have the ability to self-prune.

Silviculture practice

To date, propagation of *H. alchorneoides* has been from seed, although cuttings have been raised *in vitro* and a micropropagation methodology has recently been developed. Seeds are collected either directly from the tree or in cloths spread beneath trees to collect falling fruits. Successful germination can be achieved by washing seeds in water and lightly sanding them or by applying gibberellic acid. As fruit maturity stage affects seed germination, best germination results are achieved with seeds from purple rather than yellow or green fruits.

The preparation of stump cuttings is done using seedlings, once their stems are sufficiently lignified. Seedlings are uprooted, and the roots and shoots are pruned back to 15 and 10 cm, respectively. Stump cuttings require 5–6 months in the nursery before they are ready for planting. Plantation establishment using stump cuttings requires resprout management in the early phase of establishment.

Site preparation prior to stand establishment usually involves site clearing. Fertilizer is applied to the bottom of the planting hole, and seedlings are planted with their nursery substrate. Information about the management of *H. alchorneoides* in plantations comes from initial introductions of the species in north-east Costa Rica. Initial trials used spacings of 2 × 2 m, although a spacing of 3 × 3 m is now more commonly recommended. Most plantations, both forestry plantations and on-farm trials, have been established at 3 × 3 m spacing and are thinned after 3–4 years. Pruning is recommended at 9–12 months and thinning is recommended after 3–4 years.

Diseases and pests

H. alchorneoides is very susceptible to damping off due to fungal attack (*Phytophthora* spp., *Fusarium* spp.) at the seedling stage under very humid nursery conditions.

The majority of pests on *H. alchorneoides* are defoliators. Up to 19 insect pests have been identified on *H. alchorneoides* in Costa Rica.

Uses

H. alchorneoides is among several species that have shown potential (in Costa Rica especially) for restoring degraded areas and abandoned agricultural land, and for facilitating the regeneration of secondary forest when planted in pure or mixed species plantations. Its suitability for use in revegetation of degraded lands can be attributed in part to its ability to grow on a range of soils. Other environmental benefits of growing the species include high carbon sequestration and storage, hyper-accumulation of cobalt on contaminated sites, and facilitation of forest succession by increasing seed species richness density in plantations by attracting seed dispersal agents from nearby forests.

H. alchorneoides has reddish-brown heartwood, and sapwood with a lighter colour and pinkish hue. Its wood has low lustre, no particular odour, an interlocked grain, and a smooth texture, although fibres tend to tear while planing. Basic specific gravity is high, and ranges from 0.59 to 0.86.

In Central America, the wood has traditionally been used for heavy construction, including railway sleepers, bridges, columns and beams. Sawing is easy and it has good nail-holding characteristics after drilling. It is moderately to very durable and is fairly resistant to subterranean and drywood termites. On the basis of its durability it has also been recommended for use as fence posts, stakes, piles and in marine construction. In addition, it is used for furniture, cabinet work, flooring, and in decorative veneers, turnery and joinery.

The non-wood uses of *H. alchorneoides* are limited and highly localized. In Guyana, a decoction of the bark is used for its anti-tussive properties. Leaf extracts exhibit analgesic properties, while the oil extracted from seeds appears to be anthelminthic.

Further reading

Abdelnour, A., Aguilar, M.E. and Valverde, L. (2011) Micropropagation of pilon (*Hieronyma alchorneoides*). Editorial de la Universidad de Costa Rica, San José, Costa Rica, *Agronomía Costarricense*, 35, 9–19.

ACEEFN (1992) *Hyeronima alchorneoides*. In: Proceedings, Segundo Encuentro Regional sobre especies forestales nativas de la zona norte y atlántica de Costa Rica. Asociación Costarricense para el Estudio de Especies Forestales Nativas), Sarapiquí, Costa Rica, pp. 45–51.

Flores, E. (1993) Pilón, Bully tree. *Arboles y semillas del Neotrópico* 2, 53–74.

Jøker, D. and Salazar, R. (2000) *Hyeronima alchorneoides* Allemão. Seed Leaflet 47, Danida Forest Seed Centre, Humlebaek, Denmark, 2 pp.

Hymenaea courbaril L.

Jatoba

Importance

Hymenaea courbaril is a medium-sized Neotropical tree. The timber is hard, heavy, fine-grained with reddish-brown heartwood which, although difficult to saw, machine or plane, is strong, tough and takes a fine finish.

Although it is an important timber tree, it is in many regions better known for producing 'copal', the orange sticky resin released from the trunk, branches and the fruit pericarp when damaged, which solidifies on the tree or falls to the ground in chunks. Fossilized copal is the source of most Neotropical amber. The fresh resin has been used commercially in medicine, as an ingredient in varnish or polish, and to make incense. Large quantities are harvested from the base of living trees.

Despite its relatively slow growth, the tree's nitrogen-fixing ability and its attractiveness to bees have made *H. courbaril* a useful agroforestry and land-rehabilitation tree.

Botanical features

H. courbaril is a medium-sized, heavily branched evergreen tree with a spreading crown reaching to 40 m in height and up to 1 m dbh girth. The bark is greyish, smooth and thick. Leaves are alternate, divided in two parts, glossy with 1–2 cm long petiole. Each blade is oblong, 5–10 cm long and 2–5 cm wide, and has an asymmetrical rounded base and short acuminate apex.

The inflorescence is a 8–15 cm long terminal panicle bearing 3–4 cm long, yellow flowers with a calyx jointed at its base and red anthers. The fruit is a brown hard woody pod (5–20 cm long, 2–3.5 cm thick) that does not open at maturity. The 3–10, 2–3 cm long, ellipsoid, slightly compressed and red-brown hard-coated seeds within are embedded in a powdery, cream-coloured pulp with an unpleasant odour (hence the other common name 'stinktoe').

Several local varieties have been described.

Distribution

H. courbaril is native to central and northern South America and the Caribbean islands. It occurs in lowland forest up to 900 m altitude in high rainfall areas (1900–2150 mm annually).

The trees grow in a wide variety of soil textures, but perform best in well-drained soil, on sandy ridges or riverbanks. They can grow on clay if not waterlogged and grow well on oxisols with a pH of 5–7.

It is occasionally cultivated and frequently managed (i.e. protected when a swidden clearing is opened). It has also been planted in Peninsular Malaysia and Indonesia, and tried in southern India.

Silviculture characteristics

Considering the tree is a hardwood, seedling growth is rapid. Trees may reach a height of 8 m in 5 years and 18.5 m in 16 years.

Seedlings do not survive and establish well in open conditions. The species is intolerant to shade when mature. Planting in the open, for shade and ornamental purposes, produces attractive and spreading trees more rapidly. However, research in Brazilian forest fragments showed that *H. courbaril* is able to regenerate and to recruit in shaded environments. It also coppices well, although stumps of large trees coppice less successfully, and can maintain itself in frequently cut-over areas.

The intercropping potential of *H. courbaril* with green manure, forage and weed species is good: at its early phases of development, tree growth is not affected by the herbaceous species. In fact, a positive effect on nutrient accumulation by the weeds was observed when they were grown in the presence of *H. courbaril*.

Silviculture practice

Mature pods remain on the trees for a long time and are hard to pick when trees are tall. Pods must be collected soon after falling, to prevent them from decomposing under humid conditions and the seeds being lost. Seeds can be released from the dried pods with a gentle squeeze.

Large variation in seed weight is reported (from 250–500 seeds/kg to 800–900 dry seeds/kg according to Indian sources). This partly reflects differences in moisture content. Fresh seeds are generally more moist, and heavier. Dry seeds can be stored in an airtight container for at least 1 year, in a cool and dry place. Viability can presumably be prolonged by cold storage (3–5°C), though older sources suggest storage at 26°C and 12% humidity.

Stand establishment is by planting stock grown from seeds or from cuttings. Fresh seed with relatively high moisture content (15–20%) may germinate freely without pre-treatment. Dry seeds, however, need some pre-treatment such as manual scarification (nicking, filing, hot-wire burning), or bulk pre-treatment by dipping in boiling water, or for 30–60 min in concentrated sulphuric acid.

Seeds may be sown in pots or directly planted in nursery bags for later transplanting. Germination is usually quick; at an optimal temperature of 30°C, germination begins 20–30 days after sowing. Seedlings benefit slightly from being grown in larger root trainer-sized pots (14% in terms of height, dbh and leaf biomass). Application of phosphorus is recommended to produce good-quality seedlings.

Seedlings grow at an angle with a drooping leader, a habit that may persist for 2 or 3 years. Open planting sites and good weed control until the seedlings reach about 2 m is important for best planting stock. Seedling development is fast under controlled ideal conditions, and provides the possibility of obtaining numerous healthy and vigorous seedlings, which may be used in reforestation programmes.

Diseases and pests

H. courbaril leaves contain tannins and other complex secondary compounds, and only specialized insects predate on them. In an experimental study in Costa Rica, seedlings were infrequently affected by galls but overall impacts were low. Sometimes seedlings were partly defoliated by bees which use the leaves for building nests.

A weevil (*Rhinocherus* spp.) bores through seed-pods and eats the seeds in Costa Rica, Trinidad, and Tobago. Other insects (*Acanthoscelides* spp., *Hypothenemus buscki*, and *Myelois decolor*) feed inside seed-pods in Puerto Rico. An unidentified insect cuts twigs and branches after depositing eggs in Trinidad and Tobago. Leaf-cutter ants (*Atta* spp.) rarely harvest young leaves, and studies indicate that the leaves contain a terpenoid 'caryophyllene epoxide' that inhibits the growth of the fungus that the ants cultivate. In the Caribbean, damp wood termites (*Nasutitermes costalis* and *Nasutitermes nigriceps*) eat dead wood, and marine borers attack the wood.

Uses

Jatoba is primarily a timber species. It has a basic density of 740 kg/m^3. The wood is hard, heavy and strong and ages to a beautiful rosy-brown to deep red colour. It is resistant to white-rot fungi and termites and is used in heavy construction (hydraulic work, truck bodies and railroad ties), as well as for indoor purposes.

Fossilized resin – amber – is considered a semi-precious stone and is traded for jewellery. The ability of amber to preserve the structure in a wide variety of biological inclusions provides scientists with numerous insights concerning past fauna and flora.

The resin is used industrially in the production of varnishes. In some areas of the interior of Amazonas, Brazil, it is used in the fabrication of ceramics, mostly kitchen utensils, as a sealant to reduce or eliminate water penetration.

The bark is a source of tannin. The pods contain a thick layer of mealy or powdery, whitish or yellow, sweet pulp. Despite its unpleasant smell, the pulp surrounding the seeds is eaten in some regions, especially as a famine food. It is also mixed with water to make drinks.

The bark has many medicinal uses; tea is prepared from the bark and used as cough medicine, as a natural energy tonic, to treat urinary system problems as well as being a vermifuge and fungus cure, and general stomach cleaner. The fruit is used as a remedy for mouth ulcers, and the leaves and wood are used to treat diabetes.

Further reading

Francis, J.K. (1990) Hymenaea coubaril *(L). Algarrobo, Locust.* SO-ITF-SM-27. Institute of Tropical Forestry, Rio Pedras, Puerto Rico.

Orwa, C., Mutua, A., Kindt, R., Jamnadass, R. and Simons, A. (2009) Agroforestree Database: a tree reference and selection guide version 4.0. Available at: http://www.worldagroforestry.org/resources/databases/agroforestree (accessed 27 March 2013).

Schmidt, L. (2008) *Hymenaea coubaril* L. Seed leaflet No 139. Danida Forest Seed Centre, Humlebaek, Denmark.

Vozzo, J.A. (2002) *Hymenaea coubaril* L. In: Vozzo, J.A. (ed.) *Tropical Tree Seed Manual.* Agriculture Handbook 721, US Forest Service, Washington, DC, pp. 518–519.

I

Intsia bijuga (Colebr.) Kuntze

Moluccan ironwood

Importance

Intsia bijuga, a mangrove-inhabiting species, produces a heavy hardwood, one of the most decay-resistant woods known, with the common name 'merbau'. Its physical and mechanical properties, high natural durability and attractive appearance make it suitable for many purposes. Over-exploitation has led to the species becoming seriously threatened in many parts of its range, and it is classified as vulnerable on the Red List of Threatened Species of the International Union for Conservation of Nature (IUCN). It is almost extinct in Sabah and is uncommon in Peninsular Malaysia, rarely reaching commercial size there. It is, however, more common in Indonesia and Papua New Guinea. There have been few efforts to develop commercial plantations, so demand for the timber is met almost exclusively by removal from the wild. With bark and leaves having medicinal properties, *I. bijuga* is held to be among the three most valuable plant species for a variety of local uses by communities in Papua New Guinea.

Botanical features

I. bijuga is a large tree and may reach up to 50 m in height with a dbh of 250 cm. The bole is often straight, sometimes crooked, grey-brown, dippled (i.e. with shallow depressions where bark has scaled off). The buttresses are 2–4 m high. Stipules are intra-petiolar and united.

Leaves alternate, pinnately compound with 2–3 opposite pairs of leaflets without terminal leaflet, rarely only one pair of leaflets. Leaflets broadly elliptic, ovate or obovate, rarely almost rounded, 2–18 cm in length, and 1.5–12 cm in width, base acute or rounded, apex blunt, rounded or notched, rarely pointed, leaflet stalk 2–6 mm.

Inflorescences 10–17 cm long, glabrous, in terminal or axillary racemose panicles or in clusters. Flowers bisexual, zygomorphic, pedicel 7–12 mm long, calyx lobes 4, elliptic, 6–16 × 4–7.5 mm; petal white turning to pink, red or purple, 6–10 × 10–15 mm with claw at lower half 5 mm in length. Fertile stamens 3 with red or purple filaments, staminodes (non-fertile stamens) 4–7, ovary on a stalk, style longer than stamens, red or purple. Fruit a pod, black or brown when ripe, oblong, flattened, 12.5–25 × 4–7 cm, valves woody, splitting open. Seeds oblong or flat, large, hard, without an aril, dark brown when mature.

I. bijuga has a hypanthium 6–16 mm in length, almost as long as the calyx lobes; in comparison *Intsia palembanica* has a shorter hypanthium, 3–4 mm in length, and shorter than the calyx lobes. *I. bijuga* fruits (pods) are 8.5–28 cm in length, whereas in *I. palembanica* they are 15–40 cm in length.

Species within the genus of *Intsia* tend to shed their leaves annually and remain bare for a few days. *I. bijuga* flowers and fruits throughout the year, while *I. palembanica* tends to be more seasonal, with flowering from January to April and fruiting in February–December and August–November.

Distribution

I. bijuga is the most widespread species within the genus, and is distributed from Tanzania and Madagascar, through southern India and Myanmar, to Malesia, northern Australia, Fiji and Polynesia. Exclusive natural stands of *I. bijuga* occur in Papua New Guinea on sites destroyed by fire during natural droughts. *I. bijuga* is also indigenous to New Caledonia. It is considered endangered in many areas in South-east Asia and almost extinct in others.

Plantations of *I. bijuga* have been successful in Java and Hawaii. Attempts have also been made to establish plantations in the Philippines and Kalimantan. This species has been planted in Tanzania, Zanzibar, Calcutta and Bogor, possibly for exhibiting in living plants collections in gardens and arboreta.

Silviculture characteristics

I. bijuga is a shade bearer. It can withstand drought, salt winds and termites. Seeds are classified as orthodox, and may retain viability for a long time. Experiments to establish planting stock from cuttings have had limited success. Some seedlings are raised in the nursery, and natural regeneration is abundant.

Silviculture practice

I. bijuga seeds are protected by a hard seed coat which is impenetrable to water. Mature seeds may survive more than 3 years without any treatment due to their moisture content being less than 10%. Seeds may take up to 2 years to germinate. Scarification and then soaking in water may promote germination. Immersing the seeds in concentrated sulphuric acid for 1 h is also an effective method to rupture the seed coat. Within natural stands in Papua New Guinea, dormant seeds may germinate abundantly in the gaps remaining after logging. Germination is preferable in full sun, and is epigeal. The cotyledon's food reserve contributes to seedling development up until the first pair of leaves develop. Vegetative propagation of *I. bijuga* is limited.

Appropriate drainage and high air humidity are suitable for seedling development. Seedlings may be transplanted into the field after approximately 3 months, at a spacing of 3 × 4 m or

5 × 5 m. Initial growth of the seedlings is fast, reaching an average height of 40–50 cm after 3 months. Later, seedling growth rate is reduced to only 5–6 cm in total in the subsequent 3–10 months. *I. bijuga* seedlings require high light intensity for optimal growth rates, and growth appears more rapid in the open than under closed canopy conditions. However, experiments in Indonesia showed that growth of seedlings under shade was faster than in full sunlight.

Commercial felling of *I. bijuga* mainly occurs in Papua New Guinea, where the logs tend to be larger. *I. bijuga* is less exploited elsewhere as it is confined to mangrove belts, which are often inaccessible, difficult to log due to the muddy conditions, and logging timber of an exploitable size tends to be unprofitable. For these reasons, in Malaysia and Kalimantan *Intsia palembanica* is preferable to *I. bijuga*.

As logging companies move through Papua New Guinea, it has become clear that forest management strategies need to be developed so that local communities can benefit from income opportunities arising from sustainable harvesting of *I. bijuga* and other profitable timber species.

Diseases and pests

I. bijuga is resistant to attack by *Polystictus versicolor* (*Trametes versicolor*).

The roots seedlings can become infected by nematodes. In the Philippines, *Rotylenchulus* and *Helicotylenchus* spp. have been found in nurseries, and are a moderate problem. *I. bijuga* wood is resistant to drywood termites and subterranean termites but is reported to be susceptible to powder-post beetle attack. In Kalimantan, young plantations suffer from browsing by deer and mouse deer, and rats are also considered serious pests. Reports of diseases and pests are mainly from plantations and nurseries in the Philippines, Kalimantan and Java. No precautions are known to be taken in order to avoid these problems.

Uses

I. bijuga is a suitable tree for soil conservation in eroded gullies and in buffer strips along creeks, and can be used to reclaim land degraded by gulley erosion. It is a useful shade tree for homes and parks, and being wind tolerant is ideal for use as a windbreak.

The timber is used in construction work, especially for high-class exterior joinery, such as windows, solid panel doors, framing and weatherboarding. It is also used for high-grade flooring for both light and heavy pedestrian traffic, and for hydraulic constructions such as bridges, wharves, sheet piles and sluices. It is used for furniture making, panelling, stairs, handrails, truck bodies, fence posts, poles, turnery, carvings and musical instruments.

The timber is heavy and hard, with a density ranging from 690 to 1000 kg/m^3 at 12% moisture content. The heartwood varies from light brown to orange-brown or dark red-brown. It darkens on exposure to air, and therefore is easy to differentiate from the lighter-coloured sapwood.

Working properties vary, but generally *I. bijuga* wood cuts cleanly in most operations. The green timber may be marred by an oily exudate and the saw teeth may become covered with a gummy substance during sawing. Sawing dried wood has a moderately blunting effect on cutting edges. Generally a smooth finish is obtained without giving any problems during turning and boring. The wood is liable to split, so pre-boring for nailing and screwing is advisable. *I. bijuga* wood is not suitable for cutting and slicing into veneer even after pre-steaming, because of its high density. It glues satisfactorily with most modern glues. The wood stains black in the presence of iron, even though it does not corrode iron. Wood of *I. bijuga* contains a colouring agent, which is 80% soluble and tends to seep out very freely, forming brown stains. To avoid staining, the wood is soaked in water for some days to remove the colouring matter.

I. bijuga produces a brown to yellow dye, which is obtained from an oily substance present in the wood and bark. The leaves are used in medicinal preparations in the Philippines, and have been shown to have high antioxidant activity. The bark and leaves are widely used in traditional medicines to treat rheumatism, dysentery, diarrhoea and urinary diseases. The seeds are edible after careful preparation.

Further reading

Cornejo, A.T. (1990) Ipil, *Intsia bijuga* (Colebr.) O. Ktze. *Research Information Series on Ecosystems* 2, 1–9.

Dayan, M.D.P., Reaviles, R.S. and Moraga, M.C. (2009) Rejuvenating the environment with indigenous forest tree species. *Canopy International* 35, 6–7.

Ser, C.S. (1982) Malaysian timbers - merbau. Trade Leaflet 65, Malaysian Forest Service, Jabatan Perhutanan, Malaysia, 5 pp.

Intsia palembanica Miq.

Ironwood

Importance

Intsia palembanica, like *Intsia bijuga*, produces a heavy hardwood with the common name of 'merbau'. Trade in this species has been especially important in Peninsular Malaysia and, since the late 1980s, in Indonesia. Elsewhere in Malesia, trade in the species is less important. It is used for a wide range of purposes because of its favourable physical and mechanical properties, combined with a high natural durability and an attractive appearance. Prospects for the species are good, as long as silvicultural practices can encourage natural regeneration and growth, and if sufficiently long rotations can be applied. As it is a rather slow-growing tree, plantations of the species are not likely to be successful.

Botanical features

A medium-sized to large tree to 45 m tall, with a bole to 150 cm diameter. Bole often slightly sinuous, branchless for up to 22 m, buttresses big, steep, plank-like. Crown with large ascending limbs, domed, deciduous. Bark smooth, developing large loose-hanging scales, fresh surfaces warm brown; middle bark bright green; inner bark when cut releases a slow-flowing oily exudate.

Leaves paripinnate, 4-jugate, rarely associated with some 2-, 3-, or 5-jugate; petiole and rachis (3.5–)9.5–17.5(–26) cm, both puberulous, glabrescent. Leaflets ovate, elliptic, sometimes broad–ovate or –elliptic, or –suborbicular, (3–)10–13.5(–18.5) × (3–)5–7.5(–10.5) cm; apex obtuse or rounded, sometimes slightly notched; base obtuse, rounded, often somewhat asymmetric; nerves 6–10 per side; petiolules 2–5 mm.

Inflorescence terminal or axillary, simple raceme, often fasciculate, or paniculate; up to ca 10 cm long, pubescent to glabrescent; pedicels 3–12 mm. Flowers bisexual, zygomorphic; hypanthium cupular, narrowly infundibuliform or cylindric, 3–4 mm long. Calyx lobes 4, elliptic, ovate or slightly obovate, 6–8(–12) × 3–6.5(–8) mm. Petal only 1 fully developed, flabellate, lower half narrowed into a claw, the others rudimentary or absent 3–6(–10) × 3–6.5(–8) mm. Disk absent. Stamens 3 or very rarely 4 fertile, filaments 2–4 cm long; anthers dorsifixed ca 3 mm long; staminodes 4–7 filaments and connate at base, 4–10 mm. Ovary stipitate, puberulous, 5–6 mm long; style 2–3.5 cm long; stigma small, capitellate.

Pods oblong, flattened, glabrous, 15–19(–40) × 5.5–7(–9) cm, dehiscent, 2-valved (valves leathery or slightly woody), often 3-seeded. Seeds ovoid, oblong, discoid, flattened, 3–4.5 × 1.5–1.2 cm, 0.5–1.2 cm thick, not arillate, scurfy, exalbuminous.

I. palembanica is deciduous, shedding its leaves at least once a year and remaining leafless for several days. It flowers at rather irregular intervals and the ripe pods fall entire, not opening in the tree.

Distribution

I. palembanica is naturally distributed in the Andaman Islands (India), southern Myanmar, Thailand, Malaysia, Singapore, Brunei Darussalam, Indonesia (except Java), the Philippines, and Papua New Guinea.

Silviculture characteristics

Growth of *I. palembanica* seedlings is fast initially, then it slows down. Seedlings need a high light intensity for optimal growth, and in the open, growth is faster than under a closed canopy. However, plant performance is better under slight shade than under full sun.

Because of its slow growth, the rotation age of stands is expected to be more than 60 years, if not up to 120 years. The diameter of the largest *I. palembanica* arboretum tree (in Malaysia) was recorded as 46.9 cm dbh at 45 years old, giving a dbh growth rate of about 1 cm/yr. On average, however, a slower growth rate is expected in a normal stand; dbh growth rates of between 0.38 cm and 0.58 cm/yr have been recorded in plantation trials.

Silviculture practice

I. palembanica has orthodox seeds which can be stored for up to 3 years or more in a cool dry place without decrease in germination rate. A seed moisture content of 8–10% and a storage temperature of between 5 and −20°C have been found to be optimal. There is a tough seed coat, so if untreated, germination may take up to 2 years or more. To promote germination, scarification followed by soaking in water is necessary. The most effective scarification technique is to use a file to scrape off the small protrusion on the seedcoat located at the opposite side of the hilum. Acid scarification using concentrated sulphuric acid for about 1 h also works. Once scarified, the seed will swell in water within hours, and begin to germinate. Vegetative propagation of *I. palembanica* has been tested in the Philippines with some success.

Little large-scale planting of *I. palembanica* has been attempted. Trial plantings in the Forest Research Institute Malaysia (FRIM) used plant spacings of about 2 × 2 m, 5 × 5 m, and 3 × 1 m with varying degrees of success. Direct sowing with scarified seeds can also be done. However, the seeds must be protected against predation by rodents. Good, well-aerated soils are needed in direct-sown plantings. Seedlings perform better in high humidity areas and in soil mixed with 25% sand. Seedlings can be transplanted into the field about 3 months after sowing.

There is no record of any maintenance of *I. palembanica* plots under a plantation situation. The first few years presumably require close maintenance in order to control common weeds such as *Imperata cylindrica*, *Mikania cordata* and ferns.

No thinning regime has been established for *I. palembanica*. In general, thinning should be done to promote rapid growth and maintain wind firmness. The basis of selection is to choose the dominant leaders (i.e. straight stems with general good form). Young seedlings tend to droop and branch very low, but once they reach a height of ca 10 m, they attain a better shape.

All production of *I. palembanica* comes from natural stands. Forest inventories in Peninsular Malaysia in the 1990s showed that there was still sufficient merbau timber in natural forests. The species is rather slow growing, and cutting cycles in natural stands should exceed 60 years; a rotation length of 80–120 years is probably more realistic. Annual log production figures were more or less stable for the period 1981–1996. The standing stock in Peninsular Malaysia was estimated to range from 1.46 to 2.23 stems/ha for stems >15 cm dbh in virgin forests, and from 1.2 to 2.58 stems/ha in logged-over and regenerated forests. The management system in Peninsular Malaysia – using the Selective Management System and with the previous Malayan Uniform System for natural forest – appears to be able to sustain the production of merbau logs. To restrict the flow of illegally logged *I. palembanica* into the timber supply chain, a methodology has been developed to track the unique genetic fingerprints of single logs from logging concession to timber mill.

Diseases and pests

There are no records of diseases for *I. palembanica*, but fungal infections often occur in injured cotyledons after scarification.

Rodents are known to feed on the germinating seeds when sown directly. This can be controlled by planting the seed inside a bamboo column.

Uses

Merbau is a very good general-purpose timber. It has favourable physical and mechanical properties, combined with a high natural durability and an attractive appearance. It is classed as a heavy hardwood, with a density ranging from 515 to 1040 kg/m^3 when air dry. The sapwood is pale yellow to light buff and is sharply differentiated from the heartwood, which is yellowish to orange-brown when fresh, darkening to brown or dark red-brown on exposure. Sapwood is not durable and often is not used together with the heartwood. The timber is strong and has good working qualities. The timber seasons slowly without any degradation. Merbau is difficult to cut or slice into veneer, even after pre-steaming, mainly due to its high density. When in contact with water, water-soluble extractives create blackish-lined, dark-coloured spots on the surface which are very difficult to remove. The heartwood shows exceptional resistance to fungal and termite attacks.

Merbau is used for construction work in house building, especially for high-class exterior joinery such as windows, solid panel doors, framing and weather-boarding. It is also an excellent timber for high-grade flooring for both light and heavy pedestrian traffic. Other uses include furniture making, panelling, stairs, handrails, shop fittings, truck bodies, turnery, poles, fence posts, musical instruments and carving. It is also particularly useful for structures in contact with water such as bridges, wharves, sluices and sheet poles, and it is generally free from ship worms.

Young seeds of *I. palembanica* are sometimes eaten, after first being soaked in salt water for 3–4 days and then boiled. A brown to yellow dye is obtained from an oily substance present in the bark and wood. Bark and leaves are both used medicinally. Heartwood extracts have recently attracted the attention of the pharmaceutical industry as sources of skin-whitening and anti-acne agents.

Further reading

Grace, J.K. and Tome, C.H.M. (2005) Resistance of the Indonesian woods bangkirai (*Shorea laevis*) and merbau (*Intsia palembanica*) to Formosan subterranean termite attack. *Sociobiology* 45, 503–509.

Sasaki, S. and Ng, F.S.P. (1981) Physiological studies on germination and seedling development in *Intsia palembanica* (merbau). *Malaysian Forester* 44, 43–59.

Wong, K.N., Tan, W.L. and Chew, F.T. (2009) Identification and characterization of microsatellite loci in *Intsia palembanica* (Leguminosae), a valuable tropical timber species. *Molecular Ecology Resources* 9, 360–364.

Irvingia gabonensis (Aubry-Lecomte ex O'Rorke) Baill.

Bush mango, wild mango

Importance

Irvingia gabonensis is a tree from the wet forests of Central Africa (Guinea–Congolian region). Mature emergent trees are common in low-lying areas; however, small fertile trees are found on higher land in drier forests. It is considered rare. The tree is commonly preserved as a shade tree on farms (e.g. with cocoa and coffee).

Fruits and kernels are both eaten locally. Fruits are only traded locally, but kernels are widely traded both domestically and internationally. The purported health value of the kernels in dietary treatments for obesity has greatly increased this trade.

The wood is hard and durable, but has almost no international market.

Botanical features

I. gabonensis is unrelated to the true (Asian) mango. It is a straight, slender-stemmed, densely crowned deciduous tree reaching 40 m tall and 1 m in diameter, with narrow buttresses up to 6 m high. The bark is smooth to scaly, grey to yellow-grey.

Flowers are bisexual, with five 3–4 mm long, yellowish-white petals. The fruit is a drupe, up to 7 cm long, smooth, yellow when ripe. The orange pulp is sweet and fibrous and has a single 2–3-cm long hard seed. Two varieties have been distinguished: var. *gabonensis* which fruits in the wet season, and var. *excelsa* which fruits in the dry season.

The fruiting trees are known to attract animals such as elephant and bush-pig.

Distribution

I. gabonensis is indigenous to the humid African lowland forests, from coastal Liberia to the Central African Republic, and south to Cabinda (Angola) and the westernmost part of DR Congo; it also occurs in Uganda, and in São Tomé et Príncipe. It is sometimes planted, especially in West Africa (Cameroon through to Côte d'Ivoire). It prefers well-drained sites below 1000 m altitude and an annual rainfall from 1500 to 3000 mm, and is absent from swampy areas.

Two to three trees often grow together, and in some areas it is reported to be gregarious. The presence of *I. gabonensis* is often associated with former human habitation. Trees are sensitive to fire.

It is classified in the Red List of Threatened Species of the International Union for Conservation of Nature (IUCN) as a lower-risk species, but is close to being 'Vulnerable'. Its sustainability is mostly threatened by loss of natural forest, but this is compensated for by selective retention of the highly valued trees, as well as by domestication efforts.

Silviculture characteristics

Until recently there had been little effort towards domestication. More recently, recognizing its high local value, the International Centre for Agroforestry declared *I. gabonensis* a subject for domestication and genetic improvement and coordinated a systematic collection of good-quality seeds, with emphasis on south-east Cameroon and northern Gabon, the centre of its diversity. Research into germination and propagation for commercial development has led to standardizing of nursery practices, and successful propagation by bud-grafting and marcotting.

It is relatively tolerant of shade.

Silvicultural practices

I. gabonensis is mainly propagated by seed though transplanting of wildings; retention and protection of wildings when clearing land for agriculture are also common.

Farmers select seed from trees which bear large fruits, with good taste, high yield, regular production (every year), early maturity and easy kernel extraction.

Seeds are extracted from the fruit and dried for at least 2 days before being planted in containers. Germination takes more than 14 days and a germination rate of 80% can be reached.

Several methods of vegetative propagation have also been developed: letting leafy stem-cuttings root under mist; micropropagation; grafting; and marcotting. Plants from marcotts can fruit 2–2.5 years after transplanting.

In many areas *I. gabonensis* occurs in wild stands or is retained in plantations of cocoa, coffee or annual food crops, or in home gardens. Management tasks mostly include pruning, harvesting (gathering and picking) and fertilization.

I. gabonensis fruits are gathered from beneath each tree, or harvested by climbing shorter trees. After extraction the kernels are dried in the sun or on drying racks over a fireplace.

Diseases and pests

Seeds are infested by larvae of the merchant grain beetle (*Oryzaephilus mercator*). Preventing cracks in the seeds helps to reduce infestation.

Since 2004, the fruit fly *Bactrocera invadens*, of possible Sri Lankan origin, has infested many West African fruit trees including *I. gabonensis*. This pest poses a serious threat to the region's horticulture and, due to concerns about further spread, has led to restrictions on international trade.

Uses

The tree has been planted widely and is popular in agroforestry.

The wood is used in heavy construction and has a density of 760 kg/m^3. Local people generally value these trees more for their non-wood values, and therefore tend not to cut them down. However, the wood is occasionally used locally for heavy construction work, for decking on fishing boats, paving blocks, planking, canoes, pestles and household utensils. Young trees are used for making poles and stakes, while branches are made into walking sticks or thatched roof supports. Branches are used as firewood.

The fruit pulp of ripe *I. gabonensis* is juicy and sweet, and is eaten fresh. People prepare juice, jelly, jam and wine from the fresh fruits and the pulp has also been used as a (black) dye for cloth. Kernels of *I. gabonensis* yield an important food additive for people in West and Central Africa. The kernels are ground and crushed, and then used to thicken soups and stews. The kernels can also be made into a cake (called 'dika bread') for preservation and easy use. Oil extracted from the seed is used in cooking. It solidifies at ambient temperatures and is used as a substitute for cocoa butter in soap-making. The press cake is still suitable for thickening soup or as cattle feed.

Considerable commercial attention has been raised by trials suggesting that an extract prepared from the kernel helped obese people to lose weight and lower their blood pressure. Demand is increasing because of this, although clinical evidence data remain weak and questionable.

I. gabonensis has many traditional medicinal uses. Preparations from the bark are rubbed on to the body to relieve pains, are applied to sores and wounds and used against toothache. Shavings of bark are also used to stop diarrhoea and dysentery. In Cameroon, bark preparations are used to treat hernias and yellow fever, and as an antidote to poisoning. Powder prepared from the kernels is applied to burns.

Further reading

Atangana, A.R., Tchoundjeu, Z., Fondoun, J.M., Asaah, E., Ndoumbe, M. and Leakey, R.R.B. (2001) Domestication of *Irvingia gabonensis*: 1. Phenotypic variation in fruits and kernels in two populations from Cameroon. *Agroforestry Systems* 53, 55–64.

Ayuk, E.T., Duguma, B., Franzel, S., Kengue, J., Mollet, M., Tiki Manga, T. and Zenkeng, P. (1999) Uses, management and economic potential of *Irvingia gabonensis* in the humid lowlands of Cameroon. *Forest Ecology and Management* 113, 1–9.

Shiembo, P.N., Newton, A.C. and Leakey, R.R.B. (1996) Vegetative propagation of *Irvingia gabonensis*, a West African fruit tree. *Forest Ecology and Management* 87, 185–192.

Tchoundjeu, Z. and Atangana, A.R. (2007) *Irvingia gabonensis* (Aubry-Lecomte ex O'Rorke) Baill. Record from Protabase. Van der Vossen, H.A.M. and Mkamilo, G.S. (eds) PROTA (Plant Resources of Tropical Africa / Ressources végétales de l'Afrique tropicale), Wageningen, The Netherlands. Available at: http://www.prota4u.org/ (accessed 30 November 2012).

J

Juglans nigra L.

Black walnut

Importance

Juglans nigra is regarded as one of the most valuable hardwood species in North America and one of the world's premier timber species, internationally renowned for the manufacture of high-quality furniture. It is found in natural mixed hardwood stands, growing well on fertile sites, but the high value of the timber has prompted the establishment of extensive plantations in temperate regions around the world.

Botanical features

J. nigra grows up to 50 m tall with diameters up to 1.5 m. In forests it develops a long clear trunk with little taper and a small rounded crown, but the trunk forks low down in the open. The root system usually includes a deep taproot and several wide spreading and deeply set lateral roots. The bark is deeply furrowed. Leaves are compound, deciduous, 30–60 cm long, with 15–23 ovate to lanceolate leaflets, irregularly serrate, 6–12 cm long, glabrous above, somewhat glossy, pubescent and glandular beneath, and the terminal leaflet often very small or absent. Flowers are unisexual, appearing with the leaves in separate inflorescences on the same tree; the male in stout catkins 7–12 cm long, the female in 2–5 flowered spikes. The round fruit is solitary or in pairs, 4–5 cm across, thick shelled with a rough hull, the nut coarse and irregularly, longitudinally furrowed.

Distribution

The natural range of *J. nigra* covers much of the eastern USA and into southern Canada. It has also been established in central Europe, New Zealand, and Australia.

Silvicultural characteristics

Within the native range, the growing season is 140–280 days (from Canada to Florida), annual precipitation is 635–1780 mm, and mean annual temperatures of 7–19°C. Optimal growth is seen when mean annual temperatures are about 13°C, with 170 frost-free days and mean annual rainfall of 890 mm, with an absolute low of −43°C recorded. Although *J. nigra* is reported to survive absolute lows of −3°C, winter damage occurs at much higher temperatures, causing die-back on some branches and late flushing, and sites prone to out-of-season frost and strong winds should be avoided. Wind barriers have beneficial effects on seedling establishment by reducing wind velocity, solar radiation and raising air temperatures and can extended the growing season by 1 month. *J. nigra* is generally found as individual trees, scattered, or in small groups in mixed broadleaved stands, and the best opportunities for walnut management may occur in these stands rather than in plantations.

Silviculture practice

Poor site selection is a common cause of unsuccessful hardwood plantations, but even on suitable sites, inadequate site preparation and maintenance can lead to crop failure. Cultivation (ploughing and discing) followed by chemical weed control has provided the best height growth compared with other site preparation techniques. Both the planting of nursery raised seedlings and use of direct seeding have their advantages and disadvantages. Vegetative propagation techniques including cuttings, grafting, and budding are also commonly used.

Weed control is the most important factor after site selection. Most hardwood species of high timber value will only grow well in deep fertile soils, which are also capable of supporting dense weeds and grasses. It is also important to remember that *J. nigra* is intolerant of shade, so if planted with other trees, the growth rates should be similar to maintain the tree in a dominant or co-dominant position.

J. nigra is susceptible to weed competition and late frosts until it reaches a certain size; fertilization at the time of planting increases the early growth of seedlings on poor soils and therefore minimizes the risk period, but the growth does not equal that on good sites. Fertilizer application can also increase nut production, with treatments containing P with N and/or K being most effective. Fertilizer has also been applied to older trees.

Irrigation may be necessary where drainage is excessive, or where soil volume suitable for root development is limited by soil depth or impermeable layers. Unless other factors are limiting (and this is often the case), irrigating such sites may help to maintain tree growth during periods of inadequate precipitation and/or high evapotranspiration. In general, trees need about 2.5–4 cm of rain per week.

Pruning should start when trees are 3–4 m tall and should leave at least one-half of the tree stem with branches, with leaf area being reduced by no more than 25% in any 1 year. Eventually, pruning to a height of 6 m is necessary to maximize the production of defect-free wood, and should be carried out in winter when it can be combined with form pruning.

Initial spacing between 3 × 1.5 m and 3 × 3 m are recommended for plantations, or 10 × 10 m to 14 × 14 m in intercropped agroforestry. Planting at wider spacing in conjunction with agricultural crops allows the establishment of *J. nigra*, which helps to obtain revenues during the long rotation period, and improves growth rate and wood quality. Planting 10 times more trees than are needed for final harvest can increase the value of that harvest more than 2.5 times, with an initial stocking of 1000–1250 stems/ha decreasing to a final crop of 60–70 stems/ha; however, an initial stocking of 600 stems/ha is suggested for sheltered sites.

Growth rates in North America are similar to those in Europe, and may achieve a mean annual volume production of 8–10 m^3/ha, and an annual height growth of 0.5 m/yr over a 60–70 year rotation. However, there are reports of the production of veneer logs after only 35–40 years. Similar results are found whether in pure stands, or mixed with other hardwoods such as chestnut (*Castanea* spp.) or oaks (*Quercus* spp.). Interplanting *J. nigra* with nitrogen-fixing species can also provide competitive returns.

Diseases and pests

J. nigra is attacked by a variety of diseases and pests, including many of those that are known to attack *Juglans regia*. It has proved to be very susceptible to thousand cankers disease, a pest complex caused by the walnut twig beetle (*Pityophthorus juglandis*) and the fungus *Geosmithia morbida*.

There have been few reports of any serious disease problems in natural woodlands, although stand conditions in plantations may increase abiotic injuries and weaken trees, leading to increases in pathogen populations. Walnut anthracnose (*Gnomonia leptostyla*) is the most common foliar disease infecting *J. nigra* and is a major cause for poor and irregular nut production. Insect pests include walnut caterpillar (*Datana integerrima*), regarded as the most destructive leaf feeding insect, with fall webworm (*Hyphantria cunea*) another common defoliator. Walnut shoot moth (*Acrobasis demotella*) is considered the most destructive shoot borer; ambrosia beetles (*Xylosandrus germanus*) develop galleries in the stem; and black walnut curculio (*Conotrachelus retentus*) attacks the nuts.

Uses

As well as small-scale forestry plantings, it has also been established as amenity plantings in many temperate countries, also as *Juglans* × *intermedia*, a natural hybrid of *J. nigra* with *Juglans regia*. *J. nigra* has also been used extensively as rootstock for *J. regia* in Europe and parts of the USA.

The heartwood varies from light brown to dark brown to dark purplish-brown, getting darker with age, sometimes with purplish streaks. There are 40 different wood grain patterns recognized, which, although it is usually strongly figured, can sometimes be very plain. *J. nigra* is a medium-density hardwood with a basic density range of 479–577 kg/m^3 when dry, or 908 kg/m^3 when green. The wood is heavy, hard, strong and stiff, has good shock resistance, and is one of the most durable of any American hardwoods. *J. nigra* has a slow, variable drying rate, and it may collapse and check internally if kiln dried from green, so air-drying to 30% moisture content is recommended. It works easily with hand or machine tools, takes paints and stains exceptionally well, is readily polished and easily glued. Wood properties such as colour toughness that also influence timber value are affected by site conditions, and faster-growing plantation trees may contain approximately 20% less heartwood compared to naturally grown trees. The wood finishes beautifully with an outstandingly handsome grain colour and pattern, and is used for furniture, cabinet-making, veneers, turnery, carving, gunstocks and small turned items. Logs are exported from the USA to Europe and Asia.

J. nigra nuts are edible, oily and sweet with a full, rich and aromatic flavour, and its potential as a nut-producing tree crop in the USA has long been recognized. More than 700 nut-producing cultivars are identified. Nuts are occasionally used as a feed for pigs.

The inner bark has mild cathartic properties, sugar is made from the sap and green nut husks are used to dye cloth yellow or orange. The juice from freshly macerated unripe hulls is used in folk medicine as a treatment for localized, topical fungal infections such as ringworm.

Further reading

Jones, J.E., Mueller, R. and Van Sambeek, J.W. (2000) *Nut Production Handbook for Eastern Black Walnut*. Southwest Missouri Resources, Conservation & Development (RC&D), Inc., Missouri, USA, 147 pp.

Reid, W., Coggeshall, M. and Garrett, H.E. (2007) *Growing Black Walnut for Nut Production*. University of Missouri Center for Agroforestry, Columbia, Missouri, USA, 16 pp.

Williams, R.D. (1990) *Juglans nigra* L. Black Walnut. In: Burns, R.M. and Honkalla, B.H. (eds) *Silvics of North American Hardwoods, Vol. 2*. Agricultural Handbook 654, USDA Forest Service, Washington, DC, pp. 391–399.

Juglans regia L.

Walnut

Importance

Juglans regia is an economically important, multipurpose tree, yielding high-value walnuts and a decorative, sought-after timber. In the past, most interest was in the fruit, and orchards are still common, but plantations are increasingly grown for timber alone. *J. regia* is also suitable for mixed orchards and timber plantations, and intercropping in agroforestry systems. It was one of the most valuable trees in its native range from Europe to China, widely introduced, and is now grown in temperate regions around the world.

Botanical features

J. regia is a broadleaved deciduous tree to a height of 25–30 m and diameter of 150 cm, with a straight cylindrical stem and a large, rounded crown. Bark is smooth, white-silver, becoming longitudinally fissured with age but does not scale. Lateral roots can grow to more than 3–4 times the crown diameter. Leaves are alternate, compound, imparipinnate, aromatic, 35 cm long, each with 7–9 light green leaflets, obovate or elliptical, acute or acuminate, 6–15 cm long. Leaflet margin is almost entire, sometimes slightly dentate. Petiole without stalks, enlarged at the base. Flowers are monoecious, in catkins. Male catkins, 6–8 cm long, are pendent, greenish, many-flowered. They develop on twigs of the previous year. Female catkins develop at the top of current-year shoots, in terminal racemes with fewer flowers. The fruit is an ovoid drupe; 30–40 mm long and 25–30 mm in diameter. The epicarp (hull) is green, rich in tannins and fleshy. The woody endocarp (nut) contains the seed (kernel), rich in oils, divided into 2–4 deeply rugulose lobes.

Distribution

J. regia has been widely introduced, cultivated, and naturalized in Eurasia for thousands of years. It is difficult to distinguish the primary natural distribution, but this is generally considered to stretch from south-eastern Europe to south-western China. It includes the Balkans, Asia Minor, the Caucasus, Central Asia, the Himalayan foothills, and Xinjiang and Xizang (Tibet), China.

It is today found in most of Europe except for northern regions (Scandinavia, Baltic states and Russia); in Asia, it has been introduced and cultivated, particularly in Turkey, India and China. It has also been introduced into temperate regions of North Africa and South Africa, North America, South America, Australia and New Zealand.

Silviculture characteristics

J. regia grows in climatic conditions from dry and hot climates in North Africa to continental climate of central Russia and Canada. Rainfall during the growing season should be at least 700–800 mm, and plantations need 5 months with a mean temperature of more than 10°C and benefit from hotter summer temperatures. Resistance to low temperatures depends on tree age, ecotype, and vegetative phase. Young trees are damaged by late spring frosts, and −3°C may affect the foliage. When dormant, adult trees can tolerate low winter temperatures to nearly −30°C.

It is considered to be a fast-growing tree, and longevity is 150–200 years. *J. regia* is a light-demanding species, and young trees are sensitive to weed competition. The apex of adult trees needs to be isolated and exposed to light, or tree architecture can be affected. *J. regia* is not resistant to wind or fire. Trees regenerate by suckers and are suited to pollarding.

Silviculture practice

J. regia has been grown for both fruit and timber for centuries and has naturalized in many places. Natural and artificial selection and improvement has resulted in many varieties, so local provenances should always be used.

For propagation by seed, ripe nuts are collected, the hulls removed, and seeds air dried in the shade. Storage is possible only until the following spring if stratified at low temperatures in moist sand, peat moss or vermiculite/perlite. Seeds are sown 3.5 cm deep in well-prepared seedbeds, and carefully tended. One-year-old plants with a good taproot system are preferred, but 2-year-old and bare-rooted plants are also used. Direct sowing is promoted to ensure no damage to taproots, but losses to predators are usually considerable. Grafting is also commonly used to produce improved planting stock.

If planting in winter, soil preparation must be carried out during the previous summer, with tillage to 80–100 cm depth and fine harrowing. Planting must be carried out when the soil is not frozen or too wet, and application of manure is recommended if the soil has less than 2% organic matter. Tending during the early years after planting is important, especially weed control by tillage and mulching. Seedlings will need protecting with fences or tree guards where wild animals or livestock are present. Early pruning is very important for high-quality timber production, especially for straight, clear, knot-free veneer logs. Stand conditions dictate the timing of thinning and felling, including tree spacing, pruning history, management practices, and environmental site conditions.

Cultivation by small landowners and farmers commonly combined nut production with a subsequent timber harvest, obtained through use of a particular pruning technique. Flagpole pruning gives a rounded crown producing a large quantity of fruits, above a 2.5–4 m high branchless stem yielding high-quality timber. However, the form of modern nut varieties have much lower crowns which makes pruning and fruit collection easier, but is not so suitable for high-quality wood production.

On appropriate sites, annual girth growth rate is 3–3.5 cm/yr up to 60 years, or up to 5–6 cm in exceptional conditions. A plantation of 60 trees/ha in France, pruned up to 3–4 m of stem height, had a mean annual volume production of 1 m^3/ha of merchantable timber on a 60-year rotation.

Diseases and pests

Walnut blight caused by *Xanthomonas arboricola* pv. *juglandis* is a major disease of nuts and foliage. Deep bark canker is caused by *Erwinia rubrifaciens*; it results in large cracks in the bark which ooze a reddish-brown mixture of bacteria, dead cells, and plant sap. A soilborne bacterium, *Agrobacterium tumefaciens* (*Rhizobium tumefaciens*) causes galls that can eventually girdle a walnut tree. Walnut anthracnose is a foliar disease caused by *Gnomonia leptostyla*. At least 13 species of *Phytophthora* cause Phytophthora root and crown rot leading to the decline or death of trees. *Armillaria mellea* is the causal agent of oak root fungus disease and also causes root and crown rot. Butternut canker caused by *Sirococcus clavigignenti-juglandacearum* leads to disease and death of branches, bark discoloration and cankers on the twigs, branches and trunks. Viral diseases include blackline disease caused by a strain of Cherry leaf roll virus, leading to a lethal girdle at graft unions. Walnut bunch disease caused by a mycoplasma-like organism is characterized by clusters of wiry twigs on trunks or large branches of infected trees.

Reported nut pests include the walnut worm (*Cydia pomonella*) and navel orangeworm (*Amyelois transitella*) which make nuts unmarketable; maggots of the walnut husk fly (*Rhagoletis completa*) feed on the husk, and scale insects such as *Epidiaspis leperii*, *Diaspidiotus perniciosus*, and *Mytilococcus ulmi* promote the growth of sooty mould and the susceptibility of exposed nuts to sunburn. Damage to foliage is caused by the walnut blister mite (*Eriophyes erineus*) and aphids such as *Chromaphis juglandicola*, and roots are damaged by the nematodes *Criconemella xenoplax* and *Pratylenchus vulnus*.

Uses

Trees are used for shade, as roadside trees, and are often found in mixed stands, either with other fruit or timber trees, or intercropped with cereals.

J. regia wood is dark, often with very attractive figure, and the large burrs on occasional trees yield particularly valuable veneer. In the timber trade, a distinction is often made between 'clear' and 'coloured' walnut wood. *J. regia* wood has a density of 670 kg/m^3 at 15% moisture content; is strong without being too heavy; withstands considerable shock; is very stable, scarcely warping or checking; and after proper seasoning it shrinks and swells very little. It is easy to work and has a uniform and slightly coarse, silky texture. Both veneer and sawnwood may be treated by vaporization, a steam treatment that softens the wood and protects it against insect attack.

Straight logs with no defects, more than 2.5 m long and diameters more than 35 cm are sold as veneer logs. Logs not used for veneers, due to short lengths or defects, are used for sawn timber. This can be used for a wide range of products and is highly valued for fine furniture and marquetry work. Veneers are used for interior furnishing and furniture; solid walnut wood is also used for furniture, as well as joinery, wood carvings, small boxes, rifle butts, billiard cues and walking sticks.

The valuable nuts are eaten fresh or dried, and used in confectionery. They yield an edible oil used in cooking, and oil to make furniture polishes and specialist paints. Walnut husks have high tannin content, once used for the production of natural dyes and in the pharmaceutical industry. Leaves also have traditional medicinal properties, and many famous liqueurs are made with the nuts.

Further reading

Becquey, J. (1997) *The Cultivation of Walnut Trees for Timber [Les Noyers à Bois]*. 3rd ed. Institut pour le Developpement Forestier (IDF), Paris, 143 pp.

Germain, E., Prunet, J.P. and Garcin, A. (1999) *Le Noyer [Walnut.]*. Centre Technique Interprofessionnel des Fruits et Legumes (CTIFL), Paris, 279 pp.

McGranahan, G. and Leslie, C. (1990) Walnuts (Juglans). *Acta Horticulturae* 290, 905–951.

Taha, N.A. and Al-Wadaan, M.A. (2011) Utility and importance of walnut, *Juglans regia* Linn: a review. *African Journal of Microbiology Research* 5, 5796–5805.

Juniperus procera Hochst. ex Endl.

African pencil cedar

Importance

Juniperus procera is of exceptional interest as being one of only a few indigenous African conifers. It is illustrative of a southerly migration of northern hemisphere genera by way of the temperate rainforest belts of tropical mountain ranges, such as those of East Africa, South America, and the Indo-Malayan region, which thus form migration highways for temperate-zone genera. It is the largest juniper species in the world (up to 50 m) and produces valuable timber, being strong, easy to work, extremely durable and resistant to fungi or insect attack due to the oleoresins present. It is known commercially as the African pencil cedar because its wood closely resembles that of *Juniperus virginiana*, used in the pencil-making industry. From the late 1800s onwards, considerable quantities have been exported from East Africa (especially from Kenya) to Europe. However, *J. procera* has a genetically fixed fluting of the stem, and with frequent occurrence of heart rot and of ingrowing bark, this considerably reduces the proportion of extractable clear timber. Over-exploitation and land clearance for agriculture have reduced the original population, and the species is now threatened with extinction, but is still in great demand and the wood fetches high prices because of its excellent properties. Natural forests remain the major source of supply and there is urgent need for knowledge about the ecology of *J. procera*, to allow its conservation and/or sustainable use.

Botanical features

J. procera is a tall, evergreen tree up to 50 m in height, with a straight trunk up to 300 cm in dbh. It is usually free of branches for 15–20 m, sometimes buttressed and often with a pronounced twist. In early years the tree is pyramidal in form, becoming wide-spreading at maturity with a round-topped crown 15–20 m broad, made up of thick branches clothed with light green foliage. In pure forests trees are tall with less massive trunks and smaller crowns. Bark pale brown to reddish-brown, thin, fibrous, with thin shallow longitudinal fissures, exfoliating in thin papery strips (Plate 68). Knowledge is lacking about the characteristics of the root system. Leaves are simple, opposite–decussate and sessile. Juvenile leaves are needle-shaped, prickly, up to 2.5 cm in length; mature leaves are dark green; scale-like; triangular; sharply tipped; paired and pressed close to the branchlets with a visible resin gland on the back.

J. procera is a dioecious species, with irregular flowering and fruiting periods. It is wind pollinated and seeds are spread by birds. Male flowers are rounded or ellipsoid, 2–4 mm long, yellowish and solitary on short axillary branchlets. Female flowers are rounded, not more than 1 cm in diameter and reddish-brown to purplish-black with blue bloom when mature. Soft, berry-like cones are produced towards the end of the branches (Plate 62). They are bluish-grey at first, turning brown when ripe and with a waxy bloom. They contain one to four pyramidal, hard woody seeds embedded in a greenish pulp.

Distribution

The natural occurrence of *J. procera* begins in the western Arabian Peninsula (Saudi Arabia and Yemen), then extends southwards across the Ethiopian highlands and throughout Eastern Africa, eventually reaching Zimbabwe and the north of South Africa. The tree grows in mountainous areas, highland forests, and forest remnant areas. *J. procera* is an important component of the undifferentiated Afromontane and dry single-dominant Afromontane forests. It can also occur in transitional forests between dry single-dominant Afromontane forest and semi-evergreen bushland and thicket.

In these mainly evergreen forest types, there is commonly dense, evergreen undergrowth of shrubs and herbaceous plants, among which ferns can occur in places and the trees are sometimes covered with moss and pendulous lichens; climbers are common, epiphytic figs are occasionally found, and grass occurs in quantity in open places.

Silviculture characteristics

J. procera is a prolific seed-bearer and the seeds, though sometimes damaged by seed-boring insects, are usually very fertile. Despite this, little regeneration is observed in mature juniper forests and, under pure forest conditions, seedling establishment is hindered because the species is light-demanding and dislikes humus on the ground. Germination is, therefore, more common in open, grassed areas or among shrubs, such as in glades or forest edges where there is adequate light and mineral soils. Consequently, juniper seedlings are mainly found under the shelter of bushes such as *Philippia* spp. on forest edges, and progressively older trees are found further in the forest. Natural regeneration also occurs after a fire or in other large clearings, giving rise to tall, even-aged stands of trees with long, branch-free boles. Those nearly pure stands last longer than the type arising from forest spread.

J. procera thrives best at higher altitude and in higher rainfall areas, though it can survive in hot and/or dry conditions once established. Its adaptability and timber value make it a good plantation species. However, slow growth and poor survival in the field (due to attacks by insect borers and heart rot), has discouraged cultivation in its native range. Based on measurements in planted stands, the average growth potential is 7.5 m^3/ha/yr, considerably less than some exotic species on the same sites.

Silviculture practice

There are 42,750 seeds/kg on average, and germination in nurseries is usually around 40% after 6 weeks. Seedlings of 1–2 years old should be carefully protected and planted in dense formations, a 2 × 2 m spacing being preferred to promote self-cleaning in the extremely branchy thicket-stage. *J. procera* is suitable for propagation through vegetative means; stecklings (rooted cuttings for outplanting) with well-developed root systems establish easily and grow well, and rooting is favoured if cuttings are obtained from juvenile stock plants (1–2 years old).

Pruning can significantly increase the timber production of a stand. Thinning operations increase light through the canopy, thereby improving germination conditions and allowing natural regeneration. Early selective thinning is also recommended to enhance crown development and diameter growth.

J. procera can be grown in plantations, homesteads, or at boundaries. It is a hardy, adaptable species, with some resistance against drought periods, diseases and pests, and produces valuable high-quality timber. Weeding during early stages and pruning during establishment should be carried out. The species is not recommended for intercropping due to acidic substances in decomposing leaves.

Diseases and pests

The living wood of *J. procera* is subject to attacks of a wood-rot fungus, *Fomes juniperinus*, which is said to be identical with the species that attacks *Juniperus* spp. in North America. The fungus creates cavities of various sizes and in serious infestations a large tree may be reduced to a mere shell. Mature trees or trees growing on humid sites are particularly susceptible to the fungus. Measures to prevent damage include:

1. Strict protection from fire and other injury.
2. Maintenance of dense stands or suitable permanent mixtures to favour natural pruning while trees are still young before heartwood formation.
3. Removal of broken branches or wounds with exposed heartwood, and removal of those already attacked by the fungus.

J. procera can be affected by the cypress aphid (a species complex composed of two separate species, *Cinara cupressi* and *Cinara cupressivora*). This aphid sucks sap, causing yellowing of leaves on the affected twig; the level of damage depends upon factors such as stress and climate, but host die-back and sometimes death can result.

Uses

Replanting of *J. procera* is carried out in deforested areas for soil conservation or improvement, to protect from erosion, or even to try to convert certain poor sites into woodlands, as done in Eritrea, Ethiopia and Kenya. The species is well adapted to these purposes because it can establish in extreme conditions. However, this practice is still little used, mainly because it does not provide direct income to the countries concerned, where needs are urgent.

The wood is of light to medium weight, with a basic density varying from 450 to 650 kg/m^3. It is soft and straight-grained with growth rings not plainly marked, and a fine and even texture. It seasons well with care, is easy to work, but can split in nailing and is rather brittle, but fissile. It has a characteristic and persistent aromatic smell. The sapwood is white and the heartwood is variable in colour when freshly cut, generally light red, sometimes almost purple, turning reddish-brown on exposure. Based on a study in an Ethiopian highland forest, it was proved that *J. procera* forms annual growth rings in response to precipitation. The timber is also widely used where native for many purposes including building (both construction and lining), joinery of all kinds and flooring. The wood burns evenly when fresh, and is much appreciated as a firewood and for making charcoal.

Cedar wood oil, distilled from the wood or sawdust, is used in the cosmetic industry. Bark, leaves, twigs and buds are used locally for medicinal purposes.

Further reading

Orwa, C., Mutua, A., Kindt, R., Jamnadass, R. and Simons, A. (2009) Agroforestree Database:a tree reference and selection guide version 4.0. Available at: http://www.worldagroforestry.org/resources/databases/agroforestree (accessed 27 March 2013).

Pohjonen, V. and Pukkala, T. (1992) *Juniperus procera* Hocht. ex. Endl. in Ethiopian forestry. *Forest Ecology and Management* 49, 75–85.

Yoshikawa, K. and Yamamoto, F. (1995) *Juniperus procera* forest in Saudi Arabia. *Tropical Forestry* 34, 14–23.

Juniperus virginiana L.

Eastern redcedar

Importance

Juniperus virginiana is a common coniferous species growing on a range of sites throughout the eastern USA. On good sites it can, exceptionally, reach up to 37 m tall and 160 cm dbh. Trees over 500 years old are known. The wood is highly valued because of its beauty, durability and workability. The number of trees and volume of *J. virginiana* are increasing throughout most of its range. It provides cedarwood oil, food and shelter for wildlife, and protective vegetation for fragile soils. It is among the first tree species to invade abandoned fields and cleared land, and in many areas is considered a noxious invasive species.

Botanical features

A tree generally around 20 m in height and 100 cm in diameter, and very varied in habit. Crown on young trees is elongated ovate, with later branches horizontally arranged or nodding. Older trees have wide, fluted, buttressed boles. Bark is grey to red-brown, thin, fibrous and shreds easily.

On young plants leaves are always needle-like, 3–6 mm long, prickly acuminate, opposite, and on stout shoots and young plants in whorls of 3. Scale-like leaves decussate, 1–2 mm long with a sharp point.

J. virginiana is dioecious, reaching sexual maturity at about 10 years old. Staminate strobili or conelets develop on male trees at the tips of axillary branches of new scale-leaves. Pollen grains form in conelets having 10–12 entire-margined sporophylls. Staminate strobili turn a conspicuous yellowish-brown when they reach maturity during winter. Small green conelets begin to develop by early autumn on ovulate trees, but grow very little during the winter. They are borne terminally on axillary branches of new scale-leaves but do not become conspicuous until early spring.

Cones (fruits) are green at first but eventually turn bluish as the season progresses. Each cone contains 1–4 seeds, 2–4 mm long, often with longitudinal pits. The seed coat has a thick and bony outer layer and a thin, membranous inner layer.

Mature trees produce some seed nearly every year, but good crops occur only every 2–3 years. Cones do not open and remain on the tree throughout the winter, although many are eaten and dispersed by animals. Most remaining cones are dispersed in February–March.

Distribution

J. virginiana is the most widely distributed conifer of tree size in the eastern USA, and is found in every state east of the 100th meridian. The species extends northward into southern Ontario and southern Quebec in Canada. In its range, which has been considerably extended by natural regeneration from planted trees, especially in the Great Plains region, it is found most often growing at altitudes between 30 and 1070 m.

Silviculture characteristics

J. virginiana is shade-intolerant, but trees that have lived for decades beneath a full canopy of other trees on medium- to low-quality sites have been observed. It is a pioneer species, suited for use on mined land, old fields or pastures that are protected from fire. It grows well and faster than associated species because it is sun-adapted, drought-resistant, and has a long growing season. It is temporary on most sites and is eventually replaced by broadleaves and pines.

Thin bark and surface roots make *J. virginiana* intolerant of fire, although foliage does not burn well, affording some natural protection.

Silviculture practice

Mature fruits are usually collected in autumn by hand stripping or shaking onto canvas. Being resinous, they should be soaked in a weak sodium hydroxide solution for 1–2 days, followed by thorough washing. Seeds are extracted by running the fruit through a macerator and floating the pulp and empty seeds away.

Cleaned seeds are ready for use or can be dried to 10–12% moisture content for storage at −7°C to +4°C. The number of cleaned seeds/kg ranges from 81,570 to 121,250, averaging 96,120. If seeds are to be sown in the spring, they should be soaked in a citric acid solution (10,000 ppm) for 96 h, placed in moist-warm stratification at 24°C for 6 weeks, and finally placed in moist–cool stratification at 5°C for 10 weeks.

Germination is best if fresh seeds are used. Dry, stored seeds may be sown in mid-July for early spring germination.

Seeds are broadcast or sown in rows 15–20 cm apart in well-prepared seedbeds and covered with about 6 mm of firmed soil or sand. Stratified seeds should be sown in the spring, early enough to allow completion of germination before air temperatures exceed 21°C. Germination of stratified seed usually begins 6–10 days after sowing and is completed in 4–5 weeks. Untreated seeds may be sown in the autumn and mulched until germination during the second spring after planting. If fruits are depulped, dried and stored at −16°C, seeds germinate the first spring after summer sowing. Germination is epigeal.

Seeds that pass through digestive tracts of birds and animals, and those that remain on the ground germinate in the first or second spring; most natural germination takes place in early spring of the second year after dispersal.

Seedlings from nurseries are transplanted after 1–2 years at spacings from about 15 × 3 cm to 20 × 5 cm. Generally, *J. virginiana* is field planted as 2–0, 3–0, 1–1, 1–2, 2–1 or 2–2 stock, when stem diameter is at least 4.0 mm, preferably 5.6 mm, at ground level.

Most natural regeneration takes place on relatively poor broadleaved or pine sites, along fencerows or in pastures that are not burned or mowed. Seedlings are commonly established in rather open broadleaved stands adjacent to older

1

2

3

4

5

6

7

8

9

Plate 1. *Abies amabilis* © K. Klinka
Plate 2. *Abies lasiocarpa* © K. Klinka
Plate 3. *Acacia nilotica* © N. Pasiecznik
Plate 4. *Acer macrophyllum* © K. Klinka
Plate 5. *Adansonia digitata* © D. Gwaze
Plate 6. *Albizia lebbeck* © D. Lea/CSIRO Forestry and Forest Products
Plate 7. *Alnus rubra* © K. Klinka
Plate 8. *Alstonia scholaris* © Australian Tree Seed Centre
Plate 9. *Baikiaea plurijuga* © D. Gwaze

10

11

12

13

14

15

16

17

18

Plate 10. *Cedrus atlantica var. glauca* © V. Ordidge
Plate 11. *Cedrus libani* © V. Ordidge
Plate 12. *Chamaecyparis nootkatensis* © K. Klinka
Plate 13. *Dalbergia nigra* © S. Botelho
Plate 14. *Enterolobium cyclocarpum* © C. Hughes
Plate 15. *Eucalyptus camaldulensis* © Australian Tree Seed Centre
Plate 16. *Eucalyptus robusta* © D. Lea/CSIRO Forestry and Forest Products
Plate 17. *Gliricidia sepium* © C. Hughes
Plate 18. *Guazuma ulmifolia* © C. Hughes

19

20

21

22

23

24

25

26

27

Plate 19. *Khaya senegalensis* © D. Thies
Plate 20. *Larix occidentalis* © K. Klinka
Plate 21. *Leucaena leucocephala* subsp. *glabrata* © C. Hughes
Plate 22. *Melaleuca cajuputi* © D. Lea/CSIRO Forestry and Forest Products
Plate 23. *Melaleuca leucadendra* © Australian Tree Seed Centre
Plate 24. *Melaleuca quinquenervia* © D. Lea/CSIRO Forestry and Forest Products
Plate 25. *Parkia biglobosa* © D. Thies
Plate 26. *Peltophorum dubium* © S. Botelho
Plate 27. *Picea abies* © A. Pittaway

28

29

30

31

32

33

34

35

36

Plate 28. *Pinus ponderosa* © K. Klinka
Plate 29. *Pinus radiata* © V. Ordidge
Plate 30. *Pinus strobus* © A. Pittaway
Plate 31. *Populus tremuloides* © K. Klinka
Plate 32. *Prosopis juliflora* © N. Pasiecznik
Plate 33. *Prunus avium* © C. Teeling
Plate 34. *Pterocarpus angolensis* © D. Gwaze
Plate 35. *Quercus ilex* subsp. *rotundifolia* © C. Sousa Correia
Plate 36. *Quercus robur* © P. Robakowski

37

38

39

40

41

42

43

44

45

Plate 37. *Quercus suber* © C. Sousa Correia
Plate 38. *Samanea saman* © C. Hughes
Plate 39. *Schinus molle* © N. Pasiecznik
Plate 40. *Schizolobium parahyba* © C. Hughes
Plate 41. *Sequoiadendron giganteum* © V. Ordidge
Plate 42. *Terminalia arjuna* © D.K.N.G. Pushpakumara
Plate 43. *Terminalia superba* © D. Thies
Plate 44. *Thuja plicata* © K. Klinka
Plate 45. *Zelkova serrata* © K. Yoda

46

47

48

49

50

51

52

53

54

Plate 46. *Acacia melanoxylon* © Australian Tree Seed Centre
Plate 47. *Baikiaea plurijuga* © D. Gwaze
Plate 48. *Enterolobium cyclocarpum* © C. Hughes
Plate 49. *Eucalyptus camaldulensis* var. *obtusa* © Australian Tree Seed Centre
Plate 50. *Gliricidia sepium* © C. Hughes
Plate 51. *Leucaena leucocephala* subsp. *glabrata* © C. Hughes
Plate 52. *Paulownia tomentosa* © A. Pittaway
Plate 53. *Prosopis juliflora* © C. Hughes
Plate 54. *Schizolobium parahyba* © C. Hughes

55

56

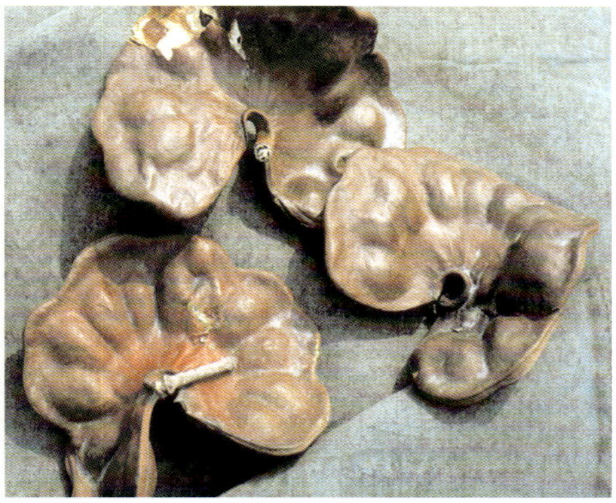

57

58

59

60

61

62

63

Plate 55. *Terminalia catappa* © A. Pittaway
Plate 56. *Enterolobium cyclocarpum* (seed pods) © C. Hughes
Plate 57. *Khaya senegalensis* (fruits are woody capsules) © D. Thies
Plate 58. *Leucaena leucocephala* (foliage, flowers and seed pods) © N. Pasiecznik
Plate 59. *Peltophorum dubium* (ripe seed pods) © C. Hughes
Plate 60. *Robinia pseudoacacia* (seed pods and foliage) © C. Hughes
Plate 61. *Cupressus macrocarpa* © G. E. Fyson
Plate 62. *Juniperus procera* © C. Couralet
Plate 63. *Sequoiadendron giganteum* © V. Ordidge

64

65

66

67

68

69

70

71

72

Plate 64. *Abies lasiocarpa* © K. Klinka
Plate 65. *Agathis australis* (single tree) © G. E. Fyson
Plate 66. *Betula pendula* © J. Barnett
Plate 67. *Fagus crenata* © T. Kamitani
Plate 68. *Juniperus procera* © C. Couralet
Plate 69. *Larix occidentalis* © K. Klinka
Plate 70. *Populus tremuloides* © K. Klinka
Plate 71. *Tsuga heterophylla* © K. Klinka
Plate 72. *Tsuga mertensiana* © K. Klinka

seed-bearing *J. virginiana* trees as a result of seed dispersal by birds.

First-year seedlings do not produce much height growth but develop a long fibrous root system. If competition from an overstorey is severe, seedlings may not survive, but if they do establish they can survive for a long time. *J. virginiana* also competes very well in shelter belts, where it is the most common natural reproduction.

Propagation by grafting, air-layering or cuttings is possible.

Growth rates of *J. virginiana* depend largely on site quality, competition from other species, and stand density, factors reflecting competition for available soil moisture. Trees 20–30 years old are generally 5–8 m tall with 6–8 cm dbh, while mature trees are usually 12–15 m tall with 30–61 cm dbh.

Growth and yield are affected by stand density and competition from broadleaves, with the highest volume growth obtained in unthinned stands from which broadleaves have been removed. Completely released stands average higher growth in dbh, basal area, and volume than stands where only crown competition is removed. The greatest mean dbh growth occurs with the lowest stand density.

On most sites *J. virginiana* grows slowly, and long rotations are required to produce conventional sawlogs. Because the wood is used for small items, however, shortening of rotations and intermediate harvesting of merchantable wood are possible. About 20–30 years are required for posts and 40–60 for sawtimber.

Diseases and pests

Important foliage diseases include the blights *Phomopsis juniperivora* and *Cercospora sequoiae*. The major stem and foliage diseases are cedar rusts in the genus *Gymnosporangium*, the most common being *Gymnosporangium juniperi-virginianae*, which attacks trees at all stages of development and also attacks apple trees; eastern redcedar as an alternate host can, therefore, be a problem for apple growers.

Several insects damage eastern redcedar trees but rarely cause serious permanent damage. Foliage is eaten by bagworm (*Thyridopteryx ephemeraeformis*) and spruce spider mite (*Oligonychus ununguis*), both of which can completely defoliate trees. Eastern juniper bark beetle (*Phloeosinus dentatus*) attacks but usually does not kill trees, except when the attack is associated with the root rot fungus *Heterobasidion annosum*, to which *J. virginiana* is very susceptible.

Mice and rabbits can damage seedlings, while deer browse *J. virginiana* and destroy many young plants in times of feed scarcity.

Uses

J. virginiana is among the best trees for protecting soils from wind erosion and reducing the desiccating effects of wind. It ranks high in the Great Plains shelter belt plantings because of its ability to withstand extremes of drought, heat and cold. The fibrous root system also helps to hold soil in place, especially on shallow soils. *J. virginiana* is important for wildlife as shelter and food. Many varieties with different form and foliage colour are grown as ornamentals.

Although *J. virginiana* is generally not an important commercial species, its wood is highly valued, with thin, white sapwood and red to dark brown heartwood. It is moderately heavy (specific gravity approximately 0.44), with a density of about 530 kg/m^3 at 12% moisture content. It is of moderately low strength and stiffness, but has high shock resistance. Texture is fine and uniform. Grain is usually straight, except where deflected by knots. It shrinks only a little during drying and has good dimensional stability. It is easy to work, splits easily, and has good nailing and gluing properties. It is very resistant to decay and insect attack. It is usually only available in small sizes and is generally knotty, so it is used for a range of smaller articles such as fence posts, chests, flooring and pencils.

J. virginiana is an important source of cedarwood oil for use in fragrance compounding. The scent of the wood is said to be a natural insect repellent. It is popular in the USA for Christmas tree production.

Further reading

Adams, R.P. (2011) *Junipers of the World: The Genus* Juniperus. 3rd ed. Trafford Publishing, Bloomington, Indiana, USA.

Lawson, E.R. (1985) *Eastern Redcedar (An American Wood)*. USDA - Forest Service, Washington, DC.

K

Khaya anthotheca (Welw.) C. DC.

African or Uganda mahogany

Importance

Khaya anthotheca is among the most important African timbers in terms of its significance to national and international trade. The timber is valued for its attractive appearance and is mainly used in furniture, panelling and decorative veneers. It is native to natural forests of Africa from Côte d'Ivoire to East Africa. There is uncertainty in the taxonomy of similar African species and some references to *Khaya grandifoliola* and *Khaya nyasica* may refer to the same taxon. *K. anthotheca* is often confused with *K. grandifoliola* (itself still an unresolved name) and the two may not be specifically distinct.

The species is heavily exploited, particularly in East and West Africa. In places where parent trees are scarce, regeneration is poor and serious genetic erosion is believed to have occurred. Various countries have now imposed felling limits and bans on the export of logs. It is listed as 'Vulnerable' on the 2002 International Union for Conservation of Nature (IUCN) Red List of Threatened Species.

There has been some planting in tropical Asia, South Africa, and the Caribbean.

Botanical features

K. anthotheca is a large deciduous tree to 50 or even 60 m tall and as much as 4 m in diameter (above the buttresses). It usually has a straight bole that is branchless to a considerable height, but often has buttresses, which may reach 6 m on larger trees. Leaves are compound, up to 40 cm long, with 6–10 leathery leaflets, each up to 17 cm long.

Flowers are unisexual, but are borne on the same tree (monoecious). It is insect pollinated. The fruit is a dry woody capsule, 4–6 cm in diameter, dehiscing by 4–5 valves with the flat, light brown wind-dispersed seeds arranged in rows around the central column.

Distribution

K. anthotheca is native to lowland rainforests and riverine forest at low to medium altitudes, up to ca 1500 m, in areas with 600–1600 mm rain/yr. It is reported from tropical Africa between 20°S and 10°N, from Sierra Leone eastwards to Uganda and Tanzania and southwards to Angola, Zambia, Malawi, Mozambique, and Zimbabwe; it is absent from the wettest of the Upper and Lower Guinea forest ecosystems.

K. anthotheca prefers moist, well-drained soils with subsoil moisture and is normally found on fertile alluvial soils, stable, gently sloping riverbanks, and adjacent alluvial slopes.

K. anthotheca is grown in plantations and used in enrichment planting. It has been successfully introduced in South Africa, Cuba, and Puerto Rico, and on a limited scale in Indonesia and Peninsular Malaysia where it has been used in taungya systems.

Silviculture characteristics

The tree is moderately shade-tolerant when young and moderately light-demanding when older. Seedlings are sensitive to competition from weeds and grass as well as to fire and frost.

Young trees are vulnerable to the shoot borer *Hypsipyla robusta*, but resistant to termites. Browsing animals can destroy young plants or slow growth. The tree coppices very poorly.

Silviculture practice

Natural regeneration is poor, as the density of mature trees is very low and seeds lose viability between 2 and 3 weeks under natural conditions or are damaged or eaten by seed-boring beetles and rodents. Experimental work has recommended keeping canopy gaps open, as well as seed supplementation to benefit regeneration of *K. anthotheca* in disturbed or managed tropical forest (to overcome dispersal limitation).

Unopened and partly opened fruits are collected directly from the tree and dried in the sun until they split, when the seeds are taken out manually. The clean seeds are then further dried in the sun to 5–7% moisture content and stored in airtight containers. Studies indicate that the seeds may be cold sensitive and that storage at 15°C is better than at 5°C or −18°C. The seeds have no dormancy and pretreatment is not necessary. Fresh seeds have a 60–90% germination success rate, usually within 3–4 weeks after sowing. The general recommendation is to sow seeds soon after collection. Even dried seeds lose viability rapidly – after 1 year viability has typically halved. There are 2000–4000 seeds/kg.

Sowing in seedbeds has shown to give better germination than sowing in containers. The seedbeds must be slightly shaded and permanently moist. Half-decomposed pine bark and compost has been successful in South Africa.

Persistent attacks by *H. robusta* have hampered efforts to restore the depleted mahogany resource base by raising seedlings in nursery plantations. Studies are ongoing to propagate superior *Hypsipyla*-tolerant genotypes. The challenge is still to improve rooting of older cuttings as the tolerance can only be recognized in older trees, while aging negatively impacted rooting ability of cuttings. Experiments have shown that cuttings with a leaf area of 30–50 cm^2 had the best rooting percentage, and cuttings with about 30 cm^2 had the most number of roots per cutting.

Propagation from wildings is also possible. The seedlings can be transferred to polypots when they are 5 cm tall. Seedlings and wildings are ready for planting at about 30 cm tall. Light shade is recommended. Reported spacings are 5 × 5 m. Periodic weeding is desirable.

A typical harvesting rotation is from 60 to 80 years.

Diseases and pests

Plantations have been discouraged by the shoot borer *H. robusta*. This impacts saplings and smaller trees in particular. Experiments have shown that medium (75%) to deep (90%) shade reduces attack but also diminishes growth.

Taller trees suffer fewer borer attacks. There has been considerable interest in exploring the genetic factors influencing borer susceptibility.

Uses

The dense crown makes it suitable as a shade tree and it is also popular as an ornamental and in windbreaks.

The timber is used in high-class cabinet work, high-quality joinery, furniture, construction, flooring, panelling, decorative veneer and plywood – all applications where attractive good-quality, medium-weight (439–620 kg/m^3) hardwood is valued. It is moderately hard, red-pink, light-coloured, but works easily and takes a fine polish and easy to saw, season and work. It weathers well, is resistant to borers and termites, but is not especially durable.

The bark is used in dyeing and tanning leather. Infusions of bark are used medicinally to relieve colds, manage malaria and repel mosquitoes. The branches provide excellent fuelwood.

Further reading

Bussche, G.H. von dem (1982) The establishment of hardwood plantations for the production of furniture and joinery timber in the Transvaal. *South African Forestry Journal* 121, 11–16, 17–23.

Palgrave, K.C. (1977) *Trees of Southern Africa*. Struik Publishers, Cape Town, South Africa.

Ward, S.E., Boshier, D. and Grogan, J. (eds) (2008) Special issue: Sustainable management of high-value timber species of the Meliaceae. *Forest Ecology and Management* 255, 265–364.

Khaya grandifoliola C. DC.

Big-leaf mahogany

Importance

Khaya grandifoliola is a deciduous monoecious tree of tropical Africa. The timber, which resembles that of true mahoganies (*Swietenia* spp.) more closely than other *Khaya* spp., is considered highly attractive.

The tree is often used as a shade tree in mixed-tree plantations, particularly grown as a standard tree with coppiced multipurpose tree species. It is planted as a roadside tree and in Uganda it is favoured to stabilize riverbanks.

Due to the high-value timber, exploitation of *K. grandifoliola* has often been heavy. Tree regeneration is also reported to be poor away from parent trees. The tree is thus rated as 'Vulnerable' on the IUCN Red List of Threatened Species. Protected subpopulations and export bans are in place in various countries.

Botanical features

The tree is deciduous, and generally smaller than *Khaya anthotheca*, with heights of 30–40 m and diameters of 1.2–2 m. Buttresses up to 3 m high are common. The bole is normally twisted or leaning near the top, with rough, scaly and dented, greyish-brown bark. The scented inner bark is red with white streaks and a viscous exudate. The leaflets are in 3–5 pairs, and have 12–15 pairs of well-defined lateral nerves and thread-like interlateral veins that appear more prominent on the lower surface.

The fruits have 5 thick valves; the seeds are flat and light, and are carried moderate distances by wind.

Distribution

K. grandifoliola is native to tropical Africa, and has been recorded from Benin, The Democratic Republic of the Congo, Côte d'Ivoire, Ghana, Guinea, Nigeria, Sudan, Togo and Uganda. This species occurs more frequently in dry semi-deciduous forest and forest outliers than *K. anthotheca*. It is sometimes considered a forest-edge species. Due to its high timber value, it has been planted on a small scale, or tested, in several Asian countries, notably India (Maharashtra) and Indonesia (Western Java).

Silviculture characteristics

Despite some plantings, *K. grandifoliola* is generally considered unsuitable for monoculture plantations, due to vulnerability to *Hypsipyla* spp. shoot borers and their infestation likelihoods being greatly increased when the trees occur at high densities. It is a plant that benefits from good light once it is well established and is generally considered to behave as a relatively shade-tolerant long-lived pioneer, and can regenerate well in heavily disturbed forests and forest edges.

Silviculture practice

K. grandifoliola is propagated by seed. Weight is 3000–5000 seeds/kg. Seeds should be carefully selected as many are already attacked by insects while still on the tree. Healthy seeds have a good germination rate (90%); seeds stored for about 5 months at −17°C retained their viability. Seeds should only be covered with a thin layer of soil, and germination takes place in 10–35 days. In general seed storage and planting advice is similar to that for *K. anthotheca*.

For *in vitro* culture of *K. grandifoliola* seedlings, optimum shoot and root length, number of nodes and number of roots was obtained on a medium supplemented with 1.0 mg/l benzyl amino purine + 0.01 mg/l naphthalene acetic acid.

In Cote d'Ivoire, trees have been planted under the shade of 2-year-old *Leucaena leucocephala*, which suppresses weeds, reduces borer attack, and fixes nitrogen into the soil. The *L. leucocephala* shade trees need regular thinning in the first years. When the *K. grandifoliola* trees have reached 15 m in height and 15 cm dbh, they are thinned to a density of

400–500 stems/ha. A second thinning to 200–250 stems/ha takes place when trees have reached 20 m height and 20 cm dbh, a third at 25 m and 25 cm dbh (to 125–150 stems/ha), and finally at 30 cm dbh down to 75–100 stems/ha.

In natural forest, rotation cycles should probably be 80–100 years, (in plantations, however, 40–60 years might be feasible if shoot borers can be controlled).

Natural regeneration is better at the savannah–forest boundary than in the logged or otherwise disturbed forest.

Diseases and pests

K. grandifoliola plantations suffer the same infestations of the wood borer *Hypsipyla robusta* as *Khaya anthotheca* plantations. Shading, mixed planting, and removal of lateral shoots can reduce damage to the wood but can also reduce growth rates.

Uses

The pinkish to reddish-brown timber is used for decorative veneers, panelling, cabinet work, superior joinery and furniture. It is of medium hardness, with a pale sapwood, easy to saw and finish and with good natural durability. Wood density is 460–800(–950) kg/m^3 at 12% moisture content, denser than that of *K. anthotheca*. The timber is generally put to similar uses to the true American mahoganies (*Swietenia* spp.) with the attractive wood being suitable for light construction, flooring, interior panelling and trim, shipbuilding, musical instruments, toys, novelties and carving. Traditionally it is used for dug-out canoes, fuelwood and for making charcoal.

Research has been conducted into the possible use of products derived from the tree to control freshwater snails and reduce schistosomes and *Fasciola* parasites. These analyses highlight the presence of alkaloids, saponins, tannins, anthraquinone and carbohydrate. The bark in particular has been investigated for its molluscicidal and trypanocidal properties, and is used as an insect repellent, anti-malarial and fever remedy in various parts of Africa. In Nigeria, concoctions derived from the plant are said to be used for the treatment of convulsions, coughs, dermatomycosis, fever, malaria, rheumatism, stomach ache and threatened abortion. Laboratory assessments have also suggested that preparations from the tree's bark have anti-inflammatory properties.

In Nigeria, the wood-ash of *K. grandifoliola* (and some other tree species) is used to preserve agricultural seeds (such as those of maize beans or melons) against fungi.

Further reading

Adio, A.F., Oyeleye, B., Gbadebo, J.O. and Jayeoba, O.J. (2005) Effects of storage period and temperature on seed germination of selected Nigerian mahogany species. *Moor Journal of Agricultural Research* 6, 30–35.

Dupuy, B. and Koua, M. (1993) Les plantations d'acajou d'Afrique: Leur sylviculture en forêt dense humide ivoirienne (The African mahogany plantations: Their silviculture in the tropical rain forest of the Côte-d'Ivoire). *Bois et Forêts des Tropiques* 236, 25–41.

Hawthorne, W. (1998) *Khaya grandifoliola*. IUCN 2012. IUCN Red List of Threatened Species. Version 2012.2. Available at: http://www.iucnredlist.org (accessed 21 March 2013).

Siepel, A., Poorter, L. and Hawthorne, W.D. (2004) Ecological profiles of large timber species. In: Poorter, L., Bongers, F., Kouamé, F.N. and Hawthorne, W.D. (2004) *Biodiversity of West African Forests: An Ecological Atlas of Woody Plant Species*, CABI Publishing, Wallingford, UK, pp. 391–445.

Khaya ivorensis A. Chev.

African mahogany

Importance

Khaya ivorensis, known as African or Lagos mahogany, is a large tree and one of the most important timber trees in West Africa. Although a host to numerous insect pests, notably the shoot borer *Hypsipyla robusta*, and therefore considered unsuitable for use in monocultural plantations, *K. ivorensis* is fairly widely grown in plantations within its natural area of distribution, as well as in tropical Asia and tropical America. Classed by the IUCN in its Red List of Threatened Species as 'Vulnerable', the species is threatened by habitat loss, degradation and selective felling in its native range.

Botanical features

K. ivorensis is a very large, high buttressed tree with a dark green, rounded crown, often with pendulous spherical fruits visible. Average height at maturity is 52.5 m, but can range from 41 to 60 m. Average dbh is about 85 cm, but can range from 70 to 115 cm.

Stem form is straight with a cylindrical bole which is branchless up to as high as 30 m above ground level and bearing well-developed plank buttresses, with one buttress more developed than the other. One side of the bole may appear as if the tree is not buttressed.

The bark is thick, brownish, and distinctly pitted by shallow shot hole depressions due to scales flaking off in discs. The slash is red with a pink underlayer, scented and bitter to taste. There is a delayed exudation of a papery gum from wounds. The sapwood is white and the heartwood is pinkish-brown.

The leaves are paripinnate, with usually 6 pairs of opposite leaflets and existipulate. In saplings the leaves are much bigger and may be 61 cm long, with about 20 pairs of leaflets. The leaflet is oblong to oblong–elliptic, about 8.7 cm long and 14.3 cm broad, glabrous, glossy, entire, acuminate, especially on the young plant, but broadly apiculate on old ones, almost rounded at the base, with a short petiolule. The midrib is fine and raised below; the nerves are very fine.

Flowers are pentamerous, yellow and scented, and occur in profuse panicles borne on the crown of the tree. The fruit is a globose, woody capsule, when open about 5–7 cm in diameter, on a woody stalk 5 cm long or more. It opens by 5 valves and the winged seeds are packed above each other against the 5 faces of the vertical columella. The seed is brown, flat, irregular, oblong or triangular depending on its place against the columella, about 1.8 cm long and 3.1 cm broad, surrounded by a narrow wing.

K. ivorensis is deciduous between September and November, with fresh leaves appearing before the old leaves fall. Flowering is from July to January, with most of the tree flowering from September to December. Fruits soon appear (January–March) and are conspicuous as they stick out from the crown of the tree. They are dispersed by wind from about February to May, although they do not travel far from the mother tree. The empty capsules may remain on the tree for months.

Trees aged 30 years can produce good seed crops; although seeds are produced annually, good seed crops are produced only every 3–4 years.

Distribution

K. ivorensis occurs in West Africa, chiefly in Côte d'Ivoire, Ghana, Togo, Benin and Nigeria, between latitudes 3° N and 10° N; its range also extends east to Cameroon and south to Angola.

Silviculture characteristics

K. ivorensis is a light-demanding species, although it will survive under shade during its first 2 years. In the sapling and young pole stage, it does best if not exposed to too much light. It is very susceptible to attack by shoot borer *H. robusta* if subjected to open conditions.

Silviculture practice

Seeds, which are classified as orthodox, do not remain viable for long and are attacked by grubs on the ground. The germination period is 8–21 days and the seedlings need to be shaded slightly. Germination frequency of fresh seeds is around 90%, but this is maintained for 2 weeks at most, falling off quickly afterwards. Seeds should be sown 5 cm apart in rows 30 cm apart.

Seedlings in nurseries respond well to the application of fertilizers, and nursery stock of 15–18 months old and about 90–150 cm high is suitable for planting. *K. ivorensis* is not really a suitable monocultural plantation crop as it is very susceptible to shoot borer attack under open conditions. Because of this, growth in plantations is uneven.

K. ivorensis can be propagated vegetatively using stumps or stripped plants. To avoid the risk of *Hypsipyla* attacks, *K. ivorensis* should only be established in mixed stands. It is recommended that it is grown in mixtures with other species, such as *Gmelina arborea*. The companion species should be added 2–3 years after planting *K. ivorensis*. Close spacing is required, since it has been shown that the incidence of attack increases with increased spacing. Older stands should be fully underplanted.

Diseases and pests

Several fungal diseases have been recorded on *K. ivorensis*, including *Phellinus noxius* and *Rigidoporus lignosus* (*R. microporus*) (Basidiomycetes) in Peninsular Malaysia, which attack the roots and are associated with poor land preparation and a previous history of root rots, and *Uredo tesoensis*, which attacks the leaves. In Brazil, where the species is increasingly preferred for reforestation due to its high resistance to *Hypsipyla grandella* (Pyralidae), attacks by pathogenic fungi are increasing. Target leaf spot caused by *Rhizoctonia solani* (*Thanatephorus cucumeris*) has been observed in the nursery and in the field, causing numerous zonate lesions on mature leaves and up to 100% leaf fall in seedlings. In young leaves, small brown spots surrounded by a purple margin characterize attacks. In older leaves, the lesions are light brown and exhibit concentric rings that resemble a target. Other disease pathogens that cause serious damage to *K. ivorensis* include *Phanerochaete salmonicolor* (the cause of pink disease), *Ceratobasidium stevensii*, and *Cylindrocladium parasiticum*.

As with other African mahoganies (*Khaya* and *Entandrophragma* spp.), *K. ivorensis* is readily attacked by *H. robusta* (Pyralidae), which feeds on the soft tissue inside the terminal stem. *H. robusta* also lays eggs on the seeds, and the larvae bore through the seeds. The mature larvae move from fruit to fruit during feeding. Trees are attacked in their second or third year. Generally, shoot borer attack weakens the trees and predisposes them to attack by other insects and fungi. Mortality can occur with repeated infestations. Economic loss is usually due to reduced height growth and poor form of infested trees. Planting *K. ivorensis* in association with insect-repellent companion species, particularly *Cedrela odorata*, mitigated shoot borer damage, lowering the incidence of attack, and consequently of epicormic branching.

Catopyla dysorphnaea (Pyralidae) lays eggs on the fruit and at first the larvae feed within individual seeds, then move from fruit to fruit. *Xyleborus compactus* (*Xylosandrus compactus*), *Xyleborus perforans*, *Xyleborus semiopacus* (*Xylosandrus crassiusculus*), *and Xyleborus sharpi* (Scolytidae) attack temporarily stressed or injured trees, but also attack transplants that have yet to recover from transplant shock.

Gyroptera robertsi (*Eugyroptera robertsi*) (Pyralidae) mainly attacks old trees. Other pyralids, *Prasinoxena monospila* and *Doloessa viridis*, have been found to cause severe but superficial bark necroses. *Pseudophacopteron zimmermanni* (Psyllidae) produces numerous galls on *K. ivorensis*, and the citrus pest *Aleurocanthus woglumi* has been observed infesting *K. ivorensis*.

Saplings suffer from damage by porcupines, which eat the bark.

Uses

K. ivorensis is suitable for use in taungya and in mixed plantations with other species such as *G. arborea*. It is recommended for planting for gully erosion control, and has also been used for rehabilitating tin mining sites.

K. ivorensis sapwood is creamy-white and is usually removed in squaring the logs to prevent attack of ambrosia beetles which cause pin holes. The heartwood is pale red in young trees but darkens on exposure. It often has an interlocked grain which, while producing figuring, makes warping more likely. It is easily air or kiln dried and, with care, splitting or checking will not occur. It is quite strong and moderately resistant to decay. It is easily worked with tools, but has a tendency to pick up in planing. It nails, screws and glues well. It is not resistant to termites. The wood is of medium density (500 kg/m^3 at 12–15% moisture content) and is considered very versatile. It has a moderate durability, low or medium strength, and medium shrinkage. It has easy sawing machining, and finishing properties.

The wood is used for making canoes and implement handles. It is very useful for furniture and interior decoration, and is a good quality joinery wood suitable for ships, cabins and railway coaches. It is excellent for plywood and veneering, and has economic potential for use in traditional carving in Nigeria. It is probably the most widely used mahogany at present, as it is the cheapest and most readily available.

K. ivorensis has traditional medicinal uses and laboratory investigations have found that bark extracts have anti-fungal, anti-bacterial, and anti-malarial properties. The stem, bark and leaves are used as traditional food preservatives in Nigeria, and seed oils have potential industrial uses.

Further reading

Jeyanny, V., Rasip, A.B. and Ghani, A.B. (2008) *Khaya ivorensis* A. Chev - early response to nutrient deficiencies in forest plantation nursery. *Planter* 84, 593–600.

Lok, E.H. and Ong, T.H. (2002) *Khaya ivorensis*. A manual for forest plantation establishment in Malaysia. *Malayan Forest Records* 45, 215–229.

Ward, S., Boshier, D. and Grogan, J. (2008) Special issue: Sustainable management of high-value timber species of the Meliaceae. *Forest Ecology and Management* 255, 265–364.

Khaya senegalensis (Desr.) Juss.

Dry zone mahogany

Importance

Khaya senegalensis (Plate 19) is an important multipurpose tree in tropical and subtropical Africa, and is particularly valued for timber, fuelwood and medicinal purposes. A medium-sized to large tree (up to 30 m), with a wide crown, it has been introduced into Indonesia, China, Thailand, Cuba, Guyana, the Antilles, Brazil and parts of India.

K. senegalensis is generally a light-demanding species (although it will tolerate up to 50% shade when immature), and one of the most drought-tolerant *Khaya* species, but is susceptible to fire when young. Although it grows best on deep, fertile soils, savannahs and rocky sites, it will tolerate dry or lateritic soils. Natural regeneration is good, and seedlings can easily be transplanted. In Africa, it is usually planted in mixed stands, sometimes under a shade tree and it should not be planted in monoculture to avoid serious attack by *Hypsipyla* shoot borers. It is also used for enrichment and improvement planting and has some potential for use in agroforestry.

The timber is resistant to termite attack, moderately durable, easy to season and has good machining properties. It has a very decorative grain, and is utilized mainly for sawtimber, furniture, veneers and light construction purposes.

The bark, roots, seeds, foliage and flowers have a very wide range of important medicinal uses. The seed oil is rich in oleic acid. *K. senegalensis* provides a relatively poor fodder which is used in Africa at the end of the dry season. Due to its various uses in pharmacopoeias and as high value timber, the tree is often preserved when fields are cleared. However, high-intensity harvesting of bark and foliage for medicinal and fodder purposes can lead to declines in local *K. senegalensis* populations, although where bark harvesting is done sustainably, complete wound recovery by edge growth can occur in 2 years.

Botanical features

A deciduous tree, reaching 15–20 m in savannah, and on fertile and deep soils up to 35 m high and up to 1.5 m in diameter, with a clean bole 8–16 m high; buttresses absent or not prominent. The bark is dark grey, with small, thin, reddish-tinged scales. The slash is dark pink to bright crimson, exuding a red sap. The crown is wide, well-developed and may be multibranched. *K. senegalensis* develops a taproot from the first year, but may creep if a soil horizon blocks taproot development. Large horizontal roots are sometimes found near the soil surface.

Leaves are alternate, compound, stipules absent; petiole and rachis 13–33 cm long; leaflets 3–6(–7) usually opposite pairs, oblong to narrowly oblong–elliptic, 7–12 cm long, 3–5 cm wide, apex acute to shortly acuminate, base rounded, margins entire, pale green, lateral nerves 8–16, petiolules about 3.5 mm long.

Flowers are borne on a lax, much-branched axillary panicle up to 17(–20) cm long. Flowers are tetramerous, monoecious, but with obvious vestiges of the opposite sex, resulting in male and female flowers closely resembling each other externally. The calyx is pale green, lobed almost to the base, lobes subcircular, about 1 × 1 mm, imbricate; petals cream, free, oblong–ovate, 4 mm long, 2.5 mm wide, contorted in bud; eight stamens or staminodes, united into a tube; disc orange-red, cushion-shaped, fused to the base of the ovary, but free from the staminal tube; ovary 4-celled.

The fruit is an upright, almost spherical, woody capsule, 4–10 cm in diameter, opening by four valves from the apex (Plate 57). Seeds are brown, broadly transversely ellipsoid to flat, paper thin (about 25 × 18 mm) with narrowly-winged margins.

Leaves fall in the dry season and are directly replaced. Flowering occurs in Burkina Faso from November to December, in Guinea from November to February, and in Côte d'Ivoire biannually, principally from August to December. *K. senegalensis* is insect pollinated. Fruiting occurs in Guinea from February to July, in Côte d'Ivoire mostly from January to April, with a second fruiting period from July to September. Trees begin to produce seed after 20–25 years.

Distribution

K. senegalensis occurs naturally in Africa between 15° N and 8° N in a strip extending from the Atlantic to the Indian Ocean across Mali and Senegal, northern Cameroon and northern Uganda, as well as south Sudan.

The natural phytogeographical distribution corresponds to the 'Sudanian Regional Centre of Endemism' and the 'Guineo–Congolia/Sudania Regional Transition Zone'. The species is often found individually dispersed in natural vegetation and secondary forest, in association with *Detarium senegalense*, *Parinari excelsa* and *Erythrophleum suaveolens*.

Plantations have been established within its area of natural distribution and are also well established in Vietnam, India, Java, Cuba and Guadeloupe. More recent plantings have occurred in Australia, Brazil, Puerto Rico, Guyana, Thailand, Cape Verde, Reunion and South Africa.

Silviculture characteristics

K. senegalensis is light-demanding, but tolerates light shade when young and can be found in the undergrowth of northern tropical deciduous and humid forests. Regeneration is favoured in clearings; the species develops stool sprouts and root suckers. Saplings are browsed by wildlife and are very sensitive to bush fires, whereas adult trees are relatively resistant and can withstand passing fires. Under natural conditions, initial growth is slow, about 12–25 cm after 2 years, increasing to 1.2–1.5 m/yr, but slowing after 5 years, when borer attacks become more likely.

Silviculture practice

Dissemination of *K. senegalensis* is anemochorous and natural regeneration is abundant. Seed is collected from mature fruits (4500–7000 seeds/kg). Before sowing, the ligneous capsule should be completely removed before soaking for 24 h in warm water (20–30°C) to facilitate germination. Germination occurs after 10–18 days, seedlings reaching 10 cm high with roots 25 cm long after 3 months.

Seeds are sown at a distance of 7–10 cm in rows spaced 20–25 cm apart and at a depth of 2 mm. Pricking out can be performed after 3 months (3–4 leaves) at a spacing of 20 × 20 cm to obtain healthy plants of 0.5–1 m, or after 1 year, to promote secondary rooting. Stump planting is possible in areas with regular precipitation after planting, but striplings or seedlings are preferable. Plants may be transplanted after 6–7 months, but longer is preferable. Bare-rooted plants have difficulties surviving when planted in the dry Sudan climate zone.

Soil preparation is mainly done to increase water availability. Sandy soils do not have to be prepared, but it is necessary to loosen vertisols, planosols, and sometimes ferruginous soils, by loosening the subsoil, constructing small dykes or preparing water retention ditches around normal planting holes.

Pure stands are not recommended as they increase the risk of damage by *Hypsipyla* spp.

Various mixed plantations have been tested, with *Dalbergia sissoo*, *A. indica*, *Senna siamea*, *Albizia lebbeck*, *Milicia regia*, *Triplochiton screroxylon*, and *Gmelina arborea*, but care is required to ensure *K. senegalensis* is not dominated by the faster-growing species. *Leucaena leucocephala* has been assessed as ground cover under *K. senegalensis* to reduce *Hypsipyla* attacks, with some success.

Pests and diseases

The most important problem that has a significant influence on the silviculture of *Khaya* spp., as well as other Meliaceae, is borer attack, principally by *Hypsipyla robusta*. Seedlings and trees are attacked in the absence of tree cover. The main shoot wilts and is replaced by an axillary bud that becomes dominant, resulting in a multi-branched stem form if attacks continue each year. This considerably reduces the growth of the tree and value of the timber. As a consequence, *Khaya* spp. are usually planted under light cover and in mixed stands, or utilized for enrichment planting.

Other stem borers include *Eugyroptera robertsi*, *Apate terebrans*, and *Apate monachus*. The twig borer *Xylosandrus compactus* has been observed. *K. senegalensis* is also attacked by defoliators such as *Bourgognea microcera*. The sapwood is susceptible to attack by longhorn (Cerambycidae) and powder-post beetles (*Lyctus* spp.). *Pseudophacopteron zimmermanni* has been reported as a gall-forming pest while the termite *Mastotermes darwiniensis* damages seedlings and young trees.

One important problem is the browsing of young *K. senegalensis* plants by animals.

Uses

K. senegalensis has some use in silvopastoral agroforestry systems. It has also been investigated for agroforestry use. However, trimming and pollarding, current practices for producing forage, can have disastrous effects on the tree. *K. senegalensis* has also been intercropped with low-growing field crops in taungya systems. As *K. senegalensis* is resistant to *Hypsipyla grandella*, it has been used successfully as a rootstock to confer resistance to scions of the susceptible species *Cedrela odorata* and *Swietenia macrophylla*. Due to its varied medicinal uses and high timber value, the tree is often left after fields have been cleared. *K. senegalensis* is commonly grown as an ornamental shade tree and as a roadside tree in its native range. It is considered a suitable species for urban forestry.

Although the wood is of medium weight (600–700 kg/m^3 when air-seasoned) and is only moderately hard, it is rated as one of the best African mahogany woods. Its heartwood is moderately resistant to attack from insects, fungi and termites. Its sapwood is occasionally affected by longhorn beetles. The wood can be seasoned quickly and easily and worked with all types of power or hand tools without any problems. Sometimes, rather woolly surfaces result from sawing, so it is always best to keep cutting edges sharp. Planks or strips can be successfully nailed, screwed or glued together to form stable joints. Finally, the surface can be varnished or stained without difficulty to produce a very attractive finish. Wood density and strength properties are moderate, processing and seasoning are good, dimensional stability is moderate and surface finishing and decorative appearance are good. Unpreserved heartwood has a medium life expectancy of 8–15 years.

Because of its attractive appearance, the wood of *K. senegalensis* is one of the most favoured timbers for making furniture in the Sudan zone. Its solidity and durability make it a much sought-after material for the construction of pre-fabricated buildings, boatbuilding, turned wood products, decorative interior constructions, windows and doors, prayer slates and mortars. Because of its irregular stem form, however, it is rarely used for veneers.

Only small logs and branches are used as firewood, because the wood is difficult to saw. The ash is used in the conservation of millet seed.

K. senegalensis has a number of non-wood uses, such as the use of leaves for livestock forage, and bark, root and seed oils for medicinal products. A decoction of the leaves is used as a febrifuge, giving rise to the common name 'quinquina du Sénégal'.

Leaf extracts, seed powder and bark extracts have been shown to protect crops from stored-product pests, soft rots and nematodes.

Further reading

Belewu, M.A., Olatunde, O.A. and Giwa, T.A. (2009) Underutilized medicinal plants and spices: chemical composition and phytochemical properties. *Journal of Medicinal Plants Research* 3, 1099–1103.

Hung, C.D. and Trueman, S.J. (2011) *In vitro* propagation of the African mahogany *Khaya senegalensis*. *New Forests* 42, 117–130.

Shukla, N.K. and Singh, R.S. (1995) Physical and mechanical properties of *Khaya senegalensis* (khaya) from Wada, Maharashtra. *Journal of the Timber Development Association of India* 41, 9–16.

Koompassia excelsa (Becc.) Taub.

Tualang

Importance

The leguminous tree *Koompassia excelsa* is among the world's tallest rainforest angiosperm species. It is an important commercial timber for Peninsular Malaysia, where it is also planted.

K. excelsa trees are also logged in Sabah and Thailand, although in the past they were not logged in Sabah and Sarawak due to their large size, heavy timber and tendency to split when felled. The heaviness of *K. excelsa* made hauling the timber out of the forest very difficult. Therefore in logged-over forest, *K. excelsa* trees were left standing, forming a striking feature in the landscape. However, with modern logging equipment and machinery, and increased demand for the timber, it has become economically viable to log *K. excelsa* in the forests of Sabah.

K. excelsa trees provide nesting places for honeybees, the honey being collected by local villagers for sale to supplement their incomes. In fact, the value of the honey in many cases exceeds the timber value of the tree. Because of this, in the Malaysian and Indonesian parts of Borneo, *K. excelsa* trees are protected by law from felling.

The bark has medicinal properties and is used locally against fever.

Botanical features

K. excelsa is one of the tallest broadleaved rainforest trees, with a maximum height of 85 m and stem diameter of up to 290 cm recorded in Sabah. The bole has huge, steep buttresses up to a height of 3 m or more, with branching occurring at above 30 m. The bark is smooth, shiny, purplish-grey, often with a delicate fluorescent green tinge, especially towards the top. It is slippery, making the tree difficult to climb, thus creating a safe habitat in the crown for honeybee colonies. The crown is bright green to pale green and made up of many smaller sub-crowns.

The leaves are alternate, pinnately compound, with alternate leaflets and a terminal leaflet; there are 7–12, but up to 17, leaflets with 0.25 cm long petioles. The leaflet blade is small, 3–4.2 × 1–1.7 cm in size, and elliptic, with a slightly recurved margin; it is shiny dark green, glabrous above but with fine velvety hairs below, and has faint venation.

The inflorescence is in axillary or terminal panicles with many small, bisexual flowers, which are sessile or stalked. The calyx is narrowly imbricate with 5 lobes. There are 5 petals, sub-equal with an indistinct claw, and 5 stamens, the anthers opening by 2 apical and 2 basal pores. The ovary is sessile with one ovule. The fruit pod is flat, oblong, 8–12.5 × 2–3.5 cm in size, including the broad, strongly veined, papery wing which surrounds the central flat seed; the fruit does not split open until the seed germinates.

K. excelsa is deciduous annually between February and June and remains leafless for a few weeks. Flowers emerge immediately after the new leaves appear, although flowering occurs at irregular intervals of 5–6 years, often during mast flowering periods. The flowering period is May–June, with fruiting in August–September. Fruit dispersal is effected by wind.

Distribution

K. excelsa has been recorded in southern Thailand, Peninsular Malaysia, north-east Sumatra, Borneo, and the Philippines. It is locally common but not abundant, and is the dominant upper-storey tree where it occurs. It is not recorded south of Kuala Lumpur and Kuantan in Peninsular Malaysia.

Silviculture characteristics

K. excelsa is a nitrogen-fixing, shade-tolerant, self-pruning species, which can regenerate rapidly.

Seedlings of *K. excelsa* are common in the forest, but there is a marked absence of saplings, poles and small trees.

Silviculture practice

K. excelsa seeds are not protected by a hard seedcoat, so in order to ensure a high germination rate, they should not be stored for a long period. They are susceptible to fungal infection, and storage at low humidity is advisable. The germination period is 1–3 weeks and germination percentage is usually high.

In the hill forests of Peninsular Malaysia, the stocking of *K. excelsa* with a diameter of 15 cm and above ranges from 2.1 to 6.3 m^3/ha. *K. excelsa* trees planted at Kepong in Malaysia reached a diameter of 36 cm after 40 years.

Diseases and pests

No serious diseases occur on *K. excelsa*, although the pest *Xyleborus fallax* has been recorded.

Uses

The trees of *K. excelsa* provide ideal conditions for the giant honeybee *Apis dorsata* to build its huge parabolic honey combs which hang from the bottom of branches. The height at which branches occur, often at 30 m or more above ground level, and the slipperiness of the bole, make the combs inaccessible to predators such as the sun bear (*Helarctos malayanus*). Accessible however to humans, the combs, which can sometimes be 2 m across in size, yield abundant honey and beeswax, and are important sources of income for local people. With a single tree capable of supporting over 100 bee colonies, the value of the honey produced exceeds the potential value of the tree's timber.

K. excelsa produces a medium-hard wood, with a density ranging from 800 to 865 kg/m^3 at 15% moisture content. The sapwood is yellow-brown and is sharply differentiated from the heartwood, which is reddish-brown to deep red-brown when fresh, darkening to chocolate-brown on drying. Wood texture is coarse, with interlocked grain, making it easy to re-saw and crosscut when green but slightly difficult to re-saw when dry. Planing is easy, giving a smooth surface. Nail-holding properties are good. The timber dries moderately slowly, with slight end-checks, surface checks and insect attack problems. *K. excelsa* timber is moderately durable.

Among Malays, the bark is used traditionally in medicinal baths for the treatment of fevers. The tualang is the subject of many myths and legends among local peoples of the region.

Further reading

Payne, J. (1990) Bears, bees and tualang trees. In: *Wild Malaysia: The Wildlife and Scenery of Peninsular Malaysia, Sarawak and Sabah*. MIT Press, Cambridge, Massachusetts, USA, pp. 26–27.

Soerianegara, I. and Lemmens, R.H.M.J. (1994) *Koompassia* Maingay ex Benth. In: Soerianegara, I. and Lemmens, R.H.M.J. (eds) *Plant Resources of South-East Asia No. 5 (1) Timber Trees: Major Commercial Timbers*. Backhuys Publishers, Leiden, The Netherlands, pp. 270–274.

Koompassia malaccensis **Maingay ex Benth.**

Kempas

Importance

Koompassia malaccensis is a common and widely distributed species in Peninsular Malaysia; it is less common or rare in southern Thailand and the Indonesian islands. It occurs in a wide variety of habitats in primary tropical rainforest, from peat and freshwater swamps to low undulating country and ridges on hills up to an altitude of 700 m. Uniform high temperature and rainfall are required for its successful growth. It grows in a wide range of soil types. Very limited information is available on silviculture and management. Stand establishment can be done using natural regeneration or planting stock.

It is a very important commercial timber species for Peninsular Malaysia, the timber being a medium- to heavy-weight hardwood suitable for structural usage. In Sarawak and east Kalimantan, it is protected by law as these trees provide nesting places for honeybees, the honey and wax being collected by local villagers for sale to supplement their incomes.

Botanical features

K. malaccensis is a very tall emergent tree, reaching 50 m tall and 210 cm in diameter. Stems clear to a height of 24–27 m to the first branch. The crown usually a little less than half the total height and spreads to about 30 m. Bole with tall, steep buttresses to 3–4 m high. Bark grey to almost black and brittle, with narrow longitudinal fissures or cracks; rarely the bark is scaly except for trees growing in swamp forests.

Leaves spirally arranged, pinnately compound with 5–9 (rarely up to 14) leaflets arranged alternately, petiole 12–20 cm long. Leaflet blade ovate to elliptic, 5.5–12.5 × 2–4 cm with pointed, blunt, or notched apex, finely velvety hairy below, venation inconspicuous; terminal leaflet stalk 5 mm long.

Inflorescences in terminal panicles. Flowers small with 5 regular sepals, 5 petals, and 5 stamens. Fruit is a pod, flat, with twisted base, oblong 10–13 cm long and 3 cm broad, with broad veined papery wing all round a central, single, flat seed.

K. malaccensis is deciduous, shedding leaves annually during February to June, remaining bare for a few days and flowering along with new leaves. The fruits ripen several months later. Dispersal of the fruits is effected by wind.

Distribution

K. malaccensis is found between 7° N and 5° S of the equator in southern Thailand, Peninsular Malaysia, and the Indonesian islands of Sumatra, the Riau Archipelago, Bangka, Belitung, and Kalimantan. In Peninsular Malaysia it is a common and widely distributed species, and is considered the third most common large tree. It is less common to rare in other countries where it is reported to occur. In certain nutrient-poor undulating land in Peninsular Malaysia, *K. malaccensis* and kedondong (species of the family Burseraceae) dominate the forest canopy in terms of size and number of stems so that it is possible to recognize a distinct forest subtype known as 'kempas-kedondong forest'.

Silviculture characteristics

K. malaccensis seedlings are slightly shade-tolerant, very slow growing when small, and can survive and grow for 10 years under suppressed conditions with heavy mortality at first. Growth is faster if the canopy is opened up to allow more light to reach the young seedlings. Generally, there is abundant regeneration in the forest, and seedlings, saplings and mature trees are well distributed in forests.

Silviculture practice

K. malaccensis seeds have a moisture content of 15% and an average single seed weight of 1.2 g. In order to ensure a high germination rate, freshly collected seeds should, if possible, be sown in nursery beds without delay. When the seeds are stored they are susceptible to fungal infection, and low humidity should be maintained. A mixture of topsoil, sand, and sawdust is used as a sowing medium.

Seed germination is epigeal. The cotyledons are large, flat, and leaf-like. Seeds germinate after 6–54 days. Germination percentage is high.

Where there is adequate light, sapling growth is fairly rapid, and height growth of 1.5 m in the first 2 years is not unusual. Where there is insufficient light, height growth can be slow.

No specific information is available on silviculture and management, except that *K. malaccensis* is categorized as a preferred species in regeneration sampling by the Malaysian

Forest Department, which treats it as a commercial timber crop. *K. malaccensis* is susceptible to drought.

The harvesting procedure from primary forest is similar to that for other commercial timber trees. *K. malaccensis* is the third most abundant timber tree species in Malaysia; an enumeration survey in Peninsular Malaysia indicated that it makes up 3.7% of the total number of trees and 6.1% of the timber volume in a forest. In hill forests in Malaysia, the stocking of large *K. malaccensis* (>15 cm diameter) ranged from 2.1–6.1 m^3/ha. In Kalimantan, it is of scattered occurrence, with an average density of 1.1 trees/ha. Studies in secondary kempas-kedondong forest in Malaysia indicated that climber cutting and removal of unwanted tree species increased the diameter growth of *K. malaccensis*.

Pests and diseases

No serious pests or diseases have been reported. Ambrosia beetles attacking living trees or freshly felled logs are rare. Sapwood is susceptible to attack by powder post beetles.

Uses

The sapwood is pale yellowish, whereas the heartwood is orange-red or red-brown, darkening on exposure, with numerous yellow-brown streaks. Density varies from 670 to 1290 kg/m^3 at 15% moisture content. The grain is interlocked, sometimes wavy; the texture is coarse but even.

Zones of hard stone-like tissues, known as included phloem, occur irregularly in the wood and can be several centimetres wide tangentially and several metres long. They are a source of weakness in structural timber and may cause damage during seasoning. Brittle heart is also a common defect. However, recent studies indicated that timber samples with heavy included phloem could be used for construction. The natural resistance of included phloem against rot fungi showed superior natural durability when compared with that of sapwood and heartwood. The heartwood is readily attacked by termites.

K. malaccensis timber in Peninsular Malaysia requires 2–6 months' air seasoning to achieve 18.1% moisture content. Radial and tangential shrinkage from green to air dry averages 2.0 and 3.0%, respectively. It takes about 8 days to kiln dry 25 mm thick boles from about 50 to 10% moisture content. Kiln drying Schedule E is used in Malaysia. Apart from springing and splitting of boards containing included phloem, the timber is free from defects in drying.

K. malaccensis timber is somewhat difficult to machine due to the fibrous texture and interlocked grain. The silica content is about 0.1%. Blunting of saw teeth is moderate to severe. To avoid tearing during planing, the cutting angle should be reduced to 20°. The wood can be satisfactorily stained and polished. Pre-boring is advisable before nailing. Good veneer of 1.5 mm thick can be made at a peeling angle of 92° without pretreatment. Wood without included phloem is suitable for plywood that meets Indonesian, German and Japanese standards.

The heartwood of *K. malaccensis* is durable in temperate climates but in the tropics it is only moderately durable. Graveyard tests in Indonesia gave an average service life of 2 years in contact with the ground. The sapwood is liable to attack by powder post beetles and fungi. *K. malaccensis* timber has shown a large and unexplained variability in natural resistance to marine borer attack.

Untreated wood can be used for structural purposes under cover such as parquet and strip flooring (especially when resistance to acids and chemicals is needed), panelling, vehicle bodies and heavy-duty furniture. The wood also produces high-quality charcoal.

Trees of *K. malaccensis*, like those of other *Koompassia* species, provide ideal conditions for honeybees, in particular the giant honeybee *Apis dorsata*, to build their huge parabolic honey combs. The combs, which can sometimes be 2 m across, yield abundant honey and beeswax, which are important sources of income for local people. Because of their value for honey production, which exceeds the potential value of the timber, *K. malaccensis* and *Koompassia excelsa* trees are protected from felling by law in Sarawak and east Kalimantan. Kempas extracts also exhibit antimicrobial properties.

Further reading

Chow, P., Pearson, O. and Heuvel, H. van den (1990) The effects of included phloem on some mechanical properties in kempas *Koompassia malaccensis* wood. *IAWA Bulletin* 11, 121.

Lee, C.T., Lee, S.L., Faridah, Q.Z., Siraj, S.S., Ng, K.K.S. and Norwati, M. (2008) Genetic diversity assessment of *Koompassia malaccensis*. *Pertanika Journal of Tropical Agricultural Science* 31, 127–133.

Ser, C.S. (1981) *Malaysian Timbers - Kempas*. Malaysian Forest Service Trade Leaflet No. 44. Malayan Timber Industry Board, Kuala Lumpur.

L

Larix decidua Mill.

Common larch

Importance

Larix decidua is a deciduous conifer, native and widely cultivated in Europe, but also gaining popularity in North America. It is a pioneer, highly adaptable, light-demanding species which achieves very good yields under a range of climatic conditions and soil fertility levels. Its wood is durable, resistant and suitable for many uses, both as timber and as pulpwood. International tree-improvement efforts have led to the identification of superior provenances, as well as the development of high-performing interspecific hybrids.

Botanical features

An elegant tree with a broad pyramidal crown, becoming wider and irregular as it gets older. Its straight stem sometimes shows some tendency toward basal sweep. Branchlets are somewhat drooping. The bark of young seedlings is whitish-grey or yellowish-grey. A mature tree can reach a height of over 20 m, with a dbh of 60 cm or more. The crown can reach a width of 15 m. The tree has a lifespan of up to 200 years, although it can be harvested for timber at around 40 years old.

Needles, grouped in fascicles of 30–40, measure 10–38 mm long and 0.5 mm wide, are keeled and have several rows of stomata. Before being shed from the tree in the autumn, the needles take on a vivid golden-yellow colour.

Mature female cones of *L. decidua* are ovoid, 15–60 mm long and 15–24 mm wide, with 40–50 scales. During the flowering season, cone bracts are crimson-red and much longer than the cone scales. On mature cones, however, the bracts are hidden by the scales. The seeds measure 1–4 mm long by 2.5–3 mm wide, with a wing 8 mm long by 4 mm wide.

Depending on local climatic conditions, flowering begins between late March and early May, before leaf bud opening. Within a given tree, pollen is usually shed from the male cones 1–2 days before the female flowers become receptive. Leaves appear during or just after flowering, and usually fall around October. Seeds are mature in September or October. Cones open in the autumn and seeds are dispersed throughout the winter and following spring. The empty cones usually persist on the tree for 2–3 years.

Distribution

L. decidua is the only species of *Larix* native to Europe. A subalpine species, it occurs naturally in four discontinuous areas: the Alps, the Carpathian Mountains, the Sudety Mountains and Poland. It is found both in pure and mixed stands in Austria, Czech Republic, Slovakia, Slovenia, south-eastern France, southern Germany, Poland, northern Italy, Romania, Switzerland and north-western Ukraine.

Silviculture characteristics

L. decidua forms pure stands or is mixed with pine (*Pinus cembra* and *Pinus uncinata*), spruce (*Picea abies*), alder (*Alnus viridis*) or beech (*Fagus sylvatica*). It is a pioneer, transition species which colonizes open terrain after natural disturbances. It aids soil retention on steep slopes.

Like all larches, *L. decidua* is a light-demanding, shade-intolerant tree, requiring abundant light and space, which is associated with relatively low leaf mass and area. It is quite tolerant of winds, especially at altitude, but does not tolerate continuous, humid winds. Though very adaptable and tolerant of urban conditions, the species is not well suited for industrial areas because its foliage is sensitive to air pollution during the summer.

Larch has a very high transpiration rate compared to other conifers. Its water consumption relative to the fresh weight of its foliage is 3–5 times higher than that of Norway spruce (*Picea abies*). This is reflected by the productivity of the foliage: the mass of needles necessary for the production of 1 m^3 of wood is much less than for other conifers. In winter, on the other hand, common larch has a low transpiration capacity. Its deciduous habit apparently confers a significant advantage by reducing susceptibility to winter desiccation damage.

Silviculture practice

L. decidua is readily propagated by seed, but seed supply is scarce due to irregular flowering and low filled-seed percentage. Seeds remain viable for at least 3 years. Cold seed stratification for 1 month accelerates germination and renders it more uniform. Seeds may be sown in outdoor seedbeds at the end of winter, and seedlings transplanted either in individual containers immediately after germination or as bare-root stock the following year. Seedlings are usually planted out as 2+0 or 1+1 stock.

L. decidua can be propagated vegetatively, either by rooting softwood or hardwood stem cuttings, or by tissue culture. Grafting is also used in horticulture and to establish clonal seed orchards.

Seedlings establish themselves only on cleared sites. Natural regeneration can thus be favoured by various intensities of regeneration cuttings, both for even-aged forests and selection forests. Various kinds of soil scarification are recommended to ensure that young seedlings obtain enough light. Care must be taken not to reduce soil water availability, as it is essential for stand development.

Successful plantation establishment requires matching the correct provenance to site conditions and preparing the site in order to ensure that trees get ahead of competing vegetation. Plantation density varies according to the management programme. In all cases, it is important to allow sufficient space for proper tree development.

For a plantation of 2500 stems/ha, eliminating one out of every three stems every 20 years, on average, is recommended. Wide initial spacing is desirable to delay the need for thinning until the trees are of commercial size. An initial stocking of 1000 stems/ha (approximately 3 × 3 m) appears optimal, giving lower planting costs and shorter rotations, accompanied by an increase in average branch and stem diameter compared to lower spacings.

For long rotations on fertile sites, a low planting density, such as 1600 stems/ha, is also recommended, provided a broadleaved sub-canopy is created later to maintain good soil quality. Heavy thinning must then be done early to leave approximately 350 stems/ha at age 70 years.

In American plantations, 2 × 3 m spacing is recommended to provide adequate growing space and ample working room for thinning, and to reduce establishment costs compared to 2 × 2 m spacing. It also eliminates the need for one pre-commercial thinning.

Plantations must be kept free of competing vegetation, but because of the rapid height growth of young trees, care of young plantations is relatively easy, with cleaning only required during the first 2–3 years.

Natural self-pruning is slow in *L. decidua* and plantation trees, especially those set wide apart, may develop fairly heavy branches which will persist for many years. Pruning increases timber value, especially for the production of clean, easily peeled, easily logged stems and veneer logs on a short rotation.

Thinning is necessary as soon as crown closure begins in order to prevent mortality and reduction of growth increment, which can occur rapidly as competition increases. In successful larch plantations, thinning should begin early. Even at an initial spacing of 2 × 3 m, the canopy of vigorous stands will close within 6–8 years. Within 15 years, keen competition between individual trees will begin. The first thinning, at about 15 years, is aimed at removing overtopped and defective trees, most of the intermediates and some of the poorer co-dominants, and will leave about 1000 of the most promising, best-distributed stems/ha. Subsequent thinnings will be crown thinnings based on basal area, taking out those trees that too closely compete with the best-formed, best-placed stems which will make up the final crop.

If the plantation was established at a good initial spacing, the first thinning should yield marketable stems of 12–20 cm dbh. Alternatively, the entire stand could be harvested as pulpwood prior to age 30 years.

The yield of natural larch stands can be described according to four basic site categories:

1. High-altitude stands, near the tree line, giving 1 m^3/ha/yr of marketable wood, with a maximum rotation of 100–130 years.
2. Alpine stands, at altitudes of 1600–1800 m, giving 1–3 m^3/ha/yr.
3. Medium-altitude stands at 1000–1600 m, giving 3–6 m^3/ha/yr.
4. Low-altitude stands, giving 6–10 m^3/ha/yr, or more.

Diseases and pests

Larch canker (*Trichoscyphella willkommii*), considered the most destructive disease of *L. decidua* in Europe, is present throughout the tree's natural range and causes significant tree mortality and damage, particularly at low altitudes and on inadequate, badly drained sites. It was introduced into North America in the 1920s. Susceptibility to the disease varies among provenances and is increased by frost injury. Selecting provenances with low canker susceptibility and which are well adapted to the chosen site, and reducing plantation density, help prevent the disease.

Mycosphaerella laricina, a serious defoliating needle blight, reduces productivity and can kill highly susceptible trees. Until recently it was found only in Europe, but it has also been observed in north-central USA. Alpine seed sources are most susceptible, while those from the Sudety Mountains and Poland are more resistant. The disease appears to be propagated mainly through contaminated nursery stock. No control measures have been reported.

In forest nurseries, *Meria laricis* needle cast causes significant seedling mortality, both in Europe and North America. Seedlings are also vulnerable to grey mould (*Botrytis cinerea*). Reducing seedling density and allowing foliage to dry between water applications are recommended control measures.

Larch sawfly (*Pristiphora erichsonii*) and larch casebearer (*Coleophora laricella*) reduce tree growth through defoliation but rarely cause tree death. In eastern Canada, the greatest killer of larch trees is the eastern larch beetle (*Dendroctonus simplex*), a phloem-boring insect which attacks stems in early spring, boring through the bark and creating tunnels underneath, in which eggs are laid.

The main mammal pest of larch is the North American porcupine (*Erithizon dorsatum*), which causes damage to plantations by feeding on the inner bark, thus killing branches and producing multiple leaders.

Uses

L. decidua is aesthetically appealing with its airy foliage, so has value as an ornamental. It can be used on abandoned land to assist other woodland trees in becoming established, and also as an effective windbreak and shelterbelt.

The wood of *L. decidua* is hard and strong, and particularly appreciated for carpentry and naval construction. The heartwood is pale red to brick-red, whereas the resinous sapwood is pale yellow. Growth rings are well defined. Average wood density at 15% moisture content is 590 kg/m^3.

Long tracheids and high density make larch wood especially useful for pulping and papermaking. Larch pulps produce papers which are superior in tear and fold, but inferior in burst and tensile strength, to those produced from spruce or pine. *L. decidua* is used in mixture with other species to produce high-quality printing papers and newsprint. However, larch is unique among softwoods in that its heartwood contains fairly high levels of water-soluble extractives, mainly arabinogalactans. These interfere with the pulping reaction in ordinary acid sulphite processes, and have caused prejudice against the use of larch for pulpwood. Nonetheless, larches are compatible with sulphate (kraft) and certain modified sulphite pulping processes.

The bark is widely used in Europe both for its high tannin content and for its medicinal properties. A good-quality resin is obtained by tapping the trunk and can be used as a wood preservative or varnish. The resin and the turpentine obtained from it also have various medicinal uses.

Further reading

Pâques, L.E. (ed.) (2002) Improvement of larch (*Larix* sp.) for better growth, stem form and wood quality. Proceedings of an International Symposium, Gap (Hautes-Alpes) - Auvergne & Limousin, France, 16–21 September 2002. INRA - Unité d'Amélioration, Olivet, France.

Perks, M., Harrison, A., McKay, H. and Morgan, J. (2006) An update on nursery propagation and establishment best practice for larch in Britain. Forestry Commission Information Note 080, Forestry Commission, Edinburgh, UK, 6 pp.

Larix gmelinii (Rupr.) Rupr.

Dahurian larch

Importance

Larix gmelinii is a fast-growing, medium-sized timber tree that can tolerate temperatures as low as −70°C. An adaptable deciduous conifer, it tolerates waterlogged as well as dry conditions, favouring sites in the middle and lower parts of gentle slopes which are moist, fertile, and have good drainage. In Russia, the main geographical range of *L. gmelinii* coincides with the continuous permafrost area.

It is the principal plantation species in north-eastern China due to its characteristics of light-dependency, strong seed reproduction, and easy natural regeneration. Its wood has a wide range of uses. In addition, it is a good species for shelter forest and landscape planting because of its tall, straight trunk with a uniform, beautiful crown shape, well-developed root system, and high smoke tolerance.

Botanical features

L. gmelinii reaches a maximum height of 30–40 m and a dbh of 60–90 cm, but it is shrubby in the northern part of its range (north of the polar circle and in mountain areas). It lives to 250–300 years old, although trees over 500 years are known. Crown is cone-shaped, becoming irregular with age. Under permafrost conditions many older trees have a columnar crown. Stem is straight, with good natural pruning. Root system is shallow.

Branches are long, horizontal, and spreading. Shoots and branchlets are pendulous, mostly glabrous, yellowish, often reddish or darker in winter.

Bark is dark brown and split into scales in young trees, and grey, brown, and greyish-brown with scaly, longitudinal checks, which become purplish-red after scaling off in mature trees.

Winter leaf buds globose, resin-free; bud scales distinct at the tips. Needle-like leaves are light yellow-green to bright green, turning bright yellow to orange before falling in autumn. They are linear, flat, 1.5–3 cm long, 0.7–1 mm wide, and have 1–2 lines of stomata on lower surfaces either side of the central vein.

Male strobili globose, yellow, about 3–4 mm long. Female strobili pale pink or red, about 9–12 mm long. Cones 2–3.5 cm long, purplish-red when young, ovate or elliptical when mature, with seed scales nearly flat, glabrous, glossy brown, usually erect. There are 11–20 seed scales per cone, 6–15 mm long and 5–13 mm wide. Bract scales are about one-third shorter than seed scales. Seeds are small, triangular oval with seed wing about 1 cm long. The weight of 1000 filled seeds is 1.5–5.0 g.

Shoot buds expand around mid-April in the southern part of its range and from mid-May in the north. Leaves open between May and mid-June, flowering generally occurs from May to June, and cones ripen from mid-August to November depending on locality. Seeds are shed from very early spring to summer the following year.

Distribution

L. gmelinii is native to eastern Siberia and adjacent north-eastern Mongolia, and northern and eastern China, in the Greater Xing'an and Xiao Xing'an Mountains. It is also found in the high mountain region of northern North Korea.

In Russia, *L. gmelinii* covers 2.6 million km^2, extending east from the Yenisei River valley to Kamchatka and the Primor'e Territory. The northern limit is controlled by the short growing season and heat deficit, and is the same as the bioclimatic timberline. Within central and eastern Siberia this limit very closely coincides to the July isotherm of +10°C and with isolines of air temperature sum 300° for the period with the temperature more than +10°C during the 80-day long vegetative season. The world's most northerly forests are the open forests of *L. gmelinii* in northern Krasnoyarsk Territory. Dwarf, shrub-like forms are found in the River Popigai valley in the far north-east of this Territory (latitude 72°50' N).

In southern Siberia, the eastern limit of *L. gmelinii* is close to Amazar on the border between Chita and Amur provinces (122°E), extending south to the central part of the Greater Xing'an in China.

In the northern part of its range in Russia, *L. gmelinii* occurs in monospecific forests, whereas in southern Siberia it is found in both pure stands and mixed with pine, spruce, birch, and aspen. In the mountain area of southern Siberia, *L. gmelinii* is found at altitudes from 300 to 1600 m. Dwarf, shrubby forms occur in montane regions of Mongolia at altitude 2000–2200 m. Dahurian larch in the Xiao Xing'an Mountains in China is mainly distributed in valley swamp areas and is often found mixed with other conifers.

Outside China, *L. gmelinii* has been planted in several exotic provenance trials or hybridization trials, and has also been planted in arboreta, mainly in Europe and North America.

Silviculture characteristics

L. gmelinii is a light-demanding species with good sapling growth under intense light conditions. It tolerates side-shading until age 20 and more years, which is good for advanced growth before selective thinning. It is an important species for barren land afforestation in China, and has also been introduced to arid and semi-arid areas because of its strong adaptability to a continental climate and its flexible moisture requirements. It is suitable for the establishment of timber production forests, and fast-growing, high-yield forests, since it grows well in deep, fertile and moist sandy or loam soil, but it is not suitable for planting under forest canopies. It is moderately fire-tolerant.

Silviculture practice

To ensure *L. gmelinii* seed quality, most seed is collected in dedicated seed production areas or seed orchards. Cones are collected in autumn as soon as they ripen, spread out in thin layers to dry in the sun or in well-ventilated sheds, and then opened by solar heat, heating in a cone kiln, placing in a heated room, or by mechanical separation. Seeds are removed from opened cones (using a shaker), dewinged, and then cleaned with a blower or fanning mill. Seeds can be removed most easily from cones that have been stored under natural conditions for stratification over a winter season. Under favourable environmental conditions, a seed orchard can produce up to 120 kg of seed/ha. Heavy seed crops occur every 3–8 years.

Good-quality 2–3-year-old nursery stock is used for planting. Containerized seedlings, or plants produced from grafting or cuttings are also used.

Afforestation is usually carried out in spring, and also in autumn particularly in mountainous areas covered with deep snow in winter. Care needs to be taken to avoid root damage due to soil lifting by frost. Site preparation should be carried out before planting. Timely tending is necessary for young trees during the first few years. Clearing weeds and earthing-up to prevent root exposure is carried out annually, twice in the first year.

The timing of first thinning and thinning intensity vary according to initial planting density, site quality, and management purpose. In a forest stand with an initial planting density of 3300–4400 trees/ha, initial thinning is by a low-thinning at around age 15 years at an intensity of 25–30%. Subsequent thinning intervals range from 5 to 8 years.

Rotation of plantations is determined by the intended end use of the timber: 40 years for large-sized logs and about 20–30 years for small-diameter timber.

Clear-cutting is mainly used for plantations, but selective thinning may be adopted in some areas, such as on steep slopes, heavy soils, areas eroded by water, and swamps. Selective thinning is frequently used in mixed forests to improve both natural and artificial regeneration.

Diseases and pests

Seedlings are seriously attacked by *Rhizoctonia* and *Fusarium* spp.; control is through ensuring good soil and drainage conditions.

Siberian pine moth (*Dendrolimus superans*) is the main insect pest in China, and can seriously damage young and mature forests. In Siberia, the main pests of seeds are *Strobilomyia laricicola*, *Eucosma impropria*, and *Dioryctria abietella*. In central Yakutia, larvae of *Hylemyia* sp. can destroy about 80% of seeds in cones.

The most serious needle pests are *Dendrolimus superans*, *Zeiraphera diniana*, and particularly *Cosmotriche lunigera*.

Stem pests include *Melanophila guttulata*, *Ips subelongatus*, and *Urocerus gigas taiganus*, with *Xylotrechus altaicus* the most damaging in the southern part of the *L. gmelinii* distribution area. Weakened trees may be attacked by secondary xylophagous pests.

Pest control measures include releasing natural enemies such as *Trichogramma* and *Anastatus* at the ovum stage, or spraying *Bacillus thuringiensis* at the larval stage. Light traps can be used to kill adults.

Uses

In the cryolithic zone of Siberia, *L. gmelinii* forests prevent solifluction on slopes and thermokarst development on flat watersheds. In China, it is planted for soil and water conservation and watershed protection because of its well-developed root system and deep understorey litter layer.

L. gmelinii produces quality timber with a heavy, medium-hard, and easily split wood. Sapwood is light yellow, and the heartwood is yellow-brown or red-brown, with a straight grain and fine texture. Wood specific gravity ranges from 0.32 to 0.52, volume shrinkage is 0.588–0.619%, crushing strength along the grain 527–576 kgf/cm^2, bending strength 1106–1133 kgf/cm^2, compressive strength parallel to the grain 1299–1314 kgf/cm^2, and hardness of cross section 377–415.

The *L. gmelinii* wood is traditionally used in Russia and China for house and engineering construction, electricity poles, train sleepers, vehicles, ships and carpentry. Branches and young thinned trees are used for pulp and papermaking.

The wood is rich in resin with an important economic value, being used in the manufacture of rubber, paint and dyes. The proportion of bark in natural stands is 23–27% of stem volume; tannin extracted from the bark is used in the leather industry.

Further reading

Bao, Y., Zhang, Q.-L., Wang, L.-M. and Wu, J. (2011) Characteristics of biomass and productivity in *Larix gmelinii* natural forests with different stand structures. *Journal of Zhejiang Forestry College* 28, 52–58.

Osawa, A., Zyryanova, O.A., Matsuura, Y., Kajimoto, T. and Wein, R.W. (eds) (2010) *Permafrost Ecosystems: Siberian Larch Forests*. Ecological Studies 209. Springer, Dordrecht, The Netherlands.

Larix kaempferi (Lamb.) Carrière

Japanese larch

Importance

Larix kaempferi is a deciduous conifer widely cultivated in Japan and introduced into neighbouring regions of Asia, as well as into Europe and North America. It is a pioneer, highly adaptable, light-demanding species, which achieves very good yields under a great range of climatic conditions and soil fertility levels. Its wood is durable, resistant and suitable for many uses, both as timber and as pulpwood. Superior provenances can be identified, and high-performing interspecific hybrids have been developed.

Botanical features

L. kaempferi is a graceful tree with a straight stem, a broad, irregular, pyramidal crown and pendulous branching. Mature trees can reach over 27 m, with a dbh of 50 cm or more. The crown can reach a width of 8 m. Trees have a lifespan of up to 200 years, although they can be harvested for timber at around 30 years old.

Significant variation in habit occurs between individual trees. Various foliage types have been identified, related to the character of secondary branches, ranging from 'fine-slender' to 'coarse-stiff', the former being more vigorous but with a higher proportion of crooked stems than the latter, which contains a higher percentage of straight stems.

Needles of *L. kaempferi*, grouped in fascicles of approximately 60, are 30–35 mm long and 0.5–1 mm wide, keeled, and have several rows of stomata. Young shoots are a reddish colour. Before being shed in autumn, the needles take on a golden-yellow colour.

Mature female cones are ovoid to ovate, 15–35 mm long and 20–30 mm wide, with 25–40 scales. Bracts are hidden by very distinctive, recurved, round to ovate cone scales. The seeds are 3 mm long by 2 mm wide, with a wing 4 mm long by 4 mm wide.

Depending on local climatic conditions, flowering of *L. kaempferi* begins between March and May, before the opening of the leaf buds. On the same tree, pollen is usually shed 1 or 2 days before the female flowers become receptive. Leaves appear in April or May and usually fall around October. Seeds mature in September or October. Cones open in autumn and seed dispersal occurs throughout the winter and following spring. Empty cones usually persist on the tree for 2–3 years.

Distribution

L. kaempferi has a very limited natural distribution range in Japan, where it forms pure stands or grows mixed with various other conifers and broadleaved species. It is found on volcanic soils on the slopes of the central mountains of Honshu Island, at altitudes reaching 2600 m near the summit of Mount Fuji.

Silviculture characteristics

L. kaempferi is shade-intolerant and requires abundant light and wide spacing, preferring full sun or only light shade. Japanese and hybrid larches tolerate exposure quite well on good soils; they withstand salt spray remarkably well, and are also better than *L. decidua* at tolerating pollution.

Silviculture practice

L. kaempferi is readily propagated by seed, but seed supply is scarce due to irregular flowering and low filled-seed percentage. Seeds are orthodox and remain viable for at least 3 years. Cold seed stratification for 1 month accelerates germination and renders it more uniform.

L. kaempferi can also be propagated vegetatively via stem cuttings or by tissue culture. Grafting is also used in horticulture and to establish clonal seed orchards.

Seedlings are usually planted out as 2-0 or 1-1 stock, either bare-root or container-grown. Particular care needs to be given to container stock to avoid root spiralling and other malformations. Depending on the climate, seedlings can be transplanted even after budbreak. Seeds may be sown in outdoor seedbeds at the end of winter, and seedlings transplanted either in individual containers, immediately after germination or as bare-root stock the following year. Shading of young seedlings is recommended.

Thinning is necessary as soon as crown closure begins in order to prevent mortality and growth reduction, which can occur rapidly as competition increases.

Plantations established with a 2.4 × 2.4 m initial spacing can be row-thinned at 15 years, with all harvested material chipped for pulp and residual trees retained for sawn logs. At 3 × 3 m spacing, thinning can be delayed to 20 years. Alternatively the plantation can be clear-cut at 20–25 years, with logs sorted for pulp or sawn timber.

For plantations established at 2000 stems/ha, the first thinning is recommended when mean tree height in the stand reaches approximately 8 m, which corresponds to an age of 10 years on very good soil or 15 years on average sites. On very fertile sites, stem density should be 930 stems/ha at age 20 years, of which 620 stems/ha remain at age 30 years, and 450 stems/ha at age 40 years. If the initial spacing is 1.5 × 1.5 m, a strong pre-commercial thinning (at least 50%) is recommended before dominant height reaches 6 m. This way, both diameter and height growth increments will continue without any reduction.

Plantations must be established carefully to ensure that young trees quickly outgrow potential competition. Care of young plantations is relatively easy. Normally, if the site has been properly prepared and the plantation was established adequately, cleaning is only required during the first 2–3 years.

Natural pruning of *L. kaempferi* is good. Adventitious branches frequently occur on branch-free stems after severe pruning or severe thinning.

The target diameter for *L. kaempferi* at the final cut is 30–50 cm. With these diameters, final yield is 300–400 trees/ha and the final cutting age is 70–100 years. Annual total volume increment at age 40 years varies between 4.6 and 13.0 m^3/ha, with stem density varying between 400 and 840 stems/ha.

Diseases and pests

In forest nurseries, larch seedlings are often infected by grey mould (*Botrytis cinerea*). Reducing seedling density and allowing foliage to dry between waterings can correct the situation.

In Japan, *L. kaempferi* plantations are very much affected by *Mycosphaerella larici-leptolepis*, which affects the current season's shoots, causing broom-shaped deformations in the crown and slowing height growth. The disease is mainly spread by wind.

In Europe and North America, *L. kaempferi* is considered more or less resistant to larch canker (*Trichoscyphella willkommii*), needle blight (*Mycosphaerella laricina*), and needle cast (*Meria laricis*).

Larch sawfly (*Pristiphora erichsonii*) and larch casebearer (*Coleophora laricella*) reduce tree vigour and growth through defoliation, the latter causing tree mortality if attacks are prolonged. In eastern Canada, the greatest killer of larch trees is *Dendroctonus simplex*, a phloem-boring beetle which attacks stems in early spring, boring through the bark and creating tunnels underneath, in which eggs are laid.

The North American porcupine (*Erithizon dorsatum*) damages larch plantations by feeding on inner bark, buds and young shoots.

Uses

The rapid growth of *L. kaempferi* makes the species useful as a windbreak and in shelter belts. In some areas where pine (*Pinus*) trees are grown, *L. kaempferi* is planted as a firebreak. It is popular as an ornamental and in bonsai.

L. kaempferi wood is hard and strong. The heartwood is reddish-brown and contains high amounts of water-soluble arabinogalactans and flavone phenols, giving it an appreciable durability and resistance to microorganisms. Average wood density at 15% moisture content ranges between 530 and 590 kg/m^3. The wood is easy to carve, but surfaces may dent easily. Bending strength when air-dry is fairly high. Pre-boring is recommended in nailing, since the wood is prone to splitting.

L. kaempferi wood is in demand for outdoor constructions such as fences, playground equipment, window frames, decking and roofs. Roundwood is used for pit props and building poles. Sawn timber is suitable for pallets and various types of container. Due to its strength, surface hardiness and decay resistance, it is much appreciated in naval construction. In the past it was also used, untreated, for railway sleepers.

Long tracheids and high density make larch wood especially suitable for pulping and papermaking. Larch pulps produce papers which are superior in tear and fold, but inferior in burst and tensile strength, to those produced from spruce or pine. Although the fairly high levels of water-soluble extractives in heartwood interfere with the pulping reaction in ordinary acid sulphite processes, larches are compatible with sulphate (kraft) and certain modified sulphite pulping processes.

The bark of *L. kaempferi* contains tannin and other phenolic substances, and it is also used commercially to make carpenter's glue. Resin obtained by tapping gives an excellent turpentine.

Further reading

Adams, W.R. and Hutchison, G.O. (1961) Total and merchantable volume growth of Japanese larch. *Bulletin - Vermont Agricultural Experiment Station* 620, 7 pp.

Hattemer, H.H. (1974) *Geographic Variations in Japanese Larch Analyzed in Various Planting Sites*. Arboretet, Copenhagen.

Koga, S., Oda, K. and Tsutsumi, J. (1996) Effect of thinning on intrinsic wood properties of Japanese larch. In: Donaldson, L.A., Singh, A.P., Butterfield, B.G. and Whitehouse, L.J. (eds) Recent Advances In Wood Anatomy. Proceedings of the Third Pacific Regional Wood Anatomy Conference, Rotorua, New Zealand, 20–24 November 1994. New Zealand Forest Research Institute Ltd, Rotorua, New Zealand, pp. 305–307.

Larix laricina (Du Roi) K. Koch

American larch

Importance

Larix laricina, American larch or tamarack, is a deciduous conifer distributed widely across North America, up to the northern tree line. It is a pioneer light-demanding species, highly adaptable to various climatic conditions, tolerant of extremely wet soils and poor fertility levels, and achieving very good yields on well-drained, moderately fertile sites. Although it is not widely exploited commercially, its wood is durable, resistant, and suitable for use as timber and pulpwood. Superior provenances with improved growth, survival and stem form have been selected.

Botanical features

L. laricina has a straight, tapering trunk and a narrow, pyramidal crown. Its bark is thin, only 1.3–2.0 cm thick at the base of mature trees. Branches are relatively sparse, horizontal or slightly ascending. Twigs are orange-brown and glabrous. The species is self-pruning, and boles of 25- to 30-year-old trees are often clear of branches for half their height or more. Mature trees are 15–23 m tall, with a dbh of 36–51 cm, but a few reach 91–102 cm. In the southern part of its range, where it grows best, it can reach a height of over 30 m. On marshy ground and in the extreme north, growth is stunted, with adults often measuring only 3 m high, with a dbh of 8 cm. Lifespan is 180 years or more.

Needles are light or bluish-green, grouped in fascicles of 10–30, measure 10–35 mm long and 0.5 mm wide, are keeled, and have several rows of stomata.

The root system is shallow, without a taproot. On sandy sites it has a plate-like rooting habit, and on dry sites the roots bend sharply from the trunk, forming characteristic 'knees'.

Mature female cones are erect, narrow ovoid, 12–19 mm long and 8–10 mm wide, with 12–20 scales. Young cones are dark red to violet, turning pale brown as they mature. Bracts are hidden by thick cone scales. Seeds are 3 mm long by 2 mm wide, with a wing 5 mm long by 3 mm wide.

Flowering begins from mid-April to early May, just before the opening of the leaf buds. Pollen is usually shed 1–2 days before female flowers become receptive. Leaves emerge from mid-April to mid-May. It is the last of the northern deciduous trees to lose its leaves, between mid-September and early November, depending on region. Seeds mature around September. Cones open in late summer and seeds are rapidly dispersed. Empty cones persist on the tree for 2–5 years.

Distribution

L. laricina has the widest range of any North American coniferous tree, extending from northern Alberta to central

Alaska, across Canada into Newfoundland, and south into the Great Lakes region of the northern USA. The central Alaskan population is isolated geographically from the rest of its continuous range.

L. laricina forms extensive pure stands in the boreal region of Canada and in northern Minnesota, USA. In the rest of its natural range, it is found locally in both pure and, as a dominant species, in mixed, open stands.

In Europe, *L. laricina* has been included in test populations for exotic and hybrid larch breeding programmes.

Silviculture characteristics

It is shade-intolerant and requires abundant light and space, so should not be planted among other fast-growing species as it must become dominant to survive. Because *L. laricina* casts only light shade, its stands tend to have a dense understorey of shrubs.

Strong winds will often uproot large trees growing in swamps or wet sites. Compared with *Picea mariana*, however, it is fairly windfirm. In urban environments, it is relatively salt-tolerant.

Silviculture practice

L. laricina can be propagated by direct sowing in outdoor seedbeds at the end of winter in a warm, moist, mineral or organic soil, with a light cover of grass or other herbaceous vegetation.

For seed extraction, sometimes it is necessary to soak the cones and re-dry them. Seeds are orthodox and can be kept at least 4 years in frozen storage after proper desiccation. Dormancy is mild to non-existent, but a 1-month cold stratification improves germination. Seeds require good moisture to germinate, but will not germinate under water as might be expected of a swamp tree. Germination is possible at temperatures as low as 12°C, but is optimal at 24°C.

At the northern limit of its range, *L. laricina* frequently reproduces vegetatively through natural layering. Roots can also produce sprouts. Propagation by rooted softwood stem cuttings or tissue culture is also possible. Grafting is used in horticulture and to establish clonal seed orchards.

Seedlings grow best when they are provided with abundant light and a constant water level. The best planting stock is begun in the greenhouse and transplanted for 1 year. Seedlings can also be planted out as 2-0 stock, either bare-root or container-grown. Shading of young seedlings is recommended. Seedlings may be autumn-lifted and overwintered in cold rooms at −3°C.

The shade intolerance of *L. laricina* dictates the use of wide spacing and even-aged management. Some adaptation of clear-cutting or seed-tree cutting is generally considered to be the best silvicultural system, because seeds germinate better in the open and seedlings require practically full light to survive and grow well. Satisfactory re-establishment, however, often requires some kind of site preparation, such as slash disposal.

Competition must also be controlled, the first 2 years after planting being critical. Growth rate declines after 12–15 years if trees are planted at close spacings. Increment is better for the first 25 years at spacings up to 2.4 × 2.4 m.

Thinning is necessary to prevent mortality and growth increment reduction as soon as crown closure begins. With good spacing, the first thinning should yield merchantable stems of 13–20 cm dbh. The financial rotation age for kraft pulping is 25–30 years, depending on site class, with optimum spacing ranging from 2 × 2 m to 3.5 × 3.5 m.

The growth rate of established *L. laricina* saplings depends largely on moisture conditions. In wet, stagnant swamps, the tree grows very slowly, reaching an average height of 2 m in 55 years. By contrast, on well-drained sites it maintains height increments averaging 0.5 m for 20–30 years. When the crowns close, however, growth drops to a very low level. In natural stands, annual growth varies between 1.9 and 2.5 m^3/ha. In 70–100-year-old stands, mean annual increment averages 3.8 m^3/ha on well-stocked plots and 1.9 m^3/ha on poorly stocked ones.

In short rotations, intensively tended plantations produce 2–3 times more biomass than conventionally tended stands. At age 7, stem-branch biomass averages 33–36 t/ha in stands initially spaced at 0.61 × 0.61 m to 0.23 × 0.23 m.

Diseases and pests

The most common foliage diseases are rusts, where *L. laricina* acts as a secondary host. European larch canker (*Lachnellula willkommii*) has been identified in some *L. laricina* stands in eastern Canada, causing a reduction in wood quality and, occasionally, resulting in tree mortality.

Larch sawfly (*Pristiphora erichsonii*) causes periodic epidemics in which entire stands can be defoliated. Although it is present over most of the natural range of *L. laricina*, its impact is most severe at more southerly latitudes. It can be controlled by biological means.

Larch casebearer (*Coleophora laricella*), introduced from Europe at the end of the 19th century and now widespread in the eastern half of the natural range of *L. laricina* as well as in British Columbia, is another serious defoliator, causing growth reduction and eventual tree mortality. Introduction of natural predators has been effective in controlling outbreaks.

In their natural habitat, *L. laricina* seedlings suffer from high mortality rates, mainly due to damping off, mechanical injury, drought, waterlogging and insect attacks. Seedlings and saplings are damaged or killed by browsing mammals, such as the porcupine, *Erethizon dorsatum*.

Uses

With rapid growth, good response to hedging and spectacular golden autumn foliage, *L. laricina* has high ornamental value.

The wood is heavy and durable, somewhat coarse, with good strength and decay resistance. It is generally without odour but has a greasy feel. The heartwood is yellowish to brown or russet-brown in colour. Slow-grown trees develop wood with high resin content, making it decay resistant.

Because of its density, which ranges from 440 to 480 kg/m^3, and sometimes spiral grains, the wood is moderately difficult to work since it can be hard to saw, chip and plane. While it dries more slowly than other softwoods, it can be kiln dried with good results.

Because of its strength, surface hardness, and decay resistance, *L. laricina* wood is suitable for use in heavy construction (poles, piles, posts and decking), as well as in lighter work (framing, pallets, boxes and laminated timber). The wood is also used for pulp and paper production via sulphate (kraft) and certain modified sulphite pulping processes, despite the presence of arabinogalactans and other water-soluble extractives in the heartwood.

In the past, root 'knees' were used in wooden ship construction, while long, straight stems were made into water pipes.

Native Americans used the fine roots for cordage, and the needles for stuffing pillows and mattresses. The bark is used locally both for its high tannin content and for its various medicinal properties. A good-quality resin, obtained by tapping the trunk, serves as a wood preservative or varnish; it also has medicinal uses.

Further reading

Davis, A., Puettmann, K. and Perala, D. (1996) *Site Preparation Treatments and Their Effects on Establishment and Growth of Tamarack*. Research Paper. North Central Forest Experiment Station Research Paper No. NC-328, USDA Forest Service, St Paul, Minnesota, USA.

Duncan, D.P. (1954) *A Study of Some of the Factors Affecting the Natural Regeneration of Tamarack* (Larix laricina) *in Minnesota*. Brooklyn Botanic Garden, New York, USA.

Eckenwalder, J.E. (2009) *Larix laricina* (Du Roi) K. Koch. In: *Conifers of the World: The Complete Reference*. Timber Press, Portland, Oregon, USA, pp. 326–327.

Johnston, W.F. and Carpenter, E.M. (1985) *Tamarack (An American Wood)*. USDA Forest Service, Washington, DC.

Larix occidentalis Nutt.

Western larch

Importance

Larix occidentalis (Plate 20) a shade-intolerant, deciduous conifer, is the fastest-growing and largest of the larches in the Pacific North-west forest, and the most important timber species of the genus. It is used for lumber, fine veneer, poles, ties, mine timbers and pulpwood. Stands are valued for their multiple resource values. Thus, the presence of the species in pure and mixed-species stands is valuable where multiple resource use is the major management objective.

Botanical features

L. occidentalis is a large tree, occasionally reaching over 60 m in height, 200 cm dbh, and 900 years of age. The trunk is usually branch-free over much of its length. The crown is short, wide and pyramidal. The principal branches are horizontal. The bark is reddish-brown, scaly when young, becoming thick and deeply furrowed with flat flaky ridges with age (Plate 69). The twigs are hairy at first, becoming hairless and orange-brown during the summer.

L. occidentalis develops a deep and extensive root system. Fibrous roots under young stands extend up to 100 cm in depth. Windfallen mature trees usually have their roots infected by root rots. Roots are associated with both ecto- and endomycorrhizae.

Needles are deciduous, flattened above, keeled below, triangular in cross section, 3–5 cm long, shiny pale green, and 10–30 per fascicle.

L. occidentalis is monoecious; both staminate and ovulate flowers develop throughout the crown. Vegetative buds are smaller than flower buds, and which be detected early in the autumn, about 1 year before subsequent cone crops mature. Methods of sampling buds and conelets have been devised for forecasting larch seed crops on individual trees, as well as stands.

Pollen and seed conelets appear several days before vegetative buds open, usually from about mid-April to mid-May. Cones complete their development in one season and mature by mid- to late August. Cones usually begin to open by early September, with more than 80% of seeds dispersed by mid-October. Cones open when they have dried to a moisture content of 35–40% and usually fall from the tree during the following winter, although many may stay attached through the following summer.

Initiation of phenological events varies greatly with latitude, altitude, aspect and weather.

Distribution

A western North American (predominantly Cordilleran) species, *L. occidentalis* has a relatively small native range. It grows in south-eastern British Columbia, the Upper Columbia River Basin of north-western Montana, northern and west central Idaho, north-eastern Washington; along the eastern slopes of the Cascade Mountains in Washington and north-central Oregon; and in the Blue and Wallowa Mountains of south-eastern Washington and north-eastern Oregon.

L. occidentalis grows in pure stands but is more frequent in mixed-species stands. Old-growth western larch stands are rare; despite being a long-lived species, it is replaced by shade-tolerant species in the absence of fire. Thus, it is present in early, mid- and late stages of fire-driven, secondary succession. Depending on climate and soil moisture, the most common associates are *Pseudotsuga menziesii* var. *glauca* and *Pinus ponderosa* in drier temperate climates; *Abies grandis*, *Tsuga heterophylla*, *Thuja plicata* and *Pinus monticola* in wetter temperate climates; and *Picea engelmannii*, *Abies lasiocarpa*, *Pinus contorta* and *Tsuga mertensiana* in subalpine boreal climates. *L. occidentalis* forests typically have a species-rich understorey vegetation with dense herbaceous and less dense shrub layers.

Although introduced to Europe and Asia, *L. occidentalis* has not been grown as a timber crop species outside its native range. It is grown in arboreta and parks.

Silviculture characteristics

L. occidentalis is characterized by rapid growth, shade-intolerance, and windfirmness.

Reproduction capacity is high, with seed production beginning at <25 years old. The northern population, however, is a poor seed producer. Seed dissemination capacity is intermediate, with seed dispersed some 250 m from the parent tree.

Potential for natural regeneration in low light is practically nil, but high in the open in the presence of exposed mineral soil or burnt forest floor. Partial shade is beneficial for seedling establishment. Potential initial growth rate (to 5 years old) is intermediate, being low (<5 cm/yr) in the establishment period and high (>50 cm/yr) thereafter.

Advance regeneration does not develop in the absence of adequate light and seedbeds.

Self-pruning capacity in dense stands is high. With wide crowns, crown spatial requirements are high. Light conditions beneath closed-canopy, mature stands are high, and associated with well-developed understorey vegetation.

Potential productivity is high, *L. occidentalis* being the most productive deciduous conifer. Longevity is high, probably even greater than the reported maximum of 900 years.

Silviculture practice

Depending on the site and management objective, clear-cutting, patch-cutting, strip shelterwood and seed-tree systems are practised. Considering its shade-intolerance and fast growth rate, *L. occidentalis* is a desirable component in mixed-species stands which include shade-intolerant or shade-tolerant tree species.

Propagation by seed is the only contemporary method for regenerating *L. occidentalis*. Natural regeneration and/or planting of containerized stock are used for stand establishment. Burning or scarification is required for successful natural regeneration.

Trees with the largest crowns produce the most cones. Cone crops differ substantially by year and location. Long-term records of seed production show that good seed crops are produced at about 5-year intervals, with fair to poor crops in the intervening years. Larch starts bearing abundant cone crops from 40 to 50 years and continues bearing heavily for 300–500 years. Only dominants and co-dominants produce significant numbers of cones.

A mature cone may have as many as 80 filled seeds, but the average is about half that number. Seed viability is related to cone crop size, ranging from 5–10% in poor crops to 70–80% in good crops. Seeds are small and lightweight, averaging 20,000 per kg.

L. occidentalis has the potential for rapid diameter growth, but overstocking, insects and dwarf mistletoe often prevent full realization of this potential. Stand density affects height and diameter growth very early in the life of the stand. Heavy overstocking is common in young stands, with densities sometimes exceeding 86,500 trees/ha. Thinning such stands relieves growth suppression, but even the dominant trees in unthinned stands perform well below their growth potential. Thinning in young stands, preferably before age 20, enhances diameter and height growth during the juvenile years when response potential is greatest.

A positive diameter growth response is obtained by fertilization with nitrogen, but the effects last only about 3 years and height growth is unaffected.

Soil moisture and nutrients account for over 80% of the variation in site index (50 yr @ bh), which ranges from 10 m on low-productivity sites to nearly 30 m on high-productivity sites. Basal area increases rapidly to about age 40 years, decelerates and nearly levels off after age 100. At age 100, the basal area of *L. occidentalis* stands approaches 69 m^2/ha on high-productivity sites and about 46 m^2/ha on low-productivity sites. Projected yields for larch forests at age 100 range from 308 to 813 m^3/ha on low- and high-productivity sites, respectively. With full stocking (but not overstocking), 544 m^3/ha is a reasonable objective by age 100 on intermediate sites.

Diseases and pests

Important fungal diseases include *Hypodermella laricis*, *Fomitopsis officinalis*, and *Porodaedalea pini*.

Dwarf mistletoe (*Arceuthobium laricis*) is the most damaging parasite of *L. occidentalis*. It can infect seedlings as young as 3–7 years old and continues throughout the life of the tree. In addition to killing tree tops, reducing seed viability, creating conditions suitable for other diseases and insects, and causing burls, brashness and some mortality, it decreases height and diameter growth. Infected residual trees left after harvesting or fires promptly infect other trees, as mistletoe seed can be ejected as far as 14 m.

Larch casebearer (*Coleophora laricella*) and western spruce budworm (*Choristoneura occidentalis*) are the most serious insect pests of *L. occidentalis*; they can sever the terminal leader, resulting in an average loss of about 25–30% of the height growth for that year.

Uses

L. occidentalis forests are appreciated for their multiple resource values. The seasonal change in hue of foliage from light green in the spring and summer, to gold in the autumn, enhances the beauty of the montane forests. Because it is a successful pioneer species, it quickly regenerates on disturbed sites and provides protection for watersheds. *L. occidentalis* forests provide ecological niches for a wide range of birds and animals.

L. occidentalis is used extensively for lumber, fine veneer, long-straight utility poles, railroad ties, mine timbers and pulpwood. The wood is heavy, hard and strong with an air-dry density of 593 kg/m^3. The heartwood is brown and moderately resistant to decay, while the sapwood is yellowish-white and narrow.

The wood of *L. occidentalis* contains about 4–23% arabinogalactan. It is the best domestic source of this water soluble gum, used for offset lithography and in the food, pharmaceutical, paint, ink and other industries.

Further reading

Parent, D.R. and Mahoney, R.L. (2008) *Western Larch: A Deciduous Conifer in an Evergreen World*. Station Bulletin - Idaho Forest, Wildlife and Range Experiment Station No. 90, Idaho, USA.

Rehfeldt, G.E. and Jaquish, B.C. (2010) Ecological impacts and management strategies for western larch in the face of climate-change. *Mitigation and Adaptation Strategies for Global Change* 15, 283–306.

Schmidt, W.C., Shearer, R.C. and Roe, A.L. (1976) *Ecology and Silviculture of Western Larch Forests*. Technical Bulletin, Forest Service, United States Department of Agriculture, Washington, DC, No. 1520, 97 pp.

Shearer, R.C. (1999) *Coram Experimental Forest: 50 Years Research in a Western Larch Forest*. General Technical Report - USDA Forest Service, Rocky Mountain Research Station No. RMRS-GTR-37, Ogden, Utah, USA.

Larix sibirica Ledeb.

Siberian larch

Importance

Larix sibirica, a deciduous conifer distributed widely across north-western Asia up to the northern tree line, is a pioneer species highly adaptable to various climatic conditions, tolerant of extremely wet soils and poor fertility levels, and achieving very good yields on well-drained, moderately fertile sites. Its wood is durable and resistant, suitable for many uses as both timber and pulpwood.

Botanical features

L. sibirica is an elegant tree with a remarkably narrow crown and straight stem. Branches are irregularly arranged, slender, and nearly horizontal, with somewhat drooping branchlets. The bark is typically scaly, with vivid reddish-purple inner bark.

In northern Russia, *L. sibirica* is only a small tree, exhibiting stunted growth in the extreme north. Under more favourable conditions further south, it can attain a height of over 40 m, a diameter of about 1.5 m, and a crown width of 7.5 m. Trees live about 200 years.

Needles are grouped in fascicles of 10–50, measure 5–60 mm long and 0.5 mm wide, are keeled, and have several rows of stomata. Young shoots are usually straw-yellow or whitish-grey, and hairless. Young trees are most easily recognized and differentiated from *L. decidua* by a sweet scent which becomes apparent under dry conditions. Before being shed in the autumn, the needles take on a spectacular golden-yellow colour.

Female cones of *L. sibirica* are usually wide–ovoid, borne erect on dwarf twigs, light brown, 24–45 mm long, 20–30 mm wide, and have 22–40 scales. During the flowering season, cone bracts are usually red, very seldom green, and much longer than the cone scales. On mature cones, however, the bracts are hidden by the cone scales. Seeds are 4–5 mm long by 3–4 mm wide, with a wing 4–6 mm long by 3 mm wide. Mean weight of 1000 seeds varies from 8.3 to 9.2 g. The percentage of filled seeds varies from 17% to 80%.

Flowering begins at a temperature sum of 18–20°C (April or May), and needle flushing at 50–60°C. Within a given tree, pollen is usually shed from the male cones 1–2 days before the female flowers become receptive. Leaves appear during or just after flowering, and usually fall in early autumn. Seeds are mature in October or November. Cones open in autumn and seeds are dispersed throughout the winter and following spring. Empty cones usually persist on the tree for 2–3 years.

Distribution

L. sibirica is a dominant tree in northern Russia and western Siberia, forming unbroken stands from Lake Baikal in the east to the White Sea, north to the mouth of the Yenisei River on the Arctic Sea, and from there south to the Altai Mountains. The eastern boundary of its range is contiguous with the limits of Dahurian larch (*Larix gmelinii*), overlapping along a north-south boundary. *L. sibirica* is also found in the Xinjiang Province of China, and in northern Mongolia.

It is an important commercial timber species in Eastern Europe. Plantations have also been established in the UK and in the Canadian Prairies, mostly for shelter belts.

Silviculture characteristics

L. sibirica is shade-intolerant, requiring abundant light and space, although light demand decreases with age.

Under favourable growing conditions, it forms more or less pure stands, but is also found scattered in forests among other conifers. As a result of forest fires and clear cuts in the far north, it is often supplanted by birch (*Betula* spp.), with such secondary stands being naturally replaced by *L. sibirica* only after 100–150 years.

L. sibirica forests are characterized by a relatively high natural fire hazard, although on the cold permafrost soil of northern Siberia trees survive fire well due to thick bark. In certain regions, fire helps natural regeneration by eliminating the dense understorey of companion species and increasing the depth of thawed soil in the summer.

L. sibirica is sensitive to air pollution and is a highly effective bioindicator of SO_2 pollution.

Silviculture practice

L. sibirica is readily propagated by seeds, but seed supply is scarce due to irregular flowering and low filling percentage. Seeds are orthodox and remain viable for at least 3 years. They are stratified by being kept under snow for 2–3 months, then treated with potassium iodide solution. Germination frequency is low, around 40%. Seeds are very susceptible to diseases, and seedlings show high mortality.

L. sibirica can be propagated vegetatively, mainly by softwood or hardwood stem cuttings from juvenile trees, or by tissue culture. Grafting is also used in horticulture and to establish clonal seed orchards. Various *Larix* species can be grafted onto *L. sibirica* rootstocks to produce rapidly growing trees.

For *L. sibirica* plantations, 2–4-year-old bare-root seedlings (15–60 cm high) are generally used. The main reasons for *L. sibirica* seedling mortality during the first year are groundwater deficit and high temperature in the upper soil layer, and also late spring and early autumn frosts.

Plantations are established on fertile, moderately wet or fresh and aerated soils. Abandoned arable lands can also be used. Clear-cut areas are regenerated in a regular way, while enrichment planting is used for areas with well-maintained advance growth, or which are already covered by successful second growth. Some 3300–4400 seedlings/ha are planted when establishing pure cultures. Commercial plantations of mixed species, in which up to 25% of trees may be larch, are established by alternating larch seedlings with seedlings of shade-tolerant species in the first row, and leaving the second row exclusively for the shade-tolerant species. In commercial plantations in eastern Siberia, seedlings of *L. sibirica* create a closed canopy 5–10 years after planting. At this stage, the number of saplings varies from 900 to 2800/ha, with height ranging from 0.7 to 1.8 m.

The planting period in spring is very short, since larch begins to grow very early. Ordinary soil preparation involves making strips or furrows. Complete soil preparation takes place only on bare places, in gaps, and on abandoned arable land. Soil loosening is important in artificial plantations as larch growth is negatively affected by sod formation and soil compaction. Despite high competitiveness against weeds, regular weeding operations result in a much higher growth rate.

Clear-cutting is the main felling system in commercial forests. In even-aged larch and mixed forests on plains and on gentle slopes which are sufficiently supplied with advance growth, the maximum width of clear-cut areas should not exceed 500 m for an area no larger than 50 ha. On steep slopes in Siberian mountain ranges, two-step shelterwood felling is recommended, while in uneven-aged forests on plains or mild slopes, the long-term shelterwood system is preferred. In central and southern Siberia, harvest age varies from 120 to 140 years.

In natural, uneven-aged stands, wood volumes up to 1000 m^3/ha are produced, with average tree heights of 36–38 m and diameters of 45–50 cm. Mean growing stock yield can reach 400–500 m^3/ha, but under severe conditions in the north, in swamps and other poor sites, this decreases to 50–80 m^3/ha.

Diseases and pests

Pathogenic fungi recorded on *L. sibirica* include *Phaeolus schweinitzii*, *Heterobasidion annosum*, *Lachnellula willkommii*, *Spadicoides atra* and *Zalerion arboricola*.

Meria laricis needle cast causes seedling mortality in *L. sibirica* both in Europe and North America. Young seedlings can also be infected by grey mould (*Botrytis cinerea*). Reducing seedling density and allowing foliage to dry between water applications can correct the situation.

L. sibirica is the least susceptible of the larches to defoliation by larch sawfly (*Pristiphora erichsonii*). Unless intense defoliation is repeated over several consecutive years, no mortality is observed, but growth is reduced and trees are weakened. *Dasyneura laricis* sometimes affects *L. sibirica* stands, but there is considerable intra- and interpopulation variability in the level of tree damage. In Russian seed orchards, *L. sibirica* can also be attacked by *Dasyneura rozkovi*, causing bud death.

Uses

A narrow, pyramidal form, fast growth, and relatively non-competitive root system make *L. sibirica* well suited for windbreaks, and for farmstead and field shelterbelts. It is used for amenity plantings, largely due to its autumn foliage colour.

L. sibirica wood is hard and strong, among the toughest and most durable of softwoods, with high values for physical and mechanical properties and natural resistance to wood-destroying fungi and insects. Average wood density ranges from 685 kg/m^3 in the northern taiga to 620 kg/m^3 in the forest steppe zone. Slow kiln drying is recommended, since the wood is liable to check, split and warp. Timber is used for construction beams, sawn timber, chipboard, furniture, railway sleepers, fuel and pulp. *L. sibirica* wood is in demand for outdoor wooden constructions, such as park and playground equipment. Roundwood is used for pit props and building poles, while sawn timber is suitable for pallets. Low-quality wood is used in particleboard production.

Despite having a heartwood containing fairly high levels of water-soluble extractives, mainly arabinogalactans, which interfere with the pulping reaction in ordinary acid sulphite processes, it is possible to get bleached and unbleached pulp for paper and cardboard production using sulphate cooking of *L. sibirica* wood.

Dulcitol, a hexahydric alcohol, is a by-product of larch pulping. Other valuable chemical products are obtained from the complex processing of *L. sibirica* wood and bark, including tanning extracts, food dyes, colophony and turpentine. Traditionally, gloves are made from the inner bark.

Hydrolysis of *L. sibirica* biomass residues gives fodder sugars and syrup. Stem waste can be used to produce charcoal, while the foliage is a raw material for obtaining chlorophyll-carotene paste, chlorophyll sodium and essential oils.

Further reading

Abaimov, A.P. (1998) *Variability and Ecology of Siberian Larch Species*. Rapporter - Institutionen för Skogsskötsel, Sveriges Lantbruksuniversitet No. 43, Swedish University of Agricultural Sciences, Umeå, Sweden.

Osawa, A., Zyryanova, O.A., Matsuura, Y., Kajimoto, T. and Wein, R.W. (eds) (2010) *Permafrost Ecosystems: Siberian Larch Forests*. Ecological Studies vol. 209. Springer, Dordrecht, The Netherlands.

Shuman, J.K., Shugart, H.H. and O'Halloran, T.L. (2011) Sensitivity of Siberian larch forests to climate change. *Global Change Biology* 17, 2370–2384.

Leucaena leucocephala (Lam.) de Wit

Leucaena

Importance

Leucaena leucocephala (Plate 21) is a versatile, fast-growing, nitrogen-fixing tree well suited to management in agroforestry systems. Its value lies in its combination of multiple products, exceptional fodder quality, fast growth, and ease of propagation and management by farmers. It is one of the foremost tropical fodder trees described as the 'alfalfa of the tropics' and was one of the first species to be grown for green manure in alley farming systems. Its uses and benefits led to pantropical promotion and planting, with *L. leucocephala* heralded as a 'miracle tree' by national and international development agencies.

Botanical features

A small tree, typically 3–15 m tall and 10–35 cm in diameter, sometimes reaching 20 m tall and 50 cm in diameter. Form varies from shrubby and highly branched in the 'common' or 'Hawaiian' type of subsp. *leucocephala* to upright with a short clear bole to 5 m and an open, rounded crown in the 'giant' or 'Salvador' type of subsp. *glabrata*. Bark is mid grey-brown with shallow rusty orange-brown vertical fissures. Leaves are bipinnate with 4–9 pairs of pinnae per leaf and 13–21 pairs of leaflets per pinna. Leaflets are small, 9–21 mm long, 2–4.5 mm wide, linear–oblong or weakly elliptic, tip acute, base rounded to obtuse, with a concave, cup-shaped, elliptic petiole gland. White flowers are arranged on compact globose heads in groups of 2–6, with each head 12–21 mm in diameter with 100–180 flowers (Plate 51). Hairy anthers (visible with a hand lens) distinguish *Leucaena* from all other mimosoid legume genera. Pods are 9–19 cm long, 13–21 mm wide, linear–oblong, and flat with papery pod walls, mid- to orange-brown, glabrous and slightly lustrous (subsp. *glabrata* and subsp. *ixtahuacana*), or densely covered in white velvety hairs (subsp. *leucocephala*), arranged in clusters of 3–20, and occasionally up to 45, per flower head (Plate 58).

Distribution

L. leucocephala originates from Mexico, but its introduction and cultivation since prehistory means that its precise natural distribution is extremely difficult to ascertain. It has been very widely introduced and in probably present in every tropical country. It may have been introduced into the Philippines as early as the 1600s, and by the 1800s it was already present throughout Asia and Africa.

Silviculture characteristics

L. leucocephala is essentially a tropical species, requiring warm temperatures of 25–30°C for optimum growth, but with poor cold tolerance and significantly reduced growth during cool winter months in subtropical areas. It grows best below about 1500 m altitude and 15–25° N or S of the equator, although it can survive over a much wider range to about 2000 m and to 30°N or S. Trees shed their leaves even after light frosts, and heavy frosts kill all above-ground growth although trees often resprout after die-back. It prefers sub-humid or humid climates (650–3000 mm) with a moderate dry seasons of up to 6–7 months, although it can survive with rainfall of as little as 300–500 mm.

L. leucocephala is a fast-growing, light-demanding species adaptable to a wide range of production systems at a wide range of spacing and with regular and intensive lopping. It resprouts vigorously after coppicing or pollarding. Trees are self-fertile and produce prodigious quantities of seed from the first year, and more or less continuously throughout the year as moisture permits. It is slightly or completely deciduous depending on the length and severity of the dry season. *L. leucocephala* has the ability to form a symbiotic association with *Rhizobium* root nodule bacteria which are able to fix atmospheric nitrogen. With wider planting, a number of important limitations have become apparent including lack of cold and drought tolerance, poor growth on acid soils, heavy pod production resulting in weediness, low wood durability and susceptibility to the defoliating psyllid Leucaena.

Silviculture practice

Precocious and virtually year-round seed production means it is not difficult to obtain large quantities of seed, with easy collection, extraction, processing and storage. Seed requires pretreatment, and soaking in hot water is most widely used. Despite application of a wide range of propagation techniques including rooted stem cuttings, tissue culture, grafting and air-layering, successful vegetative propagation has been limited to experimental rather than commercial operations.

Nursery procedures are straightforward and although establishment normally relies on container-grown nursery seedlings, direct sowing and even aerial sowing have also been used successfully. Site preparation is important, as slow early growth makes young transplants susceptible to weed competition, and the success of direct seeding depends on timing, land preparation and weed management.

For fuelwood plantations where small-diameter material harvested on short (2–3 year) rotations is required, maximum total yields are obtained at close spacing (10,000–20,000 trees/ha). For the production of larger poles, wider spacing (5000–10,000 trees/ha) and longer (3–6 year) rotations are required. High wood yields can be maintained over a wide range of rotation lengths, stand densities, and systems of management. For fodder production, high plant densities (50,000–150,000 plants/ha) are coppiced at above 60 cm high every 2–4 months to produce highly branched shrubs that regrow quickly.

On favourable sites when moisture is not limiting, exceptionally high annual yields of 40–80 t/ha fresh leaf are possible, and 20–50 t/ha in seasonally dry or subtropical climates. Wood production from short (3–5 year) rotations compares favourably with other species, with annual growth rates of 3–4 m in height and 10–60 m^3. However, on less favourable sites such as in cooler tropical highland areas, on acid soils, or under high psyllid pressure, yields have often been disappointing.

Diseases and pests

L. leucocephala is susceptible to a leaf spot disease caused by the fungus *Camptomeris leucaenae* causing die-back in areas that receive >2000 mm precipitation. Gummosis is another serious disease, being a canker caused by a fungus in the genus *Fusarium*.

The most important pest is the Leucaena psyllid (*Heteropsylla cubana*) causing wilting, defoliation and tree mortality. It was first detected in Hawaii in 1984, and infestations suddenly appeared over large areas of the Pacific Ocean, Australia, and South Asia by 1988, and the Indian Ocean and East Africa by 1991. In Latin America, large numbers of seeds are eaten by five species of bruchid beetles, including three species of *Acanthoscelides* and two species of *Stator*, and are also hosts for other seed beetles.

Uses

L. leucocephala has been used for soil improvement, soil conservation and erosion control in diverse agroforestry combinations and systems including alley farming, live barriers on terrace boundaries, shelter belts or windbreaks, or simply as dispersed trees over crops. It has also been used in wider land rehabilitation and revegetation efforts, sometimes using aerial seeding. However, weediness can be a problem.

The wood is of medium density (500–700 kg/m^3), with light reddish-brown heartwood and pale yellow sapwood, machines easily, and dries without splitting or checking. It is used for domestic and industrial fuel, poles, posts, sawn timber, furniture, parquet flooring, particleboard and pulp. However, its use as a sawn timber is limited by commonly small dimensions, often less than 30 cm in diameter, knotty and a high proportion of juvenile wood, though advances in timber processing means there is increasing use of small-dimension sawn wood. Use of poles is limited by lack of durability and susceptibility to termites and wood borers. Fuelwood and charcoal is of acceptable, although not highest, quality. It has not become a common plantation forestry species.

L. leucocephala produces one of the highest-quality and most palatable fodders of all tropical trees, comparing favourably with alfalfa (*Medicago sativa*) in feed value except for its higher tannin content and mimosine toxicity to non-ruminants. Leaves have high nutritive value (high palatability, digestibility, intake and crude protein content), resulting in impressive animal production with 70–100% increases in animal live-weight gains compared with pure grass pastures. In addition, it is very persistent over several decades of cutting or grazing, is very productive, recovers quickly from defoliation, combines well with companion grasses and can be grazed with minimal losses due to trampling or grazing. Leaf meal is also milled, pelleted and sold internationally for incorporation in commercial poultry and pig feed.

The undesirable (toxic) effects of feeding *Leucaena* fodder are largely attributed to the non-protein amino acid mimosine, which is converted to DHP (3-hydroxy-4-1(H)-pyridone) in the rumen. Toxicity can be avoided by limiting feed intake to less than 30% of the diet, but for ruminants, a much more effective solution is the introduction of the rumen microbe *Synergistes jonesii* which is capable of detoxifying mimosine and DHP. Mimosine remains a significant problem for non-ruminants, however, and is the main disadvantage of *Leucaena* in leaf meal production compared with alfalfa.

Leaves of *L. leucocephala* are widely used for green manure, with widely recognized benefits in terms of soil fertility. They decompose quickly, providing a very rapid influx of nutrients, and maize yields can be maintained in *Leucaena* alley farming in the absence of fertilizer inputs.

In Mexico, *L. leucocephala* is an important food plant and is widely cultivated in backyards, gardens and small orchards for the production of edible unripe pods and seeds which are consumed and widely marketed throughout the country. In Asia, unripe pods and seeds are also used, as well as young leafy shoots, fermented seeds and sprouted seeds.

Further reading

Brewbaker, J.L. (1987) *Leucaena*: a multipurpose tree genus for tropical agroforestry. In: Steppler, H.A and Nair, P.K.R. (eds) *Agroforestry: A Decade of Development*. International Council for Research in Agroforestry, Nairobi, pp. 289–323.

El-Nor, S.A.H.A. (2003) *Leucaena leucocephala* as a feed for ruminants in tropical and sub-tropical areas. *Egyptian Journal of Nutrition and Feeds* 6, 189–208.

Hughes, C.E. (1998a) *The Genus* Leucaena*: A Genetic Resources Handbook*. Tropical Forestry Paper 37. Oxford Forestry Institute, Oxford, UK.

Hughes, C.E. (1998b) Monograph of *Leucaena* Benth. (Leguminosae: Mimosoideae). *Systematic Botany Monographs* 55, 1–244.

Lithocarpus spp.

Stone oaks

Importance

Lithocarpus is a large Asian genus in the beech family Fagaceae. The Plantlist (available at: http://www.theplantlist.org (accessed 4 April 2013)) accepts 334 distinct taxa while other sources recognize as few as 100 species. The Californian 'tanoak' used to be included in *Lithocarpus*, but recent genetic evidence led to its reclassification and renaming as *Notholithocarpus densiflorus*. It is excluded here.

The species do not have well-known English vernacular names, though the generic term 'stone oak' has been proposed. In the equatorial rain forests of South-east Asia, *Lithocarpus* species are important timber trees, second only to the Dipterocarpaceae.

Important timber species include *Lithocarpus coopertus*, *Lithocarpus elegans*, *Lithocarpus encleisacarpus*, *Lithocarpus lampadarius*, and *Lithocarpus sundaicus*.

Stone oaks are an important element of old-growth forests in South-east Asia and reach the highest densities and diversities in the forested highlands.

Several of the species are attractive ornamental trees and are planted in parks and large gardens in warm temperate and subtropical areas.

Due to heavy losses of lowland and hill rainforest habitat, three *Lithocarpus* species (*Lithocarpus platycarpus*, *Lithocarpus kostermansii*, and *Lithocarpus crassinervius*) are listed as 'Endangered' and another seven are listed as 'Vulnerable' on the Red List of Threatened Species of the International Union for Conservation of Nature (IUCN) .

Botanical features

Lithocarpus are usually evergreen trees, with solitary terminal buds and erect flower spikes. The trees have leathery, alternate leaves, which may be either entire or toothed. Flowers are solitary or in clusters of 3 or more, with cupular calyx.

The seed is a nut similar to an (oak) acorn, but with a very hard woody nut shell (hence the genus name, from Greek *lithos*, stone, and *carpos*, seed). In some species the nut kernel is edible (e.g. *Lithocarpus edulis*) and some may be scatter-hoarded by spiny rats and/or terrestrial squirrels.

Distribution

All *Lithocarpus* are native to tropical Asia. They occur from sea-level up to over 2600 m altitude. The centre of evolutionary diversification is situated in western Malaysia, with the highest numbers of species and endemism occurring in Borneo. The genus is a major component of mid-montane Fagaceae–Myrtaceae forests in China (1800–2400 m above sea level), with four species forming around half of the total basal area. In the Bird's Head area of Irian Jaya (West Papua) it is one of the three most frequent genera found, and is especially common in secondary forest.

The genus is not found in the lesser Sunda Islands, and is restricted to southern tropical and semi-tropical parts of Indochina and coastal East Asia, indicating an intolerance of drier and more seasonal climates.

Silviculture characteristics

There has been little research on the behaviour of these trees aside from some ecological work on particular species.

Symbiosis with myccorhizae is required for successful establishment of *Lithocarpus* seedlings.

Seeds are hoarded by seed-caching rodents. A study in south-west China monitored tanoak (*Lithocarpus harlandii*) seedlings established from caches in the primary stands over a 12-month post-dispersal period. Some nut traits in tanoak species, such as large seed size, hard nut husk, lower tannin, and mast seeding, were important characteristics for seed dispersal by scatter-hoarding rodents, compared with oak species with higher tannin content.

Study of the effect of disturbance on natural regeneration of *Lithocarpus dealbatus* (and others) in India was carried out in three stands, differing in degree of disturbance, during the years 1988–1991. Seed production increased with increasing disturbance as well as with increase in dbh of the trees. The trees produced heavier seeds in the disturbed stands than in the undisturbed stand. Seed predation decreased and germination increased with distance from the parent tree in all three stands, suggesting that distance-related seed predation was not influenced by disturbance. The greatest germination of seeds was observed on the moss layer, irrespective of species and forest stand. The presence of litter inhibited germination of seeds to a great extent and germination was better in the undisturbed stand than in the disturbed stands.

Despite being recognized as a valuable timber species, *L. sundaicus* is not cultivated.

Attempts to use *Lithocarpus garrettianus* as one of 20–30 so-called 'framework species' to restore degraded tropical forestland in northern Thailand failed when dormancy time was revealed to be as long as 7 months, and seedling emergence was very slow at 11 months.

Coppicing experiments in the Philippines have shown that regeneration of local *Lithocarpus* species from sprouts is feasible.

Silviculture practice

Little is known about the behaviour of these species in production forests. In experimental plantations in Java it was found that the bark of *L. sundaicus* trees as young as 5 years already contained large amounts of tannin (16.5–22% on a dry-weight basis). A tree 8 m tall, with a trunk dbh of 12 cm, has an average yield of 3.5 kg dry bark. The bark can easily be removed from the trunk. In general, bark is dried in the open air, and normally it is used in tanneries as it is and not made into extract.

Diseases and pests

The poplar and willow defoliator caterpillar (*Lymantria obfuscata*) may attack the forestry species 'Indian oak' (*L. dealbatus*) in Kashmir and the north-west Himalayas. Parasites (including *Anastatus kashmirensis*, *Rogas indiscretus* and *Dibrachys cavus*) may attack the pest.

From a silvicultural point of view it is sometimes viewed as a problem that many animals like the fruits of *Lithocarpus*, thus limiting natural regeneration; however, the contrasting argument is that animal dispersal is key to the natural ecology of these species.

Uses

Lithocarpus spp. produce good timber, valued for a wide range of applications in construction as well as for furniture and tool handles. Wood densities of the important timber species are: *L. coopertus*, 830–910 kg/m^3; *L. elegans*, 720–960 kg/m^3; *L. encleisacarpus*, 520–720 kg/m^3; *L. lampadarius*, 850–1030 kg/m^3; and *L. sundaicus*, 500–700 kg/m^3, at 15% moisture content.

Before reliance on rice cultivation emerged, societies in the Lower and Middle Yangtze river basin have long relied on the seasonal collecting and storage of acorns including *Lithocarpus*.

The bark of Sunda oak (*L. sundaicus*) has been used occasionally in Indonesia (Java) for tanning hides. The tannin has excellent properties, giving a fine leather of light colour. Although it was already being recommended some 70 years ago as a good source of tanning material (air-dried bark contains 15–22% tannin), it has never been used extensively.

The seeds of several species are ground and eaten as starchy food, and *Lithocarpus hancei* seeds are used for brewing alcohol in Yunnan. Young shoots of *Lithocarpus polystachys* are dried for tea.

Further reading

Cannon, C.H. and Manos, P.S. (2003) Phylogeography of the South-east Asian stone oaks (*Lithocarpus*). *Journal of Biogeography* 30, 211–226.

Lemmens, R.H.M.J. (1991) *Lithocarpus sundaicus* (Blume) Rehder Record from Proseabase. In: Lemmens, R.H.M.J. and Wulijarni-Soetjipto, N. (eds) PROSEA (Plant Resources of South-east Asia) Foundation, Bogor, Indonesia. Available at: http://www.proseanet.org (accessed 17 January 2013).

Martawijaya, A., Kartasujana, I., Kadir, K. and Prawira, S.A. (1986) *Indonesian Wood Atlas. Vol. 1.* Forest Products Research and Development Centre, Bogor, Indonesia, pp. 111–115.

Soepadmo, E. (1972) Fagaceae. In: van Steenis, C.G.G.J. (ed.) *Flora Malesiana, Series 1. Vol. 7.* National Herbarium of The Netherlands, Leiden, The Netherlands, pp. 375–376.

Lovoa trichilioides Harms

African walnut

Importance

Lovoa trichilioides, the only West African species of the genus *Lovoa*, is a large monoecious forest tree growing to 40 m. It is associated with acid, base-poor soils. On suitable sites, growth is fast and stem form is good. It has valuable timber, and is less susceptible to tree borers and other insects than other *Lovoa* species. *L. trichilioides* is suitable for establishment of plantations and for enrichment planting to improve degraded forest. The seeds are relatively short-lived. It is one of the two main timber species in the Congo and exploitation rates are high. Plantations have been established in Cameroon, Nigeria, Uganda, and Côte d'Ivoire.

Botanical features

L. trichilioides, a large emergent evergreen forest tree, grows to a height of 40–46 m, with a dbh of up to 180 cm. It has a straight, cylindrical bole of about 27.5 m, with short, thick ascending buttresses. It has a dark compact crown which is not very dense but which tends to open out a little when older.

The bark is thin and smooth, has numerous lenticels and darkens with age from grey to almost black. Old trees are scaly towards the base. The slash is thick, fibrous and red with vertical streaks and is highly cedar-scented. There is a gummy exudation from wounds.

The leaves are alternate, glabrous, and paripinnate, with about 4–6 pairs of opposite or subopposite leaflets, exstipulate. The rachis is about 14 cm long and is slightly winged. The leaflet is oblong–elliptic, about 14 × 5 cm in size, entire with a short, broad tip which is often retuse, cuneate at the base, dark shining green above and dull lighter green below. The leaflet is not flat and is often twisted slightly towards the apex and base. The midrib is prominently raised below. The nerves are conspicuous on both sides. The petiolule is about 0.6 cm long, flattened above and channelled slightly. The leaf of the sapling is about 60 cm long and its leaflets are oblong and not retuse.

Flowers are small, white, scented and paniculate. The fruit is a pendulous, elongated, 4-sided, semi-woody capsule about 5 cm long. The valves are papery and separate at the base. There are often only 2 fertile seeds on each face of the central columella and they are inserted by their wings. The wing is terminal and is about 3 × 0.75 cm. The seeds are wind dispersed.

L. trichilioides is evergreen. Flowering is most profuse in December–January, although flowers occur on the tree most of the year. Fruiting occurs in March and April, but is irregular.

L. trichilioides forms vesicular arbuscular mycorrhizal associations with some fungi.

Distribution

L. trichilioides is naturally distributed in the Guinea–Congolian region, from 10°N to 10°S, from Sierra Leone east to western Uganda, and south to north-western Tanzania and northern Angola.

Although it is an evergreen forest emergent, it also occurs in thickets, gallery forest and mixed tropical forests, and is extremely common in lakeside forests in Uganda.

Silviculture characteristics

L. trichilioides is very shade-tolerant when young, but is light-demanding later. It is more shade-tolerant when young than *Entandrophragma* and *Khaya* spp.

It prefers deep humid soils and can grow at altitudes up to 1200 m.

Silviculture practice

Mature seeds of *L. trichilioides* are classified as orthodox, and there are about 4400 seeds/kg. The species seeds profusely, and seeds are gathered from the forest floor. As they are highly vulnerable to insect attack, it is advisable to store seeds in sealed containers in a cool place. Seeds germinate in about 16 days with a germination frequency of about 90%, but seed viability is lost rapidly and germination declines to about 30% after 2 months, so storage should be limited. Seeds should be sown with wings pointing upwards and three-quarters covered with soil. Young seedlings require shading in the nursery, but are relatively resistant to insect attacks. They can attain a height of 1 m after 2 years.

Rooted stem cuttings, seedlings and wildings are used in propagation. *L. trichilioides* grows well when outplanted in groups or lines under thinned natural forest. Trees tend to branch early and pruning should be done to improve tree form. In plantations, trees can take up to 30 years to mature.

Thinning to about 300 trees/ha by year 15, and to 100 trees/ha by year 30 to enable the trees to attain a diameter of about 90 cm in 60–70 years is recommended.

Average heights of 9 m over the first 7 years, and an annual diameter increment of 1.0–1.8 cm have been recorded. Initial growth rate is slower than in *Khaya ivorensis*, but growth rates increase with age. A very low average annual yield of only 1.8 m^3/ha was obtained when *L. trichilioides* trees were planted in unfavourable areas where the trees were not maintained and were subject to shade.

L. trichilioides in plantations with *Nauclea* and *Khaya* attained a diameter of about 25 cm after 25 years; trees with heights of 30 m and diameters of 30 cm after 21 years have been recorded.

Minimum diameters for felling range from 60 to 80 cm. Caution should be exercised during felling as logs can suffer from heart shake and brittle heart.

Diseases and pests

Wood deterioration by basidiomycetes such as *Gloeophyllum sepiarium* and *Pleurotus* spp. is common.

Large-scale destruction of seedlings by shoot borers has been recorded. Longhorn beetles, borers, scale insects and defoliators are the main pests observed in young plantations, and collar decay also occurs. Antelopes and rodents eat the bark of seedlings and young trees. Fruits and seeds are subject to heavy predation by insects such as *Catopyla dysorphnaea*.

Uses

L. trichilioides is suitable for establishment of plantations, for enrichment planting to improve degraded forest, for afforestation and for intercropping with crops such as coffee and banana. It is also used as a shade and shelter tree, and is planted as an ornamental avenue tree.

L. trichilioides sapwood is white and lustrous. The heartwood, which sometimes contains holes of 1 cm diameter which are probably caused by insects, is golden-brown, often with dark brown or black lines caused by gum veins. It is this colour which gives it the trade name of African walnut. The grain is normally interlocked. The wood is of medium weight, with a density of 450–610 (but up to 680) kg/m^3 at 12% moisture content, and has a basic specific gravity of 0.45. It is fairly strong and hard, though liable to attack by pinhole borers and longhorn beetles. It resists fungal attack. It works fairly easily, although it is liable to pick up on quarter-sawn material. It nails, glues and stains well and if well-sanded and filled it takes a fine polish. It has a tendency to split or warp and is resistant to impregnation. It is generally fairly easy to season. *L. trichilioides* wood is highly valued for furniture, cabinet work, flooring, panelling, veneers, carving and decorative work. It is suitable for gun stocks, railway sleepers, pulpwood, plywood and in the construction of railway carriages.

Locally, *L. trichilioides* also provides fuelwood and charcoal.

Further reading

Nyemb, N. (2008) *Lovoa trichilioides* Harms. In: *Timbers 1 (Plant Resources of Tropical Africa, volume 7)*. PROTA Foundation, Wageningen, The Netherlands, pp. 349–353.

Styles, B.T. and White, F. (1991) *Lovoa trichilioides* Harms. In: Polhill, R.M. (ed.) *Flora of Tropical East Africa: Meliaceae*. A.A. Balkema, Rotterdam, The Netherlands, p. 59.

Tchoundjeu, Z. and Leakey, R.R.B. (2001) Vegetative propagation of *Lovoa trichilioides*: effects of provenance, substrate, auxins and leaf area. *Journal of Tropical Forest Science* 13, 116–129.

M

Mansonia altissima (A. Chev.) A. Chev.

Mansonia

Importance

Mansonia altissima is generally a medium-sized tree with great potential for plantation establishment. It is moderately shade-tolerant in the first few years, requiring light shade as a sapling, but becoming more light-demanding with age. The species is moderately drought-tolerant and prefers relatively fertile soils. It produces an attractive, walnut-brown wood also sometimes known as African black walnut, which is used for furniture manufacture and other quality products.

Botanical features

M. altissima is a medium-sized to fairly tall tree with a small crown, made up of short, slender horizontal branches. The average height is about 42 m, but ranges from 40.8 to 44.2 m, with a dbh ranging from 90 to 100 cm. The bole is clear, slender and cylindrical, but tapering and buttressed when mature. The buttress may reach 3.7 m in height.

The bark is light brown, fairly smooth, becoming furrowed vertically; later, it is greyish-brown with more or less circular but irregular scales. The blaze is a very light brown, often with white, vertical lines, darkening almost immediately on exposure, and thick. The sapwood is white, with prominent ripple marks. The heartwood is dark brown, sometimes with a purplish tinge.

The leaves are simple, alternate, obovate or orbicular, about 12.5 cm long and 7.5 cm broad, entire, slightly acuminate, cordate, often unequal sided, 7-nerved at the base and with numerous, fine, tomentose hairs on the underside of the lamina and some stellate hairs. The midrib, nerves, and veins are raised below and the nerves extend to the leaf margins without looping. The petiole is 2.5 cm or more long, tomentose, and swollen at the base and apex. The leaf on young trees and saplings may be 30 cm long and markedly dentate. The stipules are triangular and ciliate, about 1.8 cm long and 0.5 cm broad at the base, persisting for at least 1 year.

Axillary cymose inflorescences occur near the ends of branches. The flowers are white, scented, and hermaphrodite. The fruit is a samara about 7.5 cm long, with 1–5 on a common peduncle. The wing is unilateral, straight on one side and bulged on the other.

M. altissima trees are deciduous from February to early April. Flowering begins in June and is usually over by the end of August, with the flowers being showy. The fruits are large, pendulous, and conspicuous, and ripen from December to February. However, some observations have shown earlier and more prolonged flowering and fruiting periods, depending on location and forest type. Seeds are dispersed by wind but do not travel far from the mother tree.

Distribution

The natural distribution of *M. altissima* is within the Guinea Region in West Africa between 3°N and 10°N. Two botanical varieties are recognized: var. *altissima* occurring from Liberia to Nigeria, and var. *kamerunica* from Nigeria east to the Central African Republic. The latter is characterized by glabrous branchlets.

M. altissima is characteristic of dense semi-deciduous forest in areas with about 1600 mm annual rainfall and a pronounced dry season.

Plantations have been established in Cameroon and Nigeria.

Silviculture characteristics

M. altissima is moderately shade-tolerant in the first few years; it requires light shade as a sapling, and becomes a light demander upon maturity. It is also moderately drought-tolerant. It shows good regeneration, especially in areas where there is a high, light canopy.

Silviculture practice

The seeds are classified as orthodox, and have a 1000-seed weight of about 330 g. They have a short viability and should be sown as soon as they are collected from the forest floor beneath the trees. The germination period is about 16 days and germination is 70–85%, although it is often considerably less due to insect attack. After 6 months, seedlings are approximately 40 cm tall, and are 90 cm tall after 1 year. *M. altissima* may be planted out at 15 months, and can reach 4–5 m in 3 years. In tropical shelterwood plots, saplings reach 1–2 m in 4 years.

M. altissima can also be planted out as striplings or stumps. After planting out, a light overhead shade is beneficial until the plants are properly established. Plantains (*Musa* spp.) are useful shade plants. Seedlings show a preference for drier forests. In wet forest zones, plantations can be established, but trees trend to be short-lived.

M. altissima has great potential as a plantation species, although it suffers in plantations due to pests. Trees attain 6 m in height in 4 years in old logged forests. Mean annual increments over the first 21 years have been recorded as 5.5 cm in Nigeria. In plantations, height increments of 2 m/yr after 2 years or 1.3 m/yr after 10 years have been recorded in

Cameroon. A tree 40 cm in diameter yields about 1.8 m^3 of timber, while trees 50, 60, and 70 cm in diameter yield 2.8, 4.1, and 5.6 m^3, respectively.

Diseases and pests

Godasa sidae will skeletonize leaves of *M. altissima* in the nursery, and may also cause occasional defoliation of plantations. Ambrosia beetles attack the boles of standing trees. Wood-boring caterpillars of *Eulophonotus* spp. may cause damage to the sapwood. Larvae of the polyphagous grasshopper *Zonocerus variegatus* may seriously attack the foliage, especially of older trees, and in plantations they may become serious pests.

Uses

M. altissima can be used for plantation establishment in degraded and disturbed areas.

M. altissima wood is purple-brown when freshly cut, fading to an even walnut-brown. The sapwood is pale. The wood is straight grained with a fine even texture, dries easily, and has very high natural durability, medium to heavy strength, a density of 650 kg/m^3 and shows little shrinkage. It has good resistance to abrasion, easy sawing and machining properties, and excellent finishing properties. It is a well-known, attractive wood used for high-quality furniture, joinery, cabinet work, flooring, sliced and rotary veneer, turnery and boatbuilding. It is also used in construction for doors and windows, in railway coaches and shop fittings and for boxes and crates.

Wood waste is used as a substrate for the cultivation of the edible fungus *Lentinus tuber-regium*, and preliminary observations indicate that this substrate when fermented has some value as cattle feed.

The bark of *M. altissima* is very poisonous and in many parts of its range is used as a component of arrow poisons. Although both the bark and heartwood contain cardiac poisons, which can cause heart disorders, decoctions of the bark are used as traditional medicines in some regions, particularly against leprosy.

The sawdust is an irritant and, in fact, *M. altissima* is one of the worst wood species in terms of toxicity and commonness of allergic reactions. *M. altissima* has been reported as a sensitizer, and though the most usual reactions simply include eye and skin irritation, the wood dust can also produce a wide range of other effects, including nausea, giddiness, sneezing, headaches and asthma. Shavings used as litter can also cause serious veterinary medical problems in livestock.

Further reading

Akinnagbe, A. (2008) *Genetic Variation in* Mansonia altissima *A. Chev. and* Triplochiton scleroxylon *K. Schum Under Different Regimes of Human Impact in Akure Forest Reserve, Nigeria*. Cuvillier Verlag, Göttingen, Germany.

Ohene-Coffie, F. (2008) *Mansonia altissima* (A. Chev.) A. Chev. In: Louppe, D., Oteng-Amoako, A.A. and Brink, M. (eds) *Timbers 1 (Plant Resources of Tropical Africa, volume 7)*. PROTA Foundation, Wageningen, The Netherlands, pp. 356–360.

Phongphaew, P. (2003) Bete/mansonia (*Mansonia altissima*). In: *The Commercial Woods of Africa*. Linden Publishing Inc., Fresno, California, USA, pp. 28–29.

Melaleuca cajuputi Powell

Swamp tea-tree

Importance

Melaleuca cajuputi (Plate 22) is a long-lived and moderately fast-growing tree particularly adapted to waterlogged soils in the lowland tropics. It is the primary source of the medicinal cajuput oil, but is also useful for fuel, posts, poles, household products, construction materials, shipbuilding, honey production, shade, shelter and ornamental purposes.

Botanical features

M. cajuputi is usually a medium-sized tree up to about 30 m tall with a single stem. In some situations it may be reduced to a multi-stemmed shrub, but in the Northern Territory of Australia it can be over 40 m tall and 1.2 m in diameter. The bark is grey to white in colour, and papery, and the crown has a somewhat silvery appearance. It has an extensive root system, and on soils subject to prolonged waterlogging it develops aerial adventitious roots, which can form buttresses on the lower trunk. It will coppice, and has the ability to reproduce from root suckers.

The dense, dull green leaves are alternate, lanceolate–elliptic, straight or curved, often hairy, and 5–7 nerved on compressed petioles 3–11 mm long. Oil glands usually obscure. Young shoots are covered in dense, silky, spreading hairs about 0.3–2 mm long. Leaf size and shape varies between subspecies.

The dense, many-flowered terminal or upper-axillary inflorescence consists of 1–3 spikes, 3.5–9 cm long, 2–2.5 cm wide with white, greenish-white or cream-coloured flowers in triads. The rachis is villous, growing out at anthesis; bracts ovate, striate, villous and caducous, without bracteoles; calyx tube turbinate to sub-cylindrical, up to 2 mm long, and pubescent; lobes are semicircular to triangular, 0.5–0.7 mm long, pubescent with a glabrous thin margin; petals are broadly obovate, clawed, up to 2.5 mm long; stamens white, glabrous; claw is 1–3.5 mm long, with 4–15 filaments attached to the upper margin of each claw, free part to 8 mm long; style 6–9 mm long, glabrous; stigma small; ovary approximately 1 mm long, pubescent above. The fruits are cup-shaped to globose capsules about 3 mm long and 4 mm wide, orifice 1.5–2 mm diameter with thinner valves than most *Melaleuca* species. The subspecies vary in numbers of stamens per bundle and length of bundle claw.

As with all species in the genus, *M. cajuputi* does not develop resting buds and grows whenever conditions are favourable.

Melaleuca spp. have hermaphrodite, protandrous flowers and are pollinated mainly by insects, but also by birds and small mammals. Young trees set their first flower buds as early as 13–14 months. Where water is in abundance throughout the year (e.g. in the higher rainfall areas of Indonesia, Papua New Guinea and northern Australia), flowering can occur throughout the year. At locations in Australia where there is a distinct dry season, *M. cajuputi* flowers between March and June and from August to December. Mature seeds have been collected in Australia in October–January.

Distribution

M. cajuputi is the only species of about 250 in the genus to occur naturally west of Wallace's Line. The exact limits of the natural range of *M. cajuputi* are not known, as the species has been cultivated in Asia for more than 100 years. The approximate boundaries cover a latitudinal range of 18° S to 12°N. *M. cajuputi* has been planted since 1926 for oil production in central Java, Indonesia. It has also been widely planted in Malaysia and Vietnam.

Silviculture characteristics

M. cajuputi is a long-lived, moderately fast-growing tropical tree adapted to both waterlogged and well-drained soils. It is tolerant of inundation, and seedlings have been known to survive complete submersion. *M. cajuputi* is resistant to fire, and tolerates exposure to salt-laden winds. It is also tolerant of acid sulphate soils and the aluminium toxicity associated with them. It regenerates by seed, coppicing and root suckers.

Silviculture practice

Propagation is usually by seeds, which germinate readily in moist, warm conditions with no pretreatment. There is an average of 2,700,000 viable seeds/kg. Viability of seed stored dry (5–8% moisture content) in refrigerated, airtight containers at 3–5°C can be maintained for several years.

Plantations have been established with small stump plants, and the species can be reproduced vegetatively from stem and branch cuttings.

Seed should be sown under shade (optimum temperature for germination is 30°C) on a free-draining and sterilized medium, and covered very sparingly with inert material (e.g. sand). Germination should be complete after 15 days, and then shade can be reduced. Germinants are transplanted at the second leaf-pair stage to containers filled with sterilized potting mix. Shade cover is needed for the first week after transplanting, after which time plants should be fully exposed. After germination the tiny seedlings are easily damaged by overhead watering or rain, or may be killed if the sowing mix dries. The 'bog' technique of watering has been adopted in Vietnam to avoid these problems. This involves standing the base of the germination tray permanently in water so that moisture soaks up to the surface, which is constantly moist but not flooded. Seed is sown evenly over the surface at a density of about 7000 viable seeds/m^2. An inflated plastic bag is fitted over the germination tray to maintain a moist environment. Once the germinants are sturdy enough to withstand overhead watering (approximately 4 weeks), the container is removed from the water and handled normally. The risk of fungal disease is high, so good hygiene is essential. After germination the tiny seedlings can be slow to develop at first, presumably while the roots establish. Once under way, however, they grow quickly and in 3–6 months seedlings are a suitable size for planting.

Melaleuca spp. form symbiotic mycorrhizal associations with various fungi. Ectomycorrhizal and vesicular-arbuscular mycorrhizal (VAM) associations both occur in the genus. Nurseries growing *Melaleuca* spp. should attempt to introduce appropriate mycorrhizae to the nursery soil, especially where the soils are deficient in phosphorus.

Diseases and pests

In the nursery, *M. cajuputi* is susceptible to damping off caused by soil-borne pathogens such as *Rhizoctonia*, *Pythium*, and *Phytophthora*.

The pests *Spodoptera litura* and *Trabala vishnou* have been reported as causing damage to *M. cajuputi* trees in the Mekong Delta, Vietnam.

Uses

M. cajuputi is an attractive ornamental tree. It also provides excellent shade and shelter, and is one of the few trees that can grow successfully in brackish soil in exposed situations. It often plays a significant environmental role in estuarine habitats, where it provides habitat for birds, fish, shrimps and other wildlife.

The wood of *M. cajuputi* has a yellowish sapwood, merging gradually into pinkish-grey heartwood. It has a high silica content (0.2–0.95%) which blunts saws. It is hard and heavy, with a green density of approximately 1070 kg/m^3 and an air-dry density of approximately 750 kg/m^3. Although often difficult to split, it makes good fuelwood and is used for this purpose in Indonesia, Malaysia, Thailand, and Vietnam. The timber is generally used in the round or roughly fashioned, and is suitable for posts, poles, piles and general construction materials. It is durable in contact with fresh- or saltwater. *M. cajuputi* is traditionally used for making dugout canoes. The bark is still used today in parts of Malaysia as a luting material in boatbuilding, and in the construction of traditional houses in Papua New Guinea.

A major use of *M. cajuputi* is as a source of cajuput oil, produced by steam distillation of the leaves and terminal branchlets. Cajuput oil has medicinal properties and is used in traditional medicine in Vietnam, Indonesia and China. The Australian Aborigines used preparations from the leaves of this species in the treatment of aches and pains, and inhaled the aromatic vapours from crushed leaves to reduce nasal or bronchial congestion. It was one of the first products imported to Europe from South-east Asia by the Dutch (an activity described in several early 18th-century publications) because of its reputation as a treatment for many diseases. *M. cajuputi* is also a good source of nectar and pollen for honey production.

Further reading

Doran, J.C. (1998a) *Melaleuca cajuputi* Powell. In: Dung, N.X. and Oyen, L.P.A. (eds) *Plant Resources of South-East Asia No. 19. Essential Oil Plants.* Backhuys Publishers, Leiden, The Netherlands.

Doran, J.C. (1998b) Cajuput oil. In: Southwell, I. (ed.) *Medicinal and Aromatic Plants - Industrial Profiles - Melaleuca.* Harwood Academic Publishers, Amsterdam, The Netherlands.

Tanaka, K., Masumori, M., Yamanoshita, T. and Tange, T. (2011) Morphological and anatomical changes of *Melaleuca cajuputi* under submergence. *Trees: Structure and Function* 25, 695–704.

Melaleuca leucadendra (L.) L.

Long-leaved paperbark

Importance

Melaleuca leucadendra (Plate 23), known as weeping or long-leaved paperbark on account of its drooping leaves which are over seven times as long as they are wide, is a fast-growing tree that will tolerate waterlogged soils including those with acid sulphate and salinity problems in the lowland tropics. It is useful for posts, poles, fuelwood, household products, construction materials, shade, shelter, ornamental plantings and honey production. The essential oils of some provenances have commercial potential.

Botanical features

On favourable sites *M. leucadendra* is a tall tree, 25–45 m, with a straight stem of up to 1.5 m diameter. Elsewhere it may be a small- to medium-sized tree with a crooked or twisted bole. The bark is white-layered and papery. The long, slender branchlets usually droop, giving the crown a weeping appearance.

The adult leaves are alternate, light green, 5-nerved, thin, narrowly lanceolate, mainly 10–19 cm long, 1–2 cm wide, on a petiole 6–12 mm long. The leaves are widest below the middle. There is usually a complete absence of hairs except on the very youngest shoots.

The inflorescence consists of 1–3 spikes, terminal in the upper leaf axils, 6–15 cm long and 2–3 cm wide, bearing many white or creamy-white flowers with stamens 5–7 per bundle and 10–13 mm long. Fruits are sessile, thin-walled cylindrical capsules about 3–4 × 3–4 mm, loosely clustered along the branchlets.

M. leucadendra is a long-lived, fast-growing tropical tree. As with all species in the genus, it does not develop resting buds and grows whenever conditions are favourable. In *M. leucadendra*, first flower buds appear on young trees after 2 years of growth. Flowering in natural stands occurs throughout the year, but chiefly during (March–)June–September. Mature capsules have been observed on trees throughout the year, but the most reliable time to collect *M. leucadendra* seed in northern Australia is just prior to or during the wet season, December–February.

Distribution

M. leucadendra occurs in Australia, mainly in coastal or subcoastal areas of tropical Queensland, the Northern Territory, and Western Australia, but extends inland for up to 350 km along major rivers. It is also found in southern Papua New Guinea and eastern Indonesia, where it is recorded in Irian Jaya (West Papua) and the Maluku Islands including Aru, Ambon, Ceram, Kai and Obi.

Silviculture characteristics

M. leucadendra is moderately fast-growing and has the ability to tolerate acidic, infertile swampy conditions. In waterlogged and flooded areas it forms aerial adventitious roots. It has reasonable coppicing ability, but is variable in this attribute. It is fire-tolerant and moderately to highly salt-tolerant.

Silviculture practice

Propagation is usually by seeds, which germinate readily in moist, warm conditions with no pretreatment. There is an average of 1,825,600 viable seeds/kg. The viability of seed stored dry (5–8% moisture content) in refrigerated airtight containers at 3–5°C can be maintained for several years. *M. leucadendra* can also be propagated vegetatively from stem and branch cuttings.

Nursery techniques are the same as for *Melaleuca cajuputi*. Seed should be sown under shade (optimum temperatures for germination of *M. leucadendra* are in the range 25–30°C) on a free-draining and sterilized medium, and covered very sparingly with inert material (e.g. sand). Germination should be complete after 15 days, and then shade can be reduced. Germinants are transplanted at the second leaf-pair stage to containers filled with sterilized potting mix. Shade cover is needed for the first week after transplanting, after which time plants should be fully exposed.

After germination, the tiny seedlings are easily damaged by overhead watering or rain, or may be killed if the sowing mix dries. The 'bog' technique of watering has been adopted in Vietnam to avoid these problems in propagation. This involves standing the base of the germination tray permanently in water so that moisture soaks up to the surface, which is constantly moist but not flooded. Seed is sown evenly over the surface at a density of about 7000 viable seeds/m^2. An inflated plastic bag is fitted over the germination tray to maintain a moist environment. Once the germinants are sturdy enough to withstand overhead watering (approximately 4 weeks), the container is removed from the water and handled normally. The risk of fungal disease is high, so good hygiene is essential.

Seedlings can be slow to develop at first, presumably while the roots establish. Once under way, however, they grow quickly and the 3(–6) months it takes for plants to reach plantable size is similar to other fast-growing species such as eucalypts.

M. leucadendra is used for a range of land-care, wood, and non-wood purposes. The silvicultural system adopted will depend very much on the end use of the planting, although it is clear from the lack of literature on the subject that little is known of optimum stand establishment, tending, and management systems for this species.

Although optimum husbandry of the species is not well known, application of a 100 g starter dose of a balanced NPK fertilizer is likely to aid in establishment and early growth. *M. leucadendra* can tolerate prolonged flooding and a fluctuating water table. The species can compete successfully with weeds, but early weed control is likely to improve growth rates. Form pruning has been advocated for garden specimens.

Diseases and pests

M. leucadendra appears to be relatively free of major damage by diseases and pests. Damage by grasshoppers and leaf-rolling caterpillars can be severe during the dry season in northern Australia. In the Mekong Delta of Vietnam, damping off, *Spodoptera litura*, and *Trabala vishnou* were identified as pathogens and pests of *M. leucadendra*, *M. cajuputi*, and *Melaleuca viridiflora*. *Eriophyes leucadendrae*, a newly described species of eriophyoid mite from Australia, forms galls on the leaves.

Flies of the family Fergusoninidae (Diptera) and nematodes of the genus *Fergusobia* (Tylenchida: Neotylenchidae) together form galls on various Myrtaceae, this being the only recorded mutualism between flies and nematodes. New species of *Fergusonina* gall-forming flies and *Fergusobia* nematodes from the *M. leucadendra* species-complex in Australia have been described.

Uses

M. leucadendra is grown in agroforestry-based plantation systems in Java (Indonesia) known as *tumpungsari*, a form of shifting cultivation. The highest leaf-oil production under this system was obtained in western Java when the companion crop was rice, whereas in eastern Java maize was better than cassava in its effect on oil production. *M. leucadendra* is used for shelter belts, in coastal plantings, and as a street or avenue tree. Its white trunk, graceful habit and sweet-smelling flowers make it a popular ornamental. It is also a good shade tree. The species is used in the rehabilitation of bauxite mining sites in the tropical north of Australia and has potential for use in rehabilitating sites contaminated with manganese. *M. leucadendra* is valuable for wildlife, the nectar from its flowers attracting bees, birds, and flying foxes (*Pteropus* spp.).

The yellowish sapwood merges into a pinkish-grey heartwood, which is hard, medium- to fine-textured, of moderate strength, difficult to season, but durable in the ground and in water. The wood has a high silica content and interlocked grain. Air-dry density is 725–800 kg/m^3. It is mainly used in the round or roughly fashioned. The wood is suitable for a wide range of purposes, including railway sleepers, house and fence posts, mine timbers, boatbuilding, joinery, poles and piles. It makes good firewood although it is often difficult to split.

The bark of *M. leucadendra* was used by Australian Aboriginal peoples to make shelters, containers, fire-tinder, bedding and wrapping. They also used decoctions of the leaves and inner bark externally as a wash or internally in the treatment of colds, fever and to alleviate general malaise.

As with other *Melaleuca* species, *M. leucadendra* is known as a tea tree due to the fact that an infusion of its leaves can be used as a tea substitute. Several authors have highlighted the commercial potential of the eugenol-rich foliar essential oils of *M. leucadendra* in the flavour and fragrance industry. The nectar of the flowers of *M. leucadendra* produces a light amber-coloured honey with a strong flavour and low density and the pollen is a good source of protein for bees.

Further reading

Blake, S.T. (1968) A revision of *Melaleuca leucadendron* and its allies (Myrtaceae). *Contributions to the Queensland Herbarium* 1, 1–114.

Brophy, J.J. and Lassak, E.V. (1988) *Melaleuca leucadendra* L. leaf oil: two phenylpropanoid chemotypes. *Flavour and Fragrance Journal* 3, 43–46.

Budiadi Ishii, H.T. (2010) Comparison of carbon sequestration between multiple-crop, single-crop and monoculture agroforestry systems of *Melaleuca* in Java, Indonesia. *Journal of Tropical Forest Science* 22, 378–388.

Melaleuca quinquenervia (Cav.) S.T. Blake

Broadleaved paperbark

Importance

Melaleuca quinquenervia (Plate 24) is a moderately fast-growing, fire-tolerant tree capable of growing on nutrient-deficient sites with continuous or periodic flooding in the humid and subhumid tropics. It is moderately salt-tolerant. This species is most likely to be grown on relatively short coppice rotations that maximize the production of small-sized logs. *M. quinquenervia* makes excellent fuelwood, provides useful timber, is a good source medicinal niaouli oil and is important in the production of honey.

Botanical features

A small- to medium-sized tree, commonly 8–12 m tall, but ranging from 4 to 25 m depending on local growing conditions. The stem is moderately straight to crooked, the crown is narrow and open, or fairly dense. The thick, pale-coloured bark is made up of many papery layers that split and peel, and on large trunks becomes rough and shaggy.

Young shoots are densely hairy and therefore silvery in appearance, with hairs 0.25–2 mm long, which are appressed on the leaves and ascending on the twigs. Mature leaves are alternate, dull green, stiff, leathery, lanceolate to oblanceolate, 4–9 by 2–3.5 cm, with entire margins, and 5 (rarely 3 or 7) prominent parallel veins from base to tip, on a petiole 6–24 mm long.

The white or creamy-white (rarely greenish or reddish) flowers are produced in thick, fluffy spikes. The conspicuous part of each flower consists of five bundles of stamens 10–20 mm long. Spikes are solitary or 2–3 together, terminal, sometimes solitary in the uppermost 1–3 axils, and 4–8.5 × 2.5–3.5 cm in size. The spike grows out into a leafy twig beyond the fruits. Each inflorescence results in 30–70 densely packed woody, stalkless capsules. The capsules are short, cylindrical, 3–4 × 4–5 mm, grey-brown, hard, and persistent. The seeds are pale brown, very small, about 1 × 0.3 mm, tapering from the dorsal end. The seeds are shed through 3–4 slits positioned horizontally below the capsule rim.

M. quinquenervia is a long-lived, moderately-fast growing tropical tree. As with all species in the genus, it does not develop resting buds and grows whenever conditions are favourable. In Florida, seedlings may grow throughout the year, but growth is most rapid in summer.

In *M. quinquenervia*, flowering in young trees commences as early as 3 years, and trees less than 1 m tall may bloom. The usual peak flowering time in Australia is March–July (autumn and winter), but flowering may occur at any time. Flowering is influenced by the weather, particularly rainfall and soil type. In more tropical environments the species tends to flower more often and flowering events are longer and later. Fruits ripen in spring and summer; in Australia mature seed has been collected during November–March.

Distribution

M. quinquenervia is a native of the coastal region of eastern Australia, from near Sydney (New South Wales) to Cape York (northern Queensland). It occurs mainly in a belt within 40 km of the sea. This species extends into southern Papua New Guinea and Irian Jaya (West Papua), Indonesia, and has an extensive distribution in New Caledonia, especially on the north-west of the island. In Australia and Papua New Guinea, *M. quinquenervia* is generally confined to the lowlands (below 100 m), but in New Caledonia it forms extensive stands in uplands to an altitude of 900–1000 m.

The species has been planted in India, Mexico, Nigeria, Philippines, Puerto Rico, Southern Africa, West Indies; and in California, Florida, Hawaii and Texas (USA).

Silviculture characteristics

M. quinquenervia demonstrates successful establishment and moderately fast growth on sites in the humid and subhumid tropics, which may be inundated with water for months at a time. In waterlogged and flooded areas it forms aerial adventitious roots. The best prospects for widespread use of this species appear to be on swampy sites, as more productive species are available for better-drained habitats. Utilization of *M. quinquenervia*, however, may be limited by the extent of swampy environments available for reforestation, and the relative performance of this species in comparison with some closely related, similarly adapted *Melaleuca* spp. such as *Melaleuca leucadendra* and *Melaleuca cajuputi*.

M. quinquenervia is highly fire-tolerant, except during the early seedling stages. The species can successfully compete with weeds, but early weed control will improve growth rates. *M. quinquenervia* is moderately salt-tolerant; it is also tolerant of aluminium toxicity. The species has the ability to coppice readily, but root suckers are not commonly produced.

Introduction of this species and its close relatives should always be treated with caution, given its potential to become an invasive weed.

Silviculture practice

Propagation is usually by seeds, which germinate readily under favourable conditions, and do not require pre-germination treatment. In soil which is seasonally or permanently flooded, seeds lose their viability in around 1.5 years, while in non-flooded soils they can survive for 2–2.3 years. Germination is epigeal. There is an average of 2,660,000 viable seeds/kg. Viability of seed stored dry (5–8% moisture content) in refrigerated, airtight containers at 3–5°C can be maintained for several years. As many species within the genus are vegetatively propagated by stem and branch cuttings, these techniques should also be successful with *M. quinquenervia*.

Nursery techniques are the same as described for *M. cajuputi*. As with other species in the genus, the seeds of *M. quinquenervia* are very small, so care is needed to ensure that the sowing mix does not dry out, diseases are avoided or controlled, and that seeds or small seedlings are not damaged or washed away by careless watering. Watering from below may be used in preference to overhead systems to avoid this last problem. When the germinating seedlings are 2–3 cm tall (4–8 weeks after sowing), they are ready for transplanting from germination trays to containers. Conventional containerized nursery procedures can produce healthy, vigorous seedlings suitable for planting in about 4–7 months from sowing. Growth rate is relatively high on sites where water is abundant and soils are deep, but rates are lower on dry or shallow soils.

Diseases and pests

The rust fungus *Puccinia psidii* can severely damage *M. quinquenervia* and other Myrtaceae species.

In Australia, over 400 herbivorous insects were found in association with *M. quinquenervia* and its close allies, but damage appears localized. The species suffers only slight damage from sawflies. The heartwood lacks resistance to damage by termites, fungi and marine borers. *Eriophyes quinquenerviae*, a newly described species of eriophyoid mite from Australia, forms galls on the leaves, and *M. quinquenervia* has been identified as a new host for larvae of the longicorn beetle *Agrianome spinicollis* in south-east Queensland.

Outside the native range of *M. quinquenervia*, it has been reported that 40 insects, 4 nematodes, and 22 plant pathogens were among the pests collected or isolated in Florida. However, none was serious to the species' survival and overall it was regarded as relatively free of diseases and pests. The stellate scale (*Vinsonia stellifera*), has been observed on *M. quinquenervia* in the Bahamas, while *Retithrips syriacus* was recorded causing damage to leaves of trees planted as ornamentals in Puerto Rico in 1994.

Uses

M. quinquenervia can be used for windbreaks and may have a role in erosion control on degraded and poor soils. It has high potential for use in constructed wetlands to provide buffering for sewage spills; such wetlands proved excellent sediment traps, with up to 98% removal of suspended solids when receiving primary settled sewage. It was also thought possible to combine this use with leaf oil production.

M. quinquenervia is a popular ornamental tree, particularly for damp conditions, and birds are attracted to the fragrant blossoms. A red-flowering form makes an excellent feature plant for use in landscaping. Remnant *M. quinquenervia* forests near Brisbane, Australia, are important habitats for the conservation of bird species, and in Queensland it is one of the key tree species utilized by koalas.

The sapwood is pale yellow to pink. The heartwood is pink to reddish-brown, rippled with light and dark tones. It is hard, fine-textured and diffuse, porous, tough, tending to warp and difficult to season. The wood contains 0.2–0.95% silica that rapidly blunts saws and planes. The wood requires care in drying to minimize checking and warping. Collapse is slight. Shrinkage is about 3.5% radially and 7% tangentially. The basic specific gravity is generally within the range 0.49–0.55 and air-dry density is 700–750 kg/m^3. The wood has been used for a wide range of purposes, including mine timbers, fence posts and rails, flooring, house timbers and pulp. It is an excellent fuel and makes good-quality charcoal.

Mulch is prepared from wood chips of *M. quinquenervia* in Florida. The bark is resistant to mildew, and consequently it is used to insulate the walls of cold-storage rooms and to stuff pillows. Its water resistance makes it useful for caulking boats.

M. quinquenervia is a good source of nectar and pollen for bees, and it is a major source of honey in Australia and Florida. Natural stands of *M. quinquenervia* on New Caledonia are the source of niaouli oil, which is produced by hydrodistillation of leaves. Niaouli oil is mainly used to treat pulmonary infections, especially colds and bronchitis. Foliar essential oils also have potential for the control of mosquito (*Aedes aegypti*) larvae, house flies (*Musca domestica*) and crop diseases and pests.

Further reading

Geary, T.F. (1988) *Melaleuca quinquenervia* (Cav.) ST Blake. In: Burns, R.M. and Mosquera, M. (eds) *Useful Trees of Tropical North America*. North American Forestry Commission Publication No. 3. North American Forestry Commission, Washington, DC.

Watt, M.S., Kriticos, D.J. and Manning, L.K. (2009) The current and future potential distribution of *Melaleuca quinquenervia*. *Weed Research (Oxford)* 49, 381–390.

Wrigley, J.W. and Fagg, M. (1993) *Bottlebrushes, Paperbarks and Tea Trees, and all Other Plants in the Leptospermum Alliance.* Angus and Robertson, Sydney, Australia.

Mesua ferrea L.

Ceylon iron wood, penaga

Importance

Mesua ferrea produces a very heavy hardwood which is used for heavy construction work. However, because of its slow growth rate, it is not suitable as a plantation tree for timber production, and so is more often grown as an ornamental tree. The fragrant flowers are used in cosmetic products and the seeds are used in traditional medicines in India, Malaysia and elsewhere in the region. The oil extracted from seed kernels has industrial uses. *M. ferrea* is considered a sacred tree in India, and is the national tree of Sri Lanka.

Botanical features

M. ferrea is a medium-sized tree, reaching heights of up to 36 m, with a dbh of up to 95 cm. The crown is dense, narrowly conical when growing in the open, with flushes of bright red leaves. The stem is straight or slightly fluted, unbranched up to 20 m, with small buttresses. The bark is irregularly fissured and scaly, light brown to grey, with a purplish tinge. The inner bark is reddish-brown or pinkish, with sparse droplets of whitish to pale yellow latex which darken on exposure.

Leaves simple, opposite, decussate, elliptic to lanceolate, 4.5–12.5 × 1–4 cm, apex pointed, base acute, white-waxy below, venation indistinct on both surfaces.

The flowers are solitary from leaf axils, large, 2.5–10.2 cm in diameter, bisexual, very fragrant, white, with a central bunch of yellow stamens. Sepals 4, decussate, persistent, and enlarged in fruits; petals 4, white; ovary superior, 1–2 celled, each cell with 1–2 ovules, style slender with 4-lobed stigma. Fruit is a capsule, ellipsoid to globose, pointed, 2.5–6.5 cm wide and 3.5 cm long, splitting into 2–4 segments, seated on the persistent sepals, often exuding resinous droplets. Seeds (1–4) 2.5 × 1.8 cm, with dark brown shiny testa.

In Assam, India, it is reported that the old leaves are shed during the cold season and that the new leaves appear towards the end of the cold season. The shedding of leaves, the appearance of fresh leaves, and the timing of flowering, fruiting and fruit ripening vary between regions.

M. ferrea starts flowering when the trees are fairly old (15–20 years). It flowers during the dry (cold) season, with the flowers often opening at 3–4 a.m. and closing again around sunset. The flowers remain open for one day only, the petals remaining within the calyx for 1 week to 10 days before dropping off.

Distribution

M. ferrea is recorded in Bangladesh, Bhutan, Cambodia, India, Indo-China, Myanmar, Peninsular Malaysia, Singapore, Sri Lanka, Thailand, and Vietnam. It usually occurs scattered but is occasionally gregarious and is found in a wide variety of habitats, such as lowland and hill areas in evergreen and semi-evergreen forests, up to altitudes of 900 m.

It is planted as an ornamental or shade tree in Malaysia and Indonesia.

Silviculture characteristics

M. ferrea is tolerant of shade; under deep shade it outperforms *Dipterocarpus* species. When grown in the open, it is seriously affected by drought; the seedlings are very sensitive to drought and suffer severely in exposed situations.

Two important factors for successful natural regeneration are good condition of the seed-bearing trees and an absence of suppression of seedlings by weeds and climbers.

M. ferrea is a slow grower. In a plantation in Peninsula Malaysia, growth of trees under full exposure was contorted, but became straight later.

The coppicing ability of *M. ferrea* varies depending on environmental factors; it appears to coppice better in drier climates, and good light seems to be a decisive factor in promoting coppicing.

Silviculture practice

M. ferrea produces abundant quantities of seed, and so natural regeneration can be profuse in the absence of weeds and climbers. Propagation is usually carried out from seed, and fruits can be collected from underneath the trees or directly from the branches; 1 kg of fruits contains 300–500 seeds. Propagation can also be carried out by grafting *M. ferrea* onto *Calophyllum inophyllum* rootstocks. A 75–90% germination success rate was recorded within 11–24 days, with germination being hypogeal. Germination of seeds originating from shaded parts of the tree is better than in seeds from parts of the tree in direct sunlight, leading to better seedling survival and growth. Seeds lose their viability within 2–3 months. Viability is directly correlated with seed weight. In India, the best storage results were obtained using perforated polythene bags stored at 5°C; storage for 4 months reduced the germination success rate to 27%.

A high degree of moisture and shade are essential for good germination and early growth. Low light intensity has been shown to give the best seed germination. Seeds are best sown in raised nursery beds at a depth of 1–1.5 cm, with a spacing of 5 × 5 cm. Application of superphosphate fertilizer is beneficial for 7-day-old seedlings, whereas ammonium-based nitrogen fertilizer is detrimental. Seedlings in natural forests can tolerate prolonged and intense shade, and can also grow well under full exposure.

In West Bengal, seedlings have been planted in 30 cm^3 pits at a spacing of 1.8 × 1.8 m. *Tephrosia candida* was sown as a nurse crop between the lines of *M. ferrea* in Tamil Nadu. Experimental plantations have been successfully raised in Kerala in deciduous, clear-felled and burned forest areas, with *T. candida* as the nurse crop and *Trema orientalis* as the shade crop. In south-west Sri Lanka, late successional tree species such as *M. ferrea* have been established successfully in rainforest by creating openings in *Pinus caribaea* plantations (by removing 3 rows of trees) and planting within and adjacent to these gaps. *M. ferrea* is suitable for growing as an understorey with fast-growing, light-demanding species, such as *Tectona grandis*.

Regular weeding and thinning are necessary with artificial regeneration until the plants are established; establishment may take up to 7 years. Regular thinnings at intervals of 4–5 years are highly beneficial.

Diseases and pests

Ganoderma lucidum, causing root and butt rot, has been recorded on *M. ferrea*.

The larvae of the insects *Phenacaspis dilatata* (*Pseudaulacaspis cockerelli*) and *Toxoptera aurantii* feed on the sap of *M. ferrea* leaves but do not cause serious injury. The buprestid borer *Chrysochroa* sp., which attacks living trees and forms large galleries in the heartwood, and a cerambycid borer, can both cause serious damage.

Uses

M. ferrea is widely cultivated as a shade tree throughout its range, and has been successfully used as such in an intercropping system with turmeric crops. It is a suitable species for underplanting in teak plantations. Because of its dense, conical crown, bright red flush of new leaves, turning vivid green later, and showy, fragrant flowers, *M. ferrea* is a popular ornamental tree grown in parks and along roads.

M. ferrea produces a heavy and very hard wood. The standard Malaysian and Association of South-east Asian Nations (ASEAN) name for the timber is penaga. When fresh, the heartwood is reddish-brown with a purple tinge, becoming dark brown on drying, differentiating it from the pale yellow sapwood. The density of the wood is 940–1195 kg/m^3 at 15% moisture content. The wood is slightly difficult to re-saw and cross cut. Planing and boring is easy, but nailing properties are poor. Being a heavy and hard wood, *M. ferrea* is used for heavy construction, including posts (pit props and transmission poles), beams, rafters, joists, columns, heavy duty flooring and furniture. It is also used for cabinet work, pallets, staircases, tool handles, agricultural implements, vehicles and boat-building. In India, it has been used for railway sleepers. *M. ferrea* timber is also suitable for gun-stocks, walking sticks and musical instruments.

All parts of *M. ferrea* are used in traditional medicine throughout its native range. Extracts have antidiarrhoeal, anticonvulsant and anti-inflammatory properties. Leaf-powder smoke can be used against adult mosquitoes (*Culex quinquefasciatus*) and leaf extract has larvicidal action. A mixture of pounded kernels and seed oil is used in poultices for wounds. Seed oil is used for treating skin itches and eruptions, dandruff and rheumatism, and has been shown to have disinfectant properties. Seed extracts have antibacterial and anti-arthritic properties. Flower extracts have potent antibacterial and antioxidant activities. The fragrant flowers are used in some cosmetic products and are used to stuff pillows and cushions. Flower, bark and leaf extracts also have potent antiviral activity against plant viruses.

Seeds contain up to 70% oil, which is dark yellow or brown with an unpleasant pungent odour. The oil has industrial uses, including the manufacture of polyurethanes for use in paints, resins and surface coatings. Interest has grown in the use of the seed oil for the production of thermostable, biodegradable nanocomposite biomaterials. De-oiled seed meal has been shown to have feed value for pigs and poultry.

Further reading

Dennis, T.J. and Kumar, K.A. (1998) Constituents of *Mesua ferrea*. *Fitoterapia* 69, 291–304.

Kadambi, K. (1954) *Mesua ferrea*, Linn., its silviculture and management. *Indian Forester* 80, 431–450, 3 photos & 1 map.

Soejono, R. (1978) The Nagasari tree (*Mesua ferrea* L.). *Buletin Kebun Raya* 3, 145–148.

Metrosideros polymorpha **Gaudich.**

'Ohi'a lehua

Importance

Metrosideros polymorpha is a highly variable, slow-growing, broadleaved tree, with an annual diameter increment of <1–2 mm, making it unsuitable as a commercially viable silvicultural species. It is, however, the most abundant and widespread tree in the Hawaiian Islands. It is the earliest tree colonist on new lava flows and forms near-monospecific stands or mixed stands with *Acacia koa* in the oldest montane wet forests. Therefore, *M. polymorpha* is important in both primary succession and in stability of mature forest ecosystems. *M. polymorpha* trees are critical as food sources for endemic nectivorous and insectivorous birds, and form critical specific habitats for endemic *Achatinella* and *Partulina* tree-snail species. Although *M. polymorpha* trees are extremely abundant, the forest area is in decline and is threatened by numerous alien invasive species.

Botanical features

Morphologically, *M. polymorpha* is highly variable. Growth form ranges from small erect or prostrate shrubs to tall trees. Crowns may be compact or multibranched. The height of the crown base may be low when growing in open habitat, but begins at mid-stem or higher in closed canopy forest. On moist soils, *M. polymorpha* grows to 20–24 m in height with a dbh of 45–90 cm. Trees to 30 m in height and 120 cm dbh have been recorded. The bole may be straight, cylindrical and smooth, or twisted and deeply fluted. Shrub growth forms predominate on poorly drained, dry or rocky soils, and at high altitudes.

Leaves are opposite, dorsiventral, clustered and highly variable, ranging in form from obovate to orbicular, elliptic or ovate. They are 1–8 cm long and 1–5.5 cm wide. Several pairs of raised veins radiate from the base. The upper surface is glabrous and the lower surface glabrous, woolly or pubescent. Margins are revolute to flat, the apex rounded or sometimes obtuse to acute, and the base cuneate to cordate. Petioles are 1–16 mm long and 1–3 mm wide.

Flowers are bright to medium red and in inflorescences of 2–5 pairs of cymules. Peduncles are 7–18 mm long, pedicels 2–8 mm long, bracts ovate to suborbicular and 5–10 mm long, sepals rounded to triangular and 1.5–4 mm long, and petals obovate to orbicular and 2.5–5 mm long. The style is 13–30 mm long and the stamens are 10–30 mm long; masses of stamens extend from the flower, giving the flowers their characteristic pompom shape. Fruiting capsules are 3–8 mm long and wide, glabrous or pubescent.

M. polymorpha is an evergreen species. Vegetative flushing varies by population, occurring seasonally with season dependent on population, or throughout the year in some populations. Flowering generally peaks after vegetative flushing.

Distribution

The natural distribution of *M. polymorpha* is limited to the six largest islands in the Hawaiian Archipelago, namely Hawaii, Maui, Molokai, Lanai, Oahu and Kauai. It grows from sea level right up to the tree line at an altitude of 2500 m, and is dominant in cloud forest above 400 m. It is common in moist and dry forests, and in high shrublands.

Silviculture characteristics

A bright, sunlit environment is necessary for the survival and maturation of most *M. polymorpha* germinants. The species is windfirm in hurricanes, and tolerates a wide range of soil, temperature and rainfall conditions. Growth form is highly dependent on environmental conditions. Unfortunately, mature stands experience canopy decline, leading to synchronous dieback of age cohorts.

Silviculture practice

Silviculture is generally not practised, except in a few ecosystem restoration experiments. Seed viability is variable, ranging from <1% to 78%, but is typically low. In one study only 14% of seeds had intact embryos. Best germination results have been recorded in conditions of full sunlight and 25°C. Seeds can remain viable for as long as 9 months if stored at room temperature. In natural settings, seeds commonly germinate on tree ferns or moss-covered logs in the forest understorey. For production of ornamental flowers, it is recommended to raise plants from cuttings or air-layering, as seedlings are often not true to type.

Although high-volume, old growth stands may yield 70–84 m^3/ha, and much less on poorer sites.

Diseases and pests

Although there have been no reports of rust diseases on *M. polymorpha* in the past, *Puccinia psidii* was identified on potted plants on Oahu in 2005. Although a serious disease of eucalyptus and other plant species throughout Hawaii, any threat to natural *M. polymorpha* forests remains to be seen. Root rots caused by *Phytophthora cinnamomi*, *Pythium vexans* and *Armillaria mellea* can also be damaging, and are most often associated with weakened stands experiencing canopy decline. Canopy decline is characteristic of the species and is associated with advanced maturity, leading to the synchronous dieback of age cohorts throughout entire stands.

The Hawaiian endemic cerambycid borer (*Plagithmysus bilineatus*) is the insect pest with the greatest potential negative impact on *M. polymorpha* trees. It may become epidemic in weakened trees and is often associated with canopy dieback. Other potentially damaging borers are *Ceresium unicolor*, *Xyleborinus saxeseni* and *Xyleborus simillimus*.

Uses

Metrosideros forests and Hawaii's native and endemic species of birds, arthropods, molluscs and plants have co-evolved on the islands, making *Metrosideros* forests especially valuable as habitats and food sources for a unique and endangered biota. *M. polymorpha* provides valuable watershed protection due both to its abundance and especially to its broad environmental amplitude. In landscaping, *M. polymorpha* is used as a windbreak, hedge, shade tree and flowering ornamental.

M. polymorpha wood is very hard, heavy and dense (approximately 930 kg/m^3), has an average specific gravity of 0.81 oven-dry weight/green volume, has a pale brown sapwood that grades into the red to purple-brown heartwood, and can develop a long curl. Its hardness makes the wood difficult to work, and it is notorious for shrinking, warping, splitting and checking during drying. Modern uses include wood-strip flooring, decking, decorative posts, roundwood construction, fence posts and furniture. Historically, the wood has been used for veneer, ukulele keys, ship blocking in dry docks, bracing stakes in agriculture and irrigation ditch construction, wharf fenders and marine construction, pallets, pile-driver cushions, railroad ties and firewood. Ancient Hawaiians used the wood to carve the large sacred images known as tikis, as well as household items such as poi boards. The wood was also used in the construction of houses and temple walls, and for spears and mallets.

The flowers are an important source of nectar for honey production; 'ohi'a honey is one of the only truly native Hawaiian honeys. The flowers are used in garlands and leis, and are considered sacred to the volcano goddess Pele.

Further reading

Friday, J.B. and Herbert, D.A. (2006) *Metrosideros polymorpha* ('ōhi'a lehua). In: Elevitch, C.R. (ed.) *Traditional Trees of Pacific Islands: Their Culture, Environment, and Use.* Permanent Agriculture Resources, Holualoa, Hawaii, USA, pp. 465–490.

Gerrish, G. (1989) 'Ohi'a dieback and forest life cycles. In: Stone, C.P. and Stone, D.B. (eds) *Conservation Biology in Hawai'i.* University of Hawaii, Honolulu, Hawaii, USA, pp. 38–42.

Hodges, C.S., Adee, K.T., Stein, J.D., Wood, H.D. and Doty, R.D. (1986) *Decline of Ohia* (Metrosideros polymorpha) *in Hawaii: A Review.* Pacific Southwest Forest and Range Experiment Station General Technical Report PSW-86. USDA Forest Service, Berkeley, California, USA.

Milicia excelsa (Welw.) C.C. Berg

African teak

Importance

Milicia excelsa (syn. *Chlorophlora excelsa*) is a highly valued timber species of humid tropical Africa. It is notably similar to *Milicia regia* and therefore the trade names iroko and mvule may refer to both species, and descriptive accounts often combine them. Its timber is strong, beautiful and resistant to insect attack and decay. *Milicia* trees are revered for their shade and medicinal properties in many local cultures, and farmers often leave naturally occurring trees while clearing land. In parts of West Africa, *Milicia* trees are traditionally sacred.

Milicia is locally threatened by over-exploitation and poor regeneration.

Botanical features

Milicia species are large, deciduous and dioecious forest trees with dense crowns and heavy branches. Trees can reach 50 m tall, with 25 m or more of clear boles and diameters approaching 200 cm. The crown can show the typical wide-spreading characteristics of a forest emergent. The tree does not form true buttresses, but roots of older trees are sometimes exposed along the surface and develop into low buttress-like structures. Seedlings and saplings have large, thick taproots. Male trees tend to be more slender and less prone to forking than female trees. The thick, hard bark is brown and scaly, with numerous lenticels. Cut bark exudes a sticky white latex. The wood is yellow-brown and granular. Leaves are simple and alternate. The juvenile leaf is oblong–lanceolate, up to 4 cm long, with a variable (entire and non-entire) but often serrate margin, soft texture, and often with a lighter, more yellowish colour. Leaves on mature trees are broader, more glossy, often darker, and tougher in texture than on juveniles.

The flowers of both male and female trees are borne on individual spikes in the axils of young leaves. Male flowers are white and form slender catkins up to 20 cm long, whereas female flowers are greenish, in 2-cm wide spikes 5–6 cm long. Pollination is by wind. Fruits appear from February to April, and resemble long green mulberries. Many fruits fall beneath mother trees while others are dispersed by bats, birds and squirrels.

Distribution

M. excelsa occurs in a wide belt completely crossing the African mainland humid tropics. *M. regia* is restricted to west Africa, up to Cameroon. Where their ranges overlap (Liberia, Côte d'Ivoire, and Ghana), distinguishing the two species is difficult and hybridization may occur.

The species is a secondary tree of wet savannah, rainforest, riverine and low-altitude evergreen forests (0–1600 m above sea level). It is sometimes left in old cultivated areas. *M. excelsa* thrives in well-drained, relatively fertile soils. The tree prefers a mean annual rainfall of 1000–2200 mm, but can tolerate a dry season of up to 6 months, and can grow in areas with mean annual rainfall as low as 700 mm provided it has access to groundwater (e.g. by streams).

The distribution of *Milicia* species in Ghana was examined using data from the Forest Inventory Project of 1989. *M. excelsa* occurs at low densities in natural forests. Based on the total number of stems, *Milicia* is most abundant in the dry semi-deciduous zone (2.4 trees/ha) and least abundant (0.18 trees/ha) in the wet evergreen zone (of Ghana). The pattern of higher frequency in drier forest areas is consistent across all size classes of trees. Two possible explanations for this phenomenon are the preference for *Milicia* seedlings to grow in open environments, and the negative effect of an extended dry season on a gall-forming psyllid (*Phytolyma* spp.).

Silviculture characteristics

The species is light-demanding, has a high growth rate compared to other hardwood species, and grows best when plant competition has been removed.

Establishing effective cultural techniques for *Milicia* plantations has been hampered by gall-forming psyllids of the genus *Phytolyma*. Development of resistant varieties and revised silvicultural techniques may improve cultivation opportunities.

Some individual trees have partial resistance to *Phytolyma lata*, and therefore it would be advantageous to develop vegetative propagation methods. Also, since older leaves are more resistant to attack by *Phytolyma* spp., successful efforts have been made to graft older branches to younger tissue. *Milicia* trees appear to coppice and pollard effectively.

Silviculture practice

Fruits are pulped in water to extract the tiny hard seeds (281,000 seeds/kg when dried). Seeds are sun dried to 8%

moisture content. Viability of seeds varies greatly, even when kept in an air-conditioned and cooled environment. Germination is light-responsive and begins at approximately 16 days.

When 3 weeks old, seedlings can be pricked out and transplanted to plastic bags or other containers. Plantable seedlings (30 cm high) can be obtained in 4 months, with seedlings growing to 1 m in 6 months and to 2 m or more in 1 year. *Milicia* seedlings can be transplanted as bare-root plants or stumps, in a similar way to teak, but the most common technique used in Ghana is to plant them out in the plastic bags in which they germinated.

Milicia has fast initial growth rates, particularly in semi-shade (30–60% irradiance). Seedlings must be grown under shade cloth or screen, even in the nursery, to avoid psyllid attack. Companion shade trees also appear to reduce psyllid galls on *M. excelsa*, although they reduce tree growth at the same time. Further experimentation should focus on achieving low levels of galls and good tree growth.

Alhough *Milicia* species have been widely planted as street and shade trees, they have not been successfully established and managed in plantations, largely because of problems with *Phytolyma lata*, and all timber has been harvested from natural forest. Given the moderate growth rate (e.g. 4–5 mm diameter increment per year in Ghana), more than 40 years would be required to increase from 50 to 70 cm in diameter, and to reach 110 cm diameter (the cutting limit in Forest Reserves in Ghana for this genus) would require 220 years from seedling establishment. Concern was therefore raised that present extraction rates will lead to extinction and many efforts were made to establish plantations.

Growth of *M. excelsa* in plantations is highly variable due to extremely local differences in soil conditions, indicating that the species may be very demanding in terms of soil fertility and physical properties. The tree is common near rivers, on termite mounds, in farms and burned areas, and generally on favourable sites. In Africa, the tree is highly valued and thus often protected and cultured, so that it is difficult to distinguish if an individual tree is growing well because of site quality or human tending and intervention.

Diseases and pests

The major impediment to growing *Milicia* spp. in plantations has been the gall-forming psyllids in the genus *Phytolyma*, which attack seedlings in particular.

Development of resistant varieties, along with appropriate silvicultural techniques, may make it more viable to cultivate this valuable tree in the future. Good results in mass propagation of cuttings have been obtained with an application of 0.2% IBA (indole-3-butyric acid), with a leaf area of around 40 cm^2 and a medium with a high moisture-holding capacity (24% or less water by volume).

Uses

M. excelsa ranks among the most valued timber trees of tropical Africa. Its dense yellow-brown heartwood is highly prized for both ornamental and structural purposes. It is durable, workable and resistant to termite and marine borer attack, making it equivalent in value to teak. The wood has an air-dry density of 673 kg/m^3 and can be worked fairly easily with hand or machine tools, and has good nailing and gluing properties. It is used for joinery, furniture, boatbuilding, marine work, domestic flooring, veneer, railroad sleepers, cabinet work and shop fittings. Deposits of calcium carbonate ('stones') can blunt and damage cutting tools. Some people can suffer skin reactions to the wet wood or sawdust.

It has been suggested that the leaves are edible and many medicinal uses are reported: bark infusions are used as a purgative and to treat whooping cough, heart disease, leprosy, venereal sores, hernia and lumbago, while the latex is used to treat river blindness (*Onchocerciasis*), toothache and tumours.

Ritual sacrifices take place underneath some *Milicia* trees, particularly in West Africa, where they are considered sacred in some localities; gifts are given to them, fertility and birth are associated with them and the timber is used to make ceremonial drums and coffins. Veneration of *Milicia* is the main approach to its conservation: the trees are conserved on farms, in sacred groves, in public places and in cemeteries.

Further reading

Bosu, P.P., Cobbinah, J.R., Nichols, J.D., Nkrumah, E.E. and Wagner, M.R. (2006) Survival and growth of mixed plantations of *Milicia excelsa* and *Terminalia superba* 9 years after planting in Ghana. *Forest Ecology and Management* 233, 352–357.

Dainou, K., Doucet, J.L., Sinsin, B. and Mahy, G. (2012) Identity and ecology of Central African timber tree species: the case of *Milicia* spp., a review. *Biotechnologie Agronomie Societe Et Environnement* 16, 229–241.

Nichols, J.D., Wagner, M.R., Agyeman, V.K., Bosu, P. and Cobbinah, J.R. (1998) Influence of artificial gaps in tropical forest on survival, growth, and *Phytolyma lata* attack on *Milicia excelsa*. *Forest Ecology and Management* 110, 353–362.

Ofori, D.A. and Cobbinah, J.R. (2007) Integrated approach for conservation and management of genetic resources of *Milicia* species in West Africa. *Forest Ecology and Management* 238, 1–6.

Millettia laurentii De Wild.

Wenge

Importance

Millettia laurentii is a timber tree native to Central Africa. The species is listed as 'Endangered' on the International Union for Conservation of Nature (IUCN) Red List of Threatened Species, due to habitat destruction and over-exploitation for timber. The heavy and hard timber, wenge, has a distinctive dark colour with tan striping.

Botanical features

M. laurentii is a medium-sized tree up to 30–45 m high with a cylindrical bole, often slightly bent and branchless for up to 20 m but usually much less. The tree can reach up to 120 cm in diameter, with small buttresses, or be fluted at the base. The outer bark is greyish, rough with lenticels. The inner bark is yellowish, granular, with reddish exudate. Branches droop with glabrous twigs. Leaves are alternate, imparipinnately compound with (4–)6–7(–9) pairs of leaflets. Stipules and stipels are absent. The petiole is 4–7 cm and the rachis 10–18 cm long. Petiolules are 4–6 mm long. Leaflets are opposite, oblong to obovate, 6–15 × 3–4(–9) cm, abruptly acuminate at apex and glabrous. Flowers are borne in a terminal panicle 20–40 cm long, with short-hairy branches up to 5 cm long. Flowers are bisexual, papilionaceous. The pedicel is 3 mm long, with 2 small bracteoles near the apex. The calyx is campanulate, 6–8 mm long with the tube as long as the lobes. The corolla is pale purple to purplish-blue, glabrous, standard orbicular, 12 mm in diameter, with a 3-mm long claw at its base. Wings and keel are 16 mm long. Of the 10 stamens, 9 are fused and 1 is free; they are 15 mm long. The fruit is an oblanceolate to linear flat pod 15–28 × 3–5 cm in size, with a stiff wall. It is glabrous, finely striped, dehiscent with 2–4 seeds. The seeds are oblong to lens-shaped, flattened, 22–25 × 18–20 mm, smooth and purplish-brown in colour.

Distribution

M. laurentii is restricted to an area between the eastern part of Cameroon, Equatorial Guinea and Gabon and the western part of the Central African Republic and the Democratic Republic of the Congo. It occurs in well-drained areas but also in forest subject to regular inundation, in riverine forest and tree savanna. The distribution is not regular across the area.

Silviculture characteristics

There is little information on silviculture characteristics for *M. laurentii*. Its timber is harvested locally from sustainably managed forests. Information on growth rates and other silvicultural characteristics is insufficient. In the Democratic Republic of the Congo, *M. laurentii* is one of the species planted for reforestation of formerly cultivated land.

Silviculture practice

M. laurentii can be propagated by seed and cuttings. Seedlings have epigeal germination. In a trial in the Democratic Republic of the Congo, 48% of stem cuttings planted at the onset of the rainy season sprouted. This ability to propagate through cuttings offers potential for plantation establishment. Wildings can also be successfully transplanted.

Diseases and pests

The heartwood of *M. laurentii* is very durable and resistant to fungal, dry-wood borer, and termite attack. It is moderately resistant to marine borers. The sapwood is susceptible to attack by powder-post beetles. Logs can be left in the forest for some time without damage, although borers may attack the green timber. In the Democratic Republic of the Congo, planted trees have been completely defoliated by the caterpillars of *Rhopalocampta libeon*.

Uses

M. laurentii is an attractive ornamental and roadside tree. Stem cuttings can be planted as a live fence. The roots have nodules containing nitrogen-fixing bacteria.

The heartwood of *M. laurentii* is yellow when freshly sawn but darkens on exposure to dark brown or black-brown, with black streaks, and is sharply demarcated from the pale yellow, 2–5 cm thick sapwood. Logs may have brittle heart. The wood has a straight grain and a medium to coarse texture. It is heavy, with a density of 750–960 kg/m^3 at 12% moisture content. It air-dries slowly, with a slight risk of distortion and checking, and high rates of shrinkage. Once dry, the wood is hard and elastic, but with a tendency to split. The wood saws and works well, but force is required and tools may blunt rapidly. Due to differences in density between light and dark areas, sanding can lead to uneven results. The gluing and varnishing properties of *M. laurentii* wood are poor due to the presence of resin cells, but results can be improved by using a filler. Contact with fine sawdust produced during processing may cause occupational asthma and allergic dermatitis in workers. Several isoflavones have been isolated from the heartwood of *M. laurentii*, as well as several alkaloids from bark and seeds.

The wood of *M. laurentii* is commonly used for heavy flooring, interior and exterior joinery and panelling, cabinet work, turnery and sliced veneer. It is also suitable for heavy construction, mine props, vehicle bodies, sporting goods, toys, canes, boxes, crates and railway sleepers. Its distinctive colour pattern makes it popular for decorative furniture and parquet flooring. It is also used for high-quality musical instruments such as guitars, and popular for the production of sculptures, masks and drums; it is sometimes used in archery bows.

M. laurentii is used in traditional medicine. A bark decoction is used to treat liver complaints, diabetes, hernias, skin diseases, constipation, fever and rheumatism. The bark is also applied as an expectorant and emetic, and to treat epilepsy, smallpox, oedema, sores and abscesses. It is used as fish poison, an insecticide, vermifuge and arrow poison. Edible caterpillars feed on the leaves.

Further reading

AFPD (2008) African Flowering Plants Database - Base de Donnees des Plantes à Fleurs D'Afrique. Available at: http://www.ville-ge.ch/musinfo/bd/cjb/africa/recherche.php?langue=fr (accessed 28 March 2013).

Senterre, B., Robbrecht, E., Degreef, J. and Friis, I. (2001) The phytogeography of the rain forests of Equatorial Guinea, with special emphasis on the Nsork rain forest. *Systematics and Geography of Plants* 71, 1087–1096.

Tchinda, A.T. (2008) *Millettia laurentii* De Wild. Protabase. Louppe, D., Oteng-Amoako, A.A. and Brink, M. (eds) PROTA (Plant Resources of Tropical Africa / Ressources végétales de l'Afrique tropicale), Wageningen, The Netherlands. Available at: http://database.prota.org/search.htm (accessed 1 December 2012).

N

Nauclea diderrichii (De Wild.) Merr.

Bilinga

Importance

Nauclea diderrichii is a species native to tropical Africa. Trees are medium to large in size and produce a cylindrical, unbuttressed bole. The wood is classed as very durable and has a variety of uses. The bark and leaves have some medicinal properties. The species regenerates very well, especially in open conditions, and is used widely within its natural distribution for taungya and plantation establishment.

Botanical features

N. diderrichii is a medium to large tree with a straight, cylindrical, unbuttressed bole, and a rather horizontal, whorled bough with a comparatively small, rounded crown. Average height of mature trees ranges from 30 to 40 m and sometimes up to 50 m. Average diameter ranges from about 0.9 to 1.7 m. Multiple stems are sometimes formed when young plants have been attacked by *Orygmophora mediofoveata*.

The bark is light brown with thin scales and may become a darker colour if the bole is in the open. The slash is thick, fibrous, and saffron-coloured. The sapwood is whitish and the heartwood is yellow or golden-brown, sometimes with a pinkish tinge.

The leaves are simple and opposite, have pairs of interpetiolar stipules at the base, and are about 5.6 × 10 cm in size (but are considerably larger on young plants). The leaf is elliptic, entire, obtuse, cuneate, glabrous and with a slender petiole about 1.8–2.5 cm long, flattened above. The midrib and nerves are raised below. The stipule is broadly ovate to almost orbicular, about 1.8 cm long, glabrous.

Inflorescences are capitate, scented, and in globose heads about 2.5–3 cm in diameter. The flower is small and consists of a short-toothed green calyx, 5-lobed yellow corolla tube, five epipetalous stamens, and an inferior, two-locular ovary with an exserted white style.

The fruit is globose, about 3.7 cm in diameter, yellow-brown, fleshy and deeply pitted. The seed is very small and light brown.

N. diderrichii is an evergreen tree, with the main flowering season from May to August (but to December in some locations). Fruits have been found throughout the year, but the main fruiting season is in the first few months of the dry season from September to February, depending on location. Seed distribution is mainly by birds, although other animals, including elephants, may disperse the seeds.

Distribution

N. diderrichii is found in the moist evergreen and transitional to moist semi-deciduous forests of tropical Africa. The natural distribution ranges from Sierra Leone as far east as Mozambique, and as far south as Angola.

The species has been grown in taungya plantations within its natural distribution. It has been grown in Nigeria since 1918 as a nurse species for many valuable mahoganies (*Khaya* spp.) and as a useful species in its own right. Plantations or trials have also been established in Cameroon, Côte d'Ivoire, Liberia and Democratic Republic of Congo. It has also been planted in trials in Brazil, Fiji, Peru and Vanuatu.

Silviculture characteristics

N. diderrichii is a very strong light demander from the seedling stage and grows vigorously when sufficient overhead light is available. However, optimum growth rates occur at well below full light. It is a species that grows in large gaps and can reach the canopy in 15 years.

Silviculture practice

Seeds, which are classified as orthodox, do not require pretreatment. There are about 60,000–100,000 seeds/kg. As the seeds are small it is easier to break up the fruits in water and dry them and then sow, or else pour the water and seeds on to a seedbed. Germination occurs within 14–21 days. Seedlings can be pricked out at about 6–8 weeks and put in pots or transplant beds. It is advisable to have light screens over the potted seedlings or transplant beds. The seedlings reach plantable size in 12 months.

Planting can be carried out using potted seedlings, stumps or striplings. Young plants are liable to insect borer attack and leaf curl, and tend to develop a bushy habit; a planting spacing of not greater than 3 × 3 m is necessary.

Plantations in Sapoba, Ona and Agbodi (Nigeria) have achieved annual diameter increments of between 1.2 and 1.6 cm, whereas plantations in south Bakundu (Cameroon) have recorded annual diameter increments of 3.1 cm up to year 8. Trees attained 12 m in height and 10 cm dbh within 4 years on old logging tracks in Ghana. In Nigeria, a 26-year-old taungya plantation achieved a mean height of 16 m and a mean dbh of 27 cm.

Diseases and pests

Orygmophora mediofoveata larvae attack young shoots of *N. diderrichii*, but are controlled to some extent by natural enemies which include the parasitoid wasps *Macrocentrus* and *Pristomerus* spp.

Uses

N. diderrichii is widely used on taungya and other plantations, and is recommended as a nurse crop for valuable Meliaceae trees such as *Lovoa trichilioides* and *Khaya* spp.

N. diderrichii has an attractive versatile wood of worldwide reputation. The heartwood has a yellow to gold colour, and the sapwood is clearly defined and whitish to pale yellow. The grain is usually interlocked and of medium to coarse texture. The wood is hard and needs careful drying to avoid hair checks. Strength and density are variable. The wood has an air-dry density of 700–900 kg/m^3, with an average of about 780 kg/m^3. The wood is sometimes brittle. The heartwood is rated as very durable; it is resistant to marine borers and moderately resistant to termites. The sapwood is liable to powder-post beetle attack. It has satisfactory sawing properties, but needs a filler to give a satisfactory finish. It is used for harbour structures, bridges, decking, boat work, exterior structures, flooring, fences, panelling, drums and household utensils such as mortars. An alkaloid in the wood is said to be a cumulative cardiac poison and may be fatal.

Preparations from the bark are used in traditional medicine to treat gonorrhoea, anaemia, stomach pains, indigestion and jaundice, while preparations from the leaves are used to treat diarrhoea and measles.

Further reading

Ansel, J.L. (2005) Le bilinga. In: Ansel, J.L. (ed.) *Les Arbres Bâtisseurs*. Eyrolles, Paris, pp. 138–139.

Irozuru, G.N. (1986) *Nauclea diderrichii* (Rubiaceae): a monograph and appraisal. MSc thesis, University College of North Wales, Bangor, UK.

Opuni-Frimpong, N.Y. and Opuni-Frimpong, E. (2012) *Nauclea diderrichii* (De Wild. & T. Durand) Merr. Available at: PROTA4U http://www.prota4u.org/protav8.asp?h=M4&t=Nauclea,diderrichii&p=Nauclea+diderrichii#Synonyms (accessed 20 March 2013).

Neobalanocarpus heimii (King) P.S. Ashton

Chengal

Importance

A large dipterocarp tree reaching to over 60 m tall, *Neobalanocarpus heimii* produces a durable and heavy hardwood, which is particularly important and popular in its native range in Peninsular Malaysia, where it has been used as a standard by which other timbers are judged. It is easy to work and can be used for a wide variety of purposes. Having been overexploited and with poor regeneration, *N. heimii* is classed as 'Vulnerable' on the International Union for Conservation of Nature (IUCN) Red List of Threatened Species, with populations in southern Thailand and Singapore now believed to be extinct. The Malaysian Government has banned the export of round logs and has implemented conservation measures.

Botanical features

N. heimii is a large tree, sometimes with a height of over 60 m and a diameter of 100 cm or more in older specimens, with tall, stout buttresses. The largest specimen in Malaysia is in the Pasir Raja Forest Reserve in Dungun and is 65 m tall with a girth of 16.75 m and dbh of 5.33 m.

The bole usually has a good shape and is clear for 30 m or more, but can also be somewhat irregular and low-branched. The bark is dark coloured, with shallow, irregular, longitudinal fissures, and becomes shaggy in older specimens; it falls off in long scales, with occasional exudations of clear, colourless or golden-coloured resin. The crown is large and heavy.

Leaves are alternate and simple, leathery, elliptical–lanceolate, 7–17 × 2.3–5 cm, acuminate, rounded to cuneate at base, unequal, pinnately veined with 9–12 pairs of ascending and arched veins, slightly sunken on the upper surface, but prominent and glabrous or only sparsely pubescent on the lower; tertiary venation is reticulate–scalariform. The petiole is 5–10 mm long. Stipules are narrowly oblong, to 12 mm long, and early caducous.

The inflorescence is paniculate, terminal or axillary, up to 9 cm long, singly branched and puberulent. Flowers are bisexual and actinomorphic. Sepals 5, subequal, imbricate at base, broadly ovate, acute, outside caducous puberulent. Petals 5, elliptic, creamy-white or greenish-yellow, outside densely buff pubescent. Stamens 15, filaments slender and tapering, anthers linear–oblong, yellow. Ovary single, ovoid, glabrous, surmounted by a filiform style. The fruit is acorn-like, with an oblanceolate, cylindrical nut on a short, stout stalk, up to 5.5 × 2.5 mm, shortly apiculate, shining, embraced at the base by a cup formed by the enlarged woody, saccate sepals, the outer 2 of which are slightly smaller and thicker than the inner 3. The seed, shaped like the fruit but a few millimetres shorter, is green at maturity.

Unlike most other species of Dipterocarpaceae, *N. heimii* flowers and fruits annually. The timing of peak flowering is variable and has been observed in mid-November and early December, depending on year. Pollination is effected by insects, mainly bees. The fruits ripen about 5 months after the first appearance of flowers. Planted trees have been known to bear fruit at 10 years old or younger. Seeds are heavy (5–7 g), so dispersal (by gravity) rarely extends beyond 16 m from the mother tree.

Distribution

The species is endemic to Peninsular Malaysia, Singapore, and southern peninsular Thailand. It is widespread in mixed dipterocarp forest below 1000 m altitude, especially on well-drained friable soils on undulating land. In Peninsular Thailand, where it is now thought to be extinct, it occurred frequently in hill dipterocarp forests along slopes and in valleys, often in association with *Shorea curtisii*.

Silviculture characteristics

Of the emergent tree species present in some lowland dipterocarp forests in Peninsular Malaysia, *N. heimii* is among the most common; in the Pasoh Forest Reserve, for example, it is only second in abundance to *Shorea maxwelliana* for all stems greater than 1 cm dbh. The silviculture of *N. heimii* is totally dependent on natural regeneration. Seedlings are capable of surviving for long periods under dense shade, albeit suppressed. They respond to openings and gaps, and well-established young trees can appear without assistance once such openings are created. They respond more positively to forest gaps than saplings of some *Shorea* species.

Silviculture practice

Seeds are highly recalcitrant and do not store for long, so timing of seed collection soon after falling is critical. As seeds are wingless and heavy, most are found under the canopy of the mother tree. Short-term storage at 14°C for up to 57 days is possible without serious loss of viability. Germination takes between 9 and 45 days. Germination rates range from 68% to 100% and are related to seed size and weight.

There is little experience with propagation in nurseries. *N. heimii* seedlings are sensitive to over-exposure and drought; seedling care is necessary as fungal infections, insect predation and water stress can result in seedling mortality rates of 31–52%. Some shade is required in the early stages, but saplings and later stages benefit from greater amounts of light. Group planting trials of *N. heimii* in secondary growth forest, followed by 4 years of treatment, resulted in good growth and survival.

Although growth is slow at the seedling stage, rapid growth takes place after trees reach a diameter of 25 cm and is maintained until the tree is over 80 cm diameter. It is estimated that it would take about 110 years for a tree to reach a diameter of 40 cm in natural forests under considerable competition.

If grown under more favourable conditions, better growth can be obtained. For example, the mean annual diameter growth of the largest trees in trial plots of the Forest Research Institute Malaysia (FRIM) in Kepong was 0.95 cm in a 40 year-old stand. In arboreta, where trees receive better maintenance, a diameter growth rate of 1.18 cm per year has been observed for 45 year old trees. It is therefore possible to attain a 70 cm diameter tree in slightly over 70 years under proper maintenance and management.

There has been some success with enrichment planting in advancing secondary forest in Malaysia, as well as with underplanting in *Acacia mangium* plantations with a view to replacing them with the higher-quality dipterocarp.

Diseases and pests

Insect predation destroys 23–37% of the seed crop before the fruits fall. Fruits are often attacked by the seed beetle *Coccotrypes graniceps* and its larvae. A shoot borer, *Laspeyresia* sp., has been reported to cause damage to saplings, while a pinhole borer, assumed to be of the genus *Diapus* (Coleoptera: Platypodidae), attacks living trees.

Uses

N. heimii timber is a heavy hardwood, with an air-dry density ranging from 915 to 980 kg/m^3. The sapwood is light yellow and sharply defined from the heartwood, which is light yellow-brown with a distinct green tinge, darkening on exposure to a dark purple-brown or rust-red. Wood texture is fine and even, with shallowly to deeply interlocked grains. The timber is very strong, slightly difficult to re-saw, and easy to moderately easy to cross-cut. Planing is easy and the planed surface produced is smooth. The timber seasons slowly with moderate end-checks and surface-checks as the main sources of degrade. The wood is extremely durable and no treatment is necessary due to the presence of natural preservatives; even under exposed conditions the timber can last around 100 years. The timber is suitable for all forms of heavy construction, railway sleepers, bridges, wharves, power-line poles, heavy-duty furniture and heavy-duty flooring; it is also a favourite for boatbuilding.

Good-quality resin, known in Malaysia as dammar penak, can be obtained by wounding the tree. It has been used only on a limited scale in the manufacture of certain classes of varnish.

Further reading

Jøker, D. and Nadarajan, J. (2003) *Neobalanocarpus heimii* (King) Ashton. Danida Forest Seed Centre Seed Leaflet 70, Humlebaek, Denmark, pp 2.

Marzalina, M., Nashatul, Z.N.A. and Jayanthi, N. (2001) Ex situ conservation efforts for *Neobalanocarpus heimii* in Malaysia. In: Proceedings of the South East Asian Moving Workshop on Conservation, Management and Utilization of Forest Genetic Resources. FAO Corporate Document Repository. Available at: http://www.fao.org/DOCREP/005/AC648E/AC648E00.HTM (accessed 15 November 2012).

Okuda, T., Manokaran, N., Matsumoto, Y., Niiyama, K., Thomas, S.C. and Ashton, P.S. (eds) (2003) *Pasoh: Ecology of a Lowland Rain Forest in Southeast Asia*. Springer-Verlag, Tokyo.

Neolamarckia cadamba (Roxb.) Bosser

Common bur-flower tree

Importance

Neolamarckia cadamba is a fast-growing deciduous tree species. It is planted as a reforestation species in watersheds and eroded areas. It is used as a plantation crop for commercial purposes and its wood has a wide range of commercial uses, and it has the potential to act as a windbreak in agroforestry systems.

It is becoming one of the most frequently planted species in the tropics; in South-east Asia alone areas reaching several hundreds of thousands of hectares are planted. Predictions estimate that this species may compete with other commercial species, such as African obeche (*Triplochiton scleroxylon*).

Botanical features

N. cadamba is a large deciduous tree and has a straight cylindrical stem with or without a small buttress. It is a fast-growing species highly tolerant of sunlight, and can reach a height of 45 m and a diameter of more than 100 cm.

The crown is typically umbrella-shaped and small, with its branches spreading and arranged in tiers. In young trees the bark is greyish and smooth, but it becomes darker and longitudinally fissured in older trees. The bark peels or sheds off in small rectangular plates.

Leaves are simple, opposite, coriaceous, 20–25 cm long and 11–15 cm wide, obovate to oblong, apex acute to acuminate, base obtuse or rounded and more or less heart-shaped, margin entire to slightly repand; veins distinct underneath; dark green above and pale yellowish-green beneath; stipules oblong and covering the buds, readily falling off. The midrib is prominently ridged below, greenish-white; petioles robust, cylindrical, 2.5–6 cm long.

Orange multiple flowers, borne in globose heads, are fragrant and attract bees and insects. The fleshy orange fruit is globose and is made up of the ovaries of many small flowers, coalesced into a fleshy mass of closely-packed capsules, which later splits into four parts and contains a number of minuscule seeds.

In the Philippines, the flowering season of *N. cadamba* runs from April to June. Fruits begin to mature or ripen from September to February. The tree begins flowering at 5 years old. In other areas the season varies; for example, India: May–June with the fruit ripening and falling in January and February; Sabah: June–September with fruit ripening in September to February; Sri Lanka: fruit ripening in September. In Indonesia, trees may begin flowering at 4 years old and flowering occurs from April to August and occasionally from March to November.

Distribution

N. cadamba is widely distributed from India to the Malaysian Archipelago and grows naturally in Nepal, Assam, Sri Lanka, Vietnam, Myanmar, Indonesia, Australia, Papua New Guinea, Sarawak and the Philippines.

Silviculture characteristics

N. cadamba is a light-demanding species. On good sites it grows relatively vigorously and rapidly. It is a pioneer species common to old logging sites. It can tolerate wind, has the ability to self-prune, and is suited to coppicing.

Silviculture practice

Where there are sufficient seed-producing trees of *N. cadamba*, clearing the soil at the time of seed ripening can help effect natural regeneration. A plantation in East Kalimantan has been managed this way since 1938. Plantations established from nursery seedlings are less successful than natural regeneration, which is possibly linked to local site conditions.

N. cadamba can be propagated by seed, nursery-grown seedlings, stumps, and stem cuttings. Seeds can be stored for up to 2 years in cool conditions and for 3–6 months in dry, airtight containers in shaded areas. The seeds are very small and there are approximately 20,000 seeds/g. Special techniques are used to extract the minute seeds from the fleshy multiple fruit. In the Philippines, seed extraction takes place by rubbing the fresh fruit and macerating it in water until the seeds are exposed. In Indonesia, the fruits are dried before the seeds are removed. The mature fruits are yellow to yellow-orange and each contains approximately 8000 seeds. Fruits are readily eaten by ants, cattle, and other animals, so should be collected immediately after the fruit falls. The seeds should be extracted from the fleshy fruit and dried; no pre-germination treatment is needed.

Germination occurs in 7–14 days. After germination, the seedlings should be watered regularly for 3–4 months or until seedlings reach about 30–45 cm tall.

N. cadamba can be planted bare-rooted without a significant loss of growth. Stump plants about 1 cm diameter give good results. Planting at 3 × 3 m or 4 × 4 m spacing is recommended, although to produce relatively straight bole timber, a close spacing of 2 × 2 m is needed. Thinning should be done early and frequently, and is an easy task when the boles are straight with regular crowns.

N. cadamba has also shown to be an excellent shade tree for line planting with dipterocarp species.

Rotations in plantations can vary from 4 to 7 years according to site location and end products. High production yields of 50 m^3/yr have been reported in the Philippines. Usual production is 10–40 m^3/yr.

Diseases and pests

The fungi *Cercospora* sp., *Alternaria* sp., and *Cladosporium* sp. cause irregularly shaped dark brown or black spots, sometimes with a distinct yellow edge, on leaves. The spots enlarge to patches until the leaves wither and drop off. A fungus which causes partial or complete defoliation and dieback is *Gloeosporium anthocephali*. An infestation by *Mycena citricolor* on *N. cadamba* and coffee trees has been reported in Costa Rica. The fungus *Meliola cadambae* has been found on the leaves of *N. cadamba* in Kerala, India.

The most serious pest of *N. cadamba* is *Arthroschista hilaralis*, an insect which defoliates the tree. Other defoliators include the hornworm *Daphnis hypothous* and *Eupterote fabia*. In India, the teak sapling borer (*Sahyadrassus malabaricus*) has been observed boring into the stem of saplings. In East Java, 1-year-old mixed plantations of *N. cadamba* and *Falcataria moluccana* were attacked by grubs of *Lepidiota stigma*, *Leucopholis rorida*, and *Holotrichia hellesi*; the taproots were damaged and the root collar peeled out.

Seedling roots can become deformed and a general yellowing of leaves may appear due to attack by root-knot nematodes. This is commonly caused by *Meloidogyne incognita* and can be prevented if the soil medium is steam- or heat-sterilized.

Uses

N. cadamba is considered by some authors as a 'miracle tree' because of its relatively fast growth rate; it is commonly used to reforest denuded watershed areas. In some parts of the Philippines the species is used to prevent the invasion of creeping and climbing vines which may interfere with the favourable growth of desired species.

Because of its spreading crown and large leaves it is a suitable species for windbreaks in agroforestry systems. It is planted with *Leucaena leucocephala* in the Philippines and Indonesia, and in Kalimantan it has been intercropped with upland rice. It is also used as an ornamental tree.

The wood cannot be used in outdoor conditions as it has poor durability. Its density is 290–465 (560) kg/m^3 at 15% moisture content. It is a soft wood and if properly treated is a good material for making beams and rafters, tea chests, packing cases, shuttering, ceiling boards, toys, yokes, carvings, boxes, ceiling boards, wooden shoes, matches, pencils, chopsticks, veneer, plywood and charcoal. It is also suitable for making canoes and cheap furniture. Its most important use is for pulp, producing low- and medium-quality paper. The pulp is sometimes mixed with other long-fibre material.

N. cadamba leaves are used as a fodder for cattle. The fruit and flowers are edible and the leaves are occasionally used as plates. The bark, when dried, has medicinal properties, as do the leaves. The bark of the roots also provides a yellow dye. *N. cadamba* also yields an important essential oil.

Further reading

Choo, K.T., Gan, K.S. and Lim, S.C. (1999) Timber notes - light hardwoods VI (dedali, kedondong, kelempayan, kelumpang, kembang semangkok). Timber Technology Center, Forest Research Institute Malaysia (FRIM), Kuala Lumpur, Malaysia, *Timber Technology Bulletin* 16, 7 pp.

Jøker D, (2000) *Neolamarckia cadamba* (Roxb.) Bosser. Seed Leaflet 17, Danida Forest Seed Centre, Humlebaek, Denmark, 2 pp.

Mondal, S., Dash, G.K. and Acharyya, S. (2009) Analgesic, anti-inflammatory and antipyretic studies of *Neolamarckia cadamba* barks. *Journal of Pharmacy Research* 2, 1133–1136.

Nothofagus pumilio (Poepp. & Endl.) Krasser

Lenga

Importance

Nothofagus pumilio is an important and abundant species in southern Chile and Argentina, where it represents one of the largest undeveloped tree resources; it was identified as the second most abundant species in Chile in the National Forest Survey. The wood is a potential hardwood substitute for maple and cherry furniture, mouldings and lumber. It accepts finishes well and is versatile with respect to manufacturing. Heart rot and windthrow are potential problems. The species appears to respond well to silvicultural practices. The accepted silvicultural regime is shelterwood cutting.

Botanical features

N. pumilio is a temperate boreal tree, with a wide variety of niches, from sea level to alpine (2000 m). It prefers well-drained soils. The habitat varies from the southern to the northern range, where it is found in subalpine and avalanche zones. It is found on all aspects, but grows in uniform stands on drier sites and in mixed stands with *Nothofagus betuloides* on wetter sites.

Size varies with location, as trees are very responsive to site characteristics. On a good site height at maturity will reach 22–25 m, with a dbh of 2 m. Tree maturity depends on site and ranges from 150 to 350 years.

N. pumilio has a cylindrical straight stem, with branches and forks from approximately 20 m upwards, and an oval crown. The crown appears to be more uniform and lighter in colour than in *Nothofagus betuloides*, commonly found in association with it.

The root system is highly variable. In soils with a hardpan or high water table, the roots will fan out without a taproot structure. In deeper soils, sinkers and a taproot will develop. Mycorrhizae appear to be important for development.

The foliage is deciduous, light to medium green, turning red or yellow in the autumn. Leaves are ovate with 5–7 veins on each half, with the leaf margins serrated (two teeth between the veins). Young shoots are reddish with white lenticels.

N. pumilio produces flowers in spring (around December), and fruits mature in autumn (around April). Single seeds are produced, with about 50,000/kg. Seeds have low viability and can develop without fertilization. Reproductive and vegetative buds are staggered along the branch, with female flower buds maturing later than male buds. In general, seeds disperse naturally no greater distance than twice the height of the crown of the parent tree.

Distribution

N. pumilio is distributed in Chile and Argentina between latitudes 45°S and 57°S. In the northern part of this range it is found at altitudes between 1300 and 2000 m, while in the southern part it is found between sea level and 500 m. Lenga-type forest occupies some 0.3 million km^2, with an estimated 15,000 km^2 being in Argentina. It is well represented in many parks and reserves in Chile and Argentina.

N. pumilio has been experimentally planted in a few areas outside South America (e.g. Scotland, Germany, Faroe Islands, and Scandinavia), and is planted as an amenity tree in the Pacific North-west states of the USA. However, it is mainly present in research collections and arboreta in temperate Europe and North America.

Silviculture characteristics

The shelterwood silvicultural systems traditionally used in *N. pumilio* forests in South America suggest that the species is shade-tolerant, needing the shelterwood cut to regenerate satisfactorily. However, preliminary forest nursery studies suggest that the species may respond well to increased light regimes. *N. pumilio* is not an aggressive competitive species, and does tolerate shade better than some competing species: this may be why it has been considered to be intolerant of high light conditions. It is not resistant to fire.

It is not an aggressive colonizer; other trees, grasses and shrubs out-compete *N. pumilio* in warmer areas.

Low seed viability is a problem which could be improved by better seed handling procedures at harvest and by controlling moisture content.

Silviculture practice

Silvicultural practices in forests designated as lenga types are prescribed by regulation in both Chile and Argentina. The silvicultural system is shelterwood with a series of precommercial and commercial thinnings, commonly referred to as 'seed cuts' and 'protection cuts'.

Site preparation studies have not been conducted for *N. pumilio*. Because the climate is dry with, at times, high winds, the use of fire is inappropriate as trees are not resistant to fire damage. Ripping to control grass competition helps natural regeneration and the establishment of seedlings. Shelterwood methods currently being used keep understorey growth to a minimum. However, *N. pumilio* grows in exposed situations where winds scour the surface, and any disturbance to the soil surface may encourage soil erosion.

Mycorrhizae, such as *Russula nothofaginea*, have been found associated with *N. pumilio*.

Preliminary thinning trials show growth responses up until about 60 years old. Elaborate silvicultural thinning regimes have been developed. Planting nursery stock and/or thinning of natural regeneration (upwards of 50,000–100,000/ha when saplings are 3–4 m tall) would probably result in the greatest productivity increases and economic advantage.

Chilean Government regulations for the management of *Nothofagus* forests require that 3000 seedlings/ha be established before regeneration can be considered successful and the shelterwood overstorey is removed. Under existing forest conditions and from a practical perspective, once a seed cut is accomplished there is very little likelihood that the remaining overstorey will be harvested due to the economics and potential damage to the regeneration. *N. pumilio* appears to respond well to silvicultural manipulation.

Management of *N. pumilio* is expected to increase forest productivity by approximately 25%. Current rotation periods are between 80 and 120 years. Unmanaged forests yield from 250 to 400 m^3 of wood/ha. Growth rates are variable; height growth tends to predominate during the first 60 years, whereas diameter increment becomes significant later. In initial nursery trials, *N. pumilio* seedlings at 8 months were the size of wildings 8–20 years old.

Diseases and pests

Heart and butt rots, caused by either *Spongipellis chubutensis* or *Pyrrhoderma andinopatagonicum*, are common and may be introduced by guanaco and cattle. Well over 50% of the mature trees in an unmanaged forest may be affected by butt rot, heart rot, and/or stem rot. Stem and heart rots may be introduced after branches in the crown are broken by ice storms or by damage from livestock and guanaco. Since *N. pumilio* is harvested from unmanaged older stands of trees, there is a high incidence of rot, resulting in only 20% of wood volume being useable.

Attacks by insects are periodic and usually do not cause tree mortality, just slower growth. The damage recorded has included temporary defoliation by *Ormiscodes cinnamomea* in the Nahuelhuapi and Lanín national parks in Argentina. Geometridae have been found infesting lenga in Punta Arenas, Chile.

Misodendron mistletoes grow almost exclusively on species of *Nothofagus*, with harmful effects on trees, forest productivity and wood quality. Control is achieved by pruning infected branches and/or removing the most infected trees.

Since their introduction in the 1950s, beavers (*Castor canadensis*) have become a significant pest, destroying an estimated 1.0–1.5% of lenga forests on Tierra del Fuego, primarily due to waterlogging of the soils as a result of flooding; however, some direct cutting damage also occurs. Beavers seemed to select lenga in mixed *N. pumilio–N. betuloides* forests, particularly trees with stem diameters <10 cm. Humane trapping, a bounty system, and the development of a fur industry are recommended as the best methods of control.

Guanaco (*Lama guanicoe*), a large camelid herbivore on the Convention on International Trade in Endangered Species (CITES) endangered species list, affects lenga forests by grazing and creating forest dust beds. Damage is correlated with population density, the availability of alternative grassland habitats, and seasonal use. Standard herbivore protection is recommended for planted stock, such as tree guards for saplings, high fencing and repellent sprays.

Fire, land conversion and livestock grazing are the most serious causes of anthropomorphic damage. *N. pumilio* is thin-barked and thus not fire-resistant. Livestock damage growing shoots and buds and may also cause root damage or be a vector of heart rot. Argentinian law requires fencing to protect recently harvested areas from livestock until the trees reach sufficient height (approximately 2 m) to withstand grazing. There is less grazing pressure on the *N. pumilio* forests of Chile.

Uses

Agroforestry, while commonly practised in the regions where *N. pumilio* occurs, is not recommended until trees are well-established, since root damage from grazing animals is thought to encourage stem heart rot. It is sometimes used in shelter belts around ranch houses in Patagonia.

In the Pacific North-west of the USA, *N. pumilio* is popular as a street tree due to its yellow and red autumn foliage colours and compact crown.

N. pumilio is used locally for construction timber, firewood, and shingles. When traded internationally, the timber has many end uses, including fine and outdoor furniture, doors, mouldings, veneers and cabinet work. It is suitable for sawmilling (mouldings, lathing and carving). It accepts dyes well, and finishes well. It retains its colour in natural light, and is resistant to ultraviolet light. The heartwood is similar in colour to cherry (*Prunus*) and the sapwood colour is similar to maple (*Acer*). The wood is popular in Europe because of its favourable physiomechanical characteristics, and in Japan because of its light colour. It has a density of 470–490 kg/m^3.

Edible fungi of the genus *Cyttaria* grow on *Nothofagus* species and are used in a limited way by local people.

Further reading

Burry, L.S. and Trivi de Mandri, M.E. (2011) Historia del bosque en el centro de Tierra del Fuego, Argentina [History of the forest of central Tierra del Fuego, Argentina]. *Revista Mexicana de Ciencias Forestales* 2, 25–36 [in Spanish, with English summary].

Veblen, T.T., Hill, R.S. and Read, J. (eds) (1996) *The Ecology and Biogeography of Nothofagus Forests*. Yale University Press, New Haven, Connecticut, USA.

O

Ochroma pyramidale (Cav. ex Lam.) Urb.

Balsa

Importance

Ochroma pyramidale, balsa, is a pioneer species native to tropical America. Under ideal conditions it is capable of producing the fastest growth rates and the lowest timber densities observed among tropical forest species. It produces one of the lightest commercial timbers, which formerly had a wide range of applications for buoys and life jackets, as well as for model-making and insulation purposes. Ecuador is still the main exporting country, supplying over 90% of the volume traded, but the species has also been widely planted in tropical Asia. Timber density varies greatly, with higher-density woods, and the heartwood of older trees, having little commercial value.

Due to its fast natural growth in large clearings, *O. pyramidale* became a favourite species to plant for restoration of degraded areas (notably in Mexico and Costa Rica).

Botanical features

O. pyramidale is a deciduous or evergreen medium-sized tree with a broad crown. It can reach as much as 30 m in height and 70 cm in diameter. Usually it possesses a short cylindrical stem with short buttresses, though some older trees develop a distorted, bulging stem form, through water shortage. It is botanically a highly variable taxon and was formerly thought to consist of at least 11 distinct species (now it is the only species in the *Ochroma* genus).

O. pyramidale has large, cordate, simple leaves with deciduous stipules. The alternate leaves are arranged spirally. The presence of two foliar nectary types, along the vein and on the petioles, results in the continual maintenance of ants on the leaf undersurface, which tend to function as antiherbivore agents.

The flowers are bisexual, large, and showy. They are axillary, solitary, the calyx tubular with 5 unequal lobes. The 5 petals are whitish, with stamens united into a briefly 5-lobed staminal column bearing sessile wavy anthers. The fruit is oblong, ribbed and 5-valved. The seeds are often embedded in hairs of the fruit walls, with little or no endosperm and flat, contorted or plicate cotyledons. Seeds are dispersed by wind.

In Ecuador, the tree flowers at the end of the rainy season. Flowers open late in the day and are rich in nectar. Bats, the kinkajou and the olingo (two nocturnal arboreal mammals) are thought to be the primary pollinators. In plantations on Java and Peninsular Malaysia, *O. pyramidale* trees flower throughout the year, and are pollinated by bats.

Distribution

O. pyramidale is a native of tropical America, from southern Mexico to Bolivia, including islands of the West Indies. It has been planted widely in tropical Asia (Papua New Guinea, Indonesia, Malaysia, Thailand and the Solomon Islands), where it has locally become naturalized.

Silviculture characteristics

O. pyramidale is a typical pioneer species. The wind-dispersed seeds can travel far and are capable of sustained dormancy. Regeneration will colonize clearings or gaps, natural or man-made. It grows gregariously, and early growth can be extremely rapid (30 m tall and 30 cm diameter at breast height (dbh) in 10 years was recorded in Panama). *O. pyramidale* is a strong light demander, but tolerates some lateral shade in the first year in cases where the summer sunlight is strong. It has the ability to self-prune.

Silviculture practice

Under natural conditions, forest clearance exposes the soil to the sun, and this triggers germination of *O. pyramidale* seed.

O. pyramidale is normally propagated by seed. Viable seeds are produced when trees are 3–4 years old upward. There are about 100,000 seeds/kg; these can be stored well for 12–18 months in a closed container. One method of removing the long silky hairs from the seeds is by burning. Dry heat (96°C) for 5 min) and immersion in boiling water improves germination rate considerably.

Seeds are sown in boxes or in plastic pots, in a 1:1 mixture of sandy loam garden soil and pure fine sand, a soil media combination which gives a high germination percentage. Slight shade is given to prevent excessive moisture loss. When seedlings are about 20 cm tall (at an age of about 3–4 months), they are ready for transplanting into holes of 30 × 30 × 30 cm, at spacing of about 4 to 5 m^2.

The roots of young trees are very susceptible to damage, so bare-rooted plants cannot be used and direct sowing is often the preferred method.

Careful tending is necessary during the first year, as the seedlings are quite fragile, and can be easily damaged by wounding, trampling or browsing. Seedlings cannot survive heavy weed competition in the first year. Nurse crops are useful to guard against excessive bark scorch. Epicormic buds should be removed. Protection against fire is also necessary.

Early thinning to maintain fast growth is particularly important. Following harvest, residues can be burned, which gives rise to abundant natural regeneration.

A taungya system at 2 × 3 m spacing can be used. In Papua New Guinea, a closer initial spacing (2 × 2 m; 2268 trees/ha) is applied, with thinning to 99 trees/ha before clear-felling at age 5 years old. In India, experiments with taungya using low crops (hill paddy rice) suggested that such low agricultural crops can help to suppress weed competition, and do not, as was once thought, retard seedling growth during the early years.

Planting *O. pyramidale* in fallow land dominated by bracken fern accelerated succession to forest (southern Mexico). Seeds were broadcasted and weeding was applied to establish the trees successfully.

Early growth can be extremely rapid, and this gives rise to the most valued, lightest-weight timber. In trials in Indonesia, mean annual growth was 2–6.6 cm in diameter and 1.2–3.3 m in height. After the tree reaches the age of 10–15 years, the growth rate slows down, and from this period the tree can deteriorate rapidly. In South America, mean annual diameter growth may be as much as 10 cm, and after 10–12 years, when growth stabilizes, trees are generally 20–25 m tall and about 100 cm in diameter. In Peninsular Malaysia, growth is usually considered too slow to produce the very low-density wood which is sought commercially. In the Philippines, an annual height increment of about 3 m for the first 2 years has been recorded, with trees reaching a height of 7.5 m by age 3 years. Diameter increment is similarly rapid during early years: from 14 cm diameter at age 1 year, to 29 cm diameter in year 3, and 60 cm diameter at age 12 years.

Short rotations of 10–12 years are usually adopted in the Philippines, although in Papua New Guinea even shorter rotations are practised. In Indonesia, rotations of 7–8 years are usual, as after this time heartwood begins to form, which gives a heavier, less valuable timber.

Diseases and pests

No serious pests or diseases are reported. The World Agroforestry website mentions a wood borer, *Zeuzera coffeae*, damaging a 1.5-year-old balsa plantation in Java.

In their native habitat the trees are often associated with an ant population on the leaves, which acts as an antiherbivore agent.

Uses

Being a pioneer species, *O. pyramidale* helps to stabilize soil in areas recovering from forest loss or damage, and with its broad leaves and rapid development of a spreading crown, it could help to minimize soil erosion problems resulting from rainfall on open areas. The Lancandon people of Mexico are aware of the tree playing an important role in the succession from fallow to forest in the management of their traditional 'slash and burn' cultivation system, and purposely plant it in fallows. The trees appear to be slowing degradation of organic matter and allowing its storage in leaf litter.

The timber, though low in density, is highly variable (typically about 160 kg/m^3 at 12% moisture content but varying from 40 to 310 kg/m^3). The wood is considered part of the 'corkwood group', and contains a large proportion of large, thin-walled parenchyma cells.

Balsa wood was formerly used for components in airplane body construction, in the refrigeration industry, for insulating plates, life-belts and jackets, surf boards and buoys. Synthetic materials have largely replaced it, though for some special applications (e.g. model plane making) balsa is still the preferred material. It has potential as a raw material for pulp and paper production, and has also been used in composite sandwich board production.

The kapok-like floss from the fruits is suitable for filling pillows and mattresses. The bark yields a strong fibre that can be used for rope. The stem and root bark are reported to have medicinal properties. The Lancandon use it for treatment of back pain.

Norwegian Thor Heyerdahl, trying to prove that contact between Polynesian and South American peoples had been possible, used balsa logs to build the *Kon Tiki* raft which he famously sailed from Peru to the Tuamotu Archipelago in 1947. 'Balsa' means 'raft' in Portuguese.

Further reading

Diemont, S.A.W., Martin, J.F., Levy-Tacher, S.I., Nigh, R.B., Ramirez, L.P. and Golicher, J.D. (2006) Lacandon Maya forest management: restoration of soil fertility using native tree species. *Ecological Engineering* 28, 205–212.

Greenhouse, S. (1935) The culture of the Balsa tree in Ecuador. *Journal of Forestry* 33, 870–876.

Ibanez, A., Hernandez, C. and Sosa, M. (1983) Balsa, *Ochroma pyramidale*, and some of its applications. *Boletin Tecnico Forestal* 12, 1–6.

Wiselius, S.I. (1998) *Ochroma lagopus* Sw. In: Sosef, M.S.M., Hong, L.T. and Prawirohatmodjo, S. (eds) *Plant Resources of South-East Asia No 5(3). Timber trees: Lesser-known timbers.* Prosea Foundation, Bogor, Indonesia, pp. 414–416.

Ocotea porosa (Nees & Mart.) Barroso

Brazilian walnut, imbuia

Importance

Ocotea porosa is a major commercial timber tree from the rain forest of southern Brazil. Due to heavy exploitation and habitat loss it is classified as 'Vulnerable' on the International Union for Conservation of Nature (IUCN) Red List of Threatened Species. It is also used in reforestation of abandoned areas.

Botanical features

O. porosa is a tall evergreen or rarely deciduous tree reaching 30–40 m in height and 50–180 cm in diameter. The bole has evident nodal swellings and the crown is composed of ascending branches. The bark is grey with thin cracks, and a scaly surface. On older trees the trunk is generally thick and crooked, with large globose excrescences. The leaves are simple, alternate, elliptical, obtuse, acuminate and penninerved with two more developed ribs, shiny on top and opaque underneath. There are no stipules and the petiole is flat. The leaves are 6–10 cm long and 15–20 cm wide with entire margins and domatia on the underside, in the axils of the veins. They are shiny on top, and opaque underneath with a flat petiole. Branches are round, thin, flexible and rough. Flowers are borne in pilose axillary panicles, with few small (3–4 cm long) asynchronous white flowers, with articulate pedicels. Flowering takes place between October and February. The fruit is a globose or ovoid-shaped drupe with a calyx growing in the shape of a shallow goblet. The fruit has a thin pericarp and brown surface and is 13–17 mm in diameter.

Distribution

O. porosa is found between −23°N and −30°N and 500–200 m in altitude, in the moist *Araucaria* forests of Paraná and Santa Catarina in southern Brazil. *O. porosa* may be found associated with other species (mixed *Araucaria* and hardwood forest) or in dense populations commonly called 'imbuiais'. It also grows in isolated clusters in the coastal tropical forest as a pioneer species penetrating in more open *Ocotea pulchella* forests. It is found less frequently in the Brazilian states of Rio Grande do Sul, São Paulo, and Rio de Janeiro. The species may also occur in Argentina and/or Paraguay. *O. porosa* is found on diverse soils, from the bottom of valleys to the tops of ridges, but mainly in soils of basaltic rocks. The predominant climate in the regions where *O. porosa* is found is temperate subhumid, with average temperatures varying between 12 and 18°C with frequent frosts in the winter. The absolute minimum temperature is −6°C. The favoured annual average precipitation should vary between between 1250 and 2500 mm and be uniform.

Silviculture characteristics

In a trial carried out in Santa Rita do Passa Quatro, São Paulo, an average height of 7.81 m and dbh of 9 cm was observed in a 23-year-old stand. No significant differences in height and dbh were found in a spacing trial with this species at 7 years of age in experiments set up in Angatuba, São Paulo. In a native forest of the highlands of Santa Catarina, a volume of 59 m^3/ha was reported. Pure plantations establish well in small parcels (maximum of 150 trees) and on fertile soils. *O. porosa* can be used in reforestation by planting with pioneer species to avoid direct sunlight and frost damage. The fruits of *O. porosa* are highly appreciated by wild animals, which disperse their seeds, facilitating recolonization of degraded areas. The tree can live up to 500 years, although much older individuals have been reported.

Silviculture practice

O. porosa can be propagated by direct sowing, natural regeneration and planting stock. Seed storage is orthodox. Vegetative propagation methods include cuttings, grafting and tissue culture. The greatest seedling survival rate seems to be obtained when seeds were sown at a depth of 0.5 cm in shade conditions. Seedlings are sensitive to direct sunlight and frost.

Diseases and pests

O. porosa is resistant to rot, mildew and insects. The heartwood is reportedly resistant to attack by decay fungi.

Uses

The tree is attractive and can be used for ornamental purposes in landscape design as well as in urban areas and amenity plantings. The tree is also used in agroforestry and forest restoration.

The wood of *O. porosa* is a hard, high-quality timber. It is used for fine furniture and hardwood flooring and also for panels and interior decorations, tiles, construction, stairs, fences, musical instruments and quality veneers. Its strength and resistance make it a preferred wood for outside constructions such as decks, foot bridges and boat docks. The wood is moderately dense (600–700 kg/m^3). The heartwood is variable from yellowish or olive to chocolate-brown; either plain variegated and figured with parallel veins or dark spots. The sapwood is grey and usually distinct. The surface has an irregular shine. The wood has a medium uniform texture, with straight grain, irregular to interlocked. It is easy to work, finishes very smoothly and keeps well after manufacture. The wood has good nailing characteristics and glues and varnishes readily. It is often used as a substitute for European walnut (*Juglans regia*). The wood is reportedly fire-resistant. If used outdoors, it requires no additional treatments such as preservatives or varnish.

O. porosa is a useful tree for apiculture, as the flowers are very attractive to bees. Research has indicated that porosin, a natural compound extracted and isolated from the heartwood, has biological activity against human lung and breast carcinoma.

Further reading

Berni, C.A., Bolza, E. and Christensen, F.J. (1979) *South American Timbers - The Characteristics, Properties and Uses of 190 Species.* Division of Building Research, CSIRO, Melbourne, Australia.

Danieli-Silva, A. and Varassin, I.G. (2012) Breeding system and thrips (Thysanoptera) pollination in the endangered tree *Ocotea porosa* (Lauraceae): implications for conservation. *Plant Species Biology* 28, 31–40.

Kalil Filho, A.N., Sousa, V.A. de, Marzollo, L.G. and Hirano, E. (2004) Dynamics of seeds germination of imbuya (*Ocotea porosa* Nees et Martius ex. Nees, Lauraceae) progenies within populations of Paraná and Santa Catarina. Embrapa Florestas, Colombo, Brazil. *Boletim de Pesquisa Florestal* 48, 121–128.

Ocotea usambarensis Engl.

(East African) camphor wood

Importance

Ocotea usambarensis is an important timber tree of East African mountain rainforests where it often reaches large sizes.

The timber is valued as it works well and produces high-quality furniture. The tree is threatened by over-harvesting, leading to the implementation of controls. For example, it is a preserved tree species in Tanzania, and cutting is banned in Kenya. However, illegal harvesting continues.

The species has prospects as a plantation tree, given its high timber value, but research is needed to clarify issues of seed availability, handling, germination and artificial regeneration through vegetative means. Silvicultural issues still need more assessment.

O. usambarensis also has various local medicinal uses.

Botanical features

It is a large densely crowned evergreen tree reaching 35–45 m height and 2–3 m diameter. The bole sometimes has small buttresses and the stem is typically branchless up to 10–15 m. The bark is grey and fissured when young, aging to red-brownish and scaling in large flakes on older trees. The leaves are opposite, oval to rounded, 4–16 cm long and 2.5–9 cm wide. The leaves have a distinct scent of camphor when crushed.

Male and female flowers are separate, 8–10, each 5 mm long, yellow-white-green and hairy.

The fruit is an ellipsoid green to orange-brown berry, up to 1 cm long, and bears only 1 small seed in pulp.

Distribution

O. usambarensis is widely distributed throughout East Africa. It is common in wetter forests at altitudes between 900 and 3000 m. It grows mostly in mountain ranges between 1200 and 2600 m. The mean annual rainfall in the area of distribution ranges from 1200 mm to 2450 mm, often with 2–3 dry months, and a mean annual temperature of 12–26°C. In Kenya and Tanzania, *O. usambarensis* used to be dominant in many mountain forests. A regional mountain forest association is sometimes called 'Ocotea forest'. In Uganda it is more scattered. Due to over-exploitation it has become rare throughout its range.

Silviculture characteristics

Harvesting of this tree has been carried out mainly in natural stands, and only a small number of plantings have been established in Kenya and Tanzania (in the early 20th century).

Seeds are only available during mast years at intervals of about 4–5 years. Seeds are heavily attacked by insects, and germination is generally poor. *O. usambarensis* regenerates mainly by root suckers. After natural breakdown of an old tree, the gap is first filled by fast-growing pioneer species, in the shade of which the suckers can establish. After death of the pioneer species they can develop into new trees.

Artificial regeneration is facilitated by lifting natural root suckers produced in profusion near and around the stumps of felled trees. However, suckers are especially palatable to elephants and other large mammals, and heavy browsing on suckers in forests with large animal populations can impair natural regeneration. On good sites, relatively rapid growth rates of up to 2 m/yr in height have been reported for young seedlings and saplings, but this seems exceptional. In a 75-year-old Kenyan plantation at 2450 m altitude, trees had only reached 15–30 m height and 20–50 cm dbh (annual increment 6 mm until 18 years old and thereafter down to 4 mm).

Silviculture practice

The seed weight of *O. usambarensis* ranges from 15,000 to 20,000/kg. The surrounding pulp must be removed from fresh seeds in water, and seeds checked for damage. They may be stored only briefly and should be kept in moist sawdust. However, it is best to use fresh seeds for sowing. Germination usually starts after 30–45 days; it may take up to 90 days but the success rate may be as low as 45%.

O. usambarensis is mainly harvested from natural stands. Coppicing is also an option, as camphor trees of all ages respond well to this. Large-scale logging leads to secondary forest in which the species does not play a significant role.

A study in well-stocked plots of naturally regenerated *O. usambarensis* (about 17 years old) in Tanzania showed that thinning to 220 stems/ha significantly increased dbh and height of the trees; this could reduce the theoretical rotation age from 60–70 years to 50 years, although with possible adverse effects on wood quality.

Another study in south Kilimanjaro and west Usambara, however, concluded that a heavy first thinning after 15 years and at 20 years was recommended, reducing the stock to about 700 stems/ha, with subsequent thinnings at intervals of 7–10 years.

Diseases and pests

Standing older trees often show heart rot caused by fungi such as *Ganoderma applanatum* and *Fomes* spp.

Fruits and seeds are often attacked by gall insects. Suckers are heavily browsed by large mammals, especially elephants; in addition, the bark of young trees is often stripped by squirrels and tree hyraxes.

Uses

The wood, bark and roots of this species are all utilized. *O. usambarensis* is one of the most valuable East African timbers; the heartwood is light yellowish-brown, darkening to deep brown on exposure and with a camphor scent. Wood density is 474–630 kg/m^3. The timber resists acids and fungi, but not termites. Given its poor durability, is unsuited to outdoor uses. It is commonly used for making high-value furniture, cabinets, veneer, panelling and for general joinery. In the past, *O. usambarensis* was a good source of railway coach frames. Local people also use the wood for building poles, and offcuts provide good firewood and charcoal.

Bark and roots are used in traditional medicines. The bark is commonly used to treat stomach aches and intestinal worms. Roots, leaves, bark, and inner bark are processed in different ways to provide decoctions and dry powders useful in treating swellings, whooping cough, measles, backache, stomach ache, malaria and wounds. The bark of *O. usambarensis* contains 11% tannin, 0.5% pseudocubebin, and small quantities of the volatile oil cineol (not camphor). Further research on its pharmacological character and on sustainable bark-harvesting methods is advised.

In Tanzania (Usambaras), *O. usambarensis* is used for water divining, in house construction and for household implements.

Further reading

ICRAF (1992) *A Selection of Useful Trees and Shrubs for Kenya: Notes on their identification, Propagation and Management for use by Farming and Pastoral Communities*. ICRAF, Nairobi.

Kigomo, B.N. (1987) The growth of camphor (*Ocotea usambarensis* Engl.) in plantation in the eastern Aberdare range, Kenya. *East African Agricultural and Forestry Journal* 52, 141–147.

Mbuya, L.P., Msanga, H.P., Ruffo, C.K., Birnie, A. and Tengnäs, B. (1994) *Useful Trees and Shrubs for Tanzania: Identification, Propagation and Management for Agricultural and Pastoral Communities*. Regional Soil Conservation Unit (RSCU), Swedish International Development Authority (SIDA), Stockholm.

Mugasha, A.G. (1980) The relationship between crown diameter and breast height diameter for *Ocotea usambarensis* Engl. as a guide to a feasible stocking in second regeneration stands. Tanzania Silviculture Research Note 38, Silviculture Research Station, Lushoto, Tanzania.

Octomeles sumatrana Miq.

Binuang

Importance

Octomeles sumatrana has high potential as a very fast-growing plantation species in the lowland humid tropics for the production of light-coloured, low-density wood for non-structural uses. Under the commercial name erima, it is one of the major exportable hardwoods produced in Papua New Guinea. *O. sumatrana* may be useful for reforesting heavily logged-over forests which have become overgrown by climbers and woody weeds. Research priorities include the need to investigate genetic variation and determine genetic parameters and correlations for economically important traits, especially wood density, and the development of silvicultural management techniques.

Botanical features

A large to very large evergreen tree to 60–80 m tall and with a bole diameter to approximately 2 m above the buttresses. In mature trees the bole is straight and generally cylindrical with little taper, clear for up to 30–40 m and with prominent buttresses. Buttresses thin, wide-spreading, straight or slightly concave, 5(–10) × 6 m. Bark grey to grey-brown, fissured or irregularly cracked, thin, only 2–4 cm thick, often pustular outer bark, greenish underbark, yellowish inner bark rapidly turning brown when cut, no exudate, unpleasant smell. The crown is open, semi-globular, pagoda-like with whorled branches when young becoming broad, rounded to flat when mature.

Leaves are arranged spirally on sharply 3-angled branchlets. The leaves are simple, thin, bright mid-green, glabrous, round, broadly ovate to cordate, acuminate, more or less entire, 12–30 × 6–23 cm. Juvenile and intermediate leaves may be much larger, and the leaf margin may have a few coarse teeth. Nervation with 5–9 palmate nerves radiating from the junction of petiole and lamina, and 5-angled. The petiole is 6–30 cm long.

The species is dioecious; male flowers are bell shaped, 4–5 × 5 mm, and arranged in long (20–60 cm) spikes. The calyx lobes are ovate, triangular, acute, 2 mm long; petals triangular, 3 mm long with an incurved appendage. The stamens are 4 mm long, filament thick with 2 mm-long kidney-shaped anther. Female flowers are borne on 8–12 cm long spikes, approximately 5 mm long, filament thick; ovary 1–2 mm high, free calyx tube 2–4 mm, lobes broadly triangular, 1 × 2 mm; styles 1–2 mm long, fleshy, thick, flattened with stigma 0.5 mm high, 1 mm diameter. The fruits are barrel-shaped capsules, approximately 12 mm long, arranged on long, 15–40 cm spikes on a 10–20 cm long peduncle. The fruits split laterally from the top downwards, nearly to base, to release numerous tiny, 1 × 0.2 mm, spindle-shaped seeds.

O. sumatrana has been reported to flower and fruit throughout the year in Papua New Guinea. In Sabah (Malaysia), mature fruits have been recorded during the period from November to January. Flowering and fruiting are often abundant, but with peaks every 3–4 years. Seasonality varies with location. Eastern Indonesia has pronounced seasonality, with seed maturing in the November–January period, while in western Indonesia fruiting may occur throughout the year but with highest frequency in May–June and in December. The period from flowering to fruiting is 3–6 months. Plants have been observed to flower and fruit as early as age 5–6 years in trial plantings in Vanuatu.

Distribution

The species has a wide range in near-equatorial regions of western Melanesia and the islands of South-east Asia (but is absent from the continent). It extends from northern Sumatra, through Borneo (including Kalimantan, Sarawak and Sabah), Sulawesi, the Moluccas, widely in the Philippines (including the Provinces of Cagayan, Bataan, Laguna, Quezon, Palawan and Mindanao), and from the Vogelkop Peninsula (Irian Jaya), throughout Papua New Guinea, east to Santa Isabel Island in the Solomon Islands.

Silviculture characteristics

O. sumatrana produces masses of very light seeds (approximately 20 million seeds/kg) which are readily dispersed by wind and water. The silvicultural characteristics are typical of a pioneer species, including rapid colonization and very fast early growth in disturbed, open areas having freshly exposed soil and full light. It does not tolerate even partial shade. Rapid early growth and crown closure are advantageous characteristics in plantations, as well as its self-pruning and freedom from serious pests. However, the species does not thrive in closed plantations and the spreading crown seems to suffer from branch contact with neighbouring trees.

Silviculture practice

The fruits are collected from the tree when the capsules begin to turn brown. The capsules are dried for 1–3 days and, after splitting, vigorous shaking is used to dislodge the seed, which is then separated by sieving. Initial seed viability is variable but generally quite low, approximately 40%. Viability decreases rapidly with time and may be totally lost by 3 months, although seeds stored in the refrigerator at 2–4°C or in an air-conditioned room at 20°C can retain >80% viability for up to 16 weeks. For sowing, seeds are mixed with fine river sand and sown in germination trays which are kept moist under full shade. Damping off is minimized by sowing the seeds thinly and ensuring adequate ventilation. Seedlings can be pricked out 5–6 weeks after sowing and are ready for planting after approximately 4 months, when they are 15–20 cm tall. Efficient vegetative propagation techniques using rooted cutting and cloning have been developed. Rooted cuttings are transferred to polybags in the nursery after 2 months, with plantlets ready for planting out 6–8 months later.

When plantlets and seedlings are planted out, close spacing, (e.g. 2.4 × 4.9 m or approximately 840 stems/ha), has given more rapid growth than wider spacing. Herbaceous and woody climbers need to be controlled prior to canopy closure.

The species is amongst the fastest-growing tropical trees for wood volume production. In trial plantations in the Solomon Islands, 4–5-year-old trees had an average height of 16 m and an average diameter of 21 cm. In Sabah, trees reached a height of 10 m 2.5 years after planting, and in Bogor (West Java) a 4-year-old tree reached a height of 25 m and a diameter of 47 cm. Trees may attain 48 m height and 105 cm dbh in 60 years. Thinned and unthinned regenerated stands in the Philippines displayed an annual increment of 46 and 36 m^3/ha, respectively, over a period of 4–6 years after logging. In another trial, a 2-year improvement cutting of a 4-year-old natural stand of *O. sumatrana* in the Philippines significantly increased diameter, height, and volume per tree; nevertheless, wood yield in both treated and untreated (control) stands was similar, the annual average volume increment being 41.1 m^3/ha. For large-scale plantations in Indonesia the expected annual increment during a rotation of 30 years is 25–40 m^3/ha.

Diseases and pests

Overall, the species is relatively free of major diseases and pests. Recorded insect pests include *Characoma* moths, an unidentified shoot borer, termites and defoliators.

Uses

The species is a pioneer in bare alluvial soils and helps in riverbank stabilization. *O. sumatrana* is one of the plantation tree species that smallholders in Ranau, Sabah use for landscaping, long-term (timber) investment and soil protection purposes.

The wood of *Octomeles* has been investigated in numerous studies, sometimes with contradictory results. The wood is soft, low density, moderately coarse grained and rather brittle. Brittle heart may account for 14% of log volume and 33% of the diameter of the logs, and felling shakes are common. The broad sapwood, 7–15 cm wide, is white with a faint greyish-yellow tinge, whereas the heartwood is pale brown-grey or yellow-brown, with a pinkish hue. It has no smell when dry but gives off an unpleasant odour when wet. Wood basic density varies in the extreme from 160 to 480 kg/m^3.

The wood deteriorates rapidly and needs to be worked up soon after cutting. Rates of shrinkage are rather low, but care is required during seasoning and the wood seasons rather slowly given its low density. The timber is easily worked by hand and machine tools, and dried timber has excellent technical properties. The pulp may be used for paper-making, but its low density limits its value for this purpose. Low density and durability mainly restrict the wood to non-structural and composite uses. The timber may be used for light furniture and joinery, interior finishes, mouldings, wide shelves, louvred doors, coffin boards, sledges, concrete shuttering, packing, low-quality crates and boxes, buoys and fishing net floats, bark and core veneer, light construction chipboard or fibreboard and matchboxes. Traditional uses include large dugout canoes, rafts, drums and firewood. Untreated sapwood is susceptible to *Lyctus* attack.

In Kalimantan (Indonesia), the bark used to be combined with roots of *Morinda citrifolia* and leaves of *Symplocos* spp. to dye split rattan red. The inner bark contains a bitter purgative and yellow dye. Young leaves are eaten as a vegetable and the juice is used in local medicine to treat stomach ache. Wild bees often nest in the trees.

Further reading

Forest Products Research and Development Institute (1984) *Octomeles sumatrana* Miq. Philippine Timber Series 1. Forest Products Research and Development Institute, Binuang, Philippines, 3 pp.

Lemmen, R.H.M.J., Soerianegara. I. and Wong, W.C., (eds) (1995) *Plant Resources of South-East Asia No. 5 (2). Timber Trees: Minor Commercial Timbers*. Prosea Foundation, Bogor, Indonesia. Backhuys Publishers, Leiden, The Netherlands, 655 pp.

Shim, P.S. (1973) *Octomeles sumatrana* in plantation trials in Sabah. *Malaysian Forester* 36, 16–21.

***Olea capensis* L.**

Black ironwood

Importance

Olea capensis is an African hardwood timber tree yielding a high-quality wood. Over-exploitation from natural forests in South Africa and montane areas in East Africa led to calls for its conservation, and harvesting in South Africa is regulated. The figured timber is valued for furniture and sliced veneer, and it has become a candidate species for plantation forestry.

Botanical features

O. capensis grows 10–20 m with diameters of 70–80 cm, rarely up to 40 m tall and a diameter of 100 cm, although it also found as a small tree or bushy shrub. Trees have a small but dense crown, and a long, straight and usually clean bole, sometimes buttressed, which can be up to 15 m to the lowest branch in the largest trees. Bark is pale grey or whitish and smooth when young, but becoming dark grey with shallow fissures with age, and characteristic blackish gum is exuded from bark wounds. The main branches ascend steeply and are not usually heavy. Leaves are lanceolate–oblong to almost circular, in opposite pairs, 3–10 cm long and 1.5–5 cm across, light to dark green and glossy above, rather paler green below, occasionally with a purplish tinge, sharply tipped, margin wavy, midrib pale and clear below, with a petiole to 0.3–3 cm long. The flowers are small, white or cream-coloured, sweetly scented, in profuse many-flowered axillary or terminal heads,

3–15 cm long. The fruit is an ovoid, fleshy drupe, purplish when ripe and about 2 cm long.

Distribution

O. capensis is found in many countries in Africa south of the Sahara, from Guinea and Guinea-Bissau in West Africa, to Sudan and Ethiopia, and south to South Africa; and at the eastern end of its range, to Madagascar, Comoros, and Socotra, Yemen. It is particularly common in Afromontane regions, especially in East Africa (including Mt Elgon, Mt Meru and the Usumbara mountains) and in the southern and eastern Cape in South Africa. It is not known to have been introduced into areas outside its native range.

The taxonomy and relationships with related species have been confused, but three subspecies of *O. capensis* are certainly accepted. The most widespread is subsp. *macrocarpa* which covers most of the entire range; subsp. *capensis* is restricted to South Africa, and subsp. *enervis* to Swaziland. *O. capensis* is also generally accepted to include taxa that used to be treated as separate species, such as *Olea hochstetteri* and *Olea welwitschii*, but they are still sometimes treated separately and there remains some confusion in the taxonomic literature.

In many regions, *O. capensis* is becoming rare due to over-exploitation. This is the case in many mountain regions in Ethiopia, Kenya, Tanzania and in South Africa where it was already noted as requiring conservation in 1890 and it is now legally protected and harvesting is regulated.

Silviculture characteristics

Young plants need shade, but saplings are light-demanding and the species is generally classified as a pioneer that depends on gap creation for regeneration. It possesses a certain degree of windfirmness but is vulnerable to strong winds, tending to be shallow-rooted even though it grows on relatively soft, loamy soils. *O. capensis* is an evergreen tree. Profuse flowering takes place at irregular intervals of up to 7 years, late in the dry season, and fruit takes 6 months to ripen.

O. capensis occurs in wet tropical montane locations where temperatures are relatively low and, apart from diurnal fluctuations, fairly constant throughout the year. Mean annual temperatures of 14–18°C seem ideal, provided they occur within the natural altitudinal limits. In general, it prefers regions with an annual rainfall of 800–1500 mm and growth seems to be checked on drier sites. In West Africa, *O. capensis* is characteristic of humid mountain forest, restricted to mountains above 600 m altitude. In Central Africa, it occurs at 1000–2800 m altitude, and in East Africa it is mainly found at 1500–2700 m (but occasionally as low as 800 m or up to 3200 m). In South Africa, it is a dominant canopy tree, especially in forests along the coast. In East Africa, the tree can dominate the climax vegetation in lower- and mid-altitude montane forest. Young trees prefer well-drained, deep, loamy and fertile soils. Once established, they are quite drought-resistant and they also grow in poor soils.

Silviculture practice

O. capensis seed is collected, cleaned of all pulp after soaking in water for 2 days and sown immediately, although the germination is low (about 35%) and uneven, taking 2–3 months but even up to 1–2 years. Seed can be stored for several years after drying and kept at low temperatures. Seedlings need to attain 30 cm in height before planting out, as shorter seedlings (15–20 cm) do not survive well. Wildings are also sometimes also collected for planting. The management of natural regeneration is being attempted, and the species has also been noted as a dominant understorey species naturally regenerating in some exotic pine plantations. Line and gap planting seem to show promise, the most suitable materials being striplings (with only the apical pair of leaves left) up to 1.8 m in height, though bare-rooted or potted transplants also survive well.

For raising plantations, the use of nurse trees such as *Grevillea robusta* has proved beneficial, with *O. capensis* planted in clumps of up to ten trees at the close spacing of 1 × 1 m, and a spacing of 7–8 m between clumps. Trees must be protected from browsing, and weeding and the removal of brambles and climbers improves growth. Trees are self-pruning, but thinning is required, with five thinnings recommended in plantations, leading to a final density of 120 trees/ha. Growth is fast in young plantations and slower in older ones. In the first few years, annual height increments can be more than 1 m, and annual diameter increments can be more than 1.2 cm. This slows in later years, to an average over 30 years of 0.5–0.8 m in height and 0.5–0.8 cm in diameter. Mean annual volume increments of 20 m^3/ha have also been reported after 25–30 years and rotation lengths of 75 years are suggested for timber production.

Diseases and pests

No diseases and pests have been reported, though *O. capensis* is susceptible to browsing by livestock and wild animals. The wood is attacked by ambrosia beetles and termites, but is not susceptible to *Lyctus* beetles.

Uses

The small but dense crown makes *O. capensis* a suitable agroforestry tree, and it is managed in cultivated land in parts of Africa, and is recommended for replanting on degraded forest soils.

O. capensis heartwood is pale golden-brown to dark brown, often with variegated, irregular, dark brown, grey-black or yellowish streaks, and is distinct from the grey to yellowish-white sapwood. Grain is straight to slightly interlocked, with a fine and even texture, a surface that is slightly oily to the touch, and which takes a good polish. The wood is heavy and hard, strong and durable, with a density of 860–975 kg/m^3 at 12% moisture content, but up to 1170 kg/m^3. The wood air dries slowly with a tendency to check, split, warp and distort, and kiln drying must be carried out at low temperatures and moderately high humidity. The wood is unstable unless properly dried. Irregular grain means that machining is difficult, and the wood is liable to split when nailed; however, nail-holding power is high after pre-drilling. The wood is easy to saw when green, but difficult to saw and work when dried, and blunts saw teeth and cutting edges rapidly. It turns fairly well and good-quality veneers are obtained from slicing.

Timber is used for heavy construction work, furniture, turnery, flooring and panelling, general joinery and veneers. The heartwood, often traded as 'ironwood', is in high demand for flooring, carpentry, and panelling, and is widely used for house and bridge construction, counter and table tops, tool handles and wagon parts. It produces beautiful furniture, turnery and sliced veneer, and is often used by African artists. It is suitable for interior trim, sporting goods, toys, novelties and agricultural implements. It is also used as firewood and for charcoal production.

The wood of wild African olive (*Olea europaea* subsp. *cuspidata*) is very similar and is used for similar purposes.

The oily fruits are edible and are used for making beer and lemonade. Many parts of the tree are used in traditional medicines. The foliage serves as fodder, especially during the dry season. The flowers produce nectar for honey bees. In South Africa, the tree is also used as rootstock for grafting European olive cultivars from the Mediterranean region. The tree is considered sacred by the Maasai people and is commonly used in ceremonies.

Further reading

Aerts, R. (2011) *Olea capensis* L. In: Lemmens, R.H.M.J., Louppe, D. and Oteng-Amoako, A.A. (eds) *Prota 7(2): Timbers/Bois d'œuvre 2.* [CD-Rom]. PROTA, Wageningen, The Netherlands.

Maundu, P. and Tengnäs, B. (eds) (2005) *Useful Trees and Shrubs for Kenya.* Technical Handbook 35. World Agroforestry Centre - East and Central Africa Regional Programme (ICRAF-ECA), Nairobi, 484 pp.

Tsingalia, M. (2010) Impacts of selective logging on population structure and dynamics of a canopy tree (*Olea capensis*) in Kakamega forest. *African Journal of Ecology* 48, 569–575.

P

Pachira quinata (Jacq.) W.S. Alverson

Red ceiba

Importance

Pachira quinata, commonly known as red ceiba in English and as pochote in Spanish, is a tree of moderate growth whose timber is highly appreciated by consumers throughout its natural distribution, particularly for construction and furniture, due to its high quality and special properties. As a result, its price is one of the highest in local markets. However, this has resulted in extensive logging which has reduced its occurrence to isolated forest patches, and the species is considered as threatened at the level of some populations. Plantations have been established in a number of countries, mostly within its natural range. Survival and growth are particularly sensitive to site and soil depth. This reduces the possible sites for plantations, as the most suitable ones are commonly used for agricultural purposes. The relatively long rotations (compared to exotic species commonly planted in the region), which create long periods without any financial returns on investment, also discourage farmers from planting *P. quinata*.

Botanical features

P. quinata is a broadleaved deciduous tree, attaining a height of 25–35 m and dbh of 1–3 m. The bole is straight and cylindrical with a fluted and buttressed base. The crown is well developed but irregular. Branches are horizontal and large, and the tree is not self-pruning. Mature bark is thick, greyish-brown, and covered most frequently by stout spines up to 3 cm long. The root system is moderately deep, with a well-defined conical taproot that reaches to 2.5 m.

Leaves are digitally compound, glabrous, alternate and petiolate, with 3–7 leaflets (most commonly 5). Petioles are long (4–15 cm) and have 2 small stipules at the base. Leaflets are petiolulate, 4–17 cm long by 1.5–7 cm wide, oblong and obovate, narrowly decurrent at the base, and the apex is caudate-acuminate. The texture varies from chartaceous to subcoriaceous and nervation is reticulo–pinnate.

The flowers are hermaphrodite, arranged in cymous or pauciflorous inflorescences. Flowers are erect, 8–14 cm long. They have 5 linear petals, 7–11 cm by 0.5–0.8 cm, white or light pink inside and green-brown outside. Stamens are white and numerous, averaging 110 per flower. The style is longer than the stamens and finishes in a 5-lobulated stigma. Ovary is superior, 5-loculed.

The fruit is a dehiscent 5-valved capsule which has a wide range of shapes and colours, although mostly ellipsoidal and khaki to dark brown. Fruit size varies between 2 and 15 cm long and 2.5 and 5 cm diameter. Seeds are subglobose, 5 × 4 mm and surrounded by a white floss (kapok) that helps dispersal. Seed number per fruit is highly variable but 30–50 on average.

Trees may start flowering when only 1 year old, but usually reach sexual maturity when 2–5 years old. The flowering season starts just after leaf fall and lasts for about 3–4 months. The production of flower buds is related to the amount of light the branches receive, and shaded/lower branches hardly produce any flower buds. Flower buds reach anthesis when 22–26 days old. The flower opens around 7:00 p.m., although this timing is delayed with increase in mean relative humidity (particularly >60%) and/or decrease in mean temperature (<16–18°C). Anthesis is simultaneous (within 5 min) for all the flowers opening on a particular day, and takes 1–9 min. Flowers are pollinated the same night by the bat *Glossophaga soricina* which feeds on intrafloral nectar. Stigmas are receptive only when the stigmatic lobes are fully reflexed, and flowers last only for the night of the anthesis. The corolla and stamens fall the morning after, although the style remains. The average proportion of flowers that develop into mature capsules is 10–20%, and these are only formed on the external part of the crown. The first flowers to open on each tree are more likely to form capsules. The total number of flowers and capsules produced varies with year and age of a population. Daily flower production is not constant throughout the flowering season or within a population. Half of a season's flowers can be produced in just a quarter of the length of the season, with great variation in flower production between individuals. Capsule maturity is achieved 35–50 days after anthesis, depending on the local climate.

Distribution

P. quinata occurs naturally in Central America and northern South America, mainly in tropical dry forest, although sometimes in premontane wet forest or very dry forest in Venezuela. In Panama and Costa Rica, the species can be found also in wet forest. It grows naturally on the Pacific coast of Honduras, Nicaragua and Costa Rica, in regions with marked seasonality. In Panama, the species extends over both coasts and in Colombia throughout the dry area of the Caribbean coast.

Silviculture characteristics

P. quinata is a semiheliophyte species of moderate growth, requiring some initial shade for seedlings but full light for normal development. Juveniles are rare in logged or unlogged natural forest. In terms of natural regeneration, *P. quinata* in

Colombian Amazonia shows low recruitment due to low fruit production, high seed predation and very high seedling mortality under continuous canopies, mainly due to damping off, indicating a need for canopy gaps. Some authors maintain that the low seedling survival rate makes the species unsuitable for restoring degraded sites; however, the species has been recommended for restoring abandoned pastures under direct and moderate light conditions. Its reforestation potential was considered high in Panama, where high wood-volume indices were attained.

The species is intolerant of competition and this will affect initial spacing, weeding, and thinning regimes. Other principal limiting factors to good growth are a high susceptibility to wind, fire and soils which are either poorly drained or with very low moisture content (due to steep slopes or other factors). Planting is economically sound only on fertile soils. The species propagates vegetatively very easily, while seed storage and germination show no particular problems. The species produces a strong taproot which appears to aid survival during the pronounced dry season. The secondary root system does not spread far and thus the species is not suited to exposed areas. It is not self-pruning and requires regular pruning to obtain straight and knot-free boles.

Silviculture practice

Effective seed collection must be at fruit maturity, just before capsule dehiscence between March and May, depending on location. Mature capsules feel hard when pressed with the thumb, and are brown. Collected capsules should be spread out under shade to reduce moisture content, opened by hand when the first capsules dehisce, the floss and seeds extracted, and then sieved to remove the seed. Seeds must be dried to 6–8% moisture content. High viability can be retained if seeds are slow-dried for 48 h, followed by rapid drying for 24 h. *P. quinata* seeds are orthodox and remain viable for more than 12 months if kept in hermetically sealed containers and at low temperatures (5–6°C). No pretreatment is required prior to germination, which occurs within 8–35 days of sowing.

The most frequent plantation methods use seedlings (15–30 cm long) or stump plants (30–50 cm long, including root, and 1.5–2.5 cm diameter) which are produced in the nursery in containers or by sowing directly in beds. Stump stock needs 6–12 months to be ready for planting, while container seedlings need 3–4 months. With a strong taproot, planting of stumped plants may begin 15–30 days before the rainy season starts. Direct sowing in the field is not practised. Stakes (obtained from branches with orthotropic growth, 1.5–2.0 m long) are used to establish living fencerows, with spacings from 2 to 5 m, particularly in Costa Rica and Nicaragua.

Recommended spacings in plantations are 3 × 3 m and 3.5 × 3.5 m. Closer spacings require earlier thinnings due to earlier canopy closure, and there is rarely a market for the resulting products. Wider spacings are required if site conditions are poor or intercropping is planned. Stands for seed collection require wider spacings from the start (at least 8 × 8 m) to facilitate bat pollination and full light to the crowns. Site preparation before planting is encouraged, particularly in clay soils. Weed control during the first 3 years is essential, due to the species' light requirements. A first pruning in the first or second year is often practised in Costa Rica for crown shaping. Then, three more prunings are common when the trees reach heights of 5–6 m, 8–9 m and 11–12 m. Thinnings must be made at canopy closure, starting usually when plantations are 5–7 years old, with 2–3 more interventions required to reach final densities of 100–300 trees/ha. Thinning at ages 7, 12 and 16 years to maintain a basal area between 14 and 17 m^2/ha is recommended.

Diseases and pests

The main problems at the nursery stage are unspecific and include root collar rot and damping off (*Fusarium* and *Cylindrocladium* spp.), and damage associated with crickets (*Acheta assimilis* (*Gryllus assimilis*)), cutworms (*Agrotis* spp.), grubworms (*Phyllophaga* spp.) and leaf cutter ants (*Atta* spp.). In plantations, the most important damage is caused by insects of the genera *Atta* (defoliation at nursery and young stages), *Phyllophaga* (root destruction of seedlings and plants less than 3 years old) and *Steirastoma* (shoot borer at young and mature stages). *Arsenura armida* is an important pest in Venezuela. Some rodents (e.g. *Orthogeomys cherriei*, *Sigmodon hispidus*) are sometimes problematic in nurseries or young plantations, causing destruction of plants and young trees through debarking. Seed losses can be severe due to the perforation of flower buds (by *Eulepidotis* sp.) or seed and fruit (by *Dysdercus bimaculatus*).

Uses

Although *P. quinata* is used mainly for timber (from natural forest since its use in plantations is as recent as 25 years), it is also used for other purposes, depending on the country. In Costa Rica and Nicaragua the establishment of living fencerows from vegetatively propagated stakes is common practice, with spacings ranging from 2 to 5 m. On other occasions, the trees have been left standing to give shade in coffee plantations (Nicaragua) or, most frequently, as remnant trees in cattle pasture. There are also examples of intercropping with maize, although only for the first 2 years of the plantation, due to a quick canopy closure.

The wood is highly valued throughout its natural distribution. The timber is easy to work, saw and veneer, and produces a good finish. Most frequent uses are for non-structural interior construction (mouldings, door and window frames) and quality furniture. However, it has also been used for veneer, plywood, particleboard, fibreboard, carpentry and joinery. Heartwood and sapwood present a characteristic difference in colour. Lustre is low, with a soft figure, medium texture and velvety feeling to the touch. The grain is straight to slightly interlocked. The density is medium (350–550 kg/m^3), with low strength values and it is thus not suitable, in general, as structural timber.

With a view to developing an alternative to chemicals for controlling gastrointestinal parasites in sheep, *P. quinata* sawdust was shown to inhibit the development of parasite larvae *in vitro*. The medicinal/pharmaceutical potential of *P. quinata* has also been investigated, showing activity in one or more of the bioassay systems used.

Further reading

Castellanos, M.C. and Stevenson, P.R. (2011) Phenology, seed dispersal and difficulties in natural recruitment of the canopy tree *Pachira quinata* (Malvaceae). *Revista de Biología Tropical* 59, 921–933.

Cordero, J. and Boshier, D.H. (2000) *Bombacopsis quinata*. Tropical Forestry Paper 39. Oxford Forestry Institute, Oxford, UK.

Kane, M., Uruena, H., Dvorak, W. and Atehortua, C (1993) The potential of *Bombacopsis quinata* as a commercial plantation species. *Forest Ecology and Management* 56, 99–112.

Palaquium spp.

Nyatoh, bitis

Importance

The genus *Palaquium* contains over 100 species, many of which are important for timber production. There is some concern regarding the overuse of some species, which has led to them becoming locally rare in some areas.

Timbers from the various species are traded under the names of nyatoh (the light to heavy hardwood species) and bitis (the heavier hardwood species).

A number of *Palaquium* species, in particular *Palaquium gutta*, are important for the production of a latex known as gutta percha which has both waterproofing and insulating properties. Fruits of some species are edible, and the seeds contain a fat with multiple uses.

Botanical features

Trees of the genus vary in size up to 60 m in height and 250 cm in diameter with a columnar buttressed bole.

The smaller species include *P. gutta* which is a medium-sized to large tree about 25 m tall (occasionally up to 45 m) and 60 cm in diameter. The larger species include *Palaquium rostratum* which is up to 60 m tall and up to 210 cm in diameter.

Distribution

The genus *Palaquium* is distributed throughout Asia including Brunei Darussalam, Indonesia, Malaysia, Singapore, Thailand, Bangladesh, India, Myanmar, Sri Lanka, China, the Philippines and Borneo. Species are also found in Polynesia.

Trees are usually found in lowland forests, including freshwater swamps.

Silviculture characteristics

Trees of the genus *Palaquium* need a considerable amount of light for development, and show increased vigour when not overshadowed by other vegetation. Some species are long lived yet slow growing; the annual diameter increment of *Palaquium maingayi* is only 0.5 cm, and it will take about 100 years to attain a diameter of 55 cm. In comparison, planted trees of *P. rostratum* in the arboretum at the Forest Research Institute Malaysia (FRIM) reached a diameter of 57 cm after 40 years.

Silviculture practice

Trees may reproduce by natural regeneration or be propagated by seed or cuttings. Seedlings may be planted out when about 20 cm in height.

As a result of the increasing demand for nyatoh timber, more research is needed on silvicultural aspects, with a view to establishing large-scale plantations. Both *Palaquium polyanthum* (for timber and latex production) and *Palaquium hexandrum* (for timber production) are promising plantation species.

Diseases and pests

Trees are susceptible to the fungal pathogens *Corticium salmonicolor* (*Phanerochaete salmonicolor*) and *Phellinus noxius*.

Pests recorded on *Palaquium* species include *Achaea serva* (a leaf-feeding caterpillar), termites of the genus *Neotermes*, and the pinhole borers *Platypus gerstaeckeri*, *Xyleborus ferrugineus* and *Xyleborus perforans*.

The tea mosquito bug (*Helopeltis bradyi*) has been observed on *P. gutta* and the pentatomid shield bug (*Erthesina fullo*) has been recorded on *Palaquium formosanum*.

Nyatoh is rated as moderately durable yet prone to termite and fungal attack; treated wood is much more durable. The sapwood of nyatoh is liable to attack by powder-post beetle. Bitis, on the other hand, is more durable and is also resistant to termites.

Uses

P. gutta has been used for site restoration planted on wasteland and tin tailings. However, its main uses are for timber and the production of gutta percha latex.

The light to heavy hardwood *Palaquium* species are traded under the name nyatoh. Nyatoh heartwood is variable in colour from pale pink through to purple-brown, with a lighter sapwood. Nyatoh (which may also include species of the genus *Payena*) is regarded as a general utility wood suitable for the production of fine furniture, interior joinery and plywood, doors, veneers, panelling, strip and parquet flooring and ceilings. It is also suitable for floorboards, partitions, household appliances, musical instruments, boat decks and masts and canoes. The buttress wood is used for making oars, cartwheels and handles for agricultural implements. Most of the nyatoh wood in commercial supply varies from 600 to 700 kg/m^3 at 15% moisture content.

The heavier hardwood *Palaquium* species are traded under the name of bitis. Bitis is heavier and more durable, and more suited to heavy construction than nyatoh. Bitis has a higher density than nyatoh, varying from 850 to 1150 kg/m^3 at 15% moisture content.

The timber characteristics of the various *Palaquium* species in commercial use are variable. The timber of *P. gutta* is fairly heavy for a *Palaquium* species, with a density of 610–910 kg/m^3 at 15% moisture content; it contains silica which makes it difficult to work, but gluing is easy. Pre-boring for nails and screws is advised. The timber is used for planks and furniture and should not be exposed outdoors or on the ground.

P. maingayi is a somewhat lighter wood varying from 570 to 850 kg/m^3 at 15% moisture content. It is easy to saw, giving a smooth finish. Nailing and boring properties are good.

P. rostratum and *P. hexandrum* have similar densities of 480–760 kg/m^3 and 450–770 kg/m^3 at 15% moisture content, respectively. *P. maingayi* is easy to work and very durable when treated. It is somewhat similar to red meranti (a light hardwood from *Shorea* species).

Palaquium species, in particular *P. gutta*, are highly important for the production of latex which is obtained by tapping the bark or by extraction from the leaves. The latex from other *Palaquium* species is sometime used to adulterate gutta percha from *P. gutta*; however, the latex from *P. polyanthum* also produces a good gutta percha. The latex contains 80% polyisoprene and 20% resin. At 60°C the gutta latex can be moulded to a desired shape. The latex is well known for its insulating properties and imperviousness to water. It is therefore used in the insulation of subterranean and submarine electrical cables, as well as the making of golf balls, certain types of surgical appliances, transmission bells, and acid-resistant receptacles. Other uses include splints for fractures, supports, pipes, speaking tubes in telephone receivers, adhesives and as an ingredient for chewing gum. It is also used for protecting wounds and in dental clinics as an alternative to synthetic fillers. The popularity of gutta latex has declined with the advent of synthetic plastics.

Palaquium seeds contain a fat which is used in the manufacture of candles and soap and is sometimes also used for cooking and lighting. The fruit is edible but sour.

Further reading

Schulte, A. and Schone, D (1996) *Dipterocarp Forest Ecosystems: Towards Sustainable Management*. World Scientific Publishing Co., Singapore.

Soerianegara, I. and Lemmens, R.H.M.J. (eds) (1993) *Plant Resources of South-East Asia. Timber trees: major commercial timbers. No. 5(1)*. Pudoc Scientific Publishers, Wageningen, The Netherlands, 610 pp.

Tropical Timber Information Center (1975) *Palaquium* sp. Brief 21. Tropical Timber Information Center, State University of New York, New York, USA, 4 pp.

Parashorea spp.

White seraya, gerutu

Importance

The genus *Parashorea* contains about 14 species, of which a number are important in timber production. Timbers of this genus are traded as white seraya (the light- to medium-weight species) and as gerutu (the heavier-weight species). White seraya is an important commercial export timber, second only to light red meranti in Sabah, Malaysia. The commercial export of gerutu is much lower.

Enrichment planning projects have been carried out to restock logged areas with dipterocarp seedlings including *Parashorea* spp. A number of *Parashorea* species are considered to be endangered or critically endangered often as a result of their over-exploitation for timber.

Botanical features

Trees of the genus *Parashorea* are large evergreens characterized by a cylindrical, tall, straight, fissured bole, the presence of large buttresses and bark with large lenticels. Species range from 35 m in height up to *Parashorea malaanonan*, which is a large species, reaching a height of 60 m and a diameter of 200 cm, with large buttresses up to 4.5 m.

Distribution

Representatives of the genus are distributed across Myanmar, Thailand, Laos, Vietnam, China, Malaysia, Indonesia, Cambodia and the Philippines.

Silviculture characteristics

Parashorea species have similar silvicultural characteristics to other dipterocarps, although some species may exhibit faster growth rates and be more light-demanding.

In *P. malaanonan*, natural regeneration occurs rapidly, and it tolerates drought, fire and wind; however, it needs partial shading in its early stages of development.

Silviculture practice

Propagation is by seed or cuttings. Wildings may also be collected from the forest. Collected wildings can be lifted with a rootball attached and transferred to a lightly shaded nursery wrapped in banana stalks or ventilated plastic bags. After 3–6 months they can be planted out, *P. malaanonan* preferring a moist soil and partial shade. A suggested spacing for planting is 4 × 4 m and fertilizer is recommended.

In *P. malaanonan* seed production is erratic (every 3–4 years) and seed viability is short-lived. Planting stock production is thus a considerable problem.

The potential of using bamboo (*Gigantochloa laevis*) as a shelter crop for dipterocarp seedlings including *Parashorea* sp. has been investigated. However, it was concluded that although the bamboo would generate some income it would need to be intensively harvested to improve light conditions for the dipterocarp seedlings. *Acacia* has also been investigated as a potential nurse crop for native species including *Parashorea* spp. in Vietnam.

Diseases and pests

Seedlings of *Parashorea* spp. may be affected by root rot and damping off. Snails feed on seedlings and *Adoretus* beetles feed on all parts of the plant; logs are susceptible to the sal borer *Hoplocerambyx spinicornis*.

The gall-inhabiting coccoid *Danumococcus parashoreae* has been observed on the leaves of *Parashorea tomentella* in Malaysia. Also in Malaysia, the cerambycid borer *Cyriopalus wallacei* has been found to cause serious commercial damage to living trees of *P. malaanonan*.

The wood is prone to attack by borers and termites, and the sapwood is susceptible to powder-post beetles.

Uses

The lightweight *Parashorea* timbers, including *Parashorea macrophylla*, *P. malaanonan* and *P. tomentella*, are known as white seraya and are considered light to medium-weight hardwoods with a density of 390–670 kg/m^3 at 15% moisture content. The heartwood is yellow to light brown, occasionally with a pinkish tinge. The sapwood, although not clearly defined, is paler in colour.

The heavyweight *Parashorea* timbers, including *Parashorea aptera*, *Parashorea densiflora*, *Parashorea lucida*, *Parashorea parviflora*, *Parashorea smythiesii* and *Parashorea stellata*, are known as gerutu which are considered more medium-weight hardwoods with a density of 640–865(–1050) kg/m^3 at 15% moisture content. The heartwood is often olive-brown in the heavy *Parashorea* timbers and may contain streaks. The texture of the wood is moderately coarse with an interlocked grain.

Both white seraya and gerutu are general-purpose timbers that are used primarily for veneer and plywood production, but may also be used for construction. White seraya is better suited to veneer and plywood production than gerutu as the latter is a denser wood. However, gerutu is considered more durable than white seraya.

Other uses include wood for cabinetmaking, furniture, flooring, hardboard, fibreboard, particleboard, musical instruments, toys, turnery, planking, decking, joinery, cladding, shuttering, railway sleepers and boxes. Gerutu may also be used for pulp and paper-making.

Of all *Parashorea* species, *P. malaanonan* is perhaps considered the most commercially important, with its timber much in demand locally and for the export market. It has a density of 435–675 kg/m^3 at 15% moisture content. The wood is used as a general-purpose timber and is preferred for the manufacture of plywood and veneer.

Further reading

Appanah, S. and Turnbull, J.M. (eds) (1998) *A Review of Dipterocarps: Taxonomy, Ecology and Silviculture*. Center for International Forestry Research, Bogor, Indonesia.

Newman, M.F., Burgess, P.F. and Whitmore, T.C. (1996) *Manuals of Dipterocarps for Foresters: Borneo Island Light Hardwoods: Anisoptera, Parashorea, Shorea (Red, White and Yellow Meranti)*. Royal Botanic Garden Edinburgh, UK, and CIFOR, Bogor, Indonesia, 275 pp.

Soerianegara, I. and Lemmens, R.H.M.J. (eds) (1993) *Plant Resources of South-East Asia. Timber Trees: Major Commercial Timbers. No. 5(1)*. Pudoc Scientific Publishers, Wageningen, The Netherlands, 610 pp.

Parinari excelsa Sabine

Guinea plum

Importance

Parinari excelsa is a large canopy tree recorded from the rainforests of Africa and the American tropics. In Africa, it is a prized timber tree producing a fairly hard wood that polishes to a bright lustre used in high-value furniture and building construction materials. The dense wood also makes good firewood or charcoal.

The tree produces fruits called 'gingerbread plums' or 'Guinea plums', valued by people in several West African countries. In Guinea, for example, it is even among the ten most important food-tree species listed by rural people. The kernels are also edible.

P. excelsa is a traditional agroforestry tree and offers the potential for extracting nutrients from deeper soil and recycling them into the upper soil to be available to crops. It is often planted (or left) in coffee farms. Indeed it is one of the tree species farmers leave standing when clearing land in various parts of Africa, including for rice cultivation in Guinea.

The trees are also important for frugivorous primates, squirrels, birds and other animal species. The flowers are considered an important bee forage.

Despite the tree's wide distribution it is poorly known outside of Africa.

Botanical features

P. excelsa is a large evergreen canopy tree, up to 45 m high with a widely spreading, golden-tinged crown. Its bole is slender and straight, and may be fluted or with narrow buttresses. The bark is smooth grey with lenticels and scattered scales. The slash is reddish, with orange flecks, and smells sweetly like sugarcane. The tree has simple, alternate leaves of 10 × 3.5 cm with entire margins, acute to rounded base and an acuminate apex. They appear silver or golden underneath due to dense, short hairs. Twigs are purplish.

Flowers are a pinkish-white and grow in a leafy terminal panicle (9–12 cm). The fruits are brown to purplish with corky spots, an ellipsoid drupe of 3–4 cm with rough skin and a hard stone enclosing two seeds.

Distribution

P. excelsa is widespread. In wet tropical Africa, it is associated with mountain and upland forests, from Senegal to Uganda and Tanzania and southwards to Angola, Zambia, Malawi and Mozambique. In western Uganda, the 'Parinari forest' type is named after it, as its wide crowns are detectable from aerial photographs.

In the Neotropics, it occurs across Amazonia and locally in Central America: the species has been recorded from Bolivia, Brazil, Colombia, Costa Rica, French Guiana, Guyana, Peru, Suriname and Venezuela.

Silviculture characteristics

Seeds are difficult to germinate, but root suckers may be used to directly establish plantings. Most *Parinari* species produce a profusion of sprouts from their roots, whenever exposed or wounded. These can be cut off and planted in sand, preferably under intermittent mist. Thinning may be required to stimulate tree growth.

In some parts of its natural range in East Africa there has been a sustained lack of recruitment that remains poorly understood (e.g. since the 1960s in Kibale Forest, Uganda). Speculation is divided: it may reflect some biological problem (such as seed predation) or that *P. excelsa*, while long lived and relatively shade-tolerant, is not a genuine climax species, and is in fact unable to regenerate under its own canopy.

Silviculture practice

Root cuttings provide a key to propagating elite specimens; superior types could be quickly and easily established across much of Africa, in village gardens or scattered alongside roads and tracks on hillsides.

Producing the tasty fruit in a managed way seems feasible, but so far few attempts have been made to improve and plant the 'Guinea plum', for example by collecting representative specimens for grafting, or hybridization. However, trees are often 'semi-cultivated' in parts of Africa where farmers spare them whenever they clear land, and subsequently take care of them in some way. This has contributed to *ex situ* conservation.

There appears to be no information on the management or use of this species in the American tropics.

Diseases and pests

No references were found to pests or diseases in *P. excelsa* trees, but we note the regeneration failure mentioned above, which may or may not result from insect or fungal damage.

Uses

The attractive finish of the hard, dense timber (air-dry density of 833 kg/m^3) makes it well suited to high-value joinery, furniture and general construction. Given its durability it is especially favoured for house posts, door frames, mining and railway sleepers. The timber is considered relatively resistant to damp and fungal attack. The sapwood is yellowish-white, while the heartwood is reddish-chocolate brown. The texture is moderately coarse with an interlocked grain. Seasoning is difficult as the timber tends to split and distort; gentle air drying is favoured. Silica in the wood makes working difficult and blunts saws. Stellite-tipped saws are recommended for dry wood. It is used locally to make canoes, mortars, tool handles and building poles, as well as for firewood or charcoal.

The so called 'gingerbread plums' or 'Guinea plums' of *P. excelsa* are used in a variety of ways. They are eaten fresh or boiled with cereal. When pounded with water, they create a colourful red juice that is added to lemonades or orange crushes. This refreshing liquid is thickened with maize or cassava flour into a tangy-tasting gruel, and syrups are prepared from it as well. The Guinea plum is also the basis for some alcoholic drinks. The kernels inside the seeds are also eaten. These are slightly oily and are enjoyed roasted as snacks; mixed into cooked dishes; and a few are pressed to yield cooking oil.

The only references to a traditional medicinal use of *P. excelsa* are from Kwazulu Natal, South Africa, where it is traditionally used for treatment of respiratory infections, and from Ivory Coast where it is sometimes mentioned for treating diarrhoea. The bark has traditionally been used as a source of tannin in East Africa. African famers who clear the forest for cultivation will often leave *P. excelsa* standing due to the value of the fruit and seed, as well as for honeybees. Where they occur, the trees are an important food source for a variety of fruit-eating wildlife including elephants, chimpanzees, monkeys and hornbills.

Further reading

Götz, E. (1983) *Timber Trees of The Gambia*. Stiftung Walderhaltung in Afrika, Hamburg, Germany, 105 pp.

Hamilton, A. (1991) *A Field Guide to Ugandan Forest Trees*. Makerere University Press, Kampala, Uganda.

Szolnoki, T.W. (1985) *Food and Fruit Trees of The Gambia*. Stiftung Walderhaltung in Afrika, Hamburg, Germany, 132 pp.

Parkia biglobosa (Jacq.) R. Br. ex G. Don

African locust bean

Importance

Parkia biglobosa (Plate 25) is an important multipurpose tree in 'parklands' and savannahs of semi-arid Africa from Senegal and Guinea to South Sudan and northern Uganda. It is particularly valued for its pods and seeds, which are used in many food preparations. Its enormous crown, crimson flowers, and the high esteem in it which it is held by local people make it stand out in its native range. It is preserved when land is cleared for cropping, though there have only been limited efforts at establishing plantations.

Botanical features

P. biglobosa is a deciduous tree, 15–20 m high, exceptionally 30 m. It has a relatively short and not always straight trunk with a diameter of up to 80 cm. The crown is large, wide spreading, and umbrella-like. Old trees have large horizontal branches. The bark is greyish-brown, thick and fissured; an amber gum exudes from wounds. The root system is a taproot with lateral horizontal roots near the soil surface.

Leaves are dark green or greyish-green, bipinnate, 15–40 cm long, with 14–30 pinnae, roughly opposite with 50–70 pairs of leaflets. Leaflets smooth and oblong, 1–1.5 cm long and 3 mm wide. Capitula are red, turning to salmon-pink, biglobose, 4.5–7 cm by 3.3–6 cm, with hermaphrodite and staminoidal flowers, on peduncles up to 30 cm long. The ball-like remnant flower-heads produce clusters of pods, which grow into fruits, slightly bent, up to 45 cm long. They are smooth and brown, and contain a yellow pulp surrounding up to 23 flattened black seeds per pod.

Distribution

P. biglobosa occurs naturally in Africa mostly between 5°N and 15°N in a strip extending from the Atlantic Ocean from Senegal to Guinea, across southern Mali and Côte d'Ivoire, to Southern Sudan and northern Uganda in the East. It is also found in São Tomé where it has possibly been introduced. The natural phytogeographical distribution corresponds to the Sudanian regional centre of endemism and the Guineo–Congolian/Sudania regional transition zone, often in savannah woodlands and in cultivated parklands near villages with short fallows, where trees are conserved during land clearing. It has been introduced to tropical America and the Caribbean and is naturalized in Haiti.

Silviculture characteristics

P. biglobosa is a drought- and fire-tolerant, multipurpose, nitrogen-fixing tree. It is not grown in plantations, although individual trees are sometimes planted by farmers; mostly, trees are selectively left standing when bush is cleared for cultivation. Reports from experimental plantations in Niger and Burkina Faso suggest that mortality rates are higher in mixed plantations than in pure plantations, and group planting rather than individual planting is recommended. Plantations are difficult to establish due to the breakage of branches, especially by early rainy-season winds, with a high incidence of mortality in 3–5-year-old seedlings.

The main geographical distribution corresponds to Sudanian, Sahel and savannah-like climates, characterized by annual precipitation of 650–1300 mm with a dry season lasting 4–8 months. The mean annual temperature in this range is about 28°C, but trees at high altitudes in Guinea can tolerate minimum temperatures of at least 5°C and annual rainfall up to 1750 mm. Flowering starts at the age of 8 years, and occurs towards the end of the dry season, from December to April in West Africa, with spasmodic flowering sometimes occurring in other months.

Silviculture practice

Mature pods are picked directly from the tree or collected when they have just fallen, with 2800–5000/kg. Washed and dried, they can be stored for 2 years. Pretreatment is required, with soaking in hot water overnight a common method and germination occurs after 7–10 days. Direct sowing produces poor results due to rodents and other predators, and container planting stock using fresh seed produce the best results, planted out after 10–14 weeks in the nursery. Stumps and plants from nursery beds also produce satisfactory results.

A spacing of 5 × 5 m is recommended, thinning to 100 plants/ha after 8–10 years, and pure stands are recommended rather than mixed stands. Growth rates are variable, and 4 years after transplanting, trees may be 3 m tall on the best sites or only 1–1.5 m tall on drier sites, with average diameter of 17–20 cm after 22 years. Farmers use various pruning techniques to enhance fruit production, using different frequency, intensity and seasonality.

After 15–20 years, trees can produce 25–100 kg of pods each, or 900 kg of seed/ha, with 2.2 t pulp and 1.9 t of waste husks.

Diseases and pests

Pods may be attacked by seed-feeding beetles, though there is little information on other insect pests or fungal pathogens, with the only records being for *Cercospora* sp., *Hypoxylon rubiginosum* and *Phyllachora leonensis*.

Uses

P. biglobosa is generally conserved during clear felling and bush clearance for its multipurpose value. It grows in savannah woodlands and is abundant in so-called parklands near villages in Africa, where semi-permanent cultivation of crops such as sorghum is undertaken. It serves as a shade tree for livestock and as a windbreak.

The wood of *P. biglobosa* is generally whitish, with hardly any difference between the sapwood and the pale yellow heartwood. The wood gives off an unpleasant smell when it is newly felled. The relatively hard and moderately solid wood has a density of 554–640 kg/m^3 when air-seasoned. Seasoning does not take long and only occasionally causes distortion. The wood is not durable, however, being readily attacked by fungi and termites, and in water, marine borers rapidly destroy the wood. In contact with the ground, the wood decays rapidly. Fungi also cause ugly discolorations, considerably reducing the value of the timber. The wood works easily with hand or power tools, but can split around knots when being processed. It is easy to nail, screw or glue planks or strips together to make firm joints and it takes paint and varnish. The wood is used mainly as a light structural timber, making vehicle bodies, agricultural implements, boxes, crates and barrels, furniture seats, hoe handles, mortars, pestles, bowls, planks and carvings. Sometimes branches are lopped for fuelwood.

The major use is for food. Seeds are fermented and give a greasy extract of unpleasant odour and appearance, or a black, strong smelling, flavoursome, tasty food, rich in proteins. Dried fermented seeds keep for more than 1 year in traditional earthenware pots, without refrigeration, and small amounts are crumbled, during cooking, into traditional soups and stews which are usually eaten with sorghum- or millet-based dumplings or porridges. It is rich in protein (37%), fat (35%) and vitamin B2, and fermented beans are rich in lysine. The fat in the beans is nutritionally valuable and approximately 60% unsaturated, the major fatty acid being linoleic acid. The dry pulp is rich in sugar (about 60%) and vitamin C, with a moderate mineral content. The pulp is eaten raw, made into a refreshing drink when mixed with water, used as a sweetener and fermented into an alcoholic beverage. The young pods are cooked as a vegetable or roasted. Young leaves are occasionally eaten as a vegetable. The seeds can be used as a substitute for coffee. The seeds yield a yellowish oil.

Whole pods are eaten by livestock, and branches are sometimes looped for fodder. Leaves are rich in energy, but are only of medium feed quality owing to poor levels of protein and some minerals and so should not be the sole source of fodder.

Bark, leaves and fruits are used in many medicinal preparations. Twigs are used to clean teeth. Bark and pods are used as piscicides. Husks of pods with the bark produce a blue dye. Fibres from pods and roots are used as sponges and as strings for musical instruments. Mucilage from part of fruit is made into a fluid and used for hardening earth floors and to give a black glaze to pottery.

Local laws often prevent destruction of the tree, and it can be individually owned and thus serve for land demarcation. It also has cultural and spiritual values.

Further reading

Hall, J.B., Tomlinson, H.F., Oni, P.I., Buchy, M. and Aebischer, D.P. (1997) Parkia biglobosa: *A Monograph*. School of Agricultural and Forest Sciences Publication No 9, University of Wales, Bangor, UK, 107 pp.

Ouedraogo, M., Ræbild, A., Nikiema, A. and Kjær, E.D. (2012) Evidence for important genetic differentiation between provenances

of Parkia biglobosa from the Sudano-Sahelian zone of West Africa. *Agroforestry Systems* 85, 489–503.
Ræbild, A., Hansen, U.B. and Kambou, S. (2012) Regeneration of *Vitellaria paradoxa* and *Parkia biglobosa* in a parkland in southern Burkina Faso. *Agroforestry Systems* 85, 443–453.

Paulownia × taiwaniana T.W. Hu & H.J. Chang

Taiwan paulownia

Importance

Paulownia × taiwaniana is widely used in plantations in Taiwan due to its fast growth rate, easy propagation from root cuttings, high timber quality, and beautiful flowers. This species has been established in plantations in several countries, and the timber is largely exported to Japan. Unlike *Paulownia tomentosa*, *P. × taiwaniana* is not recorded as an invasive species.

Botanical features

P. × taiwaniana has an open, spreading crown and a straight bole.

Leaves are simple, petiolate, opposite, broadly ovate, 10–30 cm long, 8–30 cm wide, cordate at base, acute or abruptly acuminate at apex, upper leaf surface glabrescent with few glandular hairs, lower leaf surface moderately tomentose with dendroid hairs. Petioles are 10–17 cm long, furfuraceous, glabrescent.

Inflorescence is large, paniculate, 50 cm long, composed of many individual cymes which are simple or dichotomously branched, with 3–5 flowers; peduncles of the cymes 5 mm long, pedicels 8–15 mm long, brown, lanate with dendroid hairs. Flowers fragrant with light purple infundibulo-campanulate corollas 6.5 cm long, 2 cm wide. Calyx is 5-lobed, grey-brown, campanulate, 1–1.5 cm long, covered with dendroid hairs. Fruit is capsular, woody, ovoid, 3.5–4.5 cm long, 2 cm wide, with 2 valves that are loculicidally dehiscent. Seed very numerous, small (5.0–5.5 mm), winged.

In natural forest in Taiwan, *P. × taiwaniana* is leafless from the end of December until March, and flowering begins in April. The fruits mature in October.

Distribution

P. × taiwaniana is native to Taiwan. It has been established in plantations in India, Japan, Indonesia, Malaysia, the Philippines, Thailand, Argentina, Bolivia, Brazil, Paraguay and South Africa.

Silviculture characteristics

In Taiwan, the annual growing season of *P. × taiwaniana* is 5 months, beginning in late March and ending in late August. It is a light-demanding species and has a rapid growth rate. *P. × taiwaniana* will produce coppice shoots after harvesting. Strong winds tend to damage the leaves of young trees and coppice shoots, but smaller leaves on older trees are less affected.

Silviculture practice

P. × taiwaniana is easily propagated from root, stem or shoot cuttings. Large root sections such as 4 × 6 cm, 3 × 8 cm or 2 × 12 cm (length × diameter) are used in plantations. The slender root sections (1 × 12 cm or 0.5 × 14 cm) are planted in the nursery in order to obtain further root cuttings. Root cuttings can also be obtained from seedlings, but the production and size are reduced. Before root cuttings are planted in the field, all fibrous roots are removed; they are stored under shelter for 1 week and dehydrated to 10% water content. After 4 months, if stored in plastic bags at 5°C, root cuttings will sprout at a 96% success rate.

Seedlings are seldom used for plantation establishment, but rather as a source of root cuttings in the nursery, due to their slow growth and because their hairy stems are prone to being coated in soil. Mature pods are collected and air-dried for 2 weeks, producing approximately 240,000 seeds/l. The seed remains viable for several years if stored in closed plastic bags at 0–4°C. Seed germination requires exposure to light and may be enhanced if seeds are treated with 75% ethanol (1 min) then 3.5% hypochlorite plus 1 drop polysorbate 20 per 100 ml (15 min). Seeds sown in germination boxes in a sand mixture with tree-bark compost or sown in dibbling tubes filled with vermiculite and peat may obtain an 80% germination rate within 2 weeks.

Treated seeds are sown evenly in a seedbed, at a density of 10 ml of seeds/m^2 (i.e. 2400 seeds/m^2). The seedbed should be protected by a transparent polyethylene cover approximately 60 cm high, in order to protect the seedlings from the heavy rains which tend to result in their hairy stems being coated with soil. Usually, after 6 months the seedlings are approximately 30 cm tall and are replanted in the nursery and used to raise root cuttings.

For stand establishment, site preparation is recommended, including ripping along the planting line. The initial spacing is 4 × 6 m (500 trees/ha), but in areas with high weed density 3.3 × 3.3 m (900 trees/ha) is advised. The planting pits should be 60 cm in diameter and 40 cm in depth, and filled with a mixture of 20 kg compost and topsoil. In Taiwan, growth of *P. × taiwaniana* is enhanced and disease resistance increased by use of 0.4–0.5 kg of chemical fertilizer (N:P:K 20:7:7) for each tree applied twice a year, especially before truncating and after thinning. If the soil pH is below 5.5, lime is also recommended. The root cuttings are buried vertically to 1 cm beneath the soil surface in the rainy season. The soil around the root cuttings is compressed and covered with dry weeds (from weeding) in order to maintain moisture and inhibit weed growth.

Cuttings begin to sprout in 1–2 weeks. After 1.5 months they have a dominant shoot 30 cm tall which should be retained and all other coppice shoots removed. At this stage, prevention of insect damage and weed competition is important. When seedlings reach 1.5–2 m, the terminal buds should be removed in order to encourage stem and root growth. The following spring before the leaves sprout, seedlings are cut back to ground level and covered by a thin layer of soil, ash or dry weeds. After 2 weeks many coppice shoots will sprout, and only the dominant shoot 30 cm high is retained; this may reach 6 m tall after 6 months' growth. Any lateral shoots are pruned in order to obtain knot-free logs; the wounds heal quickly and there is little risk of infecting pathogens or further production of lateral buds.

At 3 years old, *P.* × *taiwaniana* begins to flower. The inflorescences should be removed before the leaf buds burst in order to increase vegetative growth. *P.* × *taiwaniana* trees are unable to tolerate shade, so thinning is important to maintain maximum growth.

Pure stands of *P.* × *taiwaniana* may be managed intensively on an 8-year rotation, or a 10–15-year rotation for high-value logs. On a good site, 4-year-old trees may reach 12 m in height with a dbh of 24 cm; high-value logs with a 30 cm dbh can be obtained after thinning. For a 10–15-year rotation, 60% reduction from the initial 500 trees/ha to 300 trees/ha is sufficient.

Diseases and pests

In the nursery, *P.* × *taiwaniana* seedlings are susceptible to damping off and root rot caused by soil-borne fungi such as *Pythium myriotylum*, *Pythium spinosum*, *Pythium splendens*, and *Rhizoctonia solani*. Nematode root knot caused by *Meloidogyne incognita* and *Meloidogyne arenaria* may seriously affect seedling roots and reduce growth and yield of root cuttings. Paulownia witches' broom is a serious disease in *P.* × *taiwaniana* of all ages. This disease is transmitted by vectors such as *Empoasca vitis*, *Cyrtopeltis tenuis* and *Halyomorpha* spp.

Uses

Various agroforestry models using *P. taiwaniana* have been developed in China. *P.* × *taiwaniana* is suitable for intercropping, as its roots are deeper than the crop plants and hence competition for moisture and fertilizer is not severe. In Taiwan, *P.* × *taiwaniana* species is intercropped with tea or grasses for grazing cows, pigs and chickens. *P.* × *taiwaniana* is an effective shade tree for many agricultural crops, including cereals, tea and grapes. Its sparse canopy allows light to penetrate to the crops, reduces wind speed, and creates stable temperature and moisture conditions. This species is used in shelter belts for rice, and along irrigation channels and roadsides.

Generally, *P.* × *taiwaniana* wood is characterized by straight grain, uniform texture, lack of shrinkage, and a low density of 250–360 kg/m^3 at 15% moisture content. It is easy to plane, saw and carve without splitting, yet its nail-holding ability is low. The wood has good resistance to fire, decay and warping. The majority of *P.* × *taiwaniana* timber grown in Taiwan is exported to Japan, where it is used for similar purposes to *P. tomentosa*. However, the grade quality of *P.* × *taiwaniana* is not as high as that of *P. tomentosa*.

Paulownia wood is used for furniture, particularly cupboards, which are smoke-proof and resistant to insect attack. It is used for model airplanes and gliders, and interior panels on airplanes, vehicles and ships. The soundboards of some Chinese musical instruments are made from *Paulownia*, due to its good resonance properties. The wood is suitable for oil drums, wine or beer barrels, tea boxes, fruit boxes and grain storage containers. It is also used for traditional handicrafts, ornaments, beehives, furniture, veneers and agricultural tools. Following removal of foliage, branches from thinnings are used as fuelwood.

The leaves, fruits, and wood of *Paulownia* species have medicinal properties, and are especially used to treat bronchitis, coughs and phlegm. Its leaves and flowers are used as fodder for pigs, sheep and rabbits. The foliage from *P.* × *taiwaniana* thinnings is used as green manure.

Further reading

Donald, D.G.M. (1990) *Paulownia* - the tree of the future? *South African Forestry Journal* 154, 94–98.

Lin, T.P. and Wang, Y.S. (1991) *Paulownia taiwaniana*, a hybrid between *P. fortunei* and *P. kawakamii* (Scrophulariaceae). *Plant Systematics and Evolution* 178, 259–269.

Rin, U.C. (1979) *Silvicultural Studies on Taiwan Paulownia (*Paulownia taiwaniana *Hu et Chang)*. Research Bulletin 178. National Chung Hsing University, Taichung, Taiwan.

Paulownia tomentosa (Thunb.) Steud.

Paulownia

Importance

Paulownia tomentosa is one of the most important tree species in China. It is a fast-growing tree with high-quality wood.

P. tomentosa can become invasive in countries where it is introduced. It is considered an alien invasive species in the USA and Australia.

The wood is used in furniture, decorative products and musical instruments. *P. tomentosa* is also suitable for use in agroforestry systems in semi-humid temperate areas.

Botanical features

P. tomentosa is a large deciduous tree with an umbrella-shaped crown. It may grow up to 18 m tall, with a dbh of at least 1 m. Its bark is pale yellow, with numerous large lenticels.

Leaves opposite, cordate to ovate, 20–30 cm long, 15–28 cm wide, acute or obtuse, base cordate, margins entire or 3–5 lobed, undersurfaces of leaves tomentose. Terminal bud absent, axillary buds sunken in bark, winter buds with several outer scales, superposed.

Cymes peduncled, the peduncles as long as or longer than the pedicels, growing on the main axis, and branches of paniculate inflorescences 40–60 cm long. Calyx deeply lobed, the lobes as long as or longer than the tube. Flowers perfect, fragrant, showy (see Plate 52). Corolla 8–10 cm long, purple, tubular; with 5 shallow, rounded, unequal lobes. Fruit ovoid, pointed, woody capsules 2.5–4 cm long, with 2 carpels and numerous tiny winged seeds attached to the 2 very large placentas. Seeds, tiny, winged, flat, 1.5 mm long.

P. tomentosa flowers in April–May, after which the new leaves emerge. Fruits mature and turn brown in the autumn, and remain on the tree throughout the winter.

Distribution

P. tomentosa is native to China, and is widely distributed in central and northern regions, especially Shaanxi, Shanxi, Gansu, Henan, Hebei, Shandong, Anhui, Hubei, Jiangsu and the Liao'ning peninsula. It occurs in Japan and South Korea, but some Japanese taxonomists believe that these are naturalized populations resulting from past introduction and cultivation of this species in these countries. *P. tomentosa* naturally occurs in deciduous and mixed forest species, and to a lesser extent in secondary forest.

Within *P. tomentosa*, the three varieties have distinct distributions. *P. tomentosa* var. *lucida* occurs mainly in northern China, var. *lanata* in the northern Yangtse River region and var. *tsinlingensis* in central and south-western China.

P. tomentosa has been introduced as an ornamental in many European countries, New Zealand and the USA. It has been introduced to Brazil, Paraguay, Argentina and Australia, where it is used mainly for timber production.

Silviculture characteristics

P. tomentosa is a strongly light-demanding species which readily colonizes exposed sites, even including rocky cliffs and scoured riparian zones. It can sprout prolifically from adventitious buds on stems and roots, allowing it to survive fire, cutting, and even bulldozing. Since *P. tomentosa* is a light-demanding species it is not suitable to plant with other fast-growing species.

P. tomentosa has a sparse crown, a deep, widely distributed root system and high adaptability, and is therefore suitable for use in agroforestry and afforestation.

Silviculture practice

The principal method of propagation in *P. tomentosa* is from root cuttings with a diameter of 0.5–1.5 cm, which are collected from 1–2-year-old seedlings or mature trees. One-year-old seedling stock produces 15–20 effective root cuttings, which may be transplanted. It is important to select vigorous root cuttings, and to enhance germination of buds by maintaining a warm temperature or covering with thin plastic film. Transplanting occurs during the spring when the soil temperature is above 4–8°C, and cuttings should be planted upright with the widest girth at the top. After 6 months' growth from April to September, stock may reach 3.5–4.5 m tall with a dbh of 5–7 cm.

P. tomentosa is also grown from seed. The fruits are collected when they begin to open, and are stored in a dry place with good ventilation until they naturally dehisce and release the yellow-brown seeds. The seed is small and easily dispersed by wind. There are 2.5–5.0 million viable seed/kg of dry seed. Good germination rates (70–90%) may be achieved by soaking the seed in warm water at 35–40°C, planting within 5 days and maintaining a temperature of 35°C once planted. *Paulownia* can be propagated using tissue culture.

In pure-stand plantations, wide spacings are important due to the fast growth of this species, and if planted too densely *P. tomentosa* shows low survival rates. Generally, *P. tomentosa* is planted at spacings of 4 × 4 m or 6 × 6 m. In agroforestry systems such as intercropping, individual trees are planted 5 m apart in rows which may be 30–50 m apart. Fertilizer application encourages fast early growth. Spot site preparation is usually necessary when stands are planted along roads, water channels and river-banks. *P. tomentosa* is often planted on terraces in hilly or mountainous areas.

Pruning after 3 years' growth is common practice in plantations. In younger stands irrigation and weed control are necessary. Good drainage is important during the wet seasons.

Generally, a 10-year-old tree on brown forest soil has an average annual increase in diameter of 0.8–1.1 cm. In mountainous soils, hilly areas and seasonally waterlogged land, it is also capable of vigorous diameter growth, such as 0.6–0.8 cm.

Where *P. tomentosa* is an unwanted invasive it can be controlled using both mechanical and chemical means. Hand pulling can be effective on young seedlings, so long as the entire root is removed, since broken fragments of root may re-sprout. Cutting of larger trees is considered an initial control that will require repeated cutting of re-sprouts.

Diseases and pests

Paulownia witches' broom (PWB) is a major disease found in *P. tomentosa* plantations, and is caused by a phytoplasma (previously described as a mycroplasma-like organism). This disease may be identified by distinctive yellow broom-like shoots, which die back in the autumn. PWB is commonly found in seedling stocks and young trees (3–6 years old), and may greatly influence their growth. It has little effect on adult trees. In its natural environment PWB is spread by the stink bug *Halyomorpha picus*. Infected roots and seedling stock aid the spread of this disease, yet seeds obtained from infected trees may grow disease-free. The effects of PWB vary depending on the *Paulownia* species. For example, *P. tomentosa*, *Paulownia catalpifolia*, and *Paulownia elongata* are easily infected; however, *Paulownia fortunei*, *Paulownia kawakamii*, and *Paulownia fargesii* show good tolerance to the disease. PWB may be controlled by trimming diseased branches. Selecting clones with good resistance and using suckers free of the disease will reduce spread of the disease.

Anthracnose disease is a major disease in saplings which injures leaves, petolies, and shoots, and causes early leaf drop. *P. tomentosa* also suffers damping off disease caused by *Rhizoctinia solani* and *Fusarium* spp. Other diseases include the nematode *Meloidogyne marioni*, which infects the roots of seedlings and results in mortality, and the fungus *Sphaceloma tsugii*, which commonly damages seedling shoots and causes dieback. Both diseases have a common occurrence throughout the distribution range *of P. tomentosa*.

The main pest found on *P. tomentosa* is *Eumeta variegata* (syn. *Clania variegata*). *E. variegata* occurs throughout the natural distribution range of *P. tomentosa*; it develops one generation in northern China and two generations in southern China. Seedling stock is the major source of spread of this pest.

Uses

In temperate zones, *P. tomentosa* is suitable for intercropping with wheat, rape and sweet potato. In the subtropics it may be intercropped with tea and bamboo. This species may also be used to form windbreaks on farmlands, roadsides or canal-banks. In some European countries, *P. tomentosa* is cultivated as an ornamental due to its beautiful flowers, and is found in arboreta, botanical gardens and parks. *P. tomentosa* is also believed to reduce sulphur dioxide pollution and grows in environments where other trees may suffer leaf loss.

Generally, *P. tomentosa* wood is characterized by straight grain, uniform texture, lack of shrinkage and a low density of 250–360 kg/m^3 (at 15% moisture content). It is easy to plane, saw and carve without splitting or warping, yet its nail-holding ability is low. The wood is resistant to fire and decay. The wood is used for furniture, particularly cupboards, which are smoke-proof and resistant to insect attack. It is used for model airplanes and gliders, and interior panels on airplanes, vehicles and ships. The soundboards of some Chinese musical instruments are made from *P. tomentosa* wood due to its good resonance properties. It is suitable for oil drums, wine or beer barrels, tea boxes, fruit boxes and grain storage containers. It is also used for traditional handicrafts, beehives, ornaments and agricultural tools. In recent years, *P. tomentosa* has been used for plywood and particleboard manufacture, and for kitchen utensils such as wooden bowls, plates and rice steamers.

The flowers and leaves contain fats, sugars and protein, and are traditionally used as fodder for pigs, goats and rabbits. The flowers of *P. tomentosa* contain 2.7–4.1% glucose and 8.9–9.2% soluble sugars; the leaves contain 4.3% glucose and 4.3% soluble sugars. The leaves, containing 3.3% nitrogen, are also used as green manure in rural areas.

In the past, *P. tomentosa* has been used in traditional Chinese medicine to treat bronchitis, coughs, asthma and high blood pressure. Currently, the flowers, leaves and bark are effective in treating chronic tracheitis.

Further reading

Akyildiz, M.H. and Kol, H.S. (2010) Some technological properties and uses of paulownia (*Paulownia tomentosa* Steud.) wood. *Journal of Environmental Biology* 31, 351–355.

Jiang, J.P. (1988) *Silviculture of* Paulownia. China Forestry Publishing House, Beijing.

Remaley, T. (1998) Princess Tree: *Paulownia tomentosa* (Thunb.) Sieb. & Zucc. ex Steud. Plant Conservation Alliance, Alien Plant Working Group: Bureau of Land Management, Washington, DC. Available at: http://www.nps.gov/plants/alien/fact/pato1.htm (accessed 22 March 2013).

Zhu, Z.H., Chao, C.J., Lu, X.Y. and Xiong, Y.G. (1986) Paulownia *in China: Cultivation and Utilization*. Asian Network for Biological Sciences, Singapore and International Development Research Centre, Ottawa, vi & 65 pp.

Peltophorum dubium (Spreng.) Taub.

Canafistula

Importance

Peltophorum dubium (Plate 26) is a moderate to fast-growing pioneer tree species used in many mixed plantations under different site conditions in Brazil and other South American countries. It is used for land reclamation and other environmental purposes. Its medium-quality wood is mainly used for fuelwood and building construction. It is widely planted as an ornamental and shade tree. *P. dubium* is commonly known as canafistula, a name also applied to *Cassia fistula*, another leguminous tree with showy yellow flowers, but it also has a number of other names in South America, including árbol de Artigas and ibirá puitá. *P. dubium* is an endangered species in its native range, so genetic conservation of base populations needs to be undertaken.

Botanical features

P. dubium is a semi-deciduous to deciduous tree that reaches a height of 10–20 m and dbh of 40–90 cm. Very occasionally it reaches 40 m in height and 300 cm dbh. The trunk is cylindrical, almost straight or slightly curved. It has dichotomic branches and a wide crown.

Leaves are bipinnately compound, alternate, up to 50 cm long and 25 cm wide. It has 12–21 pairs of pinnae, each one with 20–30 pairs of opposite leaflets, 5–10 mm long and 2–3 mm wide.

Inflorescences are in showy panicles or ferruginous terminal racemes, tomentose, up to 30 cm long. Flowers are hermaphrodite, with a yellow or orange corolla, up to 2 cm long. Fruits (legumes) are dry, flattened, indehiscent, 4–9 cm long and 1–2.5 cm wide, with 1–3 seeds per fruit (see Plate 59).

Depending on the region, flowering occurs from September to March, while fruits ripen from April to December.

Distribution

The natural distribution of *P. dubium* ranges from north-east Brazil (7°S) to Uruguay (30°S). It occurs widely in Brazil, except in the north (Amazonian rainforest) and the most arid part of the north-east. It is found in north-eastern Argentina, in eastern Paraguay and northern Uruguay.

Silviculture characteristics

P. dubium is a light-demanding species with moderate to fast growth, according to the quality of the site where it is growing. It has a variable growth habit that can be monopodial or irregular with bifurcation from the bottom. It has the ability to self-prune and coppice. *P. dubium* grows on several soil types, even those which are acid and infertile. However, it does not tolerate shallow or excessively wet soils. The species tolerates up to 5 months of dry season with strong water deficit. In the coldest regions, young trees are injured by frost but adult trees in natural forests are cold-tolerant up to −11°C.

Silviculture practice

Seeds of *P. dubium* are collected when fruits are reddish-brown or greyish. Trees have an abundant annual fruit production, with dispersion by anemochory and barochory. After collection, the fruits must be sun-dried within 10–30 days and then broken to obtain the seed. Seeds can also be separated from fruit wastes through a screen, a gravity separator or an air screen cleaner. Newly processed seeds generally have 6–8% moisture and there are 21,600 seeds/kg. Seeds packed in paper bags and stored in a dry room (air moisture 50%) show

92% germination after 25 months. Seeds have seedcoat dormancy that can be overcome by mechanical scarification with sandpaper or by chemical scarification with sulphuric acid for 10, 17 or 30 min, depending on seed lot. After dormancy is broken, germination is quick and uniform, and can reach 95%. Seed viability can be determined by the tetrazolium test, with analogous results to the germination standard test. Seedlings are easily grown in polythene bags or in small-volume (50 cm^3) plastic tubes, and a plantable size is reached in 4–5 months. *P. dubium* does not show nodulation capability but does form endomycorrhizal associations.

P. dubium is mainly planted in mixed plantations for environmental purposes. In Brazil, in these mixed plantations, it is planted as a light-demanding climax species and grown with pioneer species (*Trema micrantha*, *Senna multijuga*, *Schinus terebinthifolius*, *Lithraea molleoides*, *Guazuma ulmifolia*, *Clitoria fairchildiana*, *Croton floribundus*), other light-demanding climax species (*Tabebuia chrysotricha*, *Tabebuia serratifolia*, *Enterolobium contortisiliquum*, *Cedrela fissilis*, *Ceiba speciosa*) and shade-tolerant climax species (*Aspidosperma polyneuron*, *Aspidosperma olivaceum*, *Myroxylon peruiferum*, *Hymenaea courbaril*, *Inga marginata*).

As a species that grows well on low-quality sites, *P. dubium* is much used in mixed planting for land reclamation. In plantations based on the secondary succession concept, *P. dubium* has shown one of the best performances among the light-demanding climax species in south-eastern Brazil. In a 3-year-old plantation in dark red compacted latosol, at a spacing of 1.5 × 3.0 m, the species reached 1.3 m in height. *P. dubium* also is one of the best species for revegetation of riparian forests, reaching 6.0 m in height and 5.7 cm dbh, at the same spacing, at 64 months.

In spite of the several applications for its wood, *P. dubium* is not generally planted with this objective in mind. The commercial production of wood comes primarily from native populations. It has an irregular bole when grown in experimental pure plantations. However, the future of the species as a commercial timber-producing species appears promising. In a plantation in São Paulo State, Brazil, trees gave a mean annual increment of 23.36 m^3/ha by 7 years of age. Considering that the seedlings were produced from seeds collected in natural populations, without any genetic improvement process, the increment could probably be still greater if improvement programmes were implemented.

Diseases and pests

There is a record of total defoliation in young pure stands in southern Brazil caused by caterpillars of *Molippa sabina*. The sawyer beetles *Oncideres ulcerosa* and *Oncideres dejeani* also cause damage in plantations.

Uses

Because of its good crown characteristics, growth, and wind resistance, *P. dubium* is planted for land reclamation, revegetation of riparian forests and windbreaks. It is planted in agroforestry and silvopastoral situations. The species is also widely planted for ornamental purposes, due to its showy flowering and graceful form, and as a shade tree.

P. dubium wood is very resistant to white rot. It has a density of 690–774 kg/m^3 and is used in building construction as beams, roof-timber, laths, door-posts, windows and flooring. Externally it is used for railway sleepers, posts, stakes and telegraph poles. It is also used for furniture, turnery, marquetry, carpentry and joinery to produce truck-beds and pieces for interior decoration. It can be used in the manufacture of particleboard. The species produces medium-quality fuelwood and charcoal.

P. dubium has high levels of saponin and tannin in its trunk bark and wood. A red dye product is extracted from the wood powder. It is also recommended for planting as a green manure.

Further reading

Carvalho, P.E.R. (2002) *Canafistula*. Circular Técnica - Embrapa Florestas No. 64, Colombo, Brazil, 15 pp. (in Portuguese).

Seneme, A.M., Possamai, E., Vanzolini, S. and Martins, C.C. (2012) Germinação, qualidade sanitária e armazenamento de sementes de canafístula (*Peltophorum dubium*) (Germination, sanitary quality and storage of seeds of *Peltophorum dubium* (Spreng) Taub.). *Revista Árvore* 36, 1–6 (In Portuguese with English summary).

Siqueira, A.C.M. de F., Sebbenn, A.M., Gurgel Garrido, L.M. do A., Castollen, M. da S. and Zanatto, A.C.S. (1999) Comportamento silvicultural e genético de duas espécies arbóreas tropicais secundárias. (Silvicultural and genetic characteristics of two tree species of secondary tropical forests). *Revista do Instituto Florestal* 11, 53–64 (in Portuguese with English summary).

Peltophorum pterocarpum **(DC.) Backer ex Heyne**

Copperpod

Importance

Peltophorum pterocarpum is a fast-growing tree with multiple uses. It is of potential use for reforestation, in agroforestry farming systems and as a source of green manure; it is also used in urban forestry. The bark has been an important source of dye for batik in Java, and the leaves are used as fodder. *P. pterocarpum* is also a very beautiful and elegant ornamental tree, and as such it has a range of common names including copperpod, yellow Poinciana and yellow flame tree. Although the branches are prone to damage from high winds, it is widely planted in gardens, parks and as an avenue tree throughout the tropics. Provenance trials and research into genetic variation in the species are worth pursuing.

Botanical features

P. pterocarpum is a deciduous tree usually reaching a height of 15–24 m, although it may attain 50 m, and a diameter of 50–100 cm. It has a dense, spreading crown, and smooth, grey bark. The root system is deep, making it a very windfirm tree.

Leaves are bipinnately compound and large, 30–60 cm long, with 8–10 pairs of pinnae each bearing 10–20 pairs of oblong leaflets 0.8–2.5 cm long and 0.4–1.0 cm wide, with oblique bases.

The inflorescence is a large compound raceme (panicle) up to 50 cm long, and is brown-tomentose. The panicle is terminal with rust-coloured buds. Flowers are showy, orange-yellow, each about 2.5 cm in diameter, and fragrant, particularly at night. Fruits are 1–4 seeded pods, flat, thin, winged, 5–10 cm long, conspicuously dark red or copper-coloured when ripe, eventually turning black. Seeds are very hard and about 9 × 4 mm in size.

In the Philippines, panicles appear from May to September, with flowering in March–April. In India, the general flowering period is March–May, although sporadic flowering may occur throughout the year (particularly in young trees), and a second flush of flowers may occur in September–November. As a fast-growing species, young trees raised from seed will, under good conditions, flower from age 4 years.

Distribution

P. pterocarpum is native to the Indo-Malayan region, and is found from the Andaman Islands and Sri Lanka in the west, throughout Malesia including Papua New Guinea, north to Indochina, and south to northern Australia.

It is widely cultivated throughout its natural area of distribution, and also in the Bismarck Archipelago, India, tropical Africa, the Caribbean, Central America, and Florida and Hawaii in the USA. It is considered to be an exotic species in the Philippines.

Silviculture characteristics

P. pterocarpum can withstand the salty winds of beach and coastal locations. The species does not tolerate severe drought, but can tolerate a certain level of air pollution. Although, as a leguminous tree, it is thought to fix nitrogen, there is no detailed information confirming this.

Under natural conditions, *P. pterocarpum* is a lowland species, rarely occurring above an altitude of 100 m. It thrives best under more or less seasonal conditions, tolerating a dry season of 1–3 months. It prefers open or disturbed forest conditions, occupying coastal vegetation, rainforests and savannah woodlands, and can be found growing along beaches and in mangrove forests, especially their inner margins. In Java, it is also found growing wild in *Imperata* grasslands and teak forests.

Silviculture practice

P. pterocarpum can be propagated by seed, grafting or branch cuttings. Untreated seeds need several months to germinate. Seeds store well for a long period. Germination frequency has been recorded at 78%, with good shoot and root growth. Germination is hastened by scarifying one end of the hard seedcoat, softening the seedcoat in dilute acid, or immersing the seed in boiling water for 2 min followed by soaking in cold water for one night. Grafts or cuttings may be used to obtain better uniformity for roadside trees.

Seedlings are best raised in a nursery for 1 year before transplanting to the field. Wildings from beneath mother trees can also be successfully used for planting. Young trees are often planted in an intercropping system with mahogany (*Swietenia* sp.) or teak (*Tectona grandis*).

After the first year of establishment, little effort is needed to maintain plantations. Stands will usually survive even with thick ground cover, such as from *Imperata* or other tall grasses. *P. pterocarpum* is fast-growing and can reach a height of 9 m in 3 years. Agroforestry trials in south Sumatra with *P. pterocarpum* hedgerows spaced at 4 m intervals and cut 2 to 4 times a year showed that, over a 3-year period, an average prunings yield of 8 t/ha was possible; the umbrella-shaped crowns which developed following cutting allowed little light to reach the ground.

Diseases and pests

P. pterocarpum does not suffer much from pests or diseases although powdery mildew caused by *Oidium* sp. is reported from India. In Singapore, foliage can be severely damaged by the night-flying beetle *Autoseria rufocuprea*. *P. pterocarpum* is a marginally suitable host for the bagworm moth (*Pteroma plagiophleps*) in Bangladesh. The larvae of *Hyposidra talaca* have been recorded as defoliators.

Uses

P. pterocarpum has been used in rotational alley-cropping/ fallow systems in Sumatra, where it shows promise. It is widely appreciated as an ornamental due to its profuse flowering, as well as other characteristics such as fairly good resistance to pests and diseases. It also provides shade due to its dense spreading crown.

Wood density is 530–600 kg/m^3. The sapwood is greyish-white, turning grey-brown on aging. The heartwood is light reddish-brown or black, and is moderately hard, moderately heavy, and somewhat lustrous, with a straight to interlocking grain. Although not a commercial timber species, the wood is used locally for light construction purposes, cabinetmaking and as fuelwood.

The decorative golden-yellow flowers are used as cut flowers.

The bark of *P. pterocarpum* has been an important component of the dark or black 'soga' dye used in Java for batik work. It is also used for tanning leather, and preserving and dyeing fishing nets. In Indonesia, the bark is used for fermenting palm wine.

All parts of the tree (leaves, roots, bark, flowers) are used extensively in traditional medicine. Extracts are used as astringents to cure or relieve intestinal disorders after pain at childbirth, sprains, bruises and swelling, or as lotions for eye troubles, muscular pains and sores; and for gargles and tooth powders.

P. pterocarpum is suitable for use as a fodder for browsing livestock. In India and elsewhere, it is a source of nectar and pollen for honeybees, including dammer bees (*Trigona iridipennis*) and rock bees (*Apis dorsata*).

Further reading

ILDIS/CHCD (1994) *Peltophorum pterocarpum* (DC.) K. Heyne. In: ILDIS/CHCD (eds) *Phytochemical Dictionary of the Leguminosae.*

Volume 1. Plants and Their Constituents. Chapman & Hall, London, pp. 542–543.

Santapau, H. (1966) The rusty shield-bearer (*Peltophorum pterocarpum* (DC.) Backer). In: Santapau, H. (ed.) *Common Trees*. National Book Trust, New Delhi, pp. 112–115.

Thakur, P., Dhiman, S.R., Gupta, Y.C., Bhardwaj, D.R. and Kashyap, B. (2011) *Peltophorum pterocarpum*: a magnificent ornamental tree for mid-hills of Himachal Pradesh. *MFP News* 21, 10–11.

Picea abies (L.) H. Karst.

Norway spruce

Importance

Picea abies (Plate 27) is economically the most important coniferous species in Europe with a history of intensive management for several centuries. This has both changed natural forests and led to its introduction outside its natural distribution.

It exhibits good yield and quality performance under very diverse site conditions but attention needs to be given to the choice of suitable provenances adapted to local site conditions. In recent decades, *P. abies* has suffered extensive forest decline in central Europe, which has reduced its use in new planting both within and outside its natural distribution range.

Botanical features

At maturity *P. abies* are ca 30–40 m high with a dbh of 0.4–0.9 m. In young trees the crown is pyramidal but old trees tend to be columnar with a conical top. The stem is typically straight and cylindrical with little taper. The bark is orange-brown, finely flaking, becoming grey-brown and scaly on old trees.

The branches are slightly ascending at the top of the tree, becoming level then drooping, so that the crown usually has a radius of 4–5 m although the branches themselves slowly increase in length. The shoots are red-brown, orange-brown or golden-brown, darkening with age; they may be hairy but are usually glabrous. Buds are ovoid–conical and generally not resinous.

The yellow- to white-coloured wood shows little delineation between heartwood and sapwood.

The root system is shallow and wide-spreading with short sinker roots into the subsoil provided there is sufficient aeration. The fine roots have obligate ectomycorrhizae.

The needles are rhombic in cross-section with stomatal bands on all sides; they are 1.4–2.5 cm × 1 mm. On shoots in full sun the needles are rigid, pointed and rising but on shaded branches they project horizontally. The needles sit on a pulvinus, a leaf pillow which remains as a peg-like structure on the shoot when the needle is shed, usually after 4–6(–10) years.

P. abies starts flowering at about age 40 years in dense stands, but earlier with open grown trees. Good seed crops usually occur at 2–4 year intervals. The pollen or male flowers are located along the previous year's growths on branches in the middle or lower crown and expanding from 1 to 2.5 cm, from red to yellow. Female flowers are produced from buds at the end of the shoots before new growth is made; initially they are erect, usually dark red or green and to 5 cm. After pollination they become pendent and ripen brown, measuring 10–15 × 3–4 cm. The seed scales are rhombic, thinly woody or leathery. As they dry the cones open to disperse the yellow-brown to dark brown 4 mm seeds with a brown 13 mm wing which clasps the seed. The seeds will germinate at temperatures above 7°C with 7–10 cotyledons.

Distribution

The natural distribution is from the Balkans north to southern Scandinavia and from eastern France to Romania and south east Russia. This is thought to result from the effects of the last Ice Age in Europe when *P. abies* survived in refugia in the Carpathian and Transylvanian Alps, Dinaric Alps and the Apennines in Italy. It migrated northwards and westwards after the last glacial period. The populations in north-eastern Europe and Scandinavia show characters of *Picea obovata* and may be the result of migration from a Russian refuge; plants from this supposed introgression have been named *Picea fennica*.

The planting of *P. abies* has been centred in Western Europe (where it also occurs naturally), as there were many suitable sites where it could grow well. In Hungary, Romania, France, Luxembourg, Switzerland, Austria, The Netherlands and Norway, *P. abies* was planted outside the areas in these countries in which it had occurred naturally. However, at this time it was also introduced as an exotic into countries such as Belgium, Denmark, Iceland, Ireland and the UK.

P. abies was introduced to the USA in 1876, and to Canada in about 1914. It was also experimentally planted in Japan. In the southern hemisphere, *P. abies* plantations were established in South Africa, Tasmania and New Zealand.

Silvicultural characteristics

P. abies is found both as a pioneer species and as a climax species in boreal forests; it is a tree of cool temperate climates with a good water supply. It does not tolerate coastal exposure (salt spray) nor does it like summer drought. It will grow in some shade but is better with good light. Seedling growth is slow for the first few years, with height growth increasing after 5–10 years. If grown throughout the juvenile stage under low light or shade conditions, it is slow to adapt to increased light conditions.

P. abies is not very tolerant of waterlogged conditions or flooding, although it is found associated with grey alder (*Alnus incana*) on river plains in the boreal zone, where it tolerates occasional flooding. It is not fire-resistant.

On unsuitable sites it is often susceptible to wind damage; this risk can be reduced by having mixed stands. Bark wounding always results in timber decay. For high-quality timber, either pruning or initial high-density stands to encourage natural pruning at an early age are necessary.

Climatic conditions are important, with high temperature during the growing season favouring the development of reproductive buds. Dry conditions and moderate to high temperatures during the flowering period are necessary to obtain sufficient pollination, as male meiosis and pollen development occur after winter dormancy in late winter or early spring. Seed maturation requires high accumulated temperatures during the summer and early autumn. Such specific weather patterns during two successive years occur rather infrequently, particularly far north and in high-latitude regions, which may explain why seed crops are both rare and irregular in these areas.

P. abies can grow well under a wide range of site conditions or soil types but ideally requires low summer rainfall with plentiful soil water but not stagnant groundwater. It will tolerate low-nutrient soils.

Silvicultural practice

Both the use of planting stock and natural regeneration are widely employed. Seeds are dispersed mainly by wind but also by mammals and birds, with only a few dispersed over long distances. Regeneration will be largely influenced by the composition of the ground vegetation and will be hindered by rank vegetation or by a thick raw humus layer, particularly at high altitudes.

Vegetative propagation of selected seedlings is used, and mirrors the natural layering which can be important in alpine areas where the climatic conditions prohibit sexual reproduction, leading to scattered clonal groups of trees. Micropropagation using somatic embryogenesis can be successful in creating unlimited propagules from valuable material.

Planting stock is usually 2+1 or 2+2 transplants, because seedling growth is initially slow at first.

For high-quality veneer timber, low-density stands with pruning from a young age is practised. Site preparation is strongly recommended for sites which have a dense ground vegetation, and in the boreal zone with a cool climate to remove some or all of the accumulated undecomposed needle litter or raw humus.

Plants should not be overfertilized, and fertilizer treatments are only necessary on very poor soils. Where 'acid rain' is the cause of forest decline, attempts to control the problem have included both fertilizers and lime application.

Waterlogged sites hinder the development of the root system, thus increasing the risk of damage from windthrow, and on such sites drainage is absolutely necessary.

There is a large genetic variation in *P. abies* influenced by many factors, both natural and the result of man's efforts. The most pronounced patterns of genetic variation are related to population responses to climatic conditions, such as resistance to spring frosts or early autumn cold, and timing of bud burst, with the variability often being related to latitude and altitude of origin, or degree of continentality in the local climate. Differences in populations in the same geographic region can be large, reflecting the history of cultivation, and may mask patterns of variation of adaptive traits of a provenance. Traits which show the most variation between provenances are: time of onset of spring growth, termination of summer growth and development of frost-hardiness in late summer. Variation in bud flushing and the initiation of shoot growth are assumed to be under the control of both differential responses to accumulated temperature sums in the spring, and by conditions during acclimatization the previous year.

Resistance to late spring frost is closely related to the time that growth starts, with late-flushing provenances from eastern Europe suffering less damage than Nordic provenances during spring frost events in Scandinavia, with a similar but weaker relationship between the time of growth cessation in summer and resistance to autumn frost events.

Wood-quality traits, such as basic wood density, have been shown to vary considerably among different provenances, but large within-population variation is also reported and is the basis of tree breeding programmes.

Diseases and pests

The most important decay fungi affect standing trees, and are caused by *Armillaria mellea agg.* and *Heterobasidion annosum*, which can affect trees either through the roots or through stem wounds. Besides the economic losses due to such infections, trees weakened by these pathogens are more likely to suffer wind and snow damage.

Deer and boar can cause severe damage by bark peeling, which will allow the entry of decay fungi.

Replanted sites following clear felling may be affected by bark beetles such as *Hylobius abietis*, with thicket-aged stands liable to attack by *Pityogenes chalcographus*. Felled timber stacked in the forest can be attacked by *Trypodendron lineatum*. The most serious pests are those which tend to have large fluctuations in population, as a mass population explosion can cause severe damage, such as *Lymantria dispar* and *Ips typographus*. Nursery seedlings may be affected by sap-sucking arthropods.

Uses

In montane regions, *P. abies* is planted at low densities for protection forests and erosion control, and such forests provide considerable recreation value. Norway spruce is used for Christmas trees and for supplying decorative foliage.

It is economically the most important softwood species in Central Europe and is used as an industrial timber, for construction timbers and as pulpwood. A special quality is the occurrence of resonance wood, used for musical instruments. The timber is soft and of medium density, between 350 and 500 kg/m^3 (12% moisture content). Shrinkage is small (13–14%) and timber with regular rings exhibits very little deformation on drying. The wood is not durable, and unsuitable for uses where durability, decay-resistance and toughness are required.

Resins, oils, tannins (from bark) and honey are non-wood products.

Further reading

Niklov, N. and Helmisaari, H. (1992) Silvics of the circumpolar boreal forest tree species. In: Shugart, H.H., Leemans, R. and Bonam, G.B. (eds) *A Systems Analysis of the Global Boreal Forest*. Cambridge University Press, Cambridge, UK, pp. 13–84.

OECD (1997) *Draft Consensus Document on the Biology of Norway Spruce (*Picea abies *L.)*. Series on Harmonization of Regulatory Oversight in Biotechnology, No. 12. OECD Paper No. ENV/MC/CHEM/BIO (97)1, OECD, Paris, France.

Schmidt-Vogt, H., Vogellehner, D., Kral, F. and Gi, J. (1977) *The Spruce (*Picea*). A Manual in Two Volumes. Vol. 1. Taxonomy, Geographical Distribution, Morphology, Ecology, and Forest Communities.* (Die Fichte. Ein Handbuch in zwei Bandern. Band I. Taxonomie, Verbreitung, Morphologie, Okologie, Waldgesellschaften), Waldbau-Institut, University of Freiburg, German Federal Republic, 647 pp.

Schulze, E.-D., Lange, O.L., Oren, R. and Asche, R. (1989) *Forest Decline and Air Pollution: A Study of Spruce (*Picea abies*) on Acid Soils (Ecological Studies).* Springer-Verlag, Berlin-Heidelburg, Germany, 475 pp.

Picea engelmannii Parry ex Engelm.

Engelmann spruce

Importance

Picea engelmannii is a subalpine coniferous tree species indigenous to western North America. It is common in old-growth and second-growth, high elevation continental forest, where it is one of the largest trees. Its wood, valued more than that of its major associate, *Abies lasiocarpa*, is used mainly for lumber and pulp.

Botanical features

P. engelmannii is a large-sized tree (dwarfed, prostrate or krummholz in subalpine parkland), occasionally over 55 m tall, with a dbh of 200 cm. The stem is straight and distinct to the top of the tree, with the leading shoot upright. The crown is dense, symmetrical, narrow and spire-like. The lower branches are often sloping downward; secondary branches may hang vertically. The bark is thin, broken into large, loose, coarse, rounded, brownish scales; newly exposed bark is silvery-white and resinous. Twigs are greyish to light brown and hairy.

P. engelmannii has a shallow root system, with most roots within 50 cm of the ground surface. In deep, well-drained soils, the lateral roots may be longer than 200 cm. Roots are associated with both ecto- and endomycorrhizae.

Needles are evergreen, slightly curved, flexible, 15–25 mm long, four-sided in cross-section, bluish-green but often coated with whitish powder, and aromatic when crushed, the pungent odour being strongly characteristic of the species.

P. engelmannii is monoecious; male and female strobili open from terminal or lateral buds of the previous year's shoots after dormancy is broken, usually in late April to early May. Ovulate strobili are usually borne near the ends of shoots in the upper crown, and staminate strobili on branchlets in the lower crown. Male flowers ripen and pollen is wind disseminated in late May and early June at low elevations, and from mid-June to early July at high elevations.

Seed cones are cylindrical to narrowly ovoid, 3–7.5 cm long and shiny. New cones mature in one season and ripen in August or early September, shedding the seed up to 250 m from the parent tree. Cone scales are finely toothed, flexible, tapered at both ends, loosely fitting and with prominent bracts. The cones may fall during the following winter or may remain attached to the tree for some time. Seeds are winged, small and light, averaging about 297,000/kg.

Initiation of phenological events varies greatly with latitude, longitude, altitude, aspect and weather.

Distribution

P. engelmannii is a western North American species which is distributed predominantly in the subalpine or Cordilleran forest region. Its range extends in Canada from northern British Columbia and Alberta, south through all western states of the USA to New Mexico and Arizona. It is a major component of the high elevation Cordilleran forest.

P. engelmannii grows in post-fire even-aged or uneven-aged old-growth, predominantly mixed-species stands, typically with *A. lasiocarpa*, occasionally in pure stands. It is present in early to late stages of secondary succession on a wide range of sites, and forms a major component in the Engelmann Spruce–Subalpine Fir zone.

The major associates of *P. engelmannii* are *Abies lasiocarpa*, *Picea pungens*, *Picea glauca*, *Pinus albicaulis*, *Pinus contorta* and *Tsuga mertensiana*.

Picea engelmannii was introduced to Europe in England in 1862 and in Germany in 1891 when it was included into a network of experimental plantations by the German Forest Research Institute. Currently, it is grown only in arboreta and parks. Provenance and species trials have been established in Norway and New Zealand.

Silviculture characteristics

Reproduction capacity is intermediate; *P. engelmannii* is a moderate seed producer, with cone production beginning at age 15–40 years. Seed dissemination capacity is intermediate.

Potential for natural regeneration in low light is low when compared to *A. lasiocarpa*; regeneration develops in canopy gaps. Potential for natural regeneration in the open is high, given the presence of exposed mineral soil or burnt forest floors. Partial shade is beneficial for seedling establishment on high elevation sites.

Potential initial growth rate at 5 years or younger is low, being usually <10 cm/year when planted, and depends on snowpack.

Response of advance regeneration to release is intermediate. Self-pruning capacity in dense stands is intermediate. Crown spatial requirements are low. Light conditions beneath closed-canopy mature stands are intermediate.

Potential productivity is high, decreasing with increasing elevation; site index on intermediate sites decreases from 30 m on lower subalpine sites to <10 m on upper subalpine sites.

Longevity is high, with trees >600 years old not uncommon. The oldest recorded specimen was at least 852 years old.

Trees are moderately tolerant of shade and snowpack, but susceptible to windthrow, especially after any initial cutting in old-growth forests.

Silviculture practice

Depending on site and management objectives, clear-cutting, patch-cutting, strip shelterwood, and group-selection systems are viable for the establishment and growth of *P. engelmannii*. It can be grown in single- or mixed-species stands, preferably with *A. lasiocarpa*.

P. engelmannii can reproduce by layering, most often near the timberline, where the species assumes a dwarf or prostrate form. However, propagation by seed is the only contemporary method of regeneration. Establishment is by natural regeneration and/or planting containerized stock. *P. engelmannii* is a moderate to good seed producer, with seed production becoming significant when trees are larger and 150–250 years old. Good to bumper seed crops are generally borne every 2–5 years, with some seed produced almost every year. Most seed is shed by the end of October and nearly all is disseminated by the wind over long distances. Viable seeds that survive over winter normally germinate following snowmelt when seedbeds are moist and air temperature is at least 7°C. In undisturbed forest, seeds germinate and seedlings become established on forest floor, decaying wood or mounds of mineral soil upturned by windthrown trees. Germination and initial establishment are generally better on mineral soil and forest floor seedbeds; however, initial survival on severe, high elevation sites is higher on forest floor than on mineral soil seedbeds.

Because of its slow initial root penetration and extreme sensitivity to heat at the succulent stage, drought and heat kill many first-year *P. engelmannii* seedlings. Drought losses can continue to be significant during the first 5 years. In addition, newly germinated seedlings are very susceptible to early autumn frosts.

Understorey vegetation can be either beneficial or a serious constraint to establishment; seedlings become established more readily on sites protected by willows (*Salix*) or fireweed (*Epilobium angustifolium*) than in the open, because these plants compete less aggressively for available soil moisture than grasses and herbs.

Early shoot and root growth at high elevations is slow. After 10 years, trees grow faster, averaging about 1.2–1.5 m in height in about 20 years in full sun or light overstorey shade, and in about 40 years under moderate overstorey shade. Severe suppression of seedlings does occur at low light levels. It is not uncommon to find trees 80–120 years old only 1.0–1.5 m tall under the heavy shade of a closed forest canopy.

P. engelmannii stand diameter varies from 38 to 76 cm and dominant height from 14 to 40 m, depending on site quality and density. Trees have the capacity to grow well at advanced ages, growing steadily in diameter for 300 years, long after the growth of most associated tree species has slowed down.

Diseases and pests

The most common diseases of *P. engelmannii* are caused by wood-rotting fungi that result in loss of volume and predispose trees to windthrow and windbreak. These include *Phellinus nigrolimitatus*, *Flammula alnicola*, *Polyporus tomentosus*, *Gloeocystidiellum radiosum* and *Coniophora puteana*. Trunk rots are associated with *Porodaedalea pini*, *Stereum sanguinolentum*, *Echinodontium sulcatum*, and *Amylostereum chailletii*. Spruce broom rust (*Chrysomyxa arctostaphyli*) is also common, causing bole deformation, loss of volume and spike tops, and increases susceptibility to windbreak and infection by decay fungi.

The spruce beetle (*Dendroctonus rufipennis*) is the most serious insect pest of *P. engelmannii*. It is restricted largely to mature and overmature trees. The western spruce budworm (*Choristoneura occidentalis*) is another potentially dangerous insect attacking both *Picea engelmannii* and *Abies lasiocarpa*.

Dwarf mistletoe (*Arceuthobium microcarpum*) can cause tree mortality.

Uses

P. engelmannii–A. lasiocarpa forests occupy the greatest water-yielding areas in the Rocky Mountains. They also provide timber, habitats for a wide range of game and non-game wildlife, forage for livestock and are used for recreation. *P. engelmannii* is used to a limited extent for long-term stabilization of mine spoil, and is planted in snowbreaks and windbreaks.

The wood is straight grained, moderately strong, and low in resin. The heartwood is light, soft, resilient and nearly white with a tinge of red, whereas the sapwood is often not distinct from the heartwood. At 12% moisture content, the wood has a density of about 370 kg/m^3. Due to the presence of numerous knots, only small amounts of select lumber grades are produced. In the past, *P. engelmannii* was used principally for mine timbers, railroad ties and poles. Much of the timber is used in home construction where great strength is not required, and for prefabricated wood products. In recent years, rotary-cut veneer has been used in plywood manufacture. Other uses include speciality items such as violins, pianos and aircraft parts.

The pulping properties of *P. engelmannii* are excellent. Long fibres, light colour and absence of resins permit trees to be pulped readily by the sulphite, sulphate or groundwood processes.

Further reading

Alexander, R.R. and Shepperd, W.D. (1984) *Silvical Characteristics of Engelmann Spruce*. USDA - Forest Service, Rocky Mountain Forest and Range Experiment Station General Technical Report RM–114, Fort Collins, Colorado, USA.

Grossnickle, S. (2000) *Ecophysiology of Northern Spruce Species: the Performance of Planted Seedlings*. Canadian Science Publishing, Ottawa, Canada.

Rehfeldt, G.E. (2004) *Interspecific and Intraspecific Variation in Picea engelmannii and its Congeneric Cohorts: Biosystematics, Genecology, and Climate Change*. USDA - Forest Service, Rocky Mountain Research Station General Technical Report RMRS-GTR-134, Fort Collins, Colorado, USA.

***Picea glauca* (Moench) Voss**

White spruce

Importance

Picea glauca is one of the most economically important tree species in North America, especially in Canada where it is found in all forested regions except the Pacific Coastal belt. It is used primarily for wood pulp and general-purpose lumber.

Botanical features

It is a medium-sized tree, generally 25–35 m tall with boles of 60–100 cm. The crown is broadly conical, ragged, irregular and densely foliated, but spire-like in northern parts of the range. The root system is shallow, with many tough, pliable, wide-spreading branch roots. The bark is grey-brown, smooth, becoming darker grey and scaly with age.

The leaves are blue-green with stomata on all four sides, stiff, 15–22 mm with a pointed or blunt tip and unpleasantly pungent when crushed. Branchlets are shiny and glabrous, light greenish-grey, often tinged with orange or purple. The buds are ovoid, 3–6 mm, non-resinous, and orange-brown.

The development of the reproductive structures follows a 2-year cycle typical of most conifers in the northern hemisphere. Bud scales are initiated at the terminal apex and at newly initiated axillary apices within the enlarging vegetative buds, from about late April. Apices differentiate as vegetative, pollen cone, or seed cone buds around mid-July, as shoot elongation stops. Pollen cone bud development is complete by early October when they become dormant.

Reproductive and vegetative buds break dormancy at about the same time, in response to photoperiod, while subsequent development is regulated by temperature. Flushing of reproductive buds precedes that of vegetative buds, and pollen is released over a 1-week period. Fertilization occurs 3–4 weeks later, and embryo development is completed in early to late August.

The cones are slender cylindrical, 2.5–6 cm long, blunt tipped, and stalkless with the seed scales: fan-shaped and broadest near rounded apex, 10–16 × 9–13 mm. The seeds are 2–4 mm with a wing of 4–8 mm long. The mature cones open from August onwards as they lose moisture, reclosing during wet periods.

Distribution

The natural range of *P. glauca* extends from the Atlantic coast to within 100 km of the Pacific Ocean, and from the northern tree limit across North America south into northern New England, New York, and the Lake States, with outlying populations occurring south to the Black Hills in Wyoming and South Dakota. It is found at altitudes from sea level to 1520 m.

Hybridization where sympatric with Englemann spruce (*Picea englemannii*) in western Canada, Montana and Wyoming gives *P. glauca* var. *albertiana* and with Sitka spruce (*Picea sitchensis*) in north-western British Columbia and Alaska, gives *Picea* × *lutzii*.

P. glauca has been planted on a small scale or tested throughout much of the cool northern temperate zone, although it is not an important reforestation species outside of its natural range. It has been planted in many towns and cities outside its natural range in North America.

Silvicultural characteristics

P. glauca is classified as intermediate to tolerant of shade and may persist as a suppressed understorey component. Establishment is facilitated by disturbance and thus the age distribution in natural stands is not continuous. *P. glauca* grows on a wide range of soils and site conditions but with podzols the most common. Although *P. glauca* can occupy extremely harsh site conditions, it is generally regarded as more demanding than associated conifers, requiring higher moisture and fertility to achieve best development on moderately well-drained soils. Soil pH values range from (4.0) 4.7–7.0(–8.0).

Mean daily temperatures for January throughout much of the species range are ca −29°C with those in July ca 13°C, with precipitation of only 250 mm and a growing season of only about 60 days, and may be as short as 20 days. At its southern limit it roughly follows the 18°C July isotherm (with maximum summer temperatures to 43°C recorded in Manitoba) and mean annual precipitation up to 1270 mm in Nova Scotia and Newfoundland.

Silvicultural practice

P. glauca seeds exhibit varying degrees of dormancy that may be broken by cold stratification. At its northern limit vegetative reproduction by layering is common as the climate limits seed production. Most planting stock is now produced from seed in containerized systems but cuttings taken from seedlings can be rooted.

Natural regeneration of *P. glauca* can be difficult to predict and is not easily established under most harvesting systems. As a shade-tolerant species, *P. glauca* is able to regenerate under mature stands of spruce and early successional species, but advance regeneration stocking is often poor. Freshly disturbed areas with exposed mineral soils offer the best conditions for germination and establishment. Survival of advance regeneration following logging on thick moss sites is often poor, as seedlings are replaced by more aggressive species. A prolonged period of very little growth ('planting check') may following planting, possibly due to nutrient stress, but allelopathic effects of *Cladonia* lichens may inhibit the establishment of regeneration. Direct sowing has given erratic results.

As *P. glauca* is adapted to a wide range of sites, yields and rotation lengths in natural stands vary widely. Mean annual increments of 6.3–7.0 m^3/ha have been attained on the best loam soils with maximum MAI at about age 55, while maximum volume was achieved when basal areas were maintained at 23–32 m^3/ha.

P. glauca has been used in provenance research with interesting results. Although local seed sources are generally recommended, some provenance transfers have been demonstrated to be particularly promising. Within-stand variation can be as important as population origin. Heritability estimates on one long-term trial were much higher when only the crop trees were considered, as would be the case during selection in older stands. Pest resistance, such as to white pine weevil (*Pissodes strobi*) varies substantially among individuals within a population. Strong inbreeding depression has been reported.

Breeding programmes are now well established throughout the planted range, with most seed orchards established by grafting cuttings from plus-trees. These include using the introgressive hybridization with Englemann spruce, and with Sitka spruce (often in the hope of imparting the resistance of *P. glauca* to the white pine weevil. The degree of cold hardiness of the hybrid *P.* × *lutzii* is related to the proportion of *P. glauca* germplasm.

Diseases and pests

Spruce cone rust (*Chrysomyxa pirolata*) can cause abnormal development of the cones, reduced seed production and decreased viability of seeds.

Damping off fungi (species of *Fusarium*, *Pythium*, *Rhizoctonia*, *Phytophthora* and *Cylindrocladium*) commonly affect emerging seedlings, particularly in nurseries. Young seedlings may also suffer from Sirococcus blight (*Sirococcus strobilinus*) and infestation by the nematode *Xiphinema bakeri*.

Spruce needle rust (*Chrysomyxa ledi* and *Chrysomyxa ledicola*) is common wherever the alternate host, Labrador-tea (*Ledum* (or *Rhododendron*) *groenlandicum* and associated species), is found, but extensive damage from the fungus is rare. Spruce broom rust is common, causing abnormal proliferation of shoots to form witches' brooms, but rarely causes death. Witches' brooms on *P. glauca* are sometimes caused by eastern dwarf mistletoe (*Arceuthobium pusillum*), although black spruce is more susceptible to this parasite.

Cytospora canker (*Valsa* sp.) can be a damaging factor in plantations. Massive tumour-like growths are commonly observed on stems and branches, particularly near coastal areas, but their aetiology is not known.

Root rot results in significant losses. *Armillaria ostoyae* and *Inonotus tomentosus* are both capable of outright mortality, and some 10–15 other fungi cause root and butt decay that result in cull and contribute to windfall. Root rots are most prevalent in off-site plantations, such as light-textured, dry or impervious soils.

A great number of insects are a natural component of *P. glauca* forest types but few are responsible for large losses. The eastern spruce budworm (*Choristoneura fumiferana*) is the most destructive, and massive epidemics of this defoliator occur periodically resulting in heavy mortality and loss of growth, particularly where *P. glauca* is associated with balsam fir. Several other defoliators cause damage or weaken trees on a smaller scale, including the yellowheaded spruce sawfly (*Pikonema alaskensis*), European spruce sawfly (*Diprion hercyniae*), needleminers, needleworms, loopers, tussock moths and the spruce harlequin. Other groups of insects attack buds and shoots, including gall-forming adelgids (*Adelges* spp.), spruce bud moths (*Zeiraphera* spp.) and the white pine weevil (*P. strobi*).

Several species of bark beetles feed and breed in galleries between the bark and wood. The spruce beetle (*Dendroctonus rufipennis*) may attack trees of normal vigour, particularly those that are of large diameter and slow-growing, and has killed large areas. Generally though, bark beetles are considered secondary pests, attacking trees weakened by other means such as budworm epidemics, and may be thought to be beneficial in that their feeding hastens the return of wood to the humus.

Warren's collar weevil (*Hylobius warreni*) causes significant damage in scattered areas, girdling smaller trees and making larger trees susceptible to root rots such as *I. tomentosus*. The strawberry root weevil (*Otiorhynchus ovatus*) and the root-collar weevil (*Hylobius congener*) can cause significant mortality to seedlings.

Many insect species inhabit or feed on spruce cones and seed as part of their life cycle, and significant losses can be caused by the white spruce cone maggot (*Strobilomyia neanthracina*), the spruce seed moth (*Cydia strobilella*), and to a lesser extent by the spruce budworm and the spruce coneworm (*Dioryctria reniculelloides*).

Uses

P. glauca is an important component of northern forests, maintaining soil stability, watershed values, and recreational opportunities. *P. glauca* stands are a source of cover and food for wildlife, and the species is planted as an ornamental and used in shelter belts.

P. glauca is a primary species for production of lumber and wood pulp from the boreal forest. The wood is light (with a density of 330 kg/m^3), soft, resilient, straight grained and white, with little contrast between sapwood and heartwood.

While of minor importance today, in times past the bark was used to cover summer dwellings, the roots for lashing canoes and baskets, the boughs for bedding and pitch and extracts used for medicinal purposes.

Further reading

Beaulieu, J. (1996) *Breeding Program And Strategy for White Spruce in Quebec*. Information Report - Laurentian Forestry Centre, Quebec Region, Canadian Forest Service, No. LAU-X-117E, Sainte-Foy, Quebec, Canada, v & 25 pp.

Li, P., Beaulieu, J., Corriveau, A. and Bousquet, J. (1993) Genetic variation in juvenile growth and phenology of white spruce provenance-progeny test. *Silvae Genetica* 42, 52–60.

Magnussen, S. (1993) Growth differentiation in white spruce crop tree progenies. *Silvae Genetica* 42, 258–266.

Nienstaedt, H. and Zasada, J.C. (1990) *Picea glauca* (Moench) Voss. White spruce. In: Burns, R.M. and Honkala, B.H. (eds) *Silvics of North America. Vol. 1, Conifers.* Agriculture Handbook 654. Forest Service, USDA, Washington, DC, USA.

Park, Y.-S., Weng, Y.-H. and Mansfield, S.D. (2012) Genetic effects on wood quality traits of plantation-grown white spruce (*Picea glauca*) and their relationships with growth. *Tree Genetics and Genomes* 8, 303–311.

Picea obovata Ledebour

Siberian spruce

Importance

Picea obovata is an important tree species in the Siberian forests and in the Altai Mountain area of Xinjiang, China, covering many millions of acres. It forms a large tree with a straight trunk and has a high-quality wood.

Botanical features

At maturity *P. obovata* is ca 30–35 m tall and has a bole diameter of 40–60 cm. The crown is columnar, or conical. The bark is grey or dark grey, finely flaky but becoming irregularly scaly in old trees. The branches are slightly pendulous. The twigs are pale brown or a buff-yellow in colour, thin and covered with short, usually glandular hairs. The winter buds are conical, somewhat resinous, with the basal scales shorter and somewhat cuspidate.

The dull green leaves are 1.5–2.3 cm long with acute tips. In section they are rhombic or flat rhombic with 5–7 rows of stomata on the upper surface and 4–5 on the lower. They are set rather appressed on the upper side of the branch and directed forward, and somewhat parted below with some spreading downwards.

Flowering occurs in the spring (May) with the ovuliferous cones carried at the tips of the branches in the upper crown. The mature cones are pendulous, drooping, cylindric–ovoid, 5–11 cm × 2–3 cm in diameter. They are purple or dark purple with some green when young, yellow-green with some purple before maturity and brown at maturity. The seed scales are triangular obovate, rounded, and moderately wood; they are about 2 cm long and 1.7 cm wide. The scales are densely arranged and begin to open 2–3 weeks after maturity. The seed is black, 5 mm long (1.5 cm long with wing).

Growth occurs in the short summer months, mainly in a period of ca 6 weeks in June and July, with the seed ripening in September.

The spruce populations in north-eastern Europe and Scandinavia show characters of both *P. obovata* and *Picea abies* and may be the result of migration from a Russian refuge; plants from this supposed introgression have been named *Picea fennica*, or as *P. obovata* var. *fennica*. In *Flora Europaea*, *P. obovata* is treated as a subspecies of *P. abies* but this treatment is generally rejected on the basis of the considerable differences in cone and foliage characters between *P. obovata* and *P. abies* and the closer similarity of *P. obovata* to various Chinese species, such as *Picea wilsonii*.

Distribution

P. obovata is widely distributed in Eurasia from the Pacific coast of the Russian Far East, into Siberia, Mongolia, Xinjiang in China, Kazakhstan, and through to central and northern Russia where its range overlaps with that of *P. abies*. It is one of the important species in cold/temperate, coniferous forests. *P. obovata* var. *fennica*/*P. fennica* is found in northern Sweden and Finland. In China, *P. obovata* only occurs in the Altai Mountains in north Xinjiang, where its distribution gradually narrows from the mountainous area of Habahe and Buerjin in the north-west to the Qinghe mountainous area in the east. It occurs from sea level in the north, and from 1300 to 2200 m in the Altai range.

P. obovata may form small areas of pure forest but is generally mixed with Siberian larch (*Larix sibirica*) in the first forest storey, with *P. obovata* occupying the second storey on alluvial sites. In low-altitude forests, *P. obovata* can either be dominant or be mixed with Siberian fir (*Abies sibirica*) and Siberian pine (*Pinus sibirica*) where it often occurs in forests with a strong grass layer. In lower and damp valleys, it is often mixed with *Populus tremula* or *Betula pendula*.

Silvicultural characteristics

P. obovata is moderately shade-tolerant with the ability to adapt to environmental conditions. It mainly grows on mountainous brown coniferous-forest soil and grey forest soil, and not on strong podzols. The soil pH is acid to neutral.

In the Altai mountains in China *P. obovata* occurs in one of four climatic zones:

1. the Taiga Forest zone in the north-west, which has a long, cold winter. There is a period of 5–7 months where the average temperature is below 0°C with the average temperature in January below −20°C. There is a short, cool summer where the average temperature is 12–18°C in July, giving a growing season of ca 100–120 days with over 600 mm of annual precipitation;
2. the sub-moist temperate climate in the middle mountainous area, which is dominated by grey forest soil. The annual average temperature is above 0°C but the extreme maxima is 26°C and extreme minima of −7°C, with annual precipitation of 400–600 mm;
3. the grassland zone, which has average temperatures of about −10°C in January, and 23–25°C in July, with annual precipitation 200–300 mm;
4. the foothills of the Altai mountains, where the annual average temperature is about 3°C, with average temperature in January is −20°C and in July is 21°C, although extremes of −45°C and 38°C are recorded, with annual precipitation of 150 mm.

Silvicultural practice

P. obovata is mainly propagated by seed which is collected in September from trees with outstanding characters such as a good trunk, vigorous growth and freedom from diseases and pests.

P. obovata is a shade-tolerant and moisture-demanding species, and is best grown in a nursery with shade, and where either annual precipitation of 600 mm is available, or irrigation. Sowing density is dependent on seed quality and standard seedling quantity. Thousand-seed weight, germination rate, germination capacity and viability should be checked before sowing. Seeds can also be washed in water, screened and cleaned to obtain a purity of over 90%. Pretreatment is necessary to kill bacteria and insects on the seed surface; it is recommended that seed should be dipped in either 0.3% potassium permanganate or 0.5% ferrisulphate solution for 4 h until the seedcoat becomes black.

Careful bed preparation and use of fertilizer are necessary for raising vigorous seedlings. Drilling with row widths of 20–30 cm is used. The seeds are covered to their own depth with an even covering of sieved soil, and the seed bed shaded to reduce strong light and excessive evaporation, and to protect from birds. Watering is required depending on weather and soil conditions.

P. obovata is commonly planted either in spring or summer, normally at a spacing of 2 × 2 m. Seedlings are transplanted when 2–3 years old. *P. obovata* can also be vegetatively propagated, most commonly by rooted cuttings from seedling trees. Good rooting can be achieved when cuttings are grown on coarse sand with a daily average relative humidity of 55–90% from May to July, when the monthly average temperature of soil 5 cm underground is 14–18°C. Grafted stocks are usually employed to establish clonal seed orchards. *P. obovata* grows slowly in its early stages and is generally less than 1 m tall at 10 years old. After that, its growth increment gradually increases and the rotation is usually 40–50 years on good-quality sites in the Altai region of China, where a stand volume of over 600 m^3/ha can be achieved, dropping to 400–600 m^3/ha on lower-quality sites.

Diseases and pests

Botrytis cinerea, *Sclerotinia sclerotiorum*, and *Truncatella angustata* can infect leaves and tender tips of spruce seedlings and saplings, especially in the nursery. *Fomitopsis pinicola* is recorded as decaying the timber. *Thekopsora areolata* is a rust fungus with cherry as the alternate host.

Recorded pests include beetles such as *Dendroctonus micans*, *Hylurgops palliatus*, *Ips duplicatus*, *Ips typographus*, *Tetropium castaneum*, and *Pityogenes chalcographus* and the weevil *Pissodes piniphilus*.

The green spruce aphid (*Elatobium abietinum*) has been reported, as have *Dendrolimus superans*, *Gilpinia hercyniae*, *Plemeliella abietina*, *Monochamus saltuarius*, *Monochamus sutor* and *Sphinx pinastri*. *Megastigmus strobilobius* attacks the seeds.

Uses

P. obovata plays an important role in watershed and environmental management in the Altai region of northern Xinjiang, China.

The wood has a density of 380 kg/m^3, with a fine grain, and is of good quality and toughness. It is traditionally used in construction, for machinery and implements and in wood processing.

Resin and tannin are the main non-timber products.

Further reading

Freiberg, I.A., Terekhov, G.G., Surnin, I.V. and Nikitin, A.M. (1986) Silvicultural requirements for growing *Picea obovata* plantations in the Central Urals. *Lesnoe Khozyaistvo* 1, 39–41.

Krussmann, G. (2009) *Manual of Cultivated Conifers.* Timber Press, Portland, Oregon, USA (transl. Epp, M.E. from the 1983 German edition; ed. Warda, H.-D.), 526 pp.

Popov, P.P. (1996) Hybrid spruce in northeastern Europe. *Lesovedenie* 2, 62–72.

Schmidt-Vogt, H., Vogellehner, D., Kral, F. and Gi, J. (1977) *The Spruce (Picea). A Manual in Two Volumes. Vol. 1. Taxonomy, Geographical Distribution, Morphology, Ecology, and Forest Communities.* (Die Fichte. Ein Handbuch in zwei Bandern. Band I. Taxonomie, Verbreitung, Morphologie, Okologie, Waldgesellschaften.), Waldbau-Institut, University of Freiburg, German Federal Republic, 647 pp.

Picea sitchensis (Bong.) Carrière

Sitka spruce

Importance

Picea sitchensis is an ecologically important species of the north-temperate coastal rainforest of western North America. It is a valuable species for both pulp and lumber. The wood offers unique qualities for use in many musical instruments, and has an outstanding strength-to-weight ratio. It is a keystone species in some of the most productive ecosystems of North America and has played an important role in plantation forestry in Northern Europe.

Botanical features

It makes a large tree, often 55 m with a 2 m bole diameter but at 95 m tall for the 'Carmanah Giant' on Vancouver Island. The crown is narrowly conical and rather open. The bark is thin, broken into large, loose, reddish-brown scales, with newly exposed bark a rusty-grey.

The rigid sharply pointed leaves are 1.5–3 cm long, flattened or broadly triangular in cross section and blue-green to light yellow-green with prominent stomatal bands. The branches are spreading with somewhat drooping branchlets. The twigs are stout, light grey to yellow-brown and glabrous. The buds are conical to dome-shaped, 5–10 mm and resinous.

The cones are broadly cylindrical, 5–10 cm long with elliptic to narrowly diamond-shaped flexible papery scales 1.5–2.2 × 1.2–1.6 cm which have an irregularly toothed wavy margin. The reddish-brown seeds are 2–3 mm with a wing 5–8 mm.

The development of the reproductive structures follows a 2-year cycle typical of most conifers in the northern hemisphere. Bud scales are initiated at the terminal apex and at newly initiated axillary apices within the enlarging vegetative buds, from about mid-April. Apices differentiate as vegetative, pollen cone, or seed cone buds around mid-July, as shoot elongation stops. Pollen cones typically develop from small axillary apices on vigorous distal shoots or terminal apices on less vigorous, proximal shoots. Seed cones usually develop on distal axillary positions on vigorous shoots or from smaller terminal apices on less vigorous shoots. Pollen is released over a 1-week period, in late April on Vancouver Island and by mid-May in Scotland. Fertilization occurs 4–5 weeks later, and embryo development is completed in mid-August.

The mature cones open as they lose moisture, re-closing during wet periods. Seed dispersal begins in the autumn, with >70% of the seeds dispersed within the first 6 weeks, 90% by February, and the remainder released over the next growing season.

Distribution

The natural range spans a narrow strip on the north Pacific coast of North America, extending for 2900 km from 61°N in south-central Alaska to 39°N in northern California with an isolated population in Mendocino County, California. Throughout this range, *P. sitchensis* is a coastal species rarely occuring more than a few kilometres from the coast or higher than 500 m. Outside its natural range, *P. sitchensis* has played an important role in plantation forestry, particularly in Northern Europe.

P. sitchensis occurs most commonly in mixed stands, usually with *Tsuga heterophylla*. *Alnus rubra* and *Populus trichocarpa* are associated throughout the range but other associates vary with latitude: with *Pseudotsuga menziesii*, *Chamaecyparis lawsoniana*, *Pinus monticola*, *Sequoia sempervirens* and *Acer macrophyllum* in the south; *Pinus contorta* subsp. *contorta* and *Thuja plicata* into south-east Alaska; and with *Cupressus* (*Chamaecyparis*) *nootkatensis*, *Tsuga mertensiana*, *Abies lasiocarpa* and *Alnus sinuata* in northern sites and at higher elevations in the south. In Alaska and the Skeena, Nass, and Bulkley river valleys of British Columbia, *P. sitchensis* is associated with *Picea glauca*, and hybrid populations are named *Picea* × *lutzii*.

Pure stands of *P. sitchensis* are common on tidewater areas that receive salt spray, and in early succession situations following disturbance.

P. sitchensis is a component of several ecological associations, characterized by available moisture and nutrient regimes. On the Queen Charlotte Islands, the *Tsuga heterophylla*–Sitka spruce forest is typically associated with *Vaccinium* and several mosses. On exposed coastal locations, salt tolerance limits the associates to *Calamagrostis nutkaensis*, *Gaultheria shallon*, and *Stokesiella oregana*. In sheltered alluvial sites it occurs with *T. heterophylla* and *A. rubra* in grass meadows.

On the drier inland valleys and eastern slopes of coastal mountains *P. sitchensis* is most likely to be sympatric with white spruce (*P. glauca*).

Silvicultural characteristics

With only limited shade tolerance, stands with its usual associate *T. heterophylla* tend towards *T. heterophylla* climax, although rarely to pure *T. heterophylla*, as *P. sitchensis* is physically large, long-lived and able to invade small openings resulting from windthrow.

P. sitchensis will produce epicormic shoots along the stem, originating from either dormant or adventitious buds, in response to light intensity. Thus, thinning of stands will stimulate epicormic branching, affecting the future quality of the trees.

Among abiotic factors, windthrow is the most seriously damaging agent, particularly in plantations of *P. sitchensis* established in the UK. Elsewhere in Europe on deeper soils it has proven more windfirm than conifers such as *Picea abies*. In North America, *P. sitchensis* is considered less wind-resistant than *P. menziesii* and *T. plicata*, but more so than *T. heterophylla* and *Abies amabilis*.

P. sitchensis is not a fire-resistant species. Autumn frost is a particular problem in the UK, where faster-growing southern provenances are particularly susceptible.

P. sitchensis grows on soils derived from a range of parent materials with best development on deep, moist, well-aerated soils, generally with pH values varying from 4.0 to 5.7.

The natural range is restricted to an area of maritime climate with abundant moisture throughout the year, relatively mild winters and cool summers.

Silvicultural practice

Outside its natural range, *P. sitchensis* has played an important role in plantation forestry, particularly in Northern Europe. In the UK, the species now accounts for almost 70% of the annual conifer planting stock and plantations cover over 20% of the forest/woodland area. *P. sitchensis* is also a primary plantation species in Brittany, northern France.

P. sitchensis seeds exhibit weak dormancy and both the rate and total amount of germination can be increased by cold stratification.

Planting stock can be produced by vegetative propagation, as juvenile cuttings are easily rooted and performance is comparable to conventional transplant stock. Although *P. sitchensis* planting stock has traditionally been produced in bare-root nurseries as 2+0 seedlings or 1+1 transplants, an increasing proportion is now produced in containerized growing systems.

P. sitchensis tends to be disturbance-dependent, regenerating naturally on landslides, newly exposed alluvial sites and openings created by windthrow. For natural regeneration there must be sufficient ground exposure and disturbance. Thinning in mixed stands with *T. heterophylla* will normally favour regeneration of the hemlock over spruce.

P. sitchensis is relatively nutrient demanding, particularly at young ages prior to crown closure. It is most productive on nitrogen-rich soils, but also requires relatively large amounts of calcium, magnesium and phosphorus; nitrogen and phosphorus have been found to be limiting on sites in both British Columbia and the UK.

P. sitchensis grows slowly for the first few years but much more rapidly thereafter. Very old *P. sitchensis* eventually assume a dominant position in old-growth hemlock–spruce stands. Stands tend to be dense, and yields are high.

Clinal variation patterns are expressed for phenological traits such as cessation of growth and is greater among than within provenances. Although variation among provenances is important in determining the risks and benefits of transferring seed sources, genetic improvement from mass selection relies primarily on variation within populations as the source of genetic gains. For productivity traits, 40% of the genetic variation for *P. sitchensis* is at the provenance level, while 60% is within provenance.

Perhaps the most precise estimates of genetic variance for *P. sitchensis* have been reported for a 7-tree diallel cross, planted at two test sites in Scotland and Wales. In this experiment, genetic variation for height growth, although under some additive genetic control, was predominantly controlled by non-additive effects, whereas diameter was only controlled by additive effects.

The ease of vegetative propagation of *P. sitchensis* offers opportunities to capture additional gains earlier in the breeding cycle. Clonal selection has been demonstrated to be highly effective for characters such as planting stock phenology, frost resistance, root-growth potential, early height and diameter growth and branching habit.

A provenance test in British Columbia demonstrated significant differences in susceptibility to white pine weevil, but unfortunately a study of variation within a resistant provenance found that taller families were more likely to be attacked.

P. sitchensis seedlots from British Columbia have been certified under the Organisation for European Co-operation and Development (OECD) scheme for sale in Europe. In the UK, southern provenances (below 47°N) grow fastest, but are susceptible to spring and autumn frosts, although clonal testing has demonstrated substantial variation in frost hardiness within provenances.

Breeding strategies for *P. sitchensis* generally utilize a system of progeny testing and recurrent selection for generation advancement, combined with clonal seed orchards for production of improved seed. Flowering of grafts can be stimulated by means of various cultural treatments, shortening breeding cycles. The vegetative multiplication of small quantities of seed by rooted cuttings optimizes the process.

P. sitchensis naturally crosses with *P. glauca* and can be hybridized with several other species.

Diseases and pests

Although many pathogens can attack *P. sitchensis* at various stages of its life cycle, only a small number of decay fungi cause serious commercial damage. *Heterobasidion annosum* causes butt rot and can result in significant loss of yield and quality. *Armillaria spp.* and *Phellinus weirii* are likely to kill young trees. *Rhizina undulata* is associated with fire, either as site preparation or following thinning, affecting both young seedlings and pole-sized trees.

The most serious insect pest of *P. sitchensis* in North America is the white pine weevil (*Pissodes strobi*). Adults overwinter in the litter, emerging in the spring to mate and lay eggs in feeding punctures made on the tree's leader. The larvae tunnel down the inner bark, killing the leader. Putative resistance is complex and appears to be genetically inherited. Maintaining partial shade by close spacing until the trees are at least one log-length in height will discourage attack and reduce the economic impact of the weevil on timber value.

The economic impact of other insect pests in North America is low although a number have the potential to do damage. In Great Britain, the diversity and impact of insect pests have been limited by the relatively short period since introduction. *P. sitchensis* has a low resistance to the sap-sucking green spruce aphid (*Elatobium abietinum*), owing to low concentrations of volatile plant compounds.

Several species of bark beetles feed and breed in galleries between the bark and wood. The spruce beetle (*Dendroctonus rufipennis*) in North America, and the great spruce bark beetle (*Dendroctonus micans*) in Europe will both attack trees of normal vigour, particularly those which are large in diameter and slow-growing.

Uses

P. sitchensis is an important component in maintaining soil stability, watershed values and recreational opportunities.

The wood of *P. sitchensis* is light, soft, resilient and relatively strong; the heartwood is light pinkish-brown with a gradual transition into a creamy-white sapwood, with a density of about 450 kg/m^3 at 12% moisture content. High strength-to-weight ratio and resonant qualities of clear sawn timber are attributes that have traditionally made *P. sitchensis* wood valuable for speciality uses.

Further reading

Fletcher, A.M. (1992) Breeding improved Sitka spruce. In: Rook, D.A. (ed.) *Super Sitka for the 90s*. Proceedings of a meeting held in Elgin, Scotland, 2–4 October 1990, organized by the UK Forestry Commission Research Division. Forestry Commission Bulletin 103. HMSO, London, pp. 11–24.

Lines, R. (1987) Seed origin variation in Sitka spruce. In: Henderson, D.M. and Faulkner, R. (eds) *Sitka Spruce, Proceedings of a Symposium*. Royal Botanic Garden, Edinburgh, 3–6 October 1986. *Proceedings of the Royal Society of Edinburgh, B Biological Sciences* 93, 25–39.

Mason, B. and Perks, M.P. (2011) Sitka spruce (*Picea sitchensis*) forests in Atlantic Europe: changes in forest management and possible consequences for carbon sequestration. *Scandinavian Journal of Forest Research* 26, 72–81.

Spittlehouse, D.L., Sieben, B.G. and Taylor, S.P. (1994) Spruce weevil hazard mapping based on climate and ground survey data. In: Alfaro, R.I., Kiss, G. and Fraser, R.G. (eds) *The White Pine Weevil: Biology, Damage and Management*. Proceedings of a symposium held 19–21 January 1994 in Richmond, British Columbia. FRDA Report No. 226. BC Ministry of Forests and Lands, Victoria, British Columbia, Canada.

Pinus banksiana Lamb.

Jack pine

Importance

Pinus banksiana is a fast-growing, extremely hardy, short-lived, fire-adapted species which is an important source of pulpwood, lumber and round timber in the northern forests of the USA and Canada. It grows further north than any other American pine and is the most widely distributed pine species in Canada. It is a shade-intolerant, early seral species invading areas where mineral soil has been exposed by major disturbances such as fires. It usually grows in pure, even-aged stands on less fertile and drier soils than those required by other native species in its range.

Botanical features

P. banksiana is a small-sized tree, occasionally over 25 m tall, with a dbh of 35 cm and longevity of 200 years. In open-grown trees, the trunk is tapered, the crown is conical and open, and the branches are ascending and arching. In forest-grown trees, the trunk is slender and straight with little taper and the crown is short. The bark is thin, reddish-brown to grey when young, becoming dark brown, flaky; with age furrowed into irregular thick plates. The twigs are slender, yellowish-green, becoming dark greyish-brown during the second season, ridged and grooved. The wood is moderately hard, heavy, weak and light brown.

P. banksiana develops a taproot as a seedling and maintains it to maturity. In deep, well-drained soils the roots may penetrate below 270 cm. The bulk of the roots consist of laterals confined largely to the upper 46 cm. Roots are associated with both ecto- and endomycorrhizae.

The needles of *P. banksiana* are evergreen, in bundles of two, 2–4 cm long, straight or slightly twisted, stiff, sharp-pointed, light yellowish-green and spread apart.

P. banksiana is a monoecious species. Ovulate cones are usually borne on primary and secondary branches in the upper tree crown and staminate cones are usually borne on tertiary branches lower in the crown. Seed cones are variable in shape, oblong to conical, asymmetrical, straight or curved inward, 3–7 cm long, yellowish-brown when mature, stalkless, usually pointing forward often in clusters of two or three at nodes; usually remaining closed and persistent on the tree for 10–20 years. Cones are opened abundantly by fire, this being the source of dense, even-aged stands of the tree in fire succession.

Initiation of phenological events varies greatly with latitude, longitude, and weather. Bud initiation for the following year's shoot systems begins in late June or early July. Fertilization occurs about 13 months after pollination. Cones mature and the seeds ripen late in the growing season of the year after pollination.

Distribution

P. banksiana is a North American species which is distributed predominantly in the Atlantic and Central regions and marginally in the Cordilleran region. Within its range, *P. banksiana* is widely but not continuously distributed. In Canada it is most abundant in Ontario, and in the USA, the largest areas are in Minnesota, Wisconsin and Michigan.

The major associates of *P. banksiana* in order of presence are: *Quercus ellipsoidalis*, *Quercus macrocarpa*, *Pinus resinosa*, *Populus grandidentata*, *Populus tremuloides*, *Betula papyrifera*, *Quercus rubra*, *Pinus strobus*, *Acer rubrum*, *Abies balsamea*, *Picea glauca*, *Picea mariana*, *Larix laricina* and *Populus balsamifera*.

Silviculture characteristics

P. banksiana has a high reproduction potential, with cone production commencing before 10 years of age. Seed dispersal is poor, typically <100 m from the parent tree. The cones are predominantly serotinous and open below 27°C. Under low light levels, the potential for natural regeneration is very low or practically nil, as *P. banksiana* is extremely shade-intolerant. Natural regeneration is favoured in the open, especially after wildfires.

Initial growth rates (<5 years) are high, about 5 cm in the first growing season and up to 90 cm in the fifth growing season. Advance regeneration does not develop in the absence of adequate light and seedbeds.

The potential productivity of *P. banksiana* is medium, with a site index (50 years at bh) of <20 m and timber yields close to 250 m^3 after 50 years on the most productive sites. Growth declines after about 80 years.

Silviculture practice

Depending on site and management objectives, clear-cutting and patch-cutting systems are viable for the establishment and growth of *P. banksiana*, which can be grown in single or mixed stands.

Propagation by seed is the only contemporary method used to regenerate *P. banksiana*. Natural regeneration (where applicable) and/or planting, using containerized stock, is also used to establish the species.

In naturally regenerated stands, *P. banksiana* typically begins to flower after 5–10 years under open-grown conditions, but not until later in closed stands. Seed production differs from year from year, but some seed is usually produced every year and total crop failures are rare. Seed yields per cone range from about 15 to 75.

Over much of its natural range *P. banksiana* bears predominantly serotinous cones, but in the southern part of its distribution cones are non-serotinous. Total seeds stored on the trees in serotinous cones can reach more than 14.6 kg or 4 million seeds/ha in well-stocked mature stands. As viability after 5–10 years may be significantly reduced, only cones 6 years old or less should be collected. Commercially cleaned seed numbers are between 156,600 and 551,000/kg.

P. banksiana exhibits epigeal germination. Under forest conditions, seedling growth is slow in the first 3 years, but increases rapidly beginning in the fourth and fifth years. Seedlings attain a height of about 5 cm in the first year, 15 cm in the second year, and 30–90 cm in the fourth year. If favourable moisture conditions prevail in late summer, *P. banksiana* frequently has a second period of shoot elongation and produces lammas and proleptic shoots.

Overstocked *P. banksiana* seedling and sapling stands with 4950 or more trees/ha should be weeded or cleaned (pre-commercial thinning) to improve growth and development, otherwise such stands may stagnate because natural thinning is slow except on the best sites.

In well-stocked stands, *P. banksiana* develops into a short to medium-tall, slender tree with a narrow, open crown covering 30–45% of the stem. Crown ratios from 10–20% are not uncommon in dense stands. Open-grown *P. banksiana* develops a stocky stem of poor form and a wide, spreading crown with persistent branches, often to the ground. Overstocked stands produce weak, spindly stems that are susceptible to breakage by wind, ice and snow.

During the first 20 years, *P. banksiana* is the fastest-growing conifer in its native range apart from *L. laricina* (tamarack). Growth is somewhat slower, but maintained longer, in Canada than in the Lake States of the USA. The annual height growth on medium sites in the Lake States averages from 33 cm at age 30 to 23 cm at age 50. At age 80 years, annual height growth is only 13 cm. On typical sandy plain sites, growth is about one site quality better where the water table is 1.2 –1.8 m from the soil surface than where it is deeper.

Normally, mature trees are about 17–20 m tall with a dbh of 20–25 cm. Stand basal areas seldom exceed 37 m^2/ha. *P. banksiana* stands begin to disintegrate after 80 years on the best sites and after 60 years on poor sites.

Diseases and pests

P. banksiana is susceptible to a number of rust fungi, with the pine-oak (eastern) gall rust (*Cronartium quercuum*) and pine-to-pine (western) gall rust (*Endocronartium harknessii*) being most prevalent. The major root-rot fungi include the shoestring fungus (*Armillaria mellea*) and annosum root rot (*Heterobasidion annosum*).

The species is subject to many damaging agents, with seedlings being affected by root borers (*Hylobius pales*,

Hylobius radicis, *Hylobius rhizophagus*, and *Hylobius warreni*), shoot and stem borers (*Pissodes approximatus*, *Pissodes strobi*, *Formica exsectoides*), leaf feeders (sawflies of the genus *Neodiprion*, *Diprion similis*, *Anomala oblivia*, *Tetralopha robustella*), needle miners (*Argyrotaenia tabulana*, *Exoteleia pinifoliella*, *Zelleria haimbachi*), root feeders (*Phyllophaga* spp.), sucking insects (*Cecidomyia reeksi*, *Cecidomyia piniinopis*), scale insects (*Nuculaspis californica*, *Chionaspis pinifoliae*, *Toumeyella parvicornis*) and spittlebugs (*Aphrophora parallela*, *Aphrophora saratogensis*).

Young trees are especially susceptible to early spring fires. Several sawflies, the pine tussock moth (*Parorgyia plagiata*) and the jack pine budworm (*Choristoneura pinus*) are the most important defoliating pests.

Uses

P. banksiana stands can be an important part of the visual resource for recreation areas; they stabilize watersheds, produce areas for blueberry picking and provide food and shelter for wild game species.

The wood has a density of 400 kg/m^3 and a whitish sapwood with conspicuous growth rings and light orange to light brown heartwood, which is late in forming (after 40–50 years). The grain is generally straight, but may be somewhat uneven, and the timber has a medium and uneven texture. The timber is reported to have little natural resistance to decay. It is used for framing and as pulpwood. As new developments such as structural particleboard become practical, the rapid juvenile growth of *P. banksiana* will be an advantage where gross cubic volume becomes important. The lumber is reported to be often graded with other spruce timber and marketed in the spruce–pine–fir group.

Further reading

Kashian, D.M., Corace III, R.G., Shartell, L.M., Donner, D.M. and Huber, P.W. (2012) Variability and persistence of post-fire biological legacies in jack pine-dominated ecosystems of northern Lower Michigan. *Forest Ecology and Management* 263, 148–158.

Leitch, M., Shahi, C. and Jackson, K. (2011) Impact of a regeneration method and vertical position on juvenile wood properties of jack pine in northwestern Ontario. *Wood and Fiber Science* 43, 32–40.

Rudolph, T.D. and Laidly, P.R. (1990) *Pinus banksiana* Lamb. In: Burns, R.M and Honkala, B.H. (technical coordinators) *Silvics of North America, Vol. 1*. USDA Agriculture Handbook 654, Washington, DC, USA, pp. 280–293.

Pinus brutia **Ten.**

Brutian pine, Calabrian pine

Importance

Pinus brutia is one of the most important low elevation Mediterranean conifers. Its ecological importance is mainly due to the fact that it is composed of subspecies that, on the whole, have a large ecological adaptability. Its tolerance to moderate frost, subhumid environments, high temperatures and drought make it suitable for reforestation in many regions, such as Italy, France, the former Yugoslavia and southern Australia. In many areas *P. brutia* var. *eldarica* is a better option than *Pinus halepensis* (the Aleppo pine) in terms of adaptability. Usually, *P. brutia* will present a straighter trunk and faster growth than *P. halepensis*.

Botanical features

P. brutia has a wide, spreading, open and rounded crown, growing taller and straighter than *P. halepensis*, reaching up to 25 m in height. It is not very long-lived (around 120 years). The trunk is usually straight, with spreading branches becoming erect near the top of the tree. Shoots are thick (4–5 mm) and yellow-reddish. The bark is thick, silvery-grey in young individuals and reddish-brown in old ones, regularly and longitudinally fissured. Buds are oblong, 10–20 mm long, and not resinous.

The needles are grouped in bundles of two; dark green (12–18 cm long × >1 mm wide), straight, and rigid. They have three layers of hypodermal cells and peripheral resin canals.

Cones are short, broad, slightly shiny, brown (6–9 × 5–7 cm), ovoid. They usually grow in a whorl of 3–4, spreading horizontally, sessile or on a very short stalk. Some cones have a curved top when closed; apophysis is convex; umbo is flat.

Seeds are 7–9 mm long, darkly mottled on the upper surface, lighter on the lower one. Trees begin producing seed at around 7 (up to 10) years of age. Ripening is biennial; 1 kg contains 16,700–25,000 seeds. The annual crop of seed is abundant, but cone production is lower than in *P. halepensis*.

Flowers appear between March and May. Cones ripen from January to March, but are serotinous and remain closed on the tree for many years.

Distribution

P. brutia occurs naturally in the eastern Mediterranean region: Greece (including Crete), Cyprus, Turkey, Syria, Lebanon, Iraq, Pakistan, Ukraine and southern Russia (Crimea and Georgia).

The species has been planted outside its natural range in Italy and Croatia, and is undergoing evaluation and breeding in South Australia. Seeds were imported from the Mediterranean Basin by Muslim pilgrims, leading to its naturalization in Iran, Pakistan and Tajikistan. In Italy, the species was imported during the Roman era and its presence in Calabria (Italy) was considered natural enough for Tenore to give it the specific name 'brutia'.

Studies on the genetic structure and pollen of pine populations appear to show that *P. brutia* var. *eldarica* probably originated from a single mountain called Eliar-Ugi in Azerbaijan and was transported along ancient trade routes used during the days of Alexander the Great.

Silviculture characteristics

P. brutia requires full illumination and is less tolerant to drought than *P. halepensis*, but is less specific with regard to soil requirements and can tolerate colder environments. The provenance of the seed is crucial, due to the considerable genetic variation between its botanical varieties. Self-pruning capacity is low (the same as in *P. halepensis*).

P. brutia has a medium tolerance of fire, similar to that of *P. halepensis*. Adaptations of *P. brutia* to fire include thick bark, seed production at a young age, a retention period of the seeds in the cones over several years (during which time the seeds retain their viability) and opening of the cones at high temperatures. It frequently represents the paraclimax species in humid and subhumid climate areas that are frequently burned.

Silviculture practice

Seeds can be stored for 5–6 years at 4°C at a moisture content of 7–8%. There is great variation in the viability of seeds and growing characteristics of seedlings are related to provenance. Stratification or pretreatment of seeds before sowing is not essential, but may promote more rapid and uniform germination. *P. brutia* is characterized by low levels of germination in the dark, with an optimal temperature range of 10–20°C, as is *P. halepensis*. In addition, its germination is photosensitive (white light promoted and far-red inhibited).

P. brutia is extremely easy to raise in nurseries. Seeds are sown at a depth of 1–2 cm at a spacing of 0.2 m before the end of March. Suitable seedling densities can be obtained by using about 25 g seed/m^2. Mulching with manure and sand is often used. Irrigation is necessary during germination and the initial growing stages. Weeding should be carried out every 3–5 weeks. Roots can be pruned at 18–20 cm after about 6 months if necessary.

As damping off is a frequent occurrence in this species, care should be taken not to seed too densely and to use soil with a pH between 7 and 8.

Both container and naked root seedlings are used to establish plantations. Plantation activities are performed using both mechanized and manual plantation methods. Typical spacing on well-prepared soils is 3 × 3 m. If terraces are used, seedlings are planted at 2 m intervals on terraces that are arranged at 3 m intervals. Planting space does not have significant effects on tree height, but diameter is enhanced significantly with increasing spacing up to 6 m^2 of growing area.

Removal and control of competing weeds and woody plants is normally carried out for 3–5 years after the establishment of a plantation. At these times, replacement plantings are also carried out. It is also normal practice to fence *P. brutia* plantations in order to reduce the damage from grazing of livestock and wildlife.

Uniform shelterwood cutting methods over the entire stand have proved to be the most effective in ensuring natural regeneration. Other methods applied include strip shelterwood cutting and strip clear-cutting. Heavy thinnings should be carried out at 15-year intervals and removal of lower branches should be carried out every 20–25 years to reduce fire hazard. In Israel, light thinning treatments (removal of 15%) gave much higher volume production and therefore larger volume removal and better financial return (more sawn timber for pallets) than removing 40–50% of the standing volume.

The age class management system appears to be the most suitable for *P. brutia*. Rotation ages are 50 years for good sites and 60 years for poorer ones. Shorter rotation ages (30–40 years) are adopted for specific products such as mine props. Maximum mean annual increments are between 5.5 and 7.5 m^3/ha, but on exceptional sites, this can reach 10–12 m^3/ha.

Diseases and pests

In good edaphic conditions, *P. brutia* is relatively free of pests and diseases.

A complex of fungi was isolated from *P. brutia* trees exhibiting needle blight and die-back in Syria. *Alternaria alternata*, *Ascochyta* spp., *Aureobasidium pullulans*, *Leptostroma pinastri*, *Lophodermium pinastri*, *Mycosphaerella* sp., *Phoma glomerata* and *Septoria* sp. were associated with needle yellowing and blight. *Coniothyrium pini*, *Pleospora* sp. and *Diplodia pinea* were associated with die-back.

The scolytid *Orthotomicus erosus* feeds on the stem, bark and cambium and can cause serious damage in plantations on poor sites. It can be controlled by using trap trees or chemical traps. *Thaumetopoea pityocampa* (pine processionary moth) causes severe defoliation and can reduce annual increments by up to 60%. Infestations are cyclic and are often linked to severe climatic conditions. Nests can be removed manually and burned. Preventive applications of *Bacillus thuringiensis* preparations can cause a reduction in egg hatch rate and increased larval mortality. Among the most frequent abiotic causes of damage are fire and grazing. The impact of both of these can be greatly reduced by using appropriate management practices such as employing firebreaks, thinning, removal of undergrowth and fencing.

Uses

P. brutia can be used as an ornamental species and as a Christmas tree.

The wood of *P. brutia* is of medium density, ranging from 468 to 530 kg/m^3, with a high resistance to compression and a medium resistance to flexing. Its production of fibre is medium to high and is best suited for kraft pulping. The timber is used to for fencing posts, telephone and transmission poles, railway sleepers, sawlogs for box making, general construction and joinery. The wood is also suitable for the production of pulp and for the manufacture of fibreboard. If *P. brutia* is not pruned at an early age, knots may appear.

P. brutia can produce significant amounts of resin for turpentine production. The resin of *P. brutia* has a high acid number and low percentage of unsaponified matter, which are indicators of good quality. Resin tapping does not have adverse effects on wood properties or seed yield.

Further reading

Gezer, A. (1987) Silviculture of the Brutian pine (*Pinus brutia* Henry) in Turkey. *Doga, Tarim ve Ormancilik* 11, 139–151.

Kurt, Y., González-Martínez, S.C., Alía, R. and Isik, K. (2012) Genetic differentiation in *Pinus brutia* Ten. using molecular markers and quantitative traits: the role of altitude. *Annals of Forest Science* 69, 345–351.

Shater, Z., Miguel, S. de, Kraid, B., Pukkala, T. and Palahí, M. (2011) A growth and yield model for even-aged *Pinus brutia* Ten. stands in Syria. *Annals of Forest Science* 68, 149–157.

Pinus caribaea Morelet

Caribbean pine

Importance

Pinus caribaea, especially *P. caribaea* var. *hondurensis*, is widely planted throughout the American, Asian and African tropics and subtropics.

Worldwide interest has been arising in the use of fast-growing selection populations and provenances of *P. caribaea* var. *hondurensis*. It has been found that *P. caribaea* var. *hondurensis* crosses naturally with *Pinus oocarpa* var. *ochoterenai* and their progeny have shown better tree formation and faster growth than the parents. *Pinus elliottii* crossed with *P. caribaea* var. *hondurensis* has made a major contribution to the success of commercial forestry in Queensland, Australia. These successes have also stimulated interest in hybrids in other countries such as South Africa, Brazil and Fiji.

Botanical features

A tree of moderately large size, *P. caribaea* is seldom higher than 30 m with a diameter from 25 to 75 cm on sites of average quality. Under particularly favourable conditions the diameter can swell to 135 cm and the tree can grow to be as tall as 45 m. It forms a straight, clean bole sometimes free of branches for the first 12 m when mature.

Lower branches are horizontal to drooping. Upper branches are often ascending. At maturity *P. caribaea* has a thin, rounded, slightly spreading crown while younger trees have more dense crowns.

The bark on old trees is thick, reddish-brown, and forms in rough plates with deep vertical and horizontal fissures. Younger trees have a reddish bark.

Buds are composed of many linear acuminate, white-ciliate, reddish-brown scales.

P. caribaea is capable of adapting the form of its root system to local circumstances.

Leaves are needle-like, crowded, and spreading at the ends of twigs, remain attached for 2 years and are often in fascicles of 3 (sometimes 2, 4 or 5). Leaves are mostly 15–25 cm long and 1.5 mm or less broad, serrulate and rigid, dark or yellowish-green, slightly shiny with stomata in whitish lines on all surfaces. There are 2–8 resin ducts internal or rarely medial. The hypoderm is biform, thick, and has 3–5 layers of cells. Sheaths are 10–12 mm long, light brown becoming brown or blackish, and are persistent.

Strobili appear before the new leaves in January or February, when pollination occurs. The male strobili are many, sessile, and in whorled, short, crowded clusters near ends of twigs mostly in the lower part of the crown. Male strobili are 20–32 mm long, 5 mm broad with 12–18 ovate, scarious margined with reddish-brown bracts at the base. Female strobili are mostly in upper part of the crown and appear near the apex of elongated twigs. However, female strobili become lateral, (1–)2–4(–8) in a whorl, and 1–3 whorls form/yr. Year old cones are erect to reflexed on scaly stalks 1–1.5 cm long. Ellipsoidal cones are 1.5–2 cm long and 1 cm broad with shiny tan scales and a minute prickle which is curved upwards toward the apex of the cone.

Mature cones are usually reflexed, symmetrical (4–)5–10(–12) cm long. When cones are closed conically they are more than twice as long, and as broad as 2.5–3.5 cm in diameter. Cones are borne singly or in groups of 2–5 on stiff, stout penduncles 1–2 cm in length. The penduncle remains attached to the branch after the cone ripens in the late summer months. Cones are deciduous but in some areas are persistent for 1 year or longer. Cone scales are reflexed or wide spreading, thin, flat, weak, often pyramidal and recurved, transversely keeled, with a dark chocolate-brown colour on inner surfaces. Apophysis with a transverse ridge, tan coloured, and shining. Umbo is small, slightly raised, grey, and ending in an inconspicuous, small, weak prickle, variable in size, but less than 1 mm long.

Female flowers are produced over many months and as late as June with *P. caribaea* var. *hondurensis*. Var. *bahamensis* has a main flowering time from November to January in the Bahamas. The time lapse between the receptive stage of the flower and ripening of the cones averages about 18–21 months.

Seeds are narrowly ovoid and about twice as long as broad, pointed at the ends, and have three angles; on average less than 6 mm in length, 3 mm in width and black to mottled-grey or light brown colour. Seeds have a membranous brown wing less than 20 mm long.

Distribution

P. caribaea has a wide though very disjunct distribution in the Caribbean basin. The tree varies a good deal, both morphologically and silviculturally, in the various parts of its range. *P. caribaea* var. *caribaea* is confined to Cuba and the Isla de la Juventud (formally Isla de Pinos). The island is sufficiently distant from the main island to have a genetically distinct population. *P. caribaea* var. *bahamensis* is indigenous to certain islands of the Bahamas and the Caicos groups. *P. caribaea* var. *hondurensis* can be found in the eastern half of Central America south-east from the Yucutan Peninsula.

On Cuba, *P. caribaea* var. *caribaea* occupies low, rolling hills and mountain slopes in the western part of the island. It ascends to an altitude of 330 m and forms open forests to a limited extent. On the Isla de la Juventud, pure stands on low coastal flats can be found in the north-west end of the island. On the rest of the island it grows mixed with other species.

In the Bahamas, *P. caribaea* var. *bahamensis* is restricted to the four north-westerly islands of the group. Here, at a mean altitude of approximately 6 m, it forms extensive, pure, well-stocked forests. The extent of the same variety in the Caicos group is limited to the Islands of North Caicos and Pine Cay, the most easterly extension of the species' range.

On the continent *P. caribaea* is found in Belize, Guatemala, the Republic of Honduras (including the island of Guanaja or Bonaca) and in Nicaragua. A small group of trees has been reported in Quintana Roo in Mexico.

P. caribaea is planted in many lowland tropical sites, with Australia, South Africa, Tanzania, Trinidad and Suriname examples of more successful sites. It is particular promising for high-volume production and is fast growing. These attributes have been exploited in other tropical and sub-tropical countries in Africa, Asia and South America.

Silviculture characteristics

P. caribaea is grown easily in nurseries, and has a good rate of seed germination. The best growth is obtained on fertile sites with good subsoil drainage and an annual rainfall of 2000–3000 mm. The tree is moderately resistant to drought. However, shallow, dry, and excessively cold sites should be avoided. *P. caribaea* is best suited to low latitudes and altitudes.

Although *P. caribaea* grows on a wide variety of soils, bole form is liable to be adversely affected by poor drainage, while the rapid growth induced by deeper, fertile soils may cause an increase in the incidence of branchless leaders.

Silviculture practice

Seeds need to be dried to 8% moisture content or less before placing in low-temperature storage (−10–0°C). Seeds can then be stored for 8–9 years without stratification.

Seeds can also be stored for shorter periods in airtight containers. The humidity should be below 10% and the temperature below 4.5°C. Fumigation is not necessary and can be harmful.

P. caribaea seeds will germinate within 8–21 days. Var. *hondurensis* can be transplanted after a period of 5–6 months. Var. *caribaea* may take as long as 8 months to be ready for transplanting, while var. *bahamensis* lies between the other two species, with approximately 7 months needed for growing to a size suitable for plantation.

A suitable sowing medium is sand:forest soil in a ratio of 3:1, the pH of the soil should not be less than 6.

P. caribaea requires full sunlight. Weeding might be necessary. Root competition can become an important factor in less fertile sites, especially with var. *hondurensis*, which will develop a very strong root system in unfavourable conditions. All varieties of *P. caribaea* are frost tender. The trees are moderately wind and fire-resistant.

Rates of growth of *P. caribaea* can vary considerably. On suitable sites in South Africa, *P. caribaea* var. *hondurensis* reached heights of 27 m and dbh of 47 cm at age 34 years old, with an MAI of 22 m^3/ha. On the best sites in South Africa, it has attained an annual increment rate of 35–42 m^3/ha at age 14 years. In South Africa, var. *hondurensis* has displayed more crooked stems than the other two varieties.

Diseases and pests

Charcoal root-rot caused by *Macrophomina phaseolina* has been reported to cause considerable damage to *P. caribaea*. Losses of trees to the root pathogenic fungi *Armillaria mellea*, *Phytophthora cinnamomi*, *Heterobasidion annosum*, and *Cylindrocladium* spp. have been reported. When injured by hail *P. caribaea* may, rarely, be attacked by *Diplodia pinea*.

Indigenous leaf-cutting ants (*Atta* spp.) have caused damage. Other pests include bark beetles (*Dendroctonus* spp.), the pine tip moths (*Rhyacionia* spp.), shoot weevils (*Pissodes* spp.), defoliators from the family Tortricidae and cone weevils (Curculionidae).

P. caribaea wood is not susceptible to attack by *Lyctus* spp. (powder-post beetles). Low resin content will increase susceptibility to the dry-wood termite (*Cryptotermes brevis*). *P. caribaea* wood has little resistance to marine borers.

Uses

P. caribaea will grow on virtually all soil types, hence the tree is particularly suitable for soil conservation, erosion control and soil improvement. Problems can arise, however, in the case of warm salt winds that lower the performance of the tree. *P. caribaea* is used as an ornamental shade tree.

P. caribaea timber has a medium to coarse texture with reported wood densities ranging from 439–680 kg/m^3. The grain is straight, the wood has a medium lustre, and compression wood is frequently present. Machining is very easy, though resin can clog and gum up cutting edges and sliding surfaces. This is particularly so with var. *hondurensis* more than 12 years old, which can have heavy infiltration of resin into the heartwood.

As the annual rings are relatively uniform or indistinct, and knots often few, the wood can be worked with greater ease than that of many temperate conifers.

Lumber is the greatest single use for the species. The wood from naturally grown stands is well suited for most outdoor and indoor applications. The wood takes nails particularly well and is suitable for joinery and moulding if free from resin.

P. caribaea is more widely used for low-stress products due to a high variability in wood strength properties, especially of timber from trees less than 15 years old. However, due to the resin infiltration of the heartwood after 15 years, marketing of the wood before the age of 15 is advisable if this is not desired.

In the round, *P. caribaea* var. *hondurensis* is too weak to meet the specifications for telephone poles. Other uses include boxes, pallets, treated posts, poles, inexpensive furniture and toys. Resin-soaked wood has high durability and is very popular for boat decking. The wood is also suitable for plywood, wool-cement board and particleboard.

P. caribaea chips are suitable for pulp, except for dissolving-grade pulp. Var. *bahamensis* and var. *caribaea* yields are slightly lower than that of var. *hondurensis*. Pulping chips are produced from plantations in Australia, Brazil, Central America and Africa.

Fibres of foxtailed trees (those with long terminal shoots, a morphological feature which this species, and other pines, is prone to develop) are in general up to 13.5% longer than the fibres of normal trees growing on the same sites. Thus foxtails can be utilized for pulp manufacturing instead of the usual disposal of foxtails from plantations. In South Africa, pulp yield at a constant kappa number was found to be highest for var. *bahamensis*, followed by var. *caribaea* and then by var. *hondurensis*.

Wood from *P. caribaea* is also used to some extent for fuelwood and charcoal.

When tapped, *P. caribaea* yields a good-quality oleoresin, distilled to produce turpentine and other products.

Further reading

Arias, D., Calvo-Alvarado, J., Richter, D. de B. and Dohrenbusch, A. (2011) Productivity, aboveground biomass, nutrient uptake and carbon content in fast-growing tree plantations of native and introduced species in the Southern Region of Costa Rica. *Biomass and Bioenergy* 35, 1779–1788.

Dos Santos, P.E.T., Kageyama, P.Y. and Ferreira, M. (1996) Genetic conservation and breeding strategy of *Pinus caribaea* Morelet: results of progeny trials established in Brazil and in Argentina. QFRI-IUFRO Conference, Caloundra, Queensland, Australia, 27 October–1 November 1996. In: Dieters, M.J., Matheson, A.C., Nikles, D.G., Harwood, C.E and Walker, S.M. (eds) *Tree Improvement for Sustainable Tropical Forestry Volume 2*, Queensland Forestry Research Institute, Queensland, Australia, pp. 338–341.

Pinus contorta Douglas ex Loudon

Lodgepole pine

Importance

Pinus contorta is a fast-growing, short-lived and fire-adapted pine species with a very wide ecological amplitude and tolerance. It is an important and valuable timber tree in western North America, with forests dominated by *P. contorta* covering some 6 million ha in the USA and 20 million ha in Canada. Because of its rapid growth rate, small taper, and thin bark, it produces a higher volume of wood than many of its associates of the same diameter and height. It is a low nutrient-demanding species and easy to regenerate and grow. A common problem of regenerating *P. contorta* is overstocking, which may result in growth stagnation during early stand development, especially on water-deficient, nutrient-poor sites.

Botanical features

P. contorta is a medium-sized tree, occasionally reaching 35 m in height with a dbh of 60 cm. The trunk is straight with little taper, especially in dense stands where the live crowns are small. The crown is narrowly conical; branches are slender and short. The bark is relatively thin (<2 cm thick) and orange-brown to grey with fine scales.

The root system of *P. contorta* is generally shallow, but a taproot and vertical sinkers develop on well-drained sites. Because of its shallow root system, *P. contorta* is susceptible to windfall, particularly after stands are opened by harvesting. Roots are associated with both ecto- and endomycorrhizae.

The evergreen needles occur in bundles of two and are 3–6 cm long. They are usually twisted, stiff, very sharp pointed, and a dark to yellowish-green.

P. contorta is a monoecious species, with female flowers most often at the apical end of main branches in the upper crown, and male flowers on older lateral branches of the lower crown. The reddish-purple female flowers grow in whorls of 2–5 and are 10–12 mm long. The pale yellow to yellowish-orange male flowers are crowded clusters of catkins at the base of new shoots and are 8–14 mm long.

Cones are short cylindrical to ovoid, 3–6 cm long, purplish-brown, stalkless in small clusters at the nodes, and usually closed on the tree for 10–20 years.

Pollen generally matures during mid-May to mid-July and seed cones in August, September or October, more than 1 year after pollination. Inland forms and high elevation stands apparently mature earlier than coastal forms or low elevation stands. Cone maturity is indicated by a change in colour from purple-green to light brown.

Distribution

P. contorta is a western North American species distributed in both Pacific and Cordilleran regions. Owing to its wide ecological amplitude, it is a ubiquitous species that grows throughout the Rocky Mountain and Pacific coast regions, extending north to about latitude 64° N in the Yukon Territory, Canada, and south to about latitude 31° N in Baja California, USA, and west to east from the Pacific Ocean to the Black Hills of South Dakota, USA.

P. contorta was first introduced to Europe in 1832, being planted in arboreta, parks and on a minor scale in forests. In 1950, *P. contorta* became a major species for afforestation of peatlands in Britain, Ireland, Sweden and Finland. *P. contorta* is planted throughout Europe and plantations exist in the Netherlands, Denmark, Iceland, Norway, Germany, Poland and the former Soviet Union. France probably has the largest share of plantations of this species outside North America. In the UK, planting of *P. contorta* has declined, owing to poor tree form and susceptibility to insect defoliation, notably by *Panolis flammea*.

P. contorta grows predominantly in even-aged, post-fire forests in pure stands. It is a primary succession species and is present in early, mid- and late stages of secondary succession on water-deficient and waterlogged sites. It is a major component in the hypermaritime forest and fire-disturbed communities in boreal and temperate climates.

Silviculture characteristics

P. contorta is a shade-intolerant and exposure-requiring species, with a high reproduction potential. Viable seed is produced very early (5–10 years) and seed set is prolific, although dispersal is poor, typically <100 m from the parent tree. Good seed crops can be expected at 1–3-year intervals. Cones are persistent, and serotinous (closed) cones accumulate for decades. The serotinous cone habit varies over wide geographic areas as well as locally. In Oregon, USA, where the non-serotinous cone habit is prevalent, seedfall ranged from about 35,000 to over 1.2 million/ha.

Under low light levels, the potential for natural regeneration is very low (only in dry and cold climates in canopy gaps). Natural regeneration is favoured in the open, especially after wildfires. Initial growth rates (less than 5 years) are high, more than 50 cm/yr after the third growing season on productive sites. The response of advance regeneration to release is very slow (more than 10 years).

In dense stands, *P. contorta* has a high self-pruning capacity and its crown spatial requirements are low. Light conditions beneath closed-canopy, mature stands are high and associated with well-developed understorey vegetation.

The potential productivity of *P. contorta* is medium, with a site index (50 years at bh) of less than 30 m. Growth declines after about 80–120 years. It generally lives for less than 300 years.

Silviculture practice

Depending on site and management objectives, clear-cutting and patch-cutting systems are viable for the establishment and growth of *P. contorta*, which can be grown in single- or mixed-species stands, preferably with either shade-intolerant or shade-tolerant species.

Propagation by seed is the most suitable method for regenerating *P. contorta*. Natural regeneration (where applicable) and/or planting using containerized stock is used for its establishment.

The ability of *P. contorta* to regenerate at the expense of other species is due not only to cone serotiny, but also to high seed viability, early rapid growth and the ability to survive a wide variety of microsite and soil situations. Germination is epigeal and the best germination occurs in full sunlight and on bare mineral soil, free of competing vegetation. Both shading and competition inhibit germination and survival.

Serotinous cones do not open at maturity because of a resinous bond between the cone scales. The bonds break at temperatures between 45 and 60°C, and cone scales are then free to open hygroscopically. Large quantities of seeds are thus available for regenerating a stand following fire. Closed cones at or near the soil surface (less than 30 cm) are also subjected to insulation temperatures sufficiently high to open them and may provide seed in harvested areas.

Drought is a common cause of mortality among first-year seedlings; losses vary with soil type and seedbed conditions. Disturbed mineral soil seedbeds generally produce the best germination and survival. Drought losses usually decline considerably after the first growing season. *P. contorta* seedlings are poor competitors, and competition from grass is often detrimental.

A common problem of regenerating *P. contorta* stands is overstocking, which results in stagnation at early ages. If trees are well distributed, stocking should not exceed 1200–2000 stems/ha between 5 and 20 years of age. Proper distribution and full utilization of the site, however, may require establishment of 2470/ha and thinning to obtain proper spacing. Control of stand density offers the greatest opportunity for increasing productivity of any readily available management practice.

P. contorta shows good response to thinning at an early age, with poor sites or overstocked stands requiring thinning after only 10 years. Diameter growth acceleration is usually greatest in heavy thinnings; cubic volume and basal area growth/ha are usually greatest in light thinnings. Although mechanical thinning, as with bulldozer strips, is a convenient alternative, obtaining a proper response is difficult. At older ages, the growth response of *P. contorta* is strongly correlated with crown size, vigour and amount of release provided.

Growth and yield of *P. contorta* is greatly affected by stand density as well as by environmental factors. For example, the mean annual growth in old-growth unmanaged stands can be as low as 0.4–0.6 cubic m/ha in the Rocky Mountains because of large numbers of small trees and a high incidence of dwarf mistletoe (for example, *Arceuthobium americanum*). Annual net growth may be increased to 2.1–5.6 m^3/ha by controlling stand density and reducing dwarf mistletoe infection. In the UK, plantations can achieve annual growth rates of 4–14 m^3/ha over rotations of 4–80 years.

Maximum yields in the Rocky Mountains are reported to be 280 m^3/ha at a density of 1980 trees/ha, but only 21 m^3/ha at a density of 4450/ha. In extreme cases, 70-year-old stands with 247,000 trees/ha averaged only 1.2 m in height and less than 2.5 cm in diameter at ground level. Yields of 168–224 m^3/ha can be found in old-growth stands in the Rocky Mountains. Yields of >336 m^3/ha can result from a combination of favourable initial stocking, good site quality, and the absence of *Dendroctonus ponderosae* (mountain pine beetle) and dwarf mistletoe.

Control of stand density offers the greatest opportunity for increasing productivity of any readily available management practice, as culmination of total cubic volume occurs as early as 40 years in severely stagnated stands and between 50 and 80 years for overstocked, but not greatly stagnated, stands. Thinning of young overstocked and stagnating stands can restore growth potential and redirect it into merchantable-size products. A stocking control thinning can achieve most of the yield increase possible from thinning if more complete timber utilization (lower merchantable dbh and top diameters) is an option.

Because of its relatively thin bark, lodgepole pine is more susceptible to fire than *Pseudotsuga menziesii* (Douglas fir) and many other associates. It is less susceptible than Engelmann spruce (*Picea engelmannii*) or subalpine fir (*Abies lasiocarpa*).

Chinook winds following extremely cold weather occasionally cause red belt injury, particularly in Canada, and Montana, USA. Defoliation of trees is common and mortality can occur over large areas. Heavy snow can break or bend trees, particularly in dense stands with narrow crowns and intense root competition. Thinning can contribute to snow breakage, particularly if previously dense stands are opened suddenly.

Diseases and pests

One of the most serious diseases is a stem canker caused by *Atropellis piniphila*. Another serious stem canker is comandra blister rust (*Cronartium comandrae*). The western gall rust (*Endocronartium harknessii*) is especially damaging. Other fungi that may cause serious losses in wood production are needle casts (such as *Elytroderma deformans and Lophodermella concolor*); root rots (such as *Armillaria mellea* and *Heterobasidion annosum*); and wood decays (such as *Phellinus pini* and *Peniophora pseudo-pini*).

The mountain pine beetle (*D. ponderosae*) is the most severe pest of *P. contorta*. Adult beetles attack in July or August, introducing bluestain fungi (*Grosmannia clavigera*). Mortality from beetle epidemics often creates large amounts of jackstrawed fuel, which ignites easily from lightning and other sources, and hampers fire-control efforts.

The pine beauty moth (*P. flammea*) has become a major defoliating pest. Dwarf mistletoe (particularly *A. americanum*) is the most widespread and serious parasite affecting *P. contorta*.

Uses

P. contorta is not only an important timber species but is also a major tree cover in many scenic and recreational areas, and on critical watersheds. It provides many acres of wildlife habitat and is associated with many grazing allotments throughout its range.

The wood is soft and light to moderately hard with a density of 380–465 kg/m^3, with a generally straight grain and medium fine texture. The timber is reported to be the strongest of the pines native to the western USA. The narrow sapwood is nearly white to pale yellow in colour, and is not clearly differentiated from the heartwood. It has very little resistance to decay when in contact with the ground and is also susceptible to attack by wood-destroying insects. It has good moulding, nailing, painting, boring, screwing and mortising characteristics. Plantation-grown material can be used for oriented strandboards (OSB) and other board composites.

P. contorta wood is used for framing, panelling, posts, corral poles, utility poles, railroad ties and pulpwood. As new developments such as structural particleboard become practical, the rapid juvenile growth of the species will be an advantage where gross cubic volume becomes important.

P. contorta is usually included in spruce–pine–fir species combinations for structural sizes and grades. It is also marketed separately or in a mixture with Engelmann spruce (*Picea engelmannii*).

Further reading

Dhar, A. and Hawkins, C.D.B. (2011) Regeneration and growth following mountain pine beetle attack: a synthesis of knowledge. *BC Journal of Ecosystems and Management* 12, 1–16.

Klinka, K., Worrall, J., Skoda, L. and Varga P. (1999) *The Distribution and Synopsis of Ecological and Silvical Characteristics of Tree Species of British Columbia's Forests*. Canadian Cartographics Ltd, Coquitlam, British Columbia, Canada.

Parchman, T.L., Gompert, Z., Mudge, J., Schilkey, F.D., Benkman, C.W. and Buerkle, C.A. (2012) Genome-wide association genetics of an adaptive trait in lodgepole pine. *Molecular Ecology* 21, 2991–3005.

Pinus densiflora Siebold & Zucc.

Japanese umbrella pine

Importance

Pinus densiflora, one of the most dominant forest conifers in Japan and Korea, occurs naturally as pure stands in various sites on a variety of soils and climatic conditions and holds an important position in forest ecology and forestry. It provides high-quality wood for light construction purposes, furniture and different valuable by-products. It has also been one of the popular ornamental pines in Korea and Japan since ancient times. However, the natural stands of *P. densiflora* in Japan and Korea are being reduced due to diverse natural disturbances, human-induced environmental changes, and severe insect damage by the pine gall midge (*Thecodiplosis japonensis*). Extensive research on gene conservation and breeding for resistance to the pine gall midge has been carried out in Korea.

Botanical features

Tree stem may be straight or contorted (the latter particularly in coastal sites) with an open, oval or umbrella-shaped crown. In optimum conditions, it reaches a height of 36 m and a dbh of 60–70 cm at maturity. Young bark peels in thin scales to reveal a red-brown colour, but older bark on the lower stem is grey and typically has a turtle-shell pattern. Current year twigs are green.

Needles are arranged in pairs and clustered 2 needles per fascicle, often twisted and roughly 5–8 cm long. Needles are bright green in colour.

P. densiflora is monoecious with staminate and ovulate strobili borne separately on the same tree. The female flowers are dark-purple and look like tiny cones. They are scattered singly or in small bundles at the top of the current shoot. Male flowers are positioned near the base of the new shoot, usually below the female flowers. They are pale yellow when full bloom, and disperse a large amount of pollen.

Cone, conic–ovoid, colour tan to golden-brown, 4–7 cm long, in whorls of 2–5 at branch nodes, remaining closed and attached for several years, on a somewhat flexible peduncle 1–3 mm long. The scale is cuneate, but the exposed part is flattened; the concealed part is dark red-brown. Seeds with attached wing are 10–17 mm long.

P. densiflora is an evergreen species; new leaves remain on the tree for about 2 years. Flowering begins in April, and continues till May; pollen dispersal and pollination occur during this season. However, fertilization does not actually happen until the next year. It takes place in June about 13 months after pollination, and the cone matures in September of the second growing year.

Distribution

P. densiflora is widely distributed in China (Shandong, Jiangsu), Korea, Japan (Honshu, Shikoku, Kyushu) and southern Ussuriland (Russian Far East) and Lake Khanka, on rocky slopes at 0–500 m altitude. In Japan, its altitudinal range varies from 0 to 2300 m. Its altitudinal range in Korea is from 0 to 1300 m, but is mainly concentrated at around 500 m.

P. densiflora has been planted in Azerbaijan, India, North America and Europe. It was introduced to the USA and Europe as an ornamental.

Silviculture characteristics

P. densiflora requires a well-drained site and full sun. It is quite fast growing when very young, but growth rates decrease and may become very slow. This species has an invasive adaptation to fire; fires typically kills the tree, but it can quickly establish and attain dominance. This invasive potential also makes it a good example of a species that has been favoured by human disturbance.

Silviculture practice

This species is easily grown from seed and transplants well. Fresh seeds require no pre-sowing stratification. No loss of viability occurred in seeds stored in sealed containers at room temperature after 2 years. Storage at 5% moisture content is detrimental to seed stored for long periods.

In nurseries, non-stratified seeds are sown in the autumn or spring, at a density to obtain 550–650 seedlings/m^2. Seeds should be sown at a depth of 10–15 mm. A short stratification period of 6 weeks at 4°C improves the germination of stored seed. Seedlings have a very sparse root system, and trees should be planted in permanent positions when they are still small, between 30 and 90 cm tall. Larger plants can be transplanted into the garden or yard, but it is not recommended to move trees over 1 or 2 m tall unless they have been transplanted every 2 or 3 years in a nursery or they have had their roots trimmed 1 year before transplanting. Planting should be done in early autumn, or in spring just before the growth period.

Propagation by cutting only works for very young trees less than 10 years old, and it is more effective to use single leaf fascicles with the base of the short shoot. Removing buds from the shoots several weeks before taking the cuttings helps. Cuttings usually grow slowly. Nursery-grown seedlings are commonly field-planted as 1+1 seedlings.

P. densiflora matures at about 40–50 years of age. In intensively managed pure stands in Kangwon province, Korea, an average height of 15 m and dbh of 20 cm have been recorded at 30–40 years. It may be possible to produce trees of more than 50 cm in diameter on rotation periods of 60–80 years in Korea.

Diseases and pests

Where *P. densiflora* has been planted as an exotic species, it is usually pest-free with occasional scale; needle blights and rusts have been reported occasionally. In order to keep trees healthy, it is important to prune out the infected branches. Needle cast is also common on small trees and plantation or forest trees and the infected needles turn yellow and fall off. *P. densiflora* is susceptible to pine wilt disease.

Pine needle gall midge (*T. japonensis*) is an endophytic gall-forming insect which occurs in Korea and Japan. It has caused severe damage to *P. densiflora* and *Pinus thunbergii* stands in the Korea Republic since the 1960s. Once its larvae are introduced at the base of pine needles, they form the gall which causes defoliation and frequent death of individual infested trees. Use of natural enemies has been proposed to control pine needle gall midge, and partially put into effect. However, because of the insect's life cycle (the larvae develop in the interior of the galls formed after egg laying by the adults), an effective measure to manage the insect population has not yet been developed.

Uses

In the view of its pioneer characteristics and ability to tolerate infertile soils, *P. densiflora* is suitable for erosion control. However, this tree is extremely intolerant of shade, thus *P. densiflora* stands are often substituted by faster-growing broadleaved species (e.g. *Quercus* spp. and *Betula* spp.) unless fire or human disturbance occurs. It takes considerable and continuous efforts to maintain *P. densiflora* forest stands.

It is also one of the most popular ornamental pines in Korea and Japan, and now in Europe and North America.

P. densiflora wood has been one of the most important materials used for palace and temple construction in East Asia. Today it is also used extensively for general construction purposes, veneer and furniture manufacture, pulping and other timber uses. Heartwood is reddish-brown in colour and sapwood is pale reddish to yellow-white. Wood density is 410 kg/m^3 and the texture of the wood is coarse, straight grained, and light and soft. This wood is judged to have moderate grade properties of shrinkage, water absorption and strength.

The characteristics of drying, adhesion and paintability are excellent, and those of sawing, cutting with a plane and decay resistance are moderate. However, bending characteristics are poor. The manufacturing characteristics of particleboards, fibreboards, wood cement board and pulp are excellent. The manufacturing characteristics of plywood and charcoal are moderate.

Leaves and pollen of *P. densiflora* are edible; they are often used as a natural medicine. The needles, which contain some antibacterial compounds, were also used in the preparation of traditional Korean cakes, to extend the storage period. The needle extract is used as an ingredient for some 'health' drinks. The leaves and branchlets can also be processed as a feedstuff for ruminants through anaerobic fermentation. The resin and terpene oil contained in leaves are commercially valuable components which are used in aromatherapy and as an antiseptic.

The Matsutake mycorrhizal mushroom, named 'Songyi' in Korea, is the most valuable mushroom produced in *P. densiflora* forests. 'Songyi' has a very distinct smell and flavour, and is very expensive in Korea and Japan. Therefore, 'Songyi' mushroom collecting is a significant income source to some rural people in Korea, along with the pine species associated with it.

Further reading

Huo, Y.L. and Li, G.W. (1990) Investigations on decline and death of *Pinus densiflora* in the Yantai region, Shandong Province. *Forest Pest and Disease* 3, 21–22.

Kim, Z.S. and Lee, S.W. (1998). Genetic diversity in East-Asian *Pinus* species. In: Puri, S. (ed.). *Tree Improvement: Applied Research and Technology Transfer.* Science Publishers, Inc., Enfield, North Carolina, USA, pp. 207–219.

Okada, K., Okada, S., Yasue, K., Fukuda, M. and Yamada, A. (2011) Six-year monitoring of pine ectomycorrhizal biomass under a temperate monsoon climate indicates significant annual fluctuations in relation to climatic factors. *Ecological Research* 26, 411–419.

Pinus echinata Mill.

Shortleaf pine

Importance

In the USA, *Pinus echinata* grows in over 22 states. It tolerates a wide range of soils and temperatures and has stronger wood than that of the widely grown *Pinus taeda*. *Pinus echinata* has an initially slow growth rate compared with *P. taeda*. In

Southern Africa, *P. echinata* has not been rated as highly as *P. taeda* and *Pinus elliottii*, because of its relatively slow growth and poorer stem form.

Botanical features

P. echinata has a multinodal growth habit, with small and horizontal branches. It reaches heights of 25–30 m and a dbh of 60–90 cm on good sites. Trees with heights of up to 40 m and diameters of 120 cm have also been recorded. The stem and crown form of this species are generally better than most of the southern USA pines, and it is self-pruning. The bole is typically well formed and has a relatively short, pyramidal or rounded crown. Young seedlings have a characteristic 'J'-shaped crook in the stem which is a result of initial prostrate growth which turns upwards to form the crook. The stem form is generally good and shows wide variation. The bark of young trees is very dark and scaly, becoming reddish-brown on older trees and broken into irregular, flat plates. The bark is about 25 mm thick and typically has resin pockets which appear as small indentations surrounded by oily spots on the bark flake.

P. echinata has two needles per fascicle, but occasionally produces three needles. The needles are usually 75–125 mm long, slender, flexible, slightly twisted and persist for 2–5 years. The basal sheath is persistent and only about 6 mm long.

P. echinata is monoecious, the cones are borne in whorls of 2–4 and reach sizes of 35–50 mm long, making them the smallest among the four major southern USA pines. The scales of the cones are thin and flattish, ending in small, straight or curved prickles which are sometimes deciduous before the cones ripen.

Cones and seed of *P. echinata* are much smaller compared to other southern USA pines, with the number of seed/kg ranging from approximately 70,000 to 160,000. *P. echinata* reaches peak cone production after 20 years.

In the north-eastern part of its natural range in USA, flowering begins in mid- to late April to June. In the southern range, flowering starts in March and peaks around late March to mid-April. Mast years occur every 3–10 years in the north-eastern part of its range and every 3–6 years in the south.

Distribution

P. echinata is reported to be the most widely distributed of all pines in southern USA, although its extension southwards is not as far as the other southern USA pines. A major break in the distribution of *P. echinata* along the Mississippi valley effectively creates an eastern and western population group. It grows best at elevations between 185 m and 455 m in the Piedmont and from 45 to 305 m in Louisiana and Arkansas, USA. Although *P. echinata* is widely distributed, this only represents about half of the total volume of *P. taeda*.

Silviculture characteristics

P. echinata is a large tree, with smaller branches compared to the widely grown *P. taeda*, but with better self-pruning characteristics. In the USA, the species has generally been used to reforest poor sites where the terrain is very steep and the soils shallow. On good sites, it tends to be replaced by more competitive hardwood species and is more competitive on thin, rocky, and nutrient-poor soils. It is shade-intolerant and does not survive or grow well when suppressed. The growth rate of *P. echinata* is slower than other pines, such as *P. taeda*.

P. echinata is moderately fire-resistant, but wild fires in young plantations are very damaging and larger trees can be killed by severe fires. If the crowns of young seedlings are killed, sprouting from the base can form new stems. Fire-damaged trees are also more susceptible to insect damage. *P. echinata* is generally resistant to wind although trees may be uprooted where root systems are shallow.

Silviculture practice

Seed production is rarely a limiting factor. The seed is orthodox and can be stored for up to 5 years (at 4°C; relative humidity of 12% in airtight bags) with a minimum loss in viability. A broadcast sowing rate of approximately 0.56 kg/ha, with half this amount being suitable for seedbeds, has been recommended.

The species is easily propagated from cuttings and can also be propagated by air-layering or grafting, usually for the establishment of clone orchards. Although it has been successfully propagated through tissue culture no commercial application of this technique has been reported.

Both bare-rooted and containerized seedlings can be used. As with other pines, soil collected from under pine stands is preferred for nursery soil, as it usually contains mycorrhizae which support seedling growth. The soil can be mixed with either pine bark, loam soil or any other available potting media. Seeds can be sown directly into nursery beds for pricking out into containers or for production of bare-root seedlings. In warm environments seedlings reach plantable sizes (15–30 cm) in 7–9 months.

In the USA, seeds are stratified for 60 days (1–5°C) before spring sowing at a rate of 430/m^2, being mulched with burlap or chopped pine needles. Seedlings are usually ready for lifting and outplanting in the autumn.

In areas with distinct dry seasons it is normal to plant out seedlings just at the onset of the rains to ensure high survival rates. In the mid-south of the USA, the best time to plant *P. echinata* is after the beginning of December until March or when the danger of frost-heaving is over, but before new growth starts in the late spring. In areas receiving rainfall throughout the year and without freezing winters, planting may be done throughout the year. Recommended seedling spaces range from 2.4 × 2.4 m to 3 × 3 m.

It is also possible to sow seed directly either into spots or by broadcasting. Establishment from regeneration is also possible and can be achieved in 2–3 years. A minimum of about 7–9 mature fruiting trees/ha is required. Regeneration is usually good if the area has been logged, as this process exposes the soil, bringing it into contact with seed. Where logging has not taken place it may be necessary to plough or disc to bury the seed. Another way of encouraging natural regeneration is the prescribed burning of litter before seed fall.

Stocking rates for *P. echinata* usually vary from 1111 to 4940 trees/ha depending on management objectives and site quality. The lowest stocking of 1111 stems/ha gives a spacing of 3 × 3 m which is the standard spacing in many countries for sawtimber. The between-row spacing depends on whether weed control is mechanical or manually performed. The planting holes are usually made to a depth of 15–20 cm depending on seedling sizes. It is common practice to clear the land of other hardwoods that would compete with the young seedlings. If the land is flat, land preparation is sometimes performed by ploughing, discing or ripping; otherwise on extremely sloping ground, there is very little land preparation.

Thinning is important to ensure a good final tree crop, with wet and fertile sites usually requiring thinning after as early as 5 years depending on initial stocking. As a rough guide, first thinnings can be expected to be performed after 6–10 years, depending on growth and initial spacing, with subsequent thinnings every 3–6 years. A rough guide to deciding when to thin is when the stand's live crown depth is less than 40% of tree height. To maximize returns, first thinning is usually done when the trees can be sent to a pulp mill, but where this is not possible, thinning to waste is sometimes the norm. Besides pulp, the thinnings may also be used as fence posts.

Pruning to enhance log quality is also important. The height of pruning depends on the end use, and where clear logs are in high demand, pruning is often taken up to 9 m. Pruning is usually done in stages, the first being usually performed when the trees are 10–20 cm in diameter.

Growth and yield of a given stand depend on the quality of the sites, stocking, and silvicultural practices administered. *P. echinata* slowly covers sites and is not overly competitive, but after 6–7 years, total exclusion of ground vegetation is sometimes achieved. Pulp rotations of 12–15 years and sawn timber rotations of 25–30 years can be achieved.

In the states of Alabama, Tennessee and Georgia, USA, the mean annual increment for total volume culminates at about 30 years on poor sites and 20 years on good sites. The total volume yield (outside bark) after 40 years at a density of 3090 stems/ha was 180 m^3/ha. For the same age and planting density, but higher site index (18.3 m), the yield was 451 m^3/ha. The basal areas for these different stocking rates were 31 and 39 m^2/ha, representing 1870 and 875 trees/ha, respectively.

Natural stands of *P. echinata* in Missouri, USA, showed significantly higher net volume yields when thinned to about 21 m^2/ha or above, after 51 years. The presence of hardwoods reduced growth and yield of the stands by 8–12%. Uneven-aged shortleaf-loblolly pine (*P. taeda*) stands in south Arkansas have been recorded to have achieved an annual sawtimber growth of 6 m^3/ha.

Diseases and pests

The most important disease in its native range is 'littleleaf disease', which is prevalent in heavy soils with poor internal drainage and reduced aeration. Other factors implicated in the complex include nematode damage, drought stress, and low fertility, which increase the susceptibility of the tree to root-rotting fungi (primarily *Phytophthora cinnamomi*) and together act to impede mineral absorption. Seedlings and older trees of *P. echinata* are moderately susceptible to root rot caused by *Heterobasidion annosum*. Red heart (*Phellinus pini*) can occur in older trees (>80 years). *P. echinata* is more tolerant of the fusiform rust *Cronartium fusiforme*, which attacks *P. taeda*.

In the nursery, seedlings are subject to damping off and root rots in wet conditions and soils with a pH above 6. Needle rusts do not normally become serious, although brown spot (*Scrirrhia acicola*) can be a problem and the rust *Cronartium comandrae* can damage seedlings. In evaluation trials in Zimbabwe, *P. echinata* suffered moderate losses from *Rhizina inflata*.

Young trees are attacked by Nantucket pine tip moth (*Rhyacionia frustrana*), the red-headed sawfly (*Neodiprion lecontei*), and the loblolly pine sawfly (*Neodiprion taedae linearis*). Other more general pine pests include pales weevil (*Hylobius pales*), pitch-eating weevil (*Pachylobius picivorus*), *Dendroctonus frontalis*, pine engraver beetles (*Ips* spp.) and the black turpentine beetle (*Dendroctonus terebrans*). In Zimbabwe, it is prone to attack by pine woolly aphid (*Pineus pini*).

Uses

It has been used for erosion control although it does not produce much litter, and it is sometimes used as an evergreen ornamental.

The wood of *P. echinata* is heavy, strong and moderately resinous and cannot be easily distinguished from that of other southern USA pines. The grain is coarse, straight and uneven. The heartwood is orange or yellowish-brown and the sapwood is creamy-yellow. It is easily worked and is used for a variety of purposes including light structural work, interior finish, panelling and sleepers. It has moderate resistance to decay.

Figures for the specific gravity of *P. echinata* wood range from 0.31 to 0.8, but are usually in the range of 0.47–0.51. In the first 10–15 years, its wood has been reported to be about 15% higher than that of *P. taeda* in basic wood density. Trees planted at lower altitudes show higher wood densities compared to trees grown at higher elevations.

P. echinata is an important raw material for kraft pulp and is also a common constituent of dissolving pulp. *P. echinata* is a poor source of commercial oleoresin.

Further reading

Budhathoki, C.B., Lynch, T.B. and Guldin, J.M. (2010) Development of a Shortleaf Pine Individual-Tree Growth Equation using Non-Linear Mixed Modelling Techniques. General Technical Report - Southern Research Station, USDA Forest Service SRS-121. In: Stanturf, A. (ed.) Proceedings of the 14th Biennial Southern Silviculture Research Conference, 26 February–1 March 2007, Athens, Georgia; USDA Forest Service, Asheville, North Carolina, USA, pp. 519–520.

Elliott, K.J., Vose, J.M., Knoepp, J.D. and Clinton, B.D. (2012) Restoration of shortleaf pine (*Pinus echinata*) – hardwood ecosystems severely impacted by the southern pine beetle (*Dendroctonus frontalis*). *Forest Ecology and Management* 274, 181–200.

Guldin, J.M. and Shelton, M.G. (2010) *Effects of Pre-Commercial Thinning in Naturally Regenerated Loblolly Shortleaf Pine Stands in the Upper West Gulf Coastal Plain: Results after Two Growing Seasons.* General Technical Report - Southern Research Station, USDA Forest Service SRS-121. In: Stanturf, A. (ed.) Proceedings of the 14th Biennial Southern Silviculture Research Conference, 26 February–1 March 2007, Athens, Georgia; USDA Forest Service, Asheville, North Carolina, USA, pp. 55–62.

Stewart, J.F., Tauer, C.G. and Nelson, C.D. (2012) Bidirectional introgression between loblolly pine (*Pinus taeda* L.) and shortleaf pine (*P. echinata* Mill.) has increased since the 1950s. *Tree Genetics and Genomes* 8, 725–735.

Wittwer, R.F., Lynch, T.B., Huebschmann, M.H. and Waldrop, T.A. (1998) Stand density index for shortleaf pine (*Pinus echinata* Mill.) Natural stands. General Technical Report - Southern Research Station, USDA Forest Service, SRS-20. Proceedings of the 9th Biennial Silvicultural Research Conference, 25–27 February 1997, Clemson, South Carolina; USDA Forest Service, Asheville, North Carolina, USA, pp. 590–596.

Pinus elliottii **Engelm.**

Slash pine

Importance

Pinus elliottii var. *elliottii* is an important timber species native to the lower coastal plain within the south-eastern USA. Because of its rapid early growth and its production of highly valuable wood products, it has been widely introduced into other countries. As an exotic, it is used in Africa, especially in Southern Africa, and in Australia and South America, for various products ranging from lumber to pulpwood. In Brazil, it makes an important contribution to the resin production industry.

Botanical features

P. elliottii var. *elliottii* is characterized by a long, clear bole. Self-pruning of the branches is common and this results in a relatively short crown.

P. elliottii var. *elliottii* varies from 18 to 30 m in height and averages about 0.6 m in diameter. Large trees are sometimes 36 m tall and 0.9 m in diameter. The trunk is straight and erect with a narrow ovoid crown with spreading and ascending limbs.

In good soil conditions, *P. elliottii* develops an extensive lateral root system and a moderate taproot.

The needles are 18–25 cm long in fascicles of 2 and 3 on the same tree, dark glossy green, and tufted at the ends of tapering branches. They extend back some distance along the branch and are persistent until the end of the second year. Resin canals number 2–10 and are mostly internal. The fascicle sheath is usually 1.25 cm or less in length.

P. elliottii is monoecious and wind pollinated. The cones are 10–15 cm in length, ovoid–conic, and sessile. They usually persist on the tree until the following summer. Cones are reddish-brown, lustrous, and are armed with a sharp spine. The seeds are about 0.6 cm long, dark brown mottled with black, with wings about 2.5 cm long that are thin and translucent.

A 3-year period is required from the time strobili are initiated until the time the seeds mature. The primordia of new strobili are detectable in late spring. These cone initials overwinter as buds the first year. Early in the second year the male strobili are purple and 5 cm long, pollen is shed in late January or February. The female strobili appear on stalks in the upper crown and are about 2.5 cm long and red to purple at the time of pollination. Soon after pollination, the pollen tube ceases to grow and appears to remain in a quiescent state for the summer and winter. During the third growing season fertilization occurs some 12–14 months after pollination. The cones enlarge and seeds mature during the third summer.

Distribution

The range of *P. elliottii* var. *elliottii* extends over 8° latitude and 10° longitude. The latitude covered in the south-eastern USA ranges from 27° to 35°N. *P. elliottii* var. *densa* occurs only in the southern portion of Florida. The naturalized range of *P. elliottii* has been extended by planting into northern Alabama, Georgia and southern Tennessee. Its western range now includes western Louisiana and eastern Texas. It regenerates well naturally in these areas. It will grow along streams and edges of swamps, and invade abandoned fields wherever soil moisture is high.

Silviculture characteristics

P. elliottii is a subclimax species that without human intervention and with the absence of fire or other catastrophic event will proceed to a mixed hardwood forest. It is considered intermediate in tolerance to shade by some authors and intolerant by others. It will reproduce naturally in small openings and will invade poorly stocked *Pinus palustris* stands. However, competition from overstorey and understorey vegetation reduces growth and causes mortality. The two varieties of *P. elliottii* differ in their patterns of growth. *P. elliottii* var. *elliottii* makes excellent early height growth, whereas *P. elliottii* var. *densa* has a grasslike, almost stemless stage, that lasts from 2 to 6 years. Moreover, *P. elliottii* var. *densa* lacks the straight axis or stem characteristic of *P. elliottii* var. *elliottii* and often develops forked boles with large branches and an open, spreading, irregularly shaped crown. In varietal trials, *P. elliottii* var. *elliottii* averaged 3–6 feet (~1–2 m) taller at 10 years than *P. elliottii* var. *densa*.

P. elliottii is slow to express dominance in dense, even-aged stands. As a consequence, height growth is slower in very dense stands than in moderately or lightly stocked stands.

Silviculture practice

P. elliottii var. *elliottii* produces good seed crops about every 3 years and *P. elliottii* var. *densa* every 1–5 years. Trees may reach seed-bearing age in 7–8 years, but do not reach their reproductive capacity until many years later.

P. elliottii cones mature during September and shed most of their seeds in October. More than 90% of the seeds fall within 48 m of their source, but some may be carried by wind as far as 75 m. The seed yield is about 29,760 seeds/kg in average and viability is usually good.

Germination normally occurs from November to April. Seeds kept in cold storage for 1 year or more benefit from stratification or cold prechilling. For storage, seeds should be dried to 7–10% moisture content and held at sub-freezing temperatures. *P. elliottii* seeds store well with good germination being maintained after 50 years of cold storage.

Germination is epigeal and newly germinated seedlings have an average of 7.2 cotyledons, 30 mm long.

The presence of mycorrhizae has an important beneficial influence on survival and early growth of bare-root seedlings.

P. elliottii seeds are usually sown in the nursery in April and seedlings lifted for outplanting in the following December through February.

Vegetative reproduction of *P. elliottii* rarely occurs naturally. The most widely used technique of vegetative reproduction in seed orchards is the use of dormant scions from mature trees that are grafted onto seedling rootstocks which are just beginning active growth in early spring.

Air-layering has also been more than 85% successful in some tests.

Cuttings from branches are difficult to root, especially when taken from older trees.

The preferred method of regenerating *P. elliottii* among large forest landowners is clear-cutting followed by chemical or mechanical site preparation, then direct seeding or planting of bare-root or container nursery stock. With the availability of genetically superior stock, planting should continue to be the preferred method of regeneration of *P. elliottii*. Survival and early growth of seedlings are frequently stimulated by intensive site preparation treatments such as flat discing, roller chopping or bedding. Discing and roller chopping are effective on deep, dry, sandy soils where they control competing vegetation, incorporate organic matter into the topsoil and may alter nutrient availability. On sites with a shallow water table, bedding provides increased rooting space, improves aeration and may increase growth. Surface drainage may produce a similar response. On drought-prone sites, weed control and irrigation may be effective. Many *P. elliottii* sites are low in available phosphorus and nitrogen.

Fire may be used for site preparation, hazard reduction and control of understorey vegetation. Fire also improves forage production for livestock and habitat for wildlife.

P. elliottii may also be regenerated naturally in uneven-aged stands by the group-selection method and in even-aged stands by either the seed-tree or shelterwood system. Even-aged silvicultural systems are preferred, as the species is comparatively shade-intolerant. Another reason for favouring even-aged systems is that they permit the use of prescribed fire to control stand composition and understorey development. Even-aged management, accomplished by leaving 15–25 trees/ha after a cut of 62–99 shelterwood trees/ha, appears to be the most successful silvicultural system for *P. elliottii* var. *densa*. If a stand is cut, direct seeding in the autumn will provide good regeneration.

Young stands respond promptly to thinning or release. Older stands whose height growth rate has declined and those in poor growing condition may respond slowly or occasionally fail to respond to thinning. For these reasons, pre-commercial thinning before stand age 5 is recommended if densities exceed 4940 trees/ha, especially if the objective of management is the production of sawtimber, veneer logs, or naval stores. Residual densities should be about 1235–1730 trees/ha. In sawtimber production and when high initial density is not a problem, or if a pre-commercial thinning has been made, commercial thinnings will maintain good stand vigour, promote diameter growth, shorten the time required for individual trees to reach sawtimber-size and concentrate volume on trees to be carried to rotation age. These thinnings should be initiated as early as economically feasible, and at the latest by age 20. Commercial thinnings will increase diameter growth of residual trees. They will not increase total yields of stands managed for pulpwood on a 20- to 25-year rotation, or those managed for sawlogs, if the rotation is long enough for a majority of trees to reach sawtimber size and competition for growing space has not caused excessive mortality.

The optimum pulpwood rotation may be about 25 years, as this age is intermediate among those at which mean annual cubic volume culminates. In unthinned plantations, this age increases to 30 as density increases and site index declines. Available information suggests that on good sites, stands averaging 41 cm dbh can be grown in 45–55 years if initial dense stocking is adjusted before age 5 to about 1235–1730 trees/ha and if commercial thinnings at about 5-year intervals reduce residual basal areas below 19.5 m^2/ha.

Plantations and natural stands have similar rates of growth on sites of a given quality if they are comparable in genetic makeup, are maintained at comparable densities and are given the same level of management.

Diseases and pests

Fusiform rust (*Cronartium fusiforme*) is the most serious threat to *P. elliottii*. Other important diseases of *P. elliottii* are annosus root rot (*Heterobasidion annosum*) and pitch canker (*Fusarium sacchari*).

Reproduction weevils, the pales weevil (*Hylobius pales*), and the pitch-eating weevil (*Pachylobius picivorus*) are attracted by the odour of pine resin to recently logged areas where they lay their eggs in lateral roots of fresh pine stumps.

Older stands are subject to attack by engraver beetles (*Ips* spp.), the black turpentine beetle (*Dendroctonus terebrans*), and the southern pine beetle (*Dendroctonus frontalis*). A recently recorded pest of *P. elliottii* is the beetle *Hylobitelus xiaoi*. Another agent that can sometimes cause damage with considerable losses under certain conditions is senna seymeria (*Seymeria cassioides*).

Uses

P. elliottii may be planted to stabilize the soil on eroding slopes and strip-mine spoil banks, where its rapid early growth is an advantage. The species has been frequently used in agroforestry because of its relatively short tree crown which permits light to reach the forest floor. Moderate grazing by cattle does little damage to trees past the seedling stage and may be beneficial by preventing the build-up of a dense and highly flammable understorey. In addition, *P. elliottii* types provide habitat for several threatened species.

The sapwood is a yellowish-white, whereas the heartwood is a reddish-brown. It is very strong and heavy, very stiff, hard and moderately high in shock resistance. The density (at 12% moisture content) was reported as 657 kg/m^3. It has straight grain, is medium in texture and relatively difficult to work with hand tools.

The straightness of the bole makes *P. elliottii* particularly suited for poles, pilings and solid wood products. However, because *P. elliottii* has rapid early growth it is well adapted for pulpwood rotations for paper production.

P. elliottii is among the most productive of some 26 species of pines that are worked for oleoresin or gum naval stores worldwide.

Other products such as bark, needles or straw are widely used in mulching.

Further reading

Gurgel Garrido, L.M. do A. and Kageyama, P.Y. (1993) Changes in estimates of genetic parameters of resin production for *Pinus elliottii* var. *elliottii* as a consequence of thinning. *Revista do Instituto Florestal* 5, 123–131.

Harley, G.L., Grissino-Mayer, H.D., Franklin, J.A., Anderson, C. and Köse, N. (2012) Cambial activity of *Pinus elliottii* var. *densa* reveals influence of seasonal insolation on growth dynamics in the Florida Keys. *Trees: Structure and Function* 26, 1449–1459.

Zhao, D.H. and Kane, M. (2012) Differences in growth dynamics of loblolly and slash pine plantations in the southeastern United States. *Forest Ecology and Management* 281, 84–92.

Pinus halepensis Mill.

Aleppo pine

Importance

Pinus halepensis, a native species to the Mediterranean region, is a drought-tolerant, fast-growing conifer. It is not used for commercial forestry anywhere in the world, mainly because of its poor stem form and low-quality timber. It is, however, very widely planted for restoration of degraded sites, especially within its large natural range. Its tolerance of poor, degraded or stony soils makes it very suitable for this purpose. Although, *Pinus pinaster* is also used for this purpose, *P. halepensis* is preferred for drier sites and calcareous soils. In parts of the Mediterranean region, Aleppo pine is planted to improve water infiltration on hill slopes.

The plantation area of *P. halepensis* has expanded dramatically in the last century. This trend was especially marked in parts of Europe where wide agricultural holdings were abandoned following the imposition of set-aside policies in the European Union.

Botanical features

A 2- (occasionally 3-) needled, monoecious pine species, usually forming a slender single-stemmed tree up to 20 m tall (occasionally reaching 25 m) with deep roots. Not long-lived, with a life span up to 150–200 years.

Needles are pale green (usually 6–12 cm long, sometimes up to 15 cm; less than 1 mm wide), and are borne on silvery-grey branches, usually abscising after the second year; with marginal resin ducts and leaf sheath 8 mm long. Cones (moderately to highly serotinous) are conical (5–12 cm long), turned downwards, grey to reddish-brown, usually opening in the third year. Seeds are dark brown, mottled, 6–7 mm long with wings 18–28 mm long. Bark is greyish, smooth at first, but finally fissured and exposing the reddish-brown or orange inner bark.

Distribution

P. halepensis has a large 'naturalized' distribution in the western part of the Mediterranean region. Like most other Mediterranean pines, *P. halepensis* has been widely planted for many centuries.

P. halepensis is often a dominant component of coniferous forests at lower altitudes in the Mediterranean region. Forests dominated by *P. halepensis* comprise about 10% of the maquis shrubland/forest biome complex in the Mediterranean region. In conjunction with *Pinus brutia*, these two pines cover about 6.8 million ha in this region.

Silviculture characteristics

P. halepensis has poor tree form for sawtimber production, especially when grown in closed stands where it tends to lean, resulting in eccentric piths and a high proportion of compression wood.

Among the most importance ecological features of *P. halepensis* are its high tolerance of drought, nutrient-poor soils, frost (provenance-specific) and salt spray. On the other hand, *P. halepensis* is completely intolerant to waterlogging.

Adaptations of *P. halepensis* to fire include its early reproductive maturity and the high ability of the seeds to germinate and establish themselves in the immediate post-fire environment. There is usually prolific seedling recruitment after a fire, leading to dense, even-aged stands. However, measures to protect seedlings from browsing and grazing mammals are sometimes required. Partly because *P. halepensis* shows a low degree of self-pruning compared to other pines, forests of this species are highly flammable.

Silviculture practice

Plants are generally raised from seed in the nursery, though direct sowing is also successful in some areas. Seed germination is rather prolonged, but can be improved by stratifying the seed. Damping off caused by *Pythium* spp. might become problematic at this stage.

In terms of stand regeneration, good results have been obtained with spot sowing under shelter. In poor sites, good results are achieved with bare-root seedlings, or containerized 3–5-month-old greenhouse-grown seedlings. In hot and dry sites, suitable transplantation spacing for seedlings are 1.8–2 m, planting 1–3 plants in each hole and thinning after 6–8 years. Planting density can be reduced as seedling quality and methods of site preparation and weed control improve.

In terms of creating single-row shelter belts, the bushy habitat of *P. halepensis* determines the need of plenty of growing space; approximately 9–10 m need to be left on each side of the row.

Mycorrhizal inoculation (together with the application of urban compost) improved survival and growth of *P. halepensis* seedlings in semi-arid conditions and poor soils, but showed less benefit on richer soils.

There are records of negative impacts from competition with annual weeds and shrubs on poor sites.

When *P. halepensis* is planted for shade, it may be desirable to deliberately encourage side-branch development by cutting out the leader at a height of 2.4–4.6 m when it is 3.2–5.1 cm in diameter. When single trees are grown in the open for shade, trees are often pruned to 3–4.5 m.

Fire is the dominant factor to consider in the management of *P. halepensis* stands in the Mediterranean region. Biomass control, involving grazing, prescribed burning and chipping of forest thinning residues, have become fundamental in reducing the hazard of fire in these forests. However, in limestone areas, prescribed burning might not be advisable because of the extreme flammability of *P. halepensis*, *Quercus ilex* and *Juniperus oxycedrus*.

In terms of *P. halepensis* annual growth, assessments undertaken in several parts of the Mediterranean region showed a variation from 2.5 to 3.3 m^3/yr.

Diseases and pests

Among the fungal diseases known to affect *P. halepensis*, the following are probably the most widespread and damaging: *Coleosporium inulae* (a heteroecious foliar rust), *Cronartium flaccidum* (blister rust), *Crumenulopsis sororia* (a canker), *Gibberella fujikuroi* var. *subglutinans* (pitch canker), *Phellinus pini* (red ring rot), *Phellinus torulosus* (white decay), *Sclerotium bataticola*, *Sirococcus strobilinus*, *Sphaeropsis sapinea* (canker/crown wilt), *Thyriopsis halepensis* and *Diplodia pinea*. *P. halepensis* is also known to exhibit knot disease possibly caused by bacteria associated with *Pissodes castaneus*.

Among the insect pests of *P. halepensis*, the following are probably the most widespread and/or damaging: *Leucaspis pusilla*, *Leucaspis pini*, *Matsucoccus josephi* (Israeli pine bast scale), *Monochamus galloprovincialis*, *Orthotomicus erosus*, *Pityogenes calcaratus*, *Thaumetopoea pityocampa*, *Hylurgus destruens*, and *Cenopalpus wainsteini*. Secondary pests include borers (*P. castaneus*, *Tomicus piniperda*, *Orthotomicus erosus*, and *Monochamus galloprovincialis*) that usually cause problems only after trees have been weakened by environmental stresses.

In many areas of the current range of *P. halepensis* in the Mediterranean region, large-scale dieback in stands has been attributed to complex interactions between various factors including the use of inappropriate provenances. Dieback starts with progressive desiccation from the lower branches and finally death. Dieback is also aggravated by ozone damage, which, in some regions, predisposes *P. halepensis* to attack by *M. josephi*. The combination of dead trees and bush encroachment in *P. halepensis* stands increases the hazard of wildfires.

Uses

P. halepensis is not a species of major commercial importance, but it is a very important species in its natural range where its main uses are in agroforestry, soil conservation, erosion control, site restoration, and land reclamation. In its natural range and in the many parts of the world where it is planted as an exotic, it has considerable value as an amenity and ornamental tree.

The amenity purposes include its use as fuelwood and charcoal, windbreaks, shelter belts and shading. In the semi-arid parts of South Africa, it is used in community forestry projects to restore degraded areas and/or provide woodlots. In many temperate regions, *P. halepensis* is grown as an ornamental tree.

P. halepensis has a reported density of 460 kg/m^3 and yields a yellowish-brown, coarse-grained, resinous, moderately dense wood of poor quality which is sometimes used for rough constructional purposes, low-grade joinery, railway sleepers, box shooks, telephone poles, mine props, firewood and charcoal.

Formerly, *P. halepensis* was often tapped for resin, especially in Portugal and Spain, but this is now only performed in some parts of Greece. The bark has been used in tanning leather.

Further reading

Dahman, M. (1974) *Pinus halepensis*: its technological characteristics and utilization. *Bulletin d'Information, Institut National de Recherches Forestieres, Tunisia* 17, 9–12.

Girard, F., Vennetier, M., Guibal, F., Corona, C., Ouarmim, S. and Herrero, A. (2012) *Pinus halepensis* Mill. crown development and fruiting declined with repeated drought in Mediterranean France. *European Journal of Forest Research* 131, 919–931.

Maestre, F.T. and Cortina, J. (2004) Are *Pinus halepensis* plantations useful as a restoration tool in semi-arid Mediterranean areas? *Forest Ecology and Management* 198, 303–317.

Poirazidis, K.S., Zografou, K., Kordopatis, P., Kalivas, D.P., Arianoutsou, M., Kazanis, D., Korakaki, E. and Rigolot, E. (2012) A GIS-based integrated approach predicts accurately post-fire Aleppo pine regeneration at regional scale. *Annals of Forest Science* 69, 519–529.

Prévosto, B., Amandier, L., Quesney, T., Boisgelin, G. de and Ripert, C. (2012) Regenerating mature Aleppo pine stands in fire-free conditions: site preparation treatments matter. *Forest Ecology and Management* 282, 70–77.

Pinus jeffreyi Grev. & Balf.

Jeffrey pine

Importance

Pinus jeffreyi is an important Pacific Coast forest tree, with wood characteristics very similar to those of *Pinus ponderosa*. This species is an important forest resource for the lumber industry.

Botanical features

P. jeffreyi is a large forest tree, commonly reaching heights of 25–50 m, and rarely 60 m. Diameters of mature trees often reach 1.2–2.3 m, and there are older reports of trees over 3 m in diameter. The canopy growth form of *P. jeffreyi* is conical in younger trees. Mature trees form a more cylindrical and open crown, tending to form a flat top with age as height growth slows.

The bark of *P. jeffreyi* is dark reddish-brown in colour, with narrow plates separating deep fissures. Bark fissures and exposed newly exposed bark areas have a distinctive smell described as vanilla, pineapple or banana. This character distinguishes this species from the similar *P. ponderosa* which has little or no odour in shoots, needles or between plates of the bark. This odour in *P. jeffreyi* results from the presence of heptanes rather than terpenes in the foliage. The root system of *P. jeffreyi* typically has both a relatively deep taproot as well as extensive lateral roots. As a result, uprooting by wind is rare.

The foliage of *P. jeffreyi* consists of glaucous, bluish-green needles 12–25 cm in length, arranged in bundles of three. Needle retention is commonly 4–6 years, but less under stress from oxidant air pollution.

The oblong female cones of *P. jeffreyi* are 15–30 cm in length and 10–12 cm wide. There are numerous coarse scales on the cones with long prickles pointing inward such that open cones held in the hand have little or no sharpness. The seeds are 10–13 mm in length, with a large wing reaching to 30 mm.

The vegetative and reproductive growth phenology of *P. jeffreyi* varies between populations in relation to latitude and altitude. Branch elongation in the southern Sierra Nevada begins in mid-May and is over in early June. The elongation of new needles begins in June and continues through July and August into early September. New male and female cones are formed and pollination takes place in June or July, depending on the habitat. Fertilization takes place about 13 months later, and female cones reach maturity in the summer of the second season. Seeds are commonly shed from September to October.

Distribution

The natural range of *P. jeffreyi* extends from south-west Oregon south through much of California (mainly in the Sierra Nevada) in USA, to northern Baja California in Mexico. It is a high-altitude species; in the north of its range, it grows widely at 1500–2100 m altitude, and at 1800–2900 m in the south of its range.

P. jeffreyi has not shown good success as a plantation tree outside of its natural range, and thus introductions are limited. It was introduced to England in 1853, but there are reports that some trees were planted as early as 1838. If true, these may well have been originally introduced as *P. ponderosa*. In New Zealand, *P. jeffreyi* has grown more slowly in trials than *P. ponderosa*. It has also been introduced in trials in Europe, South Africa and Japan.

Silviculture characteristics

P. jeffreyi is the dominant forest tree species east of the crest of the Sierra Nevada and Cascade Range, and in the Transverse Ranges of southern California and Sierra San Pedro Martir of Baja California.

West of this crest, it is more commonly a co-dominant climax forest species of upper montane forests growing with *Abies magnifica*, *Abies concolor*, *Pinus monticola*, *Pinus contorta* subsp. *murrayana*, and *Pinus lambertiana*.

P. jeffreyi is tolerant of ultramafic soils, and is often dominant in these conditions, even on dry sites at fairly low altitude. *Calocedrus decurrens* is the most common associate on ultramafic soils, but a number of other conifer species may also be present.

On other soils, it only becomes dominant at higher altitudes where the usually faster-growing ponderosa pine does not thrive.

Silviculture practice

P. jeffreyi has not been widely planted and there is little information available on its silvicultural traits outside of its natural range.

P. jeffreyi typically begins to bear seeds when it reaches a height of 18–55 m, and produces a large seed crop at intervals of 2 to 8 years. Seeds germinate quickly in the spring after seedfall, but seedlings are shade-intolerant and thus subject to intense competition from a variety of understorey species. Variable degrees of dormancy are present depending on the provenance of the seed source. Some seed populations from east of the Sierra Nevada-Cascade crest apparently do not require stratification, but most sources germinate best after 60 days of cold, moist stratification. Although seeds are adapted for wind dispersal, a secondary dispersal by vertebrates caching seeds is also significant.

P. jeffreyi does not reproduce vegetatively by either sprouting or layering under natural conditions.

Seeds are commonly sown in forest tree nurseries in April to allow a complete growing season for development. Stratified seeds are sown at a depth of 6–8 mm with a density of 270–320 seedlings/m^2. Appropriate irrigation and fertilizer application aids in obtaining sufficiently rapid seedling growth to allow planting out after just a single growing season. Seedlings from provenances east of the Sierra Nevada-Cascade crest commonly require two seasons of nursery growth. Nursery soil management to control nematodes and root rots is usually carried out.

Competition with understorey vegetation makes natural regeneration of *P. jeffreyi* inefficient in commercial stands. Nursery-grown seedlings are planted out during the spring when soils warm sufficiently to allow root growth and water uptake. Survival rates above 90% have been attained in the Sierra Nevada on appropriate sites cleared of competing species.

Invasions of either shrubs or herbaceous perennials can provide a strong negative impact on the growth and survival of planted seedlings. Suppressed saplings of *P. jeffreyi* require 3–7 years after thinning before they begin to accelerate height growth.

P. jeffreyi grows more slowly than *P. ponderosa* during its sapling stage, but attains more rapid growth once it reaches pole size at 5–10 years of age. Managed stands of *P. jeffreyi* in the central Sierra Nevada reached a height of 7 m and a diameter of 16 cm after 20 years of growth. Slightly lower growth rates have been reported for a plantation in north-eastern California with trees 5 m in height and 15 cm in diameter at an age of 30 years. Old-growth stands of *P. jeffreyi* may include individuals 400–500 years in age, with diameters in excess of 1.5 m. Growth rates on ultramafic soils are relatively low. Although yield tables have not been produced for *P. jeffreyi*, observations support the suggestion that existing yield tables for *P. ponderosa* will provide good estimates for pure stands of this species.

Diseases and pests

Fungal pathogens are also a major problem for *P. jeffreyi*, with this species subject to two needle diseases, a limb canker, five different rust diseases, three major root diseases and various heart rots. Elytroderma disease caused by *Elytroderma deformans* is a needle-cast disease, which causes premature death of 1-year-old needles and has reached epidemic proportions in specific environmental sites, most notably lake basins and stream valleys with cold air drainage. Medusa needle blight (*Davisomycella medusa*) is a serious pest in local areas, particularly on poor sites in drought years, and may cause a significant decline in tree growth. Rust disease may also have serious outbreaks in stands of *P. jeffreyi*. These include stalactiform rust (*Peridermium stalactiforme*) which infects lower limbs and eventually crowns of young trees, filamentose rust (*Peridermium filamentosum*) in the crowns of mature trees, sweetfern rust (*Cronartium comptoniae*) which may kill young trees, tarweed rust (*Coleosporium madiae*) causing heavy defoliation in wet years and western gall rust (*Endocronartium harknessii*) forming large galls and trunk cankers leading to mortality on both young and mature trees. Although major outbreaks of this last species may be decades apart, this rust is ubiquitous through the range of *P. jeffreyi*.

Fungal root diseases of *P. jeffreyi* include annosus (*Heterobasidion annosum*), armillaria (*Armillaria ostoyae*) and black stain (*Leptographium wageneri*). Heart rot fungi infecting *P. jeffreyi* include species of *Lentinus*, *Fomes*, and *Polyporus*. Red rot (*Dichomitus squalens*) attacks *P. jeffreyi* in southern California.

P. jeffreyi is subject to a large number of serious pests and pathogens that impact growth rates and survival. Infections by western dwarf mistletoe (*Arceuthobium campylopodum*) are a serious problem throughout the range of *P. jeffreyi*. Heavy infections of this parasite cause severe witches' brooms and severely reduce tree growth, eventually leading to death. Infected older trees pass on the infection to understorey saplings. Dwarf mistletoe also predisposes many stands to bark beetle attack and it has been held responsible for 60–80% of *P. jeffreyi* mortality in drought years.

The most damaging insect pest on *P. jeffreyi* is the Jeffrey pine bark beetle, (*Dendroctonus jeffreyi*) which causes severe damage throughout the range of the species. The susceptibility to attack from the Jeffrey pine bark beetle is increased significantly by any conditions that reduce tree vigour. Such conditions include annosus root disease, mistletoe (*A. campylopodum*), and Elytroderma disease, as well as drought stress, fire damage and overstocking. Other species of bark beetle whose attacks precede or accompany the Jeffrey pine bark beetle include the California flat-headed borer (*Melanophila californica*), the red turpentine beetle (*Dendroctonus valens*) and two pine engravers, the emarginate ips (*Ips emarginatus*) and the Oregon pine engraver (*Ips pini*). The maintenance of stand health and vigour is the most effective control from Jeffrey pine bark beetle damage.

Other insect pests impacting *P. jeffreyi* include twig and needle scales, various defoliators, borers and tip moths, as well as a diversity of cone and seed feeders. The most destructive of the scale insects is the ponderosa pine twig scale (*Matsucoccus bisetosus*) which feeds on branches and trees of all ages. Two serious defoliators are common. Larvae of the pine needle sheath miner (*Zelleria haimbachi*) may destroy more than 75% of new needle production in local outbreak areas. Larvae of the pandora moth (*Coloradia pandora*) cause heavy defoliation in outbreaks lasting 2–4 years that occur with a frequency of 20–30 years. The western pine shoot borer (*Eucosma sonomana*) stunts needle growth, thereby retarding height growth of infected trees by as much as 30% annually. The ponderosa pine tip moth (*Rhyacionia zozana*) kills current shoots of saplings and young trees, chronically retarding growth. The fir coneworm (*Dioryctria abietivorella*) kills the terminal buds of saplings and poles, causing a forked main stem. Larvae of the pine reproduction weevil (*Cylindrocopturus eatoni*) may significantly increase mortality of saplings and young trees under conditions of water stress. Jeffrey pine seed worm (*Laspeyresia injectiva*) and ponderosa pine seed worm (*Cydia piperana*) are major seed predators in immature cones. Mammalian herbivores are a significant cause of damage and mortality in seedlings and saplings of *P. jeffreyi*. These include deer, rabbits, snowshoe hares, pocket gophers, porcupines and domestic livestock.

Uses

P. jeffreyi provides important forest cover for montane habitats throughout its range, where it may occur either in relatively pure stands or in mixed conifer associations.

P. jeffreyi has a wood density of 370 kg/m^3 and is a general-purpose wood with widespread use for construction grade lumber. Its characteristics are virtually identical to those of the more widespread *P. ponderosa*, and no distinction is made between the two species in commerce.

P. jeffreyi produces a natural heptane, rather than the terpenes characteristic of most pines. Pure heptane distilled from the tree was once used to develop the octane scale for gasoline. In the past, these extracts were occasionally utilized as a remedy for pulmonary ailments and tuberculosis.

Further reading

Burdon, R.D., Miller, J.T. and Knowles, F.B. (1991) Introduced forest trees in New Zealand: recognition, role, and seed source. 10. Ponderosa and Jeffrey pines: *Pinus ponderosa* P. Lawson et Lawson, *Pinus jeffreyi* Grev. et Balf. *FRI Bulletin* 124, 23 pp.

Salverson, W.G., Walker, R.F., Fecko, R.M., Frederick, W.B., Miller, W.W. and Johnson, D.W. (2011a) Influences of mechanized thinning and prescribed fire on natural regeneration in an uneven-aged Jeffrey pine stand. *Journal of Sustainable Forestry* 30, 654–676.

Salverson, W.G., Walker, R.F., Fecko, R.M., Frederick, W.B., Johnson, D.W. and Miller, W.W. (2011b) Influences of mechanized thinning and prescribed fire on understory vegetation in an uneven-aged Jeffrey pine stand. *Journal of Sustainable Forestry* 30, 823–849.

Pinus kesiya Royle ex Gordon

Khasya pine

Importance

Pinus kesiya is becoming a very important timber-wood producer, particularly in Southern Africa and other tropical regions. Wood properties of straight trees are often superior to those of other pine species. The problems associated with the species, such as poor form and coarse branches, can be reduced by selective breeding. Because the tree grows in a variety of soils and is tolerant to various pests and climatic conditions (wind, frost and drought) it is possible to grow the tree widely in subtropical and tropical areas. The production rate is moderately high, due to its rapid growth rate, vigorous germination and propagation in favourable conditions. *P. kesiya* is an important pioneering species and will colonize bare sites rapidly.

Botanical features

P. kesiya is a large tree, with roughly whorled branches and an oval crown in young trees, with a flattened or rounded crown when mature. The tree has a bole free of branches for the first 15–20 m when mature. Stems can be forked and show some sweep. Apical dominance is strong, and the young trees display marked phototropism. Under favourable conditions, *P. kesiya* can reach a height of 30 m and a girth greater than 1 m. Adult trees have 2.5–4.5 cm thick bark, deeply fissured and pinkish to reddish-grey, flaking in small, thick, irregular, sometimes silvery, plates.

Branchlets and young shoots are glabrous, often pruinose with a waxy bloom. Buds are terminal and pointed, oblong–conical to cylindrical and not resinous. Bud scales are narrowly acute, free at the tips with a scarious, lacerated margin. There are often two whorls of branches produced in 1 year, the spring internode being much longer than the summer internode.

Needles of adult trees are from 10–15 to 21–25 cm long, 0.5–1 mm broad, bright green, margin finely serated, stunted on shorter trees, slender, in fascicles of three (rarely two or four). The fascicle base is enclosed in a persistent sheath 1.5–2.5 cm long and each possesses 3–5 resin ducts.

The male strobili are 3.75–5 cm long, cylindroid to oblong, bright yellow or light brown, about 0.5 cm broad, brown scales which adhere to them when they fall soon after ripening. The female conelets are greenish-purple with scales mucronulate. Mature female cones are small, bright, lustrous, glossy brown, pendulous, often strongly reflexed, strongly persistent, ovoid to ovoid–conical, more or less symmetrical to asymmetrical, and 4–5 to 8–10 cm long and 4–5 cm in diameter. The cones are subsessile or borne or a short peduncle 0.5–1.0 cm long. The cones are usually solitary or borne in whorls of three. Cone scales thicken toward the apex, up to 1.8 mm long. Apophysis is pyramidal and convex.

The seeds are small with a short wing 1.5–2.5 cm long.

Distribution

P. kesiya is widespread in South-east Asia between 10°N and 30°N and a longitudinal range between 26°E and 119°E in a large, disjunct area. It grows in India, Myanmar, China (Xizhang and Yunnan), Laos, Vietnam, Thailand and in northern Philippines. The tree grows at altitudes between 350 and 3000 m above sea level, though concentrated at levels between 700 and 1200 m. *P. kesiya* is the only tropical pine that grows in the Eastern Himalayas.

P. kesiya has been introduced to Africa, South America, Central America, Australia and some Oceanic islands for timber production. In Zambia and Madagascar, it is cultivated particularly successfully. *P. kesiya* is also planted in India outside its natural range, including Andhra Pradesh, Kerala, Orissa and Tamil Nadu.

P. kesiya grows on a variety of soils and geological formations, though it chiefly grows in areas with sandstone covered with reddish clay, and quartzite and shale. It is even frequently found on exposed ridges and spurs and in gorges, and along watercourses among trees from the broadleaved species such as *Alnus nepalensis* and *Quercus serrata*.

Silviculture characteristics

P. kesiya is a fast-growing species. Natural regeneration is profuse on abandoned cultivated land with loosened and exposed soils. The tree is very light-demanding and will not even regenerate in its own shade. Reproduction will not occur beneath a leaf canopy, but if gaps are sufficient and if soil conditions permit, the young plants will increase in number.

Young seedlings are able to push their way through a moderate growth of scrub, particularly on dry and sunny aspects. Generally, however, the tree is not very tolerant of competition, and in these situations the base of the stem will often produce whorls, which gives the tree a candelabra-like form. In a dense crop, the suppressed trees will die off and the crop will thin itself out. On sites near the equator the species shows great variability; the reasons for this are still unknown.

The young crop needs protection from fire, but older trees are more fire-resistant, due to the thick bark that develops with age. Because the tree has an adaptable root system, the tree is moderately resistant to wind. *P. kesiya* will develop a long taproot in well-drained soils, while on compact shallow soils the root system spreads out near the surface. Once the tree is established it is fairly drought and frost-resistant.

Silviculture practice

Seeds kept in dry, sealed tins will last 1 year. Fresh seeds have reportedly had a fertility rate up to 95%, while seeds stored in dry tins have a rate of 65%.

Seeds are soaked for 24 h, during which time the viable seeds swell up and begin to germinate. Germination can be hastened by covering the beds with mulching. The seeds will germinate after approximately 8–10 days and the process is completed within 20 days. Germination percentage varies between 30–80%, and partial shading may be needed after germination. Pre-sowing seed treatments are not normally undertaken, as the benefits are small.

P. kesiya responds readily to grafting. In general there was a 70% success rate where grafting was done. Grafting is normally undertaken either under shade in a nursery or in the field, field grafting being particularly superior. Grafts in the field showed at best a success rate of 82–97% and reached an average height of 3.8 m in 26 months. Humidified cotton, vermiculite and polythene bags to protect graft unions from drying out did not significantly increase the percentage of success.

Propagation by air-layering has proved to be successful but can really only be expected in cool stations at high altitudes.

P. kesiya is one of the easiest species of pines to treat in this way.

Sowing into beds and then pricking out trees or direct sowing into pots are the two most common techniques of raising *P. kesiya*. Trees tend to grow much better in soils consisting of simple media, such as sandy topsoils, fertilizers being added only when necessary. Shading of plants is only needed when plants are in a delicate condition or in extreme heat, and is not widely done otherwise. The seedlings are sensitive to flooding and very dry soils.

Inoculation with mycorrhizae can be done through introducing pure cultures, duff from already successful trees, and planting mother trees with mycorrhizal roots with seedlings in the nursery.

Root pruning or wrenching is commonly done to control the root/shoot ratio. Two weeks before transplanting hardening off trees usually begins.

The seedlings can be transplanted after 4–6 months, when they are about 7–10 cm high. The seedlings are transplanted into polythene pots after about 30–45 days.

Site preparation ranges from minimum input to the more intensive and sometimes costly site preparation for industrial plantations. Intensive site preparation starts with the removal of existing vegetation, destumping, charcoal burning, soil levelling, ploughing to a depth of 25 cm and discing before planting. Bare-root planting is possible if conducted under favourable conditions. Planting bags should be removed to prevent root deformities. Planting should start the moment the first 30 cm of soil is moist in the early rainy season. Direct sowing is not widely practised, though is successful in areas with favourable conditions.

Weeding is very important during seed establishment, both on sites that have a relatively harsh dry seasonal climate, and on those with favourable conditions where weeds are more prolific. In industrial plantations weeding is done up to 8–12 times in the first 3 years. Mechanical weeding has proved to yield the best results.

The risk of fire damage decreases once the tree has grown to a height of 3 m and the bark is thick enough to withstand fire. Fertilizers need to be added in places where the soil conditions are deficient.

The spacing required for the species depends on the produce that is desired in the end. For maximum volume of fibre per unit area, a close spacing is needed (2 × 2 m). Sawlog production will need greater spacing or more thinning as the trees grow. The width of the row should ensure that weeding machines can enter in between the trees (2.7 × 2.7 m). Closer spacing will inhibit branch growth on the lower part of the stem. Tree spacing of 1 × 1 m has been practised in the Philippines, to ensure straight stem formation, while 1.8 × 1.8 m is recommended in Zimbabwe. Generally, however, spacing closer than 2 × 2 m is not recommended. In general it would appear that the closer the spacing, the greater the height, and the wider the spacing the greater the diameter. Because the pine has a heavy whorl formation of coarse branches in the lower part of the tree, early live pruning is necessary for access and better timber quality. Pruning ensures higher-graded timber and can be started once the tree is about 6–7 m tall.

Diseases and pests

Needle blight, stem disease, root disease, and heart and butt rots may be encountered in plantations. Damping off in the nursery can also cause problems.

Few insects attack *P. kesiya*. However, in South Africa, it has been heavily attacked by the pine woolly aphid (*Pineus pini*). Weevils, aphids, termites and moth larvae all attack this species. The species also needs protection from dense grass competition, frost and some mammals and birds while the plants are still young.

Uses

The species is used in site restoration and control of soil erosion; in the Philippines it is used as shade trees for coffee plants in agroforestry systems. Reported wood densities for this species range from 420 to 561 kg/m^3 and it is an excellent timber that seasons very well and, once dry, is relatively stable. As a rule the wood is clearly differentiated into hardwood and sapwood, hardwood being yellowish- to reddish-brown in colour and sapwood a cream-white to white colour. Generally the wood is straight grained and moderately fine in texture, though the wood can often be knotty and very resinous.

The wood is fairly heavy and relatively strong and hard. It saws without difficulty, makes good joints, works to a smooth finish with either hand-held or machine tools and takes paint and varnishes well. The sapwood is not very durable when it comes in contact with the ground, though the hardwood is resistant to termite attack and resists weathering fairly well when exposed to the elements. The timber is used for house and bridge construction, flooring, ceiling, panelling, joinery, furniture, telephone poles and mining timber. It has been used for fuelwood and charcoal manufacture. It is growing in importance for pulpwood.

In South Africa, relatively large resin pockets have been observed in the timber. An exudation of resin is usually expected with kiln seasoning. It should be converted soon after felling and open-stacked in the shade.

P. kesiya also produces resin of a high quality, but it is not widely tapped because the tree does not yield very freely. The oleoresin is rich in pinenes, which comprise 21% turpentine oil and 79% rosin. The main limitation to commercial exploitation of resins is the fact that *P. kesiya* grows best at high, often inaccessible altitudes making collection uneconomic.

Further reading

Baishya, R. and Barik, S.K. (2011) Estimation of tree biomass, carbon pool and net primary production of an old-growth *Pinus kesiya* Royle ex. Gordon forest in north-eastern India. *Annals of Forest Science* 68, 727–736.

Lumbres, R.I.C., Lee, Y.J., Seo, Y.O. and Calora, F.G. (2012) Development of site index curves for *Pinus kesiya* in the Philippines. *Southern Forests: a Journal of Forest Science* 74, 115–120.

Lung, N.N. (1988) Modelling the growth and productivity of tropical pine forests (*Pinus kesiya*). *Izvestiya Vysshikh Uchebnykh Zavedeniĭ, Lesnoiĭ Zhurnal* 6, 8–13.

Pinus koraiensis Siebold & Zucc.

Fruit pine

Importance

Pinus koraiensis is one of the most important and valuable timber tree species in north-east China. A five-leaved pine, it has a straight trunk, good wood quality and high wood productivity. It is hardy but slow growing. The timber is resistant to decay. The resin, bark and seeds also have varied uses. It is planted not only for soil and water conservation, but also for ornamental purposes due to its long lifespan and attractive, exuberant and verdant crown.

Botanical features

An evergreen pine tree with a straight and full trunk, pyramidal in shape, up to 40 m high and 2–2.5(–3.5) m dbh, with stout, spreading or ascending branches. Branchlets stout, green, pubescent (yellowish-brown), glabrous after 2 years; winter buds cylindric–ovoid; apex with a short sharp point, 10–18 mm long, resinous; scales appressed, lanceolate, acute, reddish-brown, the upper ones oblong–acuminate, tips free. The bark is thin, grey or greyish-brown, peeling off in irregular flakes and reddish-brown beneath. The bark can be divided into coarse bark and fine bark, with the former coarse and thick, dark grey-brown with rectangular checks and uniform edge, whereas the latter is smooth and thin, dark greyish to red-brown, with small and shallow scaly or stripe checks. The root system is shallow without a taproot; the lateral roots expand over a large area horizontally.

Leaves occur in bundles of five needles, 6–15 cm long and over 1 mm wide, rather stiff, apex slightly blunter and obtuse, triangular, margins very rough, outside dark green, dorsal side without stomata, inner side white-green, with 6–7 stomatic lines; resin ducts medial; leaf sheath deciduous.

P. koraiensis is monoecious; male flowers are red to yellow, 2 cm long, globular; female flowers are green, or with a purple tinge, borne at the ends of new branches. Cones are sub-terminal, solitary or with a few together around the head-leader, short-stalked, cylindric–conical, with a blunt apex, 9–14 × 5–6 cm, bright brown when ripe, resinous, opening when ripe and falling with the stalk; scales broad and wedge-shaped, leathery and woody, outside convex, longitudinally streaked, wrinkled, margin sharp and slightly wavy, apex obtuse, recurved. The seeds are oblong–obovate, 15–17 × 9–11 mm, greyish-brown, edible, not winged and free of spines.

Flowering period occurs in June and fruits ripen in the September of the next year. The sap of *P. koraiensis* begins to flow generally in the latter part of April, when the average temperature is above 4.5°C and soil temperature above zero. Buds begin to expand in the middle of May when the average temperature is 10–14°C and new branches sprout from the middle of May to the start of July when the temperature is 14–17°C. Flower buds appear in early June when the average temperature is 14.8°C and open in late June. Pollen dispersal lasts for 3–5 days and needles fully expand at the same time when the temperature is over 18°C. Fertilization is not fully completed until May of the following year, with cones ripening during September. In the winter, *P. koraiensis* is normally dormant from the beginning of October when the temperature is less than 10°C.

Distribution

In China, *P. koraiensis* is mainly naturally distributed throughout the Changbai, Wanda and Lesser Xing'an Mountains, in the Provinces of Heilongjiang and Liaoning between latitudes 49°28' N and 40°26 'N.

In Russia, it is mainly distributed in the south of Boli and along the coast, concentrated in the east and west of the Xihete mountains and the south of the outer Xing'an mountains to the north of Heilongjiang branches on the west of Heilongjiang River (eastern Manchuria and south-eastern Siberia). In the North Korean Peninsula, it occurs on the Gaima Plateau, Langlin and Xianjing Mountains, the Motian Ridge in the north and sub-mountain zones in the south. In Japan, it grows mainly in central Honshu, with a disjunctive population in Shikoku.

The altitudinal distribution of *P. koraiensis* is mostly subalpine, ranging from about 500 to 1300 m in China, 600–1500 m in Korea, and 1050–2600 m in Japan. It occurs from 500 to 1200 m in the Changbai Mountain region and 500–1050 m in Xiao Xing'an Mountain forests.

P. koraiensis has been introduced to the Shangdong Peninsula, Taishan Mountain, Inner Mongolia (Nei Menggu), and Beijing areas of China. It has also been introduced to Ontario (Canada) and grown as an ornamental in regions of Canada with climates similar to its native Asian environment. It performed too poorly in Southern Africa to merit further attention as a plantation tree.

Silviculture characteristics

P. koraiensis has a degree of shade-tolerance when juvenile, but is more light-demanding when mature and grows best in full sunlight. It is best practice to plant *P. koraiensis* in mixed forests with broadleaved tree species in order to ensure optimum conditions of temperature (warmer) and moisture (higher air humidity) in the tree stand.

Silviculture practice

P. koraiensis is mainly propagated by seedlings. Seed is collected in clonal seed orchards and seed production stands, to ensure good quality. Trees start producing seed after 15 years, bearing abundant seed crops after 30 years, at intervals of 2–5 years. Cones should be stored in a ventilated and dry place after collection until seed moisture content falls to 9%, when seeds are placed in cold storage at 4°C in a sealed container.

Nursery beds should be made using moist, fertile and acid sandy loam soil with good drainage. Stratification of the seed in damp peat moss for 3 months at 3°C, exposing the seed to at least 1500 h of chill interval, either in a refrigerator or a stratification pit outdoors is recommended. Indoor natural stratification can enhance germination rates up to 64%.

In spring, when soil temperatures at a depth of 5 cm underground reach above 8°C, sowing can be done at a rate of 300–350 g/m^2. Old pine land is desirable for planting to ensure that *P. koraiensis* develops good mycorrhizal roots and therefore develops properly. Normal nursery practices can then be followed, such as irrigation, bird repelling, weeding, soil loosening and fertilizer applications. Seedlings can be transplanted 2–3 years after sowing. Care should be taken to prevent root damage when lifting. Afterwards, seedlings (1.0–1.5 m) can be stored in cool, well-ventilated conditions before planting out. Afforestation is mainly carried during the spring after soil has thawed to a depth of 15 cm. Afforestation methods include pot planting in sites with a thin, hard soil layer and poor water supply, and slit planting in areas with deep, loose soil and sufficient rainfall.

Saplings grow slowly, resulting in late canopy closure. Tending is generally performed for 5–7 years after planting. Some weeds or other shading environment should be maintained in the planting year and frost prevention methods such as covering grass or grass curtains should be employed in winter.

P. koraiensis is mainly planted in mixtures with *Larix* (larch) species, and deciduous broadleaved tree species, such as *Betula* spp., *Quercus mongolica*, *Fraxinus mandshurica*, and poplars (*Populus* spp.). Broadleaved trees usually grow faster than *P. koraiensis* during the juvenile stage, and help to shade young seedlings. In addition, moisture in the mixed stand is improved, disease and pest problems are reduced, and the litter of broadleaved trees can improve the fertility, physical and chemical properties of the soil, as well as avoiding soil acidification. The total production of mixed stands has been found to be greater than that of pure *P. koraiensis* plantations by up to about 28 m^3/ha after 21 years. The most viable way to produce larger-sized pine timber from *P. koraiensis* is to plant it in mixtures with natural broadleaved trees or other cultivated valuable tree species rather than in pure stands.

After canopy closure, thinning and pruning should be performed on lower branches during November–March. The initial period and thinning intensity of *P. koraiensis* are determined by planting density and growth vigour. In artificial pure forests, with an initial density of 4400 trees/ha, the first thinning is best made 14–15 years after planting. Selective cutting, including breeding selective cutting and management selective cutting, is mainly employed for multi-function and sustainable use of *P. koraiensis* forest.

Diseases and pests

P. koraiensis is infected by the fungal pathogen *Cronartium ribicola* on sites with poor drainage. This disease usually causes the canopy to grow in the shape of a broom, and is controlled by improving hygiene status after canopy closure, clearing infected plants and alternative hosts (*Ribes* spp.) in the autumn and selection of appropriate sites. *Armillaria mellea* is also frequently observed in natural and secondary forests, and in pine plantations. This pathogen usually causes serious tree die-back, and is controlled by clearing infected trees and avoiding root damage by frost. *Rhizoctonia* spp. (for example, *Thanatephorus cucumeris*) and *Fusarium* spp. can attack seedlings in the nursery, but fungicide applications to nursery beds or sowing rows can prevent serious damage.

Other fungal pathogens recorded for *P. koraiensis* include *Matsucoccus koraiensis*, *Bursaphelenchus xylophilus* (the causal agent of pine wilt), *Bursaphelenchus pinophilus*, *Bursaphelenchus mucronatus*, *Tympanis* spp. *Leucostoma* spp. and *Lophodermium maximum*.

Important insect pests include the yellow-spotted pine weevil (*Pissodes nitidus*) and *Pineus* spp., which usually cause serious damage to seedlings and juvenile trees. Other insect pests recorded for *P. koraiensis* include the black-tipped sawfly *Acantholyda parki*, *Cinara pinikoraiensis* and *Dendrolimus superans*.

Uses

P. koraiensis is one of the most important timber species in the north-east of China. It is mainly used in afforestation of felled areas and waste land, in secondary forest improvement and for soil and water conservation.

P. koraiensis has good-quality timber with a light and soft heartwood, straight grain and medium mechanical strength and structure. It is easy to process, resistant to decay and has good painting properties. High-temperature drying might be used to reduce drying times in seasoning *P. koraiensis*. The specific gravity is 0.44 and shrinkage volume 0.46%. It is widely used for construction, telegraph poles, sleepers, bridges, vehicle bodies and ships. It is also suitable for carving, plywood, sports equipment and musical instruments.

P. koraiensis is rich in resin. Tannin can be extracted from its bark, and pine needle oil from its needles. The oil content of the seed is about 25% and above 70% for the seed kernel. The oil is of high edible and nutritive value, and is used both for food and in industry.

Further reading

Choi D.S., Kayama M., Jin H.O., Lee C.H., Izuta T. and Koike T. (2006) Growth and photosynthetic responses of two pine species (*Pinus koraiensis* and *Pinus rigida*) in a polluted industrial region in Korea. *Environmental Pollution* 139, 421–432.

Qi, H.R., Zhang, A., Zhang, F. and Tan, X.R. (1991) Research on *Pinus koraiensis* plantations. In: Shi, K.S., Deng, J.F., Ding, Y.Y., Li, W.C. and Chen, Z.D. (eds) *Development of Forestry Science and Technology in China*. China Science and Technology Press, Beijing, pp. 137–143.

Wang, H., Shao, X.M., Jiang, Y., Fang, X.Q. and Wu, S.H. (2013) The impacts of climate change on the radial growth of *Pinus koraiensis* along elevations of Changbai Mountain in northeastern China Forest. *Ecology and Management* 289, 333–340.

Pinus massoniana Lamb.

Masson pine

Importance

Pinus massoniana is a fast-growing tree producing high-quality wood. This species is the basis of plantation programmes in southern and central China. It naturally regenerates very easily and is often found as a pioneer species of wastelands. It has multiple uses, such as the production of wood, pulp and resin, making it one of the most important coniferous tree species in China.

It has two main disadvantages: it can become an aggressive colonizer of uncultivated land and it is particularly susceptible to certain insect pests or plant diseases that can seriously affect its growth.

Botanical features

A tree with a spreading or umbrella-shaped crown. On good sites in natural stands, *P. massoniana* typically achieves a height of 30–35 m and a dbh of 70–80 cm, although it may exceed 100 cm. The bole is typically straight with some twists, cylindrical, and is often clear of branches for 50–60% of its total bole height, even when grown in the open.

Bark is red-brown in the upper stem, and grey-brown in lower parts, and rifts into irregular scale-like pieces. Large branches are slanting or almost level in older trees. It produces one flush of younger twigs each year (or sometimes two flushes at the southern distribution margin); 1-year-old twigs are light brown, with short hairs giving a whitish appearance.

There are 2 (rarely 3) needles per fascicle, 12–20 cm long, about 1.5 mm wide, slender, drooping, with thin-odontoid margins and 4–7 marginal gum ducts.

Flowers partheno-monoecious. Male strobili yellow, clustered at base of new twigs, 2 anthers per stamen, pollen grain with pneumathode. Female strobili purple, ovulate, one or more at top of new twig, with spiral ovuliferous scales (2 ovules per scale). Cones are ovoid–cylindrical, 4–7 cm long, green (when 1 year old), becoming brown (at 2 years old). Seeds are elliptical, 7–8 mm long, 2.5 cm wide, with wing.

Flowering begins in March in southern areas of its distribution, and about a month later in northern areas, and lasts for about 1.5 months. Cones mature in July–August of the following year.

Distribution

P. massoniana is widely distributed in central and southern China, with a latitude range of 12° and a longitude range of 17°. It is also native to northern Vietnam.

There are few successful plantations outside China, although it has been tried in many countries, mainly as part of species trials, for example in Africa, Jamaica, Australia, Sabah, Japan and India. In northern Vietnam, *P. massoniana* plantations have been established on a commercial scale.

Silviculture characteristics

P. massoniana is a light-demanding species, and even seedlings need adequate light conditions. The root system is usually well developed (both main and lateral roots) and long. Natural regeneration is high, even on less fertile sites. The life-span of *P. massoniana* can exceed 250 years.

Silviculture practice

P. massoniana seed is collected from mature (brown) cones. Mature seed is classified as orthodox and stores well under appropriate conditions.

In the northern zone of its natural area of distribution in China, *P. massoniana* seedlings are raised in nurseries, with seed sown at a rate of 120–150 kg/ha and a growth period of 8–10 months. In contrast, in southern China container-grown seedlings reach a plantable size in 5–6 months, sometimes in 3 months.

P. massoniana roots form ectomycorrhizal associations with several species of fungi.

The species cannot compete well against natural vegetation under good site conditions, so site preparation is advantageous in nursery and later in the forest stand.

P. massoniana may be planted in pure stands, or in mixed plantations with other coniferous or with broadleaved species such as *Quercus acutissima*, *Quercus variabilis*, *Schima confertiflora* and *Cunninghamia lanceolata*.

In China, *P. massoniana* is also grown in plantations, primarily for resin and essential oil production, especially in areas along the southern margin of its distribution zone.

On good sites in southern China, average heights of 15–28 m and dbh of 28 cm have been recorded at 30 years, 28 m and 45 cm at 50 years, and 29 m and 60 cm at 60 years. Rotations of 30–35 years are generally practised.

Diseases and pests

In most areas of China, especially moderately hilly or lowland areas, *P. massoniana* plantations are often damaged or even destroyed by the pine caterpillar, *Dendrolimus punctatus*. The margarodid scale *Matsucoccus matsumurae* and the pyralid *Dioryctria splendidella* (*Dioryctria sylvestrella*) often damage twigs, cones or stems, and may even be responsible for stem twisting.

The plant parasitic nematode *Bursaphelenchus xylophilus* (the pine wilt nematode), and its insect vector (*Monochamus alternatus*), which are widespread along coastal areas of China, have caused epidemic damage to *P. massoniana* in some areas of this region.

Uses

P. massoniana is an important component of many natural plant communities in China. Its easy natural regeneration and high survival rate even on infertile sites make *P. massoniana* suitable for establishment in many marginal or mountainous areas within its natural range of distribution in China.

P. massoniana wood has a density of 390–490 kg/m^3 (at 15% moisture content) and is widely used for construction, as fuelwood and to supply the pulp and wood fibre industries. Recently, the wood has been used mainly for pulp production; the fibre is also used to produce rayon materials. The wood is also used for a range of purposes, including boatbuilding, containers and packaging, posts, mine poles and railway sleepers. The timber is susceptible to termite attack, and is not resistant to wood decay fungi. *P. massoniana* wood is also used for charcoal and activated charcoal manufacture.

In China, pine resin is a very important export chemical product, as it is the raw material from which colophony (a gum rosin) and resin oils are produced.

Needles are used for the production of a chlorophyll–carotene pine needle oil, for needle paste (used as a feed additive), tannin extraction and fodder.

Additionally, the resin, needles, pollen, bark, cones and seeds are all used in traditional medicine in China. *P. massoniana* bedlogs are used for edible mushroom production.

Further reading

Yuan, W.K., Sun, F.F., Wen, D.Z., Guo, Y.Z. and Zhao, P. (2008) Tree-ring growth patterns of Masson pine (*Pinus massoniana* L.)

during the recent decades in the acidification Pearl River Delta of China. *Forest Ecology and Management* 255, 3534–3540.

Zhang, L.Y., Deng, X.W., Lei, X.D., Xiang, W.H., Peng, C.H., Lei, P.F. and Yan, W.D. (2012) Determining stem biomass of *Pinus massoniana* L. through variations in basic density. *Forestry (Oxford)* 85, 601–609.

Zhu, X.C. (1980) *Pinus massoniana*. China Agriculture Press, Beijing, 254 pp.

Pinus montezumae Lamb.

Montezuma pine

Importance

Pinus montezumae is one of the most important tree species in Mexico. It is distributed from the southern part of Coahuila state to the southern states of Mexico, Oaxaca and Chiapas. In central Mexico it is important for saw timber and pulpwood. *P. montezumae* tolerates a wide range of environmental conditions, and so it is used for recovering eroded areas and site reforestation.

Botanical features

P. montezumae is a large (20–30 m) tree with an irregular, rounded crown and long stout branches spreading to drooping needles. Bark is reddish-brown, rough and irregular, fissured. Twigs are uninodal shoots, of stout diameter with a very rough, brown surface. Buds are long, ovoid, reddish-brown, and resinous.

Leaves usually occur in groups of five in a cluster; crowded on the branchlets, lasting for about 3 years, mid-green to grey-green or bluish-green, erect or spreading, 17.5–25 cm long. Margins minutely toothed, apex a horny point; resin canals median; basal sheaths 3–5 cm long.

Cones are 9–15 cm (rarely up to 20 cm) in length, 1–3 per cluster, spreading or drooping position; mostly short persistent; open at maturity; slightly oblique and curved; long ovoid to conic; dull brown colour; stalk very small up to 1.5 cm, usually persistent on twig with basal cone scales; umbo dorsal and slightly pyramidal in shape; apophysis keeled; prickle small, usually deciduous.

Seeds are 6–7 mm long, with dark wings (20 × 7 mm). Cleaned seed averages 46,000/kg.

Flowering occurs from about 14 years old. Male flowers are produced at the base of twig buds, and female flowers near the tip; these can vary from 1 to 5 per bud. Female flowers sprout about 1–2 weeks after the staminate cone sprouts and they stop sprouting 2 weeks before pollen shedding ends. After the female flower is pollinated, the scales close, remaining at a size of 13–22 mm until fertilization the following spring. In Mexico, male flowers sprout in late February and pollen shedding occurs during March and into the first week of April. Conelets start growing the spring following pollination, reaching full size during July–August and maturity in December to mid-January.

Distribution

P. montezumae is one of the most widespread commercial timber trees in Mexico. The natural range of distribution for this species is from the southern part of Coahuila State to the southern states of Mexico. It also occurs in Guatemala in Central America.

P. montezumae grows in subtropical, warm-temperate and cool-temperate climates, growing fastest under warm-temperate conditions at altitudes of 2400–2700 m. In southern Mexico, in Chiapas state, it is found between 1900–3300 m. In central Mexico, in Morelos state, its altitudinal range is between 2300 and 3350 m. In north-western Mexico, in Durango state, it is found from 2300 to 2700 m. Within Mexico, *P. montezumae* can be found between latitudes of 16° to 26°N.

P. montezumae has been introduced in South Africa with different degrees of success, probably because of its several forms. It has also been successfully introduced in Queensland (Australia), Kenya, Malawi and the Central Africa Republic.

Silviculture characteristics

P. montezumae is a mostly shade-intolerant species, although it may withstand light shade. It is a sub-climax species and it requires fire to regenerate and maintain its stands otherwise invaded by species such as *Pinus ayacahuite* and *Abies religiosa*. It also tolerates moderate frost.

Silviculture practice

P. montezumae cones are mature from December to mid-January. Seed dispersal in natural forest stands occurs from February to June, reaching a peak in the middle of March. *P. montezumae* does not reproduce by vegetative means.

Propagation for forest plantations is usually carried out under nursery conditions for 12 months before transplanting at the plantation site. Root pruning intensity and frequency is recommended every 3 months at a depth of 10–15 cm. In studies comparing container and bare-root systems, container-grown plants produced more biomass than bare-root seedlings.

P. montezumae can be planted as pure forest stands for saw timber production, producing a mean annual increment in volume of 5–10 m^3, depending on site quality. Commercial rotation age can be from 20 to 30 years in forest plantations, and from 50 to 60 years for natural stands.

P. montezumae only reproduces via seed. In natural forest stands, it grows from 1.9 m in height and 5.5 cm dbh at 10 years of age, to approximately 30 m in height and 46.5 cm dbh after 70 years. The approximate tree density at 10 and 70 years is 6700 and 63 trees/ha, respectively.

In central Mexico, *P. montezumae* site quality at 50 years old varies from 22 to 25 m for poor and rich sites, respectively.

A competition index developed for *P. montezumae* showed that lateral surface area of the crown was the best indicator of competitive interactions between trees, and it is highly correlated to the periodic annual increment in volume of the tree.

P. montezumae forest management can be achieved by applying forest management methods for even-aged stands, using a seed tree regeneration system and a commercial rotation age from 60–70 years. Thinning can be performed every 10 years, starting at 20 years old and continuing until 40 or 50 years of age, depending on the final rotation age.

In *P. montezumae* forest plantations the commercial rotation age is 20–30 years, depending on whether the production objective is for cellulose or saw timber. The ideal initial number of trees is 2500/ha.

Diseases and pests

P. montezumae serves as host for several insects. These pests attack the cones, seeds, leaves, stem and branches of the tree. Some diseases, such as damping off, are common in this species during the nursery stage.

Insects attacking the cones and seeds include *Conophthorus ponderosae* and *Megastigmus albifrons*, whereas defoliating insects that have been reported to attack *P. montezumae* include *Zadiprion townsendi*, *Neodiprion gillettei*, *Halisidota alternata* and *Chionaspis pinifoliae*. However, damage from these species has not been evaluated. Other agents causing damage to cones and seeds of *P. montezumae* are *Cronartium conigeum*, *Laspeyresia* spp., *Cecidomyia* spp., *Asynapta* spp., *Dioryctria* spp., and Torticidae. There are currently no controls available for these pests due to the small effect they have on cone and seed production.

There are several bark beetles that attack *P. montezumae*, such as *Dendroctonus mexicanus* and *Synanthedon cardinalis*, the exact species depending on the altitude. While they usually kill trees that are already weakened, healthy and vigorous trees are sometimes attacked. Bark beetle control is usually performed using the fell-and-abandon method, which consists of felling the attacked tree, debarking it, burning the bark and abandoning the tree or selling it as lumber.

Chrysobothris inaequalis is a borer that attacks dead *P. montezumae* trees as well as the branches of trees weakened by fire. No control is usually applied.

In central Mexico, dwarf mistletoe (*Arceuthobium*) is commonly found in *P. montezumae* forest stands. It causes a decrease in growth increments, but does not kill the trees and is not controlled.

Uses

In central Mexico, *P. montezumae* is one of the most important pine tree species, growing in 39 of 43 counties in the State of Mexico. This region is characterized by its high demographic concentration, which exerts strong pressure on natural resources through uncontrolled grazing, forest fires, illegal timber harvesting, compacting of soil, soil erosion and decreased water infiltration. *P. montezumae* has therefore been extensively used in reforestation programmes in these states in order to recover degraded soils, prevent erosion and conserve soil.

P. montezumae produces strong, resinous, white wood of fine texture. It is classified as a light pine with an average specific gravity of below 0.45. Tracheids are 20–44 μm in diameter (mean value of 35 μm).

Wood obtained from *P. montezumae* has a medium contraction (mean tangential contraction of 8.66–8.22% and mean radial contraction from 4.24% to 4.08%). Due to this stability, it can be used in building construction and for parquet flooring and furniture. It is one of the Mexican pine species used for plywood manufacture.

P. montezumae is one of five Mexican pine species used for tapping in both Jalisco and Michoacán States, with a mean value of 3.5 kg of gum produced per tree per season.

Further reading

Blanco-García, A., Sáenz-Romero, C., Martorell, C., Alvarado-Sosa, P. and Lindig-Cisneros, R. (2011) Nurse-plant and mulching effects on three conifer species in a Mexican temperate forest. *Ecological Engineering* 37, 994–998.

Rodríguez Franco, C. (1997) La silvicultura de *Pinus montezumae* Lamb. en la región central de México. *Ciencia Forestal en México* 22, 91–115.

Viveros-Viveros, H., Sáenz-Romero, C., López-Upton, J. and Vargas-Hernández, J.J. (2007) Growth and frost damage variation among *Pinus pseudostrobus*, *P. montezumae* and *P. hartwegii* tested in Michoacán, México. *Forest Ecology and Management* 253, 81–88.

Pinus nigra Arnold

Black pine

Importance

Pinus nigra is one of the most important silvicultural species in central and Southern Europe, particularly on account of its adaptive and pioneer qualities. It is extremely useful in reforestation projects, including those in difficult soil conditions.

Botanical features

Young *P. nigra* trees are tall and slender, gradually becoming rounder with age. The species grows to a height of 15–35 m, except in the case of the Corsican pine (*P. nigra* subsp. *laricio*) which frequently reaches heights of 40(–50) m. Stem dbh often reaches 100 cm.

The branches grow horizontally, slightly bent upwards in Austrian pines (*P. nigra* subsp. *nigra*) and slightly bent downwards in Turkish varieties (*P. nigra* subsp. *pallasiana* var. *caramanica*). In Corsican pine (*P. nigra* subsp. *laricio*) the stem is very straight and the branches are very thin and sparse (airy crown effect), whereas in Austrian pines (*P. nigra* subsp. *nigra*) the stem often forks fairly near the ground and branches are thick and nodose.

Current year branches are glabrous and shiny; the colour varies both throughout the growth season and among the different subspecies. Branches turn from orange to reddish-brown in *P. nigra* subsp. *salzmannii*; from ochre-yellow to brownish-red in *P. nigra* subsp. *laricio*, and from grey to yellowish in *P. nigra* subsp. *nigra*. In Crimean pines (*P. nigra* subsp. *pallasiana*) the branchlets range from yellow to reddish-brown.

Needles are diploxyl with a persistent sheath, growing in pairs. They are (4–)8–16(–24) cm long and (1–)1.2–1.8(–2.1) mm thick; have a semi-circular cross-section and the tip is always pointed, although the needle is not always sharp. The base of the needle is protected by a persistent sheath about 10–20 mm in length.

P. nigra is a monoecious species. Microsporophylls are located at the base of the current year's branch and appear at the beginning of the bud's lengthening stage. The macrosporophylls are contained in erect structures in the subapical part of the current year's branch, so that they become visible at the end of the lengthening phase. They are clustered in groups of 2–4 on a short pedicle. Mature cones are subsessile or slightly pedicled. Cones are shiny and not very large: (3.5–)5–8(–12) cm long and 2–4(–4.5) cm wide. The scale is 2.5–3 cm long and well lignified; its apophysis, slightly raised, is dorsal, and the umbo has a short mucro.

Flowering occurs in spring, between April and early June. Pollination is anemophilous. Cones ripen in the autumn of their second year and are shed the following spring after having scattered their seeds, but – usually – are still intact. Needles persist on the branch on average about (3–)4(–8) years.

Distribution

This very long-lived mountain pine is found around the northern Mediterranean basin (and in western North Africa) and spreads as far east as Anatolia (Turkey) and the Crimea (Ukraine). Currently the distribution area of *P. nigra* is highly fragmented and involves almost exclusively mountainous areas.

The westernmost taxon, *P. nigra* subsp. *mauretanica* (or *P. nigra* subsp. *salzmannii*), is found in Morocco and Algeria. In central and eastern Spain, the Pyrenees and the Cevennes area of France, *P. nigra* subsp. *salzmannii* is found.

P. nigra subsp. *laricio* grows in the central and southern part of the overall *P. nigra* distribution area (Corsica, Sicily and Calabria). *P. nigra* subsp. *nigra* is the most widespread subspecies in south-east Austria, central and north-east Italy and in the Balkan Peninsula (including Greece). In the easternmost part of the distribution area we find Crimean pine (*P. nigra* subsp. *pallasiana*), which forms several forests in the south-east Balkans, in the Crimea (Ukraine), in Anatolia and on Cyprus.

It is used for reforestation projects throughout the area of its natural distribution, and is utilized over the whole of its ecological range. Such afforestation is mainly for watershed and soil protection, and the restoration of degraded lands. It is also planted in Western, central and Eastern Europe (Irish Republic, UK, the Netherlands, Belgium, Germany, Switzerland and Romania) for protection and production purposes.

Outside Europe, it has been introduced into New Zealand and the USA.

Silviculture characteristics

Over large areas of central and Southern Europe and the Balkans, *P. nigra* is the main species used in reforestation. It is planted in pure stands.

Silviculture practice

The seeds are not very large (5–7 mm long), slightly cuneiform and compressed, the wing several times longer than the seed. The weight of 1000 seeds ranges from 20 to 30 g. Among Corsican pine (*P. nigra* subsp. *laricio*) varieties, var. *calabrica* is the one with the smallest seeds (80,000 seeds/kg), whereas var. *corsicana* has the largest (60,000 seeds/kg). The yield of 100 kg of cones is 2–3 kg of seed.

P. nigra reaches sexual maturity at about 15–20 years of age (40–50 years in dense stands or in difficult stand conditions). Generally, fructification occurs annually, with a large seed crop every 2–5 years.

Germinative capacity is fairly high, 80–90%. Seeds are gathered from October to January in trees 30–90(–120) years old. Seeds are orthodox and may be conserved for up to 5 years at 5°C.

P. nigra propagates essentially by seed. Spring sowing is possible after a period of 30–45 days of vernalization (but pre-refrigeration is not necessary). On average, 8–12 kg of seed/ha is needed for direct sowing.

Sowing is carried out in April with 2.5–3.5 kg of seeds in a seedbed, sown at a depth of 0.5–1 cm. Transplanting is normally carried out after 1–2 years. No shading is needed. Protection from hard frosts, however, may be necessary depending on the provenances.

Afforestation programmes usually use 1-year-old seedlings to avoid the risk of damage to the taproot. If transplants are used, they must be planted out within 2 years of transplanting (1+2 or 2+2) in order to avoid the risk of transplant stresses and growth problems. In conditions of not very severe water stress, transplants will be almost 100% successful. Normally, planting density is 1500–2500 plants/ha. Roots are normally well-colonized by mycorrhizal fungi. Shelter, watering or addition of fertilizer are not necessary.

Since *P. nigra* does not tolerate too much lateral shade, it is advisable to thin out stands when the trees are aged about 25–30 years, and thinning should be repeated every 10 years. Thinning in plantations usually begins when a height of 10 m is attained. Under this regime, a final stand density of >500 plants/ha at age 100 years will be achieved. More intense thinning (leaving only 140 trees/ha) will result in the development of *P. nigra* trees of large dimensions and also the regeneration of broadleaved species.

Selective felling or clear-cutting with the retention of seed trees (50–100 trees/ha) are not recommended systems for *P. nigra*. Better results may be obtained with clear-cutting of smaller areas, normally on strips no wider than 30–40 m, occasionally also as holes or squares. Excellent results have been achieved with consecutive felling of groups of Corsican pine (*P. nigra* subsp. *laricio*). Stand regeneration may be artificial (by direct sowing or use of planting stock) or natural. In the latter case, natural regeneration will be more effective if performed on ploughed soil and with no weed competition. In such conditions densities of 30–40,000 seedlings/ha can be reached.

The most common form of management is that of even-aged high forest. *P. nigra* subsp. *laricio* is the most important subspecies for timber production. Calabrian pine (*P. nigra* subsp. *laricio* var. *calabrica*) has, at age 100 years, a considerable stand volume (450–900 m^3/ha) with stand densities of 350–700 trees/ha, exceptionally yielding as much as 1000 m^3/ha. Annual increment is normally about 4–6 m^3/ha, although in the best stands it can reach 7–10 m^3/ha. Corsican pine (*P. nigra* subsp. *laricio* var. *corsicana*) has similar increment values.

P. nigra subsp. *nigra* natural stands have a lower stand volume at the same age: between 150 and 450 m^3/ha, at stand densities of 700–1000 trees/ha.

Rotation length for *P. nigra* subsp. *laricio* in Calabria (Italy) and for *P. nigra* subsp. *nigra* in natural stands is between 80 and 100 years, occasionally even as much as 140 years. *P. nigra* subsp. *laricio* var. *corsicana* is cultivated on 70–80 year rotations both in natural stands and in continental France. The peak mean annual increment is reached at about age 50 years old on better stands, and at around 65 years old on poorest site conditions.

Diseases and pests

In the nursery, most common fungi are *Fusarium*, *Phytophthora*, *Pythium* and *Rhizoctonia*.

Under field stressful conditions, young plantations can succumb to *Gremmeniella abietina*. Needles and twigs can be attacked by rust fungi such as *Melampsora populnea* f. sp. *pinitorqua* and *Cronartium flaccidum*; they can be affected by needle blight caused by *Mycosphaerella pini*; and by needle cast diseases such as *Lophodermium pinastri* and *Naemacyclus minor*. *Heterobasidion annosum* is an agent of root decay. *Diplodia pinea* appears to attack trees that are already weakened by atmospheric pollution. *Sphaeropsis sapinea* also attacks trees already subjected to stress caused by other abiotic factors, particularly to excess nitrogen deposition.

P. nigra appears to be susceptible to the plant parasitic nematode *Bursaphelenchus xylophilus*.

P. nigra plantations might be attacked by insects such as *Rhyacionia buoliana*, *Diprion pini*, *Thaumetopoea pityocampa*, *Matsucoccus pini*, *Pissodes castaneus* and *Leucaspis pini*, particularly when trees are too dense, or are growing in unfavourable ecological conditions. Larvae of *T. pityocampa* also cause great problems.

Uses

P. nigra is one of the species most widely used in reforestation of difficult soils (e.g. calcareous, rocky, shallow) in the mountains of the Mediterranean region. It is planted to provide windbreaks in coastal areas as it is more resistant to salinity than most other pine species. As an ornamental, *P. nigra* subsp. *laricio* might be chosen for its size and habit, and *P. nigra* subsp. *nigra* for its hardiness. *P. nigra* is being increasingly used in urban and industrial environmental improvement because of its rapid growth and protoplasmic insensitivity to salt spray and to industrial dust, dry soil and smoke containing sulphur dioxide and other pollutants.

In adult *P. nigra* the sapwood (alburnum) is fairly thick and pale (yellowish-white), whereas the heartwood (duramen) is darker and very resinous; the texture is medium and the fibres tend to be fairly straight. Resin pockets are to be found frequently. The air dry-density at a moisture level of 12% is between 550 and 600 kg/m^3. The compression parallel to the grain strength is on average 43 N/m^3; bending strength is 90 N/m^3. Modulus of elasticity is 13,000 N/m^3.

Usually *P. nigra* wood is resinous, with good mechanical properties. The quality of the wood varies considerably between the subspecies. *P. nigra* subsp. *laricio* has the best-quality timber, whereas subsp. *nigra* is full of nodes.

Timber is not very durable, and is easily attacked by insects and wood-decaying fungi, but for the most common uses no treatment is carried out. However, workability is good.

Generally, *P. nigra* does not produce top-quality timber. It is used for poles and posts, low-grade carpentry, packaging and, after treatment with antiseptics for railway sleepers. The subsp. *laricio* is used as sawtimber; moreover, being a larger tree, it can be used for making beams.

P. nigra is used for the production of plywood panelling or cellulose pulp or paper. In some areas of Europe, *P. nigra* subsp. *laricio* was formerly used for resin extraction. Trees aged over 60 years could produce 2 kg of resin/yr.

Further reading

Espelta, J.M., Retana, J. and Habrouk, A. (2003) An economic and ecological multi-criteria evaluation of reforestation methods to recover burned *Pinus nigra* forests in NE Spain. *Forest Ecology and Management* 180, 185–198.

Mataruga, M., Haase, D., Isajev, V., Orlovic´, S., *et al.* (2012) Growth, survival, and genetic variability of Austrian pine (*Pinus nigra* Arnold) seedlings in response to water deficit. *New Forests* 43, 791–804.

Ordóñez, J.L., Retana, J. and Espelta, J.M. (2005) Effects of tree size, crown damage, and tree location on post-fire survival and cone production of *Pinus nigra* trees. *Forest Ecology and Management* 206, 109–117.

Pinus oocarpa Schiede ex Schltdl.

Ocote pine

Importance

Pinus oocarpa is a species of pine tree native to Mexico and Central America. It is the national tree of Honduras, where it is known as ocote.

This species is highly desirable for sawtimber production. The wood of *P. oocarpa* is of better quality than that of many other exotic and indigenous pines including *Pinus caribaea*. However, the growth rate of *P. oocarpa* is slower than that of *P. caribaea*, making it the less desirable of the two species.

P. oocarpa is also used for the production of wood for the paper industry in several countries such as Ecuador, Kenya, Zambia, Colombia, Bolivia, Australia, Brazil and South Africa.

P. oocarpa has shown high potential for breeding, in particular var. *ochoterenai* which produces progeny with better form and faster growth than the plus trees used as parents.

Therefore, the combination of *P. oocarpa* wood quality, stem form, branching habit and crossability with *P. caribaea*, has raised its importance as a species for future forest plantation after breeding.

Botanical features

The tree is medium to large, 15 m to about 37 m high. The stem normally reaches 50–70 cm and occasionally 90 cm in diameter. The lower branches are more horizontal whereas the top ones are more ascending, forming a thick, rounded crown. The bark is 2–4 cm thick and is a dark greyish-brown colour when the tree is older and mature. The bark also has shallow vertical and horizontal fissures, which gives the bark rough, longitudinal, geometric-shaped plates. Young trees also have a rough, but thin and reddish-brown bark. Branchlets are stiff, upright, rough and scaly. The bark of branchlets is reddish-brown and the bases of the leaf bracts are decurrent.

The leaves of the tree are in fascicles of 5 and occasionally 3 or 4. Needles are usually 20–25 cm long, thick and stiff and are occasionally slender and flexible. The margins are finely serrated and the stomata are present on the dorsal and ventral surfaces. There are 4–8 resin canals, which are mostly septal. The exterior wall of the endoderm is not thickened. Two contiguous but distinct fibrovascular bundles are present. Fascicle sheaths are brown and about 25 mm long.

Conelets are subterminal, borne singly and in pairs on long, 2–3 cm, scaly peduncles. The cone scales are small, thick and rounded with a very deciduous prickle. The cones are very variable in form and size, ranging from globose (almost round) to ovoid and somewhat conical. Cones taper off to the apex in general and are mostly symmetrical, oblique and reflexed on 3–4 cm long strong penduncles. The peduncles can be slender and weak in some forms, though short, thick, and very tenacious in others. Cones are between 6–10 cm long and pale yellowish-brown to polished ochre in colour.

The cones ripen from January to November and can be closed for very long periods only to release seeds during long dry periods. The cones are persistent for long periods on the branches. Cone scales are hard, strong and stiff, and the apophysis is almost flat with a well-defined transverse keel.

In Guatemala, Honduras and El Salvador, the apophysis may be raised to a pyramid with a small, recurved umbo. In most cases the umbo is small and almost flat to depressed, with a very small, early deciduous prickle. When the scales are open they form a wide open symmetrical rosette pattern which contrasts with the cones of *Pinus patula* or *Pinus pringlei*.

Seeds of *P. oocarpa* are small, dark brown, and 4–7 mm long. The wings are 10–12 mm long, articulated and thickened at the base where they join the seed. Cotyledons are 5–7, usually 6. About 120,000 seeds weigh 1 kg.

Distribution

P. oocarpa grows naturally in Mexico, Guatemala, Belize, Honduras, El Salvador and Nicaragua. Most of the trees are concentrated on the south-western (Pacific) half of the axis running through these six countries. Small concentrations are found in the north-eastern half, in Belize and the Mexican states of Veracruz, Hidalgo, northern Puebla and Tlaxcala.

In Mexico, the tree grows in abundance on the lower slopes of the sierras of the western and southern parts of the country. It is plentiful in the mountains of southern Guatemala and central Honduras, northern El Salvador and north-western Nicaragua.

Preferred elevations are 900–2400 m above sea level. In El Salvador, Honduras and Nicaragua it grows above 800 m. *P. oocarpa* var. *trifoliata* grows between 2000 and 2400 m above sea level.

Silviculture characteristics

P. oocarpa is a colonizer and therefore, dry, burnt-over, bare and eroded terrain all provide excellent seed beds for its saplings. Tree growth is rapid and mean annual height increment over the first 10 years sometimes averages 1.5 m. *P. oocarpa* seems to produce higher wood densities at lower altitudes. *P. oocarpa* does not withstand hurricane – including rain and wind – damage to the same degree as *P. caribaea*.

Trees aged 2–4 years old have the ability to resprout at ground level after a fire.

Tree stems tend to become more fire-resistant with age.

Silviculture practice

P. oocarpa is easily grown in nurseries. Plants germinate (50,000 plants/kg) within 14–21 days and can be transplanted after 6–8 months. In nurseries where survival rates were low, practice alterations, most importantly the use of larger-diameter planting tubes (about 5 or 6.5 cm) greatly improved seedling survival.

In the nurseries, the trees tend to grow better in less fertile soil (20% topsoil, 80% sand) where the use of fertilizer had little effect on the survival rate of the seedlings.

If using fertilizer, higher survival rates are achieved when it is applied after growth has started with exception for N fertilizer, which must be added to the plants in stages, as the trees grow.

Although this species is considered a strong light-demander, the height of the trees improves by the use of shading at the nursery stage.

In common with most pine species, *P. oocarpa* seems to survive better when transplanted than when sown directly. Trees that are kept in the nursery for a longer period (9–11 months) survive much better after transplanting than younger trees (after 7 months). Inoculation with mycorrhizal fungi such as *Cenococcum geophyllum*, *Rhizopogon luteolus* or *Suillus bovinus* results in better growth at the nursery stage.

Spacing in plantations is normally 2.7 × 2.7 m, and depending on the products required, the stand will be subjected to several pruning and thinning treatments.

For pulpwood, pruning to about 2 m of stem height will be applied when trees are 5–6 m tall. For sawtimber production, thinnings of 35–50% would be applied at intervals of 5–6 years starting from about 8 years old. Pruning could also be done in three lifts up to 10 m. These treatments are modified depending on the site index and management objectives. Clear felling takes place when suitable size timber can be harvested and the rotation age could vary from 23–30 years.

The total volume of wood production is good, comparing well with the North American Southern pines or *P. patula*. In South Africa, mean annual diameter increments of 1.3–1.5 cm and annual height increments of up to 85 cm have been reported in various plots.

Diseases and pests

Seedlings are prone to damping off while growing in the nursery.

Mexican pine bark beetle, (*Dendroctonus mexicanus*), can destroy a substantial number of trees. Other Central American insects that can cause damage to trees are leaf-cutting ants (*Atta* spp.), *Rhyacionia* ssp. (pine tip moths), *Pissodes* spp. (shoot weevils), Tortricidae (defoliators), *Dendroctonus frontalis* (a pine bark beetle), and Curculionidae (cone weevils).

In Africa, wild pigs have been known to gnaw at the roots and uproot *P. oocarpa* trees growing in the field. Resin content will determine whether the dry-wood termite (*Cryptotermes brevis*) will attack the trees.

Uses

P. oocarpa grows well on eroded soils, and can therefore be a valuable species to use in revegetation.

Wood from *P. oocarpa* has a density of 450–600 g/m^3. Its natural durability is moderate, and preservation, sawing and seasoning are easily achieved. The wood is less prone to splitting and warping than that of most other pines. The heartwood is pale yellowish to reddish-brown. It is fairly strong, stiff, tough, hard and moderately durable. Resin extraction does not significantly change the ranking of density, though the absolute density values are significantly reduced. Due to the density of the wood, it can be used in light construction work, framing, flooring, joinery, railway sleepers, box shooks, harbour piles, poles and posts. The tree is also used for transmission poles, fence posts and long-fibre pulp. The wood can be treated with a preservative to avoid attack by insects or fungi.

No special silviculture is practised for resin production as it is normally a by-product. Timber is still the primary product and trees are only tapped for 2–3 years, shortly before clear felling.

Further reading

Dvorak, W.S., Potter, K.M., Hipkins, V.D. and Hodge, G.R. (2009) Genetic diversity and gene exchange in *Pinus oocarpa*, a Mesoamerican pine with resistance to the pitch canker fungus (*Fusarium circinatum*). *International Journal of Plant Sciences* 170, 609–626.

Sáenz-Romero, C., Guzmán-Reyna, R.R. and Rehfeldt, G.E. (2006) Altitudinal genetic variation among *Pinus oocarpa* populations in Michoacán, Mexico: Implications for seed zoning, conservation, tree breeding and global warming. *Forest Ecology and Management* 229, 340–350.

Pinus patula Schiede ex Schltdl. & Cham.

Mexican weeping pine

Importance

Pinus patula is native to eastern Mexico. It is the most important pine species in Southern and East Africa and is commercially planted in South America and, to a limited extent, in Asia. It has a light coloured timber with medium density, useful for lumber. Although pulp yield is lower than that of *Pinus elliottii* and *Pinus taeda*, *P. patula* is favoured for groundwood pulp because it is less resinous. This species has been subjected to intensive genetic improvement programmes, especially in South Africa and Zimbabwe.

Botanical features

In its native range, *P. patula* reaches the height of 12–20(–30–40) m with dbh up to 120 cm. In plantations outside its native range, tree heights of 35–50 m have been recorded at ages ranging from 20 to 50 years.

The crown shape usually depends both on spacing as well as inherent variability of the trees. Open-grown trees tend to have spreading crowns, while closely spaced trees tend to have narrow crowns with good suppression of lower branches. Some trees will still have deep, narrow crowns, even when growing in the open. Although branching is said to be irregularly placed on the stem, in exotic environments branching tends to be regular with on average four branches per whorl, but the distance between whorls is variable. Foxtailing occasionally occurs, but the majority of trees tend to be multi-nodal with average distance between whorls of about 0.4 m. Within the native distribution range, branches have been reported to be pendent, whereas in exotic environments, particularly Africa, the branches are borne at an angle above the horizontal. Branch size is very variable.

Stem form in *P. patula* is also very variable. Although atrociously crooked trees can be found in this species, stem form is generally moderate to good. Good specimen trees with straight cylindrical stems that retain apical dominance to the age of 30 years or more are quite common. Common defects include forking, sometimes caused by mechanical damage such as bird perching; butt sweep; bow; sinuosity; and nodal swellings. Such stem features are usually found in trees growing in warm, lower-altitude environments.

Bark is thick, rough and scaly with large elongated plates and deep, longitudinal fissures especially on the lower part of the bole; bark colour is dark grey-brown on lower trunk, and higher up and on younger trees bark is reddish-brown or orange, thin and papery. The age at which mature bark appears is not yet known, but appears to vary with site.

The crown may be rounded or spire-like with needles borne on pendent secondary branches. Branching may be from the ground. Young shoots are more noded, glaucous-green, becoming reddish-brown in the second year.

Buds cylindrical, 15–25 mm long, covered with long pointed, fringed scales, which are not resinous and spreading at the tips. Needles are borne in fascicles of 4(3 or 5), in drooping or spreading tufts, 15–30 cm long, very thin, and of a shiny light to yellowish-green in colour. Stomata lines on all needle surfaces; resin ducts 2–3, usually medial, sheaths slender, persistent, 10–25 mm long. Needles may persist for 2–3 years.

Conelets are usually lateral, rarely subterminal, persistent, somewhat oblong, ovate–conical, curved, oblique at base, tapering towards apex, sessile, rarely subsessile, single or in clusters of 2–5, up to 10 or even more. Cone size 4–9 cm long, pale brown, hard, strong, scales oblong with raised apophyses on which there is a minute prickle. Seed triangular, grey mottled with black, 5 mm long, wing 12–18 mm long.

P. patula flowers as early as the age of 2 years, with female flowers appearing first and male flowers appearing in the fourth year. Five-year-old trees produce viable seeds, and cone and seed production is prolific by the eighth to the tenth year.

Distribution

The natural range of distribution of *P. patula* is entirely confined to Mexico. The two known varieties of this species are *P. patula* var. *patula* and *P. patula* var. *longipedunculata*.

The main distribution of *P. patula* var. *patula* is on the Sierra Madre Oriental and the high mountains of east-central Mexico. The altitudinal distribution ranges from 1400 m to 3300 m, but is concentrated between 1800 and 2800 m. This variety is generally restricted to humid, subtropical to warm temperate sites with an annual precipitation of 1000–2200 mm per annum.

The ecology of *P. patula* var. *longipedunculata* is very similar to var. *patula*, but its ecological amplitude is however narrower.

P. patula is naturalized in Southern Africa (South Africa and Zimbabwe) where it regenerates naturally and has taken over as the dominant species on some grasslands, as well as in some forests in cool, wetter areas.

Silviculture characteristics

P. patula is best adapted to high altitudes with cool climates. It is frost-tolerant. It is an invasive species of both fire-climax grasslands and woodland in exotic sites where climatic conditions are close to those found in its natural range. It is a self-pruning tree that regenerates easily.

It responds readily to genetic improvement for increased growth rate, stem form and branching.

Silviculture practice

P. patula is propagated from seed, collected from best trees on a stand, seedling seed orchards or clonal seed orchards.

Clonal orchards are established using clones propagated by grafting and, in a few cases of incompatibility, air-layering has been used to propagate the clones.

Nursery production of *P. patula* seedlings is best achieved on warm sites between the altitudes of 600 and 1200 m, where plants will be ready for planting (15–20 cm height) in less than 7 months.

Percentage germination may be improved through seed pretreatment to break dormancy. Soaking in cold water for 8 days has been reported to break dormancy in seed 1–7 years old. Soaking in hydrogen peroxide (1 kg of seed in 80 l of 1.5% hydrogen peroxide for 4 days) was found to increase germination. Scarification was found to be ineffective for *P. patula*.

Although *P. patula* regenerates easily, stands have not been established from natural regeneration, mainly because most countries use improved seed to produce planting stock. Planting stock remains the only method that is used for field establishment of *P. patula*. The commonest spacings vary from 2.4 × 2.4 m to 3 × 3 m.

There are currently no conclusive experiments on fertilizer use of *P. patula*. Some studies have indicated that forking in *P. patula* may be caused by boron deficiency. Weeding may be practised, depending on weed competition and resources.

Thinning for sawlogs aims to get a final crop or stand of more than 400 trees/ha with mean dbh of about 45 cm at rotation, which in Zimbabwe and South Africa may be 25–35 years, depending on site quality. For pulpwood, dbh of about 30 cm after 15–25 years is considered normal.

In Zimbabwe, for example, from an initial stocking of 1111/ha (at 3 × 3 m) a first thinning (40% removed) is usually done between age 6 and 8 years, to leave 666 trees/ha. A second and final thinning (40% removed) at age 12–15 years reduces stocking to about 400 stems/ha. At higher initial stockings, a third thinning is often necessary at about age 15–18 years.

Although *P. patula* is self-pruning, additional pruning might be applied. Generally, high pruning is for sawn timber while trees destined for pulpwood receive low pruning.

Timber yields from *P. patula* vary with site quality. Mean annual increments (MAIs) reported vary from 9.9 m^3/ha to as much as 40 m^3/ha. In Southern Africa, the MAIs range from 18 m^3/ha to 28 m^3/ha. The high MAIs are in East Africa where the dry season is very short.

Diseases and pests

Damping off diseases caused mainly by *Rhizoctonia solani*, *Fusarium* spp., and *Pythium* spp. can affect nursery stock. *P. patula* is also threatened by the pitch canker fungus, *Fusarium circinatum*, which infects nursery plants and, if transferred to the field, can cause severe mortality of young trees. In the field, *P. patula* may be attacked by *Dothistroma pini* after the trees have been damaged by hail. Other fungal pathogens are *Diplodia pinea*, *Sphaeropsis ellisii* and *Armillaria mellea*.

Pests of *P. patula* in its natural range are poorly documented. In exotic environments, many pests have been recorded but most of these do not appear to cause economic damage. They include the grasshopper *Mecostibus pinivora* in Zimbabwe, several coleopterans, larvae of several lepidopterans and aphids. The wood wasp *Sirex noctilio* and its associated fungus *Amylostereum areolatum* have infested plantations of *P. patula*.

Uses

P. patula produces a large amount of litter which might be useful for soil conservation and erosion control. It is not wind-tolerant, and thus not used for shelter belts or windbreaks. As an aggressive colonizer it might be useful in revegetation and land reclamation.

The average density of green wood is 880 kg/m^3 and at 10% moisture content the average density is 450 kg/m^3 (range 350–610). The wood of *P. patula* can be sawn easily when both wet and dry. The sawn timber may be seasoned by air or kiln drying. The wood is soft, light, and reasonably strong. The wood is white to yellowish-white in colour, with a pinkish heart which is difficult to distinguish from the sapwood. It takes nails without splitting and also takes glue easily. The wood is too soft for high-class furniture or joinery manufacture unless used in thick sections and coated with a hard finish. The wood is suitable for composite board manufacture and can be used for panelling and decorative purposes. The wood is attractive in appearance because of the colour contrast between early and late wood. It can also be used for wood-wool cement slabs and particleboard manufacture, and for pulp by the kraft process.

P. patula wood is not resistant to fungus, insect and termite attack, and in external conditions the wood deteriorates rapidly. Logs and sawn timber are susceptible to bluestain. To avoid this, logs have to be quickly transported and sawn as soon as possible.

The major limitation of *P. patula* species is the large size of the juvenile core wood which is blamed on the fast growth rate of the young trees; its reduction has been one of the most important breeding targets.

Further reading

Mitchell, R.G., Coutinho, T.A., Steenkamp, E., Herbert, M. and Wingfield, M.J. (2012) Future outlook for *Pinus patula* in South Africa in the presence of the pitch canker fungus (*Fusarium circinatum*). *Forests: a Journal of Forest Science* 74, 203–210.

Perry Jr, J.P. (1991) *The Pines of Mexico and Central America*. Timber Press, Portland, Oregon, USA, 231 pp.

Salaya-Domínguez, J.M., López-Upton, J. and Vargas-Hernández, J.J. (2012) Genetic and environment variation in two progeny tests of *Pinus patula*. *Agrociencia (Montecillo)* 46, 519–534.

Pinus pinaster Aiton

Maritime pine

Importance

Pinus pinaster plantations have long been cultivated in the Mediterranean basin region (both Europe and North Africa), particularly to stabilize sand dunes, in coastal areas and on low-fertility soils. The tree has been a major source of resin as well as a general-purpose timber. Its major disadvantages are its high susceptibility to fire, and the low quality of the resin compared to that from other pine species. Major research is needed into thinning procedures, especially in the Mediterranean region.

Botanical features

P. pinaster is a relatively large tree, 20–40 m tall, with an average dbh at maturity of 35–40 cm. Crowns of old trees are wide and flat, and the bole is clear over most of its length. Plantation trees have long, clean cylindrical stems (in contrast to those of open-grown individual trees, in which trunks are large at the base), and a pronounced taper with increased height. It has a deep taproot, with secondary roots well developed. Bark is thick, deeply fissured, dark red-brown.

Needles are spiny, stout, stiff but not rigid, frequently twisted, occurring in pairs, with a vivid green colour.

Flowers appear between late winter and mid-spring. Male flowers are usually abundant, clustered in shallow rings beneath the leaves; pollen is shed in early spring. Female flowers occur on the tips of expanding shoots, dull red in colour. Cones are nearly sessile, very oblique at base, slightly curved ovoid–conic; shiny light brown colour with scales and a broad transverse ridge rising to a central, small, upcurved prickle. Cones ripen between late summer and autumn, persisting closed on the tree for several years.

Flowering from age about 6 years old, becoming regular from age 10–15 years. Full seed crops occur at 3–5-year intervals.

Distribution

The centre of natural distribution is the Mediterranean area of southern France, Corsica, Sardinia, Italy and Spain, as well as Morocco to Tunisia (for subsp. *pinaster*), and a more westerly region, centred around the Atlantic areas of France, Spain and Portugal (subsp. *atlantica*).

P. pinaster range has been widely extended by cultivation, and it is planted in the Landes region and in the Atlantic coast of France, as well as throughout the Mediterranean region, for timber and resin production, dune conservation and, very often, for pulp production.

It was introduced into the UK by the 1600s, and has been extensively planted in Europe outside its natural distribution area, for example in Belgium, Turkey, Greece and countries of the former Yugoslavia (subsp. *pinaster*). *P. pinaster* was introduced to the Western Cape province of South Africa at the end of the 18th century, and to New Zealand in the 1830s; and subsp. *atlantica* was first planted in trials in Western Australia in 1896.

Silviculture characteristics

P. pinaster is fast-growing until it reaches age 60 years; the growth rate then usually declines. The species is often cultivated in even-aged stands due to the fact that it does not tolerate shade, and it is not usually possible to manage *P. pinaster* in uneven-aged monoculture stands, although it is possible to include *P. pinaster* in uneven-aged mixed species stands.

Silviculture practice

Following fire, natural regeneration of *P. pinaster* is usually possible and, on the best sites, it occurs even after clear felling. The regeneration originates from seed already on the ground, as well as from seed released from the cones of felled trees. This type of propagation is favoured by exposure to light, the usually abundant seed production of the species, and the loose sandy nature of the soils on most of the sites where *P. pinaster* subsp. *atlantica* is dominant.

Although regular seed production commences at 10–15 years of age, seeds are collected for propagation from trees aged at least 20–30 years old, to ensure their fertility. Seeds are easy to conserve, and maintain their germination capability for a very long time (after 10 years of conservation, more than 50% of total seeds stored may still be viable). The seeds can be successfully stored in boxes, taking care to shake the contents periodically.

Vegetative propagation is rarely used on *P. pinaster*, although it has been successfully performed in genetic breeding programmes, mainly in France.

In nursery, line sowing is used, with a distance of about 8 cm between lines and seed planted at a depth of 1 cm. Seedlings in transplant beds should be at a density of approximately 150 seedlings/m^2 (8 × 8 cm spacing). Usually, *P. pinaster* needs about 1 year in the seedbed and another year in the transplant bed. Container-grown planting stock can also be produced, especially suitable for afforestation in hot and dry climates; for production of potted seedlings, time of growth in the nursery must be about 6 months, or a little longer.

The silvicultural system most used for *P. pinaster* is the clear-cutting system, in various forms: in alternate clear-cut strips; in progressive strips; or patch clear-cutting. The seed-tree method can also be used, but this method is not frequently used in timber production forests.

Nowadays, the most frequent method for stand establishment is through plantation of nursery-produced seedlings. However, full and partial direct sowing, line sowing, and natural regeneration methods were often used in the past, and the latter is still used in the oldest stands. For plantations in the Mediterranean region, stand densities are most usually 1670 trees/ha (2 × 3 m), 1330 trees/ha (2.5 × 3 m), and 1110 trees/ha (3 × 3 m). If the plantation is initially at a high density (1600–2000 plants/ha), cleaning must be done when the trees are 8–10 years old, in order to reduce stand density to 1000–1200 trees/ha. This can be achieved with the elimination of the worst-quality trees, dead and non-vigorous trees.

Site preparation operations can be divided in three phases: brush cleaning, soil preparation and plantation. The need for extensive soil preparation increases in degraded soils, high slopes and the influence of Mediterranean climate types.

During soil ploughing it is common to use a phosphate or NPK fertilizer. Because *P. pinaster* does not have a high-density crown, understorey shrubs may grow quite rapidly especially during the early stages of stand establishment. To avoid weed competition, as well as to help prevent forest fires, brush understorey must be eliminated, either by cleanings or by the use of controlled burning. These cleanings are more often performed mechanically between plantation lines, two or three times over the 10-year period after planting, when spontaneous vegetation starts to compete with the young seedlings. If the understorey vegetation grows as high as the lower part of the *P. pinaster* crown, it is advisable to carry out supplementary brush control operations, between the lines, in order to reduce the risk of fire.

It is also advisable to provide an infrastructure to minimize the risk of fire spreading (fire breaks, roads, fuelbreaks) and facilitate fire fighting (water points, roads).

If *P. pinaster* is planted for good-quality timber production (e.g. furniture), it is important to establish a pruning plan in order to avoid knots and nodes in the timber. During a first phase (that starts when trees are between 5–7 m tall, average dbh 10–15 cm) all branches up to a stem height of 2 m are pruned, in all trees of the stand. After that, when average dbh is about 18 cm, the best 400–600 trees/ha are pruned up to a stem height of 3 m. When pruning living branches, the operation must be performed between August and February; pruning of dead branches can be undertaken at any time of the year.

The classic thinning system for *P. pinaster* is low thinning. It should be performed when the crowns of the trees start to touch each other. Normally, this happens on three occasions during the rotation, usually starting at about age 15 years; on each of these occasions a maximum of 30–40% of the trees should be removed. It is important to point out that these values may be different in areas of the world other than the Mediterranean region. In thinned stands usually only 300–500 trees/ha reach the end of the rotation. On lower-quality sites, however, it is convenient to maintain higher densities until felling, therefore thinning should be undertaken on two occasions only.

In the Iberian Peninsula, annual volume increment of *P. pinaster* varies between 5 and 10 m^3/ha, and yields vary between 300 and 470 m^3/ha, depending on site quality. On good-quality sites it is possible to achieve 30-year rotations, although 35 years is the most common. However, on normal sites rotation length is 45 years, and on poor-quality sites rotations of 60–80 years are the norm.

Diseases and pests

The species is affected by *Armillaria mellea*, *Lophodermium* spp., and *Cyclaneusma niveum*. *Heterobasidion annosum* was reported in Spain.

Cone and seed pests include *Pissodes validirostris* (the pine cone weevil) and *Dioryctria mendacella* (the cone worm) which tend to attack more significantly when there is a shortage of food supply (years following a heavy cone crop).

Bud-, shoot-, and twig-feeders including *Rhyacionia buoliana* (the pine shoot moth) tend to be more problematic in young plantations exposed to site and stand conditions that affect the vigour of seedlings and saplings.

Defoliator outbreaks are often associated with older stands, overstocked stands, or stands growing on poor sites. In addition, the weakened trees often become susceptible to tree-killing insects, such as bark beetles. Defoliators include *Thaumetopoea pityocampa* (the pine processionary moth, the most serious pest in the Mediterranean region) and *Neodiprion sertifer*.

Bark beetles can be very destructive and may introduce pathogenic fungi such as *Ceratocystis* spp. The first adult beetles to attack a tree produce powerful pheromones that, together with resins exuding from the tree, concentrate the flying beetle population in an area so that dead trees will often occur in patches.

Ips sexdentatus, *Tomicus piniperda*, *Tomicus minor*, *Orthotomicus erosus*, *Pityogenes bidentatus*, and *Hylastes ater* are all bark-boring insects, and their populations can increase greatly if a stand contains high-risk trees. *Dioryctria sylvestrella* particularly damages young trees. *Matsucoccus feytaudi* attacks weaken trees, making them susceptible to attacks from *Pissodes castaneus* and *T. piniperda*.

Root pests are not a serious problem in established *P. pinaster* stands that have well-developed and extensive root systems. They might be a problem in nurseries and newly stocked plantations where they may provide entry for root decay fungi.

The root pests *Hylastes* spp. may flourish in stands which have suffered fire; in which planting techniques were inadequate; in which there are many exposed, freshly felled stumps; or in successive drought years. *Anoxia australis*, *Vesperus xatarti* and *Vesperus luridus* may be particularly damaging in forest nurseries and in young *P. pinaster* stands.

P. pinaster is susceptible to *Bursaphelenchus xylophilus*, a nematode that causes the pine wilt disease.

Uses

P. pinaster is used for soil conservation and erosion control in hill slopes. As a pioneer species in the secondary ecological succession, it will encourage soil formation. Being resistant to salt spray, it is widely planted for dune stabilization.

Due to its fast growth characteristics and its frugality, *P. pinaster* is often used for afforestation of abandoned agricultural land and as shade trees in camping and recreational parks, despite its susceptibility to soil compaction.

Wood is the major product that is obtained from *P. pinaster*. Reported wood densities range from 360 to 839 kg/m^3 (at 12–15% moisture content). Its annual rings are very visible, creating a distinct and effective pattern to the timber. However, it is a wood susceptible to every possible defect and is very resinous. Logs usually have a high percentage of bark (20–40%) in the total volume.

The main applications include construction wood, furniture, roof shingles, shipbuilding and poles and posts for palisades and fencing. It is also used for firewood, with cones and needles (after drying) being used to start fires.

Resin is the most important non-wood product, being used, directly or indirectly (after distillation), to make turpentine, pitch, oils, varnishes, adhesives, waxes and soap. It is also used for waterproofing.

P. pinaster stands are rarely managed to optimize resin production. Therefore, the extraction of resin is usually performed in mature stands managed for wood production, starting 3–5 years before felling. However, the cuts made may have a negative effect on wood quality. On the Iberian Peninsula, a normal tree will produce 2–4 kg of resin/yr, reaching 5 kg/yr in the best sites and trees.

Essential oil has been produced from chipped crown materials. The bark of *P. pinaster* may be distilled to produce tar, or chipped and composted to produce a low-weight substrate for nursery containers. *P. pinaster* stands also are an ideal ecosystem for the development of edible fungi.

Further reading

Gaspar, M.J., Lousada, J.L., Rodrigues, J.C., Aguiar, A. and Almeida, M.H. (2009) Does selecting for improved growth affect wood quality of *Pinus pinaster* in Portugal? *Forest Ecology and Management* 258, 115–121.

Knowles, F.B. and Miller, J.T. (1989) *Introduced Forest Trees in New Zealand: Recognition, Role, and Seed Source. 8.* Pinus pinaster *Aiton - Maritime Pine.* FRI Bulletin 124, 13 pp.

Martínez-Alonso, C., Kidelman, A., Feito, I., Velasco, T., Alía, R., Gaspar, M.J., Majada, J., Alderete, A., Ammer, C., Close, D.C., Coll, L.D.R.K., Grossnickle, S.C., Landhäusser, S.M., Liu, Y., Löf, M., Pinto, J.R., Puértolas, J., Stanturf, J.A., Vallejo, V.R. and Villar-Salvador, P. (2012) Optimization of seasonality and mother plant nutrition for vegetative propagation of *Pinus pinaster*. *New Forests* 43, 651–663.

Pinus ponderosa Douglas ex Lawson & C. Lawson

Ponderosa pine

Importance

Pinus ponderosa (Plate 28) is one of the most widely distributed pines in western North America. Besides being a major source of timber, *P. ponderosa* forests are also important as wildlife habitats, for recreational use and for their aesthetic value.

Botanical features

A large-sized tree, occasionally over 60 m tall, 200 cm in dbh, and life span up to 600 years. The trunk is straight, with little taper, and often branch-free for most of its length. The crown is narrow and conic when young, becoming wider and irregularly cylindrical with age, and flat-topped in old trees. The branches are stout with lower ones often drooping; upper ones on old crowns are ascending. The bark is dark grey, rough and scaly when young, with age becoming orange-brown and deeply fissured into large, flat and flaky plates. The wood is uniform in texture, moderately strong and hard; heartwood is yellowish to reddish-brown and moderately decay-resistant; sapwood is very wide and nearly white to pale yellow.

P. ponderosa develops deep taproots, up to 50 cm in watered soils in the first growing season. Mature trees have roots to depths of more than 200 cm in coarse-textured soils but less than 100 cm in fine-textured soils. In open-canopy stands lateral roots may extend 46 m. The main root mass is concentrated within the top 60 cm. Roots are associated with both ecto- and endomycorrhizae.

Evergreen needles usually in fascicles of 3 (occasionally 2 or 5), 12–25 cm long, straight, stiff but flexible, very sharp, dark yellowish-green, edges sharply toothed. Var. *scopulorum* has a higher proportion of 2-needled fascicles.

P. ponderosa is monoecious. Flowering generally begins at the beginning of May; pollen is shed from the end of May to mid-June, and cones reach a full size of 8–15 cm from mid-July to mid-August of the next year. Seed is ripe from mid-August to the beginning of September, cones begin to open in the first half of September, and most seed is shed in early November.

Medium seed crops are borne every 2–3 years, on average. The average interval between heavy cone crops is 8 years. The species bears cones as early as 7 years and continues to produce good seed from 60 years to at least 350 years of age. In a good seed year as many as 852,050 seeds/ha may reach the ground.

Distribution

P. ponderosa is distributed in the Pacific, Cordilleran and central regions of North America. The range of *P. ponderosa* extends from southern Canada into Mexico, and from the Plains States of Nebraska and Oklahoma to the Pacific Coast.

P. ponderosa var. *ponderosa* ranges from latitude 52°N in the Fraser River catchment of southern British Columbia, south through the mountains of Washington, Oregon and California, to latitude 33°N near San Diego. *P. ponderosa* var. *scopulorum* extends east of the Continental Divide from latitude 48°N in north-central Montana, south-easterly into North and South Dakota, eastern Wyoming, and as far east as north-central Nebraska. *Pinus arizonica* (syn. *P. ponderosa* var. *ponderosa*) is found primarily in the mountains of extreme south-western New Mexico, south-eastern Arizona, and northern Mexico.

P. ponderosa was introduced to New Zealand in 1865, and its ease of establishment and hardiness resulted in its widespread use. In the mid-1960s, *Dothistroma pini*, which causes needle blight, arrived in New Zealand, and *P. ponderosa* was one of the most severely affected species. Its role as a major plantation species has declined; however, it is still an important plantation species on South Island, and performs well if properly sited. In 1990, there were over 5000 ha of *P. ponderosa* plantations in New Zealand. It has also been used in plantations in Argentina.

It was first introduced to Europe after 1825. *P. ponderosa* has also been tested at the Xiongyue Arboretum, Liaoning, China, and it was concluded that it was well adapted to the conditions; suitable areas for *P. ponderosa* in Liaoning were suggested. It has also been planted in arboreta, parks and on a minor scale in European forests. In 1880, it was included into a network of experimental plantations by the German Forest Research Institute. It has also been reported that it has been trialled in Australia, Cyprus, the UK, Kenya, Borneo, Malawi, Zimbabwe and South Africa.

Silviculture characteristics

A fire-resistant, windfirm tree adapted to warm and dry climates.

Predominantly a shade-intolerant species, natural regeneration under low light is reduced. Potential for natural regeneration in the open is high, especially after wildfires. Light conditions beneath closed-canopy, mature stands are high and associated with well-developed understorey vegetation. Self-pruning capacity in dense stands is intermediate.

Silviculture practice

Propagation by seed is the only contemporary method for regenerating *P. ponderosa*. A variety of containerized stock is used for planting. Throughout its range, natural regeneration of *P. ponderosa* is sporadic.

P. ponderosa can be grown in single- or mixed-species stands, preferably with shade-tolerant species.

Depending on site and management objectives, clear-cutting, patch-cutting, seed-tree, strip shelterwood, and group-selection systems are all appropriate for the establishment and growth of *P. ponderosa*.

Potential initial growth rate (at 5 years and younger) is high, reaching an annual height growth greater than 1.5 m after 5 years on very productive sites.

P. ponderosa has the capacity for root growth in relatively dry soil. It is a strongly light-demanding species, and the significance of competing vegetation as a deterrent to early survival and development of young seedlings has been demonstrated clearly. Height growth can be significantly reduced by overcrowding, therefore clearing of wood vegetation, especially during the juvenile stages, is very important.

Diameter growth can be rapid and remain fairly constant for long periods provided trees are given adequate growing space. On productive sites in California, free-growing trees can reach 66 cm dbh in 30 years. Vegetative competition can restrict diameter growth markedly whether it is from neighbouring trees or understorey shrubs. Stagnation in diameter, and often in height growth, represents a serious problem in densely stocked stands throughout the species' range, especially on poor sites.

Height growth is most rapid in the pole stage to about 60 years; growth rates decline gradually at older ages. Height growth increases with site productivity and is more sensitive to stand density than was once believed. Old-growth *P. ponderosa* produces clear, high-grade lumber, but young trees typically are limby as natural pruning develops slowly.

In New Zealand, on favourable sites, the mean annual increment of stands reached a peak at about 45–50 years and reached values as high as 30 m^3/ha.

Diseases and pests

Leptographium wageneri (black stain root disease) causes a dark staining of root wood and root death. *Heterobasidion annosum* causes a lethal root disease. Both these fungi can cause group mortality in *P. ponderosa* stands. *Armillaria* spp. are causing increased mortality in young plantations and thinned stands. The most damaging heart rot in the southern Rocky Mountains and the Black Hills is western red rot caused by *Dichomitus squalens*; *Phellinus pini* is the major heart rot in the Pacific Coast States. A needle cast, *Elytroderma deformans*, found throughout the range of *P. ponderosa*, is the most serious foliage disease. Several rusts of the *Cronartium coleosporioides* complex are damaging to *P. ponderosa*. *Cronartium comandrae* causes scattered mortality in some stands; in thinned stands it may cause substantial damage. The western gall rust (*Endocronartium harknessii*) attacks *P. ponderosa* from the Black Hills to the Pacific Coast.

In the mid-1960s, *D. pini*, which causes needle blight in many pine species, arrived in New Zealand. *P. ponderosa* was one of the most severely affected species, and the disease imposes considerable restrictions on the sites where it can be grown. At one time *P. ponderosa* had been established throughout much of New Zealand; now the best sites are the relatively dry *Dothistroma*-free uplands of central South Island. Other diseases affecting *P. ponderosa* in New Zealand include *Sphaeropsis sapinea* and *Armillaria*. *P. ponderosa* is attacked by a large number of insect pests. The most damaging of the tree-killing insects are several species of *Dendroctonus*. *Dendroctonus brevicomis* is a common cause of mortality in overmature stands in much of the range of *P. ponderosa*; however, apparently healthy trees may also be killed during epidemics. *Dendroctonus ponderosae* is the most destructive and aggressive enemy in the central and southern Rocky Mountains. *Dendroctonus adjunctus*, *Dendroctonus approximatus* and *Dendroctonus valens* are other species of the genus that often kill *P. ponderosa*.

Among bark beetles, *Ips* species are second in destructiveness only to *Dendroctonus*. Eleven species of *Ips* have been found attacking *P. ponderosa*. Of these *Ips latidens*, *Ips emarginatus*, *Ips pini*, *Ips lecontei* and *Ips paraconfusus* have the most impact.

Defoliating insects, such as the pine butterfly (*Neophasia menapia*) and the pandora moth (*Coloradia pandora*), periodically cause damage over extensive areas. *Sirex noctilio* has caused damage in older, over-crowded stands in New Zealand.

Dwarf mistletoe (*Arceuthobium vaginatum* subsp. *vaginatum* in the south-west USA, and *Arceuthobium campylopodum* in California and the north-west) are widespread and have a significant effect on mortality and production.

Uses

In New Zealand on *Dothistroma*-free sites, *P. ponderosa* is seen as having advantages (hardiness, drought-tolerance, narrow crowns and straight stems) as a shelter belt species.

In *P. ponderosa* forests, timber production, livestock grazing and recreation are the principal land uses. The forests offer year-round recreation and frequently border forest highways where aesthetic values are high. They provide habitats for various wildlife species. Snags in the mature pine forest provide a large number of species with nesting and roosting sites. Big game, such as deer and elk, also use the pine forests for food and shelter.

The wood is uniform in texture, moderately strong and hard. The wood has a density of about 450 kg/m^3 at 12% moisture content. New Zealand-grown *P. ponderosa* timber had a density of 350 kg/m^3. The heartwood is yellowish to reddish-brown and moderately decay-resistant. The sapwood is very wide and nearly white to pale yellow. The wood is used for window sashes and frames, siding, door mouldings, panelling, patterns, cabinetwork, boxes and crates.

Further reading

Burdon, R.D., Miller, J.T. and Knowles, F.B. (1991) Introduced forest trees in New Zealand: recognition, role, and seed source. 10. Ponderosa and Jeffrey pines: *Pinus ponderosa* P. Lawson et Lawson, *Pinus jeffreyi* Grev. et Balf. *FRI Bulletin* 124, 23 pp.

Gyenge, J., Fernández, M.E. and Varela, S. (2012) Short- and long-term responses to seasonal drought in ponderosa pines growing at different plantation densities in Patagonia, South America. *Trees: Structure and Function* 26, 1905–1917.

Oliver, W.W. and Ryker, R.A. (1990) *Pinus ponderosa* Dougl. ex Laws. In: Burns, R.M. and Honkala, B.H. (technical coordinators) *Silvics of North America, Vol. 1.* USDA Agriculture Handbook 654, Washington, DC, USA, pp. 413–424.

Perrakis, D.D.B., Agee, J.K. and Eglitis, A. (2011) Effects of prescribed burning on mortality and resin defenses in old growth ponderosa pine (Crater Lake, Oregon): 4 years of post-fire monitoring. *Natural Areas Journal* 31, 14–25.

Pinus radiata **D. Don**

Radiata pine

Importance

Pinus radiata (Plate 29), although quite an obscure conifer in its native California, is more extensively grown as an exotic than any other forest tree. It occupies some 4 million ha as fast-growing plantations, of which over 90% are located in the southern hemisphere. Its very fast growth, together with ease of establishment, considerable site tolerances, and the versatility of its wood, make it the utility softwood of choice almost wherever it can be grown satisfactorily. It is well naturalized in several countries where it has been introduced.

Botanical features

Crown in young trees pointed, usually somewhat irregular in outline reflecting complex branching pattern, varying from narrow and spire-like on many infertile sites to broad and coarse under high fertility; in old trees rounded or even flattened at top. Growth habit in the juvenile stages relatively untidy, the adult habit typically being attained only after 6–12 years.

The branching pattern is very complex and variable, the number of branch clusters (so-called 'whorls') on an annual growth stage of the leader stem ranging from one to six, depending very largely on the individual genotype. The variability of branch habit has been of major interest in breeding programmes.

Tree height in mainland California and Guadalupe Island generally reaches 20–35(–40) m, on Cedros Island it tends to be less. Mature-tree dbh in native stands tends to be extremely variable. In dense stands they typically reach 40–75 cm, but isolated trees can be far larger, well over 100(–200) cm. In exotic plantings height tends to be greater, generally ranging from 30 m to 50–55 m; rare trees reaching 60 m, when dbh would typically be 50–85 cm, with isolated trees ranging to over 250 cm.

Strong natural taproot in seedlings might be modified by preparation for planting in the nursery and the use of cuttings for propagation.

Needles, grouped in fascicles, in threes in var. *radiata*, and in pairs in var. *binata* and var. *cedrosensis*. Dimensions usually 10–18(3–20) cm long. Colour dark green, but may be yellowish or bluish-green.

Pollen strobili are cylindrical, 7–17 mm long. Female strobili are reddish-purple at emergence, with the noteworthy features of occurring on the more vigorous shoots including the leader stem, and never occurring in the cluster of branch buds occurring at the end of an annual growth stage on the shoot. Cones can occur singly but are very often in clusters encircling the shoot. They take fully 2 years from pollination to mature and ripen. Freshly ripened cones are pale or medium brown, usually 7–17(5–20) cm long and usually 4–7(–9) cm wide. Shape distinctive, if variable, ovoid–conical, usually with strong asymmetry. Large cones tend to be more asymmetric, and with greater length:breadth ratio. Scales often very large and thick with raised umbos on the outer side. Mucros very small or absent. Cones are serotinous, capable of remaining unopened for a number of years in absence of hot, dry conditions, and can remain attached long after opening.

A notable chemotaxonomic feature is the predominance of ß-pinene in wood terpentine, which is rare in pines and not shared with its close relatives.

Distribution

P. radiata occurs naturally in just five discrete populations, at Año Nuevo (or 'Swanton'), Monterey and Cambria on the Californian mainland; and Guadalupe Island and Cedros Island, off the coast of the Baja California Peninsula (Mexico). The total extent of natural forest, prior to European impacts, was slightly under 10,000 ha, of which about 5300 ha remain.

All natural habitats represent a special and highly localized variant of the dry to mesic Mediterranean climate. Nowhere do natural stands extend beyond 8 km inland. Altitude ranges from sea level to 420 m on the mainland, and from 300 to 1200 m in the more southerly island populations. Within its range, *P. radiata* is generally the sole high-forest species.

P. radiata has been planted on a massive scale in New Zealand, Chile, Australia and Spain. Significant areas of plantation exist in western Cape Province of South Africa, Argentina, Italy (Sardinia), Ecuador, Kenya, Colombia and Peru; with lesser plantings in various Mediterranean countries, France and the British Isles, and in some African and South American countries.

Silviculture characteristics

P. radiata is light-demanding, but less so than most pines. It is relatively tolerant of weed competition. Even dense natural regeneration tends to be strongly self-thinning, particularly in moister climates, with strong assertion of dominance. Notwithstanding, *P. radiata* can be highly responsive to heavy artificial thinning. Delayed thinning, however, can cause vulnerability to wind damage. Since it is not self-pruning, knots are prevalent in timber from unpruned logs. *P. radiata* has a long growing season, which accounts for its high growth rate.

Ground fires are tolerated by old trees and by pole-stage trees if intensity is low. Crown fires will kill and release abundant seed from unopened cones, often leading to dense natural regeneration. Similar regeneration can follow clear felling.

Silviculture practice

Seed collection and storage are particularly easy, with the serotinous cones that can store cumulative seed crops and make long-term seed storage a natural phenomenon. Seed extraction is best achieved in a kiln, high temperatures (55°C) being readily tolerated. There is no stratification requirement, except in the Guadalupe provenance and its hybrids, and germination rate is high.

Commercial propagation is still mainly from seed. However, a significant fraction is now done by vegetative propagation by shoot cuttings in the nursery, to multiply stocks that have been intensively improved genetically.

Bare-rooted planting stock, produced in open nursery beds, are used almost exclusively, although container stock is an option. Good planting stock can be produced in 6–8 months on warm sites and up to 18 months on cold sites. Cuttings can be set in the nursery from late autumn to early spring. Root pruning and topping are widely used to condition stock for outplanting and help control stock size. Precision-sown seed can be spaced at 5–6 cm within rows in nursery beds, cuttings at about 8 cm. Fertilizers, including N and P, are routinely used in permanent nurseries.

Establishment is now almost exclusively by planting, in order to control both stocking and genetic quality, although natural regeneration and direct seeding have been used on a large scale in the past.

Soil preparation is limited to managing the woody vegetation and the weeds on drier sites.

Mycorrhizal symbionts are needed, but special measures to introduce them are seldom required except when a nursery is started in open grassland.

Initial stockings have traditionally been around 2000 stems/ha but, with improvements in both establishment technique and genetic quality of the stock, they are now usually 1000/ha or less.

Fertilizers are often used, as basal dressings at planting and/or general topdressings of established stands. Acute micronutrient deficiencies are usually corrected readily.

Tending regimes are very varied. Zero tending can be a successful option, and worked remarkably well in the past in much of the plantings in Chile and New Zealand; it is better than active-tending regimes that are ill-conceived and/or badly executed. At one extreme, as in many plantations in Australia, tending regimes have involved a combination of low pruning primarily as a fire-control measure, a number of light commercial thinnings and long rotations (up to 50 years), where the main market is for light structural timber, and maximum wood volume production was sought. Systematic thinnings, entailing removal of entire rows of trees, have often been practised on easy terrain in Australia where site conditions have produced minimal stem malformation. At the other extreme, as in large areas of plantation in New Zealand, intensive regimes are based on a combination of early, heavy thinning, often without extracting the thinnings, and aggressive pruning to about 5 m height, to produce good yields of clear timber from the butt logs on short rotations (down to 25 years or less, with final stockings down to 200–250 stems/ha) and thereby maximizing the rate of economic return. Such 'radical' regimes were driven by a combination of factors, including: calculations of opportunity costs for the final crop of delaying the first thinning until the thinnings are saleable, a desire to minimize risks of wind damage, steep terrain, a high price premium for clear timber and a willingness to accept some loss of volume production if the economic returns are improved. However, some of the adverse effects of the more radical regimes on wood and timber quality are proving greater than anticipated, which is causing some rethinking of the regimes and a focus on genetic improvement of wood properties.

Height growth generally takes 2–5 years to reach the 'grand' growth phase, during which height growth can often reach 2 m/yr. Site indices' mean top height at 20 years averages 20 m in commercial plantations, but can exceed 30 m and range up to nearly 40 m. Stem volume mean annual increments of fully stocked stands is probably slightly under 20 m^3/ha but can exceed 30 or 40 m^3/ha.

Diseases and pests

Mycosphaerella pini (needle cast disease) and *Endocronartium harknessii* (syn. *Peridermium harknessii*) (western gall rust) are fungal pathogens which have limited where *P. radiata* can be grown in wetter conditions north of the species' natural range on the California coast.

Sphaeropsis sapinea (syn. *Diplodia pinea*) is often the immediate cause of failure in summer rainfall areas characterized by damp heat. Pitch canker, caused by *Gibberella circinata* (teleomorph) (*Fusarium circinatum*, anamorph), is a disease that has recently appeared in the native populations of mainland California. Root-rot fungi of the genus *Armillaria*, notably *Armillaria mellea*, are locally important. Their impact can be greatly increased by the presence of significant *Mycosphaerella* infection. The root pathogen *Phytophthora cinnamomi*, is locally important where there is seasonal waterlogging, especially in Western Australia. The needle-cast fungus, *Cyclaneusma minus*, is notable for its almost ubiquitous occurrence, rather than acute impacts.

The wood wasp *Sirex noctilio* can kill trees very quickly by attacking the boles. The European pine shoot moth (*Rhyacionia buoliana*) has long caused varying amounts of leader stem damage in plantings in Europe. It has spread to Argentina and then to Chile, where it is a major pest. The bark beetle *Ips grandicollis* is a significant pest in Australia.

The processionary caterpillar (*Thaumetopoea pityocampa*), is a significant defoliator in warmer parts of Europe, where it limits the use of *P. radiata*. Another defoliator is the pine emperor moth (*Imbrasia cytherea*), troublesome in South Africa. The pine woolly aphid (*Pineus pini*), is a minor pest in many parts, but has recently caused concern in South Africa. Various other insect pests are present in native stands but have historically been in ecological balance with *P. radiata*. Various scolytid beetles, and some other insects, have become significant primarily for their role as vectors for pitch canker.

Uses

P. radiata might be used in silvopastoral systems where tree density is kept low enough to allow pasture growth. It has been used in soil conservation and rehabilitation of large eroded areas in the Coast Range of Chile. Sand dune stabilization has been achieved in New Zealand with the establishment of high-quality timber crops.

P. radiata forms the basis of major wood-processing industries, the wood being used for veneers, joinery, light structural timber, both chemical and mechanical pulping and various categories of fibreboard. The wood is easily sawn and dried, but lacks natural durability. It is easily worked and very easy to nail, while it takes paint and varnish readily and is easy to glue. Relatively low stiffness, even in relation to its density, is an important limitation, along with the readiness with which large knots occur, although lamination can be used to produce large structural members. The wood is not suitable for steam bending, and requires skill to stain satisfactorily.

The wood properties are greatly influenced by ring number from the pith, wood density showing an initially rapid rise from the pith outwards to a fairly stable value at about ring 16. The first ten rings are usually viewed as 'corewood', the next five as 'transitional' wood, with 'outerwood' thereafter. Inner corewood (first five rings) often has a basic density of around 330 kg/m^3, rising asymptotically to about 450 kg/m^3 in the outerwood. Corewood is for most purposes highly inferior, but it is favoured for mechanical pulping. The effect of site on basic density can be very large, especially in the outerwood, where the stand average can range from around 400 kg/m^3 to well over 500 kg/m^3; warm, moist, low-fertility sites tend to give high density, while low temperatures, high fertility and summer drought appear to be conducive to low density.

Bark is widely burned as an energy source at large wood-processing plants. It is also used in horticulture, for mulches and for potting mixes. Although it has quite a good tannin content it appears not to have been used significantly for tannin extraction. Cones are popular as a domestic fuel.

Further reading

Bown, H.E. (2009) Representing nutrition and genotype in forest productivity models: a process-based approach for *Pinus radiata*. PhD thesis. University of Canterbury, Canterbury, New Zealand.

Burdon, R.D. (2000) *Pinus radiata*. In: Last, F.T. (ed.) *Ecosystems of the World, Vol. 19, Tree Crops*. Elsevier, Amsterdam, The Netherlands, pp. 99–161.

Burdon, R.D. and Aimers-Halliday, J. (2003) Risk management for clonal forestry with *Pinus radiata* - analysis and review. 1: Strategic issues and risk spread. *New Zealand Journal of Forestry Science* 33, 156–180.

Pinus resinosa Aiton

Red pine

Importance

Pinus resinosa was once a major component of extensive forests in mixture with eastern white pine (*Pinus strobus*), but extensive felling in the 19th century reduced it to a few inaccessible stands. During the 20th century, *P. resinosa* became an important reforestation species in the Lake States of the USA and in south-eastern Canada, because of its ease of planting, excellent stem form and desirable wood properties for paper, lumber, timbers and poles. It is valued as a cabin log and for small transmission poles, while thinnings find a ready pulpwood market.

Botanical features

P. resinosa is a medium-sized tree, typically attaining heights of 22–30 m and diameters of 50–70 cm in natural stands and plantations. Older specimens can exceed 125 cm diameter and 40 m height, but these are rarely encountered. Stems are usually very straight and mostly well-pruned, with little or no taper, especially when grown in closed-canopy situations. When open-grown, a *P. resinosa* crown develops as a long cylinder with a somewhat coarse texture owing to the 'tufted' appearance of its needles, borne on mostly stout, horizontal branches.

Needles occur in fascicles of two, about 10–18 cm long, which are flexible yet brittle, persisting for 3–5 years. *P. resinosa* is considered shade intolerant, leading to rapid upward crown recession when grown at close spacing. Trees at high density in unthinned plantations develop narrow, oval-shaped crowns and may have as little as 10–15% of their total height in live crown. Generally, only a single whorl of new growth is produced each growing season, making it relatively easy to infer tree age from the number of whorls. Shoot initiation slightly precedes needle elongation during

April–June of each year, with shoot extension generally complete by June–early July.

P. resinosa is monoecious, with strobili appearing between April and June depending upon climate. Female strobili 3–5 mm long (4–6 cm at maturity, unarmed) are borne singly or in clusters (up to 6–8) near the tip of the developing shoot in the upper portion of the crown, while male strobili (4–10 mm long) are borne in dense clusters near the base of the current shoot in the lower two-thirds of the crown. Open-grown trees may have considerable overlap between zones of male and female strobilus production. Flowering and seed production may begin by the tenth year on open-grown trees in favourable sites, but production beginning at age 15–20 is more typical. Some female cones may be produced each year, but good to excellent crops occur at intervals of 3–7 years; periods of 7–10 years or more without appreciable cone production are not unusual. Cone production improves as stand density decreases, with most trees in dense plantations seldom producing a single cone. A typical mature tree may produce from 200 to 1000 cones under favourable conditions. Cones mature the second year, between August and October, and shed seed soon after. A typical cone may produce 15–50 viable seeds. Seed dispersal is by wind, with dispersal distances typically less than 20 m, although longer distance dispersal is possible with strong winds and steep topography. Small mammals may cache seed, increasing dispersal distance slightly.

P. resinosa produces a wide-spreading and moderately deep (up to 1 m) root system, with taproots occasionally extending as deep as 4 m. On older trees, numerous stout lateral roots radiate to form an irregular oval shape and penetrate the soil at acute angles to form a windfirm root system, especially on loamy sands and coarser soils.

Distribution

P. resinosa occurs in a rather narrow range, between 42–52°N, generally in parallel with the USA–Canada border from Maine–Nova Scotia west to Minnesota–Manitoba. Although it is widely planted within its natural range, especially in Ontario and the Lake States, it is rarely used as an exotic. Limited trials of *P. resinosa* south of 42°N have typically failed, and its somewhat slower growth relative to some other pines stimulates little interest in using it outside its natural range.

Silviculture characteristics

P. resinosa is synonymous with plantation forestry across much of its natural range. It is considered a shade-intolerant species, and so more suitable to grow in even-aged silvicultural systems. At recommended stocking levels it produces a very symmetrical tree, well-pruned with little or no taper. *P. resinosa* does not readily hybridize with any other pine, making it unattractive in hybrid breeding programmes.

Silviculture practice

Natural regeneration using seed tree and shelterwood systems was widely practised prior to 1960, with varying degrees of success. However, seed tree and shelterwood systems are now considered 'unreliable' in the absence of appropriate fire regimes. Recent interest in restoration ecology in much of the northern forests of the Lake States may make shelterwood systems practical on public lands where mixed species and multi-aged stands are desired for aesthetic reasons.

Modern strategies include clear-cutting followed by intensive site preparation (including slash disposal, discing or other mechanical vegetation control) and plantation of nursery seedlings is now the most widely promoted regeneration practice for *P. resinosa*.

Propagation by seed has been a standard nursery practice for more than 50 years and forms the basis for major artificial regeneration programmes throughout much of the natural range. *P. resinosa* seed has orthodox storage requirements, remaining viable for decades under conditions of low moisture (5–10% moisture content) and temperature (3–4°C). Normal nursery practice includes late summer sowing to satisfy stratification requirements by overwintering seed in the nursery bed. Germination the following spring occurs when temperatures exceed 20°C for several weeks. Protection of seedlings from drought is accomplished with irrigation and cloth or snow fence shade. Most seedlings are 2–3 years from seed prior to distribution for planting. Nursery stock production schedules have generally reduced the time required to produce a plantable seedling from 3 to 2 years, primarily by decreasing nursery bed density. Limited containerized seedling production has been successful, but 95% or more of all nursery stock produced today is bare root.

As older plantations established at close spacing (2 × 2 m) during the 1930s and 1940s are harvested, they are replanted at lower initial planting densities (3 × 3 m) to reduce establishment costs, shorten rotations, and minimize the need for pre-commercial thinning. High initial stocking helps minimize weed competition.

Thinning *P. resinosa* can increase net productivity, with some recommended thinning regimes as frequent as every 8–10 years during an 80–110 year rotation. *P. resinosa* outyields many other species because it remains quite vigorous at relatively high stocking levels. Maximum annual growth (5–6 m^3) on sites of average fertility occurs at 40–60 years of age for stocking levels between 27–40 m^2/ha. Recommended stocking levels for production of larger diameter trees for dimension lumber are somewhat lower (20–34 m^2/ha). Thinning of pole stands (13–23 cm diameter) is recommended when stocking exceeds about 32 m^2/ha. Rotation lengths decline with increasing site quality, and vary from about 60 years of age for stands that receive frequent and heavy thinning to more than 140 years for infrequently thinned stands.

Diseases and pests

P. resinosa enjoys a reputation of being relatively pest free, but occasional pest problems occur, often associated with plantations established on former hardwood sites. *Armillaria* root rot is a frequent problem in the Lake States following thinning of *P. resinosa* planted on sites formerly occupied by oak stands. Plantations established in low-lying areas, or frost pockets, are susceptible to canker caused by *Gremmeniella abietina* following frost injury in late spring. Several foliar/shoot diseases also pose periodic problems, especially the shoot blights caused by *Sirococcus conigenus* and *Sphaeropsis sapinea* (*Diplodia pinea*).

P. resinosa roots are reported to die back in seasonally saturated soils, and root penetration is restricted by hardpans and compacted soils, reducing root growth and susceptibility to wind throw.

Saratoga spittlebug (*Aphrophora saratogensis*) can cause large losses of young *P. resinosa* planted on sites with large amounts of sweet fern (*Comptonia peregrina*), an alternate host for this insect. Several native and exotic sawflies, and pine tussock moth (*Dasychira* sp.), cause periodic outbreaks in plantations.

Uses

Mature stands of large *P. resinosa*, especially in mixtures with eastern white pine (*Pinus strobus*) are frequently retained for amenity value, especially along roads and lakeshores throughout the north-eastern USA. It is also grown in plantations as a Christmas tree.

The sapwood is nearly white to yellow, while the heartwood varies from red to reddish-brown. It is straight, even-grained, medium textured and moderately heavy. It has a density of 545 kg/m^3 (at 12% moisture content). The timber is used for piling, poles, cabin logs, railway ties, posts, mine timbers, box boards, pulpwood and fuel.

Further reading

Cheng Z., Chui, Y.H. and Meng, G. (2012) Within-stem variation in wood properties of red pine (*Pinus resinosa* Ait.). *Wood and Fiber Science* 44, 412–421.

Mack, T.J. and Burk, T.E. (2002) User's manual for Resinosa. *Staff Paper Series – Department of Forest Resources University of Minnesota* 158, 1–25.

Ostry, M.E., Moore, M.J., Kern, C.C., Venette, R.C. and Palik, B.J. (2012) Multiple diseases impact survival of pine species planted in red pine stands harvested in spatially variable retention patterns. *Forest Ecology and Management* 286, 66–72.

Pinus roxburghii Sarg.

Chir pine

Importance

Pinus roxburghii is one of the main coniferous species in the Himalayan region and is of great ecological, silvicultural and economic importance. It is a very successful pioneer on exposed sites, and is planted on bare land as the first stage in reforestation, where it withstands drought, frost and snow, and is free from serious pests and disease. The timber is used for many purposes including pulp production, where its long fibres are used to upgrade other species.

Botanical features

P. roxburghii is a large long-lived evergreen tree, sometimes deciduous in dry locations or dry seasons. It commonly attains a height of 45–55 m, with a dbh of 110–120 cm. The branches are very large, whorled, and the crown is elongated up until reasonable maturity. Later the crown becomes either pyramid-shaped or umbrella-shaped. The bark thickness varies from 1 cm in young saplings to approximately 5 cm in mature trees, or may often reach a thickness of 7.5 cm on the lower trunk.

It develops an extensive root system consisting of a deep taproot and radial surface roots.

The primordial needles of *P. roxburghii* are solitary, 1.5–6.3 cm in length. Adult needles are found in fascicles of 3, and are 20–32 cm in length, triquetrous, with one side rounded, and densely punctate in several longitudinal lines of stomata. The needles are long on vigorous trees, yet are short on over-mature, diseased or slow-growing trees (e.g. trees in poor localities or at high altitude). Their basal sheath is persistent, 1.5–2.5 cm long.

This species is monoecious with male and female flowers occurring on separate branches of the same tree. The male flowers are grouped on the axis of the new shoots and arranged in spirals of 8 × 5; on ripening, flowers are 1.2–1.7 cm long. These flowers are yellowish-green before ripening. After pollen release the flowers turn light reddish-brown. The male inflorescences are generally more numerous on the central and lower branches, rather than on the upper parts of the tree. Pollen is produced in large quantities, ensuring successful wind pollination. The pale green female flowers are found at or near the apices of new shoots, either solitary, in pairs, or in one or two whorls consisting of three flowers. The young cones are erect and ovoid, with scales arranged in spirals of 8 × 5. The winged seeds lie in pairs at the base of each cone-scale. The seeds, including the wing, are 2.8–4.6 cm in length and 0.7–0.9 cm in width. The seeds are ellipsoid and without the wing measure 0.7–1.3 × 0.5–0.6 cm. The number of seeds in each cone is highly variable, with a reported range of 38–704 seeds.

Only few *P. roxburghii* trees produce cones earlier than 30–40 years old. Male flowers may develop at an earlier age than female flowers. Crown development also influences seed production in this species: trees with full, rounded or umbrella-shaped crowns bear much higher numbers of cones than trees with conical crowns. Small trees (with a dbh of 20 cm) are capable of producing good fertile seed, and seed fertility declines only in over-mature trees.

In terms of phenology, in India and Pakistan, the period from the first emergence of the young cones until the mature cones shed their seed is about 26–27 months. At low altitudes, the male inflorescence is visible in early January. The blue or pink flowers ripen and pollen release occurs from February to April. After pollen release the flowers elongate to nearly 2.5 cm. The female flowers start appearing in February, and pollination takes place from February to April. Cone scales close after pollination. At the end of the first growing season (October), the cones turn greyish-brown. At the beginning of the second growing season, the cones turn green and begin to develop. By mid-April they are 7.5–10 cm in length, with curved scales, and begin to resemble the shape of a mature cone. In the spring of the third season the cones are light brown and 11–20 × 6–9 cm, and ripen in April–May. The cones open only in dry weather until June, when they close with the rains.

Distribution

P. roxburghii occurs naturally in the outer ranges and valleys of the Himalayas from about 500–2500 m altitude; and in the monsoon belt of Bhutan, Sikkim, Nepal, India, Pakistan and Afghanistan within the latitude range 26–35°N. It grows under a variety of conditions from moderate to steep hill slopes and on deep soil or bare rocks. However, large areas of these forests have disappeared over the years due to felling, lopping and burning.

P. roxburghii has been commonly planted as an ornamental tree outside its natural range in India and Pakistan. It has been introduced in Africa.

Silviculture characteristics

P. roxburghii forms extensive pure stands. The tree attains large dimensions, and natural regeneration springs up in abundance wherever sufficient light is admitted. It is a pioneer species and a strong light-demander, though it may require shade during the younger stages. An extensive root system makes this species drought-resistant and windfirm. It is completely frost-hardy, even in frost hollows. *P. roxburghii* is considered fire-resistant.

Silviculture practice

Over the last century, *P. roxburghii* has been managed either as uneven-aged forests under a selection system using single trees or groups, or as even-aged forests under compartment felling. The selection system has increasingly proved to be unsuitable due its light-demanding characteristics. In India, most *P. roxburghii* forest is managed under a shelterwood system (either uniform or irregular) with a floating periodic regeneration block.

The rotation period for this species varies from 80 to 120 years, with a regeneration period of 20–30 years. During regeneration felling 12–20 trees/ha are left standing as seed bearers, and fellings may be uniform or grouped. Uniform felling involves even spacing of seed bearers and is the most common practice in India and Pakistan; generally the area is allowed to regenerate naturally. Group fellings consist of the removal of trees over and around groups of promising advance growth, and natural regeneration is generally supplemented by direct sowing or planting of nursery seedlings. Thinning is conducted once the stand is 15 years old, and repeated at 10–20 year intervals.

The most common form of artificial regeneration used in India and Pakistan is nursery stock; direct sowing is less favoured. *P. roxburghii* has also been successfully propagated vegetatively by air-layering, grafting and rooting of stem cuttings.

Nursery seed should be collected from trees with superior growth and characteristic form. A good seed year usually occurs every 2–3 years. There are 11,000–13,000 seeds/kg of seed, which may be kept in a dry, cold and airtight store for several years. No pretreatment of seed is necessary, although some prefer to soak the seed for 24 h. Seed germination takes 4–6 days and is completed within 30–35 days. Germination is epigeal, with a success rate of approximately 65–98%. Forest soil is obtained for all new nurseries, in order to ensure mycorrhizal inoculation and good seedling development.

In India and Pakistan, many seedlings are grown in polythene tubes or root trainers, and are planted on cleared watershed areas in the mountain foothills once 1–2 years old. Planting is conducted at 2.5 × 2.5 m or 3 × 3 m spacing, during the spring and monsoon seasons.

The majority of *P. roxburghii* forests are managed under a uniform system, with a 80–120-year rotation, 20–30-year regeneration period, and thinnings commencing after 15 years. The growth rate of *P. roxburghii* is highly variable. Mean annual increment of *P. roxburghii* ranges from about 6 m^3/ha at age 80 years on quality III sites to about 15 m^3/ha at age 60 years on quality I sites (MAI of total stemwood and smallwood combined).

Diseases and pests

The most common nursery diseases of seedlings are damping off, *Fusarium* wilt and charcoal root rot. The fungus *Lophodermium pinastri* (syn. *Hysterium pinastri*) is both saprophytic and parasitic, killing any badly affected seedlings; it appears as yellow spots on the needles and turns the needles reddish. *Heterobasidion annosum* and *Phellinus pini* enter via wounds and cause wood rot; the sporophores appear as irregular brackets, usually where a branch has been lopped off. *P. roxburghii* is very susceptible to brown needle disease, which can attack seedlings in the nursery and trees up to 20 years old.

Fungi which cause heart rot include *Ganoderma lucidum*, *Polyporus gilvus* and *Polyporus ostreiformis*. *Helicobasidium compactum* is a soil pathogen favoured by warm tropical and subtropical conditions, and is responsible for purple root rot. Rust wood-rot disease is caused by *Cronartium himalayense* and results in bark blistering; it destroys young plants and infection risk is reduced as the plant matures.

Standing trees of *P. roxburghii* are attacked by various insects. However, insect pest problems are more pronounced in nurseries and plantations than in natural stands, particularly in *P. roxburghii* grown on marginal lands at low altitudes. Young seedlings suffer various forms of insect damage, such as defoliation by grasshoppers (*Poekilocerus pictus*). The larvae of *Cryptorhynchus rufescens* damage the bark of trees; *Chlorophorus strobilicola* larvae damage the cones, which remain underdeveloped, resulting in seed loss; and larvae of *Dioryctria abietella* bore into the tissue and feed until the autumn. *Anomala rufiventris* and *Popillia cyanea* grubs are found on leaves in nursery beds and flowers, respectively. Young larvae of *Eumeta crameri* feed on needles, sometimes in epidemic proportions. Other insect pests of importance include *Ips longifolia*, found on young trees; *Platypus biformis* and *Sphenoptera aterrima*.

Uses

P. roxburghii provides valuable soil protection, especially on steep slopes eroded during the monsoon. It is commonly planted as an ornamental and is also used to reclaim charcoal kiln sites.

The wood of *P. roxburghii* is non-porous and has distinct growth rings. The latewood is darker than the earlywood and becomes reddish when moistened with water. The heartwood is red, turning to reddish- or yellowish-brown with age, whereas the sapwood is white to creamy-white. The density of *P. roxburghii* wood is 327 kg/m^3. The timber is straight or spiral grained, medium fine and uneven textured, and has a highly resinous odour. The wood is moderately hard and heavy. The timber is durable when placed under cover, but not in exposed conditions. It is also easy to season and may be kiln dried without difficulty. However, resin exudes from the wood surface and the wood is liable to excessive splitting, cracking and warping during the seasoning process. It is moderately easy to saw and work, but may give raised, fuzzy and torn grain.

P. roxburghii is commonly used in construction particularly for beams, roofing, posts, boards, doors, windows frames, flooring, railway sleepers, transmission poles, matches, cheap joinery and furniture. The timber is also used for making packing cases, boxes, house fittings, bed arms and foot boards of carriages. The wood is good for building poles, fence posts, pattern making, drawing boards, boats, plane tables and cooling towers, provided the timber pieces are not too resinous. The wood is a good fuel, having a calorific value of 5015 kcal/kg. *P. roxburghii* may be used for manufacture of long-fibre pulp and gives a satisfactory yield of bleached paper-pulp for printing. Twisted fibre wood is used to produce pulp for wrapping paper.

Resin production from *P. roxburghii* is still widely practised in both India and Nepal to obtain turpentine. The oleoresin has medicinal properties and is a constituent of plaster.

P. roxburghii needle oil is used as an antibacterial agent for scenting soaps, room sprays and deodorants. Hardboard has been prepared from the needles, which are pulped semi-chemically. The seeds are edible, but are not nutritious and have an unpleasant odour. The use of the needles for animal bedding and feeding is important in Himalayan India and Nepal. The bark has high tannin content (11–14%) and is used for tanning. Ink is made from leaf charcoal. The cork is used as a cheap insulating material. Tar is obtained through distillation of the roots.

Further reading

Raj, A., Sehgal, R.N. and Sharma, P.K. (2012) Genetic parameters and gain estimates for growth traits in Chir pine (*Pinus roxburghii* Sarg.) in a 22 years old half-sib progeny test. *Indian Journal of Forestry* 35, 143–150.

Sharma, D.P. and Malik, R.A. (2011) Influence of site factors on the growth of chir pine (*Pinus roxburghii* Sargent) forests along elevational gradient. *Indian Forester* 137, 71–82.

Siddiqui, K.M. (1982) The wood and pulping properties of chir pine (*Pinus roxburghii* Sarg.) Part-I. Wood Properties. *Pakistan Journal of Forestry* 33, 45–54.

Pinus sibirica Du Tour

Siberian stone pine

Importance

Pinus sibirica is one of main forest-forming tree species in the boreal zone of northern Eurasia. Forests dominated by this species cover an area of 45 million ha, representing a wood stock of 8000 million m^3. Forests of *P. sibirica* are very important for the ecological balance of the regional landscape and represent a sustainable source of versatile timber and non-timber products. Unfortunately, these forests suffer from fire and insect pest damage, and further research into their ecology and population genetics is necessary.

Botanical features

P. sibirica is a long-lived (400–850 years), slow-growing coniferous tree with a narrow-cylindrical or elliptical crown at maturity. It often has a multiple top and candelabrum-shaped crown. On optimal sites mature trees grow up to 40–45 m high with a dbh of 1.5–2.0 m. The stem is straight and cylindrical. The root system is shallow, especially when growing on permafrost and poorly drained, moist soils, but is deeper in dry soils. The bark is grey-brown and cracked and some 3–5 cm thick at the base.

Needles are 5–14 cm long, in five-needled bundles, triangular in cross section, soft, dark green with a whitish pubescence. They persist for 5–8 years.

P. sibirica achieves the reproductive stage at the age of 15–25 years on open sites and at the age of 50–70 years in closed stands. Red microstrobiles are located in the basal part of annual shoots. Violet macrostrobiles are located in the apical part of annual shoots at the crown top. Mature cones are large, 6–12 cm long, 5–7 cm wide, and contain 80–140 large wingless seeds.

Buds open, needles start growing, and flowering begins in May. An intense growth of shoots and needles begins in June and ends late in July. Cones ripen and seeds are consumed and dispersed by the nutcracker (*Nucifraga caryocatactes*) at the end of August.

Distribution

The natural distribution of *P. sibirica* ranges from the north-east Russian plains (49°E) to the central part of eastern Siberia (126°E) and from the forest tundra in western and central Siberia (68°N) to the mountains in the central part of Mongolia (46°N) and north-eastern China. It is also found around Lake Baikal (Siberia). Its range overlaps with that of *Pinus pumila*.

The overwhelming majority of *P. sibirica* plantations have been established within its natural range, namely northern Russia, and the foothills and low mountains of south Siberia. However, this area is not significant in terms of Russian tree plantations. In Russia, Finland, Ukraine and Belarus there are experimental plantations of *P. sibirica* outside the natural range of this species, but these represent <1% of the trees planted in these areas.

Silviculture characteristics

P. sibirica is a shade-tolerant species at the juvenile stage and survives for relatively long periods under the canopy of small-leafed (*Betula pendula*, *Populus tremula*) or coniferous (*Pinus sylvestris*, *Larix sibirica* and *Larix gmelinii*) species with more open canopies. However, its growth after 10–15 years is better under exposure to full light. Natural regeneration of populations in extensive clear burns and cuts in Siberia is assisted by seed dispersal by the nutcracker (*N. caryocatactes*). Being a slow-growing and long-lived species, *P. sibirica* outlives other tree species along continuous succession stages and dominates at the final climax stage.

Silviculture practice

P. sibirica seeds can be collected in late August or early September from mature brown cones. They are classified as intermediate in germination behaviour, and for storage must be dried to a moisture content of 11% of fresh weight. Seeds require a temperature from 0 to 3–5°C for optimal storage. For optimal germination, seeds require stratification for 2–4 months at around 0°C with substrate moisture of 50–70%, which can be achieved, for example, under a snow cover. A significant increase in seed production can be achieved rapidly by selective felling in stands 81–120 years old, which creates more space for crown development.

Methods of vegetative propagation of *P. sibirica*, especially by grafting of *Pinus sylvestris*, are well developed. They are used to clone plus trees in seed orchards and are an aid to the introduction of *P. sibirica* outside its natural area.

Seedlings reach a plantable size in two vegetative seasons (during 16 months) for planting in green-moss forests and in 3–4 years for planting in other forests. The importance of proper rotation in the nursery has been stressed in various Siberian nurseries.

Partial plantations of *P. sibirica* in clear-cuts and burns are introduced by planting 2000–4000 seedlings/ha in strips (trenches). In moist habitats, deep trenches provide surface drainage of the soil. Plantations usually consist of 5000–6000 seedlings/ha on forest sites and 1500–2000 seedlings on seed farming sites. Fertilizers, mycorrhizal and bacterial agents are used in nurseries. Competing herbs, shrubs and small-leaved trees are controlled by mechanical or chemical methods for 3–4 years after planting out of seedlings.

In recent years, main cuts in *P. sibirica* stands have been prohibited in Russian forests and only sanitary or stand improvement cuts are permitted. Selective improvement cuts are used in seed farming forests in order to choose high-yield and macrocarpous trees. Clear-cuts are allowed in tree stands with less than 30% of *P. sibirica* trees. Clear-cuts are also allowed if tree stands are damaged by fire, pests or other catastrophic factors. Natural regeneration of populations of this species in clear-cuts and burns is usually assisted by the nutcracker *N. caryocatactes*.

The ban on logging *P. sibirica* forests imposed in 1989–1990 represents political recognition for the advantages of multiple-use management for this species, including the use of its edible seeds as a crop, production of flour from needles, resin-tapping, hunting and apiculture. It is acknowledged that some form of sustainable logging should be permitted for commercial management, particularly in old stands in which *P. sibirica* has stopped fruiting. Selective and shelterwood fellings are ineffective in maintaining the stability of *P. sibirica* forests.

When grown as an exotic in the non-chernozem zone (mixed-forest zone) of the European part of its distribution range, *P. sibirica* performs best on sites of the fairly fertile spruce-forest type. In pure plantations on this site type, assuming an initial density of 5000 trees/ha and in 30 years after establishment, the basal area is expected to be about 30 m^2/ha, the standing volume around 200 m^3/ha and tree mean height of 9 m.

Diseases and pests

The most widely occurring fungus diseases of *P. sibirica*, which attack needles of seedlings in nurseries or those of undergrowth and trees in forests, include *Lophodermium pinastri*, *Cronartium ribicola* (blister rust), and the black snow mould *Herpotrichia juniperi*. *Cenangium abietis* parasitizes branches of forest undergrowth and trees of *P. sibirica*. Trunks are infected mainly by *Phellinus pini* and *Stereum abietinum*. Diseases of roots are caused by *Heterobasidion annosum* and *Phaeolus schweinitzii*.

The main insect pests include *Dioryctria abietella* and *Monochamus sutor*, which attack cones and seeds, and *Pineus cembrae* and *Monochamus urussovii*, which damage shoots. *Dendrolimus sibiricus* and *Lymantria dispar* damage needles on wide areas of *P. sibirica* forests in Siberia.

Uses

Forests of *P. sibirica* are important in water protection and erosion control. *P. sibirica* is a very important agroforestry species and it can be used as an ornamental in urban environments.

P. sibirica has rose-coloured wood, soft and lightweight with a good texture. It has an average density of 420 kg/m^3, is slightly inferior to spruce wood with respect to compression strength and static bending and is much inferior to *Picea obovata* and *Pinus sylvestris*. *P. sibirica* wood serves a great variety of applications. It is used as rounds for various poles, pit props and sawlogs (light-duty constructions). Different types of sawn timber are used for production of slats, boat shells, carpentry, furniture veneer, articles of furniture and musical instruments. Small quantities of merchantable wood are used for barrel staves, food boxes and utensils. Groundwood pulp of *P. sibirica* is used in production of various types of cellulose.

The wood tar (turpentine gum) of *P. sibirica*, which is collected by tapping the trees, serves as a source of turpentine and colophony. In addition, immersion oil, which is used in optics (similar to 'cedar balsam') and other valuable starting materials for the chemical industry are extracted from the tar. Seeds are characterized by a high fat content (64.3%) and contain many vitamins; they represent an important product for the food pharmaceutical industry. The mean annual yield of seeds in *P. sibirica* forests is 10–20 kg/ha in northern taiga, 40–60 kg/ha in middle taiga, 135–160 kg/ha in southern taiga of western Siberia and 200–300 kg/ha in the mountain taiga sub-belt of the north-eastern Altai. Specialized seed-producing farms or orchards can produce as much as 800–3000 kg/ha. Other forestry enterprises produce additional types of goods that can provide sufficient versatility for all-year-round employment of workers, and sustained profitability.

Further reading

Bekh, I.A. and Savin, E.P. (1997) On the utilization of *Pinus sibirica* forests. *Lesnaya Promyshlennost'* 2, 19–20.

Chitorkin, V.V. (2008) Condition and growth of old plantations of *Pinus sibirica* in Western Siberia. *Lesnoe Khozyaĭstvo* 4, 38–40.
Dyukarev, A.G., Pologova, N.N., Krivets, S.A. and Chitorkin, V.V. (2009) Ecological/commercial zoning of *Pinus sibirica* stands near settlements. *Lesnoe Khozyaĭstvo* 3, 10–12.
Zavarzin, V.V. (1996) Management of *Pinus sibirica* forests. *Lesnaya Promyshlennost'* 4, 25–26.

Pinus strobus L.

Eastern white pine

Importance

Pinus strobus (Plate 30) is one of the most valuable tree species in eastern North America. The species responds well to nursery culture and is commonly used for reforestation, urban forestry and Christmas tree plantations.

Botanical features

A medium-sized tree up to 30(–70) m tall, 100(–180) cm in dbh, and 200 or more years old. On young trees, the crown is conical with regular whorls of branches. On mature trees, the crown is irregular, with a few long, stout branches set roughly at right angles to the trunk, and branches in the upper crown ascending, giving a broadly oval flat-topped outline that becomes one-sided due to the prevailing wind.

The root system is composed of 3–5 moderately deep lateral roots, with sinker roots often growing down from them.

Bark is thin, smooth, greyish-green when young, becoming grey-brown and deeply furrowed, with long, irregular, scaly branches.

Leaves are fascicular needles in bundles of 5, 5–15 cm long and 0.7–1 mm and triangular in cross section; straight and flexible, but slightly twisted; bluish-green with pale stomata lines evident on adaxial surfaces; edges finely toothed; persisting for 1–4 years, although the fascicular sheath scales usually shed during first season.

Twigs are slender, green and hairy, becoming glabrous and orange-brown during second season.

Buds are slender, 4–5(–15) mm long, ovoid–cylindrical shaped, sharp-pointed, with reddish-brown overlapping scales, slightly resinous.

Seed cones are clustered, pendent, cylindrical, 8–20 cm long, with 2–3 cm stalk; 50–80 scales, usually in 5 spiral rows, thin and rounded at tip, without prickles, yellowish-green becoming light brown when mature.

Pollen cones are ellipsoid, 10–15 mm, yellow.

Seeds are reddish-brown, mottled with black, 5–8 mm long; wing 18–25 mm long.

As in other pines, development of the reproductive structures follows a 3-year cycle. Pollination occurs in the spring of the second year, with fertilization delayed until the following spring, and seeds maturing in the fall of the third year. Seed cones generally develop on vigorous shoots in the upper portion of the crown. Initiation of seed dispersal is weather- and site-dependent, and may be delayed by cool, moist weather. Most of the seeds are dispersed in the autumn during a 4–8 week period.

Distribution

P. strobus has the largest range of any North American member of subgenus *Strobus*, and is the only species in the subgenus occurring on the eastern side of the continent. It extends from Newfoundland and Quebec, west to central Ontario and south-eastern Manitoba, south to Minnesota, north-eastern Iowa, northern Illinois, north-western Indiana, Ohio, Pennsylvania and New Jersey, and south in the Appalachian Mountains to western North Carolina, northern Georgia and Tennessee. Overall, the species spans a north–south range of over 1900 km, and about the same distance inland from the Atlantic coast. *P. strobus* has been planted in trials and as an ornamental in several European countries.

Silviculture characteristics

P. strobus is a pioneering and long-lived successional species and a component of climax forest types. It may form pure stands or occur as a major component in association with other conifers and hardwoods. *P. strobus* is considered intermediate in its tolerance to shade, less tolerant than eastern spruces and more tolerant than its pine associates. Due to competition dynamics, *P. strobus* will perform well when natural disturbance, such as fire, or silvicultural site treatments, allow the pine to become established well ahead of the associated hardwoods.

P. strobus is considered to be moderately fire-resistant. Frost heaving can cause severe damage. *P. strobus* is relatively windfirm, apart from when the stand has been thinned recently.

Silviculture practice

P. strobus seeds exhibit varying degrees of dormancy. The recommended treatment for nursery sowing is stratification for 60 days at 1–5°C. Germination is epigeal.

In terms of vegetative propagation, long-shoot cuttings from 2–3-year-old seedlings are preferred as they root easily. Propagation is also possible using fascicular shoots. Growth and performance of rooted cuttings are comparable to planting stock raised from seed. Micropropagation can also be used.

Following developments in nursery technology, *P. strobus* is now commonly produced from seed in containerized systems, in soil-less growing media. Optimum conditions are provided when moist mineral seedbeds have greater than 20% of full sun, but where partial shade reduces surface temperatures and provides better moisture conditions.

Potential for direct seeding as a regeneration technique is high, but seed stocking is often irregular. Feeding losses to small mammals can be over 80%, unless the seeds are covered with soil at the time of sowing.

In terms of natural and artificial regeneration, shelterwood harvesting systems provide good protection during initial establishment with sufficient light for subsequent growth of young stands. *P. strobus* regeneration under old-growth conditions is likely to create an uneven-aged and self-perpetuating stand, facilitated by the occurrence of local disturbances and continuous recruitment. Older trees have increased ability to recover from long periods of suppression.

Height growth of both planted and natural *P. strobus* is slow during the first 2–3 years. Afterwards, growth accelerates, peaking at an average annual rate of 1 m between 10 and 15 years on site index 80 (with a base of 50 years). Diameter growth can be highly variable. In fully stocked stands on average sites, the average tree diameter increases at a nearly uniform rate of 2.5 cm every 5–6 years. Yields of 500 m^3/ha at 50 years are entirely possible. The average stand in New England has an annual increment from 4.2 to 11.2 m^3/ha depending on age, site index, and stocking.

Diseases and pests

Emerging seedlings, particularly in bare-root nurseries, are commonly affected by damping off fungi, primarily *Fusarium*, but also *Pythium*, *Rhizoctonia*, *Phytophthora* and *Cylindrocladium*. Young seedlings may also be attacked by various species causing *Cytospora* dieback (*Valsa* spp.), tip blight (*Sphaeropsis sapinea*), snow blight (*Phacidium infestans*) and rhizina root rot (*Rhizina undulata*). Many fungi infect white pine foliage, including various needle casts (*Lophodermium* spp., *Hypoderma* spp. and *Cytospora* spp.), brown spot needle blight (*Mycosphaerella dearnessii*) and sooty mould (*Catenuloxyphium semiovatum*). Scleroderris canker (*Gremmeniella abietina*) sometimes infects stems of *P. strobus*, although other hard pine species are prime hosts. *P. strobus* plantations are susceptible to white pine root decline (*Ophiostoma piceaperdum*). Armillaria root rot (*Armillaria mellea* complex) also causes serious injury in plantations. Other fungi produce stem, butt and root rots, including red ring rot (*Phellinus pini*), red belt fungus (*Fomitopsis pinicola*), tomentosus root rot (*Inonotus tomentosus*), brown cubical rot (*Phaeolus schweinitzii*), fomes root rot (*Heterobasidion annosum*) and black root stain (*Verticicladiella* spp.).

In terms of large losses, the two most important organisms that affect *P. strobus* are the white pine weevil (*Pissodes strobi*) and the white pine blister rust (*Cronartium ribicola*). The sawfly (*Diprion similis*) has become a problem as a non-native species in the Lake States. Others capable of causing serious injury to foliage include the pine false webworm (*Acantholyda erythrocephala*), the white pine sawfly (*Neodiprion pinetum*) and, when growing near or with jack pine, the jack pine budworm (*Choristoneura pinus*). Shoots may be attacked by the eastern pine shoot borer (*Eucosma gloriola*) and the European pine shoot moth (*Rhyacionia buoliana*), causing forks and crooks when the leader is killed. Shoots may also be damaged by the pine leaf adelgid (*Pineus pinifoliae*) when *P. strobus* is growing near red spruce (*Picea rubens*) or black spruce (*Picea mariana*), the primary host species. The feeding of white pine aphids (*Cinara strobi*) causes flagging and mortality of smaller trees. In the eastern part of the range, the seedling debarking weevil (*Hylobius congener*) causes significant mortality to seedlings planted on recently cut softwood sites. Various root collar weevils, including Warren's collar weevil (*Hylobius warreni*), pine root collar weevil (*Hylobius radicis*), and pales weevil (*Hylobius pales*) can cause significant damage to smaller trees, as can mound ants (*Formica* spp.). Sapling stems may be damaged by the Zimmerman pine moth (*Dioryctria zimmermani*). Seed production can be seriously affected by the fir coneworm (*Dioryctria abietivorella*), the white pine cone beetle (*Conophthorus coniperda*) and the white pine cone borer (*Eucosma tocullionana*).

Uses

P. strobus is extensively used in maintaining soil stability, watershed protection, conservation of wildlife and in land reclamation of strip-mine spoils. It is used as an ornamental in urban settings.

The wood of *P. strobus* is soft, light, moderately strong, straight grained, easily worked and moderately decay-resistant. The sapwood is white to pale yellowish-white, and the heartwood is light brown or reddish-brown. The wood has an average density of about 400 kg/m^3 (at 12% moisture content). One of the most valuable lumber species in eastern North America, its easily machined, uniform-textured wood is unsurpassed for doors, windows, panelling, mouldings and cabinet work.

P. strobus is cultivated as a Christmas tree. The bark can be used as an astringent and an expectorant. Tar is produced from the wood and is utilized as an antiseptic, expectorant and protective.

Further reading

Buchert, G.P. (1994) Genetics of white pine and implications for management and conservation. *Forestry Chronicle* 70, 427–434.

Ostry, M.E., Laflamme, G., Katovich, S.A., Shaw, C.G. and Geils, B.W. (2010) Silvicultural approaches for management of eastern white pine to minimize impacts of damaging agents. *Forest Pathology* 40, 332–346.

Radu, S., Blada, I., King, J. and Sniezko, R. (2008) *Pinus strobus*: past and future in Europe. A page of silvicultural history and international scientific cooperation. *Annals of Forest Research* 51, 133–140.

Pinus sylvestris L.

Scots pine

Importance

Pinus sylvestris is one of the most commercially important Eurasian forest trees. As a pioneer species tolerating poor soil conditions, *P. sylvestris* is widely planted in temperate zones, where it is used for high-quality softwood production. It is the leading pine species grown for the commercial production of Christmas trees in northern USA and Canada. Mature stands of *P. sylvestris* are also a source of a valuable resin. Its greatest disadvantage is that *P. sylvestris* forms low-quality boles with poor self-pruning on fertile soils, and grows slowly in comparison with other species on less fertile soils.

Botanical features

Evergreen tree to 30–35 m tall (occasionally to 40 m), and to 50–130 cm dbh (occasionally to 1.5 m). Crown with spreading branches, open, conic and whorled when young, becoming more dense, flat- or round-topped and irregular with age.

Bole long, bare, straight, self-pruning in stands and often clear of branches for more than 50–60% of the total height.

Bark red, orange-red or pale grey to yellowish, thin and smooth with scaly ridges or finely scaly on the upper part of the trunk, darker to grey or almost black and fissured on the lower part of the trunk. The scaling orange-red bark on the upper stem is characteristic of *P. sylvestris*, and is seen in only two other two-needled pines, *Pinus densiflora* and *Pinus resinosa*.

Root system deep, with strong taproots where soil conditions allow.

Shoots glabrous, dull greenish-brown to greyish-yellow; strong shoots finely ribbed. Buds oblong–ovoid, brown, resinous or without resin.

Needles linear in pairs, rigid, usually twisted; longest (5–9 cm) on vigorous young trees; shorter on mature trees (2.5–7 cm). Needles semi-circular, 1.5–2 mm wide, thick, bluish-green (deep green closely and finely lined bluish-grey), in northern and eastern races more yellowish during winter. Needles usually retained on branches for 3 years, and in northern and eastern races for 4–6 years, and sometimes up to 9 years; in oceanic climates needles are retained for 2 years.

Strobili monoecious. Microstrobili clustered at base of weaker new shoots, catkin-like, bright yellow to crimson before they shed pollen. Macrostrobili in whorls of 1–5 at tips of new strong shoots, globular with scales protruding, rosy-purple, pink to yellowish-green during pollination period, then becoming green and downturned with closed scales until the next year. In the following year, they become conic, bright green. Mature cones in autumn dull grey-brown to tawny-yellow, woody, short-stalked, reflexed, conic–oblong, symmetrical or sometimes oblique, 3–7(–8) cm long; apophysis flat, pyramidal or reflexed.

New shoots grow in May to early July. Young trees may grow 1 m/yr. Radial increment of trunks lasts usually until late August. Flowering in middle May to early June. Wind-pollinated species. Cones ripen in autumn of the next year and open in February to April. Life-span of *P. sylvestris* is about 250 years, and up to 400 years.

Distribution

P. sylvestris has the most extensive geographical range of all species of the genus *Pinus*, and perhaps even of the whole Pinaceae family. The east–west range of *P. sylvestris* spreads out over a distance of about 14,000 km. The westernmost occurrences are situated in Portugal, Ireland and Scotland, and the easternmost are in the Russian Far East. The north–south range covers a distance of about 2700 km, with the northernmost occurrences at a latitude of 70°20'N in northern Norway and the southernmost ones at about 37°N in the Sierra Nevada Mountains in Spain.

P. sylvestris has been widely planted for timber beyond its natural distribution in Western Europe. It has been introduced in the arid, steppe regions of Eurasia and is widely cultivated in the northern USA and southern Canada, primarily in Christmas tree plantations (mainly var. *hamata*), as an ornamental tree and for erosion control. It has become naturalized in parts of New England and the Lake States in North America. Provenance and cultivation trials have been established in Korea, China, Mexico and New Zealand.

Silviculture characteristics

P. sylvestris is a very shade-intolerant species. It forms self-pruned trunks in stands with closed canopies. It can colonize exposed sites by virtue of its relatively far-flying winged seeds, and frequently acts as a pioneer species. *P. sylvestris* is relatively windfirm, but it suffers from snow damage in regions with mild and snowy winters. It is susceptible to fire.

Silviculture practice

P. sylvestris seeds are collected from mature (brown) cones during a period from late autumn to early spring. The seeds are extracted by drying the cones at a temperature of 40–60°C. Mature seeds are classified as orthodox. The seeds can germinate without any pre-sowing preparation. Stratification for 1–3 months at 1–10°C (ideally 2–5°C) stimulates germination. Sowing the seeds after 12–24 h in fresh water, and 3–6 h drying at room temperature without any stratification is a common nursery practice. Seedlings may achieve a plantable size in 7–8 months. Attention should also be paid to timing of nursery operations and disease control. Micropropagation and *in vitro* culture of *P. sylvestris* is also practised.

Various silvicultural systems are used for establishment of *P. sylvestris*, the main being clear felling, with seed-trees and short-term shelterwood.

There is considerable experience in the use of *P. sylvestris* as a plantation species. Both bare-root and containerized stock are used. Site preparation is an advantage, particularly on fertile soils where weed competition is a problem. Weed control is required during the establishment phase.

Planting density and further thinning schedules depend on the management objective. For sawtimber with a rotation of 80–100 years, the optimum initial density is 2500–3200 trees/ha. For pulpwood production the optimum initial density is 3200–4800 trees/ha. Where vitamin flour is made from needles (in eastern Europe and the former USSR), the initial density may be doubled or even trebled, and then reduced to the appropriate density for pulpwood or sawtimber production at ages 5–7 or 10–12 years.

Fertilizer application has increased productivity of *P. sylvestris* stands. Resin is often tapped 3–7 years before clear felling of mature stands.

In plantations, where sawtimber is the final objective (initial density of 2500–3200 trees/ha and rotations of 80–100 years), final yields of about 600 m^3/ha may be obtained on good sites. Mean annual increments peak at about 14 m^3/ha at an age of 65–70 years in the UK. In Ukraine, under optimum climatic and soil conditions, *P. sylvestris* plantations can reach average heights of 36 m and dbh of about 40 cm at 80 years old.

Natural stands can reach average heights of 30 m and dbh of about 35 cm at 80 years old, and heights of 38 m and dbh of about 50 cm at 140 years old.

Diseases and pests

High mortality of *P. sylvestris* seedlings caused by *Fusarium* spp. and *Alternaria* spp. have been observed in nurseries. Needle cast caused by *Lophodermium pinastri* causes serious damages of needles in nurseries and young (2–5-year-old) pure plantations of high density, particularly in conditions of humid summer, warm autumn and mild winter. *Phacidium infectans* can damage needles of young plants after heavy snows, particularly in the northern part of the species' distribution. *Heterobasidion annosum* causes root rot; pure stands of 25–35 years and established on post-agricultural lands are particularly susceptible to the disease. Armillaria root rot caused by *Armillaria mellea* and *Armillaria ostoyae* can also cause important losses. Stands infested by *Gremmeniella abietina* may be predisposed to secondary attack by *Tomicus* spp.

Larvae of *Amphimallon solstitialis*, *Melolontha hippocastani*, *Melolontha melolontha* and *Polyphylla fullo* can seriously damage root systems in nurseries and young plantations. Crown defoliation can be caused by needle-feeding larvae of some Lepidoptera (*Bupalus piniarius*, *Dendrolimus pini*, *Lymantria monacha* and *Panolis flammea*) and Hymenoptera (*Acantholyda* sp., *Diprion* sp. and *Neodiprion* sp.). Some methods of biological control are effective during the early stages of outbreaks, and some parasitoids and predators can contribute to this. Similar methods as well as trap logs and pheromone traps are used against stem pests such as *Ips* spp., *Dendroctonus* spp., *Polygraphus* spp., *Tomicus piniperda*, *Hylobius abietis* and *Aradus cinnamomeus*.

Uses

Due to its tolerance to poor soil conditions *P. sylvestris* is widely used in binding loose sands and land reclamation programmes in the temperate zone. *P. sylvestris* is also quite commonly used to establish windbreaks and anti-erosion belts in arid, mostly steppe regions. About 200 forms of *P. sylvestris* are used in ornamental gardening.

P. sylvestris wood is widely used for diverse wood constructions and products except in cases of permanent high humidity. It has a density of 510 kg/m^3 at 15% moisture content. The wood is exported in quantity from Russia and some other countries.

Tapping *P. sylvestris* for resin is a significant industry in eastern Europe and the former USSR. Pine oil extracted from needles and young shoots is widely used in medicine preparations to treat catarrh and bronchial infections. Flour from needles is used as a fodder additive. It is the most popular pine species grown for Christmas trees in North America.

Further reading

Cañellas, I., Martínez García, F., Montero, G., Mason, W.L., Alía, R., Forrest, G.I., Olsthoorn, A.F.M. and Sjoberg, K. (2000) Silviculture and dynamics of *Pinus sylvestris* L. stands in Spain. *Investigación Agraria, Sistemas y Recursos Forestales, Fuera de Serie* 1, 233–253.

Egbäck, S., Liziniewicz, M., Högberg, K.A., Ekö, P.M. and Nilsson, U. (2012) Influence of progeny and initial stand density on growth and quality traits of 21 year old half-sib Scots pine (*Pinus sylvestris* L.). *Forest Ecology and Management* 286, 1–7.

Karlsson, L., Bergsten, U., Ulvcrona, T. and Elfving, B. (2013) Long-term effects on growth and yield of corridor thinning in young *Pinus sylvestris* stands. *Scandinavian Journal of Forest Research* 28, 28–37.

Skilling, D. (1990) *Pinus sylvestris* L. Scotch pine. In: Burns, R.M and Honkala, B.H. (technical coordinators) *Silvics of North America. Vol. 1, Conifers*. Agriculture Handbook 654. USDA Forest Service, Washington, DC, USA.

Pinus taeda L.

Loblolly pine

Importance

Pinus taeda is the most important timber tree in the south-eastern USA and is planted for timber production in many other countries. It is a fast-growing tree with high-quality wood that is widely used for pulp and paper and for construction materials. It is heavily utilized in plantation programmes and as a resource for pulp and paper production. As an exotic it is used in Africa (particularly South Africa and Zimbabwe) and in South America for various products ranging from lumber to pulpwood. In South Africa, the species has developed abnormal wood defects and this has limited its planting although breeding programmes and silvicultural practices have reduced the problem. Originally found on moist sites, *P. taeda* is highly adaptable and commonly seeds into old fields. A great deal of research has been conducted into the species, including an active tree-improvement programme.

Botanical features

P. taeda is a medium to large tree that reaches a height of 34 m and a dbh of 75 cm that self-prunes well and develops a straight trunk and an oval, somewhat dense crown.

P. taeda tends to develop a taproot, especially on sandy soils. On wet sites or on those dominated by heavy clay the taproot is generally shorter. The bark is variable. When young, it appears brown and scaly. Older trees are ridged and furrowed, with somewhat apparent blocks. Very old trees have red-brown scaly plates.

Evergreen leaves, 15–23 cm long, with (usually) three yellow-green, stiff needles per fascicle.

Monoecious; male strobili are long, cylindrical, red to yellow, in clusters at branch tips. Mature strobili (cones) are ovoid to cylindrical and red-brown in colour, 7.5–15 cm long. The umbo is armed with a short spine. Seeds are small, 5–8 mm long, dark brown and rough. Seed wings 15 to 20 mm long and light brown.

Height and branch growth usually begin before the last frost in the spring. Young trees produce one to five growth flushes during the growing season. Resting buds between successive flushes are generally produced in response to drought. Height and branch growth usually cease in late summer. Needles are retained until the end of the second growing season. Pollen is shed in early Spring (mid-February to mid-April). Cones ripen September–October and seed is dispersed from October to December. Seed germination occurs the following spring, after overwintering on the forest floor.

Distribution

The natural range of *P. taeda* extends from southern New Jersey to central Florida and west to eastern Texas. *P. taeda* is found on the Coastal Plain, the upper and lower Piedmont, in the southern portion of the Ridge and Valley region, on the Highland Rim and on the Cumberland Plateau. Notable exclusions in its native range include the Mississippi River Valley and the very dry sandhills of North and South Carolina. The range of *P. taeda* has been extended north and west by planting. *P. taeda* has been planted for timber production in many countries around the world, and on a large scale in South Africa, Brazil, Australia, New Zealand and China.

Silviculture characteristics

P. taeda is a shade-intolerant, aggressive pioneer species that rapidly invades disturbed sites with exposed mineral soils and full sunlight. Within its native range it is known for its propensity to seed into abandoned fields. Natural regeneration of *P. taeda* is favoured by adequate seed sources, generally as seed trees or adjacent upwind stands. However, seeds can remain viable on the soil surface for several years.

Silviculture practice

P. taeda is most frequently regenerated either naturally or by the planting of 1–0 nursery stock. Container stock is generally not utilized for commercial production, but can aid establishment on harsh sites. Approximately 75% of the area in *P. taeda* in the south-eastern USA was established through natural regeneration. Stands are regenerated artificially either by direct seeding or by the planting of improved 1–0 nursery stock. When artificial regeneration is desired, cones are collected from designated seed collection areas or seed orchards in early autumn; natural seedfall usually occurs in October and November. Seeds will remain viable for several years when dried to 5% moisture content and refrigerated. Germination of the seed is hastened by a 30-day period of moist stratification. Mycorrhizal associations in the nursery bed are important for survival and growth after outplanting, and commercial products are available for mycorrhizal application. One-year-old seedlings are lifted from the nursery bed in early mid-winter and outplanted in late winter or early spring, when moisture is plentiful.

The most common silvicultural practice for the establishment of *P. taeda* stands involves clear felling of the pre-existing stand followed by the application of some combination of several site preparation techniques. Outplanting, when done, is usually by machine, frequently at spacings of 2.4 × 2.4 m, 3.0 × 3.0 m, or 2.1 × 3.0 m. When *P. taeda* stands are established naturally, stocking levels are frequently too high. Although *P. taeda* readily expresses dominance and self-thins, stand development of natural stands can be hastened with stocking control.

Whether established through planting or natural regeneration, young *P. taeda* growing stock often requires protection from fire and insect attack. Early stand growth can be greatly increased by release from competing vegetation. Once established, frequent, light under-storey fires are commonly used to remove competing hardwood vegetation and aid in subsequent stand establishment.

P. taeda responds to a wide variety of silvicultural treatments. Pre-commercial thinning is frequently used to regulate density in overstocked natural stands. Commercial thinnings are most frequently employed in stands greater than 17 years old and are most often used when managing for sawtimber. *P. taeda* grows very slowly on infertile soils and generally responds very well to fertilizer application.

In terms of growth and yield, a natural stand of *P. taeda* on an average site in the south-eastern USA reaches the peak mean annual increments at about the age 40, at 8.0 m^3/ha. At age 60, yields generally exceed 400 m^3/ha on better sites. Planted stands on average sites are expected to yield approximately 300 m^3/ha at age 30, with an MAI of 10 m^3/ha.

Diseases and pests

A wide variety of soil-borne organisms generically referred to as damping off can greatly affect *P. taeda* seedlings immediately after germination. Damping off is generally controlled with soil fumigation or seed treatments. Black root rot (*Fusarium* spp., *Macrophomina* spp.) is a major disease problem in *P. taeda* nursery beds, especially under prolonged warm conditions and in soils with pH above 6.0. Fusiform rust (*Cronartium fusiforme*) affects mature trees and seedlings, and can be a major nursery pest. Fusiform rust control generally involves a regular fungicide spray programme. Seeds are often treated for rust control. Seedlings with fusiform rust should be removed. Nearby oaks (*Quercus* spp.) can be removed to reduce the frequency of the alternate host. One of the major focuses of *P. taeda* tree improvement programmes involves breeding for increased rust resistance.

Nematodes and several species of grubs feed on the fine roots of *P. taeda* seedlings. Soil fumigants are used to control these organisms. *P. taeda* is host to bark beetles and the loblolly pine mealybug (*Oracella acuta*), which can cause tree mortality at high population densities. Two major insect pests affect seedlings in the nursery. Seedling debarking or pales weevils (*Hylobius* spp. and *Pachylobius* spp.) frequently kill seedlings by girdling. Tip moth (*Rhyacionia* spp.) often causes death of the terminal leader, but generally does not kill the seedling. Tip moth control may involve removal of adjacent stands of Virginia pine (*Pinus virginiana*), an alternate host species.

Uses

P. taeda is a commonly planted as an urban tree and has been used to stabilize soil. Pure and mixed stands of *P. taeda* provide habitat for several species of wildlife. It is frequently planted as a buffer or hedge, especially along interstate roads in the south-eastern USA.

P. taeda is an extremely important timber tree. The reported air-dry density of the wood is 577 kg/m^3. It is heavily utilized for construction material, poles and pilings, plywood and for long-fibre pulp. As long-fibre pulp, *P. taeda* is utilized in many different products. Residue from logging and pre-commercial thinnings is increasingly being used as a source of fuel.

Pine straw is often harvested and sold commercially as mulching material.

Further reading

Baker, J.B. and Langdon, O.G. (1990) *Pinus taeda* L., Loblolly pine, Pineaceae, Pine family. In: Burns, R.M. and Honkala, B. (technical coordinators) *Silvics of North America, Vol. 1. Conifers*. Agriculture Handbook 654. USDA Forest Service, Washington, DC, USA.

Campoe, O.C., Stape, J.L., Albaugh, T.J., Allen, H.L., Fox, T.R., Rubilar, R., Binkley, D. (2013) Fertilization and irrigation effects on tree level aboveground net primary production, light interception and light use efficiency in a loblolly pine plantation. *Forest Ecology and Management* 288, 43–48.

Costas, R., MacDonagh, P., Weber, E., Irschick, P. and Palavecino, J. (2001) *Pinus taeda* L.: plantation density and yield. *Revista de la Facultad de Ciencias Agrarias, Universidad Nacional de Cuyo* 33, 11–19.

Schultz, R.P. (1997) Loblolly pine: the ecology and culture of loblolly pine (*Pinus taeda* L.). *Agriculture Handbook (Washington)* 713, 494 pp.

Pinus tecunumanii Eguiluz & J.P. Perry

Tecun uman pine

Importance

Pinus tecunumanii has been described as a remarkable species with excellent phenotypic and silvicultural characteristics, with probably the best and most imposing stem form of all tropical pines. The bole is straight and can attain diameters of up to 1.4 m and a height of up to 55 m. If correctly sited, the species can be highly productive in tropical and subtropical regions. The major limitation to large-scale afforestation is lack of seed, as it only produces a light cone crop both in its native range and as an exotic. There is a need to optimize vegetative propagation in this species.

Botanical features

P. tecunumanii has an average height of 30–40(–50) m with a dbh of 0.5–0.9(–1) m. The crown is small or compact, light and conical in shape with a small number of thin lateral branches. The bole, which often has nodal and internodal swellings, is very straight with almost 30 m free of lateral branches.

The stem form is extremely straight and is said to be the best of all the pines of Mexico and Central America. In the wild, there is occasional nodal and internodal swelling, and foxtailing, particularly in Belize. *P. tecunumanii* has a well-developed root system and does not suffer from wind throw. The bark is reddish-grey, rough and fissured at the base of the bole, smoother and thinner above, exfoliating in papery scales, flakes or strips, leaving reddish-orange patches underneath.

The needles are slender, light green in colour, spreading, or more or less pendulous. There are usually 4(3–5) needles per fascicle; the needle length varies from 12 to 25 cm. The needle sheaths are papery, smooth, reddish-brown or grey and measure 10–22 mm long. The internal anatomy of the needles is very characteristic with two to three resin canals that are almost always in a medial position to the chlorenchyma.

Female flowers are borne on secondary branches. Cones are borne sparingly on the tree, either singly or in pairs and occasionally in whorled clusters of six or more. The cones are small, measuring up to 7 × 3.5 cm. The cones are shiny with a varnished appearance, narrowly conoidal with a pointed apex and a rounded base. They are weakly serotinous, pedunculate or sometimes subsessile and both forms may appear on the same tree. This species produces a small number of seeds per cone (3–5), with 70–110,000 seeds/kg. However, higher seed yields per cone (6–30 in natural stands and 9–40 in exotic plantings) have been reported in 8–16-year-old trees.

In exotic environments, male flowers are abundant but female flowers are very few. Flowering has been recorded in Zimbabwe in 4-year-old trees during June–November. In the highlands of Colombia, female flower initiation begins in July and the flowering may extend to January. Synchronization is usually good. Cones mature after 2 years and are ready for harvesting in September–October. In Mexico, within its native range, cones normally ripen between January and March, sometimes extending into April. There have been isolated reports of mature seed being collected in November in Honduras.

Distribution

P. tecunumanii has a very wide altitudinal (440–2800 m), but restricted (12–18°N) latitudinal distribution. The species extends from the two contiguous states of Oaxaca and Chiapas in Mexico through the Central American countries of Guatemala, Belize, El Salvador, Honduras, and down to central Nicaragua in a series of small, disjunct populations. The species' natural range of occurrence is largely determined by geology and rainfall, with distribution being restricted to those sites where soils are moderately deep and fertile, slightly acidic, and free draining with a mean annual rainfall ranging from 790 to 2000 mm. There are no known areas where this species has naturalized. *P. tecunumanii* has been planted in tropical and subtropical areas of many countries.

Silviculture characteristics

It is not known whether *P. tecunumanii* is truly frost-tolerant; however, trees 2 m high have been able to survive temperatures of −5°C in Zimbabwe. The species is known to be susceptible to stem breakage even in light winds if planted on exposed sites.

Silviculture practice

Seeds of *P. tecunumanii* are orthodox and can be stored for up to 5 years at a temperature of 4°C, at a relative humidity of 12% in airtight plastic bags.

P. tecunumanii can be propagated from stem cuttings and also from tissue culture. If propagating from stems, young seedling hedges up to 3 years old are preferred, as older ones do not root easily and the number of harvestable cuttings declines with age.

In the tropics and subtropics, the nursery period varies from 6 to 8 months depending on temperature. In warm sites without cold winters, the seedlings reach plantable sizes within 8 months. Seed of *P. tecunumanii* can germinate without any pretreatment. Composted soil from pine stands is preferred, as it will contain useful mycorrhizae important for optimum seedling growth. The soil is then mixed with milled pine bark or any other potting media. Seeds can be sown in nursery beds or directly into pots, plastic tubes or seedling trays. If sown directly into beds, the seedlings can then be pricked out into pots or trays when ready, or left to grow if the intention is to use bare-root seedlings.

In terms of stand establishment the norm is to do the planting during the rainy season, if in areas with distinct seasons. Field survival can be enhanced by watering if establishment is done before the onset of the rains or if the moisture level is still low.

The spacing currently being used in the experimental trials and the limited plantings to date is 3 × 3 m. It is common practice to clear other vegetation (trees and shrubs) from the area. On steep slopes it is common to dig holes (15–20 cm deep), but on flat ground, ploughing (ripping and or discing on compacted ground) is often a common practice.

There is little information on the use of fertilizers in the nursery or the field.

Pruning regimes of *P. tecunumanii* have not been rigorously established yet, but if inferences are to be made from other related tropical pines then pruning can commence after 4–6 years, depending on site quality. Although *P. tecunumanii* generally self-prunes in the wild, artificially imposed pruning is likely to be necessary.

Thinning regimes and timings have not been thoroughly investigated. Based on growth in evaluation trials on good sites, the first thinning of 50% of stocking can be performed after 6–8 years and the thinnings fed into a pulp mill. Based on thinnings in related tropical pines, a second thinning may be done after 12 years, with a final thinning at around 15–17 years. Thinnings may consist of selective felling of smaller trees, improvement fellings, or simply systematic thinnings depending on the quality and variability of seed source used to establish the stand.

In evaluation tests in Zimbabwe (1640 m and 1500 mm rainfall per annum), trees have attained heights of more than 2.5 m after 2 years of field growth, more than 9 m in height (dbh of 13 cm) after 5 years, and more than 14 m in height (dbh of 21 cm) after 8 years of field growth. Individual trees with diameters of more than 30 cm at 10 years have also been recorded. Some of the seed sources (San Jeronimo and Juquila) of this species have been shown to be just as productive as second generation-bred *P. patula* in prime sites (1650 m). In Zimbabwean trials after 8 years, the mean annual increment was estimated at 15–20 m^3/ha. It is expected that the species will attain millable diameters between 20 and 25 years, which is similar to other tropical and subtropical pines.

Diseases and pests

Due to its relatively short plantation history, only a few organisms have been reported as the cause of losses in *P. tecunumanii*; these include brown spot needle blight (*Mycosphaerella dearnessii*) and the Sitka spruce weevil (*Pissodes strobi*). It is semi-resistant in the seedling stage to the pitch canker fungus (*Fusarium circinatum*). *Rhizoctonia solani* causes mortality of rooted cuttings; and *Calonectria brachiatica* and *Calonectria brassicae* are able to cause disease in this pine species.

Costalimaita ferruginea proxima attacks the foliage of *P. tecunumanii*, and the tree is known to be damaged by the pine beetle, *Dendroctonus frontalis*, in its natural range. Bush pigs (*Potamochoerus porcus*) also damage young trees by digging and feeding on tree roots and occasionally uprooting them.

The mistletoes *Arceuthobium hondurense and Psittacanthus angustifolius* are parasites of *P. tecunumanii*.

Uses

There are no records of it being used for agroforestry, shelter belts, revegetation or land reclamation. It could be grown as an ornamental due to its imposing size and form.

In terms of wood properties, *P. tecunumanii* is comparable to other tropical and subtropical pines. The wood has an average density of approximately 465 kg/m^3, and a low resin content compared to *Pinus caribaea* or *Pinus elliottii*.

In its native range *P. tecunumanii* is currently exploited for sawn timber and other local wood products. As an exotic it is mostly intended for sawn timber and pulp. The wood is yellowish-white and when sawn does not require any special seasoning. The sawn timber can easily be dried in the open air or by kiln drying. There is a lack of information on the furniture-making properties of this species.

Further reading

Dvorak, W.S., Hodge, G.R., Gutiérrez, E.A., Osorio, L.F. and Malan, F.S. (2000) *Pinus tecunumanii*. In: CAMCORE (ed.) *Conservation and Testing of Tropical and Subtropical Forest Tree Species*. CAMCORE Cooperative. Grow Graphics, Botha Hill, South Africa, pp. 188–209.

Hodge, G.R. and Dvorak, W.S. (1999) Genetic parameters and provenance variation of *Pinus tecunumanii* in 78 international trials. *Forest Genetics* 6, 157–180.

Leibing, C., van Zonneveld, M., Jarvis, A. and Dvorak, W. (2009) Adaptation of tropical and subtropical pine plantation forestry to climate change: realignment of *Pinus patula* and *Pinus tecunumanii* genotypes to 2020 planting site climates. *Scandinavian Journal of Forest Research* 24, 483–493.

Pinus wallichiana A.B. Jacks.

Blue pine

Importance

Pinus wallichiana is indigenous to the Himalayan mountain regions, growing naturally over large areas in many countries. It is an important component of the middle- and high-altitude Himalayan forests. Its timber is extensively used for a variety of purposes including construction and railway sleepers, and *P. wallichiana* forests are important for the social and economic uplift of people living in mountain regions, as well as for the protection of steep slopes. It has gained world-wide attention for its resistance to blister rust disease.

Botanical features

P. wallichiana is a tall, straight, evergreen tree with smooth bark, which is slate-coloured on young and rough with shallow fissures on old trees. Branches are whorled, spreading, usually ascending in young trees and horizontal or drooping in old trees with upturned ends. Two whorls are often produced in 1 year. It attains a height of about 38 m and a dbh of about 1 m at maturity (120 years) under favourable conditions. The stem is curved at the base due to massive snow slides during winter. The root system of *P. wallichiana* is massive and somewhat lateral.

P. wallichiana is a five-needled pine, with 100–200 mm long, bluish or greyish-green, slender and drooping needles.

Male flowers or catkins are about 10 mm long and grow as clusters at the base of the current year's shoots. Female flowers are near the top of crown, 150–300 mm long, cylindrical in shape and only slightly woody with obtuse tips. *P. wallichiana* seeds are about 10 mm long, with wings three times the length of the seeds and 8–12 cotyledons. Good seed years are more frequent in *P. wallichiana* than is the case with any other important Himalayan conifer. Seed loses its viability soon if stored at high temperatures (>20°C).

New leaves appear in March or early April, according to location, and reach full size by August–September. The needles are shed between June and August. Male flowers appear on the lower branches of the trees in April and May, at different altitudes, before the emergence of the female cone, and shed pollen by June. *P. wallichiana* cones ripen from early September to November in most locations. The period between the first appearance of female flowers and the ripening of cones is about 16–18 months. Seed maturity in this species is reached 1 week or more before cone scales open to release seed. *P. wallichiana* trees start producing seeds at 15–20 years.

Distribution

P. wallichiana has a large natural distribution. The longitudinal and latitudinal distribution of this species is 69°–75°E and 26°–36°N, respectively. It is abundant on steep slopes and its altitude ranges from 1500 to 3600 m. This species is found extensively in Afghanistan, Pakistan, India, Nepal, Bhutan, Myanmar and China. Forests are both of pure and mixed type. This pine has been introduced to many European and North American countries.

Silviculture characteristics

P. wallichiana is more light-demanding from youth onwards than its associates. It is sensitive to fire, grazing and lopping damage. This species is typically gregarious, often forming extensive pure crops due to its capacity for regeneration in dense, even-aged masses on open hillsides and on abandoned cultivation. It is apt to suppress its associates by its rapid rate of growth on warm aspects. It does not appear to tolerate heavy rainfall and poor drainage in its natural range of distribution. It is frost-hardy at a young age. It also appears to tolerate air pollution in urban environments and montane areas. Snow causes crown breakage.

Silviculture practice

The forests are generally regenerated by natural regeneration assisted by artificial regeneration by direct sowing of seed or planting 2-year-old container-grown seedlings. Sowing of seed is carried out in autumn and planting in July–August in the forest. The rate of germination for this species is higher under medium soil pH values (pH 5, 6 or 7) than under pH 4 or over pH 8. *P. wallichiana* has maximum germination rate at 25°C, under light conditions.

Sun-dried seeds (5–10% moisture content) of *P. wallichiana*, kept in sealed glass jars at 4–5°C, have the highest rates of germination viability. Dry, cold conditions provide optimum conditions for storage. Best germination rates were achieved with stratification in moist sand for 120 days, followed by nursery sowing under a soil temperature of 15–20°C.

This species has successfully been propagated vegetatively by embryo culture and by air-layering. Mycorrhizal associations are especially important for seedling development in the nursery. Response to fertilizer in the nursery has been inconclusive.

Forests of *P. wallichiana* used to be managed under selection systems, but are increasingly managed under a regular shelterwood system with fixed periodic blocks. Since the forests respond well to regeneration felling, conversion to a uniform silvicultural system (irregular shelterwood system) with floating periodic blocks is also prescribed. The rotation varies from 120 to 200 years in different forests. Weeding and cleaning are usually not required in young *P. wallichiana* crops. Thinning, though desirable, is generally not carried out for economic reasons.

Diseases and pests

Brown-needle disease (*Mycosphaerella gibsonii*) and Dothistroma needle blight (*Mycosphaerella pini*) have been reported on *P. wallichiana*. Needle blight caused by *Lophodermium macci* was also reported. Needle-rust diseases also occur quite frequently on this pine, but are considered unlikely to cause serious damage. Severe needle cast disease of *P. wallichiana* has been recorded. The causal organism was isolated and identified as *Bifusella wallichi*. *Coleosporium himalayense* may cause needle rust. Among the 23 fungi isolated, those causing deep-seated infections were *Aspergillus funiculosus*, *Aspergillus niger*, *Fusarium semitectum* (*Fusarium pallidoroseum*), and *Rhizopus oryzae*. *Phellinus pini* may cause decay in old trees.

Dwarf mistletoe (*Arceuthobium minutissimum*) is damaging to *P. wallichiana* with infection ranging from 0 to 63%, and mortality from 0 to 3%.

The tree is not resistant to the weevil *Pissodes strobi*. Cones and seeds are seriously damaged by *Dioryctria abietella* in the north-western Himalayan region of India. Extensive mortality of this pine was also observed in western Bhutan in 1975, caused by bark beetles identified as *Ips stebbingi*.

Biston regalis is a serious pest; the larvae eat the bases of the leaves.

Uses

P. wallichiana is a multiple-purpose species and has been used in agroforestry and wasteland planting. It provides protection to the steep slopes in the Himalayan mountains on which it grows extensively.

Wood densities of 340–430 kg/m^3 have been reported. There are considerable differences between wood of trees growing in the dry and wet temperate zones. Wood from the dry zone has a larger percentage of latewood, smaller tracheid dimensions and ring width, higher density, and better strength properties, owing to the slower growth rate. It is a well-known and extensively used joinery wood in India, Nepal, Bhutan and Pakistan. In value, it ranks lower than deodar (*Cedrus deodara*) timber. After treatment, it is used for construction, joinery, house fitments, light furniture, packing cases, laminated boards, flush doors, plane tables and railway sleepers. The wood is durable under cover but non-durable in exposed conditions. Both air and kiln seasoning are successful. The timber can be sawn and worked to a smooth surface and it can be painted and polished well. In a laboratory study, wood-cement mixtures were suitable for the manufacture of wood cement boards.

Tapping of *P. wallichiana* is practised in Himachal Pradesh, India using the rill method, achieving yields of 1.154–1.488 kg/tree of resin in 6 months. Resin production can also be enhanced through the application of acid on the blaze. Different herbal preparations are made with *P. wallichiana* for the treatment of animal diseases.

Further reading

Garrett, P.W. (1992) Performance of Himalayan blue pine in the northeastern United States. *Tree Planters' Notes*, 43, 76–80.

Shrivastava, M.B. and Siaguru, P. (1997) Silvicultural system and sustainable management of cedar (*Cedrus deodara*) and blue pine (*Pinus wallichiana*) forests in Western Himalayas. *Indian Journal of Forestry* 20, 26–38.

Singh, O. and Thapliyal, M. (2012) Variation in cone and seed characters in blue pine (*Pinus wallichiana*) across natural distribution in western Himalayas. *Journal of Forestry Research* 23, 235–239.

Pinus yunnanensis

Yunnan pine

Importance

Pinus yunnanensis is a medium-sized, fast-growing tree, native to the hills and mountains of south-western China. There are approximately 6 million ha of *P. yunnanensis* forest in China, of which 275,000 ha have been planted. The timber is of great importance, being suitable for an enormous variety of uses as sawn timber or processed for industrial applications.

Botanical features

An evergreen tree, *P. yunnanensis* is generally 10–15(–40) m tall and 1 m in dbh. The branches extend and hang down pendulously. One-year-old branches are stout, brown or reddish-brown, glabrous. The scale leaves on 2–3-year-old branches fall off, exposing the reddish-brown inner bark. The outer bark is greyish-brown, becoming deeply furrowed, with thick splinters or irregular scales. The stem and trunk generally twist.

P. yunnanensis var. *tenuifolia* grows up to 35–40 m tall and 15–60 cm in dbh at 80 years old.

Needles are in groups of three, rarely two, usually remaining on the trees for 2–3 years, 10–30 cm long, 1 mm in diameter, sharp apex, with stoma lines on the under surface, and calloso–serrate on the margin, triangular or hemispheric in cross-section. There are 4–5 resin canals, usually located in the margin or in the middle, and leaf sheaths remain on the tree. *P. yunnanensis* var. *tenuifolia* is also 3-needled, needles softly pendulous, 20–30 cm long, less than 1 mm in diameter; triangular in cross section; 4–6 resin canals in the margin or rarely one canal in the middle.

P. yunnanensis is monoecious; the male cone is cylindrical, 1.5 cm long, arranged beneath cone bracts on the new branches in an aggregate spike. The cone is oval to egg-shaped, 5–11 cm long, 3.5–7 cm in diameter, turning from green to brown at maturity, with a short rachis. The scale shield is thick, flat or lumped, twisting occasionally, with ridges. The scale navel is slightly grooved or lumped, with or without short thorns. The cones of *P. yunnanensis* var. *tenuifolia* are 5–10 cm long, 4–5 cm in diameter, occasionally cuneate at the base; scale shields greyish-brown, ridged; scale navels flat or slightly lumped; thorns not apparent and usually fall early. *P. yunnanensis* leaf buds swell in the middle of March, leaves begin to sprout in the middle of April, and leaves fully grow out at the end of May. It flowers in April–May. Female cones start to develop in March of the following year; cones begin to mature in November–December. Cone-scales then open out in the following March.

Distribution

P. yunnanensis is distributed between latitudes 23°N and 32°N, longitude 98°E and 106°30'E. It is widely distributed in south-west China. The altitude range is (600–)1000–2800 (–3100) m, growing densely above 1300 m in south Yunnan, 1800–3100 m in north-west Yunnan, 1000–3000 m in Sichuan, 1000–2000 m in Guizhou and 600–2800 m in Guangxi.

Silviculture characteristics

P. yunnanensis is found in pure and mixed forests, regenerating naturally. It occurs on a wide range of soil types and can survive on shallow, rocky and infertile sites, tolerating drought, wind and severe frosts. It may become an aggressive colonizer on fertile sites. In general, regeneration is better on sunny south-facing slopes than on north-facing slopes, but this preference is reversed at low altitude and on dry soils.

Silviculture practice

Stand establishment is achieved from the abundant natural regeneration, by direct planting of planting stock, or transplanting wild seedlings.

P. yunnanensis exhibits mast years approximately every 3 years. Scattered trees on sunny hill slopes or ridges begin to produce seeds at 5–6 years old, whereas trees in stands usually do so at 7–9 years old. Seeds should be collected from vigorous and healthy mother trees 30–40 years old, with a normally developed crown, free from disease and straight trunk.

The maximum seed production observed is on quality sunny mountain slopes with 6000 cones/ha, equivalent to 6.5

kg seeds. On the lower sunny slopes and quality north-facing slopes, production is halved, and on lower north-facing slopes this decreases further, to 2000 cones/ha, or 2.1 kg seeds.

Seeds should be collected in December–March, before scales split. Seeds can be safely stored for 2–3 years when dry and ventilated. Cones weigh 53–66 g and contain 35–62 seeds; there are 62,000–64,000 seeds/kg and germination rates of 70–90%. Unsoaked seeds usually germinate 15 days after sowing.

P. yunnanensis is usually propagated by both sowing and grafting. From the end of May to the beginning of June, 2–3-year-old seedlings are used as rootstocks and the fresh shoots from superior trees are used as scions. The scions should be previously soaked in water to degrease and to absorb water. The survival rate is 60% with under-bark grafting and 78–90% by bud grafting.

P. yunnanensis forest is generally established by direct seeding with 7.5 kg seeds/ha, with a germination rate of 70–90%. Aerial sowing is also used on vast wasteland areas, of broad rows 70–100 m wide with 4–7.5 kg seeds/ha. A clonal seed orchard can be established by using shoots collected from plus trees.

The planting period in Yunnan covers the last 10 days of May; June; and the first 10 days of July during the rainy season. In the dry and lower regions (less than 1400 m in altitude) it is necessary to protect seedlings by sheltering. *P. yunnanensis* begins to wilt when soil moisture content drops to 10% and begins to die when it reaches 7.5%. After sowing, seeds are often damaged by birds, rats and insects, so some steps are required, such as dyeing seeds and using physical protection. The planting density is generally 10,000 seedlings/ha (1 × 1 m spacing) on lower hills and 6667 seedlings/ha (1 × 1.5 m) on mountainous sites. Fire breaks need to be installed. Weeds which grow vigorously after the first rains and survive after the spring drought need to be controlled.

In dense forest, trees grow a straight trunk and have few lateral branches, and pruning is not required. The canopy begins to close when trees are 8–10 years old, and at this stage they begin to self-thin gradually. When practised, thinning intervals are usually 3–5 years during the first 20 years, and then every 5 years. The clear-strip harvesting system is used in lower hill and dune land. The group system is applied to land with steep slopes and an area of 3–5 ha.

In terms of stand establishment, 3-year-old seedlings are generally 30 cm high with primary roots 60 cm long. *P. yunnanensis* starts to produce lateral branches at 5 years old, and grows slowly at first, with a mean tree height of 4.5 m and dbh of 5.2 cm after 10 years on the best site.

In plantations, the annual height and diameter of trees between 20 and 40 years old increases the fastest, with volume increments of 8.8–8.9 m^3/ha/year. This period of rapid growth may last until trees are 50 years old. At this age trees reach a mean height of 20.8 m and dbh of 21.2 cm, with a standing volume of 419 m^3/ha when density is about 1000 trees/ha.

Diseases and pests

Coleosporium solidaginis (*Coleosporium asterum*) attacks seedlings and needles on large trees and has a significant effect on young forest, even leading to tree death. *Cronartium flaccidum* attacks young shoots, extending to large branches and even to the trunk, causing slight spindle-like swellings where damaged. In addition, *Didymosporium* sp. or *Phaeomarssonia* sp. mainly damage needles on trees of around 2 years old, attacking young trees at 3–6 years old in west Yunnan and occasionally 20-year-old trees, but with little damage.

Important arthropod pests include *Dendrolimus latipennis*, *Dendrolimus kikuchii*, *Dendrolimus biundulata* and *Dendrolimus punctatus wenshanensis*, which eat needles and can kill trees. *Matsucoccus yunnanensis* damages branches and trunks in Yunnan, Guizhou, Sichuan and Xizhang. The quarantine pest *Bursaphelenchus xylophilus* can rapidly cause pine tree wilting and is very difficult to control. In addition, many other pests such as *Dioryctria splendidella* (*Dioryctria sylvestrella*), *Ips sexdentatus* and pine weevils may affect *P. yunnanensis. Tomicus minor* and *Tomicus yunnanensis* are two important shoot beetles on Yunnan pine.

Uses

P. yunnanensis has ecological functions such as erosion control, soil and water conservation; fixation of sand dunes; and protection of wild life. It is also used as a windbreak and ornamental.

P. yunnanensis wood is light, or light to medium in weight, usually soft, swelling and shrinking, and intolerant of termite damage. The heartwood is reddish-brown; sapwood is yellowish-brown or light reddish-brown (turning blue), easily distinguishable. Wood density is 470–509 kg/m^3 at a moisture content of <15% and mean density of 483 kg/m^3. The wood is easy to split, with no straight grain. The timber can be used for building, poles, pillars, sleepers and as a fuel.

P. yunnanensis is an important source of gum rosin and turpentine. One tree may produce 3–4 kg of resin during a season. The pine resin, turpentine oil, seed oil, young branches cones and pollen are used in traditional Chinese medicine.

Further reading

Liao, S.-X., Li, K., Lu, Y.-C., Li, Z.-Y. and Yu, J.-S. (2009) Analysis on growth of the dominant population of *Pinus yunnanensis* in Central Yunnan Plateau. *Forest Research, Beijing* 22, 80–84.

Xie, J.R. and Jiang, H.Q. (1986) *Forestry in Yunnan*. Yunnan Science & Technology Press, Yunnan, China.

Zhao, L.-Z., Cai,N.-H., Xia, Q.-Z., He, R.-X. and Xu, Y.-L. (2012) Effect of drought stress on seed germination and osmotic regulating substances of *Pinus yunnanensis*. *Journal of South-West Forestry University* 32, 21–25.

***Piptadeniastrum africanum* (Hook. f.) Brenan**

African greenheart

Importance

Piptadeniastrum africanum is a large tree from West and Central Africa, producing high-quality and valuable timber used for construction and furniture. It is widespread and quite common. The tree is planted or left during forest clearing as a shade tree in coffee, cocoa and banana plantations. In Uganda, the species has been planted successfully in experimental plantations.

Given its growing importance as a commercial timber tree, and partial bark removal for traditional medicine, genetic erosion is a possible concern.

West African countries in particular export the timber, mostly to Europe where it is considered comparable in quality with that of oak. Locally, the dense wood is used for dug-out canoes. The bark fibre may be used to weave mats, edible caterpillars feed on the leaves and the flowers attract honeybees. The tree, and in particular its bark, possesses a variety of traditional medicinal uses. In various countries *P. africanum* is considered a magical tree.

Botanical features

P. africanum is a large flat-crowned deciduous or evergreen large tree up to 50 m tall and with a diameter of up to 180(–300) cm. Its bole is straight and cylindrical, often branchless for up to 20(–30) m with large thin buttresses up to 5(–8) m high. The yellowish to greyish-brown bark is generally smooth but sometimes slightly fissured on larger stems, sometimes with 'ring marks' and numerous small lenticels. The leaves are alternate, bipinnately compound with 10–19 pairs of pinnae, and per pinna 25–60 pairs of sessile, linear, 1.5–8.5 × 1 mm leaflets. The inflorescence is an axillary or terminal spike-like false raceme up to 11 cm long, and often clustered at the ends of twigs. The individual flowers are small (5–6 mm across), whitish or yellowish, bisexual, regular and 5-merous; the petals are free, oblong to lanceolate. Fruits are flattened linear pods 12.5–36 × 1.5–3 cm, dark brown, transversely veined, containing up to 9 seeds. The wind-dispersed seeds are oblong, flat, 3–9.5 cm long including the papery wing surrounding the seed and glossy brown.

The *Piptadeniastrum* genus comprises a single species and seems to have a rather isolated position within its family.

Distribution

P. africanum occurs from Senegal east to southern Sudan and Uganda, and south to the Democratic Republic of the Congo and northern Angola. It occurs in lowland evergreen and semi-deciduous forest, up to 1200 m altitude. In Ghana, it shows a preference for evergreen forest. In general the tree occurs at low densities, but in some forests in Sierra Leone it accounted for 10% of all trees of over 60-cm bole diameter. In Gabon, it occurs scattered in primary forest, but may be locally more abundant in old secondary forest. In Uganda, it is a common and characteristic species of the forest in the Lake Victoria belt. Throughout its area of distribution average bole volume varies from 0.6 m^3/ha (Gabon) up to 4.8 m^3/ha (southern Cameroon).

The tree also often stands as a single tree on farmland.

Silviculture characteristics

Seedlings are often common in the forest, even in deep shade. However, they grow very poorly in shade, where they may be only 20–35 cm tall when 3 years old, whereas they may reach up to 150 cm tall after 4 years in less shaded conditions in the forest. Saplings and young trees demand good light, and are usually found in small gaps in the forest. The species can benefit from forest disturbance and can be a locally common component of older secondary forests.

Silviculture practice

Seeds for planting are collected from the forest floor (5000–6000 seeds/kg). Seeds should not be stored for more than a few weeks as they lose their viability in 1 month. They do not show dormancy, and germinate in 1–3 weeks. In a germination test, 96% of the seeds germinated within 8 days. They should be sown in shaded nursery beds. The growth of seedlings is slow and they may stay in the nursery for more than 1 year before planting. Wildings are sometimes collected for planting.

The minimum felling diameter varies from 60 cm in Côte d'Ivoire, Cameroon and the Democratic Republic of the Congo, to 90 cm in Ghana. The high buttresses mean that a platform has to be constructed for felling the large trees. Freshly harvested logs sink and thus cannot be transported by river.

Diseases and pests

No published information on pests or diseases of *P. africanum* is available. However, it is reported that *P. africanum* wood has a natural resistance to the Formosa subterranean termite (*Coptotermes formosanus*): termites experimentally exposed to the wood died within 1 week.

Uses

The wood has an air-dry density of 689 kg/m^3. The heartwood is pale brown to golden-brown, occasionally dark brown, with an attractive stripe pattern on quarter-sawn surfaces, distinctly demarcated from the 5–15 cm thick, pale pink to greyish-red sapwood.

The heartwood is moderately durable with some resistance to fungal, dry-wood borer, and termite attacks. A disadvantage limiting its marketability seems to be the high shrinkage rate. The wood is used in construction, including marine construction and bridges, flooring, railway sleepers, mine props and shipbuilding. It is also often used for interior trim, joinery, (garden) furniture, cabinet work, sporting goods, turnery, hardboard, particleboard and pulpwood. Traditionally people made dug-out canoes of the large trunks. It is also used as fuelwood and for charcoal production.

P. africanum roots, leaves and especially bark are all used in various traditional medicines. Bark decoctions are used internally to treat cough, bronchitis, headache, mental disorders, haemorrhoids, genito-urinary infections, stomach ache, dysmenorrhoea, male impotence and as an antidote; externally, decoctions are applied to treat fever, toothache, pneumonia, oedema, skin complaints, rheumatism, to expel worms, to dispel fleas and as a purgative and abortifacient. A decoction of the bark also enters in a complex treatment of leprosy. Pounded leaves and leaf decoctions are applied as an enema to treat gonorrhoea and abdominal complaints.

The bark is used in arrow poison, and as ordeal poison and fish poison by pygmy people in Cameroon and the Democratic

Republic of the Congo; mixed with rice it is used to poison mice. It can also be used as a soap substitute.

Very little is known about its pharmacological properties aside from the toxic effects of the bark.

Further Reading

Burkill, H.M. (1995) *The Useful Plants of West Tropical Africa. 2nd Edition. Volume 3, Families J–L.* Royal Botanic Gardens, Kew, Richmond, UK, 857 pp.

Jiofack Tafokou, R.B. (2008) *Piptadeniastrum africanum* (Hook.f.) Brenan. In: Louppe, D., Oteng-Amoako, A.A. and Brink, M. (eds) *Prota 7(1): Timbers/Bois d'œuvre 1.* PROTA, Wageningen, The Netherlands.

Katende, A.B., Birnie, A. and Tengnäs, B. (1995) *Useful Trees and Shrubs for Uganda: Identification, Propagation and Management for Agricultural and Pastoral Communities.* Technical Handbook 10. Regional Soil Conservation Unit, Nairobi, 710 pp.

Neuwinger, H.D. (1996) *African Ethnobotany: Poisons and Drugs.* Chapman & Hall, London, 941 pp.

Platanus acerifolia (Aiton) Willd.

London plane

Importance

Generally considered to be a hybrid between American sycamore (*Platanus occidentalis*) and oriental plane (*Platanus orientalis*), *Platanus acerifolia* is a fast-growing deciduous tree often planted in gardens, parks and urban areas throughout the temperate regions of the world. It is attractive, withstands wind and tolerates air pollution. It is also valuable for the production of timber and fuelwood, and for soil conservation and erosion control. Trees are tolerant of anthracnose but strongly susceptible to canker stain, a destructive disease that can kill a tree within 1 year. Trichomes shed from young leaves and fruits in spring and early summer cause allergies.

Botanical features

P. acerifolia has a fine, cylindrical, straight stem, wide-spreading branches and a pyramidal crown. It can grow up to 40 m high, with a dbh of 60–80 cm. The stem is often continuous into the crown and clear of branches in the basal part. The sympodial root system is wide and deep. Bark peels off in large flakes and is smoothish, with patches of brown, green, and grey. Young shoots are dark brown and tomentose.

Leaves are alternate, spiral, palmate, 12–25 cm wide, with 3–5 (rarely 7) triangular lobes which are toothed at the base, sometimes sparsely throughout, the base being truncate, obtuse to cordate. Young foliage is finely glabrous at first; trichomes soon drop off except on the underside veins. There are two fused fan-shaped stipules. The petiole is 3–10 cm long and encloses the axillary bud.

Inflorescences consist of pendulous strings of 2–6, dense, globose (2–2.5 cm), pedunculate heads of flowers; staminate and pistillate flowers occur in separate heads. Female flowers are dark red, globose, usually in catkins along branchlets, while male flowers are yellowish, soon shed, located at branchlet tips. The fruit is an elongated, chestnut-brown, single-seeded achene with a hairy tuft at the base. The frequency of full seeds is highly variable, resulting in low germination capacity.

P. acerifolia is leafless from November until the end of March. The growing and flowering periods are influenced by environmental factors, mostly temperature and rainfall. In Europe, flowering begins in March–April and finishes in May–June. Fruits ripen in autumn and break up during late winter–spring, disseminating the achenes. Collection of fruiting heads can take place any time after they turn brown, but is easiest after leaf fall.

Distribution

First planted in England around 1680, *P. acerifolia* was thought by some to have originated in Oxford in the 1670s, but as *P. orientalis* is very rare in the UK, the Oxford trees are now believed to have originated from fruits collected in France. *P. acerifolia* is more likely to have arisen in the 1650s or 1660s in France or, more probably, Spain, where both parent species would have been planted in close proximity.

Silviculture characteristics

P. acerifolia grows well singly, or in small groups in mixed woods. It is strongly heliophilous. Fast growth, suitability for coppicing and pollarding, and extension of the root system deep into the soil make it a typical pioneer species of river-banks and alluvial soils. In these sites its natural regeneration is easily effected.

Trees survive high levels of urban pollution by sloughing off bark whose lenticels are loaded with pollutants, exposing fresh open lenticels, making them suitable for urban forestry.

Silviculture practice

Seed production starts at about age 7–8 years in seedlings, but starts after only 3 years in vegetatively propagated plants. Germination capacity is low (20–30%). Seeds can be stored under cool, well-ventilated conditions in open-mesh bags or spread out on shelves. For storage longer than 1 year, seeds should be dried to 10–15% moisture and kept in containers at −5° to 3°C. Spring is the best time for sowing, with seeds broadcast or drilled in rows and covered lightly with soil or mulch. Seedlings are very fast growing, and plantable size is reached in 1 or 2 years (1+0 or 1+1).

For vegetative propagation, cuttings are taken from woody current-year shoots (length 20 cm, diameter 5–8 cm, with at least four buds) in autumn after leaf fall and immediately planted in rows 10 cm apart, with 25 cm between rows. Shading is recommended from early spring until August. The most vigorous cuttings are ready to be planted out in 1 year. Rooting ability is generally very high. *P. acerifolia* can also be propagated by *in vitro* methods.

When *P. acerifolia* is coppiced for the production of fuelwood, stakes and poles, rotations are usually very short. Cultivation in high forest is rare, although it has potential due to good wood quality. Coppice planting is usually preceded by deep working of the soil (70–80 cm). Trees are planted 2–3 m apart, with 3–4 m between rows. Planting material can be 1–2-year-old seedlings or 1-year-old cuttings. Weed control is important during the first growing season but depends on the density of planting in subsequent years. For coppices with an abundance of suckers, cutting must be carried out at the start of the second growing season. Thinning of stumps may be required at the beginning of the third year. Fertilization with nitrogen during coppice establishment helps increase root weight.

Mean volume increment is between 15 and 20 t/ha annually on a coppice rotation of 5–6 years. An average height of 40 m, a dbh of 48 cm, a density of 300 trees/ha and an annual increment of 16–18 m^3/ha was obtained in a 58-year-old stand in high forest in Italy. To ensure vigorous increments well into stand maturity, geometric stand thinnings are required.

Diseases and pests

P. acerifolia is host to canker stain caused by *Ceratocystis fimbriata* f. *platani*. This disease, which kills adult trees, produces spreading lesions in the phloem, cambium and extensive regions of sapwood. It arrived in Europe in the mid-1940s from the eastern USA, where it appeared during the 1930s. It is transmitted mainly via contaminated pruning tools and wound dressings, and infection can be prevented by sterilizing tools and immediately removing diseased and nearby trees. Anthracnose, caused by *Apiognomonia veneta*, affects buds, twigs, shoots and leaves. *P. acerifolia* clones vary in susceptibility to the disease, and resistant cultivars have been developed. Diseased shoots and fallen leaves must be destroyed. Other fungal pathogens affecting *P. acerifolia* include *Inonotus hispidus* and *Ganoderma* species causing white rot, and *Microsphaera platani*. White rots are controlled by avoiding root damage.

Noxious insect pests include wood-boring larvae of *Cossus cossus* and *Zeuzera pyrina*. *Ptilinus pectinicornis*, *Macrotoma scutellaris*, *Rhaesus serricollis* and *Cerambyx cerdo acuminatus* attack trees in decline, causing damage to the tree and impairing wood quality. *Rhamnusius bicolor* attacks dead wood. Larvae of *Aegeria mellinipennis* dig annular tunnels in the wood of larger branches and in the trunks immediately below the point where those branches emerge, stunting growth and causing proliferation of wood growth at infestation sites. *Stomacoccus platani* attacks trees in the south-western USA and often causes serious decline.

Corythucha ciliata originated in North America but made its appearance in Europe in the 1960s. This lacebug passes its entire life cycle on the leaves of the plane tree. Feeding activity causes irregular chlorotic areas on the leaf edge that expand from the leaf base outwards. The most severely affected trees show pronounced crown yellowing and precocious leaf fall. Other insect pests from North America are *Corythucha confraterna* and *Hyphantria cunea*, the latter producing two complete generations a year, the larvae from the second causing complete defoliation in many cases. Feeding by *Edwardsonia platanicola* larvae, nymphs, and adults causes a silvery-whitish spotting that is easily visible on the upper leaf surface. Infested trees show clear signs of stress.

Uses

In the 20th century, *P. acerifolia* became one of the most important shade trees for use in urban forestry, tolerating air pollution and compact urban soils. Magnificent crowns and the changing colours of the bark make the trees a very decorative feature along roads and in parks in cities and towns all over the world, not just in London where it was first popularized.

P. acerifolia has a wood density of 549–604 kg/m^3. It has a pinkish-coloured sapwood and heartwood, similar to that of beech but somewhat darker. The rings are not very conspicuous and the planed wood surface is dull. Wood texture is very homogeneous and open. Medullary rays are clearly visible, narrow, numerous and dark, giving planed wood its unmistakable and decorative appearance. It seasons easily without cracking and can be worked well after seasoning, but it is not durable. It is used in cabinetmaking, carpentry and lathe work. It can be sliced to produce attractive veneers. *P. acerifolia* also has potential for plywood production.

Paper from plane tree pulpwood is strong and has superior tear-resistance. Processing into chips prior to transformation into pulp yields a paper of good quality.

Despite its good characteristics, which make it suitable for a wide range of uses, the wood of *P. acerifolia* is hardly used by the wood industry, mainly because only small quantities reach the market.

Further reading

Landreth, J. (2010) *The Great Trees of London*. Time Out Guides Ltd, London.

Plotnik, A. (2000) Sycamore (planetree). *Platanus occidentalis*, *Platanus* × *acerifolia*, and others. In: Plotnik, A. (ed.) *The Urban Tree Book: An Uncommon Field Guide for City and Town*. Three Rivers Press, New York, New York, USA, pp. 60–67.

Podocarpus latifolius (Thunb.) R. Br. ex Mirb.

Broadleaved yellowwood, podo

Importance

Podocarpus latifolius is a slow-growing and exceptionally long-lived African forest tree. The wood, traded as 'podo', is highly valued for furniture, panelling, and shipbuilding (masts and planks). It used to be in high demand in southern Africa and was harvested at a large scale across Africa, but larger specimens of the species have become quite rare (e.g. in Zimbabwe and Rwanda) and international trade is limited. All *Podocarpus* species are protected in South Africa. It is included on the International Union for Conservation of Nature (IUCN) Red List of Threatened Species, although still as a 'Species at Low Risk'.

P. latifolius has been declared the national tree of South Africa and is planted there as a garden ornamental. Its characteristic conical shape, and bright glossy leaves make it a popular choice. The bright edible berries attract birds and other animals, which spread the seed.

Several countries have cultivated the species, but only small stands have been established. It is also planted as a shade tree in coffee, cocoa and banana plantations.

Botanical features

P. latifolius grows up to 35 m in height and reaches 1.5–3 m in diameter. It generally has a straight and cylindrical bole that can be branchless up to 10–20 m. Older trees sometimes have small buttresses. The bark is reddish to greyish-brown, and peels in long strips. The bark tannin content is high (2.7–3.6%). The simple and entire leaves may be up to 15 cm long, 5–12 mm wide, shiny and bright green when young, curved, grooved and with pointed tips. These leaves are arranged spirally to almost opposite with the brighter younger leaves standing out against the dark leaves of mature foliage.

The tree is dioecious. Male cones are axillary, single or paired, 1–5 cm × 3–5 mm, pinkish to reddish with many spirally arranged scales. Female cones are terminal and solitary, on a short branchlet, have 1–2 fertile scales at base with a swollen aril of 8–18 mm long, glaucous-green turning pink-reddish-purple and are edible and sweet. Seeds are drupe-like, globose, 7–12 mm long, greyish-green with a slightly woody coat. Birds and baboons eat the receptacle of the fruit and discard the seed undamaged.

Distribution

P. latifolius is an evergreen forest tree species which is widely distributed in mountain and highland forests in sub-Saharan Africa from Eritrea to the Southern Cape in South Africa. It occurs commonly at altitudes over 1000 m, and even up to 3500 m in Kenya and Zaire. The species prefers areas of low to medium temperatures with rainfall ranging on average from 1000–2000 mm or more and deep, fertile, well-drained soils. In harsh or exposed areas it tends to become stunted, small and dense, and may be covered in lichens. In South Africa it descends to sea level.

Silviculture characteristics

P. latifolius is a moderately shade-tolerant species and can regenerate in small and medium forest gaps in natural forests, where it outcompetes seedlings of angiosperm species. Old-growth Afromontane forests are dominated by *P. latifolius*, supporting the view that long time intervals are required for slow-growing conifers to reach the canopy.

The fruits are distributed by larger birds (turacos and hornbills), monkeys and baboons, thus providing a wide geographical distribution of the species within a certain forest ecosystem.

One kg of *P. latifolius* seeds contains 1700–3000 seeds. Freshly collected seeds germinate well (up to 80%) but loss of viability on storage is rapid and seeds for sowing should not be kept for longer than 4 days after collecting. Storage for 1 year in cold and humid conditions gave a germination rate of 72%. Germination occurs within 1–2 months. Studies in South Africa showed that natural regeneration of *Podocarpus* is abundant in small- and medium-sized forest gaps.

Growth of *P. latifolius* under natural forest conditions is slow, ranging from 0.2–0.32 cm/yr in mean diameter increment for trees with dbh of 10–90 cm. Under natural forest conditions, the stem form is generally good. In cultivation, with improved fertility and reliable water availability, this may increase to 4–13 mm/yr (30 year old trees in Rwanda).

Yield from sustainably managed natural forest is very low and despite the tree's highly valuable timber, its potential for commercial timber plantation development is limited due to slow growth rate.

Silviculture practice

Information on silvicultural establishment and management in natural and plantation stands is limited. In most countries, harvesting and exploitation of *P. latifolius* is based on natural forest stands.

The cleaned seeds should be planted within a few days, in a mixture of compost and sand, placed in the shade and watered regularly. Seedlings benefit from mulching.

Pole-sized trees may be coppiced. Larger trees often develop drooping branches, which may be pruned.

In South Africa, trees are selectively harvested under a 10-year felling cycle in which harvestable trees are not defined by minimum diameters, but according to signs of old age and viability. The advantage of this system is said to be its guaranteed sustainability, compatibility with nature conservation aims, optimal productivity and managerial simplicity and flexibility.

Diseases and pests

In South Africa, a canker and die-back disease caused by the fungus *Erythricium salmonicolor* (*Phanerochaete salmonicolor*) has been reported in *P. latifolius* as well as in other tree species. The fungus *Tripospora tripos* is reported to affect the leaves.

Insects eat and damage the seeds.

Uses

P. latifolius is used as an urban ornamental tree and as a shade tree in banana, coffee and cocoa plantations.

Timber from the species is pale yellowish-brown with a density of 469 kg/m^3. Heartwood and sapwood are not demarcated. It is used for house flooring, beams, veneers, furniture, joinery, panel framing, musical instruments, toys, food containers and handicrafts. The South African Railways used to make railway sleepers from the timber and it was previously used to make wagon boxes and coffins. It has no scent and does not chip easily, which is why it is favoured for butchers' blocks. In some countries it is used as firewood and for charcoal production and it appears to make high-quality pulpwood. The untreated wood is not very durable. It is susceptible to bluestain, powder-post beetle, termite and marine borer attacks. The wood is permeable to preservatives.

In the Eastern Cape, the bark of *P. latifolius* is used as a decoction to treat gall sickness in cattle and distemper in dogs.

The bark is also used in traditional medicine in other African countries, to treat human stomach ache; in East Africa it is used to treat cattle diseases. Studies indicate the plant has genuine pharmacological activities.

Further reading

Barker, N.P., Mullera, E.M. and Mill, R.R. (2004) A yellowwood by any name: molecular systematics and the taxonomy of *Podocarpus* and the Podocarpaceae in southern Africa. *South African Journal of Science* 100, 629–632.

Bolza, E. and Keating, W.G. (1972) *African Timbers - The Properties, Uses and Characteristics of 700 Species*. CSIRO, Division of Building Research, Melbourne, Australia, 710 pp.

Coates Palgrave, K. (1983) *Trees of Southern Africa*. 2nd ed. Struik, Cape Town, South Africa, 959 pp.

Midgley, J.J., Bond, W.J. and Geldenhuys, C.J. (1995) The ecology of southern African conifers. In: Enright, N.J. and Hill, R.S. (eds) *Ecology of the Southern Conifers*. Melbourne University Press, Carlton, Australia, pp. 64–80.

Polyscias fulva (Hiern) Harms

Parasol tree

Importance

Polyscias fulva is a deciduous, fast-growing tropical African tree, reaching up to 30 m tall with a dbh of up to 3 m, and with branches forming at a high level. It is often abundant at altitudes of 1100–2800 m in secondary forest, on forest edges and also along river valleys, and in wet upland forests and lowland forests. In closed forest, only old and mature dominant and co-dominant individuals occur. *P. fulva* quickly colonizes and covers degraded forest areas within its natural range, thus protecting the environment and creating and enabling conditions for rapid forest recovery.

The level of domestication of *P. fulva* both within and outside its natural range is low, and only in Kenya has it been grown in small-scale mixed species plantations, and in a few cases in pure stands. It is also used in shelter belts, in intercropping systems, and in agroforestry. It is increasingly being used as an ornamental tree.

The wood of *P. fulva* is white, soft and light, and used for making food boxes, crates, musical instruments and utensils. It is also suitable for veneer and plywood, and has potential for pulping and papermaking.

Further research into the species' genetic variation, silvicultural characteristics, seed storage and germination characteristics, and management in agroforestry systems is recommended.

Botanical features

P. fulva is a tall forest tree usually growing to a height of 18–25 m and 1 m dbh, although it can reach 30 m and up to 3 m dbh in some parts of its natural range. It often has a large, grey, extremely straight, unbranched, cylindrical bole reaching up to 15 m tall and 1 m diameter before the development of a whorl of branches resembling an umbrella, giving rise to its common name of parasol tree. The crown is obconical and flat-topped. The bark is smooth and grey-yellow in colour. The inner bark is dark cream, rapidly turning yellow-brown.

Leaves are pinnate, up to 80 cm long, usually with an end leaflet. Leaflets are usually in 6–7 pairs (range 3–12), ovate or elliptic, with base broadly cuneate, rounded or subcordate and apex acuminate or apiculate, 7–15 × 3–6.5 cm in size (in young leaves up to twice as large), dark green and stellate–pubescent to glabrescent above, pale yellow-brown and stellate–tomentose below.

Flowers are small, greenish-yellow to cream, and borne in large, loose axillary panicles; there are two styles, persistent in the fruit. Fruits are ovoid to obovoid, occasionally ellipsoid, shortly cylindroidal or subspheroidal, 3.5–6.0 × 3.0–4.5 mm, generally ribbed and markedly flattened, glabrous or very sparsely stellate–pubescent. Each fruit contains 2 small, light seeds.

Distribution

P. fulva is widely distributed across Africa between latitudes 12°N and 18°S. In this area it is predominantly found occurring naturally in montane forests. In West Africa, it has been reported to occur in Guinea, Ghana, Nigeria and Cameroon, and then eastward through the Central Africa Republic and South Sudan to the highlands of Ethiopia. In Eastern Africa, *P. fulva* spreads downward through Kenya, Uganda, Rwanda, Burundi, Tanzania and Malawi, reaching its southern limit in Mozambique. From Western Africa it spreads southward through Congo and the Democratic Republic of the Congo to Angola. *P. fulva* also occurs in Zambia and Zimbabwe in Southern Africa.

P. fulva occurs naturally in high-rainfall, montane areas generally receiving more than 1000 mm rain/yr and characterized by luxuriant vegetation growth. It is a dominant or co-dominant tree of the upper tree layer, and no regeneration occurs in closed forest, except in gaps and along forest edges. On higher ground in East Africa, *P. fulva* is also found scattered in the lower bamboo thicket zone.

Silviculture characteristics

P. fulva tolerates fire. It is able to regenerate rapidly and tolerates competition from weeds.

Silviculture practice

The seeds of *P. fulva* are orthodox and can be stored in an airtight container at 3°C for several years, although loss of viability after storage for only 1 year has also been reported. There are approximately 310,000 seeds/kg. Soaking seed in water for 24 h before sowing is recommended. Germination takes place after 36–45 days, the mean germination rate being 75%, and seedlings are ready for transplanting after 6 months.

Stands are mainly established using seedlings raised from seed or, on a more limited scale, wildings. In plantations, both mixed and pure stands can be established at spacings from 2.5 × 2.5 m to 3.6 × 3.6 m. Planted stands in Kenya are managed through the 'shamba' system, which involves intercropping with agricultural crops. The tree crop is therefore weeded as the crops are weeded. This practice goes on until the crown closes. However, due to the fast growth rate of *P. fulva*, its seedlings can readily withstand weed competition as soon as their roots are well established.

P. fulva grows to a considerable height before branching and therefore no pruning is necessary. As only the clear bole before branching is of economic importance, late pruning is not necessary. In the open, its crown grows very wide, but this can be controlled through spacing and/or subsequent thinnings.

One advantage of growing *P. fulva* with other long-lived hardwoods is that it can be harvested early, thus creating space for the other trees to continue growing.

As a fast-growing species, performance in plantations has been good; in Cameroon seedlings reached a height of 2–3 m after 4 years and 15 m with a bole diameter of 20–40 cm after 20 years.

Diseases and pests

No serious pathogens or pests of any economic importance on standing trees have been reported. Logs are susceptible to attack by bluestain fungi directly after felling.

Uses

As demand for its wood is low, except in forest areas under exploitation for peeler logs, *P. fulva* has been found to be an ideal species for soil conservation and revegetation in areas under high population pressure. Other, more valued species are usually exploited legally or illegally, leaving *P. fulva* and its close relatives to dominate. In both its natural habitat and where it has been introduced, it plays an important role in the conservation of catchment areas through the regulation of water flow and control of soil erosion. Its role in environmental protection is clearly seen in secondary forests where it is one of the first colonizers.

P. fulva is grown as a shelter belt in tea plantations and in Uganda it is intercropped with bananas, coffee and cocoa. It has also been a popular tree species in agroforestry trials undertaken in the highlands of Eastern, Central and Southern Africa, as the high crown permits penetration of sunlight and the leaves provide good mulch. Being fire-resistant, it has also been planted as a firebreak in plantations of species prone to fires. Where it is grown in dense populations, its leaves can be used as a green manure. It is also exploited as an attractive, fast-growing ornamental tree, and as a living fence.

The wood of *P. fulva* is white, odourless and soft, and not very durable. The grain is usually straight and texture is moderately fine to rather coarse. The wood is lightweight, with a density of 330–450 kg/m^3 at 12% moisture content. It is fairly easy to air dry, but splitting and ring shaking may occur. The wood is easy to saw and work, but often planes to a rather woolly surface. It is liable to splitting upon nailing and screwing, but glues well. It is suitable for interior joinery and for making doors. In industry it is used in the manufacture of food boxes, tea chests and crates. It is used on a large scale in veneer and plywood manufacturing, and has considerable potential for pulping, papermaking and matchsticks. Traditionally, the wood of *P. fulva* has been used to construct beehives and mole rat traps. It is also carved into various utensils and musical instruments. Due to its low density, it only serves as poor-quality firewood.

In some localities, the leaves and bark are used for traditional medicinal purposes. The flowers are a good source of pollen and nectar for honeybees.

Further reading

Coates Palgrave, K. and Coates Palgrave, M. (2003) *Polyscias fulva*. In: Coates Palgrave, K., Drummond, R.B., Moll, E.J. and Coates Palgrave, M. (eds) *Trees of Southern Africa*. 3rd ed. Struik Publishers, Cape Town, South Africa, pp. 845–846.

Maundu, P. and Tengnäs, B. (2005) *Polyscias fulva*. In: Maundu, P., Tengnäs, B., Muema, N. and Birnie, A. (eds) *Useful Trees and Shrubs for Kenya*. World Agroforestry Centre - Eastern and Central African Regional Programme, Nairobi, pp. 354.

Populus alba L.

White poplar

Importance

Populus alba, the white or silver-leaved poplar, is widely grown as an ornamental for its silvery foliage. It is a fast-growing tree utilized in afforestation and reforestation due to its resistance to drought, salt and wood borers. It is not as important commercially as the poplar species of sect. *Aigeiros*, but its adaptation to warm conditions allows it to be cultivated in Mediterranean regions and in Central Asia.

P. alba is widely used in interspecific breeding programmes to introduce valuable traits, such as high rooting ability of stem cuttings, into economically more valuable species, particularly *Populus tremula* of sect. *Populus*. *P. alba* hybridizes readily with this closely related species. The resulting hybrid, known as grey poplar (*Populus* × *canescens*), is intermediate morphologically between its parents, with a thin, grey, downy coating on the leaves, which are also much less deeply lobed than in white poplar. It exhibits marked hybrid vigour, with trees displaying vigorous growth and reaching 40 m in height and over 1.5 m in trunk diameter, much larger than either of its parents.

P. alba is suitable for propagation by tissue culture and is used as a subject for *Agrobacterium*-mediated genetic transformation.

Botanical features

P. alba is a medium-sized, deciduous, dioecious tree with a crown normally broad and rounded, but var. *pyramidalis*, the most common ornamental variety, is harshly fastigiate with vertical branching. *P. alba* can live to 300–400 years old. At maturity it can reach a height of 30 m and dbh of 100 cm.

The trunk is straight, at times sinuous, with large branches inserted irregularly, which often bifurcate. The bark is grey-white, rough at the base in older trees. The root system is lateral, shallow but expansive; it tends to produce abundant root suckers, often as far as 20–30 m from the main trunk, thus forming extensive clonal colonies.

Leaves are dark green above and white tomentose beneath. Leaves of long shoots are 6–12 cm long, relatively deeply lobed and palmate; the 3–5 lobes are coarsely dentate. The leaf base is rounded–cordate. Leaves of short shoots are smaller, oval–elliptic, and sinuately dentate. Petioles are rounded and pubescent, 3–4 cm long.

Staminate catkins are usually reddish, 5–8 cm long with 10–20 stamens, and fruiting catkins are 10–15 cm long and greyish-green. Capsules are bivalved, and pedicels are slender, 3–5 cm long. Flowering begins in early spring, before leaf emergence, and fruits mature 6–8 weeks later.

Distribution

P. alba can be found spontaneously along river valleys in its natural range, which stretches from central and Southern Europe to North Africa and Western and Central Asia. It was introduced into North America as an ornamental tree in the mid-18th century and now has become naturalized in many areas, in some cases to the extent of being considered an invasive weed.

Silviculture characteristics

P. alba is cultivated in river valleys where poplars of sect. *Aigeiros* do not perform well. It is grown in saline and calcareous soils and can be used in windbreaks for salt wind near coasts, and for dune stabilization. It is particularly suited to planting in regions with hot summers and mild or cold winters.

Silviculture practice

Following harvesting, seeds must be dried at room temperature for 2 days, after which the moisture content should be reduced rapidly to 4 or 5% using dry air at 30°C. After the removal of 'cotton fibres' (pappus), seeds can be conserved for several years at a temperature below −18°C without any appreciable decline. Germination and seedling development (usually in the greenhouse) are rapid if there is adequate moisture and the temperature is up to 20°C. Seeds should not be covered with soil. Control of algae, fungi and snails must be carried out during the initial growing period. Seedlings can be transplanted to field nurseries 6–8 weeks after sowing.

Vegetative propagation is usually by planting stem cuttings 30–35 cm long in nurseries established in early spring, after the soil has been ploughed and disc-harrowed; the optimal density is 1 tree per m^2. Phosphorus and potassium nutrients should be applied before soil preparation, while nitrogen can be applied twice during the growing season, once after sprouting and again 1 month later. Poplars do not tolerate weed competition so it is important to control weeds during the first few months after establishment and at the beginning of the second growing season. In summer, irrigation may be needed, especially in dry regions, and insects and other pests need to be controlled. When saplings have reached an average height of 3–7 m (this can be obtained by growing the trees in nurseries for 1 or 2 years), they can then be used to establish stands.

In countries where the species is cultivated, *P. alba* is planted in soil which is not suitable for other poplar species, for example, where the water table is inaccessible or the soil is poor or saline. The plantation soil must be prepared by deep ploughing, followed by one or two harrowings. Deep ploughing provides a well-aerated soil layer that promotes rapid root development in the cuttings or saplings. Plantations can be established during the dormant season, excluding very cold periods, using saplings 1 or 2 years old or cuttings 40–50 cm long. Saplings which are completely pruned, with or without roots, are inserted into holes 50–150 cm deep (depending on sapling height) and 15–40 cm in diameter.

Tree density depends on the specific wood product to be obtained: about 10,000 trees/ha for biomass for energy, and 400–1000 trees/ha for pulpwood and sawlogs. *P. alba* can tolerate a higher density than other poplar species. This higher density forces the tree to grow in height with fewer and smaller branches. During the middle rotation period, thinning is necessary. A square-shaped plantation is preferable.

Weed control is very important in stands, especially during the first years following establishment. This should be done by disc or rotary tiller and rarely by chemicals. Fertilizer application is necessary only in nutrient-deficient soils. Phosphorus and potassium fertilizers should be distributed before soil preparation, while nitrogen can be applied during the years following establishment and after sprouting.

Pruning is essential to avoid stem forking. During the years immediately after establishment, double tops should be removed and subsequently the side branches. In order to obtain valuable logs for industrial purposes, pruning to a height of 5–7 m, and thinning if the plantation is too dense, should be carried out.

The rotation of *P. alba* is 18–25 years with a final density of 300–400 trees/ha. Annual production is 10–15 m^3/ha, reaching 18–22 m^3/ha on the best sites.

Diseases and pests

Serious attacks by *Agrobacterium tumefaciens* cause cankering and damage. Among the fungi, the main agents affecting leaves are rusts of the genus *Melampsora*, and *Marssonina castagnei*. These can cause premature defoliation of trees. *Venturia* attacks leaves and young shoots, causing complete defoliation and deformation of shoots and branches early in spring.

P. alba, like other poplars, is host to a large number of insect species, but only a few of these are of any management significance. Important pests include the defoliating moth *Hyphantria cunea* and the large poplar-leaf beetle (*Chrysomela populi*). Borers tend to prefer other poplar wood to that of *P. alba*, although attacks of *Cossus cossus* and *Saperda* spp. are occasionally reported.

Uses

P. alba is used for windbreaks, erosion control along river banks and roadbeds, dune stabilization and reclamation of surface-mined lands and former seabeds.

Because of its attractive foliage, the leaves contrasting green above with white below, *P. alba* is widely planted as an ornamental tree in parks and gardens; in addition, *P. alba* var. *pyramidalis* has an attractive upright pyramidal form.

Wood from natural stands of *P. alba* is not of high quality, being rather woolly in texture, of low flammability, not durable, very light, soft and elastic; but very resistant to splitting, abrasion and wood borers. Wood density is 353 kg/ m^3. It is mainly suitable for local and artisan use, such as the production of matchboxes, punnets and trugs, and in the past, due to its resistance to borers, it was largely utilized for domestic furnishing. However, specialized plantations offer the advantage of higher wood homogeneity, necessary for industrial purposes. The main uses of *P. alba* wood today can be grouped in two categories. The first list of main uses includes semi-finished products of solid wood for carpentry, sawnwood for shutters and structural use, packaging of different types (crates and boxes), pallets, wood wool, fibreboards, particleboards and mechanical pulpwood for paper and semi-chemical cellulose. The second category is for specialist manufacture of matches, wooden frames, clogs, wooden soles for shoes and sticks for ice lollies.

P. alba and its hybrids have potential as biofuel crops due to rapid growth and biomass production.

Leaves and young shoots are collected for cattle feed. The bark yields a yellow dye and bark extracts have medicinal value.

Further reading

Angeles Bueno, M., Gomez, A. and Manzanera, J.A. (2003) Propagation and DNA markers characterization of *Populus tremula* L. and *Populus alba* L. In: Jain, S.M. and Ishii, K. (eds) *Micropropagation of Woody Trees and Fruits*. Kluwer Academic Publishers, Dordrecht, The Netherlands, pp. 37–74.

El Bassam, N. (2010) Poplar (*Populus* spp.). In: El Bassam, N. (ed.) *Handbook of Bioenergy Crops: A Complete Reference to Species, Development and Applications*. Earthscan, London, pp. 278–281.

Johnson, H. (2010) Poplars, cottonwoods, and aspens. In: Johnson, H. (ed.) *Trees: A Lifetime's Journey Through Forests, Woods and Gardens*. Mitchell Beazley, London, pp. 194–197.

Yang, J.J., Li, P.J. and Li, H. (2011) Growth characteristics of hybrid clones from *Populus alba* and *Populus alba* cv. Pyramidalis. *Xinjiang Agricultural Sciences* 48, 1988–1998 (in Chinese with English summary).

Populus balsamifera L.

Balsam poplar

Importance

Populus balsamifera, the northernmost American hardwood, is a deciduous transcontinental species that grows on upland sites but shows best development on floodplains. It is a hardy, fast-growing tree that is relatively short-lived, generally <200 years. The foliage, twigs and buds are a source of food for wildlife. Trees are planted for shade and in windbreaks and shelter belts. The wood is used for lumber, veneer and pulp and fibre products.

P. balsamifera hybridizes readily with other *Populus* species, including *Populus deltoides*, *Populus alba* and *Populus tremula*. Its hybrid with *P. deltoides* is *Populus* × *jackii*, commonly known as balm of Gilead.

Black cottonwood (*Populus balsamifera* subsp. *trichocarpa*, but also treated as a separate species, *Populus trichocarpa*) was the first tree species to have its entire genome sequenced (in 2006). It is the largest North American poplar, with trees up to 50 m tall with a dbh of >2 m.

Botanical features

P. balsamifera is a medium to large tree, 23–30 m tall, occasionally reaching >35 m, with a dbh of 130 cm. The stem is long and straight. The crown is narrow, somewhat columnar and open, with a few stout, ascending branches. The bark is smooth and greyish-green, becoming grey and separated into flat-topped ridges.

The root system is wide-spreading, with a few large roots. Lateral root spread on upland sites is at least 8–12 m. On floodplains, the root system is multi-layered, owing to the deposition of new soil by periodic flooding.

Leaves are broadly ovate, 7–12 cm long, tip long drawn-out, base usually rounded, finely toothed with many short, rounded teeth, upper surface dark green and silvery-green beneath, petiole round in cross-section. Twigs are stout, round, smooth-barked. Buds are reddish-brown to brown, long, curved, and resinous. *P. balsamifera* derives its Latin and common names from this resin and its strong aromatic fragrance, emitted in spring.

Flower clusters (catkins) are 5–9 cm long with many small flowers about 3 mm long. Male flowers have 20–30 reddish stamens. Mature female catkins are 10–15 cm long. Capsules (6–8 mm long) are a lustrous green and glabrous during development, but turn dull green at dispersal, discharging copious amounts of light, cottony, airborne seeds for dispersal by wind or water.

P. balsamifera generally reaches flowering age between 8 and 10 years old. It produces large seed crops almost every year, but significant annual variation in production can occur in individual stands and trees.

Initiation of phenological events varies greatly from clone to clone and with latitude, altitude, aspect and weather. Flowering occurs in April–May throughout most of the range, but not until June or July at northern limits and upper altitudes.

Distribution

P. balsamifera is distributed along the northern tree limit of North America, from Labrador to Alaska. From north-west Alaska, its range extends south to British Columbia and across Canada and the Lake States to the Atlantic coast. It is local in the western mountains in Oregon, Idaho, Utah, Colorado, Wyoming and in the northern plains in Nebraska and South Dakota. It also occurs as more scattered populations in Iowa, Ohio, Pennsylvania, West Virginia, Maryland and Connecticut.

P. balsamifera subsp. *trichocarpa* is distributed along the Pacific coast from Alaska to California. Its main range extends into British Columbia, western Alberta, Washington, Oregon, Idaho, Montana and northern California. It is also scattered south to Baja California (Mexico) and east to North Dakota.

P. balsamifera grows predominantly in even-aged, pure stands or with shade-tolerant conifers in later succession stages. It is a pioneer species on floodplains, and a minor or major component in many vegetation zones across North America, primarily those in boreal and temperate climates.

P. balsamifera has been used in breeding programmes in Europe (in Belgium, France and Italy) and in south Asia (Pakistan and India).

Silviculture characteristics

P. balsamifera has high reproduction potential; trees begin flowering from about 10 years of age and show good vegetative reproduction from root and stump sprouts.

Being shade-intolerant and exposure-requiring, *P. balsamifera* does not regenerate under low light levels, but in the open the potential for natural regeneration is high, particularly on mineral soils.

Initial growth rates (<5 years) are high, up to 2 m in one growing season for cuttings and stump sprouts. In dense stands, *P. balsamifera* has a high self-pruning capacity. Crown spatial requirements are high. Light conditions beneath closed-canopy, mature stands are intermediate and associated with well-developed understorey vegetation.

Young saplings are susceptible to early and late frosts, and ice storms and heavy snowfall cause considerable breakage and permanent bending. Although trees are generally windfirm, wind damage is common, especially in stands where trees are much taller than the surrounding vegetation.

P. balsamifera is well adapted to fire damage as it can produce sprouts from roots, stumps and buried branches, enabling rapid recovery. Mature trees can withstand mild to moderately intense fires. Pure stands burn less intensely than mixtures with conifers.

Silviculture practice

Depending on site and management objectives, clear-cutting and patch-cutting are viable silvicultural systems for *P. balsamifera*. It is best grown in single-species stands, but can also be grown in 2-storeyed stands with fast-growing, shade-tolerant conifers.

Seed dispersal begins in May and June throughout most of the range, and lasts for at least 2 weeks. Despite wind dispersal, large quantities of seeds fall within the stand and large numbers of short-lived germinants can be found on suitable substrates.

The seed does not exhibit dormancy, and germination occurs over a wide range of temperatures provided moisture is adequate. Germination is epigeal and, under ideal conditions, cotyledons expand in 18–24 h. Moist mineral soil surfaces are the best seedbeds. Depending on density, first-year seedlings are 5–32 cm tall.

P. balsamifera has high potential for vegetative reproduction. New stems can originate from intact or broken roots, preformed or adventitious buds on stumps or at the base of trees, and buried stems or branches in primary or secondary succession. *P. balsamifera* can be established by rooted or unrooted cuttings, which are taken in the dormant season and planted in the spring to a depth of 30–40 cm. Best establishment and growth are achieved when cuttings have healthy axillary buds, at least one of which remains above ground after planting.

Clonal differences are a major source of variation in rooting percentage (75–100%) and in the number of primary roots produced.

New shoots also form on stumps from suppressed buds and adventitious buds developed from undifferentiated inner bark. Sprouting response varies with genotype and declines with tree age. In intensive stands grown on short rotations, coppicing is used to replace the new crop after harvest of the original stand established from stem cuttings. Individual cuttings may produce 10–20 sprouts a year after harvesting; typically 4–8 sprouts will survive after 2 years.

On productive sites, *P. balsamifera* attains pulpwood size in 10–15 years and sawlog size in less than 25 years. Yields from plantations are much higher than yields from natural stands.

Diseases and pests

The most common fungal rot of mature *P. balsamifera* trees is *Phellinus tremulae*, with *Hemipholiota populnea*, *Crustomyces expallens* and *Bjerkandera adusta* also important. At least 70 other fungal species also cause decay, but few cause significant losses. *Melampsora* leaf rust occurs in young plantations, with susceptibility to it varying greatly across the geographic range of the species. Other foliage diseases include *Venturia populina* and *Taphrina populi-salicis*. Stem cankers caused by a range of fungal pathogens have also been reported, although none appears of great significance for management. Severe attacks of a bacterial canker have reportedly limited planting of the species in Europe. *P. balsamifera* is also prone to wet wood, a condition which leads to wood collapse during drying.

Although several insects attack *P. balsamifera*, none has yet been reported as a pest of economic significance; *Cryptorhynchus lapathi*, *Agrilus liargus* and *Saperda calcarata* are the most damaging, girdling or badly weakening trees >2.5 cm in diameter by tunnelling.

Uses

P. balsamifera and its hybrids are used for soil stabilization and urban forestry, although in urban situations they have several undesirable traits, such as brittle branches, catkin residues and root suckering. However, their rapid growth and fragrant scent make them popular as ornamentals where space allows. *P.* × *jackii* is used in windbreak and shelterwood plantings.

P. balsamifera wood is light (with a density of 310–380 kg/m^3), straight grained, soft and pale, and is used for pulp and the manufacture of a range of products, such as veneer, boxes, crates and brackets. Waferboard with excellent mechanical qualities can also be produced, although special procedures are needed to efficiently waferize the wood. In northern areas, *P. balsamifera* is used for structural timber and milled house logs when other species are not available.

The foliage and small woody component can be converted to an animal feed supplement. Volatile compounds from buds and other parts of the tree have biological, cosmetic and medicinal uses. Extracts from winter buds were recognized by Native Americans as having therapeutic value; balm of Gilead, a salve made by heating winter buds of *P. balsamifera* or related species such as *P.* × *jackii* in oil, was used to relieve congestion and is still used in some modern natural health ointments.

Further reading

Crist, J.B. and Schlaegel, B.E. (1979) *Balsam Poplar (an American Wood)*. USDA - Forest Service, Washington, DC.

Jansson, S., Bhalerao, R.P. and Groover, A. (eds) (2010) *Genetics and Genomics of Populus*. Springer, New York, New York, USA.

Khan, I.A. and Abourashed, E.A. (2009) Balm of Gilead buds. In: *Leung's Encyclopedia of Common Natural Ingredients used in Food, Drugs, and Cosmetics*. John Wiley & Sons, Inc., Hoboken, New Jersey, USA, pp. 64–65.

Roe, E.I. (1958) *Silvical Characteristics of Balsam Poplar (Populus balsamifera)*. Lake States Forest Experiment Station Paper No. 65, USDA - Forest Service, St Paul, Minnesota, USA.

***Populus* × *canadensis* Moench**

Hybrid black poplar

Importance

Populus × *canadensis* is a hybrid species derived from natural hybridization between *Populus nigra* and *Populus deltoides* in the 18th century following the introduction of the latter species to Europe. It is one of the most important rapidly growing trees in temperate latitudes, being widely used by the plywood industry. In central and Southern Europe, where it is much cultivated, this species is naturalized and tends to substitute for *Populus nigra*.

Botanical features

P. × *canadensis* is a deciduous dioecious tree, with a narrow to broad crown and reaching 18–30 m at maturity; older specimens can attain 40 m. Mean dbh is 25–50 cm in plantations at maturity, while older, isolated trees can reach 100 cm. The trunk is straight or sinuous, with generally ascending branches. The root system is lateral and shallow. Bark is smooth and sometimes rugose at the base of the tree, its colour varying from grey to brown. Young shoots are cylindrical to somewhat angular, glabrous, occasionally pubescent. Lenticels are round or linear.

Leaves are deltoid in shape with crenate margins, 7–10 cm long, acuminate and ciliate at first but becoming glabrous almost immediately or maintaining remnants of ciliate hairs. Teeth at the truncate leaf base are widely spaced. The petiole is flattened, usually reddish with 0, 1 or 2 glands at the leaf base.

Staminate catkins are reddish, 7 cm long with 15–25 stamens, while fruiting catkins are green and 10–25 cm long; capsules have 2–4 valves. Flowering begins early in spring, before full leaf emergence. The fruits mature a few weeks later.

Distribution

P. × *canadensis* was first found in central and Southern Europe, after the introduction of *P. deltoides* from North America, along the major river valleys where *P. nigra* was extensively cultivated, and then later in other parts of the world where both parental species were grown. At many of these sites a complex population evolved due to interspecific hybridization and backcrossing. Individuals from among these populations were selected and propagated by cuttings for cultivation in all temperate regions of the world. Some of them hybridized with the parental species and spread wherever they found optimal conditions for the germination of seed.

Silviculture characteristics

Species of the Salicaceae, including *P.* × *canadensis*, are the first woody species to colonize alluvial deposits created by flooding of great rivers in temperate zones. *P.* × *canadensis* can tolerate flooding for short periods without any appreciable damage. Spontaneous vegetative propagation occurs easily and it can be managed easily by coppicing.

Silviculture practice

After harvesting, seeds must be dried at room temperature for 2 days and then subjected to air drying at 30°C to rapidly reduce their moisture content to 4–5%. After the removal of 'cotton fibres' (pappus), seeds can be conserved for several years at a temperature of −18°C. Germination and seedling development are rapid with adequate moisture and at temperatures up to 20°C. Seeds do not need to be covered with soil.

In nurseries, vegetative propagation is usually by stem cuttings 20–25 cm long and 15–30 mm in diameter with primary buds and obtained from the basal two-thirds of lignified 1-year-old shoots. Nurseries are usually established at the beginning of spring, after the soil has been ploughed and disc-harrowed; the optimal density is 1 tree/m^2, although wider spacings are also used. Phosphorus and potassium nutrients should be distributed before soil preparation, while nitrogen can be applied twice during the growing season, once after sprouting and again 1 month later. Poplars do not tolerate weed competition so it is important to control weeds during the first month after establishment and at the beginning of the second growing season. In summer, irrigation may be needed, especially in dry regions, and insects and other pests need to be controlled. After the first year, during the dormant season, the basal portion of the tree can be pruned.

Saplings are planted out when they reach an average height of 3–7 m, usually after 1–2 years.

Poplars are usually cultivated as an agricultural crop, using intensive cultural practices. Plantation density depends on the specific wood product to be obtained: 8000–10,000 trees/ha for biomass production, about 1000 trees/ha for pulpwood, and 200–300 trees/ha for saw or veneer log. A square-shaped plantation (3 × 3 or 7 × 7 m) is preferable.

The soil in plantations must be prepared by deep ploughing followed by one or two harrowings. Deep ploughing provides a well-aerated soil layer that promotes rapid root development. Plantations can be established during the dormant season, excluding very cold periods, using saplings (1–2 years old) or cuttings (20–40 cm). Saplings are completely pruned, with or without roots, and are inserted into holes 50–150 cm deep and 15–40 cm in diameter.

Weed control is very important in stands, especially during the first years following establishment, and is normally performed by disc or rotary tillers. Fertilizer applications are only necessary in nutrient-deficient soils, but it is important to anticipate the removal of nutrients from the soil to maintain fertility. Irrigation is important in order to increase production and waste water can be used.

Pruning is essential if high-value veneer logs are desired. During the years immediately after establishment, double tops should be removed and subsequently the side branches. These must be removed before they exceed 6 cm diameter and when the stem at their base is about 10–12 cm. Pruning must be carried out to a height of 5–7 m.

Many clones of *P.* × *canadensis* increase in height and diameter faster than other woody species in temperate zones. Average annual production can vary between 6 and 20 m^3/ha. Under favourable conditions with a density of 300 trees/ha and 10-year rotations, *P.* × *canadensis* can reach a height of 30 m, a dbh of 40 cm and produce 35 m^3/yr. In the same location, under short-rotation forestry with an initial density of 10,000 trees/ha, annual biomass production can be 10–15 oven-dry t/ha.

Intensive management or cultivation is economically suitable only in rich, well-textured soils in areas with adequate climatic conditions. In other locations and situations, *P.* × *canadensis* requires a longer rotation and less intensive cultural models.

Diseases and pests

The fungal disease *Drepanopeziza punctiformis* and several species of the rust *Melampsora* can cause premature defoliation of trees. *Venturia populina* attacks leaves and young shoots, causing complete defoliation and shoot or branch deformation early in spring. Serious attacks of *Davidiella populorum* and *Chondroplea populea* cause cankers and dieback, especially in nurseries or on young saplings after transplantation. In India, other poplar fungal diseases that have been reported are *Myrothecium roridum*, *Athelia rolfsii*, *Sphaceloma populi* and *Alternaria alternata*.

Bacterial cankers caused by *Xanthomonas populi* can be very damaging, limiting the cultivation of poplars for industrial use.

Poplars are hosts to a large number of insect species, but only a few are of any significance. The most important are the moth *Cossus cossus*, the beetle *Saperda carcharias* and the weevil *Cryptorhynchus lapathi*. *Paranthrene tabaniformis*, *Paranthrene dollii* (in the USA), and *Phassus excrescens* (in Japan) attack young trees. In all these cases it is the larvae which cause the damage. The woolly poplar aphid (*Phloeomyzus passerinii*) can also damage trees.

Defoliating pests of *P.* × *canadensis* include the large poplar-leaf beetle (*Chrysomela populi*) and larvae of *Leucoma salicis* and *Hyphantria cunea*. Larvae of the poplar shoot borer (*Gypsonoma aceriana*) may cause failure of the leading shoots. In South America, ants of the genera *Atta* and *Acromyrmex* harm poplars by cutting and consuming leaves. Occasionally other insect pests, such as *Operophtera brumata*, *Leucoptera susinella*, *Lymantria dispar* and *Clostera* spp., can cause damage.

Uses

P. × *canadensis* is used in riverside plantations to control erosion of riverbanks and is planted as a shelter belt tree. It is also used in agroforestry, although some crop–tree interactions can be detrimental to crop production. The species can be successfully used for phytoremediation to regenerate lands with high contents of heavy metals or other wastes.

In urban areas, poplar hybrids are useful where fast-growing trees are needed for shade, landscaping and screening purposes.

P. × *canadensis* wood is soft, light coloured, fine textured and low in density (300–550 kg/m^3 at 12% moisture content), and has many industrial uses, the main ones being plywood, semi-finished solid wood products for carpentry, sawnwood for shutters and structural use, packaging of different types (crates and boxes), pallets, wood wool, fibreboards, particleboards and mechanical pulpwood for paper and semi-chemical cellulose. There is also potential in the future for use in manufacturing laminated veneer lumber and undried rotary cut veneers. The colour, flexibility and lightness of poplar plywood have contributed to making it almost irreplaceable for many specialist uses, such as matches, clogs, wooden soles for shoes and sticks for ice-lollies.

Leaves and young shoots are collected for cattle feed.

Further reading

Johnson, O. and More, D. (2004) Hybrid black poplars. In: *Collins Tree Guide*. Harper Collins Publishers, London, pp. 156–158.

Rathmacher, G. (2009) *Population Genetic Processes of a* Populus *Hybrid Complex: Spatially Explicit Studies on Gene Flow of* P. nigra *and* P. × canadensis *as a Basis for Conservation Measures.* Südwestdeutscher Verlag für Hochschulschriften, Saarbrücken, Germany.

Stettler, R.F., Bradshaw, H., Heilman, P.E. and Hinckley, T.M. (eds) (1996) *Biology of* Populus *and Its Implications for Management and Conservation.* Canadian Science Publishing, Ottawa.

Populus deltoides W. Bartram ex Marshall

Eastern cottonwood

Importance

Populus deltoides is a large, fast-growing tree with a very wide natural distribution range in eastern, central, and southern North America. It has been extensively planted in many countries in Asia, Africa and Europe, as well as in Australia and New Zealand. It is commonly used in block plantations and agroforestry situations (especially for fast-growing windbreaks). As a strong light-demander, it must be planted in the open at wide spacing for best results. It grows best on moist, well-drained sands or silts near streams, often in pure stands. Large quantities of wood are produced and used in the manufacture of various products. Selected clones are commonly propagated by cuttings, sets and stump plants. It is sensitive to drought, high temperature, fire and several pests and diseases. *P. deltoides* readily hybridizes with other *Populus* species and it has been extensively used in poplar hybridization programmes throughout the world, producing fast-growing clones. *Populus* × *canadensis* is the result of natural hybridization with *Populus nigra*.

Botanical features

With a large, spreading crown, *P. deltoides* is a lofty deciduous tree which under ideal conditions reaches up to 77 m in height with a dbh of 2.4 m. The bole is largely cylindrical. Branches are massive, somewhat upward-rising. Bark is pale green and appears whitish or light grey when young owing to a powdery surface bloom. Mature bark is rough with deep, irregular cracks, and appears greyish-brown to dark grey.

Leaves are bright shiny green above and slightly paler beneath, hairless on both surfaces. Leaf shape is roughly an equilateral triangle with sides of 5–10 cm, with dentate margins and a short tapering tip without teeth. Petioles are long and flattened.

P. deltoides is dioecious, with a sex ratio of 1 to 1. Floral buds form in the summer and open the next spring. Male buds develop earlier and are much larger. Flowering occurs from February to April before leaves appear, with males flowering a few days earlier than females. Male flowers, 8–13 cm long, have 40–60 stamens and are reddish in colour and more conspicuous than female flowers. Female flowers elongate to 15–30 cm. Flowering varies by as much as 1 month among trees in a stand. As a result, early–flowering trees do not have the opportunity to cross with late-flowering trees. Flowering can occur in trees as young as 4–5 years old. Northern trees flower at lower temperatures than southern trees. Seeds develop 30–60 per capsule, on short stalks on long catkins. Each capsule has 3 or 4 valves.

Seed production starts when the trees are 5–10 years old, increasing rapidly in amount as the trees become older and larger. Annual seed production in a single open-grown tree can be as high as 48 million seeds. Good seed crops are the rule. About 35 l of fresh fruit yield 1 kg or 770,000 seeds. Seeds are dispersed by wind and water from May to mid-July in the south of the range and from June to mid-July in the north. The dispersal pattern results in abundant deposits of seeds along water courses as spring flood waters recede.

Distribution

P. deltoides grows along streams and on bottomlands of North America from southern Quebec westward into North Dakota and south-eastern Alberta, south to central Texas, and east to north-western Florida and Georgia. The north–south distribution extends from latitude 28°N to 46°N. It is absent from the higher Appalachian areas and from much of Florida and the Gulf Coast except along rivers. Altitude is a primary factor determining distribution as the western boundary approaches the Rocky Mountains.

Silviculture characteristics

Trees of *P. deltoides* tolerate frost and waterlogging, but trees of all ages are very susceptible to fire. Trees coppice well, and can produce suckers.

P. deltoides is very intolerant of shade, more intolerant than any of its associates except willow. Although the two frequently seed together, pure stands of one or the other are the general rule after the first few years. Willow survives on the wetter sites and *P. deltoides* on the slightly higher, drier sites. Its faster growth allows *P. deltoides* to crowd out the willow except where prolonged deep flooding drowns the *P. deltoides* component.

Silviculture practice

Unless floating on or immersed in water, *P. deltoides* seeds must reach a favourable seedbed and germinate very soon after falling. Germination of fresh seeds may exceed 90%. Seedlings are delicate for the first few weeks and rains, very hot sunshine, and damping off fungi kill many. Very moist, exposed mineral soils, such as fresh silt deposits, are required. Germination is epigeal. Seedlings grow slowly for the first 3 weeks but rapidly thereafter. Full sunlight for a substantial part of each day is required after the first few weeks. Fully mature seeds that are dried promptly to 5–8% moisture and stored at temperatures just above freezing maintain viability for several months. Storage at −20°C may prolong viability for 5 or more years.

Satisfactory sprouting occurs on low-cut stumps of trees as old as 25 years. Reproduction by root suckers is not common. Artificial propagation normally involves using cuttings from 1-year stem growth from nursery trees. These may or may not be rooted before outplanting. Unrooted cuttings 30–50 cm long provide a satisfactory, economical means of planting. Survival rates of 70–90% are normally achieved, depending on the clone, quality of cuttings and field conditions. Rooted or unrooted long cuttings are sometimes used to reach moisture, reduce damage from deer, permit less intensive site preparation and provide greater flood tolerance. Because operational use of vegetative propagation of *P. deltoides* permits immediate and complete utilization of superior genotypes, rooting ability is of great importance.

Propagation from 1-year-old wood from older trees is often difficult but not impossible. Reproduction from the resulting material is often satisfactory.

Root growth of new seedlings is so slow that the plants are easily dislodged by rain droplets. After the first 3 weeks, root growth accelerates and lateral root growth may exceed height growth for the first year. Most of the roots are in the uppermost, best-aerated soil layer. They are nearer the surface in clay soils than in loam soils. Following siltation, roots develop on the covered portion of the stem. *P. deltoides* trees planted from conventional 20–60 cm cuttings have fewer deep roots and are not as well anchored against root lodging as those established naturally or as deep-planted seedlings or rooted cuttings.

P. deltoides responds poorly to release following crowding, and only those trees with the best crowns respond. In natural stands, uneven spacing and size permit some trees to become dominant, and natural thinning allows production of large trees. Under plantation conditions and particularly when only clones with similar growth rates are used and all trees get off to a good start, stagnation can occur quickly. Spacing and timing of thinning become critical under these conditions. Optimum growth of individual trees requires very wide, seemingly wasteful, spacing. Trees planted initially at a spacing of 3.7 × 3.7 m should be thinned by removing half of the trees at age 3 years and again at age 5 years if rapid growth rate of individual trees is to be maintained.

Growth and yield vary appreciably depending on location, site quality, clone and silvicultural conditions, including stocking density and frequency of thinning. Maximum annual increment occurs only if stands are heavily fertilized and irrigated and trees are spaced at 2 m or less. Intensively managed plantations at 10–20 years of age can achieve mean annual values for height growth of 0.9–1.4 m, diameter growth of 1.0–1.5 cm, volume increment of 7.0–24.5 m^3/ha and biomass increment of 4.5–20.2 t/ha.

Diseases and pests

At least 10 insect species and 12 diseases cause major damage to *P. deltoides* throughout its range. Butt rot, a common result of fire injury, is uncommon. Although seedlings and young trees are browsed by rabbits, deer and domestic stock, a substantial portion of the trees can recover from this damage. Beavers cut down saplings and pole-size trees for food and for dam construction. The resulting ponds may drown standing trees.

Uses

P. deltoides is frequently planted to give quick shade near homes, with male clones, which have none of the objectionable 'cotton' associated with seed, being preferred. It is also grown in windbreaks and is suitable for soil stabilization, particularly along stream and ditch banks. Deep planting permits reforesting of non-productive fields having sandy soils with available moisture beneath a dry surface layer. There is considerable interest in *P. deltoides* for energy biomass because of its high yield potential and coppicing ability.

The wood of *P. deltoides* is white (whereas the heartwood is light to medium brown), soft, light (360–599 kg/m^3 at 12–15% moisture content), and weak in bending and compression. Tension wood is often present, yielding a fuzzy surface when cut. It suffers a relatively high shrinkage on drying, but glues and paints well and does not split easily. Nail-holding ability is low.

The primary uses of the wood are for lumber, veneer, short plywood bolts, boxes, crates, food containers, interior frames for furniture, agricultural implements, wooden ware and cutting boards. Pulpwood is also produced for the manufacture of high-grade glossy paper. The sawdust can cause dermatitis.

Further reading

Cain, K. (2007) *The Cottonwood Tree: An American Champion*. Johnson Books, Boulder, Colorado, USA.

Dillen, S.Y., El Kasmioui, O., Marron, N., Calfapietra, C. and Ceulemans, R. (2011) Poplar. In: Halford, N. and Karp, A. (eds) *Energy Crops*. Royal Society of Chemistry, Cambridge, UK, pp. 275–300.

Johnson, H. (2010) Poplars, cottonwoods, and aspens. In: Johnson, H. (ed.) *Trees: A Lifetime's Journey Through Forests, Woods and Gardens*. Mitchell Beazley, London, pp. 194–197.

Stettler, R.F. (2009) *Cottonwood and The River of Time: On Trees, Evolution, and Society*. University of Washington Press, Seattle, Washington, USA.

Populus nigra L.

Black poplar

Importance

Populus nigra is a fast-growing tree utilized for afforestation and as an ornamental in all the temperate areas of the world. It is adaptable to different environments and soil types, and exhibits excellent rooting ability of stem cuttings and good resistance to pathogens. It is used in breeding programmes due to its compatibility with many other *Populus* species.

True *P. nigra* is considered on the verge of extinction in a large part of its natural range, particularly in Western Europe, due to the loss of its natural habitat through human activity and loss of its gene resources through hybridization with other species (especially *Populus* × *canadensis*, a hybrid originally obtained from crosses of *P. nigra* with *Populus deltoides*) and with its fastigiate form, var. *italica* (Lombardy poplar). Fortunately, the EUFORGEN (European Forest Genetic Resources) programme established in 1994 aims to conserve surviving genotypes in the many countries of its natural range.

Botanical features

P. nigra is a deciduous dioecious tree, with a broad and dense crown (pyramidal in *P. nigra* var. *italica*) and a longevity of 200–300 years. Old specimens can reach heights of 40 m with a dbh of 100 cm, sometimes 200 cm. Trunks are more or less straight with thick knots and stout branches, sometimes with suckers at the base. The root system is lateral, shallow or deep, depending on soil layer and water table depth. Bark is dark brown and, in older trees, deeply furrowed with large burrs.

Leaves are rhombic, long-acuminate, the lateral edge rounded and finely crenate, not ciliate, glabrous and lighter green beneath. Leaves of long shoots are 4–8 cm wide and 5–8 cm long, cuneate at the base, while those of shorter shoots are smaller and more or less rhomboid. Petioles are thin and flat.

Staminate catkins are usually reddish, 4–6 cm long with 20–30 stamens, while fruiting catkins are 10–15 cm long. Capsules are bivalved. Pedicels are slender and 3–5 cm long. Flowering, which begins when trees are around 6 years old, starts in early spring, before full leaf emergence and the fruits are mature 6–8 weeks later.

Distribution

The natural range of *P. nigra* extends from central and Southern Europe to Western and Central Asia, reaching the

Yenisei River in Siberia. It is also found in isolated localities in North Africa. It reaches its northern limit at 63°N (Archangelsk) along the Severnaya Dvina River in Russia and at 64°N in Siberia. The southern limit of its range is at 30°N in Pakistan. It is naturalized in North and South America.

The cultivation of *P. nigra* was once common in some European and West Asian countries for its easy vegetative propagation and fast growth. Ornamental varieties deserve particular consideration: var. *italica* is largely used for windbreaks, and other varieties can be found along roads all over the temperate regions of the world. The species is cultivated in India between 26 and 29°N latitude in agroforestry systems, whereas *P. nigra* var. *thevestina* is much cultivated in West and Central Asia and in North Africa.

Silviculture characteristics

P. nigra requires high illumination and does not tolerate shady areas. It is a typical pioneer species and easily colonizes open areas near riverbanks. It tolerates submergence during river flooding. Spontaneous vegetative propagation is common, particularly at the juvenile stage; fallen trees, broken roots and branches can root easily and root suckers are occasionally found. Black poplar can be managed easily by coppicing.

Silviculture practice

For storage purposes, harvested seeds must be dried at room temperature for 2 days, followed by treatment with dry air at 30°C to reduce the moisture content rapidly to 4–5%. They can then be conserved for several years, without any appreciable deterioration, at −18°C.

In nurseries, vegetative propagation usually involves stem cuttings 20–25 cm long and 15–30 mm in diameter with primary buds and obtained from the basal two-thirds of lignified 1-year-old shoots. The optimal density is 1 tree/m^2, although wider spacings are used. Nurseries are usually established at the end of winter or beginning of spring, after the soil has been ploughed and disc-harrowed. Phosphorus and potassium nutrients should be distributed before soil preparation, and nitrogen can be applied after sprouting and again 1 month later. Poplars do not tolerate weed competition, so it is important to control weeds during the first month after establishment and at the beginning of the second growing season. In summer, irrigation may be needed, especially in dry regions, and insects and other pests need to be controlled. After the first year, during the dormant season, the basal portion of the tree can be pruned.

When the saplings have an average height of 3–7 m (after 1–2 years) they can be used for stand establishment.

In order to obtain raw materials for industrial use, poplar plantations are cultivated as an agricultural crop, using intensive cultural practices. The density of the plantations depends on the specific wood product to be obtained: 8000–10,000 trees/ha for biomass production, about 1000 trees/ha for pulpwood and 200–300 trees/ha for saw or veneer logs. A square-shaped plantation (3 × 3 or 7 × 7 m) is preferred. The soil in plantations is prepared by deep ploughing, followed by one or two harrowings. Deep ploughing provides a well-aerated soil layer that promotes rapid root development in cuttings and saplings. Plantations can be established during the dormant season, excluding very cold periods, using saplings (1–2 years old) or cuttings (20–40 cm). Saplings are completely pruned, with or without roots, and are inserted into holes 50–150 cm deep and 15–40 cm in diameter.

P. nigra is intensively managed in Eastern Europe and Western Asia, particularly *P. nigra* var. *thevestina*. In these areas most of the wood produced is destined for local use, generally for construction purposes and for fuel. Stand density varies between 200 and 1000 trees/ha, rotations from 6 to 20 years and annual production from 6 to 20 m^3/ha. Where *P. nigra* is cultivated in single or double rows along roads, canals, and field edges as windbreaks or shelter belts to protect crops, the distance between trees is 2–3 m and between rows 4–10 m.

Diseases and pests

The fungal pathogens *Marssonina populi-nigrae* and several species of *Melampsora*, such as *Melampsora larici-populina*, can cause premature defoliation. *Venturia populina* attacks leaves and young shoots, causing complete defoliation and deformation of shoots and branches early in spring. Serious attacks of *Cryptodiaporthe populea* cause cankering and dieback, especially in nurseries and on young saplings after transplantation. *Mycosphaerella populorum* causes leaf spot and cankers. Fungal disease may be controlled by planting resistant clones.

Poplars are hosts to a large number of insect species, but only a few are of any significance. The most important are the moth *Cossus cossus*, the beetle *Saperda carcharias*, and the weevil *Cryptorhynchus lapathi*. *Paranthrene tabaniformis* in the natural range of the species, *Paranthrene dollii* in the USA and *Phassus excrescens* in Japan attack young trees. Damage is caused by larvae of these pests boring into the wood. *C. cossus* and *S. carcharias* attack old trees, whereas *C. lapathi* only damages young trees. The woolly poplar aphid (*Phloeomyzus passerinii*) can also damage trees; only genotypes of *P. nigra* collected in xeric and warm habitats are resistant.

Defoliating pests include *Chrysomela populi*, *Leucoma salicis* and *Hyphantria cunea*. Larvae of the poplar shoot borer (*Gypsonoma aceriana*) may cause failure of the leading shoots. In South America, ants of the genera *Atta* and *Acromyrmex* cause harm by cutting and consuming leaves. Occasional damage is also caused by other insect pests such as *Operophtera brumata*, *Leucoptera susinella*, *Lymantria dispar* and *Clostera* spp. *Hyphantria cunea* can be controlled by products based on *Bacillus thuringiensis*.

Uses

The compact crown and fastigiate form of *P. nigra* var. *italica* make it particularly suitable for use in windbreaks and shelter belts. *P. nigra* is planted as single or double rows to protect orchards and horticultural crops from winds and to reduce wind erosion of cultivated soil. It is also used to control erosion along riverbanks and roadbeds. In recent years, it has been reintroduced along several rivers in Hungary and Italy in protected floodplain areas.

P. nigra is utilized in agroforestry systems, and is widely planted as an ornamental and landscape tree, particularly var. *italica*.

The wood of *P. nigra* is soft, light coloured, fine textured and of low density (300–550 kg/m^3 at 12% moisture). Natural stands produce low-quality wood which is suitable only for local and artisan use. Plantations offer the advantage of greater wood homogeneity, the main uses of plantation wood being plywood, semi-finished products of solid wood for carpentry, sawnwood for shutters and structural use, packaging (crates and boxes), pallets, wood-wool, fibreboards, particleboards and mechanical pulpwood for paper and semi-chemical cellulose. It is also used in making matches, wood frames, clogs, wooden soles for shoes, sticks for ice lollies and coloured and reconstituted sliced veneers.

The wood is also used to produce high-quality, porous, light charcoal for use in medicine as an absorbent and in industry for bleaching.

The second-century physician Galen described the preparation of a vulnerary ointment called *acopon* from black poplar buds. A modern, similarly derived product is used as the base for veterinary ointments. Propolis exuded by poplar trees to protect themselves from microorganisms is used as an ecologically benign fungicide.

Further reading

Cooper, F. (2006) *The Black Poplar: Ecology, History and Conservation*. Windgather Press/Oxbow Books, Oxford, UK.

Koselka, J., de Vries, S.M.G., Kajba, D. and Wühlisch, G. von (eds) (2004) *Populus nigra* network. Report of the seventh (25–27 October 2001, Osijek, Croatia) and eighth (22–24 May, 2003, Treppein, Germany) meetings. IPGRI, Rome.

Rathmacher, G. (2009) *Population Genetic Processes of a* Populus *hybrid Complex: Spatially Explicit Studies on Gene Flow of* P. nigra *and* P. × canadensis *as a Basis for Conservation Measures*. Südwestdeutscher Verlag für Hochschulschriften, Saarbrücken, Germany.

Populus tremula L.

Common aspen

Importance

Populus tremula, the common, European or quaking aspen, is the most widely distributed broadleaved tree in the world, found throughout temperate Eurasia, from Iceland to Siberia, and beyond to Eastern Asia. It is a valuable commercial timber tree in northern and boreal areas, particularly Scandinavia, Russia and Poland. It is a tall, rapidly growing pioneer species, which can easily colonize open, wet swampy, and barren lands. It has a large water demand, and is sensitive to competition from other tree species as it is very light-demanding.

Multiplication of *P. tremula* using scions is not possible, which has limited its use in poplar breeding and in plantations. The wood is used for veneers, pulp and charcoal.

P. tremula has been hybridized with other *Populus* species to produce hybrids with high timber production and high disease resistance. Grey poplar (*Populus* × *canescens*), a natural hybrid of *P. tremula* with *Populus alba*, is a large, vigorous tree grown in plantations in Europe and Central Asia.

P. tremula, *P. alba*, *Populus* × *canadensis*, and their hybrids were among the first trees used in experimental genetic engineering for improved forestry characteristics.

Botanical features

P. tremula is a medium-sized deciduous tree growing to 20–35 m tall, with a maximum dbh of about 60 cm, with a clear slender trunk, compact crown and relatively short branches. In western Europe, it is possible to distinguish two forms of *P. tremula*, sometimes considered to be the same species or two subspecies:

1. a mountain form with a straight and slender tapered stem, and green-yellow bark; and
2. a lowland form in which the stem is more curved, and the bark is grey-green with lozenge-shaped slits.

In both forms, the bark is smooth, becoming dark grey and fissured with age.

P. tremula produces coppice shoots freely. It is short-lived, generally living less than 100 years.

The root system penetrates deep into soil, with a taproot and well-developed vertical root system. Mature trees develop lateral roots as well, from which many root sprouts develop, sometimes up to 40 m from the parent tree, forming extensive clonal colonies. *P. tremula* has relatively few mycorrhizal associations.

Leaves are 3–8 cm long, sub-rotund, with sinuate–crenate or toothed margins, truncate or subcordate at the base, pointed or rounded at the tip, thin, glabrous and glaucescent beneath. The petiole is up to 6 cm long and strongly flattened, making leaves tremble even in the slightest breeze. Leaves on young plants and suckers are very different, being up to 15 cm long, triangular–oval, hairless above but hairy below, with petioles up to 8 cm long. Leaves usually begin to fall in the second half of October. *P. tremula* can be distinguished from the North American aspen (*Populus tremuloides*) by the leaves being more coarsely toothed.

P. tremula is dioecious. Male flowers are 8–10 cm long and have 4–12 stamens and purple anthers. Female flowers are up to 12 cm long with red stigmas. Exceptionally, hermaphrodite flowers are produced. Pollination is typically anemophilous, but insect pollination can also occur.

Flowering occurs from March to May, beginning from the crown top and proceeding downwards. At the beginning of May, shoot buds also open, beginning from the bottom of the crown. Seeds ripen in May. Seed production normally begins from age 10–15 years. Seed dissemination is by wind, and generally seeds are produced in moderate quantities. Seed germination ability is limited to a few days. Seeds from hermaphrodite trees are often sterile.

Distribution

P. tremula is native to cool temperate regions of Europe and Asia, and is the second most widely distributed tree species in the world, after Scots pine (*Pinus sylvestris*), its range extending east from Iceland and Ireland to central Siberia and beyond to Kamchatka. North to south it ranges over 30° of latitude, from inside the Arctic Circle in Scandinavia and northern Russia, down to the Mediterranean from Spain to Turkey, and further east down to the Tian Shan of Central Asia, North Korea and northern Japan. There is also an isolated population in Algeria.

Many geographical races of *P. tremula* can be differentiated. On the basis of morphology it is possible to distinguish, on the one hand, races with a northern and montane origin and, on the other, those originating from lowland and plains regions of central Europe. Forms in Eastern Asia are considered to be a subspecies (or separate species by some), namely subsp. *davidiana*.

Silviculture characteristics

P. tremula is a light-demanding, rapidly growing broadleaved tree which is cold resistant, although best growth is in areas with warm summers, such as lowland Mediterranean areas. However, it can also be planted successfully in hilly Mediterranean regions. As a pioneer species, it is often used for afforestation of barren or degraded land.

P. tremula is tolerant of frosts, shade, waterlogging, wind and weed competition. It is found growing on a wide range of soil types, including compact, clay, muddy and relatively shallow soils. Light is more important than soil conditions, although it prefers moist soils with a high organic matter content. It often grows better on acidic than alkaline soils, although subsp. *davidiana* is tolerant of extremely alkaline soils.

It can be planted in mixtures with conifers and has the ability to regenerate rapidly.

Silviculture practice

Reproduction by seed is rare, as the seed retains germination ability for only a very short time, and does not keep well, so needs to be sown within a few days of ripening. If seed is used, it is surface-sown or just lightly covered. If sufficient growth is made, it may be possible to plant seedlings out in late summer, otherwise late in the following spring. As *P. tremula* hybridizes freely, the seed may not come true unless it is collected from the wild in areas with no other poplar species present.

It is possible, albeit difficult, to reproduce *P. tremula* using hardwood and greenwood cuttings, and *in vitro* methods of propagation have been used successfully. Cuttings of mature wood should be taken in November or December and planted in sheltered outdoor beds. However, propagation by root suckers in early spring and root cuttings in winter is more successful.

Coppicing, including short-rotation coppicing and coppicing with standards (where *P. tremula* forms the underwood), are practised. *P. tremula* may be managed in coppice rotations which are cut every 20 years; at this age the tree yields a wood suitable for paper pulp. *P. tremula* should not grow beyond a dbh of 30–40 cm, as older trees are susceptible to insect pest attack.

Diseases and pests

Poplars are susceptible to attack by a range of fungal and bacterial pathogens, insect pests, nematodes and mites.

Stems of *P. tremula* are damaged by fungi such as *Cryptodiaporthe populea*, *Fusicoccum* and *Cytospora*. Leaves are damaged by *Venturia populina* and *Drepanopeziza punctiformis*, which may result in spring defoliation. Infection by *Xanthomonas populi*, the causal agent of bacterial canker, can lead to branch death and the development of perennial cankers on the trunk. The root system can be attacked by *Armillaria mellea*, the honey fungus, which leads to root rot.

Aphids (*Phloeomyzus passerinii*) may damage the stem. Xylophagous insects (*Cryptorhynchus lapathi*, *Saperda carcharias*, *Cossus cossus*, *Paranthrene tabaniformis* and *Gypsonoma aceriana*) excavate galleries in the stem and branches. Shoots and leaves are damaged by insects such as *Leucoma salicis*, *Chrysomela populi*, *Phalera bucephala*, *Byctiscus populi* and *Hyphantria cunea*.

P. tremula is bark-stripped by squirrels and browsed upon by deer, although in coppice stands trees are reputed to be unpalatable to deer.

Uses

P. tremula is fast-growing and wind-resistant so is planted as a shelter belt species. It is a good wildlife tree, providing food and habitat for a wide range of animals, including over 90 insect species as well as deer and beaver.

Due to the distinctive trembling and bright yellow autumn coloration of the foliage, *P. tremula* is attractive as an ornamental and is grown as a street tree in cooler mountain areas, and in parks and gardens; it should not be planted close to buildings as its aggressive root system can invade and damage drainage systems and dry out the soil around foundations. Some horticultural cultivars (and hybrids with *P. tremuloides*) have been developed with different tree forms, including a weeping habit.

The wood of *P. tremula* is well suited for peeling into thick veneers. It is soft, light-coloured, not odorous and does not stain. It has a low density (about 300–550 kg/m^3 at 12% moisture content) and is fine textured, but the quality varies greatly according to growing conditions. Durability and resistance to axial compression and bending are usually low. The heartwood is resistant to penetration of wood preservatives.

P. tremula was formerly much planted for the manufacture of matches, pallets and crates. Today, it is utilized for veneers and for pulp, which can be used for the production of bisulphite cellulose and high-quality paper. The wood is also used in the production of good-quality charcoal.

Due to its rapid growth, *P. tremula* has potential as a biomass crop for energy production.

Further reading

Angeles Bueno, M., Gomez, A. and Manzanera, J.A. (2003) Propagation and DNA markers characterization of *Populus tremula* L. and *Populus alba* L. In: Jain, S.M. and Ishii, K. (eds) *Micropropagation of Woody Trees and Fruits*. Kluwer Academic Publishers, Dordrecht, The Netherlands, pp. 37–74.

El Bassam, N. (2010) Poplar (*Populus* spp.). In: El Bassam, N. (ed.) *Handbook of Bioenergy Crops: A Complete Reference to Species, Development and Applications.* Earthscan, London, pp. 278–281.

Perala, D.A., Leary, R.A. and Cieszewski, C.J. (1999) *Self-thinning and Stockability of the Circumboreal Aspens (*Populus tremuloides *Michx., and* P. tremula *L.).* North Central Research Station Research Paper NC-335. USDA - Forest Service, St Paul, Minnesota, USA.

Rae, A.M., Street, N.R. and Rodríguez-Acosta, M. (2010) Populus trees. In: Kole, C. (ed.) *Genome Mapping and Molecular Breeding in Plants. Volume 7. Forest Trees.* Springer-Verlag, Berlin.

Populus tremuloides Michx.

Quaking aspen

Importance

Populus tremuloides (Plate 31), the quaking or trembling aspen, is the most widely distributed tree in North America. It grows on a range of soils and is quick to pioneer disturbed sites where there is bare soil. This fast-growing tree is short-lived (up to 200 years) and pure stands are gradually replaced by shade-tolerant species. The light, soft wood exhibits very little shrinkage, and high grades are used for sawtimber and matches. However, most wood is used for pulping and flakeboard manufacture.

Botanical features

P. tremuloides is a medium-sized tree, rarely growing over 35 m, with a dbh of 100 cm and a short, rounded crown. The stem is long, cylindrical and smooth, with little taper and branch-free in the lower part. Bark is smooth with a waxy appearance, and is pale green to almost white when young, becoming darker with age and furrowed (Plate 70).

Seedlings initially have a short taproot, while clonal ramets have a flat root system when young, but in both cases a heart root system later develops on deep, well-drained soils. If rooting depth is restricted, a flat root system develops regardless of regeneration origin. Shallow and extensive laterals produce suckers, particularly when they are close to the soil surface. Sinker roots, which occur as frequently as every metre on laterals, may descend to depths of 3 m and are capable of penetrating strongly massive soil horizons or cracks in bedrock. The roots form ectomycorrhizae.

Leaves are broadly oval to kidney-shaped, 3–7 cm long, with tips abrupt, short and sharp, the base rounded, margins finely toothed and irregular, the upper surface deep green, paler beneath, and usually hairless. Petioles are flattened, causing the characteristic trembling movement in the wind. Twigs are slender, round, and shiny. Buds are slender, conical, pointed and without any gum or fragrance.

P. tremuloides is primarily dioecious, with trees in a given clone being either all male or all female, although some clones are hermaphrodite. The pendulous catkins, 3.8–6.4 cm long, are wind pollinated. Fruiting catkins are about 10 cm long when mature, each bearing many one-celled capsules containing about 10 small seeds, each of which is surrounded by silky hairs.

Initiation of phenological events varies greatly from clone to clone and according to latitude, altitude and aspect. Catkins elongate before leaves expand, generally in April or May. Sustained air temperature above 12°C for about 6 days appears to trigger flowering. Female trees generally flower and leaf out before male trees. Catkins take 4–6 weeks to mature. Leafing out and seed dispersal usually occur from early May to June.

Distribution

P. tremuloides is a transcontinental species, ranging over 111° of longitude and 48° of latitude, which represents the widest distribution of any native tree species in North America. The range extends from Newfoundland and Labrador west across Canada along the northern limit of trees, to north-western Alaska, and south-east through Yukon and British Columbia.

Throughout the western USA it grows mostly in the mountains from Washington to California, southern Arizona, Trans-Pecos Texas and northern Nebraska. From Iowa and eastern Missouri it ranges east to West Virginia, western Virginia, Pennsylvania and New Jersey. *P. tremuloides* is also found in Mexico, as far south as Guanajuato. Introduction of *P. tremuloides* outside its natural range is limited to arboreta and parks.

P. tremuloides is a minor or major component in many vegetation zones, primarily those in boreal and temperate climates. It typically grows in even-aged, pure stands and in later successional stages with many shade-tolerant conifers. It may be a pioneer species, but is present most frequently in early and intermediate stages of secondary succession.

Silviculture characteristics

P. tremuloides is a shade-intolerant, short-lived, windfirm tree with rapid initial growth rate and high reproduction potential, predominantly from root suckers and stump sprouts.

The potential for natural regeneration is poor under low light, but good in the open, with the amount of sucker regeneration being proportional to the degree of cutting and burning. Initial growth rates (<5 years) are high, about 30 cm in the first growing season from root suckers, >50 cm/yr thereafter.

In dense stands, self-pruning capacity is high and crown spatial requirements are moderate. Light conditions beneath closed-canopy, mature stands are intermediate and associated with well-developed understorey vegetation.

Potential productivity is high, with a site index (50 years at bh) of about 35 m on productive sites. Trees generally live for <150 years; the recorded maximum is 226 years.

Silviculture practice

Good seed crops are produced every 4 or 5 years, the minimum age for large seed crops being 10–20 years, the optimum being 50–70. Seeds are very light, with 5500–8000 clean seeds/g. Seed dispersal begins within a few days of ripening and may last for 3–5 weeks, with seeds potentially being carried for many kilometres by air currents. Water also serves as a dispersal agent. The viability of fresh fertile seeds is high (>95%), but normally of short duration.

Germination is epigeal. Exposed mineral soil is the best seedbed. During the first year, seedlings may attain a height of 15–30 cm and develop a 20–25-cm long taproot and 30–40-cm long lateral roots. Saplings grow rapidly for about the first 20 years, although high mortality characterizes young stands regardless of origin. Natural seedlings grow faster than planted seedlings in the first few years, but not as fast as suckers. In both seedling and sucker stands, natural thinning is rapid and trees that are overtopped by the canopy stop growing and die within a few years.

Mature stands reproduce vigorously by suckers. Root collar and stump sprouts are produced only occasionally by mature trees, but commonly by saplings. Root suckers are produced from meristems on lateral roots within 2–10 cm of the soil surface. Soil temperature is the most critical environmental factor affecting suckering, with optimum initiation and development being at 23°C. Excess soil moisture or severe drought inhibit sucker production. In general, sucker regeneration is proportional to the degree of cutting, with most suckers arising after a complete clearcut. Typically, about 25,000–75,000 suckers are regenerated/ha. Light burning on heavily cut areas increases the number of suckers and stimulates their initial growth, although hot slash fires diminish sucker vigour. Repeated burning increases stand density.

The intolerance of *P. tremuloides* to shade dictates an even-aged silvicultural system. Management can be intensive. Clear-cutting regenerates fully stocked sucker stands and maximizes growth. Extensive management, involving complete clearing for site preparation and no thinning, is used when trees are grown for fibreboard, pulpwood, flakeboard and some sawtimber. When grown as a biomass fuel crop, *P. tremuloides* should be grown in dense sucker stands on biological rotations of 26–30 years. Merchantable wood production by thinned stands over a 50-year rotation, including thinnings removed at ages 10, 20, and 30 years, is about 10.2 m^3/ha annually.

Diseases and pests

P. tremuloides is susceptible to a large number of diseases, which are usually of minor concern. Stain and decay have the greatest direct impact on wood production. Bacteria and yeast organisms are commonly associated with wet wood, a condition that leads to wood collapse during lumber drying.

Phellinus tremulae causes white rot of the heartwood, which can eventually invade the entire stem. It causes the greatest volume of decay. More fungal species cause butt and root rots than trunk rots, representing as much as one-third of the decay volume. Stem cankers caused by *Ceratocystis* spp., *Encoelia pruinosa* and *Entoleuca mammata* are common and can persist for decades or kill a tree within a few years.

P. tremuloides is host to a wide variety of defoliators, borers, and sucking pests, but only a few cause severe damage. *Malacosoma disstria* and *Malacosoma californicum* caterpillars are the most common defoliators.

Since the 1990s, the onset and rapid progression of branch dieback leading to tree mortality has been widely reported in *P. tremuloides* in western North America. Known as sudden aspen decline, it is strongly associated with recent changes in climatic conditions, especially the increase in droughts.

Uses

P. tremuloides forests provide habitats for a wide range of wildlife, produce abundant forage for browsers (1100–2800 kg/ha), are the preferred source of food and building material for beavers, and allow more groundwater recharge and streamflow than do coniferous forests. Due to low fuel accumulations, stands are low in flammability and make excellent firebreaks.

The trembling leaves and bright golden-yellow autumn foliage make individual trees and stands audibly and visually attractive.

P. tremuloides has a wide sapwood which is whitish to cream in colour and merges gradually into the heartwood, which is more yellowish-brown. The timber is typically straight grained, with a fine and even texture. The heartwood is not resistant to attack by decay fungi and other wood-destroying organisms. The wood is not hard, and can be dented and marred easily, but wears well relative to its low density (air-dry density 417 kg/m^3). Nailing characteristics, planing properties, and painting, turning and screwing qualities are all satisfactory. Wood products include pulp, flakeboard, particleboard, sawn timber, studs, veneer, plywood, shingles, novelty items, wood flour and, due to the surface not splintering, sauna benches and playground structures.

Whole-tree chips can be processed into nutritious animal feed or biomass for energy purposes.

Further reading

Hanna, P. and Kulakowski, D. (2012) The influences of climate on aspen dieback. *Forest Ecology and Management* 274, 91–98.

Petersen, D. (1991) *Among the Aspen: Life in an Aspen Grove*. Northland Publishing, Flagstaff, Arizona, USA.

Stettler, R.F., Bradshaw, H.D., Heilman, P.E. and Hinckley, T.M. (eds) (1996) *Biology of Populus and its Implications for Management and Conservation*. Canadian Science Publishing, Ottawa.

Strothmann, R.O. and Zasada, Z.A. (1957) *Silvical Characteristics of Quaking Aspen*. Lake States Forest Experiment Station Paper No. 49. USDA - Forest Service, St Paul, Minnesota, USA.

Prosopis glandulosa **Torr.**

Honey mesquite

Importance

Prosopis glandulosa is one of the most common tree species in northern Mexico and the south-western USA from California to Texas. In Mexico, the wood is used for rustic construction, posts, fuel and small timber items. In the USA, it is a source of low volumes of a very hard and dimensionally stable timber used for specialist flooring and furniture, and for charcoal that imparts a pleasant taste to food cooked over it ('mesquite smoked'). Trees improve soil fertility from leaf fall and nitrogen fixation, pods are forage for livestock, and the species serves as wildlife habitat, bee forage, and as a roadside, ornamental and shade tree. However, *P. glandulosa* is very invasive and is a declared weed in the USA and where introduced, especially in Australia and South Africa.

Botanical features

P. glandulosa is typically a large shrub or small tree, up to 9 m tall with a stem diameter up to 1 m, and although larger specimens are recorded, it is more often smaller. Spines are 1–4.5 cm long, solitary or paired. Leaves are bipinnate, glabrous, 1–2 pairs of pinnae per leaf, up to 15 cm long. Pinnae are 6–17 cm long, each with 6–17 pairs of leaflets, distant on the rachis, 2–6.3 cm long by 1.5–4.5 mm broad, linear or oblong, 5–15 times as long as broad, obtuse, glabrous, subcoriaceous, prominently veined below. Small yellow flowers are gathered on racemes 5–14 cm long. Pods are straight legumes, 8–20 cm long by 0.7–1.3 cm broad, rarely subfalcate, compressed to subterete, submoniliform, straw coloured or tinged with violet, short-stalked, varyingly acuminate. There are 5–18 seeds per pod, seeds 6–7 mm long, oblique to longitudinal.

Distribution

P. glandulosa is native to northern Mexico and the southern USA from the Pacific coast to the Gulf of Mexico. However, the two common varieties have mostly separate ranges, with var. *torreyana* in the west, in California, Nevada and Arizona and neighbouring Mexican states; and var. *glandulosa* in the east, in Texas and north-eastern Mexico. It has been widely introduced, and *P. glandulosa* is now a common invasive weed in Australia and South Africa, and also in southern Botswana, Namibia, Malawi and Zimbabwe. It is also reported, but in smaller numbers, from elsewhere including North Africa, the Middle East, Pakistan and north-western India.

Silviculture characteristics

P. glandulosa is a generally subtropical species, occurring in grasslands both in its native ranges in Mexico and the USA, and where introduced, especially where most widespread in parts of Australia and South Africa. It is also known to invade watercourses and roadsides, though is able to survive in areas with exceptionally low annual rainfall or very lengthy dry periods, but only if roots are able to tap groundwater within the first few years. *P. glandulosa* is known to survive severe frosts to −20°C in the USA, but can also tolerate some of the highest temperatures in the world, being one of the few tree species growing in Death Valley, California, USA, where temperatures often exceed 50°C. It has a symbiotic relationship with nitrogen-fixing bacteria and mycorrhizal associations to varying degrees.

Pods have a high sugar content and are low in anti-feedants, being widely sought after by a variety of animals. Disjunct stands of trees near to old centres of population suggest that man was also a dispersal agent in historic and prehistoric times. Invasions have often accompanied the arrival and expansion of cattle ranching and its associated effects on the environment. Cattle not only spread the seed widely after ingesting the pods, put it is also hypothesized that the excessive removal of nitrogen from savannah ecosystems that accompanies overgrazing by livestock, provides a competitive advantage to nitrogen-fixing pioneer species such as *P. glandulosa*. It can dominate in dry, or seasonally dry, watercourses or depressions, and the presence and depth of the water table is a decisive factor in tree distribution, size and growth.

P. glandulosa prefers clay soils but tolerates a wide variety of soil types. Soil nutrient status is rarely limiting, and soil improvement leads to an increase in soil fertility as trees mature. It is able to survive and even thrive on some of the poorest land unsuitable for any other tree species, and trees fix nitrogen even under conditions of high salinity and high water deficits. *P. glandulosa* is particularly tolerant of saline and alkaline soils. It has been noted by a commentator in the south-western USA in 1884 that there are few soils in which the tree cannot grow; whether the ground be rocky or flat, or the soil sandy or saline.

Silviculture practice

Freshly harvested seeds exhibit 90% germination. However, as the tree regenerates so easily (to the point of it being an invasive weed) it is rarely a managed plantation species in its native range, thus silvicultural practices in this sense do not apply. Also, where introduced, management focuses on control rather than increasing production. However, research in Texas successfully identified means to achieve both objectives simultaneously by thinning dense stands and pruning remaining trees (see Further reading). This process was used to convert weedy stands into productive 'orchards' for the production of sawlogs for timber as well as producing pods, posts and fuel. The effects of additional treatments such as cultivation (discing) and fertilizer applications were also assessed.

Diseases and pests

Beetles of the family Bruchidae destroy a substantial percentage of seeds in the native range, of which 93% are obligately restricted to *Prosopis*, showing a high degree of specialization. Herbivory is also important, with an estimated 40% of immature leaves removed by insects and 35% of shoots removed by rodents or insects. Defoliating insects vary in their severity of attack but have been found to be a major factor affecting the timing of bud break in North America. Locusts are not uncommon and twig girdlers (*Oncideres* spp.) are damaging in some areas, with adult beetles girdling small branches before ovipositing. Wood-boring beetles are also frequently found in the sapwood of fallen or cut trees but their effect on living trees is unknown. Nematodes are known to attack the roots of *P. glandulosa*, while occasionally infesting neighbouring crops.

Uses

P. glandulosa heartwood is light brown or pink when freshly cut, darkening to a dark reddish-golden-brown with age. This is very distinct from the sapwood, which is yellow, not dimensionably stable and highly susceptible to powder-post and longhorn beetles. Wood density is approximately 720 kg/m^3. Being very hard with a desirable colour and good finish, the timber is used mainly for flooring and furniture in the USA and Mexico. It compares favourably with the world's best timbers, and is exported, but only in limited volumes. It is, however, one of the most dimensionably stable timbers in the world. Volumetric shrinkage is only 4.7%, one-third of that of red oak (*Quercus rubra*) and almost half that of Indian rosewood (*Dalbergia latifolia*), with tangential and radial shrinkage values of 2.2% and 2.6%, respectively. Poles are used for fence posts and in rustic construction. The firewood is of excellent quality, and imparts a pleasant flavour to food cooked over it. *P. glandulosa* charcoal ('mesquite coals') is traded commercially throughout the USA for the barbeque industry.

P. glandulosa is a true multipurpose species, and all parts of the tree have been, or continue to be, used. The pods were once a staple food for numerous Native American tribes and are still a valuable fodder for livestock. The flowers are good bee forage, giving a high-quality honey, and the exudate gum is comparable to commercial gum arabic from *Acacia senegal*. The bark is rich in tannins; fibres were used for rope making; and every part of the tree has been used for making folk medicines.

Further reading

Geesing, D., Felker, P. and Bingham, R.L. (2000) Influence of mesquite (*Prosopis glandulosa*) on soil nitrogen and carbon development: implications for global carbon sequestration. *Journal of Arid Environments* 46, 157–180.

Patch, N.L. and Felker, P. (1997a) Influence of silvicultural treatments on growth of mature mesquite (*Prosopis glandulosa* var. *glandulosa*) nine years after initiation. *Forest Ecology and Management* 94, 37–46.

Patch, N.L. and Felker, P. (1997b) Silvicultural treatments for sapling mesquite (*Prosopis glandulosa* var. *glandulosa*) to optimize timber production and minimize seedling encroachment. *Forest Ecology and Management* 96, 231–240.

Prosopis juliflora (Sw.) DC.

Mesquite

Importance

Prosopis juliflora (Plate 32) is possibly the single most common tree and source of fuelwood in arid and semi-arid frost-free areas of the world. Wood is also used for posts, poles and charcoal, and the sweet pods are browsed by livestock. In its native tropical American range, pods are also processed into human food, and large trees are valued for timber. The foliage is generally unpalatable, so it is suitable for planting in hedges and live fences and has been used in reclamation reforestation of mining spoil, wastelands and sand dunes. This fast-growing, nitrogen-fixing tree requires very little rainfall and can survive on 70 mm/yr. However, it has become more notorious in recent years as an aggressive invasive weed where introduced, requiring improved management and utilization, especially in Africa.

Botanical features

P. juliflora is a shrub or small tree 3–12 m tall, with spreading branches and a round or a flat-topped crown, somewhat spiny with persistent, green foliage, spines 0.5–5.0 cm long, axillary, uninodal, divergent, paired or solitary, sometimes absent or not on all branchlets, largest on strong, basal shoots. Leaves are bipinnate up to 7.5 cm long plus petiole, each with 1–3 pairs of pinnae, 3–11 cm long, with 6–29 pairs of leaflets, generally 11–15, 6–23 mm long and 1.6–5.5 mm wide, elliptic–oblong, approximate on the rachis or distant a little more than their own width, emarginated or obtuse, pinnate-reticulately curved. Pendulous racemes are 7–15 cm long, florets usually greenish-white, turning light yellow (Plate 53). Legumes are straight to falcate with an incurved apex, straw-yellow to brown, compressed, linear with parallel margins, stalked and acuminate, 8–29 cm long, 9–17 mm broad and 4–8 mm thick. Each pod has 25 endocarp segments, is rectangular to subquadrate, mostly broader than long; seeds oval, brown, transverse.

Distribution

P. juliflora is native to western Mexico, Central America, the Caribbean and northern South America (Venezuela and Colombia). The native range was previously considered to include Ecuador and northern Peru, although these populations are now considered to belong to *Prosopis pallida* and/or other species. Populations on Caribbean islands may also have an anthropogenic origin. *P. juliflora* has been widely introduced over the past 200 years, with early records from the 1800s from present-day Australia, Hawaii, India, Pakistan, Senegal and South Africa, among others. It is now likely to be present in frost-free arid and semi-arid regions of almost every country in the tropics. However, misidentifications corrected only in recent years, indicate that *P. juliflora* and not *Prosopis chilensis* is the common species in Sudan and other tropical African countries, whereas the common species in the north-east of Brazil, Cape Verde and northern Senegal is *Prosopis pallida* and not *P. juliflora*.

Silviculture characteristics

P. juliflora is adapted to a very wide range of soils and site types from sand dunes to cracking clays. It is generally found in areas where water and soil fertility are the principal agents limiting plant growth, and is able to survive, and even thrive, on some of the poorest land unsuitable for any other tree species. *P. juliflora* dominates in dry or seasonally dry watercourses or depressions, and is often found in coastal flats and dunes. Although it is principally a noxious weed of natural grasslands, *P. juliflora* is also a common weed in many other habitats.

Trees survive in areas with exceptionally low annual rainfall or very lengthy dry periods, thriving in a wide range of rainfall zones, from 100 mm mean annual rainfall or less in dry coastal zones to 1500 mm at higher altitudes. The existence of two root systems, a deep taproot to reach ground water and a mat of surface lateral roots to make use of infrequent rainfall events, puts *P. juliflora* firmly in the category of phreatophytes, but trees show a variety of mesophytic and xerophytic characteristics depending on water availability. The need for rain or a high water table is reduced in coastal areas where sufficient atmospheric moisture exists with persistent trade winds or seasonal fog.

Optimum temperatures for growth are 20–30°C, and there appears to be no natural upper limit to temperature, with introduced *P. juliflora* known to tolerate day-time shade temperatures of more than 50°C. The major limitation to distribution is mean minimum temperature and frequency and duration of winter frosts. Light frosts cause dieback and more severe or longer frosts will kill mature trees, though damage is more severe on seedlings and younger trees.

Due to its tenacity and drought-resistance, *P. juliflora* has the potential to aggressively invade and colonize areas quickly, becoming a weed. This species can form thorny, impenetrable, dense stands with as many as 20,000 stems/ha, favoured by overgrazing which reduces soil fertility so legumes such as *P. juliflora* have a competitive advantage, and leads to a bare ground surface with no herbaceous cover which provides no competition for emerging *P. juliflora* seedlings.

Silviculture practice

As *P. juliflora* is today more often mainly considered as an invasive weed, silvicultural practices in this sense do not apply. However, it is still planted in several countries in South Asia and West Africa, as a plantation tree for fuelwood and fodder, and as a shelter belt and for erosion control and land rehabilitation. Various methods of vegetative propagation have been attempted with some success, but seed is the principal means of natural or artificial regeneration. Plantations may be established at a variety of spacings from as close as 1 × 2 m, to 4 × 4 m or 5 × 5 m in very arid sites, or up to 10 × 10 m for pod production. In fuelwood plantations, 50–60 t/ha can be harvested on 8–10-year rotations, and 75–100 t/ha on 15-year rotations.

Management of dense invasive naturally regenerated stands is now also being practised, the most appropriate long-term strategy being to thin and prune the seedlings to less dense spacings to accelerate the natural self-thinning process. Another benefit of such thinning is that large trees (>50 cm in diameter) help to increase soil organic carbon and nitrogen, which encourages growth of grasses requiring high soil fertility, and a vigorous ground cover reduces *P. juliflora* seedling encroachment.

Diseases and pests

Diseases include *Fusarium* root rot, *Macrophomina phaseolina* collar rot, leaf blights caused by *Colletotrichum capsici*, *Colletotrichum cymopsicola* and *Ravenelia spicigera*; a twig blight, *Diplodia prosopides*; and stem cankers caused by *Botryodiplodia theobromae*, *Lasiodiplodia* sp., and *Pestalotiopsis* sp.

Pests reported from the native range include ants, termites, psyllids, nematodes, defoliators, sap suckers, twig girdlers, flower eaters, seed feeders and wood borers.

Uses

P. juliflora is a valuable multipurpose tree, widely planted for soil conservation, in hedgerows, and as an urban and general amenity tree.

P. juliflora sapwood is distinct from the heartwood, being rather narrow and yellow in colour. The heartwood is a rich dark reddish-brown, similar to teak. The wood is heavy to very heavy, strong or very strong, extremely tough and extremely hard. It is also durable, and burns evenly. Reports of specific gravity vary widely, from 0.55 to 1.07, and basic density is reported to be 745–1250 kg/m^3 when green and 710–910 kg/m^3 when dry. It is one of the most dimensionally stable timbers in the world, with a volumetric shrinkage rate of 4–5%, and equal radial and tangential shrinkage values of about 3% each. The wood is easy to season, although timber may develop some surface cracking and end splitting, but finishes very well. However, virtually no *P. juliflora* wood is used as a fine timber and it is mostly used as a fuel. Satisfactory bleached pulp can be prepared from *P. juliflora*, with strength properties suitable for writing and printing paper production, although yields are low compared to other hardwoods and it is rarely used for pulping.

The most important non-timber products are the pods, which are a valuable livestock fodder, and serve as a security food item during famine. They contain 30–40% sugar and 7–17% crude protein. Production from good trees may be 50–100 kg/yr, although the very best individuals can produce as much as 300 kg/pods/yr. The most benefits are gained from milling and mixing the pods into complete rations. This utilizes the protein in the seeds, which otherwise pass undigested through an animal's stomach, and may lead to further invasion as a weed. Animals feeding solely on whole pods for any period of time may also become sick, especially from problems with teeth, the mouth and sometimes in the rumen. Pods are also processed into flour and used as a human food, although this huge potential as a new source of food especially in dryland Africa and South Asia has not yet been adequately investigated.

A reddish-amber gum, similar to gum arabic (from *Acacia senegal*), often exudes from the stem and older branches. Due to the profuse flowering in *P. juliflora*, this species also supports excellent honey production. Plant parts also have many medicinal applications.

Further reading

Landeras, G., Alfonso, M., Pasiecznik, N.M., Harris, P.J.C. and Ramírez L. (2006) Identification of *Prosopis juliflora* (Sw.) DC. and *P. pallida* (H. & B. ex. Willd.) H.B.K. using molecular markers. *Biodiversity and Conservation* 15, 1829–1844.

Pasiecznik, N.M., Felker, P., Harris, P.J.C., Harsh, L.N., Cruz, G., Tewari, J.C., Cadoret, K. and Maldonado, L.J (2001) *The* Prosopis juliflora - Prosopis pallida *Complex: A Monograph.* HDRA, Coventry, UK, 162 pp.

Tewari, J.C., Harris, P.J.C., Harsh, L.N., Cadoret, K. and Pasiecznik, N.M. (2000) *Managing* Prosopis juliflora *(Vilayati Babul): A Technical Manual.* CAZRI, Jodhpur, India and HDRA, Coventry, UK, 94 pp. (in English and Hindi).

Prunus africana (Hook. f.) Kalkman

African plum

Importance

Prunus africana is an African tree of increasing interest as a source of bark extract used in the international pharmaceutical trade. It is generally managed in natural forests, with plantations making little or no contribution at present, although sparse distribution rules out intensive management and makes supervision of harvesting difficult. To meet demand, it has been over-exploited in parts of its range and was added to Appendix II of the CITES listing, requiring permits for exporting harvested bark or bark products, and evidence that stands are being managed sustainably.

Botanical features

P. africana grows to 10–25 m with a diameter of 70–100 cm, occasionally to 40 m with a trunk diameter of 150 cm. In closed forests, the crown is compact, with a single slim, straight, cylindrical bole, occasionally with small buttresses. In open conditions a deep crown develops, with short boles and large spreading branches in old individuals. The bark is dark brown to blackish, with prominent longitudinal fissures in young trees, turning to coarse exfoliating scales on older trees. The leaves are simple, alternate and crenate–serrate, with a long (1–2 cm), often reddish, channelled petiole and a glabrous, subcoriaceous, elliptic–oblong lamina 6–15 cm long and 3–6 cm wide, with a prominent midrib. Inflorescences are slender axillary racemes 2–8 cm long, with 5–15 hermaphrodite, actinomorphic and predominantly pentamerous flowers, 15 mm in diameter, carried on pedicels 5–7 mm long. The fruit is an ellipsoid, slightly bilobed drupe, usually single-seeded, 5–10 mm long, purplish-brown or blackish when ripe. Bark, bruised leaves and fruits smell strongly of bitter almonds and are poisonous.

Distribution

P. africana is native to the eastern side of Africa, from Ethiopia to South Africa, with disjunct populations in Cameroon, Nigeria and Angola, and on the islands of São Tomé, Bioko (Equatorial Guinea), Comoros and Madagascar.

Silviculture characteristics

P. africana occurs in lowland and Afromontane forest types but more extensively where mean monthly temperatures are relatively low, optimally 11–19°C and 17–23°C for the coolest and warmest months, respectively. It is not present in areas subject to severe or prolonged frost. *P. africana* is associated principally with subhumid montane climates but is also present under humid and semi-arid conditions. It will grow at a mean annual rainfall of 500–700 mm with groundwater and low evapotranspiration and a dry season of up to 6 months, otherwise preferring 800–1300 mm and a 1–3-month dry season.

P. africana has recalcitrant seed and no seed bank accumulates. Seed crops are irregular but not infrequent. Fallen fruits remain close to parent trees and germinate even in heavy shade, though seedlings do not persist there, requiring large gaps to develop. When high concentrations of easily recognized seedlings develop, they are transplanted for plantation establishment or enrichment planting. Trees can flower before 10 years old and have a life span of 100–250 years, explaining the common occurrence of *P. africana* as scattered large individuals. *P. africana* is evergreen. In the equatorial zone flowering may occur every month, but tending to coincide with cool and dry periods at higher latitudes.

Silviculture practice

P. africana germinates well using fresh depulped seed not otherwise pretreated. However, unless carefully stored, very few seeds retain viability beyond a few weeks. Depulped seed from ripe, purple fruits has been stored with moderate success (35–40% viability) for 1 year, at 5°C and 15% moisture content. Transplanted wildings are widely used, though mortality tends to be high, particularly when using larger plants. Current interest is mainly in improved quality nursery stock, although direct seeding is also considered to have potential. A nursery period of 5–7 months producing seedlings to 40 cm tall has been recommended. Vegetative propagation via air-layering and shoot cuttings from juvenile plants has also proved successful.

Beyond the selection of appropriate sites within the natural area of occurrence, there are no specific stand management protocols for *P. africana*. Standard timber plantation procedures were used with the first recorded plantings in Kenya in 1913, but much subsequent planting undertaken has been in mixtures, for boundary demarcation, enrichment line planting and more recently in agroforestry systems. Due to the typically low stocking, it has never been a priority timber tree and has not attracted specially tailored management practices. However, transformation of transitional forests of the Lake Victoria basin by heavy felling initiated development of more valuable even-aged stands including *P. africana*.

The switch in interest from timber to bark has demanded more focused management to ensure resource sustainability. Immediate management aims to secure adequate protection of seed trees, create a favourable germination and survival environment around them, and free young trees from competition, complemented by increase the stocking of *P. africana* and replacing losses resulting from poor bark harvesting, with enrichment planting using both wildings and nursery stock. Doubts have been expressed, however, regarding the prospects of reversing reported population declines through conventional interventions to encourage regeneration or enrichment planting.

As an alternative, growing *P. africana* at 2.7 × 2.7 m spacing on a 12-year rotation concluding with a destructive total bark harvest has been suggested, yielding around 29 t/ha, followed by replanting. During the first 5–10 years in tended stands, annual height increment in excess of 1 m is not unusual in the tropical part of the range and annual diameter increment may exceed 2 cm. Conventional coppicing is unsuitable, as repeated removal of bark from the stumps reduces shoot survival and growth.

Schemes for repeated harvests from the same trees have been adopted in the main producing country, Cameroon, and follow a formal protocol. Bark is harvested in vertical panels removed without damage to the underlying cambium. The first crown branch determines the top of the strip and the lower end is 1 m above ground. Trees of less than 30 cm diameter are not harvested. From trees 30–50 cm in diameter, two opposing panels are detached, each one-quarter of the tree's circumference. Four panels can be removed from trees greater than 50 cm diameter, each one-eighth of the circumference, alternating with untouched panels of the same size. Trees are marked when harvested, and are not harvested again for at least 5 years. Yield estimates are 28–85 kg/tree, with an average of 55 kg. Felling and complete bark stripping yields 50–200 kg/tree depending on tree size.

Diseases and pests

Some nursery diseases have been reported, including *Colletotrichum gloeosporioides* causing leaf spot and shot hole disease in seedlings, and a powdery mildew caused by *Podosphaera leucotricha*.

Most of the reported insect pests are wood-boring beetles, with only occasional reports of defoliating insects, sap-feeders, or fungal or viral diseases of adult trees.

Uses

P. africana has been promoted as an agroforestry tree particularly through on-farm planting for eventual bark harvesting, and for fuelwood and timber on-farm as one of a suite of indigenous and exotic species. It has also been used for planted firebreaks, and makes an attractive garden shade tree.

The sapwood is pale pink and the heartwood pale red, darkening to a rich dark red on exposure. Wood density of air-dried timber is 785–1090 kg/m^3. The timber is heavy, hard, tough, and of above-average strength, making it suitable for heavy flooring and construction and, where available, widely used for lorry bodywork and bridge decking, but significant volumes have never been marketed. Use for furniture is limited by difficulties of seasoning, with a tendency to split and distort, and a need for careful kiln drying, but successfully seasoned timber has satisfactory working properties. The wood is close grained and even textured, but with a twisted grain that limits its use. It is resistant to termite and fungal attack but service life in contact with the ground is short and the sapwood is susceptible to attack by powder-post beetles. In the vicinity of forests where *P. africana* is relatively abundant, there are diverse local structural, implement and fuel uses of the wood.

A combination of active components in the bark is used to manufacture drugs for treating prostatitis (benign prostatic hyperplasia), a prostate gland condition common in older men. Apart from commercial pharmaceutical significance, macerated and infused bark has diverse uses in traditional human and livestock medicine, as do the leaves and fruits. There is a small but well-established international trade in raw and processed bark, mostly from Cameroon, Equatorial Guinea, Kenya and Madagascar, to major pharmaceutical enterprises, mainly in France, Italy and Spain; the final products are widely marketed in Europe and in North America. Capsules usually contain 12–13% extract combined with extracts from other plants, notably saw palmetto (*Serenoa repens*). 'Over-the-counter' trade in bark extract capsules globally has been estimated at US$220 million/yr.

Further reading

Breugel, P. van, Kindt, R., Lillesø, J.P.B., Sebsebe, M.B., Dudley, D.C., Friis, I., Gachathi, F., Kalema, J., Mbago, F., Minani, V., Moshi, H.N., Mulumba, J., Namaganda, M., Ndangalasi, H.J., Ruffo, C.K., Jamnadass, R. and Graudal, L. (2012) *Potential Natural Vegetation of Eastern Africa (Ethiopia, Kenya, Malawi, Rwanda, Tanzania, Uganda and Zambia): Volume 7: Projected Distributions of Potential Natural Vegetation Types and Two Important Agroforestry Species (*Prunus africana *and* Warburgia ugandensis*) for Six Possible Future Climates.* Forest and Landscape Working Papers 69. ICRAF, Nairobi, 63 pp.

Cunningham, M., Cunningham, A.B. and Schippmann, U. (1997) *Trade in* Prunus africana *and the Implementation of CITES.* Germany Federal Agency for Nature Conservation (BfN), Bonn, Germany, 67 pp.

Hall, J.B., O'Brien, E.M. and Sinclair, F.L. (2000) Prunus africana: *A Monograph*. University of Wales, Bangor, UK, 104 pp.

Prunus avium **(L.) L.**

Wild cherry

Importance

Prunus avium (Plate 33), or wild cherry, is the wild ancestor of most sweet cherry varieties. It is a valuable component of diversity in mixed broadleaf temperate forest ecosystems in Europe, and can tolerate a relatively wide range of environmental conditions. It is relatively short-lived (80–100 years); due to its capacity to reproduce vegetatively, it can easily colonize clear areas by means of root suckers, as well as spreading over longer distances by seed.

Current interest in *P. avium* wood production in natural forests and in plantations in Europe is high, as it produces an attractive patterned wood which is second in value only to that of walnut (*Juglans regia*); it is still valued today for panelling and cabinetmaking.

P. avium is increasingly planted in Europe, both in afforestation of abandoned farmland and enrichment of forests, due to its fast growth (rotation of 50–60 years). It is also utilized for landscape planting, in small groups or in linear or roadside plantings; it has attractive white flowers in spring and a reddish autumn foliage colour.

Botanical features

Tree 15–20 m in height, to 30 m on good sites; dbh 40 to >90 cm, according to age and site fertility; generally a pyramidal crown when isolated. The species can present a range of crown shapes and branch architecture according to the genotype or the local forest structure.

The stem is usually tapered in young trees, and cylindrical in mature individuals. They can be buttressed at the base in alluvial areas and curved when trees grow on slopes - in this case, and on sites exposed to winds, stems may be eccentric.

The root system is taprooted in seedlings and younger trees, while it can become shallow with age or on thin or clay soils. Adult trees produce long horizontal roots, able to produce wide clonal sucker structures.

The bark of young trees is thin, smooth, and dark grey-pink; later it becomes fissured and peeling, and is a dark purple-red or grey-red. Lenticels are prominent and stem wounds produce a resin-like, amber-coloured, odourless gum.

Leaves pendulous, with rare and scattered hairs on the lower surface along the rib. Upper surface dark green and shiny, lower surface lighter green. Leaves are oblanceolate (4–20 × 10–20 cm), rarely ovate, toothed. Petiole 2–4 cm long with 1–4 red-brownish glands near the base of the upper blade surface. Leaves arranged alternately on branchlets at the base of lateral buds. Smaller, obovate leaves accompany flower buds on brachyblasts.

Two to five hermaphrodite flowers together on brachyblasts. Flower 2–2.5 cm across, formed of five pure white petals (average 10–15 mm) surrounding a calyx tube constricted above by a circle of yellowish pollen sacs. Generally, fewer fruits than flowers ripen on the same brachyblast. Fruits are drupes, dark reddish to clear reddish, with 3–5 cm long green peduncles, containing woody stones relatively large compared to fruit size. Testa is bittersweet to sour.

Foliage development starts in March–April just before the end of the flowering period; flower buds burst in late February–March and/or later; fruit ripens (average duration, 60 days) from end of May to August.

Distribution

Archaeological and sub-fossil evidence indicate that wild *P. avium* is native to north-west and central Europe; it is found in Europe, Western Asia, and Northern Africa, and is thought to have originated from the Caucasus and surrounding area; it has been introduced into many countries. Records of cherries as cultivated plants began with Greek civilization. Cultivated cherry spread throughout the Roman Empire, including the area already covered by the wild forms of *P. avium* and *Prunus cerasus*, but it may have been domesticated in several widely separated regions.

P. avium is found in mixed coniferous forests and in temperate broadleaf forest ecosystems. Natural populations of this species still exist, although most trees occur as single individuals or in small groups.

Silvicultural characteristics

P. avium is strongly light-demanding, except when very young. It is frost-resistant and moderately tolerant of summer drought.

P. avium has strong apical dominance, which usually results in the tree developing and retaining a single straight leading shoot; therefore, initial stocking densities can be lower than for most other broadleaves. Regular pruning is necessary, and thinning is also recommended. *P. avium* does not coppice well.

Silvicultural practice

Seeds should be collected from populations of at least 20–30 individuals. Distances between mother trees should exceed 50–60 m, in order to avoid collection from suckers belonging to the same genotype. Collecting the seed can be very difficult, because birds tend to eat the cherries before they can be collected.

P. avium cultivated as a timber tree is managed in high-forest systems, or in coppice with standards in which wild cherry (standard) is sheltered by the coppice species. Following clump planting, selective thinning must be undertaken when the clumps become too dense.

The pruning of green branches predisposes *P. avium* to bacterial infections. To avoid this, pruning should be carried out in winter, but it can be combined with summer pruning (the latter period recommended by some authors) in order to avoid the development of adventitious buds on the stem.

P. avium is planted in some regions (notably parts of Italy) in pure plantations, or in mixtures in which it comprises 30–40% of trees. Some authorities recommend planting *P. avium* in small groups or clumps only, within mixed plantations: this mirrors its occurrence in natural forests, and another advantage is that it will help discourage serious damage to plantations from any of the bacterial and fungal pathogens to which *P. avium* is prone. One suggested guideline for planting *P. avium* is that clumps should not exceed about 20 trees, and the species should make up not more than about 10% of all trees in a wood.

A spacing of 2 × 1.5 m seems to be an optimum for *P. avium*, with auxiliary species planted between the *P. avium* rows; also at a spacing of 2 × 1.5 m. Lateral protection is very beneficial. In Italy, 4 × 4 m to 8 × 8 m spacings have been tested. Denser spacings should be used in pure *P. avium* plantations, while the wider spacings should be used for mixtures with other species.

Mulching may be used on drier sites or on those where grass competition is strong. Wire next to fencing or individual shelter tubes can be used to protect young *P. avium* plantations from grazing animals. Tubes were found to protect efficiently, and seedling height growth can be positively affected by the shelters, but root collar diameter does not increase correspondingly, leading to saplings that bend over when the tubes are removed.

Shelter diameter can influence height growth and the presence of branches, but not diameter growth. Root biomass can be also reduced. It has been suggested that tree shelters should be slit just before the stem fully occupies the diameter of the shelter, which can then be left in place for another year.

Diseases and pests

Pseudomonas syringae pv. *morsprunorum* causes bacterial canker resulting in gumming and dieback of infected branches. The shoots, flowers, leaves and fruits can also be damaged. *Monilinia fructicola* and *Monilinia laxa* are the agents of brown rot in the USA and Europe, respectively. This disease reduces yield by causing decay of the flowers, twigs and fruits on the tree and post-harvest. *Wilsonmyces carpophilus* causes shot hole disease, which produces lesions on the leaves and fruit. The fruit lesions become corky and rough, and the leaf lesions cause abscission. *Podosphaera clandestina* causes powdery mildew. The most significant economic losses usually result from fruit infection, but foliar mildew is most damaging in nursery plantings. *Phytophthora megasperma* and *Phytophthora cambivora* affect the tree roots, causing crown rot.

Cherry (*Prunus*) necrotic ringspot virus is one of the most dangerous sweet cherry viruses throughout the world. The symptoms of the chronic phase of the disease are yellowish-green or light green rings or bands on the leaves, with brown necrotic spots inside the rings. These necrotic spots may later break out of the leaf blade, leaving behind shot holes.

Cherry rasp leaf virus is two viruses with different symptoms. The American virus causes prominent enations on the lower surface of leaves that form leafy protuberances between the main veins or along the midrib. The European virus (Pfeffinger disease) causes large, yellowish-green flecks on the leaves, slightly wavy, distorted; secondary symptoms are similar to those of American cherry leaf virus.

Cherry leaf roll virus causes chlorotic leaf spots, leaf curling, shoot dieback, bark splitting, gumming and decline of the trees. Flowering and leaf emergence may be delayed and weakened. Leaves are rolled, rigid and sometimes discoloured. *Quadraspidiotus perniciosus* or San Jose scale appears as greyish, thin, flaky masses on the bark of branches and twigs. It decreases the vigour of the tree.

Myzus cerasi (black cherry aphid) colonies can attack the trees in the spring causing curly leaves, reduced shoot development and damage to the fruits and pedicels. They also produce honeydew which attracts ants. Larvae of *Recurvaria nanella* damage buds, shoots and flowers in early spring and enclose the vegetation by means of silky fibres, thus limiting the growth. The larvae of *Operophthera brumata* and *Erannis defoliaria* attack the shoots, flowers and leaves, damaging production.

In mid-spring, females of *Caliroa limacina* lay eggs into the leaves. The emerging larvae feed on the leaves, which then desiccate.

Uses

P. avium has a whitish to reddish heartwood and a white-yellowish sapwood. The main wood physical traits are as follows: the texture is fine, grain generally straight; average wood density is 850 kg/m^3 (fresh) to 650 kg/m^3 (dry); hardness and shrinkage are average; durability is low. Wood decay fungi and insects can easily attack the wood. Sawing is easy, but seasoning must be slow because of a tendency to split. Veneer slicing is currently used for good-quality timber.

The most important traits concern the general appearance of *P. avium* wood. Important studies are being carried out in France on the optical colour characterization of wild cherry.

The potential of *P. avium* as a source of wood for charcoal and liquid wood pyrolysis products has also been examined.

Flowers and fruits *P. avium* are a valuable food resource for birds and other animals, and help to maintain biodiversity in forest and agroforestry ecosystems; the species is also valuable honeybee forage during its flowering season throughout its range, as it flowers early.

Further reading

Díaz, R., Zas, R. and Fernández-López, J. (2007) Genetic variation of *Prunus avium* in susceptibility to cherry leaf spot (*Blumeriella jaapii*) in spatially heterogeneous infected seed orchards. *Annals of Forest Science* 64, 21–30.

Faust, M. and Surányi, D. (1996) Origin and dissemination of cherry. In: Janick, J. (ed.) *Horticultural Reviews 19*. John Wiley & Sons Inc., Oxford, UK.

Hedrick, U.P. (1915) *The Cherries of New York*. J.B. Lyon Company Albany, New York, New York, USA.

Russell, K. (2003) EUFORGEN Technical Guidelines for genetic conservation and use for wild cherry (*Prunus avium*). International Plant Genetic Resources Institute, Rome, 6 pp.

Prunus serotina Ehrh.

Black cherry

Importance

Prunus serotina is a deciduous, single-stemmed tree, native to the eastern USA and one of highest-value hardwoods in North America. It was first introduced into Europe in the 1600s as an ornamental, but has naturalized and is widespread, especially in broadleaved and mixed woodlands. Although it is valued as a source of high-quality timber, it is now much better known from European countries where it has become a highly invasive weed. The fruit is increasingly used, especially in Central America.

Botanical features

P. serotina is a deciduous, single-stemmed, medium- to large-sized tree. In mixed hardwood forest it can also be a small, contorted, short-lived understorey tree or shrub. In the eastern USA, the tree typically achieves 38 m in height and 1.2 m or more in diameter. South-western varieties are much smaller. In central Europe, *P. serotina* is mostly a shrub, rarely a tree up to 20 m in height. Leaves are oblong–ovate to lanceolate–oblong, with toothed edges, bright green above, pale green below. White flowers 8–10 mm in diameter grow in cylindrical racemes 6–15 cm long, each with approximately 30 flowers. Fruits are purple-black drupes of 8–10 mm diameter.

Distribution

P. serotina is native to most of the eastern half of the USA and parts of south-eastern Canada, from Nova Scotia to Florida, and west to a line as far south as eastern Texas, but with disjunct populations in central Texas, Arizona, Mexico and Guatemala. It has been introduced and has become naturalized in neighbouring regions in the USA, Canada and Central America. *P. serotina* was among the first American trees to be introduced to Europe, recorded in England in 1629, and was introduced and cultivated as an ornamental in gardens throughout Europe in subsequent centuries. It had become widely naturalized by the 1900s and spread invasively in many broadleaved and mixed forests in north-central Europe, with problems first noted in The Netherlands in the 1970s.

Silviculture characteristics

P. serotina is a shade-intolerant, gap-phase species in moist, temperate climates. Seedlings under a forest canopy will only survive a few years without high intensity daylight, but will grow rapidly if a canopy gap is created. Trees sprout readily from stumps, and the sprouts grow rapidly in full sunlight. Seedlings typically develop a taproot with numerous laterals during the first few years.

It grows from sea level to more than 1500 m, but optimal growth occurs at 300–800 m, such as on the Allegheny Plateau of Pennsylvania and New York, USA. Although there are significant variations within both its natural and introduced ranges, average annual rainfall is around 1000 mm, with a mean maximum temperature of hottest month of 28°C and a mean minimum temperature of coldest month of −8°C. The coldest recorded absolute minimum temperature is −40°C.

P. serotina occurs on a variety of soils except for extreme wet or dry ones, and is tolerant to various groundwater conditions, although productivity decreases on wet sites. It frequently occurs on very acid and relatively infertile soils. The soil texture ranges from sandy loam to silty clay loam. Individuals of *P. serotina* are, in natural conditions, scattered among other species or even form nearly pure stands at high elevations with impeded drainage.

Commonly emergent in mixed forests, it is more vulnerable to storm damage, and younger trees can be bent by wet snow causing loss of the leader and severe crooks that make them unsuitable for sawn timber. However, trees often make a rapid recovery after breakage, with little reduction in diameter growth. *P. serotina* is highly susceptible to injury by fire, with younger trees killed outright and even older trees killed back to ground level. Trees are intolerant of flooding.

Where introduced, *P. serotina* may invade semi-natural or managed woodland, particularly on acid sandy soils, establishing in forest clearings, margins, and inside forests, often following disturbance. It can be an aggressive colonizer, overtopping other trees species, and dense thickets of *P. serotina* considerably inhibit seedling development of other species. It is also very difficult to control because of very intensive sprouting and seeding, and suitably host-specific biological control agents are being sought.

Silviculture practice

P. serotina reproduces by seeds or vegetatively by sprouting. Seeds require cold stratification to germinate, but are a highly desirable source of regeneration; maximum sprouting occurs in trees less than 40–50 years old. Clear-cuttings of natural second growth stands resulted in third-growth stands in which more than half of the trees were of sprout origin.

P. serotina grows very fast in the seedling, sapling and pole stages, generally outgrowing and overtopping other species in mixed forests, and maintains growth advantage for 60–80 years. Diameter growth slows beyond 80–100 years and tree mortality increases rapidly. Seedlings reach 5–10 cm in height within 30 days of germination, but under dense shade they may grow only up to 3 cm/yr until they die because of lack of light. However, in full sunlight, annual height growth of young trees can be 50–100 cm, or 1.2–1.8 m if fertilized, and up to 2.4 m on the best sites. Mean annual diameter growth of dominants and co-dominant trees in natural stands are reported as being 0.65 cm at 10–40 years old, 0.5 cm at 40–70 years, and 0.4 cm at 70–100 years.

In Europe, more efforts are put into killing *P. serotina* as an invasive species than improving its silviculture for the production of high-value timber, but it appears that it could be managed by maintaining dense canopies to prevent seedling regeneration.

Diseases and pests

The most common disease is cherry leaf spot caused by *Coccomyces lutescens* which reduces tree vigour and can kill seedlings, though most other foliage diseases cause little damage. Of branch and stem diseases, black knot caused by *Apiosporina morbosa* is most common and can reduce timber value; *Cytospora leucostoma* canker can also be found. Root and butt rots include *Armillaria mellea*, *Coniophora cerebella*, *Polyporus berkeleyi* and *Tyromyces spraguei*, and many other fungi cause trunk rot.

The most important defoliating insects attacking *P. serotina* include the eastern tent caterpillar (*Malacosoma americanum*) and the cherry scallop shell moth (*Hydria prunivorata*). Gum spots in the wood are caused by the agromyzid cambium miner (*Phytobia pruni*), the peach bark beetle (*Phloeotribus liminaris*), and by the lesser peachtree borer (*Synanthedon pictipes*). *Archips* spp. and *Contarinia cerasiserotinae* are among the more important insects that can cause injury to terminal shoots

Uses

P. serotina is often planted as an ornamental tree in the eastern USA and in Europe, and is used for surface-mine spoil reclamation. It is reported to be used as rootstock for grafting on various plum varieties.

The wood is straight grained, moderately hard, and has a specific gravity of 0.53 when dry. Heartwood is a reddish-brown, darkening with age, with white to light reddish-brown sapwood. Both growth rings and rays are fairly distinct, and dark red gum streaks are sometimes present. Air seasoning is relatively quick and once dry the wood is stable and does not warp or check with normal changes in relative humidity and temperature. It is easily machined, saws cleanly, turns well, planes excellently, takes screws and glues well and takes a fine lustre on finishing.

P. serotina is an important hardwood in North America where it is used for furniture, especially suited for dining-room suites, large desks and tables. It is also used for panelling, veneers, trim for boats and buildings, handles, crafts, toys, scientific instruments, picture frames, piano actions, fixtures, woodenware, patterns and novelties. The highest-quality wood comes from the Allegheny Plateau of Pennsylvania, New York and West Virginia.

Extracts from the bark are used in the preparation of wild cherry syrup, a popular vehicle for cough medicines. The fruit can be used for making jelly or wine.

The fruit is eaten fresh, cooked and used for making juice, jam and wine; early pioneers sometimes used the fruit to flavour their rum or brandy in its native range. The bark of *P. serotina* is used in traditional cough medicines, tonics and sedatives. Many animals feed on the fruit, but the leaves, twigs and bark contain prunasin, a cyanogenic glycoside that is released during wilting and which may cause sickness or even death among some animals.

Further reading

Gatchell, C.J. (1971) *Black Cherry (*Prunus serotina *Ehrh.)*. FS-229. Forest Service, USDA. US Government Printing Office, Washington DC, USA, 8 pp.

Halarewicz, A. (2011) The reasons underlying the invasion of forest communities by black cherry, *Prunus serotina* and its subsequent consequences. *Leśne Prace Badawcze* 72, 267–272.

Marquis, D.A. (1990) *Prunus serotina* Ehrh. Black Cherry. In: Burns, R.M. and Honkala, B.H. (technical coordinators) *Silvics of North America. Volume 2. Hardwoods.* United States Department of Agriculture, Forest Service. Agriculture Handbook 654. Washington, DC, pp. 238–249.

Pseudotsuga menziesii (Mirb.) Franco

Douglas fir

Importance

Pseudotsuga menziesii is one of the world's most important and valuable timber trees. It has been a major component of the forests of western North America since the mid-Pleistocene. Although its native range has never extended beyond western North America, the species has been successfully introduced in the last 100 years into many regions of the temperate forest zone, especially Europe, where it is grown in forests, arboreta and parks. It is one of the most valuable conifers owing to its ecological and growth characteristics and wood properties. Both botanical varieties of *P. menziesii* are important timber trees in their native ranges, although the interior variety (var. *glauca*) does not achieve the same growth rates as the coastal variety (var. *menziesii*), the latter being the more commonly planted.

Botanical features

One of the largest North American conifers, trees of *P. menziesii* can reach over 80 m in height, 200 cm in dbh, and 600 years of age. Occasionally specimens may grow up to 100 m tall and 400 cm dbh, and live over 1300 years. Mature trees are characterized by a long, branch-free, cylindrical trunk and a short, columnar, flat-topped crown. Young trees have a narrowly conical crown that may extend to the ground. The principal branches are in irregular whorls at the annual nodes, lesser branches are in-between. The bark is grey, smooth, thin and resin-blistered when young, becoming deeply furrowed with irregular, broad, dark reddish-brown ridges, and up to 30 cm thick with age.

P. menziesii is deep-rooting. In deep, well-drained soils, taproots may grow up to 50% of their final depth in 3–5 years. Plate-like root systems develop when trees grow in soils which are shallow or have a high water table. Fine roots are concentrated in the upper 20 cm soil layer. Roots are associated with both ecto- and endomycorrhizae.

Needles are evergreen, flat, flexible, 2–3 cm long, grooved above, often sharp-pointed, yellowish-green in var. *menziesii*, bluish-green in var. *glauca*, and remain on the tree for 5–8 years.

P. menziesii is monoecious; trees commonly begin to produce strobili at 12–15 years old. Anthesis and pollination of var. *menziesii* occur during March and April in the warmer part of the range and as late as May or early June in colder areas. At low and middle elevations, cones mature and seeds ripen from mid-August in southern Oregon to mid-September in northern Washington and southern British Columbia. The phenology of flowering is similar for var. *glauca*; early flowering occurs in mid-April to early May in Colorado and as late as early May to late June in northern Idaho. Cone ripening varies from late July at lower elevations in Montana to mid-August in northern Idaho. Seed dispersal of var. *glauca* begins in mid-August in central Oregon and occurs as late as mid-September at higher elevations in Montana.

Mature cones are 8–10 cm long. The bracts turn brown when seeds are mature. Seed size varies with location, small seeds (about 132,000/kg) occurring on trees in British Columbia and large seeds (about 51,000/kg) on trees in California. Seedfall occurs soon after cone maturity with, generally, two-thirds of the total crop on the ground by the end of October. The remaining seeds fall during winter and spring. Germination, which is epigeal, occurs in mid-March to early April in the warmer portions of the range and as late as mid-May in cooler areas.

Seedling growth in the first year is indeterminate but relatively slow and limited generally by moisture, which triggers initiation of dormancy in midsummer. The dormant period normally extends from midsummer until April or May of the following year.

Initiation of phenological events varies greatly with latitude, altitude, aspect, and weather. Trees exhibit a long period of continuing height growth and cone production.

Distribution

The latitudinal range of *P. menziesii* extends from 19°N to 55°N. The range of var. *menziesii* extends from central British Columbia south along the Pacific Coast Ranges for about 2200 km into California (near latitude 35°N), while the range of var. *glauca* stretches from central British Columbia along the Rocky Mountains into the mountains of central Mexico over a distance of nearly 4500 km.

P. menziesii grows in even- or uneven-aged, pure or mixed-species stands, depending on site and successional stage. It is an outstanding example of a pioneering tree following fire, remaining dominant over large areas and outliving its more shade-tolerant associates. The species may be present in early, mid or late stages of secondary succession, and forms a major or minor component in old-growth stands in temperate and mesothermal climates.

P. menziesii is a major component of a number of forest communities in the Montane Spruce, Sub-boreal Spruce, Interior Douglas fir, Interior Grand Fir, Interior Western Hemlock, Pacific Madrone–Douglas fir, White Fir, Coastal Western Hemlock, Pacific Silver Fir, Sitka Spruce and Redwood zones.

P. menziesii was first introduced to Europe after 1825. It was planted in arboreta, parks and on a minor scale in forests. In

1880, it was included into a network of experimental plantations by the German Forest Research Institute. Plantations of *P. menziesii* now exist in over 20 European countries. In New Zealand, it ranks highly as an important softwood species. It has also been planted in New South Wales, Tasmania, Victoria and was trialled in South and Western Australia.

Silviculture characteristics

Reproduction capacity is high; cone production begins as early as 10 years and even old trees produce viable seed. Seed dissemination capacity is high; most seeds fall within 100 m of the source, with the maximum dispersal distance being 2 km.

The potential for natural regeneration in low light varies from low in the wettest climates to high in the driest climates. The potential for natural regeneration in the open is high, especially after wildfires, providing it is exposure-tolerant and favourable seedbeds are available. The potential initial growth rate (at 5 years or younger) is high; first-year seedlings on productive sites may develop shoots about 10 cm long.

The response of advance regeneration to release is high, with diameter response immediate and height growth response delayed. Self-pruning capacity in dense stands is high, and crown spatial requirements are intermediate. The light levels beneath closed-canopy, mature stands are intermediate.

The potential productivity is high; the site index (50 years at bh) is about 40 m on the most productive coastal sites. The growth in height of coastal trees continues for more than 200 years.

The longevity of *P. menziesii* var. *menziesii* is high, with the oldest trees nearly 1500 years old, while var. *glauca* is not as long-lived, trees rarely living more than 400 years, although ages over 700 years have been recorded.

Silviculture practice

Depending on site and management objectives, clear-cutting, patch-cutting, seed-tree, shelterwood, and selection systems are all appropriate for establishment and growth. *P. menziesii* can be grown in single- or mixed-species stands, preferably with shade-tolerant species.

Propagation by seed is the only contemporary method of regeneration. Containerized stock is used for planting when *P. menziesii* is prescribed as the major species.

Seed crops occur at irregular intervals, with one heavy and one medium crop every 7 years on average. Trees 200–300 years old produce the greatest number of cones.

P. menziesii can produce lammas shoots, but this habit is confined to either the wetter portion of the range or to years with abnormally heavy summer rainfall. First-year seedlings on better sites in the Pacific North-west may develop shoots 6–9 cm long. Growth in subsequent years is determinate and gradually accelerates so that when saplings are 8–10 years old, terminal growth may consistently exceed 100 cm/yr on the more productive sites.

First-year seedlings survive and grow best under light shade, especially on southerly exposures, but older seedlings require full light, particularly in humid climates. Successful regeneration of var. *menziesii* often depends on weed control. Microsites with adverse moisture and temperature conditions frequently limit establishment of seedlings of both varieties on southerly aspects. Seedlings and saplings respond satisfactorily to release from competing brush or overstorey trees if they have not been suppressed too severely or for too long. Trees of pole and small sawtimber size generally respond very well to thinning. Application of a nitrogen fertilizer in combination with thinning gives better growth responses than either treatment alone.

When *P. menziesii* develops in a closed stand, the lower limbs die rapidly, as they are increasingly subjected to overhead shade. However, natural pruning is exceedingly slow because even small dead limbs resist decay and persist for a very long time. Usually *P. menziesii* is not clear to a height of 5 m until 77 years old, and not to 10 m until 107 years old.

Natural stands of var. *menziesii* are usually densely stocked, with more than 2500 trees/ha; planted stands generally have an initial stocking of between 750 and 1500/ha. The largest height increment is attained between 20 and 30 years old, but the ability to maintain a fairly rapid rate of height growth is retained over a long period. On a medium site at low elevations, height growth – which averages 61 cm annually at age 30 – continues at a rate of 15 cm/yr at age 100, and 9 cm at age 120.

Information about yields of var. *menziesii* under intensive management for an entire rotation in natural stands is still limited. It is therefore necessary to rely either on estimates based on yields from unmanaged stands, on yields from intensively managed stands in regions where *P. menziesii* has been introduced as an exotic, or on growth models. Depending on site quality, mean annual net increments at age 50 years vary from 3.7 to 13.4 m^3/ha in unmanaged stands. Managed stands of var. *menziesii* can produce mean annual increments of 7 m^3/ha on poor sites and more than 28 m^3/ha on the most productive sites under rotations of between 50 and 80 years.

P. menziesii var. *glauca* does not usually attain the growth rates, dimensions or age of var. *menziesii*, but is recommended for growing in mountainous Mediterranean zones.

Diseases and pests

P. menziesii is host to hundreds of fungi, but relatively few cause serious problems. *Pythium*, *Rhizoctonia*, *Phytophthora*, *Fusarium* and *Botrytis* species can cause significant seedling losses in nurseries, whereas *Rhizina undulata*, *Armillaria mellea* and *Inonotus weirii* can cause significant damage in plantations, the last two representing a serious threat to management of young stands, especially west of the Cascade Mountains. Trees die or are so weakened that they are blown over. Effective control measures are not available. Of the many heart rot fungi (over 300), the most damaging and widespread is *Phellinus pini*.

Needlecast caused by *Rhabdocline pseudotsugae* is mainly a disease of younger trees, reaching damaging proportions only after prolonged periods of rain while the new needles are appearing. Needlecast caused by *Phaeocryptopus gaeumannii* also occurs on *P. menziesii* throughout its natural range, but causes little damage there. However, it causes significant reductions in growth of trees planted in eastern USA, Europe and New Zealand.

The mistletoe *Arceuthobium douglasii* is a significant parasite throughout most of the natural range of *P. menziesii*.

Orgyia pseudotsugata and *Choristoneura fumiferana* are the most important destructive insect pests of *P. menziesii* in its native range. Both insects attack trees of all ages at periodic intervals, often resulting in severe defoliation of stands. The Douglas fir beetle (*Dendroctonus pseudotsugae*) is a noxious pest in old-growth stands.

The cones and seeds are host to a number of destructive pests, including *Megastigmus spermotrophus*, which matures in the developing seed and gives no external sign of its presence; *Barbara colfaxiana* and *Dioryctria abietivorella*, whose larvae bore through developing cones; and *Contarinia oregonensis* and *Contarinia washingtonensis*, which destroy some seeds and cause galls that prevent normal opening of cones.

Otiorhynchus ovatus and *Chrysoteuchia topiaria* can cause significant damage to seedlings in nurseries.

Uses

P. menziesii is used for erosion control, as a shelter belt species and as an amenity tree. In its native range it is grown as a Christmas tree on rotations ranging from 4 to 7 years. Trees are sheared each year to obtain a pyramid-shaped crown.

The wood is moderately heavy, hard and exceptionally strong, with a density of 480–540 kg/m^3 (at 12% moisture content). Heartwood is reddish-brown and sapwood yellowish-white, and there is a marked contrast between earlywood and latewood. The wood provides pulp and sawnwood, and is used for structural purposes, in shipbuilding and in the production of items such as laminated beams and interior and exterior finishing, boxes and railway ties.

Further reading

Dirks-Edmunds, J.C. (1999) *Not Just Trees: The Legacy of a Douglas fir Forest*. Washington State University, Pullman, Washington.

McDonald, P.M. and Fiddler, G.O. (1999) *Ecology and Development of Douglas fir Seedlings and Associated Plant Species in a Coast Range Plantation*. Pacific Southwest Research Station Research Paper PSW-RP-243. USDA - Forest Service, Albany, California, USA.

Prudham, W.S. (2005) *Knock on Wood: Nature as Commodity in Douglas fir Country*. Routledge, New York, New York, USA.

Pterocarpus angolensis DC.

Bloodwood

Importance

Pterocarpus angolensis (Plate 34) is common in woodlands and wooded grasslands across dry habitats in Southern Africa and produces one of the most valuable hardwoods in the region. The timber is preferred for flooring, furniture, veneers, panelling, musical instruments and utilitarian and decorative wood carvings. Due to its high value as a timber species, wild stocks of *P. angolensis* have declined in many regions through over-harvesting, and native woodlands in which it occurs have been cleared for farming. Nursery propagation, cultivation, and establishment of plantations have been attempted in Malawi, Mozambique, South Africa, Tanzania and Zambia, but these have not increased the natural range of the species to any extent.

Botanical features

P. angolensis grows to 12–15 m tall with diameters of 20–40 cm, and up to 25 m high with a diameter of 1 m on good sites. Trees are generally single-stemmed, unarmed and deciduous, with an umbrella-shaped crown usually wider than the height of the tree. Bark is dark brown to black, rough and corky, deeply fissured vertically and horizontally, standing out in uneven, rectangular blocks, and exuding a dark red gum if wounded. The compound leaves are large, 20–50 cm long and 7–12 cm wide, with 5–25 pairs of leaflets arranged alternately or subopposite. Leaflets are elliptic to obovate, 2.5–7.5 cm long and 2–4.5 cm wide, with a tapering, bristle-tipped apex and entire margins. The orange, pea-shaped flowers, 1.2–2 cm long, are gathered in large panicles, 10–20 cm long. Distinctively large and circular fruits, 5–15 cm diameter are covered in dense, coarse bristles up to 1.5 cm long and surrounded by a stiff, membranous, wavy wing up to 3 cm wide, glossy straw-yellow when dry and fading to dull beige with age. Fruit usually contain only a single seed, occasionally two, though 20–50% of fruits do not contain any seed. Old fruits are retained on the tree for several months, even after leaf fall.

Distribution

P. angolensis is native to Southern Africa, from Lake Victoria in Tanzania to Lake St Lucia in South Africa. Its core distribution is associated with miombo woodlands, but extends beyond the miombo at its fringes. It is common in Tanzania, Mozambique and South Africa along the Indian Ocean, but does not reach the Atlantic coast in the west, and is absent from drier southern regions of Botswana and Namibia.

Silviculture characteristics

P. angolensis is frequently a dominant species by virtue of its large size, average densities, and resistance to fire. Its coppicing ability is weak, though variable, and may decline with age. Seed viability and germination are variable. Survival and growth rates through artificial regeneration are poor, and so it may be a poor candidate for plantation forestry. It proliferates on old cultivated lands.

P. angolensis is found across a wide variety of climatic zones throughout Eastern and Southern Africa, ranging from sea level to approximately 1650 m altitude. It is strongly associated with seasonal, deciduous savannahs, woodlands and dry forests, largely subject to subtropical or tropical climates. Its southern distribution is probably defined by cold winter temperature and increasing aridity. It is reasonably drought-tolerant.

Densities of mature *P. angolensis* in the wild range are highly variable (1–20 trees/ha), reflecting soil characteristics and factors that inhibit recruitment such as frequent fires or anthropogenic impacts. Typically, over 65% of the wild population exist as shrubs. In some regions people use truncheons to establish fences. In Tanzania, *P. angolensis* has been observed to dominate in abandoned fields, indicating a strong preference for disturbed sites, reduced competition and high light.

Two factors in the management of *P. angolensis* requiring further research is why such large proportions of wild populations exist as shrubs, and what cues are required to enter a period of rapid vertical growth as a sapling. A number of cues have been speculated, but none adequately substantiated, including fire, browsing, above-average rainfall, protection from fire or browsing, release from plant competition, or adequate below-ground biomass and tapping into the water table.

The species is strongly deciduous. Leaf flush in spring/summer is coincidental with the arrival of the rainy season, thus, leaf flush may be delayed if the onset of the rains is late. Fruits may remain on the tree for most of the year. Leaf fall coincides with the end of the rainy season, and is usually rapid, being completed within a few weeks.

Silviculture practice

P. angolensis can be propagated from seed or via truncheons. Seed storage is orthodox, but viability decreases with long-term storage. Seedling growth is typically slow, and the entire above-ground biomass usually dies back in the winter which may continue for several years in the nursery or field. Nursery plants benefit from inoculation with vesicular-arbuscular mycorrhizae.

Cut poles and truncheons of *P. angolensis* frequently produce shoots, which suggests it would be a good candidate for vegetative propagation. However, success is often poor (<30%) because they fail to generate roots or only produce a small root biomass; 50–80% of field plantings typically die within the first 2 years, and surviving plants often remain small with annual winter mortalities for 10 years or more. Similarly, mortality of truncheons planted out directly is frequently over 80%. Keeping the area weed free results in faster growth.

Trees may start fruiting when 8 cm in diameter, corresponding to a tree age of 26–39 years, with all trees fruiting when above 25 cm in stem diameter, corresponding to 80–92 years old. Early annual diameter growth can be 1–1.5 cm/yr, declining with age, with an average annual diameter increment of 0.5 cm. Given the relatively slow growth of *P. angolensis*, harvest rotations would have to be long, probably greater than 100 years. The minimum size for harvesting of 35 cm diameter, commonly adopted by several countries in which *P. angolensis* grows, is well above the size of reproductive maturity. Hence, if the regulation cutting size is enforced, there should be continued recruitment via seed. Seed production is positively related to tree size in a linear fashion.

The timber is highly valued by commercial and subsistence users, and the species has been heavily exploited throughout much of its range. It is protected in most countries in which it occurs, and thus cannot be felled without a permit and meeting of the minimum exploitable stem diameter. However, enforcement seems weak. This is compounded by the absence of recruitment at many sites and so there is a narrowing size class profile. However, at other sites recruitment seems vigorous, thus portraying highly episodic and spatially variable recruitment.

Diseases and pests

A serious decline leading to dieback and death of *P. angolensis* has caused concern in Southern Africa for many years, and is thought to involve *Fusarium oxysporum* as the causal organism. In addition, *Pseudofusicoccum* spp., *Diplodia* spp., *Fusicoccum* spp. and *Lasiodiplodia* spp. may also be implicated. After bark is harvested for medicinal purposes, the cut area generally heals, but is susceptible for a time to infection by pathogens, and subsequently progressive damage through repeated fires.

Uses

Being a tall tree, *P. angolensis* is suited to silvopastoral systems, providing shade and browse.

The wood of *P. angolensis* is medium to coarse grained, with an attractive colour greatly admired by furniture-makers and wood carvers. It is also one of the most stable timbers in the world. The heartwood is generally light to dark brown in colour, heavily streaked, often with a reddish tinge. This is starkly contrasted to the sapwood that is pale yellow or beige. Air-dried density is 400–700 kg/m^3, but is reported to be denser from South African populations than from those further north. The heartwood is resistant to termites and wood-rotting fungi, whereas the sapwood is susceptible to borers. It seasons well and rarely warps; drying is slow but with minimal shrinkage, and relatively little tendency to split. The timber is flexible and relatively light, making it easy to work, glue and turn. It polishes and paints well, and has good retention of nails and screws.

The timber is by far the major and most valuable product of *P. angolensis* and non-wood uses are in comparison relatively negligible. The bark of *P. angolensis* has medicinal uses, and extracts are used by a number of African groups to treat ailments such as headaches, chest pains, stomach ache, ulcers and burns. Only a few of these have been formally tested, with promising results pertaining to anti-inflammatory and anti-schistomomal properties. When injured the tree exudes a dark red sap that has been used as a dye for cloth, hair and facial or body paint. Both the bark and the sap can act as a fish poison. The distinctive and attractive fruits may be used as decorations, either natural or painted. It is also a useful species for beekeeping.

Further reading

Amri, E. and Mamboya, F. (2012) Genetic diversity in *Pterocarpus angolensis* populations detected by random amplified polymorphic DNA markers. *International Journal of Plant Breeding and Genetics* 6, 105–114.

Therrell, M.D., Stahle, D.W., Mukelabai, M.M. and Shugart, H.H. (2007) Age, and radial growth dynamics of *Pterocarpus angolensis* in southern Africa. *Forest Ecology and Management* 244, 24–31.

Vermeulen, W.J. (1990) *A Monograph on* Pterocarpus angolensis. Southern African Regional Commission for the Conservation and Utilisation of the Soil (SARCCUS), Pretoria, South Africa, 133 pp.

Pterocarpus erinaceus Poir.

African rosewood

Importance

Pterocarpus erinaceus is a tree of the Sahel and some neighbouring areas, yielding a very high-quality rose-red-brown furniture wood also used for carving and musical instruments, especially xylophones. It is also a valuable fodder tree, and the resin is used to make a dark purple dye and contains kinotannic acid (kino), a strong astringent. Extracts exhibit anti-malarial, anti-oestrogen, and wound-healing activity. High demand for the high-quality fodder for livestock keeping has led to a shortage of supply in some areas. The species has high potential as a multipurpose tree, but is limited by the need for 4–5 years protection from browsing.

Botanical features

P. erinaceus is a small- to medium-sized tree, 12–15 m tall at maturity, rarely to 35 m and with stem diameters sometimes in excess of 1 m, with an open spreading crown giving little shade, and often with low, pendulous branches. Under good conditions the bole can be straight and clean up to 6–8 m in length, but in drier areas multiple stems are typical and the bole becomes fluted with buttresses. The bark is dark grey to black, rough, scaly with scale edges turned up, exuding a red translucent gum when damaged, hardening rapidly. Branches are light and smooth and the tree is unarmed.

Leaves imparipinnate, 30 cm long with 10–15 alternate or subopposite leaflets, 6–11 cm long and 3–6 cm wide, elliptic to oblong, shortly pubescent below with many parallel lateral nerves; tip shortly acuminate or retuse. Masses of golden-yellow flowers cover the canopy with a very attractive appearance from December to February when the tree is leafless and most other trees are not in flower. Fruits are rounded or saucer shaped, indehiscent, pale green when young, looking like the leaves, becoming brown at maturity, papery, up to 7.5 cm in diameter, persistent on the tree, with prickly hairs, surrounded by membrane-like wings. Pods contain 1–3 oblong or somewhat curved seeds.

Distribution

P. erinaceus is native to open dry forests of the Sahel throughout West and Central Africa from Senegal to Chad and south to the Central African Republic and Gabon. The northern limit of its distribution is approximately 14°N, where it is a small stunted tree limited by rainfall. Southward it extends to the edges of the West African tropical humid forest. In drier areas to the north it is replaced by *Pterocarpus lucens*, and in wetter areas to the south it is replaced by *Pterocarpus santalinoides*. Along with *Parkia biglobosa*, it is thought to be a surviving species of the former, dense, dry forest of the Sudanian zone. It is not thought to have been introduced outside of its native range.

Silviculture characteristics

P. erinaceus is a deeply rooted nitrogen-fixing species and regenerates well in fallow, regenerating easily from seed, and it suckers vigorously. It grows in open mixed dry woodland in farmland, but is slow growing. *P. erinaceus* occurs in semi-arid and subhumid low altitude dry woodlands of the tropical West African Sahel, with annual rainfall of 600–1200 mm and a very long dry season up to 8–9 months. Once established, it is tolerant of drought, and survives annual bush fires. Mean annual temperature in the native range is 15–32°C, but *P. erinaceus* will also tolerate temperatures above 40°C. It prefers shallow, gravelly, well-drained soils.

Silviculture practice

P. erinaceus produces large quantities of seed and the fruit must be cracked open to release these. There are about 3500 unshelled seeds/kg and 19,800 shelled seeds/kg. Seeds require pretreatment to germinate, the preferred method being soaking in water for 18–24 h. Germination then occurs within 6–10 days. Seedlings can be grown in pots or nursery beds and develop a deep taproot. Planting out can be done after several months, either as potted plants or bare-root seedlings. The survival rate of planted seedlings can be high, but they are slow growing and need to be protected from grazing for 4–5 years after planting.

P. erinaceus is cut back every year by farmers in the Sahelian parklands, lopped to provide dry season browse by pastoralists and is also collected for sale in local markets and urban areas as livestock fodder. It is generally over-exploited for fodder during the dry season. Annual height increment has been recorded as 11–15 cm in the first year and 25–42 cm in the second year. Plantations have been established in Ghana and Senegal. The recommended spacing is 5 × 5 m for woodlots and 1 × 2 m for fodder, although the management in woodlots and intensively coppiced fodder banks is not well documented. *P. erinaceus* coppices well when cut at a height of 50 cm at the end of the rainy season and remains green in the dry season, but trees do not coppice well when cut at ground level, and results indicated that a 1 ha fodder bank could feed a herd of 24 tropical livestock units (equivalent to 1000 kg animal live weight).

Diseases and pests

There are very few records of any pests and diseases attacking *P. erinaceus*. However, a tar spot disease caused by *Phyllachora pterocarpi* has been reported though the severity of any impacts is not known.

Uses

P. erinaceus is an important agroforestry tree, particularly for provision of fodder. It is suitable for soil conservation, erosion control, soil improvement and re-vegetation.

The wood is highly attractive, fine grained, dark brown to rich rose in colour, with some dark mottling. It is very hard, termite resistant, durable and easy to work, and annual growth rings are clearly visible. Density is 640–770 kg/m^3. It is known as African kino or African rosewood (but should not be confused with other 'rosewoods' obtained from several species of *Dalbergia*). It is used for furniture, interior building work such as door and window frames, decorative panelling, parquet flooring, musical instruments such as the keys for the traditional xylophones or carved into masks. It has many other uses including posts and poles, fencing, household items and agricultural implements. The wood is also an excellent fuel with a high calorific value (5000 kcal/kg), and is used to make very high-quality charcoal which is highly valued by blacksmiths.

P. erinaceus is an important fodder tree; its dry leaves which fall to the ground are browsed by livestock. Large quantities of leaves are harvested near villages for livestock, particularly when these are tied up during crop-growing seasons. However, supply of the sought-after fodder falls far short of demand. A study in Mali found that over 1400 tonnes of fresh *P. erinaceus* leaves were sold annually to feed urban livestock in the capital city Bamako, when 7500–8000 tonnes was thought to be required. Due to over-harvesting, it was increasing difficult to find the tree in the periphery of the city where it used to be readily available, and collectors were forced to travel up to 50 km to find trees that still had leaves to harvest. The feed value of the leaves is very good, containing approximately 15% (10–19%) crude protein, an energy value of 5.2 MJ/kg dry matter, and 0.15% (0.08–0.27%) phosphorus.

The gum dries to a red resin commercially known as African kino that contains kinotannic acid (kino), a strong astringent. This is pulverized and used to satinize cloth, and mixed with shea butter (from the fruit of *Vittelaria paradoxa*) to make a dark purple dye. Dyes are also made from roots and stemwood and the bark contains tannin. The leaves, bark and roots are also widely used medicinally, and plant extracts also have anti-malarial, anti-oestrogen and wound-healing activity. The leaves are used against fever, the bark against tooth and mouth complaints and the bark and resin as an astringent against dysentery.

During December to February the tree is covered in very attractive yellow flowers, making it suitable as an ornamental.

Further reading

Adjonou, K., Ali, N., Kokutse, A.D., Novigno, S.K. and Kokou, K. (2010) A study of the dynamics of overexploited natural stands of *Pterocarpus erinaceus* Poir. (Fabaceae) in Togo. *Bois et Forêts des Tropiques* 306, 45–56.

Nacoulmak, B.M.I., Traoré, S., Hahn, K. and Thiombiano, A. (2011) Impact of land use types on population structure and extent of bark and foliage harvest of *Afzelia africana* and *Pterocarpus erinaceus* in Eastern Burkina Faso. *International Journal of Biodiversity and Conservation* 3, 62–72.

von Maydell, H.J. (1986) *Trees and Shrubs of the Sahel: Characteristics and Uses*. Schriftenreihe der GTZ, No. 147. GTZ, Eschborn, Germany, 525 pp.

Pterocarpus indicus Willd.

Redwood

Importance

Pterocarpus indicus produces a high-quality timber valued for fine furniture, decorative flooring, cabinet work, and musical instruments. It is a commercial timber species in its native range from South-east Asia to the Pacific, especially in Indonesia, Papua New Guinea, the Philippines and Thailand, but it has been over-exploited in some regions. It exudes a gum or resin which is a powerful astringent; the boiled, shredded bark is used for tanning as well as for medicinal preparations. It can be planted as a species for agroforestry, reforestation, and as an ornamental and avenue tree, but it is relatively slow-growing, which also discourages planting on a commercial scale.

Botanical features

Usually a deciduous tree growing up to 30 m tall, with an irregular fluted trunk up to 2 m in diameter, usually with a short bole, high buttresses which may be board-shaped, and a wide spreading crown, dome-like in open areas, sometimes with lower branches drooping and touching the ground. Leaves are alternate, imparipinnate, with 6–9(–11) alternate leaflets ovate to oblong–ovate, blunt acuminate, shiny and 5–12 cm long and 3–7 cm wide with entire margins. The terminal leaflet is usually largest, with the smallest leaflets being the lowest pair on the rachis. The flowers are numerous, bright deep yellow with an orange tinge, borne in axilliary racemes 5–7 cm long, flowers about 1.5 cm long, mostly single, very showy and fragrant. The glabrous, indehiscent pod comprises a smooth orbicular fruit, about 6 cm in diameter, with or without spines, and surrounded by a broad undulating wing. It contains 4–5 seed chambers with generally only 1–2 (rarely 3) seed.

Distribution

P. indicus is native to East and South-east Asia to southern China, Indonesia and western Pacific Islands. It is particularly important as a commercial timber species in the Philippines, Papua New Guinea and Thailand. *P. indicus* has also been widely introduced to Central America, the Caribbean islands, Florida, northern South America, East and West Africa, South Asia, and elsewhere in the Pacific.

Silviculture characteristics

P. indicus prefers climates with an annual rainfall of 1300–4000 mm and no pronounced dry season. Although it can adapt to sites with a dry season up to 6 months long, the timber quality will be reduced due to stem forking and it will adopt a deciduous habit. It is light-demanding, and in natural conditions is often found as a dominant upper storey species. It is found in evergreen and semi-deciduous forest, river banks, tidal creeks, and rocky shores, immediately behind mangrove swamps, and occasionally on drier slopes, but is also highly adaptable to other conditions. *P. indicus* thrives best in moist, sandy loam or clay loam soil, along gullies and stream banks up to 1300 m, but more commonly in seasonal forests to 600 m. It is expected to be severely damaged or killed by low temperatures of less than 5–8°C.

P. indicus can be planted in partly shaded or open areas. It has the ability to regenerate rapidly from seed and is nitrogen-fixing. It is windfirm with a surface lateral root system, but foliage is damaged by salt-laden winds. Being thin-barked, it is also susceptible to fire injury, but recuperates easily when damaged. It has good coppicing and pollarding ability and can tolerate some drought. Time to first fruiting is 5–7 years, or only 2–3 years if plants are raised from cuttings.

Silviculture practice

The most common technique is to use bare-rooted planting stock, but stump planting, direct sowing and cuttings are also used. *P. indicus* seeds are orthodox. Seed pods are collected where fallen and can be stored in containers for 1 year or more, with approximately 1200–1300 seeds/kg. Germination is improved if seeds are extracted from the pods before sowing, but seeds are usually sown in pods into prepared nursery beds or in individual plastic bags, with germination rates of 24–40% after 4–15 days. For outplanting as bare-rooted plants, seedlings are transplanted to a seedbed at 5 × 20 cm spacing for 3–4 months. Partial shade should be provided, and the seedbed mulched. The desirable height of planting stock is 75–100 cm if bare-rooted and 22–30 cm if potted seedlings. If direct sowing is desired, 2–4 pods should be placed in each planting hole.

P. indicus can be vegetatively propagated with relative ease using a range of techniques, often used as the germination rate is relatively low and seedling survival is vulnerable to weed competition. Large shoot cuttings are prepared from branches 1–2 m long, 2–2.5 cm (up to 6 cm) in diameter with all leaves and side branches removed, with reported establishment rates of 72–85%. Stump plants taken from seedlings or wildings are also used, with recommended stem length of 10–20 cm, root length of 20–40 cm, and stem diameter of 1.5–2 cm.

Areas to be planted should be cleaned, with an initial spacing of 1 × 1 m to encourage the development of a long, straight stem, with gradual thinning to about 2.5 × 2.5 m or 3 × 3 m to increase diameter growth, and thinned individual plants may be transplanted elsewhere. In the Pacific, low initial densities of about 400 trees/ha or a spacing of 8 × 3 m is recommended, with one or more selective thinning to a final density of 100–150 trees/ha. Regular weeding around trees for 3–4 years after planting will improve the survival rate and growth.

Little information is available about management of *P. indicus* in natural stands. The trees often occur scattered in dipterocarp forests, where harvesting is governed by diameter limits (usually 60 cm). However, as *P. indicus* wood is in great demand, trees less than this size are often cut illegally, particularly in the Philippines.

On fairly good sites, height growth may reach 2 m/yr in the first 3–4 years, before slowing to about 1 m/yr or less. Average dbh of seedlings is 15 cm at 12 years after planting, and 24 cm diameter after 17 years, with the tallest trees being 34 cm and 40 cm diameter, respectively. However, highest diameter growth of 2.5 cm (up to 4 cm)/yr is reported to occur with trees raised from large stem cuttings. The volume yield of a sawlog from a 30-year-old *P. indicus* tree is approximately 8 m^3 under favourable conditions, but is often limited by the trees having only short boles and large, spreading branches.

Diseases and pests

Angsana wilt disease caused by *Fusarium oxysporum* was first reported from Malacca in 1870 and spread to South-east Asia in the following decades, causing some significant tree mortality. Typical symptoms are yellowing and dropping of leaves, followed by infestation by pinhole borers and eventual tree death. Studies in Singapore indicated that lightning damage frequently appears to provide the stress that attracts ambrosia beetles (*Platypus* spp.), and if they are carrying *F. oxysporum* spores, infection is likely to follow. Secondary spread is by *F. oxysporum* which has entered the soil from the infected tree. Tar spot disease caused by *Phyllachora pterocarpi*, leaf spot disease caused by *Aldona stella-nigra*, attack from a leafminer (*Hyloconis* sp.), and galling and dieback caused by the psyllid *Padaukia quadrimaculata* have also been reported.

Uses

P. indicus is commonly planted on roadsides, in gardens and along field boundaries, as a live fence, windbreak and as a shade tree for crops such as coffee. As a nitrogen-fixing tree it can help improve the soil, and is recommended for use in improved fallows, reforestation of denuded areas and for stabilization of stream banks. As a multipurpose tree, it has been recommended for use in agroforestry systems and on small farms, but has limited potential for interplanting with crops due to its large size and heavy shade. It is one of the most promising multipurpose tree species in the Pacific for reforestation, woodlots, live fencing and amenity planting.

The wood of *P. indicus* has beautiful figure and is very attractive. It is of medium weight, moderately hard to hard, tough and resilient. Heartwood is light yellowish-brown to reddish-brown with dark irregular streaks, distinct from the generally pale straw-coloured to light grey sapwood. Wood density is 550–900 kg/m^3 at 15% moisture content, although due to the wide variation in environmental conditions under which *P. indicus* is found, wood properties can also vary widely. Heartwood is usually resistant to termites, but sapwood is susceptible to powder-post beetles. It is very durable when used for interior work in the tropics and subtropics.

P. indicus wood has been widely utilized for furniture, including table-tops using buttress wood. Physiologically diseased trees yield a famous strongly figured wood known as 'Ambonese gnarl wood' or 'amboyna' in the Philippines. Other high-value uses include cabinet work, veneers, radio cabinets, piano cases, interior finishes and musical instruments.

The heartwood is used locally as a red dye and the bark has tanning properties. The very young leaves and flowers are sometimes eaten. Various parts of the tree have medicinal properties, including root juices used to treat syphilitic sores; young leaves applied to boils, ulcers and prickly heat; and gum used in Malay medicine for diarrhoea and dysentery.

Further reading

Ang, L.H. (1988) A note on the growth of *Pterocarpus indicus* in a sixty-year-old plantation in Malaysia. *Journal of Tropical Forest Science* 1, 188–189.

Soerianegara, I. and Lemmens, R.H.M.J. (eds) (1993) *Plant Resources of South-East Asia No. 5(1). Timber Trees: Major Commercial Timbers.* PROSEA Foundation, Bogor, Indonesia, 610 pp.

Thomson, L.A.J. (2006) *Pterocarpus indicus* (narra), ver. 2.1. In: Elevitch, C.R. (ed.) *Species Profiles for Pacific Island Agroforestry.* Permanent Agriculture Resources (PAR), Hōlualoa, Hawaii, USA, 17 pp.

Q

Quassia simarouba L. f.

Mountain damson

Importance

Quassia simarouba is a multipurpose tree native to Central and South America and the Antilles. It is one of the whitest woods known, and commercially exploited in its native range; it is useful for making fine boxes for documents and herbariums. There is increasing interest, particularly in India, in growing *Q. simarouba* in plantations as a biofuel crop, because of its oily seeds. Extracts from *Q. simarouba* have good insecticidal properties, and it is also a traditional medicinal plant.

Botanical features

Q. simarouba is a large dioecious evergreen tree, 28–30 m (up to 35 m) in height and 50–90 cm (up to 125 cm) in diameter, with a large crown. The trunk is regular, straight and cylindrical. The bark is 1 cm but more usually 2 cm thick. The external bark is light grey, and furrowed with longitudinal fissures.

Leaves are compound, alternate, 30–40 cm in length, odd in number or almost imparipinnate, with 3–6 leaflet pairs (sometimes 9) on short petioles 0.3–0.5 cm long, not always opposite, oblong, with a defused base and apex usually obtuse, 5–10 cm in length by 3–3.5 cm in width, semi-coriaceous. The upper leaf surface is shiny green, the underside pale, light green; whole margin deflexed, with parallel secondary veins. Petiole long, 6–10 cm. New foliage is red.

Inflorescence in the form of terminal panicles with dense flowers, multi-branched, 20–30 cm in length, rarely surpassing the leaves. Flowers, white or yellow, small, 6–8 mm in diameter, clustered; 5 petals and 5 sepals with a prolonged calyx. Rudimentary gynoecium in male flowers, hairy staminodes in female flowers. Male flowers with 10 stamens, 3 mm long. Ovary 5-lobed with 5 carpels and 5 ovules, and a 5-lobed stigma. Fruit an egg-shaped drupe formed by 3–5 capsules, larger at the base, glabrous, 1 cm long by 4–5 mm wide, mature fruits dark in colour.

Being dioecious, *Q. simarouba* has separate male and female trees, with females producing an abundance of small fruits. Flowering generally occurs between August and September in the southern hemisphere and between March and April in the northern hemisphere. Trees start to produce flowers and fruits after 5 years. Pollination is effected by bees. Fruits are ripe between November and December in the southern hemisphere. Seed dispersal is primarily by vertebrates (birds, monkeys, fruit bats), but leaf-cutter ants (*Atta colombica*) also play a minor role. Trees are relatively short-lived, up to an estimated maximum of around 120 years.

Distribution

Q. simarouba is indigenous to the Amazon rainforest of South America and other tropical areas in Central America and the Caribbean. It is generally quite rare throughout Central America, the Lesser Antilles and the Guianas, Brazil, Colombia and Venezuela in South America. Some authors state the species is also native to southern Florida, USA.

Silviculture characteristics

Q. simarouba establishes well, grows rapidly, and tolerates weed competition. The species grows in full sunlight or partial shade on almost any well-drained soil. *Q. simarouba* will grow quickly on rich soils high in organic matter and should be protected from frost in southern Florida, USA, where large trees are reported to be difficult to establish from containers. It is a good species for carbon sequestration and it also has the potential to be grown in plantations on degraded land, including areas degraded by gold mining and wastelands.

Silviculture practice

Propagation is by seed, which germinates easily and rapidly; plantlets have also been produced via somatic embryogenesis. Young plants and seedlings are easily transplanted. A single tree can produce 10–20 kg of fruit every year, with 800–1600 seeds/kg. Collection of dark-coloured mature fruits should be done from the tree or from the ground. Seeds are extracted by pressing the fruits after they have been in the sun for 1–2 days, and seeds should be washed and sun-dried for 1 day. Seeds stored for 3 months show significant losses in viability. Seed pretreatments are not necessary with fresh seeds but they may be soaked in water for 48 h. After sowing, germination takes 7–21 days. When planting seedlings in nurseries, the recommended substrate is 50% black soil, 40% organic material, and 10% sand. Seedlings remain in the nursery for 12 months before reaching the ideal size for planting.

The recommended size for planting holes is 40 × 40 × 40 cm. For timber production, the spacing between trees should be 3 × 3 m and for foliage production, 6 × 6 m. Rotations of 20–25 years have been suggested on good, deep, well-drained sites to obtain trees taller than 25 m. In some plantations in Trinidad, trees reached 25 m in height after 11 years. In Suriname, plantations totalling 150 ha established since 1954 had few problems with a 40-year rotation period.

The species needs direct sunlight for best development, and

is suitable for enrichment planting in gaps created by logging. Trees require pruning early in the rotation, as no natural pruning occurs; response to thinning is slow in natural forests.

Diseases and pests

The defoliator *Atteva pustuella* seems to be the only major pest recorded in *Q. simarouba* plantations. Control can be achieved by adequate management and pruning. *Onchoscelis germari* (Coleoptera, Curculionidae) has been reported attacking *Q. simarouba* in Mato Grosso, Brazil. Laboratory tests showed *Q. simarouba* wood to be susceptible to attack by the fungi *Lenzites trabea* (*Gloeophyllum trabeum*), *Pycnoporus sanguineus* and *Polyporus fumosus* (*Bjerkandera fumosa*), and also to attack by some termites (*Nasutitermes* spp.). Pure culture tests indicate the wood to be somewhat durable with regard to attack by white-rot and brown-rot fungi; however, actual graveyard evaluations show the wood to be readily attacked by decay fungi and insects. The wood is also prone to bluestain.

Uses

In agroforestry, *Q. simarouba* is used as the first layer in the system. It creates conditions for other species to develop due to shading and litter deposition, and is grown as a shade tree for coffee plantations and in pastures. *Q. simarouba* has potential for soil conservation in degraded areas. The flowers are very attractive to pollinating insects such as bees and flies, which makes the species very desirable for pollen and honey production.

The wood is used for interior construction and furniture, and is suitable for the production of pulpwood, paper, flake-type particleboard and matches. Other uses include carpentry, cabinetmaking and finishing, boxes, crates, veneers, plywood and musical instruments. It is a soft, yellow-white, light timber, with a density of 500 kg/m^3 (at 12% moisture content). The wood has insecticidal properties and is therefore used for boxes and reinforced boxes for storing precious documents and herbarium samples, to protect material from insect attack over many years. It seasons easily, drying out quickly, practically without defects.

Aceituno oil is extracted from the seeds of *Q. simarouba* and contains palmitic, stearic and oleic acids, with a low level of linoleic acid. This seed oil can be refined to produce oil suitable for human consumption and for biodiesel. The fruits and leaves are consumed by cattle, although the foliage is considered a poor animal fodder due to low nitrogen and high fibre content.

Q. simarouba has a long history in herbal medicine in many countries. Roots are used in the treatment of bloody diarrhoea. Bark and leaves are recommended for dyspepsia and fever treatments. The bark and, occasionally the roots, are boiled in water to yield a powerful astringent and tonic used to wash skin sores and to treat many different diseases. Leaf extracts show antibacterial activity against food-borne and spoilage microorganisms, and triterpene quassinoids are the main active group of phytochemicals. Some of them, such as ailanthinone, glaucarubinone and holacanthone, are considered the plant's main therapeutic constituents and are documented to have antiprotozoal, antiamoebic and anti-malarial effects, and are even regarded as toxic to cancer and leukaemia cells. *Q. simarouba* extracts have also shown activity against plant pathogens and pests. The sawdust is used in substrates for the cultivation of edible mushrooms.

Further reading

Gilman, E.F. and Watson, D.G (1993) *Simarouba glauca*: Paradise-Tree. Fact Sheet ENH-748. Institute of Food and Agricultural Sciences, University of Florida, Gainesville, Florida, USA, 3 pp.

Pradhan, K. (2004) Promising prosperity: the oil yielding *Simarouba glauca*. *Wastelands News* 19, 54–56.

***Quercus cerris* L.**

Turkey oak

Importance

Quercus cerris is indigenous to Southern Europe and Western Asia. It occurs on a wide range of soils, and has greater drought resistance than the other oak species of the region. A sustained growth rate under suitable site conditions can provide high stocking levels for valuable fuelwood production. In the eastern part of its range, where wood quality exhibits its best features, *Q. cerris* can be managed as high forest for timber production.

Its high regeneration ability and pioneer characteristics make *Q. cerris* a suitable species for use in the afforestation of degraded areas and in agroforestry systems. It is also a valued ornamental tree.

Botanical features

Q. cerris is a large, broadleaved deciduous, and at 120–150 years is a fairly long-lived tree. When grown in dense, high-forest structures, the crown is oval-shaped, high-inserted, and medium-textured. When open-grown, it develops a wide, multi-branched, and dome-shaped crown, typical of isolated oak trees. When mature it can attain a height of 30–35 m and a dbh of 60–90 cm.

The stem form of a stand-grown tree is usually straight and does not fork below the crown base; apical dominance is evident in mature trees. The root system consists of a taproot and deep lateral root branches. Bark is smooth and grey in young trees, thick and deeply fissured with a vertical, cross-linked rib structure in adult trees.

Leaves are simple, alternate and extremely variable both in shape and size at individual tree level: length 5–10(15–18) cm, width 3–5(7–8) cm; usually oblong to obovate, rounded or subcordate at the base, the leaf blade being larger at the median length.

Male catkins are 5–8 cm long, slender, and pendent; flowers are isolated, small, sessile or subsessile. Female spikes show 1–5 flowers on a tomentose, cylindrical axis. The fruit-bearing axis is short and normally bears one fruit (an acorn). The cupule, covering one-third to two-thirds of the acorn, has ash-coloured tomentose scales, curved or uncinate at the tip. Acorns vary from 20 to 35(–40–50) mm in length and from 12 to 18(–22) mm in diameter.

Foliage opens in April–May, leaf fall is late and gradual in the autumn; depending on local or seasonal climatic conditions it may continue into the following winter. Flowering occurs in May–June. Both bud burst and flowering are relatively late. Pollination is anemophilous. Acorns ripen during the second year (August–September), about 14 months after pollination. *Q. cerris* reproduces both sexually and agamically.

Distribution

Q. cerris is found in Southern and Eastern Europe. Its present distribution may be considered to be both natural and naturalized, as this species was introduced into many areas in the northern part of its geographical range, particularly during the 19th century, as a fuelwood species. It is indigenous to most of Italy. The western limit of its distribution range is probably France. The northern distribution limit is Germany, continuing through Austria and Switzerland, eastern Czech Republic, Slovakia and Hungary.

Q. cerris is also naturally distributed in Romania, Bulgaria, and eastwards to the Crimea, Anatolia, Syria and Lebanon. It is found throughout the Balkan peninsula.

Q. cerris has been introduced to a limited number of countries outside its natural range including Ukraine, Romania, the UK, Argentina and New Zealand.

Silviculture characteristics

Q. cerris is a fairly drought-resistant, light-demanding species. It can withstand very low air temperatures in winter but it is susceptible to frost damage. It is resistant to windthrow and snow loading.

It is not very resistant to fire, but is able to regenerate rapidly from the stump following cutting. Rapid seedling growth enables *Q. cerris* to compete with herbaceous weeds, although excessive development of grass and shrub weed competition will hamper natural regeneration. *Q. cerris* is one of the tree species best suited to coppicing, because of its capacity to resprout, high shoot growth rate, and stump tolerance of repeated cutting cycles. Coppice shoots originate from both dormant and adventitious buds. *Q. cerris* will not usually regenerate from root suckers. When grown in sufficiently dense stockings, trees will maintain a highly inserted crown form, and self-pruning will remove lateral branches.

Epicormic branches can develop following overcutting, or may be due to weakening of the tree because of attacks by pathogens or pests. It is suitable for pollarding, a practice which was especially linked, in the past, to the provision of fodder for domestic animals.

Silviculture practice

Seed production is optimal between the ages of 40 and 120 years. Natural dissemination occurs in October. The interval between mast years (high acorn production) is 3 or 4 years. Seeds have a high moisture content (about 40%) and a good germinative ability (more than 80%). Seeds are recalcitrant and germination is hypogeal.

In forest stand establishment, both high-forest and simple coppice or coppice with standards systems are widely practised.

In high-forest regeneration, the shelterwood system (uniform or strip cuttings) is usually used; in mixed-oak forests group cutting is also applied. Control of shrubs and weeds is necessary when undergrowth is present, before the seed fall; cleaning the forest floor and harrowing the soil may promote seedling establishment. Additional broadcast sowing or localized direct sowing may also be used to supplement natural regeneration where needed.

The period for regeneration to establish is short, as *Q. cerris* is a light-demanding species and seedlings have rapid early growth. The shelterwood removal cutting occurs between 5 and 7 years later, when half of the trees are removed. The final cutting concludes the regeneration period after an equal interval of years. Combined group and selection fellings are also carried out in mixed-oak forests where both productive and aesthetic/protective functions are required.

Silvicultural treatments throughout the shelterwood cultivation cycle up to the regeneration period should keep the stand not too dense (moderate to heavy thinnings), but the permanent opening of canopy should be avoided.

In mixed forest where other, more valuable oak timber species (*Quercus robur*, *Quercus petraea*, *Quercus frainetto*) are present, intermediate cuttings can be used early to promote the growth conditions of individual trees selected as final crop trees.

The establishment of new *Q. cerris* stands is usually carried out by direct sowing, broadcast sowing, or group sowing of acorns. Soil tillage is necessary to clear the ground of natural vegetation, reduce weed competition, and prepare a suitable rooting environment for oak seedlings. The average amount of acorns used for sowing ranges from 200 to 800 kg/ha depending on the sowing method used.

Use of planting stock is more expensive than direct sowing. Seedling size should be at least 50 cm to overcome early weed competition. Density varies from 1000 to 2000 trees/ha, depending on site quality, forest management objectives, and the silvicultural system to be applied.

In terms of coppicing, *Q. cerris* produces a good quality fuelwood with high calorific value. All types of coppice were traditionally practised: simple coppice, coppice with standards, and coppice with standards of multiple age classes.

The ability of *Q. cerris* stumps to resprout after several coppice cuttings, and the rapid early growth of shoots, make it possible to manage simple coppices with periodic clear-cuts. The number of *Q. cerris* seed trees (standard trees) in a coppice forest should be low (40–50 standards/ha, with a stool density of 2000–3000/ha). Only high-quality standard trees should be released, originating from seed rather than from coppice shoots.

Different techniques have been used to undertake coppice conversion to high forest. Coppice stand structural changes are implemented by means of thinnings repeated every 10–15 years, starting with 20-year-old coppice stands (i.e. a few years older than the traditional oak coppice rotation).

The first two thinnings drastically modify the coppice structure into a physiognomy similar to that of a high forest, releasing selected shoots on each coppice stool, at a tree density ranging from 1600 to 2400 stems/ha. The transitory crops originating from these operations exhibit sustained incremental growth trends, up to at least age 50–60 years.

Another classical method converts coppice into high forest by progressively releasing a number of standards at each consecutive coppice clear-cutting until the time for a seed felling is reached. Under this method, coppice harvesting may be continued throughout the coppice conversion period.

Coppice regeneration in degraded stands or sites can be improved by cutting coppice stools to below ground level, and covering the cut surface with a layer of soil. More vigorous and healthy coppice shoots can be achieved this way.

In high forest management, rotations vary from 80 to 120 years in pure and mixed stands throughout the range of *Q. cerris*. Standard yields based on southern Italian data range from 445 to 657 m^3/ha at age 120 according to site class and tree density (731–241 stems/ha), with average dbh varying from 25 to 41 cm.

In coppice forests of *Q. cerris*, yields of between 100 and 150 m^3/ha at age 15 years in hilly areas of central Italy are used as reference values on average site classes in the Mediterranean region.

Diseases and pests

Many pathogens or pests exhibit recurrent outbreaks, which can seriously damage infested stands. In spite of this, *Q. cerris* trees are usually able to overcome such periodic attacks. In this sense, pathogens and pests may be considered as mere ecological factors controlling the forest ecosystem.

Oak decline is a complex disease that involves abiotic (water and/or temperature stress) as well as biotic agents. Insect infestations are associated with oak decline (*Lymantria*, *Agrilus* sp.), but many fungal species, whether or not pathogenic, also contribute, as they can damage roots (*Armillaria mellea*, *Ganoderma lucidum*, *Phytophthora quercina*), or stems and branches (*Diplodia mutila*, *Hypoxylon mediterraneum*).

Ciboria batschiana damages acorns, mainly during storage, and can be controlled by treatment in warm water. *Microsphaera alphitoides* and *Rosellinia quercina* are important pathogens in nurseries and control may be necessary.

The gall aphid *Phylloxera quercus* is very common in many Mediterranean countries, where it can be injurious to oak species. *Diphyllaphis mordvilkoi* is mainly harmful to young plants and seedlings. Among lepidopteran defoliators attacking *Q. cerris*, the most important are *Lymantria dispar*, which attacks oak foliage during the summer, and *Tortrix viridana*, which feeds on buds and young leaves in spring. Biological control using *Bacillus thuringiensis* has been successfully implemented against the moth outbreaks.

Some phloem- and wood-boring insects frequently cause damage, mainly to mature trees weakened by unfavourable climatic conditions. Other pests and diseases recorded for *Q. cerris* include *Aphelonyx persica*, *Apiognomonia errabunda* (anthracnose) and *Chilaspis mayri*.

Uses

Q. cerris has been much used for afforestation of degraded and shallow soils, because of its pioneer characteristics of rapid early growth and tolerance to a wide range of soil types. Therefore, it has a role in soil conservation, erosion control, and revegetation of bare soils.

Q. cerris is a hardwood species; wood is pinkish-brownish with evident multiseriate rays; it has a coarse texture and medium hardness. Grain is usually straight, but logs are often affected by heavy internal stresses which may produce end splits and cracks at cutting or just after. Average density at 15% moisture content varies between 800 and 880 kg/m^3; shrinkage is high. Impact resistance is good. Durability for external use is low, as wood may be attacked by wood-destroying fungi and by many insect pests, including *Lyctus* sp.

The wood saws easily; kiln-drying must be slow. It has an adequate glue strength which may, however, be reduced by the nature and concentration of wood extractives present. Planing, painting and staining properties are adequate.

Products and markets in decreasing order of importance are, or have been: fuelwood and charcoal; composite and panel board production, both pure and mixed with other hardwood species; pulp and paper; railway sleepers; parquet boards; plywood; and vehicle bodies. Bark and wood are also utilized for tanning. The prospects for utilization of *Q. cerris* wood are being improved by advances in wood processing technology.

Q. cerris trees are usually scattered in cultivated land (in hedges and along roads) and in rangelands, to provide shelter for use by farmers and shade for livestock. Acorn production and young coppice shoots were used as a source of fodder and browse in Mediterranean agrosilvopastoral systems. An annual acorn production of 4–5.4 kg/tree (corresponding to 170–230 feed units/ha) has been estimated in a *Q. cerris* oak coppice with 80 standard trees in the Apennines of central Italy.

At the regional scale, the use of oak forests for animal husbandry or for hunting produces additional revenues which may exceed those from wood harvesting. Furthermore, locally important benefits for rural populations may be provided, such as the production of edible mushrooms and truffles, typically found in pure and mixed-oak stands.

Further reading

Amorini, E., Bruschini, S., Cutini, A., di Lorenzo, M.G. and Fabbio, G. (1996) Treatment of Turkey oak (*Quercus cerris* L.) coppices. Structure, biomass and silvicultural options. *Annali dell'Istituto Sperimentale per la Selvicoltura* 27, 121–129.

Bussotti, F. (2001) *Quercus cerris*. Sherwood. *Foreste ed Alberi Oggi* 7, 41–44.

Cutini, A. (1997) Drought effects on canopy properties and productivity in thinned and unthinned Turkey oak stands. *Plant Biosystems* 131, 59–65.

Meshinev. T. and Nikolov, V. (1990) Biological productivity of the *Quercus cerris* + *Q. frainetto-Carpinus orientalis*-mixtoherbosa association in the region of the Elena Fore-Balkan mountains. *Nauka za Gorata* 27, 34–40.

Quercus ilex L.

Holm oak

Importance

Quercus ilex (Plate 35) is one of the most important species in the Mediterranean area, where it is capable of growing both as a tree (representing 'climax' Mediterranean vegetation) and as a shrub in more degraded areas. It has a compact, dark green, and rounded crown, with tough, long-lasting leaves. It can tolerate a wide range of soil types and both drought and frost. The wood is used for firewood and to produce good-quality charcoal. The acorns and foliage of *Q. ilex* subsp. *rotundifolia* are also collected and fed to animals, or directly browsed by them. *Q. ilex* is a key species of the traditional Mediterranean agroforestry systems.

Botanical features

Q. ilex typically grows up to 20 m in height with a dbh of 80 cm, although it can grow up to 30 m, with a dbh of over 2 m, and live for 1000 years or more. The plant has a straight stem with low branches and grey, tomentose tips; the bark is dark, greyish-brown, smooth when young and then splitting into small square plates; buds are small, rounded, and tomentose. The roots are strong, divided into a deep taproot and a lateral root system which anchors the tree firmly to the substrate.

The leaves are simple, alternate, 3–7 cm long, thick, variable in form, but usually elliptic to ovate or ovate–lanceolate (smaller and more rounded in *Q. ilex* subsp. *rotundifolia*), base round to cuneate, either entire or more or less dentate, dark green and smooth on the upper surface, grey-tomentose beneath; persistent for 3–5 years on the tree. The petiole is 6–15 mm long with brown stipules.

Q. ilex is monoecious, male flowers in greenish-white cylindrical aments or catkins, pendulous (short-stalked), with a tomentose perianth and six stamens. Female flowers are borne on erect spikes, with a white, tomentose stem bearing six to seven flowers. Fruits ripen in the first year, borne at the end of a short peduncle and within a light grey cupule with appressed, flat scales. The acorns are dark brown, 1.5–3 cm long, long, narrow, and tapering.

Flowering begins in May in the central part of its natural range, at the same time as leaf drop and new leaf growth. In autumn, after the dry summer months, *Q. ilex* can also grow new leaves while fruit ripens.

Distribution

The natural distribution of *Q. ilex* includes the Mediterranean basin, where it is more abundant in the west than in the eastern part, with a latitude range of 14° in the northern hemisphere. Among the two subspecies, *Q. ilex* subsp. *rotundifolia* occurs in Portugal, south and south-east Spain, and Morocco; *Q. ilex* subsp. *ilex* occurs throughout the remaining area, which includes northern Spain, France, Italy, Greece and the Aegean islands, Corsica, Algeria and Turkey. It is found sporadically in Tunisia, Israel and the Balkans.

Silviculture characteristics

Generally a hardy and resistant tree, *Q. ilex* is very resistant to maritime exposure, but is slow growing. The seed quickly loses viability if allowed to dry out and so is best stored in moist and cool conditions over winter, or sown in an outdoor seed bed when ripe. Young plants tolerate reasonable levels of side shade.

Silviculture practice

Q. ilex seed is classified as recalcitrant and can be kept only until the following spring. This species is suitable for coppicing, so vegetative propagation is performed using stump plants, whereas stand establishment is usually by natural regeneration, wildings, or direct sowing. The growth rate is slow and because of the deep taproots, transplanting from the nursery is difficult. When direct planting is required, site preparation can be advantageous.

Q. ilex can be cultivated both as mature forest, with a harvesting age of 100–130 years, and as coppice which is cut after 30 years or less. For example, in Tuscany, Italy, a dense stand ('forteto') with closely planted trees (for example, 30,000 shoots after 10 years) is grown and cut after 18 years when trees are 6 m high. In southern France, it is typically harvested every 15–20 years. An intermediate cut after 5 years can also be advantageous. Annual yields of 1–2 m^3/ha can be expected from dense plantings and <1 m^3/ha from more open stands.

In Morocco, after 25 years, mean annual biomass production for *Q. ilex* coppice was 21 t/ha in unthinned plots, 42 t/ha for lightly thinned plots, and 19.5 t/ha in heavily thinned plots.

Diseases and pests

Under conditions of water stress in Spain and Portugal, heavy damage has been attributed to the fungal pathogen *Phytophthora cinnamomi*, and cankers caused by *Cryphonectria parasitica* have also been reported. *Phellinus erectus* and *Fomes fomentarius* are responsible for wood decay, whereas *Ganoderma* sp. attacks the roots. *Q. ilex* is reported to be resistant to honey fungus (*Armillaria mellea*).

Apart from other common borers of oak trees, *Q. ilex* is particularly susceptible to *Coraebus florentinus*, a wood-eating beetle which can occur in large numbers after periods of drought. Another damaging boring beetle is *Cerambyx cerdo*, which attacks old or weak trees. Severe attacks by the kermesid *Nidularia pulvinata*, a cochineal insect, have been described in urban areas of Tuscany and south Italy. Defoliating Lepidopteran pests include *Tortrix viridana*, *Malacosoma neustria*, *Lymantria dispar*, *Catocala nymphagoga* and *Ephestia nymphaea*.

Frequent fires can degrade *Q. ilex* forest to low shrubs as the foliage is flammable all year round. Controlled fires should be undertaken with caution.

Uses

Q. ilex forms beautiful forests, composed of large, tall trees, with both protective and recreational functions. The litter acts as a soil improver. Like many other oak species, *Q. ilex* is used as an ornamental tree in gardens and parks, where it can reach a large size.

Q. ilex wood is very hard and difficult to carve and to dry, because it easily warps and breaks. Its density is 960 kg/m^3 (12% moisture content) and calorific value is 4700 kcal/kg. It is used for railway sleepers, stakes, tools, and tool handles, and benches that undergo heavy usage (for example, carpentry and butchers' benches). Experiments have demonstrated the possibility of obtaining plywood with a sheet thickness of

0.2 mm. When coppiced, *Q. ilex* can yield good-quality firewood and excellent charcoal, which is obtained from small shoots.

In Spain and in Portugal, acorns and foliage of *Q. ilex* subsp. *rotundifolia* are used for animal feed, whereas in Sardinia, *Q. ilex* subsp. *ilex* is grown for the same purpose. The bark was used to produce tannins until a few years ago, for example, in southern France. Coppice management of this species was therefore largely dictated by demand for high-quality tannins, which were extracted from younger shoots rather than older stems. This had a significant effect on rotation period and intercropping techniques.

Further reading

Andivia, E., Márquez-García, B., Vázquez-Piqué, J., Córdoba, F. and Fernández, M. (2012) Autumn fertilization with nitrogen improves nutritional status, cold hardiness and the oxidative stress response of holm oak (*Quercus ilex* ssp. *ballota* (Desf.) Samp) nursery seedlings. *Trees: Structure and Function* 26, 311–320.

Rodríguez-Calcerrada, J., Pérez-Ramos, I.M., Ourcival, J.M., Limousin, J.M., Joffre, R. and Rambal, S. (2011) Is selective thinning an adequate practice for adapting *Quercus ilex* coppices to climate change? *Annals of Forest Science* 68, 575–585.

Quercus mongolica Fisch. ex Turcz.

Mongolian oak

Importance

Quercus mongolica is a deciduous tree, and one of the major broadleaved tree species in north-east China and eastern Russia. It grows to 30 m in height, regenerates easily, and is highly productive, producing about 10 m^3/ha on medium-class sites, with a high wood production percentage (the stemwood makes up 70% of the total biomass).

In north-east China, *Q. mongolica* mainly occurs in Helongjiang, Jilin and Liaoning Provinces, predominating in broadleaved forest. In hilly and mountainous areas of eastern and central Jilin, it accounts for the largest standing volume of timber (63,072,000 m^3) and comprises 20% of the total provincial natural forest. In Liaoning Province, *Q. mongolica* plays an important role in the local forestry industry, with about 1,060,000 ha of oak stands, making up 43.2% of the total provincial natural forests and producing 40,800,000 m^3 of timber.

Q. mongolica is managed not only for timber, but also for the silkworm industry. In Liaoning Province, for example, around 90% of stands in Haicheng County are currently managed for feeding silkworms.

Botanical features

Q. mongolica grows up to 30 m in height, with a dbh of 100 cm. The bark is dark grey-brown initially, but darker with deep longitudinal stripes when old. Young branches are purplish-brown and glabrous when young, turning chestnut brown later; lenticels light brown. Buds are brown, oval and arranged in groups of three or four.

The simple, alternate leaves are obovate or obovate–oblong, 8–20 cm long and 4–10 cm wide, with an ear-shaped leaf base; upper surface dark green and undersides greyish-green; scattered hairs or glabrous at the margins. Leaf margins have 8–9 pairs of sinuate teeth and 7–11 pairs of lateral veins.

Flowers are unisexual; male catkin arises from leaf axis of new branches. Fruits (acorns) are oval or elliptic, with a very short stalk and hard skin.

Buds begin to swell in early May in Helongjiang Province, the most northern province of China. After flowering in late May, fruits ripen during September and October. Leaves become yellow and fall from mid-October.

Distribution

Q. mongolica naturally occurs in China, eastern Russia, Mongolia, the Korean Peninsula and Japan. In China, it is mainly distributed in the south of north-eastern China (the Da Xing'an (350–800 m), Xiao Xing'an, and Changbaishan (1300 m) mountain areas), Inner Mongolia (Nei Menggu), Hebei (800–2000 m) and Shandong and Shangxi (800–2000 m).

Silviculture characteristics

Q. mongolica is light-demanding and normally found on slopes or in sunny and dry areas. It is very frost-hardy in winter and can tolerate minimum temperatures of −40°C, as in north-east China. It is adaptable to a wide range of conditions because of its drought, cold and fire-resistance. Although *Q. mongolica* can be observed occasionally growing alongside rivers and streams or in poorly drained areas, its poor growth there indicates that it is intolerant of waterlogging.

Q. mongolica is considered to provide considerable ecological benefits, superior to other forest types in nutrient recycling, water conservation and air quality adjustment. When growing on dry, barren sites, it can still form stable, productive stands. It has a long and well-developed root system, whose uptake of water and nutrients is two to three times of that of birch (*Betula* spp.). The high leaf area index of *Q. mongolica* is advantageous, allowing for this tree to capture more light than most other trees. The litter has high nutritional value, with an annual return of N, P, and K of 12.8, 8.9, and 21.5 kg/ha, respectively.

Silviculture practice

Direct sowing of *Q. mongolica* is used in some locations, and trials have shown good results in terms of decreased labour costs and higher survival (up to 96.4%). The following key practices for direct sowing should be carefully considered: pretreatment of seed; the correct date for collecting seeds; seed storage; control of mice outbreaks; and field protection at seedling stage.

During seed shortages, horticultural procedures may be carried out for the vegetative propagation of *Q. mongolica* in order to produce planting stock for silkworm rearing. Generally, the highest rooting (56.5–86.7%) was obtained from spring shoots with one-third of the leaves removed, planted in vermiculite and dung, and soaked in 100 ppm IBA (indole-butyric acid) or NAA (naphthaleneacetic acid) for 6 h.

Although *Q. mongolica* is mainly regenerated naturally in the Korean Republic, grafting propagation has also been studied. When scions from *Quercus acutissima*, *Quercus variabilis*, *Quercus serrata* and *Q. mongolica* were grafted onto rootstocks of *Q. acutissima*, survival rates were 48, 47, 57 and 59%, respectively. In addition, graftings showed juvenile characteristics in leaf morphology, such as shortened petioles and increased leaf size. The selection of clones with good survival rates after serial grafting may allow this method to be scaled up for commercial production, once the problem of graft incompatibility is reduced.

The stands explored for silkworm (*Antheraea yamamai*) production are mainly distributed in secondary forestry areas of mountain regions in the eastern part of north-eastern China. The choice of crown shape is an important aspect of tree management for silkworm feeding, as different crown shapes might be produced by different pruning strategies. This is because forest canopy structure directly affects solar interception, and consequently the photosynthetic efficiency and biomass production of the entire stand. The 'ladder crown shape' provided a leaf-area coefficient as high as that of natural crown. However, the 'stair' (conical) crown form (canopy at about 2.5 m) provided a greater leaf area index, a more equitable leaf-distribution pattern, and a greater leaf-biomass production than those achieved by the other two crown forms.

In terms of dry matter/biomass production, a 5-year-rotation cycle has been found to yield the best productivity, which has been recorded in the region of 227 t/ha.

Natural forests of *Q. mongolica* are decreasing rapidly and more research is needed on artificial regeneration and nursery techniques. *Q. mongolica* can be sowed/planted in most parts of north-eastern China, particularly in dry and barren sites that are not suitable for other tree species.

Diseases and pests

The main diseases affecting *Q. mongolica* are rust (*Cronartium quercuum*) and stem rots (*Polyporus* spp.).

Q. mongolica is attacked by some insect pests, notably *Mallambyx raddei*, *Galerucida bifasciata*, *Oupyrrhidium cinnabarinum* and *Lachnus tropicalis*.

Other pests and diseases recorded to attack this species include *Lymantria mathura* (pink gypsy moth), *Ophiostoma longicollum*, *Platypus quercivorus* (oak ambrosia beetle) and *Raffaelea quercivora* (Japanese oak disease).

Uses

In natural forests, coppiced trees have short stems and produce low economic returns, but provide good ecological benefits which mainly include soil and water conservation, decrease in surface runoff, and protection from soil and water erosion. The species has high values of rainfall interception compared to other broadleaved species, including birch (*Betula* spp.), in its natural range. In addition, leaf litter enriches soil organic matter and loosens the soil structure, ensuring that approximately 90% of rainfall directly permeates the soil, and can provide a good environment for the regeneration of other species by soil improvement. For example, coppice will tend to change gradually to Korean pine (*Pinus koraiensis*) and Korean spruce (*Picea koraiensis*) by strip or patch soil improvement. In man-made forests, thinning and other silvicultural practices are required to produce timber of medium or large diameter.

Q. mongolica sapwood is pale brown, whereas the heartwood is light yellowish-brown and paler than that of *Quercus alba* or *Quercus petraea*. The timber is very hard and heavy, with a typically straight grain and coarse texture. Reported wood densities range from 603 to 689 kg/m^3. It is one of the hardest timbers in China, with medium strength properties in bending and crushing, and low stiffness and resistance to shock loads. It is somewhat weaker than European oaks, however, and generally weighs less than other oaks on the commercial market. The wood is very durable and resistant to decay, although the large proportion of sapwood makes it susceptible to attack by beetles. Gluing, nailing, screwing and staining properties are all rated as good.

It can be used for mine and telegraph poles, stakes, sports equipment, boats, vehicles, bridges, agricultural tools, casks and wine vats.

The acorns can be used for brewing, and tannins are extracted from the bark and hulls. Sawdust and chips of *Q. mongolica* are regarded as a good medium for cultivating mushrooms.

The leaves are traditionally used for cultivating silkworms and for animal fodder in north-east China, and the bark is used in traditional Chinese medicine for the treatment of diarrhoea. *Q. mongolica* provides a good environment for the growth of understorey plants and fungi, some of which are valuable ingredients in traditional Chinese medicine. For example, Yishan ginseng (*Panax ginseng*) has been grown in *Q. mongolica* forests in Liaoning Province, China, producing 300 g/m^2 after 8 years.

Further reading

Li, X.-W. (1993) Ecological effects, economic value analysis, management strategies for Mongolian oak stand. *Forestry Economy Research* 4, 44–47.

Wang, C. (1994) Some experience of growing Mongolian oak for silkworm. *Journal of North-East Forestry University* 22, 83–87.

Yang, Y. and Yi, X. (2012) Partial acorn consumption by small rodents: implication for regeneration of white oak, *Quercus mongolica*. *Plant Ecology* 213, 197–205.

Quercus pubescens Willd.

Downy oak

Importance

Quercus pubescens, commonly known as downy or pubescent oak, is a small to medium-sized tree (occasionally shrubby) with deciduous leaves, found in temperate Europe. It presents great intraspecific genetic variability, with a large number of subspecies, varieties, forms, geographical races and hybrids having been described. This wide-ranging genetic variation allows it to flourish in very different climatic and edaphic environments, and to adapt to extreme ecological conditions.

It frequently grows on shallow and calcareous soils; it can withstand summer heat and drought as well as very cold winter temperatures. It grows in coppices with a low productivity and is used almost entirely as fuelwood. The acorns possess high nutritional value and are usually used as animal feed. *Q. pubescens* woodlands are often in poor condition because of excessive felling and pasturing of animals in the past, and frequent forest fires.

Botanical features

The stem is rarely straight and branching begins low down near the ground. Branches are sinuous and form an ample, hemispherical or irregular crown, not very dense. Young branchlets are densely pubescent and preserve their down until about 1 year of age, after which they become glabrous, rigid, greyish-brown, with visible lenticels. The bark is grey-brown and fissured with longitudinal and transverse grooves forming small, very hard, trapezoid-shaped scales, which are thicker than they are wide. The root system consists of a deep taproot and dense lateral roots.

Q. pubescens can reach 10–20 m in height, occasionally to 25 m. In some cases it grows as a shrub. Stem diameter can reach 2–2.5 m.

The simple leaves are arranged in a spiral manner. They are rather thick, very hard and stiff, elongated and ovoid in shape, to 9 × 4 cm in size but their size can vary considerably, acute or obtuse, base round to cordate, margins coarsely dentate or incised, the teeth often acute and with a mucro. The leaf lamina is bright green and glabrous above, whitish and pubescent or tomentose beneath. Leaf flushing period varies considerably, depending on site conditions. Leaves die in October–November but are retained on the tree until the end of winter.

Flowering can occur before or after foliation according to the ecological conditions of the stand. Male catkins are numerous, pendulous, 5–10 cm long, with a tomentose axis bearing several flowers at its tip. Male flowers are sessile, with perianth having 5–6 deep lobes; there are normally 8 stamens. Male catkins develop between March and May from buds on the previous year's branches. Female flowers are less numerous and arranged in spikes, short, growing in isolation or in clusters of 2–5 in the axils of the upper leaves, at the tip of the young current year's branchlets. Female flowers are ready for fertilization when male flowers, which are beneath them on the same tree, begin to wither. Pollination occurs in spring; fertilization in midsummer. The fruits ripen in the autumn of their first year. The fruit-bearing pedicel is short (5–10 mm), stiff and hairy; it bears 3–4 fruits. The fruit is an ovoid to subellipsoid acorn (25–30 mm long and 15–20 mm in diameter), with longitudinal striations when fresh. The cupule is hemispherical and covers between one-third to one-half or more of the acorn. The cupule has many closely appressed scales, triangular in shape and pale grey.

Distribution

The distribution area of *Q. pubescens* is wide, extending through most of the southern part of Europe. It can be found in northern Spain, but not in Portugal. It is present throughout most of France, except for the northernmost regions, and occurs throughout Italy, both on the mainland and on islands, all the way to the southern edge of the Alps. It is frequent in Switzerland, Austria and in southern Germany, but it does not occur in central or northern Germany. It also grows sporadically in central and Eastern Europe, from Bohemia (Czech Republic), the Balkan states and Turkey, as far east as the southern Crimea and the Transcaucasian region.

Silviculture characteristics

Q. pubescens can grow on poor sites and is tolerant of drought and cold. It will survive under light shade, but full overhead light is necessary for full vigour. *Q. pubescens* produces suckers around the coppice base from dormant or adventitious buds.

Silviculture practice

Ripe acorns can be collected in autumn from the ground, or flailed or shaken from branches onto canvas or plastic sheets after ripening. The most suitable tree age for seed gathering ranges from 40 to 100 years. The interval between good seed crops (mast years) usually varies from 1 to 3 years. Mature seed is classified as recalcitrant. The mean weight of 1000 seeds ranges from 1500 to 1800 g. At the time of collection, seeds of *Q. pubescens* will usually still retain some green colour and will have a moisture content in excess of 40% fresh weight.

Seed may be stored in layers up to 30 cm deep on a wooden floor, in a place where the temperature will not be too high and where ventilation is good. The seeds germinate readily and need no pretreatment. Germination occurs at around 20°C, if the substrate is kept moist, and needs no light. Germination is hypogeal. The seedling has a very sturdy main taproot.

Q. pubescens is often managed in coppices to produce firewood and charcoal. *Q. pubescens* woods can be used for forest grazing. The best type of stand management for this purpose are open woods with tall trees. Recently, the species has been inoculated with edible fungi, such as *Tuber melanosporum*, in order to produce edible truffles.

Q. pubescens woods are frequently very open, growing on thin soils with rocky outcrops. As result, increment rates (diameter and biomass) are very low: 2–3 m^3/ha in simple coppices with a 15–16-year rotation. In the past, to increase the productivity of *Q. pubescens* coppices and to improve the quality of the soil, coniferous trees were added to the stands, using mainly Mediterranean species such as *Pinus pinea*, *Pinus pinaster*, *Cupressus sempervirens* or *Cedrus atlantica*.

Diseases and pests

Microsphaera alphitoides can be a serious pathogen if it reappears year after year, as it inhibits photosynthetic activity.

Q. pubescens suffers from oak decline, attributed to the joint action of environmental (e.g. water stress) and secondary pathogens or parasites that weaken the tree. The symptoms of this phenomenon are the production of epicormic branches, leaf yellowing, bark cracking and stem exudations. Fungi associated with this pathology are *Armillaria* spp., *Cephalosporium* spp., *Diplodia mutila*, *Hypoxylon mediterraneum*, *Phoma cavae* and *Phomopsis quercini*.

The pine processionary moth (*Thaumetopoea processionea*) and the defoliating moth *Lymantria dispar* defoliate

Q. pubescens. Biological methods can be used to control these two insects.

Other pests and diseases recorded as attacking this species include *Agrilus sulcicollis*, *Apiognomonia errabunda* (anthracnose) and *Tortrix viridana* (green oak leaf-roller).

Q. pubescens woods deteriorated due to excess animal grazing and over-logging in the past.

Uses

Q. pubescens woods are important in traditional Mediterranean agroforestry systems, because their acorns are valuable feed for pigs, and because the understorey also includes edible browse or fodder plants. Due to its deep root system, *Q. pubescens* is an important species for erosion control and in protection forests.

Q. pubescens wood is characterized by the presence of many curved fibres, by considerable specific gravity (0.9–1.25 when fresh; 0.7–0.92 at 12% moisture content) and by a tendency to warp. A wood density of 790 kg/m^3 has been reported; however, this is considered to be in the high range for this species. These characteristic features, as well as the short stem, make this species of little interest as a source of construction timber, despite the fact that histologically and anatomically, this wood is not substantially different from that of other, more highly valued, oaks such as *Quercus petraea* and *Quercus robur*.

The bark can be used to produce tannins. Acorns may be used as animal fodder.

Further reading

Alias, S., Bianchi, L., Calamini, G., Gregori, E. and Sioni, S. (2010) Shrub facilitation of *Quercus ilex* and *Quercus pubescens* regeneration in a wooded pasture in central Sardinia (Italy). *iForest* 3, 16–22.

Arend, M., Kuster, T., Günthardt-Goerg, M.S., Dobbertin, M. and Abrams, M.D. (2011) Provenance-specific growth responses to drought and air warming in three European oak species (*Quercus robur*, *Q. petraea* and *Q. pubescens*). *Tree Physiology* 31, 287–297.

Benko, M., Novotny, V., Vrbek, B. and Szirovicza, L. (2000) Volume tables for downy oak (*Quercus pubescens* Willd.). *Radovi - Šumarski Institut Jastrebarsko* 8, 68 pp.

Debussche, M., Debussche, G. and Lepart, J. (2001) Changes in the vegetation of *Quercus pubescens* woodland after cessation of coppicing and grazing. *Journal of Vegetation Science* 12, 81–92.

Quercus robur L.

Common oak

Importance

Quercus robur (Plate 36), together with *Quercus petraea* (with which it forms hybrids), are extremely important forest trees in Europe and provide high-quality timber for construction and furniture manufacture. *Q. robur* is dominant, or is a valuable admixture, in various forest vegetation types. It is most suitable for planting on moist fertile sites, characteristic of lowland deciduous forests in Europe, but sometimes it is also used for the revegetation and afforestation of degraded sites.

Botanical features

Q. robur is typically a big tree with an open, large, often irregular crown. The main branches are long and thick. Young shoots are not pubescent. The stem is relatively short, with a large circumference. Open-grown or solitary trees may have stems warped, with frequent burrs and epicormic shoots. The bark of mature trees is thick, pale or dark grey, strong and deeply fissured in short, narrow, vertical plates. A deep taproot system forms when oak grows on fertile, deep soils, but a root system with more lateral roots will form on shallow or free-draining soils.

Q. robur oak typically achieves a height of 35–40(–50) m and a dbh of 150–200 cm. It is a long-lived species which can attain an age of 400–500 years, although some individuals (which may have a stem circumference up to 14 m) are estimated to be more than 1000 years old.

Shoots are green-brown with pale buff lenticels. Buds ovoid–conic, pointed, light brown. Leaves are obovate–oblong, 5–10(14) cm long by 7–8 cm wide, with 5–18 pairs of veins; 4–5 rounded lobes on each side, cut half way to midrib by narrow, round-ended sinuses; lobes are wavy-edged with occasional teeth; leaf base is ear- or heart-shaped. Colour dull dark green above, pale green beneath. Petioles are short, 4–8 mm long. Leaves of *Q. robur* have shorter stalks than those of *Q. petraea* and are generally less regular in shape, and their basal lobes usually overlap the stalk. The secondary veins reach the top of the lobes and the edges between them.

Q. robur is monoecious and anemochorous. Male flowers are borne on long slender green catkins, clustered in thin bunches (2–3 catkins); 2–4 cm long, brownish-yellow until shedding pollen in early May when they become yellow-green; catkins are carried on previous year's shoots. Female flowers are less visible, globular, pale brown, 2 mm with dark red stigmas, growing in groups of 2–5 on long peduncles (2–5 cm) at the end of current year's shoots. Fruits (acorns) are carried on (4)5–8(12) cm-long peduncles; scales of the shallow cupules are short and appressed; acorns are 1.5–4 cm long with dark stripes when fresh.

Q. robur flowers appear at the end of April or in May together with the new foliage, and earlier than *Q. petraea*, which is therefore more tolerant of late spring frosts. *Q. robur* starts producing acorns at about 60 years old in forest stands, but before 40 years if an open-grown tree.

Distribution

Q. robur is widely distributed in Europe. It is an important component of temperate deciduous and mixed forests found from Atlantic Western Europe to continental central Russia. It is also present in more hilly or montane zones of some Mediterranean countries (Italy and the Balkan region). The natural range of common oak stretches from about latitude 6° N (in southern Norway and Sweden) to approximately 38° N (in Sicily and Portugal).

Compared to *Q. robur*, *Q. petraea* has a more restricted distribution in Asia, and Eastern and Northern Europe.

Q. robur has been introduced into south-eastern Canada and the north-eastern USA for timber production, and to provide habitat and food for wildlife and for other environmental purposes. Introductions of *Q. robur* have also been reported from Kazakhstan.

Silviculture characteristics

Q. robur and *Q. petraea* share several characteristics typical of pioneer tree species: seeds are dispersed widely by birds and mammals (through caching); they are large, with reserves that enable seedlings to compete with most grasses; they can germinate quickly in autumn; they can withstand summer drought; and they can persist in grass until roots are developed enough to allow rapid shoot growth. Also, unlike most late-successional tree species, *Q. robur* is not dependent on shade from other forest trees; in natural conditions, oak regenerates through species alternation.

Q. robur is considered to be more light-demanding that *Q. petraea*; however, side shade is considered to be favourable until the high polewood stage has been reached (Polish foresters have a saying: 'The oak likes to grow in a fur coat but with its head uncovered.'). The species grows best on warmer, sheltered sites, and can suffer from severe frost damage. However, it also grows well under the relatively cold, continental climatic conditions in central Europe. *Q. robur* is eutrophic and requires fertile soils. The best conditions for its growth are moist, well-drained sites, in broadleaved or mixed forests and floodplain forests. It can tolerate periodic flooding and compacted soils.

Q. petraea (and *Q. robur*) can be found on many site types, although growth is best on deep, fertile, fine-textured, slightly acid soils. *Q. petraea* is characteristic of lighter, sandy or silica soils, and is not found on heavy or very wet soils, which *Q. robur* can tolerate to a greater extent. In contrast, *Q. petraea* can withstand dry conditions better than *Q. robur*.

Oak coppices and pollards easily. Both species are fairly resistant to frost damage, mainly because they flush very late in spring: newly developed foliage will be killed by temperatures below −3°C. Both species are also very windfirm, due to very deep and penetrating taproot systems.

Silviculture practice

Economic rotations are longer for *Q. petraea*, at about 160 years, than for *Q. robur*, at about 130 years.

Q. robur can be propagated by seeds, or vegetatively by sprouts from the root collar, cuttings, stem explants or tissue culture.

Q. robur commences acorn production at about age 60 years in forest stands, but before 40 years of age when growing as a single tree. Mast years are irregular, every 2–5 years in Western Europe and 3–8 years in eastern France and Poland. Flowers may be damaged by late spring frosts and abundant rain. Acorns can be damaged by autumn frosts and summer drought. Acorns are often harvested from the ground as soon as possible after they have fallen, usually in early to mid-October. They are extremely sensitive to shock and heat.

Planting can take place in autumn or spring, and 1-year-old seedlings are most often used. If herbaceous cover is thick, the use of 2- or 3-year-old transplants is recommended. *Q. robur* seedlings are planted in slits or holes at close spacing (e.g. 1.4 × 0.6 m). In mixed forests and mixed pine forests, plants may be arranged in groups. In coniferous forests, *Q. robur* may be introduced either as single individuals, or in admixtures with other broadleaves such as small-leaved lime (*Tilia cordata*) and/or hornbeam (*Carpinus betulus*).

Diseases and pests

Oak decline is an important problem in Europe. Several factors are thought to be associated with oak decline including drought and frost damage, defoliation by *Tortrix viridana*, and attack by *Microsphaera alphitoides*, *Pezicula cinnamomea*, buprestid beetles and, in later phases, *Armillaria mellea*.

Air pollution is thought to be a major cause of outbreaks of oak diseases. The cause of oak dieback is unknown, but several hypotheses have been proposed including the effects of acid rain and attacks by species of *Ceratocystis*, *Valsa* and *Botryosphaeria*. Several fungal pathogens (*Cytospora* sp., *Fusicoccum quercus*, and *Dermea* sp.) are found to be associated with secondary necrosis of oak. *Ceratocystis* sp. has been isolated from the bark beetle, *Xyloterus domesticus*, collected from sapwood beneath secondary bark necrosis and from necrotic bark tissue colonized by *Dryocoetes villosus*. It is thought that the beetles transmit *Ceratocystis* sp. to oaks that are already declining.

Both *Q. robur* and *Q. petraea* suffer from acute oak decline which has symptoms of dark weeping from the stem and results in dieback and mortality.

Oak is a host for many pests. *Tortrix viridana* is one of the most damaging insect pests of oak. Oak processionary moth (*Thaumetopoea processionea*) is widely distributed in Europe and has become an increasing problem as a major defoliator of oak. Other defoliators of oak are *Lymantria dispar* and *Operophtera brumata*. The germination ability of acorns can be significantly reduced following attack by weevils such as *Curculio venosus*. *Q. robur* is also attacked by rare insects such as the great capricorn beetle (*Cerambyx cerdo*) and the stag beetle (*Lucanus cervus*).

Uses

Q. robur may be planted for land reclamation, soil improvement, in watershed protection forests and in windbreaks, and is also an ornamental tree in parks, gardens and estates. In many forests, mature specimens of *Q. robur* are extremely important for nature protection and conservation, not only due to the high aesthetic value, but also because it provides an essential habitat for a wide range of flora and fauna.

The timber is widely used for furniture and joinery where strength and durability are required. Average density of the wood (at 15% moisture content) is about 720 kg/m^3. High-quality logs are in demand for veneer, and such logs fetch the highest prices of all broadleaves (including decorative timbers such as walnut and wild cherry). Oak wood is susceptible to the development of shake; most often this occurs in timber grown on drought-prone sites.

The uses of *Q. robur* include almost everything that can be obtained from a forest tree: large, strong, and long-lasting building timbers, fuel, food for man and animals, tannins and medicinal products.

As well as sawnwood, oak, and oak chips are used for veneers, plywood, particleboards and other composite wood products.

Further reading

Deleuze, C. and Renaud, J.P. (2010) Silviculture of sessile oak and pedunculate oak: young stands response to thinning. *RenDez-Vous Techniques* 5, 56–65.

Forestry Commission (2010) *Managing Native Broadleaved Woodland.* Stationery Office Books, London, 524 pp.

Savill, P.S. (1991) T*he Silviculture of Trees used in British Forestry*. CAB International, Wallingford, UK, 143 pp.

Savill, P.S., Evans, J., Auclair. D. and Falck, J. (1997) *Plantation Silviculture in Europe*. Oxford University Press, Oxford, UK, 308 pp.

Quercus rubra L.

Northern red oak

Importance

Quercus rubra, formally named northern red oak, but often simply called 'red oak', is one of the largest and most important hardwood trees in eastern North America. A member of the red oak group (*Quercus* section Lobatae), it has enjoyed several decades of popularity as a cabinet-, flooring, moulding and furniture wood in North America. It also provides large amounts of pulpwood and pallet stock to local processors, especially as small-diameter bolts and poor-quality logs. *Q. rubra* is still preferred as a fuelwood in many rural areas owing to its density, high British Thermal Unit (BTU) value and relative ease of splitting. More recently, production of 'hard mast' for wildlife has gained in importance.

Botanical features

Q. rubra is a considered to be the fastest-growing of the northern red oaks, reaching heights of 25–40 m and diameters in excess of 100 cm. It has a lifespan of more than 100 years, but seldom exceeds 200 years of age. Forest-grown trees are usually tall and straight with a well-pruned bole, while open-grown trees form more massive crowns and may branch within 4 m of the ground.

Q. rubra has simple, alternate, large (12–20 cm) deciduous leaves with 7–11 toothed lobes separated by sinuses extending halfway to the midrib. Leaf margins are entire. Leaf surfaces are lustrous, dark green above and glabrous, somewhat pale green beneath.

Q. rubra is monoecious and wind-pollinated, producing male flowers in 'catkins' borne in clusters on short pedicels from the axils of the previous year's leaves, and appearing with the leaves in March–June, depending on climate. Female flowers are solitary or clustered (up to 5), in the axils of the current year's leaves, becoming an acorn. *Q. rubra* acorns require 2 years to mature, with the immature (3–5 mm) fruit overwintering on the tree.

Q. rubra begins to produce seed at about 20–25 years old. Acorn crops are variable in size and frequency, and are strongly influenced by weather patterns, with good crops occurring at 2–5 year intervals depending on site. Synchronous cycles of acorn production might occur in some regions.

Distribution

Q. rubra is distributed throughout most of eastern North America and south-east Canada.

It has been planted as a timber tree in virtually every country in Europe. Introductions from North America to France, Germany, The Netherlands and several other countries date to the 19th century. These early plantations often provided seed for distribution to additional countries in Europe, the Ukraine and other parts of the former Soviet Union.

Silviculture characteristics

It is considered a 'mid-tolerant' species, surviving in the partial shade of many early successional species such as birch (*Betula papyrifera*), tulip poplar (*Liriodendron tulipifera*) and black cherry (*Prunus serotina*), but failing in the deeper shade of sugar maple (*Acer saccharum*) and beech (*Fagus grandifolia*). It requires rich, well-drained sites for optimal growth. Seedlings of *Q. rubra* are poor competitors in forest understoreys and often fail to outgrow other more aggressive species when canopies are opened or removed. However, stump sprouting is often prolific following a harvest, with sprouting more or less inversely proportional to a tree's age; beyond 75–100 years old few or no sprouts may be obtained.

In many parts of its range, *Q. rubra* can be considered a 'fire sub-climax' species that is eventually replaced by more shade-tolerant species in the absence of periodic fires. Many mature oak forests appear to owe their origin to periodic fires that created conditions suitable for oak seedling establishment, as well as the vigorous sprouting from older oak root sprouts. Fire control practices introduced in the 20th century permitted these sprouts to develop into well-stocked oak forests that now appear incapable of reproducing themselves, especially on mesic sites, in the absence of periodic fires.

Q. rubra also appears to have been an understorey component of many pine forests. Following the extensive logging of the 19th century 'pineries', *Q. rubra* 'advance reproduction' came to dominate these pine landscapes. In many present-day oak forests, eastern white pine (*Pinus strobus*) appears to be regenerating successfully, leading some to propose the existence of a 'pine–oak cycle'. In this scenario, disturbance regimes, especially logging and fire, create conditions suitable for the replacement of pine by *Q. rubra*. Subsequent management regimes that attempt to perpetuate *Q. rubra* sawtimber stands by excluding fire and extensive logging usually (unwittingly) favour regeneration of white pine, a species more tolerant of shade, under *Q. rubra* canopies.

Silviculture practice

Under specified conditions, both even-aged and uneven-aged forms of silvicultural systems have been recommended for *Q. rubra*. Clear-cutting *Q. rubra* stands often fails to perpetuate oak, either because such stands lack 'advance reproduction', or even when some seedlings are present in the understorey, they fail to out-grow competing species. However, clear-cutting can be successful with *Q. rubra* if site quality is poor, so that competition from more shade-tolerant species is minimal. Many oak stands deemed suitable for clear-cutting contain mixtures of several oaks, and resultant stands may have higher value for wildlife habitat than for timber. Very poor or infertile sites are often converted to pine plantations if commodity production is desired.

On 'average' sites, silviculture of *Q. rubra* concentrates on securing adequate 'advance reproduction' as an essential condition for perpetuating oak forests. In such situations, many foresters favour some form of shelterwood system. Some control of understorey and mid-storey vegetation is also required in order to permit oak seedlings to develop into large saplings. Underplanting seedlings may also be required if adequate *Q. rubra* reproduction cannot be obtained from seed. Shelterwoods should retain approximately 70% of crown cover in order to maintain adequate stand stocking for sawtimber production, while minimizing opportunities for shade-intolerant species to invade the site.

Single-tree or group selection is controversial with *Q. rubra*, and is considered difficult or impossible to practise on productive sites where competition from more shade-tolerant hardwoods can be intense. However, silvicultural regimes that encourage development of 'advance reproduction' by maintaining very low stocking levels (i.e. basal area of <14 m^2/ha) and approximate single-tree selection conditions can perpetuate oak stands on more xeric sites. Such treatments do, however, appear to shift species composition away from *Q. rubra* in favour of species such as pines and white oak (*Quercus alba*) that produce more frequent or abundant seed crops.

Group selection with *Q. rubra* has been advocated in situations where adequate 'advance reproduction' exists and where small groups (0.2–0.5-ha openings) can be embedded in a matrix of oak forest to permit continued sapling recruitment. As with shelterwoods, the success of group selection depends on the ability to develop adequate numbers of saplings from a combination of stump sprouts and seedlings capable of out-competing other species in the 'mini-clear-cuts' that group selection creates.

In terms of productivity, yields of 80-year-old oak stands in the central USA range from about 75 m^3/ha on poor sites to 175 m^3/ha on good sites. Appropriate thinning regimes can increase total stand yields at maturity to 103 m^3/ha on poor sites to more than 275 m^3/ha on good sites. Optimal growth in fully stocked stands of oak poles (25–35 cm diameter) on average sites can be achieved with stocking levels between 16 and 25 m^2/ha. Recommended thinning regimes reduce stocking to 60–70% of 'full stocking'. Most oak stands are even-aged, so thinning should be 'from below', with smaller suppressed and intermediate trees harvested. The remaining dominant and co-dominant trees will have larger diameters and crowns and produce better quality sawtimber.

Diseases and pests

Several diseases inflict considerable losses on *Q. rubra*, including stem cankers and root rots, as well as the vascular wilt disease caused by *Ceratocystis fagacearum*. Foliar diseases tend to cause mostly cosmetic damage, although anthracnose (*Gnomonia quercina*) can defoliate entire trees. For the most part, control measures consist of maintaining vigorous trees and practising good sanitation by removing dead and dying trees as soon as possible.

Like other common eastern hardwoods, *Q. rubra* experiences damage and mortality from a number of insect defoliators, including the gypsy moth (*Lymantria dispar*), the forest tent caterpillar (*Malacosoma disstria*), the orangestriped oakworm (*Anisota senatoria*) and several caterpillars and other Lepidoptera. The twolined chestnut borer (*Agrilus bilineatus*), the carpenter worm (*Prionoxystus robiniae*), the oak timber worm (*Arrhenodes minutus*), and several other borers and beetles attack *Q. rubra*, especially trees already weakened by defoliators, and cause considerable degrade of timber and loss of ornamental value. *Q. rubra* acorn losses to insect pests also can be high, with weevils of the genus *Curculio*, and the acorn moth (*Valentinia glandulella*), occasionally destroying 80–100% of the annual crop.

Uses

Q. rubra is occasionally planted in the eastern USA as a reforestation species because of its high timber and wildlife values, but establishment difficulties are often found. It is not a species that can be used in land reclamation or soil improvement practices. *Q. rubra* also has been widely planted as a landscape tree in parks and residential areas.

In the UK, *Q. rubra* timber is traditionally thought of as inferior to that of *Quercus robur*, but it is in demand in France for furniture, flooring and interior joinery. Untreated timber is unsuitable for exterior work because of its lack of natural durability. Wood density is 790 kg/m^3 at 15% moisture content.

Further reading

Johnson, P.S., Shifley, S.R. and Rogers, R. (2009) *The Ecology and Silviculture of Oaks*. 2nd ed. CAB International, Wallingford, UK, 600 pp.

Searle, S.Y., Turnbull, M.H., Boelman, N.T., Schuster, W.S.F., Yakir, D. and Griffin, K.L. (2012) Urban environment of New York City promotes growth in northern red oak seedlings. *Tree Physiology* 32, 389–400.

Wertin, T.M., McGuire, M.A., Teskey, R.O., Tognetti, R., Massacci, A. and Mugnozza, G.S. (2011) Higher growth temperatures decreased net carbon assimilation and biomass accumulation of northern red oak seedlings near the southern limit of the species range. *Tree Physiology* 31, 1277–1288.

Quercus suber L.

Cork oak

Importance

Quercus suber (Plate 37) is one of the most characteristic and valuable of all the Mediterranean *Quercus* species. It differs from other oaks in having a thick suberose bark (cork) on its trunk and branches and in its ability to grow a new bark or layer of cork immediately after having its bark stripped. Other Asiatic *Quercus* species have more or less suberose bark, but that of *Q. suber* is the only cork with the necessary homogeneity, quality and chemical and physico-mechanical properties for the needs of the cork industry.

Botanical features

Q. suber is a xerophytic oak species, usually growing to 15 m in height, but reaching 20–25 m in ideal conditions. Stems can reach 4–5 m in circumference. Isolated individuals usually have a short thick trunk, since, if formation pruning has not been applied, early branching is at low level and at very open angles. With time, and aided by pruning, the rounded crown of young trees spreads out, giving the tree a wide, majestic appearance.

In dense stands, a more acute branching angle results in more pronounced height growth and the tree takes on an irregular shape, with less interior foliage. This permits more light into these stands which, although dense, do allow the growth of some grassy and scrub-like vegetation, as well as the germination of acorns and other seeds.

The bark is smooth during the first 4–5 years, and thereafter takes on its characteristic appearance, with a thick cover of greyish cork, deeply grooved lengthwise and with convex surfaces between the grooves. These convex surfaces are called 'bornizo' or 'cortiça' and in wild trees can be as thick as 10–15 cm. Underneath the cork is the tanbark, which is very rich in tannin and clear yellow at first, changing to a cinnamon colour and finally to reddish-black after the first stripping. After this, the secondary cork forms, with a dark outer layer called the 'raspa'.

Leaves are simple, dark green, alternate and persistent (with a lifetime of 13–23 months), with a short (6–12 mm) petiole. Leaf blades thick, but not rigid, fairly polymorphic, 3–8 × 1.5–3 cm; leaf base entire or sinuate margin, dentate spinose. Dark green above, downy at first, then smooth.

The male inflorescence is a catkin, abundant in clusters of five or six at the ends of the first-year brachyblasts. Female flowers are short-stalked, isolated or in small clusters in annual budding. Acorns varying in shape and size even on the same tree, generally long, with a fluffy point; the cup is conical, with long, lax, greyish, downy scales that are somewhat disordered or outspread.

Q. suber develops a taproot which can go down several metres, and abundant rootlets near the surface which may give rise to shoots. This characteristic can be used to regenerate old trees by means of mouldboard ploughing, which by cutting the rootlets and shifting them nearer to the surface helps shoots to develop. These shoots are clearly distinguishable from plants from acorns, by the fact that they have whitish, downy, leaf undersides, and a much faster growth rate in the early years.

The existence of mycorrhizal associations with the roots plays an important role in the nutrition of the tree, especially important in soils poor in phosphorus and potassium.

Flowering time is diffuse, beginning in April and lasting throughout spring until summer; it is even possible to find autumn flowers which will give ripe fruit the following summer. The ripening period is from September until the end of January. *Q. suber* begins to bear fruit at 10–12 years old in good seasons, and regularly after 25–30 years. It usually bears fruit every year, but only gives high yields every 2–3 years.

Distribution

A typical Mediterranean species, *Q. suber* has many xerophytic adaptions. The species is found growing wild in Portugal, Spain, the south of France, Morocco, Algeria, Tunisia, Corsica, Sardinia and Italy. The countries with the highest area of *Q. suber* are Portugal, Algeria, Spain, Morocco and Tunisia, the densest forests being in Algeria and Tunisia.

The value of cork and the growing number of applications for this raw material has led to its cultivation in many countries outside its present native zone. Interest in introducing and reproducing *Q. suber* has been in Russia (in the 'subtropical' regions of Transcaucasia and in the Crimea), the USA (in California) and Korea.

Silviculture characteristics

Q. suber is a partly sun-demanding species, but fairly hardy in character. The adult tree needs an abundance of light, but the saplings need shelter and shade for 4–5 years on sunny slopes because of the high summer temperatures, and on shaded mountain slopes because of cold winter temperatures.

Silviculture practice

Q. suber reproduces well from acorns after a suitable after-ripening period. After gathering, acorns must be transported to the store as soon as possible, in netting bags which allow some ventilation, before storage on trays. If the store is temporary, it must be a well-ventilated cool place; the acorns must be kept in layers less than 10 cm deep, and stirred at least once or twice a day. If the final storage is carried out at temperatures below −1°C, fungal treatments are not necessary. For long-term storage, acorns should be kept between −1 and −3°C at 40–45% relative humidity, in non-hermetic containers and with permanent ventilation systems. Mixing the seeds with hygrophilous material, such as peat, is advisable when storing for lengthy periods.

Reforestation may be carried out by direct sowing of the seed in the soil, or by planting seedlings previously raised in the nursery. Seeds should be sown at a depth of 5–10 cm, with one or two acorns in each hole. Planting may be manual, in holes, or mechanical, along drills, according to soil characteristics. Initial minimum density should be approximately 625/ha at a spacing of 4 m. In plantations on flat land, with the provision of individual protectors (thereby raising planting costs), densities could be reduced to as low as 320–360/ha, as long as the usual replacement procedures are strictly carried out. On steep slopes, reforesting must be carried out at densities of over 750–800 plants/ha. Protection of seedlings and saplings is advisable in areas with rabbits or hares.

Irrigation might be necessary in summer, if spring and summer rainfall is insufficient, as is often the case in areas reforested with this species. There should be two or three significant waterings during the first summer, and eventually at the driest times in the initial years.

If the density of the plantation is high (>750–800/ha), the first thinning can take place at the time of the second formation pruning, after 12–15 years. This thinning, carried out some years after the first cork stripping, gives the selected trees time to develop their crowns, and facilitates trunk diameter growth. If the initial density is low (<600/ha), this first thinning is not usually necessary, except in the case of extremely defective trees. The stand should not reach stripping age at densities lower than 450–500/ha.

The peculiar character of silvopastoral exploitation for *Q. suber* has led specialists to consider a wide range of densities for producing optimum productivity. Stands whose main yield is cork should naturally have higher densities than those where livestock pasturing is of top priority or is also important. This results in a lower cork yield, but higher production of pasture, since the ground is less shaded. Very low densities of under 40–60 adult trees/ha, according to size, do not favour pasture production and greatly reduce cork yield.

Comparing the data from many experiments with the reality observed in many cork oak stands, optimal densities (trees/ha) for cork production are 250–500 (60–90 cm dbh), 125–200 (100–130 cm dbh), 85–125 (140–170 cm dbh), and 65–85 (180–200+ cm dbh).

Young trees branch from the base of the trunk, which means that formation pruning needs to be performed between 3 and 6 years old in order to avoid branches developing at too low a position and thus limiting cork production. The second pruning is performed at 12–15 years, pruning branches to a height of 1.5–2 m. Thinning is usually carried out at the same time as the second pruning to remove the weakest trees. A third and last formation pruning, to a height of 2.5–3 m, must be done after the first cork stripping after 20–25 years.

Cork (bark) stripping must be carried out using specialized techniques and tools so that the protective inner cambium is not damaged. The first stripping should not take place until the tree has reached a girth of 60 cm, and ideally up to a height of 1.2 m. In subsequent strippings this height can be raised to a maximum of 3 m. Stripping should be performed between mid-July and mid-August, or earlier in dry years and later in the year when June has been cooler than usual.

In terms of stand regeneration management, the method most used in *Q. suber* stands is that of the shelterwood uniform system by group, which consists in opening gaps in the stand, taking out old trees that yield poor-quality cork and those with trunks deformed by successive strippings or disease. This approach produces small, regular groups alternating throughout the blocks under regeneration, which gives an appearance of an uneven-aged stand by groups.

In cases where seed felling or secondary felling are to be carried out in more extensive stands, it is usually better not to perform the final felling, leaving some trees as an adult reserve stand, for a further two or three cork-stripping cycles. The nature of this will depend on the initial density of the stand at the start of regeneration, and should vary between 20 and 30% of the existing stock at the outset of seed felling. Adult trees which are kept should be good-quality cork producers, healthy, and, if possible, should not interfere with groups of new growth. This approach enables the block under regeneration to carry on producing a certain amount of cork from these reserves. After the first thinning of the young stand, and when it begins to produce cork, the older trees can be removed and a new production cycle then continues until the end of the rotation, when the process repeats.

The rotation period of *Q. suber* is 140–175 years, depending on the number of strippings performed, site quality and on silvicultural treatments performed.

Diseases and pests

Biscogniauxia mediterranea is the causal agent of charcoal disease of *Q. suber*. The disease is widespread in the Mediterranean region, where it affects several species of *Quercus*, although *Q. suber* is the preferred species, and the disease is closely associated with cork oak stand decline. *Botryosphaeria stevensii* causes a form of canker disease of *Q. suber*. A root pathogen, *Cylindrocarpon destructans*, is newly described as attacking *Q. suber* in Spain. Numerous pathogenic fungi were found in *Q. suber*, including *Cystodendron dryophilum*, *Dendrophoma myriadea*, *Elsinoe quercus-ilicis*, *Lembosia quercina* and *Phomopsis quercella* attacking leaves; and *Cryphonectria gyrosa*, *Nectria haematococca*, *Graphium* sp., *Ophiostoma piceae*, *Phomopsis* sp., *Phytophthora cinnamomi* and *Sporendocladia bactrospora* pathogenic on the stem.

Tortrix viridana attacks new buds and continues its attack as the leaves and flowers develop, thus causing serious fruit loss. *Lymantria dispar* is very common in Andalucian (Spain) cork woods, where it makes concentrated attacks on stands or groups. It causes devastating leaf loss, disturbing the phenological rhythm and obliging the tree to undertake a second foliation well into the summer. This extra effort drastically reduces the following spring's blossom, thus affecting two acorn crops. *Malacosoma neustria* is another defoliator of *Q. suber*. *Curculio elephas* lays its eggs in cork oak acorns in August or September, and early instar larvae feed on the cotyledons, reducing their food store. When the damage is limited to only part of the cotyledons the acorn can still germinate, although the seedling that emerges from it will exhibit weaker growth. If the damage affects the embryo, the acorn naturally loses its ability to germinate. Once acorns have fallen to the ground, if they have their cases broken or there are holes made by exiting *C. elephas* larvae, they are more liable to fungal rots caused by infection with *Fusarium* or *Ceratocystis* spp. *Dryomyia lichtensteini* causes leaf galls. The moth *Cydia splendana* also damages *Q. suber* acorns.

Other pests and diseases recorded include *Agrilus sulcicollis*, *Apiognomonia errabunda* (anthracnose), *Armillaria mellea* (armillaria root rot), *Cydia fagiglandana* (beech seed moth), *Euproctis chrysorrhoea* (brown-tail moth), *Phytophthora cinnamomi* (root disease), *Pythium spiculum*, *Fomes fomentarius* (hoof fungus) and *Platypus cylindrus* (oak pinhole borer).

Uses

Q. suber, both as an individual and in stands, is a tree of great aesthetic value. It has a protective role as a hydrological regulator, creating and conserving the soil, and is at its most important within its ecological domains, as a noble or climax species.

The wood is very heavy, hard and strong with a density of 770 kg/m^3 and has an attractive colour. The charcoal is of high quality. Pruning carried out to stimulate fruit or cork growth produces good firewood.

Cork is the main product of the tree. Good cork is compact, elastic and thick. In sites of good quality, fresh cork yield (kg/ha) can range from 3850 (1.2 m height) to 5150 kg/ha (1.7 m in height). In lower-quality sites, the fresh cork yield may range from 2210 to 2960 kg/ha (1.2–1.7 m in height). The tannin-rich tanbark is, as its name implies, the source of extracts used in tanning, which are obtained by peeling thick logs or felled trunks previously stripped of their cork. *Q. suber* mast years are also very important both for wild animals and domestic livestock, especially pigs. Acorn crops are very variable, depending on stand density, differing ecological conditions and human activities, such as stripping and pruning. On average, a tree yields 10–15 kg of green acorns, whereas a fairly sparse *Q. suber* woodland pasture can yield 800 kg/ha.

Further reading

Aponte, C., García, L.V., Pérez-Ramos, I.M., Gutiérrez, E. and Marañón, T. (2011) Oak trees and soil interactions in Mediterranean forests: a positive feedback model. *Journal of Vegetation Science* 22, 856–867.

Oliveira, G. and Costa, A. (2012) How resilient is *Quercus suber* L. to cork harvesting? A review and identification of knowledge gaps. *Forest Ecology and Management* 270, 257–272.

Torres Álvarez, E., Silva-Pando, F.J., Novo Lodeiro, N. and Álvarez Álvarez, P. (2003) Experiences on the natural regeneration of cork (*Quercus suber* L.). *Cuadernos de la Sociedad Española de Ciencias Forestales* 15, 37–47.

R

Robinia pseudoacacia L.

Black locust

Importance

Robinia pseudoacacia is a fast-growing, multipurpose tree used for land reclamation and rehabilitation in temperate and subtropical regions, and is also widely planted as an ornamental. It is nitrogen-fixing, improving soil fertility, and is also adaptable to environmental extremes such as drought, air pollutants, and high light intensities. It is a strong light-demander, is frost and drought hardy, deep-rooted, and wind-firm. It can be propagated easily by seed, coppice, and root suckers. The wood is hard, heavy, strong and durable and is used in some countries for general construction, flooring, tool handles, fence posts and fuelwood. Trees are lopped for fodder and flowers are important as a bee forage. It is an aggressive colonizer.

Botanical features

R. pseudoacacia is a medium-sized tree, generally 12–18 m high and 30–76 cm in diameter with an open, irregular crown, 30 m high and 122 cm or more in diameter on better sites. Open grown, the bole is often crooked and forked at 3–5 m, but in stands on good sites the bole is clear and straight. It has a shallow and wide-spreading root system that is excellent for soil binding but is also capable of producing deep roots (5–7 m deep). Radial root spread is about 1–1.5 times tree height. The smooth bark becomes reddish-brown and deeply furrowed with age, becoming 4 cm thick. It has sharp spines or thorns at the nodes of young branches and twigs that cause difficulties when handling seedlings, harvesting saplings and mature trees, and when feeding animals.

The leaves are alternate, deciduous, compound and odd-pinnate (imparipennate); they are 20–45 cm long and consist of 7–19 small, oval, alternate leaflets. Each leaflet is 3.8–5 cm long, 1.2–1.8 cm wide, broadest near the middle to uniformly wide, dull dark green. *R. pseudoacacia* flowers at a relatively early age (3 years). The fragrant, whitish flowers (less than 20 mm long), borne in lax to pendent inflorescences (racemes), appear after leaf emergence in May or June. The fruit is a flattened, oblong pod with a narrow wing along the ventral margin (Plate 60). The small pods contain 4–8 hard-coated seeds which can persist in the soil for many years. Seed crops occur every 1–2 years beginning at age 3; pods open on the tree and seeds are dispersed in winter and early spring.

Distribution

R. pseudoacacia is native to North America. The eastern section is centred in the Appalachian Mountains and ranges from central Pennsylvania and southern Ohio, south to north-eastern Alabama, northern Georgia and north-western South Carolina. The western section includes the Ozark Plateau of southern Missouri, northern Arkansas, and north-eastern Oklahoma, and the Ouachita Mountains of central Arkansas and south-eastern Oklahoma. Outlying populations appear in southern Indiana and Illinois, Kentucky, Alabama and Georgia. It has been widely planted elsewhere in North America.

It was first introduced to Europe in 1604 and has since become widely naturalized in many countries. Many consider this tree a weed species and it covers more than 20% of the forested land in Hungary. It has been successfully introduced into many parts of the world, even where climatic conditions are different from the native range. It is grown in temperate and subtropical regions in Africa, Asia and Australasia, and is even been grown at higher, cooler altitudes in the tropics, for example, in Java, Indonesia. It has become invasive in many countries.

Silviculture characteristics

R. pseudoacacia is a fast-growing tree which matures early; growth rate decreases rapidly after 30 years, especially on poor sites. Sprouts grow more rapidly than seedlings, and trees coppice well. It is sensitive to competition and is shade-intolerant. It does not readily develop epicormic branches, and tolerates pruning well.

The native range is classified as humid, with two local areas of super-humid climate. The range includes cool temperate moist forest, warm temperate montane, moist forest, warm temperate montane wet forest and warm temperate moist forest life zones. It occurs on upland sites in hardwood forests and along streams. It is found on a variety of soils but grows best on moist, rich, loamy soils, those of limestone origin, and soils without pronounced subsoil development. Trees do not tolerate waterlogging and are very sensitive to poorly drained or compact plastic soils, but also to dry sites.

Silviculture practice

Seed production occurs when trees are 6–60 years old. Seeds retain viability for as long as 10 years if stored in closed containers at 0–5°C. Germination is induced by pretreatment, commonly by soaking in near-boiling water for several minutes. Trees sprout readily from both stump and roots, especially after being cut or damaged, and also graft easily. Propagation by root suckers is most prevalent in natural reproduction. They can also be propagated, with difficulty, from hardwood cuttings (15–30 cm long and 1–2 cm diameter) collected in winter or early spring, and from root cuttings (5–8 cm long). The tree responds well to tissue culture and has been mass propagated by this method.

Seeds are sown at 250–300 seeds/m^2 in strips 15–20 cm apart, or in open nursery beds at 800 g/m^2. From germination to planting at 80 cm height takes about 12 months. Seedlings raised in nursery beds should be at a spacing of 25–30 cm in a 1-year bed, or 50–60 cm spacing in a 2-year bed. Direct seeding may be successful, easier, and less expensive than planting with seedlings. Direct seeding can be carried out in autumn in areas without late frosts, or in spring, seeding along strips or 20 cm holes, with the best plants released with an early thinning.

R. pseudoacacia will form nodules with a range of exotic strains of *Rhizobium*, but for effective nitrogen fixation, strains from native trees work best, and newly introduced trees require inoculation using soil from existing stands. Although it is tolerant of a wide range of soil nutrient conditions, it does respond to fertilizer, especially regarding phosphorus requirements, in intensive biomass plantations.

On the best sites, post-size trees are available after 15–20 years, and it takes 30 years to produce 20-cm diameter bolts. It is used in many countries in short-rotation, intensive culture for wood biomass production for energy purposes. In China, plantations are managed in a coppiced form to harvest forage, and it has become an important tree in the Himalayas where it is heavily lopped for fodder.

Trees can reach 3 m tall in one growing season, but average annual growth rates of 0.5–1.5 m in height and 0.2–2 cm in diameter growth are common. Trees can attain 12 m height in 10 years, 20 m height in 25 years and 26 m height and 27 cm diameter in 40 years. High-density biomass stands will generally yield 3–10 t/ha/yr, more on better sites. However, on very fertile sites under intensive management combined with genetic selection, much higher production can be expected.

Diseases and pests

The most damaging insect pest is the locust borer, *Megacyllene robiniae*, which creates holes that serve as entry points for fungi that cause extensive wood decay, and along with the heart rot fungi *Phellinus rimosus* and *Polyporus robiniophilus*, make growing the tree for timber production impractical in large areas of the USA. Several leaf miners are reported, including *Odontota dorsalis*, *Phyllonorycter robiniellus* and *Parectropa robinella*, and the locust twig borer, *Ecdytolopha insiticiana*. Many other pests and diseases are reported but few cause significant damage.

Uses

R. pseudoacacia is planted as a windbreak, a shade tree, for erosion control, amelioration of sites and reclamation of disturbed sites. It is also suitable for agroforestry and soil improvement, and grown in urban areas as an ornamental tree. *R. pseudoacacia* has potential as a nurse crop or pioneer species in stressed environments such as acid soils, dry sites or mined land.

R. pseudoacacia wood is strong and hard but fairly light, with a specific gravity of 0.68 and density of 740 kg/m^3 at 15% moisture content. The wood is usually straight-grained and fairly coarse textured, and has a tendency to warp. Wood is naturally yellowish-green in colour. It has excellent properties for steam bending, and high-pressure steaming can change the colour to golden-yellow, yellowish-brown, or light or dark brown. It is a durable, tough and high-quality wood which can be used for many purposes, including posts, poles, piling, mine props, wall panelling, flooring, fence posts, vine props, furniture, fruit boxes, pulp and paper and biomass energy.

R. pseudoacacia foliage is high in crude protein (24%) and is used as a livestock forage, but leaf protein digestibility is low and its proteins have low nutritive value because of the presence of condensed tannins. Nectar and pollen is valued as bee forage and trees are planted specifically for this use in Europe. Flowers have a very pleasant aroma and are used in the perfume industry to produce 'eau de toilette' and by herbalists as a remedy against stomach acidity.

Further reading

Barrett, R.P., Mebrahtu, T. and Hanover, J.W. (1990) Black locust: A multi-purpose tree species for temperate climates. In: Janick, J. and Simon, J.E. (eds) *Advances in New Crops*. Timber Press, Portland, Oregon, USA, pp. 278–283.

Benesperi, R., Giuliani, C., Zanetti, S., Gennai, M., Lippi, M.M., Guidi, T., Nascimbene, J. and Foggi, B. (2012) Forest plant diversity is threatened by *Robinia pseudoacacia* (black-locust) invasion. *Biodiversity and Conservation* 21, 3555–3568.

Hanover, J.W., Miller, K. and Plesko, S. (eds) (1992) *Proceedings of the International Conference on Black Locust: Biology, Culture and Utilization*. Department of Forestry, Michigan State University, Michigan, USA, 277 pp.

Huntley, J.C. (1990) *Robinia pseudoacacia* L. Black Locust. In: Burns, R.M. and Honkala, B.H. (eds) *Silvics of North America. Volume 2. Hardwoods*. Agriculture Handbook, No. 654. USDA Forest Service, Washington, DC, pp. 755–761.

S

Salix alba L.

White willow

Importance

Salix alba is a frequent and characteristic element of the landscape over much of temperate Europe. It is planted to protect riverbanks and water channels, and is managed for the production of timber for boxes, basketry, cricket bats and orthopaedic aids. It is also planted for reclaiming marginal lands, but is more common in linear plantations or along narrow riparian corridors. It is not a long-lived tree species, usually living less than 20–30 years, and has extremely short seed viability of only a few days.

Botanical features

A tall tree with a slender, well-developed (but often pollarded) trunk and greyish-brown bark, not flaking but deeply fissured when old. Branches are ascending and somewhat pendent at the apex, forming a pointed, rounded or lobed crown. The average height at maturity is 10–15 m with a diameter of 50–60 cm; it may grow to 30 m high with a diameter of more than 100 cm. Root systems are well-developed and deep, with lateral branching up to 30 m long.

Leaves are lanceolate or narrowly acuminate, 5–12 cm long, 0.5–1.5 cm wide, cuneate at the base, acuminate at the apex, minutely serrate, silvery-greyish, with the upper surface becoming dull green with age, the lower surface remaining grey-whitish. Leaf venation rather obscure, with 15–18 pairs; petiole short, less than 1 cm, usually with 1–2 pairs of sessile glands at the apex; stipules caducous, narrow, less than 5 mm long and 1.5 mm wide, with margin minutely glandular–serrate.

Catkins are cylindrical, spreading or suberect, terminal or on a very short, leafy shoot; male catkins up to 5 cm long, 0.8 cm wide, densely flowered; female catkins shorter than the male, developing at maturity. Capsule flask-shaped, about 4 mm long, with 8–10 seeds with a tuft of silky hairs that enable aerial dispersal.

Distribution

S. alba is a paleotemperate species, with a natural range extending from the Mediterranean basin (including the North African coast) to Northern Europe (UK, Germany, The Netherlands), the eastern Baltic coast (Lithuania and Latvia), to Russia – as far east as the Urals and Siberia, to the Chinese border, Iran and Asia Minor. Everywhere within and beyond its natural range it is cultivated along roadsides, on banks of reservoirs and rivers, in parks, and sometimes in forest plantations, so it is difficult to distinguish the natural distribution from the naturalized range.

It has been extensively introduced, including to China, India, Argentina, the USA, Australia and New Zealand.

Silviculture characteristics

S. alba is a typical species in temperate climate areas of central and north-western Europe, characterized by mild winters and warm summers with short summer drought, a bimodal rainfall pattern (spring and autumn rains), and sporadic snowfall that can damage and crowns. *S. alba* is found at altitudes up to 600 m in the Urals, 700–800 m in the Carpathians, 900–1000 m in the southern Alps, 1800–1900 m in the Pyrenees, Caucasus and Asia Minor, and 2400 m in the Atlas Mountains, Morocco.

As with the majority of riparian species, the presence of *S. alba* is linked to microclimatic conditions, particularly ground-water supply, rather than to the climatic pattern of the area. Many anatomical and physiological adaptations (surface leaf waxes, mucilaginous cells in the mesophyll, easy formation of adventitious roots, vigorous root and shoot growth, diffuse porous wood) show a wide tolerance for semi-arid climates and relatively cold winters. The fast growth and weakness of the wood makes it very prone to damage from wind and hail. Late and early frosts are also particularly damaging to twigs, flowers and fruits.

Like most other willow species, *S. alba* is able to colonize open areas. It is strongly light-demanding and cannot tolerate shade. New alluvial soils along rivers and creeks provide ideal sites for colonization, where both light and water supply are abundant. These environments are also favourable for the natural regeneration of this species, since seeds lack albumen and quickly lose viability and do not germinate in dry conditions. In temperate climates, *S. alba* can be coppiced successfully, and will rapidly develop a new crown with abundant suckers.

Silviculture practice

Propagation by seed is common in nature but hardly used in cultivation and seed is short-lived and very difficult to store even for a short time. Moreover, progenies are usually the result of hybridization. Vegetative propagation by cuttings is safer and more successful, and easy to collect and store for a while, even from autumn to spring, either in the open or in refrigerated storage. Nursery production of young plants is rarely implemented, and hardly convenient except for ornamental plants. Generally, cuttings are planted directly in the field, shortly after they have been collected, following soil preparation (tillage) and mechanical weed control.

Since natural woodlands in Europe where *S. alba* is present are mainly riparian woods maintained to protect riverbanks from erosion, their management guidelines are often regulated by nature reserve, park or protected area objectives. For industrial plantations in valleys and plains, the same practices that apply to poplar plantations can be implemented. In such plantations, a short rotation (10–15 years) and fairly wide planting spacings of 4 × 4 m or 5 × 5 m are used. In Argentina, more than 100,000 ha of pulpwood plantations were established in the 1950s using *S. alba* hybrids and coppiced every 8–12 years, producing 12–20 m^3/ha/yr. Pollarded trees are obtained by cutting back 10–12-year-old plants to 1.5–2.0 m in height, and managed by cutting back at regular intervals (from 3 to 12 years), depending on the purpose (wickers, wattles or poles).

Plantations can be coppiced and periodically clear felled, and trees can also be pollarded to obtain osiers for wickerwork and basketry. In the UK, *S. alba* subsp. *caerulea*, the famous cricket bat willow, is grown on 2–5-year rotations along rivers and streams with aerated water. In cultivation, trees require abundant space and light, and are therefore usually established in linear plantations. Being a short-rotation, rapid-growing species, it obviously depletes the soil of nutrients that should be reintroduced using fertilizers. To obtain good-quality wood, phosphorus and potassium should be supplied.

Diseases and pests

Marssonina salicicola and *Pollaccia saliciperda* attack *S. alba* and cause leaf, shoot and bark necrosis; *Monostichella salicis* and *Rhytisma salicinum* are responsible for leaf spots and necrosis; *Colletotrichum gloeosporioides* causes leaf spots and bark necrosis; *Ganoderma applanatum* and *Laetiporus sulphureus* are agents of wood decay. Bacteria also cause a number of diseases: *Pseudomonas saliciperda* (leaf necrosis), *Xanthomonas populi* subsp. *salicis* (stem and branch canker) and *Erwinia salicis* (wilting of leaf and shoots, and die-back). *E. salicis* is a threat to the production of wood for cricket bats in the UK, as it causes 'watermark' disease, in which the wood becomes brittle.

Among the xylophagous insects feeding on soft willow wood, *Aegeria apiformis*, *Paranthrene tabaniformis* and *Cossus cossus* cause particular damage by boring into the stem and large branches. Other notable insect pests are the curculionid *Cryptorhynchus lapathi*, the cerambicid *Aromia moscata*, and the buprestid *Dicerca ulmi*. The softness of the wood also makes the wood very susceptible to damage by many fungal pathogens and bacterial diseases. Many sap-sucking insects attack *S. alba*, such as *Aphrophora salicina*, the aphid *Tuberolachnus salignus* and the coccinid *Chionaspis salicis* cause the wilting of leaves and young twigs. Lepidopteran larvae such as *Nymphalis antiopa*, *Nymphalis polychloros*, *Cerura vinula*, *Leucoma salicis* and *Lymantria dispar* feed on green plant tissue. Similar damages to leaves and twigs may be caused by beetles *Phyllodecta vitellinae* and *Plagiodera versicolor* (Chrysomelidae), *Melolontha melolontha* (Scarabeidae), and *Polydrosus frater* (Curculionidae).

Uses

S. alba is planted as a pioneer tree for reclaiming quarries, refuse dumps and ex-industrial sites. Its fast growth allows a good crown cover to develop, and can prepare favourable soil conditions for more permanent species. In landscape planting, it can be used as a fast-growing tree which can be removed later when slower-growing species reach an acceptable size. Its environmental role is particularly important as a major component of riparian woodlands and corridors within cities, and in intensively cultivated agricultural areas.

Numerous cultivars of *S. alba* have been bred as ornamental plants. The most popular are those with a bright yellow shoot colour which is persistent until 3–5 years old (var. *vitellina*), and also those of bright orange-red (var. *brizensis* or var. *coccinea*).

S. alba is the only *Salix* species that reaches an appropriate size for timber. The wood has an air-dry density of 385 kg/m^3. It can be used for posts and beams, plywood, pallets, crates, and boxes and was also a source of firewood. Smallwood assortments can be used for matches, toothpicks, charcoal and wood pulp. Special products, such as wickers for basketry, are also obtained from pollarded trees. Industrial plantations provide stems of an appropriate size for plywood and pulpwood, but other wood products can be obtained from small farm forests.

Trees form a symbiotic association with the edible 'truffle' *Tuber magnatum* in natural conditions and *Tuber albidum* in cultured conditions, and are cultivated for such in parts of Europe. Most non-wood products from *S. alba* are, however, no longer of commercial importance, including tannin extracts from the bark, and salicin (from which aspirin was produced) from the wood, as well as other chemicals (resins, oxalate and enzymes).

Further reading

Leclercq, A. (1997) Wood quality of white willow. *BASE: Biotechnologie, Agronomie, Société et Environnement* 1, 59–64.

May, S. (1982) Willows for wood production. FAO International Poplar Commission. ENCC, Casale Moferrato, Italy.

Newsholme, C. (2003) *Willows: The Genus* Salix. BT Batsford Ltd, London, 224 pp.

Samanea saman (Jacq.) Merr.

Rain tree

Importance

Samanea saman (Plate 38) is cultivated throughout the tropics as a shade, ornamental and roadside tree. It is also a prime source of dry-season feed, and is one of the most common trees in silvopastoral systems in the seasonally dry tropical zones of northern South America and Central America. It is valued for its highly nutritious pods and also for the timber. It is notable for its large size, heavy branching, and very wide, spreading crown, characteristics which contribute to its value as a shade tree, but which largely rule out its use as a plantation species and which can limit its use for smallholder woodlots and in urban areas.

Botanical features

S. saman is a medium-sized or large tree often reaching 25–30 m and occasionally 45 m in height; a short stout bole to 2–3 m diameter; nearly horizontal branches; and a wide, low, spreading crown, often twice as wide as high. Bark is grey-brown, rough and furrowed into ridges. The leaves are large (6–25 cm long × 3–8 cm wide), bipinnate, with 3–6 pairs of pinnae per leaf and 6–9 leaflet pairs per pinna, 2.4–6.2 cm long × 1–2.5 cm wide, becoming larger towards the distal end, rhombic–oblong or elliptic, unequal at the base, dark olive-green, glabrous and slightly glossy above, dull grey-green and hairy below. Flowers are arranged in loose umbelliform heads in groups of 2–5 in the axils of leaves on actively growing shoots. The terminal or central flower on each head is sessile and enlarged compared to the peripheral flowers. The exposed and conspicuous stamen filaments are bright or pale pink and showy. Fruits are broadly linear, compressed pods, 10–22 cm long and 1.5–2.2 cm wide, dark blackish-brown when ripe. The thick mesocarp is filled with dry pitch-like, sweet, nutritious pulp. The pods are indehiscent and contain 5–10 mature seeds around 1 cm long.

Distribution

The presumed natural range of *S. saman* is Central America and northern South America: Nicaragua, El Salvador, Costa Rica, Panama, Colombia and Venezuela. The current Neo-tropical distribution has been aided by seed dispersal following the introduction of horses and cattle, and dissemination in the West Indies was followed by growth of the cattle industry, as shipments of cattle from Venezuela were often accompanied by bags of *S. saman* pods for animal feed during the voyage.

It was also widely introduced to many tropical countries in the 1800s as a shade or ornamental tree for streets, parks, roadsides and in coffee plantations, and it is now so widely cultivated, particularly in South-east and South Asia, that it is often mistaken as native to that area. The abundance, extent of naturalization, and wide distribution of *S. saman* in countries such as Papua New Guinea, Indonesia and India, with some extremely large, stately trees, are further witness to its early pantropical introduction.

Silviculture characteristics

S. saman is a truly tropical species, intolerant of frost. It thrives in both the seasonally dry and wet tropics but is faster growing where annual rainfall is more than 1000 mm. In its native range it grows at the margins of seasonally dry deciduous and semi-deciduous as well as moister evergreen woodland and savannah. It is a light-demanding species and generally has a very extensive shallow root system, but it is nitrogen fixing and promotes lush grass growth under its wide crown in extensive silvopastoral systems. *S. saman* is generally deciduous, but may be evergreen where planted in wet forest areas or as an ornamental in humid climates. Trees fruit from an early age and can live to an old age, sometimes in excess of 100 years. Flowering occurs towards the end of the dry season as the tree refoliates. Tiny immature pods persist for 8–10 months before expanding, ripening, and falling in the mid dry season nearly a year after flowering. Trees resprout vigorously after cutting.

Silviculture practice

Seeds of *S. saman* are widely available and easily collected. Extraction requires pounding of the hard, dry pods and subsequent winnowing, with 4400–7700 seeds/kg. Seeds are orthodox and are readily stored under normal seed storage conditions. Seed requires pretreatment prior to sowing, with boiling water the most widely used method. Container stock is normally used for successful establishment, in large bags, and it takes 3–5 months to produce seedlings 20–30 cm tall, though 12 month 'tall' seedlings 1.5–3 m high are favoured for ornamental and roadside plantings.

S. saman is grown on very wide spacings of 10–20 trees/ha in pasture systems, reduced to 6–8 trees/ha when older, which is ideal for the promotion of early and abundant production of pods as dry-season fodder, and good shade. It is rarely planted in conventional plantations due to its heavy branching and wide crown which demand very wide spacings (only 10–20 trees/ha in pasture systems, though 6 × 6 m is sometimes used). Wide spacings, however, can still be compatible with timber production. Given the preference for heartwood, the demand for large girths and the specialist uses of *S. saman* wood (which never produces clear sawlogs), wide spacing encourages rapid diameter growth and a thick bole with plenty of heartwood. Branch wood is also often of sufficient diameter to be used. *S. saman* can also be managed for fuelwood by pollarding in agricultural areas.

Yields of *S. saman* are extremely variable and generally low, but an average diameter of 18 cm at age 5 years and annual volume increments of 10–25 m^3/ha are reported. Trees can be grown on 25–35-year rotations, but in silvopastoral systems, rotations may advantageously be much longer as trees produce prodigious crops of pods to a great age. Under favourable rainfall and soil conditions, trees reach 50 cm diameter in 60 years and 100 cm in 100 years. In association with crops, however, the branching habit and lateral root growth require regular pruning to minimize competition.

Diseases and pests

Trees are sometimes attacked by a variety of diseases and insect pests, but which only cause limited damage, such as a canker disease caused by *Fusarium semitectum* (*Fusarium pallidoroseum*), and a cerambycid borer (*Aeolesthes holosericea*). Heavy infestations of leaf-eating caterpillars periodically defoliate trees just after new leaves emerge.

Uses

S. saman is most widely planted as a shade, ornamental and roadside tree. Its canopy, with crowns of great diameter, is unsurpassed for shade in silvopastoral systems, parks or roadsides. In addition, the pink flowers add to its value as an ornamental. It is also planted as a living fence, hedge, and as a visual and noise barrier. Although the green foliage and leaf/flower litter that collects beneath trees make an excellent nitrogen source for enriching soil, it is not pruned for such, due to the relatively slow regrowth as compared to most other trees.

The wood of *S. saman* is strong, durable or very durable, with a specific gravity between 0.42 and 0.64, a light yellow sapwood and rich dark chocolate-brown heartwood. The rich colour and beautiful but subtle grain of the heartwood, resembling black walnut (*Juglans nigra*), makes larger trees highly prized for furniture making, panelling, decorative veneers, carving, and for turning bowls, platters and other handicrafts; however, the wood is often very cross-grained, making it difficult to work unless green. Another benefit of turning when green is that the wood shrinks so little on drying that products do not warp. The wood is also used for fencing, construction timber, plywood, crates and boats. Cartwheels are made from single cross sections of the thickest trunks; these are said to be very durable and are a common sight on two-wheeled oxcarts in parts of Central America. The wood also makes good-quality fuelwood and charcoal.

S. saman pods contains a sweet nutritious pulp smelling of honey when the pods are broken. They contain 12–18% crude protein and are highly digestible (40% digestibility), and are a valuable fodder especially in the middle of the dry season when they ripen. The pods are also dried and milled into a feed meal with some international trade. The leaves are not widely used for livestock fodder, but have been reported to be of high quality with 22–28% crude protein, no detectable tannins, and high *in vitro* dry matter digestibility. The sticky sweet-flavoured pulp of the pods is often eaten by children and can be used in fruit drinks. *S. saman* is a good honey plant, is an important host plant for lac insects and exudes a low-quality gum from bark wounds.

Further reading

Cascante, A., Quesada, M., Lobo, J.J. and Fuchs, E.A. (2002) Effects of dry tropical forest fragmentation on the reproductive success and genetic structure of the tree *Samanea saman*. *Conservation Biology* 16, 137–147.

Durr, P.A. (2001) The biology, ecology and agroforestry potential of the raintree, *Samanea saman* (Jacq.) Merr. *Agroforestry Systems* 51, 223–237.

Staples, G.W. and Elevitch, C.R. (2006) *Samanea saman* (rain tree), ver. 2.1. In: Elevitch, C.R. (ed.) *Species Profiles for Pacific Island Agroforestry*. Permanent Agriculture Resources (PAR), Hōlualoa, Hawaii, USA.

Santalum album L.

Indian sandalwood

Importance

Sandalwood is the highly valued aromatic heartwood from most species of the genus *Santalum*. *Santalum album* is the most commercially important, having a fast growth rate, large tree form and the highest oil content (6–7%). Trees have significant cultural, medicinal and commercial importance in Asian, Middle Eastern and Pacific Island countries. Oil distilled from the heartwood is an important fixing agent in the perfumery and cosmetics industry, widely used for aromatherapy, and powder is used for incense sticks. Large sandalwood billets are also much sought after for furniture and carvings. There are increasing exploitation pressures on remaining trees, and most countries with indigenous *Santalum* species have established *in situ* conservation and plantation establishment programmes.

Botanical features

S. album is a small tree, sometimes only growing to 4 m tall in parts of its range, but sometimes to 15–20 m tall and a stem diameter of 70 cm. It has a straight stem, high bushy crown, and slender drooping branchlets. Bark is reddish-brown to dark brown, dark grey or nearly black, smooth in young trees, rough and fissured with age. Leaves are thin, ovate–elliptic, acute at base, acute or rarely obtuse at apex; 3–8 cm long and 3–5 cm wide, shiny green above but pale below. Flowers are faintly sweet scented or unscented, turning from green through pink to dark red or purplish-brown, in small terminal or axillary clusters. The single-seeded fruit is a round drupe, 7–8 mm in diameter, dark red, purple or black when ripe, with creamy-yellow flesh and a hard-ribbed endocarp.

Distribution

The natural range is from the tropical belt of the Indian Peninsula, through highland regions of eastern Indonesia, to coastal areas of northern Australia near Darwin. The origin of *S. album* is disputed, however. Some consider it native only to India, whereas it has also been hypothesized that it was introduced to India from West Timor, but there are references to its utilization in India from 2300 years ago. There is also debate about whether *S. album* is endemic in Australia or introduced via birds and fishermen from Indonesia.

Introduced populations occur in central and northern India, and in Java and Bali, Indonesia. Experimental introductions have been made in China, Fiji, New Caledonia, Hawaii, Tonga, Papua New Guinea, Nepal, Sri Lanka, east Indonesia and north Queensland, Australia. Commercial irrigated plantations have been established in Western Australia, and it has also been introduced to Kenya, Nigeria, Zimbabwe, Tanzania and Uganda, but with varied success.

Silviculture characteristics

Due to its hemi-parasitic nature, *S. album* grows best in association with other tree species but only when not dominated by them. *S. album* requires a host which is neither too vigorous and out-competes the tree, nor too weak so that the host is exhausted. A wide range of plants are parasitized by *S. album*, but it prefers nitrogen-fixing species. However, it can grow photoautotrophically (without a host) for up to 12 months.

S. album prefers 500–2000 mm annual rainfall and elevations of 650–1200 m, though trees have less heartwood in higher-rainfall zones. *S. album* tolerates extreme temperatures throughout its distribution from 4°C to 46°C. Fruiting begins around 2–3 years of age and natural regeneration is via bird dispersal of seeds, and root suckers. *S. album* has good coppicing ability when young, but this decreases with age. Young trees are sensitive to fire and browsing damage.

S. album is capable of growing on a range of soil types from laterite, loam, sand, clay and black cotton soils. It can tolerate alkaline and sodic soils with a pH up to 9.0 but is unable to tolerate waterlogged sites.

Silviculture practice

S. album can be propagated by dibbing seeds, transplanting nursery raised seedlings and vegetative propagation by cuttings. In India, direct sowing under potential hosts is commonly practised. For *S. album* plantations, it usually takes about 9 months to raise robust seedlings in large containers. Although it can grow without a host for several months, vigour and growth soon decline, so 2–3 months after germination, cuttings of a herbaceous pot host are planted into each *S. album* seedling container.

The parasite–host combination continues throughout the nursery phase and into the early stages of field establishment, and up to 18 months following plantation establishment. Intermediate and long-term host seedlings are raised simultaneously in the nursery and are placed strategically in the field at establishment. The long-term host, which must persist as the final host for the entire rotation length (25–30 years), is planted about 4 m from the *S. album* and pot-host combination. At the same time, an intermediate host is also planted between the *S. album* and long-term host seedling, which is parasitized for about 4–5 years until it dies. Ideal intermediate hosts are fast-growing, short-lived leguminous trees. Once parasitism is initiated, *S. album* remains highly heterotrophic and this sequence of hosts is required to maintain high plantation productivity.

S. album is dependent on its host for the absorption of water and nutrients which it is unable to absorb through its own root system, and performance varies significantly with different host species. Examples of good hosts include *Calotropis gigantea* and *Cajanus cajan* as nursery hosts, *Acacia villosa* and *Cassia siamea* as intermediate hosts, and *Casuarina equisetifolia* and *Prosopis juliflora* as long-term hosts.

S. album is often a slow-growing tree in natural stands, with annual diameter increments of 1–1.5 cm/yr, but up to 5–6 cm/yr under favourable conditions. The tree age at which heartwood formation commences varies considerably, with 10–13 years reported from India and 14–46 years in Indonesia. Assuming early heartwood formation beginning at 10 years old and increasing after about 20 years when diameters are 12–20 cm means that an 8 year-old tree with a diameter of 14 cm and height of 8 m can yield approximately 15 kg of heartwood.

The commercial value of *S. album* depends entirely on the quantity of heartwood and santalol oil content. In relatively short rotations it is important to maximize heartwood production and ensure that oil content is high. Annual global sandalwood heartwood production estimates are more than 5000 tonnes with a value of more than US$ 1 billion. Assuming a heartwood yield of 20 t/ha with a rotation length of 25 years, a plantation resource of around 6400 ha is suggested to meet current global sandalwood consumption, requiring an annual planting programme of 255 ha. India and Indonesia supply about 70% and 30% of the world's sandalwood oil, respectively. However, a period of sandalwood scarcity is looming, due to heavy pressure for clearing forested land for food production, the destruction of host trees for wood products, diseases such as spike disease, and illegal harvesting. Plantations are the only way to redress increasing problems of resource availability.

Diseases and pests

Spike disease is a serious pathogen in India, causing a shortening of the internodes, reducing leaf size, killing haustoria, blocking vascular tissue, and eventually killing trees. Nursery diseases include the pathogenic fungi *Fusarium* and *Phytophthora*. A wide range of nematodes and insect pests are reported.

Uses

The tree is useful as hedge plantings or for shade or windbreaks, and has potential as an agroforestry species on marginally productive agricultural lands.

The heartwood is close grained, very fine and even textured, hard and heavy, with a specific gravity of 0.92 and density of 897–1137 kg/m^3. It is durable and renowned for carving, religious artefacts and valuable handicrafts. The timber seasons well when dried slowly. The wood can be worked to a smooth finish and takes a satin-like polish.

Oil is widely used in the agarbathi, cosmetic, fragrance and soap industries and, in terms of volume consumed, its use as a base of fragrance far outweighs its use in medicine. It is produced commercially by steam distillation of the heartwood with an average yield between 4 and 6%. Fruits and kernels are edible and may have potential as a horticultural crop until mature trees are harvested. Seed oil has industrial applications. Leaves are used for fodder and green manure.

Further reading

Dhanya, B., Viswanath, S. and Purushothaman, S. (2010) Sandal (*Santalum album* L.) conservation in southern India: a review of policies and their impacts. *Journal of Tropical Agriculture* 48, 1–10.

Gjerum, L., Fox, J.E.D. and Ehrhart, Y. (eds) (1994) *Sandalwood Seed Nursery and Plantation Technology*. Proceedings of a regional workshop for Pacific Island countries, 1–11 August 1994. UNDP/FAO South Pacific Forestry Development Programme, Noumea, New Caledonia.

Srimathi, R.A., Kulkarni, H.D. and Venkatesan, K.R. (1995) Recent advances in research and management of sandal (*Santalum album* L.) in India. Associated Publishing Company, New Delhi, India, 416 pp.

Schefflera morototoni* (Aubl.) Maguire *et al.

Matchwood

Importance

Schefflera morototoni is a well-known pioneer species found throughout the tropical Americas. It is a timber tree but often called matchwood because the wood is used for match splints in several countries. It is suitable for restoration of degraded soils or abandoned agricultural land and is also recommended for use in 'taungya' systems, where crops are grown amongside the growing trees for as long as the shade from the trees permits.

Botanical features

S. morototoni is a medium-sized tree 8–30 m in height, with a straight cylinder-shaped bole 30–80 cm in diameter. It is sparsely branched and has a small crown with ramifications only at the apex. The bark is whitish and has a smooth surface. Leaves are large, compound, digitate, 15–40 cm long, and alternate. The petiole is 30–60 cm long, with well-developed stipules, inserted obliquely on the stem. The 10–11 folioles have elliptical lance-shaped limbs and are acuminate, undulate, sinuate on the margin, semi-coriaceous, dark green on the upper surface, and brown-red and heavily pilose underneath. Flower clusters (panicled umbels) are lateral, about 20–60 cm long and broad, with grey and finely hairy branches. The flowers are very numerous in many small, rounded clusters (umbels) less than 1.3 cm across, on spreading flower stalks 1.6–4.7 mm long. The pentamerous, finely brownish and grey hairy flower of about 4.7 mm across has a minute basal tube (hypanthium) less than 1.6 mm long enclosing the inferior ovary and bearing a minute 5-toothed calyx. The flowers have 5 petals white, pointed, and more than 1.6 mm long, 5 stamens and 2 styles. The fruit is an ovate drupe, grey, and covered with a bloom, about 4–9 mm long and 6–12 mm broad; slightly flattened, with 2 styles at the apex. It contains 2 oblong, 4.7 mm long flat brown seeds. Flowering and fruiting occurs nearly throughout the year.

Distribution

S. morototoni is found between 18°N and –31°N and 0–2150 m altitude. It is native to southern Mexico, the Greater Antilles, Central America and South America. The species was introduced to Jamaica and has been planted in southern Florida. It is quite common in Puerto Rico but nowhere else in its range. Local or regional range maps are known only for Colombia and Puerto Rico.

S. morototoni occurs in a variety of habitats such as tropical dry forests (e.g. caatinga), cerrado, and in the wet forests of tropical America. It prefers open forests. An early successional species, it is shade-intolerant and rapidly colonizes disturbed sites. Throughout its range, it is common in secondary forests, and along roadsides or riverbanks. It tolerates both sandy and clay soils, but is typically found on acid soils. It grows in areas with rainfall ranging between 1000 and 8000 mm and a mean temperature of 17–27°C (minimum mean 12–26°C).

Silviculture characteristics

S. morototoni is easy to establish using nursery stock and has rapid growth. In plantation trials in Paraná, Brazil, this species grew best in well-drained clay soils with additional fertilizer. At 3 years of age, better growth in height has been obtained at 3 × 2 m spacing and a larger diameter at a 4 × 4 m spacing. Production of 15–18 m^3/ha per year in uniform stands can be expected. *S. morototoni* demands strong light. It tolerates drought and the species is often favoured due to its fire-resistant seeds. It self-regenerates and self-prunes, but in some conditions it may become an aggressive colonizer. Wind can break branches or even uproot entire trees.

Silviculture practice

S. morototoni can be propagated from seed (seed storage is orthodox) and germination is epigeous, but also from cuttings, grafting, and tissue culture. After burning in the field the seeds germinate rapidly; but when they are collected and taken away from field conditions, the number of viable seeds is quite small and germination is poor. To increase the capacity and speed of germination fruits should be harvested when they present a purplish-green coloration. The seedlings transplant easily. Early growth is fairly rapid and best when seedlings are exposed to direct sunlight. The tree may also be regenerated through coppicing.

Diseases and pests

The wood is susceptible to decay, fungal attack and termites, and once logs are felled they should be removed from the forest quickly.

Uses

The rapid growth and physical features of *S. morototoni* give it potential as an ornamental.

The low-density wood (580 kg/m^3 at 12% humidity) is easy to work with a fine finish. It is light cream in colour with some grey. It has regular grain and a medium texture, with a shiny and smooth surface. It is used for matches in Trinidad and British Guyana, and also for light interior construction, furniture, veneers, crates, boxes, toothpicks, toys and other wood-based materials. *S. morototoni* is also used in the pulp and paper industry and as a substitute for heavier grades of balsa. The wood has a high risk of deformation and needs to be dried carefully before processing.

When the bark is cut or damaged, a sticky colourless latex is produced, but no use of this is recorded. The leaves have been known to be used as home remedies.

Further reading

Carvalho, P.E.R. (1994) *Brazilian Forest Tree Species: Silvicultural Recommendations, Potential and Use of Wood.* Especies florestais brasileiras: recomendacaes silviculturais, potencialidades e uso da madeira. EMBRAPA Centro Nacional de Pesquisa de Florestas (CNPF), Colombo, Brazil, 35 & 639 pp.

Fiaschi, P. and Pirani, J.R. (2008) Patterns of geographic distribution of the extra-amazonian Brazilian species of *Schefflera* J. R. Forst. & G. Forst. (Araliaceae). *Revista Brasileira de Botânica* 31, 633–644.

Little, E.L. and Wadsworth, F.H. (1964) *Common Trees of Puerto Rico and the Virgin Islands.* Agricultural Handbook No. 249. US Department of Agriculture, Washington, DC.

USDA Forest Service (undated) *Didymopanax morototoni* (Aubl.) Decne. & Planch. Available at: http://www.na.fs.fed.us/pubs/silvics_manual/volume_2/didymopanax/morototoni.htm (accessed 1 December 2012).

Schinus molle L.

False pepper tree

Importance

Schinus molle (Plate 39) is a common component of the vegetation in parts of western South and Central America from Argentina to Mexico, where it is exploited for timber, fuel, poles, and sawn wood, as well as various non-wood products. The common name 'false pepper tree' and its variants indicate the widespread use of the fruit as a spice, similar in appearance when dehusked and in flavour to the peppercorns of cultivated black pepper (*Piper nigrum*). The small unhusked, red-skinned fruits or peppercorns of *S. molle* have been successfully commercialized by mixing them with traditional black and white peppercorns, making colourful mixtures. The flavour is slightly sweeter than that of other peppers. *S. molle* has attractive, weeping foliage and has been widely introduced to countries with a Mediterranean climate as an ornamental, and elsewhere for soil stabilization. In some countries where it has been introduced as an ornamental, notably South Africa and the US states of Florida and Hawaii, it has become an invasive weed.

Botanical features

S. molle is a tree up to 15 m tall, often less than 10 m, with a spreading crown comprising drooping, pendulous twigs, and weeping foliage. It is evergreen and unarmed. The bark is light grey-brown to dark brown, scaly or fissured, and peels to exude a sticky latex when damaged. The trunk, although often short, may be 2–4 m high and 60 cm in diameter. Branches tend to be brittle and may break in strong winds. The species is generally shallow rooted.

Leaves are pinnate, alternate, imparipinnate, 25(–30) cm long, with a peppery smell when crushed, 12–15(–40) pairs of leaflets, linear–lanceolate, narrow, 3–4(–6) cm long.

S. molle is dioecious. Flowers are very small, green-yellow in colour, with 5 sepals, arranged in lax, hanging, open panicles at the ends of branches. The fruit is a berry containing a single, small, black seed, spherical and 3–5(–7) mm in diameter, similar to a black peppercorn, in a thin red (purple to rose-coloured) shell, with a strong, peppery smell and sweet peppery flavour. The fruit hangs in dense clusters of many dozen seeds, and may be persistent on the tree.

Flowering and fruiting may occur throughout the year, but in the northern hemisphere peak flowering is from August to November and peak fruiting from January to March. Seeds are dispersed by vertebrates, including foxes and birds, whose frugivory enhances seed germination.

Distribution

S. molle is native to Central and South America. It is a common species in the central plains of Mexico, the only tree in parts, and is recorded in some Central American countries and assumed to be native to all other mainland states in mountainous areas. It is noted in Venezuela, Colombia, Peru and Ecuador in the Cordilleras and Andes. It is common in ancient woodland in Peru, where in parts it is the only tree species. It exists in the very north of Chile, Bolivia and northern states of Argentina, with outlying populations recorded from Paraguay, Uruguay and southern Brazil, assumed to be native but may have become naturalized in prehistory. A number of publications, however, state that it is native only to the southern part of this range in central-southern South America and was introduced and naturalized in Mexico and the USA.

Silviculture characteristics

S. molle is generally described as a fast-growing species and with its shallow root system could be considered in pure or mixed plantations for both production and protection purposes. Although used in traditional agroforestry systems at altitudes below 3200 m in the Bolivian Andes, it is not generally very suitable as an agroforestry species, having allelopathic effects on companion crops. *S. molle* tolerates high temperatures and once established is extremely drought-resistant. It is also resistant to frost and temperatures as low as −10°C. Being shallow-rooted, it is susceptible to windfall. *S. molle* is considered a fire-retardant species. It has shown potential for immobilizing heavy metals when planted on disused copper and molybdenum mining sites.

Silviculture practice

Fruits do not always produce viable seeds and a cutting test is recommended to ensure that seeds are ripe and ready for harvest. Flotation can also be used to separate empty, non-viable seeds from viable ones. There are 30,000–45,000 seeds/kg depending on provenance. Seeds are collected when dry on the tree. The red husks are rubbed off and the seed sown directly into pots without any pretreatment, although soaking in cold water for 12 h is sometimes undertaken to overcome dormancy in stored seeds. Germination of fresh seeds is approximately 50% (40–80%) and tends to fall dramatically after 3 months at room temperature, though if sun-dried to 9–12% moisture content and stored in airtight containers at −10°C, it is reported that seeds can be stored for several years.

After sowing, seeds germinate in (10–)14–21(–30) days. Wide spacing of seedlings is required to reduce the incidence of *Meliola rhoina* infection where this disease is present. A nursery period of 5–6 months is generally recommended, but 3–4 months is also adequate. Young plants are susceptible to termite attack and require weeding during early establishment.

Little is known about specific management of *S. molle* as it is rarely established in plantations. In field trials in Almeria, Spain, it was established successfully in small planting holes (30 × 30 × 30 cm) in microcatchments at 4 × 4 m spacing. Mean annual rainfall in the region is 350 mm, but no post-planting irrigation was applied and survival was 85% after 2.5 years. *S. molle* had the highest survival and one of the highest growth rates of the eight species tested and was recommended as a species for further planting in the region. *S. molle* has also been established in Algeria in 40 × 40 × 40 cm holes. In Ethiopia, Yemen, Israel, Australia, and elsewhere, although of the species trialled *S. molle* was never the best performing, its survival rates were often around 50% in the first few years. It appears to prefer adequate water or high humidity during early establishment to avoid losses. Because of its low-branching habit, pruning of lower branches is recommended when the tree is young if clearance beneath is desired, and to reduce the chances of the tree being toppled by strong winds. *S. molle* trees reach maturity in less than 20 years. Coppicing, pollarding, and lopping are viable and alternative methods of management.

Diseases and pests

S. molle seedlings in nurseries were observed to be infected by *Botrytis cinerea* wilt in Chile and by *M. rhoina* in Minas Gerais, Brazil, although incidence of the latter was reduced by wider seedling spacings. A phytoplasma affecting adult trees was recently identified in Cochabamba, Bolivia.

Adult *S. molle* trees are also attacked by a number of arthropod species, both in the native range and where introduced, but damage inflicted is generally only moderate. A survey of insects present on trees growing in Quebrada de Humahuaca in Argentina obtained 679 specimens, the most notable pests being *Calophya schini* and the defoliator *Tolype guntheri* on leaves; *Thrips tabaci* and several *Frankliniella* species in flowers; and *Megastigmus* sp. and *Lithraeus* sp. attacking fruits. Beneficial predators and parasitoids were also present. *C. schini* was recently found infesting *S. molle* trees in the Lisbon area of Portugal, the first record of this pest in Europe and the Palaearctic region. *Megastigmus transvaalensis* is the main seed predator in South Africa.

Heavy infestations by mistletoes were noted in *S. molle* trees growing on the campus of Egerton University in Njoro, Kenya.

Uses

S. molle is commonly planted as an ornamental tree and for amenity situations, having an attractive appearance and giving good shade. The shallow rooting system and rapid growth rates have also been utilized by planting as a part of sand and soil stabilization projects, such as in Botswana. It is reported to have been planted as a living hedge and also in shelter belts.

The wood is moderately hard and moderately durable with a density of 540–680 kg/m^3, and is thought to be termite resistant. It makes a reasonable-quality timber for rough building work but is not a highly valued timber for any specific purposes. However, *S. molle* is predisposed to burl formation, giving a swirling grain valued in furniture making. Posts and poles are also cut and utilized for fencing and small constructions. The wood has also been used for reinforcing concrete in Mexico in place of wire netting. It is also burnt directly as a fuelwood of good quality and can be made into charcoal.

The principal non-wood product is the seed which is used as a spice, having a similar flavour to black pepper, though not as strong and somewhat sweeter. The leaves are also added to dishes to impart this peppery taste. When left in their red-rose-coloured husks, the fruits enter the European and North American markets as 'false pepper', often sold in mixtures with traditional black and white peppercorns in bags, glass jars or transparent peppermills, providing a colourful and ornamental pepper mix.

Various parts of the tree are used in traditional medicine in its native range. Plant extracts have anti-inflammatory, hypo-sensitive, anticarcinogenic, antimicrobial, antifungal, antibacterial, and insecticidal properties.

The tree produces an aromatic resin used as a mastic. The foliage is used as browse in Egypt. Tannins and fibres can also be obtained from the bark. The flowers make a honey of good quality, and the tree produces latex, though it is not clear whether this is used.

Further reading

Howard, L.F. and Minnich, R.A. (1989) The introduction and naturalization of *Schinus molle* (pepper tree) in Riverside, California. *Landscape and Urban Planning* 18, 77–95.

Richardson, D.M., Iponga, D.M., Roura-Pascual, N., Krug, R.M., Milton, S.J., Hughes, G.O. and Thuiller, W. (2010) Accommodating scenarios of climate change and management in modelling the distribution of the invasive tree *Schinus molle* in South Africa. *Ecography* 33, 1049–1061.

Zahed, N., Hosni, K., Brahim, N.B. and Sebei, H. (2011) Essential oil composition of *Schinus molle* L. fruits: an ornamental species used as condiment. *Journal of Food Biochemistry* 35, 400–408.

Schizolobium parahyba (Vell.) Blake

False tree-fern

Importance

Schizolobium parahyba (Plate 40) is a fast-growing tree, and is used widely in mixed plantations in Brazil for land reclamation, environmental purposes, and also as an ornamental and shade tree. In some South American countries the species is used in agroforestry systems. In Ecuador, it has been planted since the 1980s for timber production, and currently about 12,000 ha are being grown. Its low-quality wood is mainly used for light boxes (such as for packing fruits) and corestock veneers. It is considered excellent for pulp and short-fibre paper production, with fibre length varying from 1.10 to 1.59 mm.

Botanical features

S. parahyba is a deciduous tree and commonly reaches heights of 10–20 m with a dbh of 30–60 cm, occasionally attaining 40 m in height and a dbh of 100 cm. It presents monopodial growth, a cylindrical and straight trunk marked by petiolar scars, sometimes with prop-roots, a bole up to 15 m in height, deep roots, cymose ramification, and a very wide umbelliform crown. The external bark is almost smooth, green when young and greyish when adult, with lenticels present.

Leaves are bipinnately compound, alternate, showing 15–25 pairs of opposite pinnae, each one with 20–30 pairs of oblong leaflets 2–3 cm long. In mature trees the leaves are about 0.5 m long, but in the first few years of growth they can reach 1.7 m in length or more. Young individual trees, often unbranched and with these long leaves, are often mistaken for tree-ferns or palms, hence its common name.

In plantations, flowering begins between 6 and 8 years. The flowers are large, about 3.5 cm in diameter, golden-yellow in colour and gathered in terminal racemes up to 30 cm long (Plate 54). The fruit is ovate–oblong, dehiscent, winged, sessile, flat, coriaceous or sub-woody, 10–16 cm long and 4–6 cm wide. It generally contains one seed, sometimes two. The seeds are flat, brilliant, 2–3 cm long and 1.5–2 cm wide, with a hard integument.

In Brazil, *S. parahyba* flowers from July to December, and the fruits ripen from March to October, according to region. Leaves fall during the winter.

Distribution

The natural distribution of *S. parahyba* ranges from southern Brazil, 30°15'S, to southern Mexico, 17°N. In Brazil, it is a characteristic and exclusive species of the east coast Atlantic forest and it occurs occasionally in the semi-deciduous forests of São Paulo state.

Silviculture characteristics

S. parahyba is a light-demanding climax species; it is common in secondary vegetation and may form dense groups in large gaps. It is a widespread pioneer species in tropical and premontane forests, flourishing on well-drained moist soils on plains or hillsides. It sprouts after cutting, at any height of the trunk. It presents intense self-pruning when young and is a short-lived tree. In plantations for environmental purposes in southern and south-eastern Brazil, *S. parahyba* shows fast growth in comparison with other species; at 3 years old trees can be 7–8 m tall.

Silviculture practice

The seeds are orthodox and remain viable for a long time when stored; 1 kg contains almost 550 seeds. Mean moisture contents for intact and split seeds determined by the oven-dry method are 4.5% and 11.4%, respectively. Seeds present seed coat dormancy, which is easily overcome by mechanical scarification, boiling water, or concentrated sulphuric acid. With these treatments, germination occurs within 5–15 days and is generally high. Apart from propagation by seed, *S. parahyba* is also suited to vegetative reproduction using young branch cuttings.

S. parahyba can be established by direct sowing of seed or by planting of seedlings. Seedlings can be produced in polybags or plastic tubes, and reach 25 cm in height and 5.2 cm in diameter at the stem base in three months. *S. parahyba* shows low incidence of mycorrhizal colonization (1–19%) when inoculated with *Glomus etunicatum* and *Gigaspora margarita*.

When direct sowing is used, *S. parahyba* shows good seed germination, seedling survival and early growth rates; 4.1–4.6 cm stem diameter and 1.5–1.7 m height growth during the first 2 years have been observed with sowings on abandoned farm land in São Paulo state, Brazil.

S. parahyba is being increasingly grown in plantations on abandoned agricultural land, especially in Amazonia. In a plantation on nutrient-poor soil at a spacing of 3 × 3 m in southern Minas Gerais, Brazil, which has a 5-month dry season, *S. parahyba* reached 2.35 m in height after 3 years. In São Paulo, a plantation on dusky-latosol reached 21.5 m in height and 47 cm in dbh after 20 years. Yields of 600 m^3/ha have been obtained in Brazil at 10 years, while in Ecuador, a 15 year-old plantation exhibited a reduced growth rate and high mortality; it reached a maximum height of 24 m and a total volume of approximately 90 m^3/ha. In Venezuela, annual height growth reached its maximum in the second year, with an average total increment of 10.5 m, declining to 0.2 m/year after 5 years.

Diseases and pests

A die-back and stem rot disease was detected in *S. parahyba* in 1988, killing young trees in the Ecuadorian humid tropics, and since then thousands of trees have been dying annually. Five fungi have been isolated from infected trees: *Macrophoma* sp., *Fusarium* sp., *Graphium* sp., *Ceratocystis paradoxa*, and *Ceratocystis moniliformis*. A canker infecting trees in field plantings in Pará, Brazil, was identified as being caused by *Lasiodiplodia theobromae*, the first report of this pathogen on *S. parahyba*.

S. parahyba is susceptible to attack by *Acanthoderes jaspidea*, and in Brazil this has been reported as a major pest for this species. Attacks mainly occur during summer and the beginning of autumn, and the risk is higher during the first 4 years. Other pests include *Rhaphiorhynchus pictus*, the wood-boring beetle *Micrapate brasiliensis*, the branch and stem girdlers *Oncideres dejeani* and *Oncideres saga*, and acarid mites.

Uses

S. parahyba is suitable for restoration of riparian forests in south-eastern Brazil in mixed plantations. It is also recommended for agroforestry systems, and is grown in association with banana, cassava, coffee and pasture. It is popular as an ornamental due to its masses of golden-yellow flowers, and as a shade tree.

S. parahyba wood is very light (density 320–390 kg/m^3, increasing with age), white-yellowish and soft, of low durability under natural conditions. The surface is irregularly lustrous, the texture thick, and the grain irregular. The sapwood and heartwood are not differentiated. Although the timber has an imperceptible smell, when freshly cut it reputedly has a repulsive odour. The wood has good physical and mechanical characteristics, it dries and works well, being easy to cut and process, but does not hold nails and screws well. It has low hardness and resistance to pressure and also low resistance to attack by fungi, insects and environmental factors in general; consequently it is not recommended for use as structural wood or for external use. The wood is used for panels, doors, toys, model aircraft, matches and light boxes. It produces a good-quality veneer. *S. parahyba* is considered excellent for pulp production and short-fibre pulp, producing wood and bark with more than 60% fibre. It is also suitable for the manufacture of particleboard.

S. parahyba bark contains tannins, which are used to tan leather. The bark also has astringent therapeutic properties and is used in popular medicine. Water-soluble extracts from the leaves are effective antidotes against snake (*Bothrops* spp.) bites. The flowers produce pollen and nectar for honeybees, to give a clear and perfumed honey. Seeds are traditionally used as buttons and beads.

Further reading

Carvalho, P.E.R. (2007) *Paricá,* Schizolobium amazonicum. Circular Técnica - Embrapa Florestas No. 142, Colombo, Paraná, Brazil, 8 pp. (in Portuguese).

Condit, R., Pérez, R. and Daguerre, N. (2011) *Schizolobium parahyba*. In: *Trees of Panama and Costa Rica*, Princeton University Press, Princeton, New Jersey, USA, pp. 184–185.

National Research Council (2002) *Schizolobium* species. In: *Tropical Legumes: Resources for the Future*. Books for Business, New York, USA, pp. 204–205.

Sequoia sempervirens (D. Don) Endl.

Coast redwood

Importance

Sequoia sempervirens, the sole species in the genus, has a special significance not only economically but also ecologically and aesthetically. It represents an ancient lineage of conifers which once ranged much more widely across the northern hemisphere. Durability and resistance to attacks by pests and pathogens make the timber valuable for exterior construction and furniture. *S. sempervirens*, which forms extensive stands of massive forest trees in many north coast habitats of California, USA, grows taller than any other tree species in the world and is second only to *Sequoiadendron giganteum* in trunk wood volume.

Botanical features

S. sempervirens is a large tree of pyramidal canopy form with horizontal to drooping branches. The canopy becomes cylindrical with age, with branches absent from the lower half of the trunk. Trees are often 60–100 m tall, with the tallest known tree being 112 m. Mature trees have a dbh of 1.5 to 5 m, occasionally over 7 m. Growth form is regular, with the primary bole tapering only slowly with height. The base is often buttressed. Stem fasciation may also be present.

The bark is brownish-red and fibrous, forming a thick protective layer 15–30 cm thick around the trunk. The root system is spreading, with no major taproot. In floodplain habitats *S. sempervirens* often shows layers of new lateral root formation associated with deposition of sediments around the base of the trunk. The root system also forms a distinctive root crown or lignotuber.

The foliage is non-resinous. Most leaves consist of flat sprays of lanceolate needles, 12–25 mm long, 2–3 mm wide, glossy green above, and bearing two pale glaucous bands with stomata below. Needles commonly remain for 3 years. Another form of foliage may occur high in the canopy or at branch tips, with needles scale-like, 6–12 mm long, and sharp-tipped.

Staminate cones are ovoid and green. Female cones are red-brown, 20–35 mm long, and hang pendulously when mature. They include 15–20 oblique shield-shaped scales that are terminated in a flattened spine. There are 3–7 ovules near the margin of the scales, each with 2–5 seeds, the seeds being 5–6 mm long, with two narrow spongy wings.

Staminate cones shed pollen from November to March. Female cones are receptive to pollen during this period and mature the following autumn and winter.

Tree radial growth commonly begins in late March, reaching a maximum rate in May, declining slowly thereafter until September.

Distribution

S. sempervirens is distributed in an irregular coastal belt about 725 km long and 8–55 km wide in the western USA, from the extreme south-western corner of Oregon southward to Salmon Creek in the Santa Lucia Mountains of California. The total area of redwood habitat is about 647,500 ha. Stands are naturally distributed from sea level to an altitude of about 915 m, with the majority of stands at 30–760 m. *S. sempervirens* has been widely planted throughout the world, particularly in Europe.

Silviculture characteristics

S. sempervirens is long-lived, reaching 2200 years or more in age. Trees commonly grow in clumps of clonal stems, which means that seedlings can receive carbohydrates, water and nutrition from associated canopy trees. This characteristic allows redwoods to outcompete other conifers and to regenerate even in the deep shade beneath their own canopy. Trees survive wildfires by having high stump-resprouting potential and highly fire-resistant bark, and can produce epicormic branches in the aftermath of fire. Young trees sometimes grow more than 1 m/yr, placing heavy competitive stress on associated conifers.

Silviculture practice

S. sempervirens first bears seed at 5–15 years old, but seed viability is generally higher in older trees, peaking at about 250 years. Due mainly to undeveloped embryos, germination rate is generally <15%, but with full embryos this rises to 80–100%. Stored seed loses its dormancy rapidly, usually within 3 years. No seed pretreatment is necessary, but the speed of germination is increased by soaking seeds overnight in aerated water. New seedlings are shade-tolerant and require constant soil moisture until their roots have reached a permanent soil water supply. Once they have such roots, they are relatively tolerant of direct light.

Propagation of *S. sempervirens* from cuttings is not widely practised, although 40% rooting success can be obtained for cuttings from the tips of healthy seedlings without any special treatment. Cuttings from older trees are more difficult to root. Tissue culture has been used with good success, the plantlets obtained being generally twice the size of seedlings of the same age.

Trees resprout readily in nature from root crowns and stumps. Such sprouting from dormant meristems is visible within 2–3 weeks after logging. Each sprout develops its own root system as it grows and eventually sprouts form rings of new trees around old stumps. Sprouts initially grow much more rapidly than seedlings until ages of 40–80 years.

In the nursery, seedlings require good soil moisture but are highly sensitive to damping off. Most stand regeneration utilizes natural seedling establishment from preserved seed trees or resprouts from root crowns after logging.

Silvicultural practices with *S. sempervirens* have traditionally involved both even-age (clear-cutting) and uneven-age stand management with selective cutting. Clear-cutting provides an open, sunny environment for the establishment of pure stands via seedlings and/or resprouting. Hand planting of seedlings is generally impractical and regeneration is commonly accomplished by preserving a few seed trees until new seedlings are well established. Second-growth stands are typically harvested at 60–80 years old, although mean annual growth increment does not reach a maximum until a stand is more than 100 years old.

Young growth stands respond well to thinning. As much as 50% of basal area can be removed from stands 45–50 years old without significant loss of future volume production. Moderate to heavy thinning that opens up the canopy not only allows for increased growth of remaining trees, but also promotes vigorous resprouting of thinned individuals and thus can lead to changes toward multiple age classes and selective cutting practices. Many second-growth redwood forests are managed to create a shifting mosaic of variable patches cut on a 100–200-year rotation.

Old-growth stands in highly productive areas yield 10,500–14,000 m^3/ha of scaled logs. Young-growth stands often exhibit spectacular rates of growth. Dominant trees on good sites reach 30–45 m in height at 50 years of age and 50–67 m at 100 years. Empirical yield tables developed for young-growth *S. sempervirens* under well-stocked conditions project yields in 100 year-old stands ranging from 3576 m^3/ha on highly productive sites to 742 m^3/ha on poor sites. Maximum yields in younger stands can be obtained in mixtures with *Pseudotsuga menziesii*. Economical conversion of old-growth forests to young managed stands is difficult after clear-cutting because of slow growth in the subsequent decade. More success has been achieved with mosaics of small clear-cuts to maintain natural reproductive processes and reduce stand damage.

Over 95% of the original stands of *S. sempervirens* have been cut after 150 years of logging. Today, only about 18% of redwood habitat is in public ownership, mostly in National and State parks. Management policies vary greatly over the species' range. For old-growth stands, fire and hydrological management are of primary concern. Prescribed burning is widely practised. Since sediment deposition in alluvial valley habitats is critical for seedling establishment, protection of hydrologic basins is of major importance. Management practices in second-growth forests have focused on thinning to promote uneven-age stands and old-growth community structure, as well as fire management. For stands in private ownership, management issues have increasingly emphasized sustainable use of forest resources and protection of ecosystem biodiversity.

Diseases and pests

S. sempervirens is relatively resistant to pests and pathogens, but heart rot is widespread in old-growth stands, particularly in trees with basal fire scars or broken tops, and is the cause of most culling, the most significant heart rot fungus being *Oligoporus sequoiae*. *Ceriporiopsis albipellucida* is also widespread.

Redwood canker, caused by *Seiridium* sp., is patchily distributed in forests north of San Francisco. Most common in young trees growing on poor sites, it causes die-back in lower branches.

Of the 23 insect species identified on *S. sempervirens*, most cause insignificant damage. Twig-boring insects include *Callidium sempervirens* and *Phytamotes nitidulis*. The wood borers *Atima confusa* and *Semanotus ligneus* attack weakened and recently dead trees. The bark beetle *Phloeosinus sequoiae* attacks weakened, felled or fire-scorched trees.

Black bears strip bark in the summer in some areas; trees 10–30 years old and 15–25 cm in diameter are damaged and may be girdled and killed.

Fire plays a significant role in the ecology of *S. sempervirens*. Natural fire patterns include both severe fires every few decades and low-intensity fires at more frequent intervals. The above-ground tissues of saplings are readily killed by fire, but resprouting from root crowns is rapid. Trees over 20 years old are relatively resistant to light ground fires. Intense fires that open up fire scars at the base of mature trees promote heart rot infections.

Uses

Its size and form, and the natural beauty of old-growth stands give *S. sempervirens* a special aesthetic and spiritual quality. It is widely grown as a specimen tree in parks and arboreta.

S. sempervirens wood is highly valued for exterior construction because of its durability and resistance to decay when exposed to moisture. The wood has a density of 380–400 kg/m^3. It is used for fences, outdoor decks, exterior panelling and outdoor furniture.

Bark is used as garden mulch and occasionally in insulation.

Further reading

Barbour, M., Lydon, S., Borchert, M., Popper, M., Whitworth, V. and Evarts, J. (2001) *Coast Redwood: A Natural and Cultural History*. Cachuma Press, Los Olivos, California, USA.

Carranco, L. and Labbe, J.T. (1975) *Logging the Redwoods*. Caxton Press, Caldwell, Idaho, USA.

Noss, R.F. (ed.) (2000) *The Redwood Forest: History, Ecology, and Conservation of the Coast Redwood*. Island Press, Washington, DC.

Standiford, R.B., Weller, T.J., Piirto, D.D. and Stuart, J.D. (eds) (2012) *The Coast Redwood Forests in a Changing California*. Proceedings of a symposium for scientists and managers. USDA – Forest Service, Pacific South-west Research Station General Technical Report PSW-GTR-238, Albany, California, USA.

Sequoiadendron giganteum (Lindl.) J. Buchholz

Giant sequoia

Importance

The sole species in the genus *Sequoiadendron* (Plate 41), formerly known as *Wellingtonia*, *Sequoiadendron giganteum* is a remarkable coniferous tree that can only be described in superlatives. Individual record-breaking living specimens are amongst the tallest (over 94 m tall), oldest, and longest-lived (over 3000 years) trees in the world. Individuals with the greatest girth (up to 47 m), dbh (up to 8.8 m) and total volume (over 1480 m^3) are some of the largest single living organisms on the planet. These traits and the majestic growth form of giant sequoia trees have led to the protection of almost all native groves in designated protected areas. The species, in fact, played a pivotal role in the evolution of the American conservation movement, and continues to provoke strong emotions among most people who encounter it. In its native range, which is restricted to the Sierra Nevada in California, USA, giant sequoias are major attractions for tourists in Sequoia, Kings Canyon and Yosemite National Parks.

Botanical features

Trees of *S. giganteum* are massive, exceeding any other tree in the world in total size. The Mother of the Forest tree in the Calaveras Grove stood over 97 m tall, had a girth at ground level of over 27 m and was over 2500 years old when destroyed by bark stripping in the 1850s. The estimated bole volume of the largest living tree, the General Sherman tree in Sequoia National Park, is 1486 m^3.

Crowns on young trees are well-formed and conical, but older trees develop a rounded and irregular crown. Tall trees, particularly those with large basal fire scars, often exhibit dead tops. Mature trees generally reach heights of 45–100 m, with a mean of about 75 m. There is a range of 300 to 880 cm in dbh. Individual branches may also reach very large diameters of up to 2 m or more.

The bole is straight, and commonly unbranched up to a height of 25–40 and even 70 m. The small taper of the bole to high in the canopy is a characteristic of mature tree growth form. The base of the bole is commonly buttressed up to a height of 1–3 m or more above the ground surface.

Root systems are relatively shallow, seldom extending more than 2 m in depth, but extend laterally for broad distances of up to 30 m or more, providing a stable root platform to support the massive trunk. Thus, lateral roots commonly occupy an area of 0.3 ha or more. The soft, red, fibrous bark is as much as 15–30 cm or more thick, and frequently shows fluting or deep furrows with large rounded ridges. Branches have a very thin bark, layered in thin sheets.

Foliage is evergreen, bluish-green in colour, and consists of small awl-like scales 2–6 mm in length, spirally arranged and adherent below to the stem.

Male cones are small, 4–8 mm, nearly globose to ovoid in shape and deciduous. Female cones are woody structures that remain green and vascular on branches for up to 20 years. These cones, 5–9 cm in length, comprise 25–40 woody scales, each containing 3–12 or more ovules in double crescentic rows (Plate 63). A mature tree has about 10,000–30,000 cones at any time, with two-thirds of these green. Typically 3–9 seeds mature above each scale, with cones overall each producing about 230 seeds. Seeds are small, 3–6 mm, with about 200 seeds/g; they possess two thin lateral wings broader than the seedbody itself, and are thus well adapted for wind dispersal. Seeds are released when the cones dry from fire heat or are damaged by insects, with a dispersal distance normally of around 180 m, although more than 400 m has been recorded. It is estimated that a mature sequoia disseminates 300,000–400,000 seeds/yr.

Vegetative growth occurs in early summer.

Distribution

S. giganteum is limited in distribution to about 70 groves occupying 144 km^2 scattered over 420 km of the western slope of the Sierra Nevada in California, at altitudes between 900 and 2700 m. The great majority of groves and all of the larger groves occur in the southern one-third of this range. Nowhere does it grow in pure stands, but in mixed montane coniferous forest.

Since being encountered by Europeans during the California gold rush in the early 1850s, the species has been planted widely in Europe and elsewhere, mainly as a specimen tree on estates and in parks. Only in Germany and Belgium were large stands planted as forest trials.

Silviculture characteristics

S. giganteum is fairly shade-intolerant and relatively resistant to wildfire. Seeds are only able to germinate and grow successfully in mineral soils in full sunlight, free of competing vegetation, so stands require periodic wildfires to clear away the understorey and soil humus before regeneration can occur. Trees appear to grow on a wide range of soil types, and can tolerate exposure, frost and temperatures as low as −31°C for short periods.

Silviculture practice

Seed dormancy appears to be absent in *S. giganteum*. Freshly dispersed seeds have about 45% average viability, but lose all viability in 2–3 weeks. Bare mineral soil is a normal prerequisite for rooting of germinating seeds in the field. Most new seedlings, however, die of fungal damping off or desiccation under field conditions. With proper storage, seeds retain viability for 8–14 years.

Cuttings from saplings root well, with up to 94% success. Only limited success has been achieved in rooting cuttings from trees 30–40 years old. Cuttings of tissues from mature trees root best when taken at bud break rather than during the dormant period.

Although the tree is grown in small quantities in the nursery trade for ornamental use, there are no quantitative studies relating to nursery practice with this species. It is likely that successful nursery practice would parallel that used with *Sequoia sempervirens*.

There has been no significant commercial growing of *S. giganteum*. However, experience with experimental plantings in the UK indicate that giant sequoia is less badly affected by exposure and frost than *S. sempervirens*, and will grow on most soils except wet acid peats. Losses can be high after planting, and weed control must be thorough due to very slow growth during the first few years.

Management of *S. giganteum* populations relates to resource management of the mixed conifer forests in which they naturally occur. The primary management tool in maintaining healthy reproduction and growth of these trees is prescribed burning to expose bare mineral soil, reduce shading and competition from understorey thickets of shade-tolerant *Abies concolor*, and release seeds from cones through drying by convective heat plumes from ground fire.

Diseases and pests

Several pathogens and pests are associated with bole failure in old-growth giant sequoia trees. These include wood decay fungi such as *Heterobasidion annosum, Poria incrassata* and insects such as carpenter ants (*Camponotus modoc*).

Armillaria spp. and *Botryosphaeria* spp. are pathogens associated with disease problems in young-growth trees. *Botryosphaeria* has also been reported to severely damage giant sequoia growing in Europe. Seedlings in nurseries are vulnerable to charcoal root disease caused by *Macrophomina phaseolina* and grey mould caused by *Botrytis cinerea*.

Seedlings and saplings are highly susceptible to damage or death from fire. Although large trees are impervious due to their thick bark, long periods of repeated burning may open up fire-scars and expose flammable heartwood. If these fire scars become large enough, they can cause structural damage and bole failure. Lightning occasionally ignites fires in the crowns of tall specimens, but these rarely kill the trees.

Uses

In addition to its fame as a charismatic species in its natural habitat, there is widespread use of *S. giganteum* as an ornamental. Its grandeur has led to it being planted worldwide on estates and in parks and arboreta. In fact, growth in some areas outside its native range appears very fast, with one specimen reported to have attained a height of 22 m and a diameter of 88 cm in 17 years in Italy. Its value as an ornamental has been enhanced by the selection of a wide array of horticultural varieties differing in habit and foliage colour, including weeping, variegated, pygmy, and compact forms. It is also grown on a small scale as a Christmas tree.

At about 340 kg/m^3, the wood of *S. giganteum* is even less dense than that of *S. sempervirens*. It is resistant to decay and extremely durable but, being brittle and fibrous, it was not useful commercially in the past, although it was once used for roof shingles, matches and fence posts. Wood from plantations of immature trees, however, is less brittle and similar in quality to that of *S. sempervirens*, sparking interest in giant sequoia as a high-yielding timber crop.

Further reading

Aune, P.S. (ed.) (1994) *Proceedings of the Symposium on Giant Sequoias: Their Place in the Ecosystem and Society*. 23–25 June 1992, Visalia, California. USDA - Forest Service, Pacific Southwest Research Station General Technical Report PSW-GTR-151, Albany, California, USA.

Hartesveldt, R.J., Harvey, H.T., Shellhammer, H.S. and Stecker, R.E. (2002) *Giant Sequoias*. Sequoia Natural History Association, Three Rivers, California, USA.

Harvey, H.T., Shellhammer, H.S. and Stecker, R.E. (1980) *Giant Sequoia Ecology: Fire and Reproduction*. National Park Service, Washington, DC.

Willard, D. (2000) *A Guide to the Sequoia Groves of California*. Yosemite Conservancy, San Francisco, California, USA.

Shorea albida **Sym.**

Alan bunga, alan batu

Importance

Shorea albida is a large tree found in Borneo. It is one of the species producing the high-value alan bunga 'light red meranti', known for its beautiful and uniform colour and easy processing. Some trees also produce a darker, stronger alan batu wood.

S. albida is a characteristic component of the peat-swamp forests of Western Borneo, sometimes as almost pure stands, sometimes mixed with other species in heath forest occurring on podzolic soils. Extensive draining and clearing of peat-swamp forests is threatening the survival of this species. Brunei may now have the largest remaining undisturbed area of forest containing *S. albida* trees.

The International Union for Conservation of Nature (IUCN) includes *S. albida* on its Red List of Threatened Species as 'Endangered'. Regeneration is reported to be non-existent, though the species coppices well and does produce suckers and is suited for coppicing, but its use in plantations is limited by recalcitrant seed storage properties.

Botanical features

S. albida is a small, medium-sized or very large buttressed tree up to 70 m tall and 190 cm dbh. Leaves are oblong–elliptical in shape and 7.5–15 cm in length by 4.5–6.5 cm in width.

Distribution

S. albida is reported from Brunei Darussalam, from north-west and east Kalimantan (Indonesia), and from Kapuas swamps through to the Limbang Valley (Sarawak, Malaysia). While generally considered endemic to Borneo there are some uncertain records from nearby islands (including the Philippines). It grows gregariously on oligotrophic peat swamps, except at the margin and sometimes in the centre. It occurs mixed with other species (*Gonystylus* and *Stenonurus*) on podsol terraces and plateaux in heath forests, up to 1200 m altitude. It dominates large areas of peat swamp in western Borneo, where it often forms dense, near-monospecific stands.

Silviculture characteristics

A great deal of research has been carried out into the natural dynamics and the natural disturbance regime of dense *S. albida* forest through the analyses of aerial photographs. The intention of this work was to quantify these processes and how they might help guide management.

Like all Asian dipterocarps, seeds are recalcitrant: they germinate rapidly and cannot easily be stored. If all plants are removed there is no seed bank to assist regeneration. This is a masting species; the flowering and fruiting is irregular and unpredictable.

Seedlings appear rare in a pure forest, and it is hard for *S. albida* to regenerate naturally in closed-canopy forests. The main factors are considered to be the shortage of light intensity (less than 700 lux) and the water condition of the habitat which easily becomes dry on days without rainfall on the root system and becomes flooded after continual rainfall on the peat horizon; seedlings do not survive these combined conditions.

Damage caused by lightning and wind throw is ecologically and economically significant for the management of *S. albida* forest in Sarawak. The pure stands seem particularly vulnerable to lightning strikes as a result of their aerodynamic properties. These strikes lead to a pock-marked pattern of strikes when viewed in aerial images. The mixed forest, on the other hand, is tall with a very dense and smooth canopy and more prone to wind damage from heavy squalls during thunderstorms. One-third of the *S. albida* trees killed by insect defoliation in a peat swamp were still standing 10 years later.

Due to its formation as a pure stand, harvesting of *S. albida* normally involves clearance, after which natural regeneration fails to take place. At the same time, clear felled alan forest is not suitable for general agriculture as the peat soils make the land difficult to manage with drainage, leading to acidity. These areas are, however, now often being developed for oil-palm plantations though there is considerable conservation attention due to their high carbon values.

Irregular flowering and seeding, combined with the short viability of seeds, make sexual propagation for plantations of *S. albida* difficult. Research effort has focused on developing methods for vegetative propagation.

Silviculture practice

Experiments with planting cuttings and wildlings in Sarawak resulted in a moderate success rate: after 2 years, 57% of cuttings and 53% of wildings had survived, with average growths of 90.5 ± 47.1 cm and 140.8 ± 66.7 cm, respectively. It is recommended to use cutting beds with good water permeability and seal them with a clear plastic film to keep high relative humidity inside. Washed large-grain river sand is suitable as culture medium, due to its porosity and permeability for developing a sound root system. Irrigation frequency should be kept to a minimum, due to the sensitivity of *S. albida* against excessive water. Cutting stems should be prepared by cutting between nodes. Although the most suitable light intensity has not yet been found for rooted cuttings, the inside of the nursery house should have a bright-light intensity.

Further studies on nursing, hardening and planting experiments are still required to establish the series of reforestation techniques for rooted cuttings of *S. albida* in the field.

Diseases and pests

An unidentified caterpillar belonging to the Himantidae caused patchy mortality totalling thousands of hectares in the late 1940s. *S. albida* forests are vulnerable to lightning and wind damage, which form conspicuous gaps in the even canopy.

Uses

The wood is light brown to red-brown with straight or interlocked grain. The high-density hardwood (590–850 kg/m^3 at 15% moisture content) is durable and strong. It is commonly used for flooring and construction, vehicle flooring, exterior and interior joinery, exterior and interior panelling, furniture and shipbuilding.

The trees which grow on the periphery of the peat swamp forest are often hollow and produce very hard wood of the dark red meranti timber group. Such trees are known by the vernacular name 'alan batu' (batu meaning 'stone'). The timber is used for sawn or hewn building timbers for light construction, carpentry and joinery, wall panelling, furniture, boats and for plywood.

A test found it technically possible to construct a bridge with a 36-m span from glued-laminated (glulam) alan batu in Sarawak.

No published information was found about medicinal uses of *S. albida*, which may be due to the fact that the alan forests are mainly located in unpopulated areas.

Further reading

Ashton, P. (1998) *Shorea albida*. In: IUCN 2012. IUCN Red List of Threatened Species. Version 2012.2. Available at: http://www.iucnredlist.org (accessed 17 January 2013).

Kitaoka, A., Uchimura, Y. and Asaishi, H. (1999) Planting of *Shorea albida* Sym. on peat swamp land in Sarawak, Malaysia. In: Bista, M.S., Joshi, R.B., Amatya, S.M., Parajuli, A.V., Adhikari, M.K., Saiju, H.K., Thakur, R., Suzuki, K. and Ishii, K. (eds) *Biotechnology Applications for Reforestation and Biodiversity Conservation*. Proceedings of the 8th International Workshop of BIO-REFOR, Kathmandu, Nepal, 28 November–2 December 1999, pp. 23–26.

Kitaoka, A., Uchimura, Y. and Asaishi, H. (2006) Production of cuttings in a peat swamp species from Sarawak, Malaysia. In: Suzuki, K., Ishii, K., Sakurai, S. and Sasaki, S. (eds) *Plantation Technology in Tropical Forest Science*. Springer-Verlag, Tokyo, pp. 89–99.

Kobayash, S. (1998) Forest structure and regeneration process of peat swamp Alan forests (*Shorea albida*) in Brunei Darussalam. *PUSREHUT Special Publication Proceedings of the Second International Symposium on Asian Tropical Forest Management*. PUSREHUT-Mulawarman University, Samarinda, Indonesia, pp. 166–187.

Shorea leprosula **Miq.**

Light red meranti

Importance

Shorea leprosula is a large tree producing an important light red meranti timber which has many uses. This species is distributed from Peninsular Thailand, throughout Peninsular Malaysia and Sumatra, to north Kalimantan. It is common in the lowland dipterocarp forests and can be found in hill forests up to 610 m. It is a late-pioneer species which flourishes naturally in secondary forests where mother trees are nearby. The bole exudes yellow-crusted dammar resin.

It grows well in a diverse range of soil types with good drainage, especially sandy loams, and tolerates shade. In its natural distribution range it grows well in low-lying areas and is seldom found on ridges. It is widely planted in lowland logged-over forests as an enrichment species but does not grow in waterlogged sites, especially peat soils. The planting stock is usually raised from seedlings inoculated with mycorrhizae. It is a relatively fast-growing dipterocarp that can be managed under plantation conditions with rotations of 40–50 years. With the recent developments in wood-processing technology, which can utilize smaller-diameter logs, *S. leprosula* plantations could possibly be managed on rotations of 30–40 years.

Botanical features

A large tree with yellowish umbrella-shaped crown which flourishes naturally in secondary forests where mother trees are nearby. It is a late-pioneer species which attains an average top height of 17.5 m at the age of 15 years. In the natural forests of Peninsular Malaysia *S. leprosula* normally forms the canopy layer with a typical height of 40–50 m, and the dbh frequently exceeds 90 cm. The largest tree recorded was 175 cm dbh. Elsewhere, *S. leprosula* attains a height of 60 m with a clear bole height of up to 35 m and buttresses up to 2 m high. It has a prominent buttress and a grey-brown straight bole with regular fissures. The bole exudes yellow-crusted dammar or resin.

The leaf is normally elliptic, shortly blunt acuminate, about 12.7 × 5.1 cm, with approximately 15 secondary veins, stellate tomentum on the under surface; under scanning electronic microscope shows hair-like structure. The petiole is about 1.5–2.0 cm long. Domatial bands are located along the lower half of the midrib and the bases of the secondary veins. It is stipulated with elliptic–oblong stipule of 0.7 cm long.

About 0.7 cm long, a pale yellow corolla with revolute, narrow, and incurved petals like a clutching hand. Fifteen stamens; stylopodium sparsely hairy, style long and filiform.

Winged fruit with three outer wings and two inner wings. The dimensions of the outer and inner wing range from 1.5 to 9.0 and 2.5 to 3.5 cm, respectively. The nut is ovate–conical shaped with sharp pointed end, about 1.5 cm long. The base of sepals is minutely stellate–hairy.

Distribution

This species is distributed from Peninsular Thailand, throughout Peninsular Malaysia and Sumatra, to north Kalimantan. and is considered to be a potential dipterocarp species for domestication in Malaysia and Indonesia.

Silviculture characteristics

S. leprosula is a late-pioneer species that colonizes gap openings in disturbed forests. It has self-pruning ability. It is also a light-demander, and is suitable for line planting in secondary forests. *S. leprosula* can be established in open conditions, with about 60% survival at 9 years after planting. After seedlings reach an average height of 2–3 m, full overhead light is necessary for full vigour.

Silviculture practice

The planting stock is usually raised from seedlings. Vegetative propagation techniques for *S. leprosula* such as cuttings and budgrafting have been developed. Plantlets have also been produced successfully by tissue culture.

Good-quality seeds are usually obtained by direct harvesting from the tree to minimize insect attacks. As germination of the dipterocarp seed starts almost at once, they should be sown immediately after collection.

A single layer of *S. leprosula* seed sown in fresh sawdust has been observed to give good germination. Hardening processes are necessary for dipterocarp seedlings, especially for those to be open-planted.

Seedlings of *S. leprosula* which attain an average height of 30–45 cm are suitable for field planting. Planting must be carried out during the wet seasons to minimize mortality of *S. leprosula* planting stock comprising wildings and seedlings. *S. leprosula* seedlings treated with 100 g of rock phosphate and 40 g of 12:12:17 N:P:K granular fertilizer doubled in height and biomass compared to untreated seedlings at 6 months after planting on a log landing.

A study has shown that a more healthy planting stock of seedlings and cuttings of *S. leprosula* is normally associated with mycorrhizal fungi (a high diversity of mycorrhizal fungi is known to be associated with *S. leprosula* in natural forests).

S. leprosula can be open-planted but the survival rate is lower than when planted in partially shaded plots. The survival percentage of *S. leprosula* planted in open conditions is only 45–60%. *S. leprosula* is also suitable for line-planting in *Acacia mangium* plantations.

There is no proper documentation of fertilizer trials of *S. leprosula* to date; however, potassium is suspected to be responsible for its basal growth in natural forests. Proper documentation of pruning and thinning trials of *S. leprosula* in forest plantations is also lacking.

S. leprosula remains an important species for enrichment planting programmes in logged-over forests. However, when there are sufficient mother trees in logged-over hill forests, there is great potential for dependence on natural regeneration, with proper silvicultural treatments, to provide the subsequent crops. Studies show that *S. leprosula* is the most dynamic species in lowland dipterocarp forest in terms of recruitment, as it has a low mortality rate and a high recruitment rate. The high recruitment is probably due to consistent mass fruiting seasons of *S. leprosula* and high survival of its seedlings.

S. leprosula plantations can be managed on rotations of 40 to 50 years, with an estimation of mean annual increment of timber stock volume of 14.5 m^3/ha/yr, or a total volume of 580–725 m^3/ha. It is noted that the sawlog recovery of this species is 90%, and hence it is economically feasible as a plantation species.

Diseases and pests

A recent survey of fungi associated with dipterocarp tree species revealed a high incidence of damping off in germinating *S. leprosula* seed caused by *Physalospora rhodina*. In the same survey leaf diseases caused by *Colletotrichum* spp. and *Pestalotiopsis* spp. were frequently observed in *S. leprosula* saplings. The incidence of *Fusarium sacchari* var. *sacchari* has also been reported on *S. leprosula*.

Heart rot of over-mature trees occurs in natural forests due to attacks by unidentified fungi. This has caused considerable volume loss in severe cases. At the Forestry Research Institute Malaysia (FRIM) a group of *S. leprosula*, injured by a lightning strike, were later found to be infected with root diseases.

Uses

This species has great potential in plantations for sawlog production. It is one of the selected plantation species for Malaysia and could act as a supplement to wood production from, and thus help to reduce the depletion of, natural forests.

It is also a good species for carbon sequestration because, given the same density, long-lived trees store more carbon than short-lived, fast-growing trees (such as 7-year pulpwood rotations or other vegetation), and as the wood is used for furniture or structures the carbon would remain stored for a longer period.

Wood formation of *S. leprosula* appears to be a continuous process and not related to seasonality in rainfall or phenology. *S. leprosula* produces a valuable light red meranti timber with an average air-dry density of about 545 kg/m^3.

The sapwood is lighter coloured and distinct from the heartwood which is light red or pink-brown. The grain is interlocked and wavy with a rather coarse but even texture. Plantation-grown timber of *S. leprosula* has similar fibre wall percentage but differs slightly in longitudinal and radial variations compared to timber from natural forests.

When it is adequately protected, it is an excellent timber for joinery, utility furniture, shop and office fittings, showcases, counter tops, panelling, ceiling, shelving, cabin fittings, fences, boxes, fabricated coffins, light-duty flooring and interior partitions. It has also been used for plywood manufacture and in the production of wood cement board.

S. leprosula produces a resin known as damar daging, which is used in traditional medicine. The bark is used for tanning production.

Further reading

Jøker, D. (2002) *Shorea leprosula* Miq. Seed Leaflet - Danida Forest Seed Centre 64, Humlebaek, Denmark, 2 pp.

Pande, P.K. (2008) Wood density variations in Meranti timbers of *Shorea* species of Malay Peninsula. *Journal of the Timber Development Association of India* 54, 10–19.

Schulte, A. and Schöne, D. (1996) *Dipterocarp Forest Ecosystems: Towards Sustainable Management*. World Scientific Publishing Company, Singapore, 680 pp.

Shorea macrophylla (de Vriese) P.S. Ashton

Kawang

Importance

Shorea macrophylla is a medium to large tree that produces both nuts and timber. Dammar resin also exudes slowly from cuts made in the bark. It is a locally abundant species on periodically flooded alluvium and river banks. It is the principal source of illipe nuts on Borneo Island, and is sometimes planted for this purpose. Generally, it is protected from felling by logging licences and concessions, in order to preserve supplies of illipe nuts.

S. macrophylla has great potential as a plantation species, as it is one of the fastest growing *Shorea* species, and flourishes even on sites of only average fertility. It is a popular species for enrichment planting in Sarawak.

Botanical features

S. macrophylla is a medium-sized tree, or a large tree in favourable sites, with a straight, clean bole. In its natural habitat, a mature tree attains an average size of 50 m in height with a dbh of 120 cm with prominent buttresses up to 2 m in height. On favourable sites it may reach an average dbh of 89 cm after 40 years' growth. The branches are sparse, and often spread to form a large irregular crown. *S. macrophylla* has a greenish-grey smooth bark, which is smooth in saplings or irregularly flaky in mature trees. The outer bark is thin, hard and brown. The sapwood is pale yellow with a dark yellow heartwood. Dammar (a resin) exudes slowly from cuts made in the bark.

The leaves are simple, generally large, elliptic–oblong, base obtuse or subcordate, approximately 17 × 35 cm in length, 10–14 cm in width, with approximately 11–20 pairs of well-spaced, prominent veins. The dried lamina is dull mauve-brown above and richly shiny rust-brown below. The petiole is short and less than 3 cm in length.

The flowers are arranged in a raceme approximately 17 cm in length, with whitish-pink petals which are densely tomentose and broadly ovate.

The fruit is a nut which measures 6 × 4 cm, which tapers abruptly to a small short acute style. The fruit calyx is glabrous with 3 long lobes and 3 short lobes, and is horizontally elliptic with a deeply saccate, thick base. The mature nuts weigh approximately 36 g each.

Distribution

This species is indigenous to Borneo, and is especially widespread in west and central Sarawak, Kapuas Valley, Tidung and Burau. *S. macrophylla* is commonly found in lowland tropical rainforest, and seldom occurs above 600 m in altitude. In Brunei, this species is confined to damp clay soils on near rivers and streams, and has less favourable growth in hillsides, ridge tops or upper slopes. Due to its fast-growth characteristics, *S. macrophylla* has been introduced as a plantation species in Peninsular Malaysia and Indonesia.

Silviculture characteristics

S. macrophylla is a climax species which requires full sunlight, although it may tolerate partial shade such as growing under pine plantations. The form quality of the stem is generally straight and cylindrical when established using an under-planting technique. This species may tolerate moderate drought which often results in leaf loss. However, seedlings are unable to survive a severe drought.

Reduced growth rates of *S. macrophylla* are reported on less fertile soil and steep sites. The species prefers a high moisture environment and has been successfully established along the riparian forests. It is an excellent species for moist sites where drought is unlikely to occur.

Silviculture practice

Generally, planting stock is raised from seedlings and cuttings, using selected quality trees. Seeds are collected 2–3 weeks after flowering, from fruits either on the tree or the ground. However, the best-quality seeds are obtained from ripe fruits with greenish-brown wings which are harvested directly from the tree. Seeds with approximately a 58% moisture content have high germination rate (94%) if they are sown directly after harvesting. However, seeds may be stored for 2–3 weeks without losing their viability, if kept in an air-sealed plastic bag at 25°C. Seeds of *S. macrophylla* are sown and potted using methods similar to other *Shorea* species.

Nursery techniques have been developed for *S. macrophylla* at FRIM. Seedlings are transplanted into a potting mixture with good aeration and drainage, consisting of organic matter, sand and topsoil at a ratio of 1:3:2, respectively. Potted seedlings are grown in the nursery with a monthly application of fertilizer (1 or 2 grains of NPK 15:15:15) until they reach 25–30 cm in height. Inoculation with ectomycorrhizal fungi may also improve growth, which is obtained from the topsoil used in the potting mixture.

S. macrophylla has been propagated using leafy stem cuttings with a single node, taken from 8–15-month-old stock plants. Successful planting of *S. macrophylla* involves good site preparation and includes opening up the existing vegetation. Initial spacings depend on the site, for example line planting in a secondary or degraded forest is conducted at a distance of 3.3 × 10 m. Line planting is often used in establishing plantations in secondary forest, pine plantations, degraded sites resulted from ex-shifting cultivation, and poorly stocked riverian forest.

Generally, transplanting is conducted in the early rainy season for all dipterocarp seedlings, in order to ensure good survival and establishment. Weeding and cleaning are very important silvicultural practices for enhancing survival and growth of *S. macrophylla*. The canopy above also requires opening as sapling growth declines. Little is known about its nutrient and fertilizer requirements. No thinning practices have been documented for *S. macrophylla*. Final stand density may vary from 100 stem/ha in natural forests which have undergone improvement felling, to 300 stem/ha in a plantations established in secondary forest.

Compared to other high-quality timber dipterocarp species, *S. macrophylla* may be managed on a relatively short rotation of 25–30 years. Plantations may even be managed on a rotation of less than 20 years, if there is a demand for smaller-diameter logs, for which processing has recently been developed.

S. macrophylla has a relatively high growth rate in comparison with other dipterocarps and selected non-dipterocarp species. At FRIM, an annual increase in dbh of 1.2–2.1 cm was recorded in a 40-year-old stand. The highest yield (8.8 m^3/ha) of *S. macrophylla* was recorded in a 60-year-old stand at FRIM.

Diseases and pests

Seedlings of *S. macrophylla* are affected by damping off in the nursery, which results in decay and death. It is caused by fungi which occur naturally in the soil, such as *Fusarium* spp. and *Pythium* spp. This species also suffers from wilt. Heart rot is another form of decay, whereby the heartwood loses its strength and may crumble.

A deformed stem is likely to be developed in the species if the apical shoot is destroyed either by insects, climbers or heat stress, as is found in other dipterocarps such as *Shorea platyclados*.

Uses

S. macrophylla is suitable for agroforestry, and for enrichment planting in regions with a high soil-moisture content or degraded sites resulting from shifting cultivation.

S. macrophylla produces a lightweight hardwood non-siliceous timber which is generally traded as 'light red meranti'. It is, however, markedly less dense than the general run of light red meranti, and this, together with its light colour, causes it to be known in parts of Sarawak as 'white meranti'. This naming is confusing, however, as genuine 'white meranti' (the *Anthoshorea* section of the genus) species yield a very different timber which is difficult to work.

Wood density is between 270 and 625 kg/m^3. The timber has good working properties including peeling, gluing and seasoning, and also has high finishing properties. Treated timber is excellent for interior finishing, mouldings, panels, veneer, rotary-cut veneer, plywood, planking, door frames and general carpentry. The timber quality of *S. macrophylla* may be reduced by a condition known as 'brittle heart', a condition thought to be related to growth stresses and swaying by the wind of pole-sized trees.

S. macrophylla is the most important source of the illipe nut, and the majority of plantations are established for nut production. Collection of illipe nuts in natural forests is well established, and an inventory has been conducted to determine the density of this species and potential of nut production. The famous illipe nuts are used as a substitute for cocoa butter, because the chemical composition of fatty acid and triglyceride is similar.

Further reading

Ahmad Zuhaidi, Y., Zakaria, I., Rosdi, K. and Krishnapillay, B. (2002) *Species for Timber Plantations. A Manual for Forest Plantation Establishment in Malaysia*. Forest Research Institute Malaysia, pp. 13–24.

PROSEA (1993) *Shorea macrophylla*. In: Soerianegara, I. and Lemmens, R.H.M.J. (eds) *Timber Trees: Major Commercial Timbers (Prosea 5.1)*. PROSEA - Plant Resources of South East Asia. Pudoc Scientific Publishers, Wageningen, The Netherlands, pp. 397.

Schulte, A. and Schöne, D. (1996) *Dipterocarp Forest Ecosystems: Towards Sustainable Management*. World Scientific Publishing Company, Singapore, 680 pp.

Shorea negrosensis Foxw.

Red lauan

Importance

Shorea negrosensis is one of the largest trees to be found in rainforests in the Philippines. The high-quality timber is traded internationally as red meranti and red lauan, and is the most common general-utility timber of western Malesia and one of the most important timber trees in the Philippines.

Large-scale felling has led to growing concern for the protection and re-establishment of this species in the Philippines. *S. negrosensis* is listed as 'Critically Endangered' on the IUCN Red List of Threatened Species, and is also on the Red List of the Philippines.

Botanical features

S. negrosensis is a large tree with a height reaching up to 50 m and diameter reaching up to 250 cm. It is characterized by a widely spreading and strongly branched crown, large prominent buttresses, and grey-brown bark.

Distribution

S. negrosensis is native to the Philippines, occurring in the remaining evergreen and seasonal dipterocarp forests, up to 1200 altitude.

Silviculture characteristics

S. negrosensis is a shade-tolerant species.

Silviculture practice

Stand establishment is usually by natural regeneration and also by direct sowing, or planting stock or wildings.

For direct sowing, methods to maximize seed germination have been investigated due to growing concern regarding the re-establishment of this species. Experiments showed that complete removal of the seedcoat gave the highest germination percentage and a significantly shorter duration of germination. Seeds should be collected during the first week of seedfall and kept for not more than 3 days before sowing. The seedcoat should be removed carefully. Experiments with different germination media have shown that the best medium for *S. negrosensis* is probably sand, or sand mixed with soil.

Seeds should have a light covering of grass mulch to prevent excessive moisture loss, and 75% shade. Watering should be carried out early in the morning and twice on sunny days.

If it is weeded regularly, it is a promising plantation species because it grows quickly and produces high-quality timber.

Diseases and pests

S. negrosensis is regarded as moderately durable to fungi, durable to dry wood borers, and moderately durable to termites. Heart rot has not been detected in living trees of this species.

Uses

S. negrosensis is planted for revegetation, soil conservation and erosion control.

It is mainly used for furniture and construction. The timber is traded, along with *Shorea ovata* and *Shorea polysperma*, as red lauan. It has a density of 420–805 kg/m^3 at 15% moisture content. Woods of species of the red lauan group have a red-brown to dark red-brown heartwood, sometimes tinged with purple; white resin canals can be seen. The sapwood is clearly marked and cream to light red in colour. Black holes in the timber may be frequent.

The green logs float and can be transported by water. The wood is used for transmission poles, sawn or hewn building timbers, for light construction purposes such as carpentry/joinery, flooring, wall panelling, shingles, containers, furniture, veneers, boats, fencing, plywood and laminated wood. It is not durable unless treated.

The bark has medicinal applications and is a source of tanstuffs. The flowers are a source of pollen for insects.

Further reading

Newman, M.F., Burgess, P.F. and Whitmore, T.C. (1996) *Manuals of Dipterocarps for Foresters: Philippines*. Royal Botanic Garden, Edinburgh, UK, pp. 124.

Soerianegara, I. and Lemmens, R.H.M.J. (eds) (1993) *Plant Resources of South-East Asia. Timber Trees: Major Commercial Timbers. No. 5(1)*. Pudoc Scientific Publishers, Wageningen, The Netherlands, 610 pp.

UNEP-WCMC (2007) *Strategies for the Sustainable Use and Management of Timber Tree Species Subject to International Trade: South East Asia*. UNEP World Conservation Monitoring Centre, Cambridge, UK, 150 pp.

Zabala, N.Q. (1993) *Mass Vegetative Propagation of Dipterocarp Species*. FAO, Rome, 15 pp.

Shorea robusta C.F. Gaertn.

Sal

Importance

Shorea robusta, a slow-growing but gregarious tree species, is highly valued for its hard, naturally durable wood. The species is adaptable to a variety of growing conditions, but on account of the poor viability of its seeds, and on account of the better alternative, teak, *Tectona grandis*, being available in most localities, it has not been much investigated as an exotic timber production species until very recently.

The recent interest in choosing *S. robusta* as species for afforestation and agroforestry is on account of its multiple uses. As well as the timber, it yields sal butter (fat from seeds), and a commercially important dammar resin, used as an incense. The bark yields tannins and fibres. It is an elegant ornamental tree.

Botanical features

A large tree with a spreading round crown at maturity. Young trees tend to have a conical crown. The tree attains a height of

20–40 m, and a dbh of 50–135 cm, depending on the site. The bole is typically straight, cylindrical, and clear of branches. The bark of older trees is thick, dark and longitudinally furrowed. *S. robusta* produces a long taproot of great length, 2.0–4.3 m, reaching down to strata moist enough to meet the water requirement of the tree.

Leaves are simple, petiolate, alternate, 10–40 cm long and 5–20 cm wide, ovate or ovate–oblong, rounded or cordate at the base, coriaceous, shiny and dark green when mature. Petioles are 2.0–2.5 cm long, stout.

Panicles are 12.5–22.5 cm long, tomentose, and commonly pink in colour. Flowers are sub-sessile, hermaphrodite and hypogynous. Petals are 1.0–10.3 cm long, pale yellow and tapering upward. Many stamens (up to 50), which are much shorter than petals in length. Ovary is 3-celled, the style subulate. Fruits are leathery, indehiscent, usually 1-seeded, closely surrounded by the bases of the accrescent calyx segments, some developed into wings.

Distribution

S. robusta occurs virtually exclusively in India, though it is also found in the sub-Himalayan tract of Nepal and Pakistan, and in Bangladesh. Attempts to introduce *S. robusta* as an exotic have been made in Zimbabwe.

Silviculture characteristics

In moister areas of its range, *S. robusta* is a light-demanding species, but it can bear moderate shade in drier parts, especially when juvenile. The species does not produce root suckers, but coppices extremely well until middle-age. *S. robusta* is fire-resistant, but prone to severe damage from drought and frost.

Silviculture practice

S. robusta does not seed well every year. Good seed years occur every 3–5 years, depending on the locality.

For the reasons of poor seed storage and viability, and very slow growth of the species at seedling and sapling stages, *S. robusta* has not been a popular species for use in afforestation or in agroforestry until recently.

Plantations of the species have been attempted in the past only when it has not been possible to obtain natural regeneration for any reason. All possible methods of planting – direct sowing, poly-pot plants, basket plants, root-shoot cuttings and air-layering – have been tried. There has been no uniform pattern of success, and consequently standard practices which are common in particular localities have not developed. Nevertheless, the most usual method of establishing plantations of *S. robusta* is by direct sowing.

Direct sowing is generally used in the taungya system, an agroforestry system adopted over the last century in various parts of India and South-east Asia to regenerate existing forests of *S. robusta* where natural regeneration of the species could not be obtained. In the taungya system, agricultural crops are raised between sown rows of *S. robusta* seedlings for a period of 2–4 years. A suggested method for sowing consists of sowing seeds in trenches (45 cm deep and 45 cm wide) in two rows, 7–10 cm apart, 3–5 cm deep in the soil.

Weeding and cleaning are very important for the first 5 years after planting, to save young plants from adverse competition. In even-aged *S. robusta* forest stands, whether the result of planting or natural regeneration, regular thinning and cleaning in sapling and pole crops is necessary for good growth.

A well-managed *S. robusta* plantation in West Bengal attained on average a dbh of 31 cm, and a height of 29 m at age 35 years old; and a dbh of 39 cm and a height of 33 m at age 50 years old. In naturally regenerated forests, an average dbh of 48 cm and height of 37 m are attained after 100 years on good sites. A coppice crop will attain an average dbh of 37 cm and average heights of 27 m at age 50 years old.

In India, the better-quality, regular (even-aged) stands of *S. robusta* have traditionally been managed either under a clear felling system with artificial regeneration, or a shelterwood system with natural regeneration. Irregular (uneven-aged) stands are usually managed under a selection system. Relatively poor quality *S. robusta* are managed either as a simple coppice system, or an exclusively Indian system, called coppice-with-reserves. In high forests, the rotation adopted varies from 100 to 150 years, and in coppice crops it varies from 40 to 80 years.

It needs mentioning that although many economic products other than timber are obtained from *S. robusta*, this species has invariably been managed in India for timber only.

Diseases and pests

Polyporus shoreae causes root rot in conditions of excessive moisture, and *Hymenochaete rubiginosa* is a pathogenic fungus which causes heart rot in *S. robusta*.

There has been extensive research into the damage to *S. robusta* trees by insect pests, with as many as 145 insects which can damage the species being identified.

One of the most important pests is the heartwood borer *Hoplocerambyx spinicornis*, which has periodically caused extensive damage to *S. robusta*, the number of trees affected often running into thousands, though normally the occurrence of *H. spinicornis* is endemic, and limited to a small number of trees in a small tract. The sap sucking mealybug (*Drosicha stebbingi*) is also an economically important pest species.

Mortality of seedlings in natural forests was observed in association with high population densities of the needle nematode *Paralongidorus sali*. The insect pests *Dichocrocis leptalis, Pammene theristis* and *Sitophilus rugicollis* have been found to infest the seeds of *S. robusta*.

Uses

The aesthetic appearance, and the semi-evergreen nature of *S. robusta*, added to its capacity to coppice vigorously, and its multiple uses, make the species an ideal choice for agroforestry, avenue planting and for soil and water conservation. However, slow early growth and low seed viability counteract these advantages. Nevertheless, where growth rate is not of primary importance, for example where soil stabilization is the main concern, the species has been tried and has proved reasonably successful.

S. robusta is one of India's most naturally durable woods, and is not attacked either by termites or wood-destroying fungi until after many years of use. The timber is hard and heavy with a density of 730–930 kg/m^3, and of very high strength, though difficult to peel and cut. As a result, though it is much used for railway sleepers, and valued for construction purposes, it has been of limited use in veneer and plywood manufacture. *S. robusta* poles have been extensively used for transmission poles, scaffolding and as pit-props. It yields a good-quality fuelwood.

S. robusta is a multipurpose species, and in addition to timber and fuelwood, its leaves, bark, and seeds are also economically important, and the flowers provide a source of nectar. It yields an important dammar oleoresin which is prized as an incense, besides being used as a disinfectant, fumigant and in the preparation of ointments for skin diseases. The resin has many names, including sal dhupa, ral, guggilamu, kabba, maramaram and seral.

S. robusta leaves provide a medium-quality fodder; the leaves are also used in bidi (cigarette) manufacture. The bark is used to a limited extent in the manufacture of tannins and fireboards. The seeds yield a commercially important edible and non-edible oil. De-oiled seed meals are used as cattle feed and as a fertilizer. The seeds can be eaten after roasting. Sal butter is used as an adulterant of glue and in the manufacture of chocolate.

Further Reading

Chitale, V.S. and Behera, M.D. (2012) Can the distribution of sal (*Shorea robusta* Gaertn. f.) shift in the north-eastern direction in India due to changing climate? *Current Science* 102, 1126–1135.

Gautam, K.H. and Devoe, N.N. (2006) Ecological and anthropogenic niches of sal (*Shorea robusta* Gaertn. f.) forest and prospects for multiple-product forest management - a review. *Forestry (Oxford)* 79, 81–101.

Rahman, M.M., Rahman, M.M., Zhang, G.-G. and Islam, K.S. (2010) A review of the present threats to tropical moist deciduous Sal (*Shorea robusta*) forest ecosystem of central Bangladesh. *Tropical Conservation Science* 3, 90–102.

Tewari, D.N. (1995) *A Monograph on Sal (Shorea robusta Gaertn. f.)*. International Book Distributors, Dehradun, India, 277 pp.

Shorea rugosa F. Heim

Dark red meranti

Importance

Shorea rugosa is one of a number of *Shorea* species that produces timber known as dark red meranti which is exported around the world. Also traded as meranti bakau, *S. rugosa* is threatened due to habitat loss and overharvesting. The IUCN Red List of Threatened Species lists *S. rugosa* as 'Critically Endangered'; it is considered a priority species for conservation action in Indonesia and is a protected species in the national parks of Sarawak in Malaysia.

Botanical features

S. rugosa is a very large tree up to 65 m in height and 125 cm in diameter. It has a deeply fissured bark and buttresses up to 2.5 m high.

Distribution

S. rugosa is found in Brunei Darussalam, Indonesia and Malaysia, in mixed dipterocarp forests and peat swamps.

Silviculture characteristics and practice

Little information is available specifically on the silvicultural characteristics and practices of this species. It is suggested that reference should be made to the silviculture of other *Shorea* species of the dark red meranti group.

Diseases and pests

Woods of the dark red meranti group are regarded as poorly durable to durable to fungi, durable to dry wood borers and moderately durable to termites.

Uses

The wood of *S. rugosa* has a density of 565–850 kg/m^3 at 15% moisture content. Dark red merantis are characterized by a dark red timber (although variable from pink-brown to purplish-brown) with white streaks (resin canals). The sapwood is lighter in colour and distinct from heartwood. The texture of the wood is rather coarse, yet even. The grain is interlocked and wavy; it is moderately hard to cut across the grain.

Uses are numerous and include joinery, panelling, plywood veneers, furniture, flooring, light carpentry, partitioning, mouldings and skirtings. In Europe, dark red meranti is also highly favoured for both its technical and aesthetic attributes for use in the production of window frames. The timber is considered to be moderately durable either in exposed positions or in contact with the ground.

Further reading

Günter, S., Weber, M., Stimm, B. and Mosandl, R. (eds) (2011) *Silviculture in the Tropics (Tropical Forestry)*. Springer, Dordrecht, The Netherlands.

Tsumura, Y., Kado, T., Yoshida, K., Abe, H., Ohtani, M., Taguchi, Y., Fukue, Y., Tani, N., Ueno, S., Yoshimura, K., Kamiya, K., Harada, K., Takeuchi, Y., Diway, B., Finkeldey, R., Na'iem, M., Indrioko, S., Ng, K.K.S., Muhammad, N. and Lee, S.L. (2011) Molecular database for classifying *Shorea* species (Dipterocarpaceae) and techniques for checking the legitimacy of timber and wood products. *Journal of Plant Research* 124, 35–48.

UNEP-WCMC (2008) *Strategies for the Sustainable Use and Management of Timber Tree Species Subject to International Trade: South East Asia*. UNEP World Conservation Monitoring Centre, Cambridge, UK, 74 pp.

Styrax tonkinensis (Pierre) Craib ex Hartwich

Siam benzoin

Importance

Styrax tonkinensis is an economically important tree species in Laos and Vietnam. It is used as a commercial source of benzoin resin in Laos, where collection and sale of the resin is an important cottage industry among the hill people of the north. In Vietnam, it is an important plantation species for production of pulpwood.

Botanical features

S. tonkinensis is a tree up to 25 m tall and 30 cm in diameter, with a clear bole for about two-thirds of total tree height. Bark is grey and smooth when young, becoming brown and rough with longitudinal fissures with age. Trees are light branching, with branches more upright toward the top. Young trees have a dense crown which can occupy up to two-thirds of tree height. The development of a clear bole as a result of natural pruning denotes the change from juvenile to mature tree. The root system is shallow and as the tree ages, the taproot disappears.

Leaves are simple, alternate, ovate to elliptical, light acuminate, 4.5–10 cm long and 2.6–5 cm wide. Petioles are 6–10 mm long. The leaf lamina is tough and leathery appearing smooth on the upper surface and very rough and hairy on the lower surface. There are 7–9 pairs of secondary veins which are conspicuous on the upper and lower surfaces. A typical characteristic of *S. tonkinensis* leaves is the dark green colour on the upper surface and the whitish-green colour on the lower surface.

Inflorescences up to 18 cm long, in double racemes in the upper axils of the leaves or at the terminal part of the shoot. They have a leafy base covered with yellowish stellate hair; the caducous bract and linear bracteoles are very hairy. The inflorescence is comprised of many small flowers, 12–15 mm long with 3–5 pedicels. The cup-shaped calyx is 3–4 mm long, densely covered with yellowish stellate hair on the outside and whitish hair on the inside, with 5 short lobes. The 8–12 mm long corolla has a 2.5–3 mm tube, five 6–8.5 × 2–3 mm overlapping lobes, felted with yellowish stellate hair on the outside, interspersed with simple hair on the inside. The stamens reach two-thirds of the corolla lobes; their flat filaments are covered with long stellate hair. The superior ovary, slightly adherant to the base, is ovoidal and covered with stellate, hispid hair; the style is slender, bare, 10–13 mm long. Fruits ovoid, 10–12 mm long and 5–7 mm wide, covered with greyish stellate hair, dehiscent through three valves. The pericarp is thin, about 1 mm; there is an upright, three-rowed slightly hollowed single seed (very rarely two), with a warty, hard, thick, orange coloured episperm.

S. tonkinensis is a semi-deciduous tree and sheds many leaves during the cool, dry season between November and February. Old trees shed more leaves than young trees. Flowering and fruiting time varies with location. Flowering usually occurs during April–June (sometimes until July), and fruit matures from July–November. The first flowering takes place at 4–5 years of age. When fully mature, each tree produces up to 40 kg of fruit per year. Each 2–3 kg of fruit contains 1 kg of seed; there are 8000–9000 seeds/kg.

Distribution

Natural stands of *S. tonkinensis* are found in the northern parts of Laos and Vietnam.

The species has also been introduced into a number of countries including China, for both wood and resin production; to French Guinea, West Africa, where it has become naturalized in many areas; and to Java.

Silviculture characteristics

S. tonkinensis is a light-demanding species, and regenerates well in gaps, provided that undergrowth is not too heavy, as saplings are sensitive to competition in their early years. Wild seedlings cannot survive in the heavy shade under a dense canopy. Young trees (2–3 years old) have the ability to produce coppice shoots.

Silviculture practice

Mature seed is classified as intermediate, and requires special storage treatment to maintain viability. Moisture content of the seed is the most critical factor determining seed viability. Satisfactory viability is maintained by storing seed in mixture with wet sand in order to maintain its moisture content at 30%.

Germination of untreated seed is slow and sporadic, and may last up to 5 months. Germination is rapid after suitable storage. Newly emerged seedlings should receive 50% shade, and once established 70% of full sunlight is appropriate. Plantable size of 25 cm in height is generally reached within 3–4 months from germination. There have been no reports of vegetative propagation for the species, but stump plants prepared from 10–12-month-old seedlings are viable.

Stands of *S. tonkinensis* are presently established either by natural regeneration or artificial planting. At present, Vietnam is the only country known to carry out large-scale planting. It is normally planted as a pure stand; occasionally it is planted in mixture with bamboo or *Manglietia glauca*.

Most of the stands in Laos have established as a result of natural regeneration during a slash-and-burn cultivation cycle. Mature *S. tonkinensis* regrowth forest (10–15 years-old) is clear-felled and the area burnt during the dry season. The fruit, which falls before and during clear felling, germinates in the following rainy season. The *S. tonkinensis* seedlings grow in combination with crops especially upland rice. In general, agricultural crops are grown for 1 year only. After crop harvest, the swiddeners move on to clear another regrowth forest. The area is abandoned, with *S. tonkinensis* being the main fallow species. Trees are irregularly spaced; on average there are 200–400 stems/ha. The *S. tonkinensis* stand is normally left untended and as a result trees are subjected to severe competition. A large number of trees are suppressed, resulting in a marked variation in the size of trees within a stand.

Plantations in Vietnam are established by both direct sowing and planting containerized stock. Regardless of the type of planting stock, site preparation includes clearing and burning of ground vegetation. For direct sowing, 5–6 seeds are placed in each hole. Planting density varies from 1600 to 3300 seedlings/ha, depending on the soil fertility and end use of the wood product. A starting dose of NPK fertilizer promotes initial growth of the seedlings. Weed control is essential in the first 2 years, after ground vegetation is cleared on an as-needed basis. Because of high initial stocking, two thinnings are required before the plantation reaches the rotation age of 8–10 years. Pruning is not necessary owing to the excellent self-pruning characteristics of *S. tonkinensis*.

The species is planted for production of pulpwood in Vietnam and for wood and resin production in China. In mature regrowth forest in Laos, the tree is used as a commercial source of benzoin resin. Collection and sale of the resin has been an important cottage industry among the highland people for many centuries.

Under favourable conditions an annual height increment of 3 m can be attained during the first 3 years. A mean height of 18–25 m and dbh of 20–24 cm with under bark volume of 0.20 m^3 per tree are obtainable at 10 years. Plantations in Vietnam are grown on an 8–10-year rotation. Average yield is 50 m^3 per ha at 8 years.

Diseases and pests

A defoliator, *Fentonia* sp. (Lepidoptera: Notodontidae), has caused damage to thousands of hectares of *S. tonkinensis* plantations in Vietnam. The stems of young plants in newly established plantations are bitten off by crickets (*Tarbinskiellus portentosus*). Seedlings in the nursery are susceptible to damping off.

In northern Vietnam, the aphid *Tuberaphis owadai* was found to form coral-shaped galls on *S. tonkinensis*.

Uses

S. tonkinensis is a fallow species in upland cultivation systems in northern Laos. In Vietnam, it is sometimes planted to provide shade in tea plantations. It was used to restore eroded soils, and planted to provide green fire breaks in Guinea, West Africa.

The wood of *S. tonkinensis* is white, soft and light with air-dry density (15% moisture) between 413 and 450 kg/m^3. It is used to make wooden shoes, pencils, chopsticks, toothpicks and matches. Wood from thinning is generally used as firewood or poles. It is not suitable for use in construction because the wood is not durable, and easily broken, bent or buckled. In Vietnam, it is an important source of wood fibre for the pulp and paper mill located in Bai Bang, Phu Tho. Laboratory pulping tests showed that pulp yield and quality of *S. tonkinensis* was comparable with many commercial pulpwood species.

The resin produced from trees in Laos is of high quality and suitable for use by the fragrance industry. The resin is commercially known as 'Siam benzoin'. The benzoin resin is characterized by its content of benzoic acid (10–12%), coniferyl benzoate (65–70%), and more than 20 other aromatic esters. Siam benzoin is used in modern pharmaceuticals as an inhalant or as a mild antiseptic. It is used in traditional medicine in China to treat rheumatism, coughs, colds, stomach ache and heart burn; it has also been used in the treatment of apoplexy, dizziness, convulsions and glycosuria.

Young trees are browsed by cattle, which suggests it has potential as a fodder species.

Further reading

Jøker, D. (2000) *Styrax tonkinensis* (Pierre) Hartwich. Seed Leaflet - Danida Forest Seed Centre, Humlebaek, Denmark 44, pp. 2.

Kashio, M. and Johnson, D.V. (eds) (2001) Improved silviculture of *Styrax tonkinensis. Monograph on Benzoin (Balsamic Resin from Styrax Species)*. RAP Publication: 2001/21. Chapter 4. Regional Office for Asia and the Pacific, FAO, Bangkok, Thailand. Available at: http://www.fao.org/docrep/005/ac776e/ac776e00.htm#Contents (accessed 4 April 2013).

Li, Y.-G., Liu, X.-H., Xie, J.-Q., Zhao, X., Xu, L. and Liu, W. (2011) Provenance variation and preliminary selection of Styrax tonkinensis seedlings. *Journal of Zhejiang Forestry Science and Technology* 31, 37–41.

Pinyopusarerk, K. (1994) *Styrax tonkinensis: Taxonomy, Ecology, Silviculture and Uses.* ACIAR Technical Reports Series, No. 31. ACIAR, Canberra, v & 14 pp.

Swietenia macrophylla King

Big leaf mahogany

Importance

Swietenia macrophylla produces one of the world's best-known and popular furniture timbers and has been internationally traded and utilized for timber for over 400 years. It is a moderately fast-growing species with generally favourable plantation silvicultural characteristics, and can produce sawn timber in a 30–40-year rotation. Substantial areas have been planted in Indonesia, the South Pacific and Central America. Outside of the South Pacific region, the main drawback to its planting is its susceptibility to shoot borer (*Hypsipyla* spp.), which leads to plants of a poor and much-branched form. In some regions where it has been introduced, including Fiji, the Philippines and Dominica, *S. macrophylla* has become invasive in native forest ecosystems.

As *S. macrophylla* is becoming increasingly endangered in much of its native range, efforts are being made to better understand its reproductive biology and to elucidate the extent and patterns of genetic variation and diversity within and between populations, in order to develop an efficient conservation strategy. Improvement of sustainable growth and production of mahogany in native forests offers the best opportunity to conserve both the species and forest ecosystems.

Botanical features

S. macrophylla is an evergreen tree up to approximately 30–45 m, infrequently attaining a height of 50 m, and dbh of 2 m. Trees exhibit rapid early growth in height, often forming a single unbranched stem for up to 10 m before branching. Mature trees have a tall straight, cylindrical bole clear of branches for 12–18 m, often with high buttresses, but in open-grown areas it may be short-boled with massive branching. Production of buttresses begins at a size of 10–12 cm dbh, and their height increases linearly with tree size above this point, reaching 3 m on the largest trees and extending out 10 m or more from the stem. The crowns of the largest trees are up to 20 m in radius and comprise a small number of large ascending branches. Young branchlets are glabrous, slender, rich chestnut-brown or reddish-brown, finely lenticellate; scales surrounding buds prominent, brown, ovate or acute. The bark is grey and smooth when young, turning dark brown, ridged and flaky when old, vertically fissured, becoming heavily fissured in mature trees.

The large pinnate leaves are 16–40 cm in length, and are arranged alternately and clustered at the ends of branchlets. Generally, each leaf consists of 3–6 pairs of opposite or occasionally subopposite leaflets. Leaflets are typically 9–14 × 3–5 cm, usually oblong to oblong-lanceolate or ovate–lanceolate, sometimes elliptic–ovate, slightly falcate, dark glossy above and lighter below, and on slender petiolules 0.5–1.2 cm in length. The first seedling leaves are simple.

The small flowers are borne in axillary or sub-terminal inflorescences. Flowers are unisexual, with both sexes similar, and the trees monoecious. Each inflorescence is 10–20 cm in length with short lateral, spreading, glabrous branches, generally shorter than the leaves.

The fruits are large (12–15 × 6–8 cm), woody, erect capsules, oblong to slightly sub-globulus, commonly known as 'sky fruit' as they appear to hang pointing in an upward direction. The outer valves are thick, becoming woody with a coriaceous surface when mature. When dry, the 4 or 5-valved fruits split open from the base, or from the base and the apex simultaneously. The centre of the fruit is a thick, woody 5-angled columella extending to the apex from which the seeds hang pendulous by their wing, leaving conspicuous seed scars after their release. There are usually about 35–45 winged seeds per fruit. Seeds are chestnut coloured and 7.5–12 cm in length.

Flowering and fruiting takes place annually, with the timing varying between locations according to climate. Flowering usually takes place when trees are deciduous or just coming into new leaf and shortly before the rainy season. The fruits mature in 10–11 months. In Bolivia, flower and leaf production occur simultaneously in September at the onset of the rainy season. In Central America, northern parts of South America, and the Philippines, the trees flower in March–June and the fruits mature from December to March. In the southern hemisphere flowering is from September to November and fruiting in the period June–September. Under favourable conditions *S. macrophylla* trees may flower and start to produce seed after 12 years' growth. Dry, windy weather provides the most suitable conditions for opening of the mature fruit and dispersal of the winged seed, as well as for pollen dispersal. Fruit yields may be drastically reduced by tropical cyclones.

Distribution

The genus *Swietenia* has a natural distribution from 20° N to 18° S in tropical America. *S. macrophylla* is the most widely distributed species, occurring from the Atlantic regions of south-east Mexico, through Central America, northern South America, and in an arc across the southern Amazon basin.

S. macrophylla has been cultivated throughout the lowland humid and subhumid tropics. The species has been widely planted in South and South-east Asia, the Pacific Islands, the Caribbean, and tropical Africa. Substantial areas of plantation have been established in Indonesia, Fiji and parts of Central America.

Silviculture characteristics

S. macrophylla is a light-demanding species, and seedlings do not survive long under the shady conditions of the forest understorey or in felling gaps. However, young vigorous trees growing in full sunlight are very susceptible to attack by shoot borers and more likely to succumb to drought. In the adult stage, mahogany trees almost invariably maintain their crowns above the surrounding forest, and are exposed to the intense tropical sun.

S. macrophylla is a relatively windfirm species, only moderately susceptible to stem-break and leaf loss. However, sapling-sized trees, aged about 3–7 years old, which have long rather slender boles, are at risk of being blown over during moderate intensity tropical cyclones, the risk being much greater in open-grown plantings. The low mortality of trees in hurricanes has been attributed to their heavy buttresses which provide strong support, and reduced branching and sparse foliage of the crowns that decreases wind resistance.

Young trees are very sensitive to fire. Mature trees are more fire-resistant, and controlled burning may be used to reduce competing vegetation and create conditions favourable for natural regeneration. Natural regeneration occurs after major environmental disturbances, such as fires, hurricanes, flooding and on abandoned agricultural fields. Surviving adult trees disperse their seeds across the resulting gaps or clearings, the seedlings becoming established in these open areas in essentially even-aged cohorts, intermixed with other species. Seeds do not retain their viability beyond any rainy season and there is no soil seed bank.

Silviculture practice

After logging in *S. macrophylla* plantations, generally there is prolific natural regeneration, which may be thinned and managed to form the next crop. However, in order to make use of improved genetic material it is preferable to transplant potted or bare-rooted seedlings which have been raised in nurseries.

The seed should be stored at 3–7% moisture content at low temperatures (3°C ± 2°C). Untreated seed is sown in a germination bed or in pots inside a lightly shaded germination house. Under favourable conditions of moisture supply, temperature, and available oxygen, seed germination starts within 10 days of planting and continues for approximately 20 days. When the first seed leaves appear, seedlings are ready for pricking-out into open nursery beds, pre-prepared at a spacing of 15 × 20 cm. Immediately after pricking-out the seedlings require heavy shade (e.g. 70%), reducing to 30% after a week, and full sun after 3–4 weeks. Striplings are ready for planting out after about 4 months, when they are 30–50 cm in height. Conditioning is carried out by substantially reducing the leaf area and by root wrenching 14 days before planting. The nursery duration for potted seedlings is approximately 3–4 months.

Cuttings are relatively easy to establish from 3-year-old plants, and thereafter become increasingly difficult to strike, especially once 10 years old. Leafy cuttings from seedlings may be rooted in non-mist propagators, and high success rates were achieved with a high proportion of sand in the substrate. A maximum rooting success of over 60% was achieved with a 75:25 sand/gravel mixture. Propagation using *in vitro* techniques is possible, but is currently largely confined to laboratory studies. Clonal seed orchards have been developed in Fiji through grafting of selected plus trees.

In the South Pacific region and in Puerto Rico, plantings are usually established as line plantings in logged-over and degraded forests with variable cover of residual trees. During site preparation planting lines are cut 10–12 m apart and approximately 2 m wide, and any overhanging bush is removed. Biodegradable herbicides are now replacing the chemicals that were once used for this purpose. Under optimal conditions, seedlings of *S. macrophylla* may reach 3 m in 1 year, and 6 m in 2 years. In Fiji, establishing new plantings by direct seeding is thought to be both cost effective and to result in plants with better root systems, which are more resistant to cyclones and less susceptible to termite attack. The practice involves sowing 3 seeds at each planting site, with subsequent thinning to leave the best plant.

Plantations in Martinique (Caribbean) are established by planting bare-rooted seedlings at 3 × 3 m, or 1100 plants/ha, and thinned progressively at ages 15, 21, 28, and 35 years to a final density of 120–150 stems/ha. Stands are harvested at age 50 years, when diameters have reached at least 50 cm. Shoot borers cause major damage in young monocultural stands, and it is proposed to investigate manipulation of natural regeneration of this species, and other trees of indigenous origin. Major costs are also incurred in weed control and vine removal.

S. macrophylla self-prunes satisfactorily, but low crown break may be a problem in open environments. Form pruning of young trees is not recommended as new branches are likely to regrow. A more effective approach is to plant mahogany in intimate mixture with low, densely shading species such as *Flueggea flexuosa* or suitable bamboo species, in order to shade out lower limbs and induce longer boles. Singling operations are generally not necessary except following *Hypsipyla* damage. During the first 3–4 years regular weeding should be undertaken within the planting lines when required, while inter-row areas may be left.

S. macrophylla is currently being exploited throughout much of its natural range, but detailed information is often lacking on the extent of native populations. Over-exploitation threatens its continued existence in several parts of its natural range. It was placed on the Convention on International Trade in Endangered Species (CITES) Appendix III by Costa Rica in 1995 and on Appendix II by Guatemala and Nicaragua, covering logs, sawn timber, veneer, and plywood with effect from November 2003. However, despite this and the Brazilian government's ban on logging, transport and trade in *S. macrophylla* since October 2001, the illegal logging of this species has continued and indeed has increased the supply of illegal timber, which often has been smuggled out of the country labelled as 'other tropical species'.

A wide range of management options to enhance regeneration have been canvassed and include creating larger forest openings than those resulting from the felling of individual trees; clearing competing vegetation in forest openings; retaining seed trees near regeneration sites; restricting logging until after fruiting and seed release; enrichment planting; stabilizing land-tenure and increasing concession times; and reducing the proportion of trees cut, whilst encouraging the utilization of other forest timbers.

Although plantation-grown trees may attain maturity in 30–50 years, it takes at least 60–100 years for wild trees to reach designated commercial sizes.

Diseases and pests

Shoot borers (*Hypsipyla* spp.) are the major limitation to artificial establishment of *S. macrophylla* in Central and South America. In Fiji, ambrosia beetles (*Crossotarsus externedentatus* and *Platypus gerstaeckeri*) may attack living trees and cause variable, sometimes major, levels of pinhole damage. The small holes and associated staining may considerably reduce the value of the timber for decorative uses. An ambrosia beetle *Xylosandrus compactus* has also been observed to cause damage in young stands in Puerto Rico and in seedlings in the Peruvian Amazon. In Amazonia, seedlings growing close to mature trees are more likely to be attacked by caterpillars of the moth *Steniscadia poliophaea* than those further away. Termites are potentially a major damage agent in mahogany plantings in Fiji.

Uses

S. macrophylla has been widely planted throughout the tropics as a component in reforestation and afforestation programmes. It has also been used in agroforestry systems to provide shade and fuelwood. Its cultivation in agroforestry systems, where it is established at wide spacings with many other species, may reduce the incidence of *Hypsipyla* attack. *S. macrophylla* is also a component of silvopastoral systems and has shown potential in the remediation of contaminated land and in the afforestation of disused mine sites. The species has some ornamental value. High fracture angle and radial bending strength make the trees suitable for urban planting, whereas moderate tolerance of air pollution, good carbon sequestration and foliar dust retention make them popular as roadside trees.

The wood of *S. macrophylla* has a density ranging from 485 to 840 kg/m^3 (at 12% moisture content). It is one of the most valuable furniture timbers in the world due to its often highly decorative and attractive appearance combined with good technical characteristics. Although it is principally used for furniture and veneers, being easy to work and strong for its weight, it is suited to a wide range of other uses including light construction, boatbuilding, musical instruments, models and pattern making. It has high potential in the plywood industry, although colour variation, wavy grain, pin knots and pinhole borer damage may limit the potential for face-quality veneer. As trade in the wood is now subject to international control, methods of discriminating mahogany from other similar timbers are being developed, including methods based on near-infrared spectroscopy and DNA markers.

A wide range of traditional minor uses have been reported for *S. macrophylla*, including medicines, dyes and tannins. New potential uses include the production of alternative fuels (oil and char) by pyrolysis of seed waste; and using leaf and bark extracts to control crop and honeybee pests.

Further reading

Carvalho, P.E.R. (2007) *Mahogany Swietenia macrophylla*. Circular Técnica - Embrapa Florestas 140, Embrapa, Colombo, Paraná, Brazil, 12 pp.

Jøker, D. (2000) *Swietenia macrophylla King*. Seed Leaflet - Danida Forest Seed Centre 30, Humlebaek, Denmark, 2 pp.

Lugo, A.E., Figueroa Colón, J.C. and Alayón, M. (eds) (2002) *Big-Leaf Mahogany: Genetics, Ecology and Management*. Ecological Studies. Springer-Verlag, New York, USA, 433 pp.

Mayhew, J. and Newton, A.C. (1998) *The Silviculture of Mahogany*. CAB International, Wallingford, UK, 240 pp.

T

Tabebuia rosea (Bertol.) DC.

Pink poui

Importance

Tabebuia rosea is a medium to large tropical American tree, widespread from Mexico to Venezuela, Ecuador and Peru. It is a very important species in these regions; its timber is valuable, of good-quality, excellent to work and has an attractive finish. The species shows good performance in forestry plantations and enrichment planting, and has high potential to be used in afforestation projects. It is also used in agroforestry (e.g. alley cropping), as a live fence, as a shade tree and as an ornamental. It is the national tree of El Salvador.

Botanical features

T. rosea is a medium to large tree 28–37 m tall, with a dbh of 50–100 cm. It has a wide crown which may be conical or irregular with open foliage. The buttresses can reach 2–3 m in height. The bole is cylindrical and the bark is dark grey and rough, with vertical fissures.

The leaves are compound, opposite, with five small leaflets 13–30 cm long, 8–12 cm wide, the central leaflet being the largest.

The flowers are lilac or pink in colour, hermaphrodite, 5–7 cm long, 4–5 cm wide, with a bell- or funnel-shaped tubular corolla. The fruit is a cylindrical, linear capsule, 22–38 cm long, narrow at each end, dehiscent, with a rough surface; it is light green when young and dark green when mature. One fruit contains 240–300 winged seeds.

T. rosea requires a distinct dry period in order to flower. Flowering occurs in February–March in Costa Rica, February–April in Ecuador, and twice a year in some regions. The leaves are shed just before flowering, resulting in a colourful floral display, hence the use of *T. rosea* as an ornamental. The flowers are pollinated by bees. Fruiting begins once the tree is 4 years old. Fruits mature 3 months after flowering, at the onset of the wet season.

Distribution

T. rosea originates in northern South America. It is widespread from southern Mexico, through Central America and the Lesser Antilles, and as far south as Venezuela, Ecuador and northern Peru. It has been introduced in India, Sri Lanka and Thailand.

T. rosea grows in a range of habitats. It tends to dominate humid forest lowlands, gallery forest, and seasonally flooded regions. It also occurs in dry tropical forest and mountain regions, and is also found on abandoned farmland. *T. rosea* grows in association with *Cordia alliodora* and *Cedrela odorata*, and is occasionally present in pure stands.

Silviculture characteristics

T. rosea is a light-demanding species, although it may tolerate some shade in the first few years of growth. It has a deep root system and tolerates occasional waterlogging. It prefers a region with a high annual rainfall, between 1200–2500 mm, but requires a dry period to initiate and enhance flowering and fruiting.

Silviculture practice

T. rosea may be regenerated from seed or using vegetative propagation. The seeds are collected and sun-dried. Seed is classified as orthodox. Seeds remain viable for 12–14 days if untreated, yet may be stored in a dry and cool environment for up to 2 years. The seeds germinate within 2–3 weeks without any treatment. However, to improve and homogenize germination it is recommended to steep the seed in pure water for 12 h. Using a light substrate that can be disinfected to control nematodes can encourage successful germination and good root development. Shade is also recommended during germination. Seedlings reach a size suitable for planting out after 1–3 months. Once planted out it is necessary to eliminate weeds, and to clear the area two or three times in the first year.

T. rosea propagated vegetatively has performed well in preliminary trials.

The species is not commonly used in monoculture plantations due to its common tendency to fork in the first few years. In plantations, a high-density stocking rate is recommended, such as an initial planting distance of not more than 2.5 × 2.5 m (1600 trees/ha). In Venezuela, line planting at a density of 760 trees/ha resulted in a mean annual diameter increment of 2 cm. In Quintana Roo, Mexico, the species showed good results in enrichment planting trials. In Puerto Rico, *T. rosea* was used in a silvopastoral alley cropping system in a 5 ha plantation; after 14 years, trees were 14 m tall with a dbh of 13 cm. In Colombia, natural regeneration trials of *T. rosea* showed that a production of 210 m^3/ha in 25 years could be achieved from trees of 45.5 cm dbh.

Diseases and pests

In the nursery, seedlings are prone to damping off caused by *Rhizoctonia*. *T. rosea* is susceptible to attack by *Hypsipyla grandella*, and also suffers from moderate attacks by the fungus *Prospodium bicolor*, the nematode *Meloidogyne incognita* and the cicadellid *Rhabdotalebra signata*.

Uses

T. rosea has attractive lilac-coloured flowers and is used as an ornamental in many urban settings. It is also used in agroforestry systems and has great potential to be further exploited in this respect. The species is grown in association with coffee, maize and yucca. It is used in live fencing and often regenerates naturally in pasture. *T. rosea* is also used as a shade tree in cocoa (*Theobroma cacao*) plantations.

T. rosea sapwood is pale rose in colour and the heartwood is golden-brown to dark chestnut-brown. The wood has a specific density of 480 kg/m^3, and from the bole it is possible to obtain pieces of good-quality timber 12–25 m long and 25–70 cm diameter. The wood has a medium lustre and texture and straight, closely interlocked grain. It slices easily for veneer, has excellent machining and turning qualities, glues, finishes and polishes well and responds well to hand tools. It is easy to work, and has an attractive finish. The wood is also very easy to dry, and its strength has been reported as similar to that of teak. *T. rosea* wood is used extensively for furniture, light construction, boats, sport equipment, floors and plywood. The wood is, however, not durable when in contact with the ground, being susceptible to insect attack.

T. rosea has many traditional medicinal uses, such as treatment for snake bites, fever, anaemia, constipation and malaria. Infusions of the bark are used in many countries for the treatment of medical conditions, including diabetes, allergies and immune system disorders. In the late 1960s claims were made in Brazil that the bark of *Tabebuia* trees, known collectively there as pau d'arco, was a cure for cancer. A frenzy of bark stripping ensued, to the detriment of many local tree populations. The bark is today one of the most widely used immunostimulants. Its anticancer, anti-malarial, and anti-diabetic activities are attributed to the presence of naphthoquinones such as lapachol. The potential of lapachol for the treatment of some types of cancer is now considered low due to its toxic side effects.

T. rosea is a good honey tree, providing foraging honeybees with ample pollen and nectar.

Further reading

Condit, R., Pérez, R. and Daguerre, N. (2011) *Tabebuia rosea* (roble). In: *Trees of Panama and Costa Rica*, Princeton University Press, Princeton, New Jersey, USA, pp. 94–95.

Cordero, J. and Boshier, D.H. (eds) (2003) *Tabebuia rosea* (Bertol.) DC. In: Árboles de Centroamérica: *Un Manual para Extensionistas*. OFI, Oxford, UK/CATIE, Turrialba, Costa Rica, pp. 919–922 (in Spanish).

Jones, K. (1995) *Pau d'Arco: Immune Power from the Rain Forest*. Healing Arts Press, Rochester, Vermont, USA.

Taxus brevifolia Nutt.

Pacific yew

Importance

Taxus brevifolia is a conifer associated with several coniferous and broadleaved tree species on a wide range of sites in western North America. It tolerates shade, and in undisturbed stands is usually found as an understorey tree. It is an extremely slow-growing tree, with a heavy, hard wood, which is resistant to decay. Although grown and utilized on a very minor scale, it has many special uses. The bark contains a drug that is used in cancer research and treatment. Demand for *T. brevifolia* bark increased dramatically until it became possible to synthesize the drug chemically.

Botanical features

T. brevifolia ranges in size from a shrub to a medium-sized tree 10–15 m tall, occasionally over 25 m, 40 cm in dbh, up to 300 years old. The trunk is often twisted and fluted, extending into an irregular (ragged and lopsided) crown. Large limbs are common in the mid- and upper bole. The wood is heavy, strong, resilient, even-grained, fine-textured, decay-resistant and reddish to dark orange, with yellow sapwood. The bark is thin, with large scales, and dark reddish-brown. The root system is deep and wide spreading; roots are associated with ericoid mycorrhizae.

Leaves are evergreen, remaining on the tree for about 8 years, needle-like, 15–30 mm long, flattened, somewhat curved, upper surfaces dark green, paler green beneath, tip sharp, abruptly pointed. They are poisonous to horses and cattle.

T. brevifolia is dioecious, although occasional individuals can be variably monoecious or change sex over time. Male strobili are stalked, bud-like, pale yellow, and composed of 6–12 filamentous stamens, each with 5–9 anthers. Female strobili are less abundant, greenish, and composed of several scales. Pollen is dispersed by wind in the spring. The fruit is an ovoid–oblong seed about 8 mm long, partially enveloped by a fleshy, berry-like, scarlet, cup-shaped disk. The seed is about 6 mm long with a depressed hilum, bony inner coat, and membranous outer seedcoat. Seeds are poisonous, but the fleshy covering (or aril) is not. Birds eat the fleshy arils and void the seeds, which remain viable. Rodents often only take the seeds.

Initiation of phenological events varies greatly with latitude, altitude, aspect and weather, and as a result the growing season can last anything from 60 to 300 days.

Distribution

T. brevifolia is a western North American (both Pacific and Cordilleran) species. It grows from the southern tip of southeast Alaska south through the Pacific Coast region of British Columbia, and the Olympic Peninsula of Washington. It is rare in the Coast Range south of the Olympic Peninsula and north of the Umpqua River in Oregon, but occurs with greater frequency in the Coast Range in southern Oregon and northern California. Further inland, it grows on the western slopes of the Rocky Mountains in British Columbia, south into northern Idaho and western Montana, and the Lewis Range in Montana. Introductions of *T. brevifolia* are limited to arboreta and parks.

T. brevifolia typically grows scattered in isolated clumps of one or several individuals (very rarely in pure stands) in the understorey of other tree species, mainly conifers but also some broadleaves, including *Abies amabilis*, *Abies grandis*, *Abies concolor*, *Abies lasiocarpa*, *Alnus rubra*, *Larix occidentalis*, *Pinus ponderosa*, *Populus trichocarpa*, *Pseudotsuga menziesii*, *Quercus garryana*, *Sequoia sempervirens*, *Thuja plicata*, and *Tsuga heterophylla*. It is present in early, mid-, and late stages of secondary succession, and is a minor component in several vegetation zones across western North America.

T. brevifolia is found from sea level to (in the Sierra Nevada) 2440 m altitude. Associated soils include ultisols, alfisols and inceptisols.

Silviculture characteristics

T. brevifolia is shade and exposure tolerant, windfirm, and tolerates a wide range of climatic and soil conditions. Productive growth occurs on moist, nitrogen-rich soils.

Reproduction capacity is medium. It is a prolific seed producer, but the frequency of seed crops is unknown. Seed dissemination capacity is high; seeds are dispersed mainly by birds, rodents and squirrels. Potential for natural regeneration in low light is high when considering vegetative reproduction, and in the open is high, except in situations where it requires protection.

Potential initial growth rate (up to 5 years) is moderate, being about 20 cm/yr for planted seedlings. Crown spatial requirements are high, with long branches and a wide, umbrella-shaped, live crown. Dense and closed canopy stands are non-existent. Longevity is high but not determined.

Silviculture practice

Fruits ripen from August to October of the same year that flowering occurs. Fruits either drop to the ground or are taken from trees by birds or rodents. *T. brevifolia* is a prolific seeder. Seeds, which average about 33,100/kg after cleaning, germinate slowly and require stratification. For germination, 30°C day and 20°C night temperatures are desirable. Germination is epigeal, and usually in heavy organic matter. Yew seeds sown in nursery beds in late spring require mulching. Beds require shading during the summer and again in December. Some seeds do not germinate until the second spring after sowing.

T. brevifolia is capable of layering and often sprouts from stumps or rootstocks after the top has been killed or the tree cut. Layering usually occurs after branches or tree tops have been pressed to the ground for a prolonged period by large fallen trees or limbs.

Compared to other conifers, *T. brevifolia* grows very slowly, and is thus not considered a timber crop species; trees larger than 50 cm in dbh and taller than 12 m are rare. Because of the slow growth and because the species is typically found as an occasional tree in stands of other tree species, volumes and yields of *T. brevifolia* are low, with net volume being considerably lower due to the prevalence of heart rot in large trees.

Diseases and pests

No serious leaf diseases, but two root diseases – *Armillaria ostoyae* and *Phaeolus schweinitzii* – have been reported for *T. brevifolia*. *Rhizoctonia solani*, *Phytophthora cinnamomi* and *Pythium* sp. cause damping off and seedling root rot.

Although seasoned heartwood of *T. brevifolia* is extremely durable, large living trees often have heart rot or hollow boles. Heart rot fungi infecting the species include *Phellinus nigrolimitatus*, *Phellinus pini*, *Phellinus robustus* and *Fomitopsis rosea*.

T. brevifolia is sensitive to damage from fire, and exposure to the sun, wind and cold. It resists damage from sulphur dioxide and other air pollutants.

Uses

Sprouts and epicormic branches that form in response to stand disturbance are favoured by browsing animals. Although little used as an ornamental, the species has potential as a shade tree, for hedges and for topiary.

The wood of *T. brevifolia* is fine-grained, non-resinous, and very strong with a density of 600 kg/m^3, and is used for making archery bows, canoe paddles, tool handles, gunstocks, boat decking, furniture, musical instruments, carved figurines and miscellaneous novelty items. *T. brevifolia* trees were seldom taken in primary logging operations, but some *T. brevifolia* wood was later salvaged by firewood cutters and gleaners gathering wood for speciality products.

With a history of traditional medicinal and other usage by native peoples of the West Coast of North America, *T. brevifolia* is today being used primarily for pharmaceutical purposes. In the late 1960s, paclitaxel (originally known under the more familiar name 'taxol'), a complex compound extracted from *Taxus* bark, was identified as a possible therapeutic agent. The National Cancer Institute in the USA found paclitaxel to be one of the most promising of more than 120,000 plant compounds tested for anticancer properties. Paclitaxel appears to be effective against a wide range of tumours, and good responses have been obtained in the treatment of refractory ovarian cancer. It also inhibits the replication of *Trypanosoma cruzi*, the pathogenic protozoan causing Chagas disease, as well as the disease-causing flagellate *Trichomonas vaginalis*.

Paclitaxel has been found in most species of *Taxus*, but *T. brevifolia* is the only one that is considered to be a practical source of quantities sufficient for clinical use. This discovery led to increased harvesting of bark from the wild, endangering the species in some parts of its native range; 6000–7000 trees were cut down initially to produce 27,700 kg of bark. Some 9080 kg of bark are required to produce 1 kg of paclitaxel. Although intensive silvicultural management for bark production is feasible on suitable sites, with plants being propagated through tissue culture and vegetative means, synthesis in the laboratory through plant cell fermentation is currently the preferred method of paclitaxel production. This has led to some easing of pressure on wild populations.

Further reading

Hageneder, F. (2011) *Yew: A History*. The History Press Ltd, Stroud, UK.

Itokawa, H. and Lee, K.H. (eds) (2003) Taxus: *The Genus* Taxus. Taylor & Francis, London.

Tectona grandis L. f.

Teak

Importance

Tectona grandis is the principal timber tree of peninsular India, Myanmar, Indonesia and Thailand, and one of the most valuable timbers in the world. The reputation of teak timber is due to its matchless combination of qualities such as termite, fungus and weathering resistance; lightness with strength; attractiveness; and workability and seasoning capacity without splitting, cracking, warping or materially altering shape. In India, traditionally *T. grandis* has been used as the standard for comparing the quality of other timbers. It has been widely planted both within its home range and in other tropical regions, where it is important commercially and in the livelihoods of rural communities. The world's first teak plantation was established in India in the early 1840s.

Botanical features

T. grandis is a large deciduous tree up to 45 m tall and up to 2.5 m in diameter, with a straight clear bole for 20 m and an open crown. It is sometimes fluted. Bark pale brown with shallow longitudinal grooves in young trees and cracked and scaly in mature trees, 1–1.5 cm thick. Branchlets are stout, tetragonal with quadrangular prominent depressed leaf scars, nodes annular, young parts densely furfuraceous tomentellous.

Leaves simple, opposite decussate (occasionally three at a node or rarely alternate in seedlings), 15–90 × 6–50 cm, broadly elliptic, acute or shortly acuminate, base acute to attenuate, decurrent on the petiole, often clasping at base, margin entire, sometimes repand–denticulate, chartaceous, dark green and glabrous above and pale glabrous to tomentose or often silvery below; lateral nerves 8–10 pairs, lower 4–6 pairs, branched and the branches archingly joined along the margin, transverse nerves numerous, parallel, prominent.

The inflorescence is terminal and from the axils of upper leaves, forming a large panicle, 70 × 45 cm, with opposite decussate widely divaricate, many branched cymes, densely cinereous or furfuraceous tomentose; bracts foliaceous subtending each pair of cymes, sessile, to 15 × 4 mm; prophylla linear–oblong or lanceolate, 4 × 1 cm. Flowers pedicelled, pedicels 2–3 mm long, furfuraceous. Calyx 4–5 × 3–3.5 mm, globose–campanulate, densely tomentellous, lobes 6 (5 or 7), ovate–oblong, obtuse, 2 mm long, often reflexed. Corolla white, sometimes with a pale bluish shade, tube broadly cylindrical, lobes 6 (5 or 7), ovate–oblong or ovate–elliptic, 3 × 3.5 mm, glabrous, overlapping, reflexed. Stamens 6 (5 or 7), 3–4 mm long, filaments glabrous, anthers ovate or oblong, yellow. Ovary densely pubescent, 4-celled, ovules one in each cell, hemianatropus; style 4–5 mm long, pubescent towards base with branched hairs; stigma 2-lobed, lobes often adpressed. Fruit drupaceous, 1.2–1.8 cm long and wide, subglobose or subtetragonous, enclosed in the enlarged inflated calyx, pericarp densely felted-tomentose with irregularly branched pale brown hairs, endocarp bony; fruiting calyx 1.5–2.5 cm long and 1.2–2.2 cm wide, pale brown, irregularly plaited, seeds 1–4.

T. grandis is insect-pollinated, mainly by bees (*Ceratina* spp.) which travel among inflorescences and trees. It is a deciduous tree, although in prehumid climates it often carries some leaves all year round. Otherwise, it sheds leaves in summer and remains leafless throughout the greater part of the dry season. New leaves ordinarily appear from April to June, according to locality and season. Where the soil is moist, the leaf fall period is delayed. Flowering season varies ordinarily from June to August or September. The fruits ripen from November to January and fall gradually.

Distribution

T. grandis is indigenous to the South Asian and the South-east Asian regions. It grows naturally between latitudes of 9° N to 25° 30' N in most of peninsular India, large areas of Myanmar and parts of Laos and Thailand. It has been widely planted both within and beyond its natural range.

Plantations extend across South-east Asia, the Pacific, East Africa, West Africa, South Africa, the Caribbean Islands, South America and Central America.

Silviculture characteristics

T. grandis is leafless in the dry season. This period varies among localities depending upon moisture availability. It produces a long, thick taproot which, as the tree grows, may persist or disappear, but in either instance, numerous strong lateral roots are formed.

T. grandis is a pronounced light-demander, intolerant of shade and requiring complete overhead light. Its crown requires freedom on all sides for proper development. *T. grandis* coppices and pollards vigorously, and growth of coppice and pollard shoots is rapid. It is very sensitive to mutual root competition; on drier sites in particular root competition appears to limit the growth rate before crown competition does. The seedlings are sensitive to suppression by weeds and cannot grow well without repeated weeding.

It is susceptible to frost damage but it has such remarkable powers of coppicing that if a frost-bitten plant is cut back, it may send up a shoot which often gets above frost level and survives.

T. grandis is a moderately windfirm tree owing to its well-developed root system, although in exposed places it can suffer from wind damage, leading to branching. It has remarkable fire-resisting power; young plants recover from fire as the root system survives and finally produces a permanent shoot which survives fire.

Silviculture practice

Seed production in *T. grandis* generally commences after the fifth year. There are usually between 1500 and 2500 fruits/kg. Fruits are collected from clean ground by sweeping up. For transport and storage it is useful to remove the calyx by rigorous rubbing and winnowing. Fruit can be dry-stored in gunny bags and sealed tins for up to 2 years.

Fruit viability is in the range of 40–85%. However, germination under nursery conditions is relatively low, 25–35%, and sporadic, taking between 10 and 90 days and extending further. Such germination behaviour causes a large variation in both quantity and quality of seedlings in nurseries.

Although treatment of seeds to promote germination is not deemed necessary in some regions, a number of pre-sowing treatments have been shown to improve germination including fire and smoke, repeated soaking and drying and high temperature.

Artificial regeneration may be done by direct sowing, bag planting or stump planting. Direct sowing, the oldest method, is characterized by high mortality and slow growth. Seedlings kept in bags yield plants with a good root system in a short period of time, provided care is taken to prevent root coiling in the containers. Stump planting is the most popular method of regeneration as it offers several benefits. Stumps may be produced when needed and transported over considerable distances while maintaining their viability. Moreover, they are easier and quicker to plant, and subsequent growth is more rapid and vigorous. For production of planting stumps, beds are dug deeply, edged, seed sown and covered with earth and straw to prevent soil wash before the onset of monsoon showers. Alternatively, beds can be managed under irrigation. Germination takes place in 10–20 days. Stumps can be prepared from 1-year-old teak seedlings. An ideal stump should have a 5 cm shoot and 15–20 cm taproot with a root diameter (at the thickest part of the root usually close to the collar region) of 1–2 cm. Such stumps can either be planted fresh or after proper storage in planting holes.

T. grandis coppices well and early growth is rapid. Coppicing is seen as an alternative reforestation tool in Java, Indonesia, reducing regeneration time and costs, as well as demands for labour and seed. Mean height and diameter in coppiced plantations are significantly greater than in plantations derived from seed at 3, 8, and 13 years, and there is also less disease incidence. *T. grandis* also pollards vigorously. Pre-budded stumps grown in pots are used for establishing seed orchards. The establishment rate is above 80% under ideal conditions. The budded plants are hardened off before outplanting in orchards.

Tissue culture techniques and micropropagation protocols have also been developed for *T. grandis*. Clonal mass propagation of selected superior genotypes which are selected based on their visible characteristics (phenotypes) is another method of vegetative propagation. Shoots produced from intensively managed stock are trimmed and rooted under mist system facilities for 6 weeks and raised in culture containers for 2–3 months prior to field planting. Various growth regulators stimulate root formation; rooting response varies with the hormone, its concentration, and the season.

In natural forests, *T. grandis* has traditionally been managed under a selection system, in which fellings and regeneration are distributed over the whole area and the resultant crop is so uneven-aged that trees of all ages are found mixed together throughout. In the selection system, mature trees above the exploitation diameter, together with dead and dying trees, are removed.

Initial spacing varies considerably from place to place. A square spacing of 1.8 × 1.8 m is adequate for drier areas. Wider spacings are adopted for a variety of reasons. On good sites with rainfall above 1500 mm a spacing of 2.5 × 2.5 m or 2.7 × 2.7 m produces rapid rate of growth and faster canopy establishment.

As *T. grandis* plants are susceptible to competition from weeds, especially grasses, weed control is very important, particularly during the initial 2–3 years of establishment. *T. grandis* is a strong light-demander and will suffer from crown competition; it grows best in even-aged crops, regularly and heavily thinned, particularly in the first half of the rotation. Depending on the site characteristics and management practices, rotation period varies from 30 to 150 years.

Diseases and pests

In nurseries, bacterial wilt caused by *Pseudomonas* spp., leaf spots caused by *Phomopsis* and leaf rust caused by *Olivea* often cause serious problems. *Pseudomonas solanacearum* (*Ralstonia solanacearum*) causes typical vascular wilt in trees in India, Indonesia, Malaysia, and Myanmar, while *Pseudomonas tectonae* is prominent in the Philippines and is the main threat to seedlings in root trainer nurseries in Kerala. Leaf spots caused by *Phomopsis tectonae* in combination with *Colletotrichum gloeosporioides* (*Glomerella cingulata*) result in defoliation and death of seedlings. Rust caused by *Olivea tectonae* results in premature defoliation in nurseries and plantations in India, Pakistan, Sri Lanka, Myanmar, Indonesia and Thailand. In 1–3-year-old plants, *Corticium salmonicolor* (*Phanerochaete salmonicolor*) (pink disease) causes death of terminal shoots in India and Indonesia. A new seed rot pathogen *Cephaliophora irregularis* has recently been reported from India.

In Kerala, boring by the trunk borer *Cossus cadambae* leads to infection of the tunnels by the fungus *Phialophora richardsiae* (*Pleurostomophora richardsiae*), resulting in tree dieback.

Collar and root rot caused by *Fusarium oxysporum* and *Rhizoctonia bataticola* (*Macrophomina phaseolina*) are serious problems in seedling nurseries in Akola, India, especially in the August–October period. *Helicobasidium compactum* is a root rot causing die-back and mortality in high-input plantations in Madhya Pradesh, and is controlled by removal of basal branches and leaves.

The teak defoliator *Hyblaea puera* and the skeletonizer *Eutectona machaeralis* are recognized as serious pests. The sapling borer *Sahyadrassus malabaricus* is a problem in young plantations with dense weed growth. The teak beehole borer (*Xyleutes ceramica*), is a serious pest in Myanmar and Thailand. *C. cadambae* causes similar problems in the southern states of India, but the incidence is limited to pockets where the trees are subject to repeated lopping. Other invertebrate pests of *T. grandis* include nematodes, which parasitize the roots. In nurseries, white grubs (*Holotrichia* spp.) eat the seedling roots, causing wilting and subsequent death.

Teak also suffers from injury by various mammals such as squirrels, rats, pigs, deer and buffaloes. The worst offender is the elephant, which can ruin teak plantations.

The mistletoe *Dendrophthoe falcata* (Loranthaceae) is a major problem in plantations in almost all teak-growing countries, especially India, Bangladesh, Indonesia and Trinidad.

Uses

T. grandis trees are grown on farms and homesteads mainly for timber and are used as standards (supports) for growing pepper (*Piper nigrum*) vines. In Belize, it is a component of silvopastoral systems, and in Myanmar it is used in taungya reforestation. The species has shown some potential in other agroforestry situations, including being planted with cassava, lemongrass, medicinal and aromatic crops and bamboos.

T. grandis trees and teak forests have some importance from an ecological point of view. The fruits, seeds, bark and leaves are important food sources for birds and mammals, which also use large trees for shade. However, plantations have fewer ecological benefits in comparison with natural forest, tending to suppress the growth and diversity of native trees and understorey vegetation.

Teak is a well-known multipurpose timber and the most important furniture wood in India. Other common uses include shipbuilding, exterior construction, carving, turning, panelling and woodware. Compared with other industrial woods, teak is a medium-weight, strong wood of average hardness, with a heartwood that is one of the most naturally durable woods of the world. It is usually immune to termite and insect attack for very long periods and is considered by many to be the gold standard for decay resistance. However, it is not always immune to fungal damage (rot); the sapwood is perishable and is quickly destroyed by rot, termites or borers. Wood density is (480–)610–750(–850) kg/m^3 at 12% moisture content. The heartwood tends to be a golden or medium brown, the colour darkening with age.

The leaves, bark, roots, flowers, fruits, seeds and sawdust of *T. grandis* have all been found useful in many ways, especially for medicinal purposes.

The leaves contain yellow or red dye recommended for dyeing silk, wool and cotton. Other traditional and modern uses of the leaves include wrapping and packing fresh meat and fish, for thatching huts and for making medium-strong boards through thermoplasticization and processing. Bark and leaf powder have showed promise as a substrate with good sorption capacity for removing heavy metals and other pollutants from aqueous solutions. Activated charcoal can be prepared from the sawdust. Because of their high calorific value, the wood, branches and trimmings are useful as fuelwood, whereas the sawdust also has potential for use in fuel briquettes.

Further reading

Centeno, J.C. (1997) The management of teak plantations. *Tropical Forest Update* 7, 10–12.

Restrepo, H.I., Orrego, S.A., Valle, A., J.I. del and Salazar, J.C. (2012) Yield, optimal turnover and profitability in forest plantations of *Tectona grandis* and *Pinus patula* in Colombia. *Interciencia* 37, 14–20.

Tewari, D.N. (1992) *A Monograph on Teak (Tectona grandis Linn. f.)*. International Book Distributors, Dehra Dun, India, 479 pp.

Terminalia amazonia (J.F. Gmel.) Exell

Nargusta

Importance

Terminalia amazonia is a large tree found widely in lowland humid forest types in the American tropics and highly valued for its timber. It appears to be restricted in its native range to precipitation regimes greater than 2500 mm/yr. Plantations may need some pruning or thinning of multiple-stemmed trees. Given its moderate to fast growth rate even on low-fertility soils, high-quality wood, lack of serious insect or disease problems to date and apparent ability to thrive in pure stands, *T. amazonia* is good candidate for further plantings in low to moderate elevations in the humid tropics.

Botanical features

The tree grows up to 50 m in height (usually 20–35 m) and 1.5 m diameter. Young trees produce branches in whorls. Crowns within rainforests often have a cylindrical form, whereas the crowns of large emergents spread out over the general canopy. Mature trees typically have large buttresses which support clear, symmetrical boles up to 20 m. Trees with diameters over 50 cm are often hollow inside. Saplings and pole-size trees can have multiple vertical stems or terminate in a whorl of horizontal branches with no leader.

Small (2–4.5 cm long) simple leaves, with entire margins and acute tips; concentrated at the ends of branches. Flowers are small, 6 mm in diameter, greenish, and produced on spikes when trees are briefly deciduous, usually in the middle of the dry season in February in the southern Pacific zone of Costa Rica. Papery fruits, 1.5–2 cm wide, have two larger wings and three smaller ones. These fruits are golden-brown when mature (March–April in south-western Costa Rica) and each contains one seed (2 mm in size). The pattern of flowering and seed maturation follows that of many tree species in tropical regions with dry and wet seasons, in which flowers and fruits develop over the dry season and produce seed for the beginning of the wet season.

Distribution

T. amazonia is found over a very wide area from southern Mexico through Central America and into Brazil and Peru, as well as in Trinidad and Guyana. *T. amazonia* has not been extensively used as an exotic, although it has been planted in reforestation programmes in Costa Rica since the late 1980s.

Silviculture characteristics

No results from storage trials have been reported, so it is not known if the seeds are orthodox or recalcitrant. Fruits from isolated trees often do not contain seed, so that it has been considered best simply to spread out fruits for germination rather than attempt to store mainly empty fruits.

Silviculture practice

For propagation by seed, the small (2 mm) seeds which are embedded in 1.5–2 cm, asymmetric winged fruits (samaras), are collected and used for germination without attempting to separate out the seeds. The samaras are spread out thickly in seed boxes or on raised beds and covered with approximately 2 cm of sand or soil. Germination takes place over a period from 2 to 6 weeks after sowing.

T. amazonia has a natural tendency to coppice and appears to have good potential for vegetative propagation. Although propagation by cuttings has not been reported, 'stumps' grown from seed have been used for plantation establishment. In this method, seedlings are transplanted from germination areas into raised beds at a spacing of 12–20 cm. After approximately 10–12 months the seedlings attain diameters of 0.8–2.0 cm at ground level and can be lifted and cut back to stems of 10–20 cm with 10–20 cm of root. Once planted at the plantation, the stems sprout after 1 month, after which a single stem needs to be chosen.

The standard nursery practice is to produce *T. amazonia* seedlings in plastic bags. It requires 3–5 months from transplanting seedlings to bags for a seedling 30 cm tall to develop. Because seedlings may still be small 1–2 months before the beginning of the dry season in some areas, it may be desirable to keep seedlings over the dry season for planting out at the beginning of the next wet season. Seedlings can be maintained under light shade and moderate watering so that they do not outgrow their containers.

In natural forest, *T. amazonia* is a wind-dispersed gap colonizer, and can be classified as a long-lived pioneer. In many rainforest areas densities are apparently low, less than 1–5 trees/ha. *T. amazonia* could be encouraged to grow by creating gaps near bearing seed trees near the end of the dry season, during which time fruits are falling. Care needs to be taken that isolated trees are in fact producing viable seed, as it has been found that widely scattered trees set very little viable seed.

In Costa Rica, most plantations have been established at 3 × 3 m spacings. The species produces wide branches in whorls, which begin to touch neighbouring trees at 2–3 years when the plantation is well-managed and/or on a fertile site. These lower branches appear to be fairly persistent in shade and may help suppress weeds, but could require pruning to avoid the formation of large knots. As the tree is known to coppice readily, plantations could possibly be managed through coppice systems.

Several trials of *T. amazonia* in mixed-species plantations have been established. In a humid lowland site in the Atlantic region of Costa Rica, *T. amazonia* was combined with *Dipteryx panamensis*, *Virola koschnyi* and *Albizia guachapele* as well as being planted in pure plantations. At 39 months, *T. amazonia* was the tallest (8.4 m) of the four species tested, and showed no significant difference in height between trees in pure and mixed stands, but did have a significantly greater diameter in mixed plots. In the Pacific zone of Costa Rica, at 1075 m elevation, *T. amazonia* was planted in pure plantations and in mixed stands at 50:50 ratios with nitrogen-fixing trees and in plots with different leguminous cover crops. At 3 years, the tallest *T. amazonia* trees were 1.9 m when grown in conjunction with *Inga edulis*, but this height was not significantly different from that in the other treatments.

Diseases and pests

In general, *T. amazonia* appears to be capable of growing in pure stands with few insect or disease problems, although cautious optimism is warranted in regard to its potential in pure plantations. Larvae of the genus *Cossula* (Lepidoptera: Cossidae) bore underneath the bark and into the xylem and then bore upwards for 25–40 cm, making galleries 12–14 mm in diameter. Moderate defoliation damage to plantation trees has also been reported from the *Exophthalmus* spp., the beetle. More seriously, low survival rates (66%) have been observed in planted seedlings due to attacks by leaf-cutting ants.

Care should be taken when pruning to avoid wet weather, so that fungal entry into stems is minimized.

Uses

T. amazonia has been managed informally in natural forests and in farming areas in a variety of systems. In the upland volcanic region of Coto Brus in Costa Rica, some farmers have left stands of mature trees in clearing for pastures. These have not been managed for regeneration in the pasture grasses and it appears that once these forest-grown trees are cut or blown over, there will be no more *T. amazonia* on the site. Similarly, widely spaced forest trees have been left in coffee plantations in some cases, either as the only tree or in combination with nitrogen-fixing trees such as *Inga edulis* and *Gliricidia sepium*. In one case, *T. amazonia* grew 400 cm in height after 3 years, planted at 4 m spacings over a coffee plantation.

T. amazonia produces a wood suitable for many uses and has been suggested as a substitute for oak (*Quercus* spp.). The timber is hard, durable, has high strength properties, moderate shrinkage, and is attractive, often being yellowish with red streaks. Some physical properties are green density 1.14 g/cm^3, air-dry density 0.7–0.9 g/cm^3 and basic specific gravity 0.58–0.79. Machining is moderately difficult; some tearing may occur in planing due to the interlocked grain. Finishing qualities are good, but may require some sanding. It has been used for general construction, furniture, flooring, and outside construction including railroad ties and bridges. The heartwood is extremely resistant to treatment; the sapwood is variable. It has good resistance to dry-wood termites, but not to subterranean termites and other drywood insect pests.

Further reading

Carpenter, F.L., Nichols, J.D., Pratt, R.T. and Young, K.C. (2004) Methods of facilitating reforestation of tropical degraded land with the native timber tree, *Terminalia amazonia*. *Forest Ecology and Management* 202, 281–291.

Günter, S., Weber, M., Stimm, B. and Mosandl, R. (eds) (2011) *Silviculture in the Tropics*. Springer, Heidelberg, Germany, 559 pp.

Montero, M. and Kanninen, M. (2005) *Terminalia amazonia: Ecología y Silvicultura*. CATIE, Turrialba, Costa Rica, 34 pp.

Terminalia arjuna **(Roxb. ex DC.) Wight & Arn.**

Arjun

Importance

Terminalia arjuna (Plate 42) is a large and relatively fast-growing tree. It is used for heavy construction timber, fuelwood and for medicinal purposes. Under natural conditions, the species grows on the banks of streams, rivers and canals. The root mass of the species helps to reduce soil erosion on the banks. It is a useful species for agroforestry, and is widely planted for raising tasar silkworms in India. *T. arjuna* is a very common roadside species, and is also used for shade, ornamentation and sand dune afforestation in India and Sri Lanka.

A disadvantage of *T. arjuna* is that the wood is apt to split when drying and is not easy to work; however, careful drying results in fine-quality timber. More research is required to elucidate reproductive biology and genetic variation in *T. arjuna* and selection of provenances for good stem form.

Botanical features

T. arjuna is a fluted tree with a large, spreading crown and drooping branches. Under natural conditions, the tree grows to a height up to 30 m (sometimes up to 60 m) with a dbh up to 2–2.5 m. The bole is straight, in most cases clear of branches for half of the total height, and usually buttressed. It has a superficial, shallow root system, spreading radially along stream banks.

The leaves are simple, opposite to subopposite, coriaceous, oblong to ovate lanceolate, to subobovate, cuneate, rounded to occasionally narrowly subtruncate at base, obtuse, shortly apiculate or mucronate at apex. Shortly pale sericeous when young, soon becoming glabrous but not shiny on both sides. Pale dull green, minutely verrucose. The margins are shallow crenate to serrate. Lamina is 8–25 × 5–9 cm, arcuate 10–20 pairs of pellucid veins, leaves exstipulate. The petiole is short (2–4 cm long), sericeous, with 2 (or 1) prominent glands at petiole apex.

Inflorescences are short axillary spikes or small terminal panicles, 9–13 cm long with 2.5–6 cm long inflorescence branches, rachis short, white, pubescent, lower receptacle 0.8–1.5 mm long, short sericeous, upper receptacle 1.5–1.75 mm long, glabrous except at base where slightly pubescent. Flowers are small, cup-shaped, regular, sessile, polygamous, white, creamy or greenish-white, and strongly honey-scented.

Distribution

T. arjuna is native to central and southern India and Sri Lanka, and is planted throughout these countries as a roadside, shade and ornamental tree. It has also been widely planted in India to rear tasar silkworms. The species has also been planted in Indonesia, Kenya, Malaysia, Mauritius, Pakistan and Thailand, at low elevations, mainly as a roadside or ornamental tree.

Silviculture characteristics

T. arjuna is a moderately shade-bearing tree, but does not tolerate dense overhead shade.

Natural regeneration occurs in forests through self-seeding where the seeds accumulate in loose alluvial soil along stream banks of subhumid to humid areas. Natural regeneration rarely occurs in semi-arid habitats due to lack of moisture. Seed germination takes place early in the rainy season. If seeds are exposed to the sun, they do not readily germinate. *T. arjuna* is initially a slow grower, but later grows quickly.

T. arjuna produces root suckers and is suitable for pollarding. It coppices strongly if cut when smaller than approximately 76 cm dbh, yielding poles, posts and firewood.

Silviculture practice

T. arjuna is commonly propagated by seed in India. The mature seeds are viable for 1 year when stored in dry gunny bags or in open or closed glass bottles; however, germination rates are then reduced by 10–20%. Thus, seeds can be classified as orthodox.

In India, *T. arjuna* starts bearing fruits 6–7 years after planting. Ripe fruits can be collected from the trees themselves or from the ground below trees. They are usually dried in the shade for 1 month, and the fibrous ribs are usually reduced by rubbing with the hands. Seeds are soaked in hot water for 1 min followed by soaking in cold water for 48 h before sowing. Germination rates are variable, usually ranging from 50 to 60%. Germination starts after about 20 days and is completed in 2 months.

Experience in of direct sowing in India shows that fruits of *T. arjuna* can be sown in lines 3.5–4 m apart at the onset of monsoon rains. To improve germination of seeds the fruit should be sown half below and half above ground. Direct sown plants tend to tolerate mid-season drought better than nursery seedlings. However, direct seeding is not recommended in dry areas unless watering can be done to encourage germination in case of a break in the monsoon.

The species may also be propagated by air-layering, root suckers and stumps.

Roots of older seedlings can become too long and are better when made into stumps. Normally 15-month-old seedlings are used for preparing stumps. A collar diameter of about 1.2–2.5 cm is considered as the optimum size for planting stumps.

Besides the usual weeding, protection from fire and frost is needed for the first 1–2 years. Pruning is required to remove the weaker shoots when forking takes place due to frost damage.

The formation of vesicular-arbuscular mycorrhizal association in *T. arjuna* has been reported.

T. arjuna has been successfully grown in taungya plantations. *T. arjuna* plantations established in India for tasar sericulture have a spacing of 1 × 2 m, and are managed by repeated pollarding.

Information on management of *T. arjuna* is scarce and relates only to seedlings and saplings. Under favourable conditions in India, seedlings attain 2–3 m height in 3 years. A mean annual girth increment of 1.9 cm has been reported from Uttar Pradesh while a height of 11–12 m and a dbh of 59–89 cm at 16 years is also recorded.

T. arjuna yields up to 45 kg of dry bark chips for tannin on a 3-year cycle without injury, as bark is reproduced after two rainy seasons.

Diseases and pests

T. arjuna does not suffer from any serious pests or diseases. However, incidence of powdery mildew caused by *Phyllactinia terminalae* and white fibrous rot due to *Polystictus affinis* have been reported. Larvae of *Apoderus tranquebaricus* feed inside rolled leaves, and larvae of *Gelasma goniaria* and *Lymantria mathura* cause defoliation of the plant.

T. arjuna is also affected by galls induced by the hemipteran *Trioza fletcheri minor* which cause considerable damage to the leaves. *Ceroplastes ceriferus* (white wax insects) have also been reported on the leaves of this species.

Seed damage by the rose-ringed parakeet (*Psittacula krameri*) has been reported in India.

Uses

The species is commonly found and planted along the banks of rivers, streams, old irrigation channels, and the edges of tank bunds and alluvial bars in seasonally dry water courses, helping to reduce soil erosion on the banks. It is often planted on roadsides and in sand dune afforestation programmes with *Casuarina* species. It is also planted near wells, as roots of *T. arjuna* are believed to purify and cool the water there. *T. arjuna* is an agroforestry species and is intercropped with coconut and citrus. It is also an excellent shade tree, especially in coffee plantations.

The sapwood is pinkish-white and the heartwood is brown to dark brown, very hard, lustrous, strong, and heavy (specific gravity 0.74; weight 816–865 kg/mn). The wood is streaked with dark lines, is usually without odour, has irregularly interlocked grain, and is coarse textured.

The wood is not easy to saw and work, and is difficult to season or dry, but with care it can be brought to a fine finish. It takes a very good and lasting polish. For the most successful seasoning or drying, the trees are required to be felled during or just after the rains and converted into logs with the least delay possible.

T. arjuna is widely used in India and Sri Lanka for light and heavy construction timber. The timber is suitable for sleepers, bridge planks, cartwheels, axles, agricultural implements such as ploughs and water troughs, boatbuilding, mining props, staircases, beams and in well construction.

T. arjuna also provides satisfactory rayon-grade pulp in mixture with other woods. It makes excellent charcoal and firewood, with calorific values of 5030 kcal/kg and 5128 kcal/kg for the sapwood and heartwood, respectively.

In India, the leaves of *T. arjuna* are fed to tasar silkworms, one of the species of silkworm with the greatest cocoon crop performance. For this purpose, systematic plantations have been established at 1–2 m spacings. The leaves are also used as fodder (containing 9–11% crude protein, 14–19% crude fibre, and 28–43% dry matter) in Madhya and Uttar Pradesh in India.

T. arjuna has been widely used in Ayurvedic medicine for the treatment of cancer, dermatological and gynaecological complaints, heart diseases and urinary disorders.

The bark (primarily) and fruit are used as tanning and dyeing material. The bark contains 20–24% tannin and is remarkable for the large amount of lime (calcium carbonate) it contains; it is often burnt to produce lime for chewing with betel. The bark is also used to assist precipitation of mud from turbid water. *T. arjuna* trees exude a red resin.

Further reading

Khare, C.P. and Katiyar, C.K. (eds) (2012) *The Modern Ayurveda: Milestones Beyond the Classical Age*. CRC Press, Boca Raton, Florida, USA, 413 pp.

Paarakh, P.M. (2010) *Terminalia arjuna* (Roxb.) Wt. and Arn.: a review. *International Journal of Pharmacology* 6, 515–534.

Pandey, A.K. and Mandal, A.K. (2012) Sustainable harvesting of *Terminalia arjuna* (Roxb.) Wight & Arnot (Arjuna) and *Litsea glutinosa* (Lour.) Robinson (Maida) bark in central India. *Journal of Sustainable Forestry* 31, 294–309.

Terminalia brassii Exell

Brown terminalia

Importance

Terminalia brassii is an economically and ecologically important tree species in the Bismark Archipelago (Papua New Guinea) and Solomon Islands. It dominates large tracts of swampy lowland forest in Bougainville (Papua New Guinea) and New Georgia (Solomon Islands) and provides an important commercial timber, representing over 5% of log-export volume from the Solomon Islands. It is potentially a suitable plantation species, and one with potential to reafforest lowland swampy areas. Research needs include determination of seed-storage characteristics and a continued refinement of silvicultural practice and evaluation of plantation-grown material.

Botanical features

T. brassii is a large tree, sometimes reaching massive proportions, to 35–45 m tall with a diameter of up to 150(–250) cm, and characterized by huge flange buttresses to 7 m high. Stilt roots start to form in very young trees, and a ring(s) of adventitious roots can form some way up the buttresses in trees growing in standing water. The crown is broad, monopodial at first, later sympodial, then flattish at 20–30 years of age, spreading in flat layers at maturity. Bark surface light brown/fawn, scaly fissured to shaggy scaly, and peeling off in long loose strips. The inner bark is pale brown to white, pinkish-brown. Young branchlets are tomentose or nearly glabrous.

Leaves are simple, alternate or sometimes subopposite, subcoriaceous, varying from tomentose to nearly glabrous, narrowly oblong, narrowly oblong–elliptic or elliptic 5–9 by 3–6 cm, usually gradually narrowed and pointed at the apex (sometimes rounded) rounded or subcordate at the base; nerves 20–35 pairs; petiole tomentose–pubescent or glabrous 5–10 mm, with 2 conspicuous, glabrous suborbicular glands at or near the apex. Young leaves coated with fine rusty coloured hair. Petioles often with glands.

The flowers are pale green sessile in terminal and axillary panicles, 8–13 cm long; on a tomentose rachis. Lower receptacle (ovary) is tomentose, 1.5–2 mm long elongating to 5–8 mm before the wings develop; upper receptacle cupuliform 1 × 1–2 mm, persistent; calyx-lobes scarcely developed. Panicles to 14 cm long. Individual small narrow calyx tube, protruding stamens. Fruits, flat and papery, pubescent suborbicular, obovate or ovate in outline, 10–15 mm long by 8–13 mm wide, with 2 well-developed thin flexible wings and 3 rudimentary ones, crowned at the apex by the remains of the persistent apex.

Distribution

The species is naturally distributed in the Bismarck Archipelago of eastern Papua New Guinea (including the islands of East New Britain (Gazelle Peninsula), south-eastern New Ireland and Bougainville) and most of the Solomon Islands, including Shortlands, New Georgia group, Choiseul, Santa Isabel, Guadalcanal, Malaita, Ulawa and Makira (San Cristobal). Field trials have also been planted in Australia and Fiji.

Silviculture characteristics

Seeds are wind dispersed, sometimes travelling distances of more than 1 km from the mother tree and making them difficult to collect. They also float and germinate in water. A sample of seed needs to be checked prior to collection to ensure that it is not empty. There are approximately 70,000 seed/kg. Seed should be rapidly air-dried, out of direct sunlight and in well-ventilated conditions, down to a moisture content of 7–8%. Moist fresh seed is very vulnerable to chilling damage, and seed must be kept at mild or ambient temperatures until dried.

On dry coralline sites, on clay soils on hillsides of the Solomon Islands, natural regeneration occurs; the best regeneration often occurs where the soil has been disturbed rather than in moister areas. The species is light-demanding and benefits from full site clearance.

Silviculture practice

After collection the seed is dried away from direct sunlight, the moisture content of the seed is taken down to 6–8%, and it is important to avoid overheating and to keep moisture content relatively high. The most successful method in the nursery involved the seeds being sown thickly on a bed of 'swamp mixture' approximately 7.5 cm deep. A weatherproof cover should be provided at least 1.5 m above the seeds. The seeds should not be covered, just enough sand is sprinkled on top to keep the seed in place. They should be watered at least twice a day. Germination normally commences at 4–10 days, but will continue for 3 weeks. Good fresh seed is very easy to germinate, with high germination rates of up to 80%, though viability is more usually 25–35%.

Seedlings are pricked out at the cotyledon stage, or 5–7 days after germination. Young seedlings may suffer from fungal attack shortly after emergence: any dead seedlings should be removed quickly. Seedlings are placed under heavy shade which is reduced gradually until 6–8 weeks. While in the nursery, routine fertilizer applications with liquid fertilizers at 1 week and NPK at 6 and 10 weeks are recommended. Seedlings should be grown under a regime promoting air-pruning of roots, otherwise regular root pruning should be undertaken once the seedlings have reached 15 cm in height. Seedlings are ready for planting out when they are 20–30 cm tall at 3–6 months old.

T. brassii has very low coppicing ability. Site preparation practice involves the felling of all naturally occurring trees of over 10 cm diameter at least 6 months before planting. In the Solomon Islands, a line-planting technique was adopted but replaced later by close-line planting. Line planting at standard spacing was more favoured than close-line planting at 4.5 × 3 m; plants may also be planted at 6 m apart; within-line spacing can be 3 m, 3.5 m, or 4 m; and the close-spacing regimes of 5 × 3 m have also been recommended.

Regular weeding and climber cutting are required up to canopy closure. Singling is rarely required, and the species self-prunes satisfactory. A first selective thinning is carried out at canopy closure on all stands with more than 500 stems/ha. The second thinning, again at canopy closure, is performed to achieve a final stocking rate of 150–250 stems/ha. Wildings 2–7 cm high have been used most successfully.

A commercial thinning may be possible from age 10 years. Logging in natural stands in the Solomon Islands varied from 17400 m^3 in 1987 to 7300 m^3 in 1990; harvesting at age 18–20 years can be expected to yield 200–300 m^3/ha from a fully stocked stand of 150–250 stems/ha.

Diseases and pests

Generally, the species appears to be free from serious diseases and pests. In the Solomon Islands, pinhole borer damage has been seen in the living tree; butt rot has also been noted in old, possibly over-mature trees. Young seedlings suffer from damping off.

There are a number of buprestid beetles and reports of a caterpillar, *Roesilia lignifera*, eating leaves of young trees in Solomon Islands and Papua New Guinea; however, damage was not serious.

The adult beetles of *Paracupta helopioides*, *Cyphogastra* sp., and *Iridotaenia* sp. feed upon *T. brassii*. Trees standing in exposed positions seem to be especially prone to damage. Trees are also damaged by the cerambycid *Oxymagis horni* but seemed to be protected to some extent by gum exudates.

Uses

The wood has a coarse texture with open pores and an attractive pale grey-brown colour. The heartwood is pale brown to orange-brown in colour, with slightly paler sapwood that is not clearly defined. Very large trees may have very dark central cores which are rather waterlogged and may conceal brittle heart. The sapwood, of a slightly lighter shade, is usually pale, brownish-yellow, porous, straight grained, somewhat stringy, and of medium weight, and has medium to low strength properties. The timber is usually sound and knot-free but is susceptible to bluestain, pinhole borer, and *Lyctus* attack. It is not durable but is easily seasoned. The timber does not machine particularly well, is hard to saw and has a tendency to end split.

Wood density ranges widely from 300 to 600 kg/m^3 at 12% moisture content. Plantation-grown sawn timber is rather lighter and softer than mature timber, with basic densities of 511 kg/m^3 for mature trees and 370 kg/m^3 for plantation-grown timber having been recorded in the Solomon Islands, thus indicating that individual trees growing very fast in swamps may be rather lighter in weight.

The wood is suitable for light construction, mouldings, framing, cladding, desking, interior finishes, low-grade veneer and plywood. The species is not considered suitable for pulp production requiring severe pulping conditions, and has low yields and grades.

T. brassii has potential as a shade tree for grazing cattle. A light honey is produced during the flowering season.

Further reading

Gua, B. and Sandiford, M. (1990) The results and implications of a survey of the origin of artificial populations of *Terminalia brassii* in Solomon Islands. Forest Research Note 65–05/90. Forestry Division, Honiara, Solomon Islands, 11 pp.

Neumann, A.J. (1988) General volume table for *Terminalia brassii*. Forest Research Note 35-3/88. Forestry Division, Honiara, Solomon Islands, xxiv & 12 pp.

Terminalia brownii Fresen.

Subaraya

Importance

Terminalia brownii is one of the most useful species in the semi-arid areas in Eastern, Central, and Western Africa, particularly for resource-poor farmers. Trials of 58 species in semi-arid areas of Kenya revealed that it was one of the ten most promising species. It has great potential in agroforestry systems in semi-arid areas, as it is both drought- and termite-resistant, produces a heavy leaf fall that makes excellent mulch, crops grow well under it despite the dense shade, and has many uses. Application of green manure from *T. brownii* has also been reported to greatly increase soil nitrogen and organic carbon, as well as conserving soil moisture by lowering soil surface temperature.

To improve its sustainable use, further research is required into developing techniques to increase survival of seedlings after planting in the field; mechanisms of drought resistance; the rooting ability of cuttings; and genetic variation and breeding studies in relation to drought tolerance and productivity.

Botanical features

T. brownii is a shrubby or leafy deciduous tree with an attractive layered appearance 2.5–25 m tall. Young trees have horizontal, layered branches, but the layers gradually disappear and mature trees have broad, spreading crowns. The bark has been described as ash-grey and vertically fissured; young shoots tomentose.

Leaves not crowded at the ends of branches; turning bright red before falling; elliptic–obovate to broadly obovate; length variable (7–16 cm); apex usually short–apiculate; base cuneate; petiole 1–2 cm long. The undersides of the leaves and the petioles bear white hairs.

Flowers glabrous and creamy-white, unpleasantly scented, in axillary spikes up to 12 cm long. Calyx acuminate. Fruit broadly elliptic to ovate, smooth, rose-red to purple, up to 5 cm long, the length always less than three times the width.

Flowering occurs from April to June in Sudan and during March–June in Kenya. Fruit development takes approximately 5–6 months after insect pollination.

Distribution

T. brownii occurs naturally in Tanzania, Uganda, Congo, Kenya, Nigeria, Sudan, Eritrea, Ethiopia and Somalia, in deciduous woodlands, bushlands, wooded grasslands and riverine forests, in very dry areas and often on rocky outcrops. It is rarely found on moist sites. The species is often grown in semi-arid areas. *T. brownii* grows within a wide range of ecological zones, from 600 to 2400 m altitude and in areas receiving 300–1800 mm of annual rainfall.

Silviculture characteristics

T. brownii is a fairly slow-growing species. It grows on a wide range of soils, such as black lava, alluvial, vleis, sandy, gravel, loam and deep red clays; however, it performs best on deep sandy soils. Propagation is mostly by seedlings and wildings, which need support when young. It is managed by lopping, pollarding or coppicing.

Silviculture practice

Seeding is almost continuous and fruit development takes 5–6 months, yielding up to 3000 winged fruits/kg. The fruits are ready for collection if the wings break off easily. Trees may be shaken and seeds collected on the ground and dried for about 1 week to a moisture content of 10–20%. Seeds can retain viability for a long period if kept dry in airtight containers in a cool place.

Seeds are difficult to germinate since the fruits have a hard endocarp. They can be pretreated to increase germination rates by soaking in sulphuric acid for 60 min or by carefully making a V-shaped nick at the radicle end. After treatment, seeds should be sown immediately in soil or sand. Germination of fresh seeds takes 60–90 days.

Success was found with rooted cuttings of *T. brownii* at a rate of 10–15% in a non-mist propagator, which was found to be a cheap alternative to other propagators, although large clonal variation in rooting ability were reported. In studies on *Triplochiton scleroxylon*, which was also considered difficult to root, low rooting ability was attributed to the light/nutrient interaction of the stock plant prior to severance, which may affect the subsequent capacity of unrooted cuttings to photosynthesize. Rooting ability in *T. brownii* may also be affected by the same factors.

T. brownii is fairly slow growing, attaining a height of 1.5 m in 2 years, although it grows fast on moist sites and when irrigated. Studies on the water status of seedlings at the time of transplanting indicate that the seedlings experience severe transplant shock that leads to a high degree of stomatal closure and a decline in chlorophyll fluorescence ratio with decreasing relative water content. It has been reported that initial irrigation may increase the establishment rate to 62–89% as the irrigated seedlings were associated with a significant increase in stomatal conductance. However, application of synthetic water-absorbing polymers and mycorrhizal inoculation do not have major discernible effects on seedling establishment.

T. brownii has the main disadvantage of having a slow growth rate unless irrigated.

Diseases and pests

Seeds are prone to fungal damage and exhibit poor germination due to the hardness of the seedcoat.

T. brownii wood is termite resistant, and, apart from seeds which are prone to insect damage, no other pests have been reported.

Uses

The leaves can be browsed by livestock and used as a mulch. *T. brownii* is used as a shade tree and for erosion control.

T. brownii wood has a large number of uses, including fuelwood and charcoal production. The wood is yellow-brown, medium hard, light, strong, durable and termite resistant with a density of 640 kg/m^3. It is used for poles, posts, making beehives, tool handles and utensils (mortars and pestles), as well as being highly valued for construction and fencing.

The bark and fruits of *T. brownii* contain 19% tannin. A yellow dye is extracted from the roots. In addition, the leaves and bark are used to treat stomach ache and diarrhoea in both humans and livestock, pink eye in livestock, and for the treatment of malaria, hepatitis, yellow fever, coughs and colds. The bitter fruits are eaten by children.

Further reading

Mistry, J. and Berardi, A. (2006) *Savannas and Dry Forests: Linking People with Nature*. Ashgate Publishing Ltd, Aldershot, UK, 296 pp.

Schmidt, L. (2010) *Terminalia brownii* Fresen. Seed Leaflet No. 148. Forest & Landscape, University of Copenhagen, Denmark and World Agroforestry Centre, Nairobi, 2 pp.

Terminalia catappa L.

Singapore almond

Importance

Terminalia catappa is a hardy, fast-growing, deciduous multipurpose tree, reaching 25(–40) m tall and producing an edible fruit. It plays a vital role in coastline stabilization as a tree component of strandline plant communities in the western Indian Ocean, South-east Asia, and the South Pacific. Under suitable conditions it is a well-formed tree and has been widely planted throughout the tropics for shade, ornament and nuts, especially along sandy seashores. It is much used in agroforestry systems in the Philippines.

The timber makes a useful general-purpose hardwood, and is suitable for furniture manufacture and also for fuelwood.

T. catappa is considered a promising plantation species because its growth is fast and its silvicultural characteristics are fairly well known. *T. catappa* is easily propagated and flourishes with minimal maintenance. Although it can withstand coastal winds and salt spray to flourish in exposed, coastal locations, in such situations it will usually have poor, stunted growth, which can be a disadvantage for some management objectives.

Botanical features

T. catappa is a tree to 25–40 m tall and diameter of >50 cm. Younger trees display a characteristic pagoda form, with a single bole and monopodial horizontal branching in regular false whorls of 4–5 branches. Along each lateral, new branches are formed in a characteristic bifurcating pattern. The tiered crown becomes flatter with wide-spreading branches in older specimens. The bole is more or less cylindrical, straight, sometimes crooked and/or leaning in exposed coastal situations. Buttresses, when present, are up to 3 m high, variable, even, straight to curved, thick to thin, sometimes branching. Large trees may develop large, equal, occasionally branching buttresses and often have twisted leaning trunks. The bark surface is shallowly fissured and slightly flaky, grey to dark grey-brown, inner bark pinkish-brown, rough with continuous vertical fissuring and discontinuous horizontal cracks producing a grid appearance: the bark peels off in curved or straight scales along these lines.

The leaves are arranged in close spirals, often crowded towards the ends of the upturned branchlets. The leaf blade is simple, broadly obovate, 8–25(–38) × 5–14(–19) cm, with (5–)8–12 pairs of secondary veins, with a rounded blunt tip and tapering gradually to a narrowly subcordate base (the latter being a useful diagnostic feature). New leaves have a covering of soft appressed brown hairs. Mature leaves are mostly glabrous/shiny, leathery, dark green, turning bright yellow then vivid to dark red before falling.

The flowers are small (4–6 mm across), white or cream-coloured, 5-lobed, arranged on long (8–25 cm) axillary spikes, and somewhat foetid smelling (Plate 55). Within a spike the majority of flowers are male, with only a few bisexual flowers being present towards the base.

Typically 1–5 fruits are formed on the basal part of the flower spike. Fruits are sessile, laterally compressed, ovoid to ovate, smooth-skinned drupes. During maturation, fruits change colour from green, through yellow to bright red or dark purplish-red at full maturity. The kernel consists of 2 delicate and intricately entwined cotyledons enclosed in an inconspicuous cream-coloured (sometimes red) testa.

Distribution

T. catappa has a very wide natural distribution in near-coastal areas of the Indian Ocean, through tropical Asia, and into the Pacific. The rind of the fruit, comprising light, pithy/corky tissue, enables the fruits to float and be dispersed long distances by sea currents. The extent to which its range has been extended through movement by humans is difficult to determine. *T. catappa* has been widely planted throughout the tropics.

Silviculture characteristics

T. catappa seedlings are shade-tolerant during early growth (the first 1–2 years). Thereafter they should receive full overhead light for optimal growth, although the tree may grow under moderate shade. The species is considered to be a pioneer tree. The species regenerates abundantly on its preferred littoral sites. Under favourable conditions *T. catappa* is moderately fast growing, with height growth in early years averaging around 2 m/yr.

The tree has good windfirmness, but may suffer stem breakage or uprooting during tropical cyclones. Its tolerance to salt spray and drought makes it a promising reforestation species for sandy areas and abandoned mining sites, especially on coastal areas.

Cultivated saplings of *T. catappa* produce tiers of 4–5 branches; as new tiers are produced by the leader apex, some of the lowest branch complexes die back. However, *T. catappa*, unlike most other *Terminalia* spp., does not have a good capacity for self-pruning. The main leader may be cut out of the top, sometimes more than once, to create a very wide-spreading shade or specimen tree. Because of the wide-spreading branches, the tree needs much space.

Silviculture practice

In the Philippines, when the fruits turn greenish-yellow they are collected from the tree by using a light pole with a hook attached to the end. As seeds become yellowish, they fall to the ground and may be picked up directly. Seeds are reported to stay viable for a long period, but are normally sown fresh, within a few weeks of collection; little is known of the storage behaviour of seeds, although it is claimed that they may be stored for 1 year if kept dry. Seeds can be sown without pretreatment: germination commences in 3–8 weeks.

Seeds should be sown flat in moist seedbeds using fine sand and garden soil, immediately after transport from the collection site. Germinants are transplanted to containers or polythene bags. Seedlings must be watered regularly and should be hardened before planting out. Young seedlings are potted into polypots: these grow vigorously and the nursery phase is about 4–5 months.

Seedlings, height about 0.5 m, are usually outplanted in areas cleared by brushing or ploughing, usually at the onset of the rainy season to ensure a high survival percentage. Brush clearance and weeding are necessary during the early growing stage.

Spacing will vary considerably between locations and the purpose for which the trees are planted. Wide spacings are recommended for plantings for amenity purposes, such as 13–14 m apart for avenue plantings. Spacings of 10–12 m centres may be used for the production of nuts. The main axis may be cut out of young trees to encourage a low, wide-spreading form and easy accessibility of nuts, however, mature trees will not withstand severe pruning. For timber production, the species may be planted at closer spacings such as 4 × 4 m, in order to promote self-pruning, induce better stem form, and enable the trees to gain control of the site more rapidly.

T. catappa has been largely overlooked as a species for timber-plantation purposes and consequently there is minimal information on how to manage stands for wood production. For fuelwood production in India, a rotation length of 10–15 years has been suggested, with a 10-year-old plantation expected to yield 22–36 t/ha.

Diseases and pests

The fungal pathogen *Cercospora catappae* has been reported on *T. catappa* from Orissa, India, and pink disease caused by *Corticium salmonicolor* (*Phanerochaete salmonicolor*) has been reported from Kerala, India. Leaf spot, caused by a bacterial disease *Xanthomonas arboricola* pv. *pruni*, has also been reported in India.

Only a few insects infest this tree. *Aleurodicus dispersus* was reported from Hawaii. It is also suspected that this pest is a vector of a mycoplasma causing coconut lethal yellowing disease in Florida, USA. In Brazil, *Citheronia laocoon* was reported infesting *T. catappa*. The eggs are found on the leaves and the larvae feed on the foliage. The eggs hatch in approximately 8 days. In African plantations of *Terminalia* spp. including *T. catappa*, *Epicerura pulverulenta* is an important defoliator. Attack by *Agrilus* spp. has been observed in Papua New Guinea. In Brazil, the eriophyid mite *Aculus catappae* sp. n. was reported on the lower surface of the leaves. In Puerto Rico, the scale insect *Paratachardina pseudolobata* has been found infesting *T. catappa*.

In countries outside its natural range, *T. catappa* may be host to pests that can cause serious damage or loss in other economically important species. Susceptibility to termite attack and damage may differ in different parts of the *T. catappa* range.

Uses

T. catappa is often planted as an ornamental and amenity shade tree. It plays a vital role in coastline stabilization and erosion control, its well-developed lateral root system helping hold together fragile sandy soils, especially during storm surges and king tides. It is also used for revegetation plantings and is considered an important agroforestry species, with multiple uses, including as a host tree for silkworms.

T. catappa heartwood is brown to reddish-brown, with a wide, rather indistinct, band of lighter-coloured sapwood. The timber is fairly smooth, lustrous, elastic, tough, fairly hard, medium-coarse in texture, often with irregular or twisted grain. Wood density is in the moderate range, 530–540 kg/m^3 at 12% moisture content.

The timber is easily sawn but the interlocked grain tends to pick up during planing. Other machining characteristics are as follows: shaping, boring and mortising are fair, turning is poor, whereas sanding, resistance to screw splitting and nailing properties are good. The timber is not naturally resistant to insect and fungal attack.

The timber is of generally of good quality, and while not durable in ground contact, may be used for a wide range of end uses such as general construction, interior purposes such as flooring and furniture, boatbuilding, carts and planks. Larger logs are suitable for veneer and plywood manufacture. Traditional wood uses in the South Pacific include canoe hulls and paddles, kava bowls, tool handles, war clubs, walking sticks, drums and fuelwood. In Papua New Guinea, the timber (known as 'red-brown terminalia') is utilized for house and boat construction, and for furniture and cabinetmaking. In the Philippines, *T. catappa* is an important fuelwood species.

The oil-rich nuts of *T. catappa* make a tasty and nutritious food, and the outer flesh of suitable types is widely consumed by children around the tropics. The nuts may be consumed fresh after extraction from the shell or else preserved by smoking and consumed up to 1 year later. Sun-dried nuts yield a bland yellow, semi-drying oil which is edible but becomes turbid on standing; the main fatty acids are palmitic and linoleic acids.

Tannin is extracted from the fruit shell, bark, leaves and roots, and is used for leather, for dyeing cloth and making ink. Large leaves are used as wrapping material. The leaves, bark, fruits and other parts have many medicinal uses worldwide. The foliage is suitable for feeding tasar or katkura silkworms. The flowers yield a nectar which is attractive to honeybees.

Further reading

Francis, J.K. (1989) *Terminalia catappa* L., Combretaceae: Indian almond, almendra, combretum family. U.S. Forest Service, International Institute of Tropical Forestry, Rio Piedras, Puerto Rico, 4 pp.

Ivani, S. de A., Silva e Silva, B.M. da, Oliveira, C. de and Môro, F.V. (2008) Morphology of the fruit, the seed and the seedlings of chestnut tree (*Terminalia catapp*a L. - Combretaceae). *Revista Brasileira de Fruticultura* 30, 517–522.

Morton, J.F. (1985) Indian almond (*Terminalia catappa*), salt-tolerant, useful, tropical tree with 'nut' worthy of improvement. *Economic Botany* 39, 101–112.

Phulwaria, M., Ram, K., Harish, Gupta, A.K. and Shekhawat, N.S. (2012) Micropropagation of mature *Terminalia catappa* (Indian Almond), a medicinally important forest tree. *Journal of Forest Research* 17, 202–207.

Terminalia ivorensis A. Chev.

Idigbo

Importance

Terminalia ivorensis is a large, light-demanding, fast-growing African tree. It often occurs in association with *Terminalia superba*, although it is not as abundant. Natural regeneration is abundant in open sites. Generally, it is found in small stands in moist deciduous forests, but it also penetrates evergreen forests.

T. ivorensis plays a leading role in the market for tropical timbers in terms of quality and trading volume. It has good silvicultural characteristics and is suitable for plantations and enrichment planting. Enrichment planting can be done following taungya systems, or the 'limba' method, which is a tropical system for the conversion of depleted forests. As natural regeneration is abundant, *T. ivorensis* in natural stands can be promoted by selective thinning and elimination of lianas and undesirable trees.

Botanical features

T. ivorensis is a deciduous, straight-boled tree with excellent form and large size (35–40 m, 70–120 cm dbh). It does not bear true plank buttresses, but develops buttress-like formations extending out on to 6–8 thick, lateral roots. The usual stem height is 20–25 m. Overmature trees may often have a swollen and defective lower half of the bole and a brittle heart. The crown of younger trees is very wide-spreading, with an oval pagoda-like structure, forming characteristic horizontal storeys, but in mature individuals the branches become more ascending and a conspicuous storey is less evident. The foliage zone is largely peripheral and not very dense. The tree has no leaves for about 2–3 months during the dry season.

The bark is greyish, smooth, and finely textured; in young trees dark brown to black, on older trees with deep longitudinal cracks. The slash is fairly soft and fibrous, sometimes slightly brittle and brown outside, dark or yellow inside, paler near cambium, soon turning ochre-yellow or dirty-white on exposure.

T. ivorensis has one major taproot and some thick, nearly horizontal lateral roots.

Leaves are simple, alternate and elongate (6–15 × 3–6 cm), broadly oblanceolate to elliptic–obovate, growing in small bunches crowded at the extreme ends of the shoots. Young shoots as well as the young leaves are slightly pubescent and the latter are also reddish. The medial nerves are pubescent on both sides of the leaves, lateral nerves are prominent and slightly pubescent on the underside. Petioles relatively short, only 1.5–2 cm long.

The whitish pubescent flowers grow in spikes on a pubescent axis 10 cm long. The fruits are winged and variable in size (6–10 × 1.7–2 cm), narrow, emarginated and slightly decurrent. They are finely tomentose, with short reddish to orange-brown hairs and are often galled by insects or have a feeding hole due to weevil attack. The seeds are winged, arranged around a common axis (including the wings, which are about 6–7 cm long), very hard, reddish-yellow when ripe and about 2 cm wide. They persist for a long time on the tree before they are disseminated by wind.

T. ivorensis can be distinguished from *T. superba* by its grey-black unbuttressed stem and smaller leaf pattern compared to the strongly buttressed stem and light brown bark of *T. superba*, which also has glands on the petiole.

Distribution

The natural phytogeographical distribution of *T. ivorensis* corresponds to the Guineo–Congolia/Sudania Regional Transition Zone and the Guineo–Congolian Regional Centre of Endemism, both in West Africa. *T. ivorensis* has been introduced to Central America, India, Fiji, the Solomon Islands, Sarawak, Brazil (Parana) and the Antilles.

Silviculture characteristics

T. ivorensis is a fast-growing, light-demanding pioneer species that regenerates easily in clearings. The fast growth rate and early self-pruning ability makes it ideal for establishing large, even-aged plantations, which are best arranged in line plantings, because of its wide-spreading branches. It has a good ability to sucker, but is very sensitive to fire and to wind. Natural regeneration is abundant in open sites such as forest clearings, abandoned farms and fallow and logging roads. It coppices readily from stumps and can be coppiced for a considerable time.

Silviculture practice

The number of seeds/kg is about 5500–6500 with wings and 10,000 without wings. In order to reduce the risk of damage caused by seed borers, it is recommended to collect seeds from fruits that are still on the tree or have recently fallen. The germination capacity varies greatly, but is generally low (10–50%). Alternate soaking and sun-drying of seeds for 1 week generally improves germination rates to about 40%, although this can be increased to 80% if these pretreated seeds are mixed with mulch and placed in damp sacks in the sun for a further 2 weeks.

For container plants, germination can be accelerated by allowing the substrate to heat up by removing shading for some hours. Seedlings normally remain in the nursery for 4 months, with shading being removed when the plants are 2 months old to enable them to become habituated to the sun.

The seedlings are very sensitive to drought and can tolerate a small degree of shading only when very young. Germination is irregular with a variation from 15–50 days, owing to the hard seedcoat.

In nursery beds, *T. ivorensis* is often planted as a stump or stripling. The stumps used are moderately tall (with stems cut off 60–170 cm above ground), prepared from plants 12–15 months old that have a minimum diameter at the root collar of about 1.25 cm. The striplings are used between 120 and 180 cm tall; the roots may only be pruned back very lightly and all the leaves are stripped except for the terminal cluster in order to reduce transpiration losses. Both striplings and stumps must be well-protected against desiccation when planting.

Plantations established from container plants have been observed to do better than those derived from stumps or striplings. The section of the taproot induces a change in the later development of horizontal lateral roots.

T. ivorensis can be planted in pure stands, but trees may be attacked after a certain age by various insect pests that can cause serious damage. Mixed stands are preferable, for example, in equal quantities with *Terminalia superba* or in a ratio of 3:1 with *Khaya grandifoliola*, this species being protected against borer attack by the inclusion of *T. ivorensis* and increasing the value of the plantation.

Enrichment planting can be performed following taungya systems, which involve farmers preparing planting sites assigned by the State during shifting cultivation. When sowing their crops, farmers sow fast-growing timber trees, such as *Terminalia*, *Tectona*, *Gmelina* or *Pinus* species. The 'limba' method, a tropical system for the conversion of depleted forests can also be used. This method was developed in the Democratic Republic of the Congo.

As its natural regeneration is abundant, *T. ivorensis* in natural stands can be promoted by selective thinning and elimination of lianas and undesirable trees.

Suitable spacing recommendations range from 3 × 3 m to 5 × 5 m to 5 × 10 m. Up to a height of 5 m, *T. ivorensis* grows at a rate of 1.8–2.1 m/yr. On very favourable sites it can even grow at a rate of 4.4 m/yr during the first 4 years, with dbh increments of 7 cm/yr. The main growth in height is finished after 15 years, indicating that the final stand has to be more or less in place after this period.

Diseases and pests

The fungal pathogen *Aurifilum marmelostoma* has been reported in association with disease symptoms in *T. ivorensis* in Cameroon. Another pathogenic fungus, *Rostraureum tropicale*, has been reported from *T. ivorensis* in Ecuador.

Seeds are often attacked by *Nanophyes* spp. and *Auletobius kuntzeni* (Curculinoideae), which can cause considerable reductions in germination.

Defoliation by Notodontidae larvae, borer attacks by beetles of the Cerambycidae and Bostrichidae, and Lepidopteran pests (Cossidae and Megalopygidae) have all been recorded. Commercial plantations of *T. ivorensis* and *T. superba* are periodically defoliated by larvae of *Epicerura pergrisea* in Côte d'Ivoire. In Nigeria, *Epicerura pulverulenta* is also an important defoliator of both these *Terminalia* species. In Costa Rican plantations, larvae of a species of cossid in the genus *Cossula* were responsible for boring damage to *T. ivorensis*.

If thinning is not performed, trees stop growing and natural wastage and withering affect some trees, characterized by the crown drying out and attacks by insects (Scolytidae and Cerambycidae) as well as by pathogenic fungi.

Uses

No special land use, environmental or service aspects have been recorded for *T. ivorensis*, save for use as a shade tree in coffee, cocoa and banana plantations. It is probably suitable for land reclamation and revegetation, due to its abundant natural regeneration in clearings. However, it requires a fertile soil with good water availability, and natural stands have to be adequately managed.

T. ivorensis sapwood is 3–6 cm wide, differing only slightly in appearance from the yellowish to light yellow-brown heartwood. Its specific gravity fluctuates greatly, with a density of 370–540 kg/m^3 at 12% humidity. Its mechanical properties are satisfactory. The wood can be seasoned without difficulty, is easy to stain, polish and glue, and in general lends itself well to being worked. The heartwood is rated as durable and moderately resistant to termite attack. The sapwood is susceptible to powder-post beetles.

T. ivorensis produces a highly valued, light wood which is used for a wide range of applications. It is well-suited for cabinet work, and can be easily rotary cut and veneered. Mixed with other woods, it is also used for making paper.

Medicinal uses for *T. ivorensis* include treating wounds with bark preparations and reducing rheumatic muscle pain by rubbing with the pulp of the bark. The juice of the leaves and bark decoctions are taken to treat gonorrhoea, kidney pains and as an aphrodisiac. In conjunction with the leaves of *Eugenia* spp. and *Pericopsis laxiflora*, infusions of *T. ivorensis* are used externally to treat scabies.

Further reading

Aigbe, H.I., Modugu, W.W. and Oyebade, B.A. (2012) Modeling volume from stump diameter of *Terminalia ivorensis* (A. Chev) in Sokponba forest reserve, Edo State, Nigeria. *Journal of Agricultural and Biological Science* 7, 146–151.

Carvalho, A.M.X. de, Vale, H.M.M. do, Ferreira, E.M., Cordero, A.F.P., Barros, N.F. de and Costa, M.D. (2008) Microbial activity of soil and litter in areas with forest stands of *Pinus elliottii* and *Terminalia ivorensis*. *Revista Brasileira de Ciência do Solo* 32, 2709–2716.

Jose, S. and Gordon, A.M. (eds) (2008) *Toward Agroforestry Design: An Ecological Approach*. Advances in Agroforestry. Springer, Dordrecht, The Netherlands, 314 pp.

Mattos, P.P. de, Pereira, J.C.D., Schaitza, E.G. and Baggio, A.J. (1999) Physical, chemical and anatomical characterization of the wood of *Terminalia ivorensis*. Circular Técnica 31. Emprapa Florestas, Colombo, Brazil, 14 pp.

Terminalia oblonga (Ruiz & Pav.) Steud.

Sura

Importance

Terminalia oblonga is a widely utilized, economically and ecologically valuable timber tree in lowland Latin America.

Botanical features

T. oblonga is a large tree, reaching up to 45 m in height and 80–150 cm in diameter. It has a whitish to pale orange bark and a smooth, cylindrical, mostly straight trunk. In riparian areas the trunk can be irregular and twisted. Buttresses are well developed up to 1 m. The bark exfoliates in paper-thin sheets. The fallen sheets are reddish-brown; the new bark is creamy-light-yellowish, and has a smooth texture. Occasionally branching can be low. Branches are numerous and ascending and form a dense but narrow, sometimes irregular crown. Leaves are arranged in whorl-like clusters around the tips of the twigs and are smooth, light green, simple and alternate. The leaves are obovate to oblanceolate with entire margins. Foliage yellows and is shed annually, with varying timing. New leaves are produced after a leafless period of about 2 weeks. A second episode of crown renewal sometimes occurs in the same year. Flowers are borne in long and string-like axillary racemes that emerge just before the leaves. Each raceme contains over 50 small yellow-green flowers. Five sepals form the cup-shaped calyx which supports 10 stamens and a small central pistil. Fruits are double-winged, flat and papery brown samaras containing a single, elliptical seed in the center. Harvests occur mainly in the months of March and April but ripe seeds may be seen as early as January. The seeds germinate with the onset of the rainy season in May and June.

Distribution

T. oblonga is a canopy tree native to Mexico, Mesoamerica, Brazil and western South America. It grows from sea level to 900 m elevation, with annual rainfall between 1500 and 3500 mm. It is found in wet to very wet forests, often also in riparian areas. It grows well in many different soil types, including poorly drained areas, but seems to prefer well-drained alluvial soils or moderately moist sandy, loamy soils.

Silviculture characteristics

Trials in the northern zone of Costa Rica indicate that *T. oblonga* does well in monospecific plantations, but is affected by terrain. It seems to grow best in flat terrains and has been grown at planting distances of 2 × 2 m and 3 × 3 m. Growth of 2 m in height and 5 cm in diameter have been recorded for the first year. A mean annual increment of 4.6 m^3/ha has also been recorded in secondary forest in Costa Rica. The tree prefers a sunny site.

Silviculture practice

There is little information on silviculture characteristics or practice for *T. oblonga*. The species suffers from poor natural recruitment but the factors affecting regeneration are not well understood. Large quantities of seeds are produced, but few germinate. Blooming and fruit ripening are very variable, depending on individuals and on sites. Research in forest management plots in Bolivia found that soil disturbance and a thin litter layer favoured seedling emergence and that disturbance had a positive effect on seedling survival. Light availability also had a positive effect on survival.

Diseases and pests

The wood is susceptible to termites, and durability varies with place of origin. *T. oblonga* is moderately resistant to fungal attack. The species shows susceptibility to foliage fungus and ant attack. In Costa Rica, the larvae of *Cossula sp.* have been found to damage *Terminalia* species including *T. oblonga*.

Uses

In El Salvador and Guatemala, the tree is grown in coffee plantations for shade.

Timber from *T. oblonga* is used in heavy general construction, interior and exterior construction, furniture, floors, decorative veneers, bridge foundations and ships. The timber is of a high quality with good physical and mechanical properties. The fresh wood smells unpleasant but it is tasteless. When dry, the sapwood becomes yellowish-grey, whereas the heartwood intensifies its brown-greyish colour and darker veins or bands appear in a brownish-olive-green colour. The wood tends to oxidize rapidly when it has been exposed to air and light. Drying is moderately difficult, and drying properties vary with wood origin. The wood may show splitting, checking and slight warping. It has straight or slightly undulate grain which may occasionally be intercrossed. The wood has a medium and uniform texture. *T. oblonga* wood is heavy; reported wood densities range from 370 to 910 kg/m^3 in trees from South and Central American regions.

Further reading

Flores, E.M. (2002) *Terminalia oblonga* (Ruiz & Pav.) Steud. In: Vozzo, J.A. (ed.) *Tropical Trees Seed Manual*, Agriculture Handbook 721, USDA Forest Service, Washington, DC, USA, pp. 755–757.

Harmon, P. (2012) *Terminalia oblonga* (R. & P.) Steud. Trees of Costa Rica's Pacific Slope. A natural history of tropical rainforest flora. Available at: http://blogs.cds.ed.cr/patrickharmon/?page_id=34 (accessed December 2012).

Salazar, R. (1998) *Terminalia oblonga (Ruiz & Pavón) Steud.* Nota Técnica sobre Manejo de Semillas Forestales. Centro Agronómico

Tropical de Investigación y Enseñanza (CATIE), Turrialba, Costa Rica, 2 pp.
Schiøtz, M., Boesen, M.V., Nabe-Nielsen, J., Sørensen, M. and Kollmann, J. (2006) Regeneration in *Terminalia oblonga* (Combretaceae) - A common timber tree from a humid tropical forest (La Chonta, Bolivia). *Forest Ecology and Management* 225, 306–312.

Terminalia superba Engl. & Diels

Limba

Importance

Terminalia superba (Plate 43) is a large, tropical West and Central African tree, with a long straight bole. It occurs naturally in moist, seasonally deciduous forests, extending from Sierra Leone in the west to Congo and Zaire in the south and east.

T. superba is a moderately fast-growing tree well suited to plantation development in the humid, lowland tropics. It produces a highly regarded commercial timber, especially from the southern parts of its range where there is minimal pinhole borer attack and associated staining. The wood has excellent technical properties and many end uses.

Botanical features

T. superba is a deciduous tree which normally attains heights of 30 m, but occasionally can grow to heights of 50 m or more. It bears 3–8 characteristic, thin plank-like buttresses that extend well out from the tree up to a height of 2–3 m. The stem diameter just above the buttresses is between 60 and 120 cm; the maximum recorded is 150 cm. The extraordinarily straight, slender stem is conspicuously cylindrical. The crown comprises relatively short, horizontally developed branches that form a series of easily recognizable tiers.

The bark is initially smooth and greyish-white on young trees, but after 10 years it begins to develop longitudinal cracks and thin rectangular scales peel off constantly, giving the bark a patchy appearance. The slash is medium thick, approximately 1 cm, soft-fibrous, brownish outside and pale yellow inside.

Leaves are simple, alternate, in tufts at the end of the branchlets, rusty tomentose when young but soon becoming glabrescent, green, coriaceous, leaving obvious round scars on the branchlets when shed. The petiole is 2–6 cm long, flattened above, with a pair of subopposite, not always distinct, lateral glands slightly below the blade. Leaf blade obovate, 5–14 cm long, 2.5–7 cm wide, with a cuneate base, more or less unequal-sided and decurrent. The apex is rounded, abruptly and shortly acuminate; midrib and nerves flat above, prominent beneath. There are 4–7 nerves on each side of the midrib, arching upwards and not looping, finely reticulated.

The flowers are small, greenish-white to yellow, pubescent, and grow in spikes on a pubescent axis 7–15 cm in diameter.

The seeds are small, sessile, golden-brown nuts, transversely winged and arranged around a common axis. They are 1.5–2.5 cm long (including the wing), 4–7 cm wide and glabrous. The nut contains one seed and is approximately 1.5 cm long, 0.7 cm thick and a rounded triangular shape in cross-section.

Distribution

The natural phytogeographical distribution of *T. superba* corresponds to White's Guineo–Congolia/Sudania Regional Transition Zone and Guineo–Congolian Regional Centre of Endemism, both in West Africa. It is commonly found in moist deciduous forests and secondary forests, but also penetrates evergreen forests.

The largest plantations of *T. superba* are located within its area of natural distribution, but it has been introduced to numerous other countries in Africa, Asia, South America and Oceania.

Silviculture characteristics

T. superba is a moderately fast-growing, light-demanding pioneer species that regenerates easily in clearings. The fast growth and the sound commercial prospects of *T. superba* make it one of the few tree species of which large-scale plantations have been established in tropical Africa. It is recommended for industrial plantations in monoculture or in association with other light-demanding species, such as *Khaya grandifoliola*, *Chlorophora excelsa*, *Terminalia ivorensis* or *Triplochiton scleroxylon*, but not in small private plantations or enrichment planting. The most employed planting methods are the conversion system called 'limba method' (méthode limba), the 'taungya' system, and a modified version of this in conjunction with bananas. Its self-pruning capacity for a given age and height depends on the provenance being grown. For 12-year-old trees at a height of 22 m, heights of self-pruning between 10 and 15 m have been recorded. Natural regeneration is abundant in open sites such as forest clearings, abandoned farms, fallow areas, and logging roads. It is absent beneath the canopy of mature stands, and young trees are not able to survive in a state of suppression.

Silviculture practice

T. superba produces large amounts of seed every year. The number of seeds/kg is about 6000–7000 with wings and 10,000 without. It is best is to collect fruit from felled trees, which enables the wood quality of the mother tree to be inspected. After collecting, the seeds are sun-dried for a few days in order to prevent fermentation or fungal attack during storage. It is not necessary to remove the wings before storage, but they will then occupy more room. The seeds keep their full germination potential for at least 1 month, although 10–15% of seeds are usually sterile, rising to 30% if there are bad weather conditions at anthesis. Fresh, sun-dried seeds exhibit germination rates of 60–80%, but this drops to 30–50% after 1 year. If seeds are stored at 2–4°C, germination capacity will still be approximately 60% after 2 years. No special treatment is recommended before sowing, and dissection of the wings is not necessary.

T. superba is often established in nurseries from stumps that are 18 months old and have been cut down to 1 m some days before planting, and ideally with a diameter of 1 cm at the top. The taproot is pruned with a shovel and should remain 40 cm long. The stumps are transplanted as soon as the first buds appear at the top of the stump. Care has to be taken not to bend the taproot when planting the stump. Establishment rates of 80% can be obtained in favourable conditions and after careful planting. Termites tend to attack newly planted stumps.

Seeds are sown (at a spacing of 5 × 5 cm) in seedbeds that should be slightly shaded and watered daily, germination taking place after 10–20 days. After 5–6 weeks, they are transplanted into nursery beds at a spacing of 20 × 50 cm. At first, nursery beds should be slightly shaded, but after 3–4 weeks shading can be removed so that they are fully illuminated. If seed supplies are abundant, they can be sown directly in the nursery beds and thinned out at a height of 10–15 cm.

Container plants of *T. superba* are rarely used, but can be produced easily and are ready for planting out after about 3 months. Germination can be accelerated by opening the shading for some hours to heat up the substrate. Shading has to be progressively removed when the plants are 2 months old to habituate them to the sun.

A new grafting method has been developed based on the use of the herbaceous end split, using physiologically juvenile scions and young trees, as well as a chain of successive grafts situated close to each other. However, the technique has not been used for sufficiently long to give conclusive results.

Spacing requirements vary from 3 × 3, to 5 × 5, to 12 × 12 m, according to different authors. *T. superba* can easily be planted in pure stands and should yield, as a conservative estimate, about 200 m^3 of export-quality timber/ha. In mixed stands, *T. superba* can be planted in equal quantities with *T. ivorensis* or in a ratio of 3:1 with *K. grandifoliola*. This mixed stand has increased value due to the African mahogany, and will be protected against borer attack (*Hypsipyla* sp.) by the presence of *T. superba*. It is also possible to plant mixed stands of *T. superba* with other light-demanding species, such as *C. excelsa*, *T. scleroxylon* or *Harungana madagascariensis*.

Enrichment planting can be performed following taungya systems, which involve farmers preparing planting sites during shifting cultivation. While sowing their crops, farmers sow fast-growing timber trees, such as *Terminalia*, *Tectona*, *Gmelina* or *Pinus* species (this method was developed in the Democratic Republic of the Congo). A variant of the taungya system (méthode sylvo-bananier) was used in Zaire by Belgian foresters and involved the cultivation of bananas for the first 6 years. The economic objective of this system was to produce 60 stems of veneer wood/ha on a 60-year rotation. The limba method, a tropical system for the conversion of depleted forests, can also be used.

Up to the age of 4 years, *T. superba* grows at a rate of 2.5 m/yr in height, slowing to 2 m/yr after 10 years.

Diseases and pests

Fungi that can cause cosmetic damage to the valued colour of the wood may attack fresh timber.

The principal insect pests observed in Africa are ambrosia beetles (*Doliopygus dubius* and *Doliopygus paradubius*) that attack standing trees, the stem borer *Tridesmodes ramiculata*, and pinhole borers north of the equator. The larvae of *Epicerura pergrisea* periodically defoliate commercial plantations of *T. superba* and *T. ivorensis* in Côte D'Ivoire. In Nigeria, *Epicerura pulverulenta* is also an important defoliator of both these *Terminalia* species. In Costa Rican plantations, larvae of a *Cossula* sp. were responsible for boring damage to *T. superba*. Leaf galls caused by mites have been observed in Ghana.

Defoliating larvae in the Lasiocampidae, Lymantriidae and Saturniidae, and the grasshopper *Zonocerus variegatus*, attack leaves and terminal shoots. Wounds and self-pruning can cause internal black rot pockets, which are started by the installation of tree ant nests in the tufts of the leaves. Members of the genus *Eulophonotus* (Cossidae) damage the tufts of the leaves and pruning wounds. Termites are likely to attack newly planted stumps.

Uses

The tree is probably suitable for land reclamation and revegetation on good soils with sufficient water due to its abundant natural regeneration in clearings.

Depending on where it is grown, *T. superba* wood is yellowish to brownish-black, and even white in the Mayombé region (southern Gabon, Congo, Zaire and Cabinda). It has somewhat variable physical and mechanical properties, but it is generally described as soft and light, but with good strength properties, making it comparable to a mild oak. The mean specific gravity is 520–550 kg/m^3 (12% moisture content), with the density of planted wood being inferior to that of natural forests. The compression and bending strength are average, whereas the shock resistance is poor.

It has a moderately coarse texture, an interlocked, straight or irregular grain, and is slightly lustrous with a mild odour. The wood can be easily and quickly dried with no distortion, but is not durable and tends to decay and be attacked by insects or termites, especially when moist. It can be easily worked, but has a tendency to split when nails and screws are driven into it. The wood takes different types of stained finishes very well, although the grey-brown streaks of highly figured wood (bariolé) are conspicuous and not attractive.

Most of the timber is used for interior trim and panelling, or for making plywood and particleboard; for example, selected plywood is sold in the USA under the copyrighted name of 'Korina'. The wood fulfils all conditions for use in furniture making and for peeled and sliced veneers. Yields for veneer and plywood are only modest due to the small diameter of the logs, the amount of sapwood and the occasional presence of blackheart. Veneers obtained from highly figured wood, which is fairly common in *T. superba*, are not particularly favoured. It is suitable for making a wide range of pulps in papermaking, with flexible fibres, but has not been exploited for this purpose.

The roots, leaves and bark are used for medicinal purposes. The bark is sometimes used as a black dyestuff for cloth.

Further reading

Groulez, J. and Wood, P.J. (1985) *Terminalia superba*. Commonwealth Forestry Institute, Oxford, UK, 77 pp.

Kuete, V., Tabopda, T.K., Ngameni, B., Nana, F., Tshikalange, T.E. and Ngadjui, B.T. (2010) Antimycobacterial, antibacterial and antifungal activities of *Terminalia superba* (Combretaceae). *South African Journal of Botany* 76, 125–131.

Ridder, M. de, Hubau, W., Bulcke, J. van den, Acker, J. van and Beeckman, H. (2010) The potential of plantations of *Terminalia superba* Engl. & Diels for wood and biomass production (Mayombe forest, Democratic Republic of Congo). *Annals of Forest Science* 67, 1–12.

Thuja occidentalis L.

Eastern white cedar

Importance

Native to eastern North America, the eastern (or northern) white cedar, *Thuja occidentalis*, is a tree with limited but sustained commercial value. Despite a low specific gravity, structural weakness, and a tendency to split along the grain very easily, the extreme durability of the wood even when exposed to soil and water has made it important in commercial forestry over the last 300 years. The wood, stems, leaves and seeds contain pleasant-smelling oils and other volatile materials with medicinal and industrial properties. The tree has extreme plasticity in growth form and rate, making it suitable for reforestation on a variety of sites provided they are amply supplied with water.

Botanical features

Tree size in *T. occidentalis* is enormously variable: trees have been recorded 34 m tall with 175 cm dbh after 400 years, and 2 m tall with 30 cm dbh after 1890 years. A specimen bearing seed cones was found with stem diameter 1 cm, height 10 cm and 155 annual growth rings, making it the slowest-growing tree ever reported.

Trees are pyramidal in open settings, with little branch dieback. Boles can be narrow, erect and single in dense mixed forests or multiple when emerging from rock. The stem is fissured and striped in old age if growing on rock faces, often growing with inverted axes. Such trees experience vast amounts of cambial mortality and have grossly asymmetric cross sections. The wood is extremely durable and dead branches remain on trees for hundreds of years.

Stems can be straight, buttressed, eccentric, forking or multiple, with sweep, swelling or marked taper among other forms.

The root systems are extremely plastic, mainly superficial and shallow. Even in rock, the woody portions penetrate no more than 10–30 cm. Fine secondary roots proliferate in any cavity that is provided, even if no soil is visible.

The bark is mostly thin and fibrous with the cambium extremely close to the surface, making it vulnerable to disturbance.

Flowers appear in April–May. Cones are oblong, 8–12 mm long, terminal on short branchlets, at first erect, later pendulous, ripening in the first year. Cone production can begin at age 6–8 years. Male cones are small and yellow in early spring, opening and turning brown in June and July. Female flowers form in the autumn of the previous year. Wind pollination occurs in April or May. Female flowers change from pink to green in the spring. Cones, which comprise 4–8 pairs of leathery scales each enclosing a pair of seeds, mature and turn brown by September. Seeds are released annually from October to December. They are light brown, winged on two sides, 4–6 mm long and less than 0.5 mm thick. Dry seed weight is roughly 1.0–1.5 mg. Evacuated cones persist on the tree for up to 2 years.

Leaves are scale-like and form part of a continuum of structures that vary in length, width, thickness and fleshiness. Leaves are scarce on main shoots, up to 4 mm long; on lateral shoots foliage is dense, up to 2.5 mm long, obtuse–ovate. Mature leaves on branch apices tend to be longer, more pointed and more lignified than lateral leaves.

Distribution

The natural distribution of *T. occidentalis* is in eastern North America, from Nova Scotia in the east to Manitoba in the west, north to Hudson Bay, and south to northern Tennessee. In New York, Vermont, New Hampshire, Maine, Minnesota, Michigan and Wisconsin, stands can be large enough to support local forestry, but only scattered disjunct populations occur in other states. In Canada, stands large enough to support forestry occur in Nova Scotia, New Brunswick, Quebec, Ontario and Manitoba. About 80% of the current range is in Quebec and Ontario.

T. occidentalis is planted as an ornamental in Eurasia.

Silviculture characteristics

T. occidentalis is an efficient forest gap colonizer, whether the gap is in an existing forest, on a cliff face, in a swamp or in an abandoned field. It grows well under marginal habitat conditions too severe for other trees.

Colonization of gaps occurs with massive invasion, and cohorts of even-aged stems can result. Subsequent recruitment of trees into the population can be very low until the canopy itself begins to produce gaps at age 60–80 years. Seedlings are extremely tolerant of shade but only persist as small and stunted boles until the canopy is opened, at which time their growth rate increases rapidly. Roots are infected with vesicular-arbuscular mycorrhizae, making trees relatively insensitive to site nutrient conditions.

Self-pruning of boles does not occur so clear lumber is difficult to find. As the roots are not particularly strong, windthrow distorts stems in many stands. For shingle production stem asymmetry is not a problem, but it is for pole, lumber or post production.

T. occidentalis does not show natural senescence; the production of leaves, flowers and viable seed is not diminished with age. Trees are relatively immune to air pollution and water contamination.

Due to its extremely thin bark and fire-sensitive foliage, exposure of *T. occidentalis* to forest fire usually kills trees outright. Where fire is absent or rare, trees can achieve ages of over 1800 years, making *T. occidentalis* one of only seven tree species in North America that live over 1000 years.

Silviculture practice

Seed germination rates are low in freshly collected seed, but approach 90% after 60–90 days of air-dry cold storage, remaining high for several years. Hydrated seeds germinate at 10–25°C. Natural and planted seedlings are very vulnerable to disturbance and drought at the cotyledon stage. Mortality is almost 100% if deciduous litter covers them. Mortality is lower if seeds germinate on elevated logs, rocks or ground. Nurse logs of a variety of species are optimal sites for *T. occidentalis* establishment. Once seedlings survive the first winter, rates of mortality drop very quickly.

Both even-aged and very uneven-aged stands exist. Thinning of densely stocked stands can result in an instantaneous increase in productivity. Growth to sizes appropriate for harvest can occur within 30–90 years depending on site conditions. If 20-m tall trees are required for harvest, rotations of 50 years are normally considered. Basal area is usually about 50 m^2/ha, which comprises approximately 200 m^3 of saw lumber.

Most harvesting is conducted in small shelterwood belts, strips or blocks that allow for natural re-seeding by wind. This method of cutting maximizes the edges of cleared areas which provide the shade and shelter that minimize local heat and drought effects. Sapling growth in cleared areas can be double that found in uncut stands, but rapid growth produces wood of inferior durability, so the intended use of the wood must be considered at the time of clearing and re-establishment. Sapling growth is optimal at neutral pH, and fertilizers may increase growth.

Management of *T. occidentalis* depends on intended use. If stands are to be used in recreation or as wildlife habitats, management time frames should be in the order of 500–700 years. In contrast, if trees are to be grown and harvested for shingle or log production, management time frames can be as short as 50 years.

Diseases and pests

The butt rot fungi *Poria subacida*, *Tyromyces balsameus* and *Phaeolus schweinitzii* mostly occur in rapidly growing stands. Trees growing on rock outcrops very rarely show evidence of any pathogen or rot.

The most common destructive pests of *T. occidentalis* are the leaf miners *Argyresthia thuiella* and *Coleotechnites thujaella*, the latter mainly a pest in ornamental settings. Black carpenter ants (*Camponotus pennsylvanicus*) make extensive galleries in fast-growing trees, grossly reducing their commercial value.

As the foliage of *T. occidentalis* is highly digestible, it attracts browsing by deer and porcupines, thus reducing tree productivity and altering growth form, leading to a reduction in commercial value.

Exposure of *T. occidentalis* to salt causes desiccation damage.

Uses

The ability of *T. occidentalis* to rapidly colonize abandoned land, and its dependence on persistent supplies of water, make it extremely attractive for riparian forest restoration work. It also forms valuable wildlife refuges and habitats.

With over 120 named cultivars, *T. occidentalis* is popular as an ornamental. It is widely planted in cities in the north temperate zone due to its tolerance of air pollution.

The wood is light, flexible and rot resistant, but not widely used because of the relatively small diameter of mature logs and their tendency to be curved. However, as the low specific gravity of the wood (0.33 air dry) means that thermal insulation provided by whole logs is very high, straight timbers are highly prized in log-frame construction. Wood density ranges between 352 kg/m^3 (12% moisture) and 449 kg/m^3 (green wood). Shingles cut from this species are narrow but extremely durable, especially when quartersawn. The wood shrinks little upon drying, and is not prone to splitting except at branch traces.

Traditional uses of the wood by Native Americans include building, with bark used for roofing and for the ribs of cedar-strip canoes. Various parts of *T. occidentalis* are also used in native medicine for curing scurvy and other ailments.

Volatile oils present in the wood produce a pleasant scent that is emitted for decades and also offer some protection from wood-destroying insects. Thujone is present in the leaves and is extracted for the manufacture of pharmaceutical products and beverages, including absinthe.

The extreme longevity of *T. occidentalis* has led to its use in dendrochronology, a summer maximum temperature-dependent 2878-year tree-ring chronology having been produced.

Further reading

Hofmeyer, P.V., Kenefic, L.S. and Seymour, R.S. (2009) Northern white-cedar ecology and silviculture in the north-eastern United States and south-eastern Canada: a synthesis of knowledge. *Northern Journal of Applied Forestry* 26, 21–27.

Johnston, W.F. and Hyvarinen, M.J. (1979) *Northern White-Cedar (An American Wood)*. USDA Forest Service, Washington, DC.

Kelly, P.E. and Larson, D.W. (2008) *The Last Stand: A Journey Through the Ancient Cliff-Face Forest of the Niagara Escarpment*. Natural Heritage Books, Toronto, Canada.

Thuja plicata **Donn ex D. Don**

Western redcedar

Importance

Thuja plicata (Plate 44) is the only *Thuja* species native to western North America and is one of the major tree species in old-growth and second-growth forests of the Pacific Northwest. It is very productive and grows on a wide range of sites. Introduced into Europe, it is grown on a minor scale in forests. It is popular as an ornamental tree and as hedging. It is particularly valuable owing to its unique ecological characteristics, such as tolerance of alkaline soils, and to the unique colour, texture, and durability of its wood, which is extensively used in a wide variety of products.

Botanical features

T. plicata is a very large tree, often over 60 m tall, with dbh up to 250 cm. The stem tapers rapidly and is flaring and buttressed at the base. The crown is long, symmetrical, narrowly conical, becoming irregular with age. Principal branches are spreading, drooping, and upturned at the ends. Old trees may have one or more dead spire-like tops.

Open-grown trees have live foliage extending to the ground. The bark is thin, reddish-brown, shiny when young, becoming shredded with age and forming narrow flat ridges.

Roots are extensive. Taproots are poorly defined or non-existent, but fine roots develop a profuse, dense network. Root grafting is common. Mycorrhizae are of the vesicular-arbuscular type.

Leaves are evergreen, scale-like, shiny yellowish-green, 1–2 mm long. On main shoots they are ovate and oblong pointed, usually glandular, whereas on lateral shoots they are smaller, blunt and not glandular. They are very aromatic when bruised.

When grown in the open, *T. plicata* begins to produce strobili at 10 years of age and usually every other year thereafter. The species is monoecious; male and female strobili are produced on different branches of the same tree, reddish male strobili on the lower branches and green female strobili nearer the crown top and further from the trunk.

Anthesis and pollination occur during March and April in the southern part of its natural range and near the coast, and during May and June in coastal Alaska and in interior stands. Cones are oblong–ovate, 10–12 mm long, brown, ripening in the first year, and composed of 10–12 leathery scales, out of which 3 middle pairs are fertile. There are 2–3 winged seeds beneath each scale.

Initiation of phenological events varies greatly with latitude, altitude, aspect, weather and snowpack.

Distribution

T. plicata is native to western North America, where it grows in the Pacific and Cordilleran regions, with coastal and interior populations separated from each other. Its range along the Pacific coast stretches from Humboldt County, California (40° N), to the northern and western shores of Sumner Strait in south-eastern Alaska (56°30'N). Optimal growth and development are achieved near the latitudinal centre of its range, the Olympic Peninsula in Washington State. The range of the interior population extends south from the western slope of the Continental Divide in British Columbia (54°30'N), through the Selkirk Mountains into western Montana and northern Idaho. The southern limit is in Ravalli County, Montana (45°50'N).

T. plicata grows in even- and uneven-aged, mixed-species stands, less frequently in pure stands. It is present in early, mid- and late stages of secondary forest succession. It forms a major component in old-growth forests, and is a minor or major component of a number of forest communities.

Silviculture characteristics

Generally, *T. plicata* has high tolerance of low light levels, so is frequently used for underplanting and enrichment planting. It also tolerates stagnant winter water tables averaging less than 15 cm below the soil surface, and excessive and/or surplus soil water supply.

T. plicata is seed-producing from age 10 years and heavy seed crops are frequent. Seed dissemination is adequate within 100 m from source.

Potential for natural regeneration is medium or high in low light, if vegetative reproduction is taken into account, and high in the open providing there is exposed mineral soil or burnt forest floor. Potential initial growth rate (first 5 years) is high. Response of advance regeneration to release is high.

Self-pruning capacity in dense stands is moderately good providing that there is a high initial stand density. Light conditions beneath closed-canopy, mature stands are low and result in poorly developed understorey vegetation. Potential productivity is high, site index being around 35 m on the most productive sites. Longevity is frequently >1000 years.

Silviculture practice

Depending on site and management objectives, clear-cutting, patch-cutting, shelterwood, and selection systems are all appropriate for establishment and growth. *T. plicata* can be grown in stands of mixed species where uneven-aged management is practised or when poles are to be produced under normal even-aged management regimes. Pure stands are more suitable for the long rotations needed to produce large sawtimber, shingles and shakes.

Three types of natural vegetative reproduction occur: layering, rooting of fallen branches and branch development on fallen trees. The resulting 'veglings' are not desirable, but are sometimes more abundant than seedlings in old-growth stands.

Propagation by seed is the only contemporary method used for regenerating *T. plicata*. Cones should be collected as soon as they change from bright green to yellow, and the tips of the seed wings are visible and are a light brown colour. Seeds usually germinate well without stratification, and can retain viability for at least 7 years when stored at 5–8% moisture and −18°C. Germination is epigeal. Containerized planting stock is used when *T. plicata* is prescribed as the major species; bare-rooted plants are also used.

Disturbed mineral soil seedbeds are a major requirement for natural regeneration from seed. Slash burning favours regeneration by creating more mineral soil surfaces in cut-over areas.

Direct sowing in the autumn is successful where soil moisture is available, but large quantities of seed may be required to obtain adequate stocking levels. Container-grown nursery seedlings, which can be produced in 7 months, survive as well or better than bare-rooted planting stock. Lateral shoot growth is vigorous, amounting to at least 80% of terminal shoot growth in young saplings.

Pure even-aged plantations may be established at an initial spacing of 2 × 2 m (2,500 trees/ha), and intermediate crown classes should be removed in a light thinning at age 25 years to reduce side shade. Stands of 370–430 crop trees/ha at time of harvest allow maximum diameter growth without causing poor form.

Maintaining a nearly closed canopy at all times will benefit tree form because open-grown trees tend to develop excessively large limbs and multiple tops. Faster-growing trees of acceptable quality can be grown at wider spacings if their lower boles are pruned, but the percentages of latewood will be significantly reduced.

Intermediate trees are distinguishable from dominants by age 5 years on good sites. Co-dominant trees can usually be differentiated from dominants by age 25 years. Mean annual height growth of dominant trees is 0.5 m in 40–60-year-old, pure, second-growth stands on good sites.

Growth in low light is much slower than in the open. Suppressed trees that are 200 years old but with <10 cm dbh and <10 m height are not unusual. Survival for such long periods of suppression may be due to an ability to produce new root growth in full shade and/or frequent root grafting.

Pure, even-aged stands of *T. plicata* can attain volumes comparable to pure Douglas fir (*Pseudotsuga menziesii*) by age 50 years on high-quality upland sites. Volumes of 379–825 m^3/ha have been measured in 40–60-year-old pure second-growth stands on good sites.

Diseases and pests

More than 200 fungi are found on *T. plicata*, but it is less susceptible to pathogenic fungal attack than its associates. In North America, the most important root, butt and stem rots are *Phellinus weirii*, *Armillaria mellea*, *Poria subacida*, *Poria asiatica*, and *Poria albipellucida*. The major seedling disease is Keithia disease, a leaf blight caused by *Didymascella thujina* that infects 2nd and 3rd year nursery stock; seedlings and young plants (<5 years old) can be killed by severe infections.

T. plicata suffers little damage from insects, but it is a host for several economically important species, one of the most important being the gall midge *Mayetiola thujae*, which sometimes seriously damages seeds.

Seedlings and saplings are often severely browsed by deer, elk or rodents; browse damage can be the most important stand-establishment problem.

Uses

When properly trimmed, *T. plicata* trees make excellent hedges. Drooping branches, thin fibrous bark, and flat sprays of scale-like leaves make it an attractive ornamental tree. The leaves are a major winter food for big game.

T. plicata wood is light, soft, relatively weak, and straight-grained with an air-dry density of 369 kg/m^3. It has a characteristic and pleasant odour. Heartwood is pinkish to deep brown and resistant to decay. Sapwood is yellowish-brown to nearly white, narrow, and less resistant to decay. Shingles and shakes constitute the most important special use of the wood due to its attractive appearance, durability, lightness and superior insulation qualities. The wood is also used for utility poles, fence posts, piling, pulp, furniture, coffins, crates, beehives and fish-trap floats. Thinnings are commonly used for ladders, flag poles, goal posts, fencing and garden sheds.

Perfumes, insecticides, medicinal preparations, veterinary soaps, shoe polishes and deodorants are made from the leaf oil. *T. plicata* extractives and residues are used in lead refining, boiler-water additives and glue extenders.

Further reading

Bolsinger, C.L. (1979) Western redcedar: a forest resource in transition. Resource Bulletin PNW-85. USDA Forest Service, Pacific North-west Forest and Range Experiment Station, Portland, Oregon, USA.

Harrington, C. (ed.) (2010) A tale of two cedars - International symposium on western redcedar and yellow-cedar. General Technical Report No. 828. USDA Forest Service, Pacific North-west Research Station, Portland, Oregon, USA.

Lowery, D.P. (1984) *Western Redcedar (An American Wood)*. USDA Forest Service, Washington, DC.

Massie, M.R.C. (1987) *Utilization Opportunities for Western Red Cedar - A Review*. FRDA Report No. 017. Pacific Forestry Centre, Victoria, British Columbia, Canada.

Tieghemella heckelii Pierre ex A. Chev.

Cherry mahogany

Importance

Tieghemella heckelii, commonly known as cherry mahogany or makoré, is a large, impressively cylindrical, evergreen tree. It is highly tolerant as far as light conditions are concerned. It is grown in traditional agroforestry systems in West Africa and in experimental plantations for seed oil production. It produces a hard, durable, insect-resistant, fairly workable wood with a range of uses. At first sight the tree may be mistaken for *Terminalia ivorensis* because of its dark, fissured bole, but it can be differentiated by the presence of red slash with latex as opposed to the yellow slash without latex in *T. ivorensis*.

Botanical features

T. heckelii is one of the biggest and possibly one of the longest-lived trees of the high forest, with a rounded, large and spreading crown made up of heavy branches. Average height ranges from 50 to 56 m and average dbh from 1.5 to 2.2 m.

Boles are long, cylindrical and clear. There are no true buttresses, but in very large trees thickened root spurs develop which cause the base to flare somewhat.

The bark is grey to almost black, regularly fissured vertically and particularly deep on the undersides of branches near the base. The slash is thick and red, and milky white latex exudes from wounds. Sapwood is white, heartwood dark reddish-brown. The tree is liable to shattering when felled.

The leaves are simple, alternate, exstipulate, and crowded towards the ends of branches. The leaf is oblong–obvate to elliptic–obvate, about 10 cm long and 4.3 cm wide, entire, with a short broad acumen or rounded apex, cuneate, glabrous, shiny dark green above and with a narrow petiole about 3.1 cm long. The midrib is raised below, and the nerves are fine, numerous, and parallel with one another, with a white marginal nerve seen when the leaf is held against the light.

The flowers are small, scented, yellow, axillary and on slender pedicels about 2.5 cm long. There are 8 sepals, and 8 petals formed into a corolla tube with an equal number of petaloid appendages, 8 epipetalous stamens, and a superior ovary of 8 carpels. The corolla tube with its stamens is caducous.

The fruit is yellow, ovoid, about 7–8 cm long, obtusely pointed at the apex, with seeds embedded in the pulp. Seeds are heavy (30–50/kg), hard, boat shaped, about 3.2 cm long with one side brown, smooth and shiny and the other side rough. The kernel is fatty.

T. heckelii is evergreen. Trees flower at different times from January to May. Although the flowers are quite small and not visible on large trees, the corolla tubes with their epipetalous stamens are easily seen on the ground under trees after shedding. Ripe fruits are found from August to March and are produced in reasonably large quantities. Because of their size they are conspicuous on the ground. The pulp is eaten by some small animals.

Distribution

The natural range of *T. heckelii* extends between latitudes 3°N and 10°N in West and Central Africa. It is common from Sierra Leone to Nigeria, and is present as far east and south as the Democratic Republic of the Congo and the Central African Republic.

The species is more often found in evergreen forest than elsewhere, and is found on a range of wet tropical sites up to 500 m altitude.

Silviculture characteristics

T. heckelii is usually a good shade bearer, although in Sierra Leone it has been found to be a strong light-demander. Trees prefer fairly heavy but well-drained soils and do not do well on light soils.

Best growth occurs where mean annual rainfall is 1500–2500 mm and mean annual temperature is 24–30°C; it is unsuitable for areas where the mean annual temperature falls below 22°C. It tolerates a uniform rainfall regime and a dry season of 2–6 months.

Silviculture practice

Propagation can be done easily by seed or vegetatively. With seed, the germination period is 16–34 days and germination frequency can be over 90%. As viability is lost rapidly, seeds should be planted within 2 weeks of harvesting. Seeds should be placed on the surface of the soil at a spacing of 40 × 40 cm and covered only very lightly, if at all, under light shade. When underneath a very hard soil the hypocotyl is liable to break when trying to carry the heavy cotyledons above ground. Seeds cannot be stored well because of loss of viability and the presence of fatty cotyledons.

Vegetative propagation is possible by air-layering, cuttings or stumps. Cuttings of 92–153 cm give best establishment, with over 70% rooting within 6 months. In a test with air-layering, 39% of branches rooted after 8 weeks.

Seedlings can grow 42–62 cm in 6 months and at 18 months the height ranges from 92 to 214 cm. They can be lifted from about 12–15 months of age, with underlifting required due to long taproots. In West Africa, saplings about 2.5 years old and 1–1.5 m tall are planted out into the field usually at the beginning of the rainy season, but in western Côte d'Ivoire 6-month-old seedlings are used. Transplanting should be done under high shade. Seedlings and saplings are shade-tolerant, but capable of rapid growth when released. However, when shade is maintained the growth rate is slow. Although *T. heckelii* in the nursery can grow up to 2 m in 18 months, if grown in shade the growth rate is reduced to about 1.5–3.5 m in 13 years.

Natural regeneration is reasonably common relative to the adult population, even in areas in moist semi-deciduous forests where elephant disturbance is less frequent. However, natural regeneration is most abundant in undisturbed forests. Tree density in forests is generally low, ranging from 1 exploitable tree/23 ha to 1.4 trees greater than 10 cm diameter/ha. Thinning of the forest leads to better recruitment and growth.

Average timber volumes in Côte d'Ivoire were 4.1 m^3/100 ha for trees exceeding 70 cm diameter, while in Liberia they were 33 m^3/100 ha for trees exceeding 40 cm diameter.

Small-scale plantations for seed oil production have an average density of 357 stems/ha. *T. heckelii* is often planted at a density of about 120 trees/ha in association with coffee, cocoa, rubber or rice, and at field margins and in secondary forest following clearance of undergrowth.

Care is needed during felling as logs tend to split badly, and many logs of larger trees are hollow.

Diseases and pests

No information on diseases and pests is available, although *T. heckelii* trees are hosts to the semi-parasitic tree *Okoubaka aubrevillei*.

Rodents do a great deal of damage by eating seeds and the oily cotyledons of young seedlings.

Uses

T. heckelii is used in taungya systems, with trees interplanted with a mixture of food crops such as plantains, maize, yams, tomatoes, cassava and groundnuts. The system is particularly useful in terms of weed control.

With regard to the wood, *T. heckelii* sapwood is yellowish with a fine grain, while the heartwood is pink to rich red-brown of striking chequered figure, with occasional darker streaks. The grain is generally interlocked, but sometimes straight and of fine texture. The wood is dense (wood density is 510–630 kg/m^3), hard and rather heavy. Seasoning is slow, and mild kilning is best. It is durable and resistant to insect and termite damage. The wood is fairly easily worked and takes a good finish, though curly-grained logs tend to pick up in planing. It also has a tendency to split.

The wood is used for mortars and also makes durable shingles. It is useful for shaft guides in mines and for sleepers. It is used for furniture, interior decoration, flooring and the interiors of railway coaches and liners. It veneers well and provides an attractive veneer for furniture, cabinet work and interior decoration.

The kernels of *T. heckelii* seeds are rich in fat and yield an edible oil which is used in soap making. Fat yield is 20–30 kg/tree per harvest. One kg of seed kernels yields about 200 g of fat when the traditional method of extraction is used.

T. heckelii has a number of other local uses. The seeds can be used as a fish poison. Broken shells from the seeds strung together are worn by dancers during cultural festivals. A bark decoction is used as a mouth gargle for toothache.

Further reading

Anon. (1976) Makoré - douka. *Bois et Forêts des Tropiques* 168, 31–46 (in French).

Poorter, L., Bongers, F., Kouamé, F.N. and Hawthorne, W.D. (eds) (2004) *Tieghemella heckelii*. In: Poorter, L., Bongers, F., Kouamé, F.N. and Hawthorne, W.D. (eds) *Biodiversity of West African Forests: An Ecological Atlas of Woody Plant Species*. CAB International, Wallingford, UK, 442 pp.

Tilia cordata Mill.

Small-leaf lime

Importance

Tilia cordata was once dominant in ancient European woodlands, and is still one of the most widespread trees from the Atlantic to western Siberia, with 1.7 million ha grown in Belarus, Ukraine, and Russia, but is now less frequent in Western Europe. The light and soft wood is used for furniture and panelling, and is especially valued for carving and framing. It is also common as a street tree and makes a majestic ornamental tree in parks and gardens.

Botanical features

T. cordata is a large deciduous tree but often occurs as many-stemmed coppice. Crown form is site-dependent, widely spreading and hemispherical in open situations to narrowly cylindrical in mature woodland and conspicuously emergent in old mixed stands. Trees in woodland grow to 18–30 m in height and 20–35 cm diameter in 100 years, or 22–34 m height and 34–55 cm in diameter in 200 years. Bark is pale grey, initially smooth with vertical lines of lenticels, eventually with smooth ridges and longitudinal furrows with occasional anastomoses, with strongly fibrous inner bark. Twigs are hairless shiny, green or reddish, buds ovoid hairless. Shade shoots flat, the leaves forming a single layered mosaic; all green parts mucilaginous. Leaves are alternate, blades circular, 3–8 cm diameter, cordate at the base, acuminate at the apex, evenly dentate, upper surface matt green, lower surface with small patches of rust-coloured stellate hairs in the axils of the main veins, petioles long and slender. The inflorescence is a dichasial cyme of 5–11 pale yellow, sweetly scented flowers held above the leaves. Fruits are 6–8 × 4–5 mm in size, ellipsoidal, thin walled and fragile, mostly 1-seeded, pale brown or brown when ripe.

Distribution

T. cordata is one of the most widely distributed trees in temperate Europe, extending to Western Asia, the Caucasus and western Siberia, but absent from parts of the Mediterranean. It is rarely naturalized but the distinction between the natural and planted distribution is often unclear. Populations are often fragmented, in what was once a much denser distribution. The presence of *T. cordata* is indicative that woodland is long established and often historically documented as of great antiquity. *T. cordata* is planted as an ornamental tree in towns and parks beyond its natural range in Europe and Asia, and has been introduced to other temperate regions, notably in the USA, Canada and Chile.

Silviculture characteristics

The natural distribution of *T. cordata* is entirely within the temperate zone, extending from continental to oceanic climates with the exception of parts of the Atlantic fringe, requiring uniform rainfall in excess of 500 mm/yr. The northern limits are determined by summer warmth, the eastern limits by low winter temperatures, and the southern limits are determined by water relations. Natural rainfall in central and Eastern Europe is usually adequate for healthy growth but drought induces early cessation of leader growth.

The root system consists of several main axes which descend obliquely to 1.5–2.0 m or deeper and have numerous lateral branches and support dense ectotrophic mycorrhizae. Growth is best on moist but well-drained sandy loam. Seedlings require freedom from weeds. *T. cordata* responds to fertilizers in many natural woodland soils and in forest nurseries. It regrows freely, vigorously and repeatedly after cutting and was formerly much grown as coppice. *T. cordata* and its widely planted hybrids with *Tilia platyphyllos* (*Tilia* × *europaea*) are tolerant of urban pollution and rarely show severe damage, even in industrial situations. *T. cordata* is one of the least damaged of common trees when exposed to sulphur dioxide, nitrogen oxides or ozone. Trees grow on sea cliffs and will tolerate strong wind and salt spray but snow scorches the foliage and causes distortion of the crown. Trees of high quality are restricted to the shelter of mature woodland.

Silviculture practice

Much of the *T. cordata*-dominated woodland in Eastern Europe has been propagated by sprouting and trees are of poor quality, but regeneration by seeding, though erratic, gives stands of high quality. Fruits can be stored dry (8–15% moisture) for up to 3 years but rapidly develop deep dormancy. Fruits collected in September–October, placed immediately in moist stratification and exposed to 6–8 weeks at 0–5°C give about 10% germination in the first spring and 50–90% in the second spring. It is best stratified in the open, in light shade, and protected from mice and voles.

Cuttings taken in autumn and placed in open frames give a low rooting percentage. Sprouts with some root initials from cut stumps root easily. Chip budding and grafting are very effective. *T. cordata* also layers readily. Seedlings and cuttings grow readily when rowed out. They can be successfully transplanted after 2 years but stock for amenity planting is usually transplanted at 5 years old. Weeding of young plants reduces damage by rodents. In nursery conditions pruning to remove laterals is essential for stock for amenity planting.

Growth in height and stem diameter is rapid for 60–80 years, with maximum volume increment at 50–60 years reported as 3–8 m^3/ha/yr. It is generally considered to grow faster than beech (*Fagus sylvatica*) in central Europe. Estimates of total dry matter production from coppice with stools at a density of 1200/ha are 6 t/ha at 10 years and 23 t/ha over a 20-year rotation, with no information on management and rotation length for timber.

Diseases and pests

Seedlings are vulnerable to damping off in natural conditions, during cultivation on poorly drained soils and in prolonged wet weather. Saplings and mature trees on wet soils are susceptible to *Phytophthora cactorum* and *Phytophthora citricola*, causing loss of leaves and eventual tree death. *Ustulina deusta* appears to be the only other fungus causing death of mature trees. *Armillaria mellea* is common on old coppice stools in woodland but is often not aggressive. Few of the numerous mites and insects which feed on the tree cause significant damage, including *Eriophyes tiliae*, *Eotetranychus tiliarum*, *Operophtera brumata* and *Phalera bucephala*, although complete defoliation by *Lymantria dispar* is reported from central Europe.

Uses

T. cordata is recognized as beneficial for soil structure and fertility. The deep roots transfer minerals from low in the soil profile and leaves provide a mineral- and nitrogen-rich litter which is palatable to lumbricid worms. Improvements in soil fertility may also be caused by the lime aphid (*Eucallipterus tiliae*) which deposits large amounts of honeydew (up to 1 kg/sugar/m^2/yr) which can stimulate the growth of nitrogen-fixing bacteria. Trees persist and grow well on slopes, eroding banks, and rock ledges. Sprouting after stem breakage by rockfall or snow avalanche allows persistence and contributes to slope stabilization.

The main uses of the wood are based on its softness and ability to resist splitting. The timber is light (560 kg/m^3) and strong, but decays rapidly in external use. It is used extensively in Eastern Europe for furniture, framing, small wooden objects and toys, and was formerly used for saddles. Specialized uses include plain and carved picture frames, and carved mouldings applied to panelling before painting or gilding. It was much used in ornate Baroque panelling and is now used for its restoration. The wood was and is much used for carving, as it has a uniform texture, a surface that can be smoothed to a silk-like finish, and its strength allows very delicate features to be carved and undercut. It was the preferred wood used by many European carvers in the Middle Ages for religious carvings. The wood was used for the keys of pianos but is now largely replaced by imported basswood (*Tilia americana*).

The bark was formerly harvested for its fibres from coppice poles. *T. cordata* are allowed to persist in pastureland in southern Scandinavia and the Alps where leaf fall contributes to the maintenance of soil fertility, and branches used to be harvested as 'tree-hay' and used for winter fodder.

T. cordata is an important source of nectar for honeybees and is harvested by beekeepers throughout its geographical range in Europe. As flowering is restricted to the fully exposed parts of the canopy of old trees, maximum production is only achieved from edge, emergent or open-spaced trees. As much as 75% of the honey produced in southern Russia is derived from *T. cordata*. Annual production of honey is up to 30 kg/tree produced for a period of 10–14 days at the beginning of July in southern Europe to as late as early August in northern Europe. The honey is light coloured and retains the characteristic fragrance of the lime flowers. It rarely shows a narcotic effect on bees.

T. cordata is an attractive ornamental tree with fragrant flowers which has been planted in parks and gardens, especially in Eastern Europe. Being a large, long-lived tree, it forms magnificent specimens in parkland. Since the Middle Ages *T. cordata*, but more often *T. platyphyllos*, was widely planted in central Europe as a village tree. In the pagan religions, *T. cordata* was sacred, and considered to have female qualities such as love, mercy and freedom.

Further reading

Białobok, S. (ed.) (1991) Limes (*Tilia cordata* Mill., *Tilia platyphyllos* Scop.). *Our Forest Trees, Popular Monograph, Volume 15.* (Lipy (*Tilia cordata* Mill., *Tilia platyphyllos* Scop.). Nasze Drzewa Lesne - Monografie Popularnonaukowe). Dendrological Institute, Poznan, Poland, 467 pp.

Pigott, C.D. and Huntley, J.P. (1981) Factors controlling the distribution of *Tilia cordata* at the northern limits of its geographical range. III Nature and causes of seed sterility. *New Phytologist* 87, 817–839.

Svejgaard Jensen, J. (2003) *EUFORGEN Technical Guidelines for Genetic Conservation and Use for Lime (*Tilia *spp.).* International Plant Genetic Resources Institute, Rome, 6 pp.

Trema orientalis (L.) Blume

Charcoal tree

Importance

Trema orientalis is a useful multipurpose tree that is not only tolerant of many adverse conditions but is versatile in its uses. It has a near-universal distribution in tropical and warm temperate parts of the Old World. It is fast growing, is valued as a fuelwood, and is a good source of wood fibres, providing a highly suitable raw material for the pulp, paper and wallboard industries. *T. orientalis* is grown in plantations as well as in agroforestry systems.

Botanical features

T. orientalis is, depending on climatic conditions, deciduous or, more commonly, evergreen. It is a shrub around 1.5 m tall but is more usually found as a large, slender tree, generally 18 m tall but attaining heights of up to 36 m and dbh of up to 90 cm. The crown is open with spreading branches, and buttresses are sometimes present, often up to 1.2 m high.

Branchlets, stipules, petioles, and inflorescences are densely covered with appressed and matted silvery to glaucous hairs. Bark is smooth and light grey, with conspicuous lenticels.

Leaves are simple, alternate, stipulate, ovate–lanceolate, slightly asymmetrical, 6–20 × 1.5–10 cm; leathery or often brittle and rigid, dark green above and glaucous beneath; base rounded, cordate or truncate; apex acute to acuminate–caudate; margin serrate to denticulate; upper surface sparsely set with bulbous-based hairs, lower surface densely covered with matted star-shaped and simple erect hairs. Stipule is linear–lanceolate to ovate–acute, and 3–4 × 1–2 mm. Petiole is 1–1.5 cm long, often pinkish-purplish.

The inflorescence is a much-branched panicle or thyrse, either male or female and borne on separate vegetative branches; axis 1–2 mm thick, bracts ovate–acute, 2–3 × 1 mm. The male inflorescence is 2.5–5 cm long with 20–100 flowers, while the female inflorescence is 1.5–2.5 cm long with 5–15 flowers. Flowers are in five parts, tiny and greenish-cream. The male flower is 1.5–2 mm in diameter; perianth lobes ciliate, 1.5–2 × 1 mm, filaments 1–1.5 mm long, anthers 1 × 0.5 mm, pistillode ovoid–conical, compressed. The female flower is 2–3 × 1–2 mm; perianth lobes ovate, acute, 1.5 × 0.5 mm, ciliate, densely short-pubescent or glabrescent, staminodes absent, ovary ovoid–conical, 2 × 1 mm with two slender stigmatic arms. Flowers are pollinated by bees.

The fruit is a drupe, ovoid, 5–8 × 4–5 mm, glabrous, tipped by 2 short styles, seated on persistent but not enlarged calyx, usually in clusters, green in colour, turning red then black. Fruits are eaten by birds, which act as seed dispersal agents.

Flowering and fruiting occur all year round. Trees are short-lived, generally 8–10 years.

Distribution

T. orientalis is native and widely distributed in tropical and warm temperate areas from Africa to the Pacific, its range extending from South and East Africa, through the Arabian Peninsula, the western Himalayas, the Indian subcontinent and southern China, to South-east Asia, southern Japan and Australia. Plantations have been established in Peninsular Malaysia, the Philippines, Taiwan and South Africa. The species introduced to Hawaii and Reunion has become an invasive weed.

T. orientalis grows at altitudes up to 2500 m and in areas with mean annual temperature of 20–27°C and mean annual rainfall of 1000–2000 m.

Silviculture characteristics

T. orientalis seeds require high light levels for germination, although once established seedlings can tolerate some shade. Trees coppice well and have an extensive root system which enables them to withstand dry periods; they can also withstand frost and wind damage, but are intolerant of fire. *T. orientalis* can grow on a wide range of soils from heavy clay to light sand; it tolerates moderate alkalinity and salinity, but does not withstand waterlogging. Early reports on the presence of nitrogen-fixing nodules on the roots of *T. orientalis* are not confirmed.

The tree regenerates profusely through its numerous seeds and is a common colonizer of disturbed rainforest areas.

Silviculture practice

T. orientalis can be propagated from stump cuttings or from seed. Stumps of 2–3-year-old trees, 10–14 cm in diameter and cut at a height of 30 cm, yield an optimum number of cuttings. For seed propagation, seeds are extracted from ripe fruits collected directly from healthy trees. There are 370,000 seeds/kg. The seed may be air dried and stored for up to 6 months, but may also be planted immediately. No pretreatment is required, but dormancy can be broken by steeping the seeds in gibberellic acid dissolved in agar at 500 ppm, or by storage for 3–4 months at 2°C. Germination is epigeal and occurs within 10–30 days, with 70–80% of seeds germinating. Full light encourages germination. Seedlings show rapid growth, reaching 1 m within 6 weeks of germination. They are outplanted after 3 months in the nursery. As foliage is browsed by livestock and wild animals, trees require protection. Weeds need to be controlled to prevent shading.

T. orientalis is very fast growing and has the ability to coppice readily. It can also be pollarded. It reaches a harvestable size for pulpwood in 3–4 years. Sustained periodic harvest from the same area for charcoal and firewood is carried out on a 3–4 year rotation, cutting at 20–30 cm above ground level. The cut is made at a slant to help prevent rotting of stumps. The bark is harvested for its tannins.

Diseases and pests

No serious diseases and pests are known. *Pseudomonas syringae*, which causes a bacterial disease with gall formation, and a powdery mildew fungus, *Oidium udaiyanii*, are known to infect *T. orientalis*. The new fungal species *Corynespora trematicola* was first identified on *T. orientalis* in India in 2002.

Larvae of the noctuid *Donda eurychlora* and those of *Sahyadrassus malabaricus*, a sapling borer, cause some leaf-feeding damage in India. Other insect pests which infest *T. orientalis* are *Xylotrechus subscutellatus* and *Apriona germarii*, although these do not cause serious problems.

Uses

T. orientalis is commonly planted as a shade tree in cocoa and coffee plantations in Asia and Africa. It is also widely used as a fallow species in shifting cultivations in these areas. Being a pioneer species, it is suitable for planting on poor soils for reforesting degraded and disturbed areas. Its extensive root system enables it to withstand dry periods. It has also been planted to control erosion and stabilize river banks in South Africa.

The species has high ecological impact, with at least 14 species of butterfly using it as a larval food plant. Fruits are eaten by several bird species.

Although *T. orientalis* timber is considered to be poor and of only fair quality, the wood is widely used for poles, posts and firewood, and provides excellent charcoal. It is also used in the manufacture of panel products such as medium density fibreboard (MDF), particleboard and laminated wood, and is also suitable for pulp and paper production. Being a fast-growing species, trees attain a size suitable for fuelwood and pulpwood in only 3–4 years. Despite the wood having a fairly low density (415–465 kg/m^3) and a relatively low energy value (18,900 kJ/kg), it has a high heating value and negligible ash content. Paper produced from *T. orientalis* pulp has good folding properties and tensile strength; the average length of wood fibres from 7 year old trees is greater than 1.6 mm, which compares favourably with *Populus* spp., commonly used for paper manufacture in the USA.

Silage made from *T. orientalis* foliage is high in crude protein (18.9 g/100 g dry matter), comparable with that found in common concentrates, and so can be used as feed for cattle, buffaloes and goats.

The bark is used for making ropes. It yields tannins, and is usually harvested from trees with a diameter of 40–50 cm, which on average yield 7.8 kg/m^2 green mass and 2.8 kg/m^2 air-dried mass.

Leaves and bark are used in traditional medicine to treat a range of ailments.

Further reading

Bhat, D.M., Swamy, V.S. and Ravindranath, N.H. (2003) *Trema orientalis*. In: Bhat, D.M., Swamy, V.S. and Ravindranath, N.H. (eds) *Nursery Manual for Forest Tree Species*. Universities Press, Hyderabad, Andhra Prodesh, India, pp. 292–294.

Jahan, M.S., Chowdhury, N. and Ni, Y.H. (2010) Effect of different locations on the morphological, chemical, pulping and papermaking properties of *Trema orientalis* (Nalita). *Bioresource Technology* 101, 1892–1898.

Trichilia emetica Vahl

Cape mahogany

Importance

Trichilia emetica is a widespread and important multipurpose tree, growing naturally in sub-Saharan Africa from Senegal to the Red Sea, and throughout East and Central Africa to the Democratic Republic of the Congo and South Africa. It occurs frequently in riparian forests and in some types of acacia woodland. It is regarded as an indicator of areas with palatable grass species. It is widely used as an ornamental and shade tree, for fruit and in agroforestry systems. The aril is removed from the seed and crushed with water to form a suspension used in cooking, though the seedcoat is poisonous. *T. emetica* also produces a valuable but inedible seed oil. Leaves have some soapy properties and have been used as soap during times of economic depression. Various parts of the tree are used in traditional medicine to treat ailments and for their emetic, diuretic and purgative properties. The pinkish-grey-brown wood is suitable for timber, poles and the manufacture of a range of wooden items, but is very susceptible to insect attack.

Botanical features

T. emetica is a medium-sized tree growing 8–20 m tall but occasionally reaching 30 m. It is usually evergreen or briefly deciduous. *T. emetica* subsp. *suberosa* is a shrub or small tree with corky bark. The trunk of *T. emetica* subsp. *emetica* can be swollen at the base and may become fluted with age. The branches are erect or partly spreading and produce a pyramid-shaped crown when young that becomes oval to rounded when mature. Crown diameter can sometimes exceed 15 m. The bark is dark grey to brown, grey-brown or red-brown with fine, smooth to rather rough, shallow striations and smallish scales.

Leaves are compound with four to five pairs of subopposite to opposite leaflets and a terminal one. Leaflets are elliptic to oblong–elliptic, usually 12 cm long and 5.5 cm wide, dark green with glossy upper surfaces and hairy under-surfaces, with 10–16 pairs of closely spaced lateral veins. The leaf apex is rounded or notched and the base is broadly tapering to rounded, margins are entire. The petiolules are short and thick, and the petiole is 7–12 cm long and velvety. Flowers are creamy-green and fragrant, with petals around 1.6 cm long, in short, compact branched axillary heads about 5 cm long. Male and female flowers are separate but similar in appearance, and are borne on the same tree. The fruit is an almost spherical creamy brown capsule, 2.5–3 cm in diameter which splits into two or three valves with a distinct neck, or stipe, connecting the base of the capsule to the stalk. There are 2–6 seeds per fruit, each 15–20 mm long, black and almost completely enveloped by a scarlet aril, with 500–1000 seeds/kg. Trees flower from July to November, depending on location, while fruits mature from December to March. Seed production varies annually, with little seed being produced in some years.

The closely related species *T. emetica* and *Trichilia dregeana*, which both occur in Southern and Eastern Africa, are often confused, but can be differentiated on the basis of their habitat and botanical features. *T. dregeana* grows in evergreen forests in higher rainfall areas, whereas *T. emetica* grows in lower rainfall areas and is found in woodland, bush and riverine vegetation. *T. dregeana* has pointed leaflets with 8–9 pairs of widely spaced secondary veins, while *T. emetica* has leaflets with rounded or sometimes notched tips and 13–16 pairs of closely spaced secondary veins. Additionally, the absence of a stipe in *T. dregeana* is a key feature distinguishing it from *T. emetica*.

Distribution

T. emetica is widely distributed in sub-Saharan Africa, with subsp. *emetica* occurring in Eastern Africa from Sudan in the north to Botswana and Swaziland in the south, and subsp. *suberosa* confined to West Africa from Senegal in the west, extending in the eastern part of its range into Sudan and Uganda where the two subspecies may hybridize. Both subspecies grow in riverine forests and in various types of woodland, including some types of munga woodland, occasionally in swamp forest, montane forest, savannah or alluvial lowland rainforest and escarpment miombo. Subsp. *emetica* is generally found in open riverine forests and open savannah woodlands subject to grass fires, whereas subsp. *suberosa* prefers the more fertile soils of riverbanks and seasonally flooded riverbeds.

Silviculture characteristics

T. emetica regenerates both vegetatively and by seed. It is fast-growing, with annual height increments of up to 1 m in cold areas and 2 m in warm areas. When mature, an average of 64.7 kg of fresh seed is produced per tree, but seed production is highly variable from year to year. *T. emetica* can produce root suckers, and it coppices well.

T. emetica tolerates mean annual temperatures of 19–31°C but does not tolerate frost. It grows at latitudes up to 1800 m and requires an annual rainfall of at least 1000 mm. However, it is capable of withstanding long periods of drought. Alluvial soils are preferred.

Silviculture practice

Seed storage behaviour is recalcitrant and seeds are often damaged by weevils. They are sensitive to desiccation and cannot be stored dry because they lose viability within a short time. They can, however, retain almost full viability for at least 4 months when stored at low moisture content at 16°C. Capsules are collected at the start of opening, dried in the shade and then shaken to obtain the seeds. Before they can be sown, seeds require pretreatment, which involves maceration in water to remove the impermeable aril. Best results are obtained by sowing pretreated fresh seeds either directly or into nursery containers. Germination is epigeal and occurs within 10–20 days. Seedlings are ready for transplanting within 6–8 months. Cuttings of 1-year-old shoots can also be used to propagate the species.

Recommended spacing is 3 × 3 m for pure stands for fruit production. In contrast, spacing should be 6 × 6 m when *T. emetica* is grown in combination with agricultural crops. Transplanted seedlings require shade. Saplings grow up in the shade of parent trees and are generally found in small groups of different sized individuals. Therefore, 30 trees/ha of the original vegetation should be left standing to provide shade. Seedlings are susceptible to weed competition; herbaceous material should be removed from the site before planting and weeding is required for the first few years after planting.

Trees coppice well when cut. Rotation length ranges from 60 to 80 years for pure stands.

Diseases and pests

T. emetica is a preferred host of woodrose-forming mistletoes (*Erianthemum dregei* and *Pedistylis galpinii*), generalist hemi-parasites of savannah trees which are harvested as ornamental plants.

Termites attack cuttings, and the coffee bean weevil (*Araecerus coffeae*) attacks seeds. The larvae of the whitebarred charaxes butterfly (*Charaxes brutus natalensis*) feed on the leaves, as do scale insects and various mammals. The timber is very susceptible to insect attack.

Uses

T. emetica is planted as a multipurpose tree for fruit, shade and ornament. It is sometimes planted in reforestation projects and has potential for agroforestry, where it is planted to provide shade. It is widely planted as a windbreak in urban and rural areas. It is also an excellent tree for providing shade in the garden and in urban settings, because it grows fast and has a wide crown. Trees are often planted along roads as ornamentals.

T. emetica is also used for erosion control and for rehabilitating water catchments. It is important in the local ecology where it is present. It provides food for many species of insect, mammal and bird which feed on the nectar, flower buds, fruits and seeds. Animals and birds may be important for seed dispersal.

The wood is pinkish-brown to brownish-grey and darkens with age; its density varies between 560 and 597 kg/m^3. It works well and takes a good polish, but is attacked by wood-boring insects. It is suitable for timber, poles, furniture, household utensils, tool handles, shelving and dug-out canoes. It is one of the most important timbers used in woodcarving in Southern Africa. There is no distinction between sapwood and heartwood. *T. emetica* is also commonly used for chew-sticks and firewood.

A sweet, milky suspension is extracted from the arils and used in cooking. Although the seedcoat is highly poisonous, skinned seeds are edible and are eaten raw or after soaking in water and grinding. *T. emetica* seeds have a high oil content, yielding about 308 ml oil/kg of fresh seed. This oil, known as mafura butter, produces a good finish on wood surfaces and is used in making candles, soap and cosmetics. The cake left after oil extraction has a 16% protein content and is suitable for use as a fertilizer.

The bark, roots, leaves and seed oil of *T. emetica* are all used for medicinal purposes. An oil obtained from the water in which seeds have been boiled after being ground and pounded is used to treat cuts and to anoint the body generally.

T. emetica is also important for spiritual and ritual purposes.

Further reading

Baxter, D., Erdey, D., Berjak, P., Sacandé, M., Jøker, D., Dulloo, M.E. and Thomsen, K.A. (2005) Sensitivity of *Trichilia emetica* Vahl. seeds to desiccation. In: Sacandé, M., Jøker, D., Dulloo, M.E. and Thomsen, K.A. (eds) *Comparative Storage Biology of Tropical Tree Seeds*. IPGRI, Rome, pp. 126–130.

Dharani, N. (2011) *Trichilia emetica* (*T. roka*) Cape mahogany. In: Dharani, N. (ed.) *Field Guide to Common Trees & Shrubs of East Africa*. Struik Nature, Cape Town, South Africa, p. 181.

Ishengoma, R.C., Gillah, P.R. and Chihongo, A.W. (1997) Properties of lesser utilized *Trichilia emetica* (*T. rocka*) and *Pterocarpus stolzii* timber species of Tanzania. *Annals of Forestry* 5, 10–15.

Komane, B.M., Olivier, E.I. and Viljoen, A.M. (2011) *Trichilia emetica* (Meliaceae) - a review of traditional uses, biological activities and phytochemistry. *Phytochemistry Letters* 4, 1–9.

***Triplochiton scleroxylon* K. Schum.**

Obeche

Importance

Triplochiton scleroxylon is a commercially very important timber species indigenous to the humid, tropical forests of Central and West Africa. It has high potential as a plantation species. Heavy exploitation has depleted the species from most natural forests in its native range and is likely to have eroded the native gene pools.

Botanical features

T. scleroxylon is a tall deciduous forest tree up to 45 m in height and 60–150 cm in diameter. The boles of mature trees are often heavily buttressed but usually free from branches. Trees in the dry forest zone have smaller leaves, narrower crowns and shorter bole lengths. The bark is ashy-grey or yellowish-brown, usually smooth in young trees but scaly and with fissures in older ones. The slash is fibrous, creamy-white to pale yellow. Young trees have a cylindrical-shaped crown bearing foliage almost to the ground that gradually self-prunes to a high, dense, circular crown, which finally becomes flat-topped when the trees are old. Leaves are 10–20 cm long and broad, palmate with 5–7 lobes, cordate and 5–7 nerved at the base. Lobes are broadly ovate, triangular or oblong, rounded or obtusely acuminate at the apex. Leaves are glabrous with a 3–10 cm long stalk. The leaves of saplings and coppice shoots are often larger and more deeply lobed than the crown leaves. The inflorescence is a paniculate cyme, 4–5 cm long, with dichotomous branching. Flowers are saucer-shaped; with white petals, red-purple at the base, obovate, densely hairy, and about 1 cm long and broad. They bear 30–46 stamens and 5 carpels. Fruits are brown to reddish-brown, composed of 1–5 winged carpels, each carpel more or less rhombic. The fruit measure 0.8–2 cm across the diagonal (with a 4–6 cm long wing) and are 1.2–2 cm broad, oblong–obovate with a thickened margin. The mericarp may be densely or sparsely pilose, either from the point of attachment to the slit at the apex or only on the slit and at the point of attachment.

Distribution

T. scleroxylon is naturally found between 12°N and 0°N. Within the natural range, three sections are recognized: Sierra Leone to Togo, Benin to Nigeria and Cameroon to Congo. The tree is found in the semi-deciduous moist forest and in the transition zone between forest and moist savannah. Trees normally occur in clusters of ten or more and isolated trees are exceptional. *T. scleroxylon* is a pioneer species, often found in disturbed forests, and its natural distribution in West Africa may have been influenced by shifting cultivation. It prefers well-drained acidic to neutral ferruginous soils. *T. scleroxylon* is found mainly at low and medium altitudes, between 0–900 m, with the highest concentrations below 500 m, in areas with a seasonal two-peaked rainfall distribution (dry from December–April) with an annual rainfall between 1000 and 3000 mm. Optimal growth occurs in areas with annual precipitation ranging between 1100 and 1800 mm and mean annual temperatures between 24 and 26°C (range 20–35°C). The absolute minimum temperature is 17°C.

Silviculture characteristics

T. scleroxylon is fast growing, coppicing and self-pruning. It is used for plantation establishment and in enrichment planting in Ghana and Nigeria. Growth ranges between 12–20 m^3/ha/yr depending on the site.

Silviculture practice

Propagation can be by seed (seed storage is orthodox), marcotting, cuttings, grafting and tissue culture. Stand establishment is done through direct sowing, natural regeneration and planting stock. *T. scleroxylon* has irregular seed years with unpredictable quantities. The seeds are short-lived and considerable research has been conducted into vegetative propagation techniques. Successful rooting of cuttings can be obtained only from young, leafy stem material, which makes the use of high-humidity propagators essential. Juvenile leaf cuttings have been propagated successfully in conditions of high humidity in West Africa. Medium shade enhances growth more than dense shade and full light.

Diseases and pests

A die-back fungus, *Botryodiplodia theobromae*, is sometimes present.

Leading shoots are often killed by aphids. In Ghana, heavy defoliation by the caterpillar of *Anaphe venata* occurs almost annually, although this does not seem to affect growth. *Trachyostus ghanaensis* bores tunnels in standing trees. The heartwood is also susceptible to attack by termites, pinhole borers, long-horn beetles, and sap-stain fungi. The sapwood is prone to powder-post beetle attack. Freshly felled logs are extremely prone to attack by insects.

Uses

The main use of *T. scleroxylon* is timber, but the wood is also used for fibreboard and particleboard, blockboard, boat- and shipbuilding, boxes and crates, plywood, furniture, mouldings and building materials. It is fairly elastic and resilient but should not be used for purposes where strength is critical. Wood density is around 390 kg/m^3. The wood is creamy-white to pale straw. Heartwood and sapwood are not clearly differentiated. The texture usually varies from coarse to moderately fine and even with a natural sheen on the surface. The unpleasant odour of freshly cut wood disappears after seasoning. It works very easily with hand and machine tools, and does not blunt cutting edges of tools quickly. Cutting edges should be kept very sharp when working end-grain material since it has a tendency to crumble and chip at tool exits. The wood dries rapidly, with little or no degradation. Gluing is preferable over nailing and screwing for jointed work. Staining properties are reported to be satisfactory, but the surface requires careful filling.

An aqueous extract of the bark is commonly used as an anti-diabetic.

Further reading

Hall, J.B. and Bada, S.O. (1979) The distribution and ecology of Obeche (*Triplochiton scleroxylon*). *Journal of Ecology* 67, 543–564.

Howland, P. and Bowen, M.R. (1977) Triplochiton scleroxylon *K. Schum. and other West African Tropical Hardwoods.* West African Hardwoods Improvement Project Research Report, 1971–1977. Forestry Research Institute of Nigeria, Ibadan, Nigeria.

Igboanugo, A.B.I. and Iversen, P. (2004) Conservation and use of *Triplochiton scleroxylon* in moist West Africa. In: FAO Forest Genetic Resources no 31. Food and Agriculture Organization of the United Nations, Rome, pp. 50–52.

Tsuga canadensis (L.) Carriere

Eastern hemlock

Importance

Tsuga canadensis is a highly shade-tolerant, cold-hardy, large conifer distributed widely in North America, either in pure stands or associated with many forest types. It survives to an old age, >900 years having been recorded, and tolerates a wide range of soils as long as sites are reliably cool, moist and well drained. Morphologically and physiologically it is an elastic species, with several ecological races and numerous cultivars. Its greatest importance was formerly as a source of tannin for the leather industry, but it is now important for pulp, lumber, ornamental use, as wildlife food and habitat and as a component of landscape diversity. Due to previous wasteful exploitation and various other pressures, the species is much reduced compared with its former presence.

Botanical features

T. canadensis is an evergreen tree of graceful feathery appearance when young. It has a dense, broadly and pendulously pyramidal crown, often rounded near the apex and raggedly irregular with age. Trees mature at about 300 years with average heights of 25–30 m and dbh of 1–2 m. Stems are tapered and straight. Slender twigs, nodding terminal shoots and high live crown ratio are useful diagnostic features. The root system is wide-spreading and shallow. The bark is scaly brown and thin when young, becoming deeply furrowed, dark brown and thick with broad, flat ridges with age. The outer bark is a dull reddish-purple.

Needles are almost regularly 2-ranked, about 1.5 mm wide and 15–20 mm long, the shortest needles being on the upper side. They are flat, and gradually taper from the base to a rounded or notched apex. Margins are more or less serrated. Needles are glossy and furrowed above, and exhibit two stomatal bands beneath, each with 5–6 lines, the green margin usually broader than the white bands. Buds, which are about 1.5–2.5 mm long, light brown, hairy and acutely ovate, arise from evenly pubescent young shoots. Needles are retained for 3–4 years.

Trees are monoecious with light yellow staminate and pale green pistillate flowers in separate clusters on the previous year's shoots. Cones are light to purplish-brown, 10–15 × 15–25 mm, pedunculate pendulous, ovate, obtuse, with thin, leathery, roundish scales. Seeds are light brown and about 1.6 mm long, with a slightly longer terminal wing.

Buds break dormancy late in the spring, and pollen dispersal follows about 2 weeks later. Fertilization is complete in about 6 weeks. Cones mature during their first year, reaching full size in late August to early September, open mid-October, shed seed into winter, and persist into the following season before dropping. Seeds germinate in spring.

Distribution

T. canadensis occurs naturally from Minnesota and southern Ontario in the west in a broad belt to Nova Scotia, and south to Georgia and Alabama. Its distribution is characterized by a complex pattern of disjunctions at the borders of its range in Alabama, Georgia, Indiana, North Carolina, Ohio and Wisconsin. *T. canadensis* has been introduced into Europe, where its cultivars have been widely planted on suitable sites.

It is a major component of four forest cover types, including pure conifer types and mixtures with hardwoods. It is also a common associate in seven other types, and a minor species in another 18. The *T. canadensis* vegetation type is, however, declining in area faster than any other vegetation type.

Silviculture characteristics

T. canadensis is a highly shade-tolerant species, which regenerates well on exposed mineral soil and on well-decomposed litter and large woody debris. It tolerates heavy snow loads, continental conditions, and does best on reliably cool/humid/wet sites with good drainage. Although old trees may tolerate surface fires, and regeneration may benefit from this, other major disturbances often work against *T. canadensis*.

Following logging, residual trees are often subject to significant windthrow, sunscald, winter drying and drought damage. Trees of all ages are high to intermediate in their sensitivity to various air pollutants and salt damage.

Small trees are highly susceptible to wildfire, but prescribed burns are beneficial for securing natural regeneration.

Silviculture practice

T. canadensis seed is collected from first-year, newly brown cones in September–October. There are about 410,000 seeds/kg. Seed stored at 5°C in sealed containers and at a moisture content of between 6 and 9% can remain viable for 4 years. Dormancy is variable, with some seed lots requiring or benefiting from cold stratification at 1–5°C for 1–4 months. A constant temperature of 15°C is considered optimum for germination. Germination is epigeal, with germinative capacity less than 25%.

Seedlings are slow and difficult to grow because they are easily damaged by heat, drought and frost-heave. The preferred practice is to spring sow following stratification, but autumn sowing is also possible. Usually planted as 3+0 stock, under favourable conditions *T. canadensis* seedlings tend to have good growth and survival, both in the open and shade. Containerized stock is also popular.

Seeds require 45–60 days of warm, moist conditions for optimum germination. A two to three cut shelterwood regeneration system under a high density overstorey may work well if given sufficient seed and proper site preparation. This ideally requires reduction of heavy hardwood competition, and either prescribed burning or a thorough mixing of organic and mineral soil to expose partially decomposed humus layers. In the absence of these conditions, most regeneration will be restricted to rotten logs, stumps or tip-up mounds. Seedlings develop slowly, but should not be released completely from overhead competition until about 0.9–1.5 m tall. Shade tolerance allows survival of suppressed seedlings of good stem form for decades and even centuries. Once released, *T. canadensis* usually responds well at all ages.

Early in life and under suppressed conditions, growth of *T. canadensis* can be exceedingly slow. A tree with a 5–8 cm dbh can be 200 years old. Growth rates average 21.4 cm^2/yr. Gross volumes of more than 560 m^3/ha have been reported in older pure stands. Many fully stocked stands of *T. canadensis* have basal areas over 69 m^2/ha. No more than one-third of this basal area should be removed at any one time during thinnings or harvest cuts, as trees may suffer post-logging decadence on account of drastically altered stand climate and soil moisture conditions, in conjunction with injuries to the shallow roots.

Diseases and pests

Seeds of *T. canadensis* are sensitive to damage by several mould fungi, especially *Botrytis cinerea*. Young seedlings are susceptible to damping off caused by *Pythium* and *Rhizoctonia* spp. Root rots caused by *Calonectria kyotensis*, *Rhizina undulata* and *Gibberella fujikuroi* are also common.

Several fungi attack the needles and twigs, the rust fungus *Melampsora farlowii* being one of the most damaging, causing new shoots to curl and die. It also attacks cones, often resulting in cone abortion. Three other rust fungi, *Melampsora abietis-canadensis*, *Pucciniastrum hydrangea* and *Pucciniastrum vaccinii*, attack the needles. *Fabrella tsugae* causes browning of individual needles throughout the crown.

The heartwood of *Tsuga canadensis* is attacked by *Tyromyces borealis*, which leaves white flecks in the wood. *Pholiota adiposa* causes a cavity along the pith axis. *Ganoderma tsugae* is the most common agent of decay in stumps and old logs.

Several fungi infect root systems but seldom kill trees, the most common being *Armillaria mellea*, *Phaeolus schweinitzii*, *Tyromyces balsameus*, and *Heterobasidion annosum*.

The most serious insect pest of *T. canadensis* is *Adelges tsugae*, introduced to North America from Asia around 1924. Infestations by adults and nymphs, which suck the sap from young twigs at the base of needles, retard tree growth and cause premature needle drop. Defoliation and tree death can occur within several years of initial infestation.

Approximately 24 species of indigenous insects attack *T. canadensis* but few are considered economically important, the most damaging being *Melanophila fulvoguttata*, which attacks weakened trees.

Among vertebrate pests, the yellow-bellied sapsucker (*Sphyrapicus varius varius*) drills holes into healthy bark, often girdling young trees. Porcupines (*Erethizon dorsatum*) also frequently girdle trees by feeding on the bark. In parts of the geographic range of *T. canadensis*, where more preferred browse species are lacking, browsing by white-tailed deer, *Odocoileus virginianus*, interferes with seedling establishment even in locations of rich seedling recruitment.

Uses

T. canadensis is a popular ornamental, with over 200 cultivars differing in habit. Being amenable to pruning, it makes excellent hedges except in urban settings. *T. canadensis* also creates important wildlife habitat.

The wood of *T. canadensis* is moderately strong and pliable with a density of 380–410 kg/m^3. Despite its brittleness and tendency to split due to ring and star shake, demand for *T. canadensis* lumber is increasing. It is used for boxes, pallets, crates, and for rough construction or general millwork, including light framing, sheathing, roofing and subflooring. Much current production of *T. canadensis* is used in pulping for newsprint and wrapping papers.

Native Americans used teas brewed from the needles or inner bark to cure various ailments.

T. canadensis is also a source of natural dye, the inner bark yielding a dye that renders wool a brown colour and leather reddish. Root bark was used to dye basketry. The bark also contains 7–12% tannin and the species was over-exploited in the early 20th century to provide raw material for the American tanning industry. The bark is now also used as landscaping material.

Further reading

Hough, A.F. (1960) *Silvical Characteristics of Eastern Hemlock*. USDA Forest Service, Washington, DC.

McManus, K.A., Shields, K.S. and Souto, D.R. (eds) (2000) Proceedings: Symposium on sustainable management of eastern hemlock ecosystems in eastern North America, 22–24 June 1999, Durham, New Hampshire. USDA Forest Service North-eastern Research Station General Technical Report NE-267, Newtown Square, Pennsylvania, USA.

Olson, J.S., Stearns, F.W. and Nienstaedt, H. (1959) Eastern hemlock: growth cycle and early years. Connecticut Agricultural Experiment Station Circular C205, New Haven, Connecticut, USA.

Tsuga heterophylla (Raf.) Sarg.

Western hemlock

Importance

Tsuga heterophylla is a low- to mid-elevation tree that regenerates prolifically and grows well on acid, nutrient-poor soils of wet mesothermal and temperate climates in western North America. It is a fast-growing but not nutrient-demanding conifer that is present in all stages of secondary succession and is the dominant species over most of its range. It is recognized as an efficient producer of fibre, and its moderately strong, light-coloured wood provides an all-purpose raw material.

Botanical features

T. heterophylla is a large tree, occasionally over 60 m tall, with a dbh of 250 cm and an open and conical crown. The lower trunk is long and branch-free. The principal branches are coarse, spreading horizontally from the trunk. Twigs are yellow-brown and finely pubescent. The bark is smooth when young, and reddish-brown, becoming darker and deeply furrowed with flat-topped scaly ridges with age (Plate 71).

The root system is shallow, predominantly confined to the surface organic layer and intensely mycorrhizal.

Needles are evergreen, flat, finely toothed, 5–20 mm in length, blunt-tipped, arranged in two ranks, and whitened beneath. Upper surfaces are shiny, dark green, and grooved.

T. heterophylla is monoecious; male and female strobili develop from separate buds of the previous year. Cones are ovoid, 2–2.5 cm long, blunt-tipped, short-stalked and golden-brown.

Depending on location, flowering and pollination begin from mid- to late April and continue through late May and June. Cones mature at various times on the same branch. Ripe cones change from green to golden-brown. Seeds are usually fully ripe by mid- to late September, but cone scales do not open until late October.

Distribution

T. heterophylla is a western North American species that grows in both the Pacific and Cordilleran regions. Along the Pacific coast its range extends north along the Coast Ranges from central California to the Kenai Peninsula in Alaska. It is the dominant species in British Columbia and Alaska along the Coast Mountains. Inland it grows along the western and upper eastern slopes of the Cascade Range in Oregon and Washington, and the west side of the Continental Divide of the northern Rocky Mountains in Montana and Idaho north to Prince George, British Columbia.

T. heterophylla grows in even-aged or uneven-aged, pure or mixed-species stands on a wide range of sites. It is present in early, mid-, and late stages of secondary succession. In low-elevation coastal forests of Washington, Oregon and southern British Columbia it shares dominance with *Picea sitchensis*, whereas in slightly drier environments of the western Cascade Range its primary co-dominant is *Pseudotsuga menziesii*.

T. heterophylla is a major component of a number of forest communities in the interior grand fir, interior western hemlock, white fir, coastal western hemlock, Pacific silver fir, Sitka spruce and redwood zones.

Silviculture characteristics

T. heterophylla is a prolific seed producer with high dissemination capacity, seeds mainly falling within 500 m of the source, but up to a maximum distance of over 2 km. The potential for natural regeneration under low illumination is high; it is very shade-tolerant and trees are able to survive in a relatively dark understorey until death of overstorey trees allows them to continue their growth. The potential for natural regeneration in the open is also high, especially in wetter parts of the range. Potential initial growth rate (<5 years) is intermediate; 2-year-old seedlings are often <20 cm tall, but later they may grow >60 cm/yr. Response of advance regeneration to release is high; diameter response is immediate, while height growth response is delayed.

Self-pruning capacity in dense stands is high. Crown spatial requirements are intermediate. Low light conditions beneath closed-canopy and mature stands of *T. heterophylla* are associated with poorly developed understorey vegetation. *T. heterophylla* is highly susceptible to fire and windthrow.

Potential productivity is high; site index (after 50 years) is nearly 40 m on the most productive coastal sites. The infrequency of stand-destroying disturbance is readily exploited by this conifer over its lifetime, which is typically <500 years but can be >700 years.

Silviculture practice

Clear-cutting, patch-cutting, strip shelterwood, and selection systems are all appropriate for the establishment and growth of *T. heterophylla*. Natural regeneration can be obtained through various reproduction methods, although it often results in adequately stocked to overstocked stands. *T. heterophylla* can be grown in pure stands, although higher timber yields are achieved in mixed-species stands, especially with *Abies amabilis*.

Regular cone production usually begins when trees reach 25–30 years of age. Mature trees produce heavy cone crops at average intervals of 3–4 years. Each cone contains 30–40 small seeds, with less than half being viable. There are 417,000 to >1,120,000 seeds/kg. Seeds are not deeply dormant; stratification for 3–4 weeks at 1–4°C improves germination. Germination is epigeal, and seeds remain viable only into the first growing season after seedfall. Provided there is adequate moisture, seed germination and germinant survival are excellent on a wide range of substrates.

A combination of natural regeneration and planting is the contemporary regeneration method. A variety of containerized stock is used for planting when *T. heterophylla* is prescribed as the major species and where advance regeneration is inadequate and ingress is predicted to be unreliable. Container-grown stock provides better quality seedlings, less damage to roots, and better survival compared to bare-root stock.

T. heterophylla can be propagated by layering and from cuttings. Seedlings that die back to the soil surface commonly sprout from buds near the root collar. The tree also grafts readily, and growth of grafted material is better than that of rooted material.

T. heterophylla may form pure stands or be a component of mixed stands. Young stands vary in stocking, but understocking is infrequent. Natural 20-year-old stands can contain over 25,000 stems/ha. Stocking levels of 1480–1790 stems/ha at crown closure provide the best yields if commercial thinnings are part of the management regime. If thinnings are not planned, stocking levels as low as 740 well-distributed trees/ha can provide maximum yields at rotation age.

Under conditions of dense, even-aged stocking, early natural pruning occurs and tree crowns are usually narrow, with good stem development. Given unrestricted growing space, the quality of *T. heterophylla* logs is reduced because of poorly formed stems and persistent branches. Trees that develop in an understorey vary greatly in form and quality.

The response of *T. heterophylla* to nitrogen fertilizer is extremely variable. For overstocked stands, a combination of pre-commercial thinning and fertilizer often gives the best response.

T. heterophylla forests are among the most productive forests in the world. Mixed stands of *T. heterophylla* and *P. sitchensis* are especially productive. In the Sitka spruce zone of Oregon and Washington, the mean annual increment of such stands frequently exceeds 42 m^3/ha. Yields of *T. heterophylla* on productive sites can exceed 1850 m^3/ha at 100 years of age.

A study in northern Idaho showed that in *Thuja plicata–Tsuga heterophylla–Pinus monticola* stands, uneven-age management using selection cuts is the best silvicultural system for fully stocked stands. With a maximum merchantable volume objective, even-age plantation management and an uneven-age shelterwood system are both capable of producing the same high level of yield.

Diseases and pests

Phaeolus schweinitzii, *Laetiporus sulphureus*, *Inonotus tomentosus*, *Poria subacida*, *Phellinus weirii* and *Heterobasidion annosum* are the major root and butt pathogens; the latter can also spread from the roots into the bole. Other bole pathogens include *Echinodontium tinctorium*, which can cause extensive decay in over-mature stands.

Since 1917, there have been only 10 years in which an

outbreak of *Acleris gloverana* did not cause visible defoliation somewhere in *T. heterophylla* forests. *Lambdina fiscellaria lugubrosa* causes more mortality of *T. heterophylla* than any other insect pest. Outbreaks last 2–3 years on any one site.

Important insect pests not attacking the foliage include *Megastigmus tsugae*, which attacks cones and seeds, the larvae feeding inside seeds; *Steremnius carinatus*, which girdles seedlings at the ground line; and *Tetropium velutinum*, which attacks and occasionally kills weakened trees.

Dwarf mistletoe (*Arceuthobium tsugense*) reduces growth, lowers fibre quality, and provides an entry point for decay fungi. Uninfected to lightly infected trees may have a greater growth in volume (+40%) and height (+84%) than severely infected trees; in mature stands, annual volume losses can be as high as 4.2 m^3/ha.

Uses

The wood is moderately light, fairly hard, strong and whitish to dull light brown, with little contrast between the sapwood and heartwood. It has a density of 420–460 kg/m^3. It treats well and is used for pilings, poles and railway ties. Strength and nailing characteristics make it a preferred species for construction lumber. *T. heterophylla* has good to excellent pulping characteristics and is an important fibre source for groundwood, thermomechanical, kraft and sulphite pulps.

The bark was traditionally used by Native American peoples as a tanning agent, pigment and cleansing solution. The wood was carved into implements such as spoons, combs and bowls. The branches were used as bedding, and also as spawning substrate to collect herring spawn in estuaries. The pitch and outer and inner bark were formerly widely used medicinally.

Further reading

Cahill, J.M. (1984) *Western Hemlock (An American Wood)*. USDA Forest Service, Washington, DC.

Duddles, R.E. (1998) *Managing Western Hemlock Forests in the Oregon Coast Range*. Oregon State University Extension Service, Corvallis, Oregon, USA.

Ruth, R.H. and Harris, A.S. (1979) Management of western hemlock-Sitka spruce forests for timber production. USDA Forest Service General Technical Report PNW-88, Portland, Oregon, USA.

Webber, J.E. (2000) Western Hemlock: A Manual for Tree Improvement Seed Production. British Columbia Ministry of Forests Forest Science Program Working Paper 44, Victoria, British Columbia, Canada.

Tsuga mertensiana (Bong.) Carrière

Mountain hemlock

Importance

Tsuga mertensiana is a subalpine conifer that tolerates a high snowpack and nutrient-poor soils. It grows in maritime subalpine boreal climates of western North America, more in the Pacific than the Cordilleran region. It is slow growing and forms the upper timberline. *T. mertensiana* is largely inaccessible because of the high altitudes at which it occurs (up to 3500 m in the Sierra Nevada) and is unimportant as commercial timber. However, it is harvested to a limited extent near its lower limits. Uses of its moderately strong, light-coloured wood include small-dimension lumber and pulp.

Botanical features

T. mertensiana is a small to large-sized tree, 20–40 m tall, occasionally over 50 m, and 70 cm dbh, but also a low-spreading shrub (krummholz) at the upper timberline. The trunk is strongly tapered, bearing slender branches with upturned tips, almost to the ground in open-canopy stands. The crown is narrowly conical, becoming irregular and bent or twisted on old trees. The bark is dark reddish-brown, scaly, and divided into hard, narrow, flat-topped ridges (Plate 72). The wood is moderately light, fine-grained, relatively hard and strong; heartwood is light reddish-brown, not sharply distinct from sapwood.

The root system is shallow, with roots confined to the forest floor. Two-thirds to three-quarters of the net primary productivity appear to be allocated below ground. Roots are intensely mycorrhizal, with both ecto- and endomycorrhizae.

Needles are evergreen, rounded in cross section, 20–30 mm long, blunt-tipped, both surfaces dark bluish-green with faint lines of white dots, needles crowded, spreading all around the twig, especially on the upper side.

T. mertensiana is monoecious. Pollen release occurs from mid-May to late July, with fertilization occurring from about late July to early August. Reproductive buds can easily be identified in the late summer and autumn. The presence of many female strobili indicates a potentially large cone crop in the following year. Mature cones are oblong, purple or brownish-purple, and are 2–9 cm long, generally longer than in other species of *Tsuga*.

Initiation of phenological events varies greatly with latitude, altitude, aspect, weather and snowpack. Generally, cones ripen and open from late September to November. Cone production begins as early as age 20 years. Mature trees 175–250 years old produce medium to very heavy cone crops at about 3-year intervals, but crops may be complete failures in other years. Seeds are dispersed primarily by wind. Seed dissemination capacity is high, with most seeds falling within 500 m of the parent tree. Germination is epigeal, ranging from 47 to 75%, and occurs on snow, mineral soil, or organic soil if sufficient moisture is available. Young seedlings grow best in partial shade, and early development is often slow.

Distribution

T. mertensiana grows in the mountains of both Pacific and Cordilleran regions of western North America, with a range extending from Sequoia National Park in California (37°N) to Cook Inlet in Alaska (61°N). It grows along the crest of the Sierra Nevada in California; the Cascade Range in Oregon; the Cascade Range and Olympic Mountains in Washington; the northern Rocky Mountains in Idaho and western Montana; the Insular, Coast, and Columbia Mountains in British Columbia; and in south-eastern and south-central Alaska. Introductions are limited to arboreta and parks.

T. mertensiana grows predominantly in broadly even-aged and uneven-aged, mixed-species stands, less frequently in pure, even-aged stands (after fire disturbance). It is often a pioneer species on colluvial sites, and is present in early, mid- and late stages of secondary succession. Major associates include *Abies amabilis*, *Abies lasiocarpa*, *Abies procera*, *Chamaecyparis [Cupressus] nootkatensis*, *Pinus albicaulis* and *Pinus contorta*. *T. mertensiana* is a major component of a number of forest communities in the Mountain Hemlock and Engelmann Spruce–Subalpine Fir zones of western North America.

Silviculture characteristics

Potential for natural regeneration in low light is low to intermediate, with regeneration developing mainly in canopy gaps. Potential for natural regeneration in the open is moderate. The regeneration period is long, often >50 years on high elevation sites after disturbance. Potential initial growth rate (up to 5 years) is low, less than 10 cm/year in the first two growing seasons. Response of advance regeneration to release is high.

Self-pruning capacity in dense stands is high, and crown spatial requirements are low. Light conditions beneath closed-canopy, mature stands are intermediate. Being shallow-rooted, *T. mertensiana* is susceptible to windthrow.

Potential productivity is low and decreases with increasing elevation; site index on intermediate sites decreases from ±20 m on lower subalpine sites to <10 m on upper subalpine sites.

Longevity is high, typically less than 500 years, but ages of over 1000 years have been reported.

T. mertensiana is very frost-tolerant. Seedlings and small saplings tolerate heavy snowpacks well. Bent boles and branches spring erect after snowmelt. Leader replacement by axial buds is less frequent than in other North American hemlocks, and the drooping leaders take 3 or more years to become erect.

Silviculture practice

T. mertensiana is easily transplanted and propagated by seed and cuttings. Layering is an important method of reproduction on muskegs and krummholz in Alaska but is insignificant in forested areas in general.

Depending on site and management objectives, clear-cutting, patch-cutting, strip shelterwood and selection systems are all appropriate for the establishment and growth of *T. mertensiana;* many sites dominated by the species, however, are particularly difficult to reforest following clear-cutting. It may be grown in pure stands, although higher timber yields can be achieved by growing it in mixed-species stands, especially with *Abies amabilis*.

A combination of natural regeneration and planting is the contemporary method for regenerating *T. mertensiana*. A variety of containerized stock is used for planting when *T. mertensiana* is prescribed as the major species and where advance regeneration is inadequate and ingress is predicted to be unreliable.

The species is slow to regenerate after disturbances, and many burned areas do not have adequate stocking; however, some young stands 20–40 years old (some in burned areas) in southern Oregon and northern California may be pure *T. mertensiana* and quite dense (9900–24,700 trees/ha). Stocking in clear-cuts in the Mountain Hemlock zone is typically made up of post-harvest naturally seeded trees and advance regeneration. Planting is relatively ineffective in speeding regeneration on these cold, snowy sites compared with advance and post-harvest natural regeneration that slowly provide adequate to abundant stocking.

Saplings respond well to release, in both diameter and height growth. Understorey saplings with crown ratios greater than 50% and those growing fastest before release are likely to be the best crop trees.

T. mertensiana grows slowly in height and in diameter. Mature trees range in height from 15 m on poor sites to 46 m on the best sites. Depending on stocking, the diameter of old trees ranges from 30 cm on poor sites to 150 cm on good sites.

Stands at least 200 years old can have high basal areas and volumes; the highest values are in mixed species stands. In the Mountain Hemlock zone of British Columbia, stands in which *T. mertensiana* makes up more than half the volume have volumes of 125–924 m^3/ha; volumes are much higher (588–1348 m^3/ha) in stands where the species makes up less than half (9–36%) the volume.

T. mertensiana communities in the western Cascades of Oregon and Washington, which commonly have a mixture of other tree species, have an estimated annual productivity of 3.8–7.6 m^3/ha.

Diseases and pests

T. mertensiana is very susceptible to laminated root rot (*Phellinus weirii*). The fungus, spreading from centres of infection along tree roots, kills all trees in circular areas that expand radially. It moves faster through nearly pure *T. mertensiana* stands than through mixed conifer stands. Growth and coalescence of root rot pockets can produce infected areas of >40 ha. Saplings are not susceptible to reinfection for 80–120 years.

Other common diseases and pests include several heart rots, of which Indian paint fungus (*Echinodontium tinctorium*) is the most common and damaging; snow mould (*Herpotrichia nigra*); and dwarf mistletoe (*Arceuthobium tsugense*). *T. mertensiana* is an occasional host for the western spruce budworm (*Choristoneura occidentalis*).

Uses

Perhaps the most important use of *T. mertensiana* is for watershed protection and recreation.

Its wood is generally somewhat inferior to that of *Tsuga heterophylla* for timber and pulp, although that from old-growth is considered more valuable. The wood has a density of 420–500 kg/m^3 and is used for sawn or hewn building timbers, for light construction, joinery, flooring, particleboards and short-fibre pulp.

T. mertensiana is widely grown as an ornamental and specimen tree in landscapes and gardens, where it is appreciated for its bluish-green foliage and tolerance of severe weather.

Further reading

Dahms, W.G. (1958) *Silvical Characteristics of Mountain Hemlock*. USDA Forest Service, Pacific North-west Forest and Range Experiment Station, Portland, Oregon, USA.

El-Kassaby, Y.A. and Edwards, D.G.W. (2001) Germination ecology in mountain hemlock (*Tsuga mertensiana* (Bong.) Carr.). *Forest Ecology and Management* 144, 183–188.

Peterson, D.W. and Peterson, D.L. (2001) Mountain hemlock growth responds to climatic variability at annual and decadal time scales. *Ecology* 82, 3330–3345.

Schuller, S.R. (1977) *Vegetation ecology of selected mountain hemlock* (Tsuga mertensiana) *communities along the eastern High Cascades*, Oregon. Oregon State University, Corvallis, Oregon, USA.

U

Uapaca kirkiana Muell. Arg.

Wild loquat

Importance

Uapaca kirkiana is a multipurpose tree with recognized economic significance and extensive local utilization wherever natural populations exist. Among the several products, the fruit is by far the most important basis for domestication of the species. This is widely appreciated throughout its range and made into several products, including jams, juices and alcoholic drinks. The multiple products accrued from the tree make it a particularly suitable candidate for agroforestry applications in the Southern African miombo ecozone. The tree is evergreen with a dense crown, which may adversely affect crops in some agroforestry applications due to heavy shading. Its dioecious character does not allow identification of male and female individuals prior to flowering.

Botanical features

An evergreen tree with a short trunk and spreading multiple branches forming a dense, rounded crown. The trunk is stout, attaining a height of 5–12 m and a dbh of 5–25 cm at maturity. Branchlets are short and stout with prominent leaf scars and the young shoots are typically covered with creamy-brown hairs. The bark is dark grey or grey-brown, thick and deeply fissured. The root system is a mass of small laterals and central sinkers emerging from the base of the stem, with a taproot length of 50–150 cm.

Leaves are simple, 7–36 × 4–24 cm in size and usually restricted to the ends of the branchlets in a spiral disposition. They are dark shiny green with short, grey to rust-coloured woolly (curly) hairs beneath when young, but become almost glabrescent on approaching maturity. Leaf shape obovate or subcircular–obovate, with the apex rounded and the base cuneate. Each leaf has an entire margin, with a prominent midrib (especially abaxially), and 12–24 pairs of parallel secondary nerves.

Inflorescence buds globose, borne on short slender peduncles originating from axillary positions among the leaves. On male trees, inflorescences are in fascicles, whereas on female trees they are solitary. In both cases, the inflorescences are pale yellow-green, similar in size except for the peduncles of the male inflorescences which are longer than the female. Male flowers occur in dense clusters; female flowers are solitary. Male and female flowers are borne on different trees. The ovary is ovoid–subglobose, 3–6 locular, about 3–4 mm across. The fruit is globose, yellowish-brown, about 2–4 cm across, with a tough skin containing 3–6 pyrenes, about 1–2 cm long and 1–1.5 cm thick.

Flowering and fruiting are essentially wet season events throughout the range. Male and female reproductive individuals flower synchronously during October–April (with a peak period in February–March). Anthesis in female inflorescences lasts almost twice as long as in male inflorescences. Trees can remain in flower for several months and fruit development is a 5–8-month process, commencing in the rainy season but extending through the dry season into the next rainy season. The species demonstrates a unique phenological feature: an appreciable overlap between the flowering and fruiting phases.

Distribution

U. kirkiana is typically a miombo woodland species of the Zambezian Regional Centre of Endemism and adjacent transitional centres. It occurs naturally from 2 to 21°S and from 14 to 40°E, almost from coast to coast across the African continent. It has been recorded growing naturally in Angola, the Democratic Republic of the Congo, Burundi, Tanzania, Malawi, Mozambique, Zambia and Zimbabwe.

Silviculture characteristics

U. kirkiana is principally grown as a fruit tree, but also provides other secondary products, including wood and medicinal products. It readily colonizes open gaps or woodlands and establishes well under full or partial overhead light. The species coppices well after cutting or pollarding and, in the wild, regenerates easily from seed or root suckers. It is fire-resistant. Burning late in the dry season, however, has adverse effects; controlled early burning is ideal for regeneration and establishment. *U. kirkiana* is weed tolerant and can compete well against natural vegetation. It is termite resistant and can withstand some drought.

Silviculture practice

U. kirkiana fruits are usually collected when fully mature, but prior to natural fall. Seed processing involves manual removal of the fruit pericarp and seed separation from the mesocarp by hand cleaning in water, followed by air-drying. The seed is classified as recalcitrant and cannot tolerate desiccation to below 26% moisture content. Viability decreases rapidly once processed and can only be prolonged by cold storage at freezing temperatures up to 3 weeks. Germination is epigeal and prospects are improved by pretreament procedures such as manual scarification or 24 h soaking in water at ambient room temperature.

U. kirkiana roots form ectomycorrhizal associations with several fungi and these associations improve growth in the nursery and are essential for successful establishment in the field. To ensure effective ectomycorrhizal inoculation of the seedlings, nursery potting soil from natural stands of *U. kirkiana* should be used. Individual seeds should be sown into prefilled polyethylene tubes, with the hilum facing downwards at a depth of 1–2 cm. Partial shade should be provided in early nursery stages to avoid high temperatures and direct light, followed by full exposure to the sun. Seedlings are water-sensitive, requiring regular watering to avoid wilting, and are ready for field planting within 6–10 months when a height of 10–20 cm is attained.

Conventional plantations are non-existent, but there is widespread interest in the use of *U. kirkiana* as an agroforestry species. Site preparation involving removal of vegetation reduces the danger of early mortality and provides optimum conditions for early establishment. Although a slow-growing tree, the growth rate is much enhanced if complete cultivation is undertaken following field planting.

Intensive management has not been applied to *U. kirkiana*. Currently, management exclusively entails early burning to stimulate regeneration and reduce the intensity of dry season bush fires. In pure natural stands, thinning applications to open up the canopy improve yield and quality of the fruits. A population sex ratio in favour of female individuals is ideal for maximum fruit production, but male to female dispositions should be properly designed for sufficient pollination.

Trees retained or planted in cultivated fields in association with crops are subjected to pollarding or coppicing options to reduce competition with crops or to obtain wood products. Trees with good fruiting characteristics such as heavy load, regularity, large size and good flavour are retained, but male trees and female individuals with doubtful reproductive potential are removed.

In planted stands, fruiting occurs within 9–10 years, with yields of up to 3000 fruits per tree in older stands. Mature and ripe fruits fall naturally and are easily collected from the ground. Breaking off branches or branchlets to collect fruits is not recommended as it adversely affects subsequent yield (fruits are borne only on the branchwood of the previous one or two seasons).

Diseases and pests

To date, no serious diseases and pests have been reported to attack *U. kirkiana* in natural populations or planted stands.

Necrosis linked with *Pestalotiopsis versicolor*, leaf spots caused by *Cercospora* spp., and mildews and sooty moulds such as *Cladosporium cladosporioides* have been recorded on foliage.

Insect pests *Ceroplastes uapacae*, *Ledaspis* spp., *Microsyagrus rosae* and *Euphoria* spp. may cause 10–20% foliar damage. Low arthropod damage in mixed woodlands and fire-prone areas suggests that establishment of mixed woodlands and early controlled burning can be used to minimize pests.

Uses

U. kirkiana trees are retained by farmers in cultivated fields with various crops, as well as within homesteads. The evergreen dense crown provides good shade during dry and hot seasons. Its ectomycorrhizal association with fungi is important in phosphorus supply and nutrient recycling to enrich the soil. As a dominant tree species in hilly sites, it is useful in watershed management and plays an important role in erosion control.

U. kirkiana wood is durable with a density of 600–719 kg/m^3 at 12–15% moisture content. It has low susceptibility to insect and fungal attacks and is locally used for construction purposes, domestic utensils, furniture, joinery, boxes and carvings. The species is also a useful source of good-quality fuelwood and charcoal.

The fleshy pulp of the ripe fruit is edible and eaten fresh or processed into a wide range of refreshments, including juices, squashes and wines. The leaves, bark and roots are widely used in the preparation of traditional medicines to cure indigestion and dysentery. The thick, broad leaves are used locally to make containers for storage of processed food. A blue dye is made from the roots and the wood ash is used in seasoning food. The flowers are an important bee forage.

Amanita, *Cantharellus*, *Lactarius* and *Russula* are the commonest genera of fungi which typically form ectomycorrhizal associations with *U. kirkiana*, producing mushrooms that are widely collected and consumed or sold in markets during the rainy season.

Further reading

Mwase, W.F., Akinnifesi, F.K., Stedje, B., Kwapata, M.B. and Bjørnstad, Å. (2010) Genetic diversity within and among southern African provenances of *Uapaca kirkiana* Müell. Årg using morphological and AFLP markers. *New Forests* 40, 383–399.

Ngulube, M.R., Hall, J.B. and Maghembe, J.A. (1996) A review of the silviculture and resource potential of a miombo fruit tree: *Uapaca kirkiana* (Euphorbiaceae). *Journal of Tropical Forest Science* 8, 395–411.

Ulmus americana L.

American elm

Importance

Owing to its majestic size and graceful habit, *Ulmus americana* was a common landscape tree of eastern North America until Dutch elm disease, caused by the fungus *Ophiostoma ulmi*, devastated urban and rural populations. Dutch elm disease was introduced into the USA from Europe in 1930, and spread quickly throughout most of the natural range of *U. americana*, killing tens of millions of trees. *U. americana* was the principal tree planted along boulevards and in parks during the late 19th and early 20th century, often to the exclusion of other species. The rapid loss of so many elms, especially in urban areas, was especially conspicuous and led to heroic, but mostly failed, efforts to control the disease. In addition, an important timber tree of bottomland and river valley forests was lost. *U. americana* continues to regenerate profusely throughout much of its range, but trees seldom grow beyond sapling or pole size before succumbing to the now-naturalized disease.

Botanical features

U. americana is a large tree, reaching 30–38 m tall and 150 cm in diameter on good sites, but such trees are rarely encountered today. In dense forests, stems are very straight and well-pruned to 15–18 m, forming spreading crowns that emerged above the canopy. When open-grown, *U. americana* often forms vase-shaped crowns with pendulous outer branches.

U. americana has simple, alternate, medium-to-large (8–15 cm), deciduous, doubly serrate leaves, often appearing 'two-ranked' along vigorous shoots. The leaf base is asymmetrical, and leaf surfaces are smooth-scabrous above and slightly pubescent beneath. Mature shoots are brown-grey, usually glabrous, zigzag in outline, with small buds (4–6 mm) having two-tone (red-brown/chestnut-brown) imbricate scales. The 'true' terminal bud is absent, with the 'apparent' terminal bud borne adjacent to a shoot scar.

Flowers are perfect, in 3–5-flowered fascicles on long pedicels, appearing before the leaves in February–May, depending on climate. *U. americana* is wind-pollinated, with pollen shed 1–2 days before stigma receptivity. Self-incompatibility is common. The fruit is a wingless samara with ciliate margins and the ovary borne near the centre. Seeds are dispersed during April–June, with seeds germinating 3–10 days later. *U. americana* is a prolific seed producer with some seed produced every year and good–excellent crops every 1–3 years. Seeds are primarily wind-dispersed, but water dispersal is also known and may contribute to the species' widespread occurrence in seasonally flooded bottomland forests.

The root system of *U. americana* is widespread but tending to be shallow, with absolute rooting depth dependent on soil texture and water table. In bottomland forests, the root system may be within 1 m of the soil surface, but on deeper, well-drained soils, roots may penetrate to a depth of 5–6 m.

Distribution

U. americana has one of the largest natural distributions of any eastern North American broadleaf, occurring from Prince Edward Island south along the Atlantic coast to central Florida, westward to central Texas and Oklahoma and northward to eastern Montana and Saskatchewan.

U. americana has been much planted outside its natural range, primarily as an urban tree. It is likely to have been planted in every state in the USA, most north temperate countries, and Australia, New Zealand and South Africa during the past century. However, its popularity has decreased with the spread of Dutch elm disease.

It is a major component of a number of bottomland forest types and almost always occurs in mixed stands with silver maple (*Acer saccharinum*), sugarberry (*Celtis laevigata*), green ash (*Fraxinus pennsylvanica*), sweetgum (*Liquidambar styraciflua*), sycamore (*Platanus occidentalis*) and several oaks (*Quercus* spp.). In parts of the central and southern Appalachians, *U. americana* occurs in 'cove' forests up to 750 m altitude but is much more common below 400 m.

Silviculture characteristics

U. americana is considered a mid-tolerant species with respect to shade; it survives in the shade of early successional species such as cottonwood (*Populus deltoides*), but fails under the dense shade of beech (*Fagus grandifolia*) and maple (*Acer* spp.). Seedlings can become established on litter, logs and organic soils but germination and early growth appear best on mineral soil in partial shade to full sunlight. In the southern USA, *U. americana* regeneration success appears related to the timing of weather events, with seedling abundance reduced in years of heavy spring floods or early summer droughts. Seedlings tolerate flooding and poor soil aeration during the winter and early spring, but die if flooding is prolonged during the growing season. *U. americana* is also capable of colonizing old fields in much of its range, but is very sensitive to fire, a feature that probably kept it from spreading to prairies and other rich sites in the western portion of its range.

Silviculture practice

Given the enormous losses incurred by *U. americana* beginning in 1930, silvicultural practices for the species today are essentially non-existent. Foresters have often marked small sawtimber *U. americana* trees for harvest assuming that they will succumb to Dutch elm disease before the next harvest cycle. Thus, trees seldom develop large diameters, and a 'salvage' mentality dominates many silvicultural practices involving the species.

U. americana was a notable component of several forest cover types dominated by early successional species. Current silvicultural practice for regenerating these species emphasizes even-aged management, especially clear-cutting and shelterwood systems. *U. americana* regenerates well under these conditions and is easily perpetuated by practices which provide full sunlight and disturbed conditions, especially on alluvial soils.

Despite the loss of the vast majority of large, forest-grown *U. americana* trees in eastern North America between 1930 and 1990, natural regeneration remains extensive. Frequent and copious seed crops, together with large amounts of disturbed agriculture and forest land, provide numerous opportunities for regeneration. *U. americana* is a frequent colonizer of gaps created by dead elms, of old fields and of virtually any disturbed site in riparian or bottomland forests. Unlike American chestnut (*Castanea dentata*), another species devastated by an exotic pest (*Cryphonectria parasitica*) and seldom found in eastern forests today, *U. americana* remains abundant, although restricted to smaller diameter classes than formerly.

U. americana is essentially unmanaged in natural stands. As noted above, elm is salvaged, whether infected with Dutch elm disease or not, when it occurs on sites designated for selective harvest. It is easily regenerated with even-aged silvicultural systems, although seldom growing to a large size. It is discriminated against in most silvicultural activities, and is seldom a factor in management decisions for eastern North American forests today.

Some municipalities, especially at the fringes of the range of *U. americana*, continue to remove diseased and dying elms promptly and have reduced annual mortality to less than 2–3%.

Diseases and pests

Numerous methods for protecting *U. americana* from Dutch elm disease have been attempted since 1935, including insecticide sprays against elm bark beetles, trenching of root barriers to disrupt below-ground spread via root grafts, and various chemical injection protocols. In rural areas, large numbers of elms die from Dutch elm disease each year, testimony to the presence of naturalized, virulent strains of *O. ulmi* throughout eastern North America.

U. americana is host to a variety of other pathogens and insect pests, but most of these are insignificant in comparison with Dutch elm disease. Verticillium wilt caused by *Verticillium albo-atrum*, and the phytoplasma elm yellows both posed serious problems in the past, especially in urban areas. Elm yellows is characterized by chlorotic foliage, witches' broom formation, shoot and root die-back, reduced vigour, and occasional mortality. No practical control measures have been developed for these diseases, and both continue to pose problems for remaining trees.

In addition to the elm bark beetles (*Hylurgopinus rufipes* and *Scolytus multistriatus*) that serve as vectors for Dutch elm disease, and elm leafhoppers (*Scaphoideus luteolus*), the vector for elm yellows, a number of other insects cause damage to *U. americana*. The elm leaf beetle (*Pyrrhalta luteola*), introduced from Europe during the 1830s, is the most serious, continuing to defoliate American and other elms in North America each year. Additional pests such as scales (Order Homoptera), spider mites (Order Acarina), and other beetles, sawflies, and borers posed locally serious problems in the past, but many were successfully controlled with synthetic pesticides during the mid-20th century.

Uses

U. americana has fallen into disuse given the magnitude of losses from Dutch elm disease. It is seldom propagated or planted today, although a few cultivars tolerant of Dutch elm disease have become available in recent years. Its use in windbreaks is nil today as it has been replaced by Siberian elm (*Ulmus pumila*). At present, it appears unlikely that *U. americana* will ever again enjoy its wide popularity of a century ago, although efforts to develop resistant trees by conventional breeding as well as biotechnological means continue.

The sapwood of *U. americana* is nearly white, whereas the heartwood is light brown to brown with a reddish tinge. The wood is coarse-grained, heavy and strong with a density of 460–560 kg/m^3. It lacks durability, warps and splits badly in seasoning. The wood is used in the manufacture of boxes, baskets, crates, barrels, furniture, agricultural implements and caskets. Elm veneer is used for furniture and decorative panels.

Further reading

Campanella, T.J. (2003) *Republic of Shade: New England and the American Elm.* Yale University Press, New Haven, Connecticut, USA.

Dunn, C.P. (2000) *The Elms: Breeding, Conservation and Disease Management.* Kluwer Academic Publishers, Dordrecht, The Netherlands.

Ulmus minor Mill.

European field elm

Importance

In the past, every part of *Ulmus minor* was utilized for nourishment, as fodder for cattle, for artefacts and in popular medicine. The Roman system of grapevine growing, called 'marriage', used pollarded elm to furnish living support to vines while producing valuable fodder during the dry summer.

The main use of *U. minor* today is as an amenity tree because of the beauty of its canopy, its fast growth and its tolerance of a range of unfavourable urban conditions.

The wood is aesthetically pleasing and rather easy to work, making it suitable for furniture manufacture.

Because of the persistence of Dutch elm disease, *U. minor* seems destined to survive at present only as a shrub.

Botanical features

U. minor may attain 40 m in height and 150–200 cm dbh. Trees are relatively fast-growing and erect, with ascending and spreading branches and dense foliage. Twigs are slender and glabrous. Morphological features exhibit extreme variation within this wide-ranging species. Thus, the canopy varies in shape from umbrella or ball to narrow and slender. The bole may be short or extend up to half the total tree height. *U. minor* is distinguished from *Ulmus glabra* and *Ulmus laevis* by the presence of root suckers.

The bark, smooth when young, soon becomes rough and fissured into vertical ridges. Vigorous twigs and branchlets may present corky wings. The root system is well developed and extends as deep as 150–160 cm, even in compacted soil.

Leaves, which in all elm species are characterized by bilateral asymmetry and a double-toothed margin, are variable in shape, from obovate–acute to nearly orbicular, 5–10 cm long and 2.5–5 cm wide, with 9–12 pairs of lateral veins frequently forked near the margin. Pubescence is variable. The petiole is relatively long (9–15 mm).

Very small bisexual flowers with very short peduncles occur in clusters of 15–35. The perianth is cut in 4–6 ciliate lobes, with 4 stamens and an ovary presenting a prominent bifid gland-covered stigma. Pollen, which is allergenic, may be wind-borne for distances up to 1500 km. Maturing of the male and female parts of the flower is non-simultaneous. Self-incompatibility is very high. Flowers are visited by bees.

The fruit is a samara, with a short peduncle, broadly to narrowly obovate, smooth, 13–23 mm long, 8–13 mm wide with seed located in the centre.

Leaves turn yellow before falling in autumn. Flowering is at the end of winter (February–March) before the leaves appear. Seeds ripen a couple of months after pollination and germinate immediately. Germination is epigeal.

Distribution

Ulmus minor is found over most of Europe with the exception of the north, the northernmost location being the Swedish islands of Íland and Gotland. It occurs on the majority of Mediterranean islands and in Algeria, and there are detached outliers in Israel, Lebanon and Syria. In Asia, it occurs along the Black Sea coast of Turkey, extending via Transcaucasia to the forested slopes of the Elburz Mountains south of the Caspian Sea. It is more abundant and variable in the Mediterranean area, but it is also a prominent component of riverain woods flanking the large rivers of central Europe. Field elm hybridizes with *U. glabra*, particularly along its northern geographical limit, giving rise to the natural hybrid *Ulmus × hollandica*.

Silviculture characteristics

U. minor is valued for its capacity to tolerate various and often contrasting stresses, such as waterlogging and drought. It is an important pioneer species which colonizes abandoned farmland and calcareous, clay soils on slopes in the Mediterranean area. It is light-demanding and so is more often found at the forest edge than in the interior, but it can tolerate subdued light in mixed broadleaved forest. In the vegetation alliance '*Querco–Ulmetum minoris*', *U. minor*, oak (*Quercus robur*) and ash (*Fraxinus excelsior*) represent the climax stage of the successional status of alluvial forest stands.

U. minor is often found along roadsides, as large trees in hedges and along motorway and railway embankments. It contributes to soil protection and soil improvement through production of high-quality, easily-degradable litter. Because of its ability to produce root suckers even at several metres distance from the stump and to form connections between contiguous root systems, *U. minor* is often found growing in groups of several associated individuals. It is seldom found in pure stands.

Silviculture practice

Propagation of *U. minor* by seed is simple, although many seeds are empty. Full seeds exhibit over 90% germination. When sown immediately after collection, germination occurs within 15 days. It takes 2 years in the nursery to get seedlings of plantable size (over 1 m high). Viability is lost rapidly unless seed is preserved under controlled conditions (10% humidity and 2–4°C) in sealed bottles.

Vegetative propagation is generally feasible using softwood or hardwood cuttings, particularly if stump or root suckers are collected, as these exhibit better rooting ability. Plantable stock is obtained at the end of the rooting year when using winter hardwood cuttings, and at the end of the second year if softwood cuttings are used. Recalcitrant individuals may be propagated by root cuttings or by grafting. *In vitro* micropropagation is possible.

U. minor is an easy tree to transplant and tend because it recovers well from any kind of injury and damage. Plantations are generally established using bare-rooted stock. Normal procedures are sufficient in the nursery, as is field tending during the first years. Shelters or fencing are required while trees are young.

On moist, fertile soils, *U. minor* is a fast-growing tree reaching useful dimensions by 60 years. In mixed broadleaved forests, *U. minor* is a rare, high-value, fast-growing tree which is managed in the same manner as the other 'noble' hardwoods.

Unfortunately *U. minor* is rarely planted nowadays due to the persistence of the Dutch elm disease pandemic, so forest management measures are of purely academic interest.

Diseases and pests

As a result of Dutch elm disease epidemics in the 1930s and 1970s caused by *Ophiostoma ulmi* and *Ophiostoma novo-ulmi*, respectively, virtually all mature *U. minor* trees in Europe have been wiped out. As trees are not attacked by the disease until 5–20 years old and 3–10 m high, and there is a pronounced ability to multiply by both seed and root suckers, it appears *U. minor* is not threatened with extinction but will survive as a shrub for the foreseeable future. *O. novo-ulmi* is very variable, highly aggressive and transmitted by bark beetles of the genus *Scolytus* or through elm root connections. The size of the beetles, their population dynamics and the factors which attract them to elm trees are all decisive in determining the frequency of new infections. Protection measures against Dutch elm disease have only been partially successful, and appear suitable for only a limited number of individuals. Conventional breeding to incorporate highly disease-resistant parent material from Asia has proved more successful, but has focused on elm as an ornamental tree. An alternative approach being investigated is the direct introduction of resistance genes through biotechnological means.

Another fungus, *Verticillium dahliae*, causes branch dieback, while under conditions of high humidity, *Systremma ulmi*, *Mycosphaerella ulmi* and *Gloeosporium inconspicuum* cause leaf spots, and *Nectria cinnabarina* produces cankers that kill branches.

Among the nearly 100 insect species attacking elm, the most significant are the species of scolytid beetles which are Dutch elm disease vectors. In the Mediterranean basin, *U. minor* is one of the preferred hosts of *Metcalfa pruinosa*, while larvae of the leaf beetle *Xantogalerucha luteola* severely weaken and debilitate trees during their cyclical increases in population every 3–4 years. Considerable damage is also inflicted by larvae of *Cossus cossus*, although it is not a particular pest of field elm.

Uses

In Spain, Portugal and Italy, *U. minor* was traditionally heavily pollarded to provide living support for grapevines. It provides fodder during the dry summer period and produces fuelwood and smallwood for poles when coppiced regularly. Rows and belts of elm trees act as windbreaks.

The yellowish sapwood is sharply differentiated from the dark heartwood in freshly cut *U. minor* timber; reported wood density is 551 kg/m^3. Summer vessels laid down in wavy bands confer a characteristic zigzag pattern between the main rings of spring vessels. The wood is tough and rather heavy, and difficult to split when the grain is crossed. When free from knots it bends easily. Seasoning is slow (2–3 years) and difficult, partly because of a tendency to warp. The wood is attractive and easily worked, making it suitable for the furniture industry both as veneer and solid wood. Because of its decorative value and toughness, it has also been used for turnery, particularly when marbling is present. Elm timber also possesses exceptional resistance to decay when in continuous contact with water, which has led to its use for underwater piles, in shipbuilding and as water pipes. The wood holds nails well and is used for pallets and heavy cases. It is also used for parquet flooring.

Where it has survived Dutch elm disease attacks the main use of *U. minor* today is as an amenity tree. Early flowering also makes it an important source of pollen for honeybees during early spring.

Further reading

Dunn, C.P. (2000) *The Elms: Breeding, Conservation and Disease Management*. Kluwer Academic Publishers, Dordrecht, The Netherlands.

Richens, R.H. (2012) *Elm*. Cambridge University Press, Cambridge, UK.

Sticklen, M.B. and Sherald, J.L. (eds) (1993) *Dutch Elm Disease Research: Cellular and Molecular Approaches*. Springer-Verlag, New York, USA.

Wilkinson, G. (1978) *Epitaph for the Elm*. Hutchinson, London.

V

Virola surinamensis (Rol. ex Rottb.) Warb.

Banak

Importance

Virola surinamensis, which is known colloquially as banak, baboonwood and wild nutmeg, produces a timber in Amazonia which is of high economic importance, and its export volume is second only to that of big-leaved mahogany (*Swietenia macrophylla*). The species is threatened by uncontrolled logging in its native range, but shows several characteristics that would enable its management to be sustainable. Tree improvement programmes have been recommended.

Botanical features

V. surinamensis is a medium-sized tree, attaining a height of 30 m with a dbh of more than 1 m. The species has a small crown and few branches; the primary axis is orthotropic, with limited growth, spiralled phyllotaxy and lateral inflorescences. The bark is yellowish-brown, smooth, slightly vertically striate, slightly breakable, with exudate liquid transparent. Terminal branches are flexuous, brown, tomentose then glabrous, with petiole strongly caniculate, 2–9 mm long.

Leaves are coriaceous, oblong, with parallel margins, 10–22 cm long, 2–5 cm wide, subcordate, obtuse or acute at the base, acute or shortly acute at the apex; leaf underside is pubescent. There are nerves in the upper side of the leaf, with 16–30 secondary nerves on each side of the leaf.

Male inflorescence is paniculate, slightly forked, 17 mm long and 7 mm wide; peduncle generally flat, 4–5 mm long, bracts oblong and 3–8 mm long, 5–20 flowers in each fascicle; pedicels up to 4 mm long; perianth fleshy, thin or sub-coriaceous, 1.6–2.4 mm long, lobes 3, oblong, obtuse, 3 mm long. Female inflorescence 2–11 cm long at anthesis, with 3–10 flowers per fascicle; pedicels and bracts up to 3 mm long; ovary subglobose, complete or densely pubescent or brown tomentose, stigma conspicuous, strongly lobate. Infructescence glabrous, 6–11 cm long, with 3–8 ripe fruits in each, pedicelate, ovoid or subglobose, 13–21 mm long and 11–18 mm wide, apiculate in the apex, slightly stipitate in the base, slightly carinate; pericarp coriaceous, 1–2 mm wide; aril lobate, almost down to the base. Fruits coriaceous, ellipsoid or subglobose. Seeds ellipsoid.

In French Guiana, *V. surinamensis* flowers twice a year in March and October, the fruits form in May–June and November–December. In the Brazilian Amazon, it flowers from August to early November, and fruiting occurs from January to July. Seeds are dispersed mainly by birds, but also by mammals.

Distribution

V. surinamensis is broadly distributed throughout Central America, the Antilles and the northern half of South America, from Guadeloupe, Granada, Trinidad and Tobago in the Caribbean, to the Amazon, including parts of Guyana, French Guiana, Suriname, Venezuela, Peru, Colombia, Bolivia and Brazil. In Brazil, it is present in all of Amazonia including the states of Maranhao, Ceará, Goias, Pará, Amapá, Roraima and Amazonas.

V. surinamensis grows at up to 2000 m altitude. Its natural habitats are subtropical and tropical moist lowland forests, subtropical and tropical swamps, and heavily degraded former forest. Although the species is listed as globally threatened, due mainly to habitat loss and uncontrolled logging, it is one of the most common trees in lowland Suriname, being found in almost all forest types, including secondary forest. It grows in association with *Symphonia globulifera*.

Plantations have been established in Brazil and Suriname.

Silviculture characteristics

V. surinamensis is light-demanding and is able to tolerate flooding. It produces regrowth after harvesting. It is suitable for agroforestry systems in varzeas, with uniform spacing and crops of nitrogen-fixing legumes.

V. surinamensis prefers sandy or clay-loam soils which have moderate acidity, are fertile and have free drainage, in humid or very humid environments with uniformly high rainfall. It tolerates waterlogging.

Silviculture practice

Seeds are recalcitrant and rapidly lose their viability in 5–20 days; germination takes approximately 28 days. Seedlings should be planted out 60 days after germination when 40 cm in height; bare-root planting is possible, achieving 90% survival at 2 × 2 m spacing. Two methods for planting have been suggested: 'homogenization' and agroforestry. The first consists of the progressive elimination of trees with no economic value, to allow the growth of desired species, giving more room to the trees for fast growth to commercial size. The second consists of planting *V. surinamensis* at regular spacing within a field planted with annual crops, such as nitrogen-fixing legumes.

V. surinamensis grows vigorously up to 25–34 cm dbh, then stops due to shading by other species, and only grows at an adequate rate if light is supplied through management. Fertilizer application is considered necessary in order to guarantee biomass increase.

Studies of natural stands in the Amazon estuary showed that swamp forests had a lower commercial volume/ha (up to 56.8 m^3/ha) than areas of floodplain (bottomland) forest (144.3 m^3/ha). The number of trees/ha was higher in swamp forest (265/ha) than in floodplain (10–17/ha). The exploitation in the floodplain was extensive, occupying great areas, while that in the swamp forest was intensive. Manual logging was predominant, with low productivity (10 logs/day) in relation to mechanized logging (18 logs/day).

Experience with plantations showed that, in an upland plantation at 2 × 2 m spacing, after 2 years there was 37.4% mortality, and only a 19.4% increment and irregular stems. After thinning at 12 years old, trees only attained 9.5 m in height and had a volume of 66.4 m^3/ha. A plantation in a varzea floodplain with 3 × 1.5 m spacing was selectively harvested; after 15 years it only attained 3.6 m in height and had a volume of 19.7 m^3/ha. In a plantation in Suriname with 7 × 2 m spacing, after 10 years the trees attained a mean height of 16 m and a volume of 208 m^3/ha. After 12 years this plantation showed vigorous growth, and estimates showed that a yield of 440 m^3/ha could be expected after 40 years, equivalent to 11 m^3/ha/year.

Diseases and pests

The outer heartwood of *V. surinamensis* is non-resistant to decay fungi such as *Polyporus versicolor* and *Poria monticola*. Wood moulds and wood stain fungi are associated with stored logs, the main species being *Aspergillus flavus*, *Lasiodiplodia theobromae*, *Paecilomyces variotii*, *Penicilllium citrinum* and *Trichoderma harzianum*. The wood-decomposing hymenomycete fungi *Coriolopsis occidentalis*, *Pleurotus ostreatus*, *Pycnoporus sanguineus* and *Schizophyllum commune* have also been found on logs in Amazonian sawmills.

Insect pests causing damage to the wood include beetles (Coleoptera) and termites (Isoptera).

Uses

V. surinamensis is used in traditional agroforestry systems in Brazil as a source of firewood and charcoal for commercial purposes.

The wood is of low durability, lasting not more than 3 years when exposed to weather. It has a density of 550–650 kg/m^3 and is very easy to work by hand or machine. Although it has a good finish, it is difficult to glue and paint. The wood is widely used for veneer and also in the manufacture of coffins, matches, furniture, boxes, plywood and broom handles.

V. surinamensis is used in pulping to make good-quality kraft paper. The species has potential as a source of cellulose as a by-product of the lamination process. The cellulose content is similar to that of *Eucalyptus saligna*.

The bark, wood, roots and leaves of *V. surinamensis* yield chemicals such as flavonoids and lignans; these have medicinal properties, including some with anti-malarial, antifungal, trypanosomicidal, and antiulcerogenic activities. The leaves, bark, latex and resins are used in traditional medicine to heal stomach ache, intestinal pains, sunburn and swellings. Native peoples of the Amazon prepare hallucinogenic snuff and smoke from the blood-red resin of the inner bark.

A fat known as 'ucuuba' is obtained from the seeds and is used in soap- and candle-making.

Further reading

Gurgel, E.S.C., Carvalho, A.C.M., Santos, J.U.M. dos and Silva, M.F. da (2006) *Virola surinamensis* (Rol. ex Rottb.) Warb. (Myristicaceae): aspectos morfológicos do fruto, semente, germinação e plântula (*Virola surinamensis* (Rol. ex Rottb.) Warb. (Myristicaceae): morphological aspects of fruit, seed and seedling). *Boletim do Museu Paraense Emílio Goeldi, Ciências Naturais* 1, 37–46.

Jardim, M.A.G. and Mota, C.G. da (2007) Biologia floral de *Virola surinamensis* (Rol.) Warb. (Myristicaceae) (Floral biology of Virola surinamensis (Rol.) Warb. (Myristicaceae)). *Revista Árvore* 31, 1155–1162.

Neves, E.J.M., Santos, Á.F. dos and Martins, E.G. (2002) *Virola surinamensis: Silvicultura e Usos (Virola surinamensis: Silviculture and Uses)*. Documentos - Embrapa Florestas No. 80, Colombo, Brazil.

X

Xylia xylocarpa (Roxb.) W. Theob.

Irul

Importance

Xylia xylocarpa is a deciduous, medium to large-sized tree around 25–30 m tall, maximum 40 m, with a straight, cylindrical bole, sometimes fluted. Growth is fairly rapid in young trees. It produces a hard and strong timber which, even though difficult to work because of its hardness, is very durable, making it popular as a construction timber. It has long been important for the manufacture of railway sleepers. As a leguminous species, *X. xylocarpa* is able to fix atmospheric nitrogen. It has potential as a plantation tree.

Botanical features

X. xylocarpa has a dense crown and twiggy branchlets. On rich soil under natural conditions it attains a height of 25–30 m and a dbh of 60–70 cm. The main bole can be straight and cylindrical, but on drier sites it will be a smaller tree often with a crooked, forked or fluted stem, especially when grown in open areas.

Leaves compound, bipinnate, petiolate, alternate, 5–25 cm long; leaflets entire, coriaceous, cuneate–obtuse or rarely truncate at base, acute, acuminate or rarely obtuse at apex; petioles 2–6 mm long, thick, glabrous.

Inflorescences axillary, fascicled or racemose, in dense globose heads 1–2 cm in diameter. Flowers are scented and very conspicuous in the flowering season, with creamy-white, light yellow or yellowish-white corolla, 3–5 mm long and 1–3 mm wide. The tubular, campanulate calyx is 2–4 mm long and 5 lobed. Fruits (pods) are oblong–falcate, oblong or broadly falcate, woody, 10–17 cm long, 3–7 cm broad, rusty-tomentose with 3–12 oblong–ellipsoid, smooth seeds.

X. xylocarpa trees shed their leaves either partially or fully from the beginning of February until the end of April. Flowering begins by March and continues up to May. Fruits ripen by December and seeds are dispersed by elastic dehiscence of the woody pods, leaving the two woody valves on the tree, to be shed during the next season.

Distribution

The species is restricted in its natural distribution to the tropical parts of India, Bangladesh and Myanmar. It is widespread in Thailand and Indo-China. Before the segregation of *Xylia kerrii* as a separate species, *X. xylocarpa* was also reported to be naturally distributed in Malaysia.

It has been introduced to parts of East and West Africa.

X. xylocarpa occurs on a wide range of tropical sites, including moist bamboo forest, dry evergreen forest, mixed deciduous forest and dry deciduous dipterocarp forest. It occurs up to an altitude of 600 m.

Silviculture characteristics

X. xylocarpa is shade-tolerant, particularly when young, and it owes its gregariousness in part to this characteristic, which enables it to gain a footing under the moderate shade of mixed deciduous forest. Seedlings appear to be sensitive to drought and should be protected from fire, as the degree of fire-resistance is disputed. The species is also frost-tender.

It is not too exacting as to soil conditions, growing fairly well on dry, shallow or rocky soils. It attains its largest sizes on the banks of hill streams.

X. xylocarpa coppices and pollards well, with 95% of stools producing coppice shoots and 85% of pollarded trees regenerating well. Root suckers are produced in abundance, particularly in burnt areas and wherever the roots are exposed.

Silviculture practice

Under natural conditions in India, the seeds of *X. xylocarpa* germinate with the early showers of the rainy season in May. The natural regeneration of this tree is plentiful throughout the greater part of its range. Regeneration from coppice and root suckers is also profuse. Natural regeneration can be enhanced by slightly opening the canopy around individual seed-bearing trees and burning the debris; seedling establishment is favoured on loose, bare, well-drained soils under shade. Gradual opening of the canopy is necessary for further development of established seedlings; weeds will only be a problem if the canopy is opened too fast.

Seeds are available for collection in March–April. Ripe pods are gathered from the tree before they dehisce, and then dried under the sun in cloth bags, where they dehisce and release the seeds. Seed can be stored under dry conditions without loss of viability for about 3 months, possibly up to 1 year. About 3200–4000 seeds weigh 1 kg. Seeds pre-soaked in cold water germinate in 4–11 days, at a rate of 70–90% germination.

Although plantations of *X. xylocarpa* are rarely raised, stands can be established by direct sowing, planting stock or by stump planting. Direct sowing is the best method, giving higher survival rates and height growth. Bare-rooted nursery seedlings do not survive planting out well, owing to injury to the taproot, which develops very early. Stump plants aged 8–10 months old can be used successfully, but are more costly.

Direct sowing in well-loosened patches at 2 m intervals in rows 3 m apart is advocated. Weeding is necessary during the initial years of establishment.

X. xylocarpa in natural forests of India is worked under a selection system, simple coppice system, or coppice with standards. Where the selection system is followed, the exploitable girth is 1.35–1.5 m; trees reach this dimension in about 95 years. In considerable areas of its habitat, the mixed deciduous forests containing it are worked chiefly for fuelwood and charcoal under the simple coppice or coppice with standards system. Coppice reproduction comes up quite vigorously, producing 1–20 shoots per stool. Thinning of coppice shoots to one or two per stool is beneficial in improving their development. The optimum size for fuelwood is 20 cm stem diameter, and this is reached in about 40–45 years.

General tree dimensions recorded are 5.7 m tall and 8 cm diameter at 5 years old in India, and 16 m tall and 19 cm diameter at 10 years old in Bangladesh. Growth in high forests tends to be slow, whereas under plantation conditions it is fast.

Diseases and pests

X. xylocarpa is not affected by any serious disease. However, during monsoon periods, leaf blight caused by *Glomerella cingulata* has been observed in natural forests, as has seedling blight caused by *Rhizoctonia solani* in nurseries. The leaf blight starts with the appearance of brown spots on leaves, which enlarge in size within a few days and turn dark brown with a yellowish margin. Later the spots coalesce covering large areas of the lamina, resulting in blight and defoliation. In the case of seedling blight, leaves initially turn light yellow and within a week affected leaves and apical portions of plant show wilting.

Butt rot fungi (such as *Fomes fastuosus* and *Polystichus* spp.) also attack trees, entering via stem fire scars.

Over 50 species of insects including defoliators, bark feeders, root feeders and seed borers are associated with living *X. xylocarpa*. Although causing minor damage, none of them is a threat to the establishment of nurseries or plantations. *Caryedon serratus* (Coleoptera: Bruchidae) damages seeds in storage.

Uses

Due to its excellent coppicing and root-sucker formation characteristics, *X. xylocarpa* is suitable for reclamation of degraded forests through natural regeneration. Naturally regenerating *X. xylocarpa* trees are often retained in home gardens in Kerala, South India, for domestic timber production.

The wood is hard, heavy (air-dry density about 850 kg/m^3) and strong, with a medium-fine texture, interlocking grain, and a very durable, dark reddish-brown heartwood. It requires great care in seasoning to prevent cracking, splitting and warping, and slow drying schedules are recommended. The heartwood is comparable to teak in termite resistance.

Because of its strength and hardness, the timber is often used for railway sleepers and heavy construction, such as bridges, piles, beams, rafters, mine timbers and for harbour works. It is also used for construction of railway wagons and in shipbuilding, and is generally preferred for structures subject to wear and rough usage such as fenders, gangways, chutes and paving blocks. The timber is also widely used for making agricultural implements, country carts, well-curbs, canoes, tool handles and oil presses. It is also used as roundwood for transmission poles, pit props and building poles.

The wood is a very good fuel of high calorific value and its charcoal is highly prized for iron smelters. The wood is also utilized, depending on availability, for mixed hardwood pulp which is used for producing wrapping paper.

The bark and the wood yield a tanning material and the seeds an oil. The woody pods are used as fuel for cooking in southern India. Bark and fruits are also used in local medicines.

The seeds are edible and considered an under-utilized food source. They contain over 29% crude protein and over 14% crude fat, and are eaten after roasting.

The leaves are used as a green manure and as browse.

Further reading

Arora, R.K. (1960) Ecology of *Xylia xylocarpa* Taub. *Indian Forester* 86, 306–313.

Sosef, M.S.M., Hong, L.T. and Prawirohatmodjo, S. (eds) (1998) *Plant Resources of South-East Asia. No 5(3). Timber trees: Lesser-Known Timbers.* Backhuys Publishers, Leiden, The Netherlands.

Troup, R.S. (1921) *Xylia*, Benth. In: Troup, R.S. (ed.) *The Silviculture of Indian Trees, Volume II. Leguminosae (Caesalpinieae) to Verbenaceae.* Oxford University Press, Oxford, UK, pp. 402–418.

Z

Zelkova serrata (Thunb.) Makino

Japanese zelkova

Importance

Zelkova serrata (Plate 45) is an important East Asian timber tree, whose wood, with its special figured grain, has a very high commercial value. The foliage makes the species ornamental, being light green in spring and reddish-yellow in autumn. Established trees have a good form and are tolerant of heat, drought, wind and urban conditions, making them popular for landscaping and use in towns. The species is widely used as a subject for bonsai.

Botanical features

Z. serrata is a deciduous broadleaved tree growing to a maximum height of 35 m and dbh of 2 m. The crown is vase-shaped or round and nearly as wide as the tree is tall. The trunk is straight and well branched, with branches spreading in an upward direction forming a fan-shaped or umbrella-like framework in the crown. Bark is smooth and bright reddish-brown during the first few years, turning light grey to greyish-brown.

The root system is shallow and has a large lateral spread. Taproots of mature trees divide into thinner roots in the upper soil layers, forming a horizontally expanded root system.

Leaves are alternate, flat and thin; leaf blades are ovate, serrated with a crenate margin, long acuminate tip, rotund to slightly cordate base, 3–7 cm long and 1–2.5 cm wide with short petioles (1–3 mm). Hairs are scattered on the leaf surface; lateral veins are pinnate with 8–18 pairs, terminating in the serrated margin. *Z. serrata* var. *stipulacea* has dense hairs on both sides of the leaf blade and on the petiole.

Z. serrata has male, bisexual and female flowers, which are small and insignificant. Female flowers tend to bloom earlier than male flowers. Several male flowers are axillary on the basal leaves of current shoots, with 4–6 stamens; 1–3 female flowers are axillary on the upper leaves of current shoots, with one pistil without stamen; bisexual flowers possess few stamens and are located on the middle parts of current shoots. The number of flowers per shoot and the sex ratio vary among trees and by the position of shoots in the tree crown. Primordia of male flowers are formed by late July–August, while those of female flowers are formed by the following March–April. Pollen mother cells develop in late March, and pollen is mature until late April. Fruits are small drupes, distorted spherical, ripening to a dark grey colour in September–November.

New leaf flushing and shoot elongation occur in mid-April and finish within a few weeks. Flowering occurs in mid-April to late May. Intermittent shoot elongation occurs sporadically during the summer as lammas shoots, especially in the upper crown. Leaves turn yellow, then orange to reddish-brown in October. Viable seeds are dispersed from mid-October to early November. Seed fall begins in September, but most of these early-fallen seeds are not viable. Over 65% of seeds remain attached to the fruiting shoot, such shoots detaching from the tree and dropping to the ground; some may be wind-dispersed.

Distribution

The latitudinal range of *Z. serrata* is 22–41°N, being naturally distributed in Japan from Honshu to Kyushu, in southern Korea and Taiwan, and in China from the middle and lower basin of the Yellow River to Yunnan.

It was introduced into Europe, and also North America and other temperate regions and is now widely planted as an ornamental and landscape tree.

Z. serrata grows both in cool and warm temperate forests, in deciduous broadleaved secondary forests, and on fertile sites near mountain streams. Major plant associations include *Quercus*, *Celtis* and *Acer* species.

Silviculture characteristics

Z. serrata produces seeds as early as 10–15 years of age. Seed dissemination capacity is low; most seeds fall and disperse within 20 m from the mother tree. Seedlings and saplings are shade-tolerant, but more than 15% relative light intensity promotes height growth. Potential initial growth rate of seedlings is high, reaching 20 cm in the first year and 80 cm in the third year on clear felled sites. The allocation of assimilates to leaves occurs during April–May and June–September, and the allocation to non-photosynthetic parts increases during May–June.

Forest planting is better in fertile and permeable soils than in poor soils, in which underdeveloped trees might suffer from pest attack more frequently. Crown spatial requirements are wide, and an afforestation system leading to a mixed forest of *Z. serrata* and coniferous trees might be appropriate.

Silviculture practice

Z. serrata seed crops occur at 2–3-year intervals. The number of seeds produced is about 62,000/kg; viability varies among trees, and germination is not usually high (<45%). Germination is epigeal, and occurs in March and April. Appropriate temperatures for germination are 10–15°C during the day and

1–5°C at night. Seeds are intermediate in storage characteristics, and collection of seeds is best in mid- or late October. Seeds stored at 0–4°C retain viability for about 4 years. Chilling and stratification treatments for more than 1 month promote the germination of stored seeds.

Z. serrata has high sprouting potential, and epicormic shoots suitable for use as cuttings form more easily on basal branches than on upper ones; the leaf shedding season is appropriate for this treatment. Propagation by tissue culture is also possible.

Although there is little difference in germination between spring and autumn sowing, the former may be preferable to prevent seed desiccation caused by frost in winter. Seedlings grow up to 20 cm in height at the end of the first growing season with a density of 100 plants/m^2, and then, for raising nursery stock in the second year, transplanting should be done at a density of 15–20 plants/m^2, when seedlings spread four leaves with the cotyledon still remaining. First-year seedlings grow better under 100% daylight, but also survive and grow under some shade.

Forest planting of nursery-raised stock is the most popular technique for stand establishment, with planting being carried out from February to March or from October to November at a density of 3000–4000 plants/ha. As it has a widely expanding crown which needs space, and saplings are shade-tolerant, planting in gaps in coniferous forest may facilitate the growth of *Z. serrata* trees by the competitive effect against neighbouring trees; limited space may be conducive to straighter trunks by accelerating upward growth and branch shedding. In natural stands with densities of 25–37 trees/ha, average tree height of 21–24 m, dbh of >30 cm, and wood volumes of 100–390 m^3/ha have been obtained.

Although standard afforestation techniques for *Z. serrata* have not yet been fully established, efforts have been made to develop afforestation techniques using the mixed forest system of *Z. serrata* with conifers such as *Cryptomeria japonica*. In this system, the height of coniferous trees surrounding *Z. serrata* and the distance between them are influential for height, bole straightness, crown diameter, and the volume and quality of timber. Trees grown in open sites are well branched, producing timbers with very well-figured grain, some of which are very valuable.

Diseases and pests

Fungi such as *Septoria abeliceae*, *Pseudocercospora zelkowae* and *Sphaceloma zelkovae* produce grey to yellowish-brown patches on leaf blades and may cause early defoliation. Other fungi such as *Ganoderma applanatum*, *Armillaria mellea*, *Armillaria ostoyae*, *Perenniporia fraxinea*, *Helicobasidium mompa*, *Rosellinia necatrix*, *Erysiphe* sp., *Phyllactinia* sp. and *Sphaerotheca* sp. damage stems and roots.

The wood borer *Apriona japonica*, the most destructive insect pest of *Z. serrata*, lays its eggs under the bark of lower branches and the larvae eat the wood, making tunnels from branch to trunk. Although larval survival rates are low, the tunnels greatly reduce timber quality. *Batocera lineolata* is another boring pest, while *Paracolopha morrisoni*, a sucking insect, induces leaf galls. Both larvae and adults of *Pyrrhalta maculicollis* defoliate *Z. serrata*, and leaf blades attacked by this insect become discoloured. Other leaf feeders include *Callambulyx tatarinovii tatarinovii* and *Zeuzera multistrigata leuconotum*.

Uses

Z. serrata is suitable for planting in and around car parks, on wide tree lawns and on medians along major roads. It is grown successfully in urban areas as a shade and street tree where the sites suffer from air pollution, poor drainage and compacted soil. It is also widely cultivated as a bonsai tree.

The wood is typically ring-porous, hard, moderately heavy, and strong. The heartwood is yellowish-brown to reddish-brown, while the sapwood is pale yellowish-brown. Wood density is 480–880 kg/m^3 (average 720 kg/m^3 at 13% moisture content). The wood has been used for construction (including temples) since the 1700s in Japan, often in eminent positions such as the main posts, at entrances, as facings or in reliefs. Wood with narrow growth rings from slow-grown trees are good for tools and especially as material for lacquer ware ('shikki' in Japanese), while wood with wide growth rings is mechanically strong and good as building material; intermediate wood is less warped and good for furniture. Specific items made from *Z. serrata* wood include columns, beams, floorboards, stairs, tools, trays, bowls, pestles and mortars, chests, tables, desks, chairs, cupboards, altars, drums, bridges, ship fittings and sculptures.

The bark of *Z. serrata* is used for making cases, baskets and winnowing equipment.

Further reading

Iwata, Y. and Sano, J. (2005) Fundamental studies on growth traits and management of planted trees - viability and growth conditions of *Liriodendron tulipifera* and *Zelkova serrata* (in Japanese). *Hardwood Research* 11, 23–34.

Kanazawa, Y., Kiyono, Y., Fujimori, T. and Kamo, K. (1984) Seasonal changes in the above-ground biomass of a young *Zelkova serrata* stand. *Journal of the Japanese Forestry Society* 66, 167–173.

Contributors

Original Contributors to the Forestry Compendium

Emilio Amorini, Centro di Ricerca per la Selvicoltura, Arezzo, Italy
Ang Lai Hoe, Forest Research Institute Malaysia (FRIM), Kuala Lumpur, Malaysia
Simmathiri Appanah, FAO Regional Office for Asia and the Pacific, Bangkok, Thailand
Roger Arnold, CSIRO Australian Tree Seed Centre, Canberra, Australia
Juan Baluarte, Instituto de Investigaciones de la Amazonia Peruana, Iquitos, Peru
Jim Barnett, USDA Forest Service, Louisiana, USA
Giovanni Bernetti, Istituto di Selvicoltura, Università di Firenze, Italy
Anthony Beveridge, Auckland, New Zealand
Pierre Binggeli, UK
Dominic Blay, Forestry Research Institute of Ghana (FORIG), Kumasi, Ghana
Doug Boland, CSIRO Australian Tree Seed Centre, Canberra, Australia
Simone Borelli, FAO, Rome, Italy
David Boshier, Department of Plant Sciences, University of Oxford, UK
Soraya Botelho, Universidade Federal de Lavras, Minas Gerais, Brazil
Brioni Brammall, CSIRO Australian Tree Seed Centre, Canberra, Australia
Antonio Brunori, PEFC Italia, Perugia, Italy
Piero Bruschi, Dipartimento di Biologia Vegetale, Università degli Studi di Firenze, Italy
Rowland Burdon, New Zealand Forest Research Institute (Scion), Rotorua, New Zealand
Enrico Buresti, Istituto Sperimentale di Selvicoltura di Arezzo, Italy
Rafael Cadiz, Ecosystems Research and Development Bureau, Department of Environment & Natural Resources, Laguna, Philippines
Isabel Cañellas, CIFOR-INIA, Madrid, Spain
Konnathu Chacko, Kerala Forest Research Institute (KFRI), Kerala, India
Ebby Chagala, Kenya Forestry Research Institute, Nairobi, Kenya
Camille Couralet, Australian Academy of Science, Canberra, Australia
François Courbet, INRA, Avignon, France
Tim Cumine, Gwynedd, Wales, UK
John Davidson, Eucalyptus and Forestry Services, New South Wales, Australia
John Doran, CSIRO Division of Forestry and Forest Products, Canberra, Australia
Jean-Louis Doucet, Unité de Gestion des Ressources Forestières et des Milieux Naturels, Université de Liège, Gembloux, Belgium
Michael Drexhage, INRA - Nancy, Champenoux, France
CV Spektra Ducci, Arezzo, Italy
Gianni Facciotto, CRA-Research Unit, Casale Monferrato, Italy
Chris Fagg, Departamento de Engenharia Florestal, Universidade de Brasilia, Brazil
Bernhard Felbermeier, Institute of Silviculture, Technische Universität München, Freising, Germany
Robert Fenton, Auckland, New Zealand
James Gambiza, Rhodes University, South Africa
Craig Gardiner, CSIRO Australian Tree Seed Centre, Canberra, Australia
Ab. Ghani, Forest Research Institute Malaysia (FRIM), Kuala Lumpur, Malaysia
Gitehi Giathi, Kenya Forestry Research Institute (KEFRI), Nairobi, Kenya
Elisabetta Gravano, Sezione Coltivazioni Arboree, Università di Pisa, Italy
Saw Leng Guan, Forest Research Institute Malaysia (FRIM), Kuala Lumpur, Malaysia
Susanne Guehne, Freiburg, Germany
Raymond Guries, College of Forestry, University of Wisconsin–Madison, Wisconsin, USA
David Gwaze, The Center for Agroforestry, University of Missouri, USA
John Hall, School of Agricultural and Forest Sciences, University of Wales, Bangor, UK
Nazmul Haq, Engineering and the Environment, University of Southampton, UK
Chris Harwood, Ensis Genetics and Cooperative Research Centre, Tasmania, Australia
Karen Haysom, Bat Conservation Trust, London, UK
Sheila Helgath, Washington, USA
Darrell Herbert, Department of Biological Sciences, Florida International University, USA
Ankila Hiremath, Biology Labs, Harvard University, Massachusetts, USA
Cheng-Kuen Ho, Division of Silviculture, Taiwan Forestry Research Institute, Taipei, Taiwan
Drake Hocking, Petersfield, UK

Colin Hughes, Institute of Systematic Botany, University of Zurich, Switzerland
Faridah Hanum Ibrahim, Faculty of Forestry, Universiti Putra Malaysia, Selangor, Malaysia
Bambang Irawan, Forestry Department, University of Jambi, Indonesia
Li Jiyuan, The Chinese Academy of Forestry, Zhejiang, China
Tomohiko Kamitani, Faculty of Agriculture, Niigata University, Japan
Zin-Suh Kim, Department of Forest Research and Environmental Science, Korea University, Seoul, South Korea
Karel Klinka, Forest Sciences Centre, University of British Columbia, Canada
Thomas Knoke, Technische Universität München, Munich, Germany
K. Kochummen, Forest Research Institute Malaysia (FRIM), Kuala Lumpur, Malaysia
Narine Lackhan, Ministry of Agriculture, Land and Marine Resources, Trinidad and Tobago
J. Lal, IFS, Bhopal, India
Douglas Larson, Department of Integrative Biology, University of Guelph, Ontario, Canada
Peter Lovett, Tamale, Northern Region, Ghana
Alberto Maltoni, Dipartimento di Economia, Università degli Studi di Firenze, Italy
Yury Man'ko, Institute of Biology and Soil Science, Far Eastern Branch, Russian Academy of Sciences, Vladivostok, Russia
Lorenzo Mittempergher, Consiglio Nazionale delle Ricerche (CNR), Firenze, Italy
Tim Mullin, BioSylve Forest Science NZ Ltd, New Zealand
Elena Mutto Accori, Padova, Italy
Robert Nasi, CIFOR, Bogor, Indonesia
Mzoma Ngulube, Forest Research Institute of Malawi, Zomba, Malawi
Ian Nicholas, New Zealand Forest Research Institute Ltd, Rotorua, New Zealand
J. Doland Nichols, School of Environmental Science and Management, Southern Cross University, New South Wales, Australia
John Nittler, Santa Cruz, Bolivia
Ladislans Nshubemuki, Tanzania Forestry Research Institute (TAFORI), Morogoro, Tanzania
Betserai Nyoka, World Agroforestry Centre, Lilongwe, Malawi
Phanuel Oballa, Kenya Forestry Research Institute (KEFRI) , Nairobi, Kenya
Fernando Ortiz Ramirez, CONIF, Santafé de Bogotá, Colombia
William Ostrofsky, Maine Forest Service, USA
Rudy Pacumbaba, Department of Plant & Soil Science, Alabama A&M University, USA
Paolo Paiero, Dipartimento Territorio e Sistema Agro-Forestali, Universita degli Studi di Padova, Italy
Paolo Pallicca, Firenze, Italy
Nick Pasiecznik, Agroforestry Enterprises, Cussy en Morvan, France
João Pereira, Instituto Superior de Agronomia, Universidade Técnica de Lisboa, Portugal
John Peterson, Department of Forest Research and Environmental Conservation, Virginia Tech., USA
C. Pigott, Cumbria, UK
Khongsak Pinyopusarerk, CSIRO Australian Tree Seed Centre, Canberra, Australia
Helen Porter, Oxford, UK
D. Pushpakumara, Faculty of Agriculture, University of Peradeniya, Sri Lanka
Paolo Raddi, Consiglio Nazionale delle Ricerche (CNR), Firenze, Italy
Andrew Radomiljac, Western Australia, Australia
David Richardson, Centre for Invasion Biology, University of Stellenbosch, South Africa
Piotr Robakowski, Department of Forestry, Poznan University of Life Sciences, Poland
Ricardo Rodrigues, Piracicaba SP, Brazil
Carlos Rodríguez-Franco, Maryland, USA
Philip Rundel, Department of Ecology and Evolutionary Biology, University of California (UCLA), Los Angeles, USA
Ricardo Russo, San José, Costa Rica
Stanislav Sannikov, Ekaterinburg, Russia
Alberto Santini, Consiglio Nazionale delle Ricerche (CNR), Firenze, Italy
M. Sattar, Chittagong, Bangladesh
Hans Schabel, College of Natural Resources, University of Wisconsin – Stevens Point, USA
Rudi Seitz, Departamento de Ciencias Florestais, Curitiba, Brazil
Charlie Shackleton, Department of Environmental Science, Rhodes University, Grahamstown, South Africa
Khalid Siddiqui, Islamabad, Pakistan
Wojciech Stachnowicz, Department of Plant Ecology and Environmental Protection, Adam Mickiewicz University, Poznan, Poland
Andrea Tani, Dipartimento di Economia, Ingegneria, Scienze e Technologie Agarie e Forestali, Università degli Studi di Firenze, Italy
Evy von Pfeil, Deutsche Gesellschaft für Internationale Zusammenarbeit, Eschborn, Germany
Lex Thomson, Ensis, Canberra, Australia
Mário Tomazello-Filho, Department of Forestry Sciences, University of São Paulo, Brazil
Denise Tousignant, Ministère des Ressources Naturelles et de la Faune, Quebec, Canada
Robert Van Pelt, Institute for Redwood Ecology, Humboldt State University, Washington, USA
Gerrit van Wyk, Department of Forest Science, University of Stellenbosch, South Africa
Tim Vercoe, CSIRO Forestry and Forest Products, Canberra, Australia
Roman Volosyanchuk, Laboratory of Forest Tree Breeding, Ukrainian Research Institute of Forestry & Forest Melioration, Kharkiv, Ukraine
K. Yoda, School of Science and Engineering, Ishinomaki Senshu University, Miyagi, Japan
Silvia Ziller, Curitiba PR, Brazil

Index of Species